41st European Photovoltaic Solar Energy Conference and Exhibition (EU PVSEC 2024)

Vienna, Austria
23-27 September 2024

Volume 2 of 3

Editors:

G. C. Eder   **R. Kenny**
J. Bergmiller   **J. De Gregorio**

ISBN: 979-8-3313-1538-2

Printed from e-media with permission by:

Curran Associates, Inc.
57 Morehouse Lane
Red Hook, NY 12571

Some format issues inherent in the e-media version may also appear in this print version.

Copyright© (2024) by WIP – Renewable Energies
All rights reserved.

Printed with permission by Curran Associates, Inc. (2025)

For permission requests, please contact WIP – Renewable Energies
at the address below.

WIP – Renewable Energies
Sylvensteinstr. 2
81369 Munchen
Germany

Phone: +49 89 72012735
Fax: +49 89 72012791

wip@wip-munich.de

Additional copies of this publication are available from:

Curran Associates, Inc.
57 Morehouse Lane
Red Hook, NY 12571 USA
Phone: 845-758-0400
Fax: 845-758-2633
Email: curran@proceedings.com
Web: www.proceedings.com

41st European Photovoltaic Solar Energy Conference and Exhibition (EU PVSEC 2024)

Vienna, Austria
23-27 September 2024

Volume 2 of 3

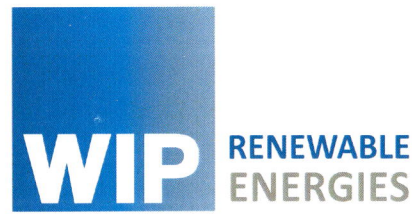

TABLE OF CONTENTS OF EU PVSEC 2024 PROCEEDINGS PAPERS

EU PVSEC 2024 Committees

Subject Index

Foreword

Oral SESSION 1AO.4 Silicon Material for Solar Cells: Growth, Stability and Reuse

1AO.4.3 Impact of High-Temperature Processing Steps on the Long-Term Stability in n-Type FZ Silicon   1

Melanie Mehler[1], Nicolas Weinert[1], Nicole Aßmann[2], Axel Herguth[1], Giso Hahn[1], Fabian Geml[1]
[1] *University of Konstanz, Konstanz, Germany;* [2] *University of Oslo, Oslo, Norway*

1AO.4.5 LeTID in Industrial Ga-doped Cz-Si with Melt Recharging   7

Joshua Kamphues[1], Axel Herguth[1], Juri Miech[1], Xueqi Bai[2], Yichun Wang[2], Giso Hahn[1], Fabian Geml[1]
[1] *University of Konstanz, Constance, Germany;* [2] *LONGI Green Energy Technology, Xi'An, China*

Oral SESSION 1AO.5 Processes for Highly Efficient Si Solar Cells

1AO.5.2 Wet-Chemically Grown Interfacial Oxide for Passivating Contacts Fabricated with an Industrial Inline Processing System   13

Byungsul Min[1], Philipp Noack[2], Bianca Wattenberg[2], Torsten Dippell[2], Henning Schulte-Huxel[1], Robby Peibst[1], Rolf Brendel[1]
[1] *ISFH, Emmerthal, Germany;* [2] *Singulus Technologies, Kahl am Main, Germany*

1AO.5.4 Local p+ Poly-Si Passivating Contacts Realized by Direct FlexTrail Printing of Boron Ink and Selective Alkaline Etching for High Efficiency TOPCon based Solar Cells   19

Berkay Uygun[1], Sven Kluska[2], Jana Isabelle Polzin[2], Jörg Schube[2], Mike Jahn[2], Katrin Krieg[2], Raşit Turan[1], Hisham Nasser[1]
[1] *ODTÜ-GÜNAM, Ankara, Türkiye;* [2] *Fraunhofer ISE, Freiburg, Germany*

1AO.5.5 Phosphorus- and Boron-doped Poly-Si/Siox Passivating Contacts via Inkjet Printing   25

Jiali Wang[1], Thein N. Truong[1], Jinlei Ren[2], Marie Adier[2], Laura Creon[2], Paula Peres[2], Rene Chemnitzer[2], Pierre-Yves Corre[2], Zhuofeng Li[1], Hieu T. Nguyen[1], Josua Stuckelberger[3], Daniel Macdonald[1], AnYao Liu[1], Sieu Pheng Phang[1]
[1] *ANU, Canberra, Australia;* [2] *CAMECA, Gennevilliers, France;* [3] *ANU, Canberra, China*

Oral SESSION 1AO.6 Highly Efficient Si Solar Cells

1AO.6.5 >24% Efficient Tunnel Back Contacted polyZEBRA Solar Cells   30

Jonathan Linke[1], Christoph Peter[1], Jan Hoß[1], Vaibhav Kuruganti[1], Saman Sharbaf Kalaghichi[1], Valentin Mihailetchi[1], Jan Lossen[1], Florian Buchholz[1]
[1] ISC Konstanz, Konstanz, Germany

1AO.6.6 Optimized Ga-Doped Cz Wafers for POLO IBC Solar Cells with High   35
Efficiency and Minimal LeTID Degradation

Thorsten Dullweber[1], Verena Mertens[1], Michael Winter[1], Sabrina Schimanke[1], M. Ripke[1], Silke Dorn[1], Yevgeniya Larionova[1], Gerrit Lange[1], Karsten Bothe[1], Jan Schmidt[1], Rolf Brendel[1], Arne K. Dahle[2], Özlem Coskun[3], Nesrin Töre Sen[3]
[1] ISFH, Emmerthal, Germany; [2] NorSun, Årdalstangen, Norway; [3] Kalyon PV, Ankara, Türkiye

Oral SESSION 1BO.1 Silicon Bottom Cells for Tandem Photovoltaics | Dielectric Layer Related Defect Characterisation

1BO.1.1 Towards TOPCon Based Bottom Cells: Current Challenges and Perspectives   40

Mario Hanser[1], Henning Nagel[1], Johannes Gry[1], Jana Polzin[1], Armin Richter[1], Jan Benick[1], Martin Bivour[1], Martin Hermle[1], Stefan Glunz[1]
[1] Fraunhofer ISE, Freiburg, Germany

1BO.1.3 Review on In-Free Recombination Junction Approaches for Two-Terminal   50
Silicon/Perovskite Tandem Solar Cells

Pia Vasquez[1], Amanda Merino Leiva[1], Perrine Carroy[1], Batiste Marteau[1], Thibaut Desrues[1], Nathalie Nguyen[1], Muriel Matheron[1], Sofia Chozas[2], Federico Ventosinos[2], Henk J. Bolink[2], Delfina Muñoz[1]
[1] CEA-INES, Le Bourget-du-Lac, France; [2] University of Valencia, Valencia, Spain

Oral SESSION 1BO.2 Advanced Silicon Solar Cell Characterisation in Laboratory and Production

1BO.2.1 Identification of Performance-Relevant Optically Detected Defects by   53
Correlative Data Analysis in Solar Cell Production

Manuel Meusel[1], A. Starke[1], Marko Turek[1]
[1] Fraunhofer CSP, Halle (Saale), Germany

1BO.2.3 Integrated Inline Characterisation Techniques for Improved Silicon   57
Heterojunction Solar Cell Production

Christian Diestel[1], Saravana Kumar[1], Alexandra Wörnhör[1], Daniel Burkhardt[1], Nico Wöhrle[1], Sebastian Pingel[1], Matthias Demant[1], Jonas Haunschild[1], Stefan Rein[1]
[1] Fraunhofer ISE, Freiburg, Germany

1BO.2.5 Expert Knowledge, AI, and Simulation: Integrative Approaches for Quality   63
Assurance in Solar Cell Manufacturing

Matthias Demant[1], Alexandra Woernhoer[1], Philipp Kunze[1], Wilkin Woehler[1], Julian Behrendt[1], Leslie Lydia Kurumundayil[1], Johannes Greulich[1], Andreas Fell[1], Stefan Rein[1]
[1] Fraunhofer ISE, Freiburg, Germany

Oral SESSION 1BO.3 Optimised Processes for the Manufacturing of TOPCon Solar Cells

1BO.3.1 Exploring the Impact and Challenges of Using Emerging Wafer Sizes in PV   76
Manufacturing

Julian Reichle[1], Hardik Gohil[1], Mehul Raval[1], Avinash Kumar[1], Wolfgang Jooß[1], Peter Fath[1]
[1] RCT Solutions, Konstanz, Germany

1BO.3.2 A Horizontal Double-Sided Copper Metallization Technology Designed for   86
Solar Cell Mass-Production

Lu Wang[1], Yusen Qin[1], Meilin Peng[2], Meixian Huang[2], ZhiPeng Liu[2], Yibo Lu[1], Guohua Zhou[1], Jingjia Ji[1]
[1] Jiangsu Xianghuan Technology, Wuxi, China; [2] Jiangnan University, Wuxi, China

1BO.3.6 Comprehensive Optimization of Glass Stencil Printing, Demonstrating   89
Ultrafine Metal Fingers Below 10 μm

Tadeo Schweigstill[1], Niko Mielich[1], Aaron Vogt[2], Malte Schulz-Ruhtenberg[2], Jonas D. Huyeng[1], Florian Clement[1]
[1] Fraunhofer ISE, Freiburg, Germany; [2] LPKF Laser & Electronics, Garbsen, Germany

Oral SESSION 1BO.4 New Concepts for the Manufacturing of IBC and HJT Solar Cells

1BO.4.3 IBC4EU: First Results of Industrialization of Low Cost, High Efficiency IBC   101
Technology

Florian Buchholz[1], Daniel Tune[1], Tobias Messmer[1], Jonathan Linke[1], Manjunath Prasad[1], Valentin Mihailetchi[1], Juras Ulbikas[2], Arne Dahle[3], Martijn Meereboer[4], Francesca Fabris[5], Erik Eikelboom[5], Tom Borgers[6], Rik Van Dyck[6], Filip Duerinckx[7], Hariharsudan Sivaramakrishnan[6], Samuel Harrison[8], Josco Kester[9], Nicolas Guillevin[10], Jan Kroon[9], Verena Mertens[11], Thorsten Dullweber[11], Ofer Shochet[12], Isaac Rosen[12], Ingo Röver[13], Wolfram Palitzsch[13], Yasmin Zaror[14], Johnnes Stierstorfer[14], Aurimas Radzevicius[15], Povilas Lukinskas[15], Julius Denafas[15], Tuomas Vanhanen[16], Tuukka Savisalo[16], Maximilian Pospischil[17], Marian Breitenbücher[17], Özlem Coşkun[18], Melodie de l'Epine[19], Philippe Macé[19]
[1] ISC Konstanz, Konstanz, Germany; [2] ProTechnologies, Vilnius, Lithuania; [3] Norsun, Årdalstangen, Norway; [4] Energyra, Westknollendam, The Netherlands; [5] Futurasun, Citadella, Italy; [6] imec, Genk, Belgium; [7] imec, Leuven, Belgium; [8] CEA INES, Le Bourget-du-Lac, France; [9] TNO, Amsterdam, The Netherlands; [10] TNO, Petten, The Netherlands; [11] ISFH, Hamelin, Germany; [12] Copprint, Jerusalem, Israel; [13] LuxChemTech, Freiberg, Germany; [14] WIP Renewable Energies, Munich, Germany; [15] UAB Valoe Cells, Vilnius,

Lithuania; [16] Valoe Cells, Mikkeli, Finland; [17] Highline, Freiburg, Germany; [18] Kalyon PV, Ankara, Türkiye; [19] Becquerel Institute, Brussels, Belgium

1BO.4.4 Self-Aligned Phase Separation for IBC Cells Using PVD Polysilicon — 106

Erik Hoffmann[1], Geoffrey Gregory[1], Massimo Centazzo[1], Muhammad Khan[1], Nabeel Khan[1], Verena Mertens[2], Philip Jäger[2], Sarah Spätlich[2], Ulrike Baumann[2], Thorsten Dullweber[2]
[1] EnPV, Karlsruhe, Germany; [2] ISFH, Hamelin, Germany

1BO.4.5 Gas Phase, Selective Etching of Poly-Silicon for Layer Patterning — 110

Laurent Clochard[1], Mingzhe Yu[2], Ruy Sebastian Bonilla[2], Paul Tierney[3], James Wright[3], Fiacre Rougieux[4], Yalun Cai[4]
[1] Nines Photovoltaics, Dublin, Ireland; [2] University of Oxford, Oxford, United Kingdom; [3] TUD, Dublin, Ireland; [4] UNSW, Sydney, Australia

1BO.4.6 Investigation of Ag-Reduction on Silicon Heterojunction Solar Cells with Different Approaches — 113

Yu Wu[1], Eric J. Kossen[1], Astrid Gutjahr[1], M. Bruggeman[2], L.J. (Bart) Geerligs[1]
[1] TNO, Petten, The Netherlands; [2] TNO, Delft, The Netherlands

Visual SESSION 1BV.5 Silicon Material: Growth, Defects and Recycling | Manufacturing of Solar Cells and Related Tools & Processes

1BV.5.5 Investigation of Oxygen and Carbon Impurities in Mono-Silicon Wafers During Rapid Thermal Annealing — 122

Nurhayat Yıldırım[1], Sertaç Eroğlu[2], Merve Çorak[1]
[1] Kalyon PV, Ankara, Türkiye; [2] Eskişehir Osmangazi University, Eskişehir, Türkiye

1BV.5.7 Increasing the Productivity of the Czochralski Process Applying Machine Learning — 125

Frank Mosel[1], Lukas Kulhavy[2], Dorra Baccar[2]
[1] PVA Crystal Growing Sytems, Wettenberg, Germany; [2] THM, Friedberg, Germany

1BV.5.8 Thermal Deactivation of Boron-Oxygen Defects in Compensated n-Type Silicon — 132

Rune Søndenå[1], Per-Anders Hansen[1], Bent Thomassen[1], Øyvind Mjøs[2], Tyke Naas[2]
[1] IFE, Kjeller, Norway; [2] REC Solar, Kristiansand, Norway

1BV.5.15 Highest Throughput Laser Processing for Thin Plated Contacts — 135

Eduardo Alvarez-Brito[1], René Haberstroh[1], Georg Hoppe[1], Keming Du[2], Florian Roessler[3], Andreas A. Brand[1], Sven Kluska[1], Fabian Meyer[1], Jale Schneider[1], Jan Nekarda[1]
[1] Fraunhofer ISE, Freiburg, Germany; [2] EdgeWave, Würselen, Germany; [3] Moewe Optical Solutions, Mittweida, Germany

1BV.5.16 Enhancement of Photocurrent Generation in Amorphous Silicon Heterojunction (SHJ) Solar Cells through the Integration of Plasmonic Nanoparticles — 139

Brahim Aïssa[1], Alessandro Sinopoli[1]
[1] QEERI, Doha, Qatar

1BV.5.21 Impact of Optimization for Mass Production PERC Solar Cell with Efficiency above 23%   144

Cheng-Wen Kuo[1], Ta-Ming Kuan[1], Yung-Chih Li[1], Chun-Wei Lee[1], Wei-Lo Chueh[1], Li-Guo Wu[1], Shih-Chieh Lin[1], Cheng-Yeh Yu[1]
[1] *TSEC, Hsinchu, Taiwan*

1BV.5.24 The Impact of Conductive Paste Composition on the LECO Process for TOPCon Solar Cells   148

Chun-Ping Lin[1], Chih-Jeng Huang[1], Han-Chen Chang[1], Sung-Yu Chen[1], Bang-Hao Wu[2], Cheng-Liang Cheng[2], Ying-Yuan Huang[3]
[1] *ITRI, Tainan, Taiwan;* [2] *TeraSolar Energy Material, Miaoli, Taiwan;* [3] *National Cheng Kung University, Tainan, Taiwan*

1BV.5.28 Realistic Estimation of Industrial TOPCon Cell Efficiency   149

Mehul Raval[1], Pirmin Preis[2], Lejo Joseph Koduvelikulathu[2], Gourab Das[1], Wolfgang Jooß[1]
[1] *RCT Solutions, Konstanz, Germany;* [2] *ISC Konstanz, Konstanz, Germany*

1BV.5.30 Optimizing the Mechanical Adhesion Properties of Plated Contacts of i-TOPCon Solar Cells   155

Christian Schmiga[1], Abdelaziz Boudellioua[1], René Haberstroh[1], Jonas Eckert[1], Sven Kluska[1], Florian Clement[1]
[1] *Fraunhofer ISE, Freiburg, Germany*

1BV.5.31 Addressing Edge Recombination Losses in Shingle Cells by Holistic Optimization of the Process Sequence   158

Alexander Göbel[1], Elmar Lohmüller[1], Dirk Wagenmann[1], Norbert Kohn[1], Marc Hofmann[1], Jonas D. Huyeng[1], Ralf Preu[1]
[1] *Fraunhofer ISE, Freiburg, Germany*

1BV.5.34 Characterization of TiOx as Electron Selective Contact Using Low-Temperature Oxidation Process via High-Pressure Sputtering   163

Franciso José Pérez Zenteno[1], Sebastian Duarte[1], Rafael Benítez-Fernandez[1], G. Godoy-Perez[1], Ignacio Torres[2], Rocío Barrio[2], Lars Rebohle[3], D. Caudevilla[1], Sari Algaidy[4], Rodgar García-Hernansanz[1], J. Olea[1], D. Pastor[1], Alvaro Del Prado[1], Eric García-Hemme[1], E. San Andrés[1]
[1] *Complutense University of Madrid, Madrid, Spain;* [2] *CIEMAT, Madrid, Spain;* [3] *HZDR, Dresden, Germany;* [4] *Polytechnical University of Madrid, Madrid, Spain*

Visual SESSION 1CV.2 Processing & Characterisation of Crystalline Si based Solar Cells | Silicon Bottom Cells for Tandem Photovoltaics | Advances in Silicon Solar Cells Characterisation and Simulation

1CV.2.2 Approaches for Reducing Metallization-Induced Losses and Cost in Industrial TOPCon Solar Cells   164

Sebastian Mack[1], Daniel Ourinson[1], Marius Messmer[1], Christopher Tessmann[2], Katrin Krieg[1], René Haberstroh[1], Sven Kluska[1], Jonas Huyeng[1], Johannes Greulich[1], Andreas Wolf[1], Florian Clement[1]
[1] *Fraunhofer ISE, Freiburg, Germany;* [2] *Fraunhofer IAF, Freiburg, Germany*

1CV.2.5 Unveiling the Synergy of Nanowires and PEDOT:PSS for Silicon Solar Cell   169
Fabrication and Leading to Mechanical Flexibility

Deepak Sharma[1], Ruchi Kumari Sharma[2], Arman Ahnood[1], Sanjay Kumar Srivastava[2]
[1] *RMIT University, Melbourne, Australia;* [2] *AcSIR, Ghaziabad, India*

1CV.2.7 Polysilicon Passivation - Tunneling Oxide Routes and Annealing Conditions   172
Effect on Passivation

Per-Anders Hansen[1], Junjie Zhu[1], Rune Søndenå[1]
[1] *IFE, Kjeller, Norway*

1CV.2.10 Selective p+ Poly-Si Fingers for TOPCon Front Contact Passivation   175

Jan Hoß[1], Saman Sharbaf Kalaghichi[1], Mertcan Comak[1], Pirmin Preis[1], Jan Lossen[1], Jonathan Linke[1], Lejo Koduvelikulathu[1], Florian Buchholz[1]
[1] *ISC Konstanz, Konstanz, Germany*

1CV.2.16 Review and Highlights of More Than 30 Years Research on Ever Improving   176
Technology for PERC Solar Cells at Fraunhofer ISE

Elmar Lohmüller[1], Sabrina Lohmüller[1], Pierre Saint-Cast[1], Johannes Greulich[1], Stefan Glunz[1], Ralf Preu[1]
[1] *Fraunhofer ISE, Freiburg, Germany*

1CV.2.17 Investigating Interfacial Phenomena in Copper-Covered, n-Type Polysilicon-   182
Based Contacts by Electron Microscopy

Reyu Sakakibara[1], Agata Lachowicz[2], Julien Hurni[1], Christophe Allebé[2], Bertrand Paviet-Salomon[2], Franz-Josef Haug[1], Christophe Ballif[1], Aïcha Hessler-Wyser[1], Audrey Morisset[2]
[1] *EPFL, Neuchâtel, Switzerland;* [2] *CSEM, Neuchâtel, Switzerland*

1CV.2.19 Robustness of Electrical Quality of Ion Implanted Black Silicon Emitters: 189
Comparison between different Ion Implantation Service Providers

Olga Morozova[1], Kexun Chen[1], Behrad Radfar[1], Ulrich Kentsch[2], Luke Antwis[3], Hele Savin[1], Ville Vähänissi[1]
[1] *Aalto University, Espoo, Finland;* [2] *HZDR, Dresden, Germany;* [3] *University of Surrey, Guildford, United Kingdom*

1CV.2.22 Excellent Passivation of Silicon Surfaces by HfO2 Layers Deposited using   193
Scalable Spatial Atomic Layer Deposition (SALD)

Jan Schmidt[1], Michael Winter[1], Floor Souren[2], Jons Bolding[2], Hindrik de Vries[2]
[1] *ISFH, Emmerthal, Germany;* [2] *SALD, Eindhoven, The Netherlands*

1CV.2.26 Simulation of Topcon/Perc Hybrid Bottom Structure for Perovskite/Silicon   197
Tandem Solar Cells Using Quokka3

Eni Muka[1], Raşit Turan[1], Hisham Nasser[1]
[1] *ODTÜ GÜNAM, Ankara, Türkiye*

1CV.2.35 A Comprehensive Analysis of the Series Resistance for Different   201
Interdigitated Back Contact Solar Cell Geometries

Telmo Isasi[1], Yeray Mateos[1], Janire Pampin[1], Vanesa Fano[1], Nekane Azkona[1], Eneko Ortega[1], Juan Carlos Jimeno[1], Eneko Cereceda[1], Alona Otaegi[1]
[1] *UPV/EHU, Bilbao, Spain*

1CV.2.37 Accuracy of Hysteresis Correction for Silicon Heterojunction Solar Cells – A   205
Simulation Study

Jonas Kern[1], Hannes Wagner-Mohnsen[2], Johannes Heitmann[1], Matthias Müller[1]
[1] *Freiberg University of Mining and Technology, Freiberg, Germany;* [2] *WAVELABS Solar Metrology Systems, Leipzig, Germany*

1CV.2.38 Contactless Carrier Lifetime Characterization of Silicon Heterojunction   209
Structures at Elevated Temperatures

Gergely Havasi[1], David Krisztián[1], Zs. Gombás[2], Zoltan Adam[2], Ferenc Korsós[1]
[1] *Semilab, Budapest, Hungary;* [2] *EcoSolifer Heterojunction, Budapest, Hungary*

1CV.2.39 Bias Light Intensity Effect on EQE Analysis for PERC Solar Cell   213

Hatice Duman[1], Özlem Coskun[1], Güven Korkmaz[1]
[1] *Kalyon PV, Ankara, Türkiye*

1CV.2.42 Improved Accuracy of Photoluminescence Images for Quality Control in   214
Solar Cell Production

Robin Wienberg[1], Jonas Haunschild[1], Saravana Kumar[1], Jurriaan Schmitz[2], Stefan Rein[1]
[1] *Fraunhofer ISE, Freiburg, Germany;* [2] *University of Twente, Enschede, The Netherlands*

1CV.2.44 Simulation and Design Optimization of Interdigitated Back Contact Silicon   219
Solar Cells with Dopant-Free Asymmetric Hetero-Contacts

You-An Li[1], Chun-Ping Lin[2], Ying-Yuan Huang[2]
[1] *NYCU, Tainan, Taiwan;* [2] *NCKU, Tainan, Taiwan*

1CV.2.45 Numerical Modeling and Design Optimization of Industrial Tunnel Oxide   222
Passivated Contact Solar Cells with Selective Passivated Contacts on the
Front

Yi-Ping Lin[1], Chun-Ping Lin[2], Jin-Cheng Chen[1], Han-Chen Chang[3], Ying-Yuan Huang[2]
[1] *NYCU, Tainan, Taiwan;* [2] *NCKU, Tainan, Taiwan;* [3] *ITRI, Tainan, Taiwan*

1CV.2.47 Modeling and Experimental Validation of Solar Cell Performance across   225
Varied Temperatures and Irradiance

Selin Cansu Gölboylu[1], Hatice Duman[1], Melisa Demir[1], Meriç Çalışkan Arslan[1]
[1] *Kalyon PV, Ankara, Türkiye*

Plenary SESSION 1EP.1 Sustainability

1EP.1.2 Copper as Cost-Effective Alternative to Silver for Si Solar Cell Metallization   226
– Status and Outlook

Florian Clement[1], Andreas Lorenz[1], Jonas Bartsch[1], Andreas Brand[1], Jonas D. Huyeng[1], Roman Keding[1], Sven Kluska[1], F. Maarouf[1], Jan Nekarda[1], Daniel Ourinson[1], Sebastian Pingel[1], J. Schube[1], Ralf Preu[1]
[1] *Fraunhofer ISE, Freiburg, Germany*

Oral SESSION 2AO.3 III-V Solar Cells & Space PV

2AO.3.2 Thermal Modeling of Triple-Junction Solar Cells Fan Out Wafer Level Packaging for Concentrated Photovoltaic   250

Konan Kouame[1], Abdul Rehman[1], Médérick Marcotte[1], Mylana Ney[1], Artur Turala[1], Corentin Jouanneau[1], Mohamed Najah[1], Serge Ecoffey[1], David Danovitch[2], Gwenaelle Hamon[1]
[1] *University of Sherbrooke, Sherbrooke, Canada;* [2] *Université de Sherbrooke, Sherbrooke, Canada*

2AO.3.3 Overview for Tandem Solar Cell R&D Activities in Japan   254

Masafumi Yamaguchi[1], Tatsuya Takamoto[2], Kyotaro Nakamura[1], Ryo Ozaki[1], Hiroyuki Juso[2], Nobuaki Kojima[1], Yoshio Ohshita[1]
[1] *Toyota Technological Institute, Nagoya, Japan;* [2] *Sharp Corporation, Nara, Japan*

2AO.3.5 Space Applications for a Variety of Solar Cell Technologies   257

Stephen Taylor[1]
[1] *European Space Agency, Noordwijk, The Netherlands*

Oral SESSION 2BO.10 New Modelling and Characterisation - Material Properties

2BO.10.1 In-depth Characterization Methodology for the Assessment of Passivation Impact in Halide Perovskite Solar Cells   260

Jonathan Parion[1], Santhosh Ramesh[1], Sownder Subramaniam[1], Henk Vrielinck[2], Filip Duerinckx[1], Hariharsudan Sivaramakrishnan Radhakrisnan[1], Jef Poortmans[1], Johan Lauwaert[3], Bart Vermang[1]
[1] *imec, Genk, Belgium;* [2] *University of Gent, Ghent, Belgium;* [3] *University of Gent, Zwijnaarde, Belgium*

Oral SESSION 2BO.8 Novel PV Material and Conversion Concepts

2BO.8.1 Pathways for Silicon Solar Cells with Molecular Singlet Fission   263

Phoebe Pearce[1], Nicholas Ekins-Daukes[1]
[1] *UNSW, Sydney, Australia*

2BO.8.2 Control of Hot Carrier Thermalization Rates in Nanowires for Advanced-Concept Photovoltaic Solar Cells   273

Hamidreza Esmaielpour[1], Nabi Isaev[1], Imam Makhfudz[2], Markus Döblinger[3], Jonathan Finley[1], Gregor Koblmüller[1]
[1] *TUM, Munich, Germany;* [2] *Aix-Marseille University, Marseille, France;* [3] *LMU, Munich, Germany*

2BO.8.5 Design and Prototyping of Spectrum-Split-Type Concentrating Photovoltaic-Thermoelectric Hybrid Power Generator   277

Kenji Kamide[1], Ryoji Funahashi[1], Tomoyuki Urata[1], Yoko Matsumura[1], Jun Sakuma[2], Hidefumi Akiyama[2], Katsuto Tanahashi[1]
[1] *AIST, Tsukuba, Japan;* [2] *University of Tokyo, Kashiwa, Japan*

Visual SESSION 2BV.1 Advances in Novel Materials, Devices and Concepts | New Modelling and Characterisation Techniques

2BV.1.1 Development of an Interdigitated Back-Contacted Solar Cell Architecture as a Platform to Assess Emerging Absorbers and New Selective Contacts — 285

Juan de Dios Castillo[1], Gerard Masmitjà[1], Pau Estarlich[1], Pablo Ortega[1], Cristobal Voz[1], Arnau Torrens[1], Oriol Segura[1], Edgardo Saucedo[1], Massoud Karimipour[2], Sonia Ruiz[2], Mónica Lira-Cantu[2], Joaquim Puigdollers[1]
[1] UPC, Barcelona, Spain; [2] ICN2, Barcelona, Spain

2BV.1.2 Annealed Phosphorus-Doped Amorphous Silicon as Electron Selective Contact for Crystalline Germanium Thermophotovoltaic Cells — 289

Gerard Rivera[1], Mansur Gamel[1], Gema López[1], Moisés Garín[2], Isidro Martín[1]
[1] UPC, Barcelona, Spain; [2] University of Vic, Vic, Spain

2BV.1.10 Sensitization of Crystalline Silicon with Organic Dye Molecules — 294

Lukáš Gdula[1], Branislav Dzurňák[1], Tom Markvart[2]
[1] Czech Technical University in Prague, Prague, Czech Republic; [2] University of Southampton, Southampton, United Kingdom

2BV.1.11 Self-Organized Films of Carbazole Derivatives on Structured Silicon Substrates for Photovoltaic Applications — 295

Sergii Mamykin[1], Daria Kuznetsova[1], Nina Roshchina[1], Petro Smertenko[1], Saulius Grigalevicius[2], Gintare Krucaite[2], Raminta Beresneviciute[2], Simona Sutkuviene[3]
[1] V. Lashkaryov Institute of Semiconductor Physics NAS Ukraine, Kyiv, Ukraine; [2] Kaunas University of Technology, Kaunas, Lithuania; [3] Lithuanian University of Health Sciences, Kaunas, Lithuania

2BV.1.12 Placement Angles for Luminescent Solar Concentrators: Simulating and Experimenting with Bifacial Photovoltaic Mosaic Devices — 299

Xitong Zhu[1], Frits Reijners[1], Michael Debije[1], Angèle H.M.E Reinders[1]
[1] Eindhoven University of Technology, Eindhoven, The Netherlands

2BV.1.20 Low Emissive Molybdenum-Doped ITO for High Vacuum Photovoltaic-Thermal Application — 302

Daniela De Luca[1], Umar Farooq[1], Paolo Strazzullo[1], Eliana Gaudino[1], Antonio Caldarelli[1], Anna Krammer[2], Andreas Schüler[2], Marilena Musto[1], Emiliano Di Gennaro[1], Roberto Russo[1]
[1] University of Naples Federico II, Naples, Italy; [2] EPFL, Lausanne, Switzerland

2BV.1.27 Optimization of a Planar Perovskite Solar Cell Layer Thicknesses: Optical and Electrical Effects — 303

Aleksi Kamppinen[1], Kati Miettunen[1]
[1] University of Turku, Turku, Finland

2BV.1.28 Photoluminescence Imaging of Perovskite Solar Cells in Full Sunlight — 308

Zhiwen Zheng[1], Felix Gayot[1], Juergen W. Weber[1], Yan Zhu[1], Ziv Hameiri[1]
[1] UNSW, Sydney, Australia

2BV.1.29 Analysis of Color Alteration as a Novel Degradation Assessment Method for Perovskite Solar Cells — 309

Rustem Nizamov[1], Aapo Poskela[1], Mahboubeh Hadadian[1], Maryam Esmaeilzadeh[1], Mikael Nyberg[1], Kati Miettunen[1]
[1] University of Turku, Turku, Finland

2BV.1.30 Statistical Model of Outdoor Perovskite Performance 315

Petra Manshanden[1], Martin Späth[1], Mark Jansen[1], Valerio Zardetto[2], Arantxa Aguirre[3], Valerie Depauw[3], Mina Heydarian[4], Juliane Borchert[4]
[1] TNO, Petten, The Netherlands; [2] TNO, Petten, The Netherlands; [3] imec, Genk, Belgium; [4] Fraunhofer ISE, Freiburg, Germany

2BV.1.32 Characterization and Degradation of Perovskite Mini-Modules 323

Rita Ebner[1], Ankit Mittal[1], Gusztav Ujvari[1], Maria Hadjipanayi[2], Vasiliki Paraskeva[2], George E. Georghiou[2], Afshin Hadipour[3], Aranzazu Aguirre[4], Tom Aernouts[4], Thommaso Fontanot[5], Sabrina Pechmann[5], Silke Christiansen[5]
[1] AIT, Vienna, Austria; [2] University of Cyprus, Nicosia, Cyprus; [3] Kuwait University, Kuwait, Kuwait; [4] imec, Genk, Belgium; [5] IKTS, Forchheim, Germany

2BV.1.34 Subcell-Resolved Electroluminescence Imaging of Monolithic Perovskite-Silicon Tandem Solar Cell for High Throughput Characterization 327

Ivanol Jaurece Djeukeu[1], Jonas Horn[1], Michael Meixner[1], Enno Wagner[2], Stefan W. Glunz[3], Klaus Ramspeck[1]
[1] halm elektronik, Frankfurt, Germany; [2] Frankfurt University of Applied Sciences, Frankfurt, Germany; [3] Fraunhofer ISE, Freiburg, Germany

2BV.1.35 A Case Study of Certainly I-V Measurement of the Perovskite Solar Cell under Dim Light Intensity for Solar/ Indoor Lighting Application 328

Yean-San Long[1], Min-An Tsai[1], Hsin-Hsin Hsieh[1], Fan-Hsuan Yeh[2]
[1] ITRI, Hsinchu, Taiwan; [2] Taipei First Girls High School, Taipei, Taiwan

2BV.1.39 Perovskite Solar Cell Light-Soaking and Relaxation Modelling for Improved Energy Yield Predictions in Indoor Environments 329

Matija Pirc[1], Špela Tomšič[1], Marko Jošt[1], Marko Topič[1]
[1] University of Ljubljana, Ljubljana, Slovenia

2BV.1.41 Modelling the Effects of Tandem Module Circuit Configurations 330

M. Ignacia Devoto[1], Daniel Tune[1], Ahmer A.B. Baloch[2], Karl Wienands[1], Rüdiger Farneda[1], Bhaskar Parida[2], Omar Albadwawi[2], Vivian Alberts[2], Andreas Halm[1]
[1] ISC Konstanz, Konstanz, Germany; [2] DEWA Research & Development Center, Dubai, United Arab Emirates

Visual SESSION 2BV.2 Compound and Organic Semiconductors

2BV.2.5 III-V Thin Films Growth by RP-CVD: Towards a Reduction of Industrialization Costs 334

Lise Watrin[1], François Silva[2], Cyril Jadaud[2], Pavel Bulkin[2], Jean-Charles Vanel[2], Kassiogé Dembélé[2], Erik V. Johnson[2], Karim Ouaras[2], Pere Roca i Cabarrocas[1]
[1] IPVF, Palaiseau, France; [2] LPICM, Palaiseau, France

2BV.2.9 Modeling and Measurement of Lumped Series Resistance with Varying Illumination and Current Condition of Low-Bandgap Solar Cells 335

Shipei Zhang[1], Xiawa Wang[1]
[1] Duke Kunshan University, Kunshan, China

2BV.2.11 Color Implementation of Cu(In,Ga)Se2 Thin-film Solar Cells with Multilayered Conductive Optical Filters   338

Yong-Duck Chung[1], Dae-Hyung Cho[1], Rina Kim[1], Woo-Jung Lee[1], Tae-Ha Hwang[1], Soyoung Lim[1], Donghyeop Shin[2], Kihwan Kim[2], Mangu Kang[1]
[1] ETRI, Daejeon, South Korea; [2] KIER, Daejeon, South Korea

2BV.2.17 Flexible Thin-Film CZTS Solar Cell based on an Electroplated Metallic Precursor Deposited on a Molybdenum/Glass Coated Stainless Steel Foil   342

Io Mizushima[1], Peter Torben Tang[1], Christoph Kammerlander[2], Andreas Zimmermann[2]
[1] IPU, Virum, Denmark; [2] Sunplugged, Affenhausen, Austria

2BV.2.23 Manufacturing, Characterisation and Stability Tests of Printed Organic Photovoltaic Devices for Indoor Applications   345

Ignacio Ballesteros Garcia[1], A. Khodr[1], Donia Fredj[2], Carmen M Ruiz Herrero[1], Hasan Alkhatib[2], O. Margeat[1], Sadok Ben Dkhil[2], Judikaël Le Rouzo[1], Jörg Ackermann[1]
[1] CNRS, Marseille, France; [2] Dracula Technologies, Valence, France

2BV.2.30 Fabrication of Highly Efficient CdSeTe/CdTe Thin Film Solar Cells with Emitter-Less Cell Structure   346

Yanbo Cai[1], Hongxu Jiang[1], Kai Yi[1], Fei Liu[1], Guangwei Wang[1], Deliang Wang[1]
[1] University of Science and Technology of China, Hefei, China

Oral SESSION 2CO.2 Triple Junctions and Advanced Concepts in Perovskite-based Tandems

2CO.2.5 Characterisation of Degradation Pathways of 3-Terminal Perovskite-Silicon Tandems After Outdoor Monitoring   349

Miha Kikelj[1], Laurie-Lou Senaud[2], Florent Sahli[2], Benjamin Lipovšek[1], Marko Topič[1], Christophe Ballif[2], Quentin Jeangros[2], Bertrand Paviet-Salomon[2]
[1] University of Ljubljana, Ljubljana, Slovenia; [2] CSEM, Neuchâtel, Switzerland

Oral SESSION 2CO.3 New Modelling and Characterisation - Device Performance

2CO.3.2 Understanding Ion-Related Performance Losses in Perovskite-Based Solar Cells by Capacitance Measurements and Simulation   363

Christoph Messmer[1], Jonathan Parion[2], Cristian V. Meza[2], Santhosh Ramesh[2], Martin Bivour[3], Maryamsadat Heydarian[3], Jonas Schön[1], Hariharsudan S. Radhakrishnan[2], Martin C. Schubert[3], Stefan W. Glunz[1]
[1] University of Freiburg, Freiburg, Germany; [2] imec, Genk, Belgium; [3] Fraunhofer ISE, Freiburg, Germany

2CO.3.4 Analysis and Modelling of Recovery and Degradation Mechanisms in Perovskite Solar Cells 379

Guillem Álvarez-Pérez[1], Arthur Julien[1], Karim Medjoubi[1], Jean Baptiste Puel[1], Jean François Guillemoles[1]
[1] IPVF, Palaiseau, France

2CO.3.5 Developments in Thermophotovoltaics (TPV) 384

Esther López Estrada[1], Alejandro Datas[1]
[1] UPM, Madrid, Spain

Visual SESSION 2CV.3 Perovskite-based Multijunctions | Perovskite Photovoltaics

2CV.3.4 Monolithic Series-Interconnected Two-Terminal Perovskite-CIGSe Tandem Solar Cells: Voltage-Matched or Current-Matched? 393

Nicolas Otto[1], Christof Schultz[1], Guillermo Farias-Basulto[2], Rutger Schlatmann[1], Eva Unger[2], Bert Stegemann[1]
[1] HTW Berlin, Berlin, Germany; [2] HZB, Berlin, Germany

2CV.3.11 Optimisation of MA-free Lead-Tin Perovskite Absorber and Interfaces in All Perovskite Tandem Solar Cells 397

Jules Allegre[1], Polyxeni Tsoulka[1], Noëlla Lemaitre[1], Baptiste Berenguier[2], Mathieu Frégnaux[3], Muriel Bouttemy[3], Philip Schulz[2], Solenn Berson[1], Kilian Alcocer[4]
[1] CEA-INES, Le-Bourget-du-Lac, France; [2] IPVF, Palaiseau, France; [3] ILV, Versailles, France; [4] CEA, Grenoble, France

2CV.3.17 Potential Induced Degradation Free Perovskite-Silicon Tandem Solar Cells 398

Kristijan Brecl[1], Matevž Bokalič[1], Gašper Matič[1], Marko Topič[1], Lisa Champault[2], Quentin Jeangros[2]
[1] University of Ljubljana, Ljubljana, Slovenia; [2] CSEM, Neuchâtel, Switzerland

2CV.3.18 Modeling of Metastability Behavior in Perovskite-based Solar Cells for Accurate Energy Yield Estimation in Realistic Operating Conditions 399

Špela Tomšič[1], Marko Remec[2], Florian Scheler[2], Mark Khenkin[2], Carolin Ulbrich[2], Rutger Schlatmann[2], Steve Albrecht[2], Marko Jošt[1], Benjamin Lipovšek[1], Marko Topič[1]
[1] University of Ljubljana, Ljubljana, Slovenia; [2] HZB, Berlin, Germany

2CV.3.24 Microstructural Analysis on the Conformity of Chemical Vapour Deposition (CVD) Perovskite Thin-Films on Silicon for Tandem PV Devices 404

Angela Chen[1], Emma Holder[1], Adrian Element[1], Yong Li[1], Kenrick F. Anderson[1], Tim W. Jones[1], Benjamin C. Duck[1], Noel W. Duffy[1], Gregory J. Wilson[1]
[1] CSIRO Energy, Newcastle, Australia

2CV.3.25 Controlling the Film Properties of SnO2 in Perovskite Solar Cells using Scalable Spatial Atomic Layer Deposition 408

Hindrik W. de Vries[1], Floor M. M. Souren[1], S. R. Ratnasingham[2], Mehrdad Najafi[2]
[1] SALD, Eindhoven, The Netherlands; [2] TNO, Petten, The Netherlands

2CV.3.33 Beyond the Lab-Scale: Perovskite Photovoltaic Fabrication and Industrial   411
Assessment with Automated Slot-Die Coater

Maurizio Stefanelli[1], Simon Ternes[1], Luigi Vesce[1], Marco Balucani[2], Aldo Di Carlo[1]
[1] *University of Rome Tor Vergata, Rome, Italy;* [2] *RISE Technology, San Martino di Lupari, Italy*

2CV.3.34 Reasoning the Change in Device Parameters with Deposition Power of NiOx   414
for Low-Dimensional Perovskite Solar Cells

Bhumika Sharma[1], Vani Pawar[1], Sushobhan Avasthi[1]
[1] *Indian Institute of Science, Bengaluru, India*

2CV.3.36 Analysis of Reverse-Bias Stability of FAPbBr3 Semi-Transparent Perovskite   417
Solar Cells

Noah Tormena[1], Alessandro Caria[1], Matteo Buffolo[1], Carlo De Santi[1], Nicola Trivellin[1], Andrea Cester[1], Gaudenzio Meneghesso[1], Enrico Zanoni[1], Fabio Matteocci[2], Aldo Di Carlo[2], Matteo Meneghini[1]
[1] *University of Padova, Padova, Italy;* [2] *University of Rome, Rome, Italy*

2CV.3.43 Enhancing Efficiency and Stability of CsPbI3 Perovskite Quantum Dots   418
Through Co2+-Doping

Pouriya Naziri[1], Naeimeh Sadat Peighambardoust[1], Umut Aydemir[1]
[1] *Koc University, Istanbul, Türkiye*

2CV.3.44 Standardized Test Routines for the Assessment of Potential Induced   422
Degradation of Perovskite Solar Cells

Beyza Durusoy[1], David Adner[2], Christian Hagendorf[3], Konrad Wojciechowski[4], Samy Almosni[4], Marko Turek [5]
[1] *METU, Ankara, Türkiye;* [2] *Martin-Luther-University, Halle, Germany;* [3] *Anhalt University of Applied Sciences, Köthen, Germany;* [4] *Saule Technologies, Wroclaw, Poland;* [5] *Fraunhofer CSP, Halle, Germany*

2CV.3.45 Evaluation of Perovskite Devices Under Real and Extreme Operating   427
Conditions - A Fundamental Step Toward Practical Applications

Marília Braga[1], Lucas Augusto Zanicoski Sergio[1], Anelise Medeiros Pires[1], Ricardo Rüther[1]
[1] *UFSC, Florianópolis, Brazil*

2CV.3.46 Enhancing Measurement Protocols for Perovskite Photovoltaic Devices: 428
Insights from the VIPERLAB Project

Eugenia Zugasti[1], Ankit Mittal[2], Lucia V. Mercaldo[3], Javier Diaz [1], Giuseppe Nasti[3], Asier Murillo Marrero[1], Natalia Maticiuc[4], Ana Belén Cueli[1], Stephan Abermann[2], Paola Delli Veneri[3], Stephane Cros[5]
[1] *CENER, Sarriguren, Spain;* [2] *AIT, Vienna, Austria;* [3] *ENEA, Portici, Italy;* [4] *HZB, Berlin, Germany;* [5] *CEA-INES, Le Bourget du Lac, France*

2CV.3.51 Solvent Engineering Driven Morphology Control of Perovskite under Air   429
Ambient Device Fabrication

Nitin Kumar Bansal[1], Shivam Porwal[2], Trilok Singh[1]
[1] *IIT Delhi, New Delhi, India;* [2] *IIT Kharagpur, Kharagpur, India*

2CV.3.55 Roll-to-Roll Printed SnO2 for Flexible N-I-P Perovskite PV   430

Thomas M. Kraft[1], Ville Holappa[1], Riikka Suhonen[1]
[1] *VTT Technical Research Centre of Finland, Oulu, Finland*

2CV.3.61 Micro Inverted Pyramid Formation in Titanium Dioxide Layer by Pulsed   432
Laser Irradiation to Improve Electron Transport in MAPBI3-based
Photovoltaic Devices

Luis Ocaña[1], Carlos Montes[1], Benjamín González-Díaz[2], Sara González-Pérez[2], Elena Llarena[1]
[1] *ITER, Granadilla de Abona, Spain;* [2] *University of La Laguna, San Cristóbal de La Laguna, Spain*

2CV.3.63 Demonstration of Industrially Scalable Chemical Vapour Deposition (CVD)   437
Process for Production of High-Efficiency Perovskite Photovoltaics

Emma Holder[1], Adrian Element[1], Yong Li[1], Faiazul Haque[1], Kenrick F. Anderson[1], Tim W. Jones[1], Benjamin C. Duck[1], Noel W. Duffy[1], Gregory J. Wilson[1]
[1] *CSIRO Energy, Newcastle, Australia*

2CV.3.64 Photoluminescence and Lifetime Stability of Pentacene and Oxide   441
Perovskites Nanoparticles Films on Nanotextured Silicon Substrate

Rémi Ndioukane[1], Diouma Kobor[1], Sergio de Armas Rillo[2], Fernando Lahoz Zamarro[2]
[1] *University Assane Seck of Ziguinchor, Ziguinchor, Senegal;* [2] *University of La Laguna, Santa Cruz de Tenerife, Spain*

2CV.3.69 Compositional Engineering of Double-cation Single-halide Perovskite for   445
Efficient Solar Cell Fabrication under Air Ambient Conditions

Mrittika Paul[1], Binita Boro[1], Amreesh Chandra[1], Trilok Singh[2]
[1] *IIT Kharagpur, Kharagpur, India;* [2] *IIT Delhi, New Delhi, India*

2CV.3.74 Interface Engineering for Perovskite Solar Cells Using Polymer-Based   446
Antisolvent Technique

Lingeswaran Arunagiri[1], Feng Wang[2], Feng Gao[2]
[1] *Linköping University, Linkoping, Sweden;* [2] *Linköping University, Linköping, Sweden*

Oral SESSION 2DO.18 Late News: Developments in High Efficiency Tandem Cells

2DO.18.3 Perovskite Record Setting Silicon Tandem Modules: Customer Expect Lower   447
LCOE

Christopher Case[1]
[1] *Oxford PV, Oxford, United Kingdom*

Oral SESSION 2DO.6 Towards Improved Understanding of Perovskite Solar Cell Device Physics

2DO.6.1 Bright Insights: Exploring Perovskite Formation Mechanisms with Combined   466
Spectral Reflectance and Photoluminescence In-Situ Data

Nasim Rezaei-Hartmann[1], Thorsten Brand[1], Adrian Adrian[1], Claudine Groß[1], M. Leyden[2], Enno Malguth[1], Aleksandra Miaskiewicz[2], Marcel Roß[2], Viktor Škorjanc[2], Lars Korte[2], Steve Albrecht[2], Christian Camus[1]
[1] *LayTec, Berlin, Germany;* [2] *HZB, Berlin, Germany*

2DO.6.5 Enhancing Crystallinity of Perovskite Materials through Rapid Microwave Annealing — 481

Syed Nazmus Sakib[1], David N. R. Payne[1], Shujuan Huang[1], Binesh P. Veettil[1]
[1] Macquarie University, Sydney, Australia

Oral SESSION 2DO.8 Scalability of Perovskite Solar Modules

2DO.8.4 Fully Printed Perovskite Solar Cells and Modules — 486

Luigi Vesce[1], Karthikeyan Pandurangan[2], Maurizio Stefanelli[1], Elena Iannibelli[1], Hafez Nikbakht[1], Maria Laura Parisi[2], Adalgisa Sinicropi[2], Aldo Di Carlo[1]
[1] University of Rome Tor Vergata, Rome, Italy; [2] University of Siena, Siena, Italy

Oral SESSION 2DO.9 Lifetime and Reliability of Perovskite Devices

2DO.9.2 TÜV Rheinland Specification on the I-V Characterization of Perovskite-Based PV Modules — 490

Giorgio Bardizza[1], Qi Gao[2], Wenhao Xu[2], Yating Zhang[2], Christos Monokroussos[2], Werner Herrmann[3]
[1] TUV Rheinland, Milan, Italy; [2] TUV Rheinland, Shanghai, China; [3] TÜV Rheinland, Milan, Italy

2DO.9.5 One-Year Outdoor Testing of 4T Perovskite/Si PV Modules — 494

Matthew Norton[1], Vasiliki Paraskeva[1], Maria Hadjipanayi[1], Elias Peratikos[1], Aranzazu Aguirre[2], Anurag Krishna[2], Santhosh Ramesh[2], Tom Aernouts[2], George E. Georghiou[1]
[1] University of Cyprus, Nicosia, Cyprus; [2] Hasselt University/Imo-Imomec, Genk, Belgium

Visual SESSION 3AV.1 PV Module Design and Manufacturing | BoS Components, Operation and Aging

3AV.1.4 Challenges for Solder Interconnection pushed by High-Efficiency Solar Cell Developments — 499

Benjamin Grübel[1], Angela De Rose[1], Achim Kraft[1]
[1] Fraunhofer ISE, Freiburg, Germany

3AV.1.5 Optimizing Sustainability: Balancing Antimony Content for Enhanced Optical Properties and Environmental Impact in Solar Glass — 505

Anika Glaubitz[1], Sven Grüttner[1], Selim Yagci[1], Oliver Pfeiffer[1], Ulf Blieske[1]
[1] University of Applied Sciences Cologne, Cologne, Germany

3AV.1.6 Photovoltaic Modules Comprising III-V Cells Encapsulated in Composite Material — 511

Francisco J. Cano[1], Werther Cambarau[1], Naiara Yurrita[1], Jon Aizpurua[1], Juan M. Hernández[1], Gorka Imbuluzqueta[1], Eduardo Román Medina[1], Oihana Zubillaga[1]
[1] TECNALIA, San Sebastián, Spain

3AV.1.11 Reliability of Aluminum-Copper Contact in PV Modules 514

Tobias Messmer[1], Dominik Rudolph[1], Gernot Emanuel[2], Andreas Nägele[2], Andreas Halm[1]
[1] *ISC Konstanz, Konstanz, Germany;* [2] *Fraunhofer ISE, Freiburg, Germany*

3AV.1.12 Lightweight Photovoltaic Modules Technologies: Reliability Evaluation and 519
Market Opportunity

Julien Dupuis[1], Christine Abdel Nour[1], J.V. Oliveira Santos[1], Paul Lefillastre[2]
[1] *EDF R&D, Moret-Loing-et-Orvanne, France;* [2] *EDF Renewables, Nanterre, France*

3AV.1.15 MgO/SiOx Adds Heat Dissipation Function to Crystalline Silicone Solar Cell 523
Modules

Eiko Shimokata[1], Yasushi Sobajima[1], Keisuke Ohdaira[2], Atsushi Masuda[3]
[1] *Gifu University, Gifu, Japan;* [2] *JAIST, Nomi, Japan;* [3] *Niigata University, Niigata, Japan*

3AV.1.16 Investigation of Temperature Homogeneity during Infrared Soldering of 527
Silicon Solar Cells using the Finite Element Method

Daniel Christopher Joseph[1], Angela De Rose[1], Dirk Eberlein[1], Onur Parlayan[1], Benjamin Grübel[1], Andreas J. Beinert[1], Holger Neuhaus[1]
[1] *Fraunhofer ISE, Freiburg, Germany*

3AV.1.17 Impact of Textured Surfaces and Cleaning on Solar Panel Glass 530
Transmittance

Aapo Poskela[1], Julianna Varjopuro[1], Tommi Jokikyyny[1], Aleksi Kamppinen[1], Heikki Palonen[1], Kati Miettunen[1]
[1] *University of Turku, Turku, Finland*

3AV.1.19 Ultra-Thin Flexible Glass as Environmental Shield for CIGS Photovoltaic 534
Modules

Nikolina Pervan[1], Sonja Feldbacher[1], Martina Harnisch[2], Tuuli Tettenborn[2], Andreas Zimmermann[2], Gernot Oreski[1]
[1] *PCCL, Leoben, Austria;* [2] *Sunplugged, Wildermieming, Austria*

3AV.1.20 Process Development and Material Evaluation of Photovoltaic Aluminum 535
Facade Element for BIPV Application

Ringo Koepge[1], Matthias Pander[1], Stephan Großer[1], Bengt Jaeckel[1]
[1] *Fraunhofer CSP, Halle (Saale), Germany*

3AV.1.22 Material Properties Requirements for Frame Sealants and Junction Box 539
Adhesives

Guy Beaucarne[1], Emmanuel Jadot[1], Dominique Culot[1], Rono Cao[2], Kayla Kenney[3], Suraj Ahuja[4], Valérie Hayez[1]
[1] *Dow Silicones Belgium, Seneffe, Belgium;* [2] *Dow (Shanghai), Shanghai, China;* [3] *Dow Silicones , Auburn, United States of America;* [4] *Dow Chemical International, Mumbai, India*

3AV.1.23 Solder Pastes in Shingled Modules 545

Karl Wienands[1], Ignacia Devoto[1], Nils Kopp[2], Carina Hallensleben[2], Rihoko Kizukuri[2], Matthias Helbig[1], Enita Kurtovic[1], Andreas Halm[1], Daniel Tune[1]
[1] *ISC Konstanz, Konstanz, Germany;* [2] *TAMURA-ELSOLD, Ilsenburg, Germany*

3AV.1.25 TiO₂/SiOx Surface Coating on Crystalline Silicon-Based-Solar Cell Module to 549
Provide Anti-Soiling Functionality

Koshiro Iwaki[1], Yasushi Sobajima[1], Keisuke Ohdaira[2], Atsushi Masuda[3]
[1] Gifu University, Gifu, Japan; [2] JAIST, Nomi, Japan; [3] Niigata University, Niigata, Japan

3AV.1.26 Performance Analysis of Different Shading-Resistant PV Module Designs under Different Partial Shading Scenarios   552

Andreas Maixner[1], Tales Siquera[1], Matthias Pander[2], Jens Froebel[2], Bengt Jaeckel[2], Hamed Hanifi[1]
[1] AESOLAR, Koenigsbrunn, Germany; [2] Fraunhofer CSP, Halle, Germany

3AV.1.33 Optimal Design for Flexible Solar Panels Attached Around Cylindrical Poles   559

Hiroki Sugimoto[1]
[1] PXP Corporation, Sagamihara, Japan

3AV.1.35 Design and Implementation of a CSI Photovoltaic Microinverter Prototype with High Frequency Switching   562

Francisco Guzman[1], Patricio Valdivia-Lefort[2], Antonio Sanchez[1], Rodrigo Barraza[1]
[1] Federico Santa Maria Technical University, Santiago, Chile; [2] Universidad de Santiago de Chile, Santiago, Chile

3AV.1.37 PV Microinverters: Balcony Power Plants, Latest Efficiency Rankings, Yield Calculation for Overpowered Mini PV Systems   563

Stefan Krauter[1], Jörg Bendfeld[1]
[1] Paderborn University, Paderborn, Germany

3AV.1.38 Aging Behavior of Polymeric Materials used in Inverter Casings   570

Eric Helfer[1], Petra Christöfl[1], Julia Petro[1], Margit Lang[1], Volker Reisecker[2], L. Heupl[3], A. Weiermair[3], Gernot Oreski[1]
[1] PCCL, Leoben, Austria; [2] Transfercenter für Kunststofftechnik, Wels, Austria; [3] Fronius International, Thalheim, Austria

3AV.1.39 Performance of Arc Fault Circuit Interrupters in Photovoltaic Inverters Connected to Long DC Cables   571

Donat Hess[1], David Joss[1], Christof Bucher[1]
[1] BUAS, Burgdorf, Switzerland

3AV.1.40 Design of the Substring MPP Tracker   578

Patrick Mader[1], Sascha Eckerter[1], Rainer Merz[1]
[1] Karlsruhe University of Applied Sciences, Karlsruhe, Germany

3AV.1.41 Testing of Electronic Interface for Diagnostic Functions of Photovoltaic Systems   583

Edoardo Celi[1], Alessandro Minuto[1], Stefano Rizzi[1], Gianluca Timò[1]
[1] RSE, Piacenza, Italy

Visual SESSION 3AV.2 PV Modules Reliability: Components, Failure Mechanisms, Testing & Modelling

3AV.2.1 Evaluation of Degradation and Impact of Climatic Conditions on PV Modules Exposed to Extreme High UV Solar Radiation   584

Patricio Valdivia-Lefort[1], Valentina Navarro[2], Rodrigo Barraza[2]
[1] Universidad de Santiago de Chile, Santiago, Chile; [2] Federico Santa Maria Technical University, Santiago, Chile

3AV.2.2 PV Module Brush Abrasion Testing 585

Gerhard Mathiak[1], Nithin Sha[1], Afra Seentakath[1], Prashanth Gabbadi[1], Yogesh Kumar[1], Mark Mirza[2]
[1] DEWA R&D Center, Dubai, United Arab Emirates; [2] Fraunhofer ISC, Würzburg, Germany

3AV.2.3 Numerical Simulation for Comparison of PV Module Designs based on Outdoor Data in Desert Climates 591

Matthias Pander[1], Bengt Jaeckel[1], Klemens Ilse[1], Amir A. Abdallah[2]
[1] Fraunhofer CSP, Halle (Saale), Germany; [2] QEERI , Doha, Qatar

3AV.2.5 Impact of Modern Cell Photovoltaic Geometries on Power and Energy Loss due to Cell Cracks 596

Ahmad Hashem[1], SL. Mortazavifar[1], Ralph Gottschalg[1]
[1] Anhalt University of Applied Sciences, Köthen, Germany

3AV.2.7 Performance Evaluation of the Custom-Made Small PV Modules after Exposure to Saudi Arabia's Climatic Conditions over 10 Long Years 600

Amir Al-Ahmed[1], Amjad Ali[1], Mohammed A. Alghamdi[2], Osama Asker[2], Ridha Ben Mansour[1], Firoz Khan[1], Atif S. Alzahrani[1]
[1] KFUPM, Dhahran, Saudi Arabia; [2] Gulf Renewable Lab, Dammam, Saudi Arabia

3AV.2.8 Analyzing the Effect of Damp Heat Test on Various PV Module Technologies, a Comparative Study 601

Ahmad Alheloo[1], Ali Almheiri[1], Baloji Adothu[1], Gerhard Mathiak[1], Vivian Alberts[1]
[1] DEWA Research & Development, Dubai, United Arab Emirates

3AV.2.9 Comparative Degradation Analysis of Emerging PV Module Technologies Undergoing Thermal Cycling 604

Ali Almheiri[1], Ahmad Alheloo[1], Baloji Adothu[1], Gerhard Mathiak[1], Vivian Alberts[1]
[1] DEWA Research & Development, Dubai, United Arab Emirates

3AV.2.10 Assessment of Critical Laminate Temperature Increase by Fast IR-based Analysis of Hot Spots on Solar Cells 607

Stephan Grosser[1], Matthias Schak[1], Stefan Eiternick[1], Bengt Jaeckel[1], Marko Turek[1]
[1] Fraunhofer CSP, Halle (Saale), Germany

3AV.2.11 Correlational Study on the Impact of Harsh Environment Stress Factors on the Ageing Effects of Several Encapsulation Materials for PV Modules 610

Tudor Timofte[1], Maria Ignacia Devoto Acevedo [1], Joachim Glatz-Reichenbach[1], Valentina Arias Reyes[2], Andreas Halm[1]
[1] ISC Konstanz, Konstanz, Germany; [2] Federico Santa María Technical University, Valparaiso, Chile

3AV.2.12 Model Calibration of Photovoltaic Modules Photodegradation in High-Radiation Environments Using UV Accelerated Exposure Testing 618

Patricio Valdivia[1], Valentina Arias Reyes[2], Rodrigo Barraza[3], Iván González Echeverria[2]
[1] Universidad de Santiago de Chile, Santiago, Chile; [2] Federico Santa Maria Technical University, Santiago, Chile; [3] Universidad Adolfo Ibañéz, Santiago, Chile

3AV.2.13 Electrical Characterization of Fresh and Degraded Photovoltaic Backsheets 619
Based on Temperature and Humidity-Dependent DC Conductivity

Anagha E R[1], Shrikrishna V Kulkarni[1], Narendra Shiradkar[1]
[1] IIT Bombay, Mumbai, India

3AV.2.14 Investigation of PV Module Degradation in Fixed Structure and Single-Axis 623
Tracker in Hot Desert Climate

*Baloji Adothu[1], Aafra Seentakath Puthiyapurayil[1], Shahzada Pamir Aly[1],
Gerhard Mathiak[1], Vivian Alberts[1]*
[1] DEWA R&D Center, Dubai, United Arab Emirates

3AV.2.17 Tackling the Fire Safety in Glass Free PV Modules 626

*Nikolina Pervan[1], Sonja Feldbacher[1], Umang Desai[2], Antonin Faes[2],
Christophe Ballif[2], Gernot Oreski[1]*
[1] PCCL, Leoben, Austria; [2] EPFL, Neuchâtel, Switzerland

3AV.2.20 Analysis and Material Modeling of Mechanical Property Degradation for 627
Simulation of Weather Exposed Polymers

*Julia Petro[1], Volker Reisecker[2], Eric Helfer[1], Gernot Oreski[1], Thomas
Antretter[3], Margit Lang[1]*
[1] PCCL, Leoben, Austria; [2] TCKT, Wels, Austria; [3] University of Leoben, Leoben, Austria

3AV.2.21 Reliability Investigation of Structural Colour Interlayers for Coloured PV 632
Modules

*Markus Babin[1], Roberto Boccardi[1], Aliihsan Bagci[1], Nanna Lysgaard
Andersen[1], Peter Behrensdorff Poulsen[1], Sune Thorsteinsson[1], Karlis
Petersons[2], Leif Yde[2], Jan F. Stensborg[2], Catarina G. Ferreira[3], Joel D. Cox[3],
Irina Vyalih[4], Jani Lamminaho[3], Morten Madsen[4]*
[1] DTU, Roskilde, Denmark; [2] Stensborg, Roskilde, Denmark; [3] SDU, Odense, Denmark; [4]
SDU, Sønderborg, Denmark

3AV.2.22 Diagnosing Potential Induced Degradation in Crystalline Silicon Photovoltaic 636
Modules

*Aysha Mahmood[1], Rodrigo del Prado Santamaria[1], Thøger Kari[1], Peter B.
Poulsen[1], Sergiu V. Spataru[1]*
[1] DTU, Roskilde, Denmark

3AV.2.25 On-Site Evaluation of Oxygen-Plasma Treated Glass Surfaces for Anti- 645
Soiling Properties

Brahim Aïssa[1], Ayman Samara[2]
[1] QEERI, Doha, Qatar; [2] HBKU, Doha, Qatar

3AV.2.27 Performance, Abrasion Resistivity and Anti-Soiling Testing of Innovative, 651
Nanostructured Anti-Reflection Coatings under Controlled and Standardized
Conditions

*Charlotte Pfau[1], Guido Willers[1], Christos Allagiannis[2], Ioannis Arampatzis[2],
Marko Turek[1]*
[1] Fraunhofer CSP, Halle (Saale), Germany; [2] Nanophos, Lavrio, Greece

3AV.2.30 Development of Encapsulant-Less Crystalline Silicon Photovoltaic Modules 656
and Their Durability Against Potential-Induced Degradation

Keisuke Ohdaira[1], Shuntaro Shimpo[1], Huynh Thi Cam Tu[1]
[1] JAIST, Ishikawa, Japan

3AV.2.35 Evaluation of the Impact of the UV Excitation Intensity on the Ultraviolet   660
Fluorescence Measurement System for Photovoltaics

Zonghan Jiang[1], Carlos Meza[1], Hugo Sanchez[1], Ralph Gottschalg[1]
[1] Anhalt University of Applied Sciences, Koethen, Germany

3AV.2.36 How to Mount PV Modules: the Effect of Different Clamping Configuration   665
on Mechanical Stresses in PV Modules

Pascal Romer[1], Andreas J. Beinert[1], Charlotte Hasselblatt[1], Cornelius Herr[1]
[1] Fraunhofer ISE, Freiburg, Germany

3AV.2.40 FMEA Based Degradation Rate Evaluation to Study Impact of Different   666
Failure Modes as Function of Mission Profiles

Bengt Jaeckel[1], Baloji Adothu[2], Vivian Alberts[3], Matthias Pander[1]
[1] Fraunhofer CSP, Halle (Saale), Germany; [2] DEWA, Dubai, United Arab Emirates; [3] DEWA, Dubai, United Arab Emirates

3AV.2.42 Numerical Simulation of the Bypass Diode Failure Resistance and those   676
Power Consumption in a Photovoltaic Solar Module with Failed Bypass
Diode

Ibuki Kitamura[1], Toshiyuki Hamada[1], Ikuo Nanno[2], Norio Ishikura[3], Masayuki Fujii[4], Shinichiro Oke[5]
[1] Osaka Electro-Communication University, Osaka, Japan; [2] Yamaguchi Gakugei University, Yamaguchi, Japan; [3] Yonago College, Tottori, Japan; [4] Oshima College, Yamaguchi, Japan; [5] Tsuyama College, Okayama, Japan

3AV.2.45 Impact of the Material Combination on the Barrier Properties and their   680
Stability in the Course of Accelerated Weathering

Daniel Schüsler[1], Patrick Wessel[1], Michael Wendt[1], Anton Mordvinkin[1]
[1] Fraunhofer CSP, Halle, Germany

3AV.2.46 Investigation of Thermo-Mechanical Behavior of Encapsulation Materials   683
used in Solar Panel Production

Umran Dilmac[1], Merve Çorak[2], Meric Caliskan Arslan[1], Yildirim Aydogdu[3]
[1] Kalyon PV, Ankara, Türkiye; [2] Kalyon PV , Ankara, Türkiye; [3] Gazi University, Ankara, Türkiye

3AV.2.49 UV Exposure of Glass/Glass Coupons with Edge Seal and Different   684
Encapsulants

Chiara Barretta[1], Lisa Meinhart[1], Andreas Brandstätter[2], Dieter Geier[3], Roland Einhaus[3], Abdulkerim Gok[4], Gernot Oreski[1]
[1] PCCL, Leoben, Austria; [2] Lenzing Plastics, Lenzing, Austria; [3] ZSW, Stuttgart, Germany; [4] Gebze Technical University, Gebze, Türkiye

3AV.2.50 Material Screening for the Development of a Photovoltaic Module Using   685
Biodegradable Materials from Renewable Raw Materials

Matthias Pander[1], Ringo Koepge[1], Bengt Jaeckel[1], Anton Mordvinkin[1]
[1] Fraunhofer CSP, Halle (Saale), Germany

3AV.2.51 Failure Mode Analysis of Austria's First Road-Integrated Photovoltaic   686
System

Alexander Erber[1], Bernhard Grasel[1]
[1] University of Applied Sciences Vienna, Vienna, Austria

3AV.2.52 Effects of Encapsulant-Backsheet Combinations on Durability of Optical   687
Properties

Jishnu Ramachandran Nair[1], Daniel Schuesler[1], Michael Wendt[1], Ralph Gottschalg[1], Anton Mordvinkin[1]
[1] Fraunhofer CSP, Halle, Germany

3AV.2.53 PID Outdoor Measurements, a New Test Setup   690

Jörg Kirchhof[1]
[1] Fraunhofer IEE, Kassel, Germany

3AV.2.54 Coatings or Tapes? Imaging Methods to Show the Successful Repair of Backsheet Cracks   694

Raffael Schifferegger[1], Yuliya Voronko[1], Anika Gassner[1], Gabriele C. Eder[1], Eric Tilly[2]
[1] OFI, Vienna, Austria; [2] ENcome Energy Performance, Klagenfurt, Austria

3AV.2.57 Effect of Weight Percent Graphene on Barrier Properties of Ethelyne Vinyl Acetate (EVA) for Improved Photovoltaic Module Packaging Reliability   695

Emeka H. Amalu[1], Oluwagbemiga A. Fabunmi[1], David J. Hughes[1], Yongxin Pang[1], Michael Short[1]
[1] Teesside University, Middlesbrough, United Kingdom

3AV.2.58 Tests beyond Standards on Bifcacial PV Modules with Transparent Backsheets   702

Alessandro Anderlini[1], Angelika Beinert[2], Ingrid Hädrich[2], Luigi D'arco[1]
[1] Coveme, Gorizia, Italy; [2] Fraunhofer ISE, Freiburg, Germany

Visual SESSION 3AV.3 PV Modules Performance: Testing, Modelling Techniques and Outdoor Performance

3AV.3.1 Enhanced Performance of PV Modules using Hierarchically Structured Glass in Different Climatic Conditions   703

Cristina Leyre Pinto[1], Jaione Bengoechea[1]
[1] CENER, Sarriguren-Navarra, Spain

3AV.3.9 A Data-Driven Calibration of the FEM Temperature Model with Wind Direction Input   709

Anastasios Kladas[1], Bert Herteleer[1], Jan Cappelle[1]
[1] KU Leuven, Leuven, Belgium

3AV.3.13 Areal Cell Temperature Monitoring Using Array of In-Laminate Integrated Sensors for Partial Shading Detection   712

Seyed Mojtaba Sadati Faramarzi[1], Georgi H. Yordanov[1], Arvid van der Heide[2], Jan Genoe[1], Jef Poortmans[1]
[1] KU Leuven, Leuven, Belgium; [2] imec, Leuven, Belgium

3AV.3.14 Maximum Power Output Predicting Algorithm of Solar Modules Based on Artificial Intelligence Technology   716

Ju-Hee Kim[1], Joonyoung Jeon[1], Yong Hyun Kim[1]
[1] KOPTI, Gwangju, South Korea

3AV.3.16 A Parametric Approach for Estimation of PV Short-Circuit Current   717

Sergiu Mihai Hategan[1], Marius Paulescu[1]
[1] West University of Timişoara, Timişoara, Romania

3AV.3.22 Comparison of Changes in the Parameters of Five PV Module Types after one Year in the Swiss Jura Mountains 721

Donat Hess[1], Fabio Panduri[1], Matthias Burri[1], Christof Bucher[1], Mauro Caccivio[2], Gabi Friesen[2]
[1] *BFH, Burgdorf, Switzerland;* [2] *SUSPI, Mendrisio, Switzerland*

3AV.3.25 Accurate Energy Performance Model for Bifacial PV Modules 738

Kristijan Brecl[1], Matevž Bokalič[1], Marko Topič[1], Antonin Faes[2]
[1] *University of Ljubljana, Ljubljana, Slovenia;* [2] *CSEM, Neuchâtel, Switzerland*

3AV.3.28 Uncertainty Assessment in the Measurement of Solar Cells under Standard Test Conditions 739

Yating Zhang[1], Wenhao Xu[1], Qi Gao[1], Giorgio Bardizza[2], Werner Herrmann[2], Christos Monokroussos[1]
[1] *TÜV Rheinland, Shanghai, China;* [2] *TÜV Rheinland Energy, Cologne, Germany*

3AV.3.29 Stabilization of Field-Aged Crystalline PV Modules Before STC Power Determination 743

Soha Essbai[1], Marcus Rennhofer[1], Ankit Mittal[1], Gusztáv Újvári[1], Thomas Weber[2], Brian Azzopardi[3]
[1] *AIT, Vienna, Austria;* [2] *PI Berlin, Berlin, Germany;* [3] *FIR Malta, Valetta, Malta*

3AV.3.30 AC/DC Electroluminescence. The War of the Currents 746

Mario Martínez[1], Sergio Suarez[1], Daniel Villoslada[1], Jose Manuel Rivas[1], Sofía Rodríguez-Conde[1]
[1] *Enertis Applus, Madrid, Spain*

3AV.3.31 Evaluation of the Contact Quality in Silicon Solar Cells and Modules Using LBIC Phase Mapping 751

Majid Salari[1], Jonas Buddgård[2], Markus Rinio[1]
[1] *Karlstad University, Karlstad, Sweden;* [2] *StickySolarPower, Sollentuna, Sweden*

3AV.3.32 Nomenclature and Description of EL Observations: Cell Cracks and Other Findings 755

Bengt Jaeckel[1], Matthias Pander[1], Paul Schenk[1], Aswin Linsenmeyer[2], Jochen Kirch[3]
[1] *Fraunhofer CSP, Halle (Saale), Germany;* [2] *Sunset Energietechnik, Adelsdorf, Germany;* [3] *Ing.-Büro Jochen Kirch, Leeder, Germany*

3AV.3.38 Daylight Electroluminescence Inspection of PV Panels On-site vs. Traditional EL Inspection with Silicon Cameras 758

Luis Alberto Carpintero[1], Diego Gónzalez-Francés[2], Kabir Paul Sulca[2], Cristian Terrados[2], Carmelo de Castro[2], Victor Alonso[2], Míguel Ángel Gónzalez Rebollo[2], Oscar Mártinez[2]
[1] *Cobra Instalaciones y Servicios, Madrid, Spain;* [2] *University of Valladolid, Valladolid, Spain*

3AV.3.39 Photovoltaic Module Array Luminescence Image Preprocessing: Heuristic Algorithms for Perspective Correction and Cell Segmentation 761

Brendan Wright[1], Ali Shakiba[1], Rama Sharma[1], Ziv Hameiri[1]
[1] *UNSW, Sydney, Australia*

3AV.3.42 Comparing Measured PV Module Power to Nameplate Values 765

Frank Weinrich[1], Stefan Riechelmann[1], Laura Stenzig[1], Stefan Winter[1]
[1] *PTB, Braunschweig, Germany*

3AV.3.44 Finding the Cell to Module Performance Values for Industrial TOPCon and HJT Technologies   768

Sraisth[1], Hardik Gohil[1], Mehul Raval[1], Wolfgang Jooss[1]
[1] *RCT Solutions, Konstanz, Germany*

3AV.3.45 The Impact of Module Degradation on the Economics of PV Projects   772

Harry Apostoleris[1], Baloji Adothu[1], Bengt Jaeckel[2], Gerhard Mathiak[1], Sgouris Sgouridis[1]
[1] *DEWA R&D, Dubai, United Arab Emirates;* [2] *Fraunhofer CSP, Halle (Saale), Germany*

3AV.3.50 Improved Sampling of IV Measurements   773

Maximilian Schönau[1], Elisabeth Schönau[2], Darwin Daume[3], Markus Panhuysen[1], Achim Schulze[4], Bernd Hüttl[3], Dieter Landes[3]
[1] *smartblue, Munich, Germany;* [2] *Catholic University Eichstätt-Ingolstadt, Eichstätt, Germany;* [3] *Coburg University of Applied Sciences, Coburg, Germany;* [4] *Rosenheim University of Applied Sciences, Rosenheim, Germany*

3AV.3.52 Enhancing Production Forecasting of Grid-Connected PV Strings Operating under Semi-Arid Climate Conditions   776

Khadija El Ainaoui[1], Mhammed Zaimi[1], Imane Flouchi[2], Said Elhamaoui[2], Yasmine El Mrabet[2], Abdellatif Ghennioui[2], El Mahdi Assaid[1]
[1] *University of Chouaib Doukkali, El Jadida, Morocco;* [2] *Green Energy Park, Ben Guerir, Morocco*

3AV.3.53 Evaluation of the Glare Function and Description of Key Measurement Procedures   780

Wolfgang Nemitz[1], Roman Trattnig[1], Jakob Zehndorfer[2], Markus Babin[3], Lukas Plessing[4]
[1] *Joanneum Research, Weiz, Austria;* [2] *Zehndorfer Engineering, Klagenfurt, Austria;* [3] *DTU, Roskilde, Denmark;* [4] *TPPV, Vienna, Austria*

Oral SESSION 3BO.11 Reliability of PV Modules: The Impact of Solar Cell Technology

3BO.11.1 Reliability of Commercial TOPCon PV Modules – An Extensive Comparative Study   786

Paul Gebhardt[1], Jochen Markert[1], Ulli Kräling[1], Esther Fokuhl[1], Ingrid Haedrich[1], Daniel Philipp[1]
[1] *Fraunhofer ISE, Freiburg, Germany*

3BO.11.3 Study and Mitigation of Moisture-Induced Degradation in Silicon Heterojunction Solar Modules   792

Lucie Pirot-Berson[1], Romain Couderc[1], Romain Bodeux[2], Frédéric Jay[1], Julien Dupuis[3]
[1] *CEA, Le Bourget-du-Lac, France;* [2] *IPVF, Palaiseau, France;* [3] *EDF, Moret Loing et Orvanne, France*

3BO.11.4 Investigation of Potential-induced Degradation and Recovery in Perovskite Minimodules   800

Junchuan Zhang[1], Haodong Wu[1], Yi Zhang[2], Fangfang Cao[1], Zhiheng Qiu[1], Minghui Li[1], Xiting Lang[1], Yongjie Jiang[1], Yangyang Gou[1], Xirui Liu[1], Abdullah M. Asiri[3], Paul J. Dyson[2], Mohammad Khaja Nazeeruddin[2], Jichun Ye[1], Chuanxiao Xiao[1]
[1] CAS, Ningbo, China; [2] EPFL, Lausanne, Switzerland; [3] KAU, Jeddah, Saudi Arabia

Oral SESSION 3BO.12 Reliability of PV Modules: The Impact of Polymers

3BO.12.4 Recent Developments in PV Module Backsheets - What Do We Really Know about Them?   809

Gernot Oreski[1], Chiara Barretta[1], Karl-Anders Weiß[2]
[1] PCCL, Leoben, Austria; [2] Fraunhofer ISE, Freiburg, Germany

Oral SESSION 3BO.14 Failure Modes and Degradation Mechanisms in PV Modules

3BO.14.2 Analyses of Glass Quality and its Influence on Mechanical Stability of Large Area PV Modules   812

Jochen Markert[1], Aditya Girish Belawadi[1], Pascal Romer[1], Frank Ensslen[1], Enzo Job[1], Ingrid Hädrich[1], Daniel Philipp[1], Tobias Rist[2]
[1] Fraunhofer ISE, Freiburg, Germany; [2] Fraunhofer IWM, Freiburg, Germany

3BO.14.3 Polarization-Type Potential-Induced Degradation in Bifacial PERC Modules in the Field   826

Peter Hacke[1], Cecile Molto[2], Dylan J.Colvin[2], Ryan Smith[3], Farrukh Ibne Mahmood[4], Fang Li[4], Jaewon Oh[5], Govindasamy Tamizhmani[4], Hubert Seigneur[2], Christopher DiRubio[6], Matthew Gardeski[6]
[1] NREL, Golden, United States of America; [2] FSEC Energy Research Center University of Central Florida, Cocoa, United States of America; [3] Pordis, Austin, United States of America; [4] ASU, Mesa, United States of America; [5] University of North Carolina, Charlotte, United States of America; [6] First Solar, Tempe, United States of America

3BO.14.5 LeTID in Real Life: The Relevance and Importance of Accelerated Tests and Treatments   834

Alison Ciesla[1], Arastoo Teymouri[1], Petra Manshanden[2], Alvin Mo[1], Astrid Gutjahr[2], Moonyong Kim[1], Li Wang[1], Catherine Chan[1], Ran Chen[1], Gianluca Coletti[1], Jakob Jan Dijksterhuis[3], Bas Van Aken[2]
[1] UNSW, Sydney, Australia; [2] TNO, Petten, The Netherlands; [3] Elsun, Roden, The Netherlands

3BO.14.6 Towards Establishing Criteria for Electrical Safety in Second-Use Photovoltaic (PV) Modules   846

Tadanori Tanahashi[1], Takashi Oozeki[1]
[1] AIST, Koriyama, Japan

Oral SESSION 3BO.15 Reliability of PV Modules: Testing and Modelling Approaches

3BO.15.2 Material Selection and Novel Reliability Testing for Floating Photovoltaic Modules   854

Nikoleta Kyranaki[1], Arvid van der Heide[1], Hamed Javanbakht Lomeri[1], Ismail Kaaya[1], Sara Bouguerra[1], Jens D. Moschner[2], Arnaud Morlier[1], Michaël Daenen[1]
[1] Hasselt University, Genk, Belgium; [2] KU Leuven, Leuven, Belgium

3BO.15.3 Outdoor Accelerated Ageing Test Using Additional Thermal and Thermomechanical Stresses   862

Ebrar Özkalay[1], Gabi Friesen[1], Alessandro Virtuani[2], Mauro Caccivio[1], Christophe Ballif[3]
[1] SUPSI, Mendrisio, Switzerland; [2] CSEM, Neuchâtel, Switzerland; [3] EPFL, Neuchâtel, Switzerland

3BO.15.4 Development of PV Module Hot Desert Test Cycle Protocol Extended Failure Modes and Effective Analysis   878

Baloji Adothu[1], Jim Joseph John[1], Gerhard Mathiak[1], Vivian Alberts[1], Bengt Jäckel[2], Ralph Gottschalg[2], Narendra S Shiradkar[3], Amir A. Abdallah[4], Juan Lopez Garcia[4], Michael Salvador[5], Bram Hoex[6], Hussein A Kazem[7], Muhammad Ashraful Alam[8]
[1] DEWA, Dubai, United Arab Emirates; [2] Fraunhofer CSP, Halle, Germany; [3] IIT Bombay, Mumbai, India; [4] QEERI, Doha, Qatar; [5] KAUST, Thuwal, Saudi Arabia; [6] UNSW, Sydney, Australia; [7] Sohar University, Sohar, Oman; [8] Purdue University, West Lafayette, United States of America

3BO.15.6 Solar Cell Crack Image Generation for Power Loss Prediction   886

Norman Jost[1], Emma Cooper[1], Benjamin G. Pierce[1], Brandon Byford[1], Ojas Singh[1], Jennifer L. Braid[1]
[1] Sandia National Laboratories, Albuquerque, United States of America

Oral SESSION 3CO.10 Materials and Processes for PV Modules

3CO.10.1 Benchmarking of Encapsulant Materials for c-Si/Perovskite Tandem Modules   892

Petra Christöfl[1], Chiara Barretta[1], Marcel Kühne[2], Frans Opden Buijsch[3], Sem Sals[3], Quentin Jeangros[3], Bernd Stannowski[4], Gernot Oreski[1]
[1] PCCL, Leoben, Austria; [2] Hanwha Q CELLS, Thalheim, Germany; [3] The Compound Company, Geleen, The Netherlands; [4] HZB, Berlin, Germany

3CO.10.2 Reliability Studies of PV Minimodules Using an Ethylene – Butyl Acrylate (EBA) Based Encapsulant and High Efficiency n-Type PV Cells   898

Ignacio Fidalgo[1], Inmaculada Campoy Felipe[2], Andreas Halm[3]
[1] Polaris Open Innovation, Oviedo, Spain; [2] Repsol Química, Madrid, Spain; [3] ISC Konstanz, Constance, Germany

3CO.10.4 Reducing Process Time of PV Module Lamination by Using Double-Side Heating System   911

Sraisth[1], Djamel Eddine Mansour[2], Aksel Kaan Öz[2], Paul Gebhardt[2], Daniel Klaus[3], Christine Wellens[2]
[1] RCT Solutions, Constance, Germany; [2] Fraunhofer ISE, Freiburg, Germany; [3] Robert Buerkle, Freudenstadt, Germany

Oral SESSION 3CO.11 Emerging Interconnection Technologies

3CO.11.2 Design Roadmap to Modules with 24 % Efficiency 920

Max Mittag[1], Christian Reichel[1], Alexander Protti[1], Dirk Holger Neuhaus[1]
[1] *Fraunhofer ISE, Freiburg, Germany*

3CO.11.3 Effect of Lowering Curing Temperature of Electrically Conductive Adhesives 924
on Ribbon Connected Solar Cells

Veronika Nikitina[1], Tim Riehle[1], Leonhard Böck[1], Torsten Rößler[1]
[1] *Fraunhofer ISE, Freiburg, Germany*

3CO.11.5 To Bypass or Not to Bypass: Integrating and Evaluating Parallel Connections 929
and Bypasses in c-Si PV Laminates

*Tom Borgers[1], Jonathan Govaerts[1], Hamed Javanbakht Lomeri[1], Apostolos
Bakovasilis[1], Rik Van Dyck[1], Bart Reekmans[1], Hariharsudan
Sivaramkrishnan Radhakrishnan[1], Jef Poortmans[1], Manuel Van den Storme[2],
Guy Van den Storme[2]*
[1] *imec, Genk, Belgium;* [2] *VdSWeaving, Oudenaarde, Belgium*

**Oral SESSION 3DO.12 Low Environmental Impact Module Design and
Technologies**

3DO.12.1 Steps Towards a 100% Renewable Material Solar Module: Evaluating 933
Material Substitutions for Encapsulation and Interconnection

Ringo Koepge[1], Matthias Pander[1], Anton Mordvinkin[1], Stephan Großer[1]
[1] *Fraunhofer CSP, Halle (Saale), Germany*

3DO.12.2 New Encapsulant for PV Modules Designed for Recycle: A Lab Scale 940
Prototype

*Margot Landa[1], Alexis Brastel[2], Eeva Mofakhami[1], Timea Bejat[1], Pierre
Piluso[2]*
[1] *CEA-INES, Le Bourget-du-Lac, France;* [2] *CEA Liten, Grenoble, France*

3DO.12.3 Laser-Assisted Delamination for Si Modules Recycling 943

*Remi Aninat[1], Maarten van der Vleuten[1], Johan Bosman[1], Henri Fledderus[2],
Anne Biezemans[1], João Gomes[1], Veronique Gevaerts[1], Ando Kuypers[1],
Mirjam Theelen[1]*
[1] *TNO, Eindhoven, The Netherlands;* [2] *TNO, Eindhoven, The Netherlands*

3DO.12.4 Innovative Design-for-Recycle for Critical Material-Free Interconnection of 953
PV Modules

Antoine Perelman[1], Vincent Barth[1], Fabien Mandorlo[2], Eszter Voroshazi[1]
[1] *CEA-INES, Le Bourget-du-Lac, France;* [2] *INSA, Lyon, France*

3DO.12.5 Bifacial Lightweight Solution without Glass 960

*Alicia Buceta[1], Ana Belén Cueli[1], Miguel Aguirre[1], Ana Linares[2], Elena
Llarena[2], Silvia Cal[2], Jaione Bengoechea[1]*
[1] *CENER, Sarriguren, Spain;* [2] *ITER, Granadilla de Abona, Spain*

3DO.12.6 Development of Novel Frontsheets with Protective Coatings to Increase the 966
Durability and Reliability of Glass-free Lightweight PV Modules

Yuliya Voronko[1], Gabriele C. Eder[1], Elisabeth Reiser[2], Markus Babin[3], Gernot Oreski[4]
[1] OFI, Vienna, Austria; [2] KANSAI HELIOS, Vienna, Austria; [3] DTU Electro, Roskilde, Denmark; [4] PCCL, Leoben, Austria

Oral SESSION 3DO.16 PV Module Assessment and Classification

3DO.16.3 Quantitative Description of the Quality of Daylight Electroluminescense (dEL) Images Against Dark Room EL Images   974

Kabir Paul Sulca[1], Carmelo de Castro[1], Diego González-Francés[1], Cristian Terrados[1], Julián Anaya[1], Victor Alonso[1], Miguel Angel González[1], Oscar Mártinez[1]
[1] University of Valladolid, Valladolid, Spain

3DO.16.4 Photovoltaic Cell Defect Classification from Luminescence Images: Embedding and Clustering with Unsupervised Machine Learning   980

Brendan Wright[1], Rama Sharma[1], Ziv Hameiri[1]
[1] UNSW, Sydney, Australia

3DO.16.5 Daylight Photoluminescence of Silicon Solar Panels in Operation by Electrical Modulation   983

Cristian Terrados[1], Diego González-Francés[1], Kabir Paul Sulca[1], C. de Castro[1], Miguel Ángel González[1], Oscar Martínez[1]
[1] University of Valladolid, Valladolid, Spain

Oral SESSION 3DO.17 Outdoor Performance and Energy Yield Estimation

3DO.17.1 PV Module Degradation in Hot Deserts: Laboratory and Outdoor Data Analysis   988

Gerhard Mathiak[1], Shahzada Pamir Aly[1], Kaushal Chapaneri[1], Baloji Adothu[1], Jim Joseph John[1]
[1] DEWA R&D Center, Dubai, United Arab Emirates

3DO.17.2 Incidence Angle Effect: Results of an Interlaboratory Comparison of Measurements on Commercial-Size Modules   995

Mauro Pravettoni[1], Min Hsian Saw[1], Giorgio Bardizza[2], Giovanni Bellenda[3], Romain Couderc[4], Gabi Friesen[3], Werner Herrmann[2], Shin Woei Leow[5], Stefan Riechelmann[6], Flavio Valoti[3], Arvid van der Heide[7], Frank Weinrich[6], Stefan Winter[6]
[1] TII, Abu Dhabi, United Arab Emirates; [2] TÜV-Rheinland, Cologne, Germany; [3] SUPSI, Mendrisio, Switzerland; [4] CEA, Le Bourget-du-lac, France; [5] SERIS, Singapore, Singapore; [6] PTB, Braunschweig, Germany; [7] imec, Genk, Belgium

3DO.17.3 Climate Specific Energy Rating (CSER) Analysis of Outdoor PV Field Data   999

Ismael Medina[1], Teodora S. Lyubenova[1], Ewan Dunlop[1]
[1] European Commission JRC, Ispra, Italy

3DO.17.4 Module Parameters Extraction for Assessing Photovoltaic Energy Yield: A Comparative Approach   1005

Ahmad Hashem[1], Hugo Sanchez[2], Frank Xu[3], SL. Mortazavifar[1], Christos Monokroussos[3], Ralph Gottschalg[4]
[1] *Anhalt University of Applied Sciences, Koethen, Germany;* [2] *Anhalt University of Applied SciencesUniversity of Applied Sciences, Köthen, Germany;* [3] *TÜV Rheinland, Shanghai, China;* [4] *Anhalt University of Applied Sciences, Köthen, Germany*

3DO.17.5 Performance and Degradation Evaluation of C-Si Modules Under Different Open-Rack and Residential Mounting Configurations   1009

Gabi Friesen[1], Ebrar Özkalay[1], Mauro Caccivio[1]
[1] *SUPSI ISAAC, Mendrisio, Switzerland*

Oral SESSION 3DO.19 Modelling Techniques for PV Modules

3DO.19.1 An Accurate Data-Driven Physical Model for Bifacial PV Power Estimation   1020

Ali Sohani[1], Marco Pierro[2], David Moser[2], Cristina Cornaro[1]
[1] *University of Rome Tor Vergata, Rome, Italy;* [2] *Eurac Research, Bolzano, Italy*

3DO.19.2 Comparative Analysis of Temperature Estimation Models in Bifacial Photovoltaic Modules   1029

Aline Kirsten Vidal de Oliveira[1], Marília Braga[1], Isadora Maciel Queiroz[1], Helena Naspolini[1], Ricardo Rüther[1]
[1] *UFSC, Florianópolis, Brazil*

3DO.19.5 Apparent Intensity Dependence of Shunts in PV Modules - Revision of the Shunt Parameterization in the De Soto Model and PVsyst   1033

Nils-Peter Harder[1], José Cano Garcia[1]
[1] *TotalEnergies, Palaiseau, France*

Oral SESSION 3DO.20 Shading and Soiling on PV Modules

3DO.20.2 Correlating Field Experimentation and Image Analysis for the Assessment of Induced Losses from Thin Object Shading on Photovoltaic Sources   1040

Matthew Axisa[1], Luciano Mule'Stagno[1], Marija Demicoli[1]
[1] *University of Malta, Marsaxlokk, Malta*

3DO.20.3 Laboratory Intercomparison on a Shading Resistance Classification of PV Modules   1046

Stefan Riechelmann[1], Hendrik Sträter[1], Laura Stenzig[1], Giorgio Bardizza[2], Werner Herrmann[2], Ebrar Özkalay[3], Gabi Friesen[3], Özcan Bazkir[4], Alexandra Schmid[5], Stefan Winter[1]
[1] *PTB, Braunschweig, Germany;* [2] *TÜV Rheinland, Cologne, Germany;* [3] *SUPSI, Manno, Switzerland;* [4] *TÜBITAK, Ankara, Türkiye;* [5] *Fraunhofer ISE, Freiburg, Germany*

3DO.20.4 Exploring Dust Particle Properties and PV Soiling Mapping: a Case Study in the Arid Landscape of a Desert Environment   1050

Brahim Aissa[1], Atef Zekri[2], Mosab I. A. Kareem Subeh[1]
[1] *QEERI, Doha, Qatar;* [2] *HBKU, Doha, Qatar*

3DO.20.6 PV Module Cleaning under Hot Desert Conditions   1055

Gerhard Mathiak[1], Afra Seentakath[1], Nithin Sha[1], Shashank Suvarn[1], Prashanth Gabbadi[1], Arumugham Muthusamy[1], Nabeel Ibrahim[1], Kaushal Chapaneri[1]
[1] DEWA R&D Center, Dubai, United Arab Emirates

Oral SESSION 3EO.1 In Field Characterisation of PV Modules | BoS Components in Operation

3EO.1.2 From Fab to Field - Quality Control with a Mobile PV Laboratory   1065

Magnus Herz[1], Hamza Maaroufi[1], Giorgio Bardizza[2]
[1] TÜV Rheinland, Cologne, Germany; [2] TÜV Rheinland, Milan, Italy

3EO.1.3 Reduction of Uncertainty of Outdoor PV Module Characterization: Test Field   1072
Experiences

Mariella Rivera[1], Christian Reise[1]
[1] Fraunhofer ISE, Freiburg, Germany

3EO.1.5 MPP Tracking Losses of Module Level Power Electronics at Partial Module   1079
Shading

Franz P. Baumgartner[1], Markus Klenk[1], Adrian Widler[1], Linus Baumann[1]
[1] ZHAW, Winterthur, Switzerland

3EO.1.6 Improvement of Tracking Algorithms using Machine Learning   1085

Sarra Ben Brahim[1], Kai Saegebarth[1], Martin Dennenmoser[2], Alsayed Algergawy[3]
[1] BayWa r.e., Munich, Germany; [2] BayWa r.e., Freiburg, Germany; [3] University of Passau, Passau, Germany

Oral SESSION 4AO.7 Advanced O&M Strategies and Methods

4AO.7.1 Best Practice Guidelines for the Use of PV System KPIs   1089

Sascha Lindig[1], Magnus Herz[2], Julián Ascencio-Vásquez[3], Marios Theristis[4], Bert Herteleer[5], Julien Deckx[6], Kevin Anderson[7], Karel De Brabandere[6], Erik Stensrud Marstein[8]
[1] UNIVERS, Munich, Germany; [2] TÜV Rheinland, Cologne, Germany; [3] UNIVERS, Redwood, United States of America; [4] Sandia, Albuquerque, United States of America; [5] KU Leuven, Leuven, Belgium; [6] 3E, Brussels, Belgium; [7] NREL, Golden, United States of America; [8] IFE, Lillestrøm, Norway

4AO.7.2 Hybrid Decision Support System: a Framework for Data-Driven   1095
Troubleshooting and Reporting

Sandra Gallmetzer[1], Mousa Sondoqah[1], Pablo Sebastian Enriquez Paez[2], Atse Louwen[1], David Moser[1]
[1] EURAC Research, Bolzano, Italy; [2] BayWa r.e., Rome, Italy

4AO.7.3 Identifying Distinct Performance Patterns in Utility-Scale Photovoltaic Plants   1104
Using an Unsupervised Machine Learning Model

Ali Shakiba[1], Brendan Wright[1], Ziv Hameiri[1]
[1] UNSW, Sydney, Australia

4AO.7.5 Enhancing Fault Diagnosis in Photovoltaic Plants: a Comprehensive 1108
Approach to Simultaneous Failures

Giosué Maugeri[1], Salvatore Guastella[1], Andrea Rossetti[1]
[1] RSE, Milan, Italy

4AO.7.6 Design and Application of Intelligent Scalable Automatic Fault Detector for 1118
Commercial Photovoltaic Systems

Mücahid Candan[1], David Melgar[1], Christian Schill[1], Mete Çubukçu[2],
Eduardo Sarquis Filho[3], Björn Müller[3], Duarte Kazacos[4]
[1] Fraunhofer ISE, Freiburg, Germany; [2] Solar Energy Institute of Ege University, Bornova,
Türkiye; [3] Enmova, Freiburg, Germany; [4] Mondas, Freiburg, Germany

Oral SESSION 4AO.8 PV Plant Performance, Analysis, Monitoring and Fault Detection in Inverters

4AO.8.1 Uncertainty-Aware Estimation of Inverter Field Efficiency Using Bayesian 1124
Neural Networks in Solar Photovoltaic Plants

Gerardo Guerra[1], Pau Mercade-Ruiz[1], Gaetana Anamiati[1], Lars Landberg[2]
[1] GreenPowerMonitor a DNV Company, Barcelona, Spain; [2] DNV Denmark, Hellerup,
Denmark

4AO.8.2 Analysis of Fault Detection and Defect Categorization in Photovoltaic 1132
Inverters for Enhanced Reliability and Efficiency in Large-scale Solar Energy
Systems

Stephanie Malik[1], David Daßler[1], Dharm Patel[1], Carola Klute[1], Robert
Klengel[1], Andreas Dietrich[2], Kai Kaufmann[3], Carsten Hennig[4], Danny
Wehnert[5], Matthias Ebert[1], Leonard Kraft[5]
[1] Fraunhofer IMWS, Halle (Saale), Germany; [2] DiSUN, Werder (Havel), Germany; [3]
DENKweit, Halle (Saale), Germany; [4] saferay holding, Berlin, Germany; [5] Leipziger
Energiegesellschaft, Leipzig, Germany

4AO.8.3 Anomaly Detection in Similarly Behaving Solar Inverters 1141

Pau Mercade Ruiz[1], Gerardo Guerra[1], Gaetana Anamiati[1], Lars Landberg[2]
[1] GreenPowerMonitor, Barcelona, Spain; [2] DNV Denmark, Copenhagen, Denmark

4AO.8.6 Towards Higher Efficiency: Data Analysis and Optimization of PV String 1146
Wiring in a Long-Running Solar Power Plant

Žiga Miklič[1], Janez Krč[1], Marko Topič[1]
[1] University of Ljubljana, Ljubljana, Slovenia

Oral SESSION 4AO.9 The Impact of Soiling on PV Systems

4AO.9.3 Qatar Dust Atlas Project: Deployment of a National Field Soiling and 1156
Environmental Parameters Monitoring Network

Brahim Aissa[1], Mohamed Abdelrahim[2], Mosab Subeh[1], Amir A. Abdallah[1],
Benjamin W. Figgis[1], Juan Lopez-Garcia[1], Veronica Bermudez Benito[1]
[1] QEERI, Doha, Qatar; [2] Bin Omran Trading & Telecommunications, Doha, Qatar

4AO.9.4 Quality Assurance from Laboratory to Field: Novel Test Solutions for 1160
Soiling-Prone PV Systems

Ioannis (John) Tsanakas[1], Rodrigo Moretón[2], Eric Pilat[1], Jorge Solórzano[3], Kévin Garcia[4]
[1] *CEA - INES, Le Bourget-du-Lac, France;* [2] *QPV, Madrid, Spain;* [3] *Entec Solar, Madrid, Spain;* [4] *CNR, Lyon, France*

4AO.9.6 Degradation Root-Cause Numerical Analysis of Around 100 PV Modules   1168
Installed in Hot and Arid Desert Environment

Shahzada Pamir Aly[1], Kaushal Chapaneri [1], Baloji Adothu[1], Jim Joseph John[1], Gerhard Mathiak[1], Vivian Alberts[1]
[1] *DEWA R&D, Dubai, United Arab Emirates*

Oral SESSION 4BO.16 Technology, Performance and Economics of PV in/on Buildings

4BO.16.1 A Systematic Approach for the Integration of BIPV Planning into the   1172
Construction Planning Process

Frank Ensslen[1], Mona Mühlich[1], Jan-Bleicke Eggers[1], Tilmann E. Kuhn[1], Bruno Bueno[1]
[1] *Fraunhofer ISE, Freiburg, Germany*

4BO.16.2 Semitransparent Bifacial PV Windows with Integrated Blinds: Experimental   1178
and Modelling Results

Simona Villa[1], Martin Hurtado Ellmann[1], Roland Valckenborg[1]
[1] *TNO, Eindhoven, The Netherlands*

4BO.16.3 Cost-Effective Energy Transition: Rooftop PV in European Union Buildings   1183

Carmen Maduta[1], Delia D'Agostino[1], Sofia Tsemekidi-Tzeiranaki[2], Luca Castellazzi[1]
[1] *European Commission JRC, Ispra, Italy;* [2] *NRB, Herstal, Belgium*

4BO.16.5 Dynamic BIPV Shading Systems: Performance Analysis for High TRL   1187
Validation and Market Transfer

Tian Shen Liang[1], Paolo Corti[1], Pierluigi Bonomo[1], Francesco Frontini[1]
[1] *SUPSI, Mendrisio, Switzerland*

4BO.16.6 Study on Improvement of Power Generation for a Window by Solar   1194
Radiation Reflected from the Low-E Coating of a Semi-Transparent
Photovoltaic Module that is Equally Arranged Linear Double-Sided Solar
Cells

Kazuhiko Umeda[1], Nobusato Kobayashi[1], Akira Yamaguchi[1], Akihiko Nakajima[2], Kengo Maeda[1], Akihiro Kuraoka[2], Naoki Kadota[2]
[1] *TAISEI, Tokyo, Japan;* [2] *KANEKA, Tokyo, Japan*

Oral SESSION 4BO.17 Characterisation, Reliability and Safety of PV in/on Buildings

4BO.17.3 Experimental Investigation of the Temperature Distribution in a BIPV Facade   1199

Nanna Lysgaard Andersen[1], Markus Babin[1], Sune Thorsteinsson[1]
[1] *DTU Electro, Roskilde, Denmark*

Oral SESSION 4BO.6 Performance and Degradation of PV Systems

4BO.6.4 Trend-Based Predictive Maintenance and Fault Detection Analytics for Photovoltaic Power Plants 1209

Demetris Marangis[1], Andreas Livera[1], George Makrides[1], George E. Georghiou[1]
[1] *University of Cyprus, Nicosia, Cyprus*

4BO.6.6 PV Module Operating Temperature: Reliable Extraction of Model Parameters from Dynamic Field Data 1214

Anton Driesse[1], Jesus Polo[2]
[1] *PV Performance Labs, Freiburg, Germany;* [2] *CIEMAT, Madrid, Spain*

Oral SESSION 4BO.7 Data Driven Field Inspection based on Imaging

4BO.7.1 From Pixels to Insights: A Software Prototype for AI-Driven Complete Diagnostics of PV Plants 1220

John (Ioannis) A. Tsanakas[1], Murielle Stepec[1], Philippe Marechal[1], Duy-Long Ha[1]
[1] *CEA - INES, Le Bourget-du-Lac, France*

4BO.7.3 Redefining Failure Detection in PV Systems: A Comparative Study of GPT-4o and ResNet's Computer Vision in Aerial Infrared Imagery Analysis 1225

Sandra Gallmetzer[1], Lukas Koester[1], Evelyn Turri[1], Mousa Sondoqah[1], Atse Louwen[1], David Moser[1]
[1] *EURAC, Bolzano, Italy*

4BO.7.4 Evaluation of Field Measurements on Hail Damage to Photovoltaic Modules 1234

Evelyn Bamberger[1], Alexandre Voirol[1]
[1] *OST, Rapperswil, Switzerland*

4BO.7.5 Evaluation of Daylight Filters for Electroluminescence Imaging Inspections of c-Si PV Modules 1241

Gisele Alves dos Reis Benatto[1], Thøger Kari[1], Rodrigo Del Prado Santamaria [1], Aysha Mahmood[1], Liviu Stoicescu[2], Sergiu V. Spataru[1]
[1] *DTU, Roskilde, Denmark;* [2] *Solarzentrum Stuttgart, Stuttgart, Germany*

Visual SESSION 4BV.3 Operation, Performance and Maintenance of PV Systems

4BV.3.1 In-Situ Maintenance-Free Measurement of Soiling-Induced Power Losses in PV Arrays 1245

Michael Gostein[1], Damien Cosme[2], Quentin Berthet-Rayne[2], Julien Chapon[3], Lluvia Ochoa[3], William Stueve[1], Dhanup Somasekharan Pillai[4], Brahim Aïssa[4], Benjamin W. Figgis[4], Juan Lopez-Garcia[4], Veronica Bermudez Benito[4]

[1] Atonometrics, Austin, United States of America; [2] TotalEnergies, Doha, Qatar; [3] TotalEnergies, Paris, France; [4] QEERI, Doha, Qatar

4BV.3.3 Improving Performance Ratio Calculations through Optimizing Front POA Irradiance Sensor Positioning — 1249

Marc A. N. Korevaar[1], Damon Nitzel[1], Shuo Wang[2], Nate Solofra[3]
[1] OTT Hydromet, Delft, The Netherlands; [2] TUAS, Turku, Finland; [3] Merit Controls, Somerville, United States of America

4BV.3.6 A Method for Detecting PV Module's Degradation due to Increased Local Resistance in Power Plant — 1253

Tohru Kohno[1], Jun Tsunoda[1]
[1] Hitachi, Tokyo, Japan

4BV.3.8 Dependence of Series Resistance on Ideality Factor and Shunt Resistance in Online Photovoltaic Module Parametric Identification — 1258

Heidi Kalliojärvi[1], Kari Lappalainen[1]
[1] Tampere University, Tampere, Finland

4BV.3.9 Predictive Maintenance and Anomaly Detection Analytics for Utility-Scale Photovoltaic Plants — 1264

Jesus Montes-Romero[1], Demeteris Marangis[2], Andreas Livera[2], George Makrides[2], Juergen Sutterlueti[3], Steve Ransome[4], George E. Georghiou[2], Nino Heinzle[3]
[1] University of Jaen, Jaen, Spain; [2] University of Cyprus, Nicosia, Cyprus; [3] Gantner Instruments, Schruns, Austria; [4] Steve Ransome Consulting, Kingston upon Thames, United Kingdom

4BV.3.12 Safety Analysis of PV Systems for Soundproof Tunnel Based on Voltage and Current Mismatch — 1269

Juhee Jang[1], Chongmin Kim[1], Sujeong Oh[1]
[1] Korea Electrical Safety, Wanju, South Korea

4BV.3.14 Improved Modelling of PV Systems with Snow Soiling for Optimized Local Energy Sharing — 1273

Ida Fuchs[1], Ole-Morten Midtgård[1]
[1] NTNU, Trondheim, Norway

4BV.3.15 Ensuring Photovoltaic Module Integrity through Electroluminescence Imaging and Machine Learning Solutions — 1279

Daniel J. Castillo Patton[1], Lucas Viani[1], Fernando García[2], Vicente Parra[1], Sofía Rodríguez-Conde[1], Jesús Cuaresma[1]
[1] Enertis Applus+, Madrid, Spain; [2] UC3M, Madrid, Spain

4BV.3.16 RACONT2050 - Reliability and Comparison of New PV Technologies — 1286

Domenico Chianese[1], Mauro Caccivio[1], Gabi Friesen[1]
[1] SUPSI, Mendrisio, Switzerland

4BV.3.18 Comparative Analysis of String IV Measurement Methods for Fault Detection in Photovoltaic Systems — 1287

Martin Bartholomäus[1], Peter Behrensdorff Poulsen[1], Sergiu Viorel Spataru[1]
[1] Technical University of Denmark, Roskilde, Denmark

4BV.3.22 AI-SafePV: An AI-Based Fault Detection Software Package to Provide Safety in Photovoltaic Arrays — 1293

Aref Eskandari[1], Jafar Milimonfared[2], Amir Nedaei[2], P. Parvin[2], M. Braga[3], Mohammadreza Aghaei[4]
[1] *Iran University of Science and Technology, Tehran, Iran;* [2] *Amirkabir University of Technology, Tehran, Iran;* [3] *UFSC, Florianópolis, Brazil;* [4] *NTNU, Ålesund, Norway*

4BV.3.23 DetectivePV: A Detection Package for Electrical Faults in Photovoltaic Arrays based on Machine Learning   1297

Aref Eskandari[1], Jafar Milimonfared[2], Amir Nedaei[2], P. Parvin[2], M. Braga[3], Mohammadreza Aghaei[4]
[1] *Iran University of Science and Technology, Tehran, Iran;* [2] *Amirkabir University of Technology, Tehran, Iran;* [3] *UFSC, Florianópolis, Brazil;* [4] *NTNU, Ålesund, Norway*

4BV.3.24 Wet Leakage and Insulation Test on String Level Through IEC 61215   1301

Mario Martínez[1], Sergio Suarez[1], Jose Cantisano[1], Jonathan Vilela[1], Jose Maria Alvarez[1], Jose Manuel Rivas[1], Sofia Rodríguez-Conde[1]
[1] *Enertis Applus, Madrid, Spain*

4BV.3.25 TALOS: Robotics and Artificial Intelligence Living Labs Improving Operations in PV Scenarios   1304

Nicolas Congouleris[1], Athanasios T. Balafoutis[1], Lisandro Puglisi[2], João Formiga[3], Daniel Albuquerque[3], Bruno Barrionuevo[1]
[1] *CERTH, Thermi, Greece;* [2] *EDP Renewables, Madrid, Spain;* [3] *EDP NEW, Lisbon, Portugal*

4BV.3.26 Harmonising Multi-Sites Measurement of Photovoltaic Systems: Comprehensive Framework for Real-Life Test Conditions in a Maltese Environment   1305

Brian Bartolo[1], Brian Azzopardi[1], Alexandre Mignonac[2], Marcus Rennhofer[3], Bernhard Kubicek[3], Rita Ebner[3], Carlos Meza[4], Melodie de l'Epine[5], Eugenia Zugasti[6], Steve Zerafa[7], Kenneth Scerri[8]
[1] *The Foundation for Innovation and Research, Birkirkara, Malta;* [2] *CEA, Cadarache, France;* [3] *AIT, Vienna, Austria;* [4] *Anhalt University of Applied Sciences, Anhalt, Germany;* [5] *Becquerel Institute, Brussels, Belgium;* [6] *CENER, Pamplona, Spain;* [7] *PIXAM, Msida, Malta;* [8] *The University of Malta, Msida, Malta*

4BV.3.27 Mediterranean Climate Impact on Photovoltaic Systems: Insights from Malta and Implications for Future European Integration   1309

Brian Bartolo[1], Brian Azzopardi[1], Alexandre Mignonac[2], Marcus Rennhofer[3], Bernhard Kubicek[3], Rita Ebner[3], Carlos Meza[4], Melodie de l'Epine[5], Eugenia Zugasti[6], Steve Zerafa[7], Kenneth Scerri[8]
[1] *The Foundation for Innovation and Research, Birkirkara, Malta;* [2] *CEA, Cadarache, France;* [3] *AIT, Vienna, Austria;* [4] *Anhalt University of Applied Sciences, Anhalt, Germany;* [5] *BI, Brussels, Belgium;* [6] *CENER, Pamplona, Spain;* [7] *PIXAM, Msida, Malta;* [8] *UoM University of Malta, Msida, Malta*

4BV.3.28 Comparison of Physical, Machine Learning and Hybrid Models of Monofacial and Bifacial PV Systems   1313

Jonas Petzschmann[1], Dirk Stellbogen[1], Manuel Heim[1]
[1] *ZSW, Stuttgart, Germany*

4BV.3.29 Quantitative Shade Detection for PV Systems Based on Clearsky Data   1314

Achim Schulze[1], Markus Panhuysen[2], Darwin Daume[3], Maximilian Schönau[2]
[1] *Rosenheim Technical University of Applied Sciences, Rosenheim, Germany;* [2] *Smartblue, Munich, Germany;* [3] *Coburg University of Applied Sciences, Coburg, Germany*

4BV.3.30 Assessing Electroluminescence Image Quality with Machine-Learning and Grey-Level Co-Occurrence Matrix Texture Descriptors — 1317

Thøger Kari[1], Aysha Mahmood[1], Rodrigo del Prado Santamaria[1], Gisele Alves dos Reis Benatto[1], Peter Behrensdorff Poulsen[1], Sergiu V. Spataru[1]
[1] DTU, Roskilde, Denmark

4BV.3.31 Forecasting the Lifetime of Photovoltaic Modules through Coupling a Physics-Based Degradation Model with 3D Heat Transfer Simulations — 1322

Timofey Golubev[1]
[1] ThermoAnalytics, Calumet, United States of America

4BV.3.32 Development of a Model to Ensure the Safety of PV Systems Using FMEA — 1328

Sujeong Oh[1], Chongmin Kim[1], Juhee Jang[1]
[1] KESCO, Wanju County, South Korea

4BV.3.35 Long-Term Monitoring of Degradation and Defect in High-Voltage Strings through Dark I-V Measurements — 1331

Samuele Chiesa[1], Gian Carlo Dozio[1], Domenico Chianese[2]
[1] SUPSI-ISEA, Lugano, Switzerland; [2] SUPSI-ISAAC, Lugano, Switzerland

4BV.3.36 Machine Learning Techniques for the Assesment of Open Circuit Voltage Losses in Photovoltaic Systems — 1335

Sandra Riaño[1], Jose Domingo Santos[1], Miguel Esteras[1], Amaia Abanda[1], Javier del Ser[1]
[1] TECNALIA, Derio, Spain

4BV.3.39 Real-Time Monitoring and Diagnostic of Rooftop Monofacial PV System Validated with Thermography — 1341

Amr Osama[1], Giuseppe Marco Tina[1], Antonio Gagliano[1], Gabino Jiménez-Castillo[2], Francisco Jose Muñoz-Rodriguez[2]
[1] University of Catania, Catania, Italy; [2] University of Jaén, Jaén, Spain

4BV.3.40 Single Image Geospatial Referencing — 1347

Evgenii Sovetkin[1], Andreas Gerber[1], Bernhard Kubicek[2], Bart E. Pieters[1]
[1] Forschungszentrum Jülich, Jülich, Germany; [2] AIT, Vienna, Austria

4BV.3.41 Outdoor Exposure Study on the Performance of Nine Different Types of Industrial PV Modules under 35° and under 90° Tilt — 1348

Carolin Ulbrich[1], Niklas Albinius[1], Luka Wernke[1], Björn Rau[1], Rutger Schlatmann[1]
[1] HZB, Berlin, Germany

4BV.3.43 Photovoltaic Output Power Modeling: a Hybrid Approach — 1354

Leticia de Oliveira Santos[1], Francisco Alexandre Andrade Souza[2], Tarek AlSkaif[3], Paulo C. M. Carvalho[1]
[1] UFC, Fortaleza, Brazil; [2] imec-NL, Wageningen, The Netherlands; [3] Wageningen University, Wageningen, The Netherlands

4BV.3.45 Estimation of Annual Power Loss of a Solar PV System due to Rise in the Cell Temperature: A Case Study for Indian Climate — 1357

Shubham Kumar[1], P. M. V. Subbarao[1]
[1] IIT Delhi, New Delhi, India

4BV.3.46 Snow Losses for Different PV Module Designs: Modelling and Validation in Southern Finland — 1361

Shuo Wang[1], Hugo E. Huerta[1], Sami Jouttijärvi[2], Aleksi Heinonen[1], Juha A. Karhu[3], Anders V. Lindfors[3], Kati Miettunen[2], Samuli Ranta[1]
[1] *Turku University of Applied Sciences, Turku, Finland;* [2] *University of Turku, Turku, Finland;* [3] *Finnish Meteorological Institute, Helsinki, Finland*

4BV.3.52 Defect Quantification System Through Aerial Inspections — 1365

Mario Martínez[1], Sergio Suarez[1], Daniel Jason[1], Daniel Villoslada[1], Jose Rivas[1], Sofia Rodríguez-Conde[1]
[1] *Enertis Solar, Madrid, Spain*

4BV.3.53 Shaping European Collaboration on Photovoltaics: A Collaborative Platform — 1369
for Simulation and Monitoring (COPLASIMON)

Simone Vitale[1], Jonathan Leloux[2], Hervè Colin[3], Eric Pilat[3], Stéphane Mollier[3], Basem Idlbi[4], Rodrigo Moretón[5], Oscar Anchorena[5], Christophe Salperwyck[6], David Melgar[7], Christian Schill[7]
[1] *LuciSun, Sart Dames Avelines, Belgium;* [2] *LuciSun, Sart-Dames-Avelines, Belgium;* [3] *CEA, Le Bourget-du-Lac, France;* [4] *Ulm University of Applied Sciences, Ulm, Germany;* [5] *Qualifying Photovoltaics, Madrid, Spain;* [6] *MyLight150, Auvergne-Rhône-Alpes, France;* [7] *Fraunhofer ISE, Freiburg, Germany*

Visual SESSION 4BV.4 Photovoltaic in/on Buildings

4BV.4.1 Performance of Vertically Mounted Bifacial Photovoltaics on High-Rise — 1375
Buildings in the Nordic Conditions

Bergpob Viriyaroj[1], Sami Jouttijärvi[2], Matti Jänkälä[1], Kati Miettunen[2]
[1] *Aalto University, Espoo, Finland;* [2] *University of Turku, Turku, Finland*

4BV.4.3 Reducing the Angular Colour Dependence of Building Integrated — 1381
Photovoltaic Modules Based on Optical Interference Coatings

Chang Chuan You[1], Ørnulf Nordseth[1], Arne Røyset[2], Tore Kolås[2]
[1] *Institute for Energy Technology, Kjeller, Norway;* [2] *SINTEF Industry, Trondheim, Norway*

4BV.4.6 Design and Optimization of Structural Colored Interlayers for Building- — 1385
Integrated Photovoltaic Applications

Catarina G. Ferreira[1], Irina Vyalih[1], Jani Lamminaho[1], Markus Babin[2], Nanna Lysgaard Andersen[2], Peter Behrensdorff Poulsen[2], Sune Thorsteinsson[2], Karlis Petersons[3], Joel D. Cox[1], Morten Madsen[1]
[1] *University of Southern Denmark, Odense, Denmark;* [2] *DTU, Roskilde, Denmark;* [3] *Stensborg, Roskilde, Denmark*

4BV.4.8 Comparative Analysis of Individual and Collective PV Integration Strategies — 1390
for a Residential Neighborhood

Qiuxian Li[1], Natasa Vulic[2], Hanmin Cai[2], Philipp Heer[2]
[1] *KU Leuven, Ghent, Belgium;* [2] *Urban Energy Systems Laboratory, Empa, Duebendorf, Switzerland*

4BV.4.10 Modelling Framework for Optimizing Hybrid Photovoltaic-Thermal Systems — 1391
in Combination with Seasonal Heat Storage

Zain Ul Abdin[1], Aron van Rossum[1], David Martinez Aguilera[1], D. N. Kanawala[1], Olindo Isabella[1], Rudi Santbergen[1]
[1] *TU Delft, Delft, The Netherlands*

4BV.4.11 Performance Assessment of Novel Solar Energy Systems for Aged Neighbourhoods and Buildings in Dutch Cities — 1392

Edward Otoo[1], Guang Hu[1], Roel C. G. M. Loonen [1], Angèle H.M. E. Reinders [1]

[1] *Eindhoven University of Technology, Eindhoven, The Netherlands*

4BV.4.12 Steel Framing/Structure as a Solution to Support BIPV Competitiveness — 1396

Simon Boddaert[1], Jean-Pierre Reyal[2], Michel Dernis[3], Philippe Alamy[4]
[1] *CSTB, Sophia Antipolis, France;* [2] *Semperstyl, Eragny Sur Oise, France;* [3] *Atrium Data, Paris, France;* [4] *EnerBim, Donneville, France*

4BV.4.13 Advanced PV and Thermal Modeling for a Feasible and Efficient BAPV-T System Design and Evaluation — 1400

Iñaki Cornago[1], Mikel Ezquer[1], Patxi Sorbet[1], Alicia Kalms[1], Gonzalo Diarce[2], Olatz Irulegi[3], Fritz Zaversky[1]
[1] *CENER, Sarriguren, Spain;* [2] *UPV/EHU, Bilbao, Spain;* [3] *UPV/EHU, San Sebastian, Spain*

4BV.4.14 PV on Green Roofs. Two Years of Comparative Measurement Data from Various System Concepts, Supplemented by Simulation Results and General Considerations — 1401

Markus Klenk[1], Roger Glarner[1], Selina Pfyffer[1], Hartmut Nussbaumer[1], Stephan Brenneisen[1], Andreas Dreisiebner[2]
[1] *ZHAW, Winterthur, Switzerland;* [2] *A777 Gartengestaltung, Seuzach, Switzerland*

4BV.4.20 PV Façades > 30 m - Fire Prevention Guidelines on High-Rise Buildings — 1409

Urs Muntwyler[1], Eva Schüpbach[1]
[1] *Dr. Schuepbach & Muntwyler, Bern, Switzerland*

4BV.4.22 Semi-Transparent CIGS Thin-Film PV Modules — 1412

Peter Borowski[1], Thomas Schutt[2], Julian Röder[1], Maik Schubert[2], Martin Hillmann[2], Kristian Herath[2], Subarna Sapkota[2], Volker Speer[2], Marko Stölzel[1], Rene Reichel[2], Thomas Dalibor[1]
[1] *AVANCIS, Munich, Germany;* [2] *AVANCIS, Torgau, Germany*

4BV.4.23 Assessing Photovoltaic-Thermal System Performance across Diverse Climates: an Economic and Environmental Comparative Analysis — 1417

Zain Ul-Abdin[1], Olindo Isabella[1], Rudi Santbergen[1]
[1] *TU Delft, Delft, The Netherlands*

4BV.4.26 A Strategic Approach to Enable Large-Scale Photovoltaic Energy Systems Deployment in Urban Areas — 1418

Joyce Arthllan Oliveira de Sousa[1], Martin Thebault[1], Lamia Berrah[1]
[1] *USMB, Annecy, France*

4BV.4.27 Performance Assessment of Colorful BIPV Facade in Norway — 1430

Junjie Zhu[1], Jørgen Young[2]
[1] *Institute for Energy Technology, Kjeller, Norway;* [2] *Isola Solar, Larvik, Norway*

4BV.4.28 Implementing Strain Relief for Improved Reliability of BIPV Modules Built on Aluminum Façade Elements — 1433

Wiebke Wirtz[1], Kevin Meyer[1], Susanne Blankemeyer[1], Thomas Daschinger[1], Henning Schulte-Huxel[1]
[1] *ISFH, Emmerthal, Germany*

4BV.4.29 CONIPHER BIPV Facades: Design and Performance Prediction — 1434

Ya-Brigitte Assoa[1], Philippe Thony[1], Emmanuel Schmitt[2], Olivier Bizzini[3], Stephane Gelibert[3], Vincent Bressy[4], Olivier Wiss[1], Alexandre Plissonnier[1], Zeina Hamam[1]
[1] CEA, Le Bourget-du-Lac, France; [2] Vicat, L'Isle-d'Abeau, France; [3] Araymond, Grenoble, France; [4] Workspaces-architecture, Grenoble, France

4BV.4.30 The Potential of Plug&Play PV in Switzerland   1435

Jan Remund[1], Anne-Kathrin Weber[1], Lukas Meyer[1], David Joss[2], Christof Bucher[2], Theo Zwahlen[2]
[1] Meteotest, Bern, Switzerland; [2] BFH, Burgdorf, Switzerland

4BV.4.32 Integration of Transparent Photovoltaic Panels into Buildings   1439

Nilşah Özar[1], Müjde Altın[1]
[1] Dokuz Eylül University, Izmir, Türkiye

4BV.4.33 Integrating FIDES Reliability Prediction into Building-Integrated Photovoltaic Systems   1443

Fereshteh Poormohammadi[1], Martijn Deckers[2], Johan Driesen[1]
[1] KU Leuven, Leuven, Belgium; [2] Energy Ville, Genk, Belgium

Oral SESSION 4CO.8 Solar Resource Assesment

4CO.8.4 Fast Horizon Algorithm – Case of Integrated PV   1444

Evgenii Sovetkin[1], Andreas Gerber[1], Bart E. Pieters[1]
[1] Forschungszentrum Jülich, Jülich, Germany

4CO.8.5 Global Patterns of Solar Resource Short-Term Variability Based on Solargis Time Series Data   1451

Juraj Betak[1], Martin Opatovsky[1], Konstantin Rosina[1], Marcel Suri[1]
[1] Solargis, Bratislava, Slovakia

Oral SESSION 4CO.9 Solar Forecasting

4CO.9.1 Can Deep Learning Replace Cloud Motion Vectors?   1456

Nils Straub[1], Steffen Karalus[1], Wiebke Herzberg[1], Elke Lorenz[1]
[1] Fraunhofer ISE, Freiburg, Germany

4CO.9.2 Skill-Driven Model Training for Solar Forecasting with Sky Images   1462

Amar Meddahi[1], Arttu Tuomiranta[1], Sebastien Guillon[1]
[1] TotalEnergies, Palaiseau, France

4CO.9.3 Ramp Rate Metric Suitable for Solar Forecasting and Nowcasting   1466

Bijan Nouri[1], Yann Fabel[1], Niklas Blum[1], Dominik Schnaus[2], Luis F. Zarzalejo[3], Andreas Kazantzidis[4], Stefan Wilbert[1]
[1] DLR, Almería, Spain; [2] TUM, Munich, Germany; [3] CIEMAT , Madrid, Spain; [4] University of Patras, Patras, Greece

4CO.9.4 Fog and Snow Detection to Improve Regional Photovoltaic Power Prediction   1476

Elke Lorenz[1], Steffen Karalus[1], Wiebke Herzberg[1], Tobias Zech[1], Babak Jahani[2], Eva Pauli[3], Jan Cermák[3], Tjade Appel[4], Merle Vespermann[4], Heidrun Misfeld[4], Jan Kühnert[4]

[1] *Fraunhofer ISE, Freiburg, Germany;* [2] *SRON, Leiden, The Netherlands;* [3] *KIT, Karlsruhe, Germany;* [4] *energy & meteo systems, Oldenburg, Germany*

4CO.9.5   Photovoltaic Power Plants as Efficient Cloud Motion Detectors   1480

Magnus Moe Nygård[1], Erling Ween Eriksen[1], Heine Nygard Riise[1]
[1] *IFE, Kjeller, Norway*

4CO.9.6   How Connected Cars can Improve Solar Forecasting - Expanding the Scale of   1491
Local Sensor Networks

Tobias Veihelmann[1], Maximilian Lübke[1], Norman Franchi[1]
[1] *Friedrich-Alexander-University, Erlangen, Germany*

Plenary SESSION 4CP.1 PV Everywhere

4CP.1.1   Dynamic Agrivoltaics: An Agronomical Tool to Protect Crops from Climate   1496
Change - Feedback from 15 Years of Research

*Damien Fumey[1], Sophie Bellacicco[1], Gerardo Lopez-Velasco[1], Jérôme
Chopard[1], Severine Persello[1], Perrine Juillion[1], Vincent Hitte[1], Yassin
Elamri[1], Isaac A. Ramos-Fuentes[1], Jean Garcin[2], Benoît Valle[2], Francis
Sourd[2]*
[1] *Sun'Agri, Paris, France;* [2] *Sun'R, Paris, France*

Plenary SESSION 4CP.2 Performance and Reliability | Thin Films and Tandems

4CP.2.3   Performance of Partial Shaded PV Generators Operated by Optimized Power   1501
Electronics an IEA PVPS T13 Activity

*Franz P. Baumgartner[1], Sara Golroodbari[2], Christof Bucher[3], Matthew
Berwind[4], Felipe Valencia[5], Ulrike Jahn[6]*
[1] *ZHAW, Winterthur, Switzerland;* [2] *University Utrecht, Utrecht, The Netherlands;* [3] *Bern
University, Bern, Switzerland;* [4] *Fraunhofer ISE, Freiburg, Germany;* [5] *ATAMOSTEC,
Atacama, Chile;* [6] *Fraunhofer CSP, Halle (Saale), Germany*

Visual SESSION 4CV.1 Solar Resource and Forecasting

4CV.1.4   Hindcasting Solar Irradiance by Machine Learning using Photovoltaic Data   1509

*Maximilian Schönau[1], Darwin Daume[2], Markus Panhuysen[1], Tristan Kreller[2],
Joseph Jachmann[2], Achim Schulze[3], Bernd Hüttl[2], Dieter Landes[2]*
[1] *Smartblue, Munich, Germany;* [2] *Coburg University of Applied Sciences, Coburg, Germany;*
[3] *Rosenheim Technical University of Applied Sciences, Rosenheim, Germany*

4CV.1.5   Climate Clustering for Photovoltaic Interest   1514

Anastasios Kladas[1], Karel Lagast[1], Bert Herteleer[1], Jan Cappelle[1]
[1] *KU Leuven, Leuven, Belgium*

4CV.1.6   Advancing Solar Resource Data: the Validation Journey of 3E's   1516
Satellite-Based Irradiation Data

Philippe Malcorps[1], Gofran Chowdhury[1]
[1] *3E, Brussels, Belgium*

4CV.1.8 Resource-Efficient PV Energy Yield Nowcasting with Sky Images: a Hybrid   1519
Global Annealing Schedule

Markos Kousounadis-Knousen[1], Apostolos Bakovasilis[2], Francky Catthoor[3], Pavlos Georgilakis[1]
[1] *NTUA, Athens, Greece;* [2] *imo-imomec, Genk, Belgium;* [3] *imec, Leuven, Belgium*

4CV.1.9 Variability of Solar Radiation in the Context of a Flat Region Highly Loaded   1524
with Aerosols

Dunia A. Bachour[1], Daniel Perez-Astudillo[1]
[1] *QEERI, Doha, Qatar*

4CV.1.10 Towards Climate-Neutral Energy: Assessing Equations for Optimization of   1528
Photovoltaic Production Estimates

Mahesh Sutariya[1], Luiz Fonseca[2], Raphael Abrahão[2], Haresh Vaidya[1]
[1] *University of Applied Sciences, Feuchtwangen, Germany;* [2] *Federal University of Paraíba, João Pessoa, Brazil*

4CV.1.14 Irradiance Transposition and Reflections in BIPV Installations   1534

Stefan Grünsteidl[1], Peter Borowski[1], Thomas Dalibor[1]
[1] *Avancis, Munich, Germany*

4CV.1.21 Irradiance Modeling for Integrated PV with OpenStreetMap   1543

Michael Gordon[1], Evgenii Sovetkin[1], Bart E. Pieters[1], Andreas Gerber[1]
[1] *Forschungszentrum Jülich, Jülich, Germany*

4CV.1.23 Availability of Solar Energy on Vehicle Roofs in German Road Network;   1544
Validation of Surface Structure Data for Shadow Loss Modelling

Christian Braun[1], Alexander Kleinhans[1], Christian Schill[1], Elke Lorenz[1], Felix Basler[1], Martin Kaiser[1], Nicolas Holland[1]
[1] *Fraunhofer ISE, Freiburg, Germany*

4CV.1.25 Physics Informed Graph Neural Networks for Multi-Site Solar Forecasting   1548

Jelena Simeunovic[1], Baptiste Schubnel[1], Pierre-Jean Alet[1], Pascal Frossard[2], Rafael E. Carrillo[1]
[1] *CSEM, Neuchâtel, Switzerland;* [2] *EPFL, Lausanne, Switzerland*

4CV.1.27 Dimensionality Reduction of Environmental Data for Long-Term PV   1552
Performance Analysis Using Graph Based Methods

Srijani Mukherjee[1], Laurent Vuillon[2], Denys Dutykh[3], Ioannis Tsanakas[1]
[1] *CEA, Le Bourget-du-Lac, France;* [2] *CNRS, Chambéry, France;* [3] *Khalifa University, Abu Dhabi, United Arab Emirates*

4CV.1.28 Statistical Methods for Monitoring Pyranometer Drift in Solar Radiation   1556
Operational Data

Lucas T. Silva[1], Rodrigo S. Queiroz[1], Nathianne M. Andrade[1], Danielle B. Cavalcante[1]
[1] *Delfos Energy, Barcelona, Spain*

4CV.1.29 Performance Evaluation of Utility-Scale Solar PV Projects in the State of   1559
Gujarat, India

Saurabh Motiwala[1], Sudarshan Kumar[1], Ashish Kumar Sharma[2], Ishan Purohit[3]
[1] *IIT Bombay, Mumbai, India;* [2] *University of Petroleum and Energy Studies, Dehradun, India;* [3] *International Finance Corporation, New Delhi, India*

Oral SESSION 4DO.1 PV System Design and Optimisation

4DO.1.1 Enhancing Bifacial Gain: Addressing Tracker Installation Challenges for 1565
Optimized Performance

*Ismail Kaaya[1], David Moser[2], Richard de Jong[1], Olivier Dupon[1], Arnaud
Morlier[1]*
[1] Imo-Imomec, Genk, Belgium; [2] Eurac Research, Bolzano, Italy

4DO.1.2 Assessing the Performance, Reliability, Economic and Environmental Impact 1576
of PV Systems Installation Parameters in Harsh Climates: Case Study Iraq

*Mohammed Adnan Hameed[1], Ismail Kaaya[2], Richard de Jong[2], Roland
Scheer[3], Ralph Gottschalg[1]*
*[1] Fraunhofer CSP, Halle (Saale), Germany; [2] Imec, Genk, Belgium; [3] MLU, Halle (Saale),
Germany*

4DO.1.3 A Techno-Economic Comparison Analysis for Optimal PV Revamping 1584
Strategies

Elina Bosch[1], Philippe Macé[1], Caroline Plaza[2], Gaëtan Masson[1]
[1] Becquerel Institute, Brussels, Belgium; [2] Becquerel Institute, Lyon, France

4DO.1.6 Innovative Setups for Photovoltaic Solar Trackers to Really Boost the 1590
Electricity Generation per Square Meter of Occupied Surfaces

Rosario Carbone[1], Cosimo Borrello[1], Ferdinando Gioia[1]
[1] University "Mediterranea" of Reggio Calabria, Reggio Calabria, Italy

**Oral SESSION 4DO.2 The Integrated Agrivoltaic Performance: Approaches,
Modelling, Experiences**

4DO.2.1 Europe's Agrivoltaic Future: Design of Four Innovative Demonstrators 1596
through Advanced Modeling in the SYMBIOSYST Project

*S Prithivi Rajan[1], Jesus Robledo[1], Jonathan Leloux[1], Christian A.
Gueymard[1], Angelo Pignatelli[2], Giovanni Borz[3], David Moser[3], Ismail
Kaaya[4], Shu-Ngwa Asaa[4], Alexandros Katsikogiannis[5], Martin Thalheimer[6],
Walter Guerra[6], Marcel Macarulla[7], Irma Roig[7], Gil Gorchs[7], Niels Groen[8],
James MacDonald[9], Giuseppe Demofonti[10], Cinja Seick[11], Giacomo Bosco[12]*
*[1] LuciSun, Brussels, Belgium; [2] EF Solare, Milano, Italy; [3] EURAC, Bolzano, Italy; [4] Imec,
Leuven, Belgium; [5] TU Delft, Delft, The Netherlands; [6] Laimburg, Laimburg, Italy; [7] UPC,
Barcelona, Spain; [8] KUBO, South Holland, The Netherlands; [9] Engie-Lab, Barcelona, Spain;
[10] Convert, Roma, Italy; [11] Aleo, Prenzlau, Germany; [12] Physee, Delft, The Netherlands*

**Oral SESSION 4DO.3 The Integrated Agrivoltaic Performance: Different
Climatic Conditions, Crops and Technologies**

4DO.3.3 A Computational Comparison and Validation Between Ray Tracing 1606
Techniques Under Special Light-Sharing Trade off Scenarios in Photovoltaics

Hugo Sánchez Ortiz[1], Roxane Bruhwyler[2], Sebastian Dittmann[1], Nicolas De Cook[2], Carlos Meza[1], Frederic Lebeau[2], Ralph Gottschalg[1]
[1] *Hochschule Anhalt, Koethen, Germany;* [2] *Liege University, Gembloux, Belgium*

4DO.3.4   Automatic Agrivoltaic Site Selection: a User-Friendly Interface powered by AHP Multicriteria Decision-Making   1610

Andressa de Sousa Cardoso[1], Alfonso López Ruiz[1], María Isabel Ramos Galán[1], Juan Manuel Jurado[1], Francisco Ramón Feito Higueruela [1]
[1] *University of Jaén, Jaén, Spain*

Oral SESSION 4DO.4 Vehicle Integrated PV

4DO.4.1   SolarMoves: The Impact on Grid Electricity Demand of VIPV   1616

Anna J. Carr[1], Ashish Binani[1], Akshay Bhoraskar[2], Oscar van de Water[2], Michiel Zult[2], René van Gijlswijk[2], Lenneke Slooff-Hoek[1]
[1] *TNO, Petten, The Netherlands;* [2] *TNO, Den Haag, The Netherlands*

4DO.4.3   Simulation and Concept Evaluation of Extendable Lightweight Photovoltaic Modules for Vehicle Integration under Wind Loads   1620

Cornelius Herr[1], Marc Andre Schüler[1], Felix Basler[1], Christopher Daniel Joseph[1], Andreas Beinert[1], Pascal Romer[1], Martin Heinrich[1]
[1] *Fraunhofer ISE, Freiburg, Germany*

4DO.4.6   VIPV: Urban Shading Effect to Solar Irradiation Estimation Method Using GIS: Case Study in Fukushima, Japan   1626

Pawita Bunme[1], Hidenori Mizuno[1], Takumi Takashima[1], Takashi Oozeki[1]
[1] *AIST, Fukushima, Japan*

Oral SESSION 4DO.5 Floating, Integrated and Hybrid PV

4DO.5.2   Exploiting the Full Performance Potential of (Offshore) Floating Photovoltaics through Thermal Approaches: an Overview of Options   1634

Oscar Delbeke[1], Jens D. Moschner[1], Johan Driesen[1]
[1] *KU Leuven/EnergyVille, Leuven, Belgium*

4DO.5.5   Performance of Zigzag Photovoltaics Noise Barrier near a Belgian Highway   1638

Sara Bouguerra[1], Richard de Jong[1], Philip Le[1], Fabio Di Giusto[1], Fallon Colberts[2], Ismail Kaaya[1], Nikoleta Kyranaki[1], Marta Casasola Paesa[1], Elke Deckers[1], Arnaud Morlier[1], Michaël Daenen[1]
[1] *IMO-IMOMEC, Diepenbeek, Belgium;* [2] *Zuyd University, Heerlen, The Netherlands*

4DO.5.6   Hybrid (Tandem?) Implementation: Solar Spectrum Splitting PV/CSP for Thermal and Electrical Energy Harvesting   1647

Jonathan Govaerts[1], Bart Reekmans[1], Patrick Choulat[1], Filip Duerinckx[1], Loic Tous[1], Bin Luo[1], Tom Borgers[1], Hariharsudan Sivaramakrishnan Radhakrishnan[1], Jef Poortmans[1], Hannes Laget[2], Qizheng Dou[3], Francis Costa[3], Lieven Stalmans[3], Ravi Kishore[4], Youri Meuret[4], Georgi H. Yordanov[4], Jens Moschner[4], Tatjana Vavilkin[5], Stefan Dewallef[5]
[1] *imec, Genk, Belgium;* [2] *Azteq, Genk, Belgium;* [3] *Borealis, Beringen, Belgium;* [4] *KULeuven, Leuven, Belgium;* [5] *Soltech, Genk, Belgium*

Visual SESSION 4DV.1 Dual Use (Floating PV, Agrivoltaics, VIPV) and other Innovative PV Applications

4DV.1.3 Bifacial Panels for Agrivoltaics and Crop Influence: Expected Benefits — 1651

Miguel-Ángel Muñoz-García[1], María Beatriz Nieto[2], Guillermo Pedro Moreda[1], Carmen Alonso-García[2], Luís Fialho[3], Fátima Baptista[3]
[1] UPM, Madrid, Spain; [2] CIEMAT, Madrid, Spain; [3] University of Évora, Évora, Portugal

4DV.1.4 Analysis of the Use of Bifacial Solar Panels in Vertical Placement and their Temporal Coupling in Agrivoltaic Irrigation — 1654

Guillermo-Pedro Moreda[1], Raúl Sánchez-Calvo[1], Luis Juana[1], Delia Rodríguez-Lucas[2], Miguel-Ángel Muñoz-García[1]
[1] UPM, Madrid, Spain; [2] Harvard University, Cambridge, United States of America

4DV.1.6 Design and Methodology for an Agrovoltaic Pilot Project in the Alentejo Region — 1658

Helena Oliveira[1], Lisa Bunge[1], José A. Silva[1], Luís Fialho[1], Paulo Infante[1], Pedro Horta[1]
[1] University of Évora, Évora, Portugal

4DV.1.8 Growing Greener. First Step on the Journey to Maximize Agri-Voltaic Potential. The SYMBIOSYST Project: Monitoring System and Platform — 1663

Giovanni Borz[1], Enrico Dalla Maria[1], David Moser[1], Maitheli Nikam[2], Gofran Chowdhury[2], Alba Perez[3], David Caballero[3], Niels Groen[4], Jennifer Porter[5]
[1] Eurac Research, Bolzano, Italy; [2] 3E, Brussels, Belgium; [3] Universitat Politecnica de Catalunya, Barcelona, Spain; [4] KUBO Greenhouse Projects, Monster, The Netherlands; [5] Above Surveying, Colchester, United Kingdom

4DV.1.9 Assessing the Agrivoltaic Potential in Hot Desert Climates — 1664

Juan Lopez-Garcia[1], Sachin Jain[1], Daniel Perez-Astudillo[1], Dunia Bachour[1], Dhanup Pillai[1], Veronica Bermudez-Benito[1]
[1] HBKU, Doha, Qatar

4DV.1.10 AgriPV in Norway: Evaluating the Initial Performance and Lessons Learned — 1669

Steve Völler[1], Marisa Di Sabatino[1], Richard J. Randle-Boggis[2], Gaute Stokkan[2]
[1] NTNU, Trondheim, Norway; [2] SINTEF Industry, Trondheim, Norway

4DV.1.12 Dual-Use Potential of Agrivoltaics in Portugal – a Case Study in Baixo Alentejo — 1675

Cláudia Fernandes[1], Jose Almeida Silva[2], Jeremias dos Santos[2], Lisa Bunge[2], André Soeiro[3], Luís Fialho[2], Pedro Horta[2], Daniel Albuquerque[1], Filipe Serra[1], Diogo Cordeiro[3]
[1] EDP NEW, Sacavém, Portugal; [2] University of Évora, Évora, Portugal; [3] EDP Generation, Lisbon, Portugal

4DV.1.16 IEA HEV TCP PVPS Task 17: VIPV Business Plan - the Long Way to the Mass Market — 1679

Urs Muntwyler[1], Eva Schüpbach[1]
[1] Dr. Schüpbach & Muntwyler, Bern, Switzerland

4DV.1.17 Cost-Competitiveness Analysis of Infrastructure Integrated PV 1682

André Penas[1], Elina Bosch[1], Philippe Macé[1], Gaëtan Masson[1], Caroline Plaza[2], Jose Maria Vega de Seoane[3]
[1] Becquerel Institute, Brussels, Belgium; [2] Becquerel Institute, Lyon, France; [3] Becquerel Institute, San Sebastián, Spain

4DV.1.22 Sierra Brava Floating Photovoltaic Plant: Real Data vs Simulation Software 1687

Dorivaldo Duarte[1], Luis Fialho[1], Sara Pereira[1], José Silva[1], Manuel Collares-Pereira[1], Pedro Horta[1], Maria Cebria[2], Nerea Vidal[2]
[1] University of Évora, Évora, Portugal; [2] ACCIONA Energía, Madrid, Spain

4DV.1.23 Numerical Model for Wave Motions and Loads of Multibody Floating Photovoltaic Structures 1691

Antonio Mikulić[1], Ivan Catipovic[1], Neven Alujević[1], Inno Gatin[2]
[1] University of Zagreb, Zagreb, Croatia; [2] Cloud Towing Tank, Zagreb, Croatia

4DV.1.24 Port of Sines Energy Transition: Photovoltaic Solutions Addressing R[4] Concept 1696

Joana Correia[1], Luís Fialho[1], José Silva[1], Pedro Horta[1]
[1] University of Évora, Évora, Portugal

4DV.1.27 Accelerate Product Development for PV in Alpine Installations 1704

Anika Gassner[1], Ebrar Özkalay[2], Gabriele C. Eder[1], Gabi Friesen[2], Markus Feichtner[3], Mauro Caccivio[2], Friedrich Bleicher[4]
[1] OFI, Vienna, Austria; [2] SUPSI PVLab, Mendrisio, Switzerland; [3] Sonnenkraft Energy, St. Veit a.d. Glan, Austria; [4] TU Wien, Vienna, Austria

4DV.1.29 Back Irradiance Measurements and Influence of the Ground Coverage on the Production of a Bifacial Agrivoltaics System 1705

Diogo Vicente[1], Dmitri Boutov[1], João M. Serra[1]
[1] University of Lisbon, Lisbon, Portugal

4DV.1.33 Assessing the Energy Yield and Irradiation Distribution in Fixed and Tracking Agrivoltaic Orchards 1709

Shu-Ngwa Asa'a[1], Ismail Kaaya[1], Olivier Dupon[1], Richard de Jong[1], Arvid van der Heide[1], Arnaud Morlier[1], Hariharsudan Sivaramakrishnan Radhakrishnan[1], Jef Poortmans[2], Michael Daenen[1]
[1] Hasselt University/Imo-Imomec, Genk, Belgium; [2] Imo-Imomec, Genk, Belgium

4DV.1.34 Economic Attractiveness of Agrivoltaics in Different Regulation Statuses – Case Study 1715

Carolina Plaza[1], Julien Van Overstraeten[2], André Penas[2], Elina Bosch[2], Melodie de l'Epine[1], Philippe Macé[2], Gaëtan Masson[2]
[1] Becquerel Institute, Lyon, France; [2] Becquerel Institute, Brussels, Belgium

4DV.1.38 Optimizing Land Productivity with Customized Tracking Algorithms for Single-Axis Trackers in Agrivoltaic Systems 1719

Gaurang Chhapia[1], Djaber Berrian[1], Johannes Linder[1]
[1] Belectric, Kolitzheim, Germany

4DV.1.39 Potential and Techno-Economic Feasibility Assessment of Utility-Scale Floating Solar Photovoltaics (FSPV) in India 1724

Saurabh Motiwala[1], Sudarshan Kumar[1], Ashish Kumar Sharma[2], Ishan Purohit[3]

[1] *IIT Bombay, Mumbai, India;* [2] *University of Petroleum and Energy Studies, Dehradun, India;* [3] *International Finance Corporation, New Delhi, India*

Visual SESSION 4DV.4 PV System Engineering | Control and Systems for Power Systems with Renewables Integration

4DV.4.1 Assessing Glare Hindrance Three Ways in Fixed Tilt PV Systems 1729

Ashish Binani[1], Antonius R. Burgers[1], Kay Cesar[1], Bas Van Aken[1]
[1] *TNO, Petten, The Netherlands*

4DV.4.3 Complementary Guide for the Electrical Design of Grid-Connected PV Systems 1733

Bruno Gaiddon[1], Marielle Perrin[1], Elika Saidi-Chalopin[2], Salomé Durand[3], David Gréau[4], Dimitri Gagnaire[5], Mathieu Mansouri[6], François Saugues[7], Olivier Verdeil[8], Gérard Moine[8]
[1] *Hespul, Lyon, France;* [2] *Consuel, Paris, France;* [3] *SER, Paris, France;* [4] *Enerplan, La Ciotat, France;* [5] *CEATECH-INES, Le Bourget-du-Lac, France;* [6] *CRER, La Crèche, France;* [7] *Stäubli, Hésingue, France;* [8] *Solarcoop, Mornant, France*

4DV.4.5 Increasing the Proportion of Winter Electricity through Design Optimisation of Photovoltaic Roof Systems 1736

Hartmut Nussbaumer[1], Roger Hiltebrand[1], Selina Pfyffer[1], Andreas Dreisiebener[2], Markus Klenk[1]
[1] *ZHAW, Winterthur, Switzerland;* [2] *A777 Gartengestaltung, Seuzach, Switzerland*

4DV.4.7 Implementation of a Sub-Hourly Clipping Correction in PVsyst 1740

Michele Oliosi[1], Bruno Wittmer[1], André Mermoud[1], Agnes Bridel-Bertomeu[1], Robin Vincent[1]
[1] *PVsyst, Satigny, Switzerland*

4DV.4.10 Highest Energy Yields per Area for PV Systems on Flat Roofs 1747

Hartmut Nussbaumer[1], Roger Hiltebrand[1], Selina Pfyffer[1], Lona Tulinski[1], Janis Preisig[1], Markus Klenk[1]
[1] *ZHAW, Winterthur, Switzerland*

4DV.4.12 Impacts of Measures to Achieve Dispatchability on the Cost of PV-BESS Power Plants 1752

Alex Renan Arrifano Manito[1], Pedro Torres[1], Marcelo Pinho Almeida[1], Gilberto Figueiredo[2], José Cesar Almeida[3], Roberto Zilles[1]
[1] *USP, Sao Paulo, Brazil;* [2] *Fluminense Federal University, Niterói, Brazil;* [3] *Mackenzie Presbyterian University, Sao Paulo, Brazil*

4DV.4.13 Analysis of Irradiation Differences on Substring Level of Modules in Solar Parks 1758

Sascha Eckerter[1], Krisztián Kerekes[2], Patrick Mader[2], Rainer Merz[2]
[1] *University of Applied Science Karlsruhe, Ettlingen, Germany;* [2] *University of Applied Science Karlsruhe, Karlsruhe, Germany*

4DV.4.14 Using Standard PV Mounting Structures with Spaced Modules in Agrivoltaic Applications 1763

Alex Renan Arrifano Manito[1], Marcelo Pinho Almeida [1], Bruno Jacomel Vieira[1], Maria Cristina Fedrizzi[1], Roberto Zilles[1]
[1] *USP, São Paulo, Brazil*

4DV.4.15 Optimization Analysis for the Best Sizing and Operation of Photovoltaic   1769
Generators in Distributed Electricity Systems

Jacopo Baldacci[1], Ciro Lanzetta[1], Antonio Piazzi[1], Nabi Taheri[2], Mauro Tucci[2]
[1] i-EM, Livorno, Italy; [2] University of Pisa, Pisa, Italy

4DV.4.19 Optimising Solar Asset Performance through Smart Module Installation using   1770
Above's Digital Twin Technology

Imke Meyer[1], Chisanupong Thawanyavitchajit[2], Inaki Perez[3], Will Hitchcock[4], Henrique Balchada[4], Jennifer Porter[4]
[1] Mott MacDonald, Brighton, United Kingdom; [2] Mott MacDonald, Bangkok, Thailand; [3] Mott MacDonald, Madrid, Spain; [4] Above Surveying, Colchester, United Kingdom

4DV.4.20 Experimental Comparison of Solar Absorption Characteristics Using   1775
Different Colors

Sedong Kim[1]
[1] KITECH, Chungcheongnam-do, South Korea

4DV.4.21 Solar Roof Potential Analysis Case Study: Test Area in South of Germany   1776

Sabrina Krähmer[1], Basem Idlbi[1], Kaouther Belkilani[1], Dietmar Graeber[1]
[1] Ulm University of Applied Sciences, Ulm, Germany

Oral SESSION 4EO.2 Planning of PV Systems | Digital PV

4EO.2.1 BIPV and PV in a Multidisciplinary Building Information Modelling (BIM)   1782
Planning and Asset Management System

Astrid Schneider[1], Karin Stieldorf[1], Christian Schranz[1], Harald Urban[1], Alfred Waschl[2], Markus Feichtner[3], Fedele Rende[4], Andreas Aiello[5], Martin Hauer[6], Kurt Battisti[7], Markus Dörn[7], Jaqueline Scherret[7], Martin Treberspurg[8], Christoph Treberspurg[8]
[1] TU Wien, Vienna, Austria; [2] buildingSMART, Vienna, Austria; [3] Sonnnenkraft Energy, Veith, Austria; [4] ACCA Software, Bagnoli, Italy; [5] ACCA Software, Vienna, Austria; [6] Bartenbach, Aldrans, Austria; [7] A-Null Development, Vienna, Austria; [8] Treberspurg and Partner, Vienna, Austria

4EO.2.2 Assessing Yield Disparities: Anticipated Versus Optimal Rooftop Solar   1790
Photovoltaic Systems and Implications for Prosumer Viability

Dominik Keiner[1], Dmitrii Bogdanov[1], Stefan Krauter[2], Christian Breyer[1]
[1] LUT University, Lappeenranta, Finland; [2] Paderborn University, Paderborn, Germany

4EO.2.3 Energy Yields and Wind Loads of Alternative PV Designs for Roofs in   1799
Snowy Climates

Maria Svedjeholm[1], Josefin Lampa[1], Anna Malou Petersson[1], Arvid Olofsson[1], Robin Andersson[2], Ehsan Fooladgar[1], Pirjo Estola[3], Mattias Lindh[1]
[1] RISE, Umeå, Sweden; [2] Luleå Technical University, Luleå, Sweden; [3] Luleå Energi, Luleå, Sweden

4EO.2.4 Digital Twin of Photovoltaic Power Plants Considering Spatio-Temporal   1806
Characteristics

Faruk Ugranlı[1], Eşref Deniz[2], Engin Karatepe[3]
[1] *Izmir Bakircay University, Izmir, Türkiye;* [2] *Entegro Enerji Sistemleri, Izmir, Türkiye;* [3] *Ege University, Izmir, Türkiye*

4EO.2.5 Fully Privacy Preserving Net-load Prediction with Federated Learning and Homomorphic Encryption   1812

Grazia Barchi[1], Mousa Sondoqah[1], Atse Louwen[1], David Moser[1]
[1] *EURAC, Bolzano, Italy*

Oral SESSION 5CO.4 PV Module Recycling

5CO.4.2 Comparative Analysis of Layer Thickness Measurement Methods for Photovoltaic Modules: a Comprehensive Study   1825

Lukas Neumaier[1], Martin De Biasio[1], Gabriele C. Eder[2], Anika Gassner[2]
[1] *Silicon Austria Labs, Villach, Austria;* [2] *OFI, Vienna, Austria*

5CO.4.4 Characterization of the Output-Fractions from Different Mechanical PV-Recycling Approaches   1828

Anika Gassner[1], Gabriele C. Eder[1], Ferozan Azizi[2], Sonja Feldbacher[3], Friedrich Bleicher[4]
[1] *OFI, Vienna, Austria;* [2] *MUL, Leoben, Austria;* [3] *PCCL, Leoben, Austria;* [4] *TU Vienna, Vienna, Austria*

Oral SESSION 5CO.5 End-of-Life PV Modules & Ecology

5CO.5.2 Comparative Analysis of Recycled Content in Metals used for Photovoltaic Applications   1835

Martina Goverts[1], Simona Villa[2], Mirjam Theelen[2]
[1] *Eindhoven University of Technology, Eindhoven, The Netherlands;* [2] *TNO Energy and Materials Transition, Eindhoven, The Netherlands*

5CO.5.3 PV Module ID: Data Driven Results to Enable PV Circularity and Address Toxicity Concerns   1841

Taylor L. Curtis[1], Ashley Gaulding[1], Ligia Smith[1]
[1] *NREL, Golden, United States of America*

5CO.5.4 Standardisation Activities on the Reuse of PV Modules in IEC TC82   1849

Arvid van der Heide[1], Serge Noels[2], Jan Clyncke[2], Rich Strömberg[3]
[1] *imec/imo-imomec, Genk, Belgium;* [2] *PV CYCLE, Brussels, Belgium;* [3] *University of Alaska Fairbanks, Fairbanks, United States of America*

Oral SESSION 5CO.6 Life Cycle Assessment of PV

5CO.6.1 Sustainability Improvement of C-Si PV Manufacturing through Technology Choices   1853

Moritz Fath[1], Mehul Raval[1], Wolfgang Jooss[1], Peter Fath[1]
[1] *RCT Solutions, Constance, Germany*

5CO.6.3 A Simplified Model to Assess the Greenhouse Gas Emissions of 1859
Perovskite/Silicon Tandem Modules

Lu Wang[1], Paula Perez-Lopez[1], Raphaël Jolivet[1], Mathilde Marchand[1], Lars Oberbeck[2]
[1] *PSL University, Sophia Antipolis, France;* [2] *TotalEnergies, Paris Saclay, Norway*

5CO.6.4 Carbon Footprint vs Reliability of Solar Photovoltaic Modules: A New 1862
Dilemma?

Alessandro Virtuani[1], Alexis Barrou[1], Bertrand Paviet-Salomon[1], Gianluca Cattaneo[1], Matthieu Despeisse[1], Christophe Ballif[1]
[1] *CSEM, Neuchâtel, Switzerland*

5CO.6.5 Are BIPV Contributing to Environmental Sustainability? An Environmental 1873
LCA Analysis of Innovative BIPV Solutions

Cristina Polacchi[1], Atse Louwen[1], Mirjam Theelen[2], David Moser[1]
[1] *Eurac Research, Bolzano, Italy;* [2] *TNO partner in Solliance, Eindhoven, The Netherlands*

5CO.6.6 The Influence of Climate Specific Degradation on the Greenhouse Gas 1885
Emissions of PV Electricity

Karl-Anders Weiß[1], Sina Herceg[1], Marie Fischer[1], Ismail Kaaya[2], Julian Ascencio-Vásquez[3], Liselotte Schebek[4]
[1] *Fraunhofer ISE, Freiburg, Germany;* [2] *EnergyVille, Genk, Belgium;* [3] *Envision Digital, Redwood, United States of America;* [4] *Technical University of Darmstadt, Darmstadt, Germany*

Plenary SESSION 5CP.1 PV Everywhere

5CP.1.2 Where Agriculture meets Energy: Assessing EU's Agrivoltaic Potential 1894

Anatoli Chatzipanagi[1], Georgia Kakoulaki[1], Nigel Taylor[1], Robert Kenny[1], Sandor Szabó[1], Ana Martinez Fernandez[1], Arnulf Jaeger-Waldau[1]
[1] *European Commission JRC, Ispra, Italy*

Oral SESSION 5DO.10 Manufacturing PV in Europe | Social Aspects of PV

5DO.10.1 Would an Increase in PV Modules Prices Impact the European PV Market? 1903

Johan Lindahl[1], Gaëtan Masson[2], Elina Bosch[2], Amelia Oller Westerberg[1]
[1] *Becquerel Sweden, Stockholm, Sweden;* [2] *Becquerel Institute, Brussels, Belgium*

Oral SESSION 5DO.11 Value and Competitiveness of PV in the Growing Market

5DO.11.1 A Snapshot of Global PV Market - 2023 1906

Gaëtan Masson[1], Melodie de l'Epine[2], Arnulf Jäger Waldau[3], Izumi Kaizuka[4], Amelia Oller Westerberg[5], Jose Donoso[6]
[1] *IEA PVPS Task 1, Brussels, Belgium;* [2] *IEA PVPS Task 1, Lyon, France;* [3] *European Commission JRC, Ispra, Italy;* [4] *RTS Corporation, Tokyo, Japan;* [5] *Becquerel Institute, Knivsta, Sweden;* [6] *UNEF, Madrid, Spain*

5DO.11.2 Driving the Quest for Reliable and Bankable PV in Europe - Status and Targets in 2030   1909

Ulrike Jahn[1], David Moser[2], Delfina Muñoz[3], Paula Sánchez-Friera[4]
[1] *Fraunhofer CSP, Munich, Germany;* [2] *EURAC, Bolzano, Italy;* [3] *CEA, Le Bourget du Lac, France;* [4] *Solkeys, Gijón, Spain*

5DO.11.3 Is the Value of (BI)PV Increasing or Decreasing Over Time?   1917

Wouter L. Schram[1], Elham Shirazi[1]
[1] *University of Twente, Enschede, The Netherlands*

5DO.11.4 The Role of Flexible Demand in Reducing the Utility-Scale PV Integration Costs: an Italian Case-Study   1921

Elisa Veronese[1], Giampaolo Manzolini[1], Grazia Barchi[1], David Moser[1]
[1] *EURAC Research, Bolzano, Italy*

5DO.11.6 Cost Analysis for a Small-Scale Hybrid, Hydrogen-Based PV Energy System   1932

Marius C. Möller[1], Stefan Krauter[1]
[1] *University of Paderborn, Paderborn, Germany*

Oral SESSION 5DO.14 Energy System Integration with Storage

5DO.14.2 Effects of the Operating Point on PV Systems Equipped with Energy Storage   1937

Kari Lappalainen[1]
[1] *Tampere University, Tampere, Finland*

Oral SESSION 5DO.15 Resilience and Security of Supply

5DO.15.1 Extraction of PV Yield Data from Smart Meter Data Disaggregation   1943

Bas van der Ploeg[1], Wilfried van Sark[1]
[1] *Utrecht University, Utrecht, The Netherlands*

5DO.15.2 Development of an Architecture for Power Interchange by Linking Photovoltaic and Electrification Vehicles   1953

Jun Tsunoda[1], Tohru Kohno[1], Issei Suemitsu[1], Kengo Kumano[1]
[1] *Hitachi, Tokyo, Japan*

5DO.15.3 Possibilities of PV Maximization for Achieving Positive Energy Districts with Respect to Building Density   1957

Helmut Bruckner[1], Maarten Verkou[2], Simon Schneider[3], Miro Zeman[2], Zain Ul Abdin[4], Rudi Santbergen[4], Olindo Isabella[4]
[1] *Sonnenplatz Grossschoenau, Grossschoenau, Austria;* [2] *PV Works, Delft, The Netherlands;* [3] *UAS Technikum Wien, Vienna, Austria;* [4] *TU Delft, Delft, The Netherlands*

5DO.15.5 Reliability Analysis of Coupled PV-Electrolyser Systems – Evaluation of Onsite Factors   1961

Stefan Niederhofer[1], Marcus Rennhofer[1], Rene Hofmann[2]
[1] *AIT, Vienna, Austria;* [2] *TU Wien, Vienna, Austria*

5DO.15.6 Grid Supporting Power Plants with 100% Energy from Wind and PV   1966

Gerhard Mütter[1], Andreas Hensel[2], Jan Winkelmann[3]
[1] Gerhard Mütter, Waldneukirchen, Austria; [2] Fraunhofer ISE, Freiburg, Germany; [3] VENSYS Elektrotechnik, Diepholz, Germany

Visual SESSION 5DV.2 Energy System Integration; Resilience and Security of Supply; Solar Fuels, Storage | PV Sustainability

5DV.2.1 Techno-Economic Analysis of Residential PV-Battery Energy System in Nordics   1969

Lauri Karttunen[1], Sami Jouttijärvi[1], Johannes Niskanen[1], Jerzy J. Jasielec[1], Hugo Huerta[2], Samuli Ranta[2], Kati Miettunen[1]
[1] University of Turku, Turku, Finland; [2] Turku University of Applied Sciences, Turku, Finland

5DV.2.2 On the Statistics of Photovoltaics in Europe   1976

Wilfried van Sark[1], Anton Driesse[2]
[1] Utrecht University, Utrecht, The Netherlands; [2] PV Performance Labs, Freiburg, Germany

5DV.2.4 Sizing of Energy Storage Systems for Different Levels of PV and Wind Power in Combined PV-Wind Power Plants   1977

Micke Talvi[1], Kari Lappalainen[1]
[1] Tampere University, Tampere, Finland

5DV.2.6 Quantitative Evaluation Method for Regional Variations in Electricity Supply-Demand Balance Fluctuation by Weather Forecast Error   1982

Issei Suemitsu[1], Tohru Kohno[1], Jun Tsunoda[1], Kengo Kumano[1]
[1] Hitachi, Tokyo, Japan

5DV.2.8 From Predictions to Profit of a Hybrid Prosumer Pilot: a Forecast-based Robust Battery Dispatch   1988

Mojtaba Eliassi[1], Anouk Hut[1], Gofran Chowdhury[1]
[1] 3E Belgium, Brussels, Belgium

5DV.2.9 Hybrid Energy Storage Systems Design Tool   1989

Ana Foles[1], Luís Fava[1], Luís Fialho[1], Pedro Matos[2], José Silva[1], Pedro Horta[1]
[1] University of Évora, Évora, Portugal; [2] Capwatt Services, Maia, Portugal

5DV.2.10 Solar PV and Battery Microgrid for Electric Cooking - Case Study Eco Moyo Education Centre in Kenya   1995

Audun Bangsund[1], Stian Rummelhoff[1], Ida Fuchs[1]
[1] NTNU, Trondheim, Norway

5DV.2.11 Integrating Bifacial PV Power Forecasting into Energy Management Systems at High Latitudes   2001

Hugo E. Huerta[1], Shuo Wang[1], Samuli Ranta[1]
[1] Turku UAS, Turku, Finland

5DV.2.12 Optimization of Vanadium Redox Flow Battery Performance for Solar PV Integrated Electric Vehicle Charging Station   2005

Ankur Bhattacharjee[1]
[1] BITS Pilani, Hyderabad, India

5DV.2.14 Challenges and Lessons in Residential Energy Storage Projects   2009

Amanda Mendes Ferreira Gomes[1], Aline Kirsten Vidal de Oliveira[1], Marília Braga[1], Ricardo Rüther[1]
[1] *UFSC, Florianopolis, Brazil*

5DV.2.17 Load Shifting in Energy Communities by Providing User-Centered Recommendations – Forecast, Optimization and Potential   2015

Lukas Gaisberger[1], Georgios Chasparis[2], Wolfgang Traunmüller[3]
[1] *University of Applied Sciences Upper Austria, Wels, Austria; [2] Software Competence Center Hagenberg, Hagenberg, Austria; [3] BLUE SKY Wetteranalysen, Attnang, Austria*

5DV.2.18 Fast Oscillations Damping Control for PV-BESS Power Plants   2021

Alex Renan Arrifano Manito[1], Pedro Torres[1], Marcelo Pinho Almeida[1], Gilberto Figueiredo[2], José Cesar Almeida[3], Roberto Zilles[1]
[1] *USP, São Paulo, Brazil; [2] Fluminense Federal University, Niterói, Brazil; [3] Mackenzie Presbyterian University, São Paulo, Brazil*

5DV.2.19 Optimal Use of Batteries on PV Systems for Solving Problems Caused by Predictable Partial Shadings   2027

Rosario Carbone[1], Cosimo Borrello[1], Ferdinando Gioia[1]
[1] *University "Mediterranea" of Reggio Calabria, Reggio Calabria, Italy*

5DV.2.20 Techno-Economic Assessment of Pumped Storage Hydro Power in Hybrid Operation with Floating Photovoltaic and Battery Energy Storage   2033

Andreas Patha[1], Sebastian Steinlechner[1], Johannes Kathan[1], Antonia Golab[2], Johann Auer[2]
[1] *AIT, Vienna, Austria; [2] Technical University of Vienna, Vienna, Austria*

5DV.2.22 Open Architecture for Battery Interfaces: Opportunities for Technological Advancements and Community Benefits   2040

Anna Ponomarenko[1], Konstantin Rozanov[2], Claudia Gutierrez Collave[2], Saif Al-Bajjali[2]
[1] *Lauder Business School, Vienna, Austria; [2] CF Energy, Vienna, Austria*

5DV.2.27 Guideline on Life Cycle Assessment of Agrivoltaic Systems   2044

Maria Anna Cusenza[1], Andrea Danelli[1], Pierpaolo Girardi[1]
[1] *RSE, Milan, Italy*

5DV.2.31 Intermediate Environmental Assessment   2045

Rene Peche[1], Karsten Wambach[1]
[1] *Bifa Environmental Institute, Augsburg, Germany*

5DV.2.37 Holistic Assessment of Scenarios for Future PV Deployment Considering Circular Economy in the EU Using PV ICE   2046

Fabian Spera[1], Andreas Schwarz[1], Robin Graeber[1], Oliver Pfeiffer[1], Ulf Blieske[1]
[1] *Cologne University of Applied Sciences, Cologne, Germany*

5DV.2.38 Development and Testing of a Thermomechanical Procedure to Assess the Disassembly Potential of a Photovoltaic Module   2052

Asier Murillo[1], Cristina Pinto[1], Alicia Buceta[1], Eugenia Zugasti[1], Antonio Urbina[2], Jaione Bengoechea[1]
[1] *CENER, Sarriguren, Spain; [2] UPNS, Pamplona, Spain*

5DV.2.40 Which is the Most Environmentally Friendly PV Technology: c-Si Solar Cell or Perovskite Silicon Tandem Solar Cell?   2056

Elisabetta Brivio[1], Andrea Danelli[1], Maria Anna Cusenza[1], Sofia Spagnolo[1], Pierpaolo Girardi[1]
[1] RSE, Milan, Italy

5DV.2.41 Considering the Environmental Consequences of the Evolution of the Risk of Extreme Natural Events on a PV Installation: a Morphological Analysis-based Prospective Method Applied to Life Cycle Assessment   2057

Alejandra Cue Gonzalez[1], Eric Rigaud[1], Paula Perez-Lopez[1], Philippe Blanc[1]
[1] PSL University, Sophia Antipolis, France

5DV.2.42 Riding the Wave: Opportunities and Constraints to Reuse and Resale of Photovoltaic PV Modules in South Africa   2062

Nicole M. Crozier[1], Jacqueline L. Crozier McCleland[2], Ernest E. van Dyk[2], Catherina Schenck[1], Palisa G. Ntsala[2]
[1] University of the Western Cape, Cape Town, South Africa; [2] Nelson Mandela University, Nelson Mandela Bay, South Africa

5DV.2.47 Circularity in the PV Industry Analysis of Environmental Impacts for Reused PV Panels   2068

Alejandra Galarza[1], Pierre-Philippe Grand[1], Nicolas Vandamme[1], Anaïs Gouabault[2], Juan Alzate[2], Nicolas Defrenne[2], Marie Lacombe[2], Lars Oberbeck[1]
[1] IPVF, Palaiseau, France; [2] SOREN, Paris, France

5DV.2.51 High Vacuum Flat Plate Hybrid Photovoltaic-Thermal Collectors: Economic and Environmental Comparison over Stand-Alone Devices   2072

Annalisa Di Napoli[1], Paolo Strazzullo[1], Roberto Russo[2], Marilena Musto[1]
[1] University of Naples Federico II, Naples, Italy; [2] National Research Council of Italy, Naples, Italy

5DV.2.53 Separation of EoL PV Modules Using Liquid-Based Methods to Achieve Better Recycling Quality   2077

Sonja Feldbacher[1], Daniel Schwabl[2], Ferozan Azizi[3], Gabriele Eder[4], Anika Gassner[4], Thomas Nigl[3], Gernot Oreski[1]
[1] PCCL, Leoben, Austria; [2] Circulyzer, Leoben, Austria; [3] University of Leoben, Leoben, Austria; [4] OFI, Vienna, Austria

Visual SESSION 5DV.3 PV Diversification Upstream and Downstream - from Industry to Applications | Costs, Economics, Finance and Markets | The Revolution of PV

5DV.3.6 The Role and Impact of Rooftop PV in the Norwegian Energy System under Different Energy Transition Pathways   2078

Stine Fleischer Myhre[1], Eva Rosenberg[1], Heine Nygard Riise[1]
[1] IFE, Kjeller, Norway

5DV.3.7 Towards a Common Strategy for Agri-PV in Europe - the Italian Perspective   2084

Celeste Mellone[1], Alessandra Scognamiglio[2], Giancarlo Ghidesi[3], Giulia Guidetti[4], Fabio Salis[5]
[1] Green Horse Advisory, Rome, Italy; [2] ENEA, Rome, Italy; [3] RemTec, Rome, Italy; [4] Green Horse Legal Advisory, Milan, Italy; [5] Iberdrola Renovables, Roma, Italy

5DV.3.9 The Impacts of Large-Scale Implementation of Solar Power in the Nordic   2087
Power Market

Dilshika Heenatigala Kankanamge[1], Jaakko Jääskeläinen[1], Sanna Syri[1]
[1] Aalto University, Espoo, Finland

5DV.3.11 Technical and Economic Analysis of the Implementation of Battery Energy   2092
Storage Systems (BESS) for Nodes in the National Electric System (SEN)
with High Concentration of Solar Energy

Fernando Flores Lizana[1], Patricio Valdivia-Lefort[2]
*[1] Federico Santa Maria Technical University, Santiago, Chile; [2] Universidad de Santiago de
Chile, Santiago, Chile*

5DV.3.13 The Role of Coupling the Heating, Cooling and Power Sectors to Achieve   N/A
100% Renewable Heating and Cooling in Europe

Olgu Birgi[1], Dominik Rutz[1], Rainer Janssen[1]
[1] WIP Renewable Energies, Munich, Germany

5DV.3.21 Fabrication Planning of Module Manufacturing Plants – Analysis of Site   2095
Parameters and Modelling Tools

Max Mittag[1], Hannah Hoffman[1], Christian Reichel[1], Dirk Holger Neuhaus[1]
[1] Fraunhofer ISE, Freiburg, Germany

5DV.3.26 Techno-Economic and Life-Cycle Assessments of Recycling Pathways for   2099
Perovskite on Silicon Tandem Modules

Lian Duan[1], Alejandra Galarza[1], George Wong[1], Lars Oberbeck[2]
[1] IPVF, Palaiseau, France; [2] TotalEnergies OneTech, Paris La Défense, France

5DV.3.27 Sensitivity of Electricity Price in the Finnish Market Conditions with   2102
Increasing Solar Energy Production

*Sami Jouttijärvi[1], Lauri Karttunen[1], Seela Tervo[2], Hugo Huerta[3], Samuli
Ranta[3], Sanna Syri[2], Kati Miettunen[1]*
*[1] University of Turku, Turku, Finland; [2] Aalto University, Espoo, Finland; [3] TUAS, Turku,
Finland*

5DV.3.30 Photovoltaic Systems and Data Centers in Africa: a Bottom-Up Analysis   2108

Marco Pittalis[1], Georgia Kakoulaki[1], Iolanda Saviuc[1]
[1] European Commission JRC, Ispra, Italy

5DV.3.33 The Benefits of a Hybrid Wind-PV Power Plant at Competitive Wholesale   2109
Electricity Market – Case Finland

Simeon Seppälä[1], Sanna Syri[1], Iraj Moradpoor[1]
[1] Aalto University, Helsinki, Finland

5DV.3.34 Profitability of Utility-Scale Photovoltaic Systems in Finland   2119

Seela Tervo[1], Sami Jouttijärvi[2], Kati Miettunen[2], Sanna Syri[1]
[1] Aalto University, Espoo, Finland; [2] University of Turku, Turku, Finland

5DV.3.35 Enhancing Energy Generation of Bifacial Photovoltaic Systems with   2124
Permeable Albedo Enhancement Composite

Filippos V. Farmakis[1], Alexandros I. Droudakis[2], George I. Tzinoglou[2]
[1] Democritus University of Thrace, Xanthi, Greece; [2] THRACE NG, Xanthi, Greece

5DV.3.40 Energy Communities-Challenge and an Opportunities for Energy   2129
Decentralization and Efficiency. A Comparison of PV based Case-Studies
with Different Control Strategies

Domenico Vito[1], Martina Bosone[2], Barbara Pirelli[3]
[1] Metabolism of Cities Living Lab, San Diego, United States of America; [2] Università degli Studi di Napoli Federico II, Naples, Italy; [3] Foro di Taranto, Taranto, Italy

5DV.3.42 Hands-On Training in Photovoltaic Reliability Assessment: A Multinational Educational Approach under the PROMISE Project   2134

Carlos Meza[1], Brian Azzopardi[2], Bernhard Kubicek[3], Aritz Legarrea Oyarzun[4], Ana Gracia-Amillo[4], Melodie de L'Epine[5], Steve Zerafa[6], Austeja Mockeviciute-Azzopardi[2], Carmel Azzopardi[2], Brian Bartolo[2]
[1] Anhalt University of Applied Sciences, Koethen, Germany; [2] The Foundation for Innovation and Research, Valletta, Malta; [3] AIT, Vienna, Austria; [4] CENER, Sarriguren, Spain; [5] ICARES Consulting, Brussels, Belgium; [6] PIXAM, Valletta, Malta

5DV.3.43 Challenges of Energy Communities at Universities – A Virtual Approach   2138

Matevž Bokalič[1], Matej Guštin[1], Marko Topič[1], Ana Belen Cristóbal[2], Marta Victoria[3], Afonso Cavaco[4], Luis Fialho[4], Alexander Gerber[5]
[1] University of Ljubljana, Ljubljana, Slovenia; [2] UPM, Madrid, Spain; [3] Aarhus University, Aarhus, Denmark; [4] University of Évora, Évora, Portugal; [5] inscico, Kleve, Germany

5DV.3.45 Developing Communication Formats for a Positive Energy Transition Focusing on Photovoltaic – A Delphi Design Sprint Approach   2139

Eva-Maria Grommes[1], Sofia Scroppo[2], Stefanie Könen[1], Laura Züll[1], Anne Karrenbrock[1], Anne-Maren Feldhof[1], Ulf Blieske[1], Thorsten Schneiders[1], Valérie Varney[1], Laura Popplow[1]
[1] University of Applied Sciences Cologne, Cologne, Germany; [2] University of Applied Science Cologne, Cologne, Germany

5DV.3.46 TRANSIT: Empowering Sustainable Energy Futures through Innovative Education and Grid-Integrated Roadmap Development   2145

Brian Azzopardi[1], Daniel Busuttil[2], Araceli Hernandez Bayo[3], Ali Ehsan[4], Eduardo Maritinez Cesenia[4]
[1] The Foundation for Innovation and Research, Birkirkara, Malta; [2] MCAST, Paola, Malta; [3] Madrid Polytechnic University, Madrid, Spain; [4] The University of Manchester, Manchester, United Kingdom

5DV.3.48 Coincidence of Photovoltaic Electric Generation During Heat Waves: An Example Analysis for Northern Italy   2149

Danny S. Parker[1], Karthik Panchabikesan[1], Delia D'Agostino[2], Dru B. Crawley[3], Linda K. Lawrie[4]
[1] Florida Solar Energy Center, Cocoa, United States of America; [2] European Commission JRC, Ispra, Italy; [3] Bentley Systems, Ismaning, Germany; [4] DHL Consulting, Pagosa Springs, United States of America

Oral SESSION 5EO.3 Challenges and Opportunities along the PV Value Chain

5EO.3.2 Comparative Global PV Manufacturing Cost and Sustainable Pricing Assessment: China, Southeast Asia, India, USA, and Europe   2152

Sebastian Nold[1], Baljeet Singh Goraya[1], Ralf Preu[1], Jochen Rentsch[1], Julian Reichle[2], Wolfgang Jooß[2], Peter Fath[2], Michael Woodhouse[3]
[1] Fraunhofer ISE, Freiburg, Germany; [2] RCT Solutions, Konstanz, Germany; [3] NREL, Golden, United States of America

5EO.3.3 Assessing the Potential of Agrivoltaic Systems in Korea through Geospatial   2160
Analysis and Multi-Criteria Scenarios

ChangYeol Yun[1], Changki Kim[1], Jinyoung Kim[1], Sangmin Jo[2], Yongil Kim[3]
*[1] Korea Institute of Energy Research, Daejeon, South Korea; [2] Korea Energy Economics
Institute, Ulsan, South Korea; [3] Seoul National University, Seoul, South Korea*

5EO.3.4 Integration of Photovoltaic Systems in the Austrian Power Plant Portfolio – a   2169
Geospatial Data Analysis

Stefan Übermasser[1], Fabian Leimgruber[1], Bernhard Kubicek[2]
[1] AIT, Vienna, Austria; [2] AIT, Vienna, Austria

5EO.3.5 Distributed Photovoltaics Provides Key Benefits for a Highly Renewable   2183
European Energy System

*Parisa Rahdan[1], Elisabeth Zeyen[2], Cristobal Gallego-Castillo[3], Marta
Victoria[1]*
*[1] Aarhus University, Aarhus, Denmark; [2] TU Berlin, Berlin, Germany; [3] Technical University
of Madrid, Madrid, Spain*

5EO.3.6 Identifying the Ecological Implications of the Repowering of Photovoltaic   2192
Systems

Karl-Anders Weiß[1], Sina Herceg[1], Marie Fischer[1], Liselotte Schebek[2]
*[1] Fraunhofer ISE, Freiburg, Germany; [2] Technical University of Darmstadt, Darmstadt,
Germany*

Plenary SESSION 5EP.1 Sustainability

5EP.1.3 Towards Reuse-ready PV: a Perspective on Recent Advances, Practices and   2200
Future Challenges

*Ioannis (John) Tsanakas[1], Gernot Oreski[2], Gabriele Eder[3], Anika Gassner[3],
Arvid van der Heide[4], Daniela Ariolli[5], Guillermo Oviedo Hernandez[5], David
Moser[6], Karsten Wambach[7]*
*[1] CEA, Le Bourget-du-Lac, France; [2] PCCL, Leoben, Austria; [3] OFI, Vienna, Austria; [4] imo-
imomec, Genk, Belgium; [5] BayWa r.e., Milan, Italy; [6] Eurac Research, Bolzano, Italy; [7]
Wambach-Consulting, Petersdorf, Germany*

5EP.1.4 Enhancing Citizens' Participation in PV Deployment   2205

*Silvia Caneva[1], Duygu Celik[1], Chiara Busto[2], Chiara Candelise[3], Alessia
Cornella[4], Letizia Bua[5], Edouard Breniaux[6], Nouha Gazbour[7], Ivan Gordon[8],
Wander Jager[9], Rudolf Kapeller[10], Gökhan Kirkil[11], Paola Mazzucchelli[12],
Osbel Almora Rodríguez[13], Marcello Passaro[14], Alessandro Sciullo[15],
Sebastien Lizin[16], Alessandro Martulli[16], Atse Louwen[4], Hanna Dittmar[17],
Thomas Garabetian[17], Rania Fki[1], Johannes Stierstorfer[1], Melanie Kern[1]*
*[1] WIP Renewable Energies, Munich, Germany; [2] Eni, Novara, Italy; [3] Bocconi University,
Milan, Italy; [4] Eurac Research, Bolzano, Italy; [5] Eni, Milan, Italy; [6] Carnot Institute Chimie
Balard Cirimat, Toulouse, France; [7] CEA, Le Bourget-du-Lac, France; [8] imec, Genk,
Belgium; [9] University College Groningen, Groningen, The Netherlands; [10] Johannes Kepler
University, Linz, Austria; [11] Kadir Has University, Istanbul, Türkiye; [12] CIRCE, Zaragoza,
Spain; [13] URV, Tarragona, Spain; [14] Sunzest Solar, Rotterdam, The Netherlands; [15]
University of Turin, Turin, Italy; [16] Hasselt University, Hasselt, Belgium; [17] SolarPower
Europe, Brussels, Belgium*

41st European Photovoltaic Solar Energy Conference and Exhibition

This presentation was selected by the Sc. Committee of the EU PVSEC 2024 for submission of a full paper to one of the EU PVSEC's collaborating peer-reviewed journals.

AREAL CELL TEMPERATURE MONITORING USING ARRAY OF IN-LAMINATE INTEGRATED SENSORS FOR PARTIAL SHADING DETECTION

Seyed Mojtaba Sadati Faramarzi [1,2,*], Georgi H. Yordanov [1,3], Arvid van der Heide [2,3,4], Jan Genoe [1,2], Jef Poortmans [1,2,3,4]

[1] KU Leuven, Department of Electrical Engineering, Kasteelpark Arenberg 10, 3001 Leuven, Belgium
[2] Imec, Kapeldreef 75, 3001 Leuven, Belgium
[3] EnergyVille, Thor Park 8310-8320, 3600 Genk, Belgium
[4] University of Hasselt, Martelarenlaan 42, 3500 Hasselt, Belgium
[*] Corresponding author; Email: seyed.faramarzi@kuleuven.be

ABSTRACT: In this work, we evaluate a novel method of measuring surface temperature distribution over a c-Si solar cell using temperature sensors integrated inside the module lamination. The target is to accurately measure thermal gradient of the cell during its operation time, evaluate cell temperature variation in real-life operation conditions, and finally detect partial shading using deviations in areal thermal readings. In contrast to measurements from the backside of the module, up to 5 °C temperature value difference has been observed which could be attributed to heat dissipation in the cell structure and wind flow. Planar temperature measurement resolution of ~ 3 cm × 3 cm has been achieved in the current work, which could be readily utilized to gain a better insight into reliability of the solar modules by continuous monitoring of the photovoltaic cell at various in-field operation conditions. As evidence for effectiveness of our method in continuous monitoring of the cell temperature, partial shading of a cell during one day of operation was evaluated, clearly showing the exact pattern of the shaded area.
Keywords: Solar Cell Thermal Profile, Continuous Monitoring, In-laminate Sensor Integration, Partial Shading

1 INTRODUCTION

Extending the lifetime of solar modules to 25-30 years while guaranteeing reliable operation in the field conditions has major financial and environmental benefits, considering high cost of recycling solar panels and long solar module investment payback time [1]. Passing the IEC certification standards cannot assure no failure or steep degradation of solar modules during their lifespan when exposed to outdoor conditions [2]. Continuous monitoring of solar modules during their operation life could be beneficial for early failure detection and timely maintenance, improved power production forecast and guaranteed lifelong operation [3-5].

Cell thermal variations and especially abnormally high cell temperatures can be the root cause of cell junction and contact degradation which eventually leads to early photovoltaic (PV) module failure. Partial shading of a cell within a PV module can force the cell into the reverse bias and load it with the entire sub-string's voltage and current. For this, a partial cell shading of as little as 5% is sufficient and can force the cell to absorb many tens of Watts, in addition to the continued sunlight heating over most of its area. The cell heating due to reverse bias is not always uniform, which can result in hotspots of different severity. These hot-spots can be detrimental in modules with double-glass structure, resulting in glass breakage. In modules with a plastic backsheet, a severe hot spot can cause backsheet burning [6]. We have seen uniform cell heating up to ~140°C.

In this study, we have integrated an array of temperature sensors inside the lamination of a single cell PV module, distributed in different locations of the cell. We used measured values of such sensors to evaluate cell surface temperature profile in different outdoor conditions. Surface thermal profile monitoring over a single cell presented herein, could help PV manufacturers to analyze and understand how their products operate when subjected to real-life conditions. Consistent thermal profile data also allows verification of physical simulation performed to analyze different solar module structures.

2 EXPERIMENTAL SETUP

In the present work, an array of 16 Negative Temperature Coefficient (NTC) temperature sensors (TT6-100KC3L-5-AUR, 100kΩ at 25°C) has been integrated inside a single monofacial mono c-Si cell laminate, measuring the temperature at the center point of a cell area of 3 cm by 3 cm. NTC sensor measurement accuracy is specified to be in the ±1 °C range.

As displayed in Fig. 1, thin Pt100 sensors (Ephy-Mess NWT-F) are carefully placed onto the backside of the glass-glass laminate while the NTC sensors are placed close to the c-Si cell and inside the laminate, between two layers of thermoplastic polyolefin (TPO) encapsulant material (Borealis BPO8828UV). The sensors are distributed over the rear area of the c-Si cell so that they have equal distance from each other. For measuring the backside temperature, 8 resistive sensors (Pt100) are taped to the backside of the module, equally spaced from each other. For Pt100 sensors, measurement accuracy is ±0.3 °C or better within our temperature measurement range.

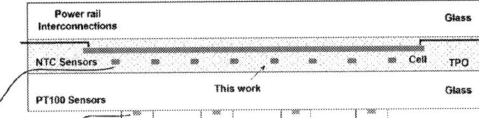

Figure 1: Cross section of single-cell module structure showing the location of integrated NTC sensors and Pt100 sensors connected to the backside. Not to scale.

The size of integrated NTC sensors is 1 mm × 1mm × 600 μm. Fig. 2 shows the cross-section view of main components of the fabricated module and their estimated thickness. Fig. 2 also includes a rough estimation of heat flux inside the module structure, together with thermal conductivities of each element in the module. Although the heat transfer is lower in the TPO layer because of high thermal conductivity, NTC sensing elements closely track the cell junction temperature because of small distance between the integrated sensors and backside of the cell.

The thickness of the TPO layer on the backside of the c-Si cell is around 900 μm and the sensors are 600 μm thick, therefore the distance between the sensor and the cell is roughly around ~ 150 to 200 μm (considering the placement error).

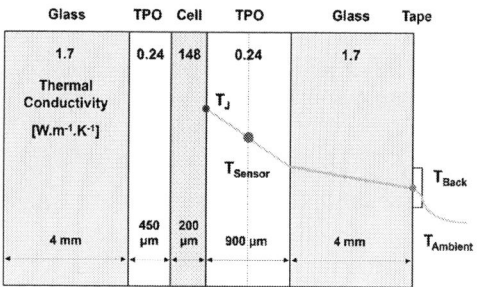

Figure 2: Schematic of the module structure and the graph of heat flux. The cell's p-n junction is on its front side.

In Fig. 3, our outdoor test setup is shown. The location of the setup was at the roof of EnergyVille 1 building in Genk, Belgium. Tilt angle is 35° from horizontal, due South. We used a DAQ6510 data acquisition system to monitor the temperature of NTC sensors and a NI compactDAQ system to measure Pt100 devices. The NTC sensors integrated inside the module have been calibrated over different temperature ranges (0-60 °C, with a 10 °C step) in a climate chamber before starting the measurements.

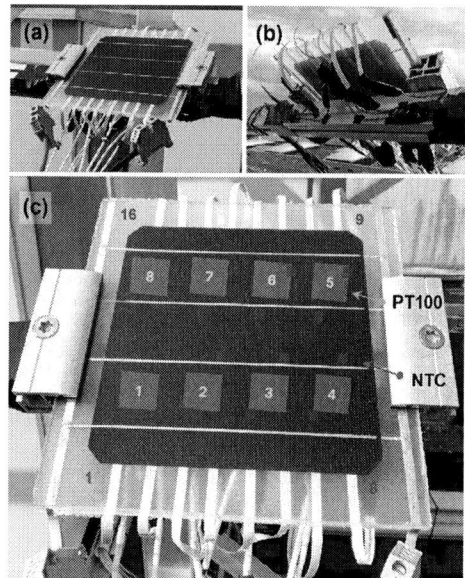

Figure 3: The measurement setup for the present work. Red dots indicate NTC sensors integrated in the module lamination and green squares (not to scale) represent Pt100 sensors taped to the back of the module.

We measured the device over a week from 16 to 23 August 2023. Single cell solar module is repeatedly cycled between open circuit (OC) and short circuit (SC) and otherwise operated at maximum power point (MPP) for the majority of the time. We also applied partial shading by covering half of the area of the cell using a black plastic cover. The effect of different environment conditions and shading was studied by analyzing the temperature sensor measurements.

3 RESULTS

The temperature difference between the ambient temperature (T_{amb}, or T_A) and cell/sensor depends on module structure, irradiance level, and wind speed [7]. As is clearly shown in Fig. 4, increasing the irradiance level leads to a rise in the cell temperature. Additionally, wind speed will cause short term variations in cell temperature value. Up to +5 °C temperature difference was observed between the cell junction and rear surface of the module, with temperature variations mostly distributed in the range of +1-3 °C under 1000 W/m^2 irradiance level.

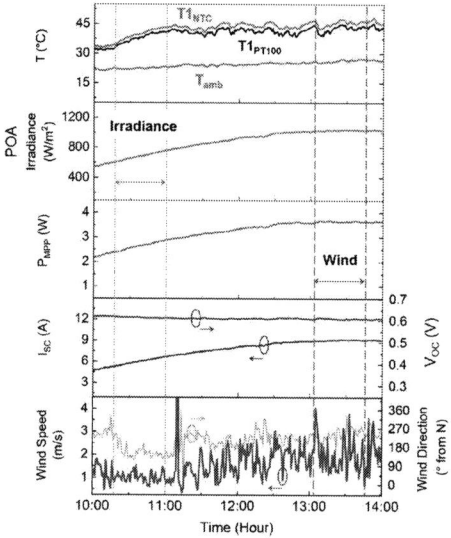

Figure 4: Effect of irradiance and wind speed on cell junction and backside temperature

Measurements during the period of 16 to 23 August 2023 (start and end time are both 7pm) are shown in Fig. 5. Having an array of sensors enables mapping of cell surface temperature variations. For three specific times shown in Fig. 5a-c subsets, this variation is demonstrated; these times are: (a) 18/08/23 3:00pm, (b) 20/08/23 10:00am, and (c) 22/08/23 00:00am. As shown in subset plots, difference between calibrated sensor values over cell area is small (< ±2 °C temperature difference), varying based on sensor location and irradiation levels.

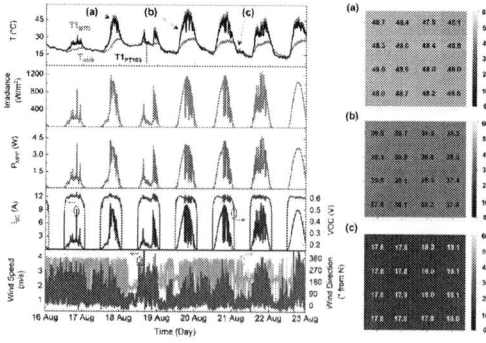

Figure 5: Cell surface temperature variations during 16 to 23 August. Starting and end time of the day are both 7pm.

In Fig. 6, the difference between the temperature values measured from the in-laminate NTC sensors and readings from the backside of the module by Pt100 sensors over one week time is displayed. At lower temperatures ($T_{Back} < 25$ °C), the delta between the two values is quite low, within ±0.5 °C. Difference values lower than zero in the plot might be originated from errors in NTC sensor readings and delay between the temperature change from the backside of the module and the cell junction. A systematic difference between the two sensors' calibration in each pair in the order of ±1°C is also possible.

As discussed earlier, the backside temperature readings fluctuate faster than the cell junction due to wind speed and direction change. These rapid changes will create negative delta values between the two sensor measurements. For the cell temperatures more than 30 °C, larger variations in ΔT values are noticeable. Most of the data are spread in 1 to 3 °C range, with a maximum temperature difference of up to 5 °C.

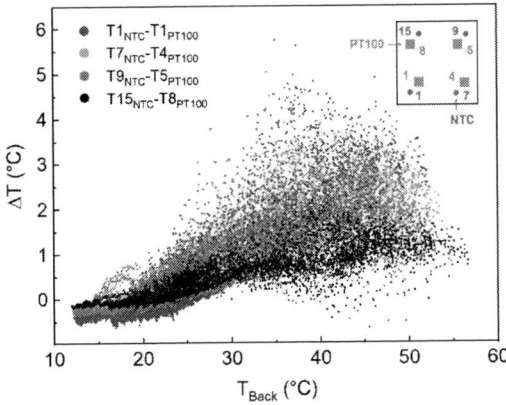

Figure 6: Temperature difference between in-laminate NTC sensors and Pt100 resistive sensing elements taped on the backside of the module

Surface temperature distribution measurement enabled by integrated sensors is beneficial for uninterrupted thermal monitoring of a cell during its operation. As an example, the effect of partial shading on the cell surface temperature has been evaluated in the current analysis, and the results are shown in Fig. 7. One half of the cell is covered using a black plastic cover. In different irradiance levels different surface temperature distributions are observed, clearly displaying the effectiveness of the method for the detection of shaded areas.

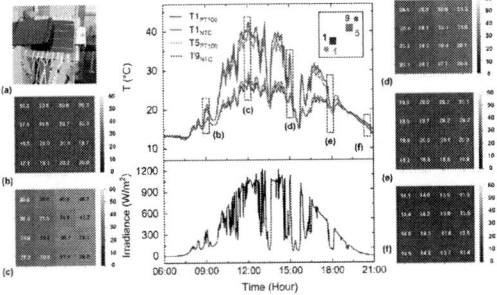

Figure 7: Partial shading detection using variations in the cell surface temperature.

Compared to other approaches for integrating sensors in solar panels such as fiber Bragg grating (FBG) sensors [3], which allow for some spatial temperature measurements over cell area, the proposed method offers much higher lateral thermal resolution with far lower cost requirements of the periphery. For FBG sensors, costly and complex interrogator devices are needed for optical signal measurement.

For accurate sensing of FBG devices, an optical fiber junction box must be added during the module assembly. However, for our suggested method all the required circuitry and wiring can be integrated close to each cell, eventually collected by a central unit located in the PV module's junction box. Table I summarizes the main advantages of presented method compared to optical sensing using fiber Bragg grating (FBG) detectors.

Table I: Comparison of in-laminate temperature sensing using integrated NTC sensors and fiber Bragg grating optical sensing method

Method	Fiber Bragg Grating (FBG) [3]	This Work
Cost	High	Medium
Implementation Complexity	High	Medium
Accuracy	± 0.3 °C	± 1 °C
Areal Meas. Resolution	Low	High

Along with temperature sensors, the same technique can be employed for integration of a matrix of thin strain sensors to detect internal strain levels inside a PV laminate. Facile integration of different thin sensors inside the laminate helps greatly in evaluating the physical and mechanical condition of the cell in operation. Such measurement data provides valuable insight on working condition of a photovoltaic device and acts as a tool for PV manufactures to improve their module fabrication methods and processes. Additionally, the temperature and strain history of the cell during its operation lifetime could aid physical simulation of solar modules by providing live cell condition statistics. Finally, research and development on solar energy would also benefit from this data to improve reliability and lifetime of solar modules.

4 CONCLUSIONS

We have proposed a novel method to map thermal distribution of a monofacial c-Si solar cell during its operation, with negligible reduction in power production. An array of temperature sensors integrated in the single cell module precisely measure cell junction temperature profile, which is closely related to irradiance levels and wind speed. This spatial temperature distribution could be used in research and development of solar modules to quantify cell degradation caused by thermal cycles or abnormally high cell temperatures.

Finally, as an example of advantage of the method, areal measurement of temperature was used to detect partial shading of the cell, showing up to 10 °C variation between hot and cold points inside a single shaded cell. Shaded area patterns could be clearly recognized by the temperature distribution. The high accuracy of this approach paves the way for implementation in some of the panels in a power plant, to enable continuous health monitoring of solar cells during in-field operation. The collected data could be used by PV manufacturers as valuable feedback to optimize the materials and technologies involved in their solar module development process, and to improve reliability and lifetime of their products.

5 ACKNOWLEDGEMENT

The authors would like to thank the WaferPV group in EnergyVille for their advice and technical support.

6 REFERENCES

[1] F. Cucchiella, I. D'Adamo, and P. Rosa, "End-of-Life of used photovoltaic modules: A financial analysis," *Renewable and Sustainable Energy Reviews*, vol. 47, pp. 552–561, Jul. 2015, doi: 10.1016/j.rser.2015.03.076.

[2] V. Sharma and S. S. Chandel, "Performance and degradation analysis for long term reliability of solar photovoltaic systems: A review," *Renewable and Sustainable Energy Reviews*, vol. 27, pp. 753–767, Nov. 2013, doi: 10.1016/j.rser.2013.07.046.

[3] P. Nivelle, L. Maes, J. Poortmans, and M. Daenen, "In situ quantification of temperature and strain within photovoltaic modules through optical sensing," *Progress in Photovoltaics: Research and Applications*, vol. 31, no. 2, pp. 173–179, 2023, doi: 10.1002/pip.3622.

[4] A. J. Beinert *et al.*, "Silicon solar cell–integrated stress and temperature sensors for photovoltaic modules," *Progress in Photovoltaics: Research and Applications*, vol. 28, no. 7, pp. 717–724, 2020, doi: 10.1002/pip.3263.

[5] K. Nishioka, K. Miyamura, Y. Ota, M. Akitomi, Y. Chiba, and A. Masuda, "Accurate measurement and estimation of solar cell temperature in photovoltaic module operating in real environmental conditions," *Jpn. J. Appl. Phys.*, vol. 57, no. 8S3, p. 08RG08, Jul. 2018, doi: 10.7567/JJAP.57.08RG08.

[6] A. Fairbrother *et al.*, "Differential degradation patterns of photovoltaic backsheets at the array level," *Solar Energy*, vol. 163, pp. 62–69, Mar. 2018, doi: 10.1016/j.solener.2018.01.072.

[7] L. de O. Santos, P. C. M. de Carvalho, and C. de O. C. Filho, "Photovoltaic Cell Operating Temperature Models: A Review of Correlations and Parameters," *IEEE Journal of Photovoltaics*, vol. 12, no. 1, pp. 179–190, Jan. 2022, doi: 10.1109/JPHOTOV.2021.3113156.

41st European Photovoltaic Solar Energy Conference and Exhibition

 **EU PVSEC**

Maximum power output predicting algorithm of solar modules based on Artificial intelligence technology

Ju-Hee Kim, Joonyoung Jeon*, and Yong Hyun Kim*

AI & Energy Research Center, Korea Photonics Technology Institute (KOPTI), Gwangju, Republic of Korea

* Corresponding author e-mail: Jyjeon@kopti.re.kr and yonghyun@kopti.re.kr

● Abstract

In this paper, we studied a measurement technology that can predict the power output of PV modules by applying constant current in a PV module using a supervised learning-based machine learning model. A total of 10 PV modules were used in the experiment, and the I-V and EL measurement data of the modules were used to compare the performance of the machine learning model and analyze the results. A total of three types of regression-based machine learning models were used, and the imbalanced data problem was attempted to be solved through SMOTE (Synthetic Minority Over-sampling Technique).

* The used PV modules had been used on the roof of a house during 2004 ~ 2022.

● Introduction

- Development of power predictive models to PV technology using supervised machine learning

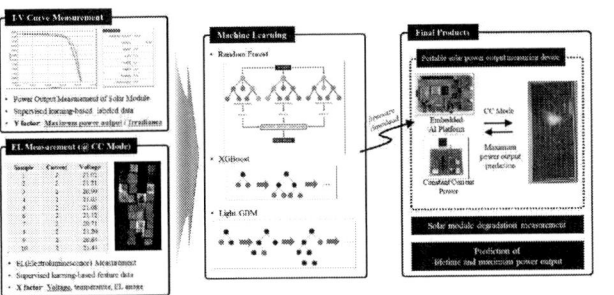

● Experimental

- PV module specification (used PV modules)

Sample	Power [W]	Imax [A]	Vmax [V]	Isc [A]	Voc [V]	Fill Factor [%]
SM-75*	75	4.35	17.3	4.75	21.8	72.7

- PV performance test :
  - EL measurement for 2, 3, 4A @ constant current(CC) mode
    : Measured the voltages with three current condition (2, 3, 4A)
    : X factor (current, voltage) of regression models
  - I-V measurement under 650, 750, 1000, 1250 W/m2 irradiance conditions
    : X factor (irradiance), Y factor (power) or regression models

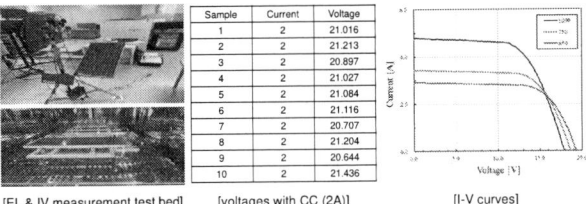

[EL & IV measurement test bed]   [voltages with CC (2A)]   [I-V curves]

- Machine learning models
  - Regression models: XGBoost, Random Forest, Light GBM

● Results & Discussions

- XGBoost

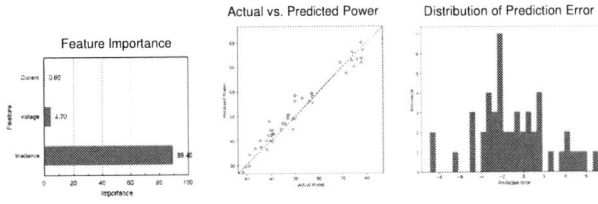

- Random Forest

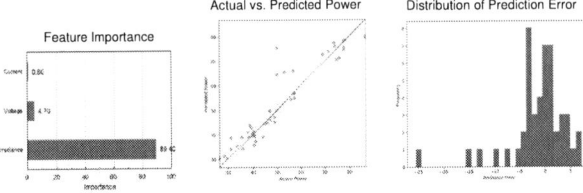

- Light GBM

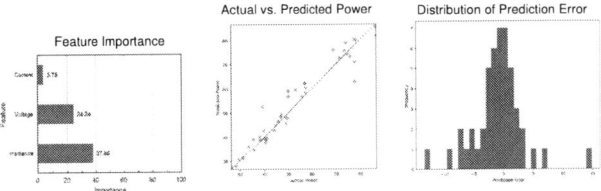

- Analysis of Regression model by MSE, MAE, and R2 Score

Model	MSE	MAE	R2 Score
XGBoost	12.453	2.837	0.955
Random Forest	32.658	3.631	0.882
Light GBM	18.458	2.903	0.933

: XGBoost model is the most suitable regression model to predict the power output (for used PV module).

● Conclusions

- The used PV modules had been installed n 18 years on the roof of a house.
- With the machine learning models, it could find out the effect of the life time and performance of used PV modules with various conditions.
- The results show that irradiance is the most important feature to the decision tree model in a feature analysis report.
- Also, it showed that the regression models(XGBoost, Random Forest, Light GBM) can predict the power of the used PV modules with CC mode.
- With the regression models, the accuracy was determined with the results of MSE, MAE, and R2 scores.
- In terms of accuracy, the XGBoost model could give very high predictive results for used PV modules(MSE(12.453), MAE(2.837), and R2 Score(0.955)).

Acknowledgement   This work was supported by the Korea Institute of Energy Technology Evaluation and Planning (KETEP) and the Ministry of Trade, Industry & Energy (MOTIE) of the Republic of Korea (No. 20213030160440, Renewable Energy-based Village-level Microgrid Demonstration).

41st European Photovoltaic Solar Energy Conference and Exhibition

A PARAMETRIC APPROACH FOR ESTIMATION OF PV SHORT-CIRCUIT CURRENT

Sergiu-Mihai Hategan[1,2,*], Marius Paulescu[1]
[1]Faculty of Physics, West University of Timisoara, Romania
[2]Institute for Advanced Environmental Research, West University of Timisoara, Romania

[*]Corresponding author: sergiu.hategan98@e-uvt.ro

ABSTRACT: Accurate modeling of short-circuit current (I_{sc}) in photovoltaic (PV) modules is crucial, predominantly impacted by solar irradiance and cell temperature. It is also significantly affected by the spectral distribution of incoming solar radiation, which is often overlooked in standard performance models. The Sandia model (SAPM) incorporates an empirically fitted spectral mismatch factor (SF) while more complex methods use spectral irradiance models that require extensive computations. This study proposes a novel alternative approach that applies parameterized atmospheric transmittances for modeling I_{sc}, aiming to maintain the accuracy of spectral methods while enhancing computational efficiency and accessibility. The model utilizes a minimal set of atmospheric parameters and a module-dependent correction factor, easily derived from the catalog sheet standard test conditions for I_{sc}. The proposed method focuses on horizontally placed crystalline silicon PV modules, under clear-sky conditions. The model yields good results, matching the spectral integration method (using SMARTS2 simulated spectra) with an R² of 0.99. Validation against data collected during a clear-sky day in June 2023 at the Solar Platform of the West University of Timisoara, Romania, showed the proposed model achieved a normalized root mean square error (nRMSE) of 1.3%, outperforming the spectral method (4% nRMSE) and the standard method (4.4% nRMSE).
Keywords: short-circuit current, parametric model, PV performance modelling

1 INTRODUCTION

Solar energy estimation and assessment is crucial for PV plants operation. I-V curve parameters are necessary for proper PV performance modelling. The short-circuit current (equivalent to photocurrent) depends on the incoming solar radiation in the plane of the cell. The impact of the spectral distribution of solar irradiance introduces uncertainty in PV performance modelling.

For a real PV cell, the most widely used model is the one diode model [1], offering an equivalent electrical circuit for a PV cell in operation. Of particular interest is the photocurrent parameter, which is the equivalent of short-circuit current I_{sc}. There is a direct relationship between I_{sc} and the spectral distribution of solar radiation.

A benchmark model for determining short-circuit current (I_{sc}) is the Sandia model (or SAPM – Sandia PV Array Performance Model) [2]. This model provides a description of the I-V curve of any given PV module. The central relationship for determining I_{sc} is:

$$I_{sc} = I_{sc,STC} \cdot \frac{G}{G_{STC}} \cdot SF \cdot (1 + \alpha(T_c - T_0)) \quad (1)$$

Here, the spectral effects are introduced by means of a spectral mismatch factor (SF). In practice, this term is often neglected, for example when solving the Shockley equation for a specific module, in order to determine its I-V curve.

The Sandia model proposes a simple polynomial fit of SF as a function of relative air mass. Other models operating on the same philosophy have expanded this procedure, by adding more variables, such as water vapor, aerosol optical depth [3] and even the average photon energy (APE) [4]. However, these all use the same type of empirical fitting procedure, which requires historical measured data for each installed module, with a low degree of generalization.

A more precise, but computationally complex method starts from the physical basis of the short-circuit current. This method also inherently includes the effects of the spectral distribution of solar radiation of PV conversion,

also ensuring a high degree of generalization. It is given by the following relationship [5]:

$$I_{sc} \sim \int G(\lambda) SR(\lambda) d\lambda \quad (2)$$

Using this method, spectral irradiance models are used to generate spectral radiation dataset, which are then integrated according to relationship (2). Some landmark spectral models that can be used are: Leckner [6], which was the first of this type, the Bird model [7], SMARTS2 [8], for clear-sky periods, or models like FARMS-NIT [9], or other radiative transfer models that can handle cloudy periods.

We propose a novel approach to this problem, using a procedure that is similar to broadband solar irradiance parametric models [10]. Employing this approach, we aim for a middle ground, preserving the accuracy and the physical basis of spectral integration methods, while also offering more accessible parametric equations for determination of I_{sc}.

The paper is structured as follows: Section 2 presents the procedures used for deriving the proposed model. Section 3 describes the synthetic and measured data used for validation of the model. Section shows the results, the accuracy of the model and discussions of future improvements.

2 METHODOLOGY

The proposed model is based on a few general assumptions. We utilized a similar procedure to parametric solar global horizontal irradiance models. Therefore, our model assumes a horizontally placed PV module. Due to the nature of the chosen spectral models, the proposed parametric method also assumes clear-sky conditions. The last assumptions are that the incidence angle of solar radiation is less than 60° and that the module is comprised of cells with a spectral response typical to crystalline silicon cells.

First, we introduce a new cell-dependent proportionality constant, to capture several factors influencing PV conversion, e.g. cell area, maximum absolute spectral response. This constant β_{cell}, can be computed in the following manner:

$$\beta_{cell} = \frac{I_{sc,STC}}{\int G_{STC}(\lambda) d\lambda} \ [A/W/m^2] \qquad (3)$$

Next, we proceed with the parameterization of the atmospheric transmittances. We used a hybrid spectral model, starting from the Leckner model atmospheric transmittances, but utilizing the high spectral resolution of SMARTS2. This is motivated by the less complex transmittance equations of the Leckner model, which lends them to easier parameterization. This model models five atmospheric extinction processes, namely Rayleigh and aerosol scattering, mixed gases, water vapor and ozone absorption.

Each atmospheric transmittance was parameterized using equation 4. The extraterrestrial spectrum (AM0 [11]) was used in this parameterization. The spectral response is the generic relative spectral response of a crystalline silicon PV cell, provided in the PVLIB library [12].

$$\overline{\tau_x}(m_x) = \frac{\int \tau_x(m_x,\lambda) G_{ext}(\lambda) SR(\lambda) d\lambda}{\int G_{ext}(\lambda) SR(\lambda) d\lambda} \qquad (4)$$

For each parametric transmittance, the following fitting equations were chosen:

$$\overline{\tau_R} = a \cdot e^{-bx} + c, x = m \qquad (5)$$
$$\overline{\tau_g} = a \cdot e^{-bx} + c, x = m \qquad (6)$$
$$\overline{\tau_{03}} = a + b \cdot x, x = m \cdot l_{03} \qquad (7)$$
$$\overline{\tau_a} = a \cdot e^{-bx}, x = m \cdot \beta \qquad (8)$$
$$\overline{\tau_w} = a \cdot e^{-bx} + c \cdot e^{-dx} + f \cdot e^{-gx}, x = m \cdot w \qquad (9)$$

The fitting procedure yielded R^2 values of at least 0.99 for each of the parametric transmittances, when compared to the hybrid SMARTS2-Leckner spectral transmittances. The corresponding coefficients are given in Table I, for each equation.

Table I: Fitted coefficient values for each transmittance.

τ	a	b	c	d	f	g
τ_R	0.54	0.127	0.44			
τ_a	0.98	1.59				
τ_o	0.99	0.028				
τ_g	0.01	0.208	0.98			
τ_w	8.76 E-1	7.95 E-4	8.05 E-2	6.57 E-2	-3.88 E-2	8.11 E-1

The beam and diffuse components are defined next, in a similar manner to broadband irradiance models. A downward scattering factor γ was introduced in the definition of the diffuse transmittance, building on the downward scattering factor introduced in the Leckner model.

$$\overline{\tau_b} = \overline{\tau_g} \cdot \overline{\tau_w} \cdot \overline{\tau_{03}} \cdot \overline{\tau_R} \cdot \overline{\tau_a} \qquad (10)$$
$$\overline{\tau_d} = \overline{\tau_g} \cdot \overline{\tau_w} \cdot \overline{\tau_{03}} \cdot \gamma(1 - \overline{\tau_R} \cdot \overline{\tau_a}) \qquad (11)$$

The downward scattering factor γ was fitted as a function of the solar elevation angle h:

$$\gamma = 0.5 \cdot (a + b \cdot \sin h + c \cdot \sin^2 h + d \cdot \sin^3 h) \qquad (12)$$

The values of the γ factor coefficient are provided in table II. They were obtained after a fitting procedure applied to the diffuse transmittance.

Table II: Fitted coefficient values for the downward scattering factor.

Coeff.	a	b	c	d
Value	0.4056	3.0056	2.4024	0.17264

The complete model is assembled in the following way, starting from the beam and diffuse parametric transmittances:

$$I_{sc} = \beta_{cell} G_{eff} \left(\overline{\tau_b}(m,\beta,w,l_{03}) + \overline{\tau_d}(m,\beta,w,l_{03}) \right) \qquad (13)$$

Here, the constant effective irradiance G_{eff} was introduced, analogous to the solar constant that is widely used for parametric solar irradiance models. We computed its value for crystalline silicon cells: $G_{eff} = 652.89$ W/m^2.

$$G_{eff} = \int G_{ext}(\lambda) SR(\lambda) d\lambda \qquad (14)$$

3 DATA

As the proposed model aims to approximate well the integrated method, a comparison with the SMARTS2 integration was necessary. Therefore, we generated a synthetic dataset of atmospheric parameters to be used as input for both models. This dataset contained 8000 points, with values in the following range: 20 values for air mass between 1 and 3, 20 values for water vapor content between 0.1 and 4 g/cm2, and 20 values for aerosol optical depth at 1000 nm (turbidity coefficient) between 0 and 0.4. Because the ozone content effect was smaller, we prescribed the standard value of 0.34 for the entire dataset.

Actual field measurements were also necessary for an appropriate comparison between the three considered models. Thus, we have the proposed parametric model, the spectral integration model using SMARTS2 and the standard Sandia model, neglecting spectral effects. Measurements were taken on the Solar Platform of the West University of Timisoara. A Cleversolar SPR-130 module was horizontally placed on the platform and connected to an amperemeter. The measurement campaign took place on 23.06.2023, around noon. We chose this period, because it was a proper clear sky day. The parametric and integrated models required atmospheric parameters as input, which were collected at the nearby AERONET station, located at a distance of less than 100 m. We ensured that the measurements were simultaneous.

For the I_{sc} measurements we applied both a temperature correction (using NOCT model) and an aging correction factor of 0.96. The catalog data sheet of the module specified an I_{sc} value at STC of 7.82 A. Using equation (3), the specific cell parameter was β_{cell}=0.0144 [A/W/m^2].

4 RESULTS AND DISCUSSIONS

The proposed model aims to approximate to a large degree the spectral integration method of estimating I_{sc} under real conditions. Therefore, a comparison of our model and the integrated method is warranted. A synthetic dataset of 8000 datapoints, described in the previous section was created, containing atmospheric input data for both models. For these conditions, the results are presented in Figure 1. With an R^2 coefficient of 0.99 and a normalized mean bias error (nMBE) of -0.4%, the

41st European Photovoltaic Solar Energy Conference and Exhibition

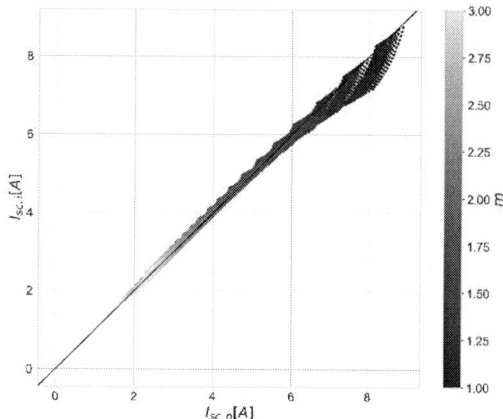

Figure 1. Agreement between the proposed parametric method (x axis) and the integrated method (using SMARTS2) on the y axis. The color of the dots represents the air mass.

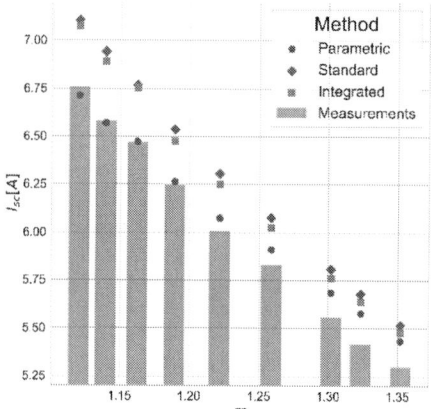

Figure 2: Comparison between the standard model (no spectral effect), integrated model and parametric model with regards to measurements. The x axis represents the air mass.

proposed model can be considered a good approximation of the spectral integration method. For relative air masses above 1.25, there is good agreement between the two models. This suggests that the proposed model could potentially be better applied in places located at higher latitudes, in temperate climate conditions.

When considering the field measurement campaign, the results are quite different, as showcased in figure 2. While the standard model, which neglects spectral effects, and the integrated model, are in good agreement, they both exhibit a large overall positive bias compared to measurements. However, the proposed parametric method deviates from the behavior of the other two models at lower air masses (below 1.25), with lower errors for these conditions.

In terms of statistical indicators, the standard method registered a normalized root mean squared error (nRMSE) of 4.8%. For all cases, the normalization was defined using the average short-circuit current for the measurement campaign. The integrated method exhibits better performance, with an nRMSE of 4%. The parametric method, however, had the best performance, with nRMSE value of 1.5%.

The atmospheric conditions for the measurement period were defined by the following values: a turbidity coefficient β of around 0.3 and water vapor content w of around 3 g/cm^2. Thus, the atmosphere was characterized by high aerosol and water vapor loading, with significant difference compared to STC.

5 CONCLUSIONS

The present study presented a novel approach in the estimation of I_{sc} for PV modules, capturing the effects of the spectral distribution of solar radiation in PV performance modeling. Conventional I_{sc} modeling methods neglect spectral effects, or they rely on empirical approaches that require ample historical measurement data. Other more complex methods are based on the integration of spectral irradiance models, being less widely accessible.

The proposed model aims at a middle ground, addressing the limitations of both conventional and integration methods, utilizing parametric atmospheric transmittances. This results in computationally efficient

and accurate means of estimating I_{sc}. The parameterization aims to accurately approximate the spectral integration method, while vastly lowering computational complexity.

The model was tested both on a synthetic dataset, and a measured dataset. On the synthetic dataset of 8000 datapoints, we tested how well the parametric model approximates the spectral integration model. We found high correlation between the models, with an R^2 of 0.99, and close to zero bias. The measured dataset comprised a small campaign during a clear sky day. On this dataset, the parametric method outperformed the standard method (which neglects spectral effects) as well as the spectral integration method.

However, the measurement sample is quite small, so more measured data is necessary for a proper validation of the proposed model, under a higher range of atmospheric conditions. Other future advancements would be employing transposition models to apply this model for modules under any possible orientation.

REFERENCES

[1] A. Sabadus and M. Paulescu, "On the Nature of the One-Diode Solar Cell Model Parameters," *Energies*, vol. 14, no. 13, Art. no. 13, Jan. 2021, doi: 10.3390/en14133974.

[2] J. Kratochvil, W. Boyson, and D. King, "Photovoltaic array performance model.," SAND2004-3535, 919131, Aug. 2004. doi: 10.2172/919131.

[3] J. A. Caballero, E. F. Fernández, M. Theristis, F. Almonacid, and G. Nofuentes, "Spectral Corrections Based on Air Mass, Aerosol Optical Depth, and Precipitable Water for PV Performance Modeling," *IEEE Journal of Photovoltaics*, vol. 8, no. 2, pp. 552–558, Mar. 2018, doi: 10.1109/JPHOTOV.2017.2787019.

[4] R. Daxini, Y. Sun, R. Wilson, and Y. Wu, "Direct spectral distribution characterisation using the Average Photon Energy for improved photovoltaic performance modelling," *Renewable Energy*, vol. 201, pp. 1176–1188, Dec. 2022, doi: 10.1016/j.renene.2022.11.001.

[5] J. Polo, M. Alonso-Abella, J. A. Ruiz-Arias, and J. L. Balenzategui, "Worldwide analysis of spectral factors for seven photovoltaic technologies," *Solar*

Energy, vol. 142, pp. 194–203, Jan. 2017, doi: 10.1016/j.solener.2016.12.024.

[6] B. Leckner, "The spectral distribution of solar radiation at the earth's surface—elements of a model," *Solar Energy*, vol. 20, no. 2, pp. 143–150, Jan. 1978, doi: 10.1016/0038-092X(78)90187-1.

[7] R. E. Bird, "A simple, solar spectral model for direct-normal and diffuse horizontal irradiance," *Solar Energy*, vol. 32, no. 4, pp. 461–471, Jan. 1984, doi: 10.1016/0038-092X(84)90260-3.

[8] C. A. Gueymard, "The SMARTS spectral irradiance model after 25 years: New developments and validation of reference spectra," *Solar Energy*, vol. 187, pp. 233–253, Jul. 2019, doi: 10.1016/j.solener.2019.05.048.

[9] Y. Xie and M. Sengupta, "A Fast All-sky Radiation Model for Solar applications with Narrowband Irradiances on Tilted surfaces (FARMS-NIT): Part I. The clear-sky model," *Solar Energy*, vol. 174, pp. 691–702, Nov. 2018, doi: 10.1016/j.solener.2018.09.056.

[10] D. Calinoiu *et al.*, "Parametric modeling: A simple and versatile route to solar irradiance," *Energy Conversion and Management*, vol. 164, pp. 175–187, May 2018, doi: 10.1016/j.enconman.2018.02.077.

[11] "2000 ASTM Standard Extraterrestrial Spectrum Reference E-490-00." Accessed: Sep. 20, 2024. [Online]. Available: https://www.nrel.gov/grid/solar-resource/spectra-astm-e490.html

[12] K. S. Anderson, C. W. Hansen, W. F. Holmgren, A. R. Jensen, M. A. Mikofski, and A. Driesse, "pvlib python: 2023 project update," *Journal of Open Source Software*, vol. 8, no. 92, p. 5994, Dec. 2023, doi: 10.21105/joss.05994.

41st European Photovoltaic Solar Energy Conference and Exhibition

COMPARISON OF CHANGES IN THE PARAMETERS OF FIVE PV MODULE TYPES AFTER ONE YEAR IN THE SWISS JURA MOUNTAINS

Donat Hess[1], Fabio Panduri[1], Matthias Burri[1], Christof Bucher[1], Mauro Caccivio[2], Gabi Friesen[2]

[1] Bern University of Applied Sciences (BFH), School of Engineering and Computer Science (TI), Institute for Energy and Mobility Research (IEM), Laboratory for Photovoltaic Systems (PV-Lab), [2] University of Applied Sciences and Arts of Southern Switzerland (SUPSI), Department of Environment Construction and Design (ISAAC), PVLab
[1]christof.bucher@bfh.ch, [2]mauro.caccivio@supsi.ch

ABSTRACT: The Photovoltaic Benchmark System Mont-Soleil (PV-Bench) pilot project was launched on Mont-Soleil at 1274 meters above sea level in the Bernese Jura in May 2023. Five PV module types, each consisting of six pieces, TOPCon, Heterojunction bifacial, Heterojunction monofacial, PERC, and PERC BIPV were operated on microinverters for 16 months. Before and after outdoor exposure on Mont-Soleil, the module performance was measured under standard conditions, and module degradation was analysed. Additionally, the time-dependent performance of the modules during the outdoor exposure on Mont-Soleil was studied. For this purpose, the IV-curve of each module was measured once every minute, and values such as current, voltage, and P_{mpp} were read and analysed from the characteristic curve. Based on these P_{mpp} and temperature measurements, the time-dependent temperature-corrected performance ratio of the modules was determined under irradiance between 900 W/m² and 1000 W/m². This paper identifies differences between the performance ratios measured indoors and outdoors and explains them. Furthermore, it outlines the challenges of determining outdoor module degradation. These challenges include natural factors, such as varying weather conditions during comparison periods, and technical factors, such as the influence of data filtering on the results, different IV curve scanning speeds indoors and outdoors, which can affect results due to capacitive effects, or software updates that may have impacted data filtering during the measurement period. Additionally, the annual and winter energy yields of the various module types are calculated, which is a relevant metric in connection with the Swiss Alpine System incentives. Based on this, estimates are made for the winter energy yield in an average year and with optimized system orientation. Mont-Soleil is also assessed for its suitability as a PV benchmark location. To this end, the spatial homogeneity of the average irradiance on the installation is examined using reference cells.
Keywords: Mont-Soleil, solar module, pv module, benchmark, degradation

1 AIM

The PV system on Mont-Soleil was commissioned in 1992 and was the largest PV system in Europe at the time. The SM-55 PV modules have a low degradation rate of approx. 0.25 % / year [1]. After 30 years of operation, the conversion of the plant into an annually growing module benchmark plant was reviewed in 2022.

With the benchmark project modern technologies are being compared for the first time at the site (see Figure 1 and Figure 2). At 1274 meters above sea level, the Mont-Soleil site is characterised as a sunny, windy and rather cool location for Central Europe. The comparison of the different technologies is one of many module comparison activity reported in solar research. Locally, it has the relevance of testing the suitability of Mont-Soleil as a benchmark plant and providing initial insights into the findings of various module types. Globally, it is another of many contributions to a better understanding of the operation of TOPCon, HJT and PERC in comparison. Also, it is the first field test of the IV curve tracer (IVCT) developed by the BFH. This device allows the performance of the modules during exposure on Mont-Soleil to be analysed. The goal is to demonstrate differences between outdoor and indoor performance measurements and highlighting the factors that make these measurements difficult to compare.

The following objectives are pursued with this paper:

1. The degradation of the different module types over the entire exposure period on Mont-Soleil is determined based on indoor measurements.
2. A detailed performance comparison of the module types TOPCon, HJT, and PERC in the climatic region of the Bernese Jura is made under irradiance levels of 900 W/m² – 1000 W/m².
3. Challenges in outdoor performance measurements and differences to indoor measurements are observed and analysed.
4. The suitability of the outdoor measuring equipment for a large-scale benchmarking project, as well as the overall suitability of the Mont-Soleil site for a benchmark installation, is examined.

41st European Photovoltaic Solar Energy Conference and Exhibition

Figure 1: Photovoltaic benchmark system on Mont-Soleil

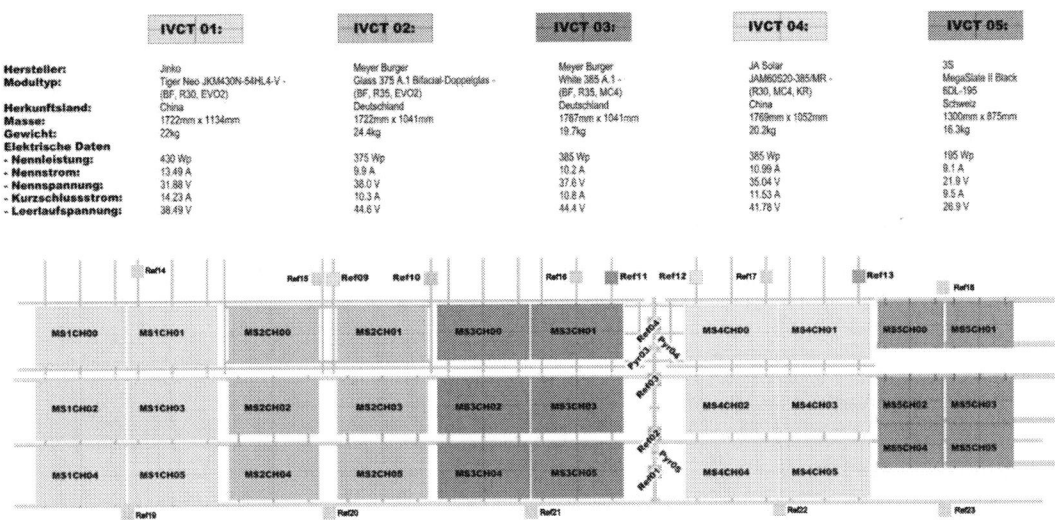

Figure 2: PV-Modules used in PV-Bench. The Jinko Tiger Neo (IVCT 01) is a TOPCon module, the Meyer Burger bifacial (IVCT 02) and white (IVCT 03) modules are Heterojunction modules, and the JA Solar (IVCT 04) and 3S Mega Slate (IVCT 05) modules are PERC and PERC BIPV modules, respectively. In the top row is no coverage behind the modules which allows backside radiation of the Meyer Burger Bifacial Module. Additionally, different reference cells as well as three pyranometers (Pyr03, Pyr04, Pyr05) can be seen. Ref 09-13 are used to detect changes in irradiance during IV curve scans. Ref 14-23 are used for inhomogeneity measurements. On the backside of each module, the module back side temperature is measured with a PT1000 temperature sensor.

2 APPROACH

The measurement process consists of a combination of indoor and outdoor measurements. Before the installation of the modules on Mont-Soleil, initial measurements were carried out. The STC power of 10 modules from each of the five module types was measured at the SUPSI PVLab. After these initial measurements, two modules per type were stabilized with light soaking to identify the initial degradation. Following these indoor measurements, six modules per type were installed on Mont-Soleil and put into operation in May 2023. The module types and their position on the installation can be seen in Figure 2. The IVCT, developed at Bern University of Applied Sciences (BFH), was integrated into the PV system between the inverter and the PV module. The IV curve is measured once per minute for each of the 30 modules. After the modules were dismounted on August 8, 2024, the indoor measurements were repeated for two modules of each type to determine the degradation of the modules over the entire exposure period. The following list provides a compilation of the project steps and the measurements conducted:

1. March 2023: Initial indoor lab measurement at the ISO 17025 accredited SUPSI PVLab:
   a. I-V measurements at Standard Test Conditions (STC) and electroluminescence imaging of 10 modules for each module type
   b. Stabilisation with light soaking according to IEC 61215-1-1 (MQT 19) of two reference modules for each type
   c. I-V measurements at Standard Test Conditions (STC) of the stabilised modules and storage in the dark
   d. Measurement of bifacility of two bifacial reference modules
2. April 26th: Installation of six modules of each type on site and start of outdoor monitoring
3. August 8th, 2024: removal of two modules of each type (lower row, MS1 to MS4, Channels 04 and 05)
4. August 2024: Repetition of indoor lab measurements of two outdoor exposed modules and one reference module for each type

To reduce errors in outdoor measurements due to system failures, missing values, and non-uniform irradiation, the data were filtered according to specific criteria. Only values that met the following conditions were considered:

- Irradiance of 900 to 1000 W/m², measured by the front-side top pyranometer (Pyr3): This ensures homogeneous conditions, as smaller spatial inhomogeneities were observed for high irradiance data. Additionally, the comparability to indoor STC performance measurements is better guaranteed due to the irradiance-dependent module efficiency.
- PDC between 0 and 600 W (125% of the highest power module installed): This excludes erroneous, unusually high IV curve readings.

The following set of filters was also applied. In most cases, these are already covered by the high irradiance filter:

- Fill factor between 0.6 and 1: This removes faulty IV readings and readings taken under partial shading conditions.
- IV readings with a maximum of one MPP: An additional filter to exclude partial shading that can occur in the morning and evening or due to soiling.
- Sunrise/Sunset filter: Dynamically considers data from one hour after sunrise until one hour before sunset.
- Reference cells' short-circuit current: Only considers IV-curve measurements when the reference cell recorded a minimum of 1 A ISC with a maximum standard deviation of 0.01 A during IV - curve scan. This ensures that the weather conditions were stable during the measurement.

The performance of the modules during the time periods considered by the filter was quantified by the temperature-corrected Performance Ratio

$$PR_{T.filt} = \frac{\sum_i E_{DC\_i}}{\sum_i (P_{STC} \frac{G_i}{G_{STC}} (1 - \delta(T_{STC} - T_i)))},$$

where E_{DC_i} is the energy over time period i, calculated based on the measured P_{mpp}. P_{STC} is the power under standard conditions (G_{STC} = 1000 W/m² and T_{STC} = 25 °C) measured in the SUPSI PV-Lab. The constant δ is the temperature correction coefficient provided by the manufacturer, and T_i is the temperature of the module over period i, determined using a PT1000 sensor on the back of the modules. The temperature correction increases the comparability of indoor measurements with outdoor measurements. Additionally, the performance of individual modules can be better compared to each other, as not all modules experience the same temperature. Differences between the module temperature are observed, for example, in the top row, where stronger cooling of the modules occurs due to the lack of coverage behind them. No correction of module back side temperature measurement has been applied to get the cell temperature. Due to the different nature of module backsides (back sheet or glass), a certain difference not accounted for is present in the outdoor measurements. The days on which the $PR_{T,filt}$ could be calculated after filtering are shown in Figure 3.

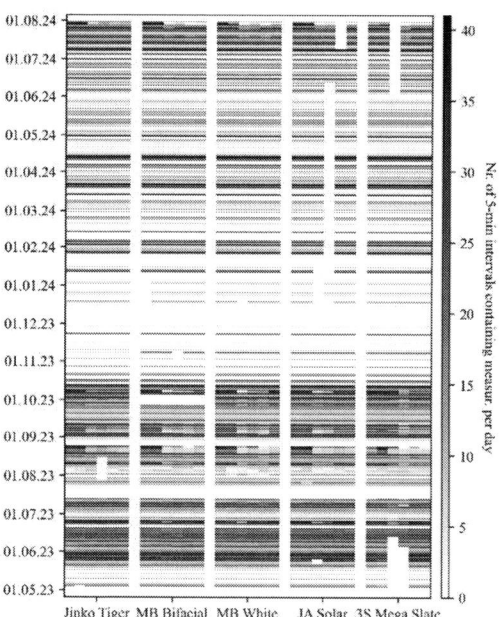

Figure 3: $PR_{T,filt}$ availability for each day and each module.

3 RESULTS INDOOR

3.1 Indoor initial power measurement and comparison to data sheet value

Initial indoor measurements conducted by SUPSI PVLab show that not all module types reach the STC power values specified by the manufacturers. Figure 4 shows the results of the STC power measurements. The values are given relative to the nominal STC power values specified by the manufacturer. For each module type ten modules were measured. The mean values and the standard deviation of the means are 96.801% ± 0.078% for the Jinko Tiger, 96.520% ± 0.099% for the Meyer Burger Bifacial, 97.431% ± 0.161% for the Meyer Burger White, 98.476% ± 0.063% for JA-Solar, and 100.091% ± 0.073% for 3S Mega Slate. The standard deviation of the mean was calculated as $\sigma/\sqrt{n}$, where σ is the standard deviation of the sample and n the sample size (n = 10). Only the 3S Solar Plus module has values in the range of 100%, close to the nominal power. Considering the measurement uncertainty of 2.7%, the Jinko Tiger and Meyer Burger Bifacial modules have STC performance values that are significantly below the manufacturer's specifications. Furthermore, it has to be considered that the modules were measured out of the box without applying any stabilisation. Two modules have been therefore light soaked under a steady state solar simulator to quantify the to be expected initial light induced degradation. The results are presented in the next paragraph.

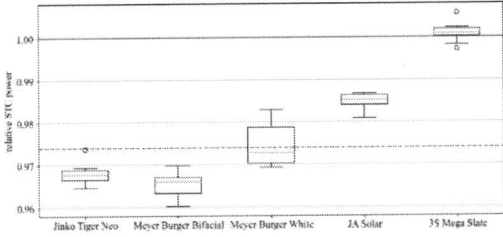

Figure 4: Initial STC power measurements of PV-Bench modules given relative to nominal values. The dashed line indicates the indoor measurement uncertainty of 2.7%.

3.2 Indoor light soaking and estimation of initial degradation

All modules exhibit initial degradations in Pmax well below 0.5% after indoor light soaking of 12 kWh/m². The PERC modules JA Solar and 3S Mega Slate modules show degradations, with measured values of 0.21% ± 0.03% and 0.24% ± 0.08%, respectively whereas the two HJT modules show a drop of 0.12% ± 0.03% for Meyer Burger Bifacial and 0.32% ± 0.00% for Meyer Burger White. For the Jinko Tiger Neo, a power increase of 0.32% ± 0.05% was detected.

3.3 Indoor measurements after 16 months outdoor exposure

Two modules for each type were re-measured indoors at SUPSI PVLab after 16 months of outdoor exposure to measure the degradation of STC parameters. Significant differences in degradation were observed across the various module types over the installation period (see Table I and Figure 5). The 3S Mega Slate module shows the lowest degradation with 0.23% ± 0.05% after 16 months which is in the range of the initial degradation measured on the reference modules exposed to a first light soaking. The JA Solar module exhibits slightly higher degradation, with 1.66% ± 0.29%. Both Meyer Burger modules show more pronounced degradation after outdoor exposure, with 3.76% ± 0.00% for the Bifacial and 4.94% ± 0.54% for the White module. The highest degradation overall is observed in the Jinko Tiger Neo module, which shows a loss of 4.99% ± 1.26% after sun exposure.

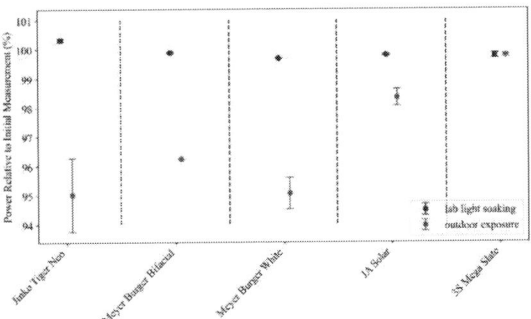

Figure 5: Overview of the indoor measurements. The averages are provided, and the error bars indicate the two measurement values.

Table I: Overview of the indoor measurements. The initial flasher power measurement before light soaking is defined as 100%. The averages of the two measurements are given, while the errors indicate the lower and higher measurement value.

STC Power	After indoor light soaking [%]	After 16 months of outdoor exposure [%]
Jinko Tiger Neo	100.32 ± 0.05	95.01 ± 1.26
Meyer Burger Bifacial	99.88 ± 0.03	96.24 ± 0.00
Meyer Burger White	99.68 ± 0.00	95.06 ± 0.54
JA Solar	99.79 ± 0.03	98.34 ± 0.29
3S Mega Slate	99.76 ± 0.08	99.77 ± 0.05

41st European Photovoltaic Solar Energy Conference and Exhibition

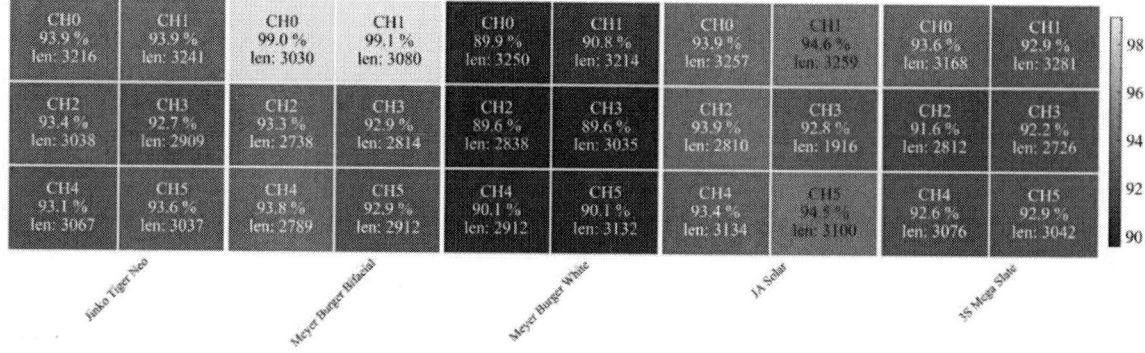

Figure 6: $PR_{T,filt}$ in % for each module over the measurement period from 01.06.2023 to 01.06.2024. Also given is the module name and the number of measurements.

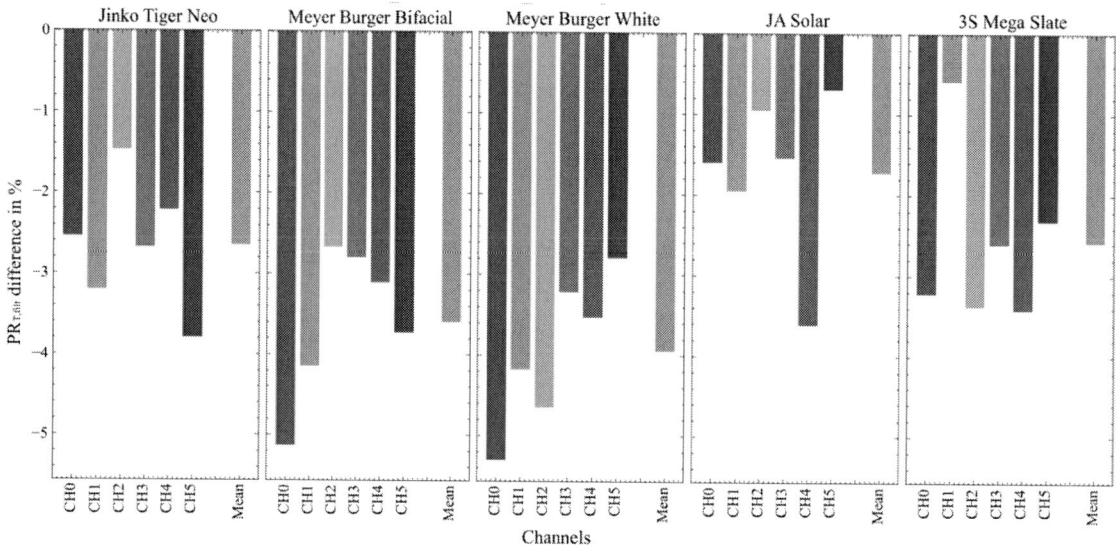

Figure 7: Relative degradation of each module over one year. The values refer to the period from 26.04 to 06.08 in the years 2023 and 202

4. RESULTS OUTDOOR

4.1 Outdoor measurement of module performance

The $PR_{T,filt}$ values measured outdoor are significantly smaller than the relative power measured indoors. Figure 6 shows the $PR_{T,filt}$ of each module over the measurement period of one year form June 1, 2023, to June 1, 2024, as well as the position of the module. The $PR_{T,filt}$ values are given relative to the initial measurements and can therefore be compared with the relative power measured indoor (Table I). The outdoor measured $PR_{T,filt}$ values of the Meyer Burger White modules are the smallest with an average of 90.01% ± 0.17% (with the standard error of the mean given as the uncertainty, see Table II). The other module types differ only slightly in $PR_{T,filt}$, with values ranging from 92.63% ± 0.27% (3S Mega Slate) to 93.83% ± 0.28% (JA Solar). Although the outdoor $PR_{T,filt}$ values are mean values over the entire measurement period, while the indoor were carried out after the entire measurement process, significantly higher degradations were measured outdoors.

For the Meyer Burger bifacial modules in the top row, values of 99.0% (CH0) and 99.1% (CH1) were observed (Figure 6). This is because only the front-side irradiation was considered in the $PR_{T,filt}$ calculation. Since the top-row modules are also exposed to rear-side illumination but the $PR_{T,filt}$ is calculated with front side irradiance measurement only, values exceeding 100% can be achieved. As a result, the top two Meyer Burger bifacial modules are not directly comparable to the others and have been excluded from the calculation of the average values.

4.2 Results of Relative Module Degradation

To determine changes in $PR_{T,filt}$ values over a year and thereby find indications of relative degradation during outdoor exposure, $PR_{T,filt}$ values over 3 months after one year are compared with those from the first 3 months (26.04–06.08). The same months were chosen to ensure similar solar irradiance angles and temperatures. The measured values of both periods and the difference in $PR_{T,filt}$ between these two periods can be found in Table II Additionally, the differences in $PR_{T,filt}$ for all modules are shown in Figure 7.

Table II: $PR_{T,filt}$ is given as mean ± standard deviation of the mean ± (standard deviation of panels) for different measurement periods and manufacturers. For the Meyer Burger Bifacial modules, the top two modules are excluded from consideration due to backside irradiation. However, for the difference between the three months of 2024 and 2023, they are included.

$PR_{T,filt}$ [%]	Jinko Tiger Neo	MB Bifacial	MB White	JA Solar	3S Mega Slate
One year (1.06.23-1.06.24)	93.44 ± 0.19 (± 0.47)	93.21 ± 0.22 ± (0.44)	90.01 ± 0.17 (±0.43)	93.83 ± 0.28 (±0.68)	92.63 ± 0.27 (±0.67)
First 3 months (1.05-06.08.2023)	91.19 ± 0.33 (±0.81)	91.06 ± 0.2 (±0.39)	87.79 ± 0.17 (±0.42)	90.83 ± 0.21 (±0.52)	89.86 ± 0.46 (±1.13)
3 months one year later (26.04-06.08.2024)	88.53 ± 0.23 (±0.56)	87.99 ± 0.3 (±0.59)	83.85 ± 0.35 (±0.87)	89.11 ± 0.57 (±1.39)	87.27 ± 0.71 (±1.74)
Diff 3 months of 2024 and 2023	-2.65 ± 0.33 (±0.80)	-3.60 ± 0.38 (±0.94)	-3.94 ± 0.38 (±0.94)	-1.72 ± 0.42 (±1.03)	-2.59 ± 0.44 (±1.08)

The indoor evaluated degradations and the outdoor evaluated relative degradations differs significantly (Table III). For the Jinko Tiger Neo, 2.3% greater degradation is observed indoors than outdoors. For the two Meyer Burger modules, degradation is also underestimated outdoors by 0.7% and 1.0%, respectively. No significant difference is observed for the JA Solar module. However, for the 3S Mega Slate, degradation is overestimated outdoors by 2.4%.

Table III: Comparison of the relative degradation measured outdoors (calculated by the difference between the $PR_{T,filt}$ values over the period from 26.04–06.08 in the years 2023 and 2024) and the degradation measured indoor.

	Rel. deg. measured outdoor [%]	Deg. measured indoor [%]
Jinko Tiger Neo	2.65 ± 0.33	4.99 ± 0.18
MB Bifacial	3.08 ± 0.23	3.76 ± 0.23
MB White	3.94 ± 0.38	4.94 ± 0.36
JA Solar	1.72 ± 0.42	1.66 ± 0.14
3S Mega Slate	2.59 ± 0.44	0.23 ± 0.17

5 DISCUSSION OUTDOOR MEASUREMENRS

5.1 Systematic measurement errors outdoor

The $PR_{T,filt}$ values averaged over a year as well as over the two 3-month measurement periods are significantly below 100% (see Table II) and, therefore, lower than the STC power measured indoors. This cannot be explained by degradation, as the outdoor measured $PR_{T,filt}$ values are lower than the relative STC power measured indoors after the entire outdoor exposition. There are many reasons that could explain systematically measuring a lower $PR_{T,filt}$ outdoors than indoors. Some of these are listed below. A combination of these causes is possible:

- Uncertainty of initial measurements (2.7%).
- The outdoor irradiance is usually not perpendicular to the plane of array (POA), which leads to losses due to the Incidence Angle Modifier (IAM). These losses are not registered by the pyranometer which has a low angular measurement uncertainty.
- Filter settings (900 W–1000 W) result in more non-clear sky data being collected in the summer than in the winter, leading to increased diffuse light in the summer.
- Temperature differences between the front side of the modules and the back side of the modules (where the module temperature is measured) Malfunction of IVCTs not correctly accounted for in the data cleaning
- The rapid scanning (100 ms) of the IV curve by the IVCT can lead to an underestimation of the $PR_{T,filt}$ in high-capacitance modules (TOPCon and Heterojunction) [2]. This is partially corrected using both SC to OC and OC to SC scans.

Systematic measurement errors in the initial measurements are possible. The initial measurements are given with a systematic uncertainty of 2.7%. Thus, the measured $PR_{T,filt}$ values are also associated with a large systematic uncertainty. While this does not affect the comparison between the modules and module types, as the same measurement error applies to all, a systematic measurement error would significantly impact the comparison between indoor and outdoor measurements.

The changing sun position throughout the year strengthens the $PR_{T,filt}$ in spring and autumn. Since the PV modules are tilted at 52°, the sunlight around midday in spring and autumn strikes the modules almost perpendicularly [3]. Since less light is reflected at a perpendicular angle [4], module efficiency is slightly increased. Indeed, slightly higher $PR_{T,filt}$ values are observed in October and February (see Figure 8), though whether this effect fully explains these increases needs further investigation. The module efficiency also depends on the air mass, which refers to the amount of atmosphere that sunlight must pass through before reaching the modules. The absorption of different wavelengths by the atmosphere affects module efficiency in varying ways [5]. Since the thickness of the atmosphere that sunlight traverses is influenced by the sun's angle, and therefore the time of year, this becomes an additional seasonal effect.

41st European Photovoltaic Solar Energy Conference and Exhibition

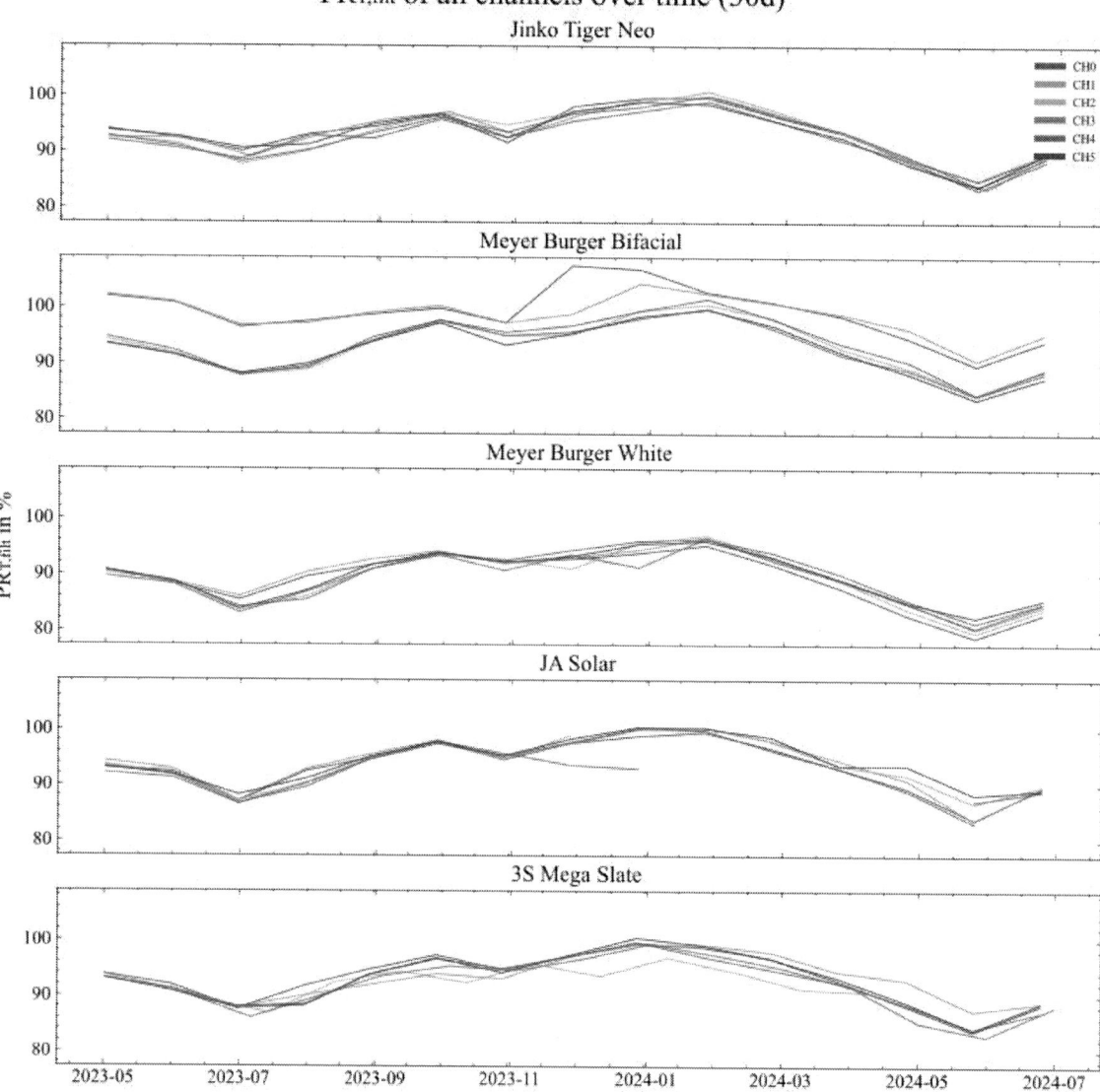

Figure 8: Monthly $PR_{T,filt}$ values over the entire measurement period with an irradiation filter of 900 W - 1000 W.

The radiation filter of 900 W/m^2–1000 W/m^2 leads to a reduction in the $PR_{T,filt}$ value during summer. In Figure 8, the $PR_{T,filt}$, averaged over each month, is shown. The $PR_{T,filt}$ values are lower in summer for all modules compared to winter. Since solar irradiance is stronger in summer than in winter, the minimum power of 900 W/m^2–1000 W/m^2, which is required for consideration due to the filter setting, is more frequently reached in summer. This means that data can be collected even with light cloud cover in summer. In such weather conditions, there is more diffuse light, which strikes the modules at a shallower angle. Shallow incident light is more likely to be reflected [4], reducing module efficiency. This also shows that missing $PR_{T,filt}$ values due to measurement errors can affect the results of individual modules (see Figure 3).

When the irradiance filter is widened to 200 W/m^2–1200 W/m^2, the seasonal fluctuations decrease. The impact of the 900 W/m^2–1000 W/m^2 filter on the $PR_{T,filt}$ value can be seen in the annual $PR_{T,filt}$ trend when using a wider filter of 200W–1200 W/m^2 (See Figure 12 in the Appendix). With this filter, the seasonal fluctuations are much smaller. Notably, all modules show a very high $PR_{T,filt}$ value in December. A possible cause of this could be snow covering the unheated pyranometer during that month, leading to an overestimation of the $PR_{T,filt}$. In the future at least one heated and ventilated pyranometer should be installed.

With a wider filter, the $PR_{T,filt}$ values are higher than with a narrower filter. When using a wider irradiance filter (200 W/m^2–1200 W/m^2), higher $PR_{T,filt}$ values are measured than with a narrower filter (900 W/m^2–1000 W/m^2). Table IV shows the annual $PR_{T,filt}$ value for different filter settings. Despite a larger proportion of shallow incident light, the $PR_{T,filt}$ values for all modules are higher with the wider filter than with the narrower filter. A possible explanation for this phenomenon is

727

higher module efficiency under irradiance bellow 1000 W/m². To ensure comparability between indoor measurements under STC and outdoor measurements, the filter will remain set at 900 W/m²–1000 W/m².

Table IV: $PR_{T,filt}$ values measured over one year for two different radiation filters.

Filter	Jinko Tiger Neo [%]	MB Bifacial [%]	MB White [%]	JA Solar [%]	3S Mega Slate [%]
900 W/m²- 1000 W/m²	93.44 ± 0.19	93.21 ± 0.22	90.01 ± 0.17	93.83 ± 0.28	92.63 ± 0.27
200 W/m² – 1200 W/m²	98.85 ± 0.19	98.1 ± 0.23	94.6 ± 0.18	98.9 ± 0.22	98.57 ± 0.34

Further temperature differences between the front side of the modules and the back side of the modules (where the module temperature is measured) can lead to an incorrect temperature correction. Especially during wind gusts, the top side of the modules can cool down faster than the underside. Since the temperature is measured on the underside of the modules, this results in an incorrect temperature correction. A solution for this would be to filter the data for rapid temperature changes and exclude measurements taken during such periods. However, no such filter was applied in the analyses conducted as part of this paper.

It is unlikely that the IVCTs are a source of error in our case. Thow the measurement accuracy of the IVCT was ensured is described in the following section. First the measurement accuracy of all IVCTs and their channels was checked by measuring static voltages and currents. These measurements were conducted once in the lab before installation and again at Mont-Soleil approximately five months after the installation of the PV system (on 27.09.2023). Static currents between 1A and 14A (up to 24A for IVCT 1) and static voltages between 5V and 65V were applied to all channels of the IVCTs. The relative deviations between the measured and applied currents and voltages for each channel are shown in Figure 14 in the appendix. The black data points refer to measurements taken before installation, while the red ones refer to measurements taken at Mont-Soleil. The individual lines correspond to the different PV modules (CH 1 - CH 6). The largest deviations (over 1.5%) were observed in the voltage measurements for IVCT 1 (CH0, CH4, CH5) at Mont-Soleil, reaching up to 3.3%. For all other IVCTs, the deviations for both current and voltage were below 1.1%. The deviations were significantly smaller at energy relevant higher currents and voltages, with values below 0.13% for currents above 5A and voltages above 20V (below 1% for voltage measurement by IVCT 1 at Mont-

Soleil). Therefore, no change in the accuracy of the IVCTs that would significantly affect the results was detected.

The low $PR_{T,filt}$ values of the Meyer Burger White modules due to possible systematic errors in IVCT 3 can be ruled out by swapping the IVCTs. Since the measurement accuracy of the IVCTs was only tested with static currents and voltages, but the IV curve in the field is measured dynamically, a reduction in measurement accuracy due to effects in the IVCT cannot be entirely excluded. Specifically, it remains to be determined whether IVCT 3 might be measuring lower P_{mpp} values as a result of such an effect. To investigate this, the IVCT of the Meyer Burger White modules (IVCT 3) was temporarily swapped with the IVCT of JA Solar modules (IVCT 4), and the behaviour of the P_{mpp} (normalised to Wp) was analysed. In Figure 13 in the appendix, the P_{mpp} values measured with IVCT 3 (orange) and IVCT 4 (blue) are shown (upper plot), along with their difference (lower plot) on October 12, 2024. At 10:30, the IVCTs were swapped, meaning that from that point, the JA Solar modules were read by IVCT 3, and the Meyer Burger White modules by IVCT 4. This swap immediately reversed the sign of the difference. This indicates that the differences in P_{mpp} are due to the differing performance of the modules, not the IVCTs.

The high capacitance of TOPCon and Heterojunction modules leads to an underestimation of their $PR_{T,filt}$. Rapid scanning of the IV curve by the IVCT can result in over- or underestimation of P_{mpp} due to the modules' capacitance [2]. The IV curves were scanned within 100 ms to make sure that the inverters do not switch off during the measurement process. This ensures that the modules were in continuous use during the time under investigation. However, at this speed, measurement errors occur in modules with high capacitance. Depending on the direction in which the curve is scanned, the curve may be measured too high or too low, leading to an over- or underestimation of the P_{mpp}. Since the IV curve is scanned alternately in both directions, the P_{mpp} is sometimes overestimated and sometimes underestimated and the typical hysteresis between the two scan directions is clearly visible. This fluctuation in P_{mpp} between individual measurements can be observed in Figure 13 in the appendix with the Meyer Burger modules (orange curve before 10:30 a.m. and blue curve after 10:30 a.m.). In [2], a study of Heterojunction modules found an underestimation of P_{mpp} by 1.2% and 1.7% despite averaging the upper and lower values. Since the measured P_{mpp} values are proportional to the $PR_{T,filt}$ of the modules, an underestimation of the $PR_{T,filt}$ value by 1% to 2% is to be expected. High capacitance is also present in TOPCon [2] and therefore an underestimation of the $PR_{T,filt}$ is also possible.

41st European Photovoltaic Solar Energy Conference and Exhibition

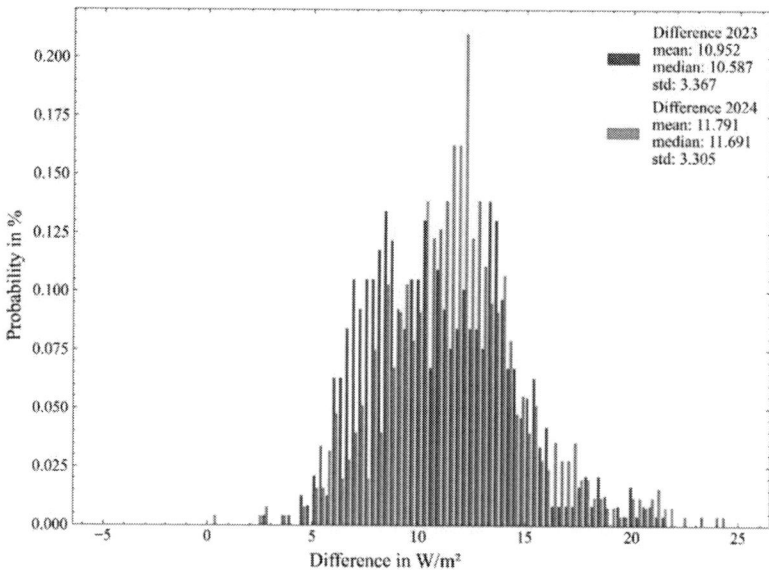

Figure 9: Difference between pyranometer Gi1 and Gi2 with the uncleaned (2023) and cleaned (2024) pyranometer Gi2. The values refer to the period from 12.06 to 05.08.

5.2 Errors of Relative Degradation Measurement Outdoor

There are reasons both for the overestimation and underestimation of the degradation by relative outdoor measurements. In Table III, it was found that the relative degradation over one year measured outdoor differs significantly from the total degradation measured indoors. Some plausible reasons for this are listed here and will be analysed in the following sections:

- Weather conditions during the measurement periods differ
- Soiling of the Pyranometer
- A shorter degradation period is investigated outdoors compared to indoors
- Software update between the measurement periods

Worse weather during the second measurement period leads to an overestimation of degradation. In the second measurement period (01.05–06.08.2024), the weather was worse than in the first period (26.04–06.08.2023). In the first period, an average irradiance of 607.8 W was measured with the pyranometer, while in the second period, it was 541.6 W. Since $PR_{T,filt}$ values are lower in poor weather due to increased diffuse light, comparing the two measurement periods leads to an overestimation of degradation. Even if the same filtering is applied in the two periods, the measurements which are compared were not measured at the same sun position.

The influence of soiling of the pyranometer on the results is minimal. Increasing soiling of the pyranometer could lead to an underestimation of the irradiance, which would result in an overestimation of the $PR_{T,filt}$ value. To check this, pyranometer Gi_2 was cleaned on 12.06.2024.

Figure 9 shows the histogram of the difference between pyranometer Gi_1 and Gi_2 for the measurement period from 12.06 to 05.08 in the years 2023 and 2024. The average difference in 2023 was 10.6 W/m², and after cleaning in 2024, it was 11.8 W/m². Cleaning changed the difference between pyranometer Gi_1 and Gi_2 by 0.8 W/m²

into the unexpected direction. Cleaning should have increased the measured irradiation on Gi_2 and thus reduced the difference. The soiling of the pyranometer is therefore a negligible effect.

Since a shorter time span is examined in the relative outdoor measurements compared to the indoor measurements, outdoor degradation should be estimated to be lower. While the indoor measurements consider degradation over the entire installation period (approximately one and one-third years), The relative degradation measured outdoors compares two periods that are only one year apart.

Between the first and second measurement periods, a software update was done on the IVCT, with the effect that more curves with overestimated P_{mpp} were considered in the second measurement period than in the first. Figure 15 in the appendix shows a scatter plot of all measured fill factors as a function of irradiance during the first measurement period (left) and the second measurement period (right) for the Jinko Tiger Neo (TOPCon) and the two Meyer Burger module types (Heterojunction). A split in the fill factor into two levels is visible in all module types, which is caused by the high capacitance of the modules. While the two levels of the Meyer Burger modules differ by a fill factor of approximately 0.05, the two levels for the Jinko Tiger Neo are almost overlapping. The two PERC modules were not plotted as there was no visible fill factor split. While the lower fill factor level was more common in the first measurement period, the higher level, and therefore higher P_{mpp} measurements, were more prevalent in the second measurement period. This results in an increase in the average fill factor between the first and second measurement periods for the TOPCon (Jinko Tiger Neo) and Heterojunction modules (Meyer Burger), while it decreases for the two PERC (JA Solar and 3S Mega Slate) modules (see Figure 16 in the appendix). Therefore, it is likely that due to this asymmetrical data selection, degradation is underestimated for the non-PERC modules.

41st European Photovoltaic Solar Energy Conference and Exhibition

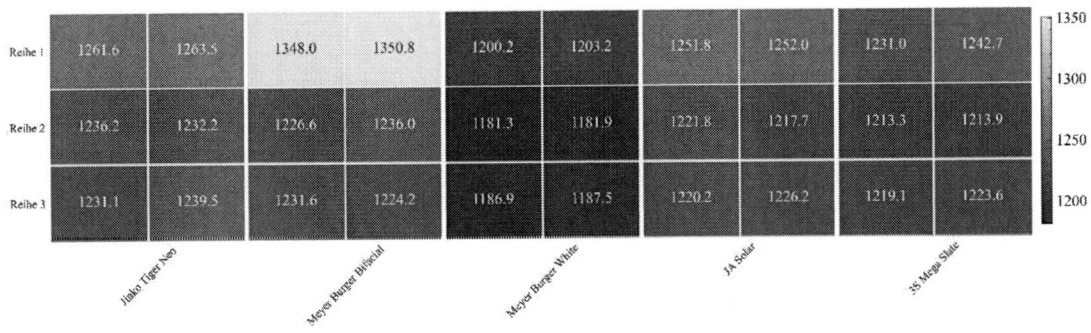

Figure 10: Estimated annual energy yield per PV module.

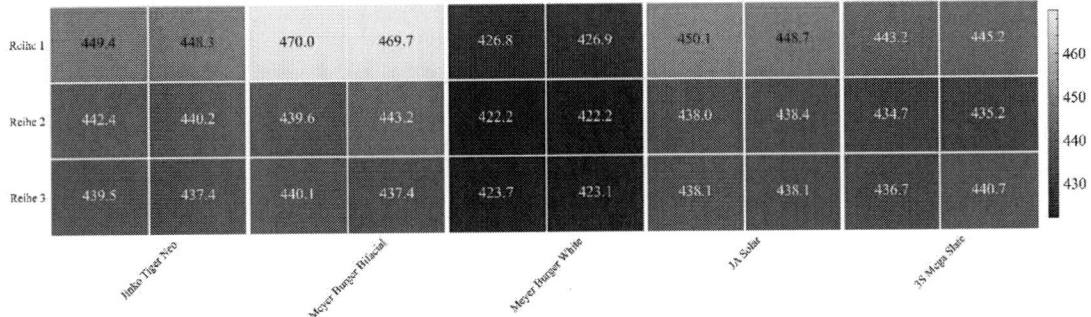

Figure 11: Estimated winter energy yield per PV module.

6. ENERGY YIELD

Compared to high alpine installations, the annual yields on the Mont-Soleil are lower. The annual energy yields are calculated based on the annual PR_{filt} (not temperature corrected with a radiation filter 200 W/m² – 1200 W/m²) and the irradiation measured by the pyranometer over the period from June 1, 2023, to June 1, 2024. The filter setting 200 W/m² – 1200 W/m² was chosen to also account for the behavior of the PV modules under low irradiation conditions. Figure 10 shows the annual energy yields per PV module, while the annual averages for each manufacturer can be found in Table V. These values are somewhere between installations in urban and alpine regions. On the Top Alp near Davos, annual energy yields between 1220 and 1530 kWh/kWp were measured for non-bifacial modules and yields of 1560 to 1800 kWh/kWp for bifacial modules, while in urban regions (Dietikon), the energy yields over the same periods were between 1180 kWh/kWp and 1230 kWh/kWp [6].

The PV system did not meet the minimum energy yield requirements for an alpine system during the winter of 2023-2024. The winter energy yield from October 1, 2023, to March 31, 2024, was estimated based on the $PR_{T,filt}$ values and irradiation measurements during this period. The resulting energy yields can be seen in Figure 11, and the averages for each manufacturer can be found in Table V. All values are below 500 kWh/kWp and therefore do not meet the minimum requirements for winter energy yield in an alpine PV system. The upper most Meyer Burger Bifacial modules (CH0 and CH1), which also absorbed backside radiation, achieved 469.9 kWh/kWp, coming closest to the target energy yield of 500 kWh/kWp.

The winter of 2023/2024 had below-average solar irradiation at Mont-Soleil. According to Meteonorm [7], Mont-Soleil (Latitude = 47.164°, Longitude = 6.990°, Elevation = 1274 m) received a total energy of 528 kWh/m² during the winter months from October 1, 2023, to March 31, 2024, given the current orientation of the PV system (tilt: 52°, azimuth: -20°). On average, from 1996 to 2015, 565 kWh/m² was irradiated during the winter under the same conditions. Therefore, an average winter energy yield 7.0% higher than what was estimated for 2023/2024 can be expected. For the top two Meyer Burger Bifacial modules, this would result in a first-winter energy yield of 502.8 kWh/kWp (see Table V).

Table V: Estimated energy yield in kWh/kWp is given as mean ± standard deviation of the mean (± standard deviation of panels). For Meyer Burger Bifacial the mean is given for modules without backside irradiation (CH2-CH5) and for those with (CH0-CH1).

	Jinko Tiger Neo	MB Bifacial	MB White	JA Solar	3S Mega Slate
Full year 01.06.23-31.05.24	1244 ± 6 (±14.7)	1229.6 ± 2.6 (±5.3) 1349.4	1190.2 ± 3.8 (±9.3)	1231.6 ± 6.5 (± 16)	1223.9 ± 4.6 (±11.3)
Winter 2024	442.9 ± 2.0 (±4.9)	440.1 ± 1.2 (±2.4) 469.9	424.2 ± 0.9 (±2.2)	441.9 ± 2.4 (± 5.8)	439.3 ± 1.8 (± 4.4)
Winter 1996-2015	473.9	470.9 502.8	453.9	472.9	470.1
Winter 1996-2015 Azim. = 0° Tilt = 70°	500.8	497.6 531.3	479.6	499.6	496.7

If the system orientation were adjusted to optimise winter productivity, a 13% higher energy yield could be achieved. According to Meteonorm, with a tilt of 70° and an azimuth of 0°, an average global radiation of 597 kWh/m² was recorded during the winters of 1996-2015. This is 13% higher than the measurement with a 52° tilt and an azimuth of -20°. For all modules except the Meyer Burger White, a winter energy yield very close to or higher than 500 kWh/kWp would be expected (see Table V). However, the annual irradiated energy would decrease from 1450 kWh/m² to 1339 kWh/m², resulting in an 8% reduction. All these values are measured on the DC side of the system. Inverter and transformer losses are not considered.

7. SUITABILITY OF MONT-SOLEIL AS PV BENCHMARK LOCATION

The homogeneity of the irradiance across the system was determined using evenly distributed reference cells located above and below each module type (see Figure 2). The I_{dc} of the reference cells, which is proportional to the irradiation, was recorded throughout the entire measurement period. Laboratory measurements of I_{dc} at irradiances of 500 W and 1000 W showed that the proportionality factors and offsets between irradiation and I_{dc} of the individual reference cells differ slightly. The module dependency of the parameters was therefore taken into account in the evaluation.

The inhomogeneity of the irradiance was investigated for different filters. Since many shadows occur after sunrise and before sunset (mainly due to trees and buildings close by), only data measured at least one hour after sunrise and one hour before sunset were considered in all measurements. Table VI shows the relative deviation to the mean for each reference cell. The values are arranged analogously to the reference cell positions. The values are given for different filters. For the first filter, no irradiance restrictions were applied (apart from twilight times); for the second filter, only measurement data with irradiance above 200 W were considered, and for the third filter, only strong irradiance between 900 W and 1000 W was considered. The evaluations in this paper are based on the third filter.

Table VI: Deviation of the measured irradiation of each reference cell from the mean in percent. The values are given for different filters and must be assigned with an uncertainty of 0.7% (1-sigma).

No restriction of irradiance	(-1.6)	0.6	0.4	0.7	-0.1
	-0.2	-0.5	-1.4	1.2	0.8

Irradiance > 200 W	(-1.7)	0.5	0.4	0.6	-0.3
	0.1	-0.4	-1.3	1.2	0.9

900W-1000W	(-1.8)	0.4	0.3	0.5	-0.4
	0.3	-0.2	-0.8	1.0	0.6

The outlier of -1.8% at the top-left of the third filter (900 W – 1000 W) is not due to an inhomogeneity of the irradiation and can also not be explained by the measurement uncertainty. The outlier in the top-left

reference cell differs from the bottom-left cell by 2.1% (third filter). If this difference were due to inhomogeneous irradiance, the $PR_{T,filt}$ of the top two Jinko Tiger Neo modules (CH0, CH1) would be lower than that of the bottom two (CH4, CH5). However, the $PR_{T,filt}$ measured for the top two modules is 93.9%, while the $PR_{T,filt}$ for the bottom two modules, CH4 and CH5, is 93.6% and 93.1%, respectively (see Figure 6). An explanation based on measurement inaccuracy is also implausible. The measurement uncertainty σ_{meas} is estimated to be 0.7%. This estimate considers the lab calibration of the reference cells and is calculated using Gaussian error propagation. The irradiance of the outlier at the top-left is 1.8% below the mean, placing it several sigma outside the measurement uncertainty $\sigma_{meas} = 0.7\%$ (and also outside the standard deviation of 0.82%). We therefore assume a measurement error in the upper reference cell.

The inhomogeneity of the irradiance σ_{irr}, excluding twilight times, is estimated to be 0.38%. In Table VII, the left column shows the relative standard deviations σ of the reference cells for the individual filters and the right column the corresponding σ_{irr}. The outlier at the top left was not taken into account in the calculations. The inhomogeneity of the irradiance σ_{irr} can be calculated as

$$\sigma_{irr}^2 = \sigma^2 - \sigma_{meas}^2,$$

since the measured standard deviation σ is composed of the measurement uncertainty σ_{meas} and the inhomogeneity of the irradiance σ_{irr}.

Table VII: Relative standard deviation σ of the measured irradiance from the reference cells (left) and the estimated inhomogeneity σ_{irr} of the irradiance after subtracting the measurement uncertainty when the outlier (top left) is excluded (right).

	σ [%]	σ_{irr} [%]
No restriction	0.80	0.38
Irradiance > 200 W	0.76	0.30
900W-1000W	0.56	0

Although no inhomogeneity in the irradiance was detected, the $PR_{T,filt}$ values are row dependent. If the average $PR_{T,filt}$ of all module rows from Figure 6 is determined, while the Meyer Burger Bifacial module is excluded (due to rear side absorption in the top row), the top row yields 92.9%, the middle row 92.0%, and the bottom row 92.5%. Thus, the top row shows values that are 0.9% higher than the middle row and 0.4% higher than the bottom row. The different $PR_{T,filt}$ values cannot be explained by an inaccurate temperature correction. On average, during the filtered time period (900 W – 1000W) the temperature is 37.6 °C in the top row, 41.7 °C in the middle row and 41.9°C in the bottom row. Thus, the lower $PR_{T,filt}$ in the middle row cannot be explained by either an overestimation or underestimation of the temperature coefficients.

One other explanation is that there is a vertical inhomogeneity in the irradiation, which was not captured by the more distant reference cells. Since higher-placed modules can absorb more scattered light from the periphery compared to the lower-placed ones, which are closer to the shadow cast by the southern PV system, it can be assumed that the upper modules have a slightly higher $PR_{T,filt}$. However, the fact that the bottom row exhibits higher values than the middle row could be due to the bright gravel path running directly in front of the PV system and is next to the bottom row. Since these influences are homogeneous in the horizontal direction, they do not affect the comparison between the different module types if they are represented in each row.

To determine the statistical significance of the differences of $PR_{T,filt}$ between the module rows, the standard deviation of the row mean, σ_m, is estimated. The standard deviation of the mean can be calculated according to Gaussian error propagation from the mean,

$$m = \frac{1}{n}\sum_{i=1}^{n} PR_{T,filt\_i},$$

as:

$$\sigma_m{}^2 = \sum_{i=1}^{n}\left(\frac{dm}{dPR_{T,filt\_i}}\sigma_i\right)^2 = \sum_{i=1}^{n}\left(\frac{1}{n}\sigma_i\right)^2 = \frac{n}{n^2}\sum_{i=1}^{n}\sigma_i{}^2,$$

where n is the number of considered modules (n=8) and σ_i is the measurement accuracy of module i. Assuming that this measurement accuracy, is identical for each module, the standard deviation of the mean σ_m simplifies to:

$$\sigma_m{}^2 = \frac{\sigma^2}{n}.$$

The measurement accuracy σ is estimated conservatively by combining the module types Jinko, Ja Solar, and 3S and determining the standard deviation of these 6 modules for each row. Since the selected modules represent 3 different module types, it is assumed that the standard deviation is overestimating the measurement uncertainty, because the variation between the module types as well as the measurement uncertainty of the modules (σ) are included in it. The Meyer Burger modules were not included in the calculation because their $PR_{T,filt}$ values differ so significantly from the other modules that they would increase the estimate of σ too much. The σ of all rows calculated in this way is on average 0.7%, leading to a standard deviation of the row mean $PR_{T,filt}$ of $\sigma_m = \sigma/\sqrt{n} = 0.24\%$. The individual rows thus differ with a significance of several sigma.

The above calculations also suggest that the differences between Jinko, JA Solar, and 3S modules are significantly smaller than the variations within the individual rows. Therefore, when comparing modules with similar $PR_{T,filt}$, it is crucial to ensure that the module types are consistently represented at the same level across the different series.

For example, the Meyer Burger Bifacial module is disadvantaged by the dependence of the $PR_{T,filt}$ values on the module row. To ensure the comparability of the different module types, the modules of the same type were arranged in pairs vertically rather than side by side (see Figure 1). Due to the chosen arrangement, the modules of all types are equally represented in all rows, so this phenomenon should not influence the comparative study. However, since the top row module of the Meyer Burger Bifacial is not considered in the average calculation (see Table II), this module type is disadvantaged. If a top module were added to the Meyer Burger Bifacial with a $PR_{T,filt}$ value 0.7% higher (average of 0.9% and 0.5%) than the average of the lower two modules, the average $PR_{T,filt}$

value for Meyer Burger would increase by $\frac{1}{3}0.7\% = 0.23$ %, resulting in an annual $PR_{T,filt}$ value of 93.44% instead of 93.21%.

8 SUMMARY AND CONCLUSION

The degradation of different module types (TOPCon, HTJ, PERC) was measured indoor and compared after being installed for a period of over one year and three months at Mont-Soleil. The TOPCon module (Jinko Tiger Neo) showed the highest degradation at 4.99% $\pm$ 0.18%, followed closely by the Heterojunction modules with 4.94% $\pm$ 0.36% (Meyer Burger White) and 3.76% $\pm$ 0.18% (Meyer Burger Bifacial) The PERC modules showed the lowest degradation, with 0.23% $\pm$ 0.17% for the 3S Mega and 1.66% for the JA Solar.

These values could partially be reproduced by outdoor measurements while the modules were positioned at Mont-Soleil. However, from the beginning of the measurement period, significantly lower $PR_{T,filt}$ values were measured outdoors compared to indoors. Moreover, the relative degradation (difference between $PR_{T,filt}$ at the end of the measurement period and $PR_{T,filt}$ at the beginning) did only partially match the degradation measured indoors.

Various reasons for the difference between indoor and outdoor measurements were identified. These include shallow incident light, which reduces $PR_{T,filt}$ outdoors due to the IAM, the irradiation which leads to more diffuse light in summer than in winter, and measurement difficulties with high-capacitance modules due to the rapid tracing of the IV curve. These observations contribute to preventing errors in future projects and ensuring that measurement data are correctly interpreted. However, although several causes for the differing results have been identified, the differences are not yet fully understood.

Additionally, Mont-Soleil was investigated as a location for alpine PV installations and as a benchmark site. Although the minimum yield of 500 kWh/kWp for an alpine PV installation was not achieved during the examined winter half-year, it was demonstrated that during an average year, and with optimised orientation and bifacial modules, the 500 kWh/kWp target can be reached. Mont-Soleil is challenging but suitable as a benchmark site due to the homogeneous irradiation. No inhomogeneity in light between 900 and 1000 W was detected by evenly distributed reference cells next to the modules. When all radiations were considered, which arise at least one hour after sunrise and one hour before sunset, the irradiation variance between the reference cells was estimated to be 0.38%. Filtering is therefore necessary on Mont-Soleil to have the same predictions as more homogeneous sites. However, a module row dependence of the $PR_{T,filt}$ values was also observed, with the lowest $PR_{T,filt}$ values found in the middle module row. The cause of this row dependence remains unclear.

9 OUTLOOK

Further causes for the differences between the indoor and outdoor measurements need to be identified. For example, the TOPCon module (Jinko Tiger Neo) shows greater degradation indoors compared to the Meyer Burger modules, while outdoors it shows less degradation than the Meyer Burger modules. It remains to be clarified whether this can be explained solely by the capacitance of the modules. Additionally, one of the PERC modules (3S Mega Slate) shows deviations in its relative degradation compared to the indoor measurements, while the other PERC module (JA Solar) matches the indoor measurements, even though the capacitance of these two modules does not affect the measurements.

The IV curve tracers (IVCT) need further optimization. The speed at which the IV curve is traced must be optimised to avoid errors due to capacitive effects while ensuring that the inverter does not shut down. It should also be ensured that, regardless of the direction in which the IVCT curve is traced, an equal number of measurements are used.

A multi-year study would be beneficial to average out weather-related phenomena. Ideally, this study would consider all radiation intensities, eliminating the unintentional weather filtering. As a result, the measurements would better represent actual field performance by accounting for weak light conditions as well.

10 ACKNOWLEDGEMENT

The authors of this paper would like to thank the project partners Société Mont-Soleil (SMS) and Espace découverte Énergie (EdE) for their support and co-operation. Special thanks also go to the Swiss Federal Office of Energy (SFOE) and the Bern Building Insurance (GVB) for their financial contribution to the project.

References

[1] L. El Boujdaini, C. Bucher, E. Özkalay, M. Burri, M. Caccivio, and F. Gabi, "Analysis of Non-Linear Long-Term Degradation of PV Systems," 2022.

[2] M. Pravettoni, D. Poh, J. Prakash Singh, J. Wei Ho, and K. Nakayashiki, "The effect of capacitance on high-efficiency photovoltaic modules: a review of testing methods and related uncertainties," *J. Phys. D: Appl. Phys.*, vol. 54, no. 19, p. 193001, 2021, doi: 10.1088/1361-6463/abe574.

[3] SunEarthTools, "Sonnenstand Rechner: Position der Sonne in Abhängigkeit von Ort und Zeit.," [Online]. Available: https://www.sunearthtools.com/dp/tools/pos_sun.php?lang=de

[4] N. Martin and J. Ruiz, "Calculation of the PV modules angular losses under "eld conditions by means of an analytical model," *Solar Energy Materials & Solar Cells 70*, 2000.

[5] W. Durisch, B. Bitnar, J.-C. Mayor, H. Kiess, K. Lam, and J. Close, "Efficiency model for photovoltaic modules and demonstration of its application to energy yield estimation," *Solar Energy Materials and Solar Cells*, vol. 91, no. 1, pp. 79–84, 2007, doi: 10.1016/j.solmat.2006.05.011.

[6] F. Carigiet, D. Grunauer, and Baumgartner, P, Franz, "Performance Analysis of PV Modules installed in the Alpine Region," *EUPVSEC*, vol. 2021.

[7] Meteotest, *Meteonorm*. [Online]. Available: https://meteonorm.com/

41st European Photovoltaic Solar Energy Conference and Exhibition

APPENDIX

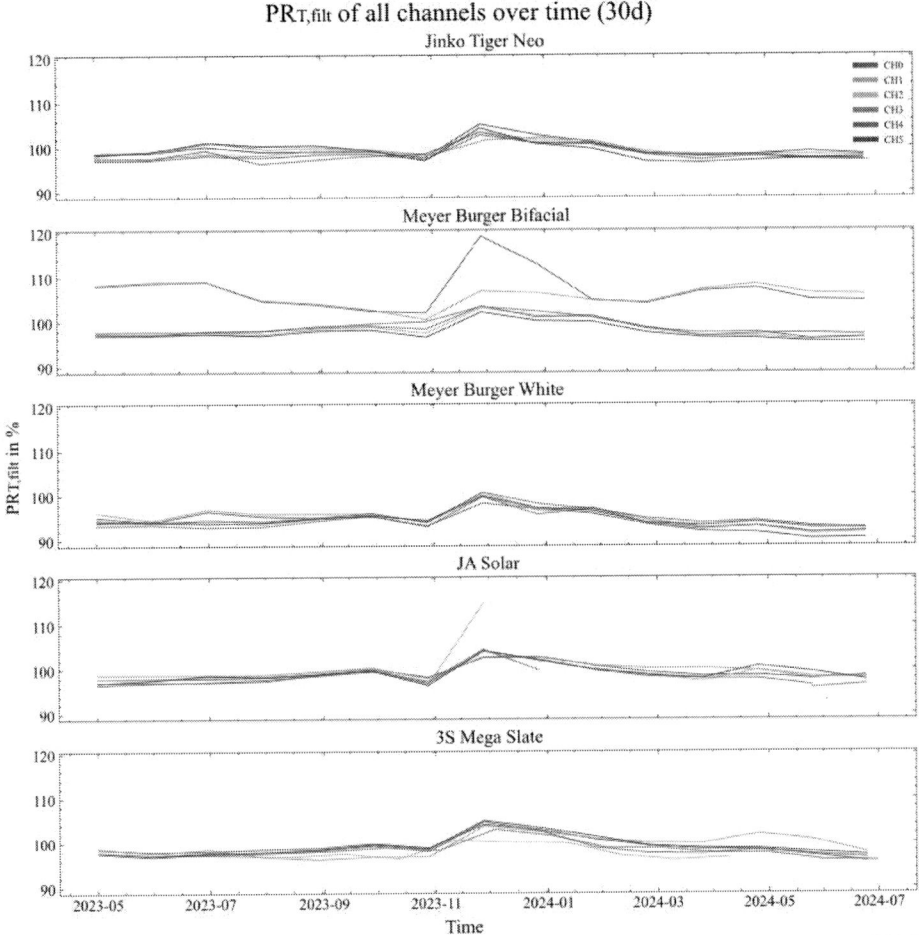

Figure 12: Monthly $PR_{T,filt}$ values over the entire measurement period with an irradiation filter of 200 W - 12000 W.

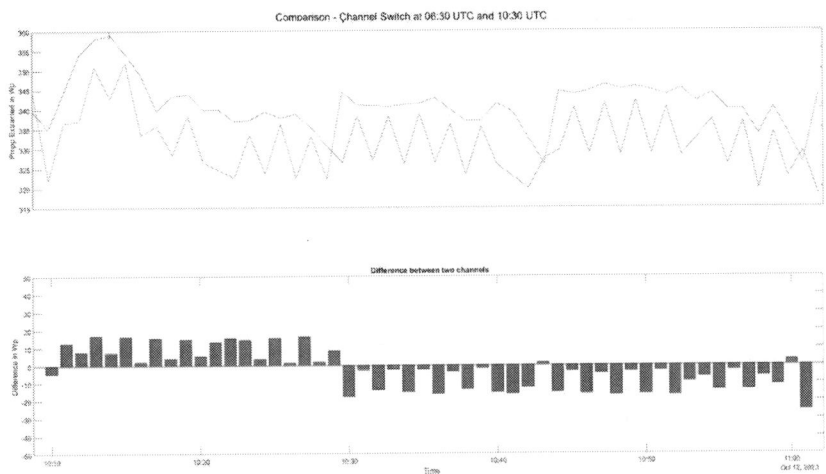

Figure 13: P_{mpp} measured by IVCT 4 (blue) and IVCT 3 (orange) during the IVCT swap at 10:30 (upper plot). The lower plot shows the difference between the measured P_{mpp}

41st European Photovoltaic Solar Energy Conference and Exhibition

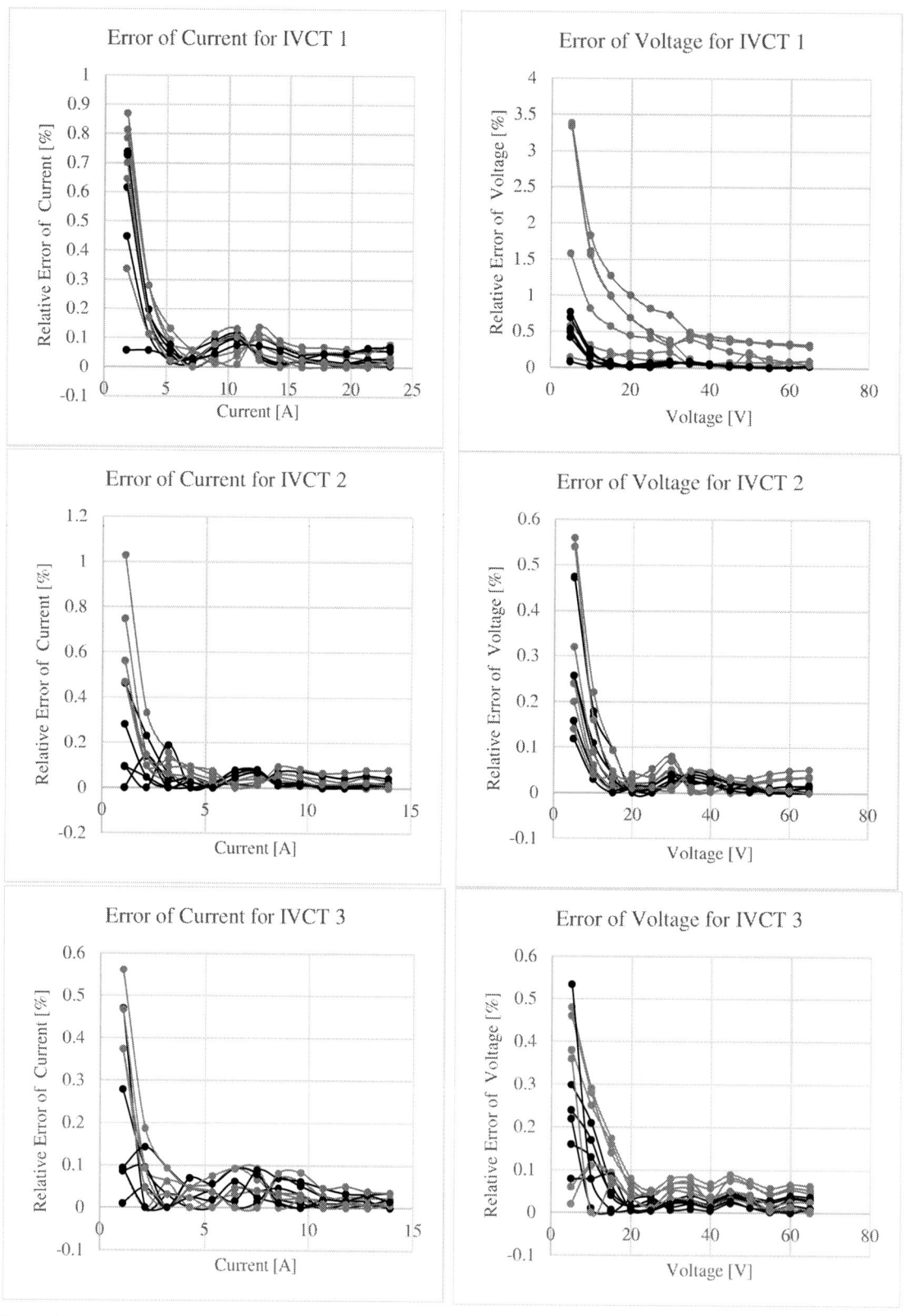

Figure 14: Relative current and voltage errors of each IVCT measured before installation (black) and after installation at Mont-Soleil on 27.09.2024 (red). The individual graphs correspond to the different modules.

41st European Photovoltaic Solar Energy Conference and Exhibition

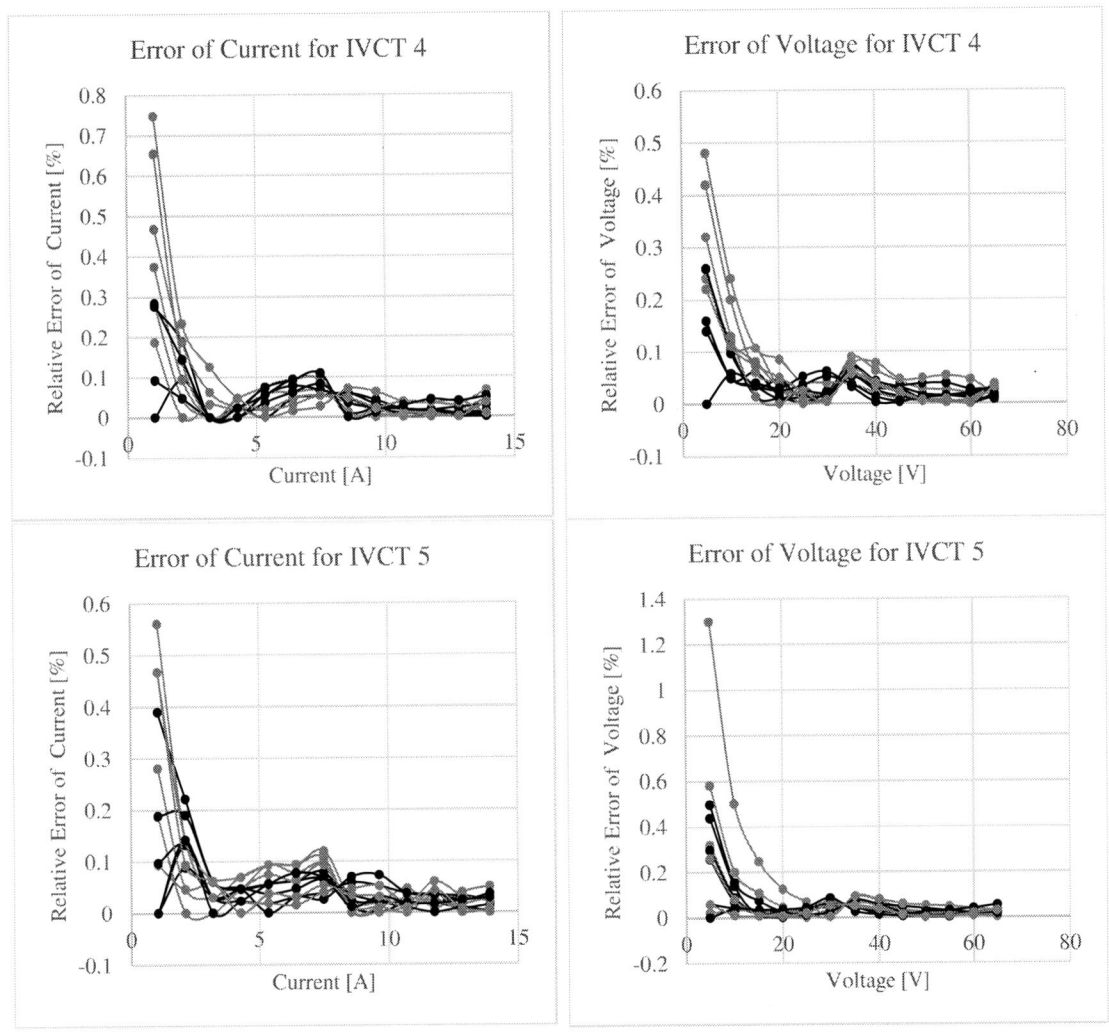

Figure 14 continued: Relative current and voltage errors of each IVCT measured before installation (black) and after installation at Mont-Soleil on 27.09.2024 (red). The individual graphs correspond to the different modules.

41st European Photovoltaic Solar Energy Conference and Exhibition

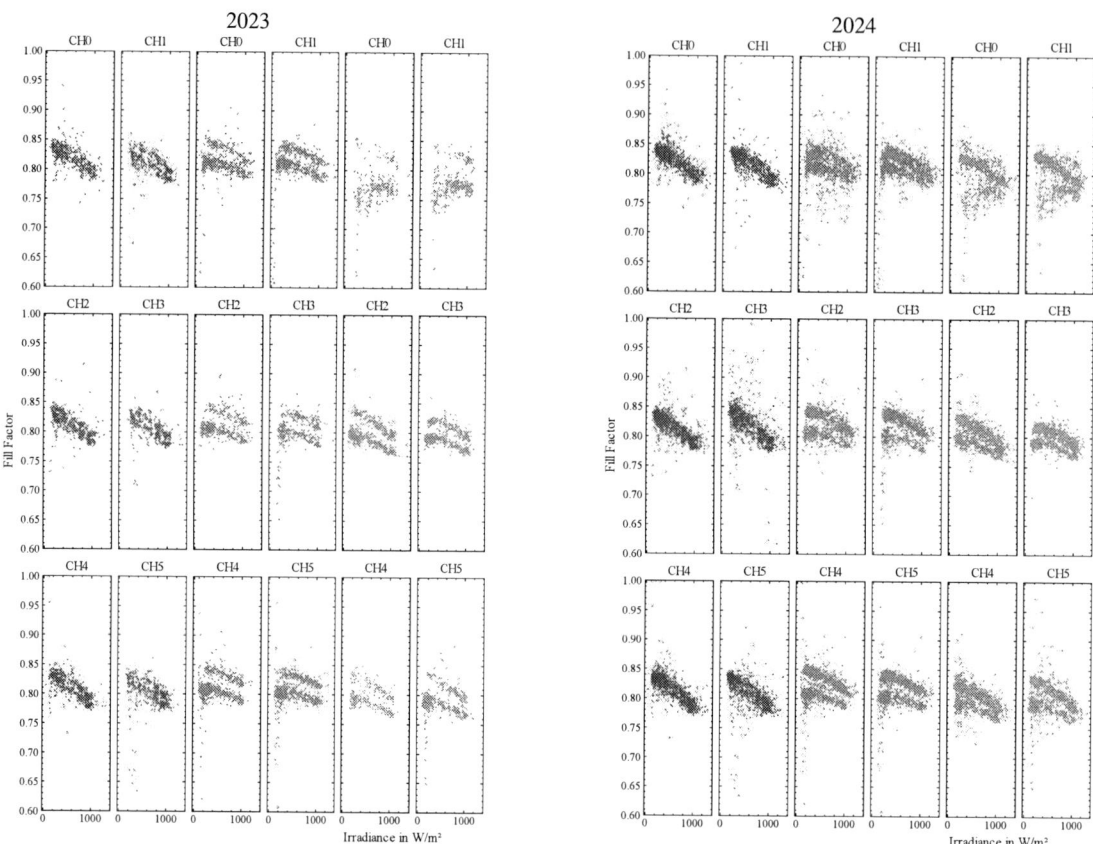

Figure 15: Scatter plot of all measured fill factors as a function of irradiance during the first measurement period (left) and the second measurement period (right). The plots are given for all modules CH0-CH5 of the module types Jinko Tiger Neo (TOPCon, blue) and the two Meyer Burger module types (Heterojunction, green and orange).

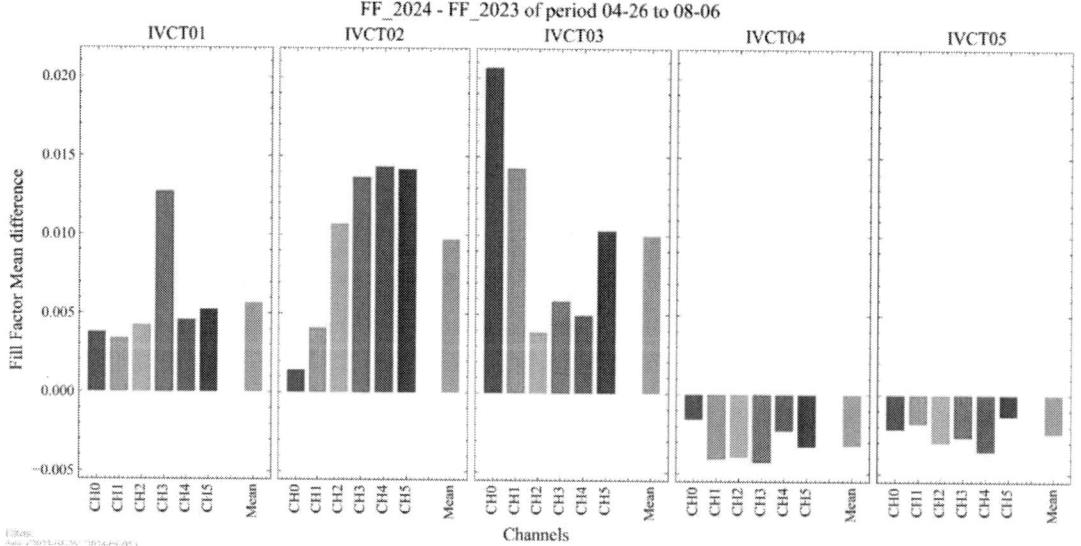

Figure 16: Change in fill factor of each module over one year for the Jinko Tiger Neo, Meyer Burger Bifacial, Meyer Burger White, JA Solar, and 3S Mega Slate. The values refer to the period from 26.04 to 06.08 in the years 2023 and 2024

41st European Photovoltaic Solar Energy Conference and Exhibition

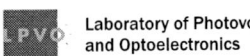

Laboratory of Photovoltaics and Optoelectronics

Accurate Energy Performance Model for Bifacial PV Modules

Kristijan Brecl, Matevž Bokalič, Marko Topič
University of Ljubljana, Faculty of Electrical Engineering, Ljubljana, Slovenia

Antonin Faes
CSEM, Neuchâtel, Switzerland

FE UNIVERSITY OF LJUBLJANA
Faculty of Electrical Engineering

:: CSEM

Abstract

In recent years, the bifacial crystalline silicon solar cells have gained the majority of the PV market. Although the PV module market is already being conquered, modelling for forecasting the energy production is not yet fully solved due to the specific variable conditions of the reflected light. We are presenting the new performance rating model for bifacial PV modules. The new model is based on separation of horizontal global and diffuse irradiance, plane-of-array irradiance, and back irradiance. The model was developed based on the measurements at the LPVO measurement site in Ljubljana, Slovenia and validated on modules of different silicon technologies (PERC, SHJ, IBC, TopCon) at different locations in Europe. The results show that the root mean square error (RMSE) of the simulated to measured performance ratio of the new model can be reduced to about 2% if all irradiance values (GHI, DHI, G_{poa} and G_{back}) and module temperature are monitored.

Irradiance
Sun's geometry

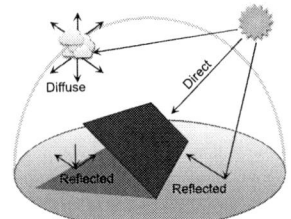

Front-side irradiance:
- direct
- diffuse
- ground reflected

Back-side irradaince:
- diffuse
- ground reflected
- direct
  (in summer months)

Model developement and verification
Test location

Bifacial PV modules of different silicon based technologies (PERC, SHJ, TopCON, IBC) have been monitored at LPVO, UL, Ljubljana, Slovenia. Front- and back-side irradiance was measured by pyranometer and reference solar cell, respectively.

Verification locations

The model was applied on measured data at five locations across Europe (Germany, Switzerland, France, Spain and Slovenia).

Different mounting system and distances from the ground were used at different locations.

Direct-diffuse performance rating model
Power calculations

Direct part of the output power with the module's angular dependence

$$P_{dir}(G_{poa-dir},\theta) = [k_{dir1} \cdot G_{poa-dir} + k_{dir2} \cdot G_{poa-dir}^2] \cdot [1 - \frac{e^{(-\frac{\cos\theta}{a_r})} - e^{(-\frac{1}{a_r})}}{1 - e^{(-\frac{1}{a_r})}}]$$

Diffuse part of the output power

$$P_{dif}(G_{poa-dif}) = k_{dif} \cdot G_{poa-dif}$$

Back-side part of the output power

$$P_{back}(G_{back}) = k_{back} \cdot F_{bifi} \cdot G_{back}$$

DDPRbifi model

$$P_{DDPRbifi} = [P_{dir}(G_{poa-dir},\theta) + P_{dif}(G_{poa-dif}) + P_{back}(G_{back})] \cdot [1 + \gamma \cdot (T_{module} - T_{STC})]$$

Results

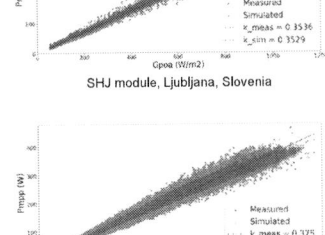

SHJ module, Ljubljana, Slovenia

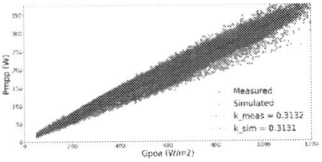

IBC module, Konstanz, Germany

TopCON module, Neuchâtel, Switzerland

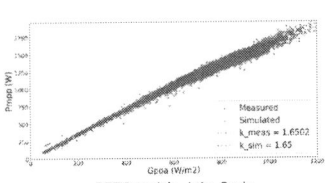

PERC module, Jaén, Spain
(only one month)

Conclusion

We developed an accurate performance rating model for bifacial PV modules. The model was tested on PV modules of four different silicon techologies (PERC, SHJ, TopCON, IBC) at five different locations in Europe. The RMS error of simulated annual performance ratio vs. measured one was below 2%. For the back-side irradiance an average share value (G_{back}/G_{poa}) for the whole year was used for each site.

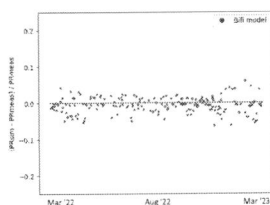

Simulated vs. measured performance ratio

More information:

K. Brecl et al., "An accurate bifacial PV module energy performance model using Direct-Diffuse Power Rating model", submitted to Applied Energy

Acknowledgments

- Slovenian Research and Innovation Agency (Research Programme P2-0197).
- This work is cofunded from the European Union's Horizon 2020 research and innovation programme under grant agreement No. 857793 (Highlite project).

41ª EU PVSEC, 3AV.3.25, 23ʳᵈ – 27ᵗʰ September 2024, Vienna, Austria

kristijan.brecl@fe.uni-lj.si

41st European Photovoltaic Solar Energy Conference and Exhibition

UNCERTAINTY ASSESSMENT IN THE MEASUREMENT OF SOLAR CELLS UNDER STANDARD TEST CONDITIONS

Y. Zhang[1], W.Xu[1], Q.Gao[1], G. Bardizza[2], W. Herrmann[2], C. Monokroussos[1]
[1] TÜV Rheinland (Shanghai) Co., Ltd, Shanghai, P.R. China
[2] TÜV Rheinland Energy GmbH, Cologne, Germany
Corresponding author: Christos Monokroussos
Telephone: +86 21 6108 1188
Email: Christos.Monokroussos@tuv.com

ABSTRACT: This study evaluates the uncertainty in current-voltage (I-V) performance testing of solar cells under Standard Test Conditions (STC) at the TÜV Rheinland solar cell testing laboratory in Shanghai, China. Several factors that could contribute to measurement uncertainty under STC are identified and analyzed, including electronic instrumental errors, temperature variability, optical effects, contact-related influences, and uncertainties arising from repeatability, reproducibility and data processing procedures. Methods are applied to analyze some sources of uncertainty, such as statistical analysis of experimental data, uncertainty budgets from instrument specifications, Monte Carlo simulations, and error propagation calculations from independent variable uncertainties. With spectral mismatch correction, the uncertainties for I_{SC}, V_{OC} and P_{MAX} are determined to be 0.93% (k=2), 0.54% (k=2), and 1.68% (k=2) respectively. Without the correction for spectral mismatch, the uncertainties are determined to be 1.46% (k=2) for I_{SC}, 0.54% (k=2) for V_{OC}, and 2.02% (k=2) for P_{MAX} respectively. The key factors influencing measurement uncertainty in solar cell I-V testing under STC conditions are identified. The outcomes of this study provide valuable guidance on optimizing measurement methods and controlling uncertainty budgets to stringent levels required by international testing standards.

Keywords: solar cell, measurement uncertainty, Monte Carlo, I-V testing, ISO/IEC JCGM 100:2008

1 INTRODUCTION

Current-voltage (I-V) characterization serves as a core method for evaluating solar cell performance through power output analysis. However, I-V testing is inevitably subjected to measurement uncertainty arising from instrumentation limitations and variability in testing procedures. A rigorous quantification of this uncertainty is therefore important to improve the reliability and credibility of reported test results. Moreover, uncertainty assessment plays a key role in laboratory accreditation compliance with standards requiring quantified confidence levels associated with output metrics. Presenting measurement results along with their respective uncertainties lays down key performance markers, providing vital insights that drive the continuous improvement of solar technology.

Solar cell measurement under STC follows a structured and methodical process. These include the calibration of reference device and traceability procedure, properly executing I-V measurements and applying corrections to match STC, as well as routine instrument calibration and maintenance. Additionally, the measurement system incorporates several key components, such as a light source, temperature control unit, electrical contacts, and spectral measurement system. The quality of testing instrumentation as well as certain empirical assumptions can considerably influence the overall uncertainty assessment.

Some organizations actively share their insights on the estimation of uncertainty in photovoltaic device measurements [1-3]. The particulars of uncertainty estimation are closely linked with the traceability chain and the specific measurement practices inherent to each lab. TÜV Rheinland's laboratory in China stands out as a certified authority, accredited by China National Accreditation Service for Conformity Assessment (CNAS), Deutsche Akkreditierungsstelle (DAkkS), and American Association for Laboratory Accreditation (A2LA) specifically for photovoltaic modules, equipped

with a comprehensive system dedicated to the measurement of PV devices. Previously, drawing from our collective expertise, we presented a refined approach to appraise the uncertainty due to spectral mismatch corrections [4]. In separate work, an uncertainty analysis was also performed on STC performance ratings of perovskite photovoltaic modules [5].

In this study, we undertake a quantitative analysis of the uncertainties associated with I-V testing under STC for solar cells at the TÜV Rheinland solar cell testing laboratory in Shanghai, China. We offer a detailed analysis of the various factors contributing to the uncertainty in measurements of short-circuit current (I_{SC}), open-circuit voltage (V_{OC}), and maximum power output (P_{MAX}). This study explores several crucial sources of uncertainty and the analytical methods used to assess them, including calibration and traceability methods, as well as thermal effect. The uncertainties for I_{SC}, V_{OC}, and P_{MAX}, following spectral mismatch correction at the TÜV Rheinland Shanghai laboratory, stood at 0.93% (k=2), 0.54% (k=2), and 1.68% (k=2), respectively. Without the correction for spectral mismatch, the uncertainties were determined to be 1.46% (k=2) for I_{SC}, 0.54% (k=2) for V_{OC}, and 2.02% (k=2) for P_{MAX}, respectively.

2 METHODOLOGY AND RESULTS

In this study, we conducted an uncertainty analysis following the guidelines in ISO/IEC JCGM 100:2008 [6]. Initially, we identified potential sources of uncertainty in solar cell measurements, which were broadly categorized as electronic, temperature-related, optical, contact-related, as well as uncertainties due to repeatability, reproducibility, and data processing. Following this, we calculated the standard uncertainty for each individual variable, employing Type A and Type B evaluations as appropriate. The concluding step involved estimating the combined uncertainty for the target measurement parameter by integrating the uncertainties of multiple variables. To elaborate, if an output estimate y is derived from several inputs $x_1, x_2,..., x_N$ as indicated in Equation (1), the equation of the combined standard uncertainty $u_c(y)$ can be linked to the standard uncertainties of these inputs as delineated in Equation (2). In instances where performing algebraic calculations based on Equation (2) proves to be complex, the Monte Carlo method presents itself as a practical alternative to address the complexity

$$y = f(x_1, x_2, \cdots, x_N) \qquad (1)$$

$$u_c^2(y) = \sum_{i=1}^{N} \left(\frac{\partial f}{\partial x_i}\right)^2 u^2(x_i) \qquad (2)$$

2.1 Electronic-related uncertainty

Voltage and current readings for the solar cell under test are simultaneously measured using a Keithley source meter. Concurrently, an integrated reference cell with a shunt resistance continuously monitors the irradiance, and the corresponding short-circuit current is measured via a digital multimeter. Based on the specifications of the datasheet, the uncertainties of voltage and current channels are calculated as 0.043% (k=1) and 0.052% (k=1), respectively. The uncertainty of irradiance measurement channel is estimated at 0.144% (k=1), accounting for the shunt resistor variation and zero offset. Additionally, thermal and cyclic drift characteristics of all instruments are also incorporated into the uncertainty analysis.

2.2 Effect of temperature

Our uncertainty analysis considers three key temperature-related aspects: 1) control and logging of the solar cell temperature, 2) instruments' thermal drift sensitivity to fluctuations, and 3) uncertainty due to temperature correction temperature coefficients. Solar cell measurements are conducted inside an environmentally controlled room maintained at a stable temperature of (23±2)°C. Additionally, a semiconductor-based temperature control stage was utilized to maintain the temperature precision of soalr cell of ±1°C. The temperature variation across different areas was verified to be within ±1°C.

In this work, we investigated the impact of non-uniform temperature distribution on solar cell performance via simulations. The solar cell was modeled as a network of 100 sub-cells connected in parallel, each randomly assigned a different temperature. A single diode model characterized the electrical behavior of individual sub-cells based on parameters such as photocurrent (I_{ph}), reverse saturation current (I_0), ideality factor (n), series resistance (R_S) and shunt resistance (R_{ph}) to replicate the performance of each sub-cell. Among these, I_0 exhibits strongest sensitivity to temperature changes and is described by the temperature dependence Equation (3) [7]:

$$I_0(T) = D \times T^3 \times e^{-\frac{E_g}{k \times T}} \qquad (3)$$

where D is a constant combining the doping and material parameters of solar cells, T is the temperature, E_g is the bandgap energy and k is the Boltzmann's constant. Other parameters excluding I_0 were treated as temperature-independent and uniformly set across all sub-cells. Leveraging Monte Carlo method, we simulated variations in I_{SC}, V_{OC}, and P_{MAX} relative to their values under uniform temperature conditions. The simulations were conducted assuming an average temperature of 25°C, generating 1000 random temperature profiles within ranges of (25±1)°C, (25±2)°C, and (25±3)°C. It was observed that, in comparison to I_{SC} and V_{OC}, P_{MAX} exhibits a higher responsiveness to temperature non-uniformity.

However, the impact of temperature non-uniformity on I_{SC} and V_{OC} is negligible when temperature fluctuations remain within the ±3°C range. Figure 1 depicts discrepancies in P_{MAX} under conditions of uniform and non-uniform temperature distributions.

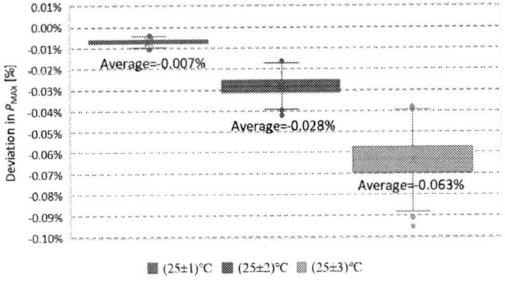

Figure 1: Deviation in P_{MAX} between uniform and non-uniform temperature conditions. It is also noted that the measurement uncertainty has been taken into accounted separately and not included in this graph.

2.3 Uncertainty on effective irradiance

The precision of irradiance adjustment is a key focus due to its direct correlation with solar cell outputs. In our laboratory, the accuracy of effective irradiance relies on regular calibration with a reference cell, supplemented by continuous monitoring with a monitor cell. Aside from electrical and temperature-related uncertainties discussed in Sections 2.1 and 2.2, the calibration uncertainty of reference cell emerges as a notable irradiance uncertainty source. The calibration uncertainty for the reference cell's I_{SC}, as calibrated by Physikalisch-Technische Bundesanstalt (PTB), is quantified at 0.200% (k=1). Annual recalibration of the reference cell is a requisite procedure, and attention must be given to any gradual shifts in performance, known as operational drift. Specifically, the drift was estimated as 0.062% (k=1), using recent calibration records from five crystalline-silicon reference cells over recent years. Another comparable factor involves spectral mismatch between the Device Under Test (DUT) and reference cell. Previously, Dr. Monokroussos outlined the method for uncertainty computation using Monte Carlo simulations [4]. Based on our analysis, the uncertainty is 0.600% (k=1) when the spectral mismatch is not considered in the irradiance adjustment. Even with adjustments for spectral mismatch, an associated uncertainty of around 0.200% (k=1) for the value of spectral mismatch remains.

The effect of non-uniform irradiance on the performance of photovoltaic modules was previously investigated in our work [4]. However, a solar cell should be modeled as comprising multiple sub-cells connected in parallel, differing from the series interconnection of sub-cells within a PV module. Based on Monte Carlo simulation, the relative uncertainties for I_{SC}, V_{OC}, and P_{MAX} were within the magnitude of 10^{-4} % under the irradiance non-uniformity of 1%. Therefore, the uncertainty caused by irradiance non-uniformity can be neglected.

Lastly, the effective irradiance for the solar cell under test should be further corrected to eliminate the shading effect of the probes. To accomplish this, a linear regression analysis was performed on data obtained with varying numbers of contact bars [8]. The associated uncertainty with this correction method arises primarily from the variability in I_{SC} measurements, as well as from the inherent uncertainty of the linear regression process itself. Table I details the specific components of uncertainty for I_{SC} considered in each category. Given that the variability in I_{SC} measurements across different number of contact bars is consistent, an ordinary least squares (OLS) model was employed for the linear regression analysis. It is worthwhile noting that during this linear regression analysis, the confidence interval indicates the degree of uncertainty in the estimated value. Here in this study, the estimated value is the true I_{SC} value after eliminating the shading caused by the contact bars. Besides these factors, all other specific components within the uncertainty on reference irradiance are illustrated in Table II.

2.4 Uncertainty due to contacting and I-V correction

Current and voltage measurements of the solar cells under test are taken by connecting the front side via probing bars and the back side on a copper stage to the measurement instrumentation. Three specific uncertainties associated with this contact method have been considered: firstly, the intrinsic uncertainty due to the contact configuration for multi-busbar solar cells [9]; secondly, the potential variability introduced by the aging of the probes and thirdly, the precision of probe alignment.

While efforts are made to maintain irradiance and temperature levels close to STC, some discrepancies are inevitable in practical measurements. In this study, the concluding stage in the measurement of solar cells under STC involves applying I-V corrections following the IEC 60891 Procedure 2 [10]. We calculate the uncertainty related to the I-V correction utilizing the error propagation method, which takes into account the known uncertainties of the independent variables (such as effective irradiance, cell temperature, measured voltage and current, along with correction parameters like temperature coefficients, internal series resistance, and curve correction factor).

Table I: Measurement uncertainty sources on measured I_{SC} of the DUT.

	N°	Uncertainty source	U($I_{SC.DUT}$) k=1	U($I_{SC.DUT}$) Total Contr.
Electronic Uncertainties	1	Current measurement channel uncertainty	0.058%	1.567%
	2	Voltage measurement channel uncertainty	0.043%	0.861%
Electronic Thermal Drift	4	Thermal drift of the current measurement channel	0.023%	0.251%
	5	Thermal drift of the voltage measurement channel	0.033%	0.500%
Optical Uncertainties	6	Uncertainty due to non-uniformity of irradiance	0.000%	0.000%
Temperature-related Uncertainties	7	Uncertainty due to non-uniformity of temperature	0.000%	0.000%
	8	Temperature controller accuracy	0.030%	0.418%
Contacting unit	9	Uncertainty related to alignment of probes	0.132%	8.192%
	10	Uncertainty related to aging of probes	0.100%	4.702%
Repeatability	11	Repeatability	0.081%	3.101%
	12	Reproducibility	0.157%	11.612%
Reference cell integration	13	Uncertainty due to reference irradiance	0.383%	68.796%
Total			0.467%	100%

Table II: Measurement uncertainty sources on reference irradiance.

	N°	Uncertainty source	U($I_{SC.DUT}$) k=1	U($I_{SC.DUT}$) Total Contr.
Electronic Uncertainties	1	Current measurement channel uncertainty	0.058%	2.278%
	2	Irradiance measurement channel uncertainty	0.144%	14.238%
	3	Voltage measurement channel uncertainty	0.000%	0.000%
Temperature-related Uncertainties	4	Thermal drift of the current measurement channel	0.023%	0.365%
	5	Thermal drift of the voltage measurement channel	0.000%	0.000%
	6	Thermal drift of the irradiance measurement channel	0.023%	0.365%
	7	Uncertainty in Temperature Measurement	0.042%	1.227%
	8	Uncertainty due to the uncertainty of temperature coefficient	0.003%	0.005%
	9	Uncertainty due to monitor cell temperature instability	0.035%	0.820%
Optical Uncertainties	10	Uncertainty on current related to misalignment of reference	0.144%	2.278%
Reference device Uncertainties	11	Uncertainty of short-circuit current calibration certificate	0.276%	27.338%
	12	Estimated drift of reference device	0.062%	2.891%
	13	Spectral mismatch uncertainty between DUT and reference cell	0.200%	27.338%
Repeatability	14	Repeatability	0.016%	0.166%
Shading correction	15	Uncertainty due to linear regression	0.174%	20.692%
Total			0.383%	100%

3 DISCUSSIONS

The measured uncertainties of I_{SC}, V_{OC}, and P_{MAX} for the solar cell characterized at the laboratory of TÜV Rheinland in Shanghai were 0.93% (k=2), 0.54% (k=2), and 1.68% (k=2) with spectral mismatch correction, and 1.46% (k=2), 0.54% (k=2), and 2.02% (k=2) without spectral mismatch correction respectively. Notably, the spectral mismatch correction has significant effect on the accuracy of I_{SC} and P_{MAX} measurements. The relative proportions of the final contributions from each uncertainty source are displayed in Figure 2. For I_{SC}, the main contributors to its uncertainty included reference irradiance, measurement repeatability and reproducibility, and the uncertainties associated with contacting. It is worth noting that the calibration of the reference cell, which significantly affects effective irradiance, is conducted externally by PTB, thus it falls beyond our laboratory's control.

When it comes to V_{OC}, the temperature was identified as the most significant influence, accounting for 65.28% of the uncertainty, with measurement repeatability, reproducibility, and electronic measurement uncertainty also having considerable impacts. As for P_{MAX}, which is derived from I_{SC}, V_{OC}, and fill factor (FF), its uncertainty is tied to the uncertainties associated with these parameters. The uncertainty in FF measurement can reveal distinct factors that predominantly affect P_{MAX}. These encompass uncertainties related to hysteresis effects, metal stabilization, and the inherent uncertainties of the contacting configurations. The uncertainty arising from the FF measurement contributes significantly to the P_{MAX} measurement uncertainty, accounting for 58.95%. The contributions from I_{SC} and V_{OC} were 30.82% and 10.23%, respectively.

41st European Photovoltaic Solar Energy Conference and Exhibition

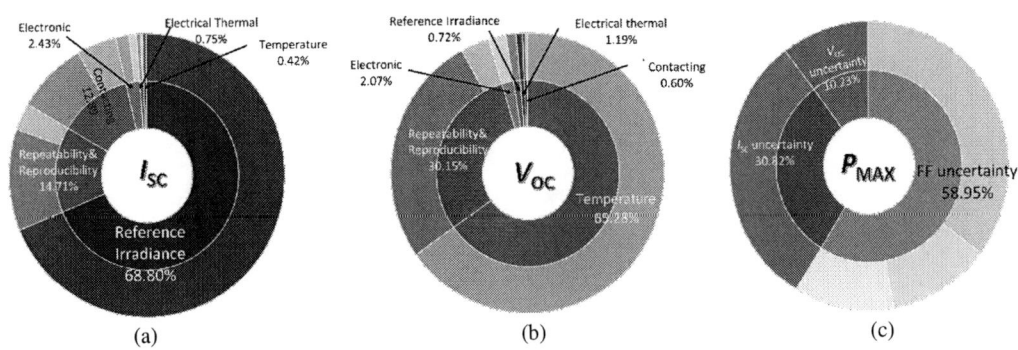

Figure 2: Proportion distribution of contribution from each uncertainty source on(a) I_{SC} and (b) V_{OC} and (c) Pmax. The inner layer of the pie graph corresponds to the major categories of the uncertainty source, while the outer layer corresponds to the subclass of each categore.

4 CONCLUSIONS

This study quantifies the uncertainties in I-V testing of solar cells. The uncertainties for I_{SC}, V_{OC}, and P_{MAX}, following spectral mismatch correction were 0.93% (k=2), 0.54% (k=2), and 1.68% (k=2) respectively. Without this correction, the uncertainties were determined to be 1.46% (k=2) for I_{SC}, 0.54% (k=2) for V_{OC}, and 2.02% (k=2) for P_{MAX} respectively. It is emphasized that the role of reference cell calibration and traceability, thermal management and the consistency of measurements demonstrated by repeatability and reproducibility in achieving precise measurements. The reference irradiance was found to be a critical variable affecting I_{SC}, while V_{OC} and P_{MAX} uncertainties were primarily influenced by the temperature and determination of FF. The outcomes not only pave the way for refining I-V measurement of TÜV Rheinland China but also align measurement practices with stringent international laboratory accreditation requirements.

5 REFERENCES

[1] Dirnberger, D. and Kräling, U., Uncertainty in PV Module Measurement—Part I: Calibration of Crystalline and Thin-Film Modules, IEEE Journal of Photovoltaics, vol. 3, no. 3, pp. 1016-1026, July 2013.

[2] Winter, S., Witt, F., Sperfeld, P., Nevas, S., Albert, H., Nagel, H.D., Ramspeck, K., Haas, F., & Plag, F., Comprehensive Analysis of a Pulsed Solar Simulator to Determine Measurement Uncertainty Components, 2014.

[3] Müllejans, H. et al., Analysis and mitigation of measurement uncertainties in the traceability chain for the calibration of photovoltaic devices, *Meas. Sci. Technol.*, 2009, vol. 20, no. 7.

[4] Monokroussos, C., Etienne, D., Morita, K., et al., Impact of calibration methodology into the power rating of c-Si PV modules under industrial conditions. 28th EUPVSEC, Paris, 2013: 2926-2934.

[5] Gao, Qi, Lau, Cho Fai, Lee, E., Monokroussos, C., Test Method of Current-voltage Characterization of Perovskite PV-module, 36th EUPVSEC, 2019.

[6] Evaluation of measurement data — Guide to the expression of uncertainty in measurement, JCGM 100 (2008).

[7] M.E.Nell, A.M.Barnett, The spectral p-n junction model for tandem solar-cell design, IEEE Transactions on Electron Devices 24, 1987.

[8] Rauer, M., Fell, A., Wöhler, W., Hinken, D., Reichel, C., Bothe, K., Schubert, M.C. and Hohl-Ebinger, J., The Impact of Measurement Conditions on Solar Cell Efficiency. Sol. RRL, 8: 2300873, 2024.

[9] Bothe, K. and Hinken, D., Precise and accurate solar cell measurements at ISFH CalTeC, PV Tech, 2019.

[10] IEC 60891: 2021 Photovoltaic devices - Procedures for temperature and irradiance corrections to measured I-V characteristics.

41st European Photovoltaic Solar Energy Conference and Exhibition

STABILIZATION OF FIELD-AGED CRYSTALLINE PV MODULES BEFORE STC POWER DETERMINATION

Soha Essbai[1], Marcus Rennhofer[1], Ankit Mittal[1], Gusztáv Újvári[1], Thomas Weber[2], Brian Appozardi[3]

[1] AIT Austrian Institute of Technology GmbH (AIT), Austria,
[2] PI-Photovoltaik-Institut Berlin AG, Wrangelstr. 100, D-10997 Berlin, Germany.
[3] The Foundation for Innovation and Research – Malta
marcus.rennhofer@ait.ac.at

ABSTRACT: This study evaluated how field-aged crystalline PV modules can be rated under standard test conditions (STC) and how they can be stabilized before that. The challenge was to compare methods with the standardized method of the IEC standard that promise to function even under the influence of reversible and metastable states or advanced, irreversible degradation. PV modules can show various anomalies due to time, temperature, voltage and light. Stabilization procedures included continuous lighting and electrical power in the dark, as well as stabilization according to IEC 16215 [1]. The research results are part of the ReliaREN-Pro and PROMISE project. One of the main goals of this projects is to achieve international harmonization and potential further development of measurement methods, especially for field-aged crystalline modules.

1 INTRODUCTION

Photovoltaic (PV) systems are a crucial part of the global shift towards sustainable energy sources. When operating photovoltaic power plants, two decisive factors are the availability of irradiation and the performance of the modules. The performance of PV modules can change over time due to degradation, and this effect varies depending on the technology [2-4]. Thin-film photovoltaics [5-7] and crystalline silicon [8] have different degradation over long periods of time. Therefore, each technology requires special methods for stabilization before STC power measurement. While IEC 61215 in the procedure instruction MQ19 provides for treatment under static lighting and at elevated temperature, methods are particularly interesting that could provide information on defects simultaneously (illumination under open circuit voltage) or which can be applied in parallel (dark current).

Basically, each method forms several possibilities for kinetic or dynamic changes in the modules and would thus often no longer be reproducible or non-destructive. Furthermore, lighting or long dunnage storage or electricity can also lead to power increases (such as by post-crosslinking of the EVA) or uncontrolled change in material (such as in different thin film technologies). All this can lead to different reactions of new or old modules during stabilization [9-10], as shown in the diagram in Fig.1, top.

2 METHODOLOGY

For the definition of stabilization in the tests according to MQT 19 of IEC 61215 [1] Δ p +/-0,01 was used. The modules were measured out-of-the-box STC (PASAN-D AAA flasher with AM1.5 and at 25 °C) and then subjected to stabilization in pairs with different methods, see fig. 2 The 15 modules were from a field installation in Malta (polycrystalline modules with 60 cells, their power ranges from 255 to 270 W). The pairs were chosen as the nearest modules for each method according to current and voltage values of the STC measurement, see Fig. 1, Right for sorting and scattering of the values. The two modules with the lowest and middle values served as a reference. The two similar modules each undergo one of the stabilizations step by step with subsequent STC measurements.

- MPP stabilization (IEC 61215) at 1000 W/m 2, 20 kWh, 50°C 1.
- Open circuit voltage (OC): (according to IEC 61215) at 1000 W/m 2, 20 kWh, 50°C 1.
- Dark current at I=0.6Isc, 50°C 1.
- Dark current at I=1.6oC, 50°C 1.

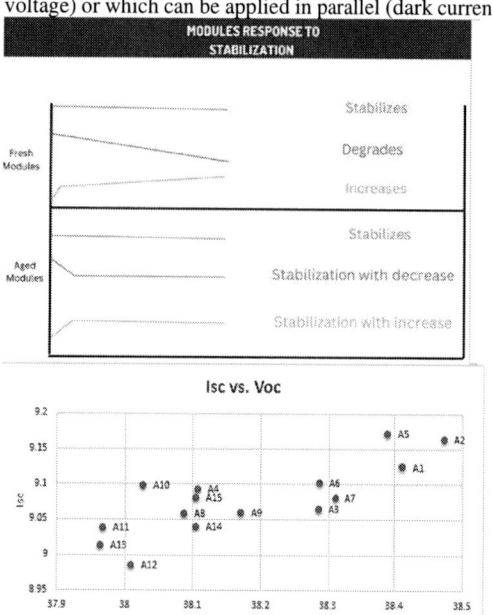

Figure 1: Top: Possible stabilization paths; Bottom: Sorting of values of field-aged polycrystalline modules.

743

41st European Photovoltaic Solar Energy Conference and Exhibition

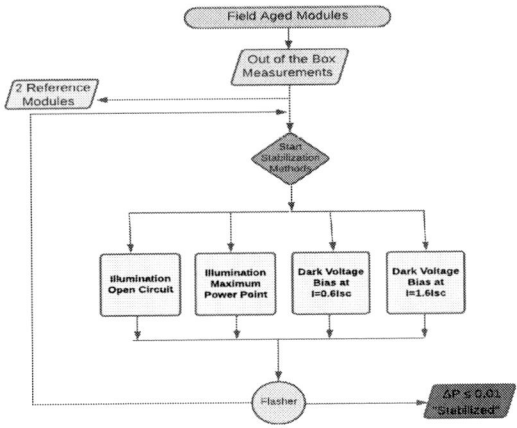

Diagram of the stabilization procedures and the sequence of measurements.

3 RESULTS AND DISCUSSION

The IV curve changes after the snap-in give basic information about the effects of various stabilization methods. Most modules show good stabilization for all procedures, some improved and some degraded. Nevertheless, all show a fundamental stabilization after three steps. 3 the changes in the entire characteristic curve are not clearly visible, you can see them clearly in the magnification around the MPP (insets).

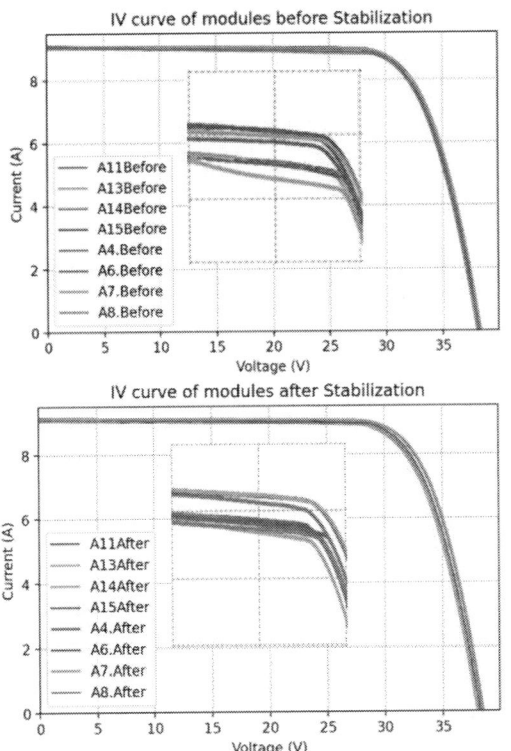

Figure 3: Characteristic of the modules in STC, left before and right after stabilization (the insets have the same scaling for each box).

Exemplarily, two time-dependent curves during stabilization are shown here: the IEC standard stabilization at MPP and lighting and the dark supply with low current (0.6 ISC).

3.1 Temporal development of stabilization in MPP

Initial stabilization slightly increased performance, while filling factors decreased slightly. After that, only slight changes were recorded within the measurement accuracy. The modules showed no further significant change in performance, indicating that they were operated stabilized in the field. The corresponding images for modules A11 and A14 are shown in Figure 4.

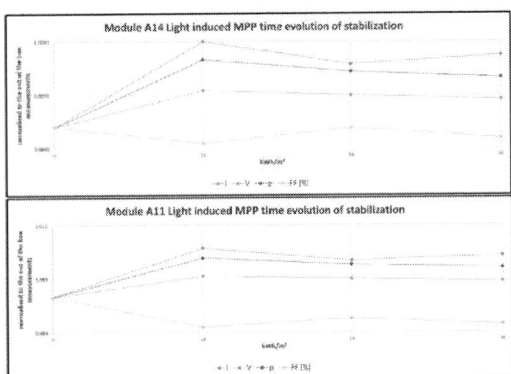

Figure 4: Time stabilization in swap steps for the modules under MPP (following MQ19 IEC 61215).

3.1 Time evolution of the stabilization at dark bias of 0.6 Isc

The stabilization at low dark current led to slightly improved current, voltage, power filling factor at the beginning. Later the filling factor decreased slightly (step 2 of the swap), while the other parameters were stable. In total, all parameters were stable with very little change after 3 steps, with almost no changes, similar to MPP. The corresponding images for modules A4 and A15 are shown in Figure 5.

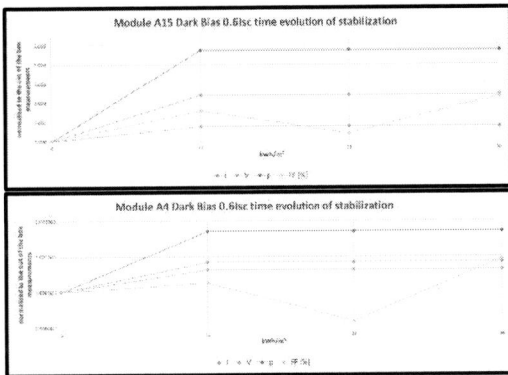

Figure 5: Time stabilization in swap steps for the modules under dark current 0.6 ISC (alternative to MQ19 IEC 61215).

The outflows at 1.6 ISC and OC showed more fluctuating results but also stabilization, only under OC a module was not stabilized (not shown here).

3.2 Comparison of stabilization methods

The results are summarized in Figure 6. Among the analyzed stabilization methods, MPP is the best stabilizing method with the lowest absolute changes, followed by dark current at 0.6 ISC. While 1.6 ISC stabilized very quickly, but led to a greater increase in performance, no stabilization within the normal value could be achieved under OC, but a very large increase in performance. It is noteworthy that 0.6 ISC showed the smallest further changes after the first step of the retrieval (except for FF).

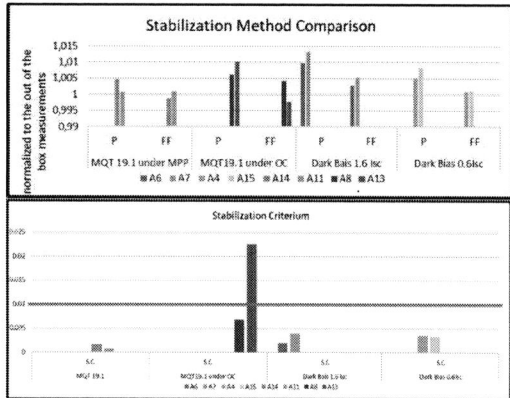

Figure 6: Stabilization values Top: electrical characteristics normalized to the input values (out-of-the-box). Below: Compliance with IEC-61215 stabilization criterion.

4 Conclusion

Photovoltaic solar modules are subject to a variety of degradation processes over time, which affect important performance characteristics such as Voc, Isc, FF and Pmpp. Understanding and managing these degradation effects is crucial to maintaining and increasing the efficiency of solar energy conversion. Various stabilization methods were evaluated, and the common standard as well as weak dark current proved to be the most effective strategies. The confirmation of MQ19 promises confidence for quality assurance in aging plants. In contrast, the good result of dark supply is a potential possibility for automated and multiplied stabilization of modules before STC power measurement. The further results of the ReliaREN-Pro project will compare several module types and collect more statistical data to confirm these first results.

The main findings may be summarized finally as:

(1) Photovoltaic solar modules are subject to a variety of degradation processes over time, which affect important performance characteristics.

(2) Various stabilization methods were evaluated and the common standard as well as weak dark current proved to be the most effective strategies. The confirmation of MQ19 promises quality assurance in ageing plants.

(3) The good result of dark bias opens a potential opportunity for automated and multiplied stabilization of modules before STC power measurement.

(4) The further results of the ReliaREN-Pro project will compare several module types and collect more statistical data to confirm these first results.

The research project was supported by the PROMISE project (Grant Agreement n°101079469) and the SES Call 2020 project ReliaREN-Pro).

References

[1] IEC 61215-1-1: Terrestrial photovoltaic (PV) modules – Design qualification and type approval - Part 1-1: special requirements for testing of crystalline silicon photovoltaic (PV) modules. Ed2 2021.

[2] Michael Gostein and Larry Dunn. Light soaking effects on photovoltaic modules: Overview and literature review. Conference Record of the IEEE Photovoltaic Specialists Conference, pages 003126–003131, 06 2011.

[3] Senthilarasu Sundaram, Katie Shanks, and Hari Upadhyaya. 18 - thin film photovoltaics. In Trevor M. Letcher and Vasilis M. Fthenakis, editors, A Comprehensive Guide to Solar Energy Systems, pages 361–370. Academic Press, 2018.

[4] P. Wu¨rfel. Physics of Solar Cells: From Basic Principles to Advanced Concepts. Physics textbook. Wiley, 2009.

[5] R. A. Sasala and James R. Sites. Time dependent voltage in CIGS and CdTe solar cells. Conference Record of the Twenty Third IEEE Photovoltaic Specialists Conference - 1993 (Cat. No.93CH3283-9), pages 543–548, 1993.

[6] Samuel Demtsu, Shubhra Bansal, and David S Albin. Intrinsic stability of thin-film CdS/CdTe modules. 2010 35th IEEE Photovoltaic Specialists Conference, pages 001161–001165, 2010.

[7] C. Deline, A. Stokes, T. J. Silverman, S. Rummel, D. Jordan, and S. Kurtz, "Electrical bias as an alternate method for reproducible measurement of copper indium gallium diselenide (CIGS) photovoltaic modules," in SPIE Solar Energy+ Technology. International Society for Optics and Photonics, 2012, pp. 84 720G–84 720G.

[8] I.L. Repins, F. Kersten, B. Hallam, K. VanSant, M.B. Koentopp, Stabilization of light-induced effects in Si modules for IEC 61215 design qualification, Solar Energy, Volume 208, 2020, Pages 894-904, ISSN 0038-092X,

[9] Jean Zaraket, Chafic Salame, and Michel Aillerie. Dark and illuminated characteristics of photovoltaic solar modules. Part II: Influence of light electrical stress. AIP Conference Proceedings, 1758(1), 07 2016. 030052.

[10] Munoz, M. A., Faustino Chenlo, and M. Carmen Alonso-García. "Influence of initial power stabilization over crystalline-Si photovoltaic modules maximum power." Progress in photovoltaics: Research and Applications 19.4 (2011): 417-422.

41st European Photovoltaic Solar Energy Conference and Exhibition

AC/DC Electroluminescence. The war of the currents

*Mario Martínez[1], *Sergio Suarez[1], Daniel Villoslada[1], Jose Manuel Rivas[1], Sofía Rodríguez-Conde[1]

[1]Enertis Applus Parque Empresarial Las Mercedes, C/ de Campezo, 1, 28022 Madrid, Spain

www.enertisapplus.com

*Phone: +34 91 651 70 21, *sergio.suarez@enertisapplus.com, *Mario.martinez@enertisapplus.com*

ABSTRACT: Solar photovoltaic energy has emerged as a pivotal solution to address global energy challenges, offering clean and sustainable power sources. However, ensuring the quality and reliability of photovoltaic modules is paramount to optimize the efficiency and longevity of solar systems. This necessitates precise characterization of modules and photovoltaic plants, with early detection of faults becoming crucial for maintenance and productivity. Among defects, cell-level anomalies can lead to significant failures and long-term repercussions on surrounding PV modules. Conventional defect detection methods, such as high-resolution electroluminescence (EL) imaging using direct current (DC) polarization at short-circuit current (I_{sc}), face limitations in detecting latent anomalies like poor soldering joints or potential induced degradation (PID). To address these limitations, an innovative approach employing alternate current (AC) polarization signals for EL imaging is proposed. This method enhances defect visibility and enables comprehensive defect analysis, including hotspots, cracks, degradation zones, and PID occurrences. The approach allows simultaneous polarization at the effective short-circuit current (I_{sc_RMS}) and its 10% equivalent, facilitating thorough defect analysis. Preliminary results, conducted at the Enertis Applus+ laboratory, demonstrate the identification of previously undetectable disconnected areas not detectable through conventional methods. Moving forward, the goal of this work will be to conduct a more in-depth analysis of other widespread defects such as PID and poor soldering.

Keywords: Photovoltaic modules, Electroluminescence imaging, Defect detection, AC polarization, PID.

1 INTRODUCTION

Solar photovoltaic energy has increasingly gained prominence as a vital solution to tackle global energy challenges, offering a reliable source of clean and sustainable power. With the global transition towards renewable energy, the role of photovoltaic systems in meeting energy demands has become more significant. However, ensuring the quality and reliability of photovoltaic modules is essential to maximize both the efficiency and lifespan of solar installations. This is particularly important as the long-term performance of solar systems directly depends on the consistent output of individual modules. In this context, precise characterization of photovoltaic modules and plants plays a critical role in maintaining system performance.

In the operation and maintenance (O&M) of photovoltaic (PV) plants, the early identification of faults has become an essential factor in preserving productivity and extending the lifespan of system components [1][2]. Among the various defects that can occur, cell-level anomalies are of particular concern, as they have the potential to cause significant failures that not only affect individual modules but can also propagate, leading to broader impacts on surrounding modules over time.

Figure 1: Technicians conducting night-EL inspection on PV plant.

Traditionally, fine-scale defects in PV modules are identified using high-resolution electroluminescence (EL) imaging [3][4] as shown in Figure 1. This method involves polarizing the module with DC at its short-circuit current (I_{sc}) [5], allowing for the visualization of defects such as cracks, multiple cracks, and finger-related anomalies, as well as the identification of disconnected regions. While EL imaging has proven effective in detecting visible defects, its capability is limited when it comes to identifying latent issues such as poor soldering joints or potential induced degradation (PID) [6][7]. These underlying defects can remain hidden during conventional inspections and may lead to long-term performance degradation if not addressed.

2 ENHANCING DEFECT DETECTION: AC VS DC POLARIZATION

To overcome the limitations of conventional defect detection methods, researchers have proposed polarizing photovoltaic modules at 10% of the short-circuit current (I_{sc}), though this approach often compromises defect visibility. In this context, our study, conducted at the Enertis Applus+ laboratory, explores an innovative approach utilizing AC polarization signals for EL imaging. This methodology offers promise in identifying a range of anomalies, including hotspots, cracks, degradation zones, and PID occurrences.

By utilizing sinusoidal AC signals, this technique allows for simultaneous polarization at the effective short-circuit current (I_{sc_RMS}) and its 10% equivalent across different frequencies. This enables a more comprehensive defect analysis by accounting for the differences in series and parallel resistances within the cells. Unlike steady-state DC polarization, which may fail to reveal certain latent defects, AC polarization dynamically interacts with the cell's electrical characteristics, leading to enhanced defect visibility.

The schematic representation of this approach is provided in Figure 2, which compares the results of DC and AC polarization. The top portion of the figure shows EL images under DC polarization, where multicracks and

PID patterns are only partially visible. In contrast, the bottom portion displays EL images captured under AC polarization at I_{sc_RMS}. The AC polarization uncovers more detailed information about multicracks and PID patterns that were otherwise not fully visible with DC. Additionally, it shows how AC polarization at both I_{sc_RMS} and 10% of I_{sc_RMS} enhances the overall characterization of defects by addressing the resistive behavior within the photovoltaic module.

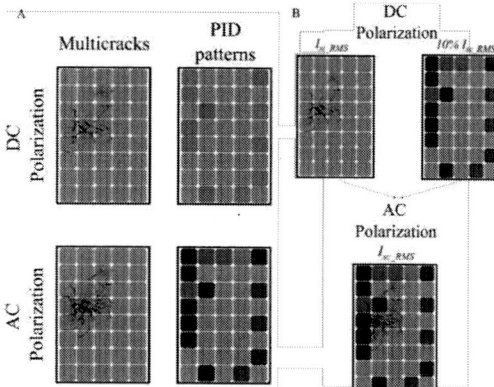

Figure 2 A): Exemplary drawing illustrating defects found in modules during electroluminescence inspection. The top section shows the expected results for multi-cracks and PID under direct polarization, while the bottom section displays the results of the same defects under alternate polarization. Additionally, an explanatory diagram of EL images obtained for different polarization currents in direct polarization (top) and alternate polarization (bottom) is provided. B) Images obtained under AC polarization would correspond to direct polarization at I_{sc_RMS} and 10% of I_{sc_RMS}.

This innovative AC polarization method is expected to improve defect detection, enabling more accurate and thorough diagnostics for PV module maintenance and performance analysis. The following sections will discuss the methodology and experimental results, which confirm the enhanced visibility of defects using AC polarization in comparison to traditional DC methods.

3 METHODOLOGY

In this study, EL imaging was performed under both AC and DC polarization to evaluate and compare defect detection capabilities in PV modules and cells. The primary objective was to assess how AC polarization at the I_{sc_RMS} enhances defect visibility when compared to traditional DC polarization at both full I_{sc} and 10% of I_{sc}.

3.1 Polarization and setup

The devices under test (DUT), which included both individual photovoltaic cells and full modules, were polarized using a programmable power supply. This power supply allowed for precise control over the polarization signal, enabling easy switching between AC and DC polarizations. For AC polarization, the signal was set to the effective root mean square (RMS) of I_{sc_RMS}, simulating the alternating current conditions under which PV modules may operate in the field. This AC signal ensures that the same total energy was injected into the module as would be under DC conditions but with the potential to reveal different defect characteristics.

In the case of DC polarization, the modules were polarized at the full I_{sc}, as is standard practice in traditional EL imaging. Additionally, to provide a more comprehensive analysis, the modules were also polarized at 10% of the I_{sc}. This allows for the identification of defects that may only manifest under lower current densities, a factor that can sometimes influence the visibility of latent issues such as poor soldering joints or weak interconnections between cells.

3.2 Image capture and processing

The EL images were captured using a high-resolution camera. To ensure consistency and accuracy in the comparative analysis between AC and DC polarizations, all images were taken under the same conditions. The exposure time, aperture settings, and camera sensitivity (ISO) were kept constant for all images, regardless of the polarization method applied. This approach ensures that any observed differences in defect visibility are solely attributed to the effects of the polarization technique, rather than variations in imaging conditions.

The images were then processed using advanced pixel-wise analysis techniques. This processing focused on identifying variations in pixel intensity that correspond to underlying defects within the PV cells and modules. By comparing the intensity variations across the AC and DC images, specific defects, including multicracks, disconnections, and PID zones, were identified and mapped.

3.3 DUT characteristics

The DUTs included both multicrystalline and monocrystalline silicon PV modules and cells. These samples were selected based on their varying susceptibility to certain types of defects, such as cracks, multicracks, and PID. Testing on both cells and modules allowed for a more comprehensive understanding of how interconnections within the module affect defect visibility under different polarization methods.

4 EXPERIMENTAL RESULTS

The experimental results presented in this section focus on the comparison of EL imaging conducted using both DC and AC polarization.

4.1 Testing on cells

The first set of tests was conducted on individual photovoltaic cells with multicracks, using both DC and AC polarization. These tests aimed to compare the effectiveness of each method in revealing common defects found in photovoltaic cells, such as multicracks and shaded regions. As shown in Figure 3, the images captured under both polarization methods, taken under identical optical conditions, present distinct results. While both methods successfully detect defects like multicracks, the extent to which these defects are visible and the overall contrast between damaged and intact regions differ significantly between DC and AC polarization.

In particular, the images captured under AC polarization exhibit enhanced contrast in areas where defects are present, suggesting that this method may provide clearer insights into certain defect types. The ability of AC polarization to reveal more detailed information about the condition of the cell is particularly

41st European Photovoltaic Solar Energy Conference and Exhibition

evident in regions where the defects are more subtle, such as small cracks or partially shaded areas. This observation suggests that the interaction between the applied polarization method and the internal electrical properties of the cells, including resistive effects and charge transport dynamics, plays a crucial role in the visibility and detection of defects.

detection, underscoring its potential advantages for identifying defects that may be overlooked or appear less pronounced when using DC polarization alone.

The pixel-wise analysis thus offers a more comprehensive comparison of the two methods, illustrating how AC polarization enhances defect visibility in key areas of the cell. This increased visibility could be particularly useful for identifying subtle or early-stage defects that might not yet be causing significant performance degradation but have the potential to do so over time.

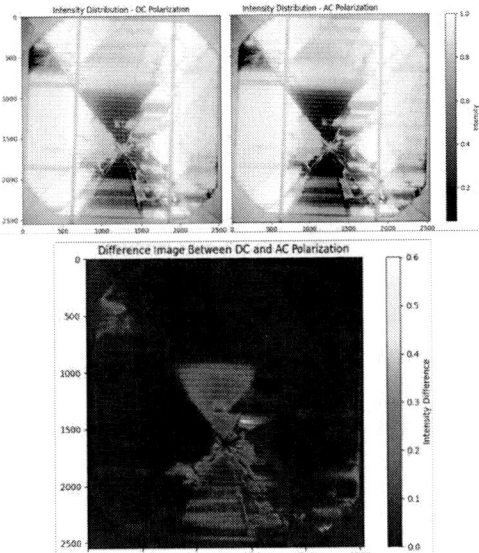

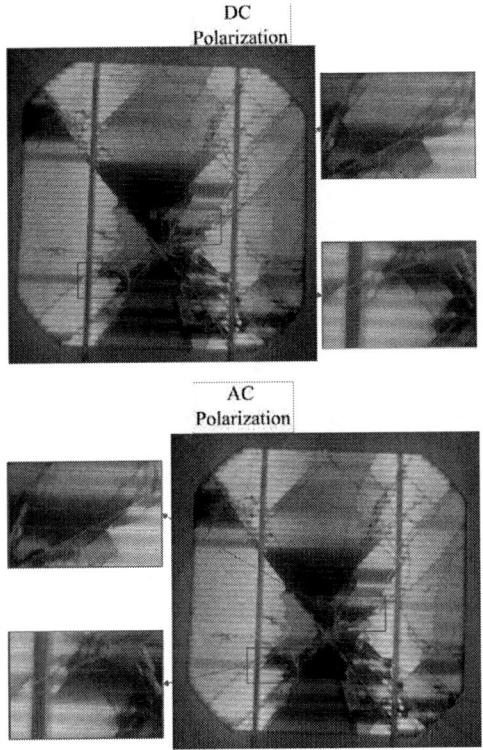

Figure 3. EL images of multitracked cell under DC and AC polarization at I_{sc_RMS}, with red boxes highlighting areas of interest.

To explore these differences further, a pixel intensity analysis was conducted to quantify the variations in defect visibility between the two polarization methods. This analysis focuses on regions affected by multicracks and shaded areas, which are particularly prone to reduced performance over time. The pixel-wise intensity distribution, shown in Figure 4, compares the images captured under DC and AC polarization. The top and bottom images represent the intensity distribution for DC and AC polarization, respectively, while the rightmost image illustrates the pixel-wise intensity difference between the two methods.

The analysis reveals that AC polarization results in a more uniform intensity distribution across the multicracked regions. This leads to improved contrast between defective and non-defective areas, making the cracks and disconnections more easily distinguishable. This enhanced contrast is particularly noticeable in the areas highlighted by red boxes in Figure 3, where the AC-polarized image reveals more detailed information than the corresponding DC image.

The intensity difference map further emphasizes the areas where AC polarization offers greater clarity in defect

Figure 4. Pixel intensity electroluminescence images of a cell with multicracks. On the left, the top image is under DC polarization, and the bottom one is under AC polarization. On the right is the pixel-wise intensity difference between the two polarizations.

4.2 Testing on modules

The second set of experiments was conducted on photovoltaic modules, focusing on the detection of multicracks using both DC and AC polarization. As shown in Figure 5, significant differences in brightness and defect visibility are observed between the two polarization methods. Both DC and AC polarization reveal the presence of multicracks; however, the areas around the busbars and certain regions of the module appear brighter and more distinct in the AC-polarized images.

The enhanced brightness around the busbars in the AC-polarized images highlights regions where cracks and electrical disconnections are more pronounced. This increased contrast provides greater clarity in identifying multicracks, particularly in areas that may not be as easily visible with DC polarization. The red boxes in Figure 5 mark areas of interest where these differences are most apparent, showing how AC polarization enhances the visibility of cracks and other defects by interacting more dynamically with the internal resistances of the module.

Additionally, the AC-polarized image shows that certain areas of the module appear more illuminated compared to the DC-polarized image, suggesting that AC polarization can better capture fine details across the module's surface. This difference in brightness is especially important for detecting multicracks that may

748

41st European Photovoltaic Solar Energy Conference and Exhibition

otherwise be obscured or less defined in DC-based images, offering a more thorough assessment of the module's condition.

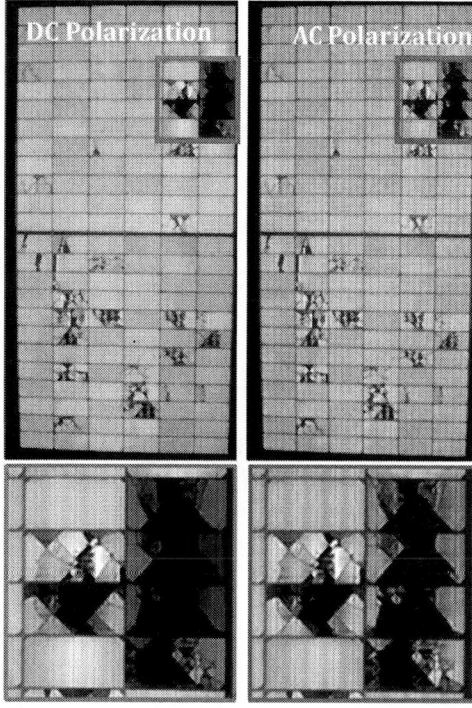

Figure 5 . EL images of multitracked solar module under DC and AC polarization at I_{sc_RMS}, with coloured boxes highlighting areas of interest.

The experiment shown in Figure 6 focuses on a polycrystalline photovoltaic module affected by PID, both experiments were conducted at the effective short-circuit current (I_{sc_RMS}). The comparison highlights differences in the visibility of PID-affected cells between the two methods.

In the DC-polarized image, certain areas of degradation are visible, but some cells remain undetected or show lower contrast. In contrast, the AC-polarized image reveals additional cells that were not visible under DC polarization, suggesting that AC polarization allows for the detection of more subtle degradation patterns. These findings are supported by the pixel-wise intensity difference map, where the dark regions indicate areas that both methods detected, while the brighter areas correspond to cells that are only visible under AC polarization.

This comparison demonstrates that, while both methods capture the degradation in some regions, AC polarization is capable of detecting additional cells affected by PID that are not visible under DC conditions, all while maintaining pixel intensity. This suggests that AC polarization may offer improved sensitivity in identifying defects in certain cells that may not be detectable with DC polarization.

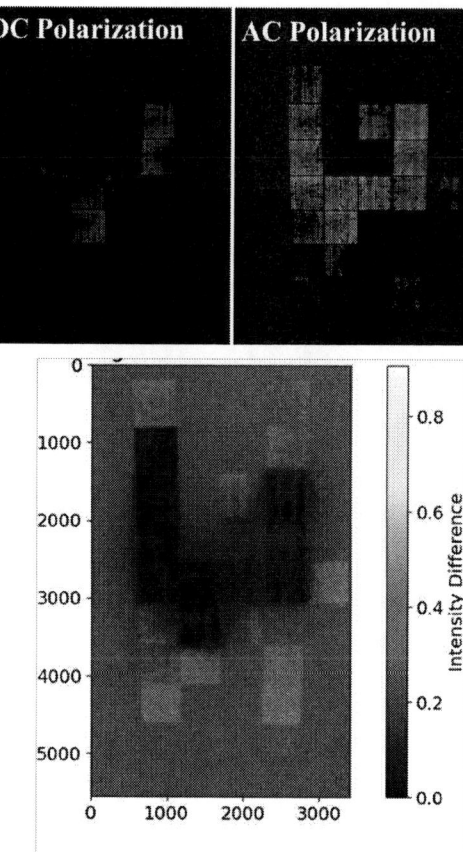

Figure 6 . Electroluminescence images of a PID-affected solar module under DC and AC polarization at I_{sc_RMS} near the edges. The bottom image shows the pixel-wise intensity difference between the two images.

5 CONCLUSIONS

The comparative analysis of electroluminescence (EL) imaging using both direct current (DC) and alternating current (AC) polarization on photovoltaic cells and modules has demonstrated key differences in defect detection capabilities. AC polarization consistently showed improved visibility of defects, particularly in areas affected by multicracks and potential-induced degradation (PID). The images captured under AC polarization exhibited higher brightness and contrast in regions around busbars and other critical areas, making defects more distinguishable compared to DC polarization.

Furthermore, AC polarization revealed defects, such as multicracks and PID-affected cells, that were not detected under DC polarization, highlighting its potential to detect more subtle or early-stage defects that may go unnoticed with traditional methods. Despite these differences in detection, both DC and AC polarization methods maintained similar pixel intensity distributions in areas where both detected the same defects, indicating that AC polarization enhances detection without compromising image quality. The results suggest that AC polarization could play a key role in improving the accuracy of photovoltaic module diagnostics, offering more detailed insights into cell-level and module-level defects and potentially enabling earlier maintenance interventions to reduce performance losses over time. In conclusion, AC polarization presents clear advantages in terms of defect detection and visibility compared to DC polarization, making it a valuable tool for enhancing the reliability and longevity of photovoltaic systems.

6 REFERENCES

[1] Haque, A., Bharath, KVS, Khan, MA, Khan, I. and Jaffery, ZA (June 1, 2019). Fault diagnosis of Photovoltaic Modules. Energy Science and Engineering. John Wiley & Sons Ltd. https://doi.org/10.1002/ese3.255

[2] Zhao, J., Sun, Q., Zhou, N., Líu, H., & Wang, H. (2020). A photovoltaic array fault diagnosis method considering the photovoltaic output deviation characteristics. International Journal of Photoenergy, 2020, 1-11. https://doi.org/10.1155/2020/2176971

[3] Kropp, T., Schubert, M., & Werner, J. (2018). Quantitative prediction of power loss for damaged photovoltaic modules using electroluminescence. Energies, 11(5), 1172. https://doi.org/10.3390/en11051172

[4] Otamendi, U., Martinez, I., Quartulli, M., Olaizola, IG, Viles, E., & Cambarau, W. (2021). Segmentation of cell-level anomalies in electroluminescence imaging of photovoltaic modules. Solar Energy, 220 , 914-926. https://doi.org/10.1016/j.solener.2021.03.058

[5] Kropp, T., Berner, M., Stoicescu, L. and Werner, JH (2017). Self-sourced daylight electroluminescence from photovoltaicmodules,IEE Journal of Photovoltaics , 7 (5), 1184-1189. https://doi.org/10.1109/JPHOTOV.2017.2714188

[6] Yamaguchi, S., Yamamoto, C., Ohdaira, K., & Masuda, A. (2018). Comprehensive study of potential-induced degradation in silicon heterojunction photovoltaic cell modules. Progress in Photovoltaics: Research and Applications, 26(9), 697-708. https://doi.org/10.1002/pip.3006

[7] Spataru, S., Séra, D., Hacke, P., Kerekes, T., & Teodorescu, R. (2015). Fault identification in crystalline silicon pv modules by complementary analysis of the light and dark current–voltage characteristics. Progress in Photovoltaics: Research and Applications, 24(4), 517-532. https://doi.org/10.1002/pip.2571

41st European Photovoltaic Solar Energy Conference and Exhibition

EVALUATION OF THE CONTACT QUALITY IN SILICON SOLAR CELLS AND MODULES USING LBIC PHASE MAPPING

Majid Salari[a], Jonas Buddgård[b], Markus Rinio[a]
a Department of Engineering and Physics, Karlstad University, 651 88, Karlstad, Sweden
b StickySolarPower, Vändvägen 10F, 192 48 Sollentuna, Sweden
majid.salari@kau.se, jonas.buddgard@stickysolarpower.com, markus.rinio@kau.se

ABSTRACT: Making good electrical contacts when interconnecting solar cells with no or very small busbars in a module can be challenging. Light beam induced current (LBIC) mapping reveals the contact quality with a high spatial resolution when evaluating the phase shift between the pulsating laser light and the current. In this research, we have fabricated solar modules with Sticky Solar Power's "The Tape Solution™" and assessed the modules using LBIC phase mapping. Enhanced LBIC phase shifts correlate with smaller series resistances between the point of illumination and current extraction. The areas with abrupt changes in the phase shift were observed under optical microscope and revealed breakage or weakening along fingers. We also developed a cross-section preparation method to investigate the physical contact between the solar cell and the taped wires at the areas of interest. This method included cutting the laminated module using rotary diamond blades followed by grinding and polishing the surface to remove the damaging effects of cutting pressure. Optical microscopic and SEM/EDS images of the cross-section near the silver pads at areas with higher resistance showed low silver content in some parts and minor gaps between the silver and solder alloy.
Keywords: LBIC, The Tape Solution, Silicon Solar Cells, Phase shift, Cross-section

1 INTRODUCTION

Silicon solar cells and modules are still the dominating photovoltaic systems in the market. The trend went towards front grid designs with less area coverage driven by the need to save silver and to increase efficiencies. Busbarless solar cells have to be interconnected in a way where the interconnection wires are directly soldered to the contact fingers. Only a few methods are currently available on the market with Sticky Solar Power's "The Tape Solution™" (TTS) being one of them [1]. Many essential characterization techniques are used to optimize cell performance, among which are photoluminescence (PL), Electroluminescence (EL) and I-V curve measurements. The materials' structure and their optical and electrical properties are analyzed with the aim to reduce defects and make the device work at its peak efficiency. Regardless of the technology, device type and research level, characterization will remain a crucial step in improving the efficiency, durability and scalability of photovoltaics.

Another powerful characterization tool in solar cell technology is Light-Beam-Induced Current (LBIC) mapping which has been used for decades to study the photoelectric properties of solar cells. Our LBIC tool results in maps of the internal quantum efficiency at a fixed wavelength, showing recombination active electrical defects in the devices. LBIC mapping has been widely used to locate different types of defects such as mechanical cracks and disconnections, extended crystal defects that reduce the minority carrier life time. It was also used to analyze the effect of different fabrication steps in single- and multicrystalline silicon solar cells [2–7]. This technique is also used for organic and novel inorganic solar cells [8–10]. There are similar and even more advanced characterization methods like Light-Beam-Induced Voltage (LBIV) and solar CELl LOcal characterization (CELLO) for studying solar cells and detecting their defects [11].

In this research, we have used an advanced version of LBIC which also measures the local phase shift between the induced current and the laser light. This phase shift reveals more information about the physical and electrical properties of the solar cell. We have laminated some small-sized solar modules and performed LBIC measurements on them. After analyzing the maps, some points of interest on the modules were selected and underwent more physical study. This included cutting the module and preparing cross-section samples to observe the physical contact between the cell and the wires. Finally, microscopic and scanning electron microscopy (SEM) images of the cross-sections were performed that provided complementary information to the LBIC measurements.

2 EXPERIMENTAL METHODS

2.1 LBIC phase mapping

In LBIC mapping, a focused light beam scans across the surface of a solar cell and the resulting current at each point is measured. At every point, also the amount of reflected light as well as the laser power is measured, leading to maps of the internal quantum efficiency. Using this tool, it is possible to identify defects and their structure. The LBIC technique can map these variations that provides insights into internal electrical activity and physical structure of the cell. This helps to optimize material quality, improve the manufacturing processes and enhance overall cell performance.

For inducing photocurrent in the cell in LBIC mapping, a laser with one specific wavelength is used. The depth that the laser penetrates the material depends on the material's absorption coefficient. Our system is equipped with two laser systems with wavelengths of 400 nm and 828 nm. It has separate optics for each laser to provide the maximum possible spatial resolution of about 3 and 6 µm, respectively. In this work, we have used 828 nm wavelength. The LBIC machine is equipped with detectors for each laser system to continuously measure the laser power and the reflected light over nearly the full solid angle above the cell at every position.

The laser is modulated by a signal with specific frequency which generates a current with the same frequency in the cell. However, there is a phase shift between the light signal and the current extracted at the contacts. This phase difference comprises of a constant phase because of the connections and measurement equipment and a phase shift during the carrier generation

41st European Photovoltaic Solar Energy Conference and Exhibition

and extraction. We have mapped this second term over the solar cell.

2.2 The Tape Solution™

TTS was developed to be able to interconnect solar cells with small or no busbars to save silver. The interconnection wires are glued to the cells using small tapes. The wires are covered with a low temperature solder. The tapes keep the wires in place and connects all cells safely in strings. The actual soldering takes place simultaneously with lamination into modules where all contacts are formed. The tapes stay in place and become a nearly invisible part of the module. The process spares the classical soldering step prior to lamination. This reduces thermal stress of the strings.

For our experiment, we cut 5-busbar M2-sized monocrystalline bifacial silicon solar cells into two halves using a 532 nm laser and cleaving to make small-sized modules. Wires were taped to both sides of these cells and two-cell small sized modules were fabricated by laminating a sandwich of frontglass, EVA, solar cell, EVA, and backsheet at our laboratory. The layer structure is shown in Fig. 1. The lamination process involved controlled change of heat and pressure applied to the stack according to the schematic plot in Fig. 2.

2.3 Physical contact evaluation

We have selected some regions of interest according to the LBIC phase shift maps for further study and physically observation of the contacts. The laminated modules were cut using rotary diamond blade cutters with as much precision and stability as possible. However, some deformity of the contact between the cell and the wires were visible at the cross-section. Therefore, we cold-mounted the samples in epoxy resin and ground the cut

surface with the details in the step 1 of Table I until the sample's cross-section was far enough from the cutting line. We continued checking the sample using an optical microscope to make sure that the contact after grinding looks intact from the pressure effects during cutting. After reaching to this point, we proceeded with the grinding steps according to Table I, followed by 2 polishing steps as listed in the recipe of Table II.

After preparing clear polished cross-sections of the samples, we were able to achieve high resolution optical microscopic images of the contacts between the neighboring layers. Furthermore, SEM and Energy Dispersive X-ray Spectroscopy (EDS) with accelerating voltage of 18 KV and a working distance of 8.6 mm revealed scans of these contacts with much higher precision at the cross-sections.

Table I: Grinding Recipe

Step		1	2	3
Grinding surface		Foil	Foil	Foil
Abrasive	**Type**	SiC	SiC	SiC
	Size	#220	#500	#800
Suspension/ Lubricant		Water	Water	Water
Round Per Minute		300	300	300
Force (N)		25	25	25
Time (min)		As needed	4	1

Table II: Polishing Recipe

Step		1	2
Polishing surface		MD-Mol	MD-Chem
Abrasive	**Type**	Diamond	-
	Size	3 µm	-
Suspension/ Lubricant		DiaPro/ Mol R 3	Iron (III)/ Nitrate
Round Per Minute		150	150
Force (N)		25	15
Time (min)		4	1

3 RESULTS AND DISCUSSION

As a first approach, we can use the model shown in Fig. 3 as the simplified equivalent circuit between the current generation and extraction point. In this model, I_1 is the generated current and I_2 is the measured current at the extraction point. Between these points, there is a series resistance R_S and a p-n junction capacitance C. The phase difference between the two currents is:

$$\varphi = \varphi_0 + \arctan(-\omega C R_s) \approx \varphi_0 - \omega C R_s \quad (1)$$

with $\omega = 2\pi f$, where f is the laser modulation frequency, and φ_0 a constant offset phase shift from different amplifiers and the light detector. According to equation (1), the change in the phase might correspond to series resistance or capacitance variations. An additional small phase change can come from the minority carrier lifetime of the silicon, which is not evaluated here.

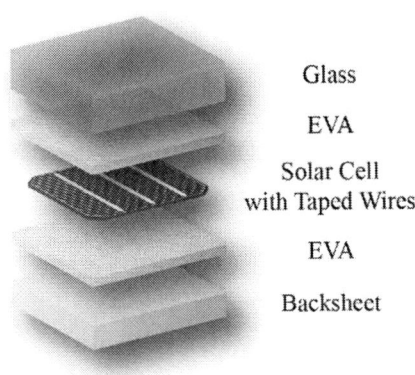

Glass

EVA

Solar Cell with Taped Wires

EVA

Backsheet

Figure 1: Layers of the laminated small-sized modules

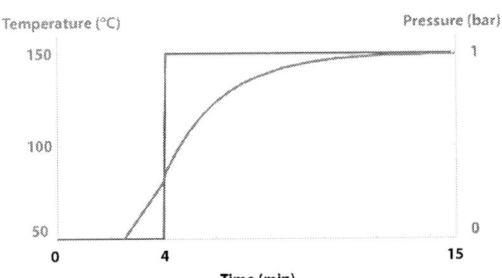

Figure 2: Schematical changes of temperature and pressure during lamination.

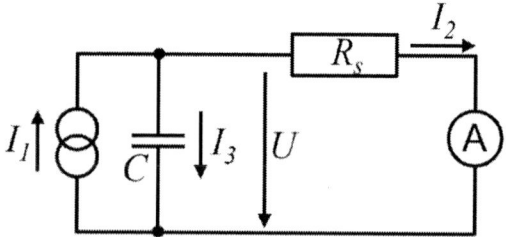

Figure 3: Simplified model of the electrical circuit between the current generation and current extraction points in the LBIC system.

The LBIC maps of the solar cell are shown in Fig. 4. As can be seen, there are some abrupt changes in the phase shift on the maps which are the areas with significant resistance changes. These areas with high phase shift were investigated with optical microscopy and the images are depicted in the right side of the figure. These images clearly show finger interruptions or reduced finger thicknesses at the marked areas. Although such defects could also be detected by electroluminescence imaging, we use these results as a proof of the dependency of the phase shift on the series resistance.

Figure 5 shows two microscopic cross-section images of small-sized modules around silver pads on the rear side of the cell. The image on the top and bottom correspond to areas around silver pads at the rear side with lower and higher resistances, respectively. It is significant that in the bottom image, there are some areas with low amount of silver paste. This might be the reason for the enhanced series resistance at the contact as indicated by the phase shift map. Figure 6 also shows SEM/EDS scans of the corresponding areas. The minor gaps between the silver

and solder alloy seen in the bottom image cause nonuniformity in the connection and the increase in resistance can be a consequence of it.

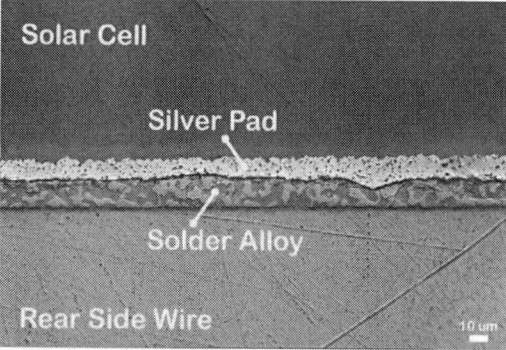

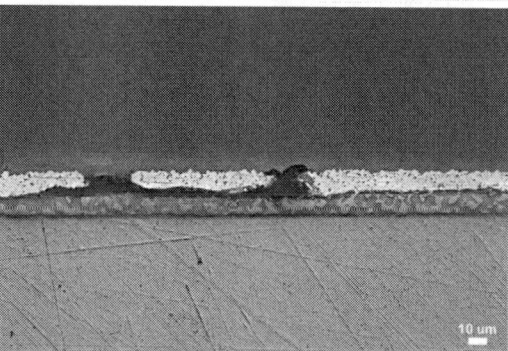

Figure 5: Microscopic images of wire-cell connection at (Top) areas with lower resistance and (Bottom) areas with higher resistance due to missing silver paste in this area.

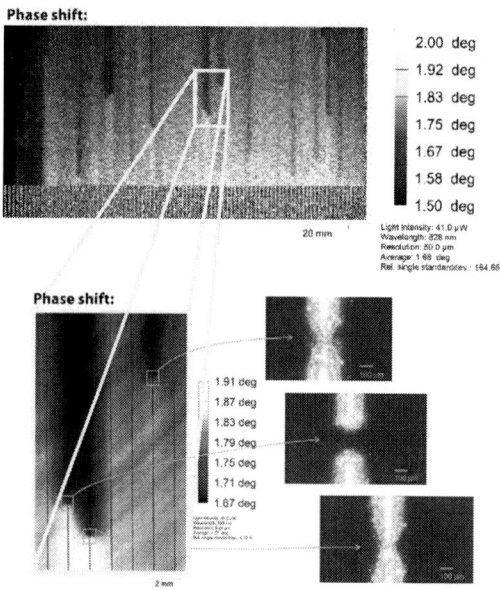

Figure 4: Top: LBIC phase shift maps on a busbarless solar cell. The current is extracted at the bottom of the image. Bottom Left: Enhanced map of the marked region on the left showing strong steps in the phase at weak finger positions. (Ring-like structures in both phase maps have a likely origin in the CZ-growth process of the wafers). Bottom Right: Optical microscope images of the three marked areas.

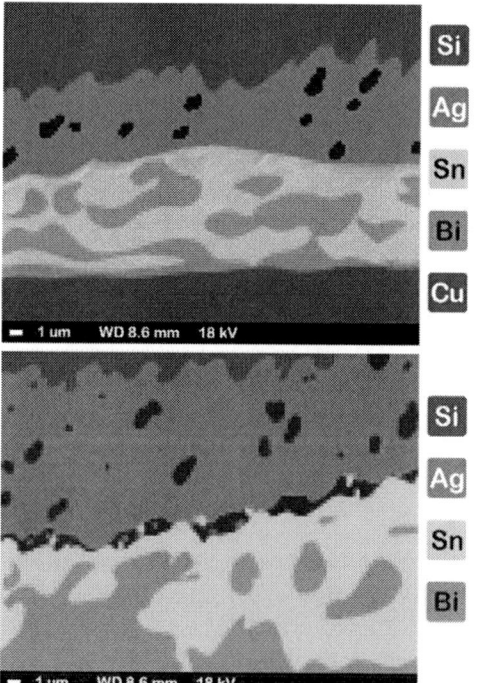

Figure 6: SEM/EDS scans of wire-cell connection at (Top) areas with lower resistance and (Bottom) areas with higher resistance due to misconnection between the silver and solder alloy seen as gaps in the image.

4 CONCLUSIONS

In conclusion, the research underscores the significance of LBIC mapping as a powerful technique to evaluate the electrical performance of solar cells and detect their defects. The ability to map the phase shift changes locally provides deeper insight into the internal electrical properties of the cells, especially issues such as higher resistance and poor connections. The complementary use of cross-sectional microscopy further assisted the diagnosis, revealing more information about the causes behind performance losses. Such findings can ease the way to optimize material quality and improve fabrication processes, thereby producing devices with higher efficiencies.

ACKNOWLEDGEMENTS

We acknowledge funding from the Solar Electricity Research Center, Sweden (SOLVE). SOLVE is co-financed by the Swedish Energy Agency (project number 52693-1) as a national centre of excellence. We also acknowledge funding from the Bussard project. The Solar-Era.Net project "Bussard" was funded in Sweden by the Swedish Energy Agency under contract number 51193-1. The authors would like to thank Gunnar Hjern for his contribution in programming of the phase-shift LBIC.

REFERENCES

[1] Buddgård J, Lagerstedt T, Machirant A. Automation of Silicone Solar Module Production with Low-Cost Tape Interconnection Method. In: 36th European Photovoltaic Solar Energy Conference and Exhibition. 2019. p. 46–9.

[2] Marquez S, Varra T, Christensen C, Rajasekharan O, Dojan C, Hobbs J, et al. LBIC Imaging of Solar Cells: An Introduction to Scanning Probe-Based Imaging Techniques. J Chem Educ. 2023 Feb 14;100(2):1011–6.

[3] Cossutta H, Taretto K, Troviano M. Low-cost system for micrometer-resolution solar cell characterization by light beam-induced current mapping. Meas Sci Technol. 2014 Aug 29. 25(10):105801.

[4] Rabha M Ben, Dimassi W, Bouaïcha M, Ezzaouia H, Bessais B. Laser-beam-induced current mapping evaluation of porous silicon-based passivation in polycrystalline silicon solar cells. Solar Energy. 2009 May 1;83(5):721–5.

[5] Pacho AP, Rinio M. A Method to Quantify the Collective Impact of Grain Boundaries on the Internal Quantum Efficiency of Multicrystalline Silicon Solar Cells. physica status solidi (a). 2020 Sep 1. 217(18):2000229.

[6] Rinio M, Yodyunyong A, Keipert-Colberg S, Mouafi YPB, Borchert D, Montesdeoca-Santana A. Improvement of multicrystalline silicon solar cells by a low temperature anneal after emitter diffusion. Progress in Photovoltaics: Research and Applications. 2011 Mar 1. 19(2):165–9.

[7] Pernau T, Fath P, Bucher E. Phase-sensitive LBIC analysis. Conference Record of the IEEE Photovoltaic Specialists Conference. 2002. 442–445 p.

[8] Feron K, Nagle TJ, Rozanski LJ, Gong BB, Fell CJ. Spatially resolved photocurrent measurements of organic solar cells: Tracking water ingress at edges and pinholes. Solar Energy Materials and Solar Cells. 2013 Feb 1;109:169–77.

[9] Krebs FC, Søndergaard R, Jørgensen M. Printed metal back electrodes for R2R fabricated polymer solar cells studied using the LBIC technique. Solar Energy Materials and Solar Cells. 2011 May 1;95(5):1348–53.

[10] Swanson DE, Geisthardt RM, McGoffin JT, Williams JD, Sites JR. Improved CdTe solar-cell performance by plasma cleaning the TCO layer. IEEE J Photovolt. 2013;3(2):838–42.

[11] Carstensen J, Schütt A, Popkirov G, Föll H. CELLO measurement technique for local identification and characterization of various types of solar cell defects. physica status solidi c. 2011 Apr 1. 8(4):1342–6.

41st European Photovoltaic Solar Energy Conference and Exhibition

This presentation was selected by the Sc. Committee of the EU PVSEC 2024 for submission of a full paper to one of the EU PVSEC's collaborating peer-reviewed journals.

NOMENCLATURE AND DESCRIPTION OF EL OBSERVATIONS: CELL CRACKS AND OTHER FINDINGS

Bengt Jaeckel[1*], Matthias Pander[1], Paul Schenk[1], Aswin Linsenmeyer[2], Jochen Kirch[3]

[1] Fraunhofer Center for Silicon Photovoltaics CSP, Otto-Eissfeldt-Str. 12, 06120 Halle (Saale), Germany
[2] Sunset Energietechnik GmbH, Industriestr. 8-22, 91325 Adelsdorf
[3] Ing.-Büro Jochen Kirch, Lindenweg 18, 86925 Leeder
*bengt.jaeckel@csp.fraunhofer.de

ABSTRACT: Crystalline silicon solar cells were, are and will dominate the Photovoltaic module market. In the past decade a lot of work was done to better understand and characterize findings utilizing Electroluminescence EL to look into the appearance of c-Si solar cells. Even standards work was done, however, a clear definition based on a standard is still missing.
Within the project PV-Riss, a group, including manufacturers, advisory experts and research facilities, is working on an overview of different findings in the literature (typically of Al-BSF and PERC cells) and bringing this into the context of today's cell technologies (TOPCon, HJT, IBC…), multi-Busbar (more then 6), large(r) cells (M6, M10(R), M12(R)) and larger module sizes with even varying constructions (glass/backsheet vs glass/glass).
The first step is to agree on a clear wording and description of certain findings. These will be presented in this contribution and a few examples are given in the "experimental results". Besides the clear nomenclature of certain findings, the location within the module is relevant for clear and comprehensive communication. Therefore, a chess-board pattern is proposed on how to clearly locate cells within the different module types, including half-cut, third-cut, and even shingled module layouts.
The final result from the project will be published as the English VDE-SPEC 90031 that is targeted to be published in Q4 2024. Depending on the results and feedback the VDE-SPEC can also serve as an IEC NWIP, e.g. within the IEC 62446-series or as a significant update to IEC 60904-13. A detailed paper is submitted to EPJ Photovoltaics.

Keywords: PV Module, outdoor, FMEA, Defects, degradation

1. INTRODUCTION

Electroluminescence (EL) imaging, first introduced in 2006 by Fuyuki [1], is widely used in the PV industry for various tests, ranging from end-of-line to field testing. Setups vary significantly, from high-resolution laboratory images to drone captures [2][3]. The focus is on PV modules with crystalline silicon cells (c-Si), which account for over 90% of the market [4][5]. Different stakeholders in the industry maintain various failure catalogs to name, group and classify observations [6]-[11]. EL images provide critical insights into the health status of PV modules. Interpretation heavily depends on when the image is captured. Key questions include: What are the observations? How are they named and classified? The assessment of criticality also varies based on the timing of the capture and the type of system. Cell cracks occur at multiple stages, and their severity assessment relies on crack type and cell technology, which is getting more and more complex due to the diversification of cell technologies.
Typically, analysis involves identifying anomalies, classifying them, and making prognoses. While cell cracks are reliably identified, qualitative assessment of this identification is lacking. Approaches to bridge this gap are discussed. The aim of this paper is to standardize image evaluation and nomenclature for existing EL images, improving communication in future discussions. Four main categories have been developed, and a preliminary classification table will be presented.

2. TERMS

To ensure a standardized observation description, it is first necessary to be able to name the exact position of the observation. Therefore, a chessboard pattern is proposed on how to clearly locate cells within the different module types (see **Fehler! Verweisquelle konnte nicht gefunden**

werden., including half-cut, third cut and even shingled module layouts (see Figure 1)
Some of the observations require further definitions to describe defects, especially in the case of cracks, it is important to know whether a crack is complete or incomplete in order to better assess the potential influence. A detailed explanation of this is given here [12][13].
Typically, c-Si cells are produced from either mono-crystalline or multi-crystalline ingots, which are then sliced into thin wafers. Their sizes can range from 4 inches to the current M12 (210 mm) [4].
Mono-crystalline silicon wafers exhibit an almost ideal crystallographic lattice structure that extends across the entire wafer. This type of material is predominantly used in today's PV modules. Multi- or poly-crystalline wafers consist of numerous adjacent areas (or grains) separated by grain boundaries. While the crystalline structure within a single grain is similar to that of mono-crystalline, the lattice orientation varies from one grain to another. Crack pattern and resulting power losses may also be different compared to mono c-Si material.

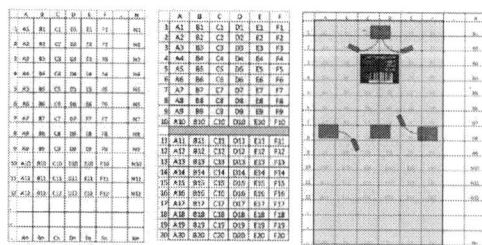

Figure 1: Location of observations within the PV Module

3. GROUPING OF EL OBSERVATIONS

EL-images are used to identify different kinds of observations. Herein the word "observation" is used to cover all observations that can be identified in an EL-image. This includes silicon solar cell cracks, wafer related structures, cell manufacturing features and things attributed to PV module manufacturing and observations after use (outdoor and indoor accelerated stress testing). Observations are not necessarily defects or per-se structures to cause increased degradation rates.

The grouping of the observation into four major categories is intended to help faster evaluation of EL-images and therefore sorting of PV modules into certain classes. Herein the observations are divided into four main groups: 1. single crack, 2. multiple cracks, 3. anomalies in electrical circuit, and 4. Miscellaneous. Within each group there are then further differentiations or groupings of certain observations. These four groups were developed after the evaluation of several failure catalogues available to the consortium and from years-long daily work with lots of EL images of different PV module designs and stages in their service life. Based on this evaluation the focus was first drawn to observations in individual solar cells followed with a potential impact in the serial/parallel connection in PV modules. Due to the fast change in cell size and number of bus bars in the last years clear conclusions for certain observations are not possible, also because long term outdoor exposure experience is missing. More background regarding the groups is published in [12][13].

The following grouping was done in detail.

Group 1: Single crack
- a) Complete
- b) Incomplete

Group 2: Multiple cracks
- c) Double
- d) Multiple crack pattern
- e) Dendritic shaped crack structures (worst case of 2b.)
- f) Tiny X-V-shape cracks

Group 3: Anomalies in electrical circuit
- a) Finger interruptions
- b) Missing cell to cross connector joint
- c) (completely) Dark cells and strings
- d) PID, LETID, UVID (module pattern!)
- e) Severe mechanical load and hail (module pattern!)

Group 4: Miscellaneous
- a) Dark spots
- b) Wafer related structure
- c) Belts structures
- d) Solar Cell "artefacts"
- e) Grayish and darker areas
- f) Scratches

It is important to note that some observations must be considered in the overall image and as a pattern. This was necessary to include some of the most relevant power loss causing effects from the past (PID) and of today (UVID – see Figure 2)

4. CLASSIFICATION

Electroluminescence imaging is a very helpful tool to investigate general quality of PV modules. However, the classifications are mostly done as company standards, e. g. for end-of-line quality control or defined in contracts between manufacturer and buyer. No general accepted classification exists, yet.

For an acceptable classification of EL observations a few key questions must be answered, followed by a severity/ priority analysis and definition.

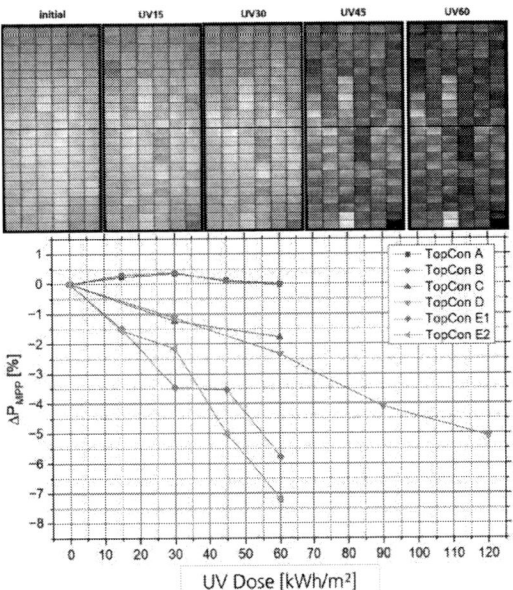

Figure 2: Top: EL pattern evolution during UV-testing per IEC 61215-2 MQT 10 and bottom corresponding power losses for various TOPCON PV modules

The first point is whether the damage is so critical that it can directly cause an electrical safety issue and/or harm person. Second point is the possible future development of the observation. Will there soon be an electrical safety issue or is there a potential power loss that does not comply with the performance guarantees from the manufacturer. Follow-up inspections might then be a consequence, and a reclassification might be necessary after a certain period of time, typical one year for larger electrical systems [14]-[16] but it also depends on the country and local grid codes.

Additionally, one of the next very important questions is to find out if it is an isolated case or a systematic fault that may lead to the full replacement of all PV modules from the entire system.

The classification is split into different points in time when the EL imaging was done, namely after manufacturing (end-of-line inspection), after shipment, after installation and during operation. Additionally, it is possible to discuss recommendations after severe weather events such as extreme snow events, hailstorms or floods.

Table 1 provides an overview of different periods during the life of a PV system and states numbers for allowed observation for discussion. The numbers after manufacturing are from best practices. The others are not stated anywhere yet and should serve as a first guide and a starting point for more clear statements along the value chain. Follow-up discussion, including running research on power loss as function of number of busbars and cell technology will hopefully bring more insights and good arguments for reasonable number of permissible observations.

Table 1: Classification proposal for selected observations; more can be found in [13].

Type of EL anomaly	after manufacturing: end of line inspection	In the field	After severe weather events
Single complete crack	≤6BB: max. 5 >6BB: max. 5	+1 per 5 years	+5
Single incomplete crack	≤6BB: max. 1 >6BB: max. 5	+1 per 5 years	+5
Dendritic crack structures	≤6BB: not allowed >6BB: not allows	Not allowed	1 per PV module, follow-up inspection required

5. SUMMARY

Electroluminescence (EL) is a powerful tool for assessing the quality of PV module manufacturing and the care taken during transportation, installation, and use. This imaging technique is effective for monitoring trends, such as crack growth. However, the industry still lacks a unified nomenclature. This paper provides a detailed overview of EL observations, explicitly naming them. The results play a significant role in the upcoming VDE SPEC 90031. Both the paper and the VDE SPEC aim to support a common nomenclature for EL images and the location mapping of observations.

The naming of observations applies to all existing and new cell types, as long as they are based on crystalline wafers. It is independent of size (ranging from the "old" 4-inch to the new M/G12), shape (full, half, third, x-cut), material (mono or multi), or silicon cell technology (Al-BSF, PERC, TOPCON, HJT, IBC, Perovskite on Si). Additionally, the design of the PV module (size, glass/backsheet, glass/glass, internal wiring of the electric circuit) does not affect the naming and localization of observations.

ACKNOWLEDGMENT

This work was funded by the Federal Ministry for Economic Affairs and Climate Action in the project PV-Riss with grant #03TN0033A.

6. REFERENCES

[1] Y. N. Fuyuki, "Quantitative Imaging of Excess Minority Carrier Density in Crystalline Silicon Cells by Luminoscopy," 21st European Photovoltaic Solar Energy Conference 2006.

[2] U. Jahn, "Review on Infrared and Electroluminescence Imaging for PV Field Applications", Report IEA-PVPS T13-10:2018

[3] VDI 2879 SPEC, "Inspection of installations and buildings with UAVs (unmanned aerial vehicles)", 2018

[4] ITRPV, "International Technology Roadmap for Photovoltaic (ITRPV)", 2024

[5] Solar Power Europe, "Global Market Outlook for solar power 2023 - 2027

[6] M. Koentges, "Elektromineszenzmessung an PV-Modulen," Photovoltaik Aktuell, vol. 7/8, pp. 36–40, 2008

[7] M. Koentges, "Review on Failures of Photovoltaic Modules", Report IEA-PVPS T13-01:2014

[8] M. Koentges, „Assessment of PV Modules Failures in the Field", Report IEA-PVPS T13-09:2017

[9] W. Herrmann, "Qualification of Photovoltaic (PV) Power Plants using Mobile Test Equipment", Report IEA-PVPS T13-24:2021

[10] B. Jaeckel, "Blick in die Zukunft", PV Magazin 1, 8 (2013)

[11] MBJ PV-Modul Bewertungskriterien, "MBJ PV-Modul Bewertungskriterien – Bewertungskriterien für die PV-Modulprüfung im Mobilen PV-Testcenter", MBJ, (2014)

[12] VDE SPEC 90031: Electroluminescence (EL) of photovoltaic modules – Terms and classification

[13] B. Jaeckel, "Nomenclature and description of Electro-Luminescence (EL) observations: Cell cracks and other observations", submitted to EPJ Photovoltaics, 2024

[14] IEC 62446-1: Photovoltaic (PV) systems - Requirements for testing, documentation and maintenance - Part 1: Grid connected systems - Documentation, commissioning tests and inspection

[15] IEC 62446-2: Photovoltaic (PV) systems - Requirements for testing, documentation and maintenance - Part 2: Grid connected systems - Maintenance of PV systems

[16] IEC 60364-7-712: Low voltage electrical installations - Part 7-712: Requirements for special installations or locations - Solar photovoltaic (PV) power supply systems

41st European Photovoltaic Solar Energy Conference and Exhibition

DAYLIGHT ELECTROLUMINESCENCE INSPECTION OF PV PANELS ON-SITE VS. TRADITIONAL EL INSPECTION WITH SILICON CAMERAS

L.A. Carpintero[1], D. González-Francés[2], K.P. Sulca[2], C. Terrados[2,3] C. de Castro[2], V. Alonso[4], M.A. González Rebollo[2], O. Martínez[2]*

(1) Cobra Instalaciones y Servicios, S.A. Calle Cardenal Marcelo Spinola, 10. 28016 Madrid (Spain).
(2) GdS-Optronlab group, Dpto. Física de la Materia Condensada, Universidad de Valladolid, Edificio LUCIA, Paseo de Belén 19, 47011 Valladolid (Spain).
(3) Solar and Wind Feasibility Technologies (SWIFT). Escuela Politécnica Superior, Universidad de Burgos. Avda. Cantabria s/n 09006 Burgos (Spain).
(4) GdS-Optronlab group, Dpto. Física Aplicada, Universidad de Valladolid, Facultad de Ciencias, Paseo de Belén 7, 47011 Valladolid (Spain).
*oscar.martinez@uva.es

ABSTRACT: This paper presents a comparative study of electroluminescence (EL) imaging of silicon photovoltaic (PV) panels in three currently operating PV plants. EL images were acquired using two methods: in a darkroom with a silicon sensor reflex camera after dismantling the panels; and during daylight without dismantling the panels using an InGaAs sensor camera (dEL). The results demonstrate that dEL can detect the same important defects in the PV panels while being not only less costly in terms of time and money but also able to prevent the production of new defects resulting from disassembly, transportation and reassembly of the modules. This way, dEL provides a more efficient and reliable procedure for quality control and maintenance.
Keywords: Electroluminescence, silicon panels, InGaAs camera, non-destructive inspection, daylight inspection

1 INTRODUCTION

Regular inspections of the solar photovoltaic (PV) plants are crucial to ensure optimal energy production and to prevent potential future failures and module degradation. Among the available inspection techniques, electroluminescence (EL) stands out as a powerful tool for detecting hidden defects and anomalies in panels [1, 2].

EL inspection involves applying a current to the panel and capturing the emitted light using specific cameras. Normally, this process is carried out with specialized silicon cameras, either at night (on-site), or by dismounting the panels and taking them to a darkroom or mobile laboratory. However, both alternatives present significant drawbacks:

Night-time inspection:
- o Requires specific weather conditions (no nearby light or moonlight).
- o Involves safety risks for personnel.
- o Higher staff costs for night working.

Dismounting panels:
- o Increases the risk of panel damage.
- o Disrupts power generation.
- o Time consuming and costly.

In the last years, on-site inspection techniques have gained much more acceptance [3]. Among them, daylight Electroluminescence (dEL) by using InGaAs cameras have been developed by several groups [4-7], including our own [8]. This innovative solution allows daylight inspection without disassembling the panels, overcoming the limitations of traditional methods. It also allows inspections to be performed regardless of the weather conditions (except for rain or high wind gusts), increasing operational efficiency, and minimizing downtime. By eliminating the need to remove panels, dEL also reduces the risk of potential damages of the modules, minimizes production losses, reduces labor costs, and ensures the integrity of the solar installation. With our compact and

lightweight design, the camera-tripod is easily portable and convenient to transport, facilitating quick setup at various locations, see Fig. 1.

Figure 1: Compact dEL system developed by our group [8].

This communication presents a comparative study of EL images of monocrystalline and polycrystalline PV panels obtained using both conventional EL with silicon cameras in a darkroom (under standar test conditions, STC), after panel disassembly, referred here as high-resolution night EL (HR-nEL), and dEL, without panel disassembly. The dEL measurements were performed by our research group, while the HR-nEL measurements were performed by two different companies specialized in the inspection of photovoltaic systems. The aim of the study was to compare the advantages and disadvantages of each method in terms of image quality, ease of use, inspection time, cost and similarity of the defects detected by both techniques.

2 MATERIALS AND METHODS

More than 1,000 mono and polycrystalline modules from three different PV power plants currently in real operation in Spain were analyzed.

In order to compare both imaging methods, a first dEL inspection was performed on-site prior to dismounting the panels for a HR-nEL inspection in a dark room. After the latter, the modules were reassembled at their original location, and a second dEL inspection was finally conducted again on-site. This way, the potential damage induced by the disassembling/handling/reassembling processes could also be studied.

The InGaAs sensor used for the dEL imaging is a Hamamatsu C12741-03 (- 640 × 512 pixel – 0.33 MPixel) camera, which works at 14-bit resolution. The pixel noise and dark current of this camera are 250 e⁻ rms and 360 000 e⁻/(px·s), respectively. The exposure time can range from 1 μs to 1 s; allowing the acquisition to be adapted to the different lighting conditions during the daytime.

Conventional EL was performed using two different silicon cameras (at least 24.2 MPixel – 6000 x 4000 pixels) in a dark room at STC conditions after dismounting the modules.

3 RESULTS

3.1 HR-nEL vs dEL

Fig. 2 shows EL images of a polycrystalline silicon panel.

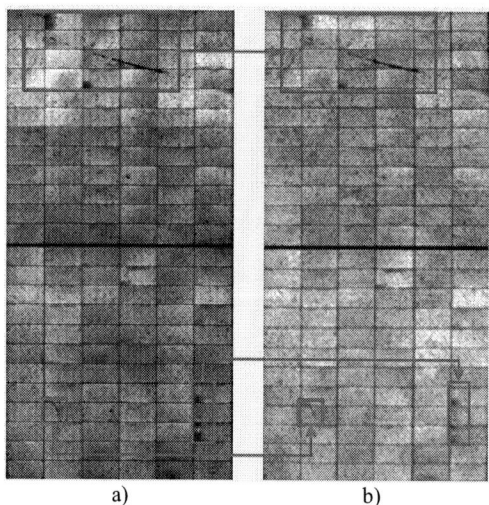

a)   b)

Figure 2: Defects in a polycrystalline silicon panel. a) High Resolution night Electroluminescence (disassembling) obtained with a Si camera; b) dEL image (without disassembling) obtained on-site with an InGaAs camera.

They were obtained by the two procedures mentioned above: using a high-resolution reflex camera after disassembling the panels (HR-nEL) (a), and using an InGaAs camera (on-site), before disassembly (dEL) (b). The results showed that despite the lower resolution of the InGaAs camera, the most significant defects [2], which can lead to a reduction in panel performance or an increase in degradation, are still detected.

Fig. 3 shows the EL images of a monocrystalline silicon panel obtained by the two methods. The small fractures surrounding the large defect are not properly detect by dEL, although their contribution to the efficiency loss is considered lower.

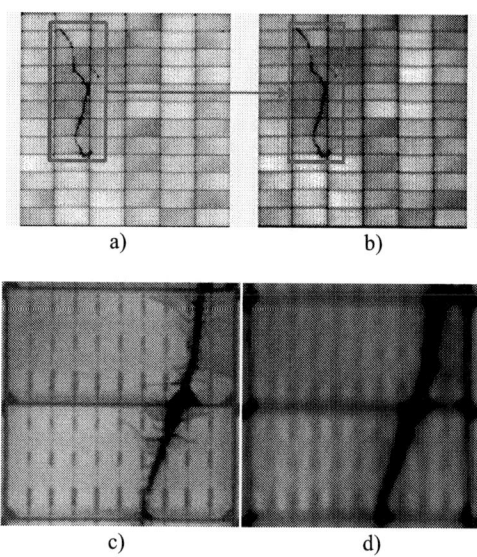

a)   b)

c)   d)

Figure 3: a) HR-nEL image of an extensive defect in a monocrystalline panel; b) dEL image of the same monocrystalline panel; c) magnification of part of the defect in the same area observed by HR-nEL; d) same magnification in the dEL image

3.2 Induced damaged

In addition to the comparative analysis of EL imaging methods, the study also investigates the potential damage caused by panel disassembly, handling and reassembly. The results show that certain defects appeared in some panels at some point during these processes. Fig. 4 presents such a case. Three EL images are shown in the figure: Fig. 4a), taken with the InGaAs camera before disassembly; Fig. 4b), taken with a Si camera after disassembly; and Fig. 4c), a new image taken again with the InGaAs camera after reassembly. Fig. 4b) shows a defect not visible in Fig. 4a). Fig. 4c) confirms the presence of the defect, which was not initially found. Fig. 5 shows the area of this defect magnified for each image. The defect should be thus ascribed to the disassembled/handling process for the HR-nEL inspection. This finding highlights the importance of using non-destructive inspection methods, such as dEL imaging on-site, to prevent damage to PV panels during EL inspection procedures.

41st European Photovoltaic Solar Energy Conference and Exhibition

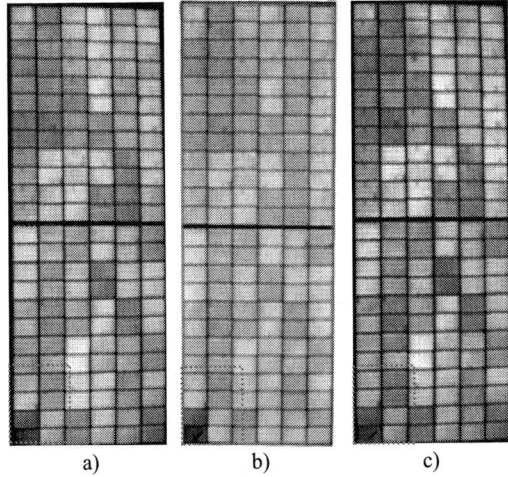

Figure 4: a) dEL image taken with the InGaAs camera before disassembly; b) HR-nEL taken with the Si camera after disassembled the panel for a darkroom inspection; c) dEL image after reassembly. A defect has appeared in the lower left corner

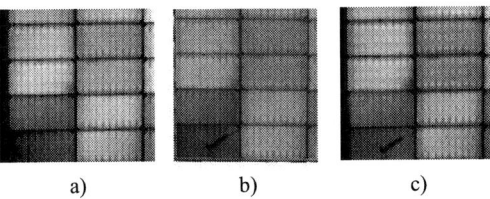

Figure 5: Magnification of the interest zone in Fig. 4 (red dotted); a) dEL image taken with the InGaAs camera before disassembly; b) HR-nEL taken with the Si camera after disassembled the panel for a darkroom inspection; c) dEL image after reassembly

4 CONCLUSIONS

In summary, dEL represents a significant advancement in EL inspection procedures, enabling daytime inspection of solar panels without dismounting them. This solution offers numerous advantages over traditional methods, including increased efficiency, safety (not only for the workers but also for the installation), cost reduction and convenience. The lower resolution of the InGaAs camera may not be sufficient to detect very small defects. However, the camera's ability to inspect panels during the day and without dismounting them makes it a valuable tool for identifying larger and more critical defects that can significantly impact panel performance. The study also highlights the inherent risks of the assembly and disassembly processes involved in performing EL inspections with Si cameras. The successful application of dEL technology in real-world settings demonstrates its potential to revolutionize the maintenance and inspection of PV systems, ensuring their optimal performance and longevity.

5 ACKNOWLEDGMENTS

This work has been financed by the Spanish Ministry of Science and Innovation, under project PID2020-113533RB-C33, and by the Regional Government of Castilla y León (Junta de Castilla y León) and by the Ministry of Science and Innovation and the European Union NextGenerationEU / PRTR under the project "*Programa Complementario de Materiales Avanzados*". C. de Castro is grateful for the funding received through the "Investigo" Programme of the Ministry of Labour of the Government of Spain. K.P. Sulca is grateful for the funding received through the predoctoral contract programme of the University of Valladolid (2022), co-funded by the Santander Bank.

6 REFERENCES

[1] Jahn U, Herz M, Köntges M, et al. Review on infrared and electroluminescence imaging for PV field applications. Int Energy Agency Rep IEA-PVPS T13-10; 2018.
[2] Review on Failures of Photovoltaic Modules. Marc Köntges, et al. Report IEA-PVPS T13-01:2013
[3] O. Kunz, J. Schlipf, A. Fladung, Y.S. Khoo, K. Bedrich, T. Trupke, Z. Hameiri, Prog. Energy 4 (2022) 042014.
[4] L. Stoicescu, M. Reuter, J.H. Werner, in 29th Eur. Photovolt. Sol. Energy Conf. Exhib. (2014) p. 2553.
[5] J. Adams, B. Doll, C. Buerhop, T. Pickel, J. Teubner, C. Camus, C.J. Brabec, in 32nd Eur. Photovolt. Sol. Energy Conf. Exhib. (2015) p. 1837.
[6] S. Koch, T. Weber, C. Sobottka, A. Fladung, P. Clemens, J. Berghold, in 32nd Eur. Photovolt. Sol. Energy Conf. Exhib., (2016) p. 1736.
[7] G.A. dos Reis Benatto, N. Riedel, S. Thorsteinsson, P.B. Poulsen, A. Thorseth, C. Dam-Hansen, C. Mantel, S. Forchhammer, K.H.B. Frederiksen, J. Vedde, M. Petersen, H. Voss, M. Messerschmidt, H. Parikh, S. Spataru, D. Sera, in Proc. 44th IEEE Photovolt. Specialist Conf. (2017) p. 2682.
[8] M. Guada, A. Moretón, S. Rodríguez-Conde, L.A. Sánchez, M. Martínez, M.A. González, J. Jiménez, L. Pérez, V. Parra, O. Martínez, Energy Science & Engineering 8 (2020) 3839.

41st European Photovoltaic Solar Energy Conference and Exhibition

PHOTOVOLTAIC MODULE ARRAY LUMINESCENCE IMAGE PREPROCESSING:
HEURISTIC ALGORITHMS FOR PERSPECTIVE CORRECTION AND CELL SEGMENTATION

Brendan Wright[1]*, Ali Shakiba[1], Rama Sharma[1], Ziv Hameiri[1]
[1]School of Photovoltaic and Renewable Energy Engineering, UNSW Sydney, Australia
*brendan.wright@unsw.edu.au

ABSTRACT: To address the significant and growing challenge of photovoltaic (PV) module end-of-life (EoL) management, characterisation techniques including infrared (IR) imaging, or more recently electroluminescence (EL) and even photoluminescence (PL) imaging are increasingly being employed to monitor the quality of installed PV modules. However, current methods for module quality assessment are often manual and time-consuming, particularly for luminescence-based imaging, performed by domain experts on individual images. These approaches cannot meet the needs of the rapidly growing PV installations, and therefore automation is required to unlock the significant potential of these imaging techniques. Existing image preprocessing methodologies typically fail to successfully handle images with the complexity of features commonly found in real-world images. This study outlines a novel methodology for robust PV module array image preprocessing, comprising a pipeline of heuristic algorithms, enabling subsequent cell-level analysis and quality assessment of individual modules. The developed preprocessing methodology has been successfully validated on multiple separate real-world image datasets, including those comprising EL, PL, and IR images of module arrays. Further, upon integration within a novel module quality assessment pipeline, we have shown the end-to-end capability to automatically identify and classify defective cells and modules at scale with high accuracy, solely from as-measured PL images of fielded PV modules. This helps to address the significant challenge of informed PV module EoL decisions for utility-scale systems.
Keywords: luminescence imaging, perspective correction, cell segmentation

1 SCIENTIFIC INNOVATION AND RELEVANCE

The challenge of photovoltaic (PV) module end-of-life (EoL) management is receiving increasing attention from the broader PV community. However, currently, there is a significant lack of adequate large-scale characterisation methods and associated analytical capabilities for automated module quality assessment. To address this challenge, characterisation techniques including infrared (IR) imaging, or more recently electroluminescence (EL) and even photoluminescence (PL) imaging are increasingly being employed to assess the quality of installed PV modules.[1] Each technique is capable of identifying various types of defects or degradation that can present operational risks and impact power generation capacity.

However, the image analysis methodology for module quality assessment is typically a manual and time-consuming one, particularly for luminescence images, performed by domain experts on individual images. This cannot scale to meet the needs of rapidly growing PV installation, and requires automation to unlock the significant potential of these imaging techniques. The key challenge hindering this automation is the first image preprocessing stage, where the raw as-measured images are prepared for further automated analysis and quality assessment.

While there exist some methods for image correction of module luminescence or infrared images, these have proven inadequate for preprocessing images featuring more significant variability in image quality typical of real-world datasets.[2]

Typical real-world images are non-ideal, featuring various combinations of imaging camera lens and perspective distortion, multiple adjacent modules or arrays, some with partial occlusion or cropped regions, and highly variable background features and contrast artefacts. Existing image preprocessing methodologies typically heavily rely on the detection of module edge or cell grid features to deskew and segment images, leading to cumulative processing errors or outright failure.

To address the limitations of prior methodologies, we have developed a novel pipeline of heuristic algorithms thatprove capable of successfully processing images with features typical of those obtained during real-world on-site measurements. In particular, those more problematic datasets with significant variability in image quality. Further, the proposed single algorithm is capable of cross-application to preprocessing of EL, PL, and IR module array images.

2 AIM AND APPROACH

The objective of this reasearch is to develop an image preprocessing methodology for automated preparation of as-measured module array images for further analysis. This included the development of a two-stage pipeline for (i) image perspective correction and (ii) cell-in-module segmentation. The resulting processed and extracted cell images can be directly utilised for subsequent quality assessment, including defect and degradation classification.

The first stage (i) of our novel preprocessing methodology comprises a heuristic perspective correction algorithm. This includes a modified Radon transform to compute a rotation map of the source module array image – see Figure 2 (a) – providing features arising from the image axis summation and corresponding alignment between image rotation angle and module in-plane angle. [3] Importantly, this method is highly tolerant to inconsistent or non-uniform cell and module grids within the image, such as variable adjacent module spacing or separated groupings of cell grids within a single module. This is particularly important in the case of typical split half-cell modules, where an offset is present at the centre of each module, breaking from the expected consistency of a uniform cell grid. Additionally, these half-cells present an aspect ratio that is closer to 2:1 rather than the standard 1:1 of full cells, requiring a method that can

41st European Photovoltaic Solar Energy Conference and Exhibition

automatically handle such variability.

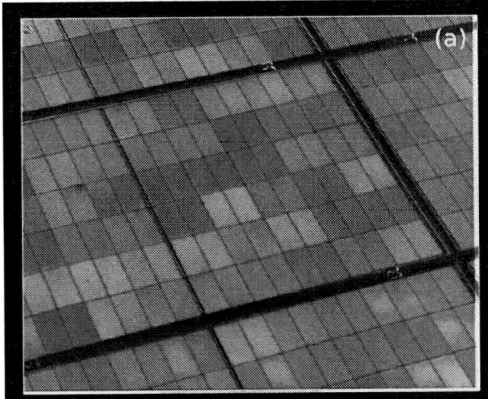

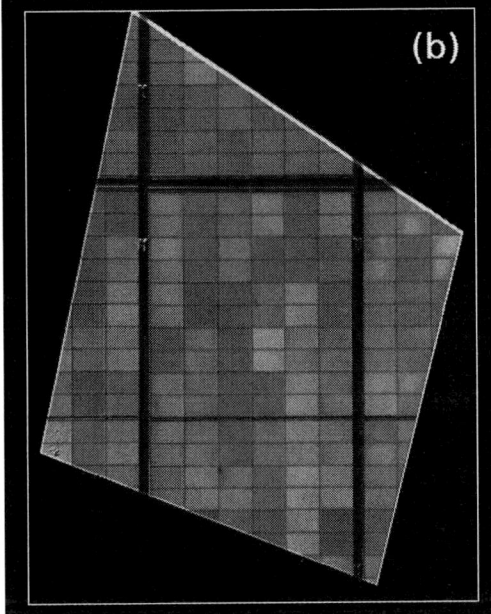

FIGURE 1: (a) An example as-measured in-field photoluminescence image of a module array containing multiple adjacent split-modules, each comprising 120 half-cells; (b) the module image after perspective correction transform yielding a de-skewed image correctly aligned to the module grid axes.

Utilising this feature map, and through identification, filtering, and fitting of these grid edge-related features to a pair of linear approximations, the module-plane angles and relative perspective offset can be calculated. This enables the computation of module plane angles and positions in image pixel coordinates – see Figure 2 (b) – facilitating a four-point perspective correction and deskew transformation. While computationally expensive, this methodology enables the consistent, accurate, and reliable (error and artefact tolerant) deskewing of arbitrary module array images – see Figure 1 for an example before (a) and after (b) – including for PL, EL and even IR images with a single algorithm.

Examples of intermediate outputs from the preprocessing stage (i) algorithms are displayed in Figure 2, including the rotation map (a) and computed in-plane

axes overlain on the source image (b). The visible sharp features in the rotation map are identified and used to compute the in-plane axes, enabling the perspective correction. Additionally, information is present within the rotation map corresponding to cell and module edge locations, relative offsets, and magnitude of lens distortion. Utilisation of this information to correct additional image distortions is possible, with further research in progress to develop such a capability.

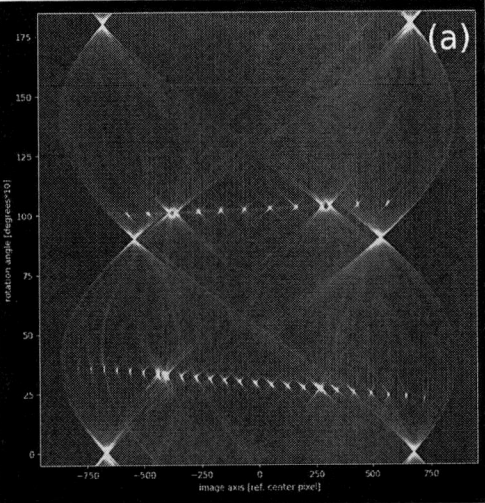

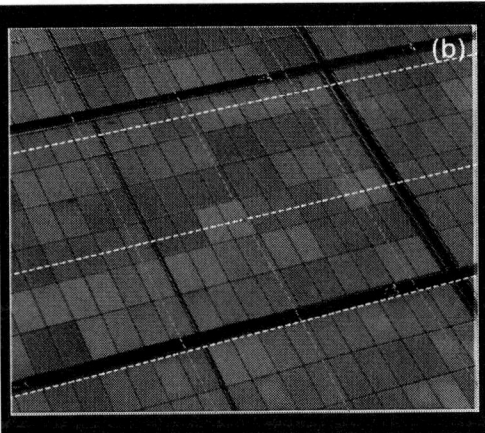

FIGURE 2: (a) Example of rotation map derived from modified Radon transform of a source module array image – Figure 1 (a) – highlighting point features that correspond to alignment of module-plane edge grid to image rotation axis, representing cell and module edges and spacing along both axes; and (b) corresponding module plane axis computation and perspective transformed overlay lines (broken), used for computation of four-point perspective correction to deskew source image, illustrated in Figure 1 (b).

The second stage (ii) of our preprocessing methodology is a heuristic cell-in-module segmentation algorithm, to identify module and cell grid edges for the extraction of individual cell images. Enabled by the preceding high-quality perspective correction, where module axes are aligned tightly to the image axes, a direct image edge feature extraction algorithm was developed. A normalised morphological gradient image is computed and Otsu thresholding is applied to generate a uniform linear binary mask that separates all consistently isolated regions of the module array.[4] This is illustrated in Figure 3 (a).

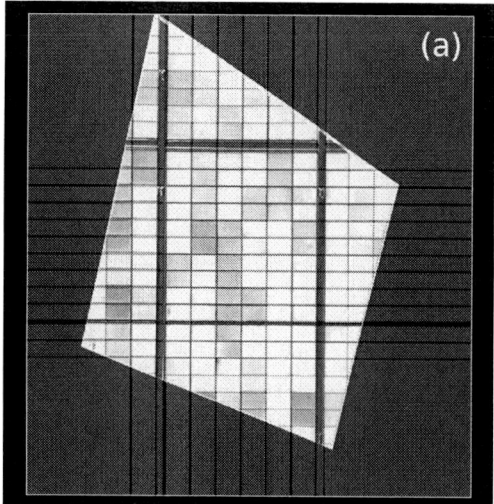

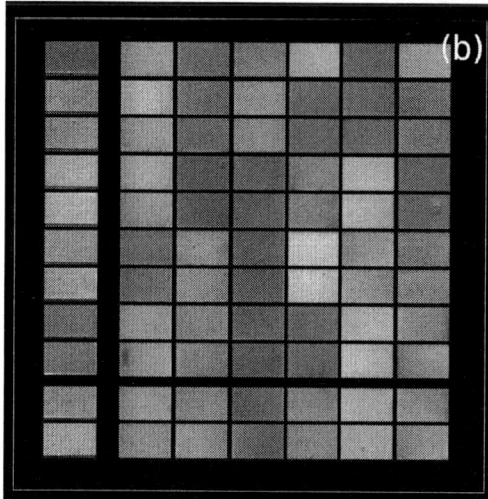

FIGURE 3: (a) A binary line mask overlay (greyscale lines) on source de-skewed module array image (Figure 1, b) illustrating the computed consistent module and cell edges, used to partition and segment regions of the image corresponding to isolated cells; and (b) the corresponding filtered and cropped image reconstruction of all individual cells at their original relative position with background features removed.

Through further filtering of the relative pixel area and aspect ratio distributions for the identified area features, a final labelled mask of individual cell images is generated and used to extract each individual cell image – see Figure 3 (b) – ready for further analysis and quality assessment. This process preserves the original sharp edges and high resolution of the source image, and the relative position of each cell within the parent module grids.

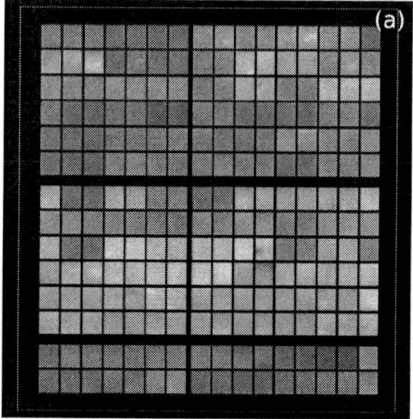

FIGURE 4: (a) The transformed source photoluminescence image of an example module array after deskew and cell segmentation; (b) the

corresponding cell-level defect classification, with defect class (text label) and severity (colour increasing from yellow to red) overlay; and (c) a close-up segment of (b), indicated by the red rectangle region, for enhanced cell-level clarity.

3 RESULTS AND CONCLUSIONS

A representative example set of results illustrating the capability and operational principles of the developed image preprocessing methodology have been presented within the previous section in Figures 1 to 3. The dataset used for these experiments was obtained through PL imaging of fielded module arrays measured on-site at a utility-scale PV installation. This particular dataset was chosen as it comprises a broad range of imaging complexities typical of real-world measurements. These include images of partial (crop or occlusion) or single modules (with split grids of 120 half cells in total), or multi-modules (adjacent within an array), all with varying degree of perspective distortion and feature complexity (background, contrast, artefacts, etc.). Importantly, the majority of these fielded operational modules exhibit clearly visible defective or degraded cells of varying severity, clearly motivating the importance of a robust and scalable automated quality and risk assessment capability.

Experimental testing was performed with a novel automated cell-level defect classification pipeline, which requires a uniform normalised map of individual cell images as input. This is used to validate the robustness and accuracy of the developed image preprocessing methodology on a subset of the module array PL imaging dataset. This was achieved through integration within a preliminary version of a developed quality assessment pipeline, already proven capable of achieving module quality and risk analysis from manually preprocessed EL images of modules (segmented to cells as herein). Figure 4 (a) displays an example source PL image of a module array after successful perspective correction and segmentation using the developed heuristic algorithms. Figure 4 (b) shows the corresponding cell-level defect classification, with defect class (text label) and severity (colour increasing from yellow to red) overlay. A close-up segment of (b), indicated by the red rectangle region, is also presented in Figure 4 (c) for enhanced clarity. This cell quality assessment is based on a statistical analysis of relative image area-residual and threshold-based defect classification. The displayed representative example highlights the capability of the developed pipeline; it has yielded comparable performance on multiple PL- and EL-based imaging datasets, as well as preliminary validation on an IR image dataset.

Importantly, the proposed methodology has been validated on particularly problematic image datasets where existing alternative methods were unsuccessful, presenting a single robust algorithm capable of preprocessing EL, Pl, and IR module array images. This will enable the development of automated quality assessment pipelines to address the significant challenge of informed PV module EoL decisions for utility-scale systems.

4 REFERENCES

[1] T. Trupke, R.A. Bardos, M.D. Abbott, P. Würfel, E. Pink, Y. Augarten, F.W. Chen, K. Fisher, J. E. Cotter, M. Kasemann, European Photovoltaic Solar Energy Conference (2007) pp22–31.
[2] E. Sovetkin, A. Steland, Integrated Computer-Aided Engineering 26 (2019) pp123–137.
[3] G. Beylkin, IEEE Transactions on Acoustics, Speech, and Signal Processing 35 (1987) pp162–172.
[4] T. Y. Goh, S. N. Basah, H. Yazid, M. J. A. Safar, F. S. A. Saad, Measurement 114 (2018) pp298–307.

41st European Photovoltaic Solar Energy Conference and Exhibition

COMPARING MEASURED PV MODULE POWER TO NAMEPLATE VALUES

Frank Weinrich, Stefan Riechelmann, Laura Stenzig, Stefan Winter
Physikalisch-Technische Bundesanstalt (PTB), Bundesallee 100, 38116 Braunschweig, Germany
Corresponding author: frank.weinrich@ptb.de

ABSTRACT: Within the frame of a market study, the nameplate values of 40 PV modules of 22 different module types from market-relevant manufacturers are compared. The PV modules are measured with an LED-based solar simulator of class A+A+A+ and an uncertainty for P_{mpp} of 0.9 % (k=2). To cross-check our results, 2 PV modules of different types are also measured using the direct sunlight method with our solar module tube. In both approaches, the respective I-V characteristic is recorded and corrected to standard test conditions in accordance with IEC 60891. The LED simulator has been used to determine the relative spectral responsivity of the modules and a spectral mismatch correction has been conducted in accordance with IEC 60904-7 for both, indoor and outdoor measurements. The determined P_{mpp} from our solar simulator shows an average deviation from the nameplate value of around -1.33 %. A cross-check with our solar module tube is in accordance with results of the solar simulator. For all measured PV modules, the power values are specified higher on the nameplate than the real power values measured by our facilities.

Keywords: PV Module, Energy Rating, Solar Simulator, Solar Module Tube, Direct Sunlight Method (DSM)

1 INTRODUCTION

In the first half of 2024, photovoltaics accounted for 13.9 % of total electricity generation in Germany [1]. According to the current Renewable Energy Sources Act (EEG), its share of German electricity generation is expected to double in the near future. To reach this goal, the government wants to triple the annual expansion from 7.5 GW in 2022 to 22 GW in 2026 to achieve the target of 215 gigawatts (GW) of photovoltaics by 2030 [2]. The focus of our working group is to reduce measurement uncertainty along the photovoltaic value chain, which provides an important basis for investment decisions for consumers, investors and manufacturers alike. Therefore, we present a market study which compares the nameplate values of 40 PV modules of 22 different types from market-relevant manufacturers manufactured between 2021 and 2024. The PV modules are measured with an LED-based solar simulator. Within the presented market study, the measurement results of the solar module tube (SMT) [3] are compared for 2 PV modules.

2 SETUP

Two measurement sites for secondary calibration of solar modules are utilized in this study. The indoor test stand used in is study is equipped with an LED-based solar simulator. The instrument is classified as class A+A+A+ according to IEC 60904-9 [4]. The measurement uncertainty of the P_{mpp} is 0.9 % (k=2). The outdoor test site is a field of view limiting setup for PV devices called solar module tube [3], applying the direct sunlight method (DSM) for calibration of PV modules. While we measured all 40 PV modules with the LED solar simulator, 2 PV-modules of two different module types are also measured with the SMT for a cross-check.

2.1 PV module samples

The characterized PV modules are either based on monocrystalline silicon (c-Si) of PERC and TOPCon technology or silicon-based heterojunctions (c-Si/a-Si) with production dates between 2021 and 2024. The PV modules range in power from 320 Wp to 545 Wp and include full-cell, half-cell, monofacial and bifacial designs. For most module types, we measured 2 to 3

modules of the same type to account for variation in power output due to variations in module production within manufacturing processes. In addition, the modules were not randomly purchased directly from the manufacturers on the market to avoid any advantageous pre-selection.

2.2 Indoor test site (LED)

The LED-based solar simulator (SINUS-3000, Wavelabs) has 26 differently colored LED channels to replicate the solar spectrum and is suitable for monofacial and bifacial modules of up to 2.4 m x 1.4 m in size. Measured I-V curves are corrected to standard test conditions using module temperature and irradiance measured with a WPVS reference cell in accordance with IEC 60891 method 1 [5]. Figure 1 show the setup during measurements.

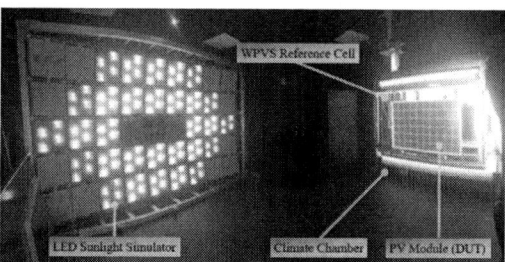

Figure 1: LED-based solar simulator during measurements with a mounted PV module inside a climate chamber.

By conducting a differential-like spectral responsivity measurement by varying the irradiance of each individual LED channel sequentially, the spectral responsivity is derived between 341 nm and 1154 nm at 26 different center wavelengths [6]. This spectral responsivity is used for a spectral mismatch correction of both indoor and outdoor measurements.

2.3 Outdoor test stand (SMT)

For the outdoor direct sunlight measurement, a large rectangular tube with a size of 4.0 m x 4.0 m x 7.2 m capable of tracking the sun is limiting the field of view of a mounted PV module under test. The tube is lined with optical black fabric and equipped with a shutter to allow well defined start/end of the module's light exposure and limit the heating of the PV module due to the direct irradiance. The shutter also divides the inside of the device in two parts. The rear area in which the module is located, is constantly cooled towards 25 °C by using an active cooling system. The module under test is mounted on an adjustable module holder inside the tube and aligned directly towards the sun during the measurement. Figure 2 shows the setup during measurements. The measurements were performed on days with clear sky and the current weather condition is monitored with an all-sky camera. The DSM measurements for the two modules are conducted in August 2024 in the region of Northern Germany. *I-V* curves are corrected to standard test conditions using module temperature and irradiance measured with a WPVS reference cell in accordance with IEC 60891, respectively. In addition, a spectral mismatch correction in accordance with IEC 60904-7 [7] is performed, based on filter radiometer data derived during the *I-V* curve measurements and module spectral responsivity measured in the lab in accordance with IEC 60904-8 [8]. An ETC-like temperature correction is conducted based on the V_{oc} values measured in the lab.

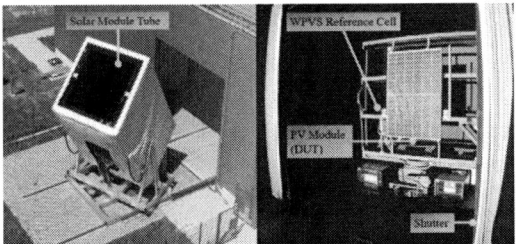

Figure 2: Photograph of the solar module tube during measurements with a mounted PV module inside.

Multiple measurements are carried out for each module, some of which are spread over several days for better statistics and to examine irradiance and temperature drifts during the measurements, to account for atmospheric/weather fluctuations. We observed variations between the corrected P_{mpp} values of up to 0.6 % per module caused by unstable irradiance conditions. The reasons for this are cirrostratus clouds and high fog, which have an influence on the light spectrum and irradiance. Another reason is an increase in temperature during the measurement, from the moment the shutter is opened.

3   RESULTS

Figure 3 show the experimental results of our comparative study. The deviation of the P_{mpp} from its nominal power of -0.05 % up to -2.42 % for the LED-based solar simulator, resulting in an average P_{mpp} 1.33 % lower compared to nameplate values. Almost half of all module types show deviations of more than -1.5 %. We did not observe any specific patterns like size or year of production yet. This might change with a larger dataset in the future. Figure 4 shows the cross-check between solar simulator and solar module tube with differences in P_{mpp} of -0.49 % (Module Type 7) and +0.65 % (Module Type 2) for the 2 PV modules compared. These cross-checks are in accordance with the results of the solar simulator.

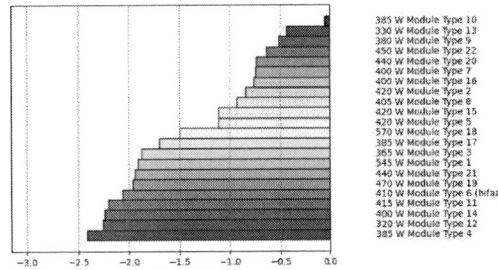

Figure 3: Deviation of the measured output values from the LED-based solar simulator from the values specified on the module nameplate.

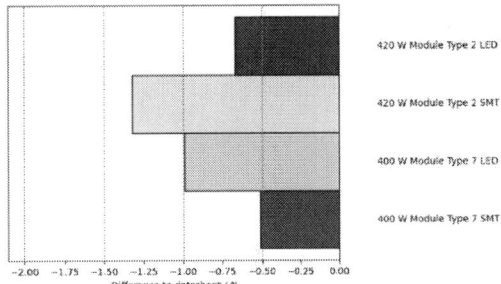

Figure 4: Deviation of the measured power values of the LED-based solar simulator (LED) and the solar module tube (SMT) from the values specified on the module nameplate.

4   CONCLUSION

For all measured PV modules, the power values are specified higher on the nameplate than the power values measured by our facilities. The deviation of the LED-based solar simulator, resulting in an average P_{mpp} 1.33 % lower compared to nameplate values. Furthermore, almost half of all module types show deviations of more than -1.5 %. To cross-check our results, 2 PV modules of different module types are also measured using the direct sunlight method with our outdoor test stand. The results of the solar module tube are in accordance with the results of the solar simulator. An explanation for the observed deviations in this study is out of our scope, since we do not know the exact traceability chain of module manufacturers. Therefore, we plan to conduct periodic market studies on this issue with a continuously updated selection of PV modules. Further improvements to our solar module tube are planned, including a better control of the PV module temperature, improving ECT module temperature measurements and characterization of irradiance non-uniformity within the measurement chamber.

5   ACKNOWLEDGEMENT

This work is developed within the project „MetroKomPV", which is funded by the Federal Ministry for Economic Affairs and Climate Action (BMWK), Germany (funding reference number 03EE1024).

6 REFERENCES

[1] https://www.destatis.de/DE/Presse/Pressemitteilunge
n/2024/09/PD24_334_43312.html

[2] https://www.bundestag.de/dokumente/textarchiv/202
4/kw17-de-eeg-photovoltaik-999570

[3] Riechelmann S, Friedrich D, Müller M et al. (2022)
Primary Calibration of Solar Modules With Direct
Sunlight. 39th European Photovoltaic Solar Energy
Conference and Exhibition; 474 - 476

[4] International Electrotechnical Commission
Photovoltaic devices - Part 9: Classification of solar
simulator characteristics (IEC 60904-9:2020)

[5] International Electrotechnical Commission
Photovoltaic devices - Procedures for temperature
and irradiance corrections to measured I-V
characteristics (IEC CD 60891:2019)[6]

[6] Sträter H, Riechelmann S, Neuberger F et al. (2019)
LED-Based Differential Spectral Responsivity
Measurements of PV Modules. 36th European
Photovoltaic Solar Energy Conference and
Exhibition; 1121-1126

[7] International Electrotechnical Commission
Photovoltaic devices - Part 7: Computation of the
spectral mismatch correction for measurements of
photovoltaic devices (IEC 60904-7:2019)

[8] International Electrotechnical Commission
Photovoltaic devices - Part 8: Measurement of spectral
responsivity of a PV device (IEC 60904-8:2014)

41st European Photovoltaic Solar Energy Conference and Exhibition

FINDING THE CELL TO MODULE PERFORMANCE VALUES FOR INDUSTRIAL TOPCON AND HJT TECHNOLOGIES

Sraisth[1], Hardik Gohil[1], Mehul Raval[1], Wolfgang Jooss[1]
[1]RCT Solutions GmbH, Line-Eid-Str. 1, 78467, Konstanz, Germany
sraisth@rct-solutions.com

ABSTRACT: This paper examines the cell-to-module (CTM) power conversion efficiency for glass-glass (GG) solar modules available on the market. Despite the impressive efficiency records reported by cell manufacturers of Tunnel Oxide Passivated Contact (TOPCon) and Heterojunction (HJT) cell technologies, a corresponding substantial increase in module power output is not consistently observed. Our analysis reveals a trade-off, where either higher CTM efficiency is achieved with relatively lower cell efficiency, or vice versa, to meet competitive module power outputs. This study highlights the critical need for accurate assessments of CTM loss and gain mechanisms to better understand their impact on overall cell and module efficiencies. Several photovoltaic (PV) suppliers were analyzed through simulations and calculations to determine the actual CTM power and efficiency conversion values. Results from four different suppliers using various cell technologies demonstrate that CTM power performance consistently exceeds 100%.

Keywords: Cell-to-module, CTM, TOPCon, HJT

1 INTRODUCTION

Cell-to-module (CTM) conversion value for any particular module type is crucial for total cost of ownership calculation during the feasibility study and investment purposes for module factory. The wrong calculation can lead to wrong values, especially when it is considered on the higher side. And if in actual production the module output is less, leading to lower production capacity compared to the planned capacity, company will go through financial losses. According to the recent NREL's market study, the CTM power conversion is somewhere in the range 99% to 101% or greater, for the new cell and module technologies like Tunnel Oxide Passivated Contact (TOPCon) and Heterojunction (HJT). However, the value varies from product to product and how they are designed. Cell interconnection plays an important role in achieving higher CTM values [2]. But such high cell efficiencies recorded in recent times in manufacturers datasheet doesn't correlate to the module power output. And CTM performance here plays an important role. Understanding the correct CTM combination with each cell technology is crucial for optimizing module power output. Accurate knowledge of CTM loss and gain mechanisms is essential to assess the true impact of solar cell efficiency on module performance. This study investigates the relationship between solar cell efficiency and CTM performance by evaluating data from different solar cells and modules datasheets. Multiple photovoltaic (PV) suppliers are examined to calculate, simulate, and determine the actual cell-to-module conversion values, with results from two specific suppliers showing CTM power performance exceeding 100%.

CTM power conversion is calculated by dividing the module power by the total cell power. And module power is calculated by adding optical gains and subtracting optical, geometrical, and electrical losses from the cell power.

$$CTMpower\ (\%) = \frac{Module\ Power}{Total\ Cell\ Power}\ x\ 100$$

Whereas,

$$Module\ Power = Total\ Cell\ Power + Optical\ Gains$$
$$- Lossed\ (Optical + Geometrical$$
$$+ Electrical)$$

The below picture, Fig. 1 and 2, explains all the CTM gains and losses. Orange and red color in the graphics depicts the losses and green color depicts the gains [3].

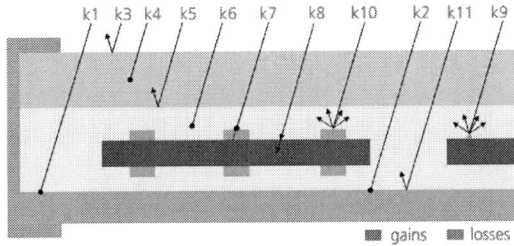

Figure 1: Schematic for CTM optical and geometrical losses and gains [3].

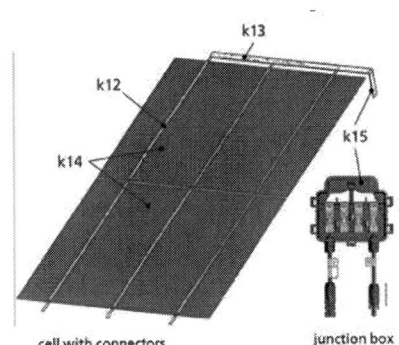

Figure 2: Schematic for CTM electrical losses [3]

2 EXPERIMENTS

To evaluate CTM performance, the following methodology was employed:
CTM Calculation Process: The cell-to-module performance was calculated by inputting solar cell electrical parameters, module layout (e.g., glass-to-glass

configuration, module dimensions, Junction Box (JB) specifics, ribbon dimensions, etc.), and the photovoltaic (PV) module supplier's parameters into the SmartCalc.CTM software developed by Fraunhofer ISE. It is important to note that the selected cell and module parameters were from the same manufacturers.

CTM Breakdown: The CTM values were categorized into electrical and optical components. These values were then input into a module electrical parameters calculation sheet to derive the cell's electrical parameters.

Data Comparison: After calculating the parameters based on CTM simulations, these values were compared to the industrial solar cell datasheet to validate the findings.

For this study four suppliers (A, B, C, D)—two for each technology, TOPCon and HJT, are selected. An important point to mention here is that the chosen suppliers have their own cell and module production. This helps in obtaining the cell type used in a module by checking the cell datasheets. A supplier with only module production is not selected because it is difficult to find the cell type and its right parameters for the CTM power conversion. Suppliers A and B are producing TOPCon, and Suppliers C and D are producing HJT. In this study only glass-glass (GG) based PV modules are investigated. Other module types are equally important, however, require extensive study and investigated, which can be considered in future.

For the CTM analysis, the advanced software, SmartCalc.Module from Fraunhofer-ISE is used [4]. Tables I, II, III, and IV show the electrical properties taken from the solar cell and module datasheets used for the simulations and calculations.

Table I: Cell Parameters provided from the datasheets for TOPCon cells

Parameters	Supplier A	Supplier B
Wafer Size	M10	G12R
Voc (V)	0.720	0.724
Vmpp (V)	0.618	0.622
Isc (A)	6.774	7.9615
Impp (A)	6.574	7.674
Power (Wp)	4.06	4.78
Eff (%)	24.6%	25.0%

Table II: Module Parameters provided from the datasheets for TOPCon cells

Parameters	Supplier A	Supplier B
No. of cells	72	66
Voc (V)	50.69	52.22
Vmpp (V)	43.57	42.84
Isc (A)	14.058	15.23
Impp (A)	13.20	14.59
Power (Wp)	575.00	625.00
Eff (%)	22.26%	23.14%

Table III: Cell Parameters provided from the datasheets for HJT cells

Parameters	Supplier C	Supplier D
Wafer Size	G12R	G12
Voc (V)	0.756	0.750
Vmpp (V)	0.686	0.680
Isc (A)	7.69	8.712
Impp (A)	7.30	8.370
Power (Wp)	5.01	5.62
Eff (%)	26.2%	25.8%

Table IV: Module Parameters provided from the datasheets for HJT cells

Parameters	Supplier A	Supplier B
No. of cells	66	66
Voc (V)	49.43	49.51
Vmpp (V)	41.32	41.69
Isc (A)	16.26	18.14
Impp (A)	15.39	17.27
Power (Wp)	635.00	720.00
Eff (%)	23.51%	23.18%

Data validation: For validating that cell parameters in the supplier's datasheet are not the same as expected from the CTM calculation, supplier A cells were tested at the ISC-Konstanz laboratory.

Additionally, quick simulations were performed for Supplier B to find the CTM power conversion values for different rear covers in a module. Supplier B rear cover was tested with three other rear covers, keeping all other information as it is. The rear covers used were glass without any grid pattern, white backsheet, and black backsheet based on the standard datasets in the software.

3 RESULTS

CTM analysis was performed using SmartCalc.CTM after adding the geometrical requirements, optical and electrical information of the PV modules components, where the cell and string distances are set to a conventional value of 1 mm and 2 mm for both HJT and TOPCon modules are considered respectively. Other parameters, for busbars, string ribbons and junction box, that can influence performance are taken from the datasheets and matched with the market trends. One more point to mention is that the same encapsulants and back cover are used for the simulation due to a lack of detailed information in the module datasheets. Extra recombination losses of 0.5% is considered for TOPCon suppliers due to cutting operation at module factory and having no passivation step like in HJT Cells production, where wafers are already half-cut [5]. All the modules have the white grid patterns at the back glass covering the cell-to-cell gaps, string gaps and string to edge gaps.

3.1 TOPCon cell-to-module performance analysis

Fig. 3 and 4 describes the optical and electrical gains and losses properties for the power output for TOPCon supplier A and supplier B.

The cell type for supplier A is M10 182 mm x 182mm. The CTM power conversion comes out 100.42%, as expected based on the market trends [1].

41st European Photovoltaic Solar Energy Conference and Exhibition

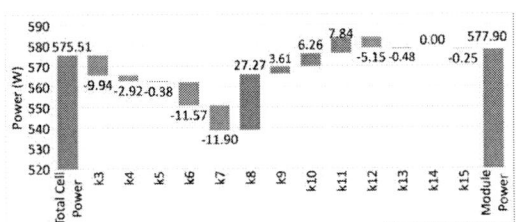

Figure 3: Waterfall diagram of the cell-to-module power performance for the supplier A (TOPCon)

It is important to note that the calculated cell efficiency using CTM parameters has approximately 0.4% difference from the datasheet value, see Table V. This cell is also measured in the lab to find out the exact cell efficiency.

Table V: TOPCon supplier A cell and module parameters comparison with calculation result.

Parameters	Module (Datasheet)	Cell (calculated)	Cell (datasheet)
Voc (V)	50.69	0.7060	0.72
Vmpp (V)	43.57	0.6133	0.618
Isc (A)	14.058	6.93	6.774
Impp (A)	13.20	6.50	6.574
Power (Wp)	575.00	3.99	4.06
Eff (%)	22.26%	24.2%	24.6%

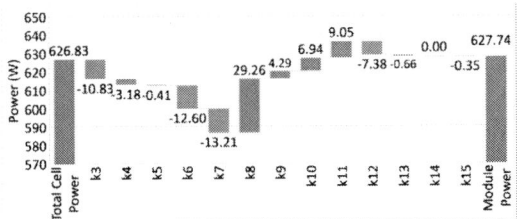

Figure 4: Waterfall diagram of the cell-to-module power performance for the supplier B (TOPCon)

The cell type for supplier B is G12R 182 mm x 210 mm. The CTM power conversion comes out 100.15%. The calculated cell efficiency using CTM parameters has approximately 0.2% difference from the datasheet value, see Table VI. This supplier has the lowest difference in CTM and datasheet values.

Table VI: TOPCon Solar supplier B cell and module parameters comparison with calculation result.

Parameters	Module (Datasheet)	Cell (calculated)	Cell (datasheet)
Voc (V)	52.22	0.7932	0.724
Vmpp (V)	42.84	0.6598	0.622
Isc (A)	15.23	7.50	7.9615
Impp (A)	14.59	7.19	7.674
Power (Wp)	625.00	4.74	4.78
Eff (%)	23.14%	24.8%	25.0%

3.2 HJT cell-to-module performance analysis

Fig. 5 and 6 describes the optical and electrical gains and losses properties for the power output for HJT supplier C and supplier D.

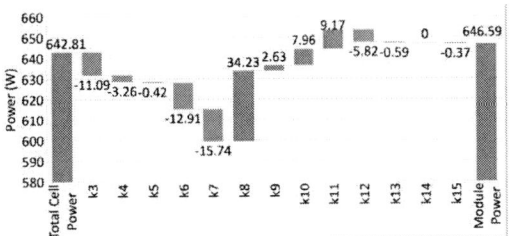

Figure 5: Waterfall diagram of the cell-to-module power performance for the supplier A (HJT)

The cell type was G12R 182.1 mm x 210 mm. The CTM power conversion comes out 100.59% The calculated cell efficiency using CTM parameters has a significant approximately 1.1% difference from the datasheet value, see Table VII.

Table VII: HJT supplier C cell and module parameters comparison with calculation result.

Parameters	Module (Datasheet)	Cell (calculated)	Cell (datasheet)
Voc (V)	49.43	0.7509	0.756
Vmpp (V)	41.32	0.6347	0.686
Isc (A)	16.26	8.00	7.69
Impp (A)	15.39	7.57	7.30
Power (Wp)	635.00	4.80	5.01
Eff (%)	23.51%	25.1%	26.2%

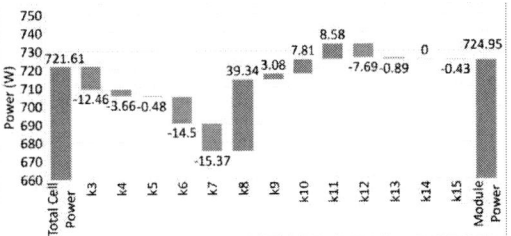

Figure 6: Waterfall diagram of the cell-to-module power performance for the supplier B (HJT)

The cell type for supplier D is G12 210 mm x 210 mm. The CTM power conversion comes out 100.46%.

Table VIII: HJT supplier D cell and module parameters comparison with calculation result.

Parameters	Module (Datasheet)	Cell (calculated)	Cell (datasheet)
Voc (V)	49.51	0.7522	0.75
Vmpp (V)	41.69	0.6416	0.68
Isc (A)	18.14	8.91	8.712
Impp (A)	17.27	8.49	8.37
Power (Wp)	720.00	5.45	5.62
Eff (%)	23.18%	24.7%	25.8%

The calculated cell efficiency using CTM parameters

41st European Photovoltaic Solar Energy Conference and Exhibition

has a significant approximately 1.1% difference from the datasheet value, see Table VIII. CTM conversion values is higher for HJT than TOPCon and so is the difference in the calculated and datasheet cell efficiency.

3.3 TOPCon Supplier A cell measurement performance and comparison

The cell from supplier A was measured in the lab of ISC-Konstanz. See the measured values in Table IX below.

Table IX: TOPCon supplier A cell measurement parameters comparison with datasheet.

Parameters	Cell (datasheet)	Cell (measured)
Voc (V)	0.72	0.731
Vmpp (V)	0.618	0.631
Isc (A)	6.774	6.70
Impp (A)	6.574	6.39
Power (Wp)	4.06	4.03
Eff (%)	24.6%	24.4%
FF (%)	83.24%	82.36%

Comparing both values, Voc is close to the datasheet value, with just a 0.01 difference. However, the main difference in efficiency is due to the datasheet's higher short circuit current density in the datasheet, a 0.074A difference. Thus, fill factor (FF) is 0.88% more in the datasheet than the measurement.

The difference in the Isc value arises due to the difference in the measurement chuck setup at the certification body (reflective chuck) and the production factory (non-reflective chuck). And this results in module power lower than what was expected from the cell parameters in the module production. Therefore, it is important that the cell from the suppliers should be measured at the module factory before feeding them to the production line, starting with the Stringer machine.

3.4 Influence of module back covers on the CTM

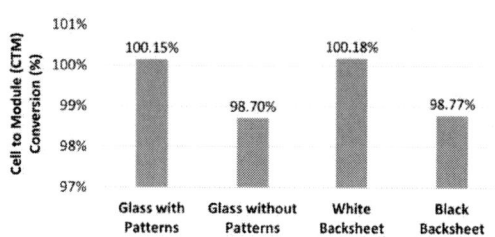

Figure 7: CTM power conversion comparison for different rear covers in Module supplier B.

Reference cell and module from TOPCon Supplier B is considered for this simulation. Standard parameters from software considered for back covers, as it is hard to get the actual data from the datasheets and industry unless it is self-measured in the lab.

From the simulation it was found out that CTM power conversion values are similar for GG with patterns and a white backsheet. Similarly, Glass without Patterns and black backsheets have similar values.

The highest CTM is for a white backsheet, thanks to

its completely white surface. There is approximately a ~1.5% difference in CTM values for different back covers. However, it can vary depending on the suppliers and types.

4 CONCLUSIONS

This study focused on understanding the impact of cell-to-module (CTM) conversion and its direct correlation with overall module efficiency for glass-glass modules. The study highlighted the need for solar cell efficiency measurement tools for module manufacturers, as discrepancies between labeled and actual solar cell efficiency can lead to elevated CTM loss values.

Simulations demonstrate that minimizing cell efficiency gaps and related losses can make achieving 100% or greater CTM feasible for future solar module development. This can be achieved by initial cell measurement without a reflective chuck during module production.

This work can be followed for other module types, such as glass-backsheet modules, including different back covers. Different back covers are also studied briefly for just one supplier. There is approximately a 1% to 1.5% difference in CTM values for different back covers.

ACKNOWLEDGEMENTS

Thanks to Fraunhofer-ISE for providing the Smart.Calc Module Simulation tool and ISC-Konstanz for doing the cell measurement.

REFERENCES

[1] J. Zuboy et al., "Getting Ahead of the Curve: Assessment of New Photovoltaic Module Reliability Risks Associated With Projected Technological Changes", IEEE journal of photovoltaics, Vol. 14, No. 1, pp. 4-22, Jan. 2024. 10.1109/JPHOTOV.2023.3334477

[2] A. Protti et al., "Analysis of Optical Coupling Gains from Cell Interconnection for the Energy Rating of PV Modules" pp. 020155-001 – 006, 2023. doi:10.4229/EUPVSEC2023/3AV.1.9

[3] M. Mittag et al., "Techno-Economic Analysis of Half-Cell Modules - The Impact of Half-Cells on Module Power and Costs" pp. 1032-1039, 2019. doi:10.4229/EUPVSEC20192019-4AV.1.20

[4] Fraunhofer Institute for Solar Energy Systems ISE, SmartCalc.Module, http://www.cell-to-module.com (accessed 10 March 2024).

[5] G. Xu, et al., "A comparative experimental study on front and back laser cutting technology for mass separation of N-TOPCon crystal silicon solar cells", Solar Energy Materials & Solar Cells 271 (2024) 112844. https://doi.org/10.1016/j.solmat.2024.112844

41st European Photovoltaic Solar Energy Conference and Exhibition

The Impact of Module Degradation on the Economics of PV Projects

Harry Apostoleris (harry.apostoleris@gmail.com)[1], Baloji Adothu[1], Bengt Jaeckel[2], Gerhard Mathiak[2], Sgouris Sgouridis[1]

[1]Dubai Electricity & Water Authority (DEWA) R&D Center, Mohammed bin Rashid Al Maktoum Solar Park, Saih al Dahal St., Dubai, UAE
[2]Fraunhofer Center for Silicon Photovoltaics CSP, Otto-Eissfeldt-Str. 12, 06120 Halle (Saale), Germany

1. Introduction

As market pressures drive solar electricity prices as well as module & hardware prices lower, developers must be alert to the potential risks posed by accelerated PV module degradation to the economic viability of solar energy projects. Here we consider the impact of module degradation on project returns. We calculate minimum sustainable PPA prices for utility scale solar projects based on different sets of initial assumptions regarding long-term module performance, and compute project internal rates of return under a range of performance scenarios with different values of linear module degradation, initial degradation and module failure rates.

2. Methodology

This study focuses on the economic impacts for PV projects of PV module degradation. We combine actual measurements of module performance over a 7-year study period with a project finance model based on annual cash flows. This hybrid approach allows technical characteristics & measured performance of modules to be directly connected to project economic indicators such as LCOE and IRR. We present this study as a general model that can also be used to consider the impact of newly-relevant degradation and failure modes that are becoming more widely observed in newer generations of modules

As a starting point we consider the measured degradation rates of modules from 3 manufacturers over 7 years. All modules were purchased in 2015. The measured linear degradation rates taken as (initial power – final power)/7 are plotted in Figure 1a for all modules, sorted by manufacturer & type (mono-multi). In Table 1 we display the rated and measured initial power, as well as the rated and measured degradation rates. We note that most modules are underrated and initial measured power is typically slightly higher than rated power. Initial power measurements were done after light soaking and should be equivalent to "year-1" performance, after any initial degradation. Based on component cost assumptions given in Table 2 (based on NREL benchmarks & current market conditions), we compute the minimum sustainable price per kWh as the LCOE for a solar plant built today using modules with the datasheet-provided characteristics. We compute a conservative and aggressive LCOE based on different assumptions about module performance. In the conservative assumption, we consider nameplate efficiency minus warranty initial degradation as the year-1 real power, with linear degradation as per the warranty in subsequent years. For the aggressive assumption, we consider linear degradation at warranty rate from an initial value assuming underrating ($P_{init} = 1.01 \times P_{nameplate}$ as per manufacturer 1 multi module data) – these assumptions could be justified as reflecting "real" module performance as past experience could indicate that such underrating is standard. The degradation trends for the different scenarios are plotted in Figure 1b.

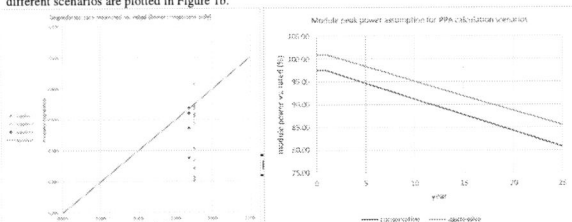

Figure 1 a) measured linear degradation rates of 2015 modules from 3 manufacturers compared to rated degradation; b) degradation

Manufacturer	cell type	Rated power (W)	Initial Power	Final Power	Initial/rated power
Manufacturer 1	mono c-Si	260	267.6	259.83	1.0292
Manufacturer 1	Multi c-Si	265	267.8	255.86	1.0106
Manufacturer 1	mono c-Si	260	267.3	263.03	1.0281
Manufacturer 1	Multi c-Si	265	267.3	254.51	1.0087
Manufacturer 1	mono c-Si	260	267.1	261.75	1.0273
Manufacturer 1	Multi c-Si	265	267.8	256.16	1.0106
Manufacturer 1	mono c-Si	260	268.6	264.67	1.0331
Manufacturer 2	Multi c-Si	255	258.2	243.19	1.0125
Manufacturer 2	Multi c-Si	255	258.4	246.31	1.0133
Manufacturer 2	Multi c-Si	255	257.8	247.41	1.0110
Manufacturer 2	Multi c-Si	255	257	250.94	1.0078
Manufacturer 3	Multi c-Si	250	255.6	243.47	1.0224
Manufacturer 3	Multi c-Si	250	255.5	245.76	1.0220
Manufacturer 3	Multi c-Si	250	256.5	245.01	1.0260
Manufacturer 3	Multi c-Si	250	255.6	249.30	1.0224

Table 1 Table 1 Rated and measured parameters for modules measured at DEWA

4. Conclusions

The data used here were collected on modules that, overall, performed in accordance with manufacturer claims. Nonetheless, these results show the importance for PV plant developers and operators of understanding the actual field performance of their modules and appropriately accounting for risk (unless a government or other entity provides backing to de-risk projects) when setting pricing agreements. Failure to do so can potentially lead to high-profile project failures and bankruptcies that impede the market for new solar projects. This will become especially critical as downward cost pressures on module and component manufacturers as well as project developers drive changes in module and system design that may cause new and possibly unexpected degradation modes to become more prominent; we illustrate this by including accelerated testing results of newer modules which show much stronger degradation. We present this study as a starting point for discussion on the critical importance of module quality and predictability of degradation rates in the solar energy industry; the risks that overconfident projections of PV module and system performance can have on the viability of solar energy projects; and the importance of continuously monitoring and verifying the performance of new module and cell technologies to avoid the risk of unexpectedly high degradation in new PV projects.

3. Results

Under the aggressive pricing assumptions, the risk to the developer is increased considerably as many conditions that would cause the project to fall short of performance targets would not be covered by the warranty, but as seen in Table 2 it allows the developer to justify a lower price which can be critical to securing large projects under the prevailing PPA auction system. In the cash flow analysis, four performance scenarios are considered with key parameters also listed in Table 2. In Figure 2 we show the influence of degradation rate on project IRR. When the conservative pricing is used, the combination of module underrating and the initial degradation term in the warranty allow IRR targets to be met even when the module significantly underperforms expectations in terms of annual degradation (based on measured initial power). However, the aggressive assumption (which still does not replicate the low PPA prices observed in some recent projects) leads to significantly below-target IRR for even small increases in degradation rate. We also consider the impact of module failure under the scenario that a certain percentage of modules fail per year and are replaced with new modules at cost. This scenario represents the risk that it may not in practice always be possible to invoke a manufacturer warranty in the case of module failure or damage. Again the aggressive PPA calculation leads to economic underperformance while the conservative assumption is far more robust.

We should also consider observed degradation in newer module technologies which can potentially present an even worse scenario. In Figure 3 we compare electroluminescence results from thermal cycling tests for modules with 5-busbar cells (technologically similar to the modules studied at DEWA), and newer modules with 16-busbar cells. The 16BB cells showed significantly more dark spots, accompanied by approximately 5% power loss, after only 200 cycles while the older modules showed minimal damage or power loss after 1000 cycles. This indicates that our degradation results from DEWA may actually paint an optimistic picture of module performance, and newer modules may see significantly higher degradation. The final manuscript will consider the effect of these "worst case" scenarios on project economics. This study is therefore presented, in part, as a warning to maintain accurate, data-based expectations regarding module performance to avoid severe economic underperformance of commercial solar projects over their lifetime.

	Pricing scenarios		Performance scenarios			
	conservative	aggressive	Measured	Warranty	5% init. deg.	1% annual module failure
capex (USD/kW)	600	600	600	600	600	600
capacity MW	1000	1000	1000	1000	1000	1000
opex %	1%	1%	1%	1%	1%	1%
IRR target %	5%	5%	--	--	--	--
1st year power % of nameplate	97.50%	101%	101%	97.5%	95%	95%
Linear degradation %/y	0.70%	0.65%	[0.5-2%]	[0.5-2%]	[0.5-2%]	[0.5-2%]
nom. kWh/kW	2150	2150	2150	2150	2150	2150
Annual module failure	0	0	0	0	0	1%
LCOE (USD/MWh)	24.78	23.75	--	--	--	--

Table 2 inputs & results for pricing and performance scenarios

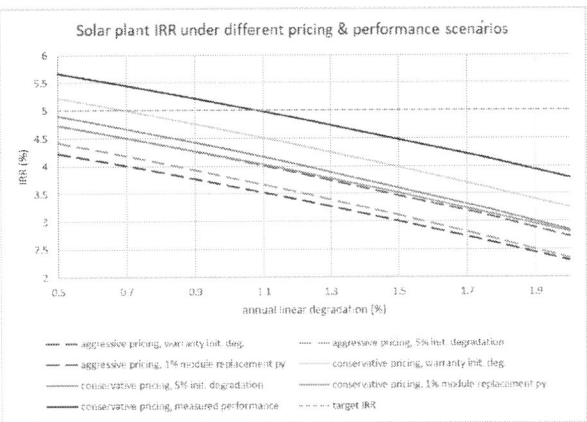

Solar plant IRR under different pricing & performance scenarios

Figure 2 Impact of degradation rate on IRR for different sets of pricing, initial power and module failure/replacement assumptions

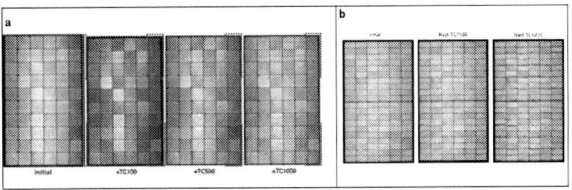

Figure 3 (a) Indoor testing results from Fraunhofer CSP of 5BB PERC cells. There was no significant change in EL appearance and power loss detectable even after extended testing. (b) newer module with 16BB cells, showing significant dark spots on EL image after 200 cycles

41st European Photovoltaic Solar Energy Conference and Exhibition

IMPROVED SAMPLING OF IV MEASUREMENTS

Maximilian Schönau[1,2], Elisabeth Schönau[3], Darwin Daume[2], Markus Panhuysen[1],
Achim Schulze[4], Bernd Hüttl[2], Dieter Landes[2]
[1]smartblue AG, Kistlerhofstraße 75, Munich, Germany
[2]Coburg University of Applied Sciences, Dept. of Electrical Engineering and Computer Sciences,
Friedrich-Streib-Straße 2, Coburg, Germany,
[3]Katholic University Eichstätt-Ingolstadt, Faculty of Mathematics and Geography
Ostenstraße 26, Eichstätt, Germany
[4]Rosenheim Technical University of Applied Science, Dept. of Appl. Natural and Social Sciences,
Hochschulstraße 1, Rosenheim, Germany
Maximilian.Schoenau@smartblue.de

ABSTRACT: The measurement of IV curves is the most important characterization technique of photovoltaic devices. This work seeks to determine a fair sampling rate for IV curves. Traditional sampling methods, such as using equidistant voltage steps, result in biased data by over- and undersampling the measurement at different voltages, which affect critical evaluations such as maximum power point (MPP) determination as well as series and shunt resistance estimation. Therefore, an adaptive sampling approach based on calculating the arc length of the IV curve is proposed, aiming to distribute data points equidistantly along the curve. This method is intended to provide a more balanced representation of the measurement data across all segments of the curve. The results indicate that using this adaptive approach, the measurement points can be significantly reduced at lower voltages while maintaining essential data density at key characteristics such as the MPP, facilitating faster and more accurate assessments.
Keywords: IV Curves, Adaptive Sampling

1 INTRODUCTION

The measurement of IV curves is the most important characterization technique of PV devices. IV curves are used for Standard Test Conditions (STC) [1], which acquire the most basic photovoltaic parameters of a device. They are used to monitor various stages within the deployment of PV, from creating PV cells to modules to strings. IV curves are also gaining popularity as an online monitoring method for operating PV strings [2], with inverters that are capable of IV measurements coming to market [3], [4], [5]. Finally, the measurement of current over voltage is used to acquire the MPP in the operation of PV.

This work seeks to determine an improved sampling rate for IV curves. Minimizing sample points in IV curve measurements provides substantial benefits across various applications: Reducing the number of sample points decreases the volume of data, which streamlines automated analysis processes. More critically, it shortens the measurement duration. In a production environment, expedited measurements enhance throughput and improves measurement accuracy by limiting exposure to variable conditions during the assessment period [6]. Similarly, for outdoor measurements and MPP trackers, a reduced measurement time minimizes losses and increases accuracy, among others by mitigating errors due to fluctuations in irradiance during data acquisition [7].

2 METHODOLOGY

Figure 1 displays an example of an IV curve with equal distant voltage steps, which is a common sampling method. The module is a Canadian Solar CS5P-220M at $1000\frac{W}{m^2}$ simulated by the De Soto model [11] as published in pvlib [11], [12]. For demonstration, the sampling is showcased with only 10 data points. The conventional sampled IV curve is compared with the proposed improved sample rate.

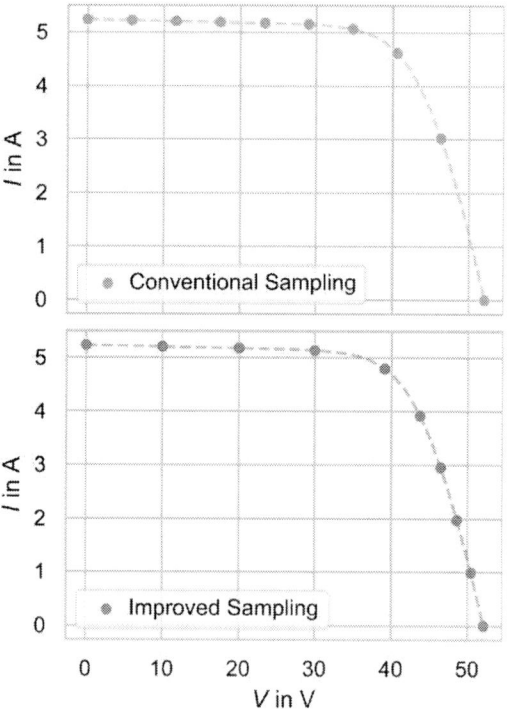

Figure 1: Conventional and improved sampled IV curve

While there are no universally optimal sample points for IV measurements, conventional sampling methods introduce biases. Equidistant voltage steps are particularly biased, by oversampling the low-voltage region of the curve while undersampling the high-voltage region. This discrepancy can lead to inaccuracies, when high-voltage characteristics are critical, such as in the evaluation of the MPP or the series resistance [8].

In addition to that, a fair sample rate is beneficial for machine learning applications. For instance, when a neural network processes IV curves, it may require fixed voltage sampling points in the input layer [9]. To maximize the effectiveness of the machine learning approach applied to an IV curve, it is beneficial to use purposefully chosen sample points, which enhance the network's ability to analyze and interpret the data accurately.

While the use of improved adaptive sample rates has been previously explored in the literature and adaptive sampling is implemented by measurement technology manufacturers [7], [10], this work proposes calculating the arc length of IV curves to establish a fair adaptive sampling rate. The voltages V_j of this improved sampling with n datapoints can be calculated with:

$$V_j = \arg\min_V |L_j - L(V)| \qquad (1)$$

Where:

$$L_j = \frac{L(V_{OC})}{n-1} \cdot (j-1), \qquad j \in \{1, 2, \dots, n\} \qquad (2)$$

$$L(V) = \int_0^V \sqrt{1 + \left(\frac{dI(V)}{dV}\right)^2}\, dV \qquad (3)$$

Eq. (1)-(3) are for the ideal case of a given continuous IV curve $I = f(V)$ with V_{OC} being the end of the curve.

The numerical implementation of the calculation of an optimal sample rate with a discrete IV curve is in the Appendix. The proposed strategy ensures a balanced information distribution across the curve, avoiding unwanted biases toward any particular segment.

3 RESULTS AND DISCUSSION

Figure 2 displays the resulting adaptive sample rate ΔV normalized to the initial sample rate ΔV_0 at $V = 0$ for the exemplary IV curve of Figure 1. The resulting graph starts with the initial normalized voltage interval of 1 and decreases to a normalized voltage interval of 0.16. Thus, the fairly sampled IV curve should contain around 6.3 times more datapoints at the end of the curve and about 1.7 more datapoints at the MPP than at the low-voltage region.

In many applications of IV curve measurements, the density of measurements is not needed to be uniformly, but instead specific regions of the curve are targeted with a higher sample rate. Determining whether and which part of an IV curve should contain more data points depends on the specific application of the measurement.

For example, areas like the MPP or other crucial segments could be targeted for increased detail. For this, the fair voltage steps ΔV may be multiplied with a factor at critical regions to have regions in the measurements with higher accuracy. This scaling of the fair sample rate ensures a higher resolution in critical areas while maintaining a progressively increasing sample rate that aligns with the typical shape of the IV curve to be determined.

For most practical applications, it is necessary to determine the sample points before measuring the IV curve, rather than adjusting them after the curve's characteristics are fully understood. As the true optimal sampling rate based on the arc length cannot be precisely known in advance, we recommend using the sample rate

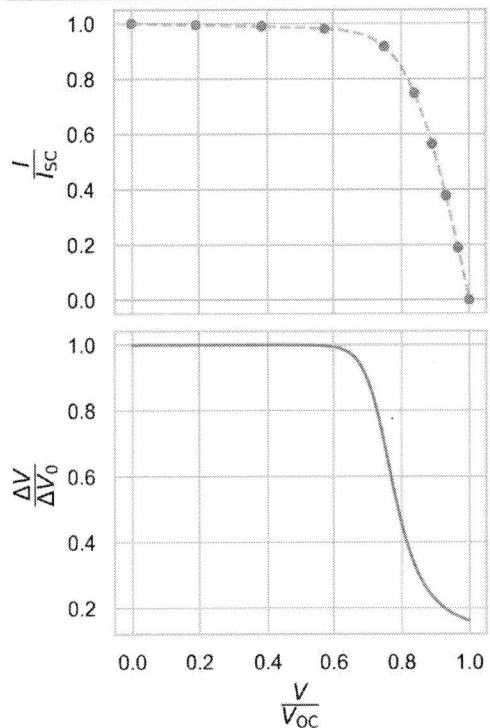

Figure 2: Plot of the increase in the sample rate of an IV curve that is necessary for equally distributed data points within an exemplary IV curve (see Figure 1)

of a typical IV curve (or the mean of a set of typical IV curves) as a reference for the used application and device characteristics. This approach allows for the estimation of a sampling rate that ensures an improved distribution of sample points on average, even though some measurements from defective devices may not follow the typical shape.

4 CONCLUSION

An adaptive sampling method for IV curve measurements is introduced that enhances the accuracy and efficiency of PV characterization. The method ensures a uniform distribution of sample points along the curve, effectively reducing biases inherent in traditional methods. The approach minimizes the number of necessary measurement points, which improves the precision of IV measurements while decreasing the data processing time and measurement duration. The fair sample rate is a first step for creating adaptive sampling strategies tailored to distinct requirements that align with the typical shape of the IV curve to be measured.

ACKNOWLEDGEMENTS

The authors gratefully acknowledge the Bavarian Research Foundation for their financial support of the project Kick-PV: "AI-based characterization and classification of PV-plants for predictive maintenance" under reference number AZ-1564-22.

Special thanks to Joseph Jachmann and Franziska Weber for their support with this work.

LITERATURE

[1] *IEC 60904-1:2020, Photovoltaic devices. Part 1, Measurement of photovoltaic current-voltage characteristics*, Edition 3.0. Geneva, Switzerland: International Electrotechnical Commission, 2020.

[2] M. Scheler *et al.*, "Precise On-Site Power Analysis of Photovoltaic Arrays by Self-Reference Algorithm," *8th World Conf. Photovolt. Energy Convers. 1070-1073*, p. 4 pages, 28190 kb, 2022, doi: 10.4229/WCPEC-82022-4DO.1.4.

[3] Huawei, "Smart I-V Curve Diagnosis." Accessed: Jun. 28, 2023. [Online]. Available: file:///C:/Users/maxim/Downloads/IV-Curve-1.pdf

[4] Growatt, "Growatt I/V Curve Diagnosis." Accessed: Jun. 28, 2023. [Online]. Available: http://www.growatt.pe/upload/file/contents/2019/06/5cf8adfa1d465.pdf

[5] SMA, "Technical Information - SUNNY TRIPOWER CORE1-US Commissioning and Servicing Commercial PV Systems with SunSpec Rapid Shutdown." Accessed: Jun. 28, 2023. [Online]. Available: https://files.sma.de/downloads/STP-CORE1-US-RSS-TI-en-11.pdf

[6] P. Kunze *et al.*, "Contactless Inline *IV* Measurement of Solar Cells Using an Empirical Model," *Sol. RRL*, vol. 7, no. 8, p. 2200599, Apr. 2023, doi: 10.1002/solr.202200599.

[7] Y. Zhu, "An Adaptive I-V Curve Detecting Method for Photovoltaic Modules," in *2018 IEEE International Power Electronics and Application Conference and Exposition (PEAC)*, Shenzhen: IEEE, Nov. 2018, pp. 1–6. doi: 10.1109/PEAC.2018.8590262.

[8] K. Mertens, *Photovoltaics: fundamentals, technology and practice*, Second edition. Hoboken, NJ Chichester, West Sussex: Wiley, 2019.

[9] M. Schönau, D. Daume, B. Hüttl, and D. Landes, "Improving IV Curve Classification by Machine Learning Methods Using Deep Autoencoders," *40th Eur. Photovolt. Sol. Energy Conf. Exhib.*, pp. 020410-001-020410–004, 2023, doi: 10.4229/EUPVSEC2023/4CV.1.53.

[10] halm, "PES-S-1-20-halm-EJ-V10-1.pdf." pessolar.com. Accessed: May 20, 2024. [Online]. Available: https://cdn.pes.eu.com/v/20180916/wp-content/uploads/2020/06/PES-S-1-20-halm-EJ-V10-1.pdf

[11] W. De Soto, S. A. Klein, and W. A. Beckman, "Improvement and validation of a model for photovoltaic array performance," *Sol. Energy*, vol. 80, no. 1, pp. 78–88, Jan. 2006, doi: 10.1016/j.solener.2005.06.010.

APPENDIX

Algorithm 1 contains the Python code for the numerical calculation of the proposed methodology of Eq (1)-(3) using NumPy.

Algorithm 1, Calculation of a fairly sampled IV-curve

```python
def fair_sample_points(
    I: np.ndarray, V: np.ndarray, n: int
) -> np.ndarray:
    """
    Returns fairly distributed voltages to
    sample the IV curve.

    Arguments
    ---------
    I: np.ndarray
        Normalized current of an IV-curve.
    V: np.ndarray
        Normalized voltage of an IV-curve.
    n: int
        Number of sample points.

    Returns
    -------
    V_new: np.ndarray
        n voltages after equal arc lengths in
        the IV curve.
    """
    # Linear upsampling of data
    n_resolution = 10000
    V_i = np.linspace(0, 1, n_resolution)
    I_i = np.interp(V_i, V, I)

    # IV arc length over i
    L_i = np.cumsum(np.abs(
        np.gradient(V_i + 1j * I_i))
    )

    j = np.arange(n)
    L_i, j = np.meshgrid(
        L_i, j, indexing="ij"
    )

    # Get the indices of equally spaced
    # arc lengths
    L_x_j = L_i[-1] / (n - 1) * j
    k_j = np.argmin(
        np.abs(L_i - L_x_j), axis=0
    )
    V_new = V_i[k_j]
    return V_new
```

41st European Photovoltaic Solar Energy Conference and Exhibition

ENHANCING PRODUCTION FORECASTING OF GRID-CONNECTED PV STRINGS OPERATING UNDER SEMI-ARID CLIMATE CONDITIONS

Khadija El Ainaoui[a, b*], Mhammed Zaimi[a], Imane Flouchi[b], Said Elhamaoui[b], Yasmine El mrabet[b],
Abdellatif Ghennioui[b], El Mahdi Assaid[a]

[a]Electronics and Optics of Semiconductor Nanostructures and Sustainable Energy Team,
Laboratory of Instrumentation of Measure and Control, Department of Physics,
Faculty of Sciences, Chouaïb Doukkali University, El Jadida, Morocco

[b]Electrical Systems and Photovoltaics Department, Green Energy Park (IRESEN/UM6P), Benguerir, Morocco

*Corresponding author: elainaoui.k@ucd.ac.ma; elainaoui@greenenergypark.ma; Tel: +212 696 321 700

ABSTRACT: Integrating photovoltaic (PV) solar energy into the electrical grid poses significant challenges due to the unpredictability of PV systems production, which introduces variability and complicates synchronization between electricity generation and consumption [1], [2]. Thus, accurate forecasting of PV power production is crucial for ensuring grid stability and efficient energy management [3]. This study focuses on modeling and forecasting the performance of grid-connected PV strings coming from different technologies, aiming to enhance the accuracy of output power prediction. The semi-arid climate, characterized by high solar irradiance and significant temperature fluctuations [4], [5], presents unique challenges and opportunities for solar energy generation. To accurately predict the performance of PV strings in these conditions, the study employs predictive approaches based on both implicit and explicit models: Single Diode Model (SDM) and Das Model (DM). The reliability of the proposed approaches is assessed by comparing I-V curves, P-V curves and peak power predicted by SDM and BM models to the actual measurements of PV strings coming from different technologies and operating at Green Energy Park (GEP) research facility in Morocco. The comparison reveals that the predicted outcomes align well with the real data, with an average of Normalized Root Mean Square Error (NRMSE) not exceeding 3.34% for SDM and 5% for DM across the three PV strings throughout the day. The results demonstrate that the predictive approaches effectively forecast the production of PV strings, accounting for climatic influences and providing insights into optimizing PV system performance, thereby contributing to improved grid stability in semi-arid regions.

Keywords: Grid-connected PV strings; Single diode model; Explicit model; PERC; I-V curves.

1 PV STRING MODELING

1.1 Single diode model (SDM)

SDM is an implicit model widely used in PV modeling. It includes five physical parameters: photocurrent (I_{PV}), saturation current and ideality factor (I_s, η), series resistance(R_s), and parallel resistance (R_P). The output current of a PV string is calculated as follows [6] :

$$I = I_{PV} - I_s\left(\exp\left(\frac{V + R_s I}{N_s N_M \eta V_{Th}}\right) - 1\right) - \frac{V + R_s I}{R_P} \quad (1)$$

Where I is the string output current, V is the string output voltage, $V_{Th} = k_B T_s / q$ is the thermal voltage, q is the electronic charge, k_B is Boltzmann constant, T_s is the string temperature, N_s is the number of cells mounted in series and N_M is the number of modules mounted in series. The resolution of SDM equation (Eq. (1)) via LambertW function, leads to [7], [8]:

$$I = -\frac{N_s \eta V_{Th}}{R_s} W_0\left(\frac{R_P R_s I_s}{N_s N_M \eta V_{Th}(R_P + R_s)}\exp\left(\frac{R_s(I_s + I_{PV}) + V}{N_s N_M \eta V_{Th}(R_P + R_s)}\right)\right) + \frac{R_P(I_{PV} + I_s) - V}{R_P + R_s} \quad (2)$$

1.2 Das model (DM)

Das proposed an explicit model to characterize the behaviour of PV modules, it incorporates two PV metrics and two shape parameters [9]:

$$I = I_{SC}\frac{1 - (V/V_{OC})^m}{1 + \alpha(V/V_{OC})} \quad (3)$$

1.3 Dependence of model parameters on solar irradiance and string temperature

1.3.1 PV metrics models

The dependence of short circuit current and open circuit voltage on solar irradiance and string temperature is given by the following formulas [10]:

$$I_{SC}(I_{POA}, T_s) = A_1(1 + A_2(T_s - T_{STC}))\left[1 + \sum_{i=3}^{6} A_i\left(\frac{I_{POA} - I_{STC}}{I_{STC}}\right)^{i-2}\right] \quad (4)$$

$$V_{OC}(I_{POA}, T_s) = B_1 + B_2(T_s - T_{STC}) + N_s V_{Th}(T_s)\sum_{i=3}^{6} B_i \ln\left(I^*_{POA}\right)^{i-2} \quad (5)$$

Where, T_{STC} is the temperature at STC, I_{POA} is the plane-of-array irradiance, I_{STC} is the solar irradiance at STC, and $I^*_{POA} = I_{POA}/I_{STC}$ stands for the normalized solar irradiance.

1.3.2 SDM parameters models

To describe variations of SDM physical parameters, the following mathematical models are used [10], [11], [12]:

$$I_{PV}(I_{POA}, T_s) = C_1(1 + C_2(T_s - T_{STC}))I^*_{POA} \quad (6)$$

$$\eta(I_{POA}, T_s) = D_1(1 + D_2(T_s - T_{STC}))\left(1 + \sum_{i=3}^{6} D_i \ln\left(I^*_{POA}\right)^{i-2}\right) \quad (7)$$

$$R_s(I_{POA}, T_s) = E_1(1 + E_2(T_s - T_{STC}))\left(1 + \sum_{i=3}^{6} E_i \ln\left(I^*_{POA}\right)^{i-2}\right) \quad (8)$$

$$R_P(I_{POA}, T_s) = F_1(1 + F_2(T_s - T_{STC}))\left(1 + \sum_{i=3}^{6} F_i \ln\left(I^*_{POA}\right)^{i-2}\right) \quad (9)$$

$$I_s(I_{POA}, T_s) = \frac{I_{PV}(I_{POA}, T_s) - V_{OC}(I_{POA}, T_s) / R_P(I_{POA}, T_s)}{e^{(V_{OC}(I_{POA}, T_s)/N_s \eta(I_{POA}, T_s)V_{Th}(T_s))} - 1} \quad (10)$$

1.3.3 DM parameters models

To describe variations of DM shape parameters of a PV string operating under real world conditions, new mathematical models are developed:

$$m(I_{POA}, T_s) = G_1(1 + G_2(T_s - T_{STC}))\left(1 + \sum_{i=3}^{6} G_i \ln\left(I^*_{POA}\right)^{i-2}\right) \quad (11)$$

776

41st European Photovoltaic Solar Energy Conference and Exhibition

$$\alpha(I_{POA}, T_S) = H_1\left(1 + H_2\left(T_S - T_{STC}\right)\right)\left(1 + \sum_{i=3}^{6} H_i \ln\left(I^*_{POA}\right)^{i-2}\right) \quad (12)$$

$A_i (i = 1 \text{ to } 6)$, $B_i (i = 1 \text{ to } 6)$, $C_i (i = 1 \text{ to } 2)$, $D_i (i = 1 \text{ to } 6)$, $E_i (i = 1 \text{ to } 6)$, $F_i (i = 1 \text{ to } 6)$, $G_i (i = 1 \text{ to } 6)$, $H_i (i = 1 \text{ to } 6)$ are determined by fitting Eqs. (4) to (12) to real-time dependent parameter values throughout a reference day.

1.4 Assessment of predictive approaches

For evaluating the accuracy of predictive approaches in forecasting PV strings production, a thorough assessment using Normalized Root-Mean-Square Error (NRMSE) is conducted [13]:

$$NRMSE = \sqrt{\frac{1}{N}\sum_{j=1}^{N}\left(I^j_{Measured} - I^j_{Calculated}\right)^2} \Bigg/ \frac{1}{N}\sum_{j=1}^{N} I^j_{Measured} \quad (13)$$

2 RESULTS AND DESCUSSION

Field experiments were conducted to assess the performance of PV strings coming from different technologies. The first PV string comprises 15 polycrystalline JKM330PP-72-V PV modules, while the second features 10 monocrystalline JKM525M-72HL4-V PV modules. The third PV string contains 12 Passivated Emitter and Rear Contact (PERC) JKM405M-72H PV modules. All three PV strings are fitted with multi-busbar technology, while the second and third PV strings specifically encompasses half-cut cell technology for increased efficiency. These strings are set in field at GEP research facility, at latitude 32.23°N, longitude -7.95°W and an altitude of 449 m. They are facing south direction with a tilt angle of 32.5°. For real-time monitoring, we employed specific tools: a calibrated PVPM1540X I-V tracer for electrical parameters measurements, a reference cell for solar irradiance monitoring, and a PT1000 thermal sensor for string temperature harvesting. Fig. 1 illustrates the evaluated PV strings (a, b, c), PVPM1540X I-V tracer (d), the reference cell (e) and the PT1000 thermal sensor (f).

Figure 1: Photographs of JKM405M-72H PV string (a), JKM525M-72HL4-V PV string (b), JKM330PP-72-V PV string (c), PVPM1500CX I-V tracer (d), reference cell (e) and PT1000 thermal sensor (f).

Figs. 2 and 3 illustrate the comparison between the measured and predicted I-V and P-V curves for JKM405M-72H, JKM525M-72HL4-V, and JKM330PP-72-V PV strings operating outdoors in Benguerir, Morocco, on June 20th, 2024, June 21st, 2024, and April 20th, 2024, respectively, under various weather conditions. The curves predicted by SDM (Fig. 2), and DM (Fig. 3) show a close match with measured data across different levels of temperature and solar irradiance. These findings affirm that the predictive approaches offer high fidelity in modeling the behavior and forecasting I-V and P-V characteristics of PV strings.

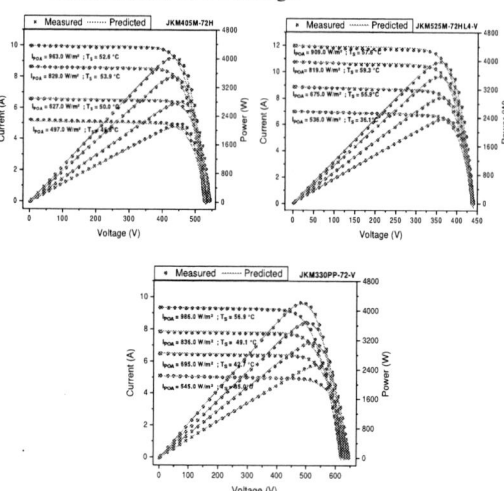

Figure 2: Measured data (garnet dots), I-V and P-V curves predicted via SDM (blue lines) for the three PV strings serving outdoors in Benguerir (Morocco) on June 20th, 2024, June 21st, 2024, and April 20th, 2024.

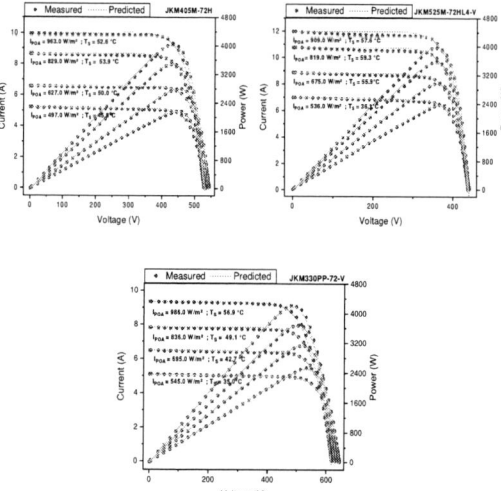

Figure 3: Measured data (garnet dots), I-V and P-V curves predicted via DM (green lines) for the three PV strings serving outdoors in Benguerir (Morocco) on June 20th, 2024, June 21st, 2024, and April 20th, 2024.

Fig. 4 presents temporal evolution of NRMSE values of current predicted via SDM and DM based approaches and corresponding daily average values for the three PV strings serving outdoors in Benguerir (Morocco). The figure shows that average value of NRMSE do not surpass 3.34 % for SDM based approach and 5 % for DM based approach for all PV strings. The evaluation demonstrates the accuracy of predictive approaches in forecasting with a high degree of confidence I-V curves, P-V curves and temporal evolution of peak power for PV strings operating under arbitrary weather conditions. It is important to note that DM offers a simpler approach, requiring fewer parameters compared to SDM. Furthermore, the new mathematical models of DM are quite accurate in determining PV string performance in real weather conditions. This accuracy enhances the credibility and precision of modeling and forecasting PV string outputs.

41st European Photovoltaic Solar Energy Conference and Exhibition

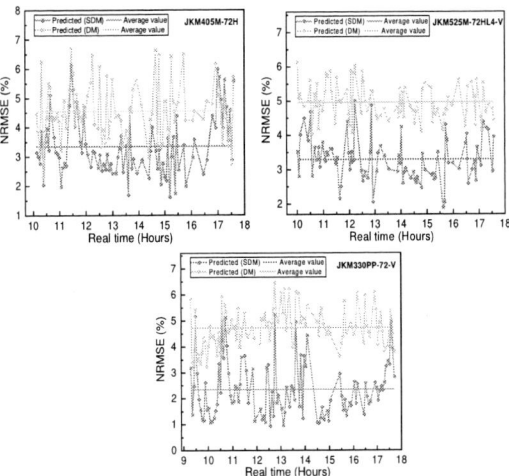

Figure 4: NRMSE values of current forecasted by SDM and DM based approaches for the three PV strings serving outdoors in Benguerir (Morocco) on June 20th, 2024, June 21st, 2024, and April 20th, 2024.

Fig. 5 shows the comparison between predicted and measured output power for the three evaluated PV strings serving outdoors under varying weather conditions. As illustrated in the figure, output power predicted by SDM and DM based approaches are in good agreement with measured data. This alignment highlights the effectiveness of both SDM and DM based methods, in accurately forecasting PV strings performance under different operational conditions.

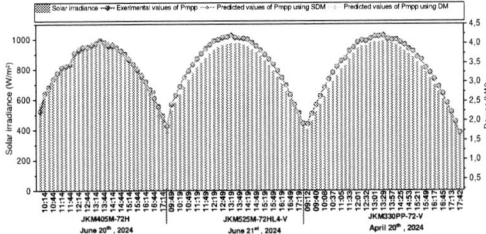

Figure 5: POA solar irradiance, measured and predicted values of peak power versus real time t for the three PV strings serving outdoors in Benguerir (Morocco) on June 20th, 2024, June 21st, 2024, and April 20th, 2024.

3 CONCLUSIONS

In conclusion, the study highlights the critical challenges of incorporating PV solar energy into the electrical grid, primarily due to the inherent unpredictability of PV systems. This variability complicates the synchronization between electricity generation and consumption, making accurate forecasting of PV power production essential for maintaining grid stability and efficient energy management. By focusing on modeling and forecasting the performance of grid-connected PV strings across various technologies, the research aims to significantly improve output power prediction precision using both implicit (SDM) and explicit (DM) models. The study introduces new analytical formulas for DM shape parameters, validated against real

measurements from PV strings operating at the GEP research facility in Morocco. The results show a strong alignment between predicted outcomes and real measurements, with an average NRMSE not exceeding 3.34% for SDM and 5% for DM throughout the day. DM emerges as particularly advantageous due to its simplicity with fewer optimization parameters compared to SDM. The newly developed mathematical models demonstrate significant accuracy in evaluating PV string performance under real weather conditions. This precision not only enhances reliability in forecasting PV outputs but also optimizes PV system performance to strengthen grid stability in semi-arid regions.

4 ACKNOWLEDGMENTS

The authors acknowledge the support of Green Energy Park and appreciate the opportunity to conduct research on its platform in BenGuerir, Morocco.

5 REFERENCES

[1] P. Gupta and R. Singh, "PV power forecasting based on data-driven models: a review," *International Journal of Sustainable Engineering*, vol. 14, no. 6, pp. 1733–1755, Nov. 2021, doi: 10.1080/19397038.2021.1986590.
[2] B. KROPOSKI, "Integrating high levels of variable renewable energy into electric power systems," *Journal of Modern Power Systems and Clean Energy*, vol. 5, no. 6, pp. 831–837, Nov. 2017, doi: 10.1007/s40565-017-0339-3.
[3] M. Paulescu, E. Paulescu, P. Gravila, and V. Badescu, *Weather Modeling and Forecasting of PV Systems Operation.* in Green Energy and Technology. London: Springer London, 2013. doi: 10.1007/978-1-4471-4649-0.
[4] A. Azouzoute *et al.*, "Modeling and experimental investigation of dust effect on glass cover PV module with fixed and tracking system under semi-arid climate," *Solar Energy Materials and Solar Cells*, vol. 230, p. 111219, Sep. 2021, doi: 10.1016/j.solmat.2021.111219.
[5] C. Hajjaj *et al.*, "Degradation and performance analysis of a monocrystalline PV system without EVA encapsulating in semi-arid climate," *Heliyon*, vol. 6, no. 6, p. e04079, Jun. 2020, doi: 10.1016/j.heliyon.2020.e04079.
[6] M. Chegaar, Z. Ouennoughi, and A. Hoffmann, "A new method for evaluating illuminated solar cell parameters," *Solid-state electronics*, vol. 45, no. 2, pp. 293–296, 2001.
[7] T. C. Banwell and A. Jayakumar, "Exact analytical solution for current flow through diode with series resistance," *Electron. Lett.*, vol. 36, no. 4, p. 291, 2000, doi: 10.1049/el:20000301.
[8] A. Jain and A. Kapoor, "A new approach to study organic solar cell using Lambert W-function," *Solar Energy Materials and Solar Cells*, vol. 86, no. 2, pp. 197–205, Mar. 2005, doi: 10.1016/j.solmat.2004.07.004.
[9] A. K. Das, "An explicit J–V model of a solar cell using equivalent rational function form for simple estimation of maximum power point voltage," *Solar Energy*, vol. 98, pp. 400–403, Dec. 2013, doi: 10.1016/j.solener.2013.09.023.
[10] M. Zaimi, K. E. Ainaoui, and E. Mahdi Assaid, "Mathematical models to forecast temporal variations of power law shape parameters of a PV module working in real weather conditions: Prediction of maximum power and comparison with single-diode model," *Solar Energy*,

778

vol. 266, p. 112197, Dec. 2023, doi: 10.1016/j.solener.2023.112197.

[11] W. De Soto, S. A. Klein, and W. A. Beckman, "Improvement and validation of a model for photovoltaic array performance," *Solar Energy*, vol. 80, no. 1, pp. 78–88, Jan. 2006, doi: 10.1016/j.solener.2005.06.010.

[12] T. U. Townsend, "A Method for Estimating the Long-Term Performance of Direct-Coupled Photovoltaic Systems," University of Wisconsin-Madison. Accessed: Mar. 14, 2023. [Online]. Available: https://minds.wisconsin.edu/handle/1793/46720?show=full

[13] K. El Ainaoui, M. Zaimi, and E. M. Assaid, "Innovative approaches to extract double-diode model physical parameters of a PV module serving outdoors under real-world conditions," *Energy Conversion and Management*, vol. 292, p. 117365, Sep. 2023, doi: 10.1016/j.enconman.2023.117365.

41st European Photovoltaic Solar Energy Conference and Exhibition

EVALUATION OF THE GLARE FUNCTION AND DESCRIPTION OF KEY MEASUREMENT PROCEDURES

Wolfgang Nemitz[1], Roman Trattnig[1], Jakob Zehndorfer[2], Markus Babin[3] Lukas Plessing[4]
[1]Joanneum Research – Materials, Weiz; [2] Zehndorfer Engineering GmbH, Klagenfurt; [3] Technical University of Denmark, Department of Electrical and Photonics Engineering, 4000 Roskilde, Denmark; [4] Österreichische Technologieplattform Photovoltaik, 1060 Wien

ABSTRACT:

The glare effect of PV systems is an increasingly discussed topic. Despite the growing concern about disturbing glare behaviour, data on building materials, windows, façade elements, roof tiles, or solar modules are often unavailable or, when provided, not comparable across manufacturers. The aim of this standardized glare measurement guide is to establish a comprehensive and repeatable procedure for assessing the potential glare from artificial surfaces and building materials, such as tiles, window glasses, façade materials, photovoltaic (PV) modules or PV cover glasses. The primary objective is to ensure the safety and comfort of people exposed to sunlight reflections by accurately quantifying and analysing glare phenomena. A secondary objective is to provide PV module manufacturers and installers with a reproducible method for assessing glare potential. Two approaches will be presented to yield results which are comparable to each other. Several PV cover materials are used as examples for the measurement methods and for the illustration of the resulting glare behaviour.

1 INTRODUCTION AND OUTLINE

Reflections of sunlight from flat surfaces, such as photovoltaic (PV) modules, can cause glare for observers in the vicinity. With the increased prevalence of PV, concerns about the glare behaviour raises increasingly. This concerns both, utility-scale PV plants, but also PV at buildings, especially in urban environments, where there is a higher risk of affecting neighbouring buildings and/or traffic areas.

Although many façade materials exceed the reflection levels of low-glare PV modules [1] or the absolute glare limit of 100,000 cd/m², it is hardly noticed. For example, textured glass like Albarino (Saint-Gobin) has been tested [2] and shows a low glare risk. Guidelines were developed, but a consistent measurement procedure is still missing.

Different approaches to the subject of glare caused by sunlight reflections can be found in the literature. Specific attention is given to the Bidirectional Reflectance Distribution Function (BRDF) and the specification of luminance in various contexts. However, a standardized description and comparison of these measurement techniques is lacking, as well as a standardized method for presentation and evaluation.

The following description outlines the measurement process in the laboratory using a gonioreflectometer on the one hand and using a luminance camera for outdoor measurements on the other hand. This serves as a basis for reproducible and comparable measurement results and sets the starting point for discussions about the evaluations concerning glare.

This guideline is based on discussions on the evaluation of the glare effect of photovoltaic (PV) systems, which is why the measurements and analyses specifically target front materials of PV modules.

However, it applies to most surfaces of building materials. The determined glare behavior provides a basis for the direct comparison of different materials as well as the evaluation of PV systems and of facades and roofs with conventional materials without the need for further measurements.

2 FUNDAMENTALS AND BACKGROUND

2.1 Glare

Glare is a disturbance of the vision caused by excessive light intensity, resulting in a reduction in visual performance. A distinction is made between direct glare, which arises from light sources with excessive luminance, and reflected glare, which is caused by reflections on surfaces. Absolute glare occurs when luminance levels are so high that the human eye cannot adapt. In the range from 10^4 cd/m² to $1 \cdot 10^5$ cd/m², adaptation is no longer sufficient and absolute glare occurs [3], which can result in a direct health risk to the eye.

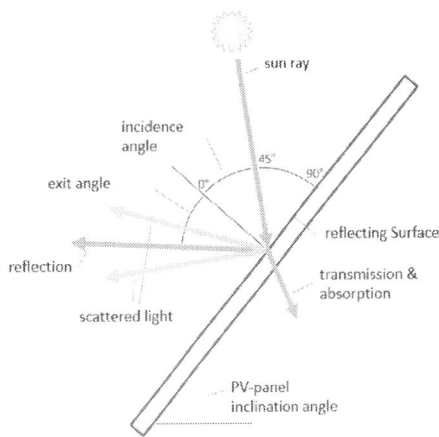

Figure 1: Illustration of optical reflections from a PV module including direct (specular) reflections and scattered (diffuse) light.

2.2 Material properties influencing glare

The glare behaviour of PV modules is primarily caused by the front sheet as most of the light transmitted at the front sheet is absorbed within the PV cell. Though, the material of the front sheet and its surface treatment determine the reflection of the incident light.

The law of reflection applies to smooth surfaces, stating that the angle of incidence equals the angle of reflection (specular reflections), as can be seen in Figure

1. Following from the Fresnel equations, the ratio of reflected light increases with the incidence angle. This increase is initially slowly, but beyond a source angle of 70° (relative to the normal of the module surface), this increase becomes higher and higher and reaches 100% at 90° (parallel to the module surface). Therefore, it is particularly important to analyse reflections generated by high source angles (flat light incidence) with precision.

For samples which have anisotropic reflections due to their surface structure, the maximum reflection value must be determined for each source angle. It is important to notice that this angle may also lie outside the measurement plane defined by the source angles. It is important to determine the BRDF, which describes the reflections as a function of both incidence and reflection angle.

2.3 Assessment techniques
BRDF measurement following the work of Nicodemus [4] are a common approach for assessing glare behaviour of different materials. Goniometer-based BRDF measurements of PV modules as well as direct luminance measurements were presented previously [5-7]. A simplified two-dimensional approach to represent the BRDF was demonstrated at the last EUPVSEC conference [8].

In Austria, a guideline [9] has been published that outlines a legal framework, based on rules for wind turbines and medical considerations [10,11] regarding light intensity limits. However, this document has been criticized as unhelpful in practice.

For the PV systems [12], the reflection characteristics of the PV modules serve as baseline. Using this data, the sun's path is simulated over the course of a year, and the glare effect is calculated for a specific observation site, as illustrated in Figure 2.

Although gloss measurements are considered in some local building regulations, they are not suitable for glare assessment due to the inherent ambiguity in the definition of gloss [13]. Thus, gloss measurements are excluded from this discussion, as they do not provide valuable contributions to a clear framework

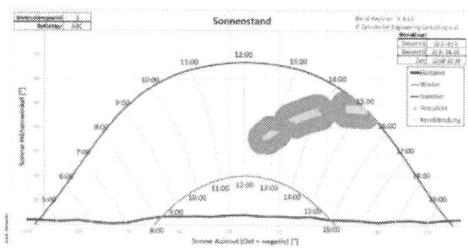

Figure 2: Example: The simulation shows a critical time window in which disturbing glare is to be expected. Illustration: Zehndorfer GmbH

3 MEASUREMENT RECOMMENDATIONS

Due to the higher reflection values at increasing incidence angles, as described in section 2.2, it is recommended to measure at least at incidence angles from 10° to 80° in 10° steps.

For the visualization, it is essential to determine the maximum reflection for each source angle and to display

this as a function of the source angle in terms of luminance [cd/m²]. This can be reached through angle-dependent reflection measurements, BRDF (Bidirectional Reflectance Distribution Function) measurements or luminance measurements. A suitable directional light source that covers the visible wavelengths of the solar spectrum must be used. If the light source do not have a continuous spectrum, it is not possible to measure the reflection values at the missing wavelengths.

If the solid angle of the reflection is smaller than that of the measuring device – indicated by the detector area not being fully illuminated – a correction factor for the measuring cone may be necessary using the luminance (cd/m²). The angle-dependent reflection data obtained from these measurements can then serve as a basis for simulating glare for a specific PV system setup and observer position (day/year).

3.1 Laboratory measurements
The advantage of laboratory measurements is an easier control of measurement conditions, resulting in reproducible measurement results. A prerequisite for this is a directional light source that covers the relevant wavelengths of the visible solar spectrum. Additionally, the reflections of light from this primary source must be isolated, either by preventing stray light from reaching the sample surface, or through lock-in techniques.

Before measuring, the spectrum of the specific light source has to be characterized to allow transfer of results to solar glare. A gonioreflectometer or an integrating sphere is typically required as the measuring device to control source and detector angles. The measurement data can also be utilized to simulate glare for a specific setup and observer position (day/year).

Another advantage is the automation of the measurement process, facilitated by the goniometer arms. However, a disadvantage is that most measuring devices record radiometric values, which must be converted into luminance for glare assessment through complex conversions.

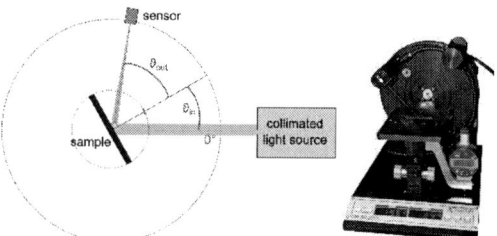

Figure 1: Illustration of the Goniometer setup, and picture of the Goniometer.

The measured reflectance values R_{meas} as a function of the wavelength λ, the illumination angle ϑ_{in} and the measurement angle ϑ_{out} lie in the value range between 0 and 1: These are converted into reflected irradiance E_{meas} using the solar spectrum with air mass 1.5 (AM1.5G), the formula is the first in the mathematical representation field below.

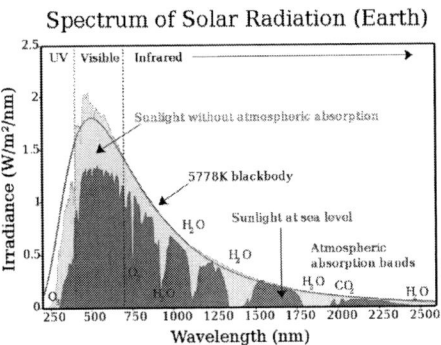

Figure 3: Full solar spectrum, including UV, visible, and infrared irradiance. Source: Wikipedia

This results in values for the irradiance as if the sample was illuminated by the sun. The eye sensitivity curve for day vision is used to convert the irradiance E_{meas} into the illuminance E_{phot}:

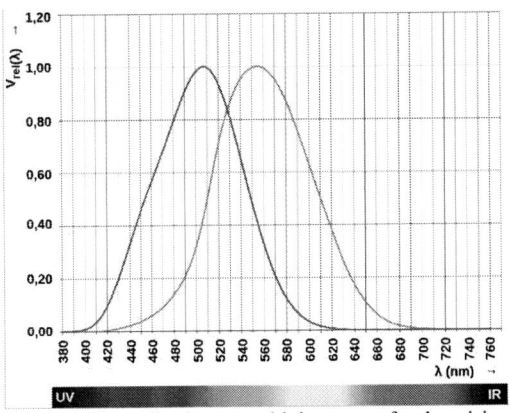

Figure 2: Normalized eye sensitivity curves for day vision (red) and night vision (blue).

Illuminance is a photometric quantity that depends on both the angle of illumination and the measuring angle. For diffusely reflecting surfaces (where the aperture angle of the reflection is greater than the solid angle of the measuring device), illuminance is converted into luminance L_{phot} using the solid angle of the measuring device Ω_{meas}.

For specularly reflecting surfaces (where the aperture angle of the reflection is smaller than the solid angle of the measuring device), illuminance is converted into luminance L_{phoz} using the solid angle of the sun $\Omega_{SUN}=$ 6.8e-5 sr.

The maximum luminance $L_{phot}(\vartheta_{in}) = \max(L_{phot}(\vartheta_{in}, \vartheta_{out}))$ [cd/m²] is determined for each illumination angle.

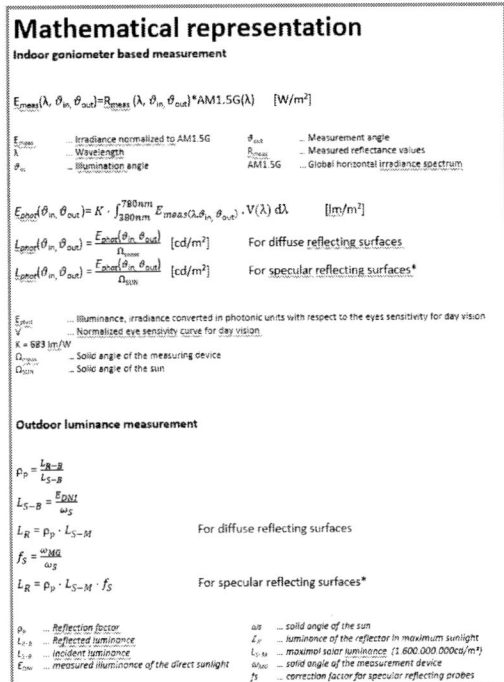

3.2 Outdoor measurements

The advantage of outdoor measurements is that the sun's rays, just like in real conditions, directly illuminate the PV module rather than relying on an artificial light source. As a result, the light spectrum, light distribution, and parallelism of the light rays are close to real-world scenarios. Special effects, such as anisotropic reflections and bundle displacement can also be observed and measured more effectively outdoors than with single-axis measurement systems indoors, making it easier to determine the absolute maximum of the reflected luminance.

Additionally, measurements are taken with a luminance spotmeter, which directly displays the luminance – the relevant value for glare – thereby reducing the need for subsequent calculations. However, the disadvantage of outdoor measurements is that they are considerably more time-consuming and hardly reproducible (in respect to laboratory measurements), as they require manual operation.

A prerequisite for the measurement is the rotatable mounting of the test specimen, allowing different angles of solar incidence to be set without having to wait for the natural course of the sun. The angle of the sun is determined using a sun angle meter, which indicates the normal angle of the sun relative to the surface of the test specimen. This setup ensures that various angles of solar incidence can be tested efficiently.

Since the luminance of the reflections may exceed the measuring range of the device, it may be necessary to use ND filters (neutral density or grey filters). The filter factors indicated on ND filters often deviate significantly from the actual values, so they must be calibrated before measurement.

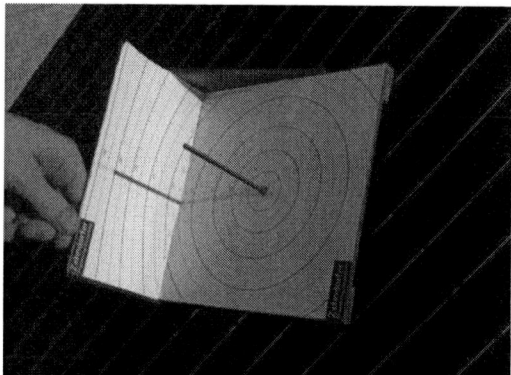

Figure 3: solar angle meter

The partial reflection factor ρ_p is determined by dividing the reflected luminance L_{R-B} by the incident luminance L_{S-B} (from the sun). For this measurement, a luminance spot meter with a small measurement angle (e.g., 1° or smaller) should be used to capture the maximum luminance. The following formula for the partial reflection factor describes the division of the calculated luminance values from the sun and the reflection:

$$\rho_p = \frac{L_{R-B}}{L_{S-B}}$$

This allows for an accurate assessment of how much light is reflected from a surface compared to the incident light from the sun.

Since the luminance of the sun depends on the solar elevation angle and humidity due to atmospheric absorption, it is necessary to capture the actual solar irradiance. In principle, this can be done with the same luminance spot meter used to measure reflections. However, there is a disadvantage: the spot meter must be precisely centered to the sun, which can be challenging to achieve by hand due to the small measurement angle and the small solar disc. Additionally, the extremely high luminance of the sun will necessitate a filter change (e.g., ND filter 40k).

A simpler approach is to use a lux meter with a shading device that blocks out the rest of the hemisphere, allowing only direct sunlight (EDNI) and excluding the rest of the sky. The luminance can be calculated from the illuminance divided by the solid angle of the sun ω_S, using the following formula:

$$L_{S-B} = \frac{E_{DNI}}{\omega_S}$$

For each incident angle (e.g., in 10° increments), the maximum reflected luminance is measured to determine, the partial reflection factor. If an ND filter is used, the reflected luminance value must be multiplied by the corresponding filter factor.

For samples that primarily reflect specularly (i.e., not diffusely), the measurement result must be multiplied by a measurement cone correction factor f_s. This correction factor is the reciprocal of the portion of the measurement cone that is actually covered by the smaller solar disc. The following formula describes the division of the solid angles of the measuring device ω_{MG} and the sun ω_s. Thus, the correct result for specular reflection is obtained by multiplying the measured value by f_s, ensuring accurate reflection measurements when the measurement cone is larger than the solid angle of the sun.

3.3 Other measurement techniques

In addition to reflectance measurements, other measurement methods and setups are also conceivable. For example, luminance camera measurements can be done indoors. It is essential to ensure that a light source with a continuous spectrum in the visible range is used, and that the angles can be set precisely and reproducibly. An integrating sphere or other measuring devices for BRDF measurements can also be used. The BRDF data must be converted into luminance data to facilitate comparison.

4 VISUALISATION, CALCULATIONS AND COMPARABILITY OF RESULTS

Once the maximum luminance has been determined for each illumination angle, it can be clearly visualized in a graph. The x-axis represents the illumination angles (0° corresponds to vertical incidence of light), while the y-axis shows the corresponding maximum reflected luminance. The maximum reflected luminance can be measured at a surface with regular reflection and corresponds to the maximum luminance of the sun on Earth (1.6×10^9 cd/m²).

The threshold for absolute glare is an individual limit. For glare assessment, a single value is required. We propose to set this limit at 100,000 cd/m² for practical reasons. The graphical presentation of the glare behaviour allows a quick visual assessment of whether this limit is exceeded at any given lighting angle. Depending on the installation situation, this enables a rapid evaluation of whether the PV modules may cause glare.

The outline for the visualization and measurement follows this:

- Recommended source angles:
  10°, 20°, 30°, 40°, 50°, 60°, 70°, 80°
- Maximum reflection for each source angle.
- Display as a function of the source angle as luminance in [cd/m²].
- One may have to account for reflection maxima at non-specular angles.
- Limit for absolute glare: 100 000 cd/m².

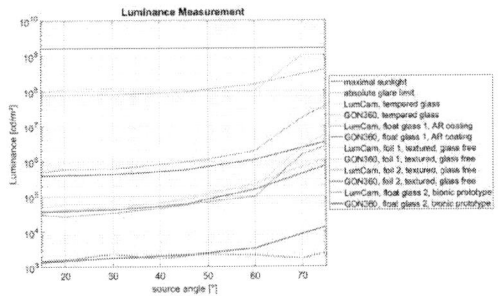

Figure 4: Visualization of the results determined by Luminance camera (LumCam) vs Goniometer (GON360).

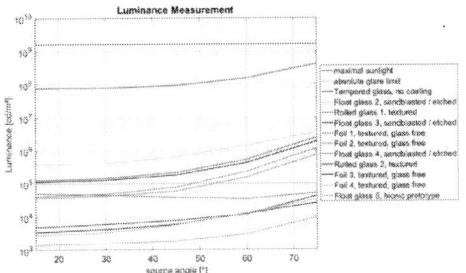

Figure 5: Visualization of the comparison tempered glass (window glass) vs PV front materials.

6. MEASUREMENT PROTOCOL

The measurement protocol for the guideline on standardized glare measurements for building materials and façade elements aims to ensure a uniform and comprehensible presentation of the assessment of the glare behaviour of surfaces. Since reflections can potentially disturb or even endanger residents, road users, and aircraft pilots, accurate measurement and assessment of glare effects are essential.

This measurement protocol provides clear instructions for conducting glare measurements, defines the measurement instruments and parameters to be used, and ensures that the results are comparable under different conditions. The goal is to identify potential glare risks at an early stage and, if necessary, recommend suitable measures to minimize glare effects, thereby ensuring the safety and comfort of all those affected.

Minimal standards for the measurement protocol
Description of the test sample (module, front-sheet, solar glass, building material): The following information must be provided:
- Photo of the sample
- Material description and application

• Surface characterization and description of the structure, particularly in the case of surface coatings

• Description of the expected optical properties, including a discussion of the possibility of anisotropy. This must be considered during the measurement if necessary.

• Manufacturer, serial number if available
Especially for solar modules:
- module type
- Module description including structure of the front of the module as far as known and possible.
- Encapsulation material, coatings and structure of the surface, including visual description of the module are essential aspects of the documentation.

Description of the measurement method:
- Measuring method
- Measurement setup
- Measuring devices must be described and documented, manufacturer, name and serial number.

Presentation of the measurement results

- Luminance at different angles of incidence (10°, 20°, 30°, 40°, 50°, 60°, 70°)
- In tabular and graphical form
- Description of the conditions under which absolute glare (> 100,000cd/m² is to be expected)

Documentation and Quality Assurance (QA):
Procedures for recording measurement data, calibration certificates, and measurement reports are established for documentation and testing purposes. The guidelines provided by the manufacturers of the measuring devices must always be followed. All relevant data and interim results must be recorded clearly and comprehensibly in the protocol or measurement report.

7 CONCLUSIONS

Emphasize values (depending on the angle of incidence) above 100,000 cd/m². Highlight their relevance in relation to guidelines and the legal framework. An application recommendation can be made based on these results, which may also include the assessment of potential discomfort and the identification of critical glare zones.

Conclusion and discussion

Required:
- Standard for glare evaluation of surfaces.
- Consistent rules for tolerable glare effects.

Proposed:
- Definition of the absolute glare limit: 100 000 cd/m².
- Legal framework for the glare effect of all façade and roof surfaces including PV-modules.

7. References

1) J. Moereke et al, 2022, LIGHT REFLECTION ANALYSIS OF PV MODULES: COMPARISON TO BUILDING FACADES AND ASSESSING THE POSSIBILITY OF GLARE; 8th World Conference on Photovoltaic Energy Conversion, Milano

2) M. Schiavoni et al, 2013, ANTI-GLARE EFFECT OF DEEPLY TEXTURED COVER GLASSES FOR SOLAR MODULES FOR INSTALLATIONS CLOSE TO AIRPORTS AND HELIPORTS,

EU PVSEC, Paris

3) SSK- Strahlenschutzkommission, 2006, Blendung durch natürliche und neue künstliche Lichtquellen und ihre Gefahren, 205. Sitzung

4) F.E. Nicodemus[2] et al, 1977, Geometrical Considerations and Nomenclature for Reflectance, NATIONAL BUREAU OF STANDARDS, US

5) M. Babin et al, 2022, Glare potential evaluation of structured PV glass based on gonioreflectometry, IEEE Journal of Photovoltaics

6) C. Bucher et al, 2021, Glare Hazard Analysis of Novel BIPV Module Technologies, International Solar Energy Society, Proceedings

7) Ruesch et al, 2016, Methode zur Quantifizierung der Blendung durch Solaranlagen- Vergleich mit anderen Materialien der Gebäudehülle, Institut für Solartechnik SPF, Rapperswil

8) C. Bucher et al, 2023, TWO-DIMENSIONAL REPRESENTATION OF THE BIDIRECTIONAL REFLECTANCE DISTRIBUTION FUNCTION OF PHOTOVOLTAIC MODULES, PVSEC, Lisbon

9) Glare from photovoltaic systems, OVE-Richtlinie R 11-3

10) Borgmann et al, 2014,Leitfaden "Lichteinwirkung auf die Nachbarschaft", Köln, Fachverband für Strahlenschutz

11) Moshammer et al, 2013, Medizinische Beurteilungsgrundlagen der Passiven Blendung; Studie im Auftrag von: Amt der Kärntner Landesregierung, Unterabteilung Sanitätswesen

12) J. Zehndorfer, 20xx, Teure Folgekosten vermeiden, Gewerbe & Kommune

13) Hanson, 2006, Good Practice Guide for the Measurement of Gloss, National Physical Laboratory, Teddington

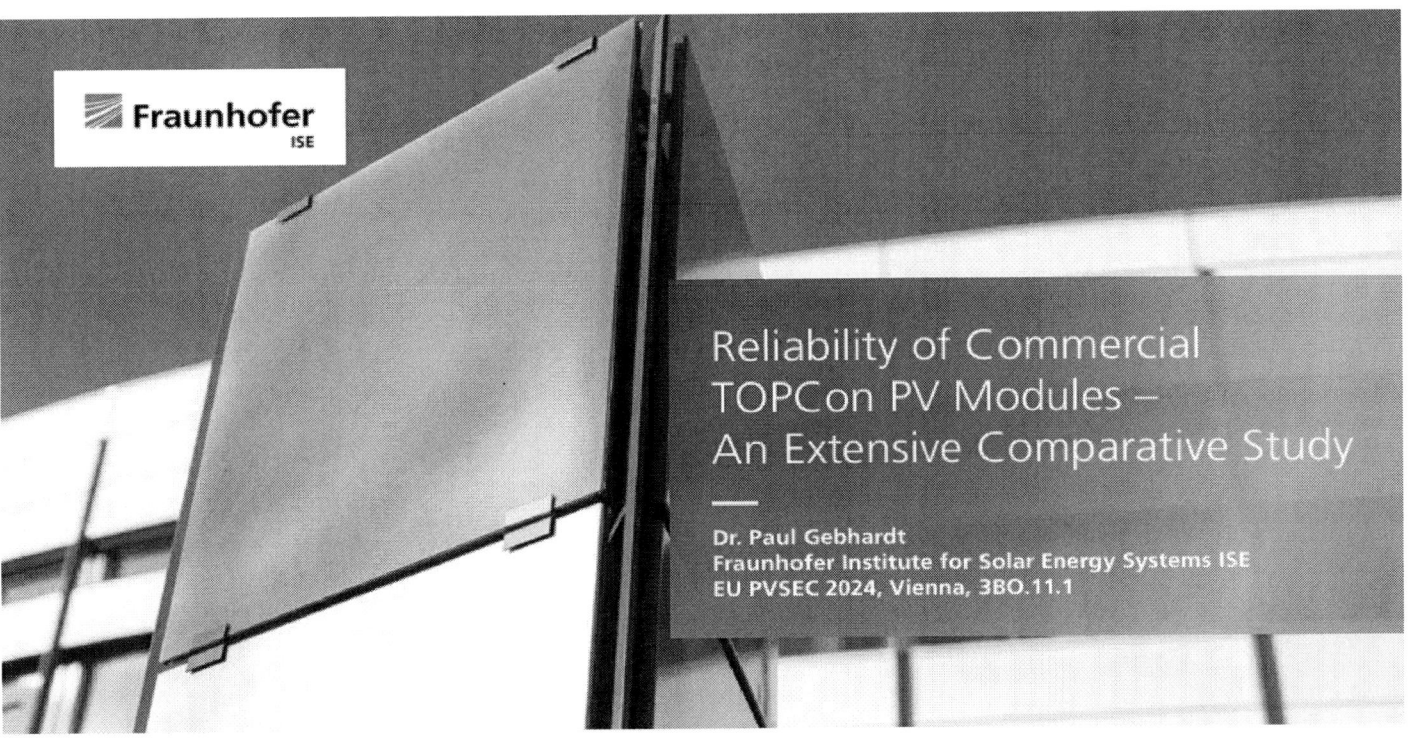

Reliability of Commercial TOPCon PV Modules – An Extensive Comparative Study

Dr. Paul Gebhardt
Fraunhofer Institute for Solar Energy Systems ISE
EU PVSEC 2024, Vienna, 3BO.11.1

Agenda

- Applied Test Program
- Corrosion/Moisture ingress
- UV irradiation
- Learnings

Test Program

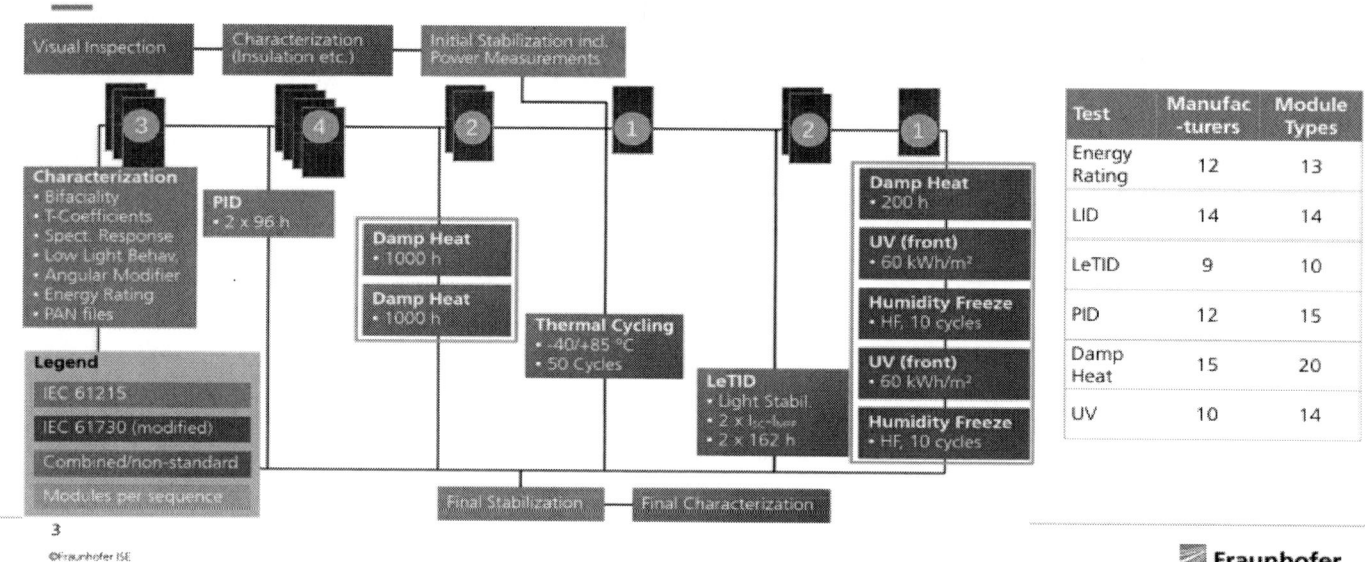

Test	Manufac-turers	Module Types
Energy Rating	12	13
LID	14	14
LeTID	9	10
PID	12	15
Damp Heat	15	20
UV	10	14

Module Selection

...and Initial Power Measurements

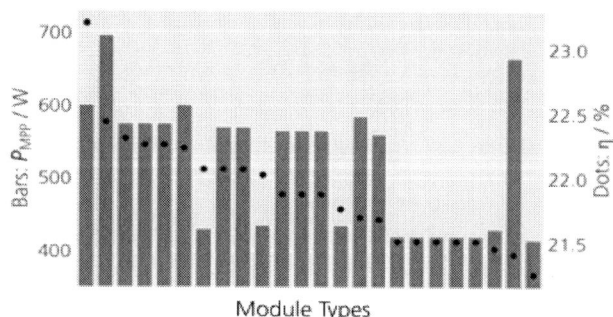

Nominal (label) values of PV modules in this study

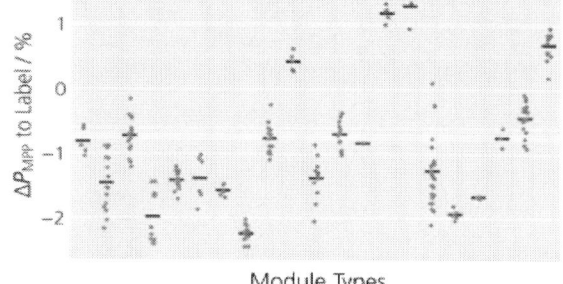

Deviation [1] of the initial power measurements (P_{MPP}) relative to label

[1] D. Philipp, U. Kräling, and M. Kaiser, "Long-Term Trends: Nominal Module Power vs. Measured Power," Bad Staffelstein, 2024.

Damp Heat Aging

- PERC glass-glass modules show usually little degradation (≤ 1 %)
- Known issue of front metallization corrosion
  - Influenced by encapsulant polymer and additives, flux, ...
- Glass-glass modules degradation pattern
  - DH 1000 h causes I_{SC} degradation
  - DH 2000 h causes FF degradation

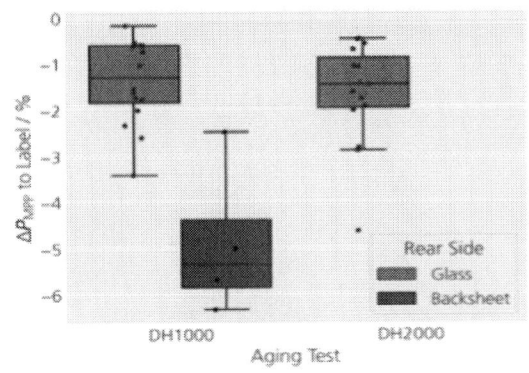

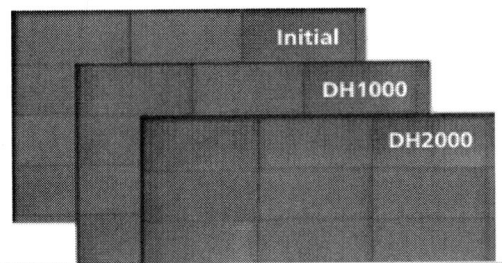

Detailed EL images (top left of PV module) initially (left), after DH1000 (middle) and after DH2000 (right).

UV Aging
Tested in modified sequence B

Test conditions
- Modified Sequence B (IEC 61730-2)
- UV-A on front, 2 x 60 kWh/m²
- Interim performance measurements (in contrast to IEC)

Observations
- Strong degradation (V_{OC}) of some types
- (Partly) regeneration during HF

→ **Relevance for field application unclear**
- Root cause analysis and outdoor tests ongoing

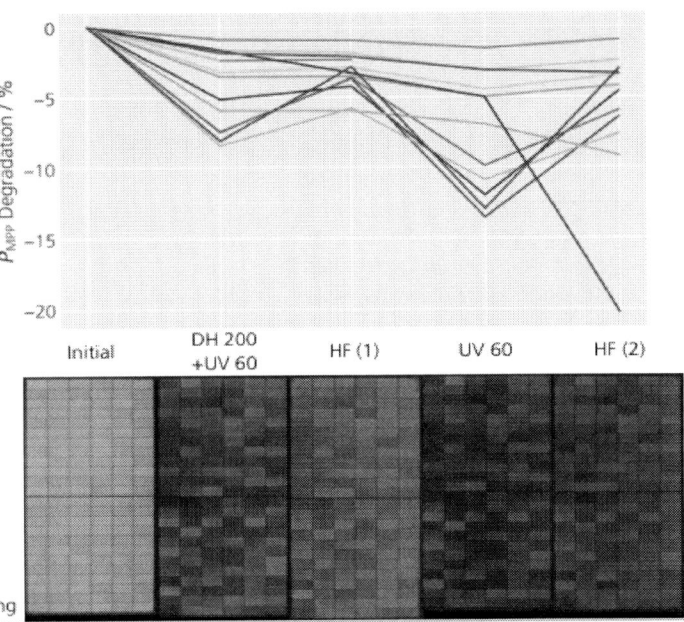

Right: Exemplary EL Images after UV aging

Further Investigation: UV-induced Degradation (UVID)
Ongoing Work

Degradation

- Influence of moisture excluded
- Relevance of operation condition
- Outdoor operation: degradation in V_{OC} and I_{SC}

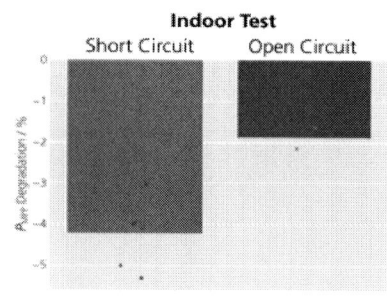

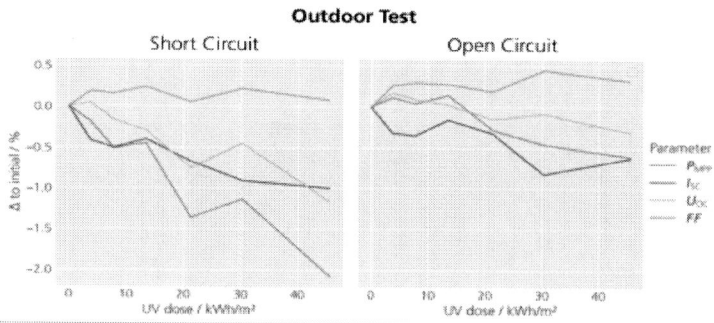

Further Investigation Recovery after UVID
Ongoing Work

Recovery

- Further degradation during dark storage
- Effect of moisture excluded
- Influence of temperature, light, operation mode, dark storage

Current Focus

- Adjustment of stardardized procedures
- Stabilization before and after UV aging

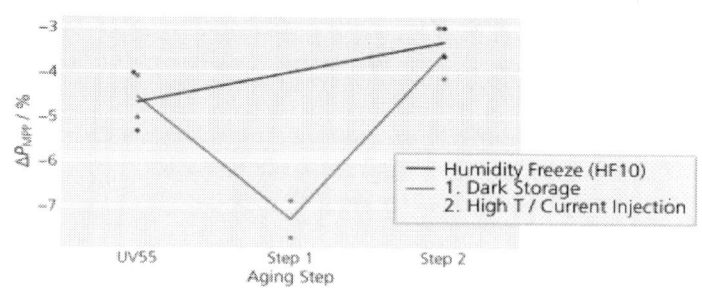

Conclusion and Learnings

Learnings

- Moisture
  - Corrosion of cell metallization

- UV
  - W-pattern (degradation/recovery)
  - Analysis and validation ongoing
  - Stabilization and correlation to outdoor performance

**Recommended testing focus
for current TOPCon Modules:**
PID | DH | UV | ML

3BO.14.2, Jochen Markert, today 3:30 PM

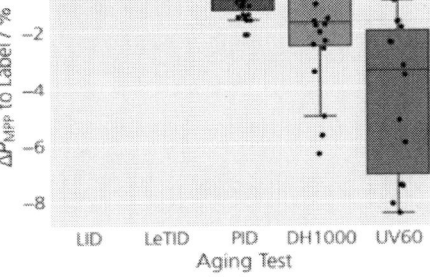

This work is under revision in
„Progress of Photovoltaics"
Preprint:

Federal Ministry for Economic Affairs and Climate Action

Funding:
03EE1149A "MiMoRisk"
03EE1133A "GagaRln"

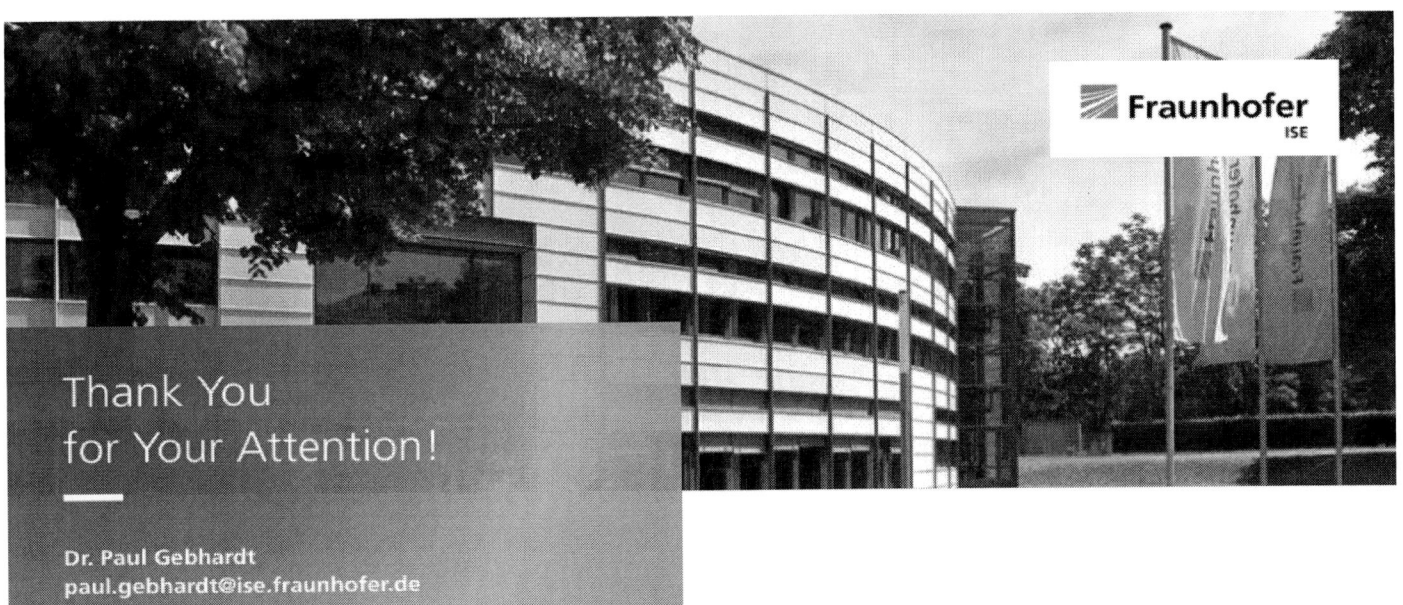

Thank You
for Your Attention!

Dr. Paul Gebhardt
paul.gebhardt@ise.fraunhofer.de

Electrical Characterization
Climate Specific Energy Rating (CSER)

Bifaciality and Thermal Coefficients

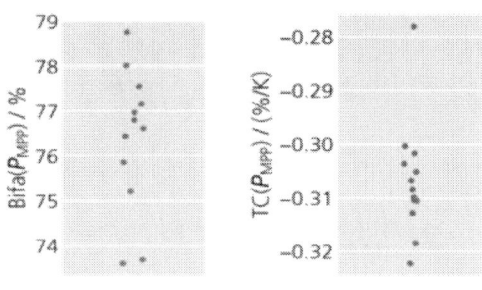

Climate-Specific Energy Rating

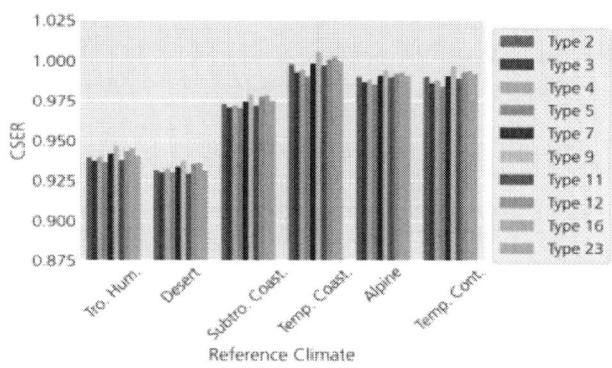

[1] M. Rivera et al. *Energy Tech*, vol. 11, no. 12, 2023
[2] L. Wang et al. Sol. Energy, vol. 238, pp. 258–263, 2022
[3] U. Kräling, et al. WCPEC-8, Milan, Italy, 2022.

LID, LeTID and PID

Negligible changes due to LID and LeTID aging
- LID test: Lightsoaking, 30 kWh/m²
- LeTID 2 x 162 h @ 2 x I_{SC}-I_{MPP}
- Similar to latest PERC module types

PID: Low to moderate losses
- Degradation slightly stronger at negative potential
- Regeneration (PID-p) not tested
- Similar to last PERC module types

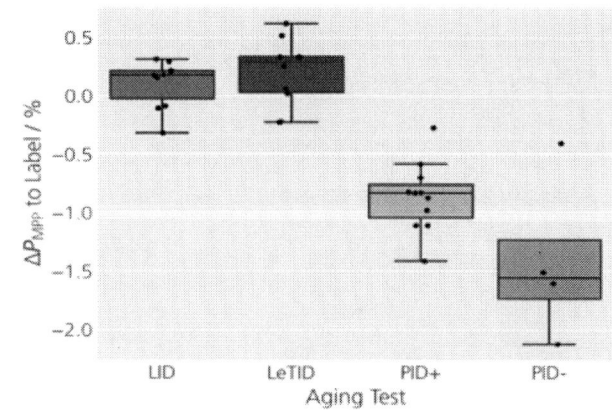

Study and Mitigation of Moisture-Induced Degradation in Silicon Heterojunction Solar Modules

Lucie Pirot-Berson[1,2,3], Romain Couderc[1], Romain Bodeux[2], Frédéric Jay[1], Julien Dupuis[3]

1 Univ. Grenoble Alpes, CEA, Liten, Campus INES, 73375 Le Bourget du Lac. France
2 EDF R&D – IPVF, 18 boulevard Thomas Gobert, 91190 Palaiseau, France
3 EDF R&D, EDF Lab Les Renardières, Avenue des Renardières, 77250 Moret Loing et Orvanne, France

---

Context

Record of crystalline Si cells
→ Silicon Heterojunction Cell
→ **27,1 % Eff**

NREL - Best Research-Cell Efficiency Chart

~20% of the world market share in 2034

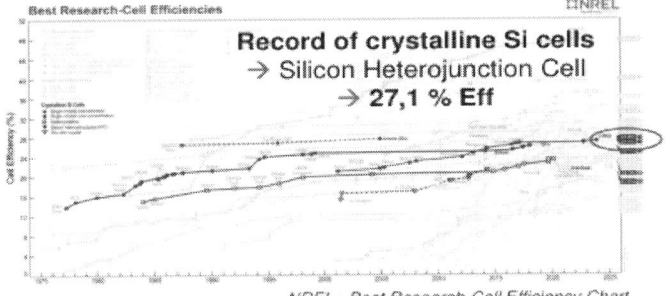

SHJ cell production limited to 37 or 95 GW depending on the scenario

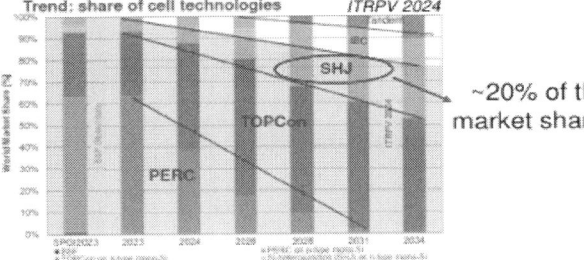

Improve SHJ reliability to maintain high efficiency

Meet the SHJ cells challenges

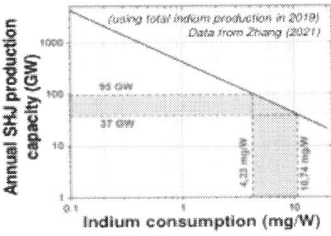

EU PVSEC 2024

DH and sodium-induced degradation of SHJ cells

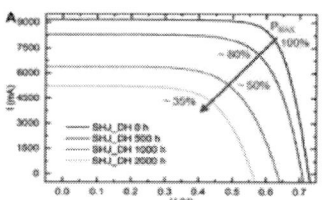

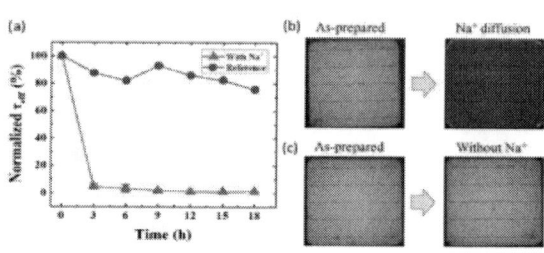

Sensitivity of SHJ modules to damp-heat (DH) environment

Gnocchi et al., Cell Reports Physical Science, 2023

Li et al., SolMat, 2021

Depassivation of SHJ cells by sodium ions from soda-lime glass

Crossing of TCO layers by sodium ions, reaching a-Si layers

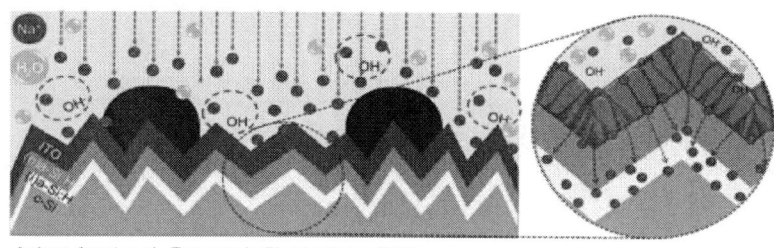

Arriaga Arruti et al., Progress in Photovoltaics, 2023

EDF EU PVSEC 2024

Experimental details

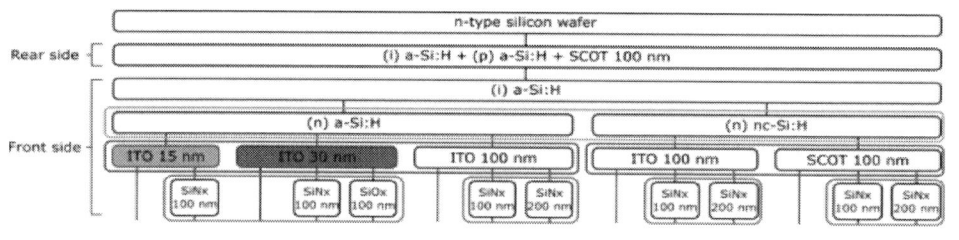

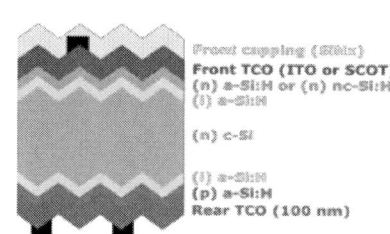

1. **Selective layer** → PECVD → Selective layer : amorphous silicon *a-Si* and nanocrystalline silicon *nc-Si*

2. **TCO thickness and morphology** → PVD → TCO : **ITO** (15, 30 and 100 nm) and **SCOT** (100 nm)

3. **Capping and their thickness** → PECVD → Dielectric capping layers : SiN_x (100 and 200 nm) and SiO_x

EDF EU PVSEC 2024

Experimental details

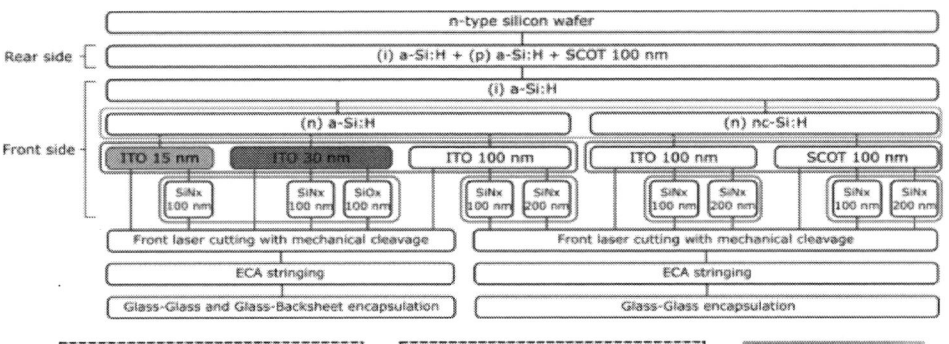

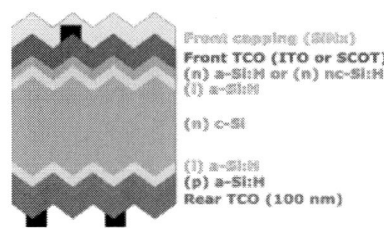

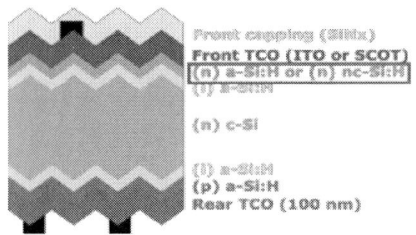

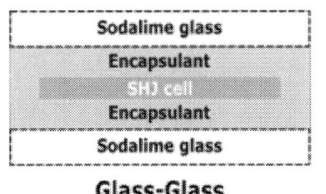

Glass-Glass

Glass-Backsheet

3 modules per configuration (20x20 cm²)

Highly permeable encapsulant

DH testing
85 °C / 85% RH
(IEC 61215)

Effect of the selective layer : *a-Si* and *nc-Si*

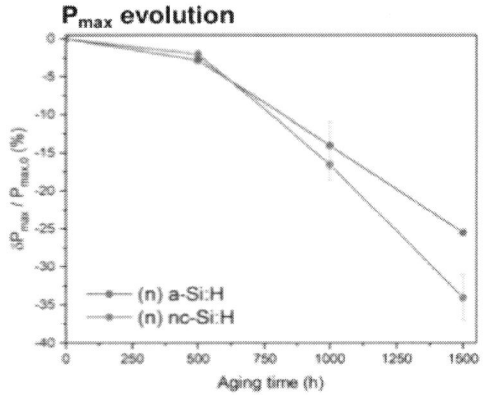

Glass-Glass

P_max evolution

nc-Si modules more degraded than a-Si

Effect of the selective layer : *a-Si* and *nc-Si*

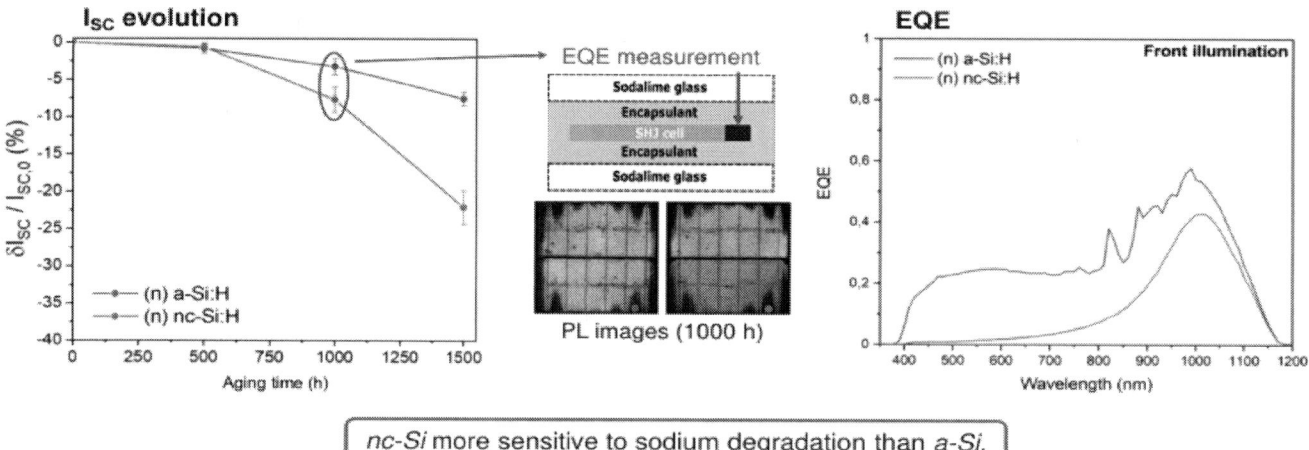

I_{SC} evolution — **EQE measurement** — PL images (1000 h) — **EQE**

nc-Si more sensitive to sodium degradation than *a-Si*,
probably due to layer density

eDF EU PVSEC 2024   *Reference : Jiang et al., SolMat, 2022*

Effect of the TCO grain size

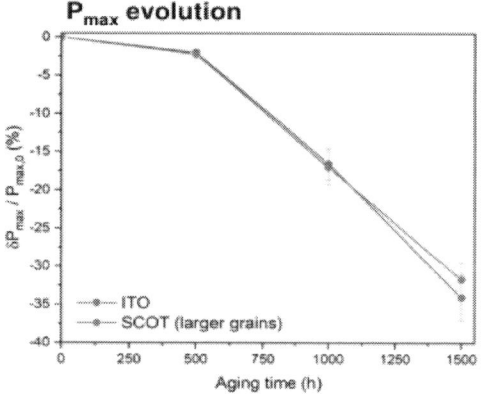 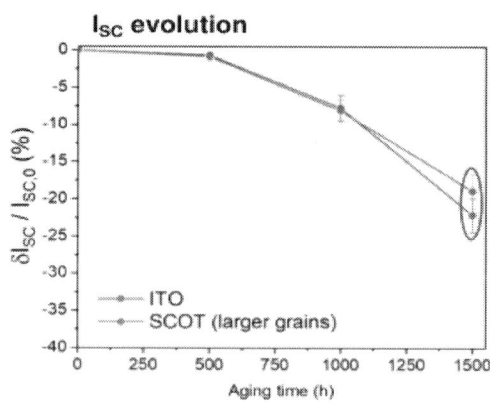

P_{max} evolution — **I_{SC} evolution**

- <u>XRD results *(nc-Si)*</u> : SCOT grains size 16% > ITO grains size → 4% difference in I_{SC} degradation
- No real difference in P_{max} degradation

Limited effect of TCO grain size

eDF EU PVSEC 2024   *Reference : Li et al., SolMat, 2021*

Effect of ITO thickness on degradation

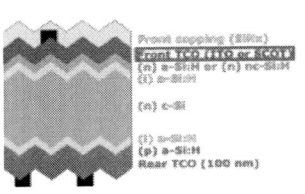

Indium consumption :
ITO 30 nm → -70%
ITO 15 nm → -85 %

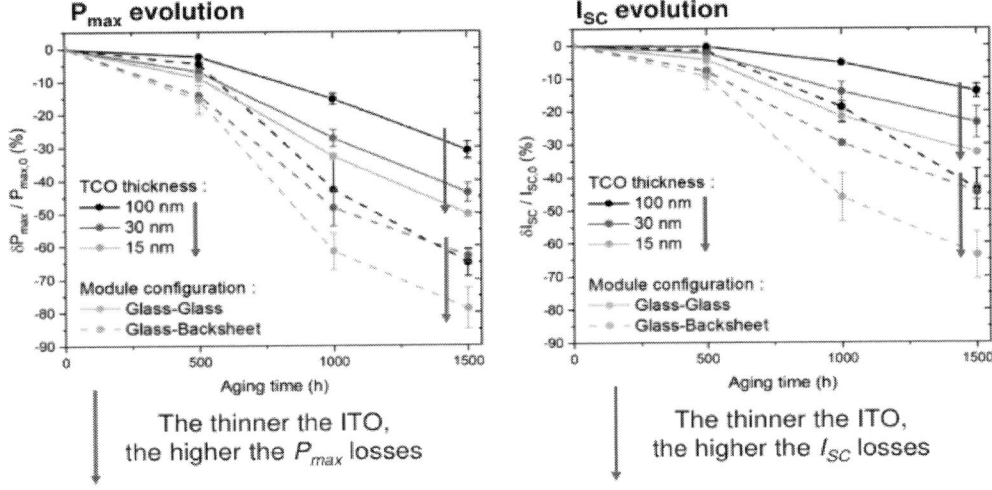

The thinner the ITO,
the higher the P_{max} losses

The thinner the ITO,
the higher the I_{SC} losses

Reducing ITO thickness facilitates sodium migration

  EU PVSEC 2024

Effect of capping layers (SiN$_X$ and SiO$_X$)

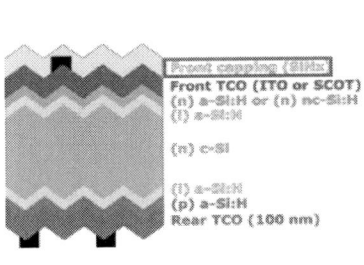

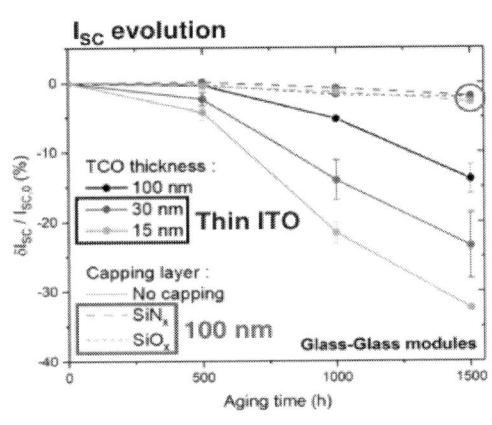

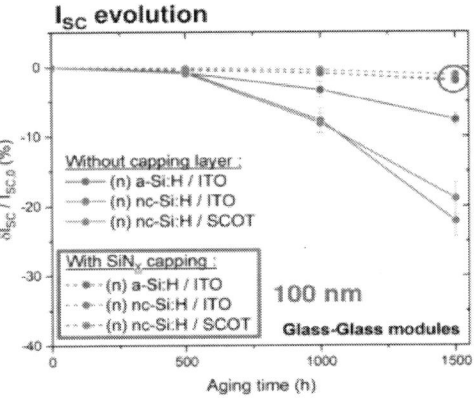

Capping layers provide protection against sodium-induced degradation
(even with thin ITO layers)

 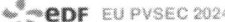 EU PVSEC 2024

Effect of capping layers (SiN$_X$) – FF losses

FF evolution

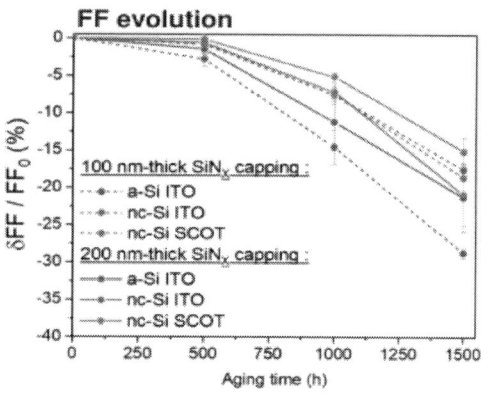

IV evolution

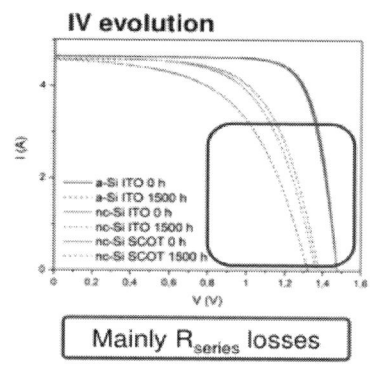

Mainly R$_{series}$ losses

ITO 100nm VS ITO 15nm capped

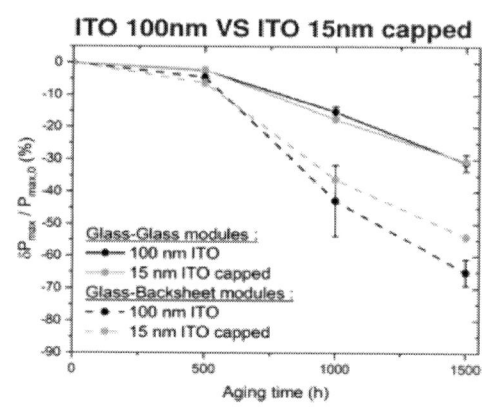

FF losses remain with capping layers

Rear capping is needed ?

Similar degradation for ITO 15 nm and 100 nm

Effect of capping layers – Glass-backsheet modules

PL images

Aging	Before aging	After 1500 h of damp heat aging		
Dielectric capping	All	No capping	SiNx	SiOx
ITO thickness	All		30 nm	
Glass-Backsheet — PL				

* Protection by capping on front side
* Beginning of degradation in laser-cut area
* SiO$_X$ provide more protection than SiN$_X$

Laser-cut area degradation hypothesis

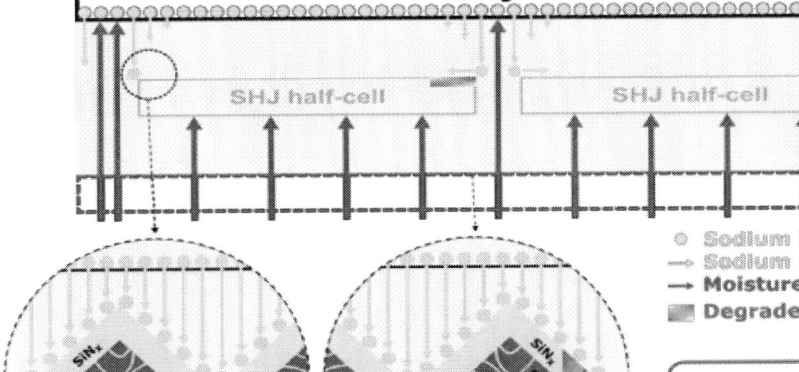

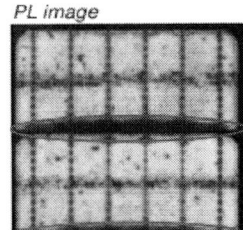

PL image

- ○ Sodium ion
- → Sodium migration
- → Moisture ingress
- ▨ Degraded capping

Front capping preserved

Front capping degraded

- • Protection by preserved capping on front side
- • Degradation of the SiN$_X$ capping due to laser cutting
- • Cell front side degradation by sodium ions
- • Laser cutting optimization required

EU PVSEC 2024

24/09/2024   13

Conclusions

→ nc-Si more sensitive to sodium degradation than a-Si
→ Limited effect of TCO grain size
→ Reducing ITO thickness facilitates sodium migration
→ Capping layers protect against degradation (SiN$_X$ and SiO$_X$)

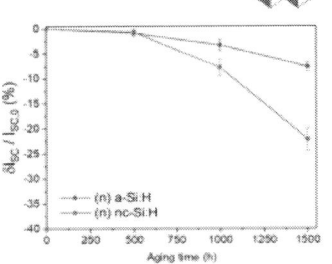

ITO 100nm VS ITO 15nm capped

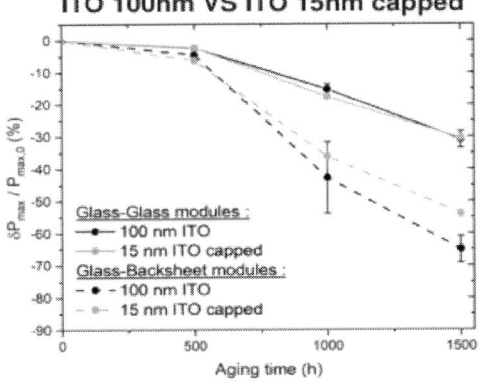

85% reduction in indium consumption, maintaining or even improving durability → TW-scale production capacity

Outlooks

→ Implementation of capping layers on the rear side and edges
→ Investigation of the effect of layer density on degradation
→ Laser cutting optimization

EU PVSEC 2024

24/09/2024   14

Lucie Pirot-Berson

lucie.pirot-berson@cea.fr

CEA INES : 50 avenue du Lac Léman,
73370 Le Bourget du lac

Investigation of Potential-induced Degradation and Recovery in Perovskite Minimodules

Junchuan Zhang[1,2], Haodong Wu[1], Yi Zhang[3], Fangfang Cao[1], Zhiheng Qiu[1,2], Minghui Li[1], Xiting Lang[1], Yongjie Jiang[1,2], Yangyang Gou[1], Xirui Liu[1], Abdullah M. Asiri[4], Paul J. Dyson[3], Mohammad Khaja Nazeeruddin[3], Jichun Ye[1], **Chuanxiao 'Nick' Xiao**[1,5]

1. Ningbo Institute of Materials Technology and Engineering, Chinese Academy of Sciences
2. University of Science and Technology of China (USTC)
3. École Polytechnique Fédérale de Lausanne (EPFL)
4. King Abdulaziz University (KAU)
5. Ningbo New Materials Testing and Evaluation Center Co., Ltd

---

Overview

- ## Background Introduction
  - Potential-induced degradation (PID)
  - Research status

- ## Failure Analysis
  - Experimental design
  - Multiscale and multimodal characterization

- ## Summary & Lesson Learnt

Introduction

- Potential-Induced Degradation (PID)
- Leakage current and ion migration caused by high potential difference between the frame and module

Reduce system output power by 40% in one year [1]

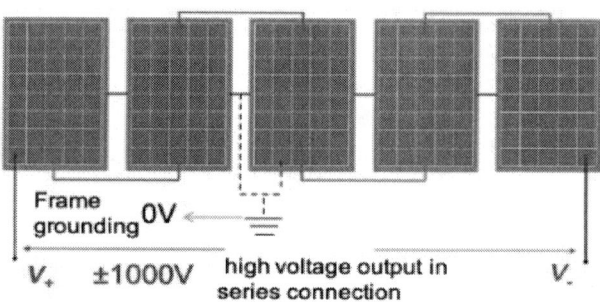

Simplified Structure Diagram of PV System [2]

The most common form of PID damage

The module (-1000 V) has a negative bias voltage relative to the outer frame (0 V)

Alkali metal ions (mainly Na^+) in glass drift into the cell due to electric field

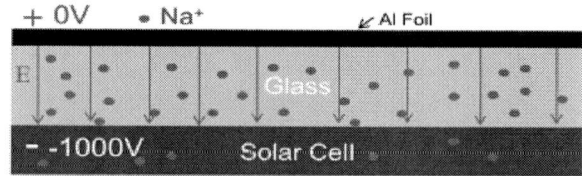

Schematic diagram of Na^+ migration causing cell damage

[1] Kumari V, et al. Materials Today: Proceedings, 2020, 30: 229-233.
[2] Luo W, et al. Energy & Environmental Science, 2017, 10(1): 43-68.

Research Status

✓ **Since 2019, researchers have been studying the PID problem of perovskite solar cells**

✓ **Most studies focused on laboratory-level small-scale experiments with an area ≤ 1.2 cm²**

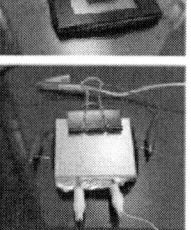
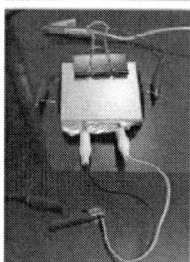

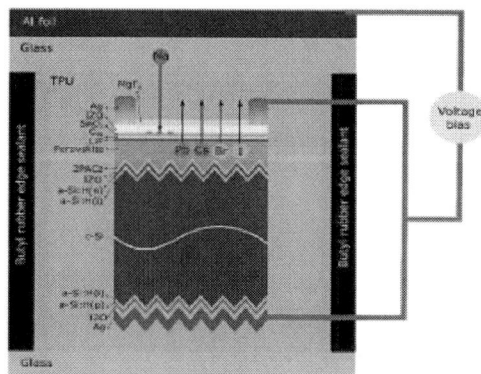

Purohit Z, et al. Solar RRL, 2021, 5(9): 2100349.   Xu L, et al. Cell Reports Physical Science, 2022, 3(9): 101026.

- **We work on perovskite minimodules with in-depth study on PID and recovery**

Potential-induced Degradation of Perovskite

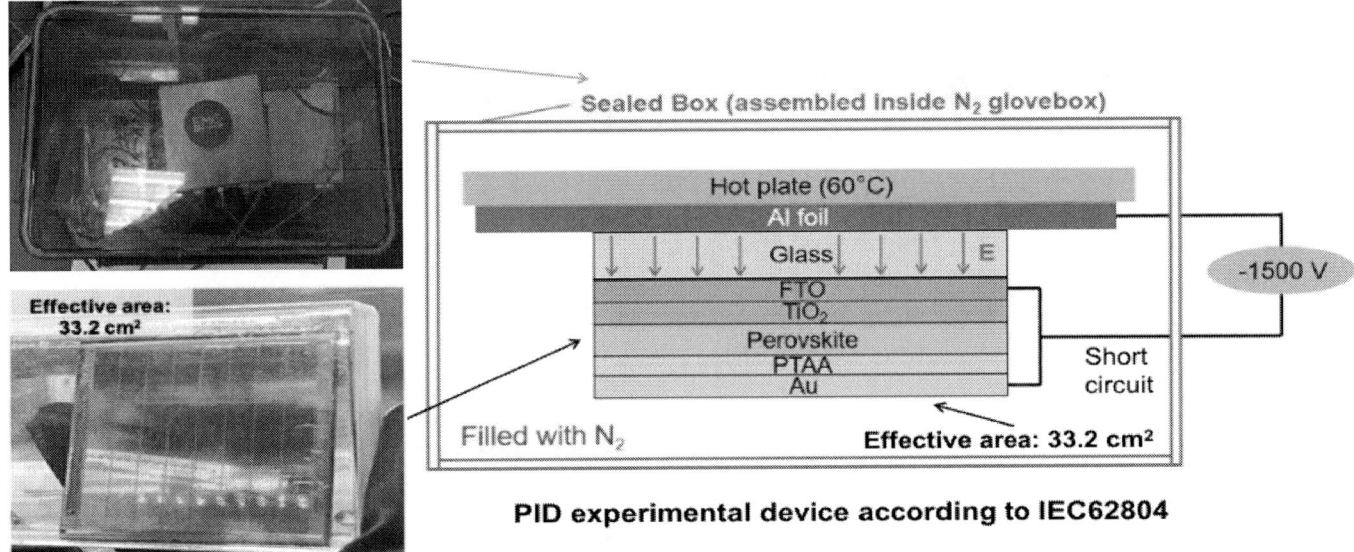

PID experimental device according to IEC62804

19.13% efficiency perovskite mini-module

I-V Degradation

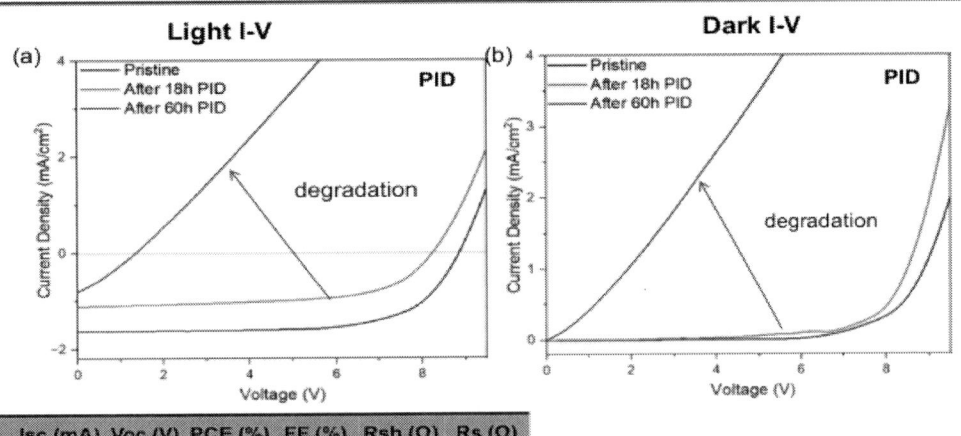

PID	Isc (mA)	Voc (V)	PCE (%)	FF (%)	Rsh (Ω)	Rs (Ω)
Pristine	53.84	8.85	16.63	68.54	2811	23.53
PID-18h	36.89	8.29	9.74	62.36	1906	35.09
PID-60h	26.77	1.34	0.57	28.75	70	41.13
Degradation degree	-50.3%	-84.9%	-96.6%	-58.1%	-97.5%	+74.8%

Severe degradation, PID polarization, shunting

I-V Recovery

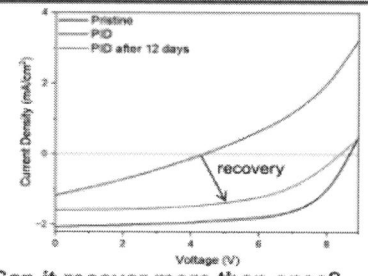

Can it recover more than once?

Recovery	I_{sc} (mA)	V_{oc} (V)	PCE (%)	FF (%)	R_{sh} (Ω)	R_s (Ω)
Pristine	61.79	8.72	14.86	62.55	1034	23.17
After PID	38.72	4.35	2.14	28.43	146	87.17
PID-after 12 days	52.17	8.41	12.13	55.37	758	41.65
recovery degree	84.43%	96.44%	81.63%	88.52%	73.31%	55.63%

80% recovery in ~2 weeks

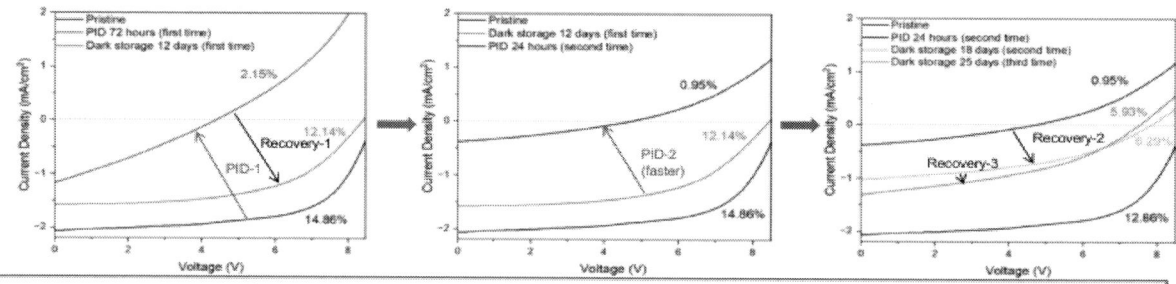

Fast recovery rate under room temperature storage, and can be restored multiple times

---

Photoluminescence Degradation

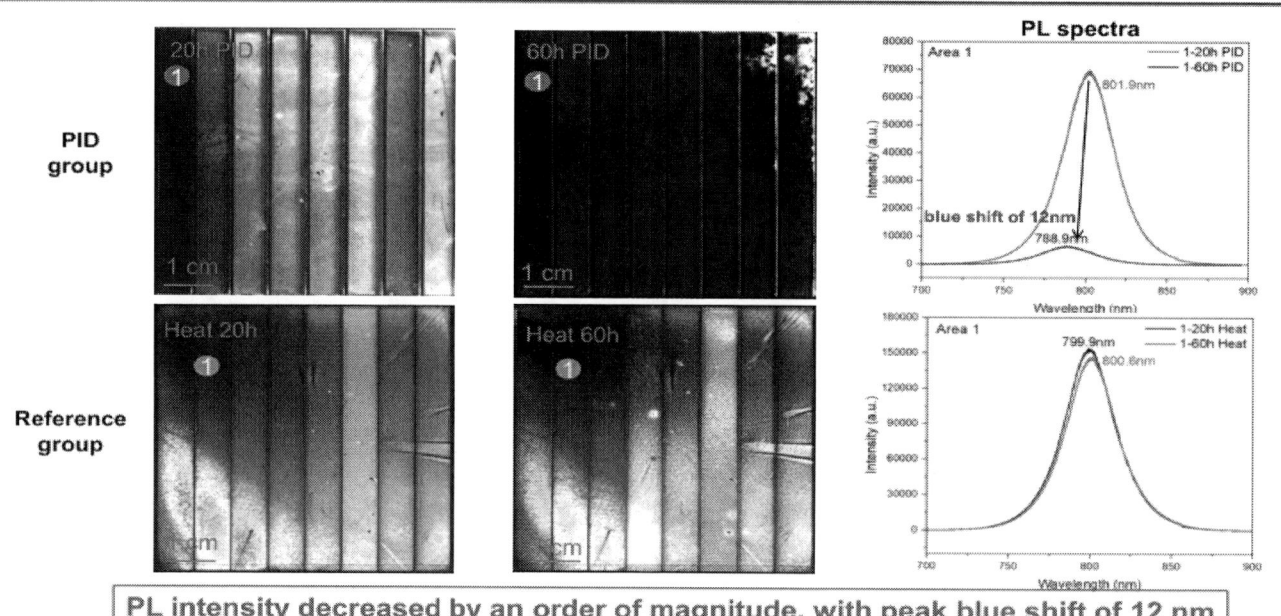

PL intensity decreased by an order of magnitude, with peak blue shift of 12 nm

Photoluminescence Degradation

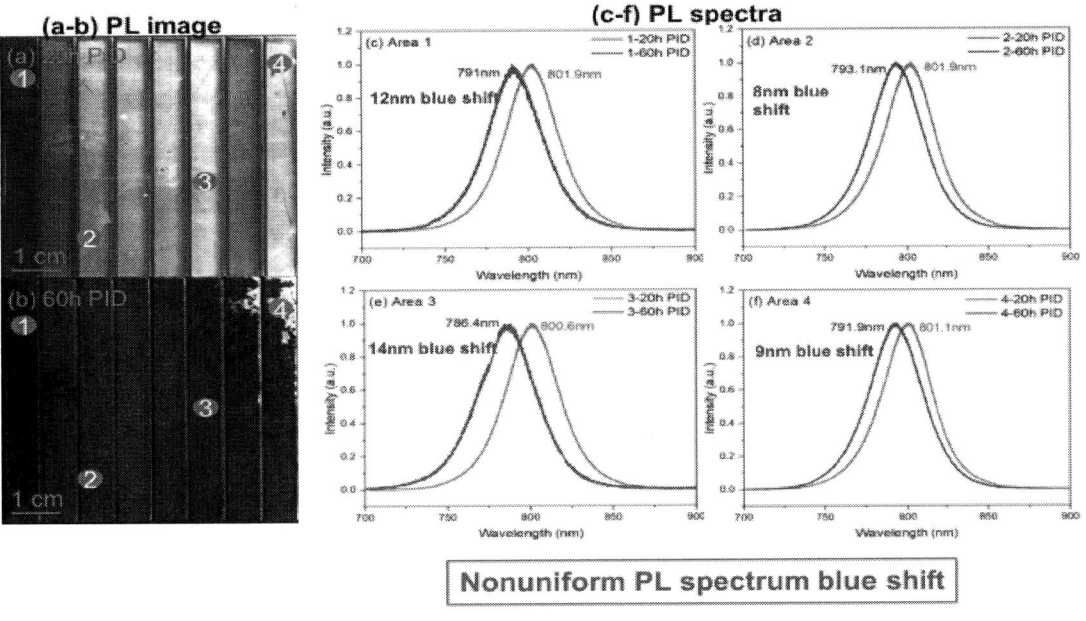

Nonuniform PL spectrum blue shift

Photoluminescence Recovery

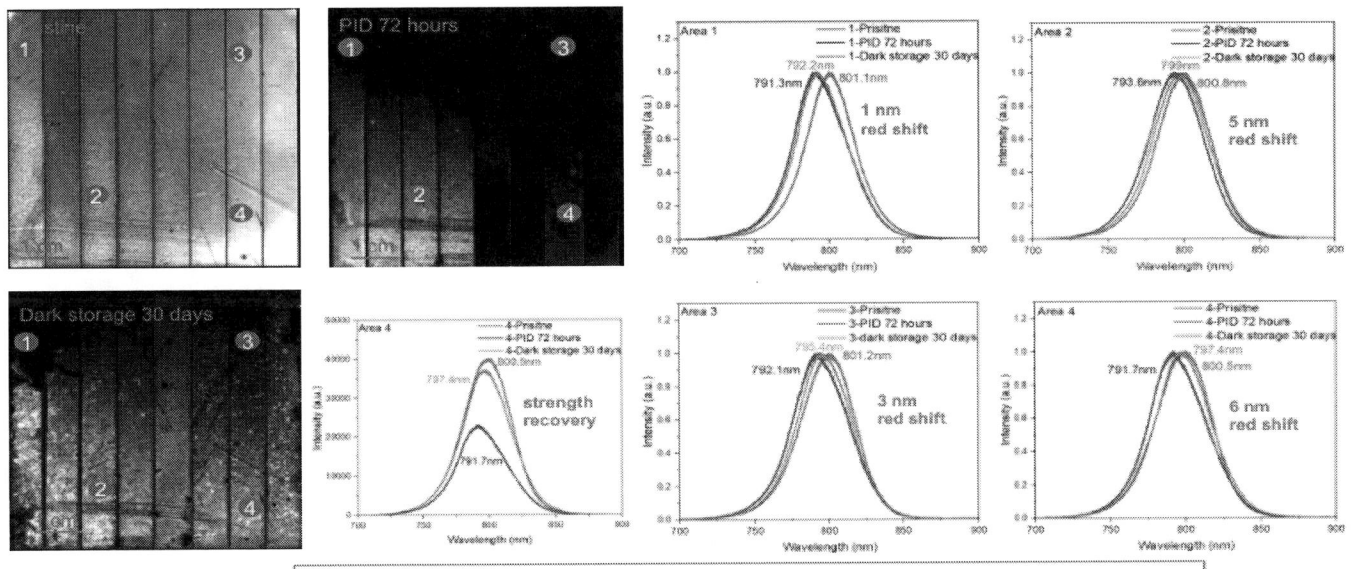

Signal strength recovery, 1-6 nm red shift, nonuniform recovery

Chemical Analysis

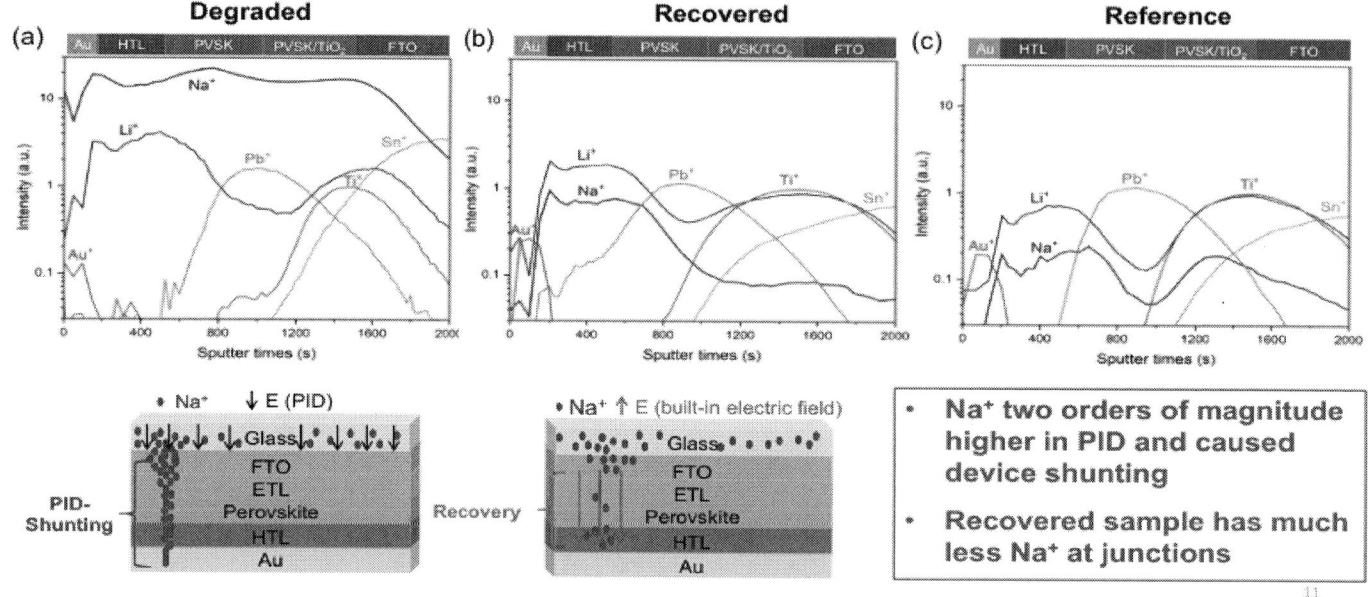

- **Na$^+$ two orders of magnitude higher in PID and caused device shunting**

- **Recovered sample has much less Na$^+$ at junctions**

Diffusion Path

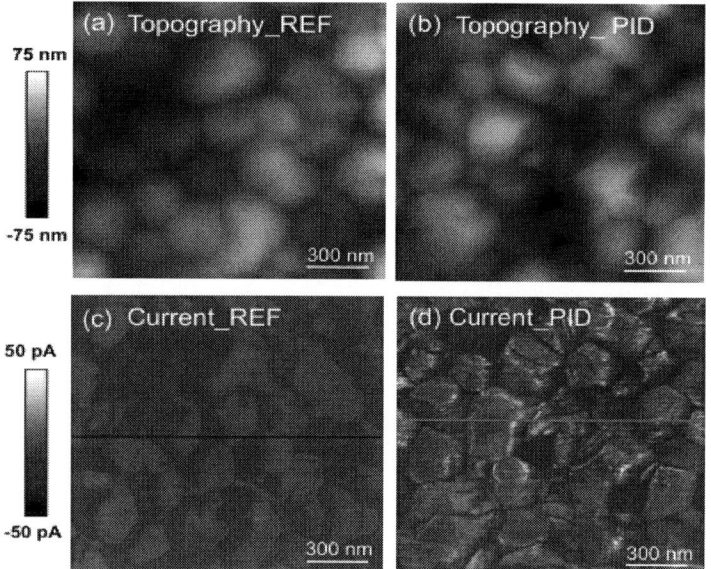

Conductive-atomic force microscopy

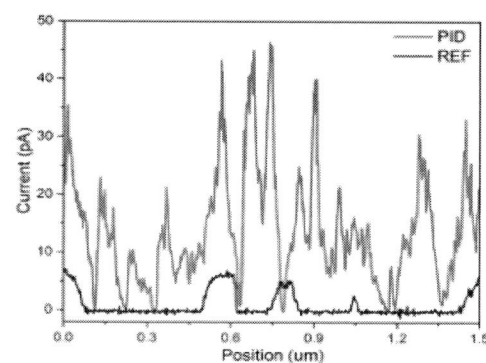

Na$^+$ tends to accumulate at the edge of grains, rather than at grain boundaries

PID Evolution

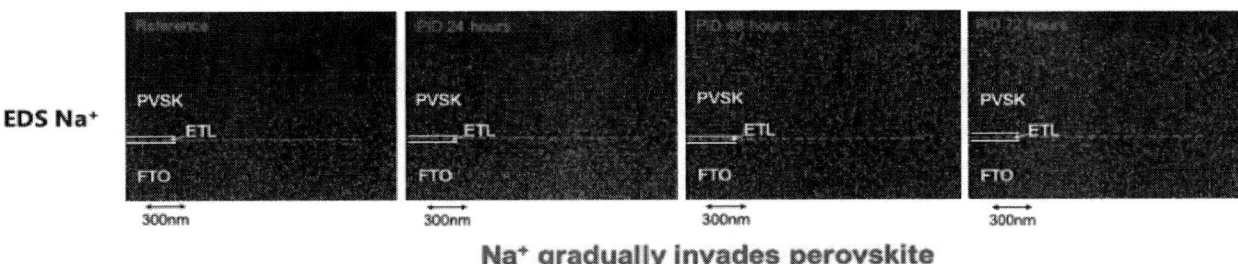

EDS Na+

Na+ gradually invades perovskite

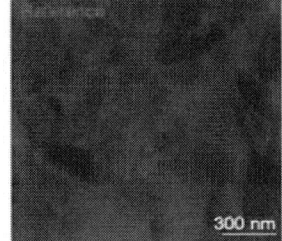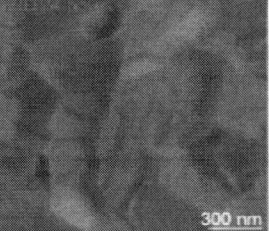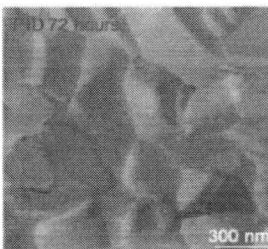

Surface potential

The potential of the perovskite layer continues to increase

Summary & Lesson Learnt

- **Failure analysis of PID, severe degradation on perovskite minimodules, but it also recover quickly**

- **Spatially nonuniform degradation, as well as the recovery**

- **Na+ (from glass) may cause PID-polarization and shunting, and the built-in electric field drift Na+ away for recovery**

- **Na+ accumulate at the edge of grains rather than grain boundaries**

- **We welcome collaboration, cxiao@nimte.ac.cn**

Note: Paper with part of the results is accepted by Progress in Photovoltaics: Research and Applications

Accurate Measurement of Photovoltaic Efficiency

High-profile Equipment

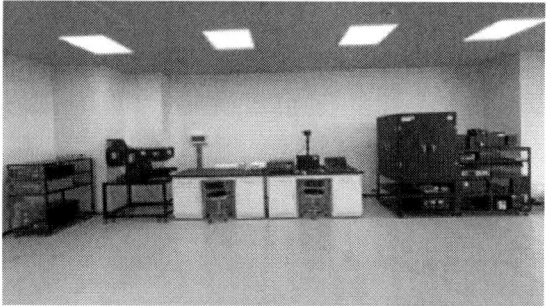

Dual-light solar simulator
Spectral responsivity measurement system

Methods: Asymptotic, Maximum Power Point Tracking

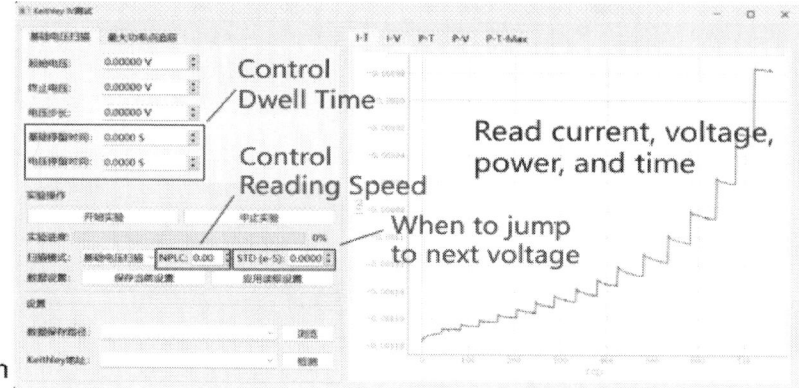

Control Dwell Time

Control Reading Speed

Read current, voltage, power, and time

When to jump to next voltage

(Software patent pending)

We can certify cell efficiency with high standard

Acknowledgements

This work was supported by Ningbo Institute of Materials Technology and Engineering (NIMTE), a research institute of the Chinese Academy of Sciences (CAS). National Natural Science Foundation of China (52302327); Deputyship for Research & Innovation, Ministry of Education in Saudi Arabia, project no. 526 and 425.

Supporting Information

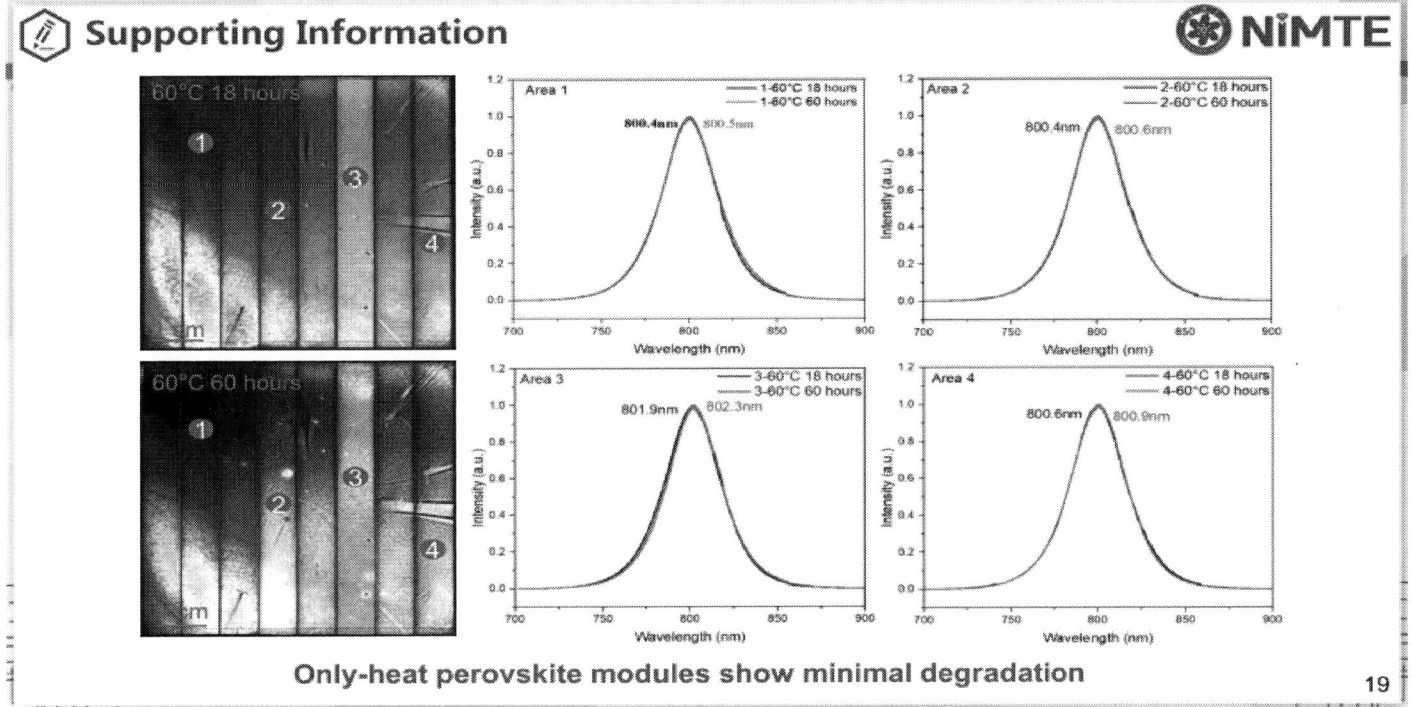

Only-heat perovskite modules show minimal degradation

41st European Photovoltaic Solar Energy Conference and Exhibition

RECENT DEVELOPMENTS IN PV MODULE BACKSHEETS - WHAT DO WE REALLY KNOW ABOUT THEM?

Gernot Oreski[1], Chiara Barretta[1], Karl Anders Weiß[2]
Polymer Competence Center Leoben GmbH. Leoben, Austria
2 Fraunhofer ISE, Freiburg, Germany
Gernot.oreski@pccl.at; chiara.barretta@pccl.at; karl-anders.weiss@ise.fraunhofer.de

ABSTRACT: This paper offers a comprehensive overview of the scientific literature focusing on the reliability of backsheet designs in photovoltaic (PV) systems, with a particular emphasis on recent material advancements. Motivated by a perceived information gap regarding the latest developments in backsheet composition and a lack of understanding concerning long-term stability and compatibility with other solar module components, the authors conducted a literature review spanning 2018 to 2023.

In recent years, there has been a significant shift in the architecture of backsheets used in the photovoltaic (PV) industry. Until 2020, backsheets made from polyethylene terephthalate (PET) laminated with polyvinyl fluoride (PVF) or polyvinylidene fluoride (PVDF) dominated the market with a combined market share of around 85%, along with fluorine-free PET laminates. Meanwhile, alternative architectures like double-sided coated PET backsheets and co-extruded backsheets held a smaller market share of 15%. However, by 2022, there has been a notable transformation, with the market share of double-sided coated PET backsheets reaching 50%. This shift signifies a move away from the traditional dominance of PVF/PVDF-laminated PET backsheets.

The Scopus based literature survey identified a total of 296 hits, with an annual publication range of 40-50 papers. Authors with American affiliation are dominating the field of backsheet reliability research. Only one Asian stakeholder was found in the top ten list of affiliations. The analysis of the backsheet research revealed a significant misalignment between research interest and market reality. Notably, the focus on materials like PVF, PVDF, and PA contrasts with limited exploration of double-sided coated backsheets. Innovative designs' reliability data primarily originates from non-established market players outside China, suggesting potential gaps in addressing industry needs.

Keywords: backsheets; reliability; BOM transparency

1 Introduction

Backsheet failures in solar modules can take various forms, including cracking, blisters, delamination, and chalking. Mechanical damage, which compromises the electrical insulation capacity of the module, poses a safety issue. It is important to note that damage does not always result in an immediate power loss. Additional backsheet issues may include yellowing, UV degradation, and moisture ingress, all of which can impact the long-term performance and durability of solar modules. The occurrence of backsheet cracks has prompted the development of test protocols designed to replicate these cracks through indoor accelerated aging tests [1–3]. This involves subjecting the backsheet to simultaneous combined or sequential stresses such as UV exposure, humidity, temperature fluctuations, and thermo-mechanical loads. Moreover, specific test sequences have been incorporated into standards, notably in IEC 63209 (Extended Stress Testing of PV Modules), focusing on backsheet cracking. A significant challenge lies in the extended time frame it takes, often several years from installation, to observe the first occurrence of backsheet failure modes and fully understand them. Complicating matters further, any alteration or variation in the Bill of Materials (BOM) can potentially impact the outcome, making it essential to account for all changes to ensure the reliability and durability of backsheet materials in photovoltaic systems.

Figure 1 shows the change of the backsheet market from 2020 to 2022 [4]. A significant change in backsheet architecture has been observed in recent years. Until 2020 the market was dominated by backsheets based on PET laminated with either PVF or PVDF. Together with fluorine free PET laminates these backsheets achieved a market share of around 85%. Double side coated PET backsheets and other architectures such as co-extruded backsheets had a combined market share of 15%. Two years later, the share of double side coated backsheets reached 50%.

Between 2020 and 2022, the global photovoltaic (PV) industry witnessed a substantial installation of approximately 560 gigawatts (GW) of solar power, marking a significant milestone as it constitutes nearly half of the cumulative PV installations to date. Asia, and notably China, played a pivotal role in this surge, producing over 90% of all PV modules, with approximately 80% of global production originating in China [5]. A notable trend within this period is the emergence of new double-sided coating backsheets, with an estimated potential deployment exceeding 100 GW of PV modules. It is worth remembering the AAA backsheet disaster, where around 12-15 GW of PV modules were affected [6]. This incident underscores the importance of thorough research and development in ensuring the reliability and durability of materials in the solar industry, especially given the substantial scale of installations and the critical role of Asia, particularly China, in PV module production. As the industry continues to expand, careful attention to the quality and reliability of materials, including backsheets, remains crucial for the sustainable growth of solar energy globally.

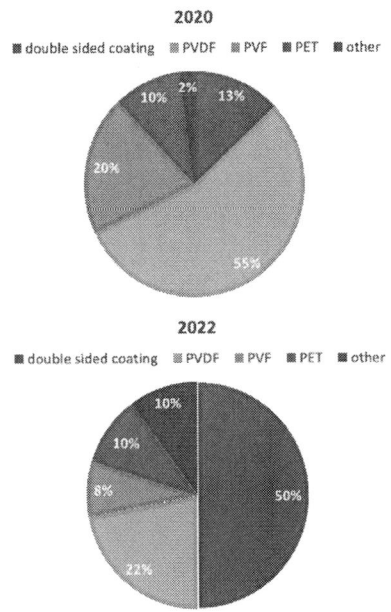

Figure 1: Comparison of PV backsheet market: 2020 (top) and 2022 (bottom) [4]

2 METHODOLOGY

The paper aims to provide a comprehensive summary of the scientific literature on the reliability of backsheet designs, with a particular focus on recent material developments. The motivation behind this research is the authors' perception of a lack of information on the most recent advancements in backsheet composition and a dearth of knowledge regarding the long-term stability and compatibility of these materials with other components in solar modules. To achieve this goal, the authors conducted a literature search covering the period from 2018 to 2023. The search was carried out using the keywords "PV" OR "photovoltaic" AND "backsheet" to ensure a broad and inclusive overview of the available literature. The content extracted from the papers was categorized into the following groups to facilitate a structured analysis:

   (1)   Review articles
   (2)   Field performance and failure mode analysis
   (3)   Modelling of backsheet properties
   (4)   Backsheet development and testing
   (5)   Module performance & reliability (no special focus given on backsheets)
   (6)   Other PV topics

Furthermore, the paper includes an analysis of the material composition of the backsheets discussed in the literature. This analysis aims to shed light on the types of materials that have been used in recent developments. Additionally, the paper identifies key players and their affiliations involved in research and development within the field of PV backsheets. Overall, the paper seeks to contribute valuable insights into the current state of knowledge on backsheet designs, emphasizing recent material developments and addressing the perceived gaps in understanding the long-term stability and compatibility of these materials in the context of solar modules.

3 RESULTS

Figure 2 summarizes the content of the papers found using the aforementioned search terms. In the literature survey conducted on Scopus covering the years 2018 to 2023 with the keywords PV or photovoltaic and backsheet, a total of 296 hits were identified, indicating an annual publication range of 40-50 papers. Notably, the top ten authors contributed significantly, accounting for a total of 175 papers, reflecting a concentrated influence within the field. Additionally, the top ten affiliations played a prominent role, collectively contributing to 202 papers. The landscape of backsheet reliability research is currently dominated by American stakeholders. Only one Asian stakeholder was found in the top ten list of affiliations doing research on PV backsheets.

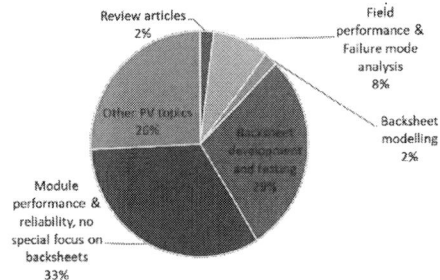

Figure 2: Classification of content of the literature search

The literature review also revealed a notable misalignment exists between research priorities and market realities, highlighting potential gaps in addressing industry needs. The predominant focus has been on backsheets containing materials like PVF, PVDF, and PA, with co-extruded polypropylene (PP) backsheets receiving considerable attention in new design studies (see Figure 3). Surprisingly, there is a limited exploration of double side coated backsheets. Published data on the reliability of innovative backsheet designs predominantly comes from non-established market players outside China (e.g. Borealis & Endurans for PP, Fuji Film for double side coated PET).

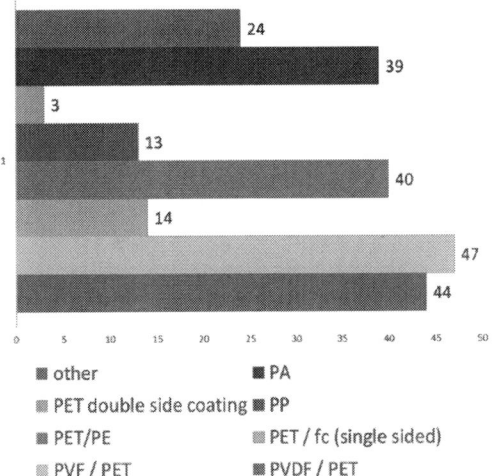

Figure 3: Reported backsheet compositions

4 SUMMARY AND CONCLUSION

Backsheets are essential for the performance and safety of photovoltaic (PV) modules. The backsheet market is highly dynamic, with a significant shift towards the use of double-side coated PET films (CPC). However, there is a concern that research on backsheet reliability does not accurately reflect market conditions, which could pose a potential risk. Much of the focus in backsheet studies has been on materials such as PVF, PVDF, PA, and PP, while there has been almost no research on double-side coated backsheets. Moreover, limited data is available on CPC backsheets from manufacturers, with important details such as the chemistry and key properties remaining undisclosed. Most of the published data on the reliability of new backsheet designs comes from non-established market players, indicating a gap in data from well-known industry leaders.

5 REFERENCES

[1] Owen-Bellini M, Hacke P, Miller DC, Kempe MD, Spataru S, Tanahashi T, Mitterhofer S, Jankovec M, Topič M. Advancing reliability assessments of photovoltaic modules and materials using combined-accelerated stress testing. Prog Photovolt Res Appl 2021; 29(1): 64–82, DOI: 10.1002/pip.3342.

[2] W. Gambogi, T. Felder, S. MacMaster, K. Roy-Choudhury, B. Yu, K. Stika, H. Hu, N. Phillips, T. J. Trout. Sequential Stress Testing to Predict Photovoltaic Module Durability. In: 2018 IEEE 7th World Conference on Photovoltaic Energy Conversion (WCPEC) (A Joint Conference of 45th IEEE PVSC, 28th PVSEC 34th EU PVSEC); 2018, pp. 1593–1596.

[3] M. D. Kempe, T. Lockman, J. Morse. Development of Testing Methods to Predict Cracking in Photovoltaic Backsheets. In: 2019 IEEE 46th Photovoltaic Specialists Conference (PVSC); 2019, pp. 2411–2416.

[4] Chunduri S, Schmela M. Taiyang News Market Report Backsheets & Encapsulants 2022/2023; 2023.

[5] Izumi K, Arnulf J-W, Jose D, Gaetan M, Elina B, Adrien vR, Melodie dl. 2023 Snapshot of Global PV Markets; 2023.

[6] Eder GC, Voronko Y, Oreski G, Mühleisen W, Knausz M, Omazic A, Rainer A, Hirschl C, Sonnleitner H. Error analysis of aged modules with cracked polyamide backsheets. Solar Energy Materials and Solar Cells 2019; 203: 110194, DOI: 10.1016/j.solmat.2019.110194.

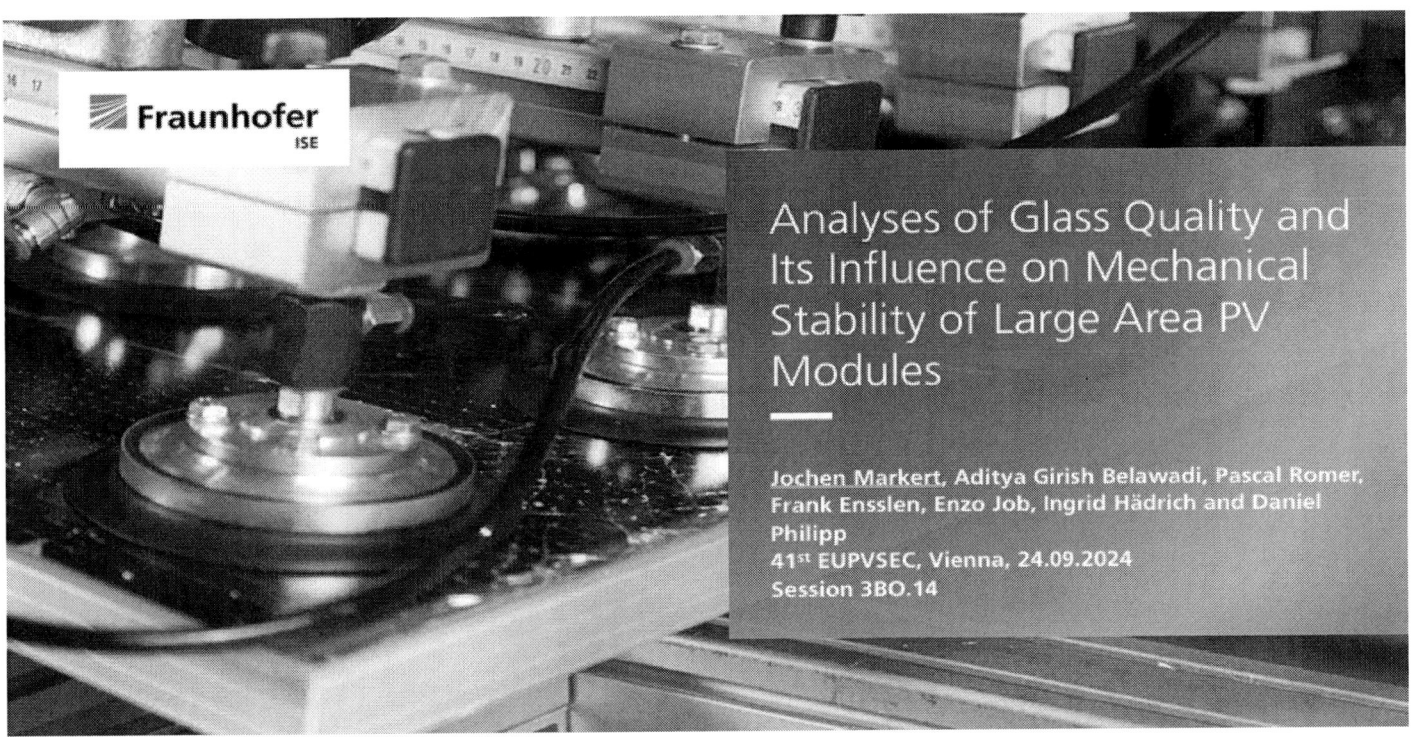

Analyses of Glass Quality and Its Influence on Mechanical Stability of Large Area PV Modules

Jochen Markert, Aditya Girish Belawadi, Pascal Romer, Frank Ensslen, Enzo Job, Ingrid Hädrich and Daniel Philipp
41st EUPVSEC, Vienna, 24.09.2024
Session 3BO.14

Scope of Work

—

Part 1	Part 2
Influence of *Glass Pre-Stress* on Failure Load	Towards *Inhomogeneous Loads* Glass/Glass vs. Glass/Backsheet

Motivation

Increased number of inquiries concerning glass breakages in PV power plants from all over the world

- **Affected are:**

  - Framed/unframed modules in different mounting configurations
  - Module area ~2.5 m²
  - Glass/glass designs with glass thickness ≤ 2 mm

Motivation

Increased number of inquiries concerning glass breakages in PV power plants from all over the world

- **Affected are:**

  - Framed/unframed modules in different mounting configurations
  - Module area ~2.5 m²
  - Glass/glass designs with glass thickness ≤ 2 mm

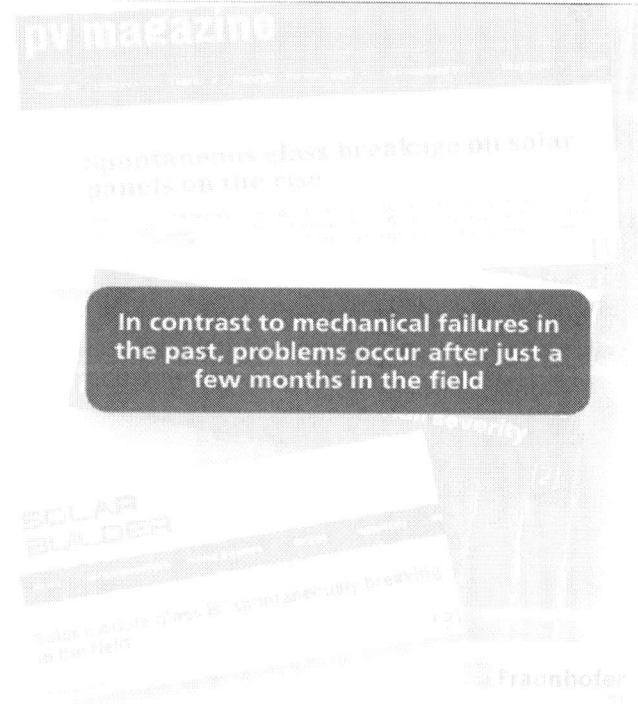

In contrast to mechanical failures in the past, problems occur after just a few months in the field

Motivation

Glass quality

* Edge quality
* Defect density
* Bending strength
* **Pre-stressing of glass**

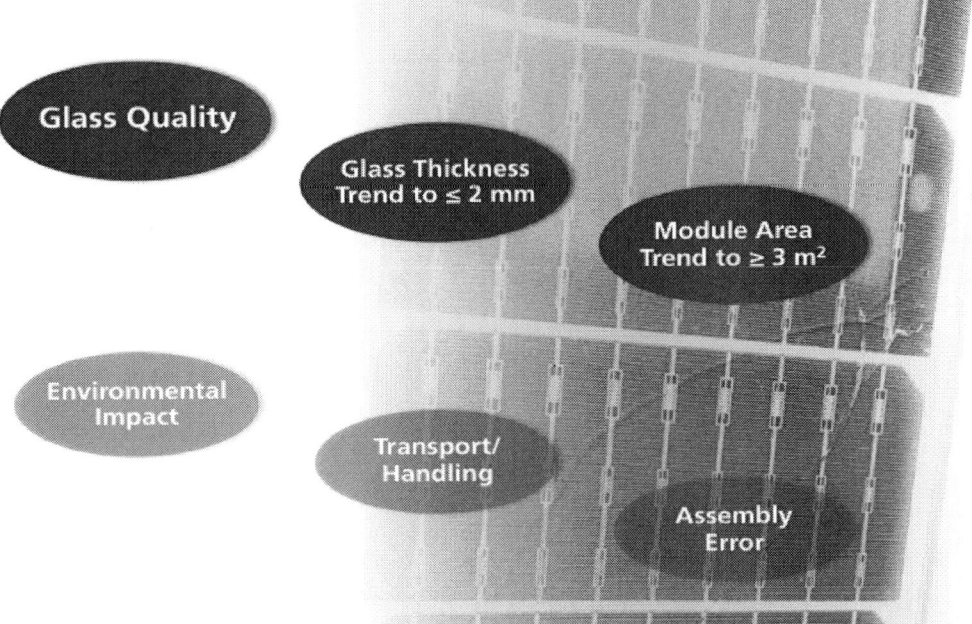

Motivation
Experimental Approach – SCALP

Scattered light polariscope:

* Non-destructive measurement of the surface pre-stress

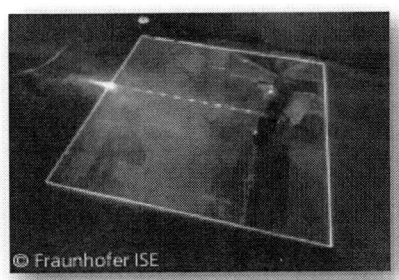

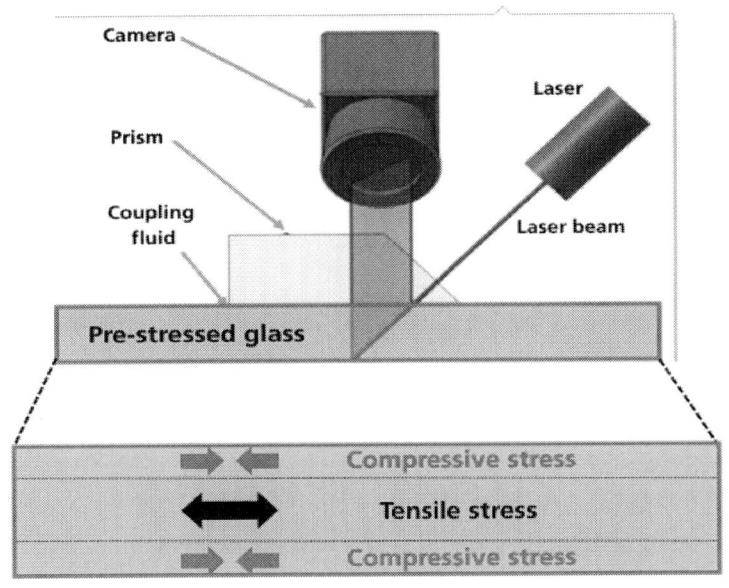

Motivation
Conditions Closer to Reality

Current approach:

- PV modules are tested with homogeneous load profiles with min ±2400 Pa acc. to IEC 61215

Aim:

- Develop test conditions closer to reality
  - Asymmetric and inhomogeneous load profiles

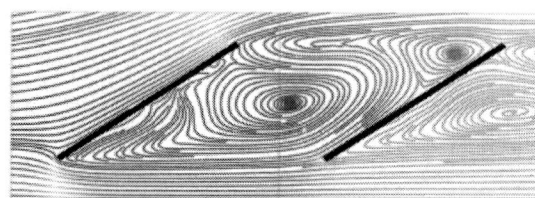
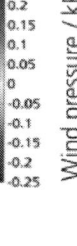

[4] Wind simulation in side view on two tilted PV module rows

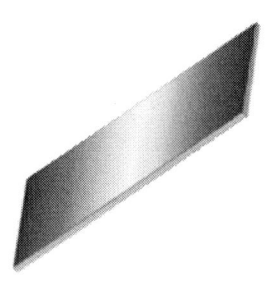

Motivation
Conditions Closer to Reality

New mechanical load test stand:

- Application of defined inhomogeneous load profiles

Motivation
Experimental Approach – Setup Conditions

Mechanical load tests in different mounting configurations:

Standard Setup
- Clamping at 20 % of long module side

Tracker Setup
- Central clamping with 400 mm clamp distance

Scope of Work

Part 1

Influence of *Glass Pre-Stress* on Failure Load

Part 2

Part 1 – Scope of Work
Continuation of Previews Investigation…

Publication on the PV Symposium 2024

- Indications for correlation between low surface pre-stress and reduced failure load were found

- **Design with 2 mm glass and lowest surface stress failed first**

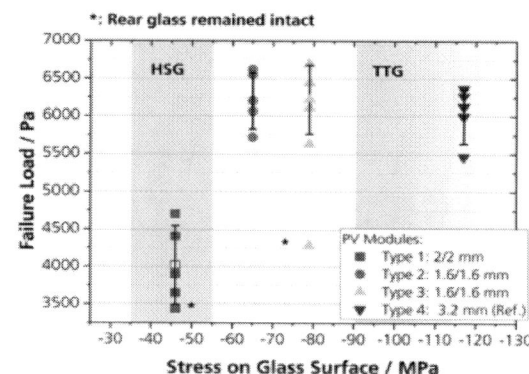

Mechanical Stability of PV Modules
Analyses of the Influence of the Glass Quality

Jochen Markert, Frank Ensslen, Tobias Rist, Andreas J. Beinert, Enzo Job, Ingrid Hädrich, and Daniel Philipp

---

Part 1 – Scope of Work
PV Modules with Different Glass Pre-Stress

Three module types provided by DMEGC Solar with defined glass pre-stress:

- 5 modules per type
- Glass:
  - Pre-stress (label): 50 MPa, 70 MPa, 80 MPa
  - Thickness 2 mm

  - Similar design: Dimesnsions and frame

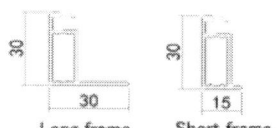

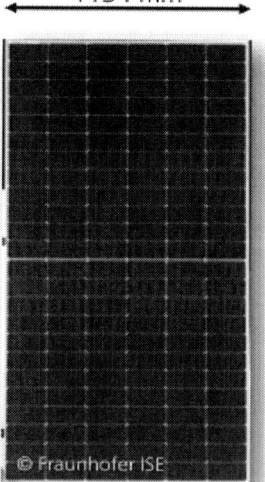

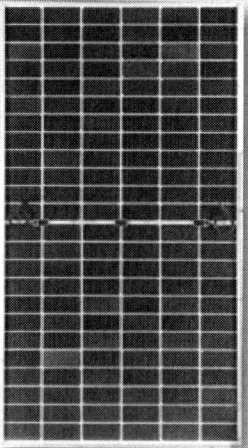

Part 1 – Results
SCALP Measurements

SCALP measurements verfied range of labeled pre-stress:

- Labeled pre-stess: 50 MPa, 70 MPa, 80 MPa

- Scattering of results due to inhomogeneity of surface pre-stress

- Shown are results from rear glass because of absence of coating

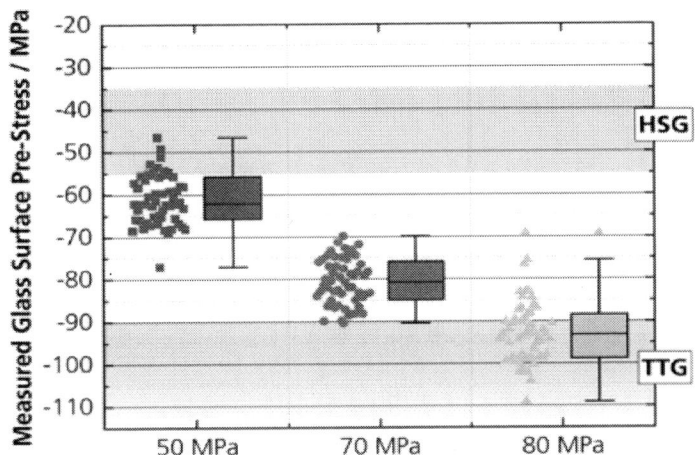

Part 1 – Results
Homogeneous Load – Pressure (Until Failure)

Mechanical load tests in different mounting configuartions:

- Relatively high failure loads in standard setup

- Significantly reduced failure loads in tracker setup

- Previously assumed significant influence of pre-stress on failure load could not be verified

  - Less pre-stressed glass -> slight decrease in failure load

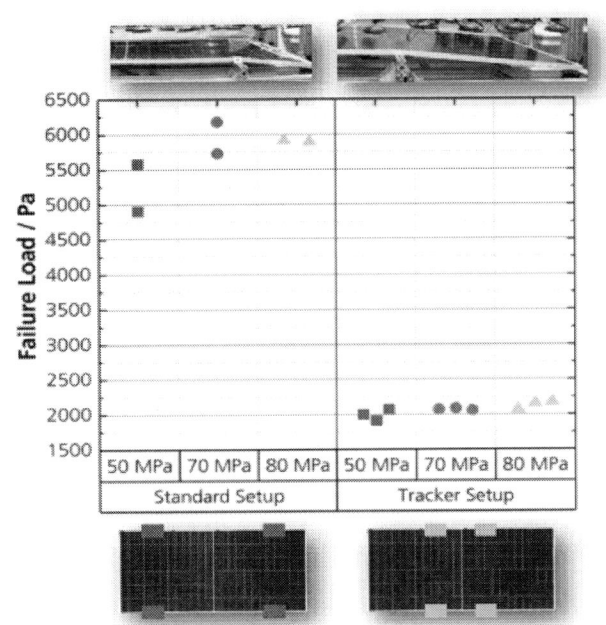

Scope of Work

Part 1	Part 2
Influence of *Glass Pre-Stress* on Failure Load	Towards *Inhomogeneous Loads* Glass/Glass vs. Glass/Backsheet

---

Part 2 – Scope of Work
Glass/Glass vs. Glass Backsheet

Two commercially available module types provided by RWE Renewables from the same manufacturer:

- Glass Thickness
  - 2 mm (Glass/Glass)
  - 3.2 mm (Glass/Backsheet)

  - Similar design: Dimensions and frame

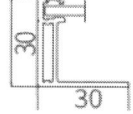
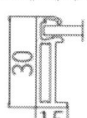

1131 mm

2383 mm

Glass/Glass

Glass/Backsheet

Part 2 – Scope of Work
Glass/Glass vs. Glass/Backsheet

Glass/glass vs. Glass/backsheet:

* **Homogeneous and inhomogeneous loads**

Part 2 – Results
Homogeneous Load – Failure Loads

Homogeneously applied load in two configurations

* Module designs in standard configuration showed comparable stability

* In tracker configuration, glass/backsheet design showed a ~2 times higher failure load

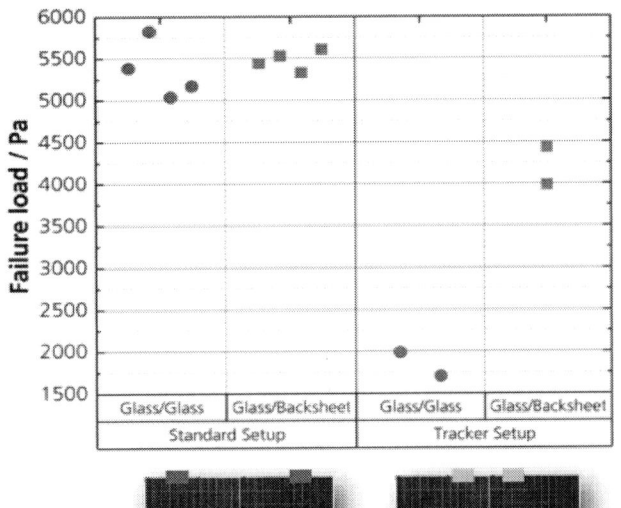

Part 2 – Results
Homogeneous Load – Module Deflection

Deflection at 2000 Pa: Significant differences in deformation behavior

- Less stiffness for glass/glass design
  - Assumption: 2 mm glass behave as two decoupled thin glass panes

- Glass/backsheet design showed additional bending component along short side

 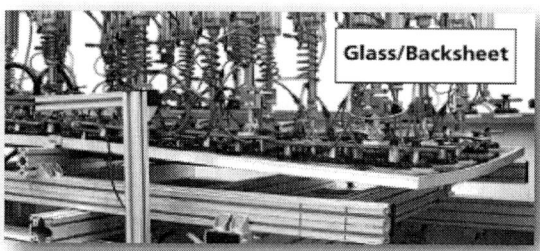

Part 2 – Results
Inhomogeneous Load – Pressure/Suction

  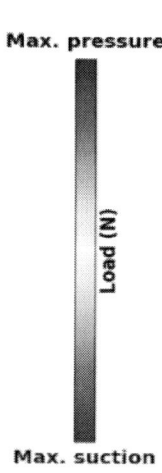

Part 2 – Results
Inhomogeneous Load – Failure Loads

Inhomogeneous/symetric load in tracker setup

- Modules passed experiments in standard setup without glass breakage

  - „Balancing effect"

- Significant difference in failure load between glass/glass and glass/backsheet design

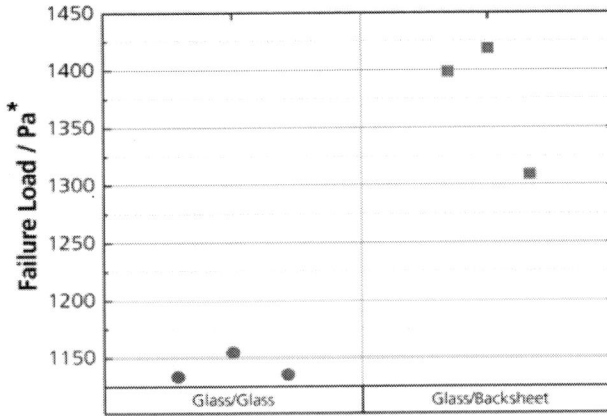

* Equivalent load

Part 2 – Results
Inhomogeneous Load – Module Deflection

Deflection at 1150 Pa: Significant differences in deformation behavior

- Less stiffness for glass/glass design

- Glass/backsheet: No deformation in short module direction

Summary

Part 1: Pre-stress

- SCALP measurements could verify labeled pre-stress from DMEGC modules

- Previously assumed significant influence of pre-stress on failure load could not be verified
  - Especially in less stable setups (e.g. tracker setup) more fundamental design issues might play a more important role

Part 2: Inhomogeneous load

- Significantly higher failure loads of glass/backsheet design in tracker setup
  - Less stiffness for glass/glass design
    - Assumption: Decoupled behavior of the two thin glass panes

- „Balancing effects" could be observed
  - Assumption: More pronounced in symmetric scenarios

Outlook

- Understanding the failure mechanisms is crucial to mitigate the glass breakages in future

  - Other glass quality parameters

  - Investigation of load profiles closer to reality:

  - Investigation on influence of frame design

- Continuing experiments to establish more statistics supported by simulations

Snow Load

Front Wind

Cross Wind
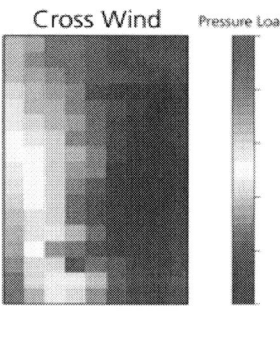
Pressure Load

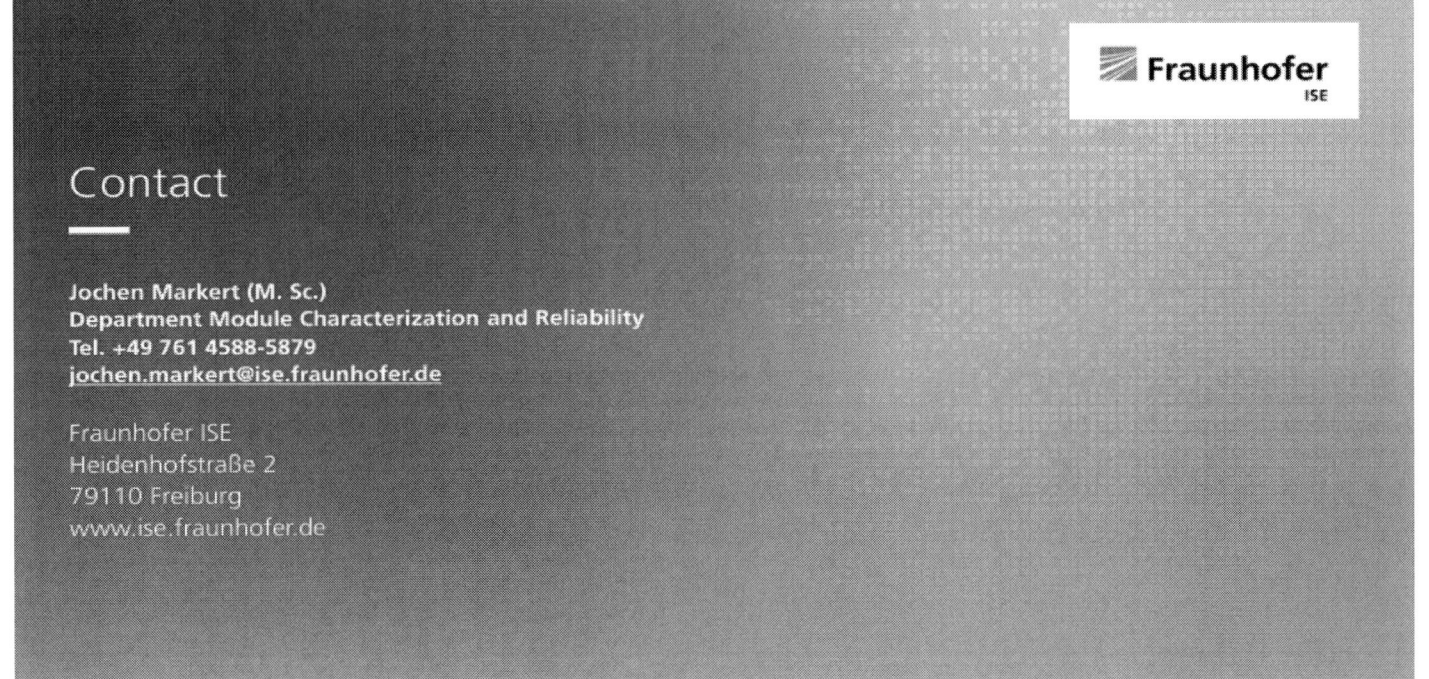

References

[1] PV magazine (2024): Spontaneous glass breakage on solar panels on the rise, https://www.pv-magazine.com/2024/06/24/spontaneous-glass-breakage-on-solar-panels-on-the-rise/, accessed 16.09.2024

[2] The American Ceramic Society (2024): Solar panel breakage on the rise as glass thickness decreases and hail severity increases, https://ceramics.org/ceramic-tech-today/energy-1/solar-panel-breakage-on-the-rise-as-glass-thickness-decreases-and-hail-severity-increases/, accessed 16.09.2024

[3] Solar Builder (2024): Solar module glass is 'spontaneously breaking' in the field, https://solarbuildermag.com/featured/solar-module-glass-is-spontaneously-breaking-in-the-field/, 16.09.2024

[4] Pascal Romer; Kishan Bharatbhai Pethani; Andreas Beinert. Effect of inhomogeneous loads on the mechanics of PV modules. *Sol. RRL* 2023

Part 1 – Results
Mechanical Load Tests until Failure

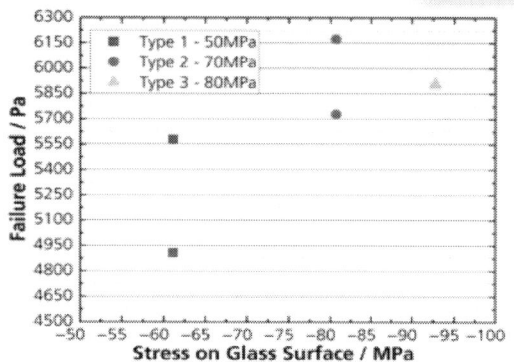

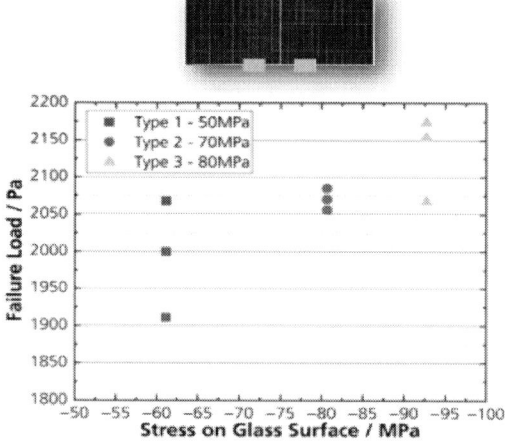

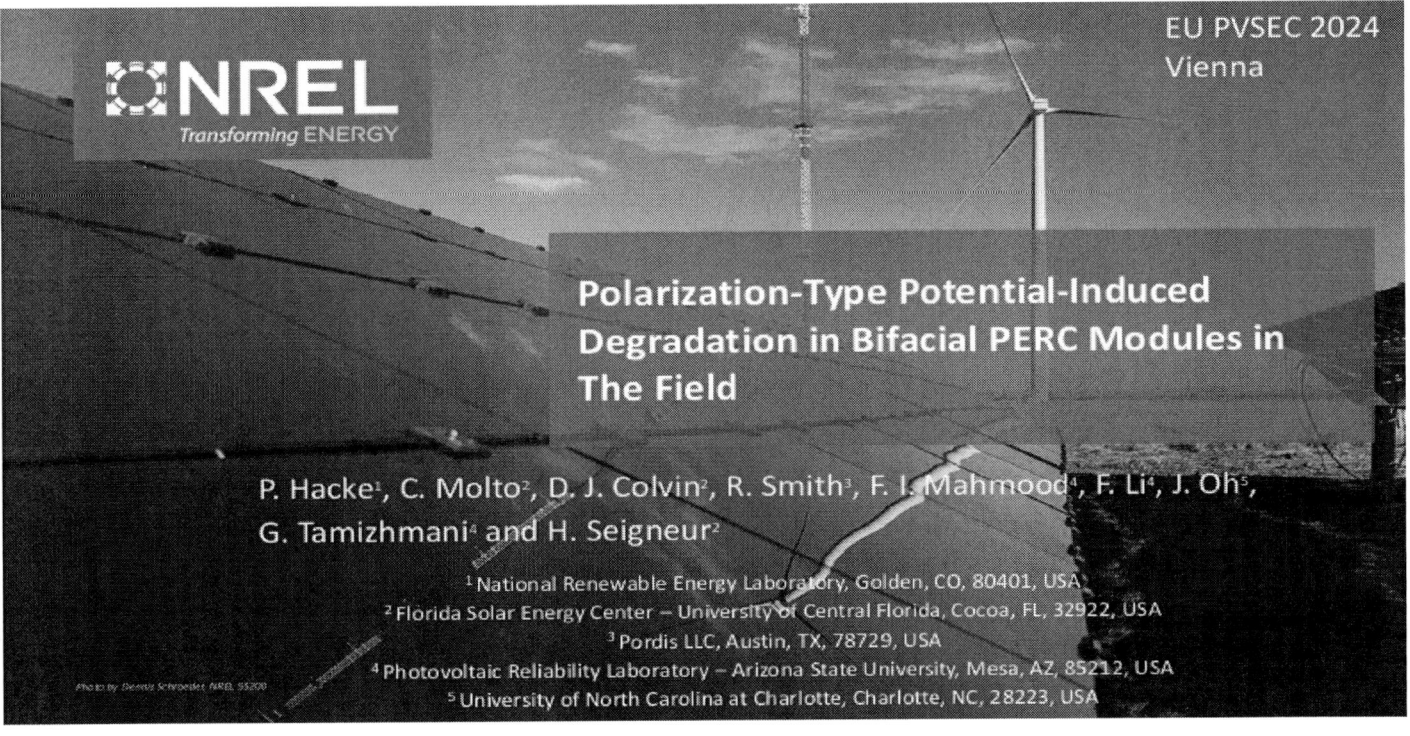

Polarization-Type Potential-Induced Degradation in Bifacial PERC Modules in The Field

P. Hacke[1], C. Molto[2], D. J. Colvin[2], R. Smith[3], F. I. Mahmood[4], F. Li[4], J. Oh[5], G. Tamizhmani[4] and H. Seigneur[2]

[1] National Renewable Energy Laboratory, Golden, CO, 80401, USA
[2] Florida Solar Energy Center – University of Central Florida, Cocoa, FL, 32922, USA
[3] Pordis LLC, Austin, TX, 78729, USA
[4] Photovoltaic Reliability Laboratory – Arizona State University, Mesa, AZ, 85212, USA
[5] University of North Carolina at Charlotte, Charlotte, NC, 28223, USA

Potential induced degradation

PID

Shunting
- n-type & p-type junction
- Na⁺ migration from glass
- **"PID-s"**

Polarization
- n-type & p-type
- Charge in passivation/ AR layer
- **"PID-p"**

Delamination
- Redox Rxn
- Stress/ pressure at cell surface
- **"PID-d"**

Other
- Penetration
- Corrosion
- Any voltage-driven degradation

PID-polarization

PID-p on n⁺/n fronts of IBC (Swanson 2005)

Field Performance Decreased 20% After Several Months Operation

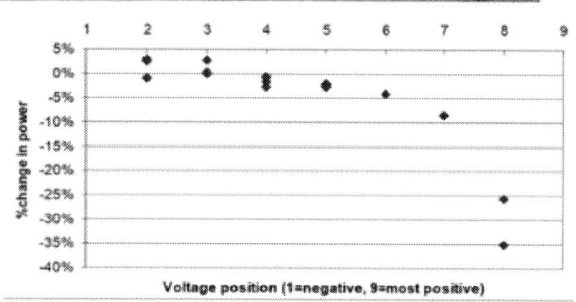

PID-p on p⁺/n fronts of PERT (Ohdaira & coworkers 2023)

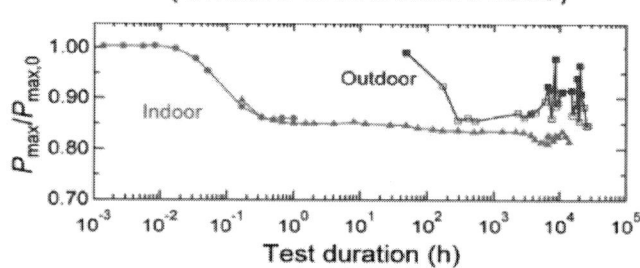

How about PERC in the field?

PID-polarization in bifacial PERC

- **Numerous studies show PID-polarization on PERC modules in accelerated testing**
- **No field studies to understand if it happens with PERC in the field.**

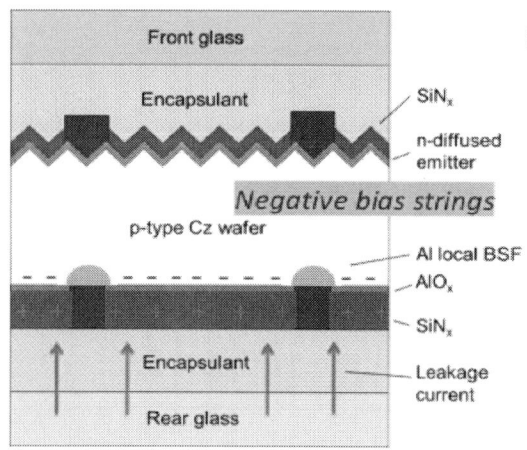

Sequence of PID-p mechanism on module back:

- Positive charge drifts through module packaging and imbeds in back dielectric
- Positive charge in dielectric attracts minority carrier electrons in the p-type base promoting rear surface recombination→ V_{oc}, I_{sc} loss
- In the extreme (more positive charge in rear dielectric), rear surface inverts, becomes effectively n-type, where the minority carrier is the hole, repelled → V_{oc}, I_{sc} increase
- Dissipation of charge, especially by photoconductivity of SiN_x, dissipation of PID-p

PID-polarization in bifacial PERC (back)

PC1D simulation
varying charge in the back dielectric

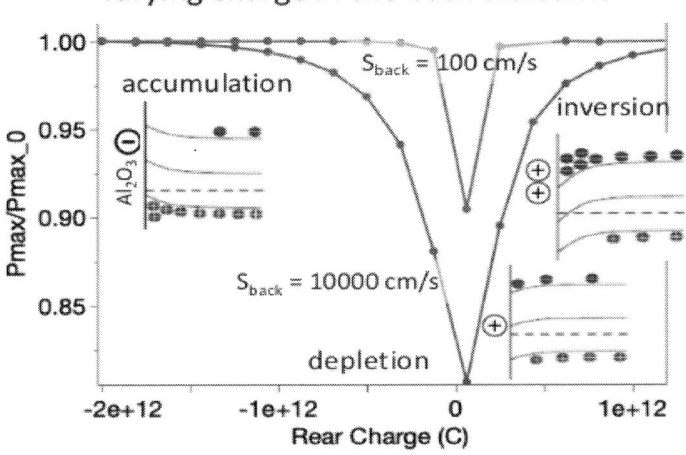

3 different PERC samples
undergoing PID accelerated stress

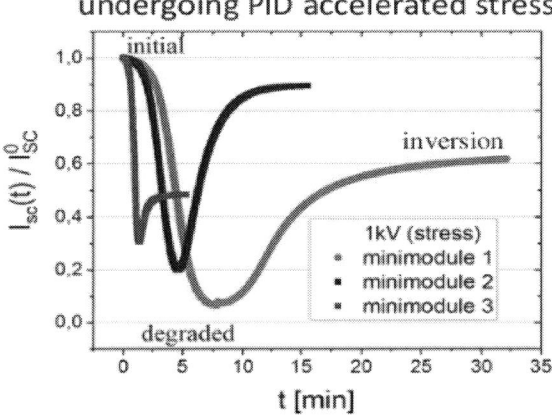

Sporleder, Nauman, Bauer and coworkers
AIP 2487,030011 (2022)

Experiment part 1
PERC module screening tests

IEC 62804-1 method b: grounded Al foil module face (front & rear), cell circuit Vsys = ± 1500 V
25 °C for 168 h; Commercial bifacial glass/glass PERC modules, 420 W

Test Step	Isc (A)	Voc (V)	FF (%)	Pmax (W)	% Pmax change from initial	Isc (A)	Voc (V)	FF (%)	Pmax (W)	% Pmax change from initial
Stress level (25 °C 168 h, Al foil)		-1500 V (rear side stress)					+1500 V (front side stress)			
Initial	10.74	48.87	79.9	419.3	0	10.69	48.59	79.9	415.0	0
PID stress test	10.50	47.82	79.4	398.8	- 4.9	10.57	47.96	79.3	402.1	-3.1
Intermediate recovery (storage)	10.11	46.78	78.8	372.6	- 11.1					
Final recovery 19 kWh/m² illum.	10.71	48.83	79.8	417.1	- 0.5	10.69	48.59	79.9	415.0	0

For more info: F. I Mahmood, F. Li, P Hacke and coworkers,
Prog Photovolt Res Appl. 2023;31:1078–1090

Experiment part 1
PERC module screening tests

IEC 62804-1 method b: grounded Al foil module face (front/rear), cell circuit Vsys = ± 1500 V
25 °C for 168 h; Commercial bifacial glass/glass PERC modules, 420 W

Test Step	Isc (A)	Voc (V)	FF (%)	Pmax (W)	% Pmax change from initial
Stress level (25 °C 168 h, Al foil)		-1500 V (rear side stress)			
Initial	10.74	48.87	79.9	419.3	0
	10.50	47.82	79.4	398.8	- 4.9
Intermediate recovery (storage)	10.11	46.78	78.8	372.6	- 11.1
Final recovery 19 kWh/m² illum.	10.71	48.83	79.8	417.1	- 0.5

Screening test result:
This module type tested in the field

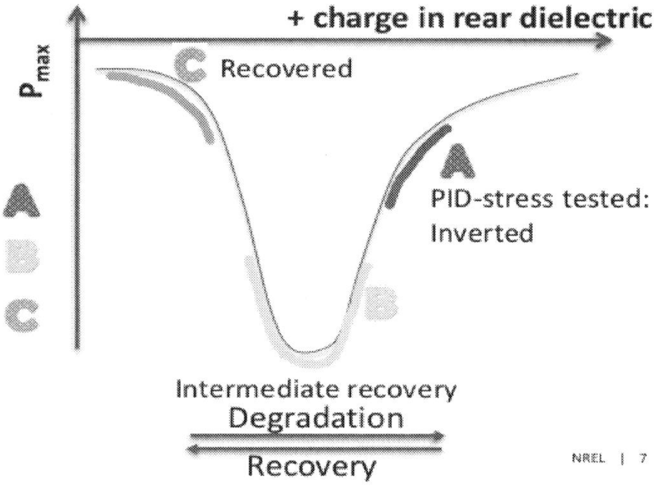

+ charge in rear dielectric

P_{max}

C Recovered

A PID-stress tested: Inverted

Intermediate recovery
Degradation
Recovery

Experiment part 2
Field Testing

6 months humid subtropical field test statistics

Rack type:	Near ground	Open rack	Open rack	Roof
Elevation (m)	0.3	2.0	2.0	0.3
Bias (V) (daytime):	-1500 V	-1500 V	+1500 V	-1500 V
Modules biased (controls):				
Averages				
Daytime T (°C) [stdev]	35.8 [11.4]	36.6 [11.5]		37.1 [12.1]
Daytime RH (%) [stdev]	46.0 [25.9]	46.2 [26.4]		45.6 [26.8]
Coulomb/ day	0.012	0.007		0.009
albedo kWh/m²	0.376	0.240		0.015

Experiment part 2
Field Testing
Power of modules undergoing field testing with ±1500 V peak system voltage

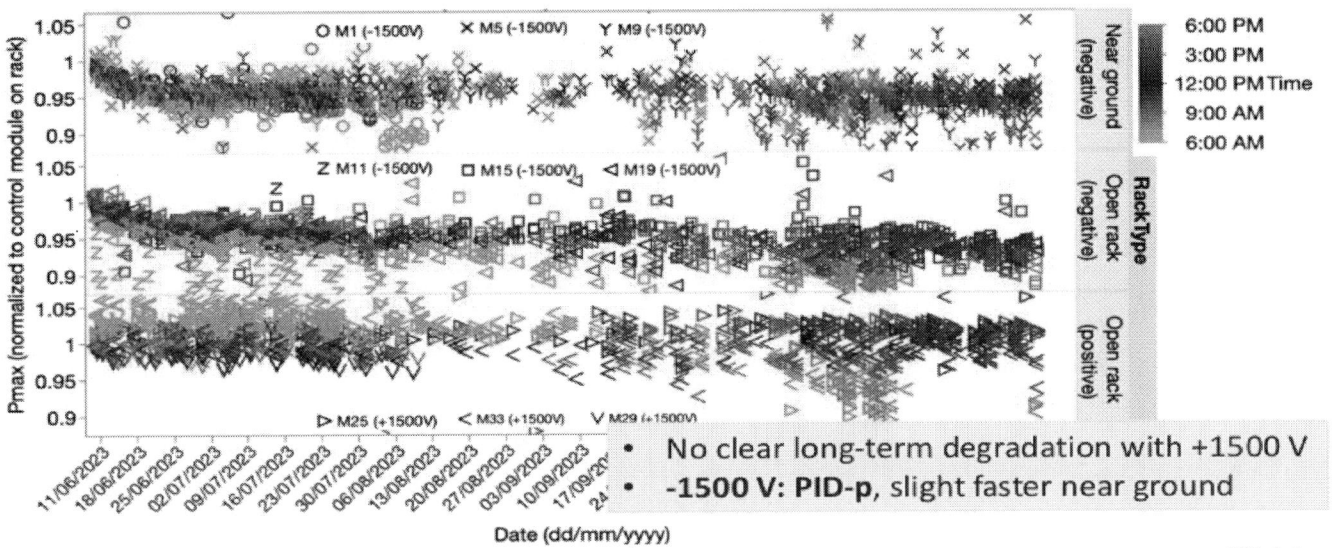

- No clear long-term degradation with +1500 V
- **-1500 V: PID-p**, slight faster near ground

Main effects (for modeling)
during first 2.5 weeks of major PID-p

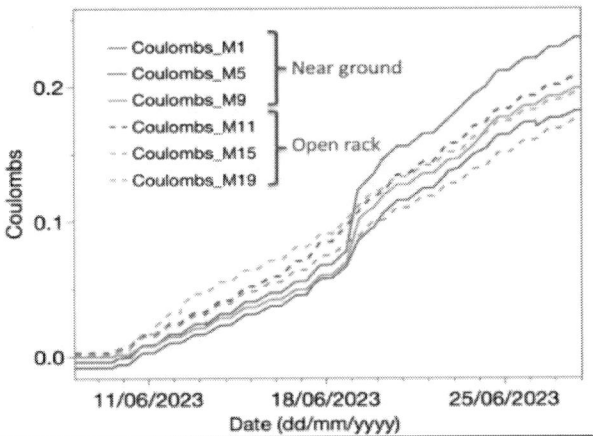

Not much difference in coulomb transfer ($Q = f$ (V, T, RH))

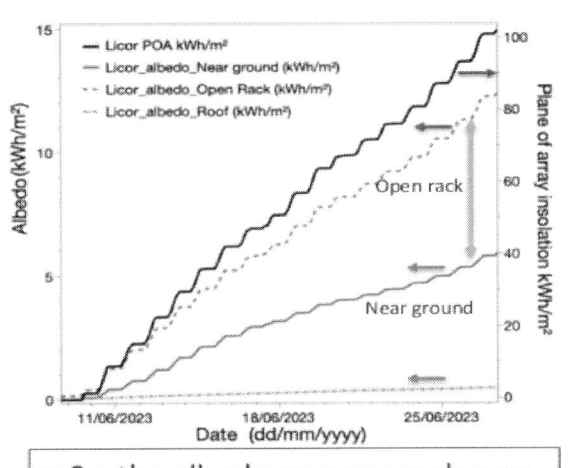

~ 2 x the albedo on open rack over near ground

Charge model for PID-p

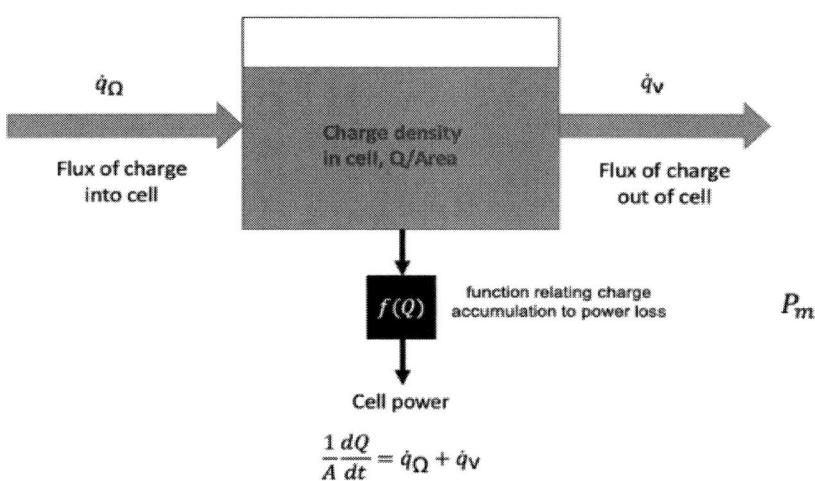

Hattendorf & coworkers (2012)

Habersberger and Hacke (2022)

$$\frac{dQ}{dt} = \frac{V}{\rho l} - kEQ$$

$$Q(t) = \frac{V}{\rho lkE}(1 - e^{-kEt})$$

$$P_{max,norm}(Q) = (1 - P_\infty)\frac{A+1}{A + e^{BQ}} + P_\infty$$

Charge model for PID-p

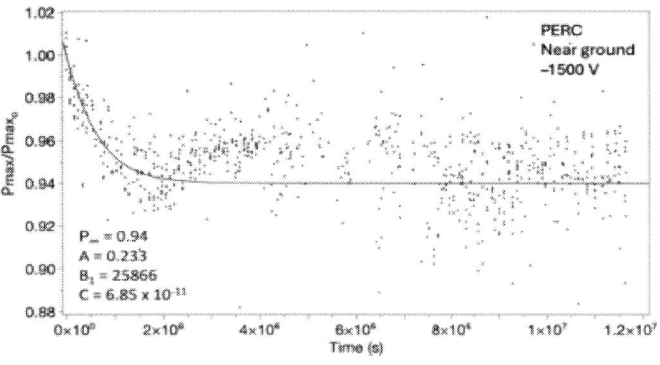

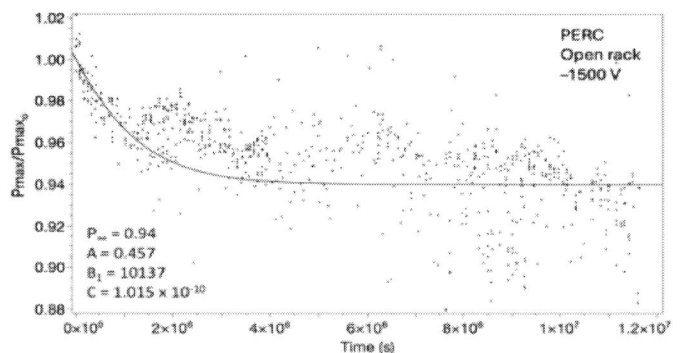

$$\frac{P_{max}}{P_0} = (1 - P_\infty)\frac{A+1}{A + \exp(B_1(1 - \exp(-Ct)))} + P_\infty$$

$$B_1 = \frac{BV}{\rho lkE}$$

$$C = kE$$

$$\frac{1}{\Big/}\frac{B_{1\ open\ rack}}{B_{1\ near\ ground}} = 2.55$$

$$\frac{C_{open\ rack}}{C_{near\ ground}} = 1.48$$

Actual albedo incident

$$\frac{E_{open\ rack}}{E_{near\ ground}} = 2.1$$

Inversion in fielded modules?

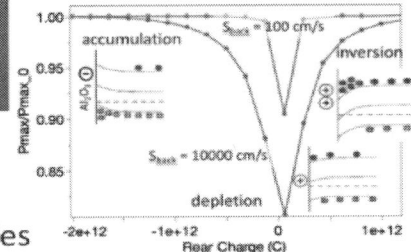

Electroluminescence at 10 weeks of open rack modules

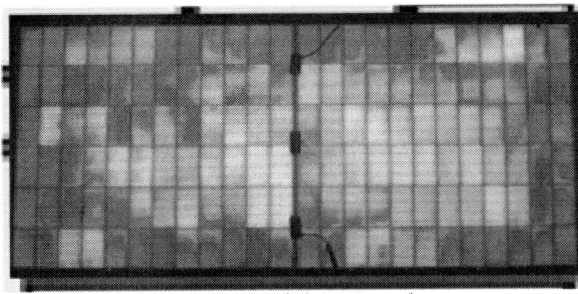

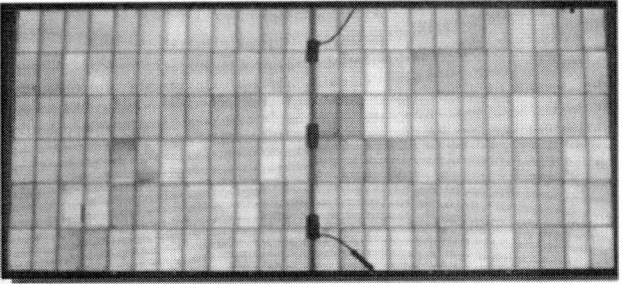

-1500 V (daytime)

0 V (outdoor control)

No, probably not

Bifacial PERC recovery rate
(reflected from concrete)

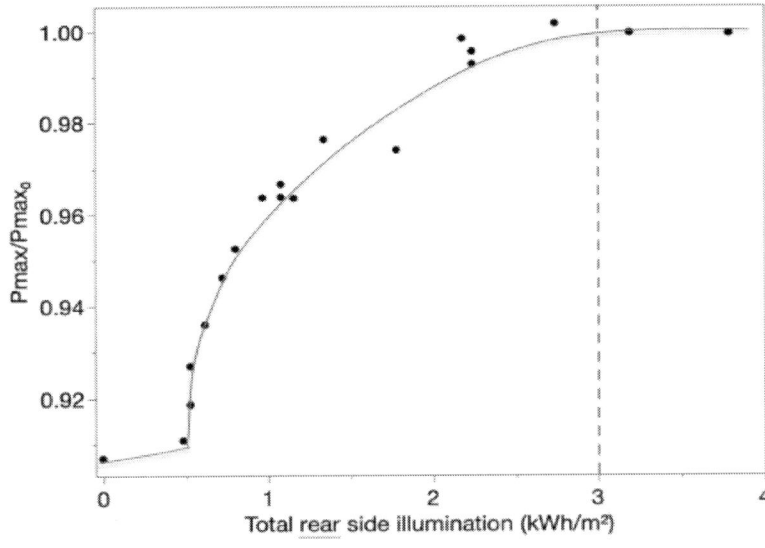

- Open rack daily average daily albedo incident on module rear: 0.24 kWh/m^2

→ Estimated 12 Days without any PID-stress to make a full recovery in the field.

Summary

- PID-p <u>does</u> happen in the field
  - n^+/n fronts of IBC (Swanson)
  - p^+/n fronts of PERT (Ohdaira)
  - p-base backs of bifacial PERC (this work)
- Bifacial glass/glass PERC modules confirmed to show PID-p in the field under V_{sys} of -1500 V in about 2.5 weeks, no matter what mounting configuration
- Charge model consistent with about 2x irradiation on <u>open rack</u> compared to <u>near ground</u> module; degradation rate similar
- Inversion at PERC back in accelerated testing, <u>not</u> in field testing: 6 % degradation in the field.

Thank you

This work is supported by the US Department of Energy Solar Energy Technology office SETO Project 38263.

Faculty of Engineering
School of Photovoltaic and Renewable Energy Engineering

LeTID in Real Life: The relevance and importance of accelerated tests and treatments

EuPVSEC Vienna, Austria
24th September 2024

Dr. Alison Ciesla, on behalf of Letitia project team

Letitia Team: Arastoo Teymouri, Petra Manshanden, Alvin Mo, Astrid Gutjahr, Moonyong Kim, Li Wang, Catherine Chan, Ran Chen, Gianluca Coletti, Jakob Jan Dijksterhuis, Bas Van Aken

Outline

- Motivation
- Letitia Project
- Cell qualification
- Treatments
- Accelerated testing
- Real life: early field results
- Conclusions & next steps

LETITIA
Light & Elevated Temperature Induced Degradation Repair Approach

Letitia team meeting at Elsun (Sept 2022)

Motivation: LeTID – what is it?

Light- and elevated Temperature-Induced Degradation [1]

An electrical defect that forms under light or heat -> caused by hydrogen

>10% power loss if untreated [2]

Degrades over several years, may not ever recover in working life especially under continental climate

Affects all silicon, especially B-doped PERC cells, but highly variable

Different thermal history [3]

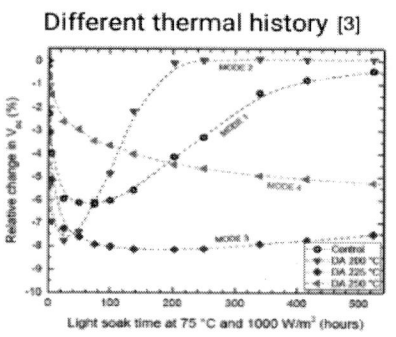

Different wafer or technology[4]

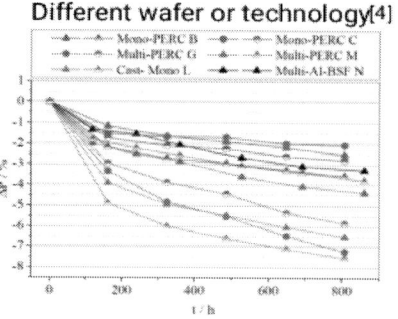

Same wafer and technology[4]

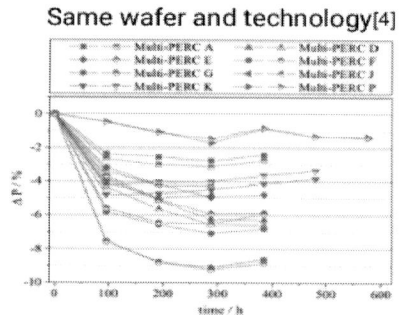

[1] F. Kersten *et al. IEEE 42nd PVSC* (2015) 1–5; [2] F. Kersten et al., Sol. Energy Mater. Sol. Cells 142 (2015) 83–86;
[3] C. Chan et al., Sol. RRL. 1, 1600028 (2017); [4] E. Fokuhl et al., 36th Eur. PV Sol. Energy Conf. Exhib. 4 (2019) 75–84.

3

Motivation: Relevant testing for the Netherlands [2]

Accelerate -> heat + increased electrical carriers (light/voltage/current)

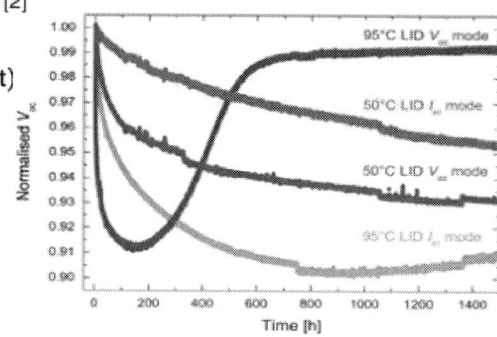

[2] F. Kersten et al., Sol. Energy Mater. Sol. Cells 142 (2015) 83–86.

Motivation: Relevant testing for the Netherlands

Accelerate -> heat + increased electrical carriers (light/voltage/current)

Can be more severe at lower temperatures, but not always predictable

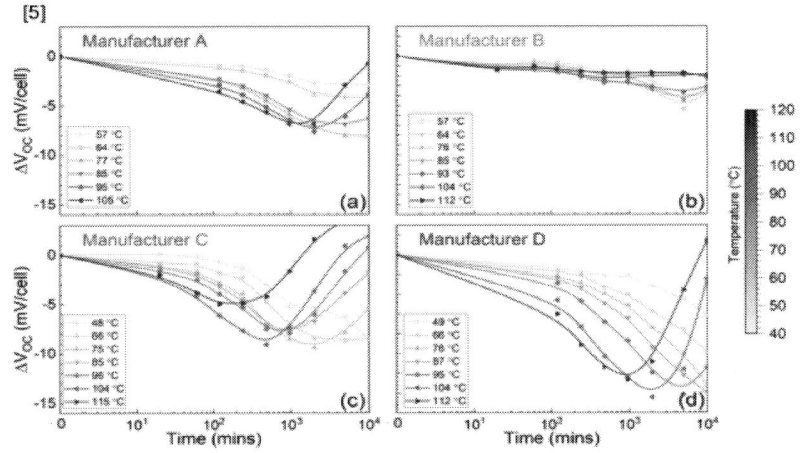

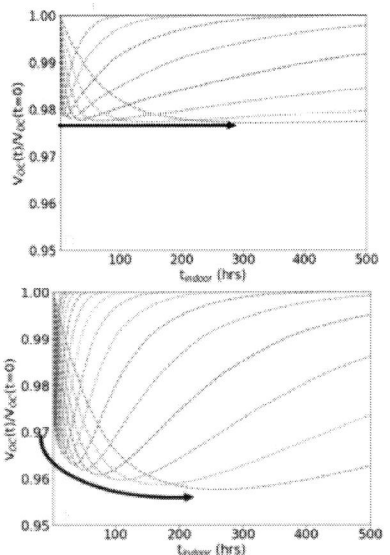

[5] A. M. Ciesla *et al.*, *Progress in PV.* **29(11)**, 1202 (2021)

Motivation: Relevant testing for the Netherlands

Accelerate -> heat + increased electrical carriers (light/voltage/current)

Can be more severe at lower temperatures, but not always predictable

Commonly tested for at ~75 °C (eg. IEC TS 63342)

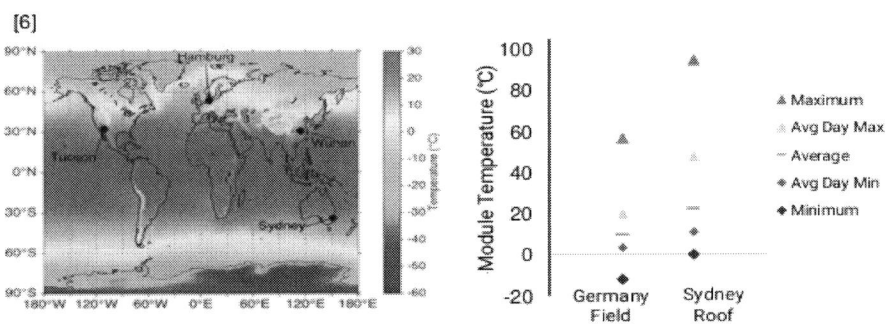

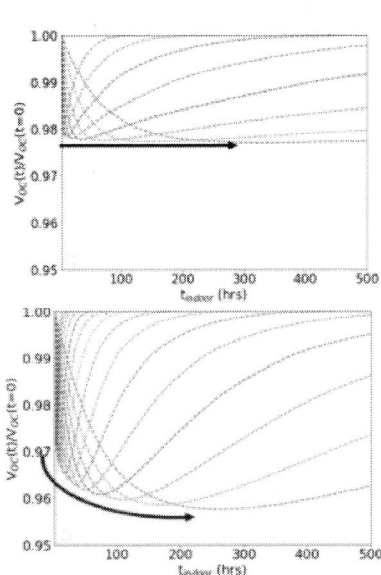

[6] A. M. Ciesla *et al.*, *IEEE J. Photovoltaics.* **10**, 28–40 (2020)

Motivation: Assured reliability for module manufacturers

- Solutions exist (eg. modify doping, structure, SiNx, firing with slow cooling).
- The amount of LeTID has been decreasing, but still significant [7].
- Commonly, module manufacturers don't know susceptibility before they buy, and many don't have capability to test and/or treat. Uncertainty on final energy yield without accurate accelerated tests.
- Solutions can be applied to finished cells

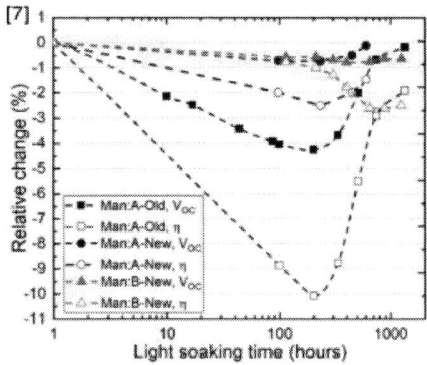 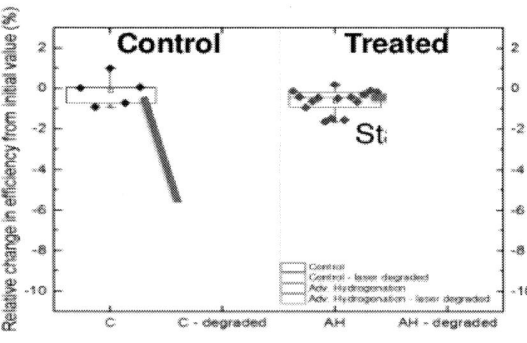

[7] Sen, C *et al.* APSRC conference proceedings 2019.

Letitia project - Partners

- Research partners:

- Industry partners:

Elsun is a premium module producer with a 100MW production line. Elsun solar modules are designed and made in the Netherlands with the best materials available. Our mission is to place as many solar panels as possible on Dutch rooftops.
http://www.elsun.nl - info@elsun.nl

SAS Sunrise is a leading solar wafer, cell and module provider in Taiwan. SAS Sunrise has presence on the complete solar value chain from manufacturing to system integration, fulfilling responsibility as a global citizen in the environmental protection of the Earth.
https://www.sas-globalwafers.com/

Letitia project – Aims

Letitia: Light and Elevated Temperature Induced Degradation Repair Approach

- Cell qualification (baseline testing) under different conditions incl. NL climate
- Optimize accelerated test conditions for NL
- Optimize treatment for selected cells
- Produce and compare treated and untreated modules (lab, field and models)

Research question: in real outdoor continental climate conditions how does degradation and recovery progress and what is the impact on the energy yield?

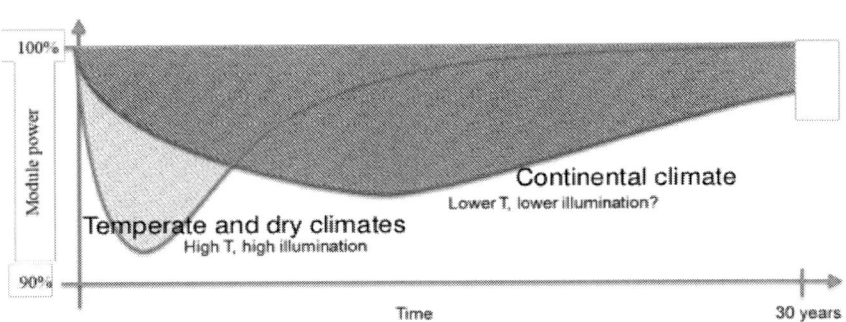

Results – Cell qualification (baseline testing)

- 4 of 8 mono-PERC cell types tested showed LeTID degradation
- 3 identified to have BO and LeTID
- 1 Ga doped
- Added in mc-PERC option (for B comparison without BO)

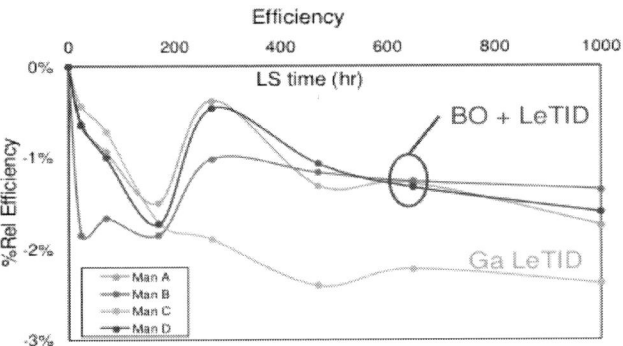

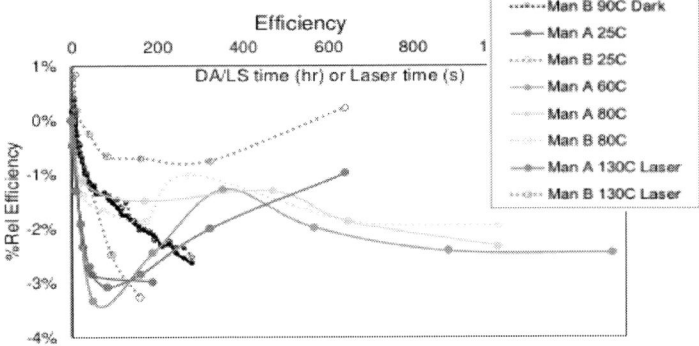

Results - Optimised accelerated laser test

- Optimized test– 130 °C, ~3suns (10% power)
- Identified LeTID in all significantly susceptible cells (B mono with BO, B mc-Si LeTID, Ga LeTID)

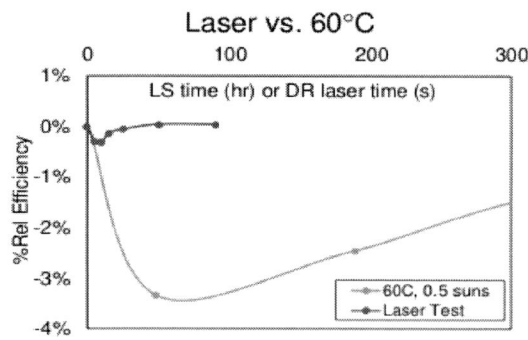 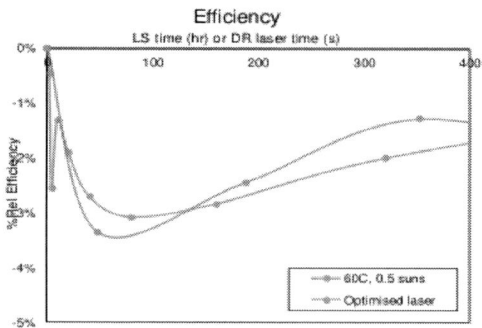

11

Results - Treatments

- Hydrogenation treatments should:
  1. **passivate defects (efficiency boost not reduction)**

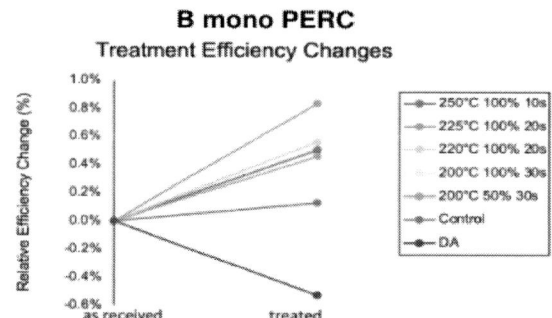

*Ruled out current injection and firing treatments due to negative effect on many cells

12

Results - Treatments

- Hydrogenation treatments should:
    1. passivate defects (efficiency boost not reduction)
    2. **clear out excess H to prevent LeTID (enhance stability)**

B mono PERC

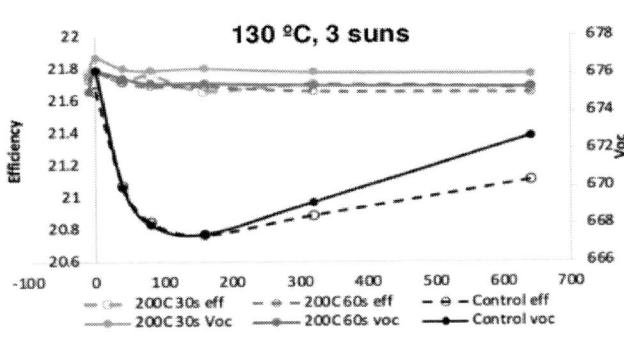

Results - Treatments

- Hydrogenation treatments should:
    1. passivate defects (efficiency boost not reduction)
    2. clear out excess H to prevent LeTID (enhance stability)
    3. **be suitable for mass manufacture (high throughput, commercial equipment, universal)**

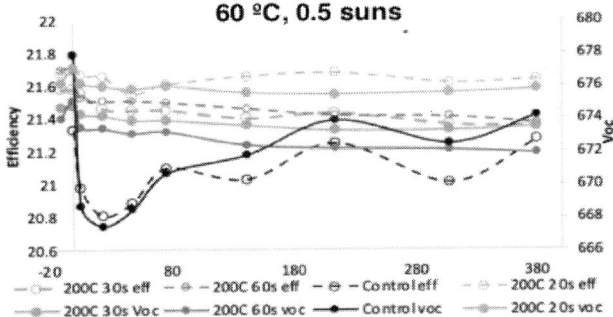

Decided on 20s process to align with Elsun throughput and overall stability.

Results – Treatments on mc-Si and Gallium cells (LeTID only)

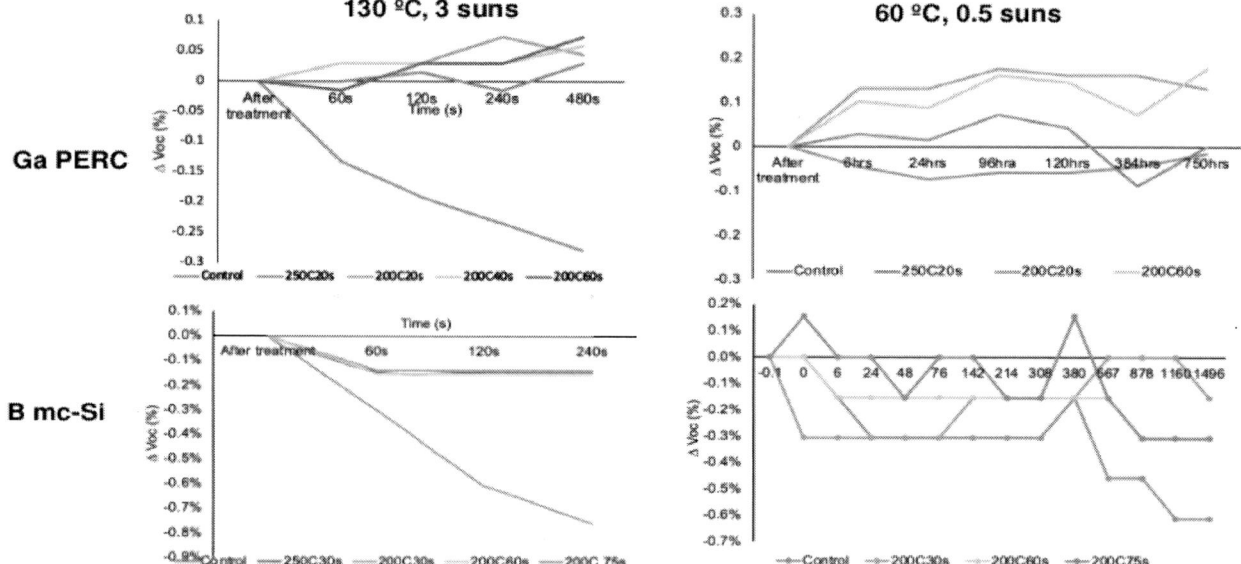

Ga PERC

B mc-Si

130 ºC, 3 suns

60 ºC, 0.5 suns

15

Results – Treatments on mc-Si and Ga

B mc-Si PERC:

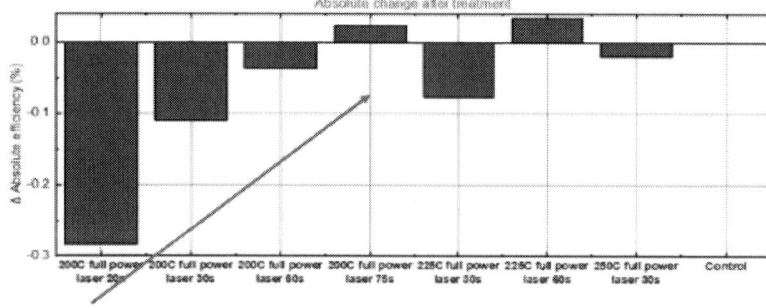

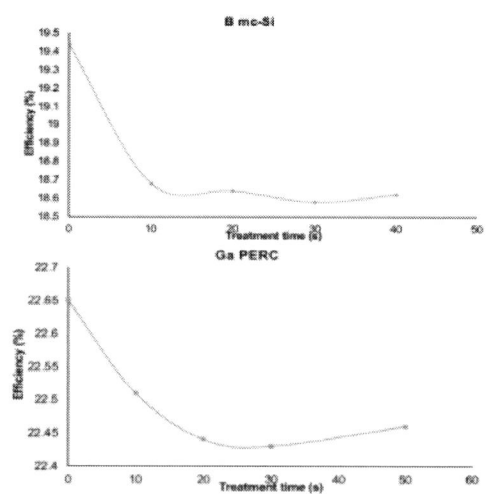

- Mc-Si and Ga: Improved stability as already degraded , but lost efficiency
- B mono-PERC: BO defects passivated gives boost and soaks up the Hydrogen

16

Results – Mini-modules in-situ illuminated climate chamber

Eternal Sun Spire
Class A+A+A+ Solar Spectrum Climate Chamber

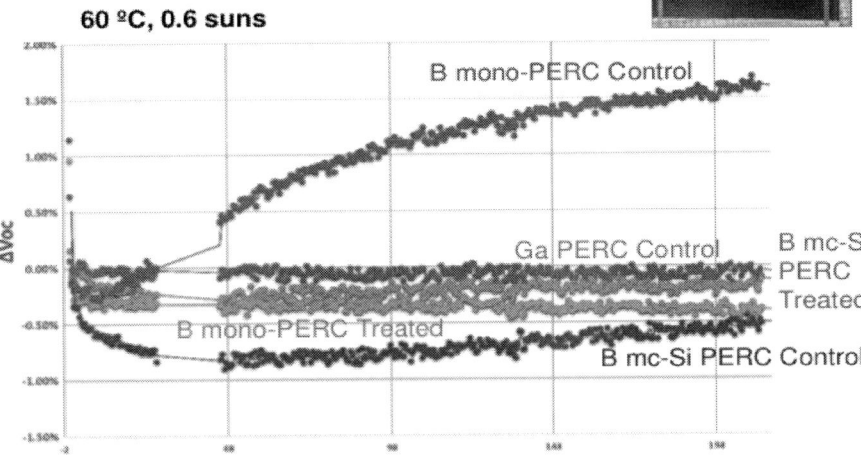

Why are we not seeing LeTID in Ga PERC in the lab?

Possible reasons?
- Suppliers to Elsun already have very good LeTID treatments
- Very slow in low temperatures
- On the border between degradation and recovery?
  - **60C, 0.6 suns**

Gallium shows different behaviour:
- Slower [8] [9]
- Higher temps exacerbates but faster (60-120 °C [10][11][12])
  - <60°C not as well known?
- Higher carrier injection -> suppresses LeTID [9][11]
  - Can be worse in lower illumination – common in NL?
- Temporary recovery appears at higher temperature with higher illumination [9][11] eg. 75°C 1 sun
  - but degrades with an n-type emitter [9] (not well studied!)

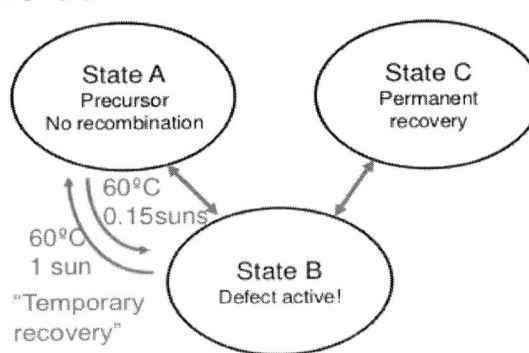

[8] Fritz *et al.*, *Silicon PV.* (2017) [10] Winter *et al.* JPV 2021 [12] Thome *et al.* SOLMAT 2024
[9] Kwapil *et al.*, *Solar RRL.* (2021) [11] Winter *et al.* Sci, reports 2022

Results – PERC modules manufactured

Modules	B mono	B mc-Si	Ga
Control cells	500	300	300
Control modules	8	4	4
Treated cells	500	250*	250*
Treated modules	8	3*	3*

*Still to be made

Elsun Roof:

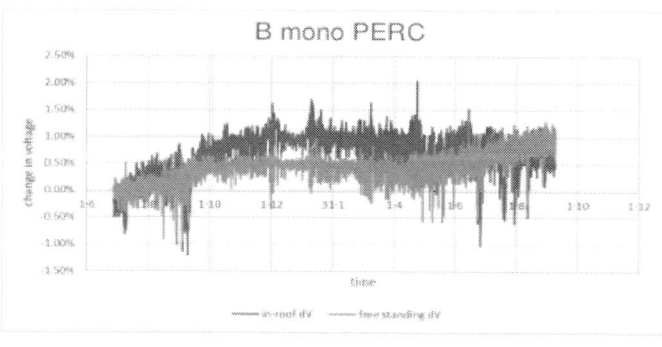

Results – Early field results (TNO)

Jun 2023 – Sept 2024:

- B mono PERC: BO recovery observed
- B mc-Si PERC: similar to reference module

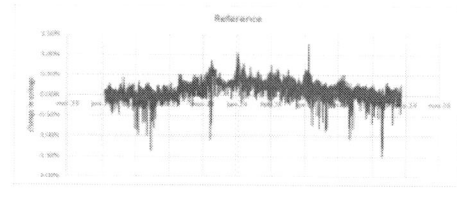

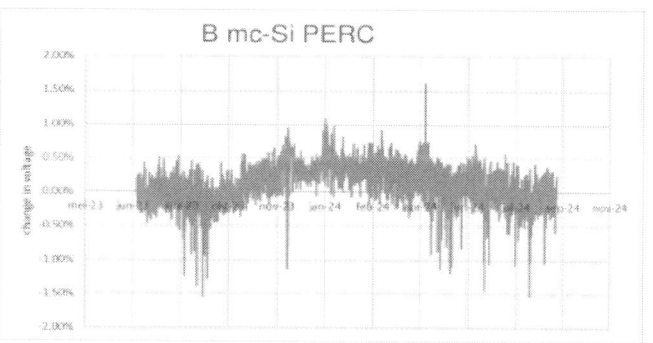

Results – Early field results (TNO)

Jun 2023 – Sept 2024:

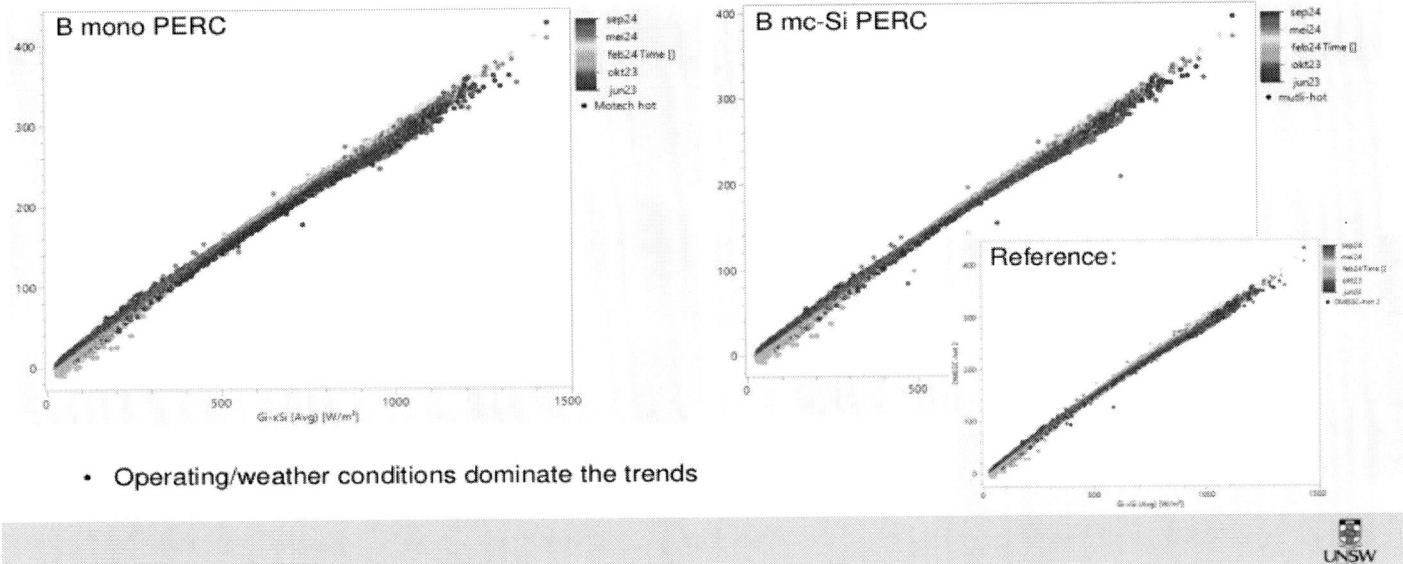

- Operating/weather conditions dominate the trends

Results – Field operating conditions (TNO)

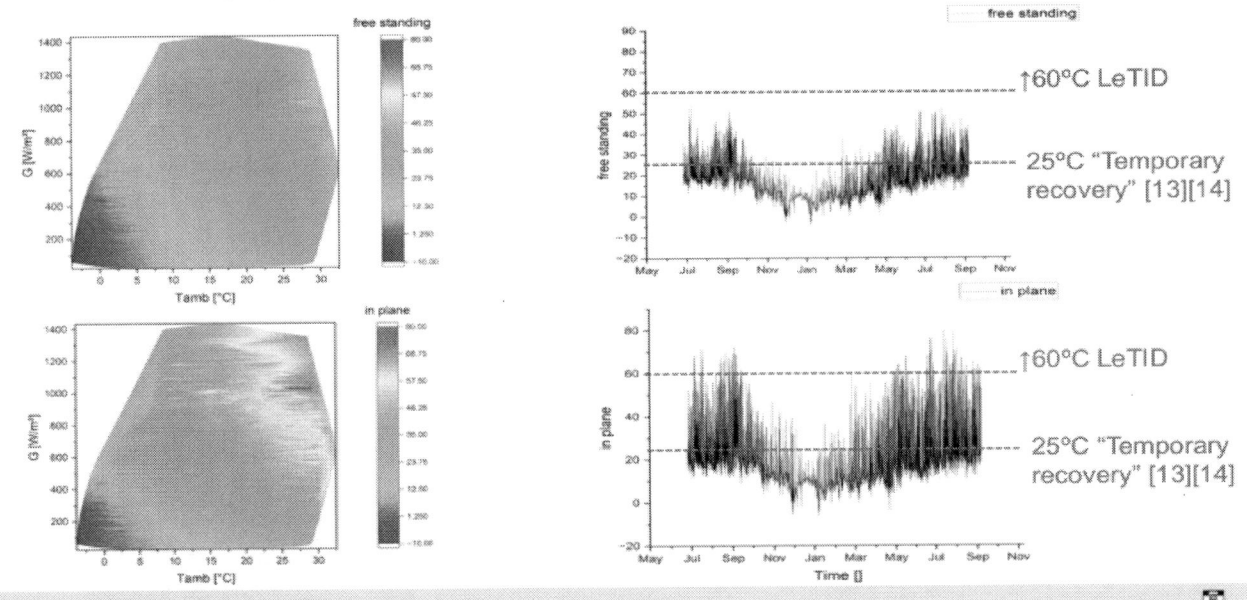

[13] Kersten, *et al.* EuPVSECm 2015   [14] Kwapil, *et al.* JPVm 2015

Conclusions

- LeTID in PERC cells is still present at reduced severity and prevalence
- A 130°C, ~3-sun accelerated laser process can identify susceptibility
- 75 °C (IEC) is not a realistic testing temperature for the Netherlands (NL)
- Consistent 60 °C, 0.6suns is also NOT accurate to simulate NL
- Variable field conditions likely shift modules between degradation and recovery modes
- Not enough known to predict field degradation accurately (esp. Ga with emitter <60°C)

Next Steps – nearing project end!

- Install mc-Si and Ga-PERC control and treated modules in the field
- Monitor for longer (past project end)
- Data analysis on distinct periods of weather conditions and with STC correction
- New project? Kinetic analysis in range of NL conditions to enable improved field predictive models

Acknowledgment
The Letitia project receives a Topsector Energy subsidy of the Ministry of Economic Affairs and Climate Policy, implemented by Netherlands Enterprise Agency (RVO) - HER20-00199097

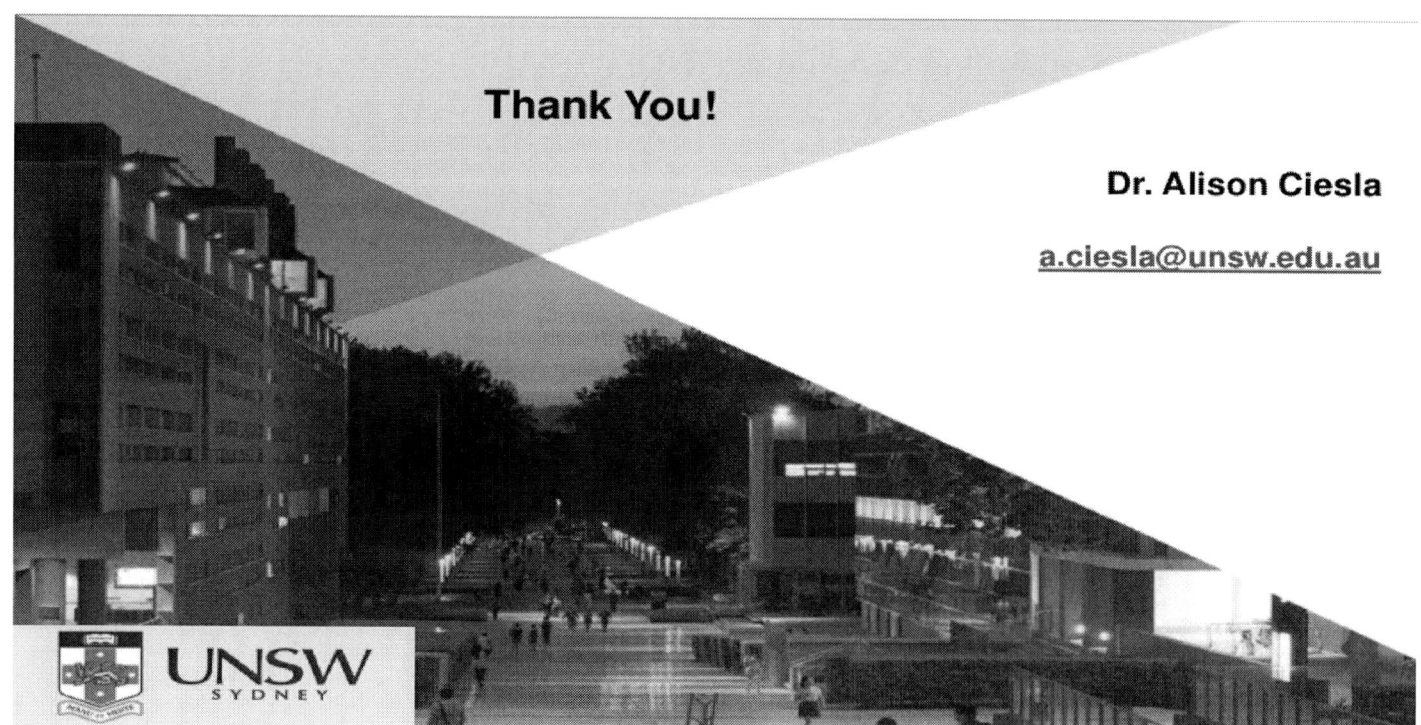

Thank You!

Dr. Alison Ciesla

a.ciesla@unsw.edu.au

3BO.14.6

Towards Establishing Criteria for Electrical Safety in Second-Use Photovoltaic (PV) Modules

Tadanori Tanahashi & Takashi Oozeki
(AIST, JP)

This presentation is based on results obtained from a project, JPNP20015, commissioned by the New Energy and Industrial Technology Development Organization (NEDO).

EU-PVSEC 2024 (Vienna, September 24, 2024)

Motivation: Re-Use of PV Modules
Higher Priority in PV Circular Economy

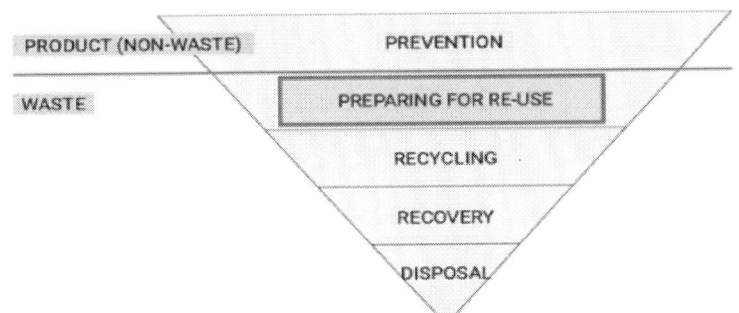

Crucial Roles
- Reducing Waste

- Supplying Spare PV Modules
  (e.g., during Refurbishment Stage)

- Providing Low-Cost PV Modules
  (for New Installations)

Potential Risks
- Electrical Safety Issues
- Performance Degradation

[1]   A. van der Heide, L. Tous, K. Wambach, J. Poortmans, J. Clyncke, and E. Voroshazi, "Towards a successful re-use of decommissioned photovoltaic modules," *Prog. Photovoltaics Res. Appl.*, vol. 30, no. 8, pp. 910–920, Aug. 2022, doi: 10.1002/pip.3490.
[2]   A. van der Heide, D. M. Godinho Ariolli, G. O. Hernandez, S. Noels, and J. Clyncke, "Re-use of PV modules: Progress in   standardisation and learnings from a real case study," in *40th European Photovoltaic Solar Energy Conference and Exhibition*,   2023, pp. 020515-001-020515–013. doi: 10.4229/EUPVSEC2023/5DO.15.6.

Challenges in This Study
Proposing a Detection Procedure for Latent Electrical Risks/Hazards

•Based on the Evaluation of Field-Aged PV Module

•Focusing on Electrical Safety Issues (i.e., Electrical Insulation)

•If Possible, Discussing Potential Applications in the 2nd-Use PV Market

Brief History of the PV Module Tested in This Study

Manufactured in 2011 (230 Wp, Conventional Al-BSF c-Si Cells)

Outdoor Exposure with Grid Connection **from Late 2011**
   10 kW PV System was installed <u>at Closed-Coastal area</u>

Decommissioned in 2019 (ca. 8 Years Work) ← Severe BS Cracks

Retrieve a PV Module from the PV System **in Mid 2019**
   Store in a Warehouse **until Late 2020** (ca. 1 Year Storage)

Indoor Evaluation **(1st Round)**
   Store in a Warehouse (ca. 8 Months Storage) ⎤ This Study
Indoor Evaluation **(2nd Round)** ⎦

Pristine Characteristics of the PV Module Tested in This Study
Determined from <u>Sister PV Modules with the Same Model Number</u>

•**Highly Reliable** against DH/TC Stressors (Extended Stress Tests)

•**Resistant to PID-s** (IEC TS 62804-1)

•**High Electrical Insulation Resistance** (Wet Leakage Current Test)

Detailed Results in the Pristine Stage have been Published

[1] JET, SAGA, AIST, PVTEC, "Report on Asia standards and conformity assessment promoting project (in Japanese)" (2014); https://www.pvtec.or.jp/data_files/view/122.

[2] S. Kawai, T. Tanahashi, Y. Fukumoto, F. Tamai, A. Masuda, M. Kondo, "Causes of degradation identified by the extended thermal cycling test on commercially available crystalline silicon photovoltaic modules." IEEE J. Photovoltaics 7, 1511 (2017). https://doi.org/10.1109/JPHOTOV.2017.2741102.

Retrieved PV Module (Rear Side View)

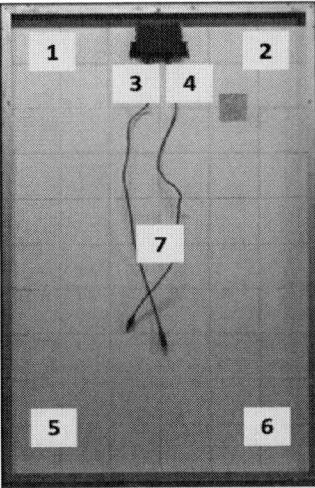

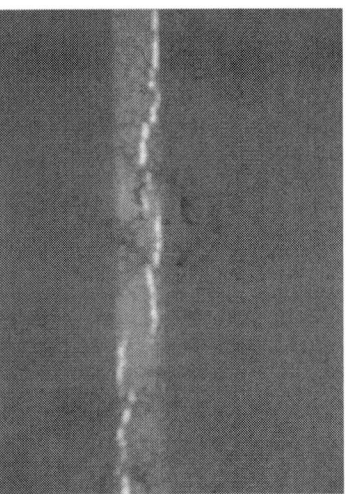

(a) (b) (c)

Enlarged photos (b and c) were captured under transmitted light for better visibility of backsheet cracks.

NIR Spectra for the Tested PV module with Backsheet Cracks.

Front Side

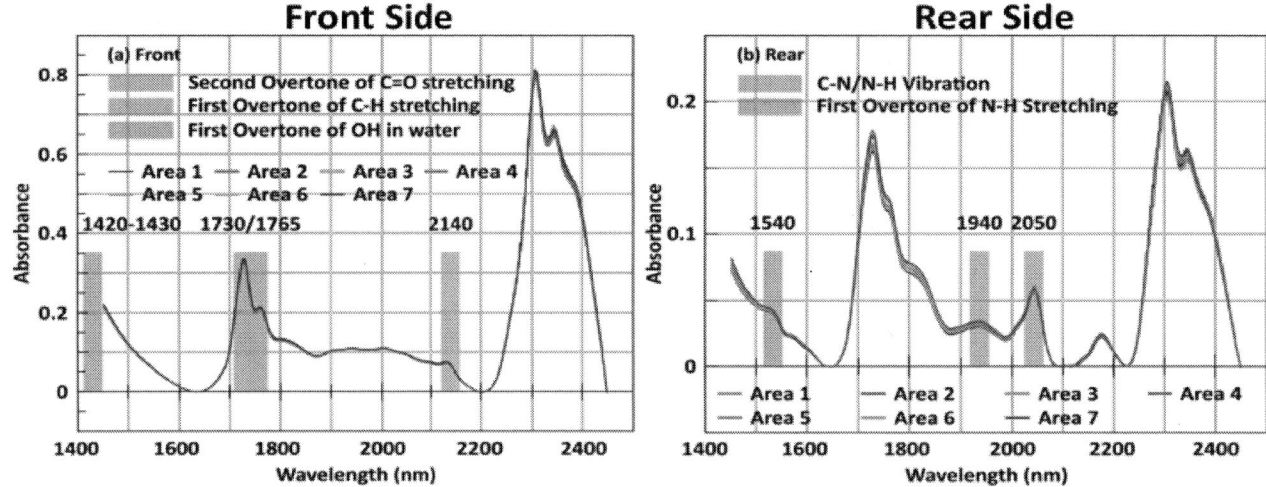

Rear Side

Encapsulant: EVA

Backsheet: PA

Electrical Safety Issue

Wet Leakage Current Test (WLCT: IEC 61215-2 MQT 15)

Electrical Insulation (WLCT*)	
Pristine Modules (2 Sister PV Modules)	> 0.12 GΩ·m²
Field-Aged PV Module (Pre-Indoor Test)	0.60 GΩ·m²

* **Pass/Fail Criterion in WLCT: > 0.04 GΩ·m²**

Even in the PV modules with severe BS cracks, the electrical insulation issue could not detect with WLCT.

High Voltage Stress Test with in-situ Leakage Current Measurement

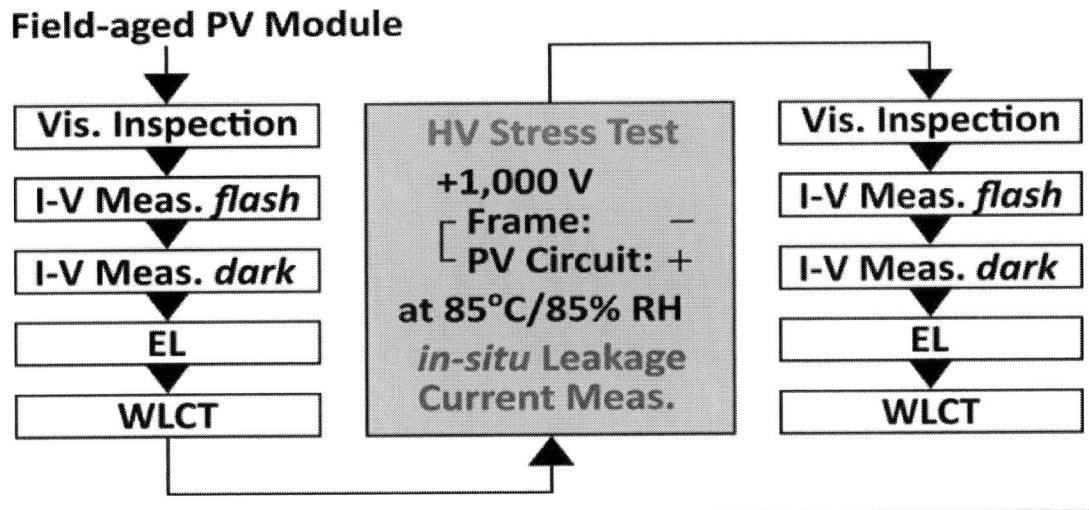

High Voltage Stress Test with in-situ Leakage Current Measurement

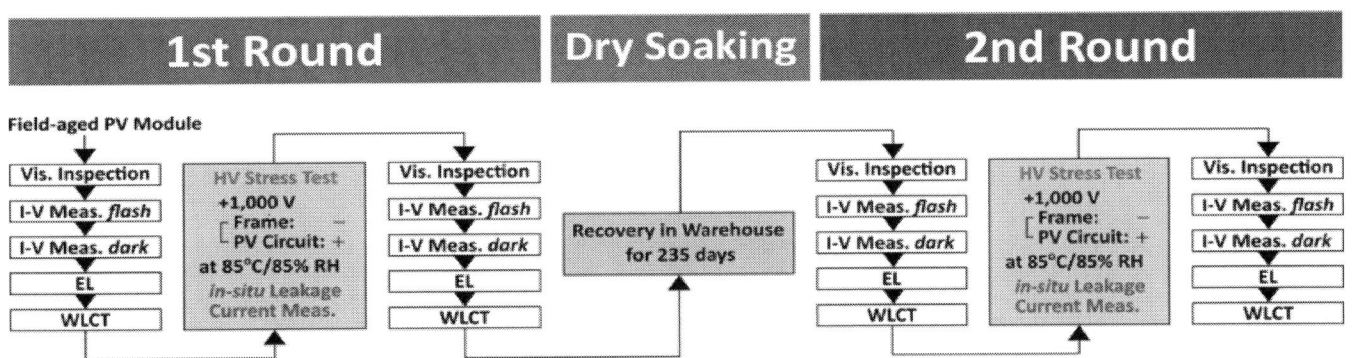

Electrical Safety Issue

Evolution of Electrical Insulation (WLCT*)			
Pristine Modules		**Field-aged PV Module (HV Stress Test)**	
		1st Round	2nd Round
> 0.12 GΩ·m²**	Pre	0.60 GΩ·m²	ND***
	Post	ND***	ND***

* Pass/Fail Criterion in WLCT: > 0.04 GΩ·m²

** Resistance in 2 sister PV modules

*** ND: Insulation resistance cannot be determined (with high current leakage)

Cumulative Transferred Charges during HV stress test

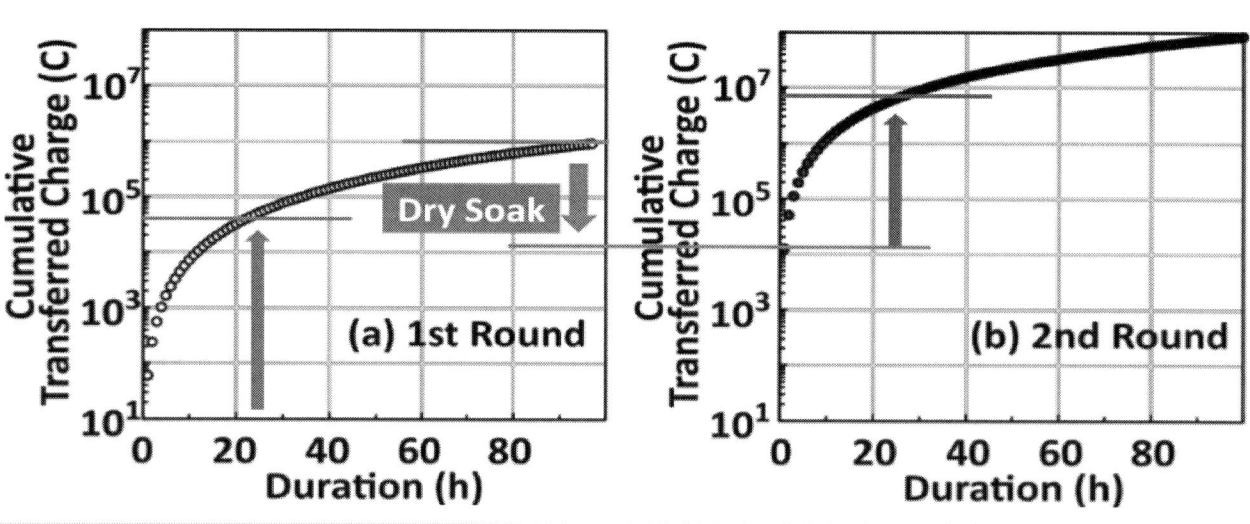

Evolution of I-V Curves with HV Stress

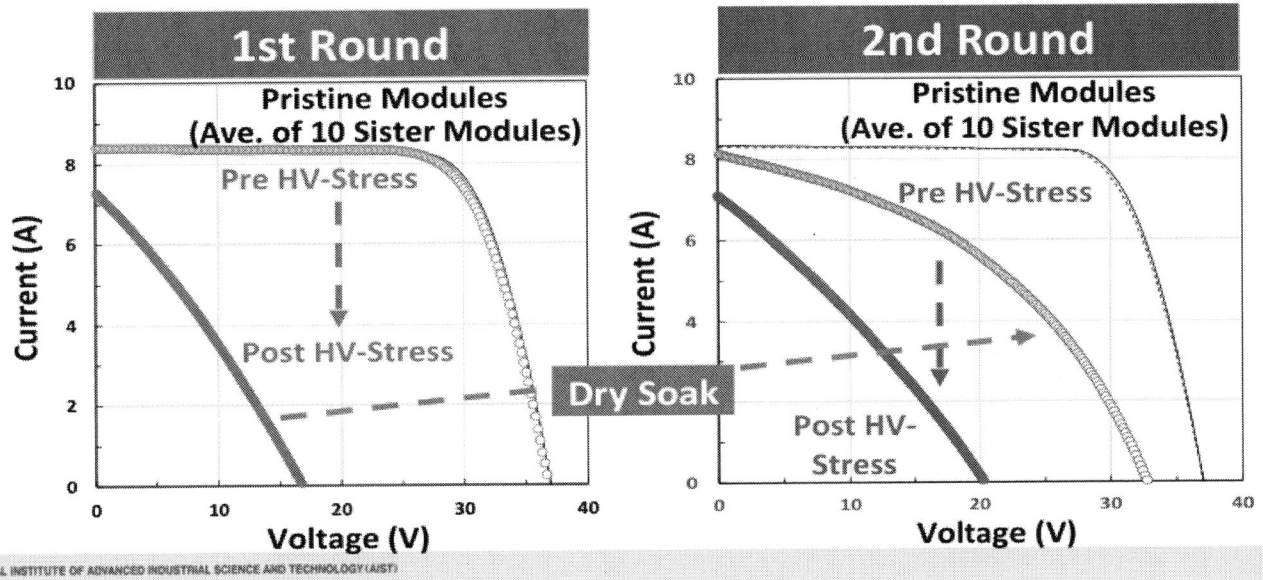

Evolution of EL Images with HV Stress

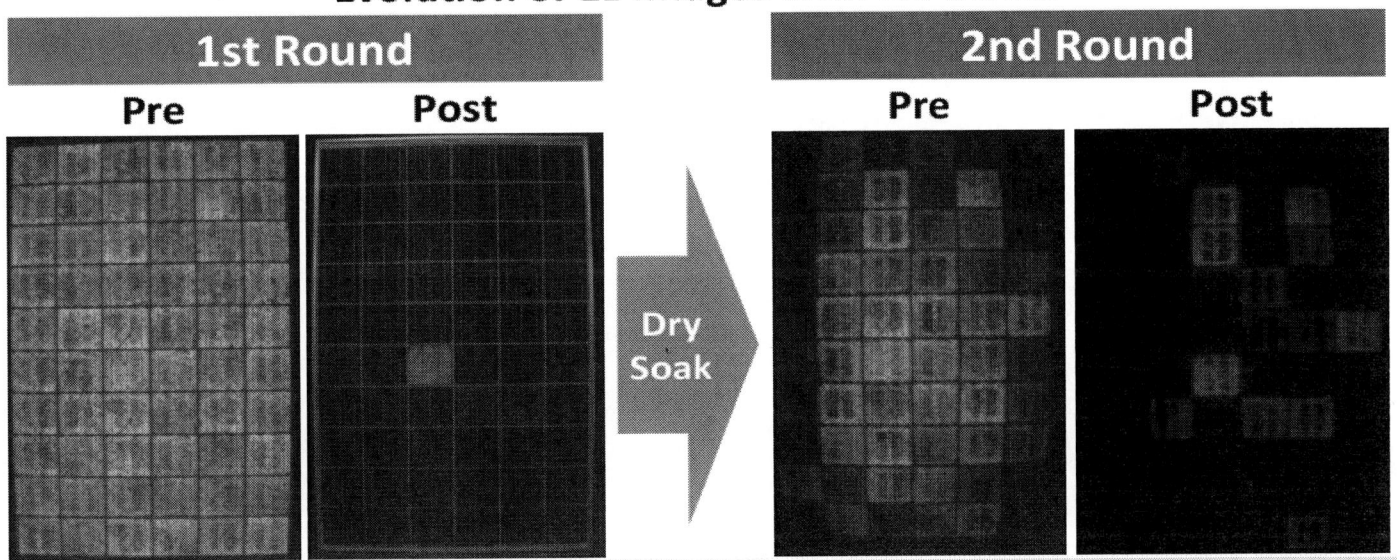

Summary:

Significant insulation issues were not detected by the Wet Leakage Current Test (WLCT), even in a fielded PV module with obvious backsheet cracks, despite previous literature recommending WLCT to identify electrical safety hazards.

- To address this limitation, we conducted a **High Voltage (HV) stress test**.
- We found the **HV stress test** to be effective for evaluating the electrical safety of insulation in PV modules, not just for detecting performance degradation under HV stress (similar to PID-s).

High cost of conducting HV stress tests for re-use inspections is a major concern.

- Introducing Acceptable Quality Limit (AQL) testing through random sampling of similar bulk PV modules (e.g., from the same PV plant with same model number) may help mitigate this issue.

Material Selection and Novel Reliability Testing for Floating Photovoltaic Modules

Nikoleta Kyranaki, Arvid Van der Heide, Hamed Javanbakht Lomeri, Ismail Kaaya, Sara Bouguerra, Jens D. Moschner, Arnaud Morlier, Michaël Daenen

Outline

- Why Floating PV?
- Which are the most critical PV module failure modes?
- Impact of mechanical load, thermal cycling and humidity
- Specific PV module design for floating applications
- Impact of thermal cycling, humidity and salt
- Conclusions
- Future Work

Why Floating PV?

- The availability of suitable land poses a limitation to further growth [1]
- Enhanced efficiency of PV modules due to reduced thermal impact [2]
- Minimal shading thanks to a more open area [3]
- Decreased evaporation loss

HOWEVER!

Different/more severe stressors:
Mechanical load due to wind and wave induced vibrations, moisture, salt (for off-shore applications)

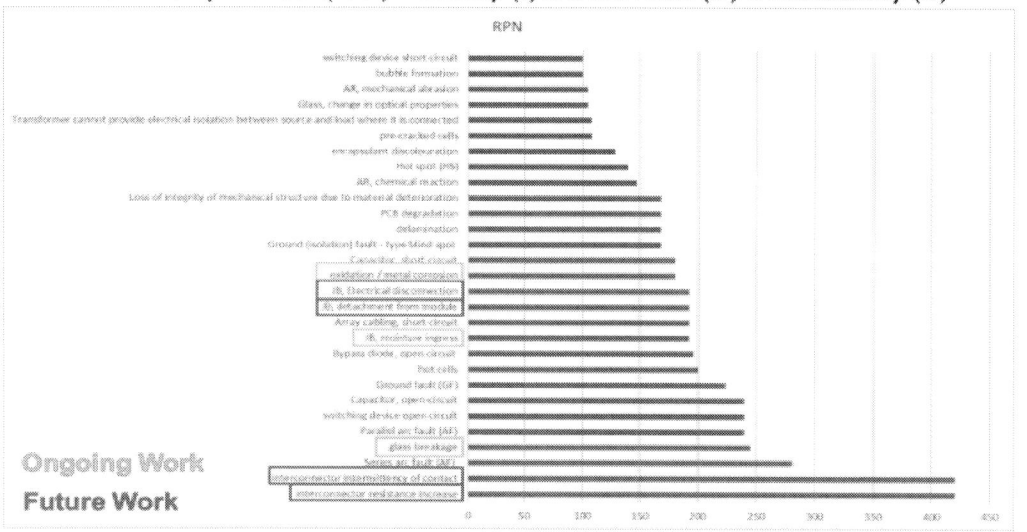

[1] 1.6 MW, Bergen, Norway. © BayWa r.e.
[2] H. Nisar et al., Solar Energy. 241, 231–247 (2022).
[3] https://www.offshore-energy.biz/dnv-and-moss-maritime-team-up-to-propel-floating-solar-technology-in-harsh-marine-environment/

Which are the most critical PV module failure modes?

Risk Priority Number (RPN) = Severity (S) x Occurrence (O) x Detectability (D)

- Interconnector failure
- Glass breakage
- Junction-box, moisture ingress, detachment and electrical disconnection
- Oxidation metal corrosion

Ongoing Work
Future Work

Glass-breakage study presented at EU PVSEC 2023 and available at:
Journal of Engineering Structures, https://doi.org/10.1016/j.engstruct.2024.118760

Impact of mechanical load, thermal cycling and humidity

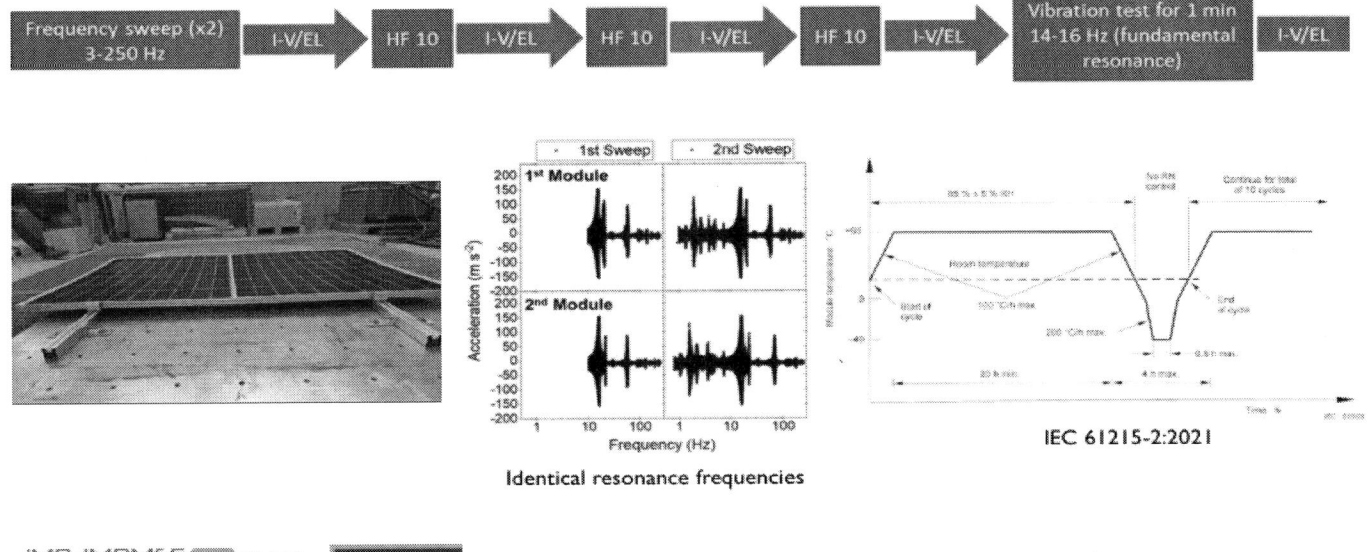

Identical resonance frequencies

IEC 61215-2:2021

Impact of mechanical load, thermal cycling and humidity: I-V curves

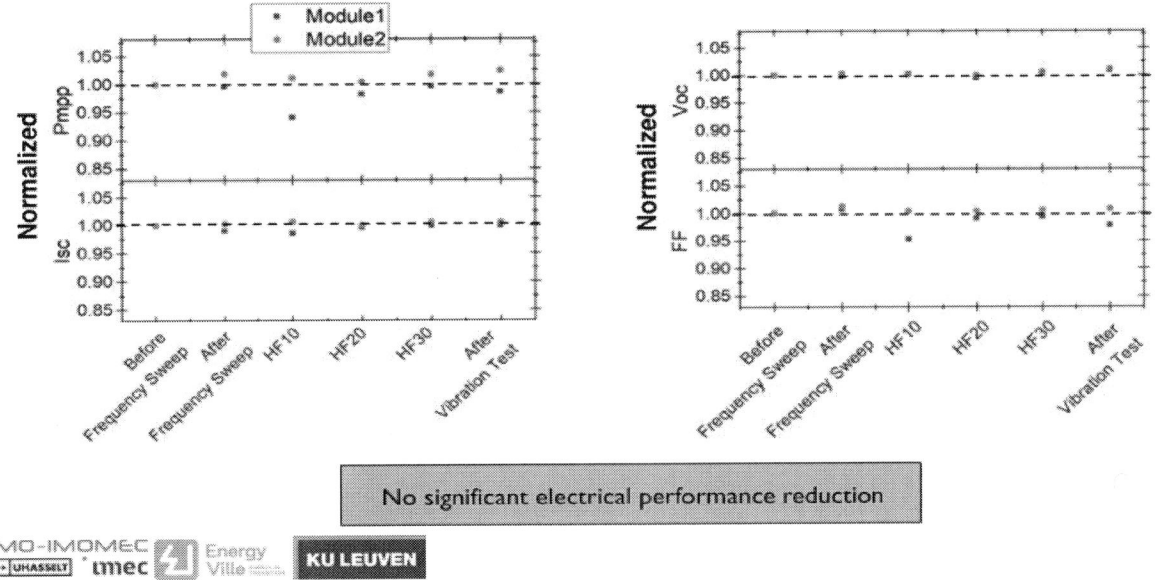

No significant electrical performance reduction

Impact of mechanical load, thermal cycling and humidity: EL imaging

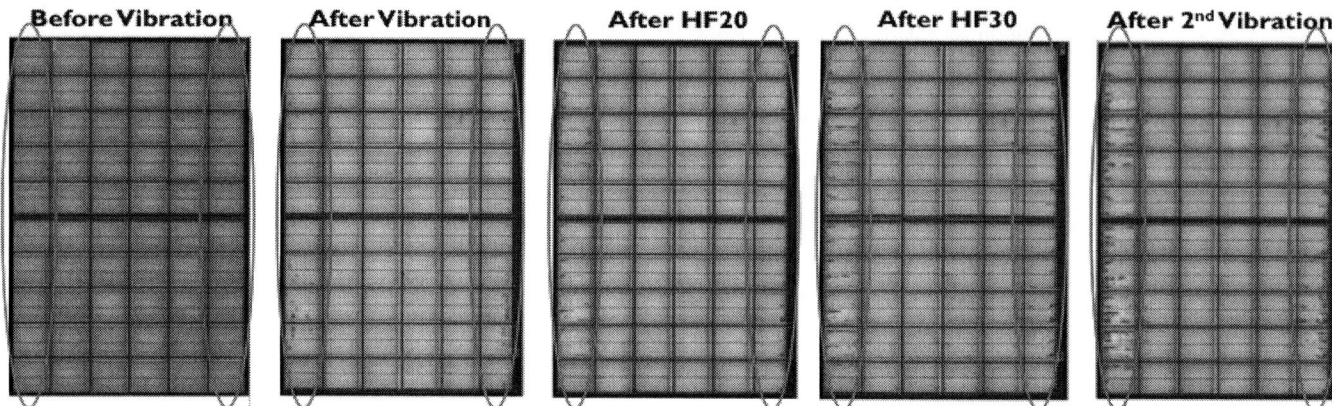

Finger detachment was observed after vibration which became more extended during the testing sequence.

Specific PV module design for floating applications

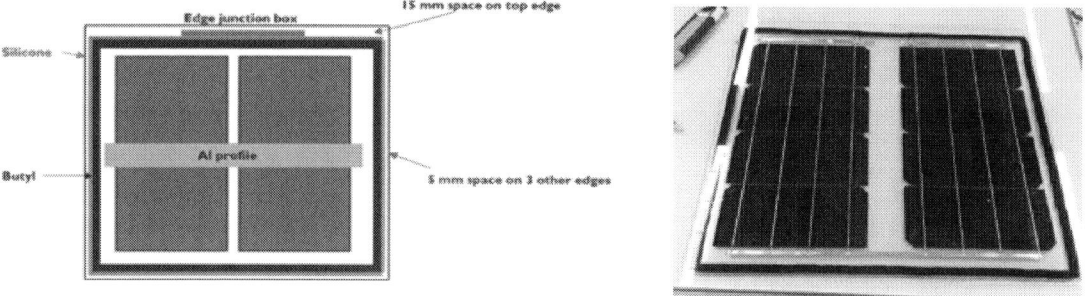

- Edge sealing of the junction box + silicone potting (IP68, no water ingress for 30 minutes up to 1.5 m depth)
- 4x2 coupons with 4BB PERC half cells
- Additional sealing of the edges of the PV modules (PVB+silicone) [1]
- Glass-glass encapsulation (40 x 40 cm²)
- No frame, addition of marine grade anodized aluminium profile glued to the rear surface
- Encapsulant with high resistivity, good adhesion, low moisture ingress and low stiffness (TPO)

[1] N. Roosloot et al. SOPHIA Workshop 2022.

Impact of thermal cycling, humidity and salt

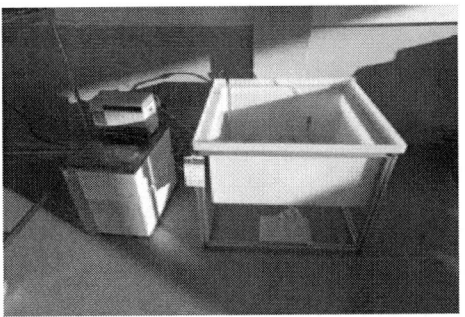

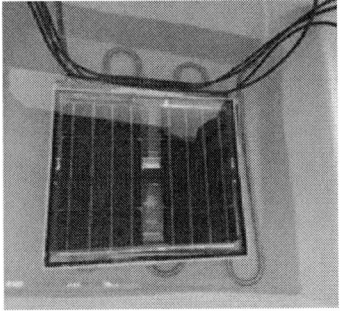

4 samples

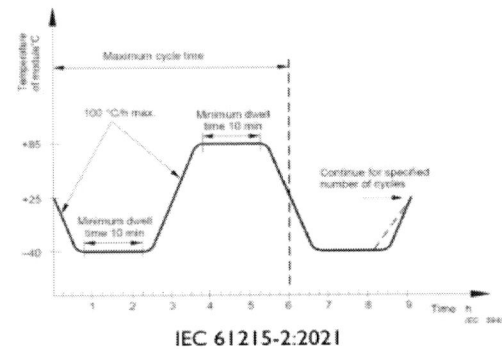

IEC 61215-2:2021

Impact of thermal cycling, humidity and salt: I-V curves

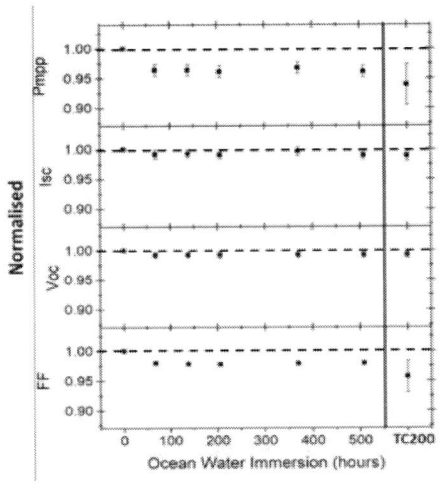

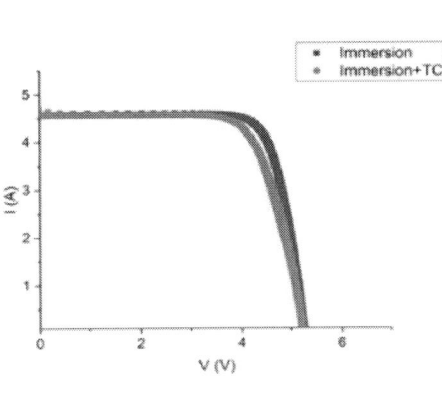

- Initial FF loss (after 69 hours of immersion) is observed because the measurement at 0 hours was conducted without a junction-box.
- Degradation of the FF by 0.043±0.026 after thermal cycling due to series resistance increase.

Impact of thermal cycling, humidity and salt: EL imaging

0 hrs

138 hrs

370 hrs

508 hrs

Before TC

After TC

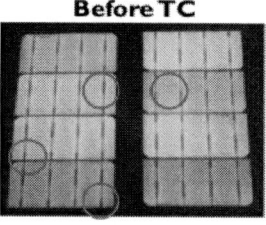
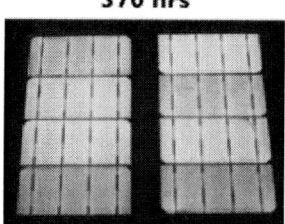
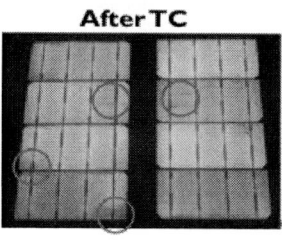

- Finger/ribbon detachment after thermal cycling which does not explain the severe FF reduction.
- Junction box failure may be occurring.

Impact of thermal cycling, humidity and salt: Visual inspection

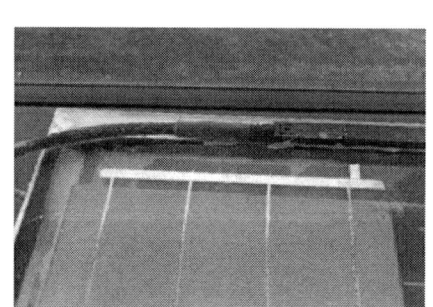

Salt accumulation after immersion but no visible corrosion on the aluminium profile

Glass crack on the perimeter after thermal cycling, not expected in commercial PV modules where the glass is usually tempered.

Conclusions

- Most severe **stressors** for floating PV: **Mechanical load** due to wind, and the waves-induced system movement, **moisture**, **salt** and **thermal cycling** allowing moisture condensation.

- Two new testing sequences:

  - **Sequence A** (commercial PV): combination of mechanical load with **humidity causes finger detachment near the clamping points.**

  - **Sequence B** (modified BoM): **thermal cycling**, **moisture** and **salt ingress/condensation** causes **finger/ribbon detachment, reducing the FF.**

- Selected **BoM**: double glass, double edge sealant, TPO encapsulation proven **adequate**

- Further **investigation** is required regarding **improvement** and the estimation of the **acceleration factor** of the test sequences

Future Work

- Comparison of the sequences to outdoor data (NauticalSunrise project), for sequence adaptation and acceleration factor estimation.

- Application of the sequences on TOP-Con and Tandem PV modules.

Acknowledgements

This work was conducted within the project "Marine Solar POtential and Technology Study" (MarineSPOTS) project and has been funded with the support of the Belgian Energietransitiefonds. Traveling to this conference is financed by FWO. Many thanks to Materials and Packaging Research & Sevices and Dimitri Adons for providing the facilities and helping with the vibration test. Additionally, we would like to thank Thijs Vandenryt for his help with the accelerometer and Jan Mertens, and Geert Doumen for designing and developing the mounting configuration for the attachment of the PV module to the vibration table, and the immersion setup.

Thank you for the attention!

nikoleta.kyranaki@uhasselt.be

  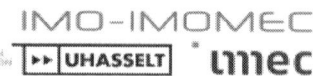

Scuola universitaria professionale della Svizzera italiana
Dipartimento ambiente costruzioni e design
Istituto sostenibilità applicata all'ambiente costruito
Laboratorio SUPSI PVLab

SUPSI

Outdoor Accelerated Ageing Test using Additional Thermal and Thermomechanical Stresses

Ebrar Özkalay[1], Gabi Friesen[1], Alessandro Virtuani[2], Mauro Caccivio[1], Christophe Ballif[2,3]

1 - SUPSI PVLab – University of Applied Sciences and Arts of Southern Switzerland, Mendrisio, Switzerland

2 - CSEM PV-Center, Neuchâtel, Switzerland

3 - EPFL – École Polytechnique Fédérale de Lausanne, Institute of Electrical and Micro Engineering, Photovoltaics and Thin-Film Electronics Laboratory, Neuchâtel, Switzerland

EUPVSEC, 23-27 September 2024
Vienna, Austria

SUPSI

Introduction

BIPV modules operate at harsher conditions

1. Elevated operating temperatures [1,2,3]
2. Larger diurnal (day-night) temperature changes [1,4]
3. More frequent partial shadow [5]

Long-term performance of BIPV modules

1. Faster photothermal degradation of encapsulant, resulting in I_{sc} loss [6]
2. Damaged cells and metallization, resulting in FF loss [6]
3. Discolouration, damaged metallization, glass breakage etc.

Outdoor Monitoring (Open-rack & BIPV)
Indoor Accelerated Ageing Tests
Outdoor Accelerated Ageing Tests

1. Shadow masks [7]
2. Heat blankets — **TODAY!**

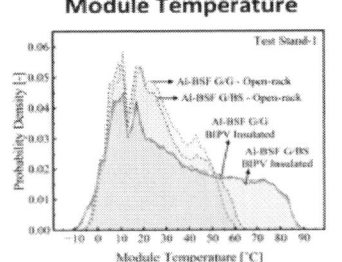

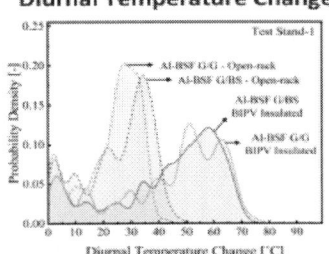

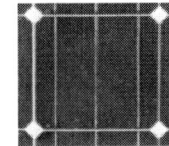

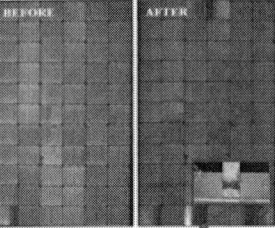

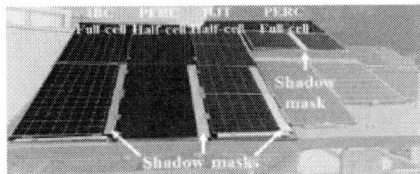

[1] E. Özkalay et al., IEEE JPV (2021)
[2] T. Nordmann et al., WCPEC (2003)
[3] T. Sample and A. Virtuani, EUPVSEC (2009)
[4] H. Hongjie et al., EUPVSEC (2018)
[5] A. Fairbrother et al. Solar RRL (2021)
[6] E. Özkalay et al. Energy and Buildings (2024)
[7] E. Özkalay et al. EPJ Photovoltaics (2024)

41st EUPVSEC: Outdoor Accelerated Ageing Test using Additional Thermal and Thermomechanical Stresses

Accelerated Ageing Test using Heat Blankets

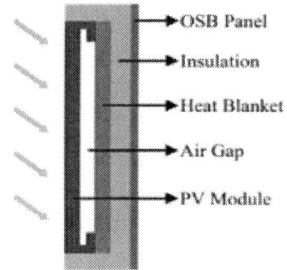

- OSB Panel
- Insulation
- Heat Blanket
- Air Gap
- PV Module

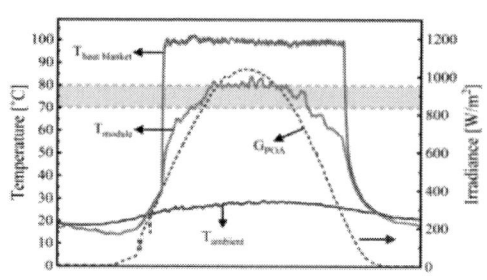

Initial Experiments of the design

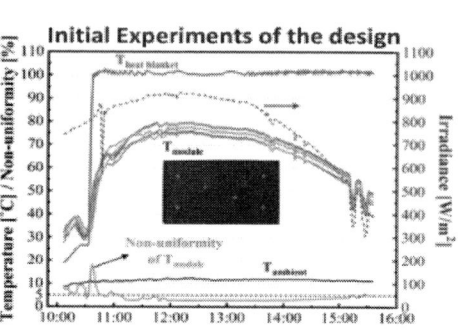

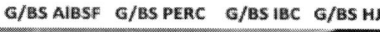

Open Rack

G/BS PERC   G/BS IBC   G/BS HJT

BIPV Insulated

G/BS PERC   G/BS IBC   G/BS HJT   G/BS AlBSF

Heat Blanket

G/BS AlBSF   G/BS PERC   G/BS IBC   G/BS HJT

41st EUPVSEC: Outdoor Accelerated Ageing Test using Additional Thermal and Thermomechanical Stresses

Temperature Analysis

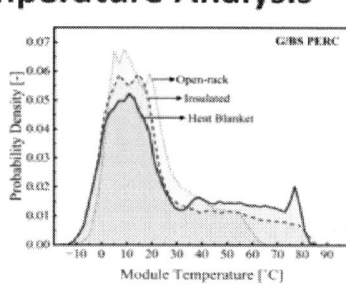

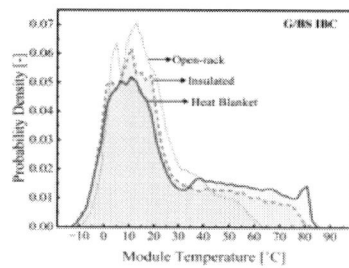

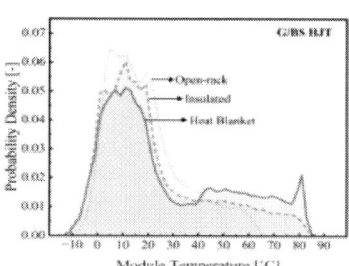

Temperature Analysis

Module Temperature

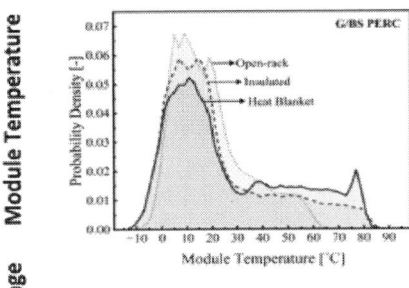
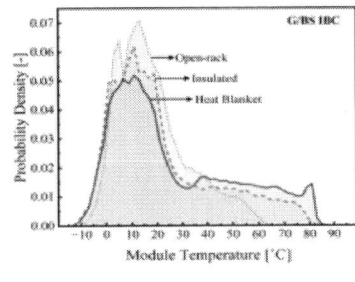
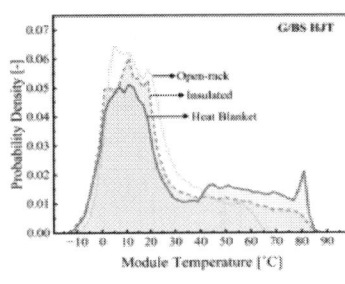

Diurnal Temperature Change

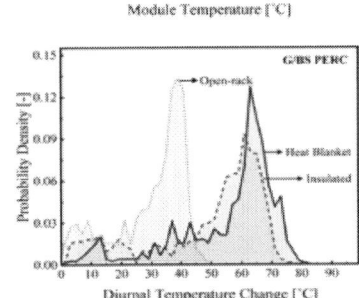
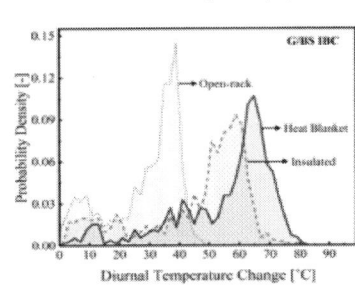
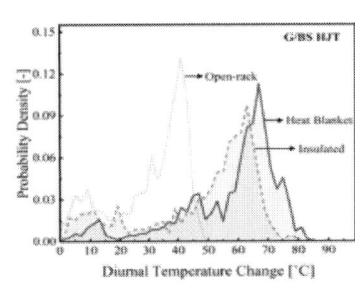

Acceleration Factor

Acceleration Factor – Thermal Stress

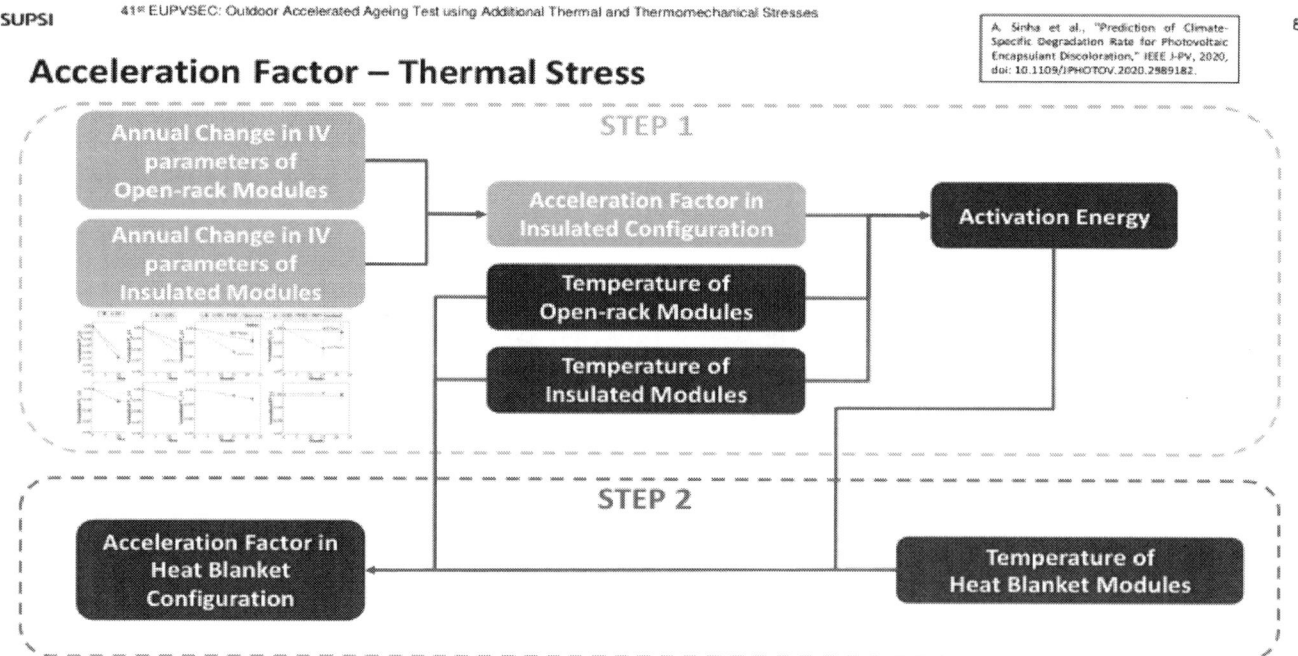

Acceleration Factor – Thermal Stress

A. Sinha et al., "Prediction of Climate-Specific Degradation Rate for Photovoltaic Encapsulant Discoloration," IEEE J-PV, 2020, doi: 10.1109/JPHOTOV.2020.2989182.

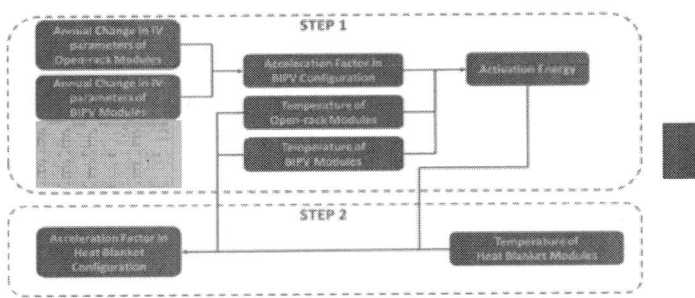

Heat Blanket versus	Module Technology	AF in P_m due to larger thermal stress (modelled) [-]
Open-rack	G/BS PERC	3.8
	G/BS IBC	1.85
	G/BS HJT	1.71
BIPV Insulated	G/BS PERC	1.42
	G/BS IBC	1.23
	G/BS HJT	1.21

- AF is **1.7 to 3.8** with respect to **open-rack** configuration
- AF is **1.2 to 1.42** with respect to **insulated** configuration

Acceleration Factor – Thermomechanical Stress

- Formula proposed by N. Bosco et al. to simulate solder fatigue damage

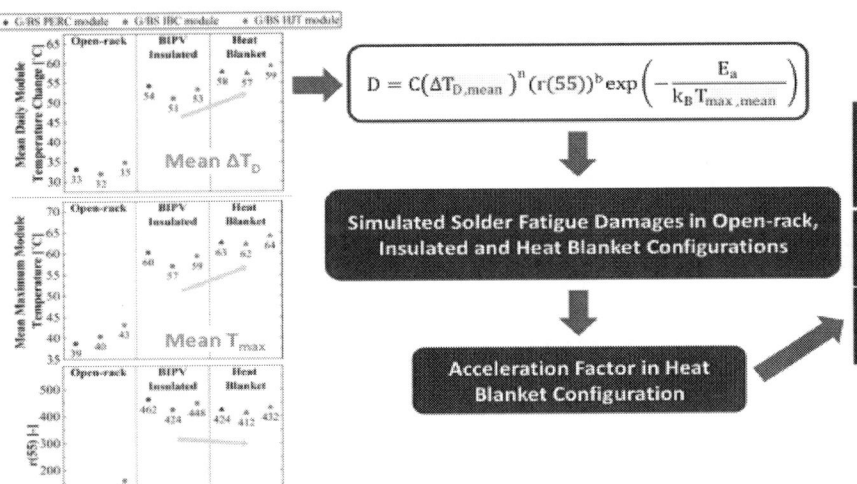

$$D = C\left(\Delta T_{D,mean}\right)^n \left(r(55)\right)^b \exp\left(-\frac{E_a}{k_B T_{max,mean}}\right)$$

Simulated Solder Fatigue Damages in Open-rack, Insulated and Heat Blanket Configurations

Acceleration Factor in Heat Blanket Configuration

Heat Blanket versus	Module Technology	AF in solder fatigue damage due to larger thermomechanical stress (modelled) [-]
Open-rack	G/BS PERC	1.9
	G/BS IBC	2.1
	G/BS HJT	1.63
BIPV Insulated	G/BS PERC	1.01
	G/BS IBC	1.05
	G/BS HJT	1.02

N. Bosco et al., "Climate Specific Thermomechanical Fatigue of Flat Plate Photovoltaic Module Solder Joints," Microelectronics Reliability, 2016, doi: 10.1016/j.microrel.2016.03.024.

RESULTS – G/BS Al-BSF Module

Indoor STC Analysis – G/BS Al-BSF

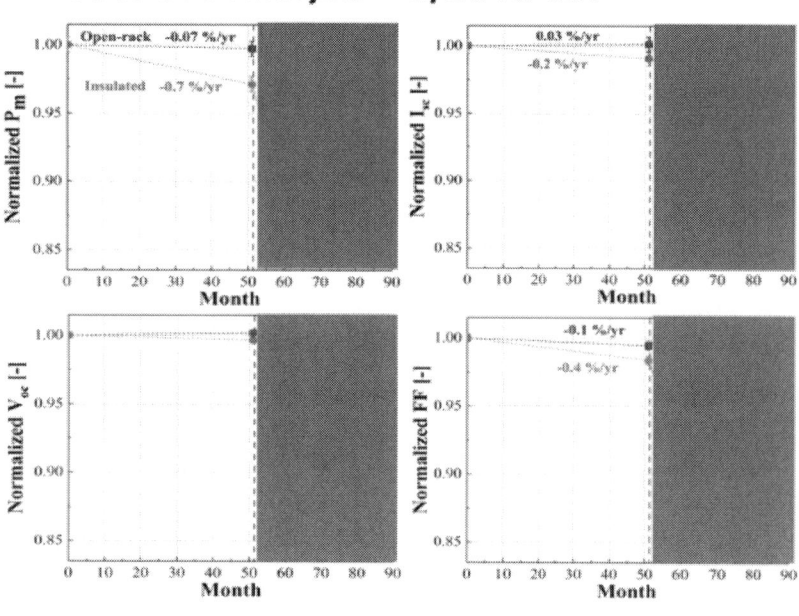

- Faster photothermal degradation of encapsulant, resulting in I_{sc} loss

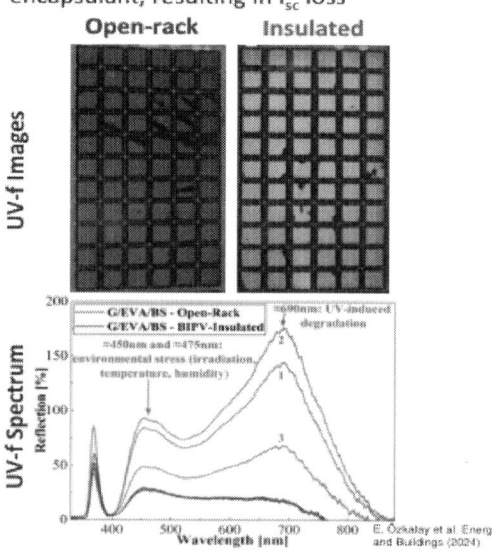

Indoor STC Analysis – G/BS Al-BSF

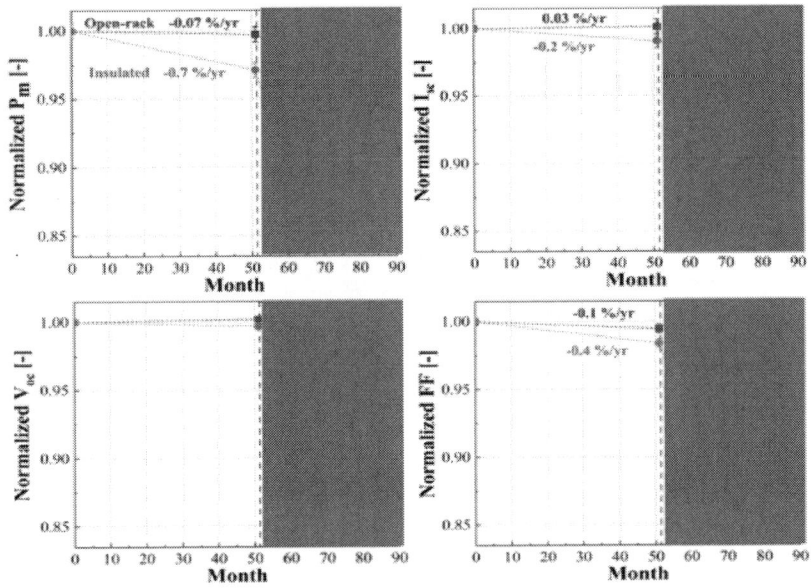

- Faster photothermal degradation of encapsulant, resulting in I_{sc} loss
- Damaged cells and metallization, resulting in FF loss

Insulated

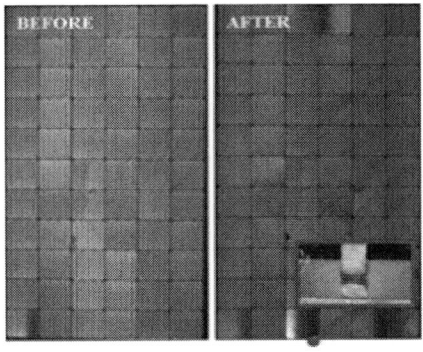

BEFORE | AFTER

E. Özkalay et al. Energy and Buildings (2024)

Indoor STC Analysis – G/BS Al-BSF

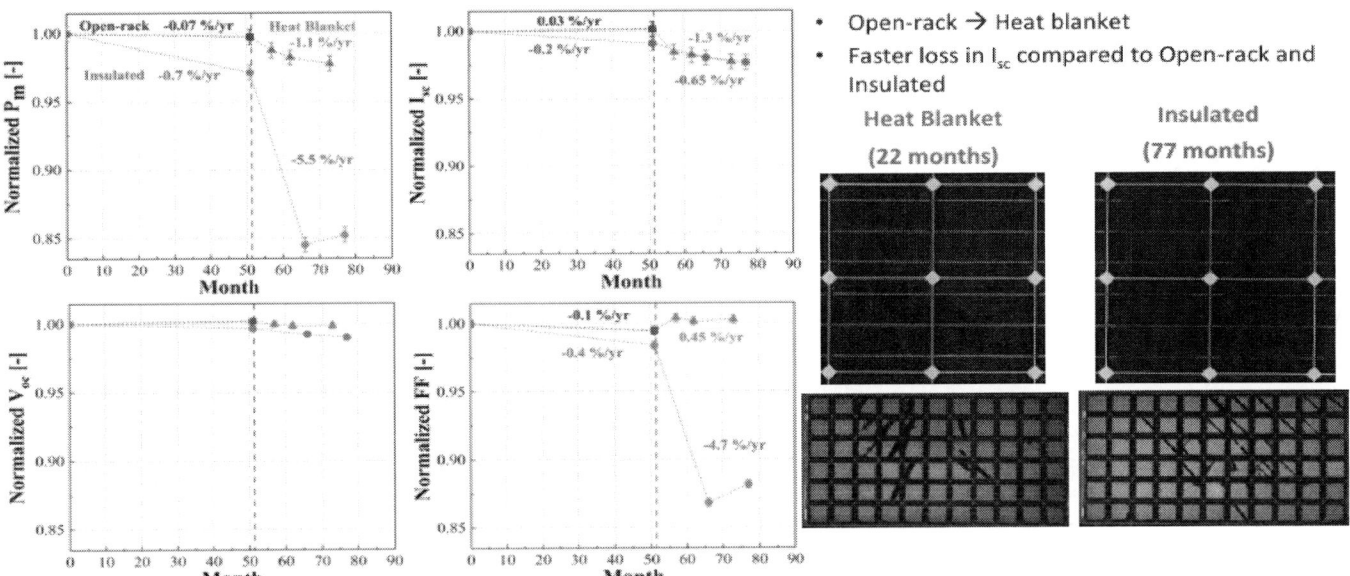

- Open-rack → Heat blanket
- Faster loss in I_{sc} compared to Open-rack and Insulated

Heat Blanket (22 months)

Insulated (77 months)

Indoor STC Analysis – G/BS Al-BSF

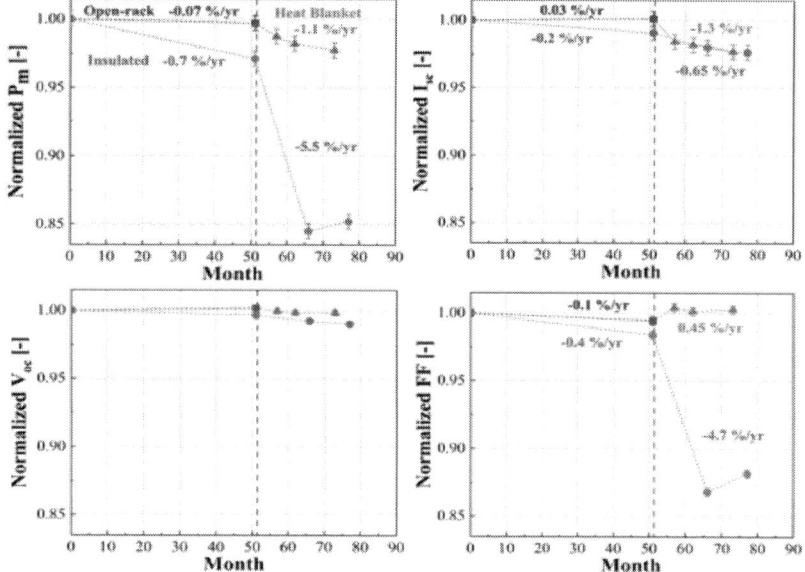

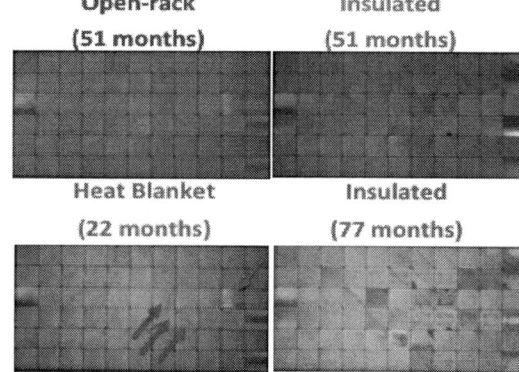

- Open-rack → Heat blanket
- Faster loss in I_{sc} compared to Open-rack and Insulated
- No loss in FF, only few new cell cracks
- Insulated module – Corrosion of contacts, faster FF loss

Open-rack
(51 months)

Insulated
(51 months)

Heat Blanket
(22 months)

Insulated
(77 months)

RESULTS – G/BS PERC Module

Indoor STC Analysis – G/BS PERC

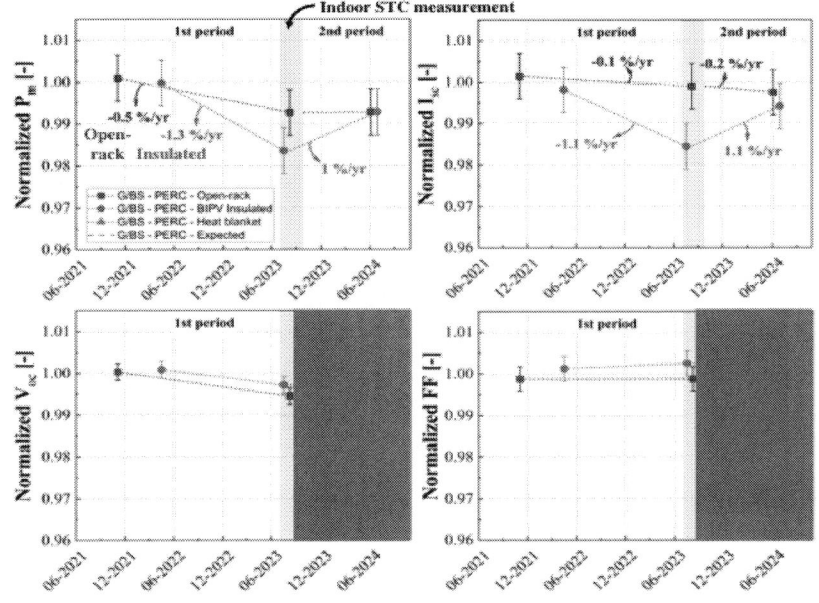

Indoor STC Analysis – G/BS PERC

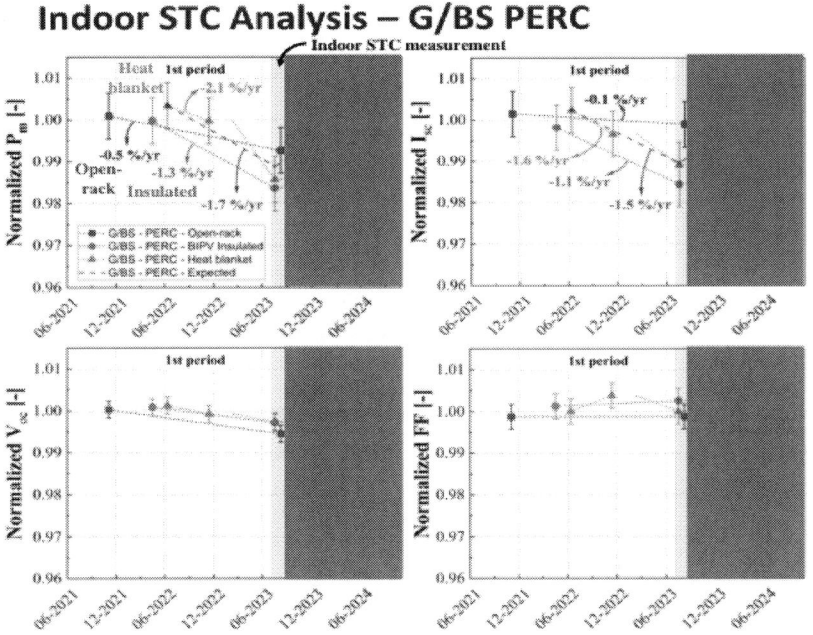

Indoor STC Analysis – G/BS PERC

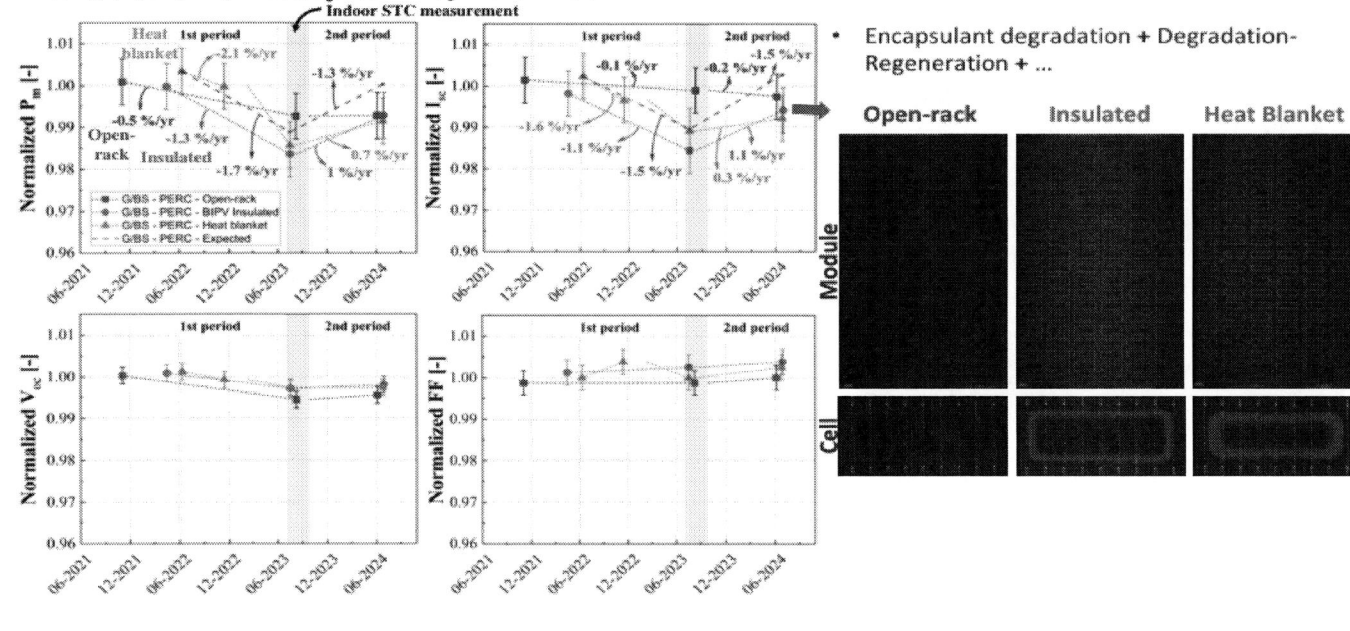

- Encapsulant degradation + Degradation-Regeneration + ...

Indoor STC Analysis – G/BS PERC

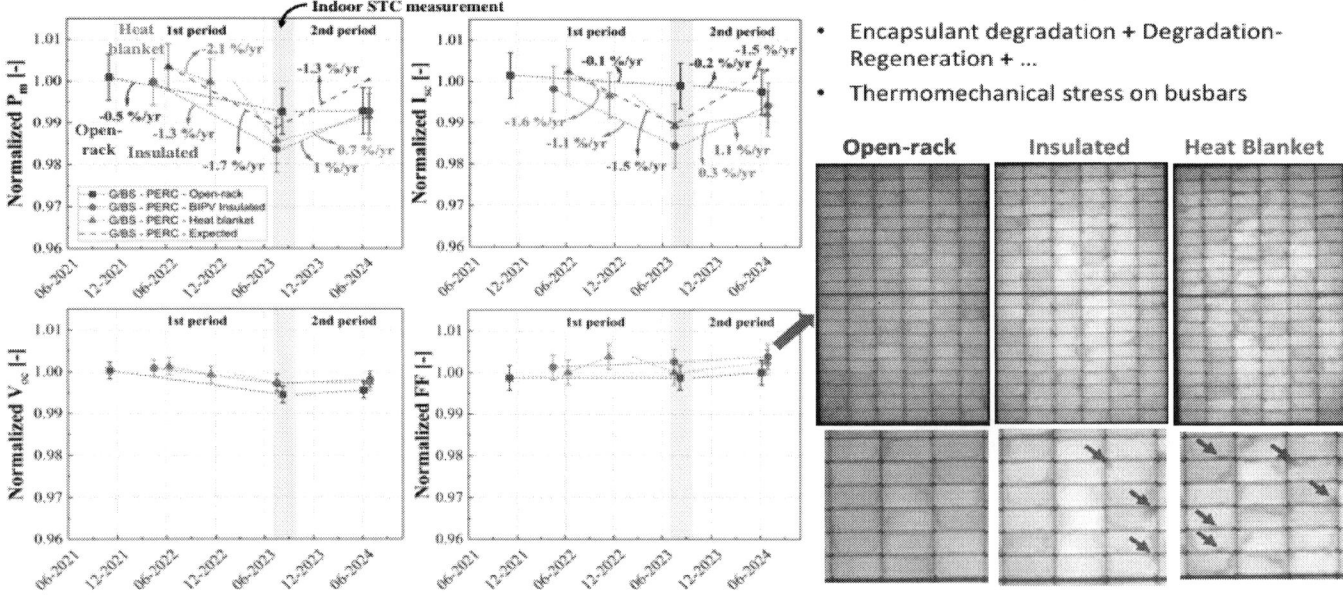

- Encapsulant degradation + Degradation-Regeneration + ...
- Thermomechanical stress on busbars

RESULTS – G/BS HJT Module

Indoor STC Analysis – G/BS HJT

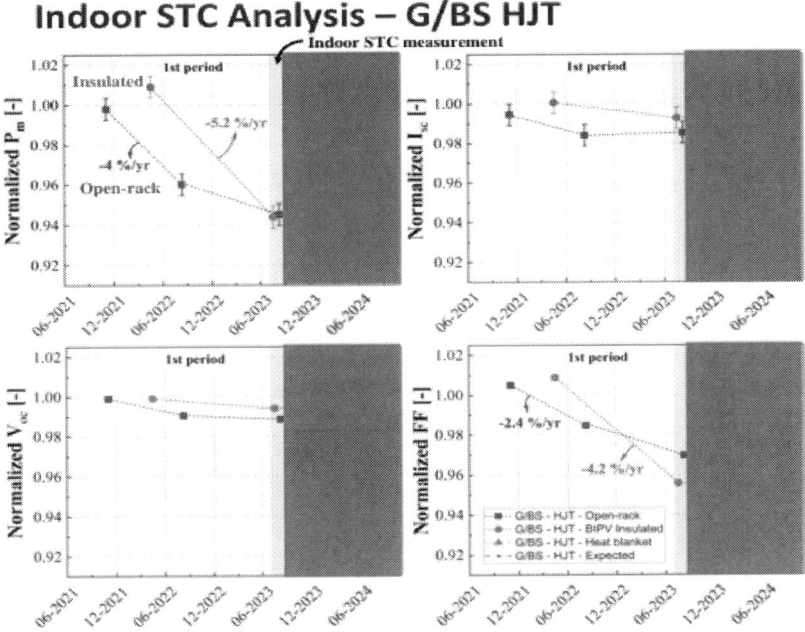

- Faster FF degradation in insulated configuration

Indoor STC Analysis – G/BS HJT

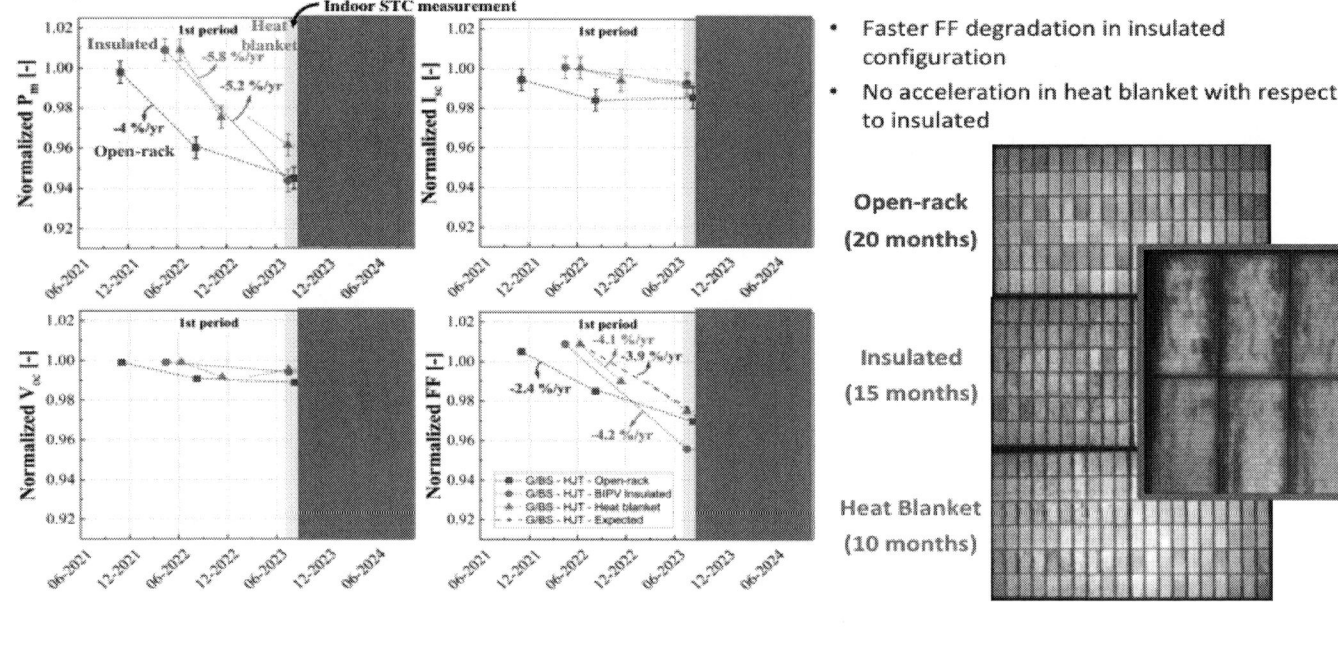

- Faster FF degradation in insulated configuration
- No acceleration in heat blanket with respect to insulated

Open-rack (20 months)

Insulated (15 months)

Heat Blanket (10 months)

Indoor STC Analysis – G/BS HJT

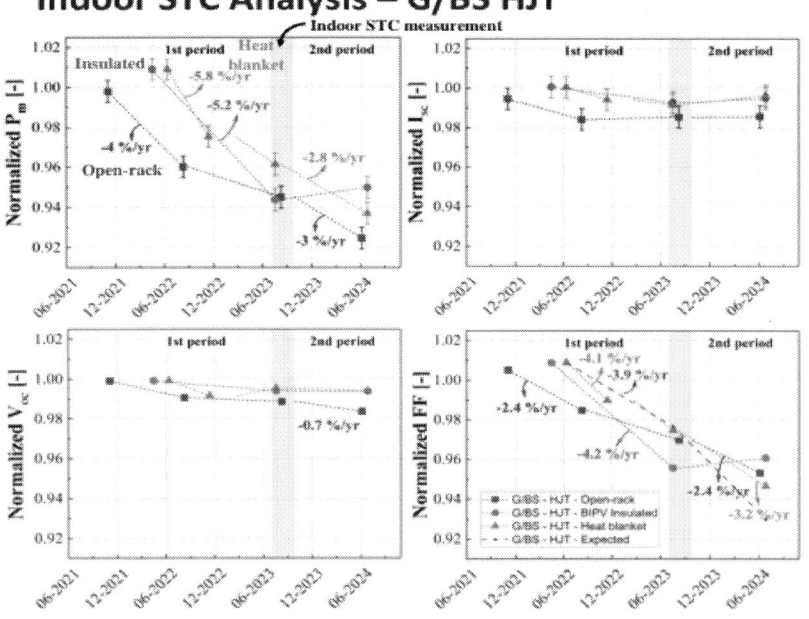

- Faster FF degradation in insulated configuration
- No acceleration in heat blanket with respect to insulated
- Stabilization of I_{sc} and V_{oc}
- Humidity ingress

Open-rack

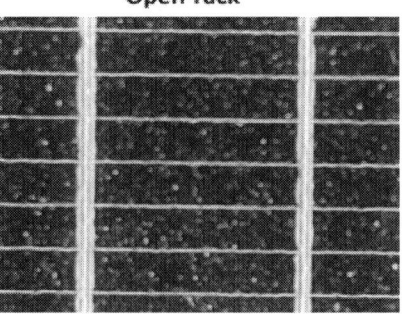

RESULTS – G/BS IBC Module

Indoor STC Analysis – G/BS IBC

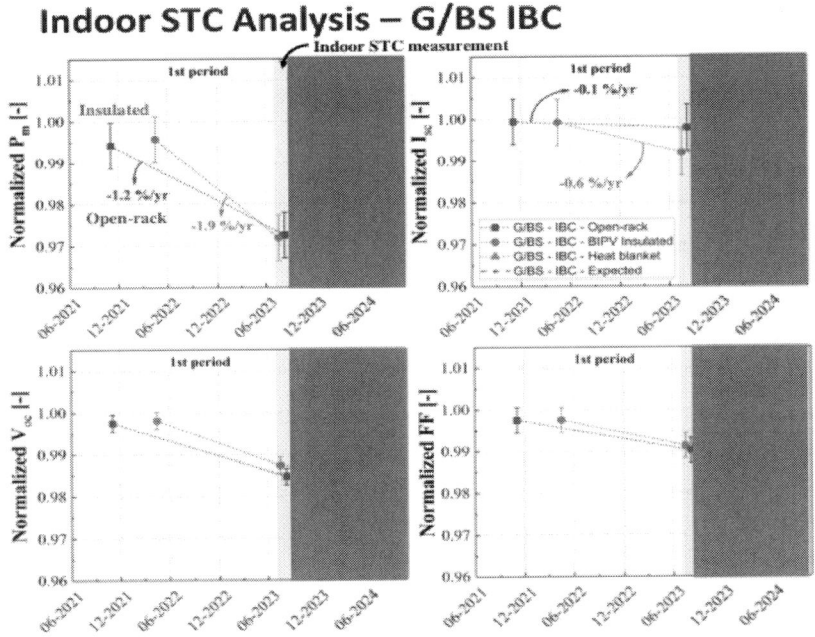

Indoor STC Analysis – G/BS IBC

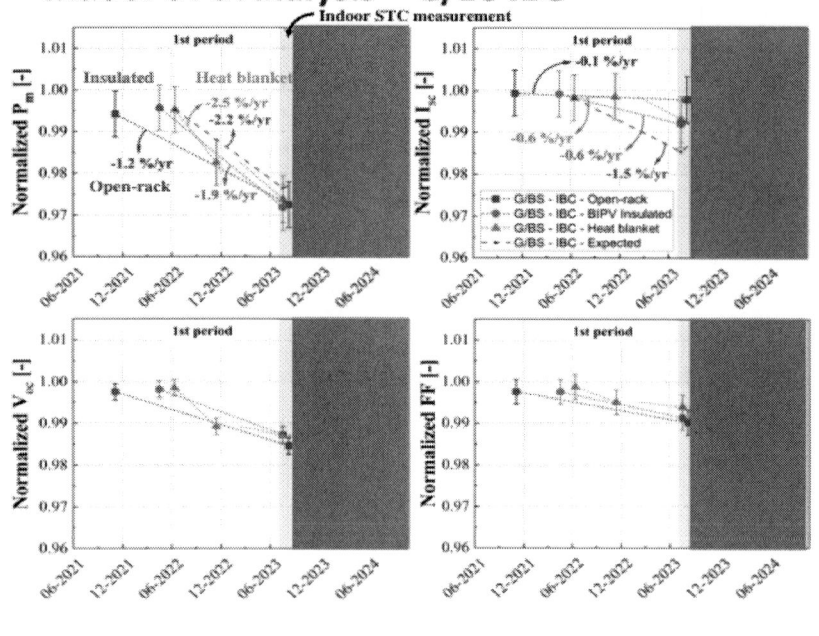

- Accelerated degradation versus open-rack module but not versus insulated module
- Slightly accelerated P_m degradation but not because of I_{sc}
- Slightly faster V_{oc} degradation – non-linearity?

Indoor STC Analysis – G/BS IBC

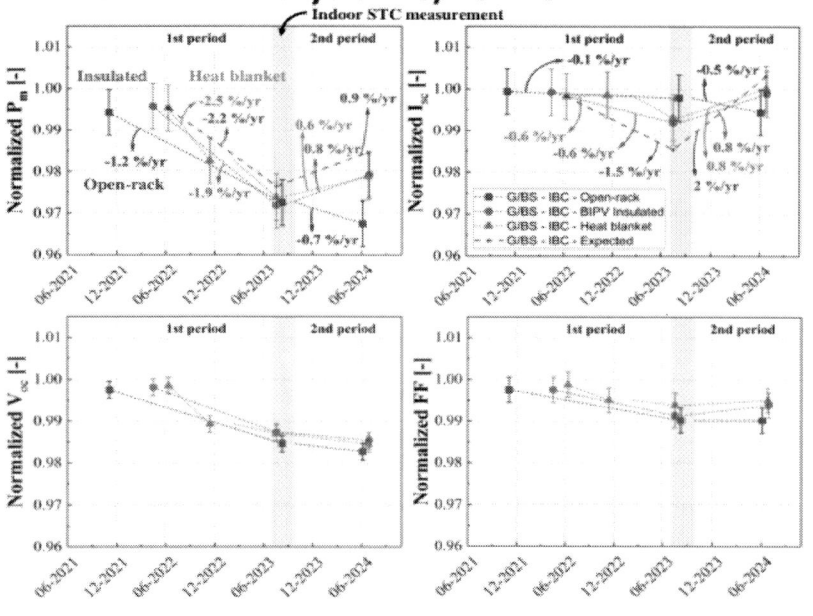

- Accelerated degradation versus open-rack module but not versus insulated module
- Slightly accelerated P_m degradation but not because of I_{sc}
- Slightly faster V_{oc} degradation – non-linearity?
- Regeneration of I_{sc} ?
- No difference between EL and UV-f images

Conclusions and Outlook

Conclusions and Outlook

Benefits

- Combined testing (i.e. irradiance, temperature, etc.)
- Increase in thermal and thermomechanical stresses with respect to open-rack, ranging from 1.7 to 3.8 and 1.6 to 2.1, respectively
- Promising results for some degradation routes (e.g. I_{sc} for Al-BSF and PERC)

Challenges

- Larger thermomechanical stress with respect to insulated modules
- Humidity ingress into the module from their back may be limited due to the heat blanket concept
- Stability of PERC, IBC and HJT technologies

Limitation

- Controlling dosage of environmental stresses like indoor testing

Outlook

- Longer monitoring
- Indoor testing for further understanding on stability of PERC, IBC and HJT
- Outdoor data analysis ➡

3DO.17.5, Thursday 10:30, Gabi Friesen, "Performance and Degradation Evaluation of c-Si Modules under Different Open-rack and Residential Mounting Configurations"

SUPSI

Thank you for your attention!

Acknowledgement:
This work is funded by the **Swiss National Science Foundation** under COST IZCOZO_182967/1 and **Swiss Federal Office of Energy**, BFE.
SUPSI PVLab and **SUPSI ISAAC Engineering team** researchers and personnel

SUPSI PVLab
Virtual Tour:

Ebrar Özkalay
ebrar.oezkalay@supsi.ch

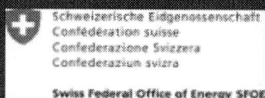

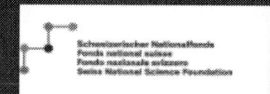

www.pvlab.solar

26/09/2024

Development of PV Module Hot Desert Test Cycle Protocol Extended Failure Modes and Effective Analysis

Baloji Adothu[1,*], Jim Joseph John[1], Gerhard Mathiak[1], Vivian Alberts[1], Bengt Jäckel[2], Ralph Gottschalg[2], Narendra S Shiradkar[3], Amir A. Abdallah[4], Juan Lopez Garcia[4], Michael Salvador[5], Bram Hoex[6], Hussein A Kazem[7], and Muhammad Ashraful Alam[8]

1DEWA Research & Development Center, MBR Solar park, P.O.Box: 564 Dubai, United Arab Emirates
2Fraunhofer Center for Silicon Photovoltaics CSP, Otto-Eissfeldt-Str. 12, 06120 Halle (Saale), Germany
3National Center for Photovoltaic Research and Education, Indian Institute of Technology Bombay, Mumbai, Powai, Maharashtra 400076, India
4Qatar Environment and Energy Research Institute, Researchery (HBKU Research Complex), Education City, PO Box 34110, Doha, Qatar
5King Abdullah University of Science & Technology, KAUST Solar Center, Thuwal, 23955, KSA
6School of Photovoltaic and Renewable Energy Engineering, University of New South Wales, Sydney, NSW, Australia
7Faculty of Engineering, Sohar University, PO Box 44, Sohar, PCI 311, Oman
8Department of Electrical and Computer Engineering, Purdue University, West Lafayette, IN, USA.

* Corresponding author: Phone +971 522578980 | baloji.adothu@dewa.gov.ae

Abstract— **More and more GW-scale PV power plants are being installed in the hot desert regions of the world. The large space, high irradiance, and low maintenance can achieve the low cost (OPEX and CAPEX). However, high temperatures, strong temperature fluctuations, and UV radiation can lead to the PV module's performance loss. The aim of the work is to develop an accelerated aging test standard protocol for the selection of PV modules for desert application. Accelerated test procedures for the reproduction of typical failures in the hot desert will be proposed and verified. To avoid different regulations in different countries, it is planned to develop an internationally recognized standard that complements the existing PV standards.**

This study presents measured degradation rates alongside a Failure Mode and Effectiveness Analysis (FMEA), including an assessment of Desert Future Risk Priority Numbers (DF-RPN) as evaluated by international global PV experts. The average degradation rate across all desert locations is approximately 1.63% per year. The top three failure modes with the highest DF-RPN are UV-induced degradation (UVID), interconnect or finger breakage, and light- and elevated temperature-induced degradation (LETID). These failure modes, which primarily affect solar cells, are triggered by environmental factors such as UV irradiation, thermomechanical stress, and light exposure at elevated temperatures. To address these issues, a hot desert testing cycle (HDTC) has been **proposed, incorporating increased UV exposure, thermal cycling, and LETID testing. Additionally, the HDTC includes assessments for mechanical load, sand abrasion, and exposure to extreme temperature variations to better simulate harsh desert conditions. This comprehensive test cycle is designed to mitigate the effects of UV irradiation, thermomechanical stress, and high temperatures in desert environments. The HDTC protocol is intended for both, present and future PV module technologies used in power plants.**

Keywords—Desert climate, Failure modes and Effective Analysis, Desert testing standards, PV modules

I. INTRODUCTION

The Middle East's hot desert climate presents ideal conditions for harnessing solar energy due to its abundant solar irradiation and vast expanses of available land [1]. Countries in the region, such as Dubai, are increasingly adopting photovoltaic (PV) technology for both residential and large-scale utility power plants. Dubai, in particular, has laid out ambitious goals through its Clean Energy Strategy, aiming to source 75% of its energy from clean and renewable sources by 2050. One of the key initiatives driving this transition is the Mohammed Bin Rashid

Al Maktoum Solar Park, which is set to become the largest single-site solar park globally.

A PV module consists of several key components, collectively known as the Bill of Materials (BOMs), which determine its durability and performance over time. However, the performance of these PV modules can degrade over time, particularly in harsh climates like the desert, where high temperatures, UV radiation, and soiling are significant stress factors [2,3]. The type of BOM used in the PV module plays a crucial role in how quickly the module degrades. Typical degradation modes include encapsulant discoloration, delamination, corrosion, snail trails, chalking, hotspots, interconnect and finger breakages, discoloration, and cell cracks [4]. These defects can substantially impact the module's efficiency and overall lifespan, especially in desert environments where stressors are more intense compared to moderate climates. As the installation of large-scale PV plants in desert regions grows, the low operational costs (OPEX) and capital expenditure (CAPEX) make these projects more economically viable. However, the extreme conditions of the desert, such as significant temperature fluctuations and intense UV exposure, contribute to accelerated failure modes and performance loss in PV modules. Current PV standards, such as the IEC 61215 series, are primarily designed for moderate climates and fall short in addressing the unique challenges posed by desert environments. Specific guidelines for selecting and testing PV modules under desert conditions are lacking, making it difficult for manufacturers and project developers to ensure long-term performance in these regions. To address this gap, this study aims to develop an accelerated aging test protocol that simulates the specific failure modes commonly observed in desert climates.

The proposed accelerated aging test procedures will reproduce typical desert-induced failures such as thermal stress, UV degradation, and soiling-related issues. These tests will help in the development of an internationally recognized standard tailored for desert environments, complementing existing PV standards like IEC 61215. This new standard will provide a unified framework for selecting and testing PV modules, helping manufacturers design more robust modules for desert applications and reducing the risk of premature failures. Furthermore, a Failure Modes and Effects Analysis (FMEA) will be used to evaluate the severity, occurrence, and detectability of different failure modes under desert conditions. This methodical approach will provide insights into how these failures affect system performance and reliability. By systematically assessing desert-specific failure modes, the study will support the formulation of the Hot Desert Test Cycle (HDTC) sequence, which will serve as a benchmark for PV module testing in desert climates.

The primary objective of this work is to develop a systematic methodology for creating the Hot Desert Test Condition (HDTC) protocol, which will address the specific challenges faced by PV modules in desert environments. The initial proposal for this protocol is founded on the collaborative input of various global solar energy institutes with extensive experience in hot desert climates. Their collective expertise of FMEA, combined with data obtained from experiments and field observations, will form the basis of the HDTC protocol.

II. METHODOLOGY

The main objective of this work is to outline a methodology for the development of the HDTC protocol. The first proposal is shown in Fig. 1. This approach is based on the views of various global solar institutes with expertise in hot desert regions. Their opinions combined with the failure modes obtained through experiments and field observations will contribute to the design of the proposed HDTC protocol. By considering the dominant degradation mechanisms that occur in hot desert regions, this collaboration aims to establish a thorough FMEA discussion and approach for developing testing standards to improve the durability and reliability of commercial and next-generation c-Si solar module technologies. Four common approaches for the discussion and development of HDTC sequences in six global hot desert areas are presented below; Observation and assessment of failure modes and degradation rates from global hot desert regions.

- Evaluation of degradation rate and FMEA/RPN for desert-related failure modes
- Discussion and correlation of RPN for the development of HDTC sequences
- Verification of the proposed round-robin tests.

FMEA evaluates each failure mode's severity, occurrence, and detectability to determine how those causes will affect the system. A methodical evaluation of severity, occurrence, and detection factors related to desert failure modes strongly underpins the formulation of the HDTC sequence proposal. This HDTC proposal suggests an accelerated aging test to target failure mode observed in the hot desert environment, specifically designed for testing in desert locations. The primary emphasis lies in assessing FMEA and RPN for failure modes induced by desert conditions. The evaluation of the RPN is obtained by multiplying the severity, occurrence, and detection rankings (RPN= S * O * D), each falling within the range of 1 to 10 [5–7]. For this purpose, various PV experts were asked to participate. They were asked for their assessment and feedback, simply in numbers (see Table 1) from 1 (e.g. low, simple) to 10 (e.g. extremely critical), on

- Severity in terms of power degradation or safety risks
- Occurrence in terms of the number of failure cases
- Detection in terms of the ability to detect a failure during the field operation, characterization, design, and production phase to minimize the impact or correct it before delivery.

- Desert relevance is the acceleration of a failure effect in a hot desert compared to a moderate climate.
- Future relevance is the importance of a failure type for future module and cell technologies. E.g. Perovskite-Tandem, TOPCON, PERC, Big-Size modules, Bifacial modules.

Generally, a higher RPN value signifies that specific defects predominantly impact system performance. For the design of the test sequences of HDTC, the extended FMEA (see Table 1) contains failure root causes, the proposal for accelerated testing, and the relevance for hot deserts.

Tab. 1 Criteria for the Evaluation of Severity, Occurrence, Detection, Desert, and Future Relevance.

Points	Severity	Occurrence	Detection	Hot Desert Relevance	Future Relevance
1	**Insignificant**	**Very unlikely**	**Simple**	**Low**	**Less**
	Barely noticeable, no impact on performance or safety	The cases are very unlikely	FM is detected in any case by common measures so that a very quick correction is possible	Less strong in the hot desert	New technologies will less prone
2 to 3	**Minor**	**Unlikely**	**Moderate**	**Similar**	**Constant**
	Aesthetic, very minor effect on performance, no impact on safety, not covered by warranty	Some modules under very specific conditions are affected	Procedures are in place to enable early detection (end of line and/or a few days of lab test)	Similar in moderate and in hot desert climates	No effect of new technologies
4 to 6	**Serious**	**Low**	**Difficult**	**Slightly more**	**Slightly more**
	A measurable influence on performance, degradation rate at warranty level, minor impact on safety.	Some modules under special conditions are affected	Detectable only with accurate and comprehensive testing	Slight acceleration due to the hot desert climate	The failure mechanism can be more relevant to new technologies
7 to 8	**Critical**	**High**	**Very difficult to detect**	**Significant**	**Significant**
	Measurable and significant impact on performance, degradation rate well above warranty, impact on electrical safety or fire.	Affects many modules	Not detectable with standard QA measures. Process monitoring provides no clues. Small changes in a recipe can cause major changes	Significant acceleration and severity in hot desert conditions	It is expected that the new technologies will be prone
9 to 10	**Extremely critical**	**Very high**	**Unlikely to detect before delivery**	**Strong**	**Strong**
	This can lead to serious injury or death or complete replacement of the modules	Affects all products of the same type	There are practically no effective early detection methods.	Hot Desert environment greatly accelerates the failure mechanism	Future technologies will be strongly vulnerable

III. RESULTS AND DISCUSSIONS

1. Degradation rate and failure modes

The findings and descriptions of different hot desert places across the world that are relevant to the FMEA analysis and HDTC sequence talks are listed below.

- The harsh conditions of the UAE hot desert, marked by elevated ambient and module temperatures, substantial daily temperature fluctuations, intense UV radiation, moderate humidity, and increased soiling rates, result in notable performance deterioration. The degradation rate of PV modules from the UAE DEWA hot desert, as determined by Rd tools, is around 1.44% annually, with a confidence interval of -1.52 to -1.40 percent annually (Fig. 1). Primary issues include discoloration, followed by interconnect defects such as thermomechanical fatigue, rusting, and discoloration of fingers and busbars. Additionally, backsheet-related problems like chalking, cracking, and delamination's contribute to the challenges[4]. PERC modules experienced significant power degradation within just one year of testing, possibly attributed to the LeTID signature, evident in the dark pattern observed in EL images[8].

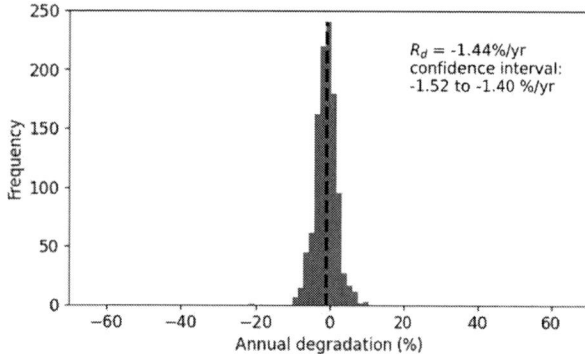

Fig. 1 RdTools degradation rate of mono c-Si PV modules from DEWA OTF facility

- In Indian hot desert climates, degradation rates (1.04%/a) surpass those in non-hot zones (0.56%/a). These elevated rates align with specific failure modes, notably an increased occurrence of encapsulant discoloration and delamination[9]. Additionally, there is a lesser but observable connection between metallization defects and snail trails. The Indian hot desert is characterized primarily by greater UV radiation and higher ambient and module temperatures.

- The harsh climate of Qatar's hot desert has a substantial impact on PV module performance, resulting in an annual loss of 1.59%/a[10]. This is attributed to harsh conditions including elevated ambient and module temperature, increased daily temperature variations, higher UV irradiance, and higher soiling rates. Predominant failure modes in modules operated in Qatar's hot desert involve encapsulant discoloration, followed by interconnect issues affecting fingers and busbars. These interconnect challenges encompass thermomechanical defects, discoloration, and corrosion[10]. Additionally, backsheet problems like chalking, cracking, and delamination further contribute to the performance degradation of PV modules in this region.

- The challenging KSA hot desert climates pose stability issues for next-generation c-Si and tandem perovskite PV modules[11,12]. The choice of encapsulant material is pivotal for stabilizing these advanced PV modules. Encapsulation has been associated with significant challenges, including pronounced yellowing and catastrophic failures[11,12].

- The PERC, TopCon, and HJT modules experience issues with robust front metallization failures, including interruptions and corrosion of fingers/busbars caused by hot climates [13]. The Australian hot desert can have a notable impact on thermomechanical failures and corrosion in novel c-Si technology, particularly affecting TopCon and SHJ modules[13,14].

- The soiling is the main issue along with other harsh weather conditions of Oman's hot desert. The daily power loss is around 0.08 to 0.1% per day[15,16]. Large variations in daily module temperature also showed an impact on power reduction. This can lead to thermomechanical defects formation and cementation of soiling on the module surface. Cementation and corrosion were found near the frame area and higher soiling area of the PV modules. The soiling is more common in Oman and other MENA hot desert regions and needs a suitable cleaning test protocol for the long-term operation of PV modules[15,16].

- The simulation study revealed that solder bond and finger failures are more dominant in hot desert climates than the non-hot climates[17,18]. The daily large temperature variation leads to the development of thermomechanical defects. This can lead to increased solder bond failure and shorten the solder bond life (20-25 N/S)[17,18]. Developing a temperature cycle test specific to hot desert conditions is essential for assessing and predicting the thermomechanical flaws.

- The degradation rates of mono c-Si PV modules in the hot desert environments of the UAE and Qatar have been compared with those in various climates, including the tropical climate of India [19], the cold desert of the Gobi [20], the Indian cold desert [21], and regions with moderate climates [22]. Additionally, they have been compared with other hot desert regions such as Algeria [19], Arizona [23], Oman [24], India [25], and Egypt, [26]. On average, the degradation rate from all these desert locations is approximately 1.63% per year.

2. Failure modes and Effective analysis (FMEA)

The global PV experts were also asked for their feedback on the following points.

- Lack of relevant failure modes of PV modules.

- Preference for the hot desert test as a design test (approx. 2 months test duration) or lifetime test (approx. 6 months test duration)
- Ambiguities in the definitions of Table 1 (severity, ...)

Most of the questionnaires were fully completed, although some failure modes were omitted for various reasons. The average values for all 18x5 fields of the questionnaire were calculated, as shown in Table 2. The DF-RPN value is determined by multiplying the factors of severity, occurrence, detection, desert, and future relevance.

The top three failure modes with the highest DF-RPN are UVID, interconnect/finger breakage, and LETID. These failure modes are associated with solar cells and are triggered by factors such as UV irradiation, thermomechanical stress, and light irradiation combined with elevated temperatures. As a result, a hot desert test should incorporate increased UV exposure, thermal cycling, and a LETID test.

Table. 2 Ranking of the Failure Modes considering the desert future RPN (DF-RPN). The Top-1 column gives the number of experts suggesting the particular failure mode on rank 1. A shared first rank is divided under the failure modes.

S.no	Failure modes	Severity	Occurrence	Detection	Desert	Future	DF-RPN	Top1
1	UVID	5.95	6.50	5.15	7.30	6.95	10101	3.5
2	Interconnects/finger breakage	5.73	6.09	5.55	6.59	5.91	7534	0.7
3	LETID	5.62	5.29	4.95	7.14	6.10	6404	3.5
4	Hot spot due to partial shading	7.14	5.14	4.41	6.09	5.55	5459	1
5	Solder bond breakage	6.32	4.95	5.09	6.32	5.36	5401	0.5
6	Solar cell cracks	6.27	5.41	4.64	6.00	5.68	5363	0.5
7	Solar glass abrasion/ARC removal	4.50	5.91	5.23	7.09	5.41	5331	1
8	Encapsulant delamination	6.23	5.14	4.68	6.43	5.43	5226	1
9	Corrosion of metallization and interconnects	5.87	5.39	5.17	5.78	5.35	5063	1.5
10	Encapsulant discoloration	4.64	5.68	4.77	7.00	5.68	5001	3
11	PID	6.50	4.95	5.05	5.41	5.55	4874	2
12	Backsheet cracking. embrittlement	5.73	5.32	4.64	6.91	4.59	4479	3
13	Cables. cable ties and connectors	5.86	5.43	4.38	5.76	5.33	4281	0.5
14	J-Box and bypass diode failures within the junction box	7.33	4.48	4.33	5.71	5.10	4142	0.2
15	Solar glass breakage	7.23	4.68	3.77	4.91	5.82	3646	1
16	Backsheet chalking	3.68	5.27	4.14	6.09	4.29	2096	0
17	Snail trails	3.45	4.27	4.59	4.95	3.95	1328	0

| 18 | Frame pieces. module clamps | 3.16 | 3.05 | 2.79 | 3.58 | 3.37 | 324 | 0 |

3. Hot desert test cycle (HDTC) sequence:

- Through a desert UV test sequence (Fig.2), the first test leg seeks to simulate the encapsulation browning/photobleaching process, and the second test leg seeks to replicate the backsheet flaws such as discoloration, cracking, delamination, and embrittlement.
- Through desert mechanical stress sequence (Fig.2), the first leg seeks to test the PV modules desert induced thermomechanical fatigue (of interconnects, fingers, and busbars) and the second test leg seeks to replicate the thermomechanical fatigue, backsheet delamination, and cracks.
- Through a desert sand abrasion sequence (Fig.2), the first leg seeks to test the PV modules blowing sand-induced scratches and abrasion. The second leg seeks to replicate the daily robot cleaning-induced damages such as the removal of ARC coating and scratches.

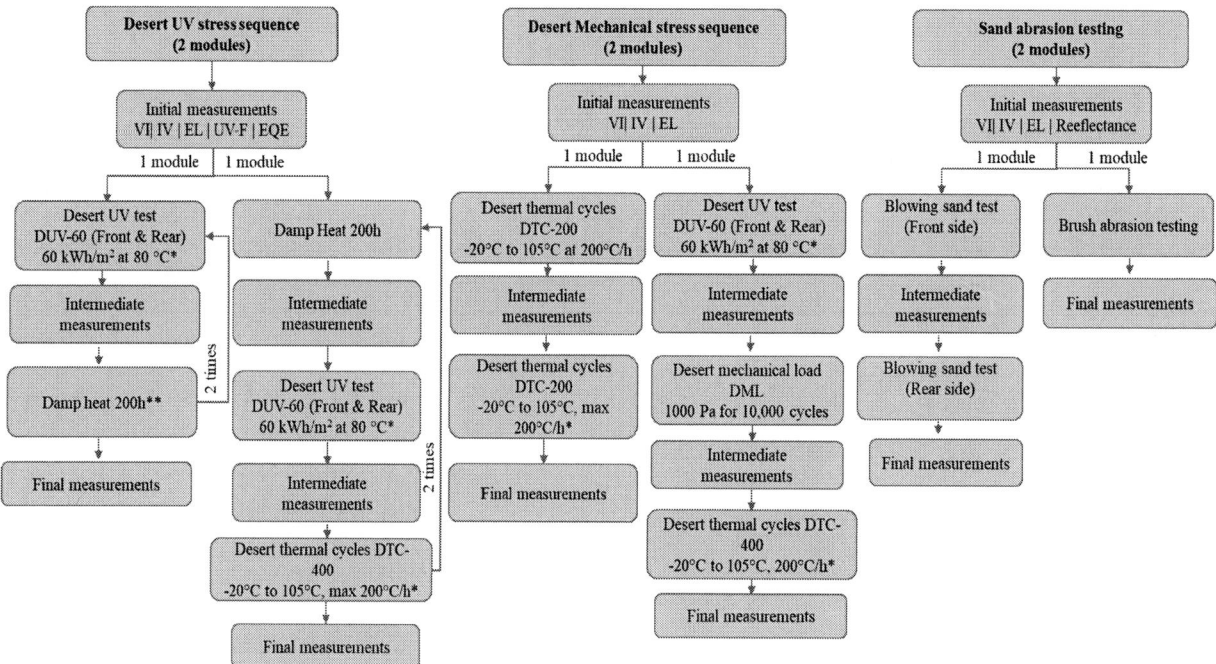

Fig.2 Hot desert test cycle (HDTC) sequence flow chart: Desert UV stress sequence (2 modules), Desert mechanical stress sequence (2 modules), and Desert sand and brush abrasion testing (2 modules)[27][28].

4. Correlation of desert future RPN for the development of HDTC sequences

Fig. 3 illustrates the suggested assessment for each failure mode, incorporating insights from expert opinions, initial observations, and experiences. A high RPN indicates the prevalence of specific defects in desert regions. The RPN values specifically highlight the dominance of encapsulant discoloration, interconnect breakages/interruptions, backsheet defects, glass abrasion, and cell degradation in the desert environment. To effectively capture and address these defects

induced by desert conditions, the HDTC protocol is put forth (Fig. 1). Proposing a prioritized approach based on RPN values, consider initiating desert UV stress testing as a primary step, followed by subsequent desert mechanical stress testing, desert sand, and brush testing in the sequence to enhance the overall performance of PV modules in desert climate.

Proposed Desert UV Stress testing aims to simulate encapsulation discoloration and backsheet issues like cracking embrittlement, and delamination. A corresponding Desert Mechanical Stress sequence is suggested to replicate thermomechanical defects, and a combination of thermomechanical fatigue, cell cracks, and backsheet delamination and cracking. The desert sand and brush abrasion sequence aims to mimic damages induced by blowing sand, such as scratches and abrasions, as well as damages from daily robot cleaning, including the removal of ARC coating and scratches.

IV. CONCLUSIONS

With the increasing installation of gigawatt-scale PV power plants in hot desert regions, these areas present several advantages, including abundant land, high solar irradiance, and lower maintenance costs. However, the harsh desert climate poses significant challenges, such as extreme temperatures, large temperature fluctuations, and intense UV radiation, all of which can contribute to performance degradation in PV modules. The average yearly degradation rate in desert regions is approximately 1.63%/year. Failure modes in the desert are different than in the moderate climate. Failure Modes and Effects Analysis (FMEA) will be employed to assess the severity, occurrence, and detectability of various failure modes specific to desert conditions. This systematic approach will provide valuable insights into how these failures impact the overall performance and reliability of PV systems. UVID, interconnect/finger breakage, and LETID were found to be the three most common failure modes in an extended FMEA on 18 failure modes pertinent to hot desert settings. These failure modes are brought on by UV irradiation, thermomechanical stress, and high temperatures. A hot desert test cycle (HDTC), which includes testing for UV exposure, temperature cycling, mechanical load, sand, and abrasion, has been proposed as a solution to these problems. This sequence is designed to improve the overall performance and durability of PV modules in desert conditions. Continued research on FMEA and the development of the HDTC test conditions and sequence is essential for ensuring long-term reliability and establishing qualification standards for PV modules in desert climates. The HDTC methodology is presently undergoing experimental validation with the goal of improving the dependability of PV modules in arid regions.

REFERENCES

[1] J. Ascencio-Vásquez, K. Brecl, M. Topič, Methodology of Köppen-Geiger-Photovoltaic climate classification and implications to worldwide mapping of PV system performance, Solar Energy 191 (2019) 672–685. https://doi.org/10.1016/j.solener.2019.08.072.

[2] B. Adothu, S. Mallick, P. Kartikay, Determination of Crystallinity, Composition, and Thermal stability of Ethylene Vinyl Acetate Encapsulant used for PV Module Lamination, in: Conference Record of the IEEE Photovoltaic Specialists Conference, Institute of Electrical and Electronics Engineers Inc., 2019: pp. 491–494. https://doi.org/10.1109/PVSC40753.2019.8981151.

[3] B. Adothu, R. Pugstaller, M. Tiefenthaler, F. Reny Costa, S. Mallick, G.M. Wallner, Crosslinking Kinetics of Photovoltaic Module Encapsulants – Investigation of Selected EVA and POE Grades, in: 38th Edition of the European Photovoltaic Solar Energy Conference and Exhibition (EU PVSEC-2021), 2021: pp. 779–783. https://doi.org/10.4229/EUPVSEC20212021-4AV.1.35.

[4] Baloji Adothu, Sagarika Kumar, Bengt Jaeckel, Neha Lyka Muttumthala, Z. Shekason, David Daßler, Kaushal Chapaneri, Prashanth Gabbadi, Yogesh Kumar, Ahmad Alheloo, Ali Almheiri, Jim Joseph John, Gerhard Mathiak, Vivian Alberts, Ralph Gottschalg, Identification and Investigation of Materials Degradation in Photovoltaic Modules from Middle East Hot Desert, in: EU PVSEC 2023, 2023: pp. 001–004. https://doi.org/10.4229/EUPVSEC2023/3AV.2.29.

[5] A. Colli, Failure mode and effect analysis for photovoltaic systems, Renewable and Sustainable Energy Reviews 50 (2015) 804–809. https://doi.org/10.1016/j.rser.2015.05.056.

[6] R. Pimpalkar, A. Sahu, R.B. Patil, A. Roy, A comprehensive review on failure modes and effect analysis of solar photovoltaic system, Mater Today Proc 77 (2023) 687–691. https://doi.org/10.1016/j.matpr.2022.11.353.

[7] Pramod Rajput, Maria Malvoni, Nallapaneni Manoj Kumar, O.S. Sastry, G.N. Tiwari, Risk priority number for understanding the severity of photovoltaic failure modes and their impacts on performance degradation, Case Studies in Thermal Engineering 16 (2019) 1–11. https://doi.org/https://doi.org/10.1016/j.csite.2019.100 563.

[8] B. Adothu, S. Pamir Aly, A. Seentakath Puthiyapurayil, K. Chapaneri, J. Joseph John, G. Mathiak, V. Alberts, Investigation of PV module degradation in fixed and single-axis tracker in hot desert climate, in: 41st EUPVSEC, 2024. abstract submitted (accessed February 5, 2024).

[9] Rajiv Dubey, Sachin Zachariah, Shashwata Chattopadhyay, Vivek Kuthanazhi, Sugguna Rambabu, Sonali Bhaduri, Hemant K. Singh, Archana Sinha, Birinchi Bora, Rajesh Kumar, O. S. Sastry, Chetan S. Solanki, Anil Kottantharayil, Brij M. Arora, K. L. Narasimhan, Juzer Vasi, Performance of Field-Aged PV Modules in India: Results from 2016 All India Survey of PV Module Reliability, in: 2017 IEEE 44th

Photovoltaic Specialist Conference (PVSC), 2017: pp. 1–4. https://doi.org/10.1109/PVSC.2017.8366143.

[10] A.A. Abdallah, K. Ali, M. Kivambe, Performance and reliability of crystalline-silicon photovoltaics in desert climate, Solar Energy 249 (2023). https://doi.org/10.1016/j.solener.2022.11.042.

[11] E. Aydin, T.G. Allen, M. De Bastiani, L. Xu, J. Ávila, M. Salvador, E. Van Kerschaver, S. De Wolf, Interplay between temperature and bandgap energies on the outdoor performance of perovskite/silicon tandem solar cells, Nat Energy 5 (2020). https://doi.org/10.1038/s41560-020-00687-4.

[12] M. Babics, M. De Bastiani, E. Ugur, L. Xu, H. Bristow, F. Toniolo, W. Raja, A.S. Subbiah, J. Liu, L. V. Torres Merino, E. Aydin, S. Sarwade, T.G. Allen, A. Razzaq, N. Wehbe, M.F. Salvador, S. De Wolf, One-year outdoor operation of monolithic perovskite/silicon tandem solar cells, Cell Rep Phys Sci 4 (2023). https://doi.org/10.1016/j.xcrp.2023.101280.

[13] C. Sen, X. Wu, H. Wang, M.U. Khan, L. Mao, F. Jiang, T. Xu, G. Zhang, C. Chan, B. Hoex, Accelerated damp-heat testing at the cell-level of bifacial silicon HJT, PERC and TOPCon solar cells using sodium chloride, Solar Energy Materials and Solar Cells 262 (2023). https://doi.org/10.1016/j.solmat.2023.112554.

[14] C. Sen, H. Wang, X. Wu, M.U. Khan, C. Chan, M. Abbott, B. Hoex, Four failure modes in silicon heterojunction glass-backsheet modules, Solar Energy Materials and Solar Cells 257 (2023). https://doi.org/10.1016/j.solmat.2023.112358.

[15] H.A. Kazem, M.T. Chaichan, Experimental analysis of the effect of dust's physical properties on photovoltaic modules in Northern Oman, Solar Energy 139 (2016). https://doi.org/10.1016/j.solener.2016.09.019.

[16] H.A. Kazem, A.H.A. Al-Waeli, M.T. Chaichan, K. Sopian, Modeling and experimental validation of dust impact on solar cell performance, Energy Sources, Part A: Recovery, Utilization and Environmental Effects (2022). https://doi.org/10.1080/15567036.2021.2024922.

[17] Reza Asadpour, Muhammed Tahir Patel, Steven Clark, Nick Bosco, Timothy J. Silverman, Muhammad A. Alam, Worldwide Physics-Based Lifetime Prediction of c-Si Modules due to Solder-Bond Failure, IEEE J Photovolt 12 (2022) 533–539. https://doi.org/10.1109/JPHOTOV.2021.3136164.

[18] R. Asadpour, M.A. Alam, Worldwide Lifetime Prediction of c-Si Modules Due to Finger Corrosion: A Phenomenological Approach, IEEE J Photovolt 12 (2022) 1211–1218. https://doi.org/10.1109/JPHOTOV.2022.3183384.

[19] M. Malvoni, N.M. Kumar, S.S. Chopra, N. Hatziargyriou, Performance and degradation assessment of large-scale grid-connected solar photovoltaic power plant in tropical semi-arid environment of India, Solar Energy 203 (2020) 101–113. https://doi.org/10.1016/J.SOLENER.2020.04.011.

[20] B.-E. Bayandelger, Degradation rate and mechanisms of PV modules in the Gobi Desert of Mongolia after 14 years operated in the field, in: 7th National Renewable Energy Forum-2017, 2017: pp. 1–5. https://www.researchgate.net/publication/338501050.

[21] S.S. Chandel, M. Nagaraju Naik, V. Sharma, R. Chandel, Degradation analysis of 28 year field exposed mono-c-Si photovoltaic modules of a direct coupled solar water pumping system in western Himalayan region of India, Renew Energy 78 (2015) 193–202. https://doi.org/10.1016/J.RENENE.2015.01.015.

[22] D.C. Jordan, S.R. Kurtz, K. VanSant, J. Newmiller, Compendium of photovoltaic degradation rates, Progress in Photovoltaics: Research and Applications 24 (2016) 978–989. https://doi.org/10.1002/pip.2744.

[23] A.P. Patel, A. Sinha, G. Tamizhmani, Field-Aged Glass/Backsheet and Glass/Glass PV Modules: Encapsulant Degradation Comparison, IEEE J Photovolt 10 (2020) 607–615. https://doi.org/10.1109/JPHOTOV.2019.2958516.

[24] M.S. Honnurvali, N. Gupta, K. Goh, T. Umar, N. Nazeema, Study of Photovoltaics (PV) Performance Degradation Analysis in Oman, International Journal of Sustainable Energy Development 6 (2017) 334–343. https://doi.org/10.20533/ijsed.2046.3707.2017.0043.

[25] R. Dubey, S. Chattopadhyay, V. Kuthanazhi, A. Kottantharayil, C. Singh Solanki, B.M. Arora, K.L. Narasimhan, J. Vasi, B. Bora, Y.K. Singh, O.S. Sastry, Comprehensive study of performance degradation of field-mounted photovoltaic modules in India, Energy Sci Eng 5 (2017) 51–64. https://doi.org/10.1002/ese3.150.

[26] D.M. Atia, A.A. Hassan, H.T. El-Madany, A.Y. Eliwa, M.B. Zahran, Degradation and energy performance evaluation of mono-crystalline photovoltaic modules in Egypt, Scientific Reports 2023 13:1 13 (2023) 1–16. https://doi.org/10.1038/s41598-023-40168-8.

[27] B. Adothu, S. Kumar, J.J. John, G. Oreski, G. Mathiak, B. Jäckel, V. Alberts, J. Bin Jahangir, M.A. Alam, R. Gottschalg, Comprehensive review on performance, reliability, and roadmap of c-Si PV modules in desert climates: A proposal for improved testing standard, Progress in Photovoltaics: Research and Applications 32 (2024) 495–527. https://doi.org/10.1002/PIP.3827.

[28] Gerhard Mathiak, Jim Joseph John, Sagarika Kumar, Baloji Adothu, Prashanth Gabbadi, Yogesh Kumar, Vivian Alberts, Bengt Jaeckel, Ralph Gottschalg, Proposal for Hot Desert Test Cycle, in: EU PVSEC 2023, 2023: p. 001-020192-001.

Solar Cell Crack Image Generation for Power Loss Prediction

Norman Jost[1,*], Emma Cooper[1,2], Benjamin G. Pierce[1,3], Brandon Byford[1,4], Ojas Singh[1,5] and Jennifer L. Braid[1,*]

[1] Sandia National Laboratories, 87123 Albuquerque, New Mexico, USA
[2] University of Colorado Boulder, 80309 Boulder, Colorado, USA
[3] Case Western Reserve University, 44106 Cleveland, Ohio, USA
[4] New Mexico State University, 88003 Las Cruces, New Mexico, USA
[5] University of Arizona, 85721 Tucson, Arizona, USA
*nrjost@sandia.gov, jlbraid@sandia.gov

ABSTRACT: Due to the extreme thermomechanical stresses imposed on photovoltaic (PV) modules during manufacture, shipping, installation, regular operation, and extreme weather, fracture of Si cells may be inevitable. As the industry moves thinner glass and larger format modules, and as the frequency of extreme weather events continues to increase, the problem of cell cracks is likely here to stay. Therefore, when cracks are observed, it is important to understand their immediate effects on PV performance. Furthermore, the associated power loss in the years of operation following the fracture event is even more difficult to forecast, as the evolution of cell crack properties is not well understood. Currently, measuring the extent of power loss due to cell cracks in fielded PV systems is costly and requires several steps such as locating modules with cracks via electroluminescence (EL) imaging, manually removing identified cracked modules, and flash current-voltage (IV) testing identified modules with cell cracks. Then those modules are either re-installed or replaced. This process is costly, so often when a large site has widespread cell cracking and associated power loss, there are issues of identifying extent of damage and liability. Therefore, methods to cost-effectively identify the current health state of modules and future risk of power loss due to cell cracks are of great interest to the solar industry, particularly asset owners.

The scope of the work presented here collects three elements to form a model chain for predicting power loss due to c-Si cell cracks from EL images. First, we modify the open-source deep learning model *pv-vision* to segment EL images for locations of cracks and other cell features such as busbars. Second, we have developed a Variational Autoencoder (VAE) to parameterize crack geometries and their spatial relationships to cell busbars and edges. Third, power loss prediction will be a feature of our fully-trained VAE. Training of this model will include associated EL and IV data from cracked cells. The final trained VAE model will be capable of co-generation of EL and single-diode model parameters for cracked cells, as well as prediction of power loss given a previously unseen EL image.

1 INTRODUCTION

As the world approaches 2 TW peak of installed photovoltaic solar energy [1] we face a challenging task of identifying when to repower PV systems due to degraded solar modules. In this study, we specifically look at power degradation associated with cell cracks. Though cracking does not typically cause major power degradation at an initial stage, power loss is observed when modules with cracked cells undergo cyclic thermomechanical stress [2].

With the large number of modules in the field, ways to predict power loss with very little effort are needed. This is important for plant valuation and commissioning, maintenance, and to forecast when repowering of a solar site will be needed. In the DuraMAT consortium funded by the U.S. Department of Energy Solar Energy Technology Office there are multiple projects focusing on how to be able to accurately model degradations in solar modules. Some degradation modes can already be modeled with open source tools [3], however for solar cell crack evolution and the related power loss there currently are no tools available.

This project aims to develop a deep learning model chain to predict crack evolution and associated performance losses in solar modules. The goal is to have a framework that applies to module level in-situ EL (electroluminescence) images from a PV site. The approach is to modify existing and proven deep learning methods, including U-Net [4] for image segmentation and variational autoencoders (VAE) for parameterization, generation, and prediction.

Previous work has used deep learning, more specifically convolutional neural networks (CNNs) to detect and extract cell cracks [5]. There are published machine learning approaches to detect power losses from full module EL images [6], [7], [8], [9], [10], [11], [12], [13]. For the scope of our work we will leverage existing approaches to further contribute and develop new open-source tools such as *pvimage* [14] for EL image cropping and planar indexing, and *pv-vision* [5] for EL image segmentation. In our approach we will be studying the crack evolution and power loss on single solar cells to limit the influence of other effects such as interconnection losses. This work will produce an image processing pipeline to be used on any EL data to determine the in-situ performance and how this performance will be affected in the future due to thermomechanical stresses acting upon the modules in the outdoor environment.

2 METHODOLOGY

The final goal of our work is to build a physics-informed, data-driven model for PV cell crack associated power loss in PV modules. Our approach requires generation of associated crack and performance data and several analysis and modeling steps detailed below.

2.1 Proposed model chain

The steps in our proposed data-driven model chain are seen in Figure 1. The first step is to use the open-source *pvimage* [14] and *pv-vision* libraries [5] for cell image cropping/planar indexing and feature extractions, respectively. The second step is to parameterize the geometric features of cell cracks using machine learning. The third step is to use the cell crack parameters to estimate power losses for the cell. We plan for this model chain to eventually predict power loss on the module level.

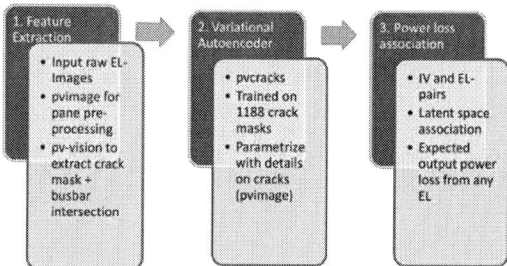

Figure 1 Steps of the data-driven modeling. Retraining pv-vision to be compatible with the new images and multi-classing, then training the variational autoencoder and lastly, associate the power loss by using IV and EL pairs.

2.2 Data + Experiment

To enable the building of our data-driven model from the cell level, we use encapsulated PV cells exhibiting cell cracks and characterize these cells with IV (current-voltage curves) and EL imaging. As little power loss occurs at the cracking event, we will age these cracked cells through thermal cycling. Additionally, experimental data from X-ray topography (XRT), alongside finite element analysis simulations of cracked cells will be used to validate our and inform the structure of our data-driven model.

In addition to our experimental data, we also used a collection of mono-crystalline Si solar cell EL images from our partners Arizona State University and Case Western Reserve University for model training. From these images we produced a dataset of 1188 binary cell crack masks, which we have made publicly available on the DuraMAT datahub [15].

2.3 Preprocessing + Feature Extraction

EL images are first preprocessed using pvimage to crop images to the cell area and planar index them into a regular square cell shape before attempting segmentation. For our needs the *pv-vision* library needed to be retrained as it would not perform well with more modern solar cells with more/thinner busbars, which were being mislabeled as cracks. In addition, we implemented multi-classification (the same pixel can be in two classes). This was needed to avoid discontinuations in the crack mask when a busbar-crack intersection was present. The retrained pv-vision gives us our crack masks, which we used for training our variational autoencoder (VAE). By using the crack features extracted with *pv-vision*, we eliminate inconsistencies arising from EL image quality, intensity, setting, camera, etc., to prevent improper power prediction associated with a particular dataset.

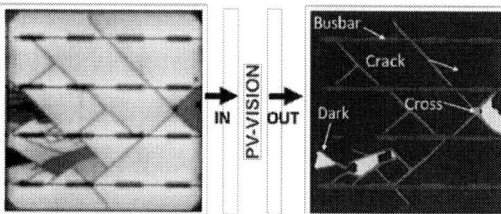

Figure 2 Schematic of how pv-vision works. Cropped EL images are used as input, the CNN extracts the features and separates them in multiple classes/layers as an output.

2.4 Variational Autoencoder (VAE)

The primary purpose of the VAE is for dimension reduction, parameterizing the crack geometry for correlation to power loss. The classical VAE structure (Figure 3) resembles a pair of convolutional neural networks (CNNs) also named U-Net, which encode an image into a latent vector, and then decode that vector to reproduce the original image. Opposed to the true U-Net structure, the layers in the CNNs are disconnected, meaning that the latent vector is a standalone representation of the input/output image. We employ the VAE technique with modified loss functions to accurately capture and reproduce cell crack geometries.

In our VAE structure, both encoder and decoder have six convolutional layers. To the typical loss functions, which are Binary Cross Entropy (BCE) and Kullback-Leibler Divergence (KLD), we also added the structural similarity index (SSIM) which is useful for quantifying similarity in continuity and shapes in images. Each loss function is weighted and added for the total loss, which is minimized during training. A secondary feature of the VAE (because of its standalone latent space) is its ability to generate artificial data. Image generation can be used to verify the quality of VAE training; an underfitted VAE would generate random shapes that do not resemble cell cracks while an overfitted VAE would only generate cracks in its training set. For testing the autoencoder we are using a modified version of a feature extractor based on image processing. This code used is mentioned in Whitaker et al. [3]. We modify the code as we do not have to extract busbars (previously done by pv-vision) and our image sizes are different. The code is available as part of the pvimage package on github [20]. The crack feature extractor is applied to our crack images and gives us the following parameters to compare: perimeter, slope, area, orientation and we can count how many features each image contains. When running a t-test to determine if the original and reproduced images and features are not significantly different, in both cases we get p-values of over the alpha threshold 0.05.

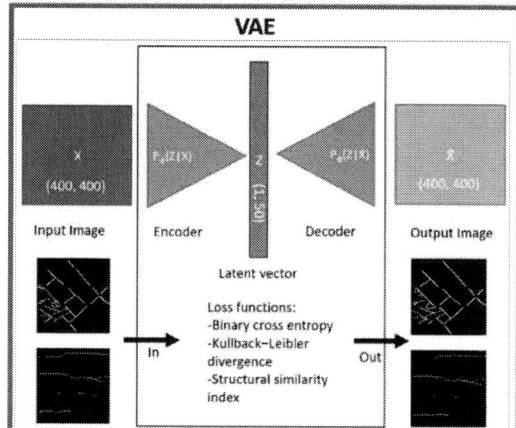

Figure 3 Structure of the variational autoencoder (VAE) showing example input/output images with their dimensions, the dimension of the latent vectors [1,50] and the weighted loss functions used.

3 RESULTS

In this section we present the current results for each

step in our proposed model chain.

3.1 pv-vision Modifications and Implementation

pv-vision is an open-source model available for feature segmentation of EL-images [5], example in Figure 2. The model is based on the U-Net architecture and is pretrained on ImageNet data. We are re-training the *pv-vision* model to accurately identify and distinguish thin busbars and cell cracks on modern cell architectures via transfer learning. Additionally, we modified the output of the U-Net to co-identify pixels as both cracks and busbars, whereas the published model could only provide single classifications for any given pixel. This capability is demonstrated in Figure 4ab. Co-classification of pixels is necessary for a) continuous segmentation of cell cracks (i.e., not introducing discontinuities in crack geometries, and b) identifying the relationship between crack intersections with busbars and power loss.

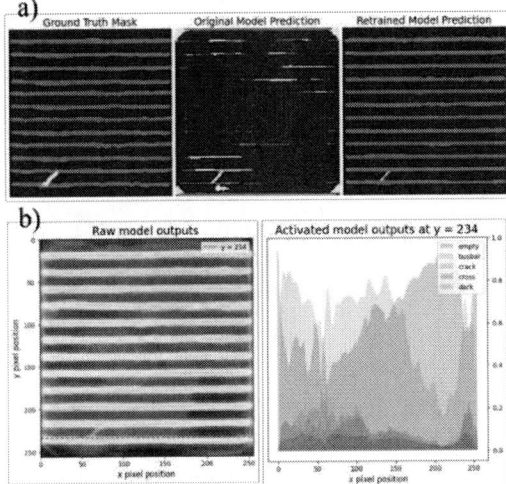

Figure 4 Results of the re-trained pv-vision for thin busbars (a) and multi-classing (b). One can see how the previous model was detecting the thin busbars as cracks or ignoring them, this has been improved. The possibility of activating the same pixels as multiple classes is a new feature not available before.

The example codes showing how the re-training was done including the validation is available online as jupyter notebooks [18]. The final weights are also uploaded to the DuraMAT datahub [19].

3.2 Cell Crack Parameterization

Our VAE reduces 400 x 400 pixel binary input images into latent vectors of 1 x 50 length. To evaluate the ability of the VAE to capture crack geometries as latent vectors, we look at the latent space of the testing batch. This is done by doing a principal component analysis (PCA) to reduce the 50 dimension to 3 and cluster using k-means. We chose 5 clusters for the binning; the number of clusters was determined by the elbow and silhouette method. In **Error! Reference source not found.** we show the latent spaces were clustered in PCA space, each color representing a cluster. Also included are sample images from each cluster. We can see that the VAE does learn crack distributions and shows clusters with little to none cracking as well as cracks with similar parameters in one cluster.

An additional feature of the VAE is the possibility to generate new images by passing new latent vectors. We use this as an additional method to test how well the VAE works. For this a t-test is performed on each cluster comparing the replicated to the original images of the testing batch. This is done to determine which cluster replicates the images best. For generating new images, we perform vector arithmetic on the chosen cluster. New latent vectors are sampled with a multivariate normal distribution around the centroid of the best cluster. Following the same previous procedure of extracting the crack features of the newly generated crack images and doing a t-tests against the testing image batch leads to the same result that the images are not significantly different.

3.3 Power Loss Prediction

The last part of our model chain will predict power loss from cell crack features. For this we will use the latent vector from the VAE and develop a model to correlate with IV curves. The model is a multi-linear-regression model.

From the experimental data we have EL-IV pairs. IV curves are processed using the single diode equation fit from pvlib [16]. We can calculate the power difference (Pmpp – power at maximum power point) and open-circuit voltage (Voc). The Voc is known to be reduced when cells shows cracking cause a decline in parallel (shunt) resistance (Rp) [2]. The here mentioned parameters can easily be fitted from the measured IV curves by using the single diode model equation fit from the open-source library pvlib [16], [17].

We extract all the diode parameters including resistances but for simplification we just look at the power loss and difference in Voc. In the cell batch we have two solar cell types, as we did not have IV curves of the solar cells before they were damaged in shipping we used the non-cracked cells as our controls. So the difference in Pmpp and Voc is the delta between the undamaged control cell and the cracked cells for each different cell type. The resulting values for each cluster in the latent space are shown in **Error! Reference source not found.**b.

From the latent space analysis we can see a correlation between the clusters and the Voc difference. The next step is to see if there is a correlation between each latent vector and the extracted diode parameters. For this we use a Pearson correlation seen in Figure 6a.

A multi-linear regression model was used to correlate the latent vector to the solar cell parameters Pmpp difference, Voc difference and Rsh. As indicated by the Pearson correlation the R^2 values of the regression are not very high for EL and IV features (Figure 6b). This result was not unexpected as initial power loss due to cell cracks is low. As these cells undergo thermal cycling in the future, we expect that these correlations will improve, particularly when associated with a quantification of thermal stress.

4   DISCUSSION

In this work, we have showcased the primary tools for developing a data-driven model chain for cell crack power loss prediction. The re-trained *pv-vision* for EL image segmentation is available on github [18] and the weights on the DuraMAT Datahub [19]. The training crack mask image set for training the variational autoencoder is available online [15] along with the current weights [18]. The code of the VAE and examples on how the validation with *pvimage* is done is on github [18].

The segmentation+VAE model was shown to be successful in reducing the dimensionality of cell cracks in EL images by a factor of 32k, while also enabling accurate

41st European Photovoltaic Solar Energy Conference and Exhibition

reproduction of the input mask through the VAE's decoder structure. Seeing the results from the latent space analysis and the multi-linear correlation in **Error! Reference source not found.** & Figure 6 we can see that the results are not entirely conclusive yet. While crack geometries cluster into groups of similar appearance, the initial power

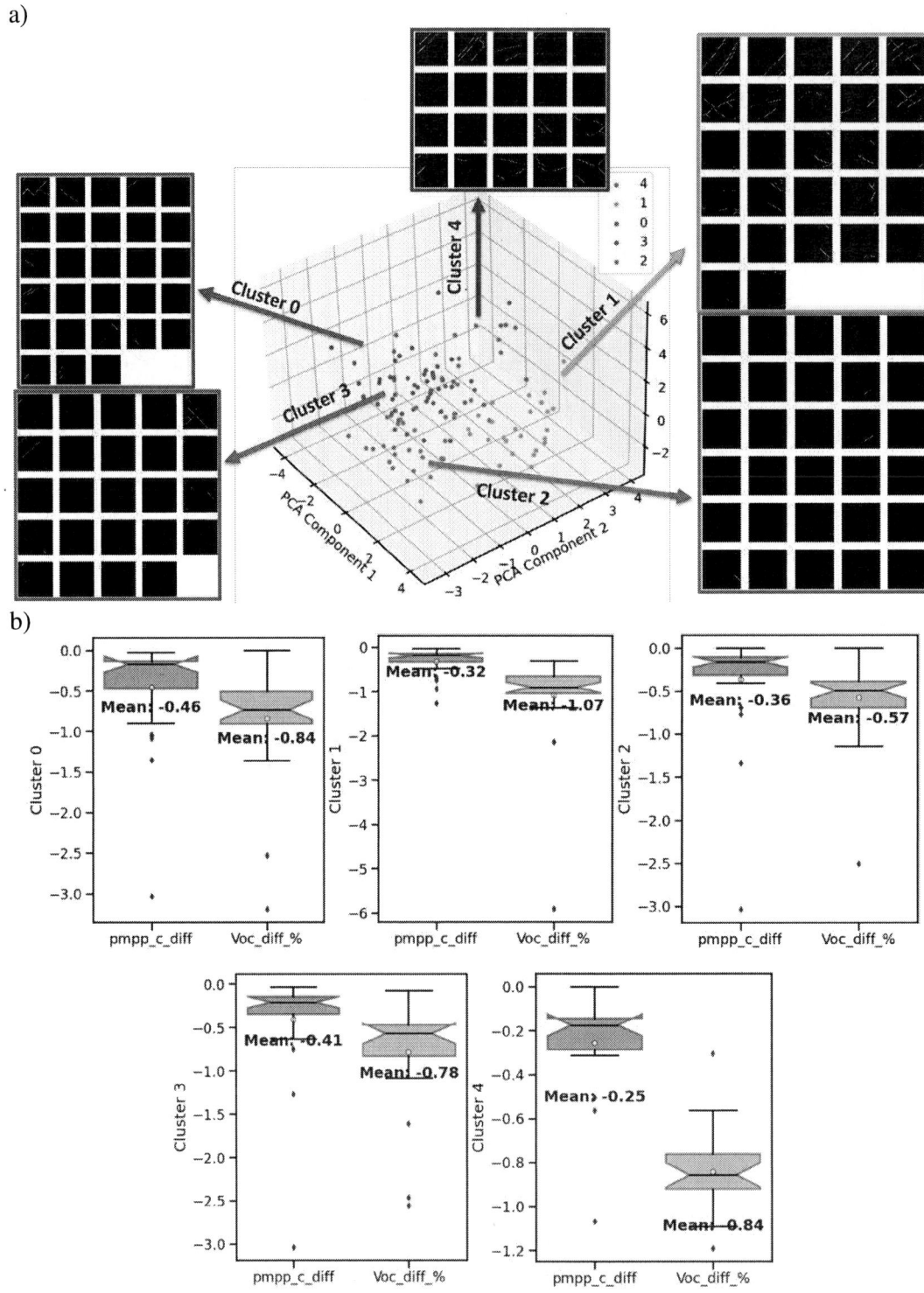

Figure 5 Latent space analysis for the EL and IV pairs; a) shows the latent space with the images from each cluster. Some clusters show a larger amount of Els with cracks and others fewer, indicating that some parametrization is happening in the latent space. b) shows the resulting Pmpp differences (in W, max. Pmpp = 3W) and Voc differences (in %). The distribution of Pmpp between the clusters makes it difficult to draw conclusion but we can see that cluster 2 with minor visible cracks has a low difference in Voc. Contrary to cluster 1 & 4 with many visible cracks having a high difference in Voc.

41st European Photovoltaic Solar Energy Conference and Exhibition

a)

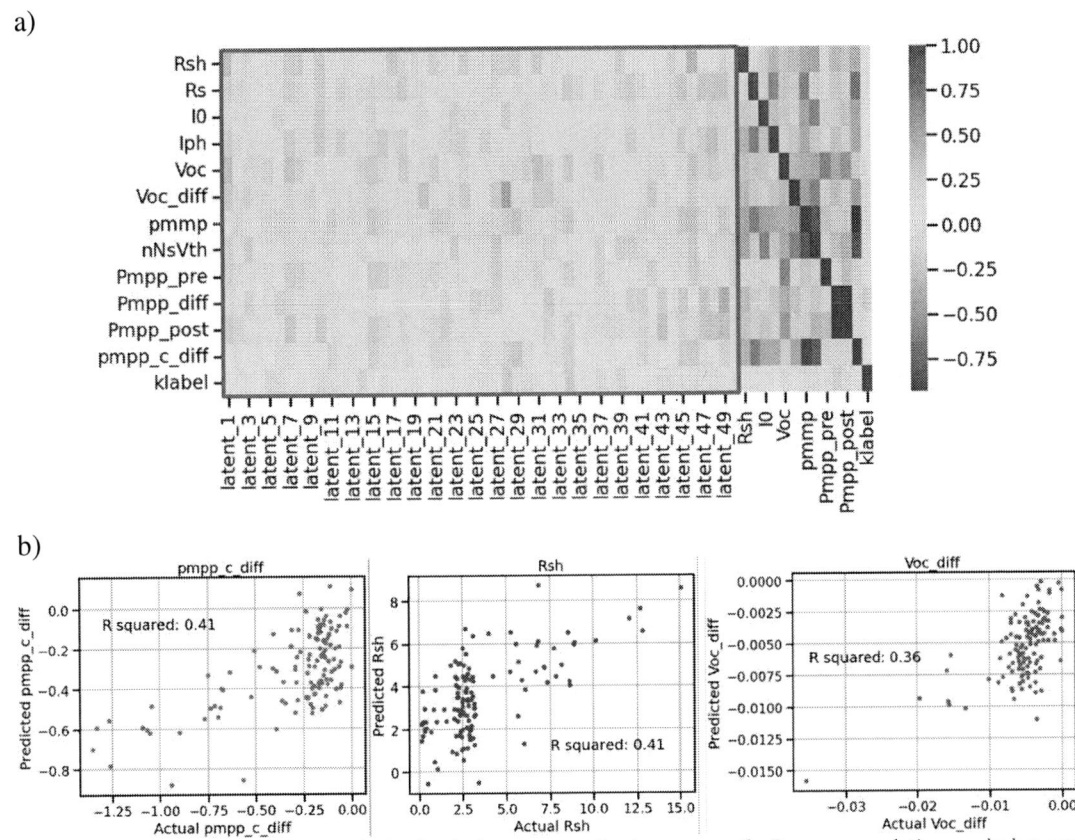

b)

*Figure 6: Results from the power loss correlation for the latent vector. In a) we can see the Pearson correlation results between the diode parameters from the single diode equation fit and the latent vectors (**blue** rectangle). We see a few latent values reaching values of ±0.3 indicating a not very high correlation. Below in b) we show the results of the multi-linear regression from the latent vectors to Pmpp difference, Rsh and Voc difference. We achieve R^2 values of around 0.4. We expect this to improve once the thermal cycling is complete and the parameter differences are taken for before and after the cycling.*

loss from these samples is too low to gauge whether the crack features captured by the VAE or summarized by PCA will be good predictors of cell power loss. Once more data in the form of the thermal cycling results are available, we will repeat this process. Here we clustered latent vectors from the VAE, then mapped the IV performance characteristics to those clusters. In the future, clustering on a combined vector of VAE features and IV features may yield more refined clusters for cell crack behavior and inform improvements to our model structure. Additionally, our final model will incorporate finite element model simulations of X-Ray topography experiments on encapsulated cracked cells, to ensure our model accurately reflects the behavior of cracks as they undergo thermomechanical stress.

CONCLUSION

This manuscript reports our progress towards a physics-informed machine learning pipeline for predicting c-Si PV cell crack evolution and the attributed power loss. We have already developed the necessary tools and re-training / tuning the machine learning models including EL image processing and planar indexing, cell feature segmentation, and crack parameterization of cell performance characteristics. Future work will include thermal cycling of encapsulated cracked cells, to enable association of power loss mechanisms with crack geometric features and classes.

ACKNOWLEDGEMENTS

Sandia National Laboratories is a multimission laboratory managed and operated by National Technology and Engineering Solutions of Sandia, LLC, a wholly owned subsidiary of Honeywell International Inc., for the U.S. Department of Energy's National Nuclear Security Administration under contract DE-NA0003525. Funding provided as part of DuraMAT funded by the U.S. Department of Energy, Office of Energy Efficiency and Renewable Energy, Solar Energy Technologies Office, agreement number 32509.

In addition, we would also like to thank the team at ASU for their contribution to this work providing images and consultation. The team at ASU are Prof. Mariana Bertoni, Dr. Ian Slauch and Isabela Schinella.

REFERENCES

[1] A. Jäger-Waldau, "Snapshot of photovoltaics À February 2024," *EPJ Photovolt.*, 2024.

[2] M. Köntges, I. Kunze, S. Kajari-Schröder, X. Breitenmoser, and B. Bjorneklett, "Quantifying the risk of power loss in PV modules due to micro cracks," in *25th European Photovoltaic Solar Energy Conference*, Valencia, Spain, Sep. 2010, pp. 3745–3752.

[3] M. Springer *et al.*, *NREL/PVDegradationTools: 0.3.2.* (May 2024). Zenodo. doi: 10.5281/zenodo.11123249.

[4] O. Ronneberger, P. Fischer, and T. Brox, "U-Net: Convolutional Networks for Biomedical Image Segmentation," *ArXiv150504597 Cs*, May 2015, Accessed: Jan. 12, 2022. [Online]. Available: http://arxiv.org/abs/1505.04597

[5] X. Chen *et al.*, "Automatic Crack Segmentation and Feature Extraction in Electroluminescence Images of Solar Modules," *IEEE J. Photovolt.*, pp. 1–9, 2023, doi: 10.1109/JPHOTOV.2023.3249970.

[6] S. Hassan, "Dual spin max pooling convolutional neural network for solar cell crack detection," *Sci. Rep.*, 2023.

[7] A. M. Karimi *et al.*, "Generalized and Mechanistic PV Module Performance Prediction From Computer Vision and Machine Learning on Electroluminescence Images," *IEEE J. Photovolt.*, vol. 10, no. 3, pp. 878–887, May 2020, doi: 10.1109/JPHOTOV.2020.2973448.

[8] W. Tang, Q. Yang, K. Xiong, and W. Yan, "Deep learning based automatic defect identification of photovoltaic module using electroluminescence images," *Sol. Energy*, vol. 201, pp. 453–460, May 2020, doi: 10.1016/j.solener.2020.03.049.

[9] M. Hoffmann *et al.*, "Deep-learning-based pipeline for module power prediction from electroluminescense measurements," *Prog. Photovolt. Res. Appl.*, vol. 29, no. 8, pp. 920–935, 2021, doi: 10.1002/pip.3416.

[10] M. Hoffmann *et al.*, "ELPVPower: A dataset for large scale PV power prediction using EL images of cells." Jülich DATA, Mar. 16, 2022. doi: 10.26165/JUELICH-DATA/TVWUUP.

[11] L. Lüer, "PV module power prediction by deep learning on electroluminescence images - Assessing the physics learned by a convolutional neural network," *Sol. Energy Mater. Sol. Cells*, 2024.

[12] Brandon Byford, Laura E. Boucheron, and Jennifer L. Braid, "Advanced Photovoltaic Panel Characterization: Using Image Transformers for Current Voltage Curve Prediction from Electroluminescence Images," *Submiss.*, Sep. 2024.

[13] Brandon K. Byford, "Transforming electroluminescence images to current-voltage (IV) curves using deep learning," in *PVSC Proceedings*, Seattle, WA, USA: IEEE, Jun. 2024.

[14] C. M. Whitaker, B. G. Pierce, A. M. Karimi, R. H. French, and J. L. Braid, "PV Cell Cracks and Impacts on Electrical Performance," in *2020 47th IEEE Photovoltaic Specialists Conference (PVSC)*, Jun. 2020, pp. 1417–1422. doi: 10.1109/PVSC45281.2020.9300374.

[15] Norman Jost, "pvcracks/VAE/crack-masks." DuraMAT Datahub. [Online]. Available: https://datahub.duramat.org/dataset/pvcracks-crack-masks-for-vae

[16] K. S. Anderson, C. W. Hansen, W. F. Holmgren, A. R. Jensen, M. A. Mikofski, and A. Driesse, "pvlib python: 2023 project update".

[17] W. F. Holmgren, C. W. Hansen, and M. A. Mikofski, "pvlib python: a python package for modeling solar energy systems," *J. Open Source Softw.*, vol. 3, no. 29, Art. no. 29, Sep. 2018, doi: 10.21105/joss.00884.

[18] N. Jost, E. Cooper, B. Pierce, J. Hartley, and J. Braid, *pvcracks*. (Mar. 12, 2024). Python. Sandia National Lab. (SNL-NM), Albuquerque, NM (United States). doi: 10.11578/dc.20240606.4.

[19] Emma Cooper, "pvcracks/pv-vision/re-train/weigths." DuraMAT Datahub. [Online]. Available: https://datahub.duramat.org/dataset/pvcracks-re-trained-pv-vision-model

[20] *cwru-sdle/pvimage*. (Jun. 25, 2024). Jupyter Notebook. CWRU SDLE. Accessed: Jul. 19, 2024. [Online]. Available: https://github.com/cwru-sdle/pvimage

BENCHMARKING OF ENCAPSULANT MATERIALS FOR c-SI/PEROVSKITE TANDEM MODULES

P. Christöfl[1], C. Barretta[1*], M. Kühne[2], F. Opden Buijsch[3], S. Sals[3], Q. Jeangros[3], B. Stannowski[4], G. Oreski[1,6]

1) Polymer Competence Center Leoben GmbH (PCCL), Leoben, Austria
2) Hanwha Q CELLS GmbH, Bitterfeld-Wolfen, Germany
3) The Compound Company, Entschede, The Netherlands
3) Swiss Center for Electronics and Microtechnology (CSEM), Neuchatel, Switzerland
4) Helmholtz-Zentrum Berlin für Materialien und Energie (HZB), Berlin, Germany
6) Montanuniversitaet Leoben, Leoben, Austria

41st EU PVSEC, 25.09.2024, Vienna

Co-funded by the European Union

*) Sauraugasse 1, 8700 Leoben, Austria
petra.christoefl@pccl.at

www.pccl.at

Introduction
Polymers: properties and their impact on PV modules

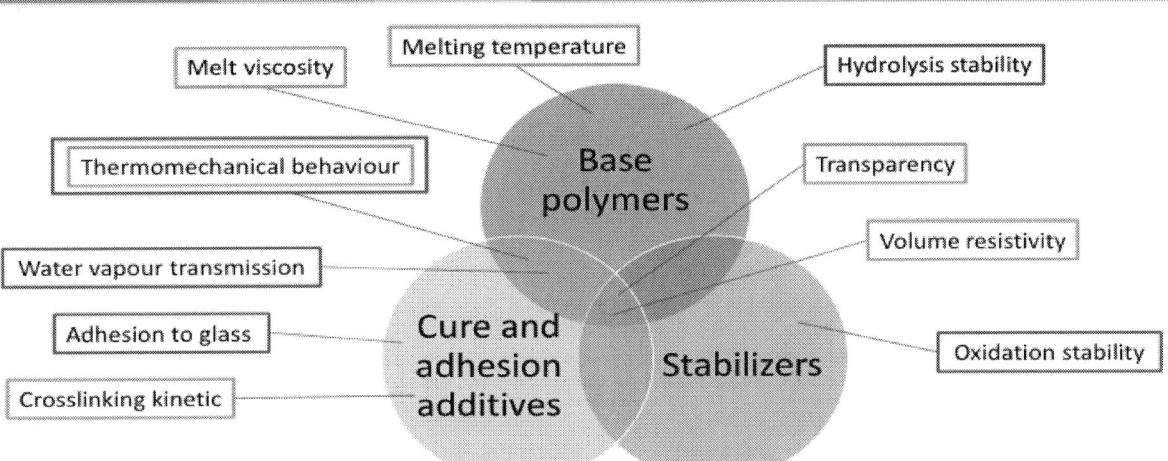

Effects on PV modules: ➤ Production ➤ Efficiency ➤ Safety ➤ Reliability

www.pccl.at

Introduction
Encapsulating c-Si/perovskite tandem cells

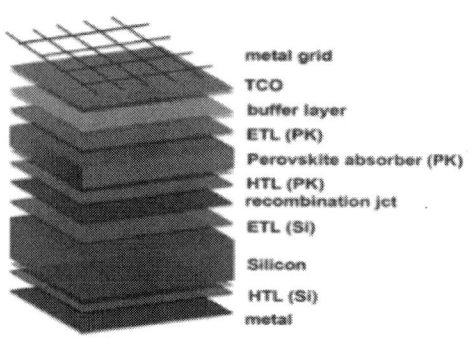

Perovskite tandem full stack

labels: metal grid, TCO, buffer layer, ETL (PK), Perovskite absorber (PK), HTL (PK), recombination jct, ETL (Si), Silicon, HTL (Si), metal

Challenge	Perovskites are flexible, fragile materials that need protection from environmental factors such as moisture and oxygen
Solution: Encapsulation with Polyolefin elastomers	**Polyolefin elastomers offer low water vapor permeability, thermal stability, and optical transparency**
Key Open Question	What is the role of melt viscosity and thermal expansion ? **→ Risk of delamination of the solar cell stack during processing**

www.pccl.at

Experimental
Approach: material selection and characterization methods

Definition of selection criteria
- Low melting point
- Low softening point
- Low anisotropy
- Low shrinkage
- Low onset T of crosslinking (or no crosslinking)

Encapsulants
- Commercially available
- 16 different types from various suppliers
- Crosslinking and not crosslinking encapsulants

Characterization

Dynamic Mechanical Analysis (DMA)

Fourier Transform Infrared Spectroscopy (FTIR)

Differential Scanning Calorimetry (DSC)

Digital Image Correlation (DIC)

Ranking/Selection

Selection of 3 encapsulants for lamination
- 2 crosslinking
- 1 non-crosslinking

www.pccl.at

Results
Thermal properties: melting and crosslinking temperatures

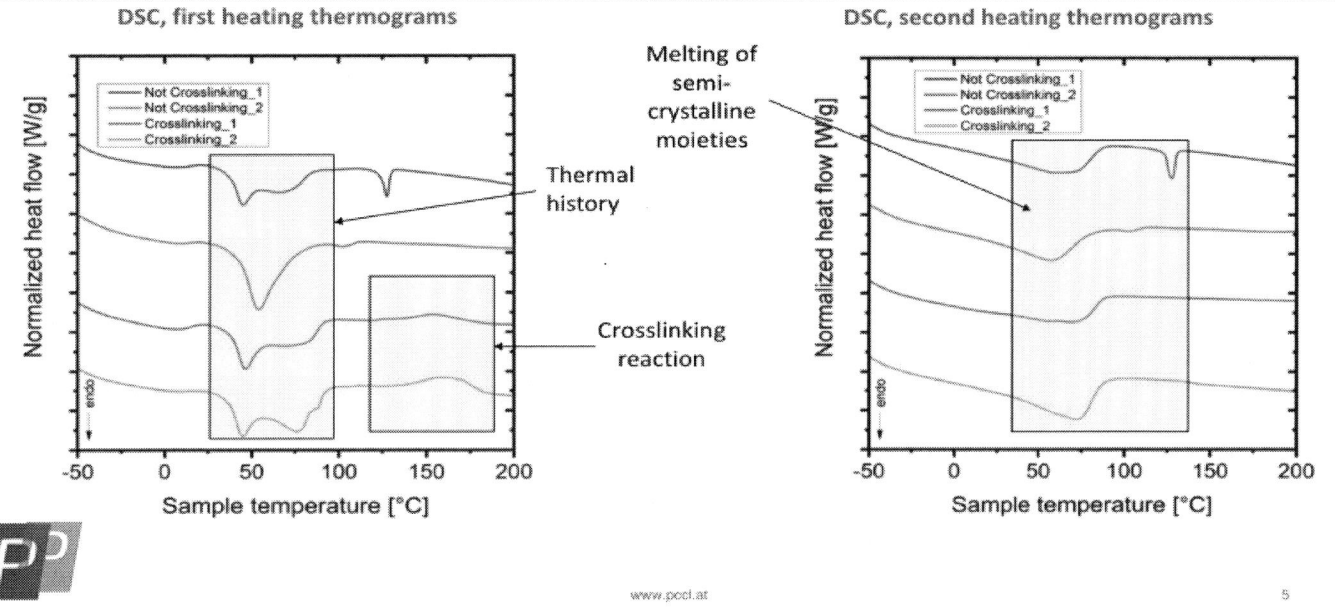

Results
Thermomechanical properties

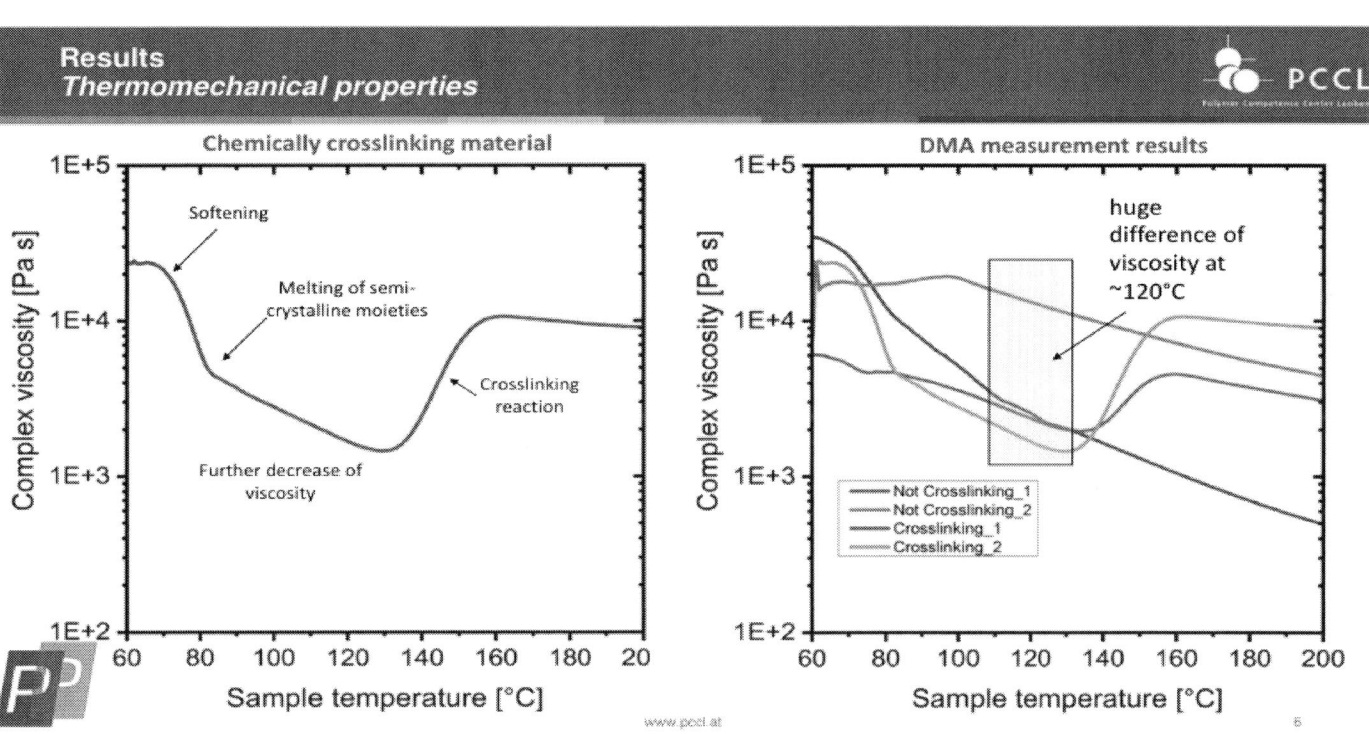

Results
combined results of non-crosslinking and crosslinking encapsulants

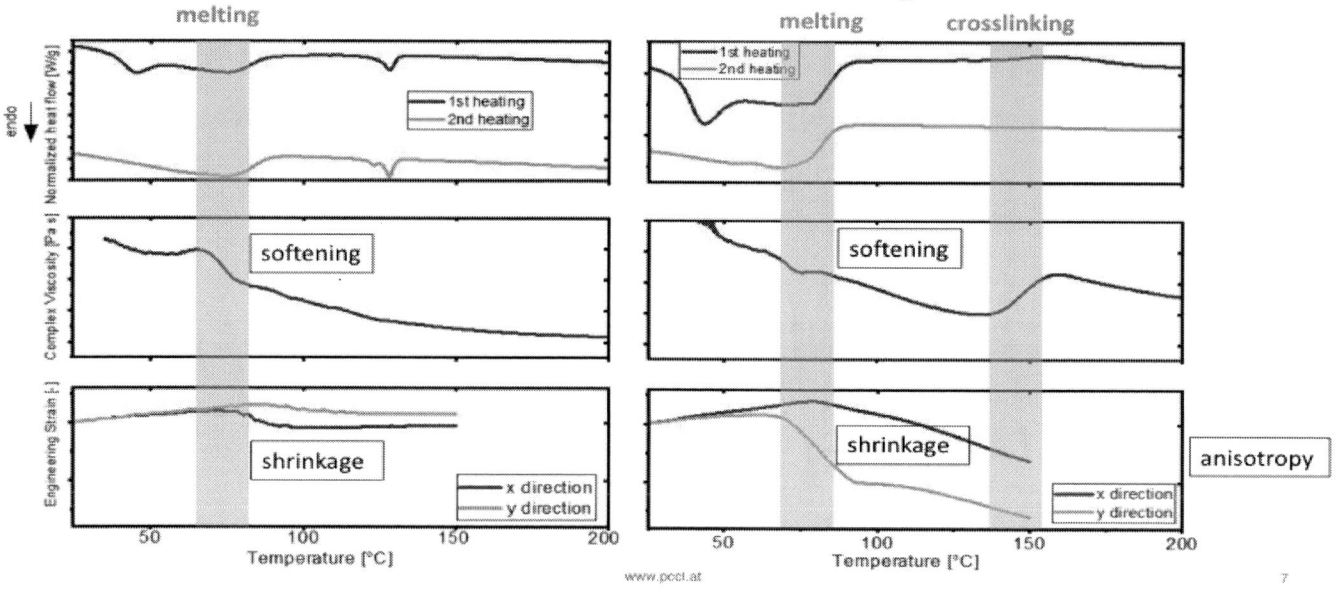

Results
Ranking and material selection

Selection criteria		low	low	low	low	low	low
Encapsulant	Polymer	Melting temperature [°C]	Softening temperature [°C]	Crosslinking onset temperature [°C]	Crosslinking peak temperature [°C]	Complex viscosity at 120°C [Pa s]	Anisotropy
A1	Ethylene α-olefin	54	-	143	161	-	
A2	Ethylene α-olefin	55	-	145	157	-	
A3	Ethylene α-olefin	55	56	135	151	2376	
B11	Ethylene α-olefin	64	60	132	151	2128	
B12	Ethylene α-olefin	60	49	137	163	2416	
B13	Ethylene α-olefin	65	63	134	153	1790	
B21	Ethylene α-olefin	69	66	136	154	1616	
B22	Ethylene α-olefin	68	65	140	158	1921	40°C, low
C1	Ethylene α-olefin	62	63	137	164	2842	
D1	Ethylene α-olefin	72	70	136	162	1669	40°C, high
G1	Ethylene α-olefin	103	65	156	166	1981	
D21	Ethylene α-olefin	70 (120)	70	-	-	2355	
D22	Ethylene α-olefin	75 (128)	71	-	-	2660	60°C, low
D23	Ethylene α-olefin	60 (128)	61	-	-	2532	
E1	Ethylene α-olefin	57 (103)	67	-	-	14633	40°C, high
F1	Ethylene α-olefin	60 (120)	57	-	-	2611	

Results
Ranking and material selection

Selection criteria		low	low	low	low	low	low
Encapsulant	Polymer	Melting temperature [°C]	Softening temperature [°C]	Crosslinking onset temperature [°C]	Crosslinking peak temperature [°C]	Complex viscosity at 120°C [Pa s]	Anisotropy
A1	Ethylene α-olefin	54	-	143	161	-	
A2	Ethylene α-olefin	55	-	145	157	-	
A3	Ethylene α-olefin	55	56	135	151	2376	
B11	Ethylene α-olefin	64	60	132	151	2128	
B12	Ethylene α-olefin	60	49	137	163	2416	
B13	Ethylene α-olefin	65	63	134	153	1790	
B21	Ethylene α-olefin	69	66	136	154	1616	
B22	Ethylene α-olefin	68	65	140	158	1921	40°C, low
C1	Ethylene α-olefin	62	63	137	164	2842	
D1	Ethylene α-olefin	72	70	136	162	1669	40°C, high
G1	Ethylene α-olefin	103	65	156	166	1981	
D21	Ethylene α-olefin	70 (120)	70	-	-	2355	
D22	Ethylene α-olefin	75 (128)	71	-	-	2660	60°C, low
D23	Ethylene α-olefin	60 (128)	61	-	-	2532	
E1	Ethylene α-olefin	57 (103)	67	-	-	14633	40°C, high
F1	Ethylene α-olefin	60 (120)	57	-	-	2611	

Summary and outlook

- **Perovskites** are **flexible, fragile** materials that need **protection** from environmental factors such as moisture and oxygen
- Appropriate encapsulant materials have to be selected to reduce thermomechanical stresses onto the ETL during lamination and to ensure long-term reliability

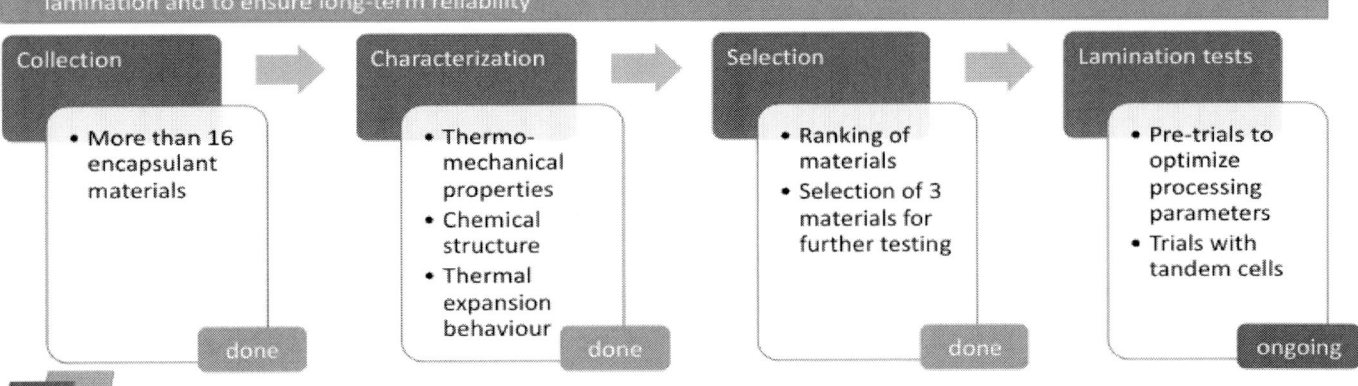

Collection
- More than 16 encapsulant materials

done

Characterization
- Thermo-mechanical properties
- Chemical structure
- Thermal expansion behaviour

done

Selection
- Ranking of materials
- Selection of 3 materials for further testing

done

Lamination tests
- Pre-trials to optimize processing parameters
- Trials with tandem cells

ongoing

Thank you for your attention! Any questions?

Petra Christöfl
Sauraugasse 1, 8700 Leoben, Austria
Petra.christoefl@pccl.at

pepperoni-project.eu

@PEPPERONI_EU

 **YPAREX**

qcells **HZB** Helmholtz Zentrum Berlin

This project is co-funded by the European Union. Views and opinions expressed are however those of the author(s) only and do not necessarily reflect those of the European Union or the European Climate, Infrastructure and Environment Executive Agency (CINEA). Neither the European Union nor the granting authority can be held responsible for them.

The project is also supported by the Swiss State Secretariat for Education, Research and Innovation (SERI).

www.pccl.at

41st European Photovoltaic Solar Energy Conference and Exhibition

Reliability studies of PV minimodules using an Ethylene – Butyl Acrylate (EBA) based encapsulant and high efficiency n-type PV cells

Ignacio Fidalgo*; Polaris Open Innovation; ignacio.fidalgo@polaris-oi.com
Inmaculada Campoy Felipe; Repsol Química, S.A.; icampoyf@repsol.com
Andreas Halm; ISC Konstanz; andreas.halm@isc-konstanz.de

Project Scope and Collaborators

	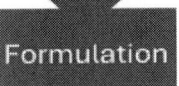 Resin	• EVA Primeva 28025S • EBA Ebantix E20020
	 Formulation	• EVA and EBA formulations based on Repsol resins
	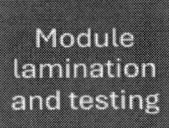 Film	• EVA and EBA encapsulant films based on PolarisOI formulations and process parameters
ISC	Module lamination and testing	• Preparation of minimodules • Aging tests

Why EBA?

- Ethylene butyl acrylate copolymers (EBA) are an interesting alternative to EVA as material for encapsulation:

  - Minimized Corrosive Potential (free of acetic acid)

EVA   Acetic acid   EBA

Why EBA?

- Ethylene butyl acrylate copolymers (EBA) are an interesting alternative to EVA as material for encapsulation:

  - Minimized Corrosive Potential (free of acetic acid)

  - Improved Thermal and Chemical Stability

  - Similar or better mechanical and optical properties

  - Better aging behaviour

  - Lower Moisture Vapor Transmission Rate (MVTR)

  - Volume Resistivity (VR) significantly improved

Why EBA?

- There are commercial grades of EBA based encapsulants available in the market[1]

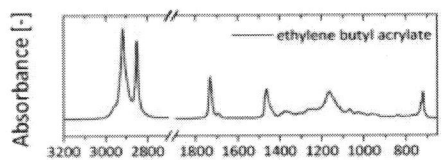

Figure 2: Infrared spectra of identified chemical structure within the monolayer encapsulants collected.

#	Manufacturer	Material Composition	Melting temperature	Curing Chemistry	Optical properties	Availability 2023
1	M1	ethylene butyl acrylate	81°C	peroxide	UV-T	no
2	M1	ethylene butyl acrylate	90°C	peroxide	UV-T	yes
3	M1	ethylene butyl acrylate	90°C	peroxide	UV-B	yes
4	M2	ethylene ethyl acrylate	92°C	silane	UV-T	yes
5	M3	ethylene ethyl acrylate	93°C	silane	W	yes
6	M4	ethylene α-olefin	62°C	peroxide	UV-T	no
7	M4	ethylene α-olefin	65°C	peroxide	UV-T	yes
18	M8	ethylene acrylic acid	90°C	none	UV-B	no
19	M8	ethylene acrylic acid butyl acrylate	97°C	none	UV-B	no
20	M9	ethylene acrylic acid	93°C	none		no
25	M22	ethylene α-olefin	106°C	none	UV-B	yes
26	M23	ethylene α-olefin		none	UV-B	yes
27	M24	ethylene α-olefin				

- Attending at the melting points[1] and FTIR spectra[2,3] published in the literature, it can be inferred that an EBA grade with 20% BA and 20MFI is being used in these commercial products and has been described in different published papers.[4,5]

1. ORESKI, Gernot, et al. Polyethylene copolymers as solar cell encapsulants: A critical overview. 40th European PVSEC 2023.
2. ORESKI, Gernot, et al. Properties and degradation behaviour of polyolefin encapsulants for photovoltaic modules. Progress in Photovoltaics Research and Applications 28(1):1-12.
3. DINCHEVA NT., et al. Durability and Performance of Encapsulant Films for Bifacial Heterojunction Photovoltaic Modules. *Polymers*. 2022; 14(5):1052.
4. REID, Charles G., et al. Contribution of PV encapsulant composition to reduction of potential induced degradation (PID) of crystalline silicon PV cells. 28th EU PVSEC 2013 (4AV.5.49)
5. SCHNEIDER, Andreas, et al. Material Developments allowing for new applications, increased long term stability and minimized cell to module power losses. 31st EU PVSEC 2015 (1BV.6.36)

Why EBA?

- Competitive price compared to EPE or POE's*

  - Reduction of monomer content down to 20%BA results in competitive pricing, even if compared to standard EVA (28% - 33% VA)

  - Prices for different encapsulants (H2 2022*)

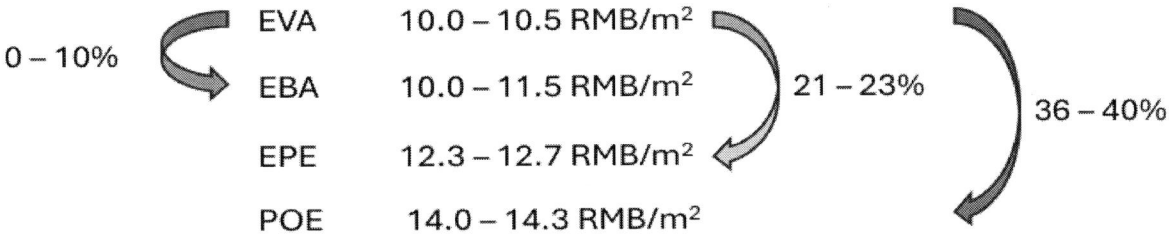

EVA	10.0 – 10.5 RMB/m²
EBA	10.0 – 11.5 RMB/m²
EPE	12.3 – 12.7 RMB/m²
POE	14.0 – 14.3 RMB/m²

0 – 10%

21 – 23%

36 – 40%

New encapsulants

- Two new encapsulants were prepared for this study in a pilot extrusion line using peroxide curing chemistry:

  ➢ **EVA based encapsulant**
    - Repsol Primeva 28025S resin **(28%VA; 25MFI)**
    - PID resistant formulation
    - **VR > 10^{15} Ω·cm**
    - Width 500mm. No embossing.

  ➢ **EBA based encapsulant**
    - Repsol Ebantix E20020 resin **(20%BA; 20MFI)**
    - PID resistance higher than EVA
    - **VR > 10^{16} Ω·cm**
    - Width 500mm. No embossing.

- These new formulations are an evolution of those previously described in the literature.

- Their performance will be checked using **high efficiency cells** (IBC and TopCon), and compared to different competitive materials (**EPE and two different POE's**) with **special focus on PID testing**

- Results and conclusions in this presentation are preliminary and valid for specific material-process combinations. Tests were only executed for one single cell batch per cell type (IBC and TopCon) as well as one single batch of each type of encapsulant (EVA, EBA, EPE, POE1 and POE2).

New encapsulants

- Properties of EVA and EBA compared to competitive materials:

Properties	Units	EVA	EBA	EPE	POE1	POE2
Melt Flow Index (190°C, 2.16Kg)	g/10'	25	20	Trilaminate		
Monomer content	%	28	20			
Melting point	°C	70	89		40 - 80	65 - 75
Density	Kg/m^3	950	925		880	879
Thickness	mm	0,45 - 0,50	0,45 - 0,50	0,45 - 0,50	0,45 - 0,50	0,45 - 0,50
%T at 500nm	%	91	90	90-91	90	91
UV cutoff	nm	<300	<300			<300
Volume resistivity	Ω·cm	>10^{15}	>10^{16}	>10^{15}	>10^{15}	>10^{16}
Adhesion to glass	N/cm	140	120-140			
MVTR 25°C, 100%RH	g/m^{2}d	18	5,2		< 5	

New encapsulants

- Lamination cycles recommended for the products used in this study:

	EVA	EBA		EPE	POE1	POE2
Module constr.	G/B	G/B	G/G	G/B	G/B	G/B
Platen Temperature (°C)	152	152	152	155	155	155
Evac. time (min)	4	5*	5	6:20	6:20	6:20
Ramp up time (min)	1	1	1	1	1	1
Cure time (min)	8	10	11	15	14	20
Top Temp. (°C)	60	60	90	60	60	60
Total cycle time (min)	**13**	**16***	**17**	**22:20**	**21:20**	**27:20**
Gel content (%)	85	70	70	>80	>60	>70

In general terms, lamination cycles are shorter for EVA and EBA compared to EPE and POE's

* Evacuation time had to be extended 5 minutes on pins in these tests for EBA in G/B modules due to the appearance of some bubbles, probably due to the absence of texture in the film and the upper heating system. Cycle optimization is pending. No bubbles were observed with the standard lamination cycle in a standard laminator without upper heating system.

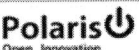 

Reliability studies

- The main goals of these reliability study can be summarized as follows:

  - Perform aging studies of minimodules manufactured with this new **EBA** encapsulant and two different high efficiency cells (**IBC** & **TopCon**)

  - Investigate the **PID protection** ability of this new **EBA** based encapsulant compared to other materials available in the market (**EVA, EPE** & **POE's**)

  - Investigate the resistance of the new **EBA** based encapsulant under accelerated **Damp Heat** and **Thermal Cycle** tests (tests in progress)

Reliability studies

PID test according to IEC 62804

- Testing matrix:

  5 encapsulants
  2 cell types (IBC & TopCon)
  <u>Glass/backsheet modules</u>

- EL/PL measurements
- IV measurements

- EBA will be tested up to 192h
  with 2 different cell types

Testing Group	1	2	3	4	5	6
Purpose	PID 192	PID 192	PID 96	PID 96	PID 96	PID 96
Cell Type (only half cells for all groups)	IBC ZEBRA M6 688	TopCon M6 988	IBC ZEBRA M6 688	IBC ZEBRA M6 688	IBC ZEBRA M6 688	IBC ZEBRA M6 688
Ribbon	0,8 x 0,24mm SnPb	0,8 x 0,24mm SnPb	0,8 x 0,24mm SnPb	0,8 x 0,24mm SnPb	0,8 x 0,24mm SnPb	0,8 x 0,24mm SnPb
Cell Matrix / Minimodule	2x Half Cell/ 2mm Cell - Cell Distance	2x Half Cell/ 2mm Cell - Cell Distance	2x Half Cell/ 2mm Cell - Cell Distance	2x Half Cell/ 2mm Cell - Cell Distance	2x Half Cell/ 2mm Cell - Cell Distance	2x Half Cell/ 2mm Cell - Cell Distance
Encapsulation Material	EBA	EBA	EVA 01 EVA from Repsol	POE 01 (VR 10*15)	POE 02 (VR 10*15)	EPE 01
Backsheet Material	PET/PET/Primer	PET/PET/Primer	PET/PET/Primer	PET/PET/Primer	PET/PET/Primer	PET/PET/Primer
Backsheet colour	transparent	transparent	transparent	transparent	transparent	transparent
Thickness Backsheet	0,26 mm	0,26 mm	0,26 mm	0,26 mm	0,26 mm	0,26 mm
Glass	220x220x2,0mm	220x220x2,0mm	220x220x2,0mm	220x220x2,0mm	220x220x2,0mm	220x220x2,0mm
Module Type	GB*	GB*	GB*	GB*	GB*	GB*
Module Stack Structure	Glass-1xEBA-Cells-1xEBA-Backsheet	Glass-1xEBA-Cells-1xEBA-Backsheet	Glass-1xEVA-Cells-1xEVA-Backsheet	Glass-1xPOE-Cells-1xPOE-Backsheet	Glass-1xPOE-Cells-1xPOE-Backsheet	Glass-1xEPE-Cells-1xEPE-Backsheet
Lamination Recipe	Recipe for EBA	Recipe for EBA	Recipe for EVA	Recipe for POE	Recipe for POE	Recipe for EPE
Count of modules / laminates	3+1	3+1	3+1	3+1	3+1	3+1

*GB = glass - backsheet

Reliability studies

PID test according to IEC 62804

- Test conditions:
  - 85% relative humidity
  - 85°C
  - 1000V
  - Polarity:
    IBC (with FSF): positive to cells, negative to foil
    TopCon: negative to cell, positive to foil

- Schematic:

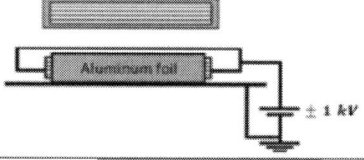

Reliability studies

IV measurements after **96h** in PID test show:

- All values within pass criteria of 5%

- EBA outperforms EVA with IBC Cells

- EBA/IBC performs similar to EPE/IBC

- IBC performed better than TopCon with EBA

- POE's outperform both EBA and EPE with IBC cells

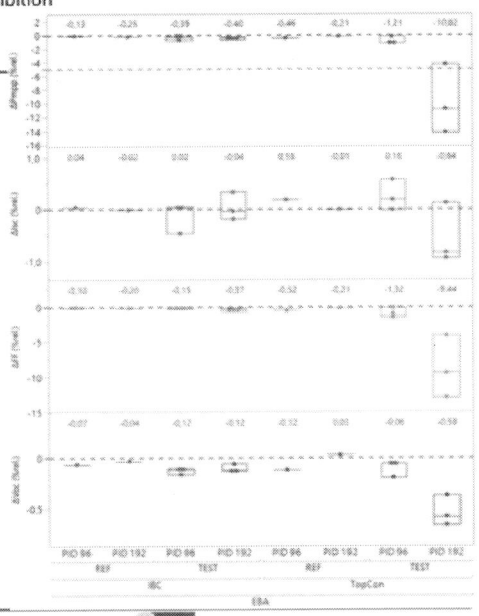

Reliability studies

IV measurements for EBA after **192h** in PID test show:

- IBC:
    - Median degradation of 0,40% for P_{MPP}
    - Related to the reference cell, this degradation becomes even less significant
    - No apparent impact of the second PID cycle

- TopCon:
    - Second PID aging cycle has a strong impact
    - Median degradation in P_{MPP} is >10% in after this cycle
    - Degradation is mainly driven by a drop in the FF, though Isc and Voc are also impacted

Reliability studies

PID aging - EL/PL measurements for EBA minimodules:

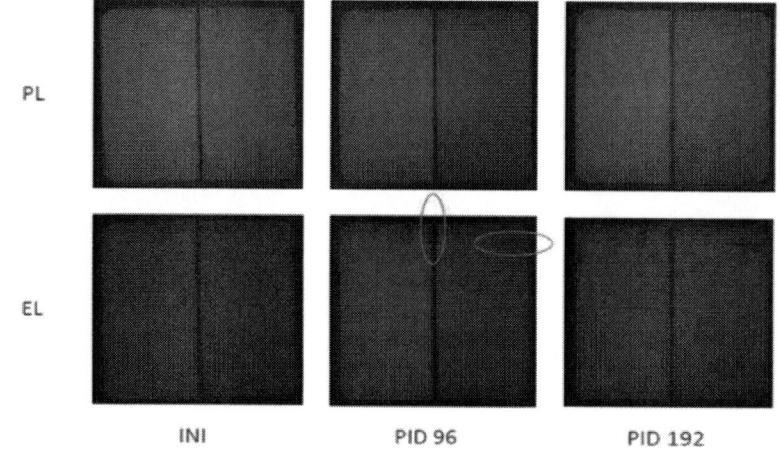

PL		
EL		
INI	PID 96	PID 192

- Example G01 M02
  - EBA
  - IBC
- EL/PL measurements after PID 96 aging show
  - One new dark patch in EL (upper right corner)
  - Electrical contact to cells intact after test
  - One additional finger interruption after PID 96 (* in the other samples of this group this was not the case)
  - Degradation on cell that seems to be linked to the structure of the cells (pattern runs parallel to busbar layout)

Polaris Open Innovation   **REPSOL**   **ISC**

Reliability studies

PID aging - EL/PL measurements for EBA minimodules:

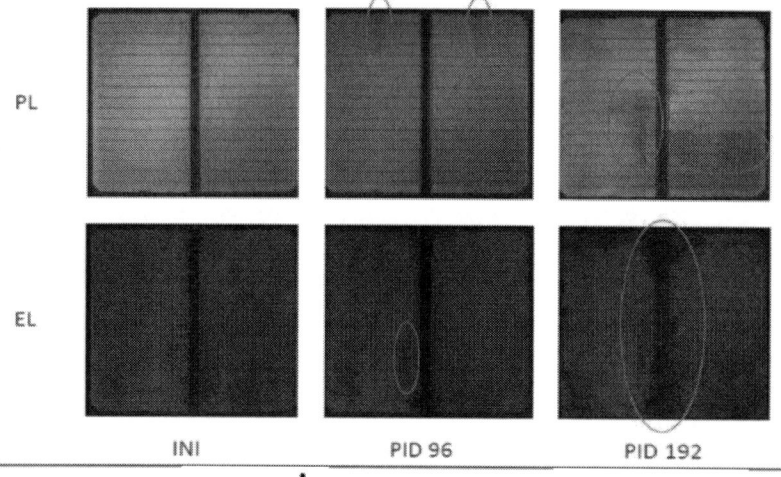

PL		
EL		
INI	PID 96	PID 192

- Example G02 M03
  - EBA
  - TopCon
- EL/PL measurements after PID 96 aging show
  - Growing degraded patches
  - Consistent to decrease in P_{MPP} in IV measurements (affecting mostly FF, but effect also visible in I_{sc} and V_{oc})

Polaris Open Innovation   **REPSOL**   **ISC**

Reliability studies

Summary of preliminary PID results:

- Performance of EBA in glass/backsheet construction:
    - Passing (loss in P_{MPP} < 5%)
        - PID 192h for IBC
        - PID 96h for TopCon

- Comparison with other encapsulants with IBC cells:

    EVA < EBA ~ EPE < POE1 ~ POE2

- Next steps:
    Test EBA with TopCon cells in glass/glass construction
    Extend comparison to EPE with TopCon in G/B and G/G construction

Reliability studies

Damp Heat aging according to IEC 62804

- Testing matrix:

    2 encapsulants (EBA & EVA)
    2 cell types (IBC & TopCon)
    <u>Glass/backsheet modules</u>

- EL/PL measurements

- IV measurements

Testing Group	7	8	9
Purpose	DHT 3000	DHT 3000	DHT 3000
Cell Type (only half cells for all groups)	IBC ZEBRA M6 6BB	TopCon M6 9BB	IBC ZEBRA M6 6BB
Ribbon	0,8 x 0,24mm SnPb	0,8 x 0,24mm SnPb	0,8 x 0,24mm SnPb
Cell Matrix / Minimodule	2x Half Cell/ 2mm Cell - Cell Distance	2x Half Cell/ 2mm Cell - Cell Distance	2x Half Cell/ 2mm Cell - Cell Distance
Encapsulation Material	EBA	EBA	EVA 01 EVA from Repsol
Backsheet Material	PET/PET/Primer	PET/PET/Primer	PET/PET/Primer
Backsheet colour	transparent	transparent	transparent
Thickness Backsheet	0,26 mm	0,26 mm	0,26 mm
Glass	220x220x2,0mm	220x220x2,0mm	220x220x2,0mm
Module Type	GB*	GB*	GB*
Module Stack Structure	Glass-1xEBA-Cells-1xEBA-Backsheet	Glass-1xEBA-Cells-1xEBA-Backsheet	Glass-1xEVA-Cells-1xEVA-Backsheet
Lamination Recipe	Recipe for EBA	Recipe for EBA	Recipe for EVA
Count of modules / Laminates	3+1	3+1	3+1
*GB = glass - backsheet			

Reliability studies

- Damp Heat testing is currently in progress

- Target: complete 3000h (x3 IEC 62904)

- Previous results in the literature[5] (2015 EU PVSEC paper) demonstrate that both EVA and EBA passed 2500h and 2000h in DH85/85 respectively with IBC cells[6]

- Comprehensive publication of these results after completion of this study

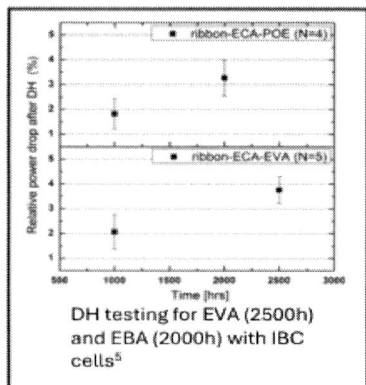

DH testing for EVA (2500h) and EBA (2000h) with IBC cells[5]

5. SCHNEIDER, Andreas, et al. Material Developments allowing for new applications, increased long term stability and minimized cell to module power losses. 31st EU PVSEC 2015 (1BV.6.36)

Reliability studies

Thermal Cycles aging according to IEC 62804

- Testing matrix:

  1 encapsulants (EBA)
  2 cell types (IBC & TopCon)
  Glass/glass modules

- EL/PL measurements

- IV measurements

Testing Group	10	11
Purpose	TCT 600	TCT 600
Cell Type (only half cells for all groups)	IBC ZEBRA M6 6BB	TopCon M10 16BB
Ribbon	0,8 x 0,24mm SnPb	0,4x 0,2mm SnPb
Cell Matrix / Minimodule	2x Half Cell/ 2mm Cell - Cell Distance	2x Half Cell/ 4mm Cell - Cell Distance
Encapsulation Material	EBA	EBA
Backsheet Material	-	-
Glass	220x220x2,0mm	250x250x2,0mm
Module Type	GG*	GG*
Module Stack Structure	Glass-1xEBA-Cells-1xEBA-Glass	Glass-1xEBA-Cells-1xEBA-Glass
Lamination Recipe	Recipe for EBA	Recipe for EBA
Count of modules / Laminates	3+1	3+1
*GG= glass - glass		

Reliability studies

- IV preliminary results after TC 200 show:

  - All material-cell combinations pass test with **less than 5% degradation in P$_{MPP}$**

  - No significant degradation with IBC cells

  - IBC cells shows better performance compared to TopCon with EBA

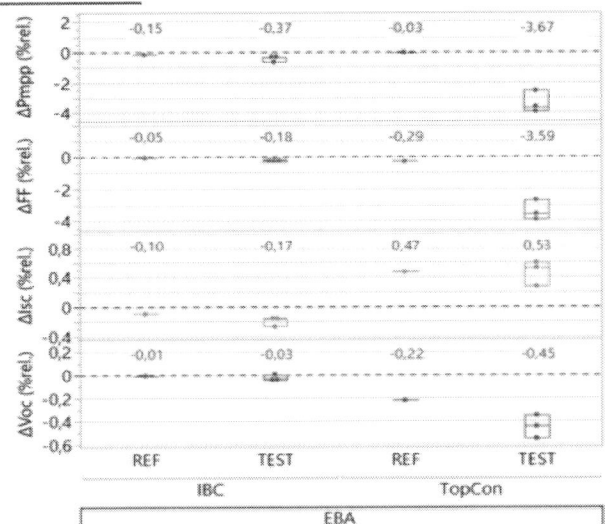

Polaris Open Innovation   **REPSOL**   **ISC**

Conclusions

- According to these preliminary results EBA is a promising candidate for the encapsulation of high efficiency solar cells

- EBA can be processed using existing extrusion technology for standard EVA encapsulants

- EBA has demonstrated good performance with IBC cells in different aging tests:
  - PID: 192h with neglectable power loss in glass/backsheet construction
  - 2000h in DH (prior art) and 200 TC (this work) (further testing ongoing)

- Performance comparable to EPE encapsulant when using IBC cells and:
  - EBA is more competitive in terms of pricing
  - Shorter lamination cycle time for EBA* compared to EPE and POE's

*Optimization pending in laminators with upper heating system

Polaris Open Innovation   **REPSOL**   **ISC**

Next Steps

- Complete accelerated aging of EBA with IBC cells:

    - 3000h in DH85/85
    - 600TC

- Extend PID testing of EBA with TopCon cells:

    - PID testing with glass/glass construction
    - Comparison with EPE

- Comprehensive publication in preparation, to include additional tests results

Bibliographic References

1. ORESKI, Gernot, et al. Polyethylene copolymers as solar cell encapsulants: A critical overview. 40th European PVSEC 2023.

2. ORESKI, Gernot, et al. Properties and degradation behaviour of polyolefin encapsulants for photovoltaic modules. Progress in Photovoltaics Research and Applications 28(1):1-12. DOI:10.1002/pip.3323

3. DINCHEVA NT., et al. Durability and Performance of Encapsulant Films for Bifacial Heterojunction Photovoltaic Modules. *Polymers*. 2022; 14(5):1052. https://doi.org/10.3390/polym14051052

4. REID, Charles G., et al. Contribution of PV encapsulant composition to reduction of potential induced degradation (PID) of cristalline silicon PV cells. 28th European PVSEC 2013 (4AV.5.49)

5. SCHNEIDER, Andreas, et al. Material Developments allowing for new applications, increased long term stability and minimized cell to module power losses. 31st European PVSEC 2015 (1BV.6.36)

6. POE in this article is Photocap 35521P from STR. Both Photocap 35521P and Photocap 35530P (STR X-31 Series) have been identified via FTIR as EBA encapsulants elsewhere[2,3]

Thank you

Muchas gracias

Vielen Dank

EUPVSEC 41st Edition 2024
3CO.10.4

Reducing Process Time of PV Module Lamination by Using Double Side Heating System

Authors: Sraisth[1], Djamel Eddine Mansour[2], Aksel Kaan Öz[2], Paul Gebhardt[2], Daniel Klaus[3], Christine Wellens[2]

[1]RCT Solutions GmbH, Line-Eid-Str. 1, 78467, Konstanz, Germany
[2]Fraunhofer Institute for Solar Energy Systems ISE, Heidenhofstr. 2, 79110, Freiburg, Germany
[3]Robert Buerkle GmbH, Stuttgarter Str. 123, 72250, Freudenstadt, Germany

25 September 2024

Contents

➢ Motivation

➢ Lamination Process and Systems

➢ Materials and Equipment

➢ Experiments and Results

➢ Conclusion

"GEPARD"

Motivation

Product Focus Glass-Backsheet Module (using PERC cell Technology)
In 2020, the market share of GB modules was 85% [ITRPV, 2019]

Glass-Backsheet Module

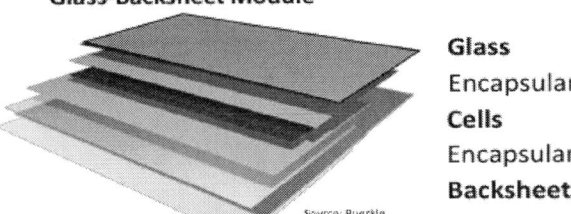

Glass
Encapsulant
Cells
Encapsulant
Backsheet

Source: Buerkle

Production Focus To avoid the laminator as the bottleneck step

Factory Space Constraint – To shorten the process
Existing Machines – To increase the throughput

Module Production Line

Source: JSG

Lamination

Lamination Process and Systems

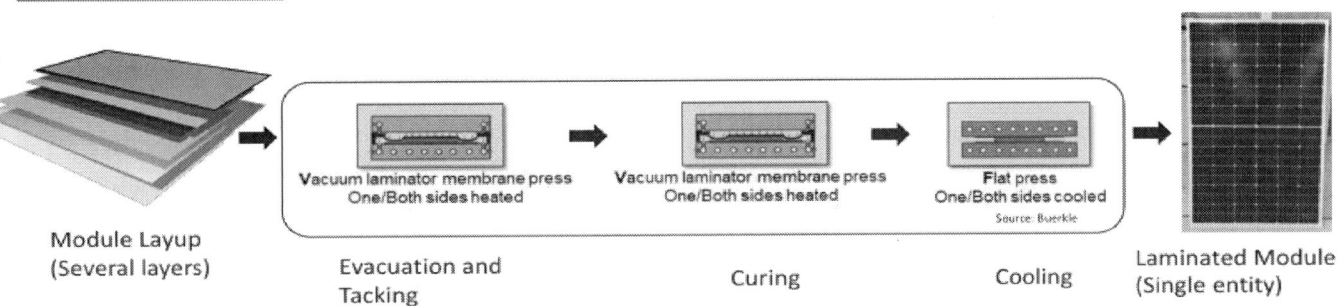

Module Layup (Several layers) → Evacuation and Tacking | Vacuum laminator membrane press One/Both sides heated → Curing | Vacuum laminator membrane press One/Both sides heated → Cooling | Flat press One/Both sides cooled (Source: Buerkle) → Laminated Module (Single entity)

Process
Temperature, Pressure, and Time.
⬇
Evacuation time, pressing time, point of pressing, rate of pressure application.

Machine
Heating System (Oil or Electrical),
Single-side heating or double-side heating,
Cooling with fan or press (cooling water), Pins system, with/without membrane

Lamination Process and Systems

Standard Lamination Process (Single-Side Heating)

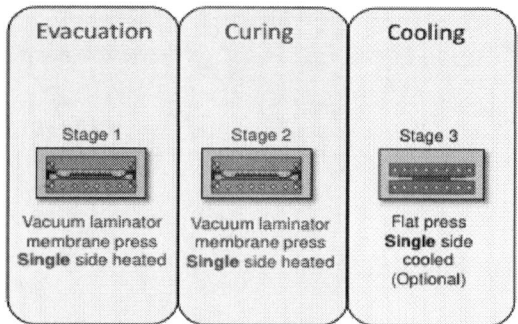

Process time = 14 min

Short Lamination Process (Double-Side Heating)

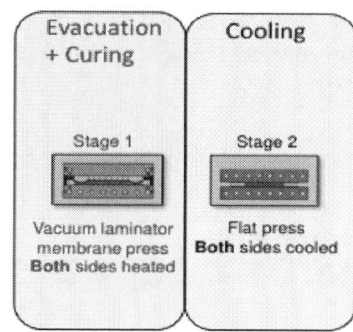

Process Time = 7 min

Process time = not including non-value added time (press closing, opening and module transfer)

Materials and Equipment

Module Type	60 cells glass-backsheet (GB) modules
Cell Technology	4 busbars (BB) mono PERC
Glass	3.2 mm thick glass, 1695 mm x 995 mm
Encapsulant	EVA F406P (front), F806P (rear) 0.45 mm
Backsheet	1000V 0.275 mm PPE structure PET-PET-Primer (Ethylene)
Ribbons	1.2 mm x 0.2 mm (Cu coated with Sn60Pb40 solder)

Buerkle Lab Laminator

Source: Buerkle

➤ Standard 60 Cells PERC Glass-backsheet Module BOM Selected

Experiment

Finding Right Temperature and Process Time	Finding the process parameters	Proving it with Full Size module lamination	Testing the modules for long term stability

➢ **DSC Measurement:** (@Buerkle)

➢ **Soxhlet Measurement** (@Farunhofer-ISE)

Adhesion Test
A.K. Öz et al., EUPVSEC 2021
DOI: 10.4229/EUPVSEC20212
021-4AV.1.5

➢ **Lamination of full-size modules** without interconnection, Visual Defect Free process (@Buerkle)

➢ **Temperature Measurement** (@Buerkle)

➢ **Process transfer** to Fraunhofer-ISE

➢ **Production of 8 full-size glass-backsheet**(GB) modules (@Fraunhofer-ISE)

➢ **Testing full-size modules** according to IEC 61215:2 and IEC 61730:2 (@Fraunhofer-ISE)

Finding Process Temperature and Time

DSC Measurement (@ Buerkle)

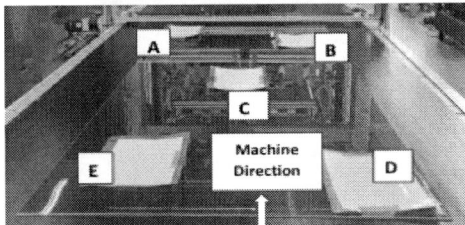

Netsch DSC 200 F3 machine
DSC: Differential Scanning Calorimetry

➢ **40 Recipes ->160 °C to 180 °C (5 °C Δ), 2 - 9 minutes process (1 min Δ)**

➢ **5 samples per recipe**

➢ **200 samples per heating system**

Soxhlet Measurement (@ Fraunhofer-ISE)

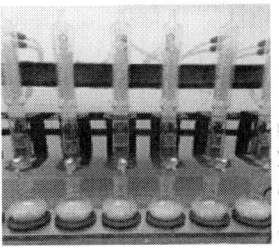
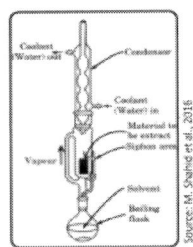

Behrotest apparatus based on the IEC 62788-1-6:2017

➢ **2 Samples per recipe**

➢ **8 hours extraction, xylene, 20 cycles per hour, no antioxidant, drying time 14 hours**

DSC & Soxhlet Measurement Results
Single-side Heating Vs. Double-side Heating

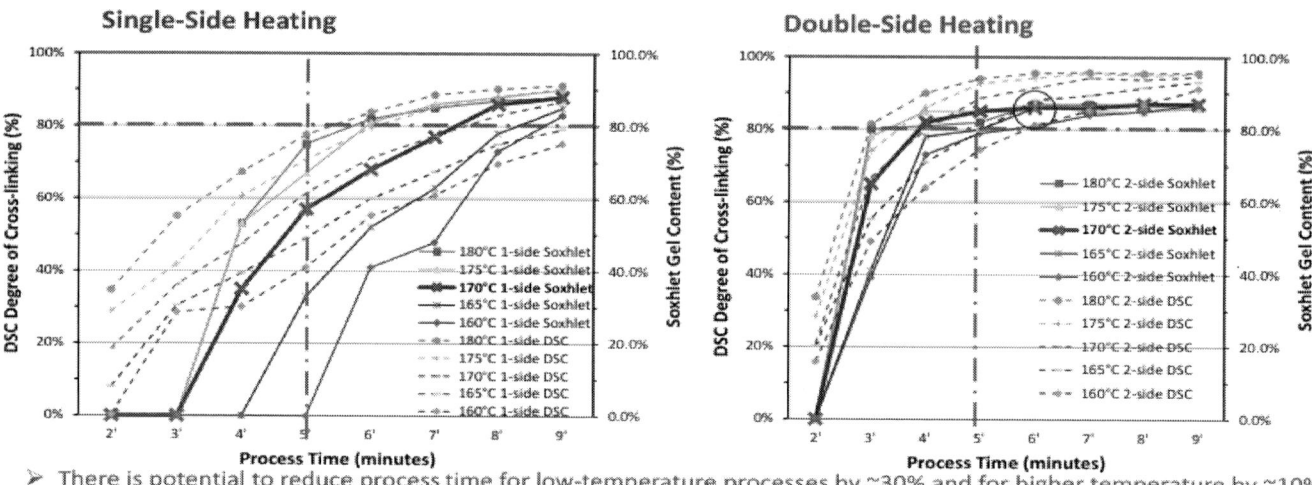

- There is potential to reduce process time for low-temperature processes by ~30% and for higher temperature by ~10%
- Processes selected: 170 °C, 6 minutes process

Finding Process Parameters

Module Lamination (@ Buerkle)

- **7 Processes tested** (Modules laminated without soldering)
- **Finding the right process parameters**
- **Criteria: No visual defects**

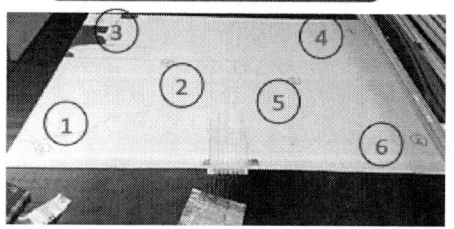

Temperature Measurement (@ Buerkle)

DATAPAQ Q18, Type K 6 Thermocouples

- **5 recipes measured**
- **To confirm the module temperature and process steps**

Temperature Profile: A.K. Öz et al., IEEE Journal of PV, 2024, DOI: 10.1109/JPHOTOV.2024.3414117

Visual Results (170 °C Set Temp.)

	6 Minutes process						5 minutes process
Recipe No. **Category**	#1	#2	#3	#4	#5	#6	#7
Backsheet optic	1	5	1	8	5	10	10
Bubbles b/w cells	10	10	10	10	10	10	10
Bubbles at the contact area	10	10	10	10	10	10	10
String movement	10	10	10	10	10	10	10
Cell breakage	10	10	10	10	10	10	10

10= Good, without defects; 5= Minor defect; 1=Bad, many defects;

Bubbles criteria: 10 = no bubbles, 1 = more than 5 bubbles
Backsheet criteria: 10: flat surface, 1 = no wrinkles and bumps

➢ Final Process selected: 170 °C, 5 Minutes Process Time (including sub-steps like evacuation time, curing time)

Bad Backsheet Optic

Good Backsheet Optic (Final Process)

Temperature Measurement (Double-side heating)

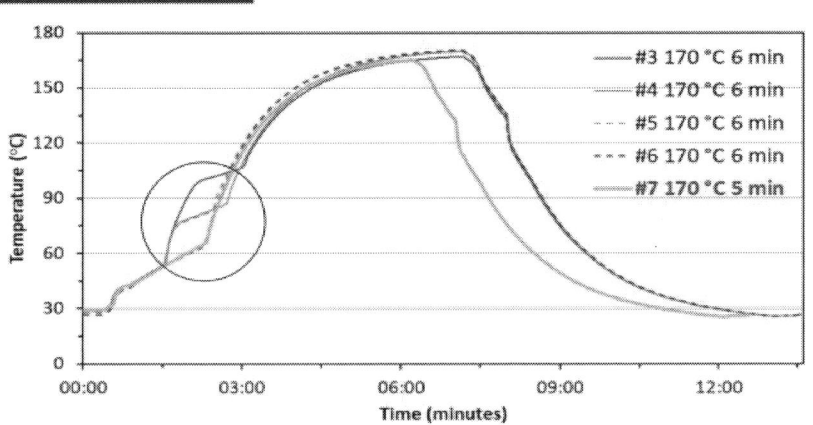

Only 5 processes were tested

The first two processes were neglected due to optical defects of the backsheet

➢The poor backsheet appearance is due to the early jump in module temperature during the evacuation phase.

➢ For 6 minutes process, the module's maximum temperature was ~168 °C, and for 5 minutes process, it was ~163.5 °C

➢ Final Process selected: 170 °C, 5 Minutes Process Time

Long-term Durability Testing of Full-Size Modules
IEC 61215-2 and IEC 61730-2

No. of modules	Testing condition	Module ID
2	**Thermal cycling tests** (TC 200, -40/+85 °C) IEC 61215-2. **Extended to 800 cycles.**	M18, M19
2	**Damp-heat test** (DH 1000 h 85 °C/85 %RH) IEC 61215-2. **Extended to 3000 h.**	M21, M23
1	**Mechanical load test after damp-heat (DH 1000)** IEC 61215-2	M22
1	**Sequence B**, IEC 61730-2	M24
1	**Sequence B1**, IEC 61730-2	M25
1	Reference	M20

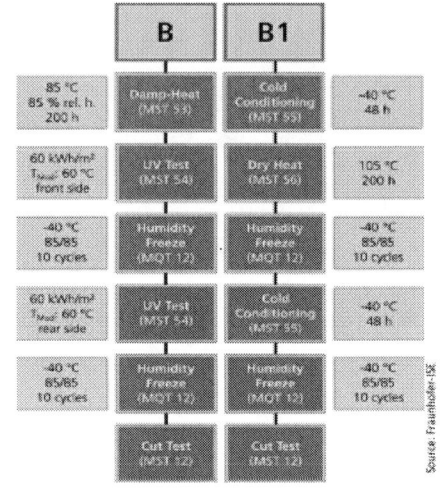

Long-term Durability Testing Results (IEC Standard)

Power loss after IEC 61215-2

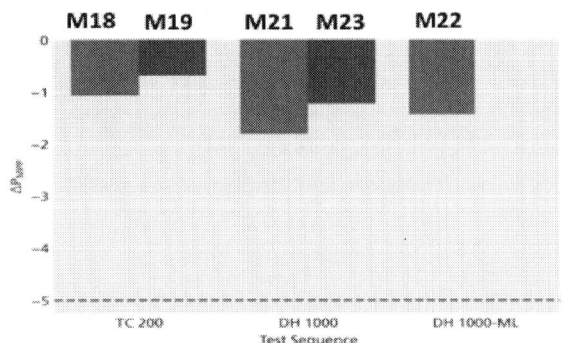

Module ID: M21
DH 1000h

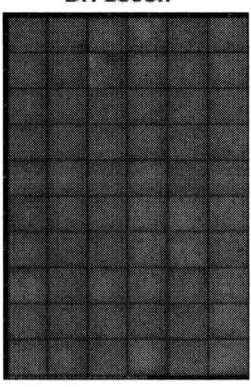

Pmpp degaradation **<-2%**

➢ Module passed the IEC 61215:2 Standard

Long-term Durability Testing Results (Extended IEC)

Power loss after extended IEC 61215-2 and IEC 61730-2

Module ID: M21

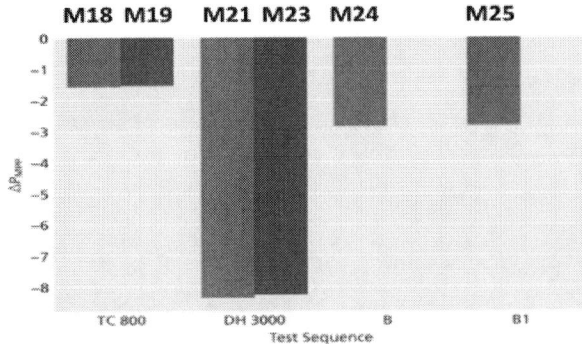

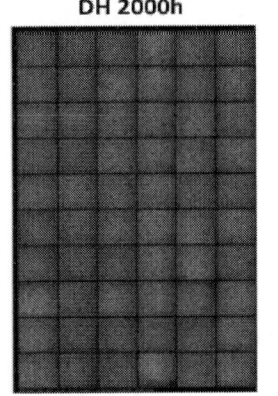

DH 2000h   DH 3000h

Pmpp degradation   Pmpp degradation
<-2.5%   **-8.4%**

Pressure Influence:
Sraisth et al., EUPVSEC 2022, DOI: 10.4229/WCPEC-82022-3DV.1.17

➤ Module passed the IEC 61215:2 and IEC 61730:2 Standard

Conclusion & Discussion

➤ Double-side heating has an **influence on the gel content** and, eventually, the lamination process.

➤ Based on the tested conditions, the lamination process can be **shortened by 10 to 30%** using double-side heating compared to the common single-side heating processes.

➤ The **short lamination process was verified** by laminating and testing full-size modules for long-term stability tests according to IEC 61215:2 and IEC 61730:2.

➤ Double-side heating system can help in **saving production space (20~25%)** and **increasing the machine's throughput**.

➤ The same procedure can be followed for **new BOMs**
   ➤ Glass-Glass (GG) Modules with new cell technologies (TOPCon, HJT, IBC) and encapsulants combination (EVA, EPE, POE)
   ➤ Glass-Backsheet (GB) Modules with new cell technologies (TOPCon, HJT, IBC) and encapsulants combination (EVA, EPE, POE)

This work was supported in part by the German Federal Ministry for Economic Affairs and Energy (BMWi) in the scope of the project **"GEPARD"** (contract number. 0324287C)

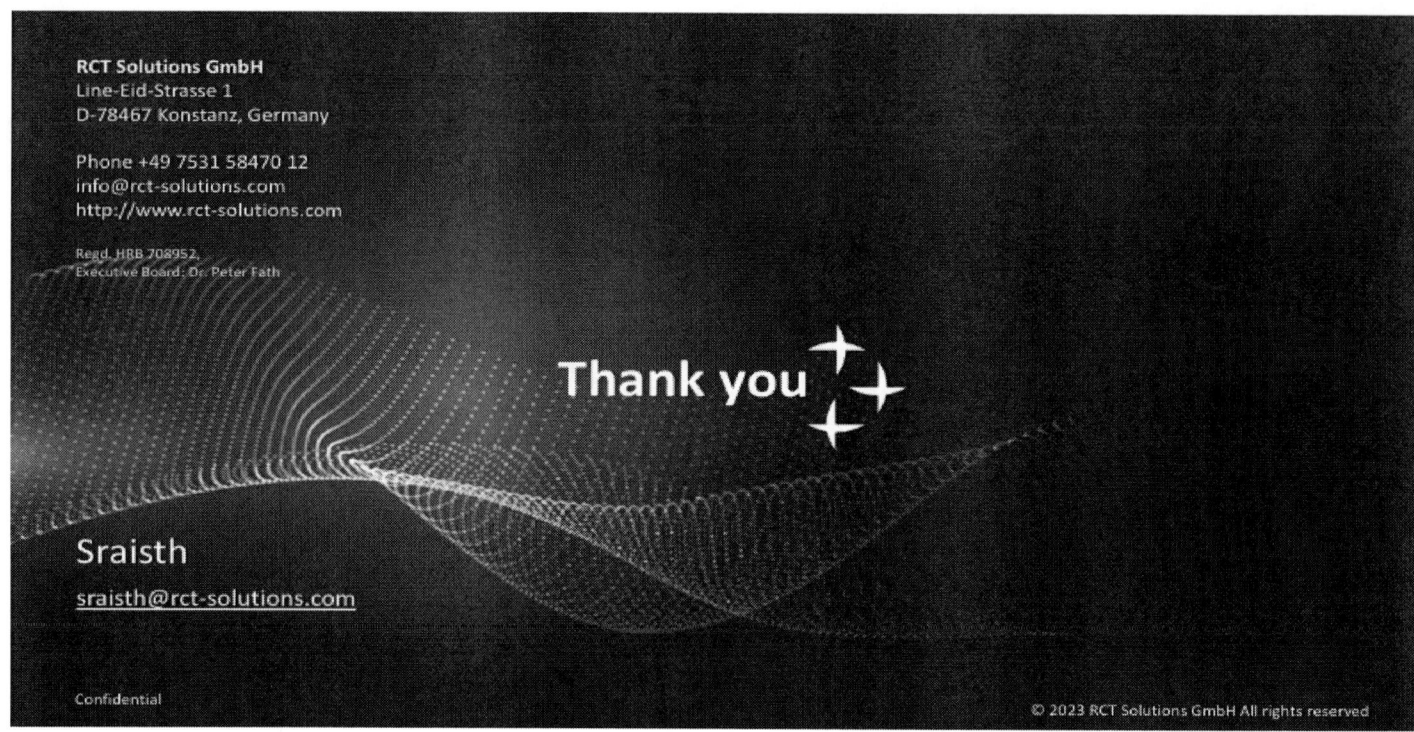

41st European Photovoltaic Solar Energy Conference and Exhibition

Design Roadmap to Modules with 24 % Efficiency

Max Mittag, Christian Reichel, Alexander Protti, Dirk Holger Neuhaus
Fraunhofer Institute for Solar Energy Systems ISE
max.mittag@ise.fraunhofer.de

ABSTRACT: We analyze loss channels in module efficiency using the cell-to-module (CTM) methodology for three different module concepts (based on conventional half cells, shingled cells and interdigitated back contact cells) and assess changes in module and component design to reduce losses and to reach 24 % in module efficiency. We quantify the improvements and calculate the necessary cell efficiency to reach 24 % in module efficiency. Our findings indicate that reducing cell spacing significantly enhances module efficiency, with additional gains observed from decreasing string and edge spacing, as well as implementing frameless designs. However, these changes require careful consideration of manufacturing processes, reliability and electrical safety standards. For shingled modules, reducing cell overlap and applying edge passivation techniques can mitigate power losses from cell splitting and edge recombination, respectively. Optimization efforts, including the use of macro-structured front and rear glass, improved anti-reflective coatings, and lower effective shading ribbons, result in substantial efficiency gains across all module types in the range of 0.7 %$_{abs}$. The study concludes that while the half-cell module exhibits the lowest efficiency, shingled modules with edge passivation achieve the highest efficiencies, potentially reaching 24 %. This research provides valuable insights into the necessary cell efficiencies for achieving 24 % module efficiency across different designs, highlighting the potential of advanced cell and interconnection technologies in future commercial applications.

I. Introduction

Module efficiency is one of the main drivers to reduce the costs of electricity. Several cell and interconnection technologies are available as promising candidates to allow for further improvement of module efficiency. We analyze modules based on conventional half cell modules with round ribbons, shingled modules with electrically conductive adhesives (ECA), and modules with ribbon-based interconnection of interdigitated back contact (IBC) cells.

We analyze these technologies using the cell-to-module (CTM) loss analysis methodology and calculate a reference module efficiency starting from comparable cell efficiency. We further analyze efficiency potentials through module component and module design improvements and quantify improvements in module efficiency. We also explore possible options to further increase module efficiency and quantify the effects.

In a third step, we apply a reversed calculation to answer the question what cell efficiency is necessary to achieve a 24 % module efficiency for the three module and cell designs?

II. Methodology

Module efficiency is calculated using Equation 1.

$$module\ efficiency = \frac{module\ power}{module\ area \cdot irradiance} \quad (I)$$

Following this, we find the optimization of module area and module power to be relevant to increasing module efficiency. Module area consists of six different area shares that need to be considered: solar cell area, pseudo-square edge area, cell spacing, string spacing, edge spacing, and frame area (Figure 1).

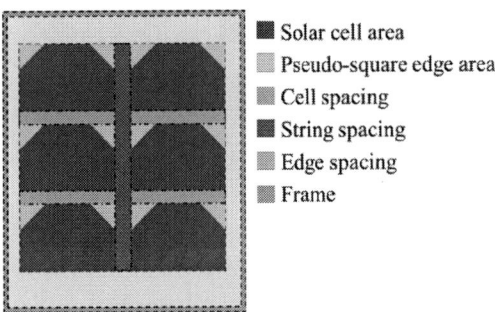

Figure 1: Module area shares (dark blue = cell, light blue = cell spacing, red = string spacing, green = pseudo-square area, orange = frame, grey = edge spacing

The goal is to maximize cell area and to reduce the other areas in order to increase module efficiency.

Module power can be increased by reducing cell-to-module loss channels and by increasing gains on module level (Figure 2). More specifically, module power can be increased by reducing optical losses (reflection, absorption and shading), the improvement of internal light recycling, component tuning and the reduction of electrical losses.

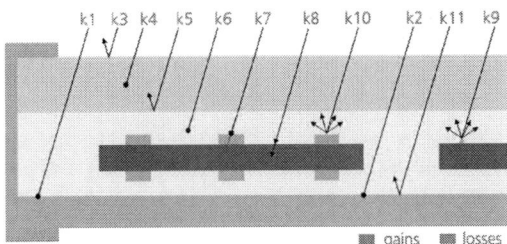

Figure 2: Optical gain and loss channels in photovoltaic modules; modified from [1].

The impact of changes in module design is quantified using SmartCalc.Module, developed by Frauhofer ISE, which is a software tool to calculate module power and efficiency based on material properties and module design information (bottom-up approach) [2].

A conventional half cell module, a module with shingled cells and a module based on interdigitated back-contact (IBC) cells are compared.

Based on reference modules design we analyze the impact of changes in module layout. The reference for the half cell is based on a commercially available utility-scale module with 144 TOPCon cells (210 mm × 91 mm based on a 295 mm ingot). It is a glass-glass module with POE encapsulant and 250 μm round wire interconnection. Cell spacing is 1.5 mm and string spacing is assumed to be 1.7 mm. The initial host cell efficiency is set to 25.3%. Module dimensions are 2.278 m × 1.304 m.

The shingled module comprises 6 strings with each 72 cells (210 mm × 32 mm). Cell overlap is 0.9 mm and edge spacing distances are the same as in the half cell module.

The IBC module has the same layout as the half cell module and is considered to also use 16 cell interconnection ribbons (3 mm × 0.15 mm).

Regarding the optimization of area shares within the module we analyze a low gap cell spacing option (0.5 mm), a no gap option (0 mm) and an overlap/paving option (-0.5 mm) (Figure 3).

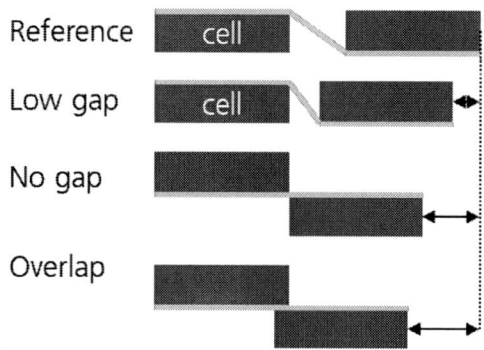

Figure 3: Cell spacing scenarios with depicted area optimization potential

In power optimization we consider the use of improved components such as macro-structured glass or reflective ribbons.

Losses on cell level are considered in the form of cell splitting losses (half cell module) and in losses from additional finger length (shingled module).

Only for shingled cells also cell level optimizations are considered by including a scenario where edge recombination losses are reduced with the application of edge passivation techniques.

III.   Results

We calculate the efficiency of the reference half cell module to be 22.5 % assuming a cell cutting loss of 0.1 %abs and a host cell efficiency of 25.3 %.

Reducing the cell spacing is found to be an option to improve module efficiency (Table 1).

Table 1: Module efficiency increase from changes in cell spacing

	Module efficiency	Δ_{abs}
Reference (1.7 mm)	22.52 %	
Low gap (0.5 mm)	22.85 %	0.23 %
No gap (0 mm)	22.97 %	0.35 %
Paving (-0.5 mm)	22.98 %	0.36 %

A reduction in string spacing from 1.7 mm to 0.7 mm increases module efficiency by 0.46 %abs and a reduction of the module edge from 14.75 mm to 10 mm increases module efficiency by 0.26 %abs. A frameless module gains 0.16 %abs with a reduction of 6 mm in module width and length (2× 3 mm frame).

A reduction in string spacing requires that the lamination process or the encapsulants allow such a change and that no movement of the strings during lamination occurs. Reducing the edge spacing needs to consider electrical safety as required in IEC 61730 or required barrier properties (e.g. diffusion of water). Omitting the frame has an impact on the mechanical stability of the module [3]. Therefore, the suggest changes are not trivial. We thus consider a less drastic improvement in the following and assume string distance of 1 mm, an edge spacing of 10 mm. We keep the module frame. Furthermore, the space required for the interconnection of strings is reduced in our analysis which leads to gains of 0.15 %abs (Figure 4). Reduction is performed by reducing the distance between the ribbons and the cells.

Those changes seem to be more feasible and improve module efficiency by 0.7 %abs.

The changes in module dimension follow the assumption that they are accepted by the market. If no such change in module dimensions is possible, then a optimization of module are shares requires changes in cell dimensions to increase their area share [4–6]. Both approaches lead to a similar increase in module efficiency.

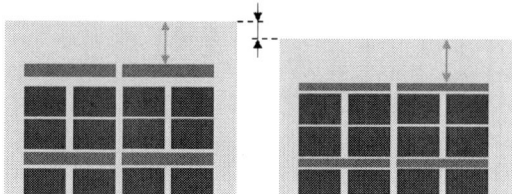

Figure 4: String spacing area reduction

The optimization of the IBC module follows the half cell module and the same changes are applied.

For the shingled module, the module area is changed by reducing the cell overlap from 0.9 mm to 0.7 mm. This increases the module area and the cell area share and reduces to power losses from cell overlap [7].

Besides the losses from cell overlap, two other losses are necessary to mention for the shingle approach. The first results from losses in the finger metallization on cell level [8]. Due to the special cell-to-cell interconnection layout losses of 0.2 %abs need to be considered for increased finger length. Also, 0.7 %abs result from edge recombination losses after cell splitting. Edge recombination losses can be reduced by applying edge passivation techniques which reduce the losses by up to 80 % [9,10]. Therefore – starting from a 25.3 % host cell – the separated shingle cell has an efficiency of 24.4 % for the unpassivated scenario and 24.9 % after application of edge passivation.

For the optimization of module power, we apply macro-structured front and rear glass [11] which increases module power by 1.9 % for the front glass by decreasing reflection losses and by 1.2 % or the rear glass through reflective gains. An improved anti-reflective coating of the glasses is assumed which results in a power gain of 0.1 %. The cell ribbon is replaced by a triangular ribbon with lower effective shading [12,13] and a power gain of 0.5 %. This leads to an increase in module power by 3.6 % for the half cell module.

For the shingled module the structured glasses are used as well. The cell overlap is reduced which results in a gain of 0.7 % and the ECA is optimized resulting in a lower contact resistance of 0.1 Ωmm^2 (compared to a reference assumption of 0.2 Ωmm^2) and a power gain of 1.4 %. The total power increase for the shingled module is 5.6 % not considering a possible gain of 0.5 % from edge passivation.

The impact of area and power optimization on module efficiency is summarized in Table 2.

Table 2: Module Optimization Results

	Half Cell	**Shingle**	**IBC**
Host cell efficiency	25.3 %	25.3 %	25.3 %
Cell level losses	-0.1 %	0.9 %	0.0 %
Cell efficiency for CTM-Analysis	25.2 %	24.4 %	25.3 %
CTM$_{efficiency}$ losses	-2.7 %$_{abs}$	-1.5 %$_{abs}$	-2.1 %$_{abs}$
Module efficiency (unoptimized)	22.5 %	22.9 %	23.2 %
Optimization	+0.7 %$_{abs}$	+0.7 %$_{abs}$	+0.7 %$_{abs}$
Module efficiency (Optimized)	23.2 %	23.6 %	23.9 %
Edge passivation		+0.5 %$_{abs}$	
Module efficiency (Optimized)	23.2 %	24.1 %	23.9 %

We find the half cell module to have the lowest efficiency. The cell level losses of the shingled cell have a significant impact on the results and low CTM$_{efficiency}$ losses for the shingled module barely compensate them. The optimization of module area shares and components is beneficial for all module concepts but they are not sufficient to reach 24 % in module efficiency. Applying an edge passivation technique results in a significant gain for the shingled module and allows for module efficiencies of 24 %.

Answering the question what cell efficiency is necessary to reach 24 % modules we find very different answers for the different module designs (Table 3). The lowest cell efficiency is necessary for the shingled cell (with edge passivation) and the half cell design requires the most efficient cell.

Table 3: Necessary cell efficiencies to reach 24 % modules

Module Design	Cell Efficiency
Ribbon-based half cell	25.9 %
Shingling unpassivated	25.7 %
Shingling passivated	25.1 %
IBC module	25.5 %

While the half cell is the most common in commercial modules [14,15], other approaches might be more suitable to reach higher module efficiencies in the future.

IV.   Summary

We analyze cell-to-module (CTM) loss channels in module efficiency for three different module concepts (based on conventional half cells, shingled cells and interdigitated back contact cells) and find an increase in module efficiency by 0.7 %$_{abs}$ to be feasible with reduced cell, string and module edge spacing.

We analyze options on how to increase module power and find that the application of macro-structured glass may increase module power by 1.9 % (if used as a front glass) and 1.1 % (if used as a rear glass). For shingled modules the reduction of cell overlap and the use of electrically conductive adhesive with a low contact resistance (1 Ωmm^2) are relevant. Cell level losses from edge recombination and increased finger lengths are major loss factors in shingle module efficiency (-0.9 %$_{abs}$). Edge passivation is therefore crucial and may reduce losses by 0.5 %$_{abs}$.

The study finds that the conventional half-cell module exhibits the lowest efficiency and shingled modules with edge passivation achieve the highest efficiencies, potentially reaching 24 %. Modules based on interdigitated back-contact (IBC) cells show a significantly higher module efficiency compared to conventional half-cells as well.

References

[1]   K.R. McIntosh, J.N. Cotsell, J.S. Cumpston, A.W. Norris, N.E. Powell, B.M. Ketola, An optical comparison of silicone and EVA encapsulants for conventional silicon PV modules: A ray-tracing study, in: Proceedings of the 34th IEEE Photovoltaic Specialists Conference, Philadelphia, Pennsylvania, USA, 2009, pp. 544–549.

[2]   cell-to-module.com. http://www.cell-to-module.com (accessed 28 February 2018).

[3]   A.J. Beinert, M. Ebert, U. Eitner, J. Aktaa, Influence of photovoltaic module mounting systems on the thermo-mechanical stresses in solar cells by FEM modelling, in: Proceedings of the 32nd European Photovoltaic Solar Energy Conference and Exhibition, Munich, Germany, 2016, pp. 1833–1836.

[4]   M. Mittag, A. Pfreundt, J. Shahid, N. Wöhrle, D. Neuhaus, Techno-Economic Analysis of Half Cell Modules - The Impact of Half Cells on Module Power and Costs, 2019.

[5]   M. Mittag, A. Pfreundt, J. Shahid, Impact of solar cell dimensions on module power, efficiency and cell-to-module losses, in: 30th PV Solar Energy Conference, 2020.

[6]   M. Mittag, C. Reichel, A. Protti, D.H. Neuhaus, Techno-Economic Analysis of Solar Modules based on Rectangular Wafers, in: 41st EUPVSEC 2024.

[7]   M. Mittag, T. Zech, M. Wiese, D. Blaesi, M. Ebert, H. Wirth, Cell-to-Module (CTM) analysis for photovoltaic modules with shingled solar cells, in: 44th IEEE PV Specialist Conference PVSC, pp. 1531–1536.

[8]   J.D. Huyeng, E. Lohmüller, B. Shabanzadeh, C. Reichel, T. Rößler, J. Weber, M. Hofmann, D. von Kutzleben, N. Abdel Latif, A. Kraft, H. Neuhaus, F. Clement, R. Preu, Challenges and advantages of cut solar cells for shingling and half-cell modules,

EPJ Photovolt. 15 (2024) 22.
https://doi.org/10.1051/epjpv/2024019.

[9] E. Lohmüller, P. Baliozian, L. Gutmann, L. Kniffki, A. Richter, L. Wang, R. Dunbar, A. Lepert, J.D. Huyeng, R. Preu, TOPCon shingle solar cells: Thermal laser separation and passivated edge technology, Prog. Photovolt: Res. Appl. 31 (2023) 729–737. https://doi.org/10.1002/pip.3680.

[10] A. Göbel, E. Lohmüller, D. Wagenmann, N. Kohn, M. Hofmann, J.D. Huyeng, R. Preu, Addressing Edge Recombination Losses in Shingle Cells by Holistic Optimization of the Process Sequence", in: 41st EUPVSEC 2024.

[11] M. Hofmann, L. Stevens, P. Hör, P. Barth, B. Bläsi, S. Riepe, S. Kalthoff, B. Kafle, M. Zimmer, M. Mittag, S. Nold, I. Sen, J. Reck, N. Schröder, L. Clochard, S. Ihlow, C. Horch, Improvement Options for PV Modules by Glass Structuring, in: 40th EUPVSEC 2023.

[12] M. Mittag, A.J. Beinert, L.C. Rendler, M. Ebert, U. Eitner, Triangular ribbons for improved module efficiency, in: Proceedings of the 32nd European Photovoltaic Solar Energy Conference and Exhibition, Munich, Germany, 2016, pp. 169–172.

[13] A. Protti, A. Welpulwar, J. Shahid, M. Mittag, A. Tummalieh, C. Reichel, Analysis of Optical Coupling Gains from Cell Interconnection for the Energy Rating of PV Modules. WIP-Munich, 40th European Photovoltaic Solar Energy Conference and Exhibition (2023). https://doi.org/10.4229/EUPVSEC2023/3AV.1.9.

[14] ITRPV, International Technology Roadmap for Photovoltaic (ITRPV): 2022 Results, fourteenth ed., 2023.

[15] China Photovoltaic Industry Association, Photovoltaic Industry Development Roadmap of China: 2023 edition, 2023.

41st European Photovoltaic Solar Energy Conference and Exhibition

EFFECT OF LOWERING CURING TEMPERATURE OF ELECTRICALLY CONDUCTIVE ADHESIVES ON RIBBON CONNECTED SOLAR CELLS

Veronika Nikitina, Tim Riehle, Leonhard Böck, Torsten Rößler
Fraunhofer Institute for Solar Energy Systems ISE, Heidenhofstraße 2, 79110 Freiburg, Germany
Corresponding Author: Veronika Nikitina | +49 761 4588 2199 | e-mail: veronika.nikitina@ise.fraunhofer.de

ABSTRACT: Electrically conductive adhesives (ECAs) can be used for the interconnection of temperature-sensitive solar cells like perovskite-silicon tandem (PVST) due to their low processing temperatures. This study investigates the impact of lowering the ECA curing temperatures ≤ 140 °C on the volume resistivity, the cell interconnection quality, and the long-term stability under thermal cycles. With ECA interconnection, additional silver in the module is introduced. This work estimates the total silver consumption in PVST modules and explores possible routes for its reduction. Replacement of Ag interconnector coating with Sn is investigated in terms of long-term stability under damp heat and joint resistance. Variation of tested ECAs included products with 17, 60, 70 and > 70 wt% silver, to realize a detailed study of the silver content effect on volume resistivity, joint resistance, module performance and interconnection reliability. Silver consumption in PVST module was estimated to be 168 mg per M6 cell, corresponding to ~24 mg Ag/W. Joint resistance with tin was observed to be higher than with silver, corresponding to –0.2 %$_{rel.}$ P_{MPP} loss attributed to series resistance in a PVST module. No increase in joint resistance with ECA interconnected tin-coated ribbons after 3000 h in damp heat was observed. Volume resistivity as well as module performance after lamination was found to profit from higher ECA silver content. ECA processing temperature was found to influence the performance of the interconnection differently for each ECA. However, longer curing time resulted in higher fill factor and better long-term stability for all tested products.

Keywords: low-temperature interconnection, electrically conductive adhesives, perovskite-silicon tandem, module technology

1 INTRODUCTION

Electrically conductive adhesives (ECAs) are increasingly used in the photovoltaic industry to interconnect temperature sensitive solar cells or in combination with the shingle [1,2] technology.

ECAs consist of a polymer matrix with conductive particles (fillers) [3], which typically consist entirely or partially of silver. Upon curing of the polymer, conductive paths are formed. In the photovoltaic industry, thermal curing of the ECAs is common, where the assembled string is heated to a processing temperature [4].

Advantages of ECAs include their versatility [5–10], the ability to tailor the product characteristics for specific applications [11–13], interconnection geometry flexibility (metallization not necessarily required), and low processing temperatures (down to 80 °C). The latter makes ECA interconnection a suitable method for perovskite-silicon tandem (PVST) solar cells. However, a drawback of using ECAs is the increased silver consumption and the limited availability of data regarding long-term stability.

PVST is a rapidly developing cell technology. Its current record for power conversion efficiency (PCE) stands at 34.6 % on a laboratory device [14] and 30.1 % on an M6 (166 mm × 166 mm) cell [15]. PVST cells are expected to enter the market after 2025 with a PCE of ~26 % [16]. Besides their high PCEs and steep learning curve, PVST cells have approximately half the current density (J_{SC}) compared to PERC, TOPCon or SHJ cells [17], which could help reducing the module power loss (ΔP_{MPP}) due to series resistance (R_S). Challenges associated with PVST cells regarding interconnection and module encapsulation are their moisture and temperature sensitivity [18]. This requires adjustments to the processing temperatures, bill of materials (BOM) and module design.

This study investigates the impact of lowering the curing temperature of ECAs on the performance and long-term stability of interconnection and the pathways to drastically reduce silver consumption in PVST modules.

2 METHODS

2.1 Silver Consumption in PVST Modules and Reduction Strategies

We estimate the silver consumption in various components of a PVST module and investigate strategies to reduce it, including the use of tin-coated interconnectors, ECAs with low silver content, and busbarless cell interconnection.

To estimate the silver consumption, we used published data and technical data sheets. The calculation is based on the input parameters provided in Table I.

Table I: Input parameters for the calculation of the silver consumption in a PVST module. It is assumed that the ECA is applied at the metallization pad positions.

	Value	Ref.
Wafer size	166 mm × 166 mm	-
Number of fingers, front	60	[19]
Number of fingers, rear	120	[20]
Finger width	25 µm	[21]
Finger height	8 µm	[22]
Number of busbars (BB)	6	[23]
Width of a line connecting BB pads	100 µm	[24]
Number of pads per BB	12	-
Mean pad area	1.4 mm^2	-
BB height	20 µm	-
ECA density	4.2 g/cm^3	TDS
ECA Ag content	17 wt%	TDS
Ribbon width	0.5 mm	-
Ribbon coating thickness	1 µm	TDS

2.3 Electrical characterization

The difference in anodic indexes between two metals (–0.15 V and –0.65 V for Ag and Sn, respectively) may lead to galvanic corrosion of a less noble one (Sn) in a joint. However, in the environment of a solar module using encapsulation materials with very low water vapor transmission rate this scenario is highly unlikely. To study the hypothesis, that the encapsulation polymer in a PVST module design acts as a sufficient barrier to isolate Ag/Sn contacts, joints (Fig. 1) consisting of Sn-coated ribbons interconnected with an ECA-2 were fabricated and aged in damp heat for 3000 h (85 % RH and 85 °C for 3000 h). Sample preparation included printing the ECA using a stencil with dimensions of 0.1 mm × 1 mm × 2 mm. The assembled joints were cured on a hot plate at 140 °C for 5 min. After curing, the samples were laminated with polyolefin (POE) foil at 150 °C for 20 min in a glass-glass setup. Two variations were prepared: with butyl-based edge seal and without one. Joint resistance was measured both after production and after aging in DH3000.

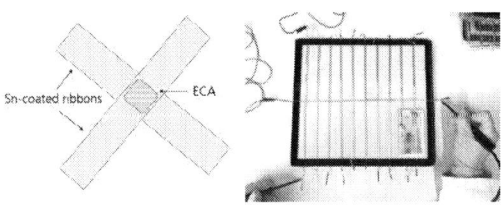

Figure 1: Sample design (left) and measurement (right) for the investigation of the Sn/Ag joint galvanic corrosion.

In order to examine how the replacement of Ag with Sn in the ribbon coating influences the joint resistance (R_{joint}), we used the method described by Böck et al. [25]. Printed circuit boards (PCB) with Ag and Sn surfaces were utilized to mimic ribbon coating. Samples with four ECAs (Tab. II) were cured in the dry convection oven for 30 min at 100 °C, 110 °C, 120 °C, 130 °C and 140 °C. Chosen ECAs have different silver content and can be cured at temperatures $\leq$ 140 °C. The range of the tested temperatures lays above the curing reaction point of each ECA and within the processing temperature range of PVST cells. Processing temperature may influence the curing behavior of the ECA (polymer contraction, evaporation of additives) and, as a result, volume resistivity.

After curing of the PCBs, joint resistances (R_{joint}) were measured. Subsequently, samples were laminated at 160 °C for 10 min with Teflon foil to simulate the solar module lamination process and measured again.

The PCBs design furthermore allows measuring volume resistivity ($R_{vol.}$) of the ECAs, which may vary dependent on the curing scheme, silver content and polymer chemistry [11].

Table II: ECAs used in this work. "DSC" refers to differential scanning calorimetry, dynamic measurement with 10 K/min ramp [26].

	ECA-1	ECA-2	ECA-3	ECA-4
Polymer matrix	A	B	B	B
Density [g/cm³]	4.4	4.2	3.0	2.4
Filler type	Ag	Cu/Ag	Ag	Ag
Ag content [wt%]	> 70	17	70	60
DSC curing peak [T °C]	85	98	89	88

2.4 Performance in the module

To evaluate the effect of curing temperature on the quality of the cell interconnection, small-scale solar modules were fabricated. The interconnection was carried out on a TT1600ECA stringer [27] with an integrated screen printer for ECA application. Strings with two half-cells were produced with industrial M6 (166 mm × 83 mm) busbarless silicon-heterojunction (SHJ) cells. Due to the absence of busbar metallization, ECAs were printed continuously to ensure proper electrical connection between fingers. The ECA print width accounted to 0.4 mm at the front and 0.6 mm at the rear side, 5 busbars were printed. The busbar amount was dictated by the constraints of the production equipment and does not represent an optimal value. This ECA application design resulted in 40 mg, 119 mg, and 82 mg silver per M6 wafer for ECA-2, ECA-3 and ECA-4, respectively. The utilized ECA printing pattern represents an unoptimized design, where the focus is laid on expected sufficient long-term stability. This was done in order to study the curing scheme effect without possible additional influences of ECA reduction. ECA-1 was not used in the module performance investigation due to processing difficulties. Severe printing defects (line interruption) were observed after 3 h on screen, worsening with time. We assume it is associated with the low pot life (6 h) of the material. The ECAs were cured using schemes considered compatible with the industrial interconnection of PVST cells, with temperatures up to 140 °C and curing times of up to 90 seconds, while recognizing that these parameters may vary depending on specific material and process requirements. Silver-coated interconnectors with a light reflective structure at the sunny side with 1.0 mm × 0.2 mm cross-section was used. The lamination was performed in a plate-membrane laminator at 150 °C for 20 minutes. The samples' stack consisted of front glass (3.2 mm), POE encapsulation foil and white polyethylene (PET) backsheet with an additional aluminum layer. Given the number of samples (180), the glass-backsheet design was chosen to simplify the sample logistics. It is not expected that the used encapsulation BOM will have a major effect on the experiment results.

To assess the long-term stability of the interconnection and its possible dependency on the curing scheme, small-scale modules were subject to 200 cycles in an accelerated thermocycling chamber (aTC200), where the ambient temperature was varied between –40 °C and +85 °C with a rate of 8 K/min and dwell times of 10 min [28]. During the accelerated thermocycling the components of the joint experience contraction and expansion according to their thermal expansion coefficients (CTE). Aging the small-scale modules with aTC helps understanding if the combination of materials and processes results in an interconnection which will be stable during operation under changing environmental conditions.

Characterization of the small-scale modules after production and after aTC200 included visual examination, current-voltage (I-V) measurement and electroluminescence (EL) imaging. The combination of these three methods allows the most efficient evaluation of the interconnection quality. Visual inspection was used to check for the macroscopic degradation features (change in color, appearance of corrosion or gas inclusions etc.). Using *I-V* measurement, electrical characteristics of the small-scale modules like short-circuit current (I_{SC}), open-circuit voltage (V_{OC}), fill factor (*FF*) and maximum power point (P_{MPP}) can be examined. I_{SC} represents optical effects, V_{OC} allows tracking the possible cell damage, *FF* reflects the series resistance losses and, as a result, interconnection quality. P_{MPP} encompasses the influence of all the above. EL images facilitate the *I-V* data interpretation by revealing local cracks, interconnection defects or cell degradation.

Figure 2: Small-scale solar module fabricated for this study, with a total of 180 modules produced (10 modules per group).

3 RESULTS AND DISCUSSION

3.1 Silver consumption in a PVST module

Considering the input parameters given in the table I, the following silver use was estimated (Tab. III).

Table III: Calculated silver consumption per PVST cell (M6, 6 busbar design) for metallization (contact fingers and busbar) and interconnection (ECA and interconnectors)

	Ag [mg/cell]
Fingers	60
Busbars	75
ECA	7
Interconnector	26
Total	**168**

For a 60-cell PVST Module of 421 W [29], total silver consumption of 168 mg/cell accounts to ~24 mg Ag/W (metallization and interconnection) of which ~19 mg Ag/W is attributed to the metallization. This corresponds to the silver use in TOPCon cells (19 mg Ag/W) and is ~5 mg Ag/W lower than for SHJ (24 mg Ag/W) [14].

3.2 Electrical characterization

After the exposure to DH3000, no significant change in the R_{joint} of Sn-coated ribbons interconnected with ECA-2 was observed (Fig. 3). Samples laminated with POE only (no edge seal) also demonstrated no increase in R_{joint}. The latter implies that POE module encapsulation either served as a sufficient barrier from the moisture ingress or that the moisture diffused through the module edges did not cause joint degradation.

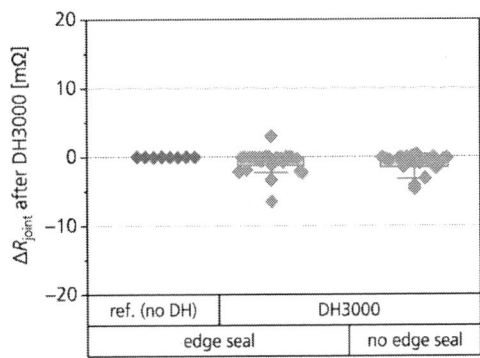

Figure 3: Difference between the initial R_{joint} (after curing and lamination) and R_{joint} after DH3000 for modules with and without edge sealing. Per sample group, 30 measurements were performed. One datapoint represents one measurement.

The results of the joint resistance measurements are presented in Figure 4. Joints with Sn-coated PCBs overall demonstrate higher resistance. However, the values with Sn are acceptable for the interconnection of the PVST cells. Taking the module design from sections 2.1 and 3.1 into account, joint resistances of 10 mΩ and 100 mΩ would result in −0.2 %rel. and −1.6 %rel. P_{MPP} loss attributed to R_S, respectively. A larger contact area would correspond to lower ohmic losses (not shown in Fig. 4).

Strong data scattering prevents from recognizing trends connected with ECA curing temperature. Values obtained after curing did not differ considerably from the presented data in Figure 4.

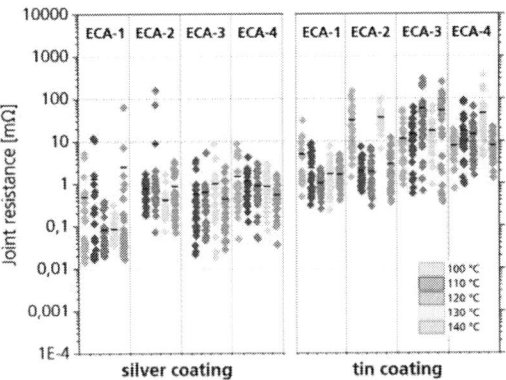

Figure 4: Results of the joint resistance measurement after curing at different temperatures and lamination of the PCBs. Per group, 19 to 57 measurements were performed. One datapoint represents one measurement.

The results of the volume resistivity measurements are shown in Figure 5. Overall, the tested ECAs exhibit different values, which correspond to their composition. Those with higher silver content show lower $R_{vol.}$ After curing (Fig. 5, transparent boxes), temperature-dependent behavior is evident with ECA-1, where samples cured at 100 °C demonstrate the median value of 2.5×10^{-4} $\Omega\cdot$cm, and the samples cured at 140 °C show significantly lower median of 3.0×10^{-5} $\Omega\cdot$cm. ECA-2 and ECA-3 show slightly higher $R_{vol.}$ for samples cured at 100 °C.

After lamination, the $R_{vol.}$ reduced strongly with

ECA-1, evening out the median values for various curing temperatures to ~1×10⁻⁵ Ω·cm. ECA-3 and ECA-4 behave alike, where groups cured at lower temperatures (100 °C, 110 °C) show $R_{vol.}$ decreasing after lamination. This may be due to the incomplete curing in the convection oven and reaction completion during lamination. The lower the oven temperature was, the more pronounced the effect appears. ECA-2 does not follow a particular trend: samples cured at 100 °C show the decreasing $R_{vol.}$ whereas with other groups the $R_{vol.}$ increases by ~2×10⁻³ Ω·cm (140 °C) to ~6×10⁻³ Ω·cm (130 °C). For all three polymer B-based materials (Tab. III) lowest $R_{vol.}$ was achieved through curing at lowest temperature with subsequent lamination. In general, all considered ECAs demonstrate volume resistivity values compatible with solar cell interconnection.

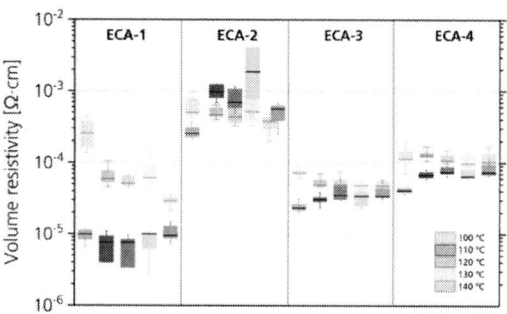

Figure 5: Results of the volume resistivity measurement after curing (transparent) at different temperatures and lamination (full color) of the PCBs. One box plot is represented by 10 measurements. Black lines refer to median values.

3.3 Performance in the module

Results of the *I-V* characterization of small-scale modules are given in Figure 6. The fill factor is shown as this parameter depends on the series resistance and can be evaluated as an indication for the interconnection performance. Initial *FF* of the cell accounted to 80.9 % (not shown in Fig. 6).

Median of the cell-to-module (CTM) loss in *FF* varies from −6.2 %rel. (ECA-3, cured 90 s at 100 °C) to −11.4 %rel. (ECA-2, cured 45 s at 130 °C). Depending on the curing scheme, each ECA behaves differently after lamination (Fig. 6, transparent boxes):

- ECA-2 shows higher *FF* if cured slower at lower temperatures (−8.6 %rel. and −10.6 %rel. for 90 s at 120 °C and 45 s at 140 °C, respectively).
- ECA-3 exhibits slightly higher median values at lower curing temperatures independent of the curing time (−6.2 %rel. and −6.9 %rel. for 100 °C and 140 °C, respectively).
- ECA-4 demonstrates a distinct increase in *FF* from 130 °C upwards (−6.4 %rel. and −10.4 %rel. for curing 90 s at 130 °C and 90 s at 120 °C).

After exposure to aTC200 (Fig. 6, full color boxes), *FF* decreases in all groups to a different extent. General trends highlighted in the previous paragraph endure. Least degradation is observed with ECA-2, in samples cured 90 s at 120 °C. They show median *FF* loss (relative to the module performance after lamination) of −1.2 %rel.aTC. ECA-3 demonstrates lowest degradation at lowest curing

temperature (Δ*FF* of −2.5 %rel.aTC, cured 90 s at 100 °C). However, higher median value for the group cured for 90 s at 130 °C than neighboring groups (120 °C and 140 °C, both for 45 s) indicates, that increasing curing time may improve the long-term stability of the interconnection. The latter is consistent with ECA-4, where increasing curing time from 45 s to 90 s results in Δ*FF* improvement from −5.7 %rel.aTC to −3.7 %rel.aTC. Longer curing results in a higher achieved degree of curing (DoC) which could have facilitated the long-term stability of the interconnection.

No differences between groups or systematic degradation features were observed in EL images or during the visual inspection of the small-scale modules.

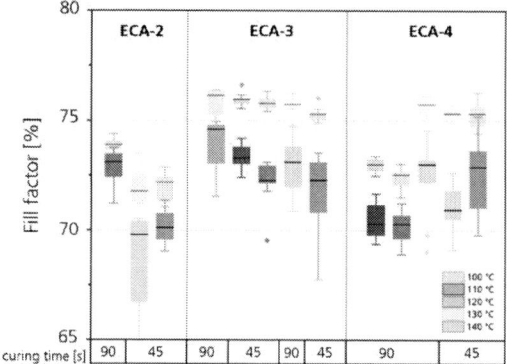

Figure 6: Fill factor of the small-scale modules measured after lamination (transparent) and aTC200 (full color). One box plot is represented by 10 samples. Black lines refer to median values.

4   SUMMARY

In this study, the use of ECAs for the interconnection of PVST cells is reviewed. The main focus lays on investigating the effect of low curing temperatures (≤ 140 °C) on interconnection quality. Additionally, the silver consumption in various components of a PVST module is estimated and three strategies to minimize Ag usage are explored: replacement of silver interconnector coating with tin, ECAs with low silver content, and busbarless interconnection. Tests were performed with four ECAs with different chemistry, including varying Ag content. The following key findings were acquired in this study:

- Silver consumption in PVST module was estimated to be ~19 mg Ag/W (metallization only) and ~24 mg Ag/W (metallization and ECA interconnection).
- No increase in R_{joint} in Sn/Ag-based ECA/Sn joints encapsulated in POE was observed after DH3000.
- R_{joint} is higher for the Sn coated surfaces. However, all values are compatible with the PVST cell interconnection: replacing Ag coating with Sn would results in ~0.2 %rel. P_{MPP} loss in a PVST module.
- ECAs with higher Ag content show lower $R_{vol.}$ After curing, ECA-1 (polymer type A, > 70 wt% Ag) exhibits lower values at higher T°C, which even out and decrease to

~1×10^{-5} $\Omega\cdot$cm after lamination. For all 3 polymer B-based materials lowest $R_{vol.}$ was achieved through curing at lowest temperature (100 °C) with subsequent lamination. In general, all considered ECAs demonstrate volume resistivity values compatible with solar cell interconnection ($<10^{-2}$ $\Omega\cdot$cm).

- The *FF* of small-scale modules with busbarless SHJ cells varied based on used ECA and curing conditions. For each ECA, different behavior after lamination was observed. Samples with ECA-2 and ECA-3 showed less *FF* loss at lower curing T °C, whereas ECA-4 demonstrated distinct improvement in *FF* from 130 °C onwards. After aTC200 the positive effect of longer curing time on long-term stability of the interconnection is observed.

5 OUTLOOK

This study highlights the impact of curing conditions on the performance of low-temperature ECA interconnection. The findings suggest that optimizing curing parameters can enhance module performance and long-term stability. Additionally, the feasibility of replacing Ag interconnector coating with Sn was demonstrated, particularly in terms of joint resistance and stability under damp heat conditions, presenting a promising route to reducing Ag consumption in solar modules. The automatic interconnection process $\leq$ 140 °C and $\leq$ 90 s was successfully implemented using busbarless SHJ cells, which further contributes to lowering silver usage.

Future research should focus on exploring the effects of combining the Sn-coated interconnectors, low-Ag ECA and busbarless metallization grid in a low-temperature ECA interconnection process.

REFERENCES

[1] N. Abdel Latif, et al., 40th European Photovoltaic Solar Energy Conference and Exhibition (2023).

[2] T. Roessler, et al., Proceedings of the 10th Metallization and Interconnection Workshop (2022) 20012.

[3] T. Geipel. Electrically Conductive Adhesives for Photovoltaic Modules, Fraunhofer ISE, Freiburg, 2017 // 2018.

[4] T. Geipel, et al. (2019).

[5] T. Geipel, et al., Photovoltaics International 21 (2013).

[6] DELO Industrial Adhesives. Product overview. https://www.delo-adhesives.com/us/products/bonding/overview-of-adhesives.

[7] Henkel Adhesive Technologies. Electricall conductive adhesives. https://www.henkel-adhesives.com/de/en/products/industrial-adhesives/electrically-conductive-adhesives.html.

[8] Panacol-Elosol. Technical data sheets. https://www.panacol.com/products/data-sheets#elecolit.

[9] Polytec Polymere Technologien. Two-component conductive epoxies. https://www.polytec-pt.com/int/electrically-conductive/products/2-component-conductive-epoxies.

[10] B. ROARTIS bvba. Electrically condictive adhesives (isotropic). https://www.iq-bond.com/electrically-conductive-adhesives-isotropic/.

[11] F. Tan, et al., International Journal of Adhesion and Adhesives 26 (2006) 406–413.

[12] M.R. Sanghvi, et al., Polymer Bulletin 79 (2022) 10491–10553.

[13] H. Derakhshankhah, et al., Journal of Materials Science: Materials in Electronics 31 (2020) 10947–10961.

[14] LONGI. 34.6%! Record-breaker LONGi Once Again Sets a New World Efficiency for Silicon-perovskite Tandem Solar Cells, 2024. https://www.longi.com/en/news/2024-snec-silicon-perovskite-tandem-solar-cells-new-world-efficiency/.

[15] LONGI. LONGi announces the new world record efficiency of 30.1% for the commercial M6 size wafer-level silicon-perovskite tandem solar cells, 2024. https://www.longi.com/en/news/is-m6-wafer-silicon-perovskite-tandem-cells-new-efficiency-record/.

[16] VDMA Photovoltaic Equipment. International Technology Roadmap for Photovoltaic (ITRPV), 2023. itrpv.vdma.org.

[17] Y. Zhang, et al., Energy & Environmental Science 14 (2021) 5587–5610.

[18] A.u. Rehman, et al., Progress in Photovoltaics: Research and Applications 31 (2021) 429–442.

[19] V. Nikitina, et al., Solar Energy Materials and Solar Cells 263 (2023) 112590.

[20] S. Pingel. Pers. communication, 2024.

[21] A. DeRose, et al., Solar Energy Materials and Solar Cells 261 (2023) 112515.

[22] S. Pingel, et al., Solar Energy Materials and Solar Cells 265 (2024) 112620.

[23] C. Messmer, et al., Progress in Photovoltaics: Research and Applications 30 (2022) 374–383.

[24] TongWei Solar Energy Co., Ltd. TW Solar Energy HJT 166 9B Silicon Heterojunction Solar Cells TW166Y209.

[25] L. Böck, et al., Proceedings of the 14th International Conference on Crystalline Silicon Photovoltaics 2024.

[26] Deutsches Institut für Normung e. V., Prüfung von Kunststoffen und Elastomeren; Thermische Analyse; Dynamische Differenzkalorimetrie (DDK). Prüfung von Kunststoffen und Elastomeren; Thermische Analyse; Dynamische Differenzkalorimetrie (DDK).

[27] https://www.teamtechnik.com/.

[28] Schiller C. et al., Proceedings of the 36th European Photovoltaic Solar Energy Conference and Exhibition (2019) 995–999.

[29] Fraunhofer ISE / Oxford PV. Oxford PV and Fraunhofer ISE develop full-sized tandem PV Module with record efficiency of 25 %, 2024. https://www.ise.fraunhofer.de/en/press-media/press-releases/2024/oxford-pv-and-fraunhofer-ise-develop-full-sized-tandem-pv-module-with-record-efficiency-of-25-percent.html.

41st European Photovoltaic Solar Energy Conference and Exhibition

TO BYPASS OR NOT TO BYPASS:
INTEGRATING AND EVALUATING PARALLEL CONNECTIONS AND BYPASSES
IN C-SI PV LAMINATES

Tom Borgers[1], Jonathan Govaerts[1], Hamed Javanbakht Lomeri[1], Apostolos Bakovasilis[1], Rik Van Dyck[1],
Bart Reekmans[1], Hariharsudan Sivaramakrishnan Radhakrishnan[1], Jef Poortmans[1], Manuel Van den Storme[2],
Guy Van den Storme[2]

[1]imec-EnergyVille-UHasselt, Genk, Belgium
[2]VdSWeaving, Oudenaarde, Belgium

ABSTRACT: In this abstract we discuss our approach on investigating the potential and a method for including electrical bypass diodes and parallel connections in a PV laminate. Such an approach could help to mitigate the impact of shading but requires an industrial viable implementation method, together with the necessary simulations to determine the effectiveness on the module performance under realistic conditions. For such an assessment it is needed to develop fabrication technology and samples for evaluation towards model validation.

1 BACKGROUND

Already early on in PV, bypass diodes were implemented in the modules to deal with partial shading and their detrimental effects, primarily to protect the cells from being exposed to reverse bias and the associated hotspots. Next to these safety concerns, bypass diodes can also help in improving performance in partial shading conditions. In the future, this topic will only grow in importance, considering the continued growth of PV deployment and strategies of implementing PV everywhere. Such integrated PV (BIPV, VIPV...), close to the consumer will invariably be exposed more to partial shading conditions.

A growing community of researchers is dealing with this challenge of partial shading of PV modules. This challenge can be targeted through the module topology, either "passive", through the use of parallel (fixed) connections or bypasses as shown in Figure 1, or more "active", with a reconfigurable switch matrix as shown in Figure 2.

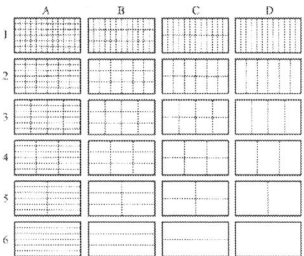

Figure 1 Different passive module topologies for dealing with shading [1]

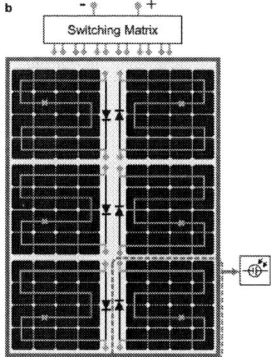

Figure 2 Active topology with switching matrix [2]

AESolar is currently commercializing an interesting approach, where integrating one bypass diode per cell allows to limit the impact of shading only to the shaded cells [3]. This approach is illustrated in Figure 3.

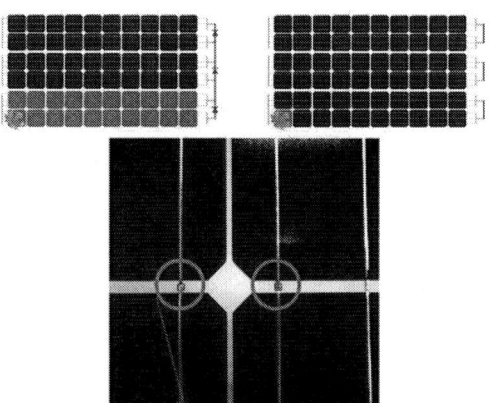

Figure 3 Schematic concept drawing illustrating the benefit of 1 bypass/cell (top); implementation in strings reported by AESolar (bottom) [3]

In this work, we look at integrating bypass diodes and parallel connections into the laminate, and evaluate performance through characterization and

41st European Photovoltaic Solar Energy Conference and Exhibition

simulation, and validate its manufacturability.

2 EVALUATION SAMPLE FABRICATION

In first instance, we prepared a mini-module with two strings, each with four M10 half-cut cells in series (10BB design). In one of the strings, we implemented a bypass diode for each half-cell, while the other (identical) string is kept as reference without bypass diodes. An impression of the mini-module is shown in Figure 4. The bypasses are implemented on commercially fabricated strings, between crossing ribbons connecting the ten wires at the back of the cell. With this approach, also indicated in Figure 4, the gap between the cells does not need to be enlarged.

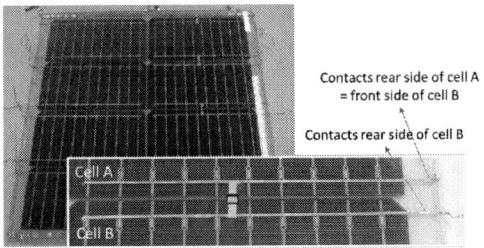

Figure 4 Fabricated mini-module consisting of 2 strings of 4 M10 half-cut cells: front (top) and rear (bottom) view illustrating the method for integrating the bypass diodes (bottom detail)

This mini-module allows to evaluate the possibility of laminating a bypass diode into a glass-glass laminate and to measure its functionality. Additionally, since the cross-connector ribbons for the bypass diodes are also accessible outside the module, the bypass functionality can be compared to a configuration where neighbouring cells of the parallel string are individually also connected in parallel (externally).

3 CHARACTERIZATION

Figure 5 shows the performance of the reference (Ref) and the bypassed (Bypass) string when they separately operate under increasing shading of one cell in steps of 25%. Without bypass diodes, the Ref string current drops linearly with the increasing

shading of one cell. With one bypass per cell in the Bypass string, a higher current is still possible, albeit only for lower voltages due to the bypassed cell and bypass diode voltage.

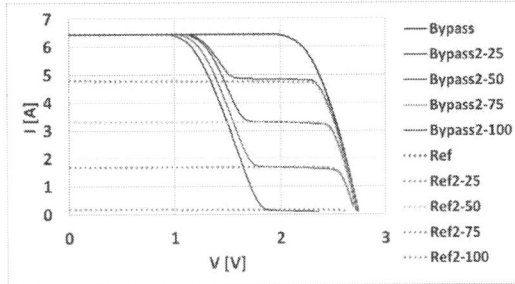

Figure 5 Performance of the separate strings under increasing shading conditions

Figure 6 demonstrates the bypass functionality when both strings are measured in parallel (module-level) and one of the cells of the Bypass string is increasingly shaded.

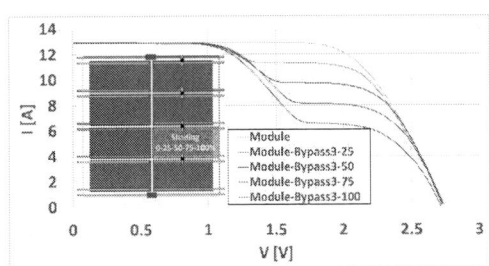

Figure 6 Bypass functionality during shading of a cell with bypass (mini-module with parallel strings)

Alternatively, we can also connect the shaded cell without bypass in parallel with the corresponding cell in the parallel string with bypass. The results are shown in Figure 7.

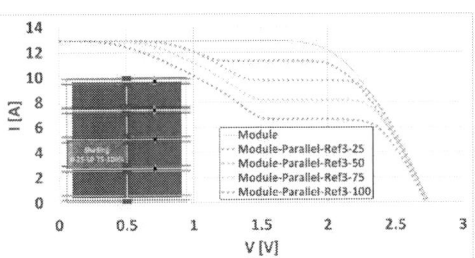

Figure 7 IV-curves of the module with parallel (external) connections between neighbouring cells and shading applied on a cell from the reference string (without cell-level bypass)

Additionally, the inverse combination was measured: with parallel (external) connections on cell-level, a cell from the bypass string (with cell-level bypass) was shaded and IV-curves recorded. The resulting graphs are shown in Figure 8.

930

41st European Photovoltaic Solar Energy Conference and Exhibition

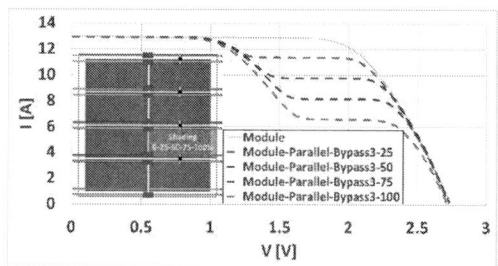

Figure 8 IV-curves of the module with parallel (external) connections between neighbouring cell and shading applied on a cell from the bypass string (with cell-level bypass)

4 OVERALL COMPARISON FOR SIMULATION

Targeting maximum power, also under shaded conditions, implies going for the global maximum power point in the P-V curve. Figure 9 shows the P-V curves calculated for the different cases in the above section.

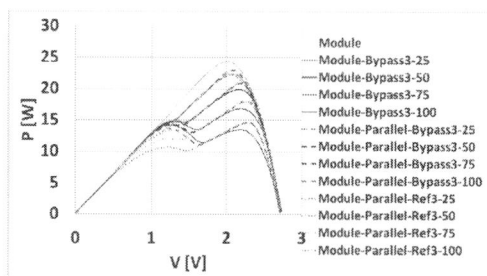

Figure 9 P-V curves comparing local and global maxima between the different configurations and varying shading levels

From this graph, it is clear that the global maximum depends on the amount of shading (in this scenario of one-cell-shading). Zooming in in Figure 10, the parallelization of cells is always beneficial for the global maximum, although for the local maximum where the bypass is active, the higher currents incur higher resistive losses for the parallelization.

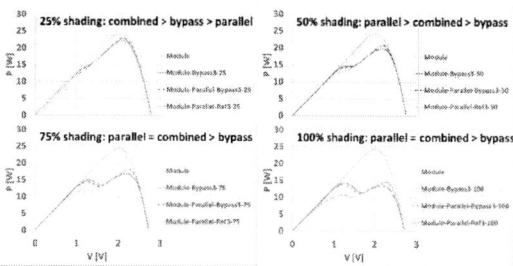

Figure 10 Distinct P-V curves focusing on the comparison of configurations for varying shading levels

This is obviously one specific case and should not be used to draw more general conclusions. Those should rather follow from extensive simulations with a model that could be validated with the above

measurements. Building such a model, we are extending the optical-electrical-thermal model [4] to be able to incorporate these configurations. As a first preliminary result, this impact due to resistive losses is confirmed by simulations with our model. Figure 11 shows the simulated impact of additional resistance in the parallel connection of cells on the IV-curve, which is mostly captured when we cover a cell from the reference string. In this case, the current will travel a longer distance, due to the external connection of the ribbons, thus generating a higher resistive path that increases losses when operating in high current.

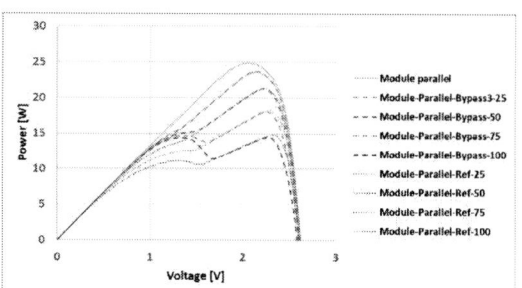

Figure 11 Simulated impact of resistance in the parallel connection

The complexity increases with an increasing number of shaded cells, creating as many additional local maxima. This is illustrated in Figure 12, where two cells have been shaded with varying intensities.

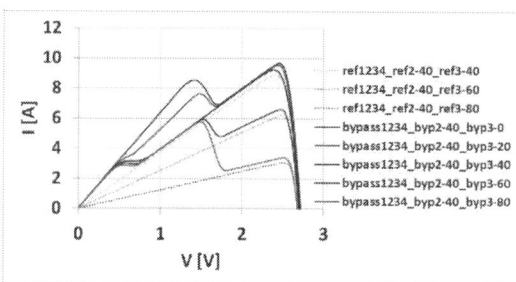

Figure 12 P-V curves under various shading levels of multiple cells in the module

5 INDUSTRIAL IMPLEMENTATION POTENTIAL

Apart from the potential energy yield gains in shaded conditions, the feasibility of fabrication (and its cost-effectiveness) is crucial for the industrial viability of the concept. To this end we designed a woven fabric combining glass fibres and metal ribbons, with the option of integrating bypass diodes inside [5]. This fabric can be aligned on the cell strings during the layup phase (with high alignment tolerance). It is designed to incorporate the parallel connections through integrated ribbons, and simultaneously allow the incorporation of a bypass diode per cell behind the cell strings.

As work in progress, such a proof-of-concept fabric has been made, as well as a first proof-of-concept module with this fabric. Figures 13-14 give some impressions of the fabrication, the resulting laminate and EL image (without cracks).

Figure 13 Layup of cells on a woven fabric (with bypass diodes)

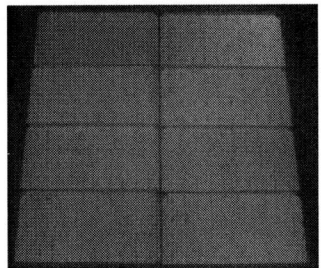

Figure 14 (Frontside) EL image of the module after lamination

First shading experiments in Figure 15 indicate the reduced resistive losses in the parallel connection due to the internally-integrated wiring in the new module. The currents are at a slightly lower level, which may be attributed to the dark, light-absorbing fabric at the rear side. The FF is lower due to the increased cross-section of the used connector ribbons.

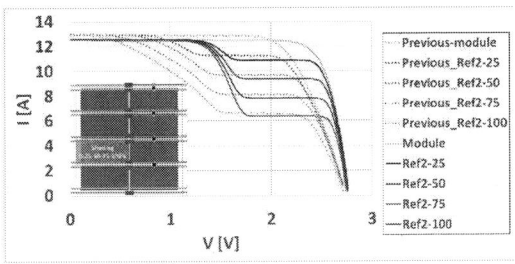

Figure 15 Performance of new module with integrated wiring compared to the previous, with identical shading conditions

CONCLUSION / OUTLOOK

This research aims to determine the practical feasibility and potential of incorporating local bypass diodes and/or parallelization of cells into the PV laminate. Ideally the simulation model is extended and can be used to estimate the impact in energy yield depending on shading conditions, cell and bypass diode characteristics, wiring scheme and

sizing. Further work is required (and planned) in tuning the measurements and simulations as well as the industrial implementation.

ACKNOWLEDGMENT

This work has been funded by the Flemish Government through the imec.icon project BIPV4ALL (HBC. 2022.0082).

REFERENCES

[1] A. Calcabrini et al., Simulation study of the electrical yield of various PV module topologies in partially shaded urban scenarios, Solar Energy 225 (2021) 726–733

[2] A. Calcabrini et al., Electrical performance of a fully reconfigurable series-parallel photovoltaic module, Nature Communications, https://doi.org/10.1038/s41467023-43927-3

[3] AESolar presentation InterSolar, 2019

[4] H. Goverde et al., Energy yield prediction model for PV modules including spatial and temporal effects, 29th EUPVSEC, Amsterdam, 2014

[5] Patent application EP 23197552.5

41st European Photovoltaic Solar Energy Conference and Exhibition, Vienna, 2024

Steps Towards a 100% Renewable Material Solar Module: Evaluating Material Substitutions for Encapsulation and Interconnection

R. Koepge, M. Pander, A. Mordvinkin, S. Großer

3DO.12.1

Content

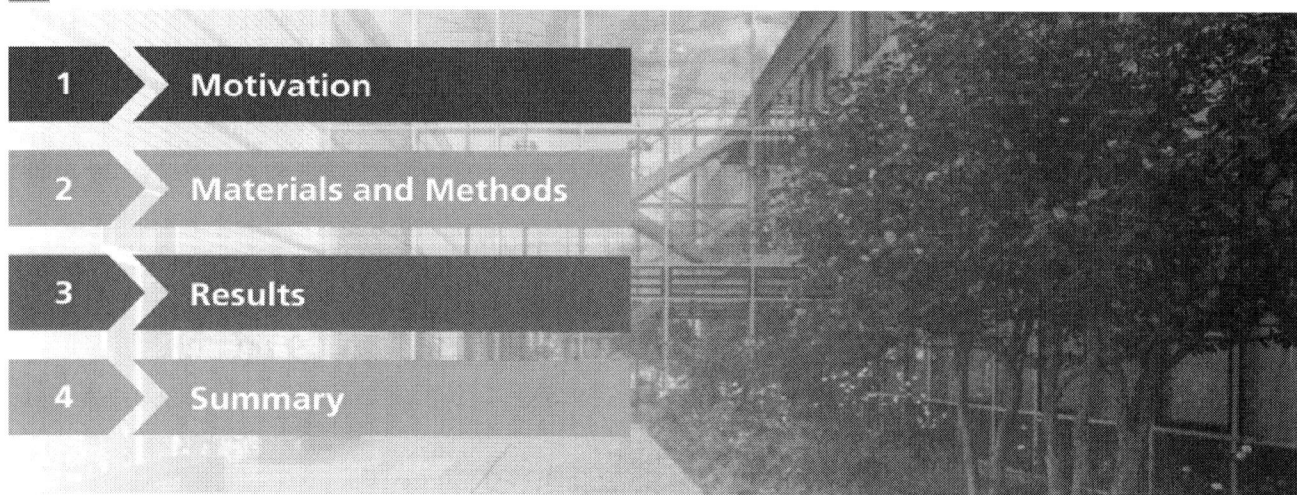

1. Motivation
2. Materials and Methods
3. Results
4. Summary

Motivation

A SOLAR MODULE MADE OF SUSTAINABLE HARMLESS MATERIALS

Reach climate goal of max. 2°C
temperature increase

> become sustainable

Restriction of Hazardous Substances (RoHS)
does not totally apply to solar panels!

> get ready to be lead-free

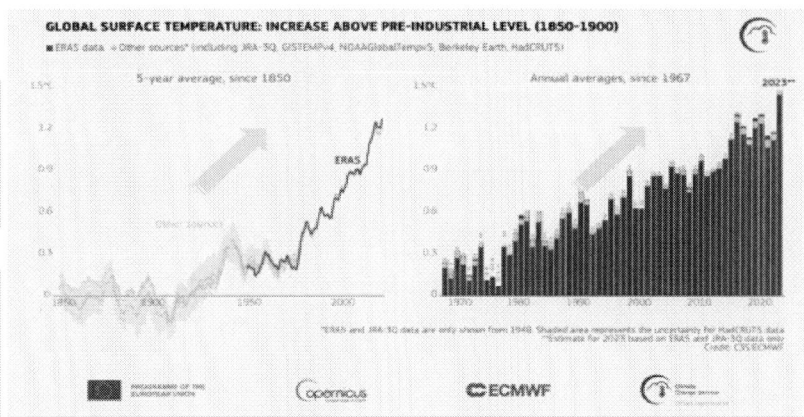

Motivation

A SOLAR MODULE MADE OF SUSTAINABLE HARMLESS MATERIALS

APPROACH – Replacement of all module components
that are not involved in power generation

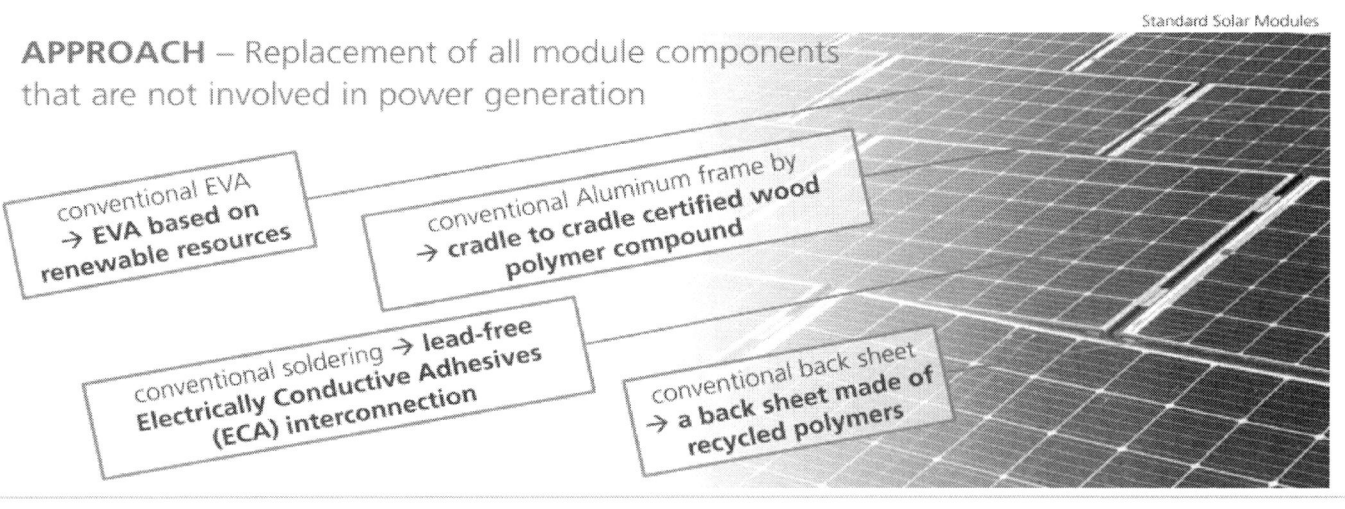

Standard Solar Modules

conventional EVA
→ EVA based on
renewable resources

conventional Aluminum frame by
→ cradle to cradle certified wood
polymer compound

conventional soldering → lead-free
Electrically Conductive Adhesives
(ECA) interconnection

conventional back sheet
→ a back sheet made of
recycled polymers

Materials and Methods
Sample Overview

3AV2.50 | "Material Screening for the Development of a Photovoltaic Module Using Biodegradable Materials from Renewable Raw Materials"

Material Batch

Please visit 3AV.2.50

	Inter-Connection	Materials				
		Front Cover	Front Encapsulat	Cell Type	Back Encapsulant	Backsheet
Batch Mat1 (Reference)	Lead Solder	1	EVA	2	EVA	PET based
Batch Mat2	Lead Solder	1	Bio-EVA	2	Bio-EVA	PET based
Batch Mat3	Lead Solder	1	Bio-EVA degradable	2	Bio-EVA degradable	PET based
Batch Mat4	Lead Solder	1	Bio-EVA degradable	2	Bio-EVA degradable	Bio-PET based
Batch Mat5	Lead Solder	1	Bio-EVA degradable	2	Bio-EVA degradable	30% recycle PET based

1 tempered glass, Optiwhite, 3mm thickness | 2 PERC Half Cell M3

Materials and Methods
Sample Overview

Interconnection Batch

ECA Connection

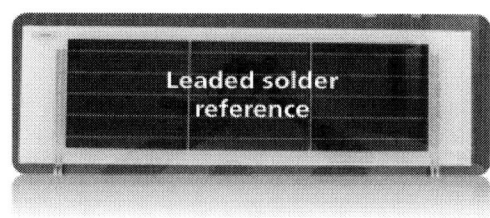

	Inter-Connection	Materials				
		Front Cover	Top Encaps.	Cell Type	Back Encaps.	Backsheet
Batch Int1 (Reference)	Lead Solder	1	EVA	PERC Full Cell M3	EVA	PET based
Batch Int2	ECA Epoxy based	1	EVA	PERC Full Cell M6	EVA	PET based

1 tempered glass, Optiwhite, 3mm thickness

Materials and Methods
Module Characterization

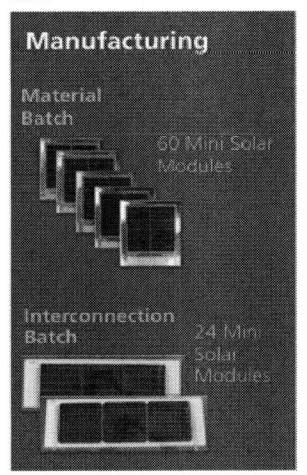

Manufacturing

Material Batch
60 Mini Solar Modules

Interconnection Batch
24 Mini Solar Modules

Testing

IEC 61370 - Sequence B
Multiple stressor sequence | Damp Heat (DH), Ultra Violette (UV), Humidity Freeze (HF)

IEC 61215 - Sequence C
Multiple stressor sequence | Ultra Violette (UV), Temperature Cycling (TC) Humidity Freeze (HF)

IEC 61215 - Sequence D
Single stressor sequence | Damp Heat (DH)

IEC 61215 - Sequence E
Single stressor sequence | Thermal Cycling (TC)

Characterization

Degradation analysis by I-V-Curve determination
@ Berger class AAA Modulflasher

- ❖ STC(standard test conditions) | (25±1)°C | Irradiation 1000 W/m² | Spectrum AM1.5
- ❖ Repeatability accuracy<0,3% (95% confidence)
- ❖ Measurement uncertainty ±2.5% without reference module
- ❖ Measurements are performed without spectral mismatch correction

Degradation analysis by EL Imaging
@ Greateyes Lumi Solar Professional

- ❖ Current supply: 10 A / 1 A
- ❖ Exposure time: 1 s / 10 s
- ❖ Aperture: 2.8

7   9/26/2024   © Fraunhofer CSP

Results
Material Batch | IEC 61730 - Sequence B | Multiple Sequences

Degradation a... ...aging

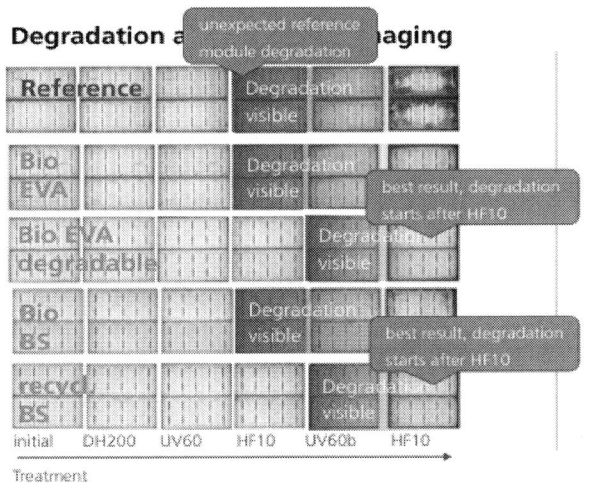

unexpected reference module degradation

best result, degradation starts after HF10

best result, degradation starts after HF10

Reference — Degradation visible

Bio EVA — Degradation visible

Bio EVA degradable — Degradation visible

Bio BS — Degradation visible

recyc BS — Degradation visible

initial | DH200 | UV60 | HF10 | UV60b | HF10

Treatment

Degradation analysis by I-V-Curve determination

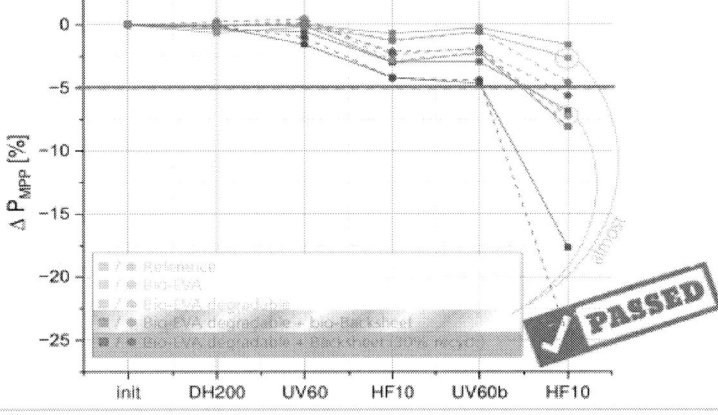

8   9/26/2024   © Fraunhofer CSP

Results
Material Batch | IEC 61215 - Sequence C | Multiple Sequences

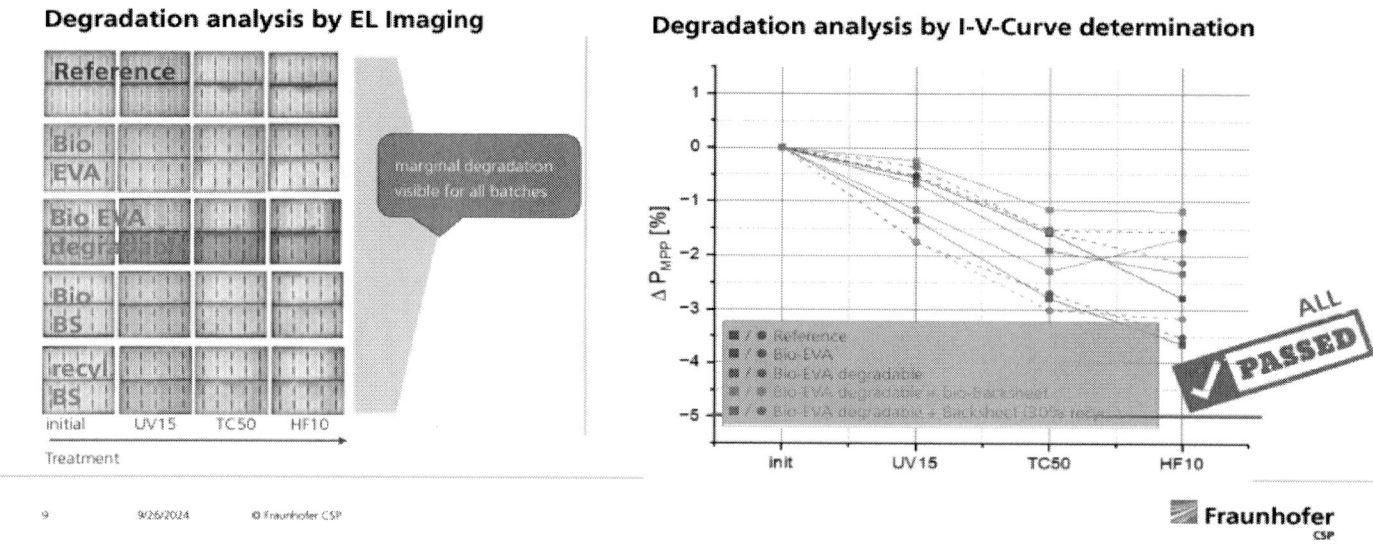

Results
Material Batch | IEC 61215 - Sequence D | Temperature Cycling

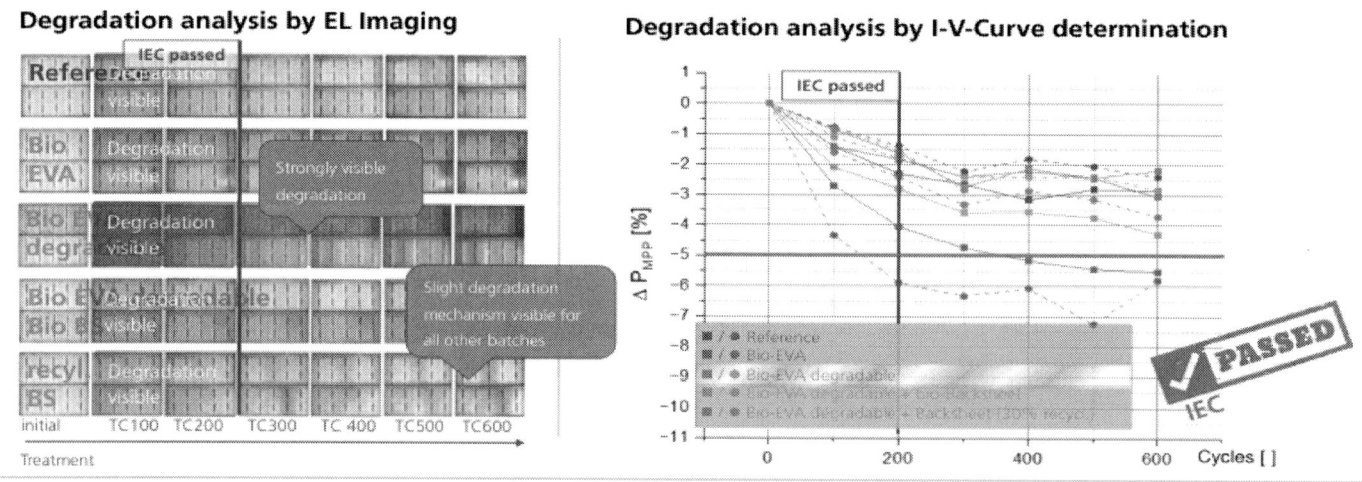

Results
Material Batch | IEC 61215 - Sequence E | Damp-Heat

Degradation analysis by EL Imaging

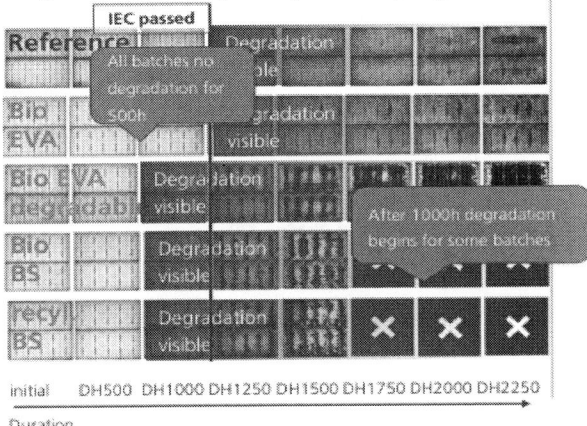

Degradation analysis by I-V-Curve determination

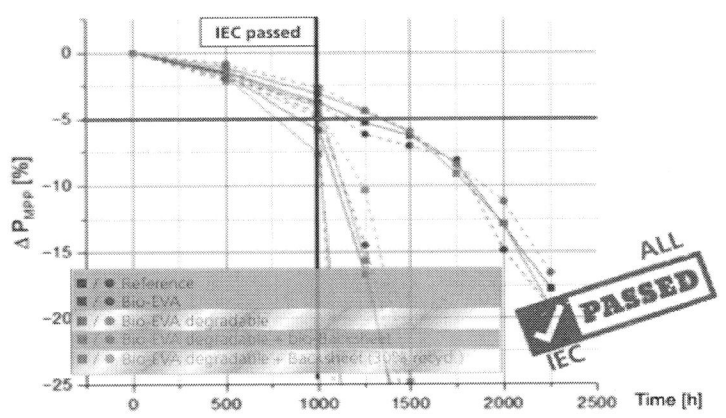

Results
Interconnection Batch | EL and I-V Analysis

Degradation analysis by EL Imaging

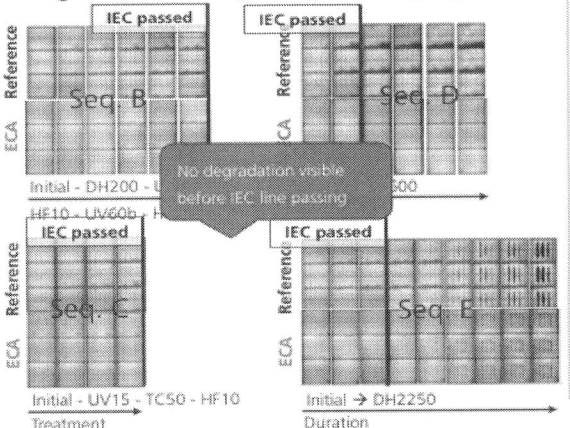

Degradation analysis by I-V-Curve determination

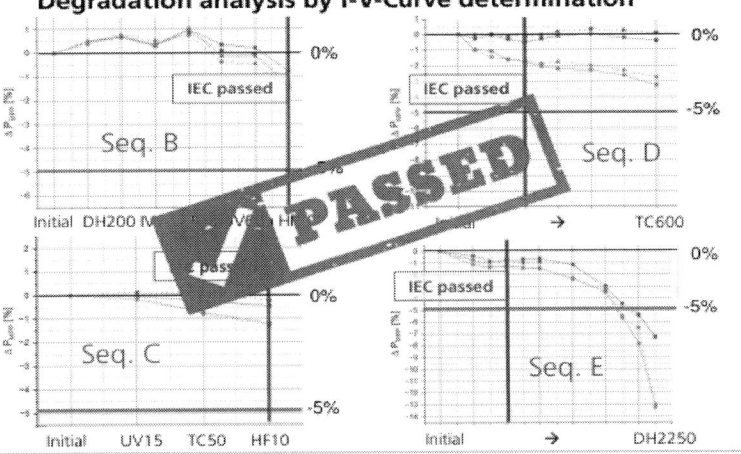

Summary
Material and interconnection evaluation

Solar Module Prototype presented @ Intersolar Munich 2024

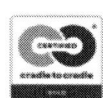

❖ Successful manufacturing of mini solar module batches with an unknown material behavior

❖ Performing accelerated aging tests according to IEC

❖ Characterization by EL and I-V-Curve analysis show weak materials, damp heat weaknesses for all Materials

> **Takeaway Message – Material Substitution Works!**
>
> ⇨ **Lead-free ECA solar cell interconnection** ⇨ Seq. B, C, D, E ✓
>
> ⇨ **EVA based on renewable resources by 60%** ⇨ Seq. (B), C, D, E ✓
>
> ⇨ **Back sheet made of 30% recycled PET** ⇨ Seq. B, C, D, (E) ✓
>
> ⇨ **Module frame made of 100% recyclable wood-plastic composite** ⇨ visit 3AV2.50 ✓

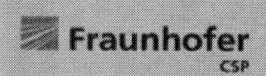

Thank you for your attention!

PUBLICATION HAS BEEN SUBMITTED

3DO.12.1 ➜ in Progress in Photovoltaics | "Solar module made of renewable raw materials through material substitutions in encapsulation, backsheet and a sustainable lead-free interconnection"

3AV2.50 ➜ in EPJ Photovoltaics | "Material screening for the development of a photovoltaic module using biodegradable materials from renewable raw materials"

Ringo Koepge

PV-Modules, Components and Manufacturing

Fraunhofer Center for Silicon Photovoltaics CSP
Otto-Eissfeldt-Strasse 12 | 06120 Halle
+49 345 5589-5311 | +49 345 5589-5999
ringo.koepge@csp.fraunhofer.de | https://www.imws.fraunhofer.de

Financial support by the Federal Ministry for Economic Affairs and Climate Action of the project "3-Quadrat" (grant no 03EET1161) is gratefully acknowledged. We thank Prof. Dr.-Ing. Mirko Feldmann head of Polymer Synthesis and Processing at Fraunhofer IMWS for support and organization for compounding and manufacturing of the bio-EVA. The Folienwerk Wolfen GmbH is gratefully acknowledged for material supply.

41st European Photovoltaic Solar Energy Conference and Exhibition

This presentation was selected by the Sc. Committee of the EU PVSEC 2024 for submission of a full paper to one of the EU PVSEC's collaborating peer-reviewed journals.

NEW ENCAPSULANT FOR PV MODULES DESIGNED FOR RECYCLE: A LAB SCALE PROTOTYPE

Margot LANDA[1], Alexis BRASTEL[2], Eeva MOFAKHAMI[1], Timea BEJAT[1]*, Pierre PILUSO[2]
[1]Univ. Grenoble Alpes, CEA, Liten, INES, 73375 Le Bourget du Lac, France
[2]Univ. Grenoble Alpes, CEA Liten, Grenoble, France
timea.bejat@cea.fr

ABSTRACT: Most of polymer families used today as encapsulants in PV modules are cross-linking polymers. This chemical reaction is an obstacle to their reuse after recycling process where a thermomechanical separation can supply encapsulant films to reuse them in new or repaired PV modules. In this document, we present the development of a dynamically cross-linked polymer resin (vitrimer) dedicated to PV applications. This material must necessarily possess properties essential to photovoltaic encapsulants such as good optical transmission and high thermomechanical properties to protect solar cells and maintain good adhesion between the different layers of a PV module. Simultaneously this vitrimer resin will make it easier to separate the various components of a PV module at its EoL (end of life), to improve recyclability and recovery of the various materials.
First material level experimental campaign concerned the optical performance assessment and its positioning compared to commercially available products and enabled a first selection to upgrade the chosen formulations up to mono module size samples. Second material level test covered the adhesion testing where special samples had been obtained and tested at initial and after accelerated aging (1000 h DH). Both interfaces glass/encapsulant and encapsulant/backsheet were measured. These preliminary results showed an exceptional material performance that allowed us to process to module integration. This resin was successfully integrated into 2 ½-cell PV modules (200 x 200 mm) in glass-glass mono modules. These modules were tested at initial stage and then underwent to accelerated aging for 1000 h of damp heat exposure.
Results showed a very good electrical performance for at least one of the three tested resins in mono-modules. The durability and recyclability of these PV modules with vitrimer encapsulant was demonstrated. These trials, which are very encouraging for further development, showed the main advantage of this new material namely its reusability as thermomechanical separation technics allow its recovery from the module and enables its reuse in a new photovoltaic module.
Keywords: vitrimer, CAN chemistry, module, accelerated aging

1 INTRODUCTION

The photovoltaic industry has been growing exponentially over the past twenty years. As a result, the quantity of photovoltaic (PV) modules reaching end-of-life is also increasing. Projections by the International Energy Agency (IEA) indicate that the cumulative volume of end-of-life PV modules could reach 60 - 70 million tons (Mt) by 2050 [1]. In anticipation of the large volume of end-of-life PV modules, and to maintain the position of photovoltaics as a clean energy technology, end-of-life management and recycling of PV modules must be taken into account. The key points that this project aims to address in order to reduce the LCA of photovoltaics are:
- Propose an alternative to EVA, the encapsulant traditionally used, which poses durability problems (yellowing, acid corrosion of interconnections due to the release of acetic acid following the degradation of the encapsulant in contact with water).
- Develop a solution with better chemical, mechanical and thermal resistance than thermoplastic polyolefins (TPO), non-crosslinkable encapsulants currently under development at industrial level.
- Promote re-use/repair by offering a solution that can be re-used several times in the event of defects (delamination, glass breakage, etc.).
- Facilitate recycling by developing a resin that enables the various PV module components (glass, solar cells, encapsulant, backsheet) to be separated by hot delamination.
- High value recycling: possibility to obtain silicon wafer without contamination and potentially without cracks after recycling.
Even today, most encapsulants used in photovoltaics

are cross-linked polymers such as EVAs or polyolefin elastomers (POEs). As a result, once they have been processed in a PV module, they can no longer be reused without being destroyed, as non-mechanical separation of the components is not possible. Residuals from encapsulants remain on separated parts. TPO encapsulants were developed to overcome this problem. Today, however, the physical performance of TPOs does not allow them to be deployed on a large scale (notably because of their high water permeability). The market share of TPOs as PV encapsulants is less than 2%, so it would be interesting to develop a product capable of meeting both the demand for reusability and durability.

The technological solution corresponds to a single-step thermoplastic polyolefin mass vitrimerization using a process that can be transposed to single-screw reactive extrusion. The dynamic cross-linking chemistry chosen was siloxane chemistry. The results of the vitrimer synthesis, the cross-linking solutions and the thermomechanical characterization of the material are described in detail in Brastel et al. [2]. This new material enables high value recycling which supplies silicon wafers from recycling process, without polymer residuals on it. It is a major achievement and marks a milestone in encapsulant development in general.

2 MODULE MANUFACTURING AND PERFORMANCE ASSESSMENT

2.1 Module manufacturing

We successfully fabricated several film samples of variable thicknesses (300 – 800 μm) of 500 cm² surface which allowed us to assemble mini modules (2 ½-cell) with the help of two vitrimer materials (size

940

200 x 200 mm) to evaluate their performance as encapsulants (Fig. 1):
- With catalyst (1 module - SPLIT04)
- Without catalyst (2 modules - SPLIT05 and SPLIT 06)

The addition of catalyst is envisaged to improve vitrimer exchange dynamics.

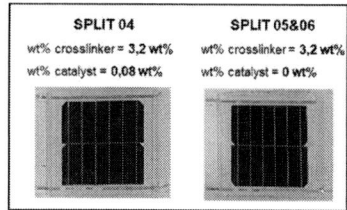

Figure 1: Tested vitrimer samples at mono modules' level.

The fabrication of PV modules with a vitrimer encapsulant went without a hitch for the two formulations presented above. PV modules after lamination showed no visual defects or cell breakage (Fig. 4). Adhesion forces between vitrimer and glass were quantified. An adhesion force > 60 N was obtained, but this result remains to be optimized as the values measured were heterogeneous depending on the location on the sample. Adhesion strength at the backsheet/ vitrimer interface has not yet been quantified, but appears to be very accurate when assessed by manual solicitation.

2.2 Electrical performance measurements
The electrical properties of the manufactured modules (**Erreur ! Source du renvoi introuvable.**) and the optical properties of the encapsulants were measured (Fig. 2). The data collected were compared with those of a commercial thermoplastic polyolefin (TPO) encapsulant.

Table I: Electrical performance of tested mono modules at initial stage

	Reference module TPO encapsulant	SPLIT 04 0.08wt% cat.	SPLIT 05 without cat.	SPLIT 06 without cat.
P_{max} (W)	5.464	5.387	5.340	5.303
I_{sc} (A)	4.830	4.787	4.776	4.765
V_{oc} (V)	1.453	1.454	1.455	1.451
FF (%)	77.8	77.2	76.9	76.7

The electrical performance of the three vitrimer modules was close to that of the reference module (P_{max} difference < 3%), while I_{sc} performance remained in the 1.35 % range, reflecting the excellent optical performance of the vitrimer encapsulant developed. These initial values enabled us to consider further accelerated aging tests. This initial performance is all the more exceptional given that the material has only been developed since the start of the project.

Optical transmission values are very satisfactory, with a total transmission value measured close to 90 %. Both materials, with and without catalyst, achieve the same level of optical transmission as the reference thermoplastic

encapsulant in the 400 – 1100 nm spectral range (Fig. 3). A difference can be observed between reference TPO and the two vitrimers in the 300 - 400 nm wavelength range corresponding to UV, which can be explained by the presence of UV absorbers in the reference's formulation that cut off part of the UV, which are not present in the vitrimers.

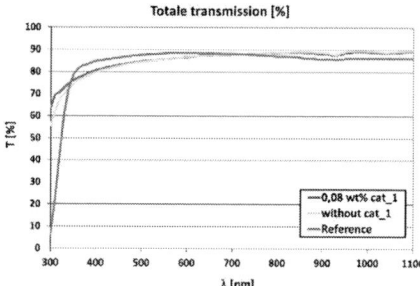

Figure 2: Optical total transmission of studied two vitrimers with reference TPO's Ttot.

3   RESULTS OF ACCELERATED AGING
The three modules were then placed in accelerated aging conditions (85 % humidity, 85 °C for 1000 h). Electrical power loss Pmax is less than 5 % for all modules at 266 and 500 h (Fig. 3). At 1000h, the limits of the materials are reached, and cell damage due to water ingress is visible in the electroluminescence images (Fig. 4).

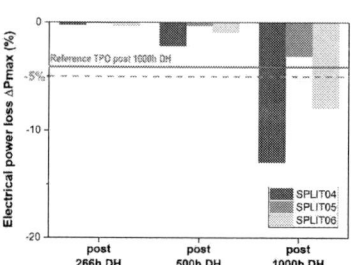

Figure 3: Pmax variation in time (DH accelerated aging conditions).

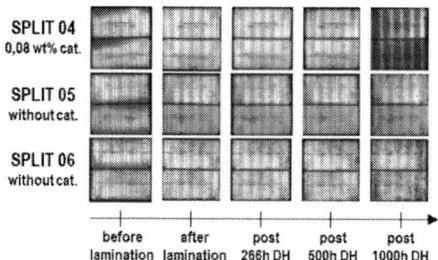

Figure 4: EL results of tested vitrimer containing mono modules during accelerated aging under damp heat conditions.

Finally, a feasibility test was carried out to show how easy it is to laminate/ delaminate/re-laminate a glass-encapsulant-backsheet assembly, confirming that a solar panel can be dismantled at its EoL, or its lifespan extended by hot relamination in case of delamination of several layers within the module (*Fig. 5*).

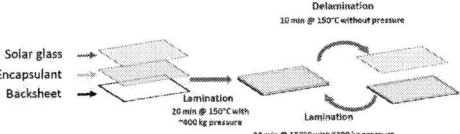

Figure 5: Lamination/delamination process

4 CONCLUSIONS

In conclusion, the completion of the first proof-of-concept modules has confirmed the value of vitrimer encapsulants in meeting the challenge of combining the advantages of thermoplastics and cross-linked encapsulants. The main advantage of vitrimers over conventional encapsulants (EVA, POE) lies in their ability to be reprocessed, which could make it possible to re-laminate a PV module in the event of delamination or glass breakage, for example. A supplementary advantage is coming from the lack of peroxides to initiate the cross-linking reaction as it could reduce the temperature and manufacturing time of a PV module.

5 REFERENCES

[1] Status of PV Module Recycling in Selected IEA PVPS Task 12 Countries, Report IEA-PVPS T12-24:2022 July 2022

[2] A. Brastel, M. Pliquet Landa, E. Mofakhami, T. Bejat, P. Piluso: Transparent vitrimer polyolefins synthesis through melt grafting of siloxane moieties, IUPAC Macro 2024.

Solar Energy at TNO

TOWARDS AFFORDABLE CUSTOMIZATION

Cell research & development	Characterization, testing & diagnostics	Module R&D	Pilots & field trials
Thin functional coatings Wafer to cell processing • Hybrid Tandem • Bifacial cells • Back contact cells • Perovskite cells	IEC level testing & characterization Advanced diagnostics • Silicon cells and tandems • Thin film • Modules • PV products	Test samples Prototyping Material development Aesthetics toolbox • Full size products • Integrated products • Special products • 2D & 3D shapes	Outdoor test facilities • Module and system level • BIPV • Floating • VIPV • Field trials at partners

Why recycle solar panels?

- Large amount of modules reaching end-of-life in the coming years

- High value materials (Ag, PV grade Si)
  - Recover
  - Reuse

- Toxic materials (Pb, Sn, Ag, PFAS)
  - Recover
  - Reuse/dispose safely

- Two timelines to consider
  - Coming 20 years ⏱ → *New recycling technology needed*
  - 20 years → ∞ ⏱ → *Design for recycling (+ new technology?) needed*

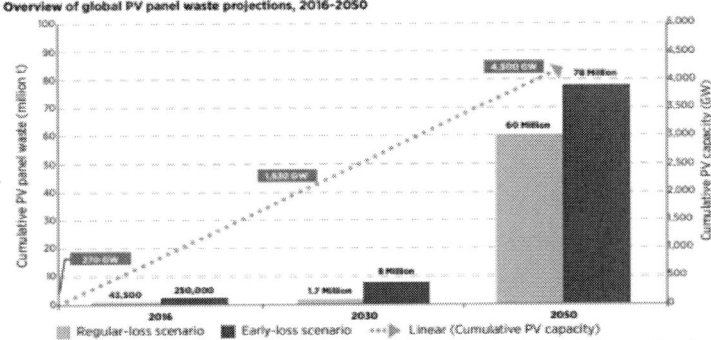

Overview of global PV panel waste projections, 2016-2050

IRENA and IEA-PVPS (2016) "End-of-life Management: Solar Photovoltaic Panels," International Renewable Energy Agency and International Energy Agency Photovoltaic Power Systems.

Delamination methods: what is already out there?

- Opening up the module is arguably the hardest step

- Downcycling
  - Grinding/shredding for reuse in mortar, road filler etc
  - Glass recovery only

- Recycling
  - Hot knife
  - Gas discharge lamps
  - Diamond wire
  - Pyrolysis
  - Chemical dissolution of the encapsulant

- We investigate NIR laser:
  - Energy efficient & fast
  - Can target the silicon surface
  - Wide parameters space for fine tuning
  - Have become more industrialised and cheaper in the past few years

Our approach

NIR laser processing

- Laser in the near infrared range (1030 nm or 1064 nm)
  - ✓ Passes through glass and encapsulant with minimum energy loss
  - ✓ Interacts strongly with the silicon

- Pulse length can be tuned
  - ✓ Keep Si as intact as possible

- The valuable components can be accessed
  - ✓ Glass in one piece
  - ✓ Wafer
  - ✓ Metallisation (Ag, Pb, Al, Sn etc)

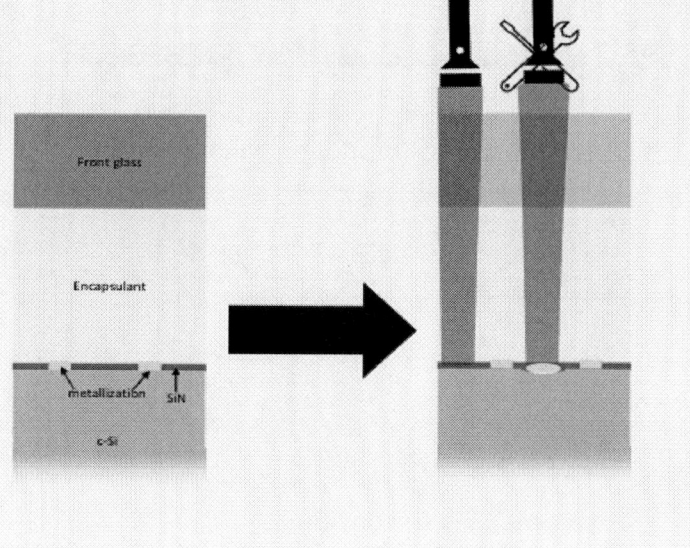

Results

Ablation type

- Basic ablation pattern: lines
- Inside laser lines, no N is detected
  - o SiNx is removed
  - o No direct ablation of SiNx by NIR is possible (laser ≈1.2 eV vs SiNx > 4 eV)

  → Indirect ablation, via the underlying silicon

- No significant groove formation in the silicon

  → sub-micron Si removal

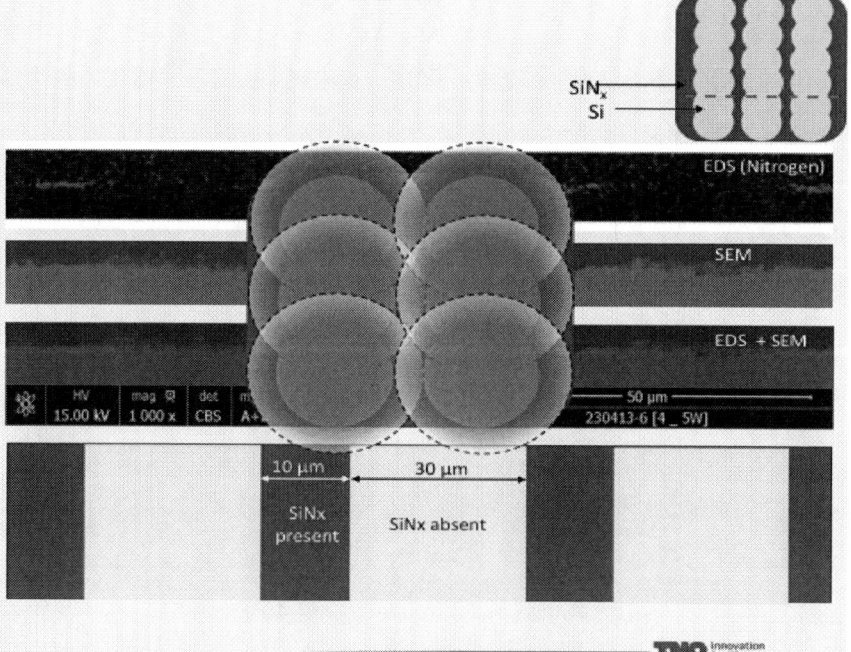

Other features of the process

- Si signal strongest in 15 J.cm^{-2} area
- Some C found inside laser lines (more at 30 J.cm^{-2})[1]
→ residual heat from Si melted encapsulant on top of Si
- Narrower N-rich strips between laser lines at 5W
→ less unablated SiNx left

❖ Knowing the ablation threshold can help optimise the process

- Minimise energy
- Minimise cross heating

[1] the clusters of C in the 30 J.cm^{-2} part are unburnt EVA

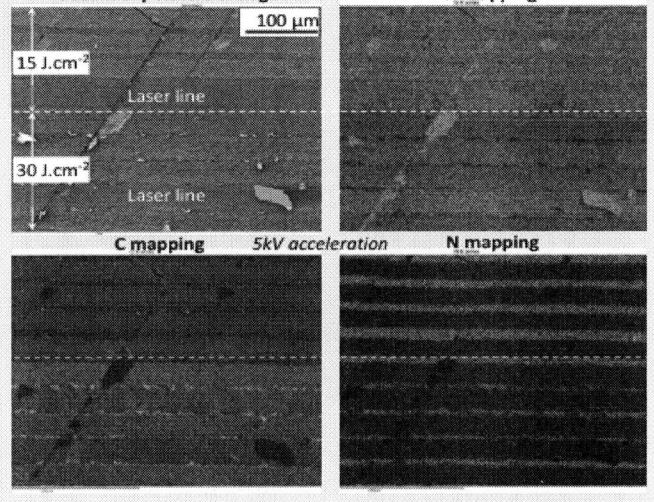

Ablation threshold determination

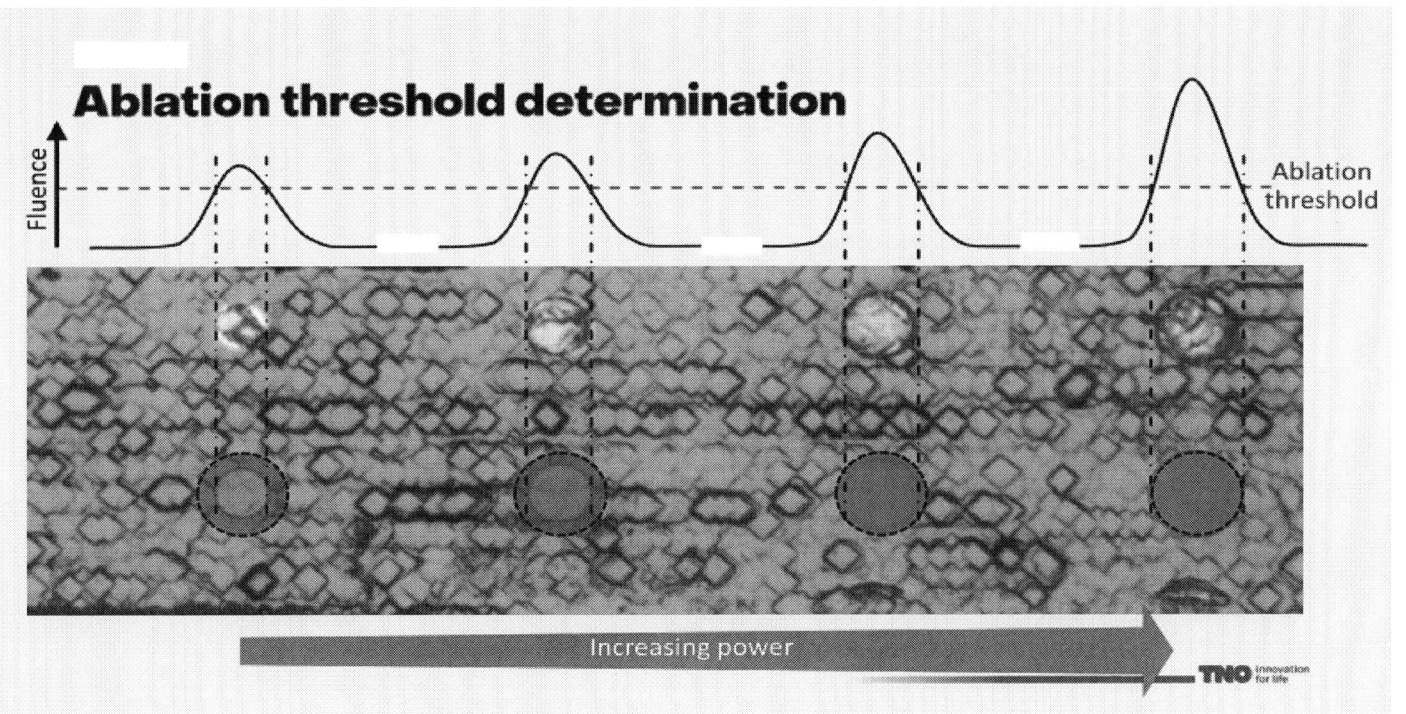

Laser pattern tuning

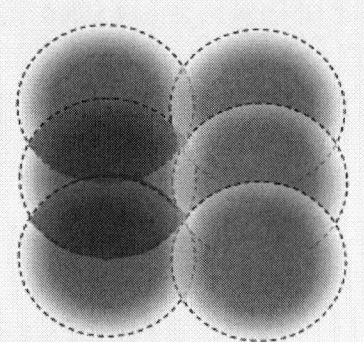
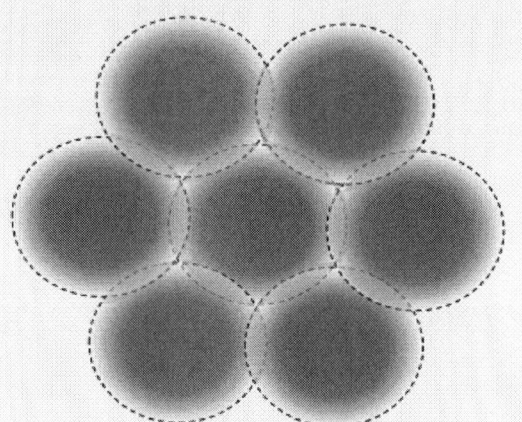

Straight lines pattern:
- Large overlap along the lines
- Lower overlap in the X direction

Hexagonal dots pattern
- Always the same overlap between dots

Clean separation

- Different fluences were tested, to find an optimum
- The best compromise was then applied to a 5x5cm² sample:
  hexagonal pattern, 25% pulse overlap
- It results in a uniform appearance and easy delamination*

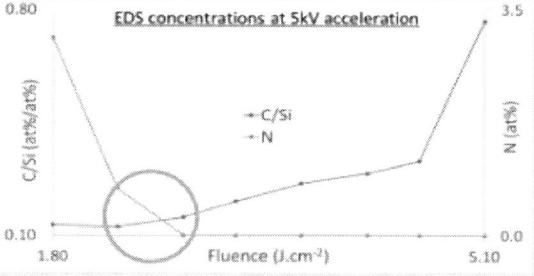

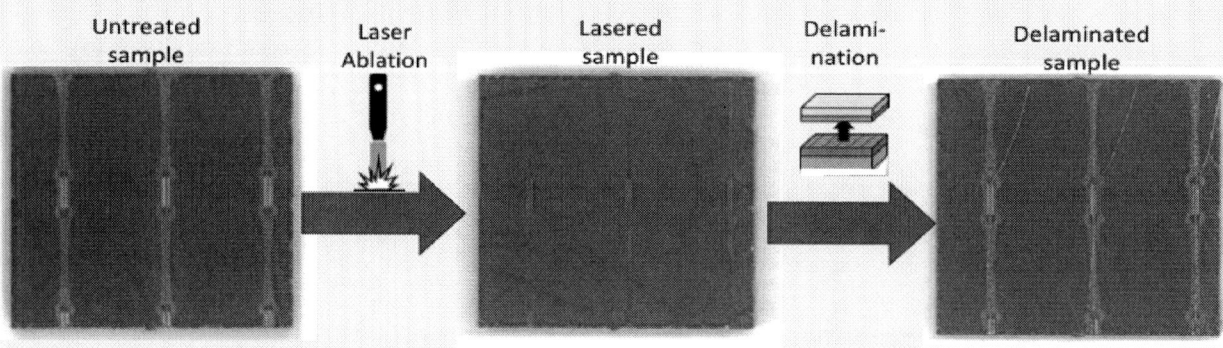

Untreated sample → Laser Ablation → Lasered sample → Delamination → Delaminated sample

* Less than 15 N.m using torque between the frontsheet and the backsheet

Clean separation

- Uniform sample surface

- Practically no N found in the treated area → SiNx is ablated uniformly

- Uniformly low C signal → minimum encapsulant melting *

→ Successful settings for this type of sample

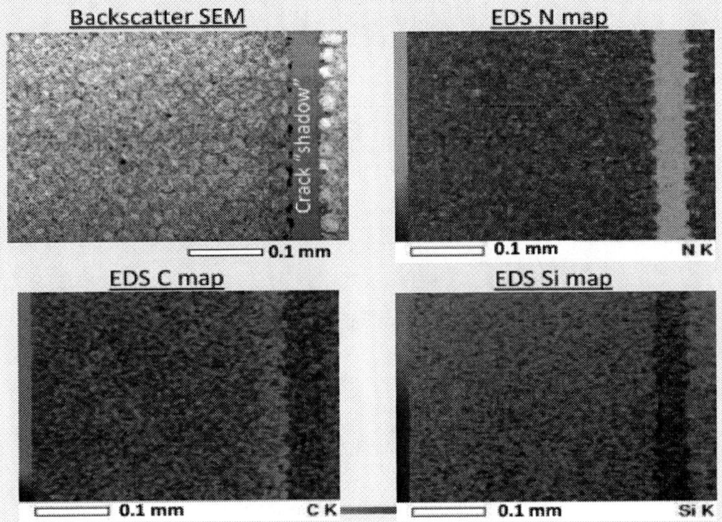

*Absolute C quantification is ongoing

Backscatter SEM

Crack "shadow"

0.1 mm

EDS N map

0.1 mm   N K

EDS C map

0.1 mm   C K

EDS Si map

0.1 mm   Si K

Conclusion

- NIR laser processing, combined with a mechanical step, is an effective method for silicon PV delamination

- It gives access to essentially unaltered metallization

- The silicon wafer has only very low superficial contamination → potential for lower energy wafer recycling

- We are studying the scale up of this technology

TNO innovation for life

Full ERP plan

Background information

Forecast PV waste

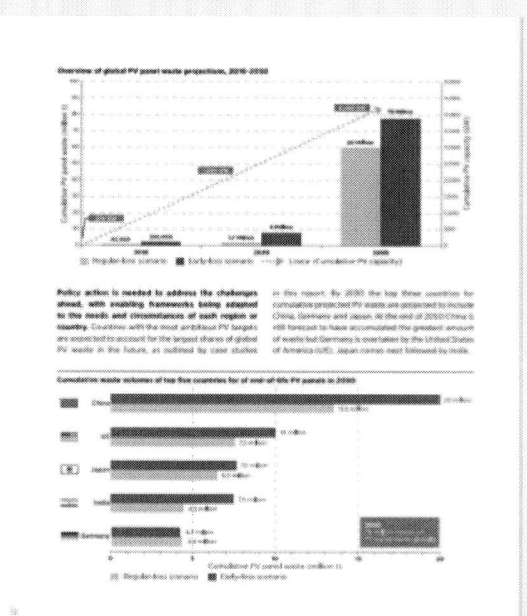

IRENA and IEA-PVPS (2016), "End-of-Life Management: Solar Photovoltaic Panels," International Renewable Energy Agency and International Energy Agency Photovoltaic Power Systems.

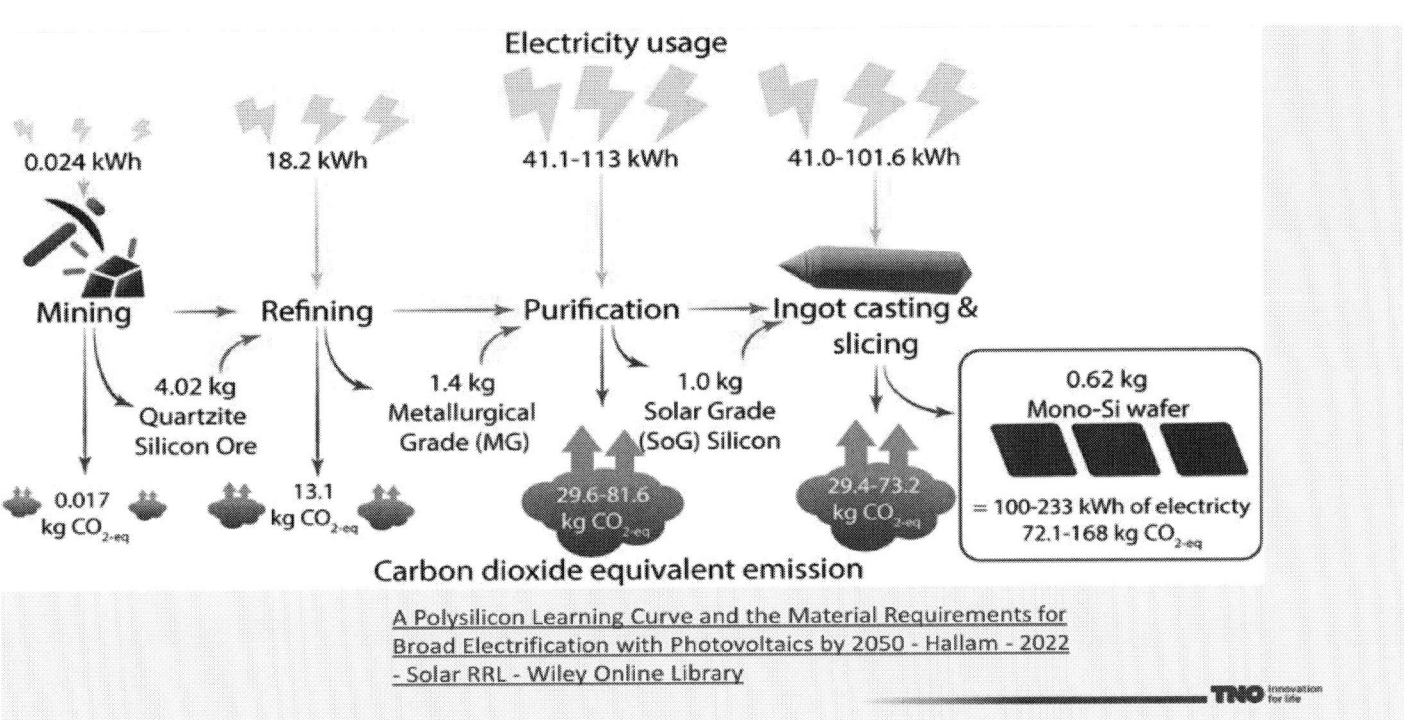

A Polysilicon Learning Curve and the Material Requirements for Broad Electrification with Photovoltaics by 2050 - Hallam - 2022 - Solar RRL - Wiley Online Library

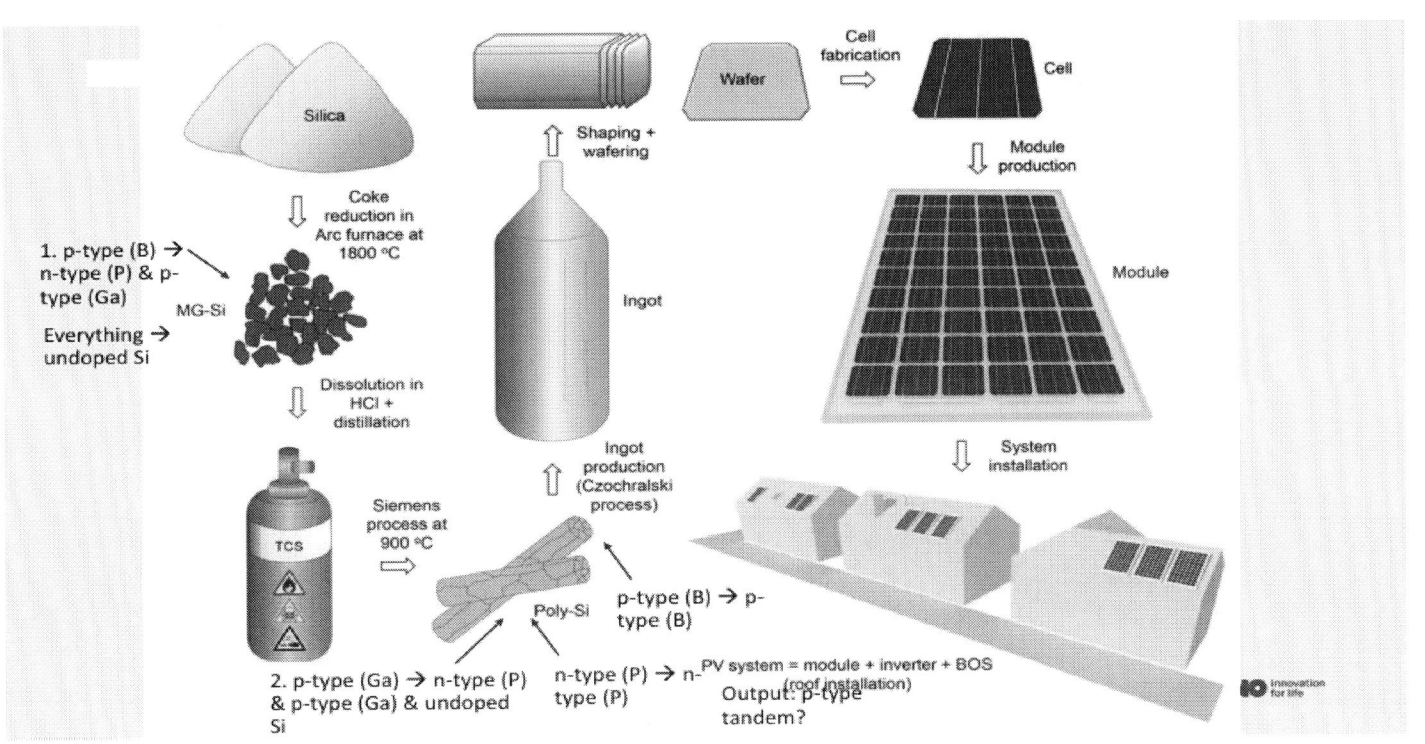

Ablation threshold determination

- Principle of the measurement:
  - Scan various pulse energies/fluences (here, x direction)
  - Measure the diameters of the ablated craters (D)
  - Plot D^2 vs $\ln(E)$ → E_{th} is the X intercept
  - Get the ablation threshold fluence

$$F_{th} = \frac{2E_{th}}{\pi w_0{}^2}$$

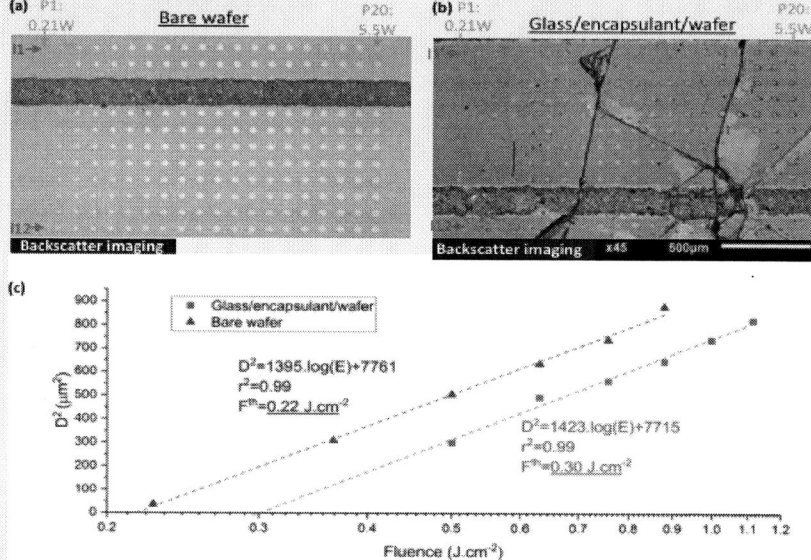

Defining the 'sweet spot'

- Settings with lowest:
  - Carbon
  - Nitrogen

Element\ power (W)	2	2.5	3	3.5	4	4.5	5	5.5
C	11.1	10.9	13.2	16.7	20	21.7	24	40.5
N	3.1	0.7	0	0	0	0	0	0
O	2.1	2.1	2.4	2.6	2.9	3.7	2.9	5.8
Si	83.8	86.3	84.3	80.7	76.4	73.9	72.3	52.9
Al	0	0	0	0	0.6	0.7	0.8	0.8
N/Si	0.04	0.01	0	0	0	0	0	0
C/Si	0.13	0.13	0.16	0.21	0.26	0.29	0.33	0.77
C/O	5.39	5.17	5.44	6.35	6.94	5.92	8.15	6.93

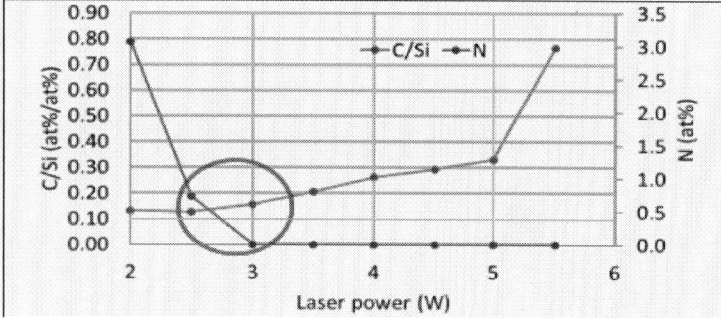

Ablation type

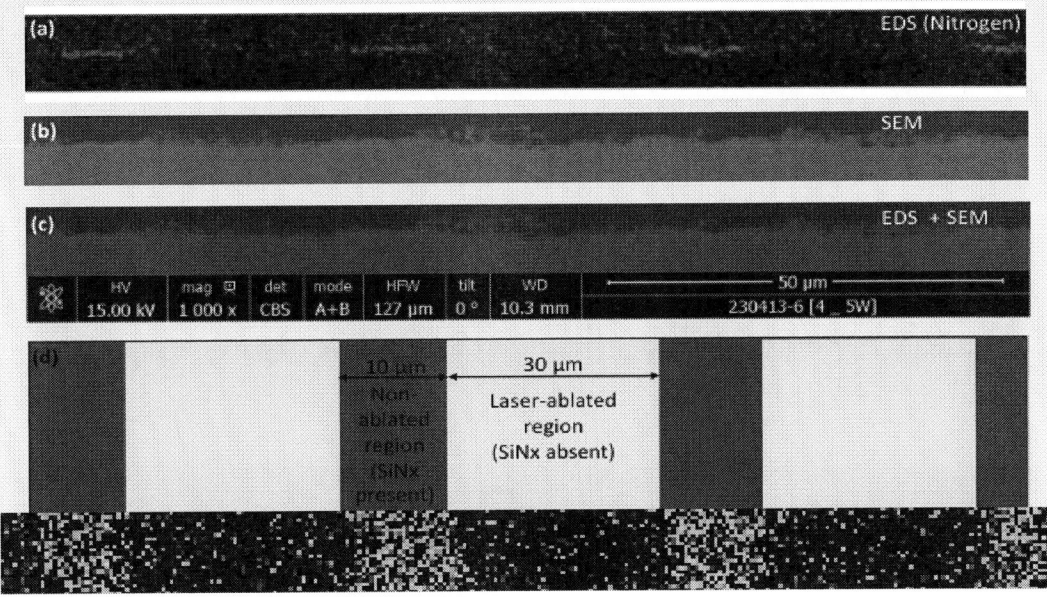

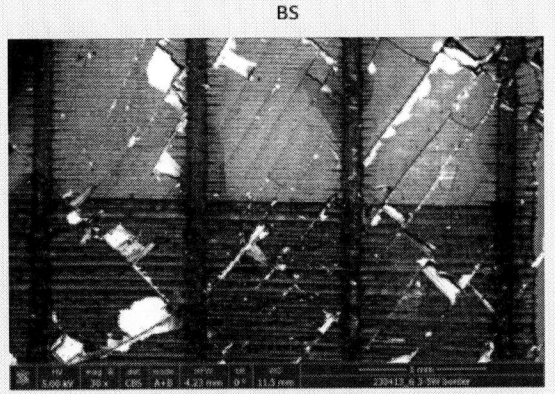

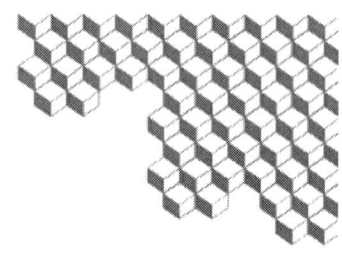

Innovative Design-for-Recycling for Critical Material-Free Interconnection of PV modules

Antoine PERELMAN, Vincent BARTH, Fabien MANDORLO, Eszter VOROSHAZI

Context

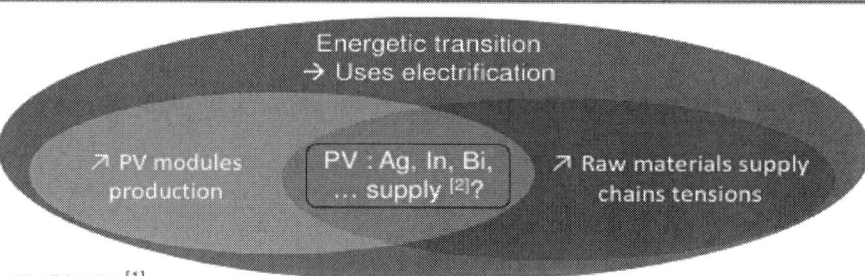

Energetic transition
→ Uses electrification

↗ PV modules production

PV : Ag, In, Bi, ... supply [2]?

↗ Raw materials supply chains tensions

Global PV Installation and corresponding PV market [1]

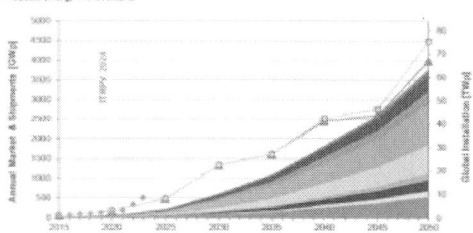

European Commission

Critical Raw Materials Act [3]

[1] VDMA, « ITRPV 2024 ».
[2] Zhang et al., « Design Considerations for Multi-Terawatt Scale Manufacturing of Existing and Future Photovoltaic Technologies ».
[3] European Commission, « Critical Raw Materials Act »

EUPVSEC 2024 - 3DO.12.4

Context

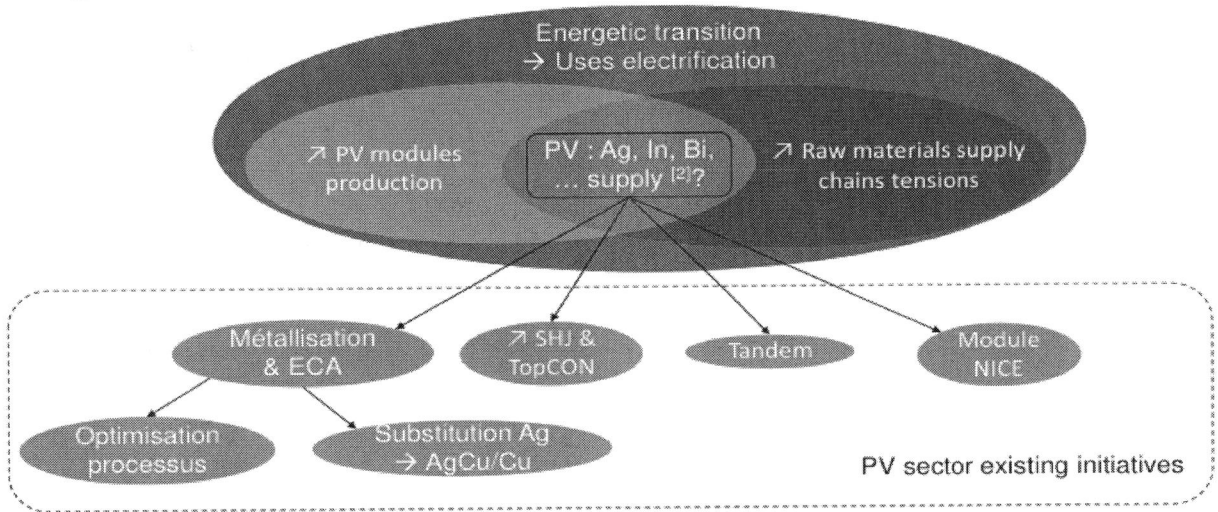

[1] VDMA, « ITRPV 2024 ».
[2] Zhang et al., « Design Considerations for Multi-Terawatt Scale Manufacturing of Existing and Future Photovoltaic Technologies ».
[3] European Commission, « Critical Raw Materials Act »

EUPVSEC 2024 - 3DO.12.4

Work objectives and chosen approach

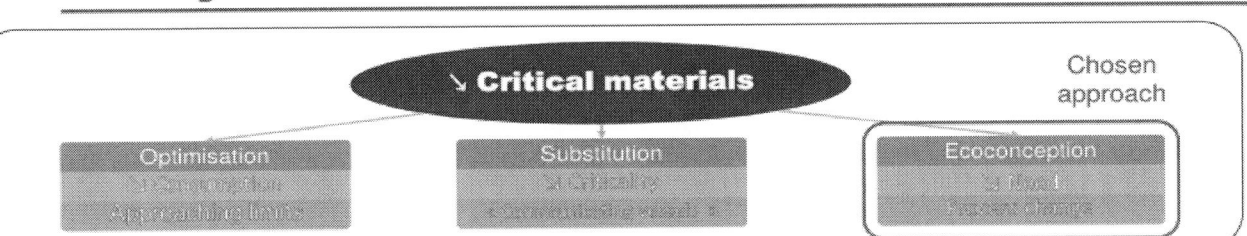

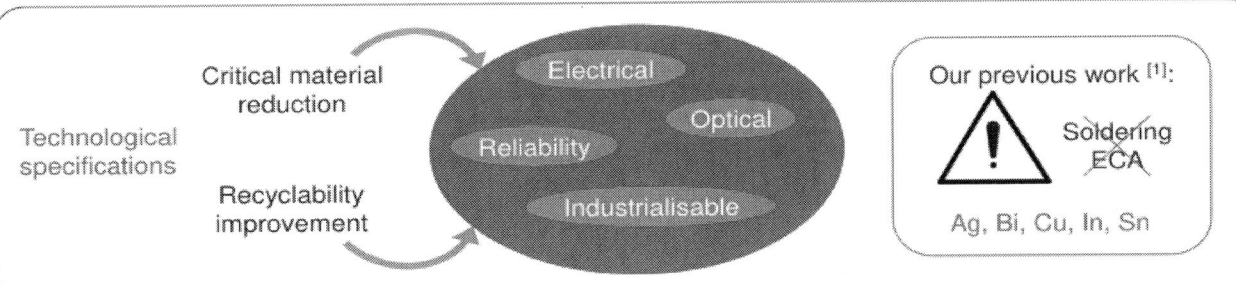

[1] Perelman et al., « Critical materials and PV cells interconnection », 10.1051/epjpv/2023034.

EUPVSEC 2024 - 3DO.12.4

Module architecture

Standard

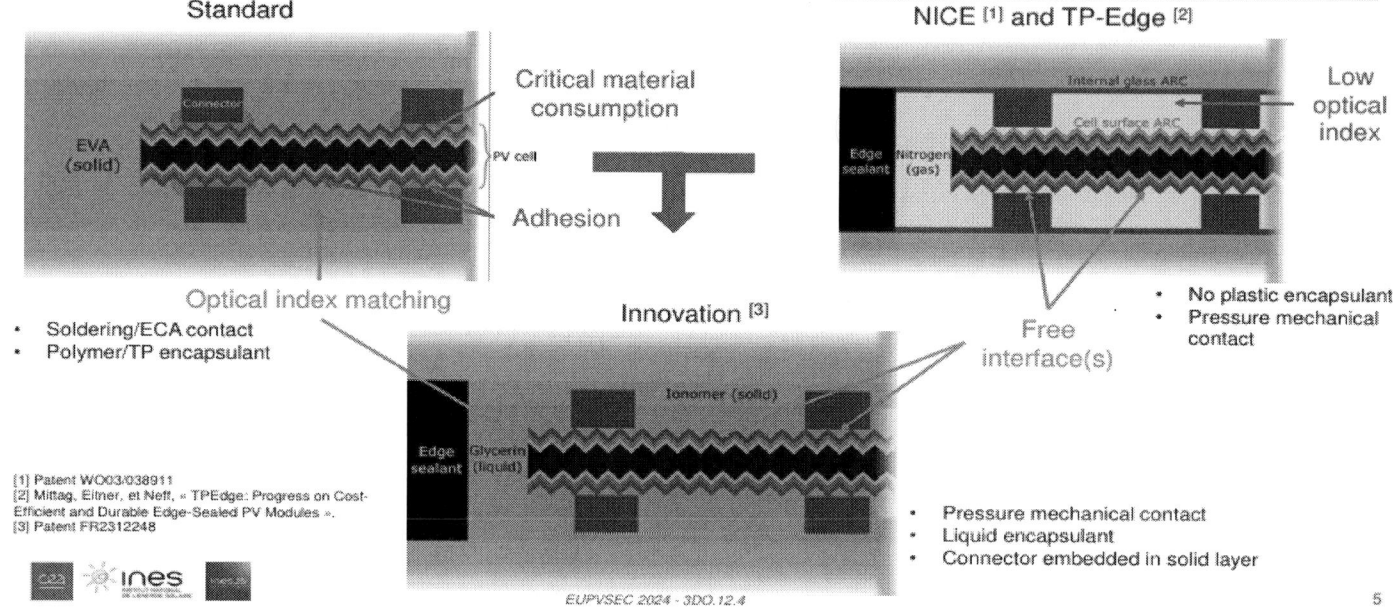

EVA (solid)

Connector

PV cell

Critical material consumption

Adhesion

Optical index matching

- Soldering/ECA contact
- Polymer/TP encapsulant

[1] Patent WO03/038911
[2] Mittag, Eitner, et Neff, « TPEdge: Progress on Cost-Efficient and Durable Edge-Sealed PV Modules ».
[3] Patent FR2312248

NICE [1] and TP-Edge [2]

Internal glass ARC

Cell surface ARC

Edge sealant

Nitrogen (gas)

Low optical index

Free interface(s)

- No plastic encapsulant
- Pressure mechanical contact

Innovation [3]

Ionomer (solid)

Edge sealant

Glycerin (liquid)

- Pressure mechanical contact
- Liquid encapsulant
- Connector embedded in solid layer

EUPVSEC 2024 - 3DO.12.4

Optics

✓ CROWM software [1]

✓ 3 architectures
- Standard
- NICE [2] / TP-Edge
- Innovation

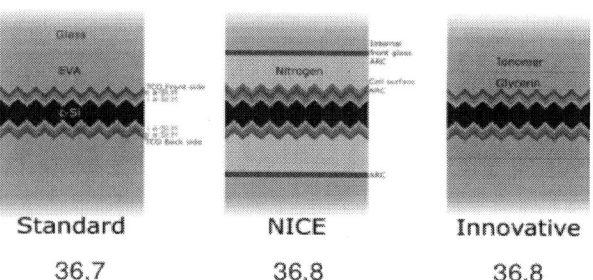

Standard	NICE	Innovative
36,7	36,8	36,8

mA.cm⁻² ➡ ✓ Optics validated

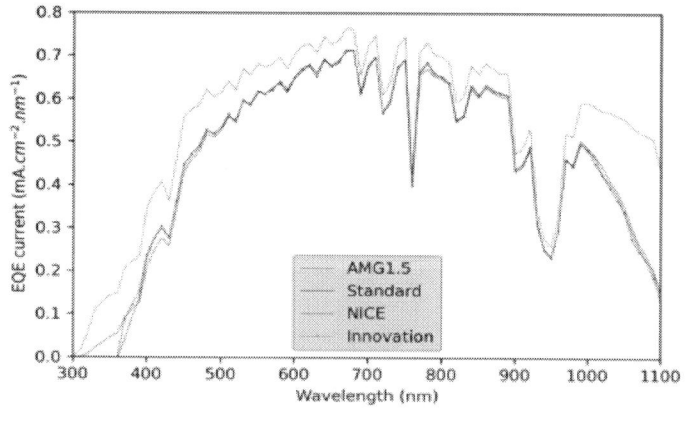

[1] Zinßer et al., « Optical and Electrical Loss Analysis of Thin-Film Solar Cells Combining the Methods of Transfer Matrix and Finite Elements ».
[2] Couderc et al., « Encapsulant for Glass-Glass PV Modules for Minimum Optical Losses ».
EUPVSEC 2024 - 3DO.12.4

Electrical results

- ✓ Feasible process
- ✓ No interconnection specific loss
- ✓ Encapsulation validated

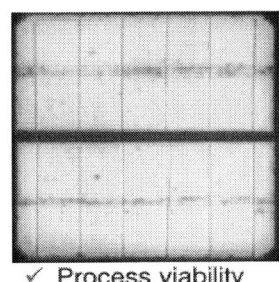

- ✓ Process viability

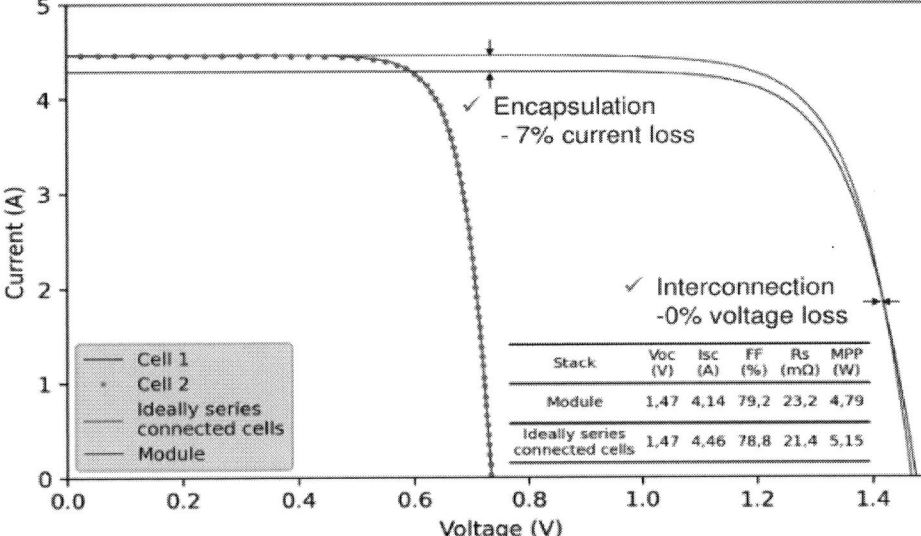

- ✓ Encapsulation - 7% current loss
- ✓ Interconnection -0% voltage loss

Stack	Voc (V)	Isc (A)	FF (%)	Rs (mΩ)	MPP (W)
Module	1,47	4,14	79,2	23,2	4,79
Ideally series connected cells	1,47	4,46	78,8	21,4	5,15

EUPVSEC 2024 - 3DO.12.4

Reproducibility

Prototyping stage
28 fabricated
valid modules

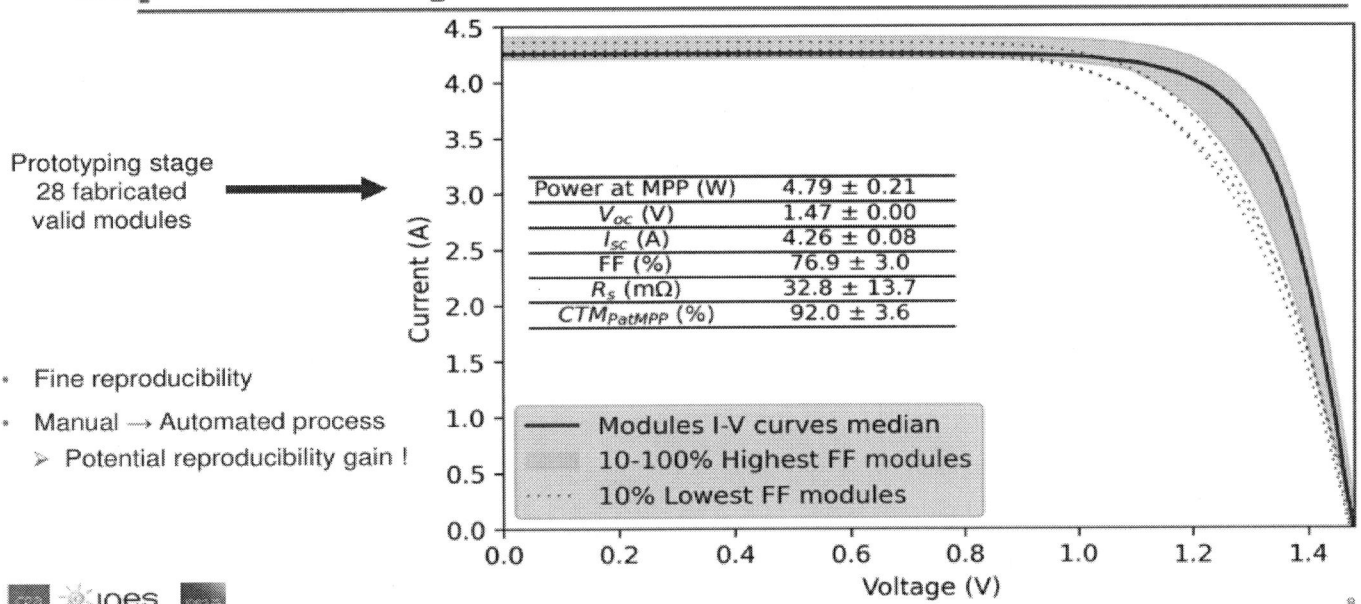

Power at MPP (W)	4.79 ± 0.21
V_{oc} (V)	1.47 ± 0.00
I_{sc} (A)	4.26 ± 0.08
FF (%)	76.9 ± 3.0
R_s (mΩ)	32.8 ± 13.7
CTM_{PatMPP} (%)	92.0 ± 3.6

- Modules I-V curves median
- 10-100% Highest FF modules
- 10% Lowest FF modules

- • Fine reproducibility
- • Manual → Automated process
  - ➢ Potential reproducibility gain !

EUPVSEC 2024 - 3DO.12.4

Thermal cycling aging

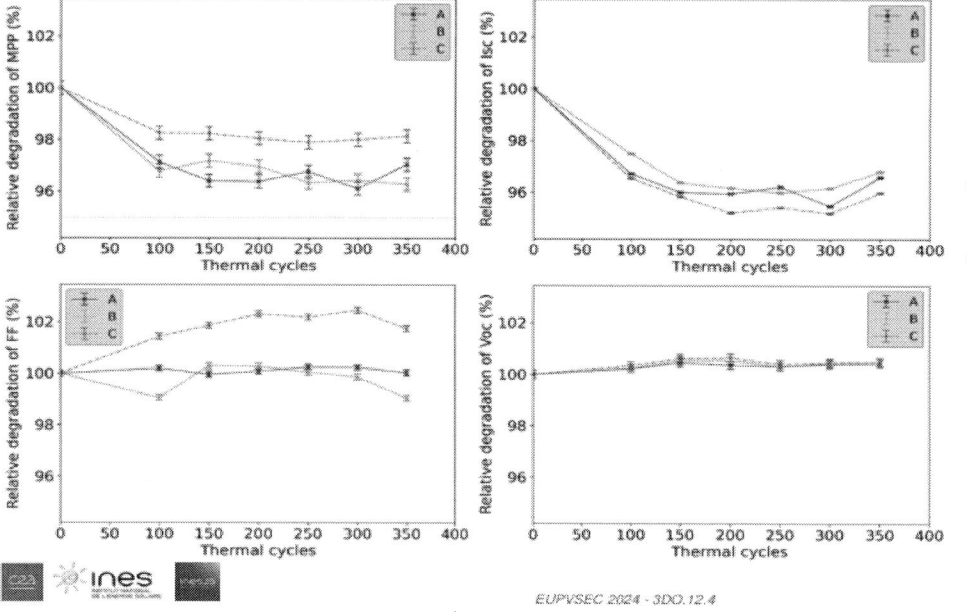

* Satisfying stability of interconnection (FF & V_{oc})

* Main driver of ↘ max power: ↘ photocurrent
  - Delamination / bubbles ?
  - Material degradation ?

Cycles [-40, +85]°C [1]

[1] NF EN IEC 61215-1-1

EUPVSEC 2024 - 3DO.12.4

Recycling

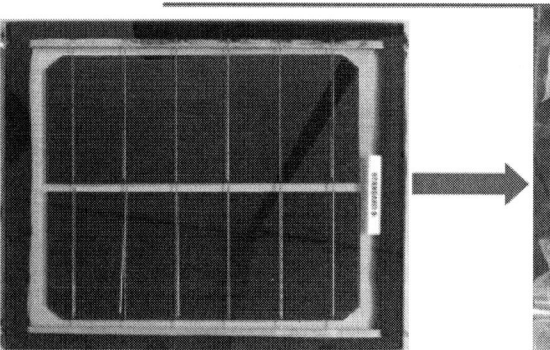

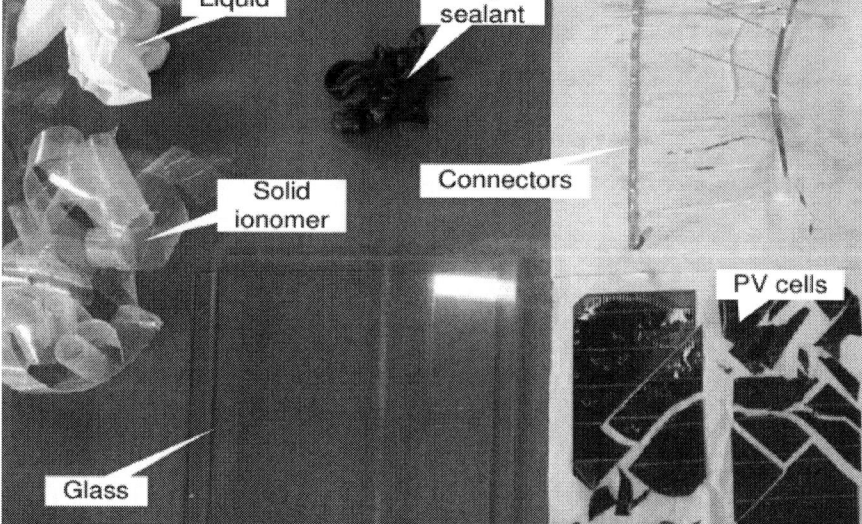

Sealant cutting

+ Connectors & ionomer stripping

= Components sorting easiness

EUPVSEC 2024 - 3DO.12.4

Conclusions & perspectives

- ✓ No connector-metallization bonding material
  - ✓ Critical material saving
- ✓ Easy module disassembling
  - ✓ Recycling
  - ✓ Repair
- ✓ High electrical and optical performances at T0
- ✓ Satisfying reliability under TC

- ➢ Process automation
  - ➢ Precision & reproducibility gains
- ➢ Material study
- ➢ Reliability: add UV aging tests

EUPVSEC 2024 - 3DO.12.4

Thank you for your attention !

This work has been realized with the participation from members of INES.2S and received funding from the French State under its investment for the future program with reference ANR-10-IEED-0014-01.

antoine.perelman@cea.fr
vincent.barth@cea.fr
fabien.mandorlo@insa-lyon.fr
eszter.voroshazi@cea.fr

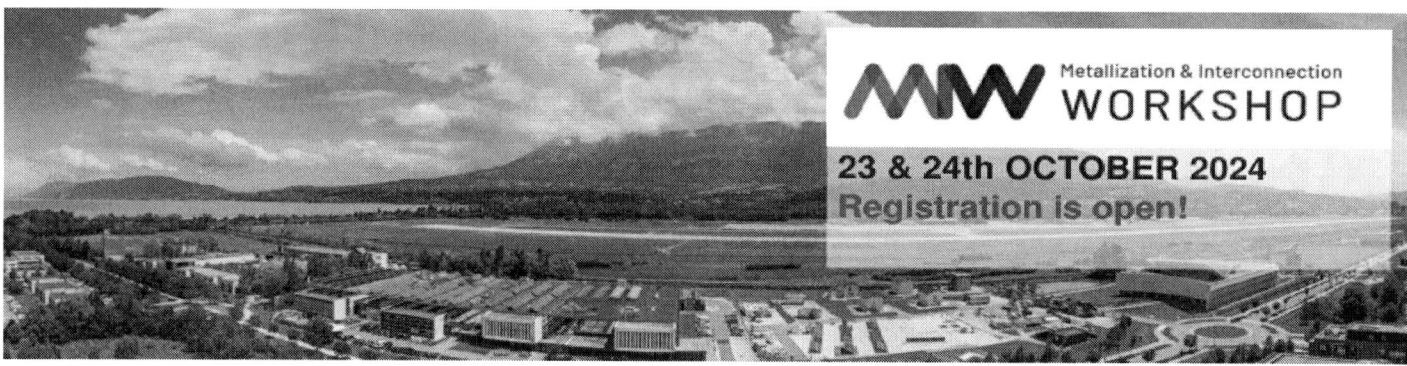

Bourget-du-Lac/Chambéry (France)

Website and programme: https://miworkshop.info/

Proceedings in partnership with Solar Energy Materials & Solar Cells

Sponsors

41st European Photovoltaic Solar Energy Conference and Exhibition

BIFACIAL LIGHTWEIGHT SOLUTION WITHOUT GLASS

Alicia Buceta[a], Ana B. Cueli[a], Miguel Aguirre[a], Ana Linares[b], Elena Llarena[b], Silvia Cal[b] and Jaione Bengoechea[a]
[a] National Renewable Energy Center (CENER)
Avenida Ciudad de la Innovación, 7, 31621 Sarriguren (Navarra) SPAIN
Tel.: +34 948 25 28 00; info@cener.com
[b] Instituto Tecnológico y de Energías Renovables, S. A. (ITER)
Pol. Industrial de Granadilla, s/n, E 38600 Granadilla de Abona (Canarias), SPAIN
Tel.: +34 922 74 77 00; difusion@iter.es

ABSTRACT: A well-known limitation of photovoltaic (PV) modules is their weight, especially with the current trend towards larger module sizes. Conventional crystalline silicon (c-Si) monofacial modules, with 3.2 mm glass as the front panel, typically weigh around 10 kg/m², which can limit their range of applications. For conventional c-Si bifacial modules, typically 2 mm glass-glass constructions, the weight is around 13 kg/m². In this work, we present the results obtained using a translucid composite in the lamination, which allows us to reduce the module weight by 73% compared to bifacial and 65% for monofacial ones. Despite the necessity for further enhancements to avoid yellowness in response to elevated temperatures and humidity, the PV modules have demonstrated the capacity to withstand hail impact without damaging the solar cells, which is a noteworthy achievement.
Keywords: lightweight, bifacial, c-Si

1 INTRODUCTION

The development of renewable energy sources is of vital importance, not only in the fight against climate change but also in securing the energy supply. In this context, solar technologies are already playing an essential role in the shift towards carbon-neutral economies, while ensuring a reliable and competitive energy supply. Seeking higher power throughput, the market has evolved toward bigger PV modules and bifacial solutions. This circumstance aggravates the already existing drawback of conventional c-Si modules high weight. Monofacial c-Si modules, with 3.2 mm glass as frontsheet, typically weigh around 10 kg/m², while conventional bifacial c-Si modules, typically 2 mm glass-glass constructions, weight around 13 kg/m². This is a limitation not only in terms of possible applications, especially when the maximum weight limit is restricted, but also in terms of the handling and transport. Therefore, finding a lightweight solution for c-Si modules is of great interest.[1]

The glass accounts roughly 60%-70% of monofacial modules total weight while in bifacial glass-glass configuration it accounts over 80% of the weight. Hence, avoiding the use of glass, or reducing its thickness, are the preferred strategies to reduce the weight of PV modules.[2], [3] One of the crucial aspects of this approach is to guarantee the mechanical integrity of the modules, which is frequently accomplished through the utilisation of composite materials that offer a combination of lightweight properties and high mechanical strength.[4], [5] However, these approaches tend to use opaque solutions as backsheet which are not valid for bifacial topologies.

In this work, this task has been addressed through the use of light transmitting materials by both sides, developing a one step process using a conventional laminator to ease the implementation in PV industry. The material of choice to replace the glass is a series of polymeric composites reinforced with woven fibber of different grammages. The results of hail impact test, damp-heat, UV irradiation and outdoor exposure will be discussed.

2 METHODOLOGY

The fabrication of the laboratory scale PV modules was performed at CENER's facility, while the scale up to a 1 m x 1.5 m size module was made at ITER's facility. The prototypes were manufactured manually, including the soldering of the solar cells.

The reliability of the produced prototypes has been assessed following the conditions laid down in PV module qualification standard IEC 61215-2,[6] with increased severity for humidity testing. More specifically, the Hail impact test, UV weathering and climatic tests have been performed on the modules, given that polymeric materials are susceptible to suffer degradation when exposed to mechanical and environmental stresses. The Module Quality Tests (MQT) were conducted at CENER's laboratory, which is accredited to perform IEC standard tests for PV modules.

The outdoor tests were conducted in both locations, at CENER, in Navarre (north of the Iberian Peninsula), and at ITER in the Canary Islands (Atlantic Ocean, west of Africa).

2.1 Lamination

For the lab scale modules, the laminator used was an automated module laminator from PEnergy for samples up to 60 cm x 60 cm and 3 cm maximum thickness.

The full-size modules were fabricated in a 2BG automated laminator, with a module size limit of 2 m x 4 m and 3cm of maximum thickness.

2.2 Hail impact test

The hail impact resistance test was performed with an ice ball launcher LBH-25 from ARIES. The experiment was implemented according to the IEC 61215-2 standard; 25 mm diameter ice balls were launched at a velocity of 23 m/s ± 2% (MQT 17).

2.3 Ultraviolet exposure

The UV irradiation was applied in a weathering chamber using as light source lamps with a dedicated filter to match the spectral distribution stated in the qualification standard of PV modules. The temperature of the samples during UV irradiation was fixed at 60 °C and a total dose of 15 kWh/m² was applied as stated in IEC 61215-2 standard.

2.4 Climatic chamber

Their resistance to temperature and humidity was tested in a C-70/200 climate chamber from CTS. The damp-heat test (MQT 13), 85 °C and 85% relative humidity (RH), was completed in accordance with the IEC 61215-2 standard, but increasing the total duration.

3 RESULTS AND DISCUSSION

The objective was to develop a bifacial lightweight solution that could be manufactured in existing industrial lines, avoiding the use of glass. Accordingly, the compatibility of all materials considered with conventional PV module lamination process was determined by visual inspection after the manufacturing process. In this work we report the results obtained using a polymeric composite in the lamination. Once the lamination process was adapted to the different configurations and the samples fulfilled the visual requirements, their resistance to hail, temperature and humidity was tested. Concurrently, efforts were made to identify potential protective measures against UV radiation, given that polymers are prone to suffer degradation, including yellowing or embrittlement under UV conditions. In the absence of glass to provide mechanical integrity, efforts were concentrated on identifying a solution capable of withstanding hail impact. When a suitable solution was identified, the samples were scaled up, initially to a 4-cell module and subsequently to a 20-cell module. The best configurations were tested outdoors in two different locations, at CENER's facilities (north Spain) and at ITER's facilities (Canary Islands).

3.1 Hail impact test

One of the main constraints in glass-free PV modules is to withstand mechanical stresses as hail impact. In order to prove the ability of the samples to fulfil the IEC 61215-2 requirements, the hail test is performed first using a methacrylate plate as back support and, if the solar cell withstands the impact without damage, then the PV sample is tested again removing the back support, with the sample free-standing position (Figure 1).

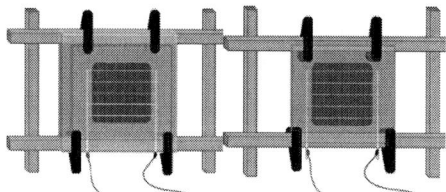

Figure 1: Schematic representation of the set up with methacrylate (left) and the sample free-standing (right) during the hail impact test.

A number of configurations were subjected to hail testing prior to identify one that would provide the required mechanical protection to the solar cells. In Figure 2, it can be observed how the solar cell breaks after the first hail impact (using methacrylate back support) when the polymeric composite shows poor mechanical protection.

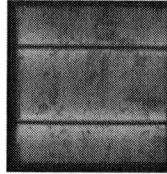

Figure 2: EL image of a polymeric composite sample before hail impact (left) and after one hail impact (right) using methacrylate as back support.

Once a reinforced polymeric composite solution able to withstand hail impact with and without back support was obtained, the sample was upscaled to a 4 solar cells module. As it can be observed in Figure 3, the upscale did not affect the module mechanical properties. The solar cells show no damage, not even the one that was already broken before lamination shows propagation of its existing cracks.

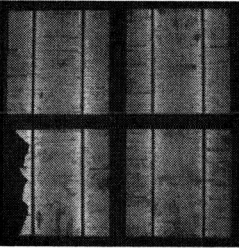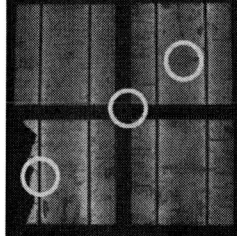

Figure 3: EL image of the reinforced polymeric composite sample before hail impact (left) and after a total of six impacts (right), three impacts with the methacrylate back support and three with the sample free-standing.

This reinforced polymeric composite configuration able to withstand mechanical stress from hail, will be considered as the reference on which optimization will be carried out.

3.2 UV irradiation

As UV radiation can be detrimental for polymers, several UV protection layers were added to the composite to evaluate its effectiveness. The yellowing index is a number calculated using transmittance or reflectance measurements and is an extended methodology to assess the colour variation of polymers due to UV radiation. To calculate it, the reflectance of the test samples before and after UV exposure was used. Negative yellowing index values indicate that the sample colour has more blue than yellow contribution while positive ones imply the opposite.

Three different approaches were tested and after being exposed to 15 kWh/m², only the solution C showed almost no colour variation (Figure 4). Hence, the sample with the UV protection C and the reinforced composite without additional UV protection (reference), were exposed to an extra UV dose of 15 kWh/m² where protection C still presented a great performance.

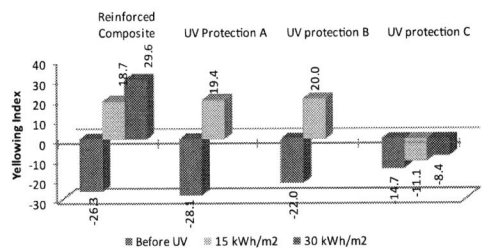

Figure 4: Yellowing index of the reinforced polymeric composite as reference and three different proposed solutions, protection A, B and C, where only C shows nearly no yellowing after a total of 30 kWh/m².

To ensure that the changes in the configuration did not affect the PV module mechanical properties, a one solar cell sample was tested under hail impact (Figure 5).

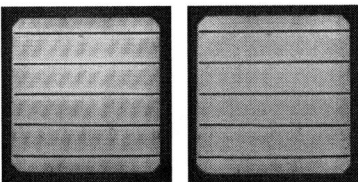

Figure 5: EL image of sample with UV protection C before hail impact (left) and after two hail impacts (right), one with methacrylate back support and another one with the sample in free-standing position.

Given that no detrimental effects were observed subsequent to the hail impact, even when the sample was in a free-standing position, Solution C was deemed to be a highly promising solution.

3.3 Damp-heat test

The two samples that had previously passed the hail impact test (a 4 solar cells reinforced composite module and a 1 solar cell module with UV protection C) were then subjected to a damp-heat test. The IEC standard stipulates a test duration of 1,000 hours at 85 °C and 85% RH. However, the aforementioned samples were subjected to a total of 2,548 hours. An examination of the EL images captured during the course of the test reveals no discernible deterioration of the active area of PV samples (Figure 6).

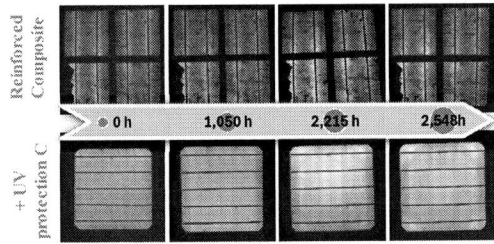

Figure 6: EL images of the samples tested under damp-heat after different times of exposure.

Nevertheless, the visual appearance of the samples indicates a distinct yellowness over time, reflected in the IV curve measurements by the decline in the I_{SC} (Figure 7).

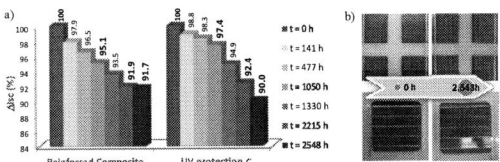

Figure 7: a) decay in percentage of the I_{SC} value over time under damp-heat, b) images before and after damp-heat test.

Although the I_{SC} descent in percentage demonstrates comparable values at the conclusion of the experiment for both samples, the decrease was more gradual at the initial stage in the sample with protection C. This phenomenon may be attributed to the potential for the protective material C to function also as a humidity barrier. This tendency is also evident in the increase of the yellowing index (Figure 8).

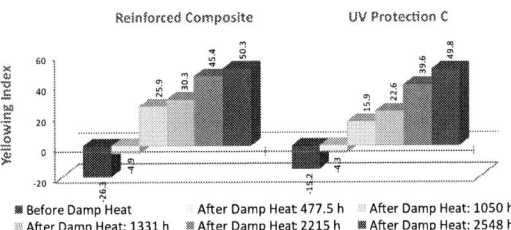

Figure 8: Yellowing index evolution over time under damp-heat. Calculated using reflectance measurements.

This result presents a new challenge, as the test is conducted in the absence of light and the yellowing is therefore not attributable to UV radiation. Consequently, an experiment was conducted in which the composite was subjected to heating at 85 °C in an environment devoid of moisture. The results of this experiment indicated that the application of heat can induce a degree of yellowing (Figure 9), although the presence of humidity appears to enhance the rate of colour change (Figure 8).

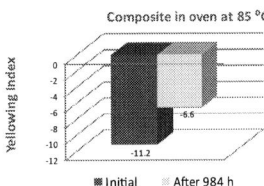

Figure 9: Small yellowing effect after 984 h at 85 °C in the oven. Yellowing calculated with reflectance measurements.

To ascertain whether the mechanical properties had been compromised during this test, the samples were subjected to further examination under hail impact. The reinforced composite module withstood a total of 13 hail impacts including tests with and without the methacrylate back support, whereas the sample with UV protection C had a total of 5 hail impacts. This subsequent analysis revealed that, while some delamination was evident in the sample with UV protection C after the impact, but no damage was observed within the solar cells (Figure 10).

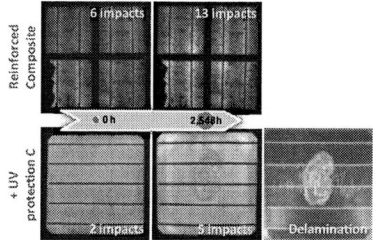

Figure 10: EL images of the reinforced composite sample and the sample with UV protection type C after the first hail impact test prior to damp-heat (left side) and after being 2,545 h under damp-heat and a second round of hail impact (centre). The picture on the right corresponds to the photograph of the delamination observed in the sample with UV protection C after damp-heat and hail impact.

Consequently, it may be concluded that the mechanical properties remain intact after damp heat tests, although the occurrence of delamination should be addressed.

3.4 Outdoor test

The configuration with UV protection A and protection C were upscaled at ITER's facilities up to 20 solar cells PV modules (1 m x 1.5 m). It should be noted that the configuration using protection A does not provide protection against UV radiation. However, an improvement in the protection against humidity is being tested. The modules were subjected to outdoor testing in two locations: CENER facilities (situated in the vicinity of Pamplona in Navarre) and ITER facilities (located in the vicinity of Granadilla in Tenerife).

3.4.1 Northern Iberian Peninsula location

At CENER facilities the samples were arranged in the rooftop at an inclination of 42° and facing south (Figure 11).

Figure 11: Installation on CENER's rooftop (left) and used pyranometers to monitor irradiance (right)

The large modules have 20 monofacial solar cells connected in series. The modules can be easily adapted to the bifacial configuration and, in principle, no differences in durability are expected between the two configurations. They were fabricated at ITER and were exposed from the beginning of June until September, a period of three months. In contrast, the lab-scale modules fabricated at CENER with bifacial solar cells were exposed from the beginning of July until September 2024, a period of two months. During this period, the climate exhibited maximum temperatures ranging from 17.6 to 39.9 °C, minimum temperatures ranging from 4.9 to 22.3 °C, and the largest thermal amplitude observed in a single day was 23.4 °C. In addition to days with sunshine and cloud cover, a few summer storms occurred, with precipitation reaching 40.5 l/m² in a single day. The storm was accompanied by

a maximum wind gust of 74.16 km/h and the occurrence of hail. The total accumulated precipitation for the whole period was 161.4 l/m² and the average wind speed was usually around 10 km/h. This data was obtained from AEMET open source from Pamplona airport.[7] The irradiance was monitored with a pyranometer situated on the rooftop at the test plane inclination (42°).

The large modules were connected in pairs in series (Figure 12) under MPPT conditions., while the laboratory scale modules were left in open circuit. Their IV curve, EL and optical measurements were performed periodically in the lab at standard conditions.

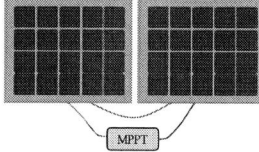

Figure 12: Large size modules connected in pairs and in series.

The laboratory scale bifacial module was outdoor a total of 62 days and no evidence of yellowing was observed during this period (Figure 13), indicating that UV protection type C is highly effective.

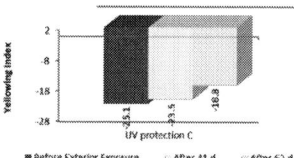

Figure 13: Yellowing index, calculated with reflectance measurements, after different outdoor exposure times of the bifacial module at CENER.

Furthermore, no clear indication of degradation was discerned when comparing the EL images, which corroborates the observation of relatively constant IV curves, exhibiting only a 1.3% decrease in P_{MAX} after 62 days outdoor. This could be attributed to the inherent degradation of the solar cells (Figure 14).

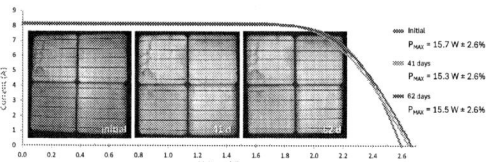

Figure 14: EL images and IV curves behaviour after different exposure times.

In the case of the monofacial PV modules, the yellowing was clearly visible in the module with UV protection type A after the first week. Therefore, it seemed clear that UV light, rather than temperature and humidity, is the main cause of yellowing. Although yellowing was not the only problem, delamination also occurred over the weeks, eventually forcing the removal of the outer layer after a couple of months. In Figure 15, the yellowing index is presented, and as it can be observed, protection C demonstrates superior performance. It should be noted that the values of the yellowing index are higher than those of the bifacial modules due to the reflection of the back side.

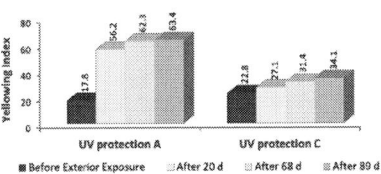

Figure 15: Yellowing index, calculated with reflectance measurements, of monofacial PV modules after being exposed outdoors at CENER.

When comparing the EL images for the module with protection A, no discernible deterioration was evident. However, the yellowing did manifest as a decline in the I_{SC} within the IV curve (Figure 16). To ensure that the enhanced module size does not compromise the durability of the hail resistance, a hail test is conducted following the outdoor exposure period. As illustrated in Figure 16, no damage is evident in the EL image, thus maintaining the IV curve's values. Nonetheless, ISC decreases around 9.5% and P_{MAX} around 11% after the outdoor exposure mainly due to yellowing.

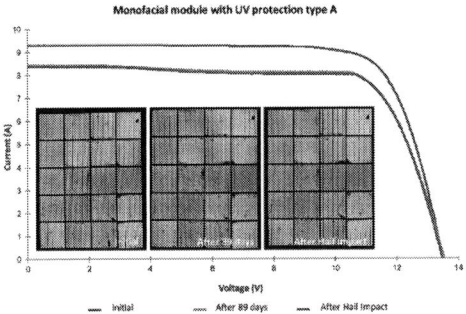

Figure 16: EL images and IV curves of monofacial module with UV protection A after outdoor exposure at CENER and subsequent hail impact test.

In the case of the monofacial module with UV protection C, not only do EL images demonstrate no evidence of degradation following the outdoor exposure and subsequent hail impact, but the IV curve also exhibits minimal variation (Figure 17).

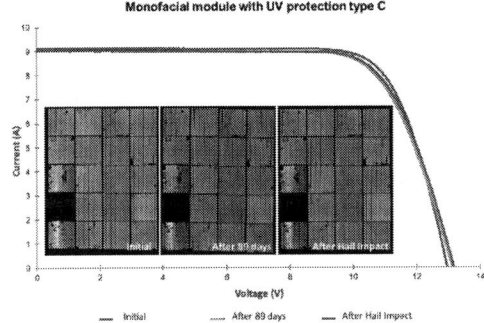

Figure 17: EL images and IV curves of monofacial module with UV protection C after outdoor exposure at CENER and subsequent hail impact test.

Given the minimal degree of yellowing observed, the decline in I_{SC} is merely 1%.

3.4.2 South of the Canary Islands location.

At ITER facilities the samples were arranged at an inclination of 20° and facing south (Figure 18). In this case, all exposed modules were 1 m x 1.5 m in size. Two of the modules have monofacial solar cells, one module with UV protection A and the other with UV protection C. The remaining two modules have bifacial solar cells, also with UV protection A and UV protection C, respectively. They were exposed outdoor from beginning of April until September 2024, a total of 5 months. During these months, the maximum temperature ranged from 22.7 to 38.3 °C while the minimum temperatures ranged from 15.7 to 23.7 °C. The maximum thermal amplitude in one day was of 15.7 °C. In general, the days were sunnier but with stronger winds. The maximum precipitation registered in one day was of 4.6 l/m² with a total precipitation during these five months of 11.5 l/m². The maximum wind gust was of 87.12 km/h although the average speed is usually around 30 km/h. This data was obtained from AEMET open source from Tenerife South airport.[7]

Figure 18: Installation at ITER's facilities(left) and the pyranometers to monitor irradiance (right).

The irradiance was quantified with a pyranometer situated on the racks at the test plane inclination (20°). The total front irradiance and the albedo (Figure 18) were monitored every 10 minutes.

Here, the large modules were individually monitored by measuring their outdoor IV curves every 10 minutes and then left in open circuit.

Visually, the same behaviour was observed. The modules (monofacial and bifacial) with UV protection A started to show yellowing after one week and delamination requiring removal of the outer layer after three months outdoors (Figure 19). Those with protection C remained apparently stable.

Figure 19: Photographs of bifacial modules after different exposure times outdoors at ITER. The upper module corresponds to the one with protection C and the lower one has protection A.

It should be noted that the I-V curve measurements of these modules were taken outdoors, so the irradiance and temperature are not stable and there are large variations from day to day. Only the data obtained at irradiances around 1,000 W/m² and 45 °C were considered for the study. In Figure 20, the decrease in efficiency due to yellowing is clear, and the differences between the monofacial and bifacial configurations are mainly related to the calculation of the total irradiance (frontal irradiance plus albedo) rather than differences in yellowing or degradation of the module.

964

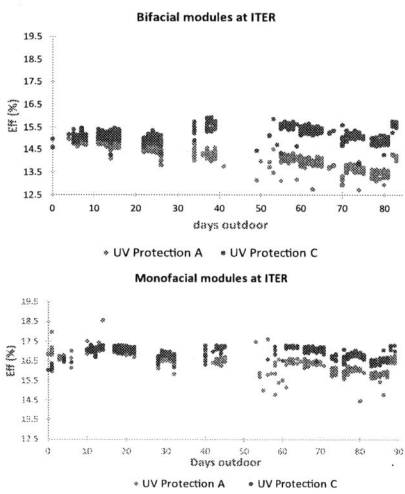

Figure 20: The graphs represent the efficiency (%) degradation along the outdoor experiment of the PV modules at ITER. The upper graph corresponds to bifacial modules while the lower graph corresponds to monofacial modules.

To compare the behaviour of the large monofacial modules located at CENER (north of the Iberian Peninsula) and at ITER (in the Atlantic Ocean of west Africa) the variation in percentage of the efficiency was considered. As it can be observed in Figure 21, the degradation at CENER's location is apparently higher than at ITER's location. While protection A loses about 11% at CENER, the trend line shows a loss of about 5% at ITER. For modules with protection C, the loss at CENER is about 3%, while at ITER the trend line shows a loss of about 1%.

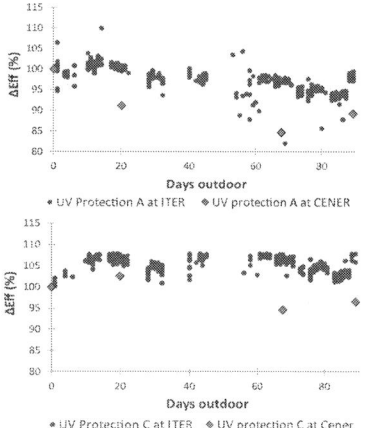

Figure 21: Comparison of the variation in efficiency of monofacial modules at CENER and ITER.

These results are an approximation due to the variations in measurement conditions, further experiments would be required to have a better understanding of how the differences in locations affect degradation.

4 CONCLUSIONS

The results showed that it was possible to produce glass-free lightweight bifacial PV modules that could withstand the hail impact test without damaging the solar cells or even propagating pre-existing cracks.

Furthermore, the one-step lamination process followed can be implemented in conventional PV module manufacturing lines.

A solution, UV protection C, was found to protect the module from yellowing caused by UV radiation. This change in configuration did not affect the module's resistance to hail impact test.

After 2,000 h under damp-heat conditions (85 °C and 85% RH), the modules suffer a severe yellowing, causing approximately 10% of the I_{SC} degradation. This yellowing is caused by a combination of high temperatures and humidity. Further work is needed to improve the polymer's resistance to moisture and heat.

The modules were tested outdoor in two different locations: CENER in Navarre (Noth of the Iberian Peninsula) and ITER in the Canary Island (Atlantic Ocean at the west of Africa). The modules without UV protection C started to yellow after the first week, and even some delamination was observed, forcing the removal of the outer layer after two to three months of outdoor exposure. Although a more thorough study is required, initial results in the IV curves indicate that the modules at CENER show greater degradation than at ITER location.

5 ACKNOWLEDGEMENTS

These results are part of project AiSoVol2,[8] extension of AiSoVol (photovoltaic generation solution as an alternative construction material). It is a project funded by the Ministry of Economy and Competitiveness, within the State Program for R&D&I oriented to meet societal challenge 3: Secure, Clean and Efficient Energy Challenge. The project, which began on 1st June 2020 and will end on 31st December 2023, is composed by the partnership of the Technological Institute of Renewable Energies, SA (ITER) as coordinator and the National Renewable Energy Centre, CENER Foundation.

Special thanks to the entire technical team for their invaluable help in carrying out the tests.

6 BIBLIOGRAPHY

[1] P. Grygiel *et al.*, Sol. Energy Mater. Sol. Cells, no. 233, (2021).

[2] A. C. Martins, V. Chapuis, A. Virtuani, and C. Ballif, IEEE J. Photovoltaics, vol. 9, no. 1, (2019), 245–251

[3] F. Lisco, A. Virtuani, and C. Ballif, Proceedings of the 37th European Photovoltaic Solar Energy Conference and Exhibition, (2020), 777–783.

[4] M. A. Schüler *et al.*, Sol. Energy Mater. Sol. Cells, vol. 277, no. April, (2024), 113048

[5] G. Oreski *et al.*, IEA-PVPS T13-13, (2021).

[6] IEC, *IEC 61215-2 Terrestrial photovoltaic (PV) modules – Design qualification and type approval – Part 2: Test procedures*, 2nd ed. Geneva, (2021).

[7] "AEMET open data." Accessed: Sep. 19, 2024. [Online]. Available: https://datosclima.es/index.htm

[8] CENER, "AiSoVol2." [Online]. Available: https://www.cener.com/en/areas/photovoltaic-solar-energy-department/outstanding-projects/aisovol-2-adaptive-photovoltaic-generation-solution-for-use-in-building-and-distributed-generation/

Development of novel frontsheets with protective coatings to increase the durability and reliability of glass-free lightweight PV modules

Yuliya Voronko, Gabriele C. Eder, Elisabeth Reiser, Markus Babin, Gernot Oreski

41st EU PVSEC, 2024, Vienna, Austria

Outline

- Lightweight PV modules – Frontsheets

- UV-curable coatings
  - key considerations for optimized performance for PV-frontsheets
  - formulation, application and curing

- Characterisation & results
  - coating quality: visual assessment & light microscopy
  - adhesion: cross-cut testing
  - light transmission: UV-VIS-NIR spectroscopy
  - chemical stability: thermal analysis

- Conclusion

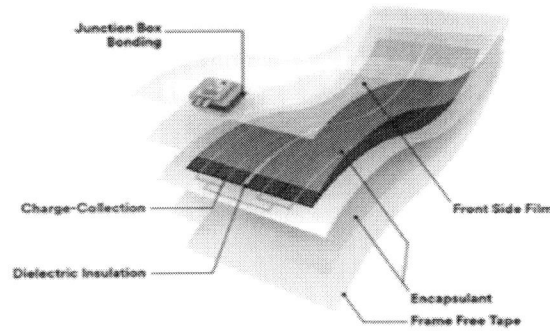

https://multimedia.3m.com/mws/media/6028100/3m-products-for-solar-energy.pdf

Lightweight PV module – Polymer frontsheet

ofi

PV frontsheets:

→ Commonly constructed as (i) single film or (ii) multi layer-composite

→ Composite mostly with polyethylene terephthalate (PET) core laminated with fluoropolymers like
  * ETFE (ethylene tetrafluoroethylene)
  * FEP (fluorinated ethylene propylene)
  * PVDF (polyvinylidene difluoride)

→ Environmental and health concerns:
European Chemicals Agency (ECHA) has proposed a ban of PFAS materials in 2023

The development of frontsheet materials with novel, solvent-free and fluorine-free coating systems which combine good barrier properties and high weathering resistance are required

Canditate front sheet materials

#	Material Description	Cost	Selection Type
1	PETFE (Monolithic)	High	PETFE (Positive control)
14	PVDF (Monolithic)	High-Med	Non-PET option
15	PC + Acrylic-coating 4	High-Med	Non-PET option
3	PET + Fluorinated layer	Medium	PET with UV filter layer
13	PET + Fluorinated coating 2	Medium	PET with UV filter coating
4	PET + Acrylic-coating 3	Med-Low	PET with UV filter coating
6	PET + UV blocker 1 (high)	Med-Low	PET formulated with UV absorbers
8	PET + UV blocker 2 (high)	Med-Low	PET formulated with UV absorbers
9	PET + Acrylic-coating 1	Med-Low	PET with UV filter coating
11	PET + Acrylic-coating 2	Med-Low	PET with UV filter coating
12	PET + Fluorinated coating 1	Med-Low	PET with UV filter coating
5	PET + UV blocker 1 (low)	Low	PET formulated with UV absorbers
7	PET + UV blocker 2 (low)	Low	PET formulated with UV absorbers
2	PET #6 but no UV blocker	Lowest	Bare PET (Negative Control)
10	PET #9 but no Acrylic-coating	Lowest	Bare PET (Negative Control)

Kempe et al.: Highly Accelerated UV Stress Testing for Transparent Flexible Frontsheets; DOI: 10.1109/JPHOTOV.2023.3249407

OFI 2024 | www.ofi.at

020267-003

UV-curable coatings

ofi

→ Advantages of UV-curable materials:
  – good chemical resistance
  – good weatherability
  – solvent-free
    → reduced environmental impact
  – short cure times
    → shorter production lines & low energy costs

* UV-curable materials nowadays are found in a number of different applications......

→ For PV application additional requirements:
  – high flexibility
  – high hydrophobicity
  – stability at low and high temperatures
  – good adhesion
  – good electrical insulation

Schematic representation of the curing process of a UV curable coatings system on substrate

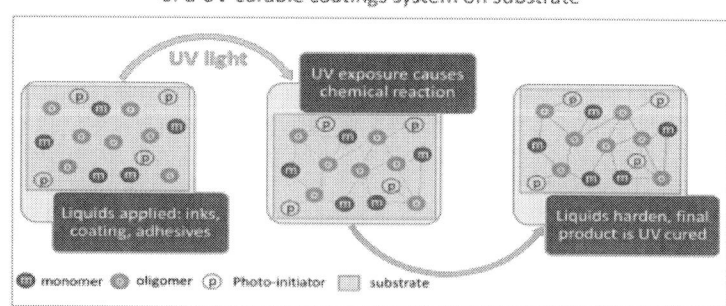

OFI 2024 | www.ofi.at

020267-004

Key considerations for optimized performance of coated frontsheet

ofi

Surface pre-treatment:
- adhesion promoters
- corona treatment
- .....

UV curing parameters:
- light source type
- light source power
- energy value
- duration

Adhesion testing
- cross-cutting test

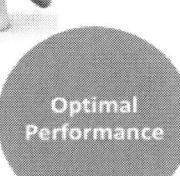

 Formulation & pretreatment

 Application & curing process

 Optical properties

 Adhesion to substrate

Aging Stability

Optimal Performance

UV curable coatings:
- Monomer
- Oligomer
- Additives
- Photoinitiator

High light transmission
→ > 90% @ 400-1100nm

- UV resistance
- moisture resistance
- chemical resistance

OFI 2024 | www.ofi.at

Coating formulation and pretreatment to PET-foil

ofi

Monomer:

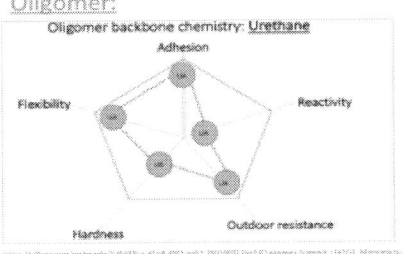

→mono-functional acrylates

Oligomer:

Oligomer backbone chemistry: Urethane

→ difunctional aliphatic hydrophobic urethane acrylate

Sample overview

Sample ID	PET Foil	Monomer A	Monomer B	Oligomer	UV-Absorber	Barrier pigment
F0		-	-	-	-	-
F1	PET	x	-	x	-	-
F2	Mylar	-	x	x	-	-
F3	204	-	x	x	x	-
F4		-	x	x	x	x
R	As reference: PET with ETFE coating (commercial frontsheet)					

Monomers A & B:
→ almost comparable in their viscosity
→ but differ in their glass transition temperature (Tg); $A_{Tg} > B_{Tg}$

PET-substrate pre-treatment: Corona treatment
→ both sides of PET foil
→ duration: 5 sec
→ at nominal power supply of 300 W

OFI 2024 | www.ofi.at

Application & Optical properties

- Coating on both surfaces
- Coating layer thickness of 30 – 50 µm

Visual characterization

→ coating homogeneously applied

→ without defects such as bubbles, cavities or delaminations

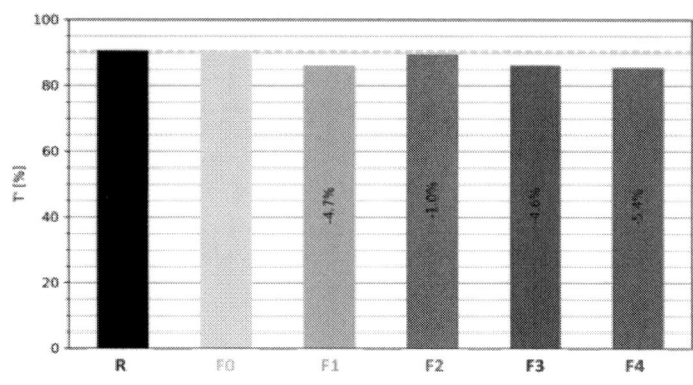

effective transmission T'

pure PET foil (F0)
coated samples F1, F2 (non-stabilized), F3, F4 (stabilized)
reference sample R (PET+ ETFE)

OFI 2024 | www.ofi.at

Adhesion to Substrate

Cross cut test:
- EN ISO 2409: Paints and varnishes
- determination of the adhesive strength and elasticity of coating

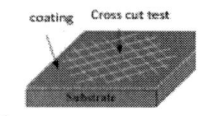

EN ISO 2409: Paints and varnishes

ISO	ASTM	TEST RESULTS	HAPPENING
0	5B	The edges are smooth and the edges of the lattice are not peeled.	
1	4B	There are small pieces peeled at the intersection of the incision, while real damage is not more than 5%.	
2	3B	The edge or intersection of the incision is peeled off, the damaged portion is between 5% and 15%.	
3	2B	Large part on the edge of the incision or the grid are totally peeled off. The peeled off area is between 15% and 35%.	
4	1B	Large part on the edge of the incision for the large part of grids are totally peeled off. The peeled off area is between 35% and 65%.	
5	0B	Over the last level.	

OFI 2024 | www.ofi.at

Aging stability: visual assessment and cross-cut test

Accelerated aging tests:

Aging test	Exposure	T [°C]	r.H. [%]	UV (300-400nm) [W/m²]	Time 1000 h
Xenon EN ISO 4892-2	102 min dry	38	65	60	✓
	18 min water spray	-	-	60	
Damp Heat ICE 61215-2/MQT13	constant	85	85	-	✓

Results of the visual assessment and the cross-cutting tests for samples F1-F4 before and after accelerated aging

	Visual assessment				Cross-cut testing			
exposure	F1	F2	F3	F4	F1	F2	F3	F4
0h	O.K.	O.K.	O.K.	O.K.	0	0	0	0
Xenon 1000h	O.K.	O.K.	O.K.	O.K.	2	2	0	1
Damp Heat 1000h	Coating detachment (edge)		O.K.	O.K.	3	3	4	4
	OK = Coating is homogeneous on both sides of the PET foil, NO bubbles or delaminations				EN ISO 2409: rating from 0 (very good) to 5 (very bad)			

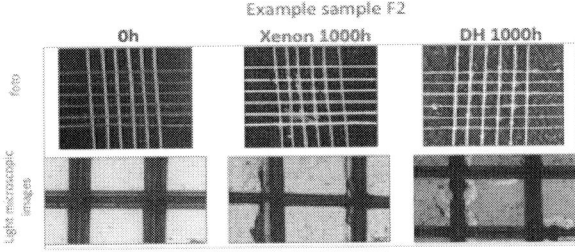

Example sample F2

→ DH 1000h: rating 3 und 4 part of the coating peeled off (15% - 50%)

Aging stability: optical properties

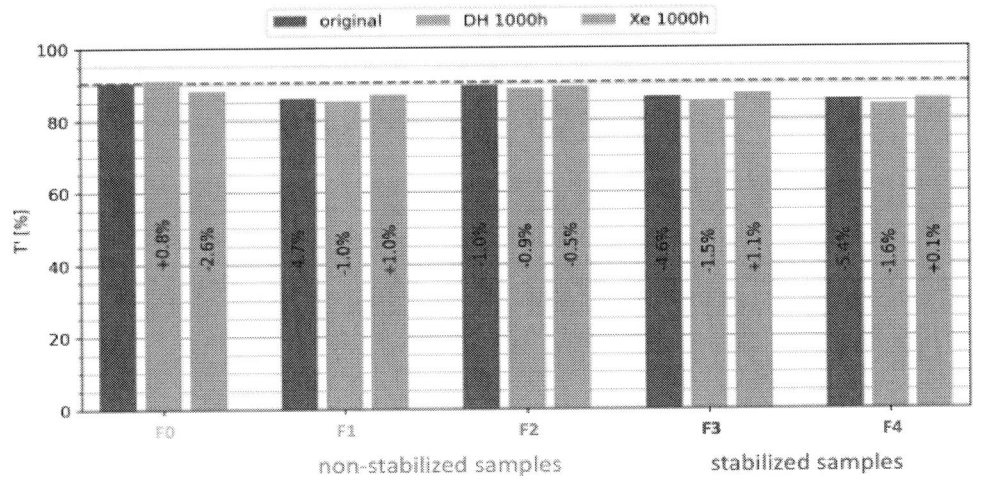

Effective transmission T'

Damp Heat:
All samples:
- optical loss
- optical changes are related to the coating

Xe-Irradiation:
F1, **F3** and **F4**:
- slight increase in transmission → a post-curing process during the Xe-exposure test

F0 & **F2**:
- optical loss

Aging stability: thermal properties of the PET core layer

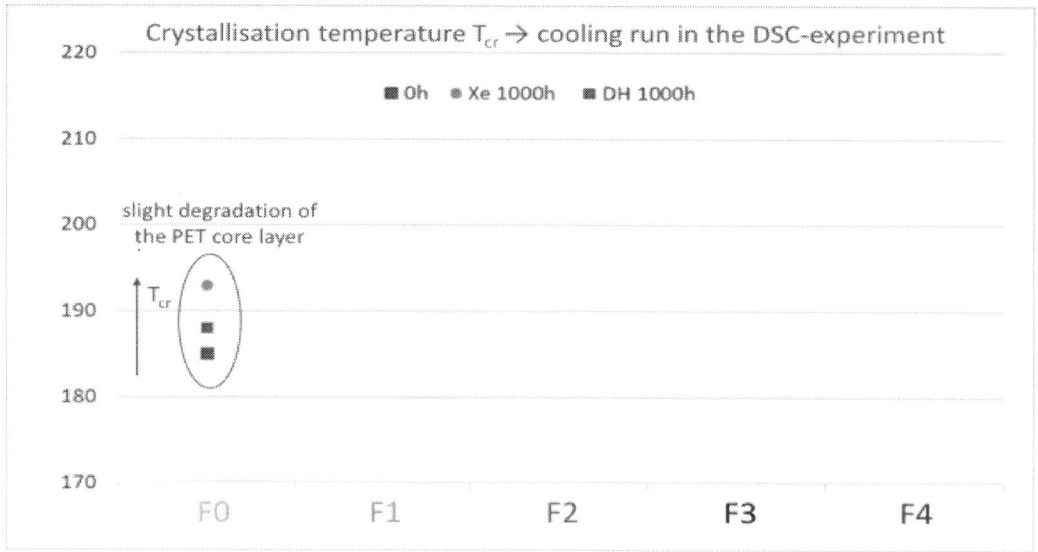

Aging stability: thermal properties of the PET core layer

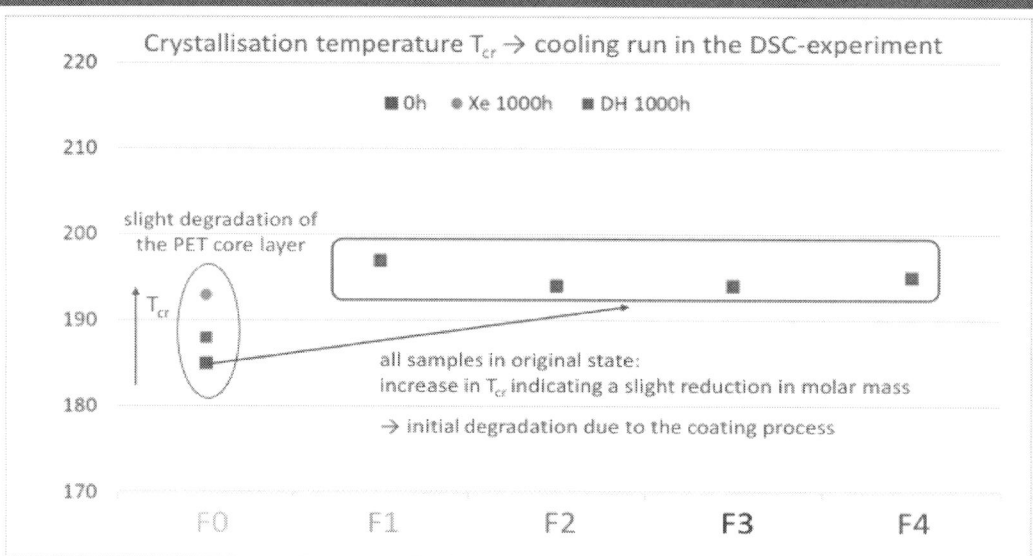

Aging stability: thermal properties of the PET core layer

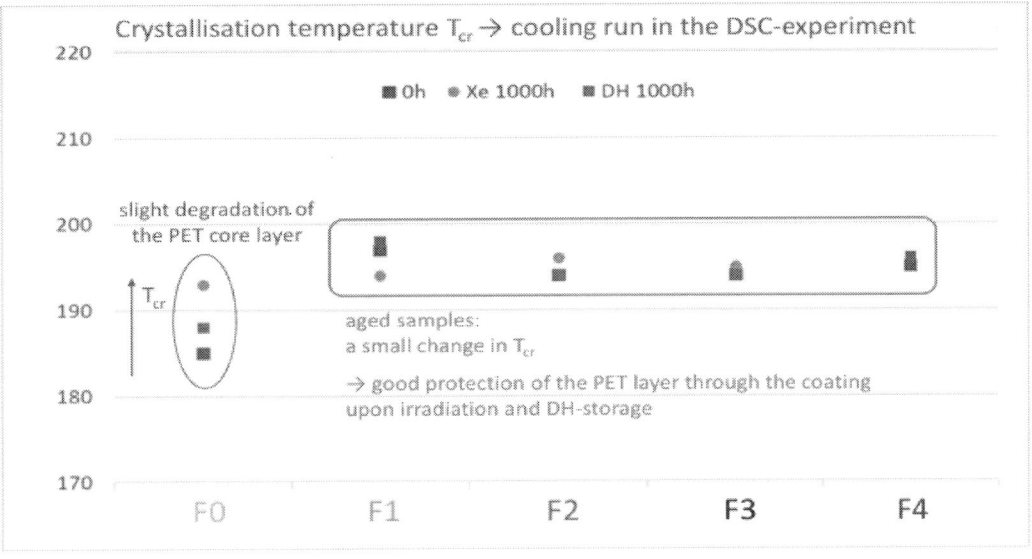

Conclusions

Aim	The development of frontsheet materials with novel, solvent-free and fluorine-free coating systems
Formulation	Four acrylate-based UV-curing coating systems were tested: Two types of acrylic monomers were combined with a urethane-acrylate oligomer
Curing & Application	Successful coating application on the PET substrate Cross-cut test revealed good adhesion to PET substrate BUT: initial degradation of PET substrate due to the coating application
Aging Stability	Xe-Irradiation: coating system with stabilizers provides good irradiation protection to the PET core substrate **Damp Heat:** developed coating systems are not <u>highly</u> moisture resistant, optimization is required

Thank you for your attention

Acknowledgement

This work was conducted as part of the Solar Era Net-Project „DELIGHT", funded by the Austrian Research Promotion Agency (FFG, Nr. 999897443) which is gratefully acknowledged.

Further presentation of results of the project "DELIGHT":
Mechanically Robust and Environmentally Stable Novel Lightweight PV Modules for Building Integration Based on Polymeric Honeycomb
4BO.17 T4.4 Dr. Umang Desai, PhD

41st EU PVSEC, Vienna, Austria. 2024

41st European Photovoltaic Solar Energy Conference and Exhibition

This presentation was selected by the Sc. Committee of the EU PVSEC 2024 for submission of a full paper to one of the EU PVSEC's collaborating peer-reviewed journals.

QUANTITATIVE DESCRIPTION OF THE QUALITY OF DAYLIGHT ELECTROLUMINESCENSE (dEL) IMAGES AGAINST DARK ROOM EL IMAGES

K.P. Sulca[1]*, C. de Castro[1], D. González-Francés[1], C. Terrados[1,2], J. Anaya[1], V. Alonso[3], M.A. González[1], O. Martínez[1]
(1) GdS-Optronlab group, Dpto. Física de la Materia Condensada, Universidad de Valladolid, Edificio LUCIA, Paseo de Belén 19, 47011 Valladolid (Spain)
(2) Solar and Wind Feasibility Technologies (SWIFT). Escuela Politécnica Superior, Universidad de Burgos. Avda. Cantabria s/n 09006 Burgos (Spain).
(3) GdS-Optronlab group, Dpto. Física Aplicada, Universidad de Valladolid, Facultad de Ciencias, Paseo de Belén 7, 47011 Valladolid (Spain).
*kabirpaul.sulca@uva.es

ABSTRACT: We present a robust comparison between Electroluminescence and daylight Electroluminescence images taken with an InGaAs camera, and "golden images" taken under optimal conditions with a Si camera in the dark, which serve as the benchmark. We study the data quality between EL/dEL correlated with the golden images. A key contribution of this work is the correlation of the SNR25 metric, with a pixel-by-pixel mathematical metric called the Similarity Index (SI) which is used to assess image similarity. We compare the similarity between a high-resolution dark EL image of a panel and a dark EL image taken with a low resolution InGaAs camera and quantify the differences showing that there are more complex than a simple resolution degradation. Then, we compare daylight EL images taken at very short exposure times with the high-resolution dark room image. We found that the SNR25 metric correlates well with the SI, allowing us to use this metric as a proxy to assess the quality of the dEL images taken on the field. The results of this research help validate the use of dEL imaging for practical applications, paving the way to define clear cut-offs for minimum quality when using this technique in the field.
Keywords: Daylight Electroluminescence, InGaAs cameras, image quality

1 INTRODUCTION

The exponential growth of photovoltaic (PV) solar energy production in recent years has established it as a leading green alternative for energy production, hence making the research in this field a priority. Ensuring the reliability and longevity of solar panels is crucial for their optimal operation and to prevent power losses. This makes the characterization and description of solar panel defects a key area of research. Electroluminescence (EL) has emerged as an invaluable technique in this context. It provides spatial information about various detrimental defects in solar modules, thereby contributing to the overall assessment of panel health and efficiency [1-4].

Traditionally, high-resolution Si cameras have been used to perform dark room EL, which, given the technical limitations of Si sensor cameras, must be performed in a dark room to remove all the light sources besides the luminescence from the PV module. However, this method requires no-light conditions, posing a challenge for inspecting PV modules in solar power plants requiring dismounting the solar panels and transport to a dark room for inspection or conducting inspections at night, which lead to production losses and/or are logistically complex. Despite its limitations, dark room EL with Si cameras remains the standard method for characterizing solar panels post-manufacturing. Recently, daylight EL (dEL) using InGaAs cameras has emerged as a promising alternative for on-site inspections [1]. The InGaAs camera is well suited for dEL thanks to its technical capabilities, the most relevant characteristics are high quantum efficiency on the spectral range of PV modules emission and fast acquisition time. Usually, SWIR optic lens and bandpass filters centered around 1150 nm are used.

However, InGaAs sensors have much lower resolution than Si counterparts, and besides dEL requires a filtration process to suppress ambient light, which can potentially reduce the quality of the obtained information.

Different parameters have been defined to quantify the quality of dEL images; here we use the SNR25 defined in [3] for asynchronous dEL schemes [3,4]. In general, this method has the potential to significantly reduce the time required for acquisition and inspection. However, and to the best of our knowledge, the impact of the lower resolution and potential quality degradation due to filtration processes in the obtained images are not fully understood.

Therefore, the aim of this work is to quantify the quality of the information recovered from dEL images obtained with our asynchronous method and to validate the use of our quality metric as a proxy for the reliability of the obtained information. For this, we first obtain "golden images" with a high-resolution Si camera under the best possible conditions in a dark room, which serve as a standard. These images are used as a reference to compare against the dEL images taken under the same conditions but with an InGaAs camera, as well as dEL images. The comparison with the "gold standard" is performed using robust algorithms to extract a metric that measures how similar the images are. We use a combined metric, called Similarity Index (SI), that can detect differences in the local structure, focus, grain and sharpness of the images. This index is then compared with the images' quality metric (SNR25), allowing us to robustly correlate this index with the quality of the information in the panel.

This work is a first step to enable robust large-scale solar panel daylight inspections by leveraging the speed of InGaAs cameras by establishing a robust methodology for quantifying the quality of dEL images.

2 METHODOLOGY

The analysis is performed in six PV modules of reflective back sheet and PERC cell technologies, the panels coming from three different manufacturers. They were all known to display cracks of several types on EL inspection. For each module, an area of four adjacent cells was selected to be imaged depending on the quantity and variety of existing failures.

Two different cameras were used for the imaging. For high-resolution EL a Sony ILCE-7SM3 12.1 MPixel CCD camera with a Sony FE 35mm F1.8 lens was used. The IR filter of the camera was removed to allow operating in the SWIR region, and instead a Hoya R72 infrared filter was used. It should be noted that although this camera has a 12 Mpixel RGB sensor, the EL images captured by this camera have only 3 MPixel. This is a consequence of operating outside the visible RGB range of the sensor, and thus not knowing the response of the green, blue and red pixels in the Bayer CCD to allow an accurate demosaicing (color reconstruction) of the raw image. Instead, the raw 14 bits intensities of each BGGR (Blue, Green, Green and Red) pixel group were normalized and averaged in a single pixel. For the InGaAs sensor we used a First Ligh C-RED 2 Lite 640x512 (~0.33 MPixel) InGaAs camera along with a Kowa LM16HC lens and a Salvo Technologies 1160 nm FWHM bandpass filter.

To ensure that both cameras were collecting the images of the exact same location of the panels they were mechanically fix atop a prism-shaped aluminum structure focalized to capture the same region of the panels. This structure also served as a frame for Musou Black IR Flock sheets to create a black box able to eliminate any SWIR illumination not originated from the panels. The flock sheets, which absorb more than 99.5% of light in near-infrared ranges, ensured optimal conditions for the dark room images. It should be noted that without this setup, the EL light emitted by the panels might illuminate the whole scene, causing non-emitting regions to display some background light. Therefore, the dark room images were taken then in the darkest room as possible to avoid these artifacts. Musou Black Paint was also applied to remaining inner elements. The structure could easily be attached to any module by means of a four-bar movable frame designed to be fixed to its long sides. This ensures that the cameras are collecting the images in the exact same position both for dark room EL and dEL, which later helps in the image comparison process.

For dark room measurements an Aim-TTi CPX400DP DC power supply was used to inject an electric current into the modules. This current was set to the short circuit current (Isc). For each configuration, the Sony ILCE-7SM3 camera was used to produce a high resolution EL image, the so called "golden image" (shutter speed of 30', F1.8 and ISO 80) by a pixel-by-pixel subtraction from the image taken at Isc the image at 0 A. The C-RED 2 Lite provided one single image which is the average of an image stack of 100 images captured at 300 frames per second (fps) (shutter speed of 3.33 ms), with its lens aperture completely open, and again an Isc and 0 A images were generated and final images created by pixel-by-pixel subtraction. Identical configuration was used to capture background images (i.e disconnect power supply from the module). Later we do, background subtraction performed during the subsequent processing. On the other hand, daylight imaging implied taking the modules outdoor and powering them. For this an EA-PSI 91500-30 power supply was used, which allows for current modulation creating custom square waves with a peak-to-peak amplitude of Isc. A square wave with a frequency of 12 Hz was applied to capture 400 images with the same configuration as in dark room. The dEL image was created by performing a frequency domain filtration as described in [3].

The first step necessary to ensure data comparability, requires crop and align both images to the same overlapping region of interest (ROI). However, since the Sony is an HD image (3 MPix) and the C-RED is a low-resolution one of 0.33 Mpix, we need to rescale the low-resolution image to high-resolution or vice versa. Then we need to morph images to correct for any small misalignments warping perspective. These two steps are performed simultaneously using the *warpPerspective* function included in the *CV2* python library, using a *LANCZOS4* interpolation algorithm.

Before comparing images, we also performed an equalization process of the images to mitigate possible illumination differences consisting on applying a contrast limited adaptive histogram equalization with a kernel of 8x8 pixels, followed by a Z-score normalization on the image and rescaling the data to the same data type (unsigned integer 16 bits).

Finally, a custom image similitude algorithm is used to calculate the similarity index *SI* between an image and its "golden image" counterpart. As mentioned before, this algorithm uses different metrics to assess similitude. First, we quantify the structural similarity of the images at the local level. For this we use a variation of the Structural Similarity Index SSIM which considers the spatial correlation and pixel intensity analysis [5]. There are several approaches of application of the SSIM, in this work we use the Multiscale SSIM (MS-SSIM) described in [6], and implemented in the *skimage.metrics* python package.

Inside the MS-SSIM parameters, we use three scales with weights [0.9, 0.08, 0.02] to skew the result towards the smaller detail. The gaussian sigma is kept minimal (0.05), to ensure small cracks or black line defects are not blur and the image kernel is 3x3 to be more sensitive to small local changes. Typically, the similarity index is calculated as the average of the MS-SSIM score for all pixels, but here we only compute the average of the pixels in the lowest decile, i.e. the worst 10% of the pixels in the image. We do this because we are more interested on detecting small detail differences with this metric, which is our *index*1.

Quality of an image is also defined by the local sharpness, i.e. how well we distinguish borders. For this we perform the Fast Fourier Transform (FFT) of the images and take the high frequency part (upper decile) of the resultant power spectra. *Index*2 is then defined as the ratio of this magnitude for the image and the golden image.

$$Index2 = \begin{cases} FFT^{LR}_{>90\%}/FFT^{HR}_{>90\%} & if < 1 \\ FFT^{HR}_{>90\%}/FFT^{LR}_{>90\%} & if \geq 1 \end{cases}$$

Another quality metric we introduce is the global focus of the image. For this we perform a relative comparison comparing the normalized Tenengrad sharpness [7]. The index is again calculated from the ratio between images' Tenengrad sharpness (T_s). Since this index is too aggressive, we use a log10 of the ratio and call it *index*3.

$$S = \begin{cases} T_s^{LR}/T_s^{HR} & if < 1 \\ T_s^{HR}/T_s^{LR} & if \geq 1 \end{cases}$$
$$index3 = 1 - \min\left(\frac{-log10(S)}{10}, 1\right)$$

Finally, to account for the relative noise of the images, i.e. the grain present in the image, we use the variance of the Laplacian of each image. We create an index (*index*4) as the ratio of this metric, but as in the previous case we kept it in log10 scale to soften its influence.

$$S' = \begin{cases} L_p^{LR}/L_{ps}^{HR} & if < 1 \\ L_p^{HR}/L_p^{LR} & if \geq 1 \end{cases}$$

$$index4 = 1 - \min\left(\frac{-\log 10(S')}{10}, 1\right)$$

The final step to assemble our similarity index (SI) is combine all the previous quality indexes:

$$SI = |index1 \times \sqrt[3]{index2 \times index3 \times index4}|$$

Note that for the quality maps we show a slightly different metric were $index1$ is replaced by the pixel MS-SSIM score:

$$SSIM(xy) = |MS\_SSIM(x,y) \\ \times \sqrt[3]{index2 \times index3 \times index4}|$$

3 RESULTS

Different datasets over 6 PV modules were obtained using the parameters summarized in table I, with the two cameras mentioned above.

Table I: Acquired datasets for the experiment

Dataset			
	Si Camera	InGaAs Camera	
	darkEL	dEL	darkEL
		12 Hz 300 fps	300fps
CANADIAN X3	X	X (780,925,590 wm²)	X
GLC X2	X	X (795,650 wm²)	X
QPLUS	X	X (780 wm²)	X

It should be noted that to perform a 1:1 comparison we define a ROI of the images that is identical for both high- and low-resolution type of images. Since the focal length and sensor geometry are not identical in both cameras, this implies adding a correction in resolution for the InGaAs case. To have images of same number of pixels in ROI, InGaAs images are equivalent to a 0.733 Mpix Si camera and not 0.33 Mpix. This happens because the ROI is not spatially the same between the two images. The ROI in the "golden image" is taking a relatively smaller portion of the image, while the InGaAs image ROI takes a bigger portion of the image.

3.1 Degradation algorithm test

Several degradation tests were first performed over the golden images to test the performance of the comparison algorithm. The following degradation test were performed:

- Resolution degradation: The original high-resolution image was downgraded to lower resolutions by averaging contiguous pixels, i.e. 1 pix, 2x2 pixels, 4x4 pixels etc.
- Noise degradation: The original high-resolution image was degraded by adding random noise. This is done by defining a random value within a range defined as ± percentage of pixel value and adding that noise to the pixel value.

- Erosion degradation: we set the pixels of the image to 0 if their value is below a given percentile value.
- Defocusing degradation: we apply a Gaussian blur to the image with a kernel size that is a percentage of the image resolution.

Note that other aberrations like barreling or image distortions will be captured by the $index1$ as these are more related to structural changes in the image.

The results of the resolution degradation test are presented in Fig.1 and shows that there is a steady reduction on the SI while reducing image resolution for all the studied panels. We first observe that the impact in each panel is not identical, as it depends on the individual features of each panel. For instance, a completely homogeneous panel will show 0 degradation but a very fractured panel with many small details will suffer a very aggressive degradation. It should be noted that we are not intending to compare different panels, but different images of the same panel against the "golden image" for each one.

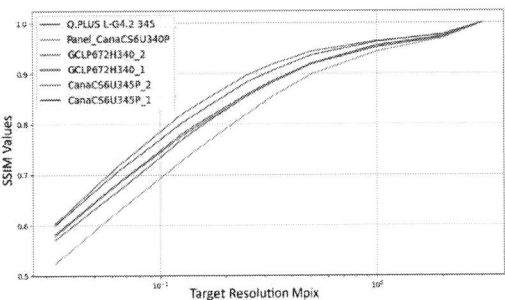

Figure 1: Similarity index algorithm test downgrading high resolution images and then testing against the original 3 Mpix image

To better illustrate this effect, we show in Fig.2 the "golden image" and the resolution degraded "golden image" of one studied portion of one panel; the colormap shows the SSIM variations. We can observe that, despite the degradation, the difference of the image quality is not evident, but the color map clearly shows regions that are different, mainly in sharp borders region like the bus bars and fracture borders.

Similar tests can be done for each image quality degradation effect, as shown in Fig.3. There we can observe in all the cases the same consistent pattern: a steady decrease in SI value when the degradation level of the image increases, which indicates that the degradation is well recognized by the SI. Again, we should remark that although the trends are the same for all studied panels, their absolute values depend on their details, and therefore they cannot be directly comparable, only images of the same panel are comparable.

3.2 Dark room EL images comparisons

The next step is an analysis of the dark room EL between the golden image and low resolution InGaAs camera image. Our main goal here is to assess the quality of the InGaAs image, for this we test if the SI compares well with the high-resolution image and against an equivalent resolution-degraded "golden image".

41st European Photovoltaic Solar Energy Conference and Exhibition

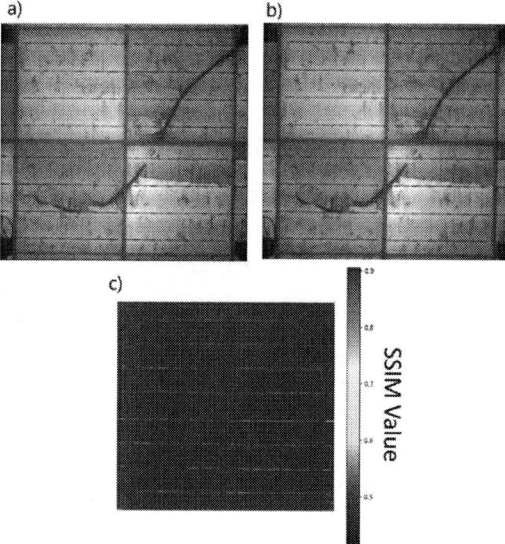

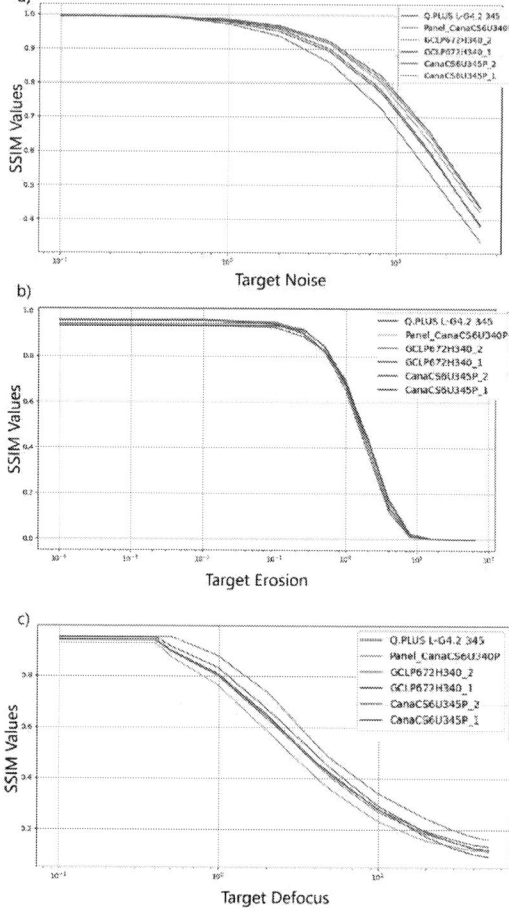

Figure 2: a) Original resolution 3 Mpx vs b) degraded 0.733 Mpx C) SSIM colormap shows that degradation affects the similarity at the borders

For this we first upscale the low-resolution image to the high-resolution and compare the images, and also the high-resolution is downscaled and compared against the InGaAs image. In Fig.4 we can observe that both processes give very similar results, having a slightly better SI with the upscaling process. However, and according to the results of Fig.1, between the high-resolution image and a resolution-degraded 0.733 Mpix image the SI is 0.94 and here we observe an SI of 0.765, evidencing that there are more differences than just the resolution. Indeed, from the color maps in Fig.4 we can observe that most of the differences arise in the darker regions of the panel, namely the inter-cell spaces and the large defects, where the InGaAs camera shows lower pixel values (i.e the blacker regions) than the Si camera.

To better illustrate these differences, we performed a more comprehensive analysis by degrading the 0.733 MPix resolution-degraded "golden image" to identify to which level of noise a 0.765 Si value corresponds, see Fig.5. Notice that where the SI shown in Fig.4c) crosses the curves gives us the amount of degradation that needs to be applied to the golden image to obtain the same value, being different for each type of degradation.

In Fig.6, we demonstrate the visual impact of applying different degradation mechanisms by showing a magnified ROI around a defect-rich area of the QPLUS panel. The effect of lowering the resolution is evident when comparing the InGaAs image to the high-resolution "golden image". When we compare the InGaAs image to the degraded 0.733 Mpix "golden image," we observe that the lower SI value arises from a combination of degradation effects, not just one. The InGaAs image exhibits a clear visual defocus, increased graininess (random noise), and darker black regions compared to the 0.733 Megapixel "golden image". Therefore, none of the single-process degraded images fully captures the loss of quality seen when using the InGaAs camera.

Figure 3: Similarity index algorithm test vs erosion percentage for different image erosion mechanisms. a) Random Noise degradation, b) Erosion degradation and c) Defocus degradation

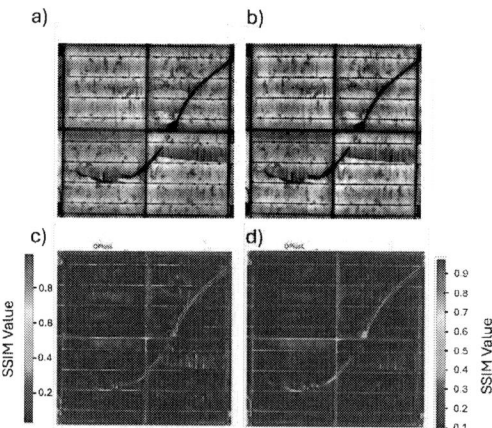

Figure 4: a) ROI of the High-resolution equalized image (QPLUS panel). b) ROI of the Low-resolution equalized image. c) Colour map showing the pixel similitude when comparing a downscaled Si image against the InGaAs image. d) Same as c) but comparing an upscaled low-resolution image against the Si image

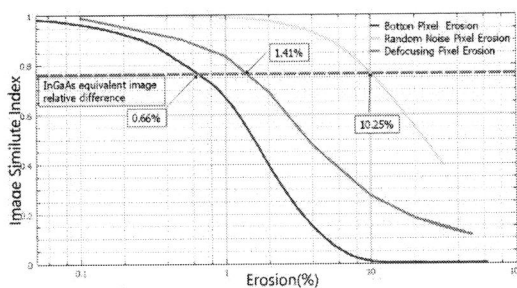

Figure 5: SI of different degradation methods vs % of degradation of the golden image (QPLUS panel)

This comparison also indicates that part of the loss of quality arises from a focus issue during image capture with the InGaAs camera, which is a technical glitch that is easy to correct. This suggests that, with proper focusing, the use of an InGaAs camera can yield information similar to that obtained with an equivalent-resolution Si camera.

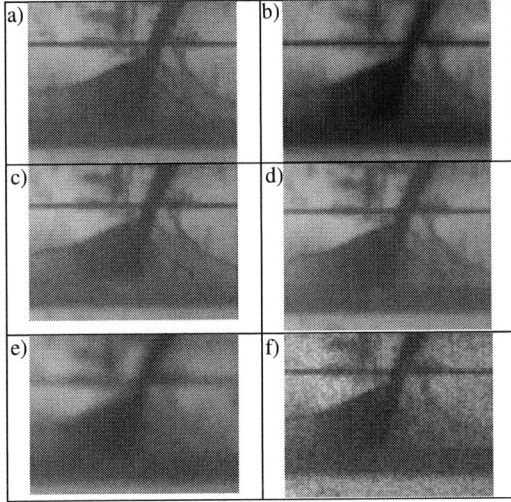

Figure 6: a) Original High-resolution Si image ROI of QPLUS panel. b) Original low-resolution InGaAs image. c) 0.733 Mpixel Si image. d) 0.66% Erosion degraded 0.733 Mpixel Si image. e) 1.41% Defocus degraded 0.733 Mpixel Si image. f) 10.21% Random noise degraded 0.733 Mpixel Si image

3.3 Golden image vs dEL comparison

Finally, we evaluated the loss of quality when acquiring EL data in daylight conditions compared to the high-resolution golden image collected under perfect dark conditions. For this study, dEL images were collected in datasets consisting of 400 images using an asynchronous scheme, and the dEL images were recovered by frequency filtering, following the process described in the Methods section. In Fig.7, we present an example of the results obtained for the QPLUS panel. Visually, the images are very similar; however, the dEL image exhibits more grain, some defocus, and shows darker regions than both the golden image and the InGaAs image collected in dark conditions. These differences are reflected in the SI value, which drops from 0.765 in the dark EL case to 0.345 for the dEL. The main differences, as shown in the colormap in Fig. 6, appear in the same regions as in the dark case, but new discrepancies also emerge. There are some dark

regions that were present in the dark EL but are not detected in the dEL (labeled as 1 and 2), and conversely, regions that are now completely dark in the dEL but were not present in the dark EL (labeled as 3 in the colormap). It should be noted that the images were collected on the same day; therefore, these different features may arise from simply moving the panel carefully from indoors to outdoors.

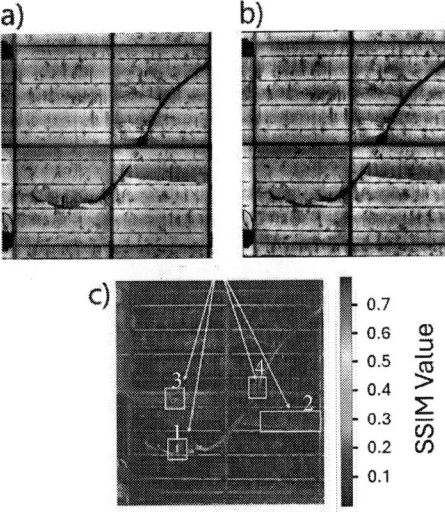

Figure 7: QPLUS panel Similarity index comparison and original images a) golden image, b) dEL InGaAs image and c) colormap similarity index

We present the remaining results for all PV modules in Fig.8, illustrating that, in general, dEL images show similar information than optimal dark EL images at a coarse level. However, when evaluating the quality using our algorithm, it becomes evident that certain regions experience more significant quality loss. These regions are mainly associated with highly defective areas featuring large black spots, where the dEL images display substantially more grain and bottom erosion. Other areas suffering quality loss include those near defective edges and smaller cracks, which are lost in the noise. This loss of quality is also captured by the SNR25 metric extracted during the frequency filtration of the images. Indeed, when we represent the relationship between the obtained similarity index for all the PV modules and the quality metric SNR25, we observe a monotonous increase of SNR25 as the SI index increases, Fig.9. This suggests that we can use the SNR25 as a reliable proxy for determining the quality of the final image. It should be noted that when processing the dark images collected with the InGaAs cameras using frequency filtration rather than subtraction, we obtain SNR25 values above 58 dB, while in these dEL images we are getting SNR25 values below 23 dB. Having established the validity of the SNR25 as a proxy for the quality of a dEL image and as a metric to quantify the quality loss of an EL image, further work is now needed to precisely determine the minimum SNR25 value necessary to ensure that a given level of detail is not lost when performing dEL compared to the information that can be retrieved under perfect dark conditions.

41st European Photovoltaic Solar Energy Conference and Exhibition

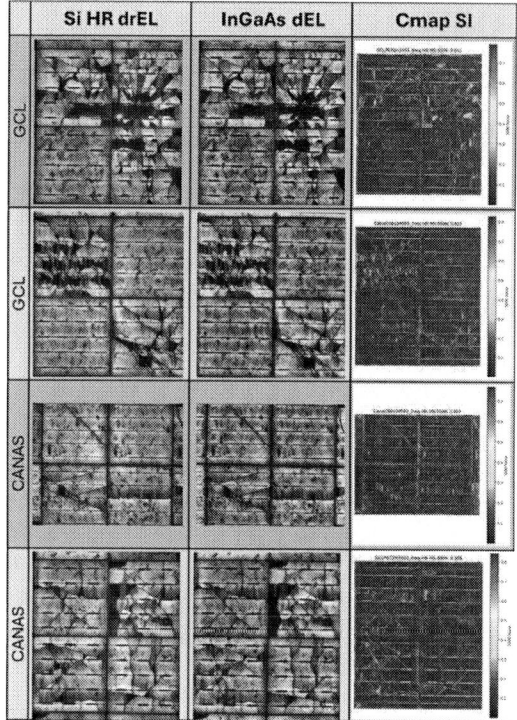

Figure 8: Results of the dEL images against golden high-resolution images for the rest of PV modules analyzed in this work

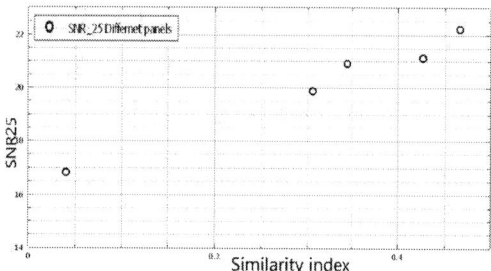

Figure 9: SNR25 vs Similarity index for all the analyzed PV modules

4 CONCLUSIONS

In this work we have created a rigorous test method for quantitatively assessing the quality of dEL images by comparing them to high-quality "golden images" obtained under optimal dark-room conditions with standard Si sensors. For this an algorithm capable of quantifying differences in fine structure and various image quality degradation effects has been developed. Our findings demonstrate that the SNR25 metric maintains a strong correlation with the similarity index SI. This correlation suggests that SNR25 can be used as a reliable proxy for assessing dEL image quality in practical applications. By providing a quantitative framework for evaluating the quality of the dEL images, our method facilitates the use of dEL imaging in the field and paves the way for establishing clear quality thresholds to ensure that critical details are not lost during image acquisition.

5 ACKNOWLEDGEMENTS

This work has been financed by the Spanish Ministry of Science and Innovation, under project PID2020-113533RB-C33, and by the Regional Government of Castilla y León (Junta de Castilla y León) and by the Ministry of Science and Innovation and the European Union NextGenerationEU / PRTR under the project *"Programa Complementario de Materiales Avanzados"*. C. de Castro is grateful for the funding received through the "Investigo" Programme of the Ministry of Labour of the Government of Spain. K.P. Sulca is grateful for the funding received through the predoctoral contract programme of the University of Valladolid (2022), co-funded by the Santander Bank.

6 REFERENCES

[1] M. Guada, A. Moretón, S. Rodríguez-Conde, L.A. Sánchez, M. Martínez, M.A. González, J. Jiménez, L. Pérez, V. Parra, O. Martínez, Energy Science & Engineering 8 (2020) 3839.

[2] G. A. dos Reis Benatto, R. del Prado Santamaría, T. K. Hass, M. Bartholomäus, L. Morino, P. B. Poulsen, S. V. Spataru, 8th World Conference on Photovoltaic Energy Conversion, Vol I (2022) 735.

[3] C. Terrados, D. G. Francés, J. Anaya, K.P. Sulca, V. Gómez-Alonso, M. A. González, O. Martínez, Proceedings of the 40th European Photovoltaic Solar Energy Conference and Exhibition, Vol I (2023) 258.

[4] G. A. dos Reis Benatto, T. Kari, R. del Prado Santamaría, S. V. Spataru, C. Terrados, D. G. Francés, J. Anaya, K. P. Sulca, V. Gómez-Alonso, M. A. González, O. Martínez, Proceedings of the the 40th European Photovoltaic Solar Energy Conference and Exhibition, Vol I (2023) 374.

[5] Sampat, M. P., Wang, Z., Gupta, S., Bovik, A. C., & Markey, M. K, IEEE transactions on image processing, 18(11) (2009) 2385.

[6] Wang, Z., Simoncelli, E. P., & Bovik, A. C, The Thrity-Seventh Asilomar Conference on Signals, Systems & Computers, Vol 2 (2003) 1398.

[7] Her, L., & Yang, X. 4th International Conference on Image, Vision and Computing (2019) 93.

41st European Photovoltaic Solar Energy Conference and Exhibition

This presentation was selected by the Sc. Committee of the EU PVSEC 2024 for submission of a full paper to one of the EU PVSEC's collaborating peer-reviewed journals.

PHOTOVOLTAIC CELL DEFECT CLASSIFICATION FROM LUMINESCENCE IMAGES: EMBEDDING AND CLUSTERING WITH UNSUPERVISED MACHINE LEARNING

Brendan Wright[1]*, Rama Sharma[1], Ziv Hameiri[1]
[1]School of Photovoltaic and Renewable Energy Engineering, UNSW Sydney, Australia
*brendan.wright@unsw.edu.au

ABSTRACT: The significant challenge of photovoltaic (PV) module end-of-life (EoL) management is receiving increasing attention from the broader PV community. However, currently there is a lack of adequate large-scale characterisation techniques and associated analytical methods for the quality assessment of fielded modules in utility-scale PV plants. This study aims to develop a methodology for (a) classifying defects and degradation using luminescence images of PV modules (obtained on-site or indoors), and (b) automated decision-making regarding the preferred EoL route. We achieve this through the development of a novel linear unsupervised machine learning (ML) methodology that provides robust and consistent cell defect/degradation classification capability. This approach involves three key stages: (a) a suite of algorithms for image feature extraction, (b) feature vector embedding with principal component analysis (PCA), and (c) k-means clustering for classification of individual cells by defect/degradation type and severity. For validation, module quality assessment is performed using the aggregate cell-level quality (identified defects or degradation). This automated quality assessment is then compared to a manual assessment performed by a representative domain expert evaluating the same images, and also against reference current-voltage measurements of each module, as well as a reference area-statistics-based method. The proposed ML-based assessment methodology achieves strong alignment with that of the expert assessment, including both the expected module performance and risk, and outperforms the reference method. Importantly, this automated quality assessment is able to perform accurately and consistently at a speed that is orders of magnitude faster than possible by manual assessment, enabling the low-cost utilisation of contactless imaging techniques for utility-scale PV system EoL monitoring and assessment.
Keywords: defect classification, luminescence imaging, unsupervised machine learning

1 SCIENTIFIC INNOVATION AND RELEVANCE

The significant challenge of PV module EoL management is receiving increasing attention from the broader PV com unity, including PV plant owners, assets management and operations and maintenance (O&M) companies, module manufacturers, and government regulatory organisations. The decision regarding the future of these modules (reuse, re-sell, recycle) is not straightforward, and a methodology for making these decisions has not been developed for utility-scale PV plants. This is due, in part, to a lack of adequate large-scale characterisation and associated analytical methods for model quality assessment sufficient to address the needs of rapidly expanding utility-scale installations. Robust (accurate, consistent) and scalable (automated, low-cost) methods for module quality assessment are vital for monitoring fielded PV modules and making informed decisions regarding their EoL.[1] This research presents a novel analysis methodology to enable the identification and classification of defects and degradation within modules at the cell-level solely from the luminescence images, supporting the subsequent EoL decisions.

Fundamentally, the core of this research problem is one of automated image classification. The traditional approach to such a classification task – using supervised ML models – would require a very large dataset (~10,000's of samples) comprising pairs of individual cell luminescence images and corresponding class labels of visible defects (inc. various faults, damage, and degradation modes).[2] Few of such datasets, if any, exist at the required scale to adequately cover the range of defect classes observed in images of real-world fielded modules. This results in difficulty training models, poor overall performance, and a lack of transferability to new technologies or operational environments.

To avoid the problems and limitations of supervised ML methods, our research pursues an unsupervised ML approach. This, rather than relying explicitly on learning to classify based on a pre-prepared labelled dataset, seeks to develop algorithms that can directly model the structure and relationships of a given system (i.e. a small unlabelled cell image dataset, typically containing only 1,000's of samples). In doing so, these data modelling algorithms provide a learnt representation of the system (cell images and defects), that can facilitate interpretable partitioning and automated classification, assisted through expert supervision and the incorporation of available domain knowledge.[3] These methods for unsupervised ML model development also present the opportunity for semi-supervised labelling of significantly larger datasets from a small training subset, providing a mechanism for the compilation of valuable large labelled datasets suitable for more optimised supervised ML model development.

2 AIM AND APPROACH

The aim of this study is to develop an automated decision-making method for EoL of PV modules. Using luminescence images obtained either through on-site imaging or indoors, defects and degradation are classified. A novel image analysis methodology was developed, comprising algorithms and unsupervised ML models to perform image feature extraction, high-dimensional embedding, and clustering for the final classification of cell images by defect type and severity (see Figure 1).

Firstly, we developed a simple baseline methodology that will be used as a reference (Auto-Stats). This baseline method involves area-based

980

statistical feature extraction, normalisation using batch-average residuals, and calibrated thresholds yielding a coarse three-axis set of defect classes with three-level severity. This approach has proven capable of basic defective cell identification and rough classification. However, it requires significant per-dataset calibration, is highly sensitive to image quality and preprocessing artefacts, and exhibits a high false-positive rate (incorrect classification) in addition to limited overall accuracy. These limitations motivated the development of an improved method, in particular to remove the requirement for manual calibration or fine-tuning for each batch of modules (with different technology, fielded duration, etc.).

The developed novel linear unsupervised ML methodology (Auto-ML) provides a more robust and consistent cell defect classification capability. This approach features an enhanced suite of image feature extraction algorithms, followed by a transformation of this feature vector through embedding with PCA, and finally the application of k-means clustering to enable improved classification of individual cells by defect type and severity.[4] This is illustrated in Figure 2. For validation, the cell-level defect classification is used to perform module quality assessment using the aggregate cell-level quality (a weighted summation of identified defective or degraded cells).[5]

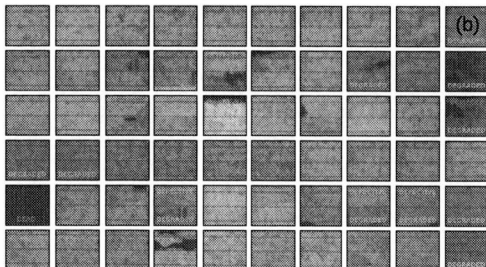

FIGURE 1: Example full-area photovoltaic module luminescence image (a), and cell-level defect assessment (b), after pre-processing and segmentation (analysis performed on individual cell images); identified defects are labelled by class (none, damaged or defective, dead or degraded) and severity (coloured by increasing risk: yellow, orange, and red).

3 RESULTS AND CONCLUSIONS

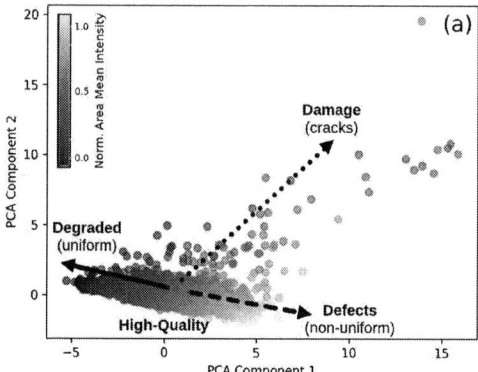

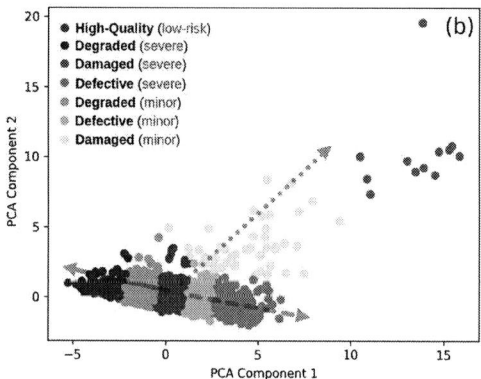

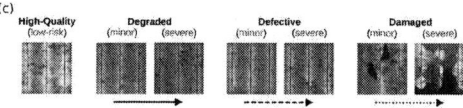

FIGURE 2: An illustrative example of Principal Component Analysis and k-means clustering methodology for cell defect classification with: (a) 2D slice of the PCA embedding from the input feature vector, with each point representing a distinct cell, coloured according to the normalised area average luminescence intensity, (b) the result of k-means clustering on the embedding to separate distinct cell groups based on feature similarity, and (c) a set of representative source cell luminescence images corresponding to the centroids of each cluster, labelled by assigned defect type and degree of severity (illustrative defect axes and direction are overlain on each plot to enhance understanding).

The developed linear ensemble pipeline was trained in an unsupervised manner on a real-world module luminescence imaging dataset comprising 60 modules of varying quality (over 3600 individual cells after preprocessing and segmentation). The details of this process are illustrated in Figure 2, showing a 2D embedding of all the individual cell images (points) after PCA on the feature vectors, where point-to-point distance represents high-dimensional image similarity. These plots are coloured by (a) cell average luminescence intensity, and (b) k-means cluster label.

Three approximate defect axes are visually represented to aid in understanding. They correspond to a cluster region of high-quality cells, and those of each degraded, defective, and damaged cells (presented with increasing severity). A set of representative source cell luminescence images corresponding to the centroids of each cluster is shown in Figure 2 (c) to highlight the strong correlation between input cell images and the corresponding classification.

In order to determine performance and accuracy, the developed cell defect classification methods were applied towards full module quality assessment, aggregating the results of individual cell-level assessments from fielded modules to obtain a single score of module quality and risk. From this score, and potentially incorporating user input regarding a desired relative distribution (i.e. a known fraction of modules to be removed), each module is assigned a quality bin from one (high-quality) to five (low-quality, significant defects or degradation). The five-bin allocation was chosen as to directly align with the quality bin assignments performed manually by domain experts for study validation.

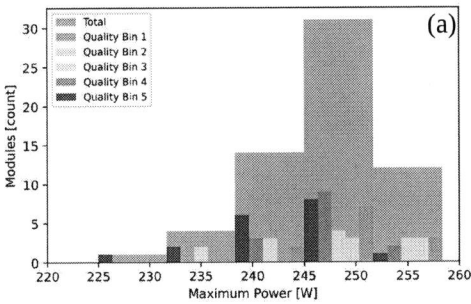

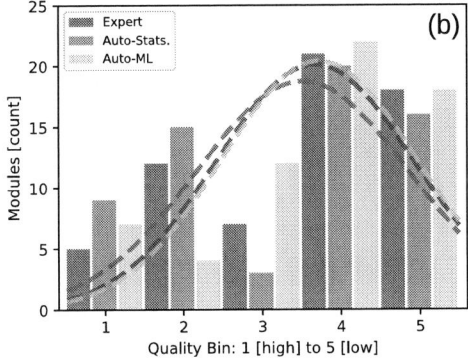

FIGURE 3: Representative example distribution of module quality bin assignment (a) from a manual assessment by a domain expert against the reference measured module maximum power. (b) Comparison of module quality bin assignment distribution between automated methods, using the improved ML-based cell defect classification (Auto-ML) compared to that of the simpler statistical residual method (Auto-Stats), with a representative reference manual expert assessment (Expert).

The distribution of quality bin assignment (coloured) as assessed by the domain expert is presented in Figure 3 (a), aligned to module maximum power output as determined through reference current-voltage measurements obtained for each PV module. This illustrates the potential for accurate manual quality assessment using luminescence images, where the assigned quality bin distribution aligns strongly with the measured electrical performance of the modules. Additionally of interest is the apparent risk bias, where a fraction of the higher-performing modules are assigned to poorer quality bins than would be expected based solely on the measured power output. A detailed analysis of these individual module images supports this trend in assignment, where clearly visible defects are present that are not yet impacting module performance, however do pose an increased risk for future degradation. As such, this quality assessment incorporates both the expected current performance and risk of modules with regard to EoL decisions.

The module quality assessment results from each automated method are compared to the manual expert assessment in Figure 3 (b), including the reference basic statistical method (Auto-Stats) and the developed linear embedding and clustering methodology (Auto-ML) for defect classification and subsequent quality bin assignment. The novel Auto-ML methodology for defect classification achieves strong alignment with that of manual domain expert assessment in module quality bin assignment, exceeding the performance of our reference Auto-Stats methodology, and more accurately captures the performance-risk bias of the manual expert assessment as discussed previously. As such, the developed automated module quality assessment methodology is capable of performing an accurate performance and risk analysis of PV modules solely from luminescence images. And, importantly, can do so at a speed and degree of consistency that far exceeds the capability of manual assessment, thereby enabling the utilisation of scalable contactless imaging techniques to monitor and assess fielded PV modules, yielding more informed EoL decisions at low-cost.

4 REFERENCES

[1] T. Trupke, R.A. Bardos, M.D. Abbott, P. Würfel, E. Pink, Y. Augarten, F.W. Chen, K. Fisher, J. E Cotter, M. Kasemann, European Photovoltaic Solar Energy Conference (2007) pp22–31.
[2] Y. LeCun, Y. Bengio, G. Hinton, Nature 521 (2015) pp436–444.
[3] Y. Bengio, A. Courville, P. Vincent, IEEE Transactions on Pattern Analysis and Machine Intelligence 35 (2013) pp1798–1828.
[4] M. Greenacre, P.J.F. Groenen, T. Hastie, A.I. d'Enza, A. Markos, E. Tuzhilina, Nature Reviews Methods Primers 2(2022) pp100.
[5] B. Wright, J. Petesic, T. Dawson, Z. Hameiri, IEEE Photovoltaic Specialists Conference (2023) pp1-3.

41st European Photovoltaic Solar Energy Conference and Exhibition

This presentation was selected by the Sc. Committee of the EU PVSEC 2024 for submission of a full paper to one of the EU PVSEC's collaborating peer-reviewed journals.

DAYLIGHT PHOTOLUMINESCENCE OF SILICON SOLAR PANELS IN OPERATION BY ELECTRICAL MODULATION

C. Terrados[1,2], D. González-Francés[1], K.P. Sulca[1], C. de Castro[1], M.A. González[1], O. Martínez[1]*
(1) GdS-Optronlab group, Dpto. Física de la Materia Condensada, Universidad de Valladolid, Edificio LUCIA, Paseo de Belén 19, 47016 Valladolid (Spain)
(2) Solar and Wind Feasibility Technologies (SWIFT). Escuela Politécnica Superior, Universidad de Burgos. Avda. Cantabria s/n 09006 Burgos (Spain)
*oscar.martinez@uva.es

ABSTRACT: Daylight Photoluminescence (dPL) has appeared in recent years as a useful tool for the inspection of solar panels, allowing for the identification of several kind of defects with good spatial resolution. The commutation between two states (On and Off) is usually necessary for the filtration of the ambient light. Several practical solutions have been implemented to do this kind of commutation, both electrically or optically. Here we explore in detail the method consisting on the electrical commutation using an electronic device connected in parallel to an adequate number of panels of a string, allowing to inspect the panels during operation, which is contactless once the electrical device is installed. The method can be also applied for the inspection of whole strings, in this case the electronic device is connected in series to the string to be inspected. The advantage of the method is the very fast commutation of the state of the string, between the MPP state and a state at/or very close to OC conditions. The values of the On and Off signals, the process and quality of the images, and the response of the inverter have been checked.
Keywords: Photoluminescence, daylight, inspection, PV panels

1 INTRODUCTION

Among the inspection techniques for defect characterization of Si solar panels, luminescence techniques provide a good spatial resolution and allows for the identification of several types of defects [1-3]. Electroluminescence (EL) inspection with Si cameras in complete dark ambient has been traditionally performed as a standard testing technique and is very well suited as an approval testing technique prior the Si solar panels are installed on a PV plant. Also dark EL (nEL) are used extensively for the inspection of the solar panels installed on a solar PV plant, but the strict dark conditions required when using Si cameras is a high disadvantage, since requires working during the night, or dismount the panels to be inspected on a laboratory or in a dark ambient in a mobile van [3-5]. Daylight luminescence techniques has been developed in recent years, with the possibility to inspect the panels on-site, and the aim to arrive to a massive inspection of the PV plants. Daylight EL (dEL) has been developed in recent years [6-10], requiring cameras with a high QE in the near IR range, as well as methods for filtering the ambient light. The dEL image is usually obtained by subtracting the signal when the panel is powered ("On" state) from when the panel is not powered (open circuit conditions –OC, "Off" state). On the other hand, daylight Photoluminescence (dPL) have the advantage of not needing a power source, since uses the sun as the light excitation source. However, it usually still requires two states ("On" and "Off" conditions) to distinguish the light coming from the panels form the ambient light, which still require electrical or optical contacts to be performed [11, 12]. The capability to made dPL in the more contactless way possible will be very beneficial for the inspection of a large amount of Si panels on-site on the PV plants.

We have previously developed a dPL system that allows to commutate the panels between the open circuit – OC ("On" state) and short circuit – SC ("Off" state) points of the I-V curve of the panel [10]. In OC conditions, the photogenerated carriers can recombine radiatively, producing a luminescence signal. In SC conditions, most

of the photogenerated carriers scape through the electrical circuit and the luminescence signal is reduced. The difference between the two states allows to extract the PL signal [10, 13]. In our system we used an InGaAs camera, specific filters to filtrate as much ambient light as possible, and an electronic device that rapidly commutes between OC and SC conditions and is also synchronized with the InGaAs camera to collect the signal in both states (dPL$_{OC/SC}$). For the practical realization of the dPL$_{OC/SC}$ measurements, the whole string was disconnected from the PV plant, and the individual Si panels from the string. The electronic device was then connected to each of the individual panels, in a contact way. The electronic device itself serves as the electric connection for the SC condition [10].

Different approaches have been also used to commutate the panels between two states, to have a large difference in the PL emission. For this purpose, electrical or optical commutations have been developed [11,12]. In the case of optical commutation, a LED can be placed to cover a solar cell, to inspect one panel, or several optical modulators (LEDs) to inspect a whole string [13, 14]. In this case, although the system can be denoted as contactless, and the optical commutation can be very quickly, the optical modulators should be putted on some of the solar cells, and then removed, which still results in an operation procedure not really simple. Also, not the complete panel or string can be inspected. Different approaches in the case of electrical commutation have been also developed in the last years, mainly through the use of the inverter to produce the change between two states. Some developments use the sweep of the I-V performed by some inverters [15]. Others, force the inverter to change between two states, which is not a quickly way [16]. More recently, a new development was performed by modifying the inverter itself to produce the change between the two states in a quickly way [11]. However, in these cases, the dPL image is obtained by forcing the inverter to operate at two voltage points or it is necessary to modify the inverter to introduce this capability.

Here, we show a modification of our dPL$_{OC/SC}$ method

by electrical commutation [10] using an electronic device to force the panels itself to work at two different points of the I-V curve, one at the maximum power point (MPP) at which the string usually works due to the maximum power point track (MPPT) function of the inverters, and thus with high Intensity (I_{MPP}) ("Off" state), and another at high voltage and thus low intensity (near or at the OC condition) ("On" state). The method can also be applied to a whole string of S panels; in this case, one string (of p strings connected in parallel to an inverter) is forced by means of the electronic device to work at these two points of the I-V curve.

In this communication we show the results corresponding to the disconnection of N panels from a string of 20 panels, and the response of two different inverters. A demonstration test was also carried out for the case of inspecting a whole string, using a micro-inverter working with two parallel strings of only one panel each.

2 METHODOLOGY

2.1 Electrical commutation

Let´s consider a string of S panels connected in series with the inverter. The electrical commutation proposed here is performed by using and electronic device which is inserted in parallel with an adequate number of panels (N), which force N panels to be disconnected from the rest of the string, such that the inverter is now working with only S–N panels. This is performed very quickly, and controlled by remote control using a computer and wireless connection. Due to the very quick change, the inverter is supposed to not be aware of the change, this means that the rest of the string (S–N panels) works at a higher voltage, and thus the operation point of the rest of the string moves to a much higher voltage in the I-V curve, very close to V_{OC} if the number of panels is properly selected, thus with a much lower current than before. This procedure is thus used to commutate the state of the panels between two points with a large difference in intensity, Fig. 1, avoiding forcing the inverter itself. By using the electronic device it is possible to make a very fast commutation, electrically connecting and disconnecting the N panels from the rest of the string, allowing for the obtention of a dPL image by subtracting the On and Off signals, which are obtained at nearly the same external irradiation conditions, thus favoring the quality of the obtained dPL image, and repeating the process as many times as desired to filtrate as much as needed the ambient light.

The method can be extended to inspect an entire string in operation. In this case, for a configuration with p strings connected in parallel to the MPPT of the inverter, it is possible to connect the electronic device in series with the string(s) to be inspected. Here, the Off state will be again the one at the MPP where the p strings normally operate, and therefore with a high current (I_{MPP}) drawn from all the strings, and which is obtained in this case when the electronic device is activated. In order to obtain an On state, the electronic device is deactivated, thus disconnecting the inspected string from the rest of the p strings, thereby forcing this string to work at OC conditions. Remotely disconnecting the string does not change the voltage in the MPPT of the inverter, while the change in the current drawn from the remaining p-1 strings connected to the MPPT of the inverter would be small, depending on the number (p) of strings connected in parallel to the inverter.

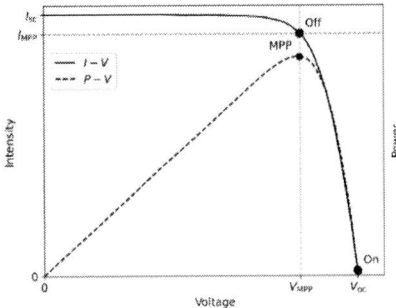

Figure 1: I-V curve showing the two points for the electrical commutation. The Off state is the normal operation point of the string (MPP point) due to the MPPT function of the inverter.

2.2 Materials and methods

We used an InGaAs camera, Hamamatsu C12741-03, with 640 x 512 pixels and 14 bits' resolution. Exposure times range from 1 μs to 1 s, which enables acquisition to be adapted to the different lighting conditions. A Kowa short-wave infrared (SWIR) optical system with 16 mm focal length was used for image acquisition. A SWIR bandpass filter – centred around 1160 nm with a bandwidth of 150 nm and a transmittance close to 90% – is used.

The electronic device used to switch the polarization states consists of a 1700 V IGBT (IXGN100N170 model), which is sufficient for a complete string operating at 1500 V and carrying 10 A.

Probes were performed using a whole string of 20 modules (mc-Si, ND-AR 330 W model from Sharp), with $V_{OC} = 45.5$ V, $I_{SC} = 9.4$ A, $V_{MPP} = 37.1$ V, $I_{MPP} = 8.9$ A (at STC). We tested two different inverters. Inverter 1 is a SUN 3Play TL-20 kW Ingeteam inverter, with an operating range of 560–820 V. Inverter 2 is a Fronius Symo 4.5-3-M model with a working range of 150–1000 V. We also used a micro-inverter (APS DS3 880W 230V model) capable of working with two panels in parallel, in order to demonstrate the method for inspecting whole strings in operation. Effective voltage and current signals at the exit of the inverter were recorded using Fluke 80K-40 and Fluke 80i-110s probes, respectively. Voltage and current measurements are shown as 5-second time segments.

2.2 Subtraction procedure

The subtraction procedure involves subtracting the On signal vis-à-vis the Off signal for each pixel, and accumulating the signal differences over a certain number of cycles (nc). Due to the presence of ambient light (background), the intensity signal (for both On and Off periods) can be very large and may even saturate the sensor, while the On–Off signal difference can be very small. To avoid saturating the sensor, it is usual to play with the aperture of the camera objective. Exposure time (t_{exp}) can also be varied. For fast switching, t_{exp} is usually chosen in the range [3-12] ms, and the aperture is modified accordingly. For the InGaAs camera used – with a resolution of 14 bits – the signal is limited to 16,384 grey levels (counts).

Our software is programed to store all the images, both for the On and Off periods, for the nc cycles. 2 x nc images are thus obtained. The software is also programed to make the difference $Signal_{(1)}{}^k = On_{(1)}{}^k - Off_{(1)}{}^k$ for each pixel k, store it as $Signal_{(accum,1)}{}^k$, and then make the difference $Signal_{(2)}{}^k = On_{(2)}{}^k - Off_{(2)}{}^k$ and add it to the previous accumulated value. A final image is obtained with the final $Signal_{(accum,nc)}{}^k$ over the nc cycles for all the pixels, giving the resulting dEL/dPL image.

In order to quantify the quality of the images obtained, the signal-to-noise ratio SNR_{avg} was calculated from the 2 x nc partial images, according to the expression given in [17].

3 RESULTS

3.1 Inspection of S–N panels of a string

For the probed string of 20 panels, according to their V_{OC} and V_{MPP} values, and to the working range of Inverter 1, the maximum number of panels that can be electrically disconnected but maintain the inverted working should be such that $Voc \times (20-N) \geq 560$, thus is N=7. This would be the case for an irradiance of 1000 W/m² and 20ºC (STC). dPL measurements on this string by disconnecting N panels from the string, with N from 1 to 6, have been performed (denoted as $dPL_{20/N}$). Figure 2(a) shows the dPL image of one defective panel obtained with the use of the electronic device in parallel with N=6 panels ($dPL_{20/6}$), for G = 1020 W/m² (the inverter is thus working with 14 panels in the Off periods). An exposure time of 5 ms and 300 On/Off cycles has been used. The On and Off signals can largely fluctuate (irradiation fluctuations due to clouds, for instance) and the On–Off signal differences are usually small (less than 15 counts in these measurements, Fig. 2(b), which makes the image obtention process not easy), but still enough to obtain a good quality dPL image, with a relatively large SNR_{avg} value of 17.3 [17]. The dPL image obtained by inspecting this isolated panel commuting between OC (On) and SC (Off) conditions ($dPL_{OC/SC}$) for the same irradiation conditions and camera parameters gives a SNR_{avg} value of 23.8, Fig.3(a), and nearly exactly the same visual information as the $dPL_{20/6}$ case. On the other hand, the dEL image obtained for the same irradiation conditions (G = 1020 W/m²) and camera parameters (t_{exp}=5 ms, nc=300), injecting a current of 9 A, give a larger SNR_{avg} value of 43.2, Fig.3(b). The same defective cells are detected, although the level of information is different. The difference between the dEL and the dPL information has been discussed in some papers [12, 18, 19]. In any case, it can be observed that the defective cells are distinguished in both cases.

Fig.2(c) shows the voltage and current measured at the output of the inverter during the first cycles of the $dPL_{20/6}$ measurement, showing the commutation between the On and Off states. A good square wave modulation is obtained for the current intensity, between 0 and ~8 A, while the voltage (that of the inverter) – which is modulated according to the electrical grid value– shows no important changes. The maximum values of the current intensity (Off periods) (~ 8 A) corresponds well to the high irradiation conditions of the measurement (I_{MPP} = 8.9 A at STC). It is also interesting to note that the current intensity modulation was observed to be constant all along the $dPL_{20/6}$ measurement, that indicating that both On and Off states are well fixed during the measurement. In fact, it can be clearly seen that the mean On–Off signal differences are fully constant, Fig. 2(b). The good quality of the obtained $dPL_{20/6}$ image is thus attributed to the large

difference in current intensity values between the On and Off periods (ΔI ~ 8 A), the fast switching between the two periods, and the perfect square modulation.

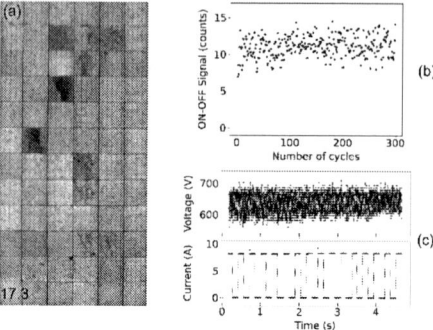

Figure 2: (a) dPL image of an individual panel of the string of S=20 panels connected with Inverter 1, with the electronic device connected in parallel with N=6 panels ($dPL_{20/6}$) (G = 1020 W/m², t_{exp}=5 ms, nc=300) (the calculated SNR_{avg} value is indicated on the bottom-left corner); (b) On–Off signal differences as a function of the number of cycles; (c) voltage and current values measured at the output of the inverter during the first cycles of the measurement

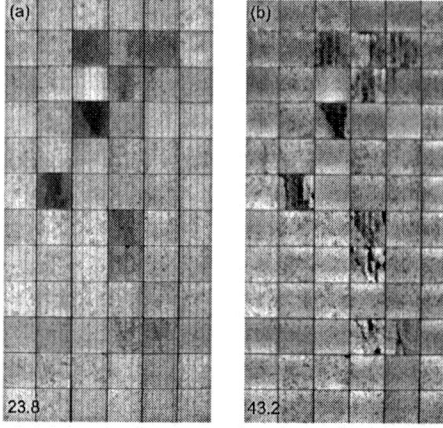

Figure 3: (a) $dPL_{OC/SC}$ and (b) dEL images of the same individual panel as shown in Fig. 2(a) (G = 1020 W/m², t_{exp}=5 ms, nc=300; $I_{current}$ = 9 A for the case of the dEL image)

Figure 4(a) shows the dPL image obtained for the same panel of this string of 20 modules, with the use of the electronic device in parallel with N=3 panels ($dPL_{20/3}$), for G = 1020 W/m². Again, an exposure time of 5 ms and 300 On/Off cycles has been used. The quality of the image has largely degraded, with an SNR_{avg} value decreasing to 13.2. The On–Off signal differences, as well as the voltage and current measured at the output of the inverter during the first cycles of this $dPL_{20/3}$ measurement, are shown in Fig. 4 (b) and (c), respectively. The current intensity modulation was observed not to be constant all along the $dPL_{20/3}$ measurement, with an Off state not fixed. This is reflected in the On–Off signal differences, which decreases monolithically to zero, Fig. 4(b). Thus, the bad quality of the obtained $dPL_{20/3}$ image is due to the small difference in current intensity values between the On and Off periods, with a large imperfect square modulation.

41st European Photovoltaic Solar Energy Conference and Exhibition

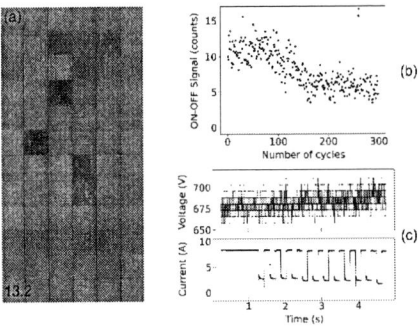

Figure 4: (a) dPL image of the same individual panel of the string of S=20 panels connected to Inverter 1, with the electronic device connected in parallel with N=3 panels (dPL$_{20/3}$) (G = 1020 W/m^2, t$_{exp}$=5 ms, nc=300); (b) On–Off signal differences as a function of the number of cycles; (c) voltage and current values measured at the output of the inverter during the first cycles of the measurement

3.2 Response of the inverter

The dPL$_{S/N}$ procedure has been checked for two different inverters, in order to study the inverter response and their effect on the obtained images. Fig. 5(a) shows the dPL$_{20/6}$ image obtained by using in this case Inverter 2, for the same conditions (G = 1020 W/m^2, t$_{exp}$ = 5 ms, nc = 300). The On–Off signal differences, as well as the voltage and current measured at the output of the inverter during the first cycles of this dPL$_{20/3}$ measurement, are shown in Fig. 5 (b) and (c), respectively. It can be seen now a quite different response of the inverter, which should be ascribed to the presence of capacitors, and the corresponding discharge processes. In spite of this, the dPL$_{20/6}$ image obtained are still good enough (SNR$_{avg}$ = 16.6) to clearly observe almost the same defective cells as with Inverter 1.

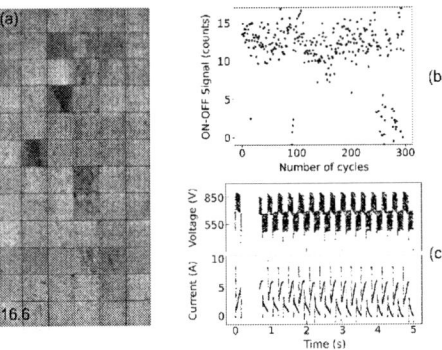

Figure 5: (a) dPL image of the same individual panel of the string of S=20 panels connected with Inverter 2, with the electronic device connected in parallel with N=6 panels (dPL$_{20/6}$) (G = 1020 W/m^2, t$_{exp}$=5 ms, nc=300); (b) On–Off signal differences as a function of the number of cycles; (c) voltage and current values measured at the output of the inverter during the first cycles of the measurement

3.3 Inspection of a whole string (dPL$_S$)

In order to test this methodology, and as a first attempt to validate it, we tested the case of two strings of just one panel (S=1) connected to a micro-inverter, with one of the panels being the defective panel shown in the previous figures. In this case, the electronic device was connected in series with the 1-panel string. Fig. 6 shows the dPL$_S$ image for the case G = 800 W/m^2, nc=300, t$_{exp}$=8 ms, the On-Off signal differences and the measured current and voltage values for the inspected panel. As can be seen, an almost perfect square wave modulation is again obtained for the current intensity drawn from the inspected panel. The image quality is still good enough (SNR$_{avg}$=13.9) to distinguish the defective cells.

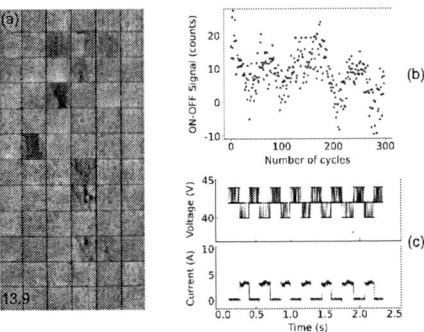

Figure 6: (a) dPL image of the same individual panel, with two panels connected to the micro-inverter, with the electronic device connected in series with the inspected panel (dPL$_S$) (G = 800 W/m^2, t$_{exp}$=8 ms, nc=300); (b) On–Off signal differences as a function of the number of cycles; (c) voltage and current values measured for the inspected panel during the first cycles of the measurement

It is also quite interesting to note that this dPL$_S$ image provides different information about the defects in the cells. The dPL$_S$ information is much more similar to the dEL image (Fig. 3(b)) with regard to the dPL$_{OC/SC}$ image (Fig. 3(a)). This result aligns with previous discussions about the information provided by dPL depending on the current drawn from the PV panels in the On and Off states, where the possibility of distinguishing a region's degree of isolation on a single dPL image was seen to depend on the level of current extraction and on the region's degree of isolation [12, 18]. This fact is now being studied in depth.

4 CONCLUSIONS

In conclusion, the electrical commutation described here does not change the state of the inverter. A very fast switching between the On and Off states can be performed by controlling the electronic device with wireless communication. The electronic device is a very cheap element that can be installed in the string and remain for the entire lifetime of the PV plant, if desired, allowing for the continuous monitorization of the state of the panels. The information provided by the dPL image is quite similar to the dEL. The filtration of the ambient light is properly performed, and the main defects of the solar cells of the panels can be detected. The method thus would allow for the inspection of solar plants, on-site, with the PV panels in operation, in a quiet contactless way, with a remarkably quality of the obtained images.

5 ACKNOWLEDGMENTS

This work has been financed by the Spanish Ministry

of Science and Innovation, under project PID2020-113533RB-C33, and by the Regional Government of Castilla y León (Junta de Castilla y León) and by the Ministry of Science and Innovation and the European Union NextGenerationEU / PRTR under the project *"Programa Complementario de Materiales Avanzados"*. C. de Castro is grateful for the funding received through the "Investigo" Programme of the Ministry of Labour of the Government of Spain. K.P. Sulca is grateful for the funding received through the predoctoral contract programme of the University of Valladolid (2022), co-funded by the Santander Bank. C. Terrados is also grateful for the financial support received under project PDC2022-133419-I00, funded by MCIN/AEI/10.13039/501100011033 and NextGenerationEU/PRTR.

6 REFERENCES

[1] L. Koester, S. Lindig, A. Louwen, A. Astigarraga, G. Manzolini, D. Moser, Renew. Sustain. Energy Rev. 165 (2022) 112616.

[2] I. Høiaas, K. Grujic, A. Gerd, I. Burud, E. Olsen, N. Belbachir, Renew. Sustain. Energy Rev. 161 (2022) 112353.

[3] M. Köntges, S. Kurtz, C. Packard, U. Jahn, K.A. Berger, K. Kato, T. Friesen, H Liu, et al., IEA-PVPS Task 13: Performance and Reliability of Photovoltaic Systems. Subtask 3.2: Review of Failures of Photovoltaic Modules. Technical report, International Energy Agency (2014).

[4] M. Navarrete, L. Pérez, F. Domínguez, G. Castillo, R. Gómez, M. Martínez, J. Coello, V. Parra, in 31st Eur. Photovolt. Sol. Energy Conf. Exhib. (2015) p. 1989.

[5] O. Kunz, J. Schlipf, A. Fladung, Y.S. Khoo, K. Bedrich, T. Trupke, Z. Hameiri, Prog. Energy 4 (2022) 042014.

[6] L. Stoicescu, M. Reuter, J.H. Werner, in 29th Eur. Photovolt. Sol. Energy Conf. Exhib. (2014) p. 2553.

[7] J. Adams, B. Doll, C. Buerhop, T. Pickel, J. Teubner, C. Camus, C.J. Brabec, in 32nd Eur. Photovolt. Sol. Energy Conf. Exhib. (2015) p. 1837.

[8] S. Koch, T. Weber, C. Sobottka, A. Fladung, P. Clemens, J. Berghold, in 32nd Eur. Photovolt. Sol. Energy Conf. Exhib., (2016) p. 1736.

[9] G.A. dos Reis Benatto, N. Riedel, S. Thorsteinsson, P.B. Poulsen, A. Thorseth, C. Dam-Hansen, C. Mantel, S. Forchhammer, K.H.B. Frederiksen, J. Vedde, M. Petersen, H. Voss, M. Messerschmidt, H. Parikh, S. Spataru, D. Sera, in Proc. 44th IEEE Photovolt. Specialist Conf. (2017) p. 2682.

[10] M. Guada, A. Moretón, S. Rodríguez-Conde, L.A. Sánchez, M. Martínez, M.A. González, J. Jiménez, L. Pérez, V. Parra, O. Martínez, Energy Science & Engineering 8 (2020) 3839.

[11] L. Koester, A. Louwen, S. Lindig, G. Manzolini, D. Moser, Solar RRL 8 (2024) 2300676.

[12] M. Vuković, M.S. Wiig, G.A. dos Reis Benatto, E. Olsen, I. Burud, Prog. Energy 6 (2024) 032001.

[13] R. Bhoopathy, O. Kunz, M. Juhl, T. Trupke, Z. Hameiri. Prog. Photovoltaics Res. Appl. 26 (2018) 69.

[14] O. Kunz, G. Rey, M. K. Juhl and T. Trupke, IEEE 48th Photovoltaic Specialists Conference (PVSC), Fort Lauderdale, FL, USA, (2021) p. 0346.

[15] M. Vuković, M. Jakovljevic, A.S. Flø, E. Olsen, I. Burud, Appl. Phys. Lett. 120 (2022) 244102.

[16] M. Vuković, I.E. Høiaas, M. Jakovljevic, A.S. Flø, E. Olsen, I. Burud, Prog. Photovolt. Res. Appl. 30 (2022) 436.

[17] C. Mantel, G.A. dos Reis Benatto, N. Riedel, S. Thorsteinsson, P.B. Poulsen, H. Parikh, S. Spataru, D. Sera, Søren Forchhammer, in Proc. IEEE 7th World Conf. Photovolt. Energy Convers. (2018) p. 3285.

[18] G. Rey, O. Kunz, M. Green, T. Trupke, Prog. Photovolt., Res. Appl. 30 (2022) 1115.

[19] C. Terrados, D. González-Francés, V. Alonso, M.A. González, J. Jiménez, O. Martínez, J. Electron. Mater. 52 (2023) 5189.

3DO.17.1

PV MODULE DEGRADATION IN HOT DESERTS: LABORATORY AND OUTDOOR DATA ANALYSIS

Gerhard Mathiak, Shahzada Pamir Aly, Kaushal Chapaneri, Baloji Adothu, Jim Joseph John
DEWA R&D Center, Dubai, UAE

DEWA Outdoor Test Facility

Data Acquisition	Rate
MPP (Imp/Vmp) and Tmod	30 secs
Irradiance and Tmod	30 secs
I-V curve tracing	10 mins

DEWA Outdoor Test Facility (DEWA OTF) is located next to MBR Solar Park, Dubai, UAE (2015 to present).

DEWA OTF-A (4 Rows with 30 modules each)

- Total Modules = 120 installed in stand-alone configuration. 30 different module types (c-Si, Thin film).

- Each module has its own MPP readings, I-V curve tracer (H&H loads 450 W for c-Si) and Tmod measurement (Pt-100 temperature sensors).

- All 4 rows have their own irradiance sensor (Pyranometer) for measuring the plane-of-array (POA) irradiance (daily cleaning).

  - Tilt angles: R1 (5 °), R2 (25 °), R3 (25 °), and R4 (90 °).

  - Cleaning frequency: R1, R2 and R4 monthly. R3 weekly.

- All chosen Modules are monofacial crystalline silicon modules (with Al-Back Surface Field)

Soiling at OTF Dubai

 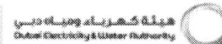

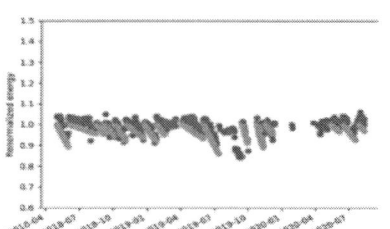 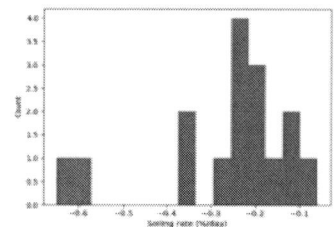

Sand and Dust
- ➤ Quartz Sand on Dunes
- ➤ Dust Devils and Shamal Dust Winds
  - Sticky dust (High calcium content)

Soiling at Dubai (Historical)
- ➤ Single Module Analysis at OTF
  - Crystalline module, 25° tilt angle
  - Fit (orange) to Renormalized Energy
  - Histogram (Mean at 0.25%/day)

Soiling at OTF
- ➤ Fixed Structure
  - No cleaning/ Monthly cleaning
  - Soiling Band at the bottom frame

Lab Measurements

 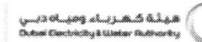

%/a	Tilt angle	Pmpp	Isc	Voc	FF
J mono	Tilt 5°	-0,41	-0,24	-0,01	-0,18
	Tilt 25°	-0,22	0,02	0	-0,25
	Tilt 25°,w	-0,28	-0,19	0,03	-0,11
	Tilt 90°	-0,20	0,01	0,01	-0,22
Average		**-0,28**	**-0,10**	**0,01**	**-0,19**
J multi	Tilt 5°	x	x	x	x
	Tilt 25°	-0,63	-0,31	-0,1	-0,25
	Tilt 25°,w	-0,68	-0,36	-0,08	-0,26
	Tilt 90°	-0,62	-0,23	-0,07	-0,32
Average		**-0,65**	**-0,30**	**-0,08**	**-0,28**

Yearly degradation in Pmpp, Isc, Voc and FF of PV modules in the OTF based on IV curve measurements in the laboratory.
- For all modules the degradation in Voc is low with <0.1%/y.
- The main difference between monocrystalline and multicrystalline modules is in Isc with -0.1%/a and -0.3%/a.

Electroluminescence

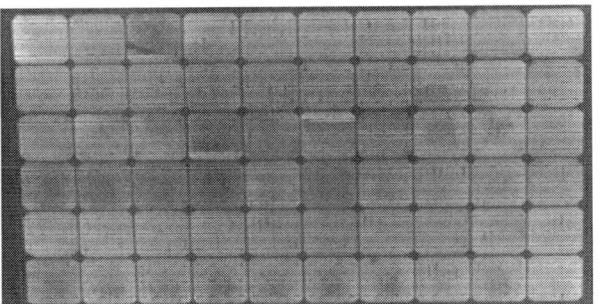

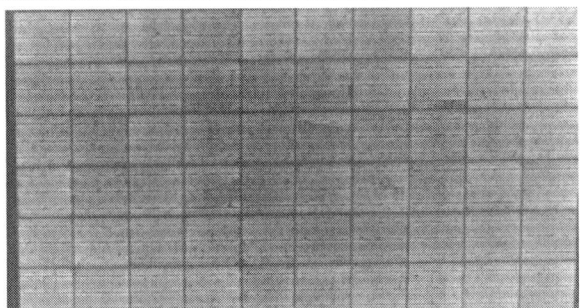

EL image of J mono module tilt 25°,m (223):
- Significant cracks (cells C6, D4, and F4)
- Significant finger breakages
- Highest loss in fill factor of the J mono modules

EL image of J multi module tilt 90° (422):
- Significant cracks
- Significant finger breakages
- Highest loss in fill factor of the J multi modules

UV-Fluorescence

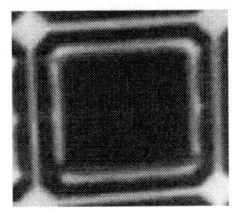

UVF image of J mono tilt 25°,m (223):
- The detailed UVF image of cell D4 clearly shows a frame structure typical for UV transparent in front of cell and UV blocking behind the cell.
- Browning is visible.

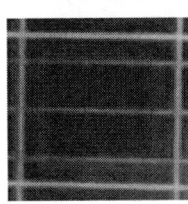

UVF image of J multi tilt 90° (422):
- The UVF signal of 90° tilted modules is relatively weak.
- The detailed UVF image of cell D4 shows a frame structure.
- Browning was not detectable.

Suns-Vmp Methodology

See more 4AO.9.6

Developed at Purdue University (USA) - Prof. Alam (2017)*

1. **Choosing equivalent circuit model**
   - 2-diode model with $n_1 = 1$ and $n_2 = 2$
2. **Extracting the pristine parameters (5 : Iph, I01, I02, Rsh, Rs)**
   - PV parameters before installation (Lab or datasheet measurements)
3. **Pre-processing the field data**
   - Module temperature(Tmod)
   - Plane of array (POA) irradiance
4. **MPP (Imp and Vmp) fitting algorithm (3-day-window)**
   - Calculate Pseudo MPP using Tmod and POA
   - Compare Pseudo MPP vs Actual Measured MPP
   - Optimize to minimize (Pseudo MPP – Actual Measured MPP)
5. **Post-processing**
   - Degradation visualization & interpretation

*Ref: X. Sun, H. Chung, R.V.K. Chavali, P. Bermel, M.A. Alam, Real-time Monitoring of Photovoltaic Reliability Only Using Maximum Power Point-the Suns-Vmp Method, in: 2017 IEEE 44th Photovoltaic Specialist Conference (PVSC), IEEE, 2017: pp. 1904–1907. https://doi.org/10.1109/PVSC.2017.8366225.

Suns-Vmp Degradation

5 Parameters (Iph, I01, I02, Rsh, Rs) Fitting (J mono 25°, weekly cleaning)
- Time Series of 5 parameters Fitting using actual I-V curve measurements every 10 minutes.
- Data loss is due to COVID and bad server migration.
- The parameters are primarily filtered (Irradiance >400 W/m²)
- Rsh which is often a function of irradiance shows a large variance

5 Parameters (Iph, I01, I02, Rsh, Rs) Degradation
- Suns-Vmp Time-Series Fitting of Pseudo IV curves
  - Boundary condition: I01, I02, Rsh, Rs monotonous Degradation
  - Iph is flexible (soiling and cleaning)
- Maximum degradation due to series resistance (Rs), followed by light current (Jph) and then shunt resistance (Rsh).
- More degradation observed in the first year, then degradation trend becomes less and almost linear.

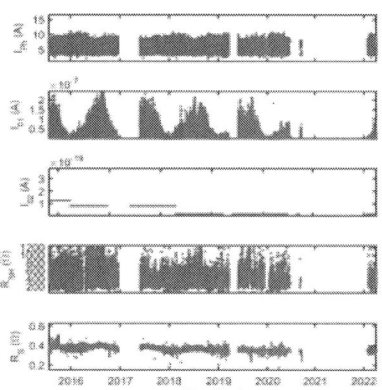

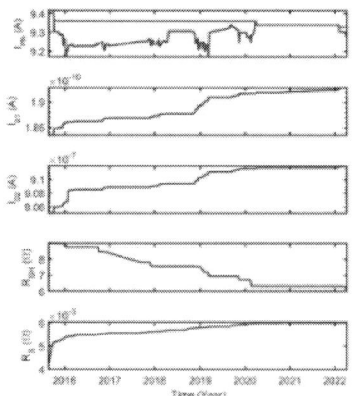

Suns-Vmp Degradation

EL image of J mono module tilt 25°,m (223):
- Significant cracks (cells C6, D4, and F4)
- Significant finger breakages
- Highest loss in fill factor of the J mono modules

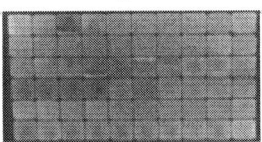

EL image of J multi module tilt 90° (422):
- Significant cracks
- Significant finger breakages
- Highest loss in fill factor of the J multi modules

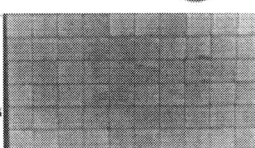

5 Parameters (Iph, I01, I02, Rsh, Rs)

Degradation

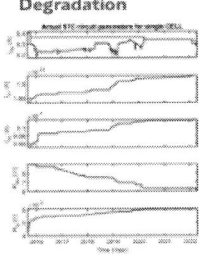

Deconvolution

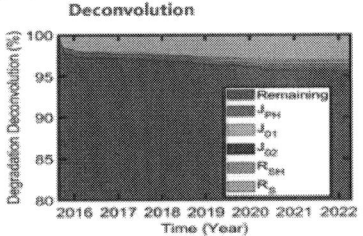

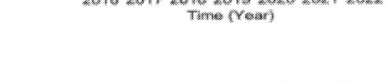

Degradation

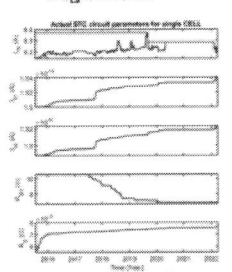

Deconvolution

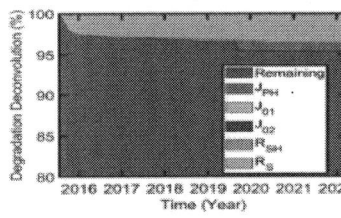

- Maximum degradation due to series resistance (Rs), followed by light current (Iph) and then shunt resistance (Rsh).
- More degradation observed in the first year, then degradation trend becomes less and almost linear.

- Maximum degradation due to series resistance (Rs), followed by light current (Iph) and then shunt resistance (Rsh).
- More degradation observed in the first year, then degradation trend becomes less and almost linear.
- Except around 2019 mid there seems to be sudden drop in efficiency due to apparently due to light current (Iph) reduction.

Suns-Vmp Degradation

%/a	Tilt angle	Pmpp	Isc	Voc	FF
J mono	Tilt 5°	-0,11	0,01	-0,02	-0,09
x23	Tilt 25°	-0,16	0,02	0	-0,15
	Tilt 25°,w	-0,07	0,01	0	-0,07
	Tilt 90°	-0,01	0	0	-0,01
Average		-0,09	0,01	-0,01	-0,08
J multi	Tilt 5°	-0,54	0,16	-0,03	-0,5
x22	Tilt 25°	-0,32	-0,03	0	-0,31
	Tilt 25°,w	-0,25	0,02	0	-0,24
	Tilt 90°	-0,13	0	0	-0,13
Average		-0,31	0,04	-0,01	-0,30

Yearly degradation in Pmpp, Isc, Voc and FF of PV modules in the OTF calculated with Suns-Vmp
- For all modules the degradation in Voc is low with <0.1%/y.
- Suns-Vmp shows the main degradation in the fill factor.
  - This can be caused by the decrease in the shunt resistance Rsh or by the increase of the serial resistance Rs.
  - The cracked modules have also the highest degradation rate in the fill factor by excluding the multicrystalline module with tilt 5°

992

OTF-Degradation: Suns-Vmp

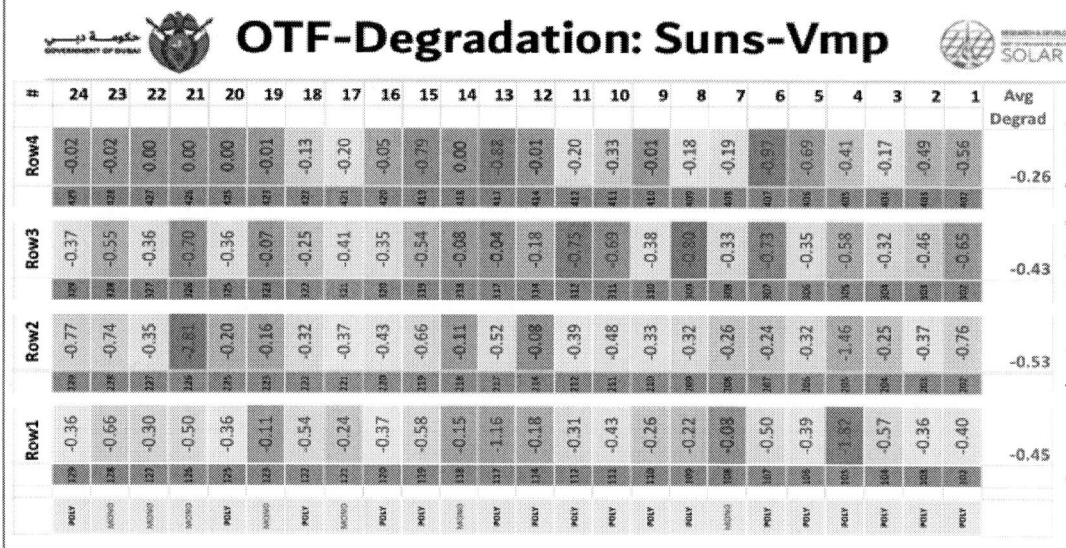

96 Monofacial C-Si modules in the OTF

The matrix shows linear degradation (%/year) for 2016-2023.

- Degradation between 0.00 and 2.81 %/year
- For modules below 0.2 %/year the low value could be biased due to missing data. (E.g row 4 19 to 22)
- 2.81 (226 in row 2) and 1.82 (105 in row 1) are outliers due to poor data quality (e.g. only one year of data)

- On average, Monocrystalline show 0.21%/year and polycrystalline show 0.49%/year linear degradation.
- Row 4 (90° tilt) shows lowest average degradation
- J Mono: Column 19 (x23), J Multi: 18 (x22), V multi: 3 (x04), P multi: 8 (x09)

Comparison Lab and Suns-Vmp

%/a	Tilt 5°		Tilt 25°		Tilt 25°		Tilt 90°		Average		Warranty
Cleaning	monthly		monthly		weekly		monthly				
Method	Lab	SVmp	Lab	SVmp	Lab	SVmp	Lab	SVmp	Lab	SVmp	
J mono	-0,41	-0,11	-0,23	-0,16	-0,29	-0,07	-0,21	-0,01	**-0,29**	-0,09	-0,7
J multi	x	-0,54	-0,64	-0,32	-0,68	-0,25	-0,62	-0,13	**-0,65**	-0,31	-0,7
V multi	-0,83	-0,57	-0,67	-0,25	-0,58	-0,32	-0,35	-0,17	**-0,61**	-0,33	-0,67
P multi	-0,66	-0,22	-0,49	-0,32	-0,42	-0,80	-0,34	-0,18	**-0,48**	-0,38	-0,7
Average	**-0,63**	-0,36	**-0,51**	-0,26	**-0,49**	-0,36	**-0,38**	-0,12	**-0,50**	-0,28	-0,69

Yearly Pmpp degradation of PV modules in the OTF of Typical 2015 modules (Al-BSF, 3 Busbars)
The Lab degradation calculation is made with the initial IV curve measurements at the contractor's laboratory before sending the modules to Dubai in 2016 and a measurement in the new lab in Dubai in 2023
- The Lab rates of the mono modules are clearly below the warranty, while the multi modules are close to it
- The degradation rates of the Suns-Vmp assessment are around 50% of the laboratory measurements

Summary

15 modules from Tier-1 manufacturers show relative low degradation (<0.7%/year) in the lab

- The monocrystalline modules in particular, show relatively low degradation and low browning compared to other modules installed in the OTF.
- The degradation rates of all four monocrystalline Al-BSF modules are significantly lower than the warranty rates, despite the harsh weather conditions in the hot desert.
    - This shows that there are high quality desert modules that can operate in hot deserts with low degradation rates.
    - An accelerated test protocol to select high quality desert modules is urgently needed.
- The laboratory measurements show consistent results in terms of tilt angle and mono or multi cell type.
- The degradation rates of the Suns-Vmp assessment are around 50% of the laboratory measurements.

Continuous outdoor measurement can be improved

- Irradiance sensors, especially module temperature sensors, and IV curve measurements
- The spectral irradiance data can be integrated into the evaluation.
- A shorter PV module cleaning period (weekly) is necessary

New measurements with PV modules with high efficiency concepts are started (in OTF row 3)

- Electronic loads with higher power (1000W) were purchased for this purpose.

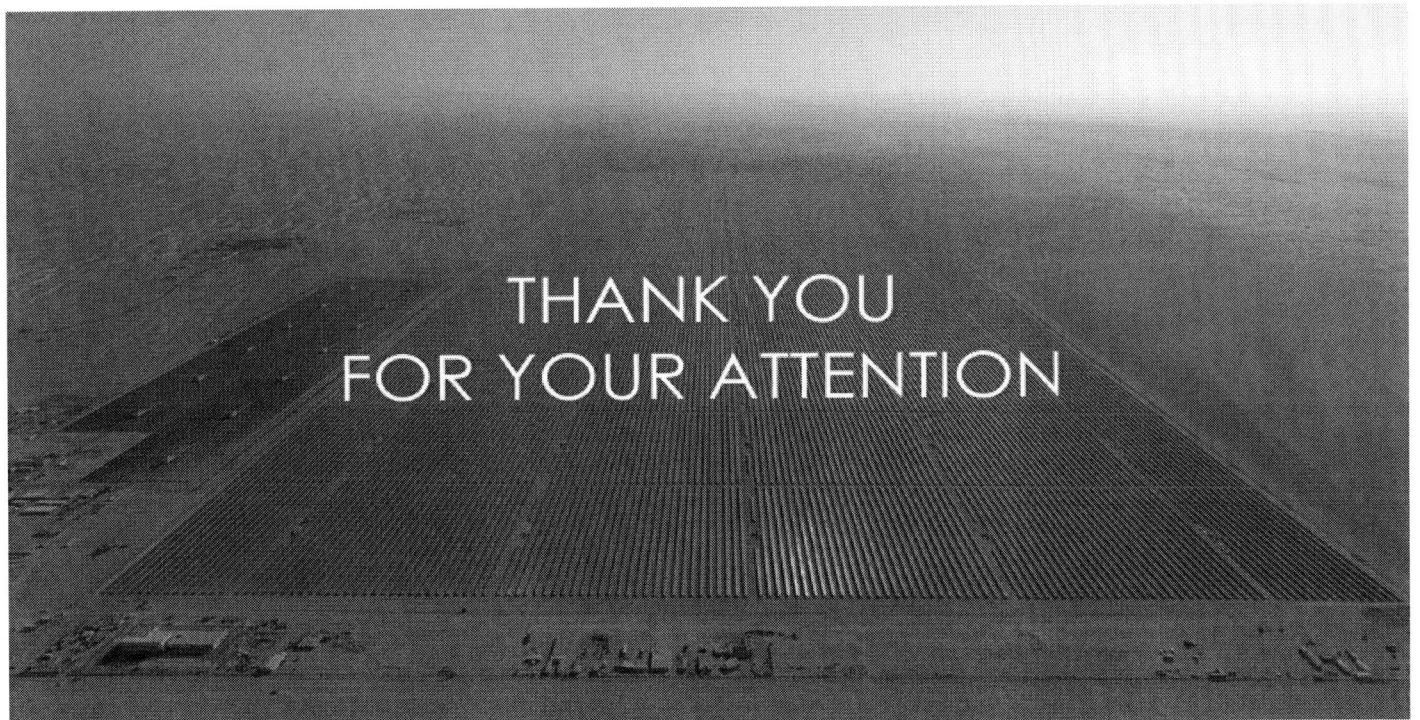

41st European Photovoltaic Solar Energy Conference and Exhibition

This presentation was selected by the Sc. Committee of the EU PVSEC 2024 for submission of a full paper to one of the EU PVSEC's collaborating peer-reviewed journals.

INCIDENCE ANGLE EFFECT: RESULTS OF AN INTERLABORATORY COMPARISON OF MEASUREMENTS ON COMMERCIAL-SIZE MODULES

Mauro Pravettoni[1], Min Hsian Saw[1,2], Giorgio Bardizza[3], Giovanni Bellenda[4], Romain Couderc[5], Gabi Friesen[4], Werner Herrmann[3], Shin Woei Leow[2], Stefan Riechelmann[6], Flavio Valoti[4], Arvid van der Heide[7], Frank Weinrich[6] and Stefan Winter[6]

[1]Technology Innovation Institute (TII), Renewable and Sustainable Energy Research Center, Abu Dhabi, UAE
[2]National University of Singapore, Solar Energy Research Institute of Singapore (SERIS), Singapore
[3]TÜV-Rheinland, Cologne, Germany
[4]Scuola Universitaria Professionale della Svizzera Italiana (SUPSI), Istituto di Sostenibilità Ambientale Applicata all'Ambiente Costruito (ISAAC), Mendrisio, Switzerland
[5]CEA, Institut National de l'Energie Solaire (INES), Le Bourget-du-lac, France
[6]Physikalisch-Technische Bundesanstalt (PTB), Braunschweig, Germany
[7]IMEC, IMO-IMOMEC / Energyville, Genk, Belgium

ABSTRACT: The incident angle effect causes a decrease in the photogenerated current of PV modules when they are subject to incident irradiance at wide angles: its relevance should be quantified for accurate energy yield purposes and has recently gained significance due to the raising interest in innovative vertically integrated PV applications (e.g. in urban structures, in agrivoltaics, and in vehicles). The international standard IEC 61853-2 presents both an outdoor and an indoor measurement method: however, the indoor measurement method for commercial-size modules is often impractical due to irradiance uniformity limitations on the volume spanned by the tested module upon rotation. In recent years, new solutions have been proposed to overcome these limitations and allow a wider adoption of this standard, but method validations and interlaboratory comparisons have been conducted so far only at small-area samples and a real validation on commercial-size modules is still to be performed: in this work we aim at fulfilling this goal, reporting the results of an interlaboratory comparison conducted within the international Project Team that is currently working at the new edition of IEC 61853-2. Result shows remarkable agreement between different measurement methods, thus validating more options for the evaluation of this important effect.
Keywords: Antiglare Treatment, Energy Rating, Experimental Methods, Incidence Angle Modifier, Relative Angular Transmittance

1 INTRODUCTION

By incidence angle effect we mean the decrease in the photogenerated current of photovoltaic (PV) modules when they are subject to direct incident irradiance at wide angles. The relevance of this effect may be quantified for energy rating purposes in terms of an Incidence Angle Modifier (IAM). It has recently gained significance due to the rising interest in novel integrated PV applications, where vertical or non-optimal tilt are favoured (e.g. in agrivoltaics, integration into vehicles, buildings, etc.), or where antiglare is of interest (e.g. in buildings, urban infrastructures, airport installations, etc.).

Among the International Electrotechnical Commission (IEC) standards, measurement methods for the incidence angle effect are described in IEC 61853-2 (in Ed. 1.0 at the time of writing [1]). The two procedures described in it, which are detailed in the following section, has proven during time to have various challenges [2] that have driven various authors to propose alternative methods [2-11]. These new methods were included in the draft for the Ed. 2.0 of IEC 61853-2.

The objective of this work is to demonstrate the reproducibility of the new methods with reference to the existing outdoor procedure of IEC 61853-2 via an interlaboratory comparison on commercial-size modules.

2 NEW METHODS FOR MEASUREMENTS OF THE ANGLE OF INCIDENCE EFFECT

2.1 IEC 61853-2 (Ed. 1.0)
Table I lists the two existing methods for the measurement of IAM that at the time of writing are present in IEC 61853-2: an indoor method, requiring a solar simulator with minimum Class B spatial uniformity of irradiance upon full rotation of the test device (i.e. ±5%, according to IEC 60904-9 [12]); and an outdoor method, requiring a two-axis tracker to provide rotation of the test device with respect to normal incidence and means of subtracting the diffused component of irradiance from the direct one.

For a module with short-circuit current $I_{SC}(\theta)$ when subject to irradiance at incidence angle θ, IAM is defined as *relative angular transmittance*

$$\tau_{rel}(\theta) = \frac{I_{SC}(\theta)}{\cos\theta I_{SC}(0)}.$$

Table I: The methods for measurements of IAM in Ed. 1.0 of IEC 61853-2, with main features

Method	Scheme	Main features
Indoor (full irradiation)		With solar simulators; Requires a simulator with uniformity over the volume of rotation; challenging for large area samples and stray light
Outdoor (absolute method)		1 or 2-axis tracker; subtraction of the diffuse component; it approximates the ideal conditions

2.2 Draft of Ed. 2.0

Table II lists the 6 methods that are currently present in the circulating draft for the Ed. 2.0 of IEC 61853-2. Indoor method 1 and outdoor method 2 are the same of indoor and outdoor methods of Ed. 1.0, respectively. The indoor method 2 (with a solar simulator that fully irradiates a module with a partially shaded cell) was originally proposed by W. Herrmann et al. [3-4]. In the indoor method 3 the light source is driven by optical tools (e.g. fibre optics, a collimator and a rotating stage) to irradiate the target cell at various angles: this method was proposed for cells by M. H. Saw et al. [2] and later upgraded to the module level. Indoor method 4, originally proposed by Plag et al. [7] can be adapted to any of the indoor methods 1 to 3 and requires the incident irradiance to be monochromatic, scanning the entire spectral responsivity band of the target device, to obtain its spectrally resolved IAM. Outdoor method 2, proposed by van der Heide et al. [11] is a relative method, in which the IAM of the testing sample is calculated relatively to the known IAM of a reference devices that is measured side by side and has been previously calibrated with any of the other methods.

Table II: The methods for measurements of IAM in the draft of the Ed. 2.0 of IEC 61853-2 with their main features

Method	Scheme	Main features
Indoor 1 (full irradiation)		Same as Ed. 1.0
Indoor 2 (full irradiation on a partial-shaded cell)		Partial shading of the target cell; can be more easily applied to tunnel simulators than indoor 1; stray light and uniformity may be critical
Indoor 3 (partial irradiation on a target cell)		Requires lock-in technique and bias light & voltage; module is fixed while the beam moves (optical fibres); effects of busbars may be critical
Indoor 4 (spectrally resolved method)		Can be applied to any of the indoor method above; requires monochromatic incident beam.
Outdoor 1 (absolute method)		Same as Ed. 1.0
Outdoor 2 (relative method)		Requires a calibrated reference device; It can be faster than outdoor 1

3 THE INTERLABORATORY COMPARISON

3.1 Participants and test samples

The six laboratories (SERIS-NUS, TÜV-Rheinland, SUPSI, CEA-INES, PTB and IMEC-Energiville) were labelled randomly as Lab 1 to Lab 6. Each laboratory adopted one of the methods of Table II: no laboratory adopted indoor method 1.

Table III shows the three test samples that were distributed to the participants in the interlaboratory comparison. These were provided by the Solar Energy Research Institute of Singapore (SERIS) and are: a 72-cell silicon heterojunction (HJT) module by Sanyo, a 120-half-cut-cell polycrystalline silicon (poly-Si) module by REC and a 60-cell poly-Si module by Gintung. All modules were commercially available at the time of production and represent a selection of a variety of products in the field: in particular, Gintung has a peculiar antiglare treatment, which can be visually observed when a spot of light is shone on top of it, as illustrated in the last row of Table III.

Table III: The test samples

	Sanyo	REC	Gintung
Series	HIP	TP2	WG
Cell type	Si HJT	poly-Si	poly-Si
Configuration	72 cells	120 cells, half-cut	60 cells
Dimensions	1580 × 798	1675 × 997	1640 × 992
Antiglare	No	No	Yes
Glare effect			

3.2 Statistical design

The three samples of Table III circulated in a sequential scheme (commonly referred to as "round-robin"), in which all participant laboratories measured all samples: no homogeneity requirement is required in this type of exercises for the test samples.

The assigned values (i.e. the "true" values attributed to the relative angular transmittance as a function of the incidence angle) were determined by *consensus* as the median of the non-outlying measurement results after all participants have completed their measurement. The results of participant laboratories are compared considering their claimed measurement uncertainties via assignment of E_n scores, defined as [13]

$$E_n = \frac{x_i - x_{ref}}{\sqrt{U_i^2 + U_{ref}^2}},$$

where x_i and x_{ref} are the i-th laboratory's measurement and assigned value, respectively; U_i and U_{ref} are the expanded uncertainties (confidence level of approximately 95%) of the i-th laboratory and the assigned value, respectively. U_{ref} was calculated by doubling the scaled median absolute deviation [13].

The stability of the relative angular transmittance properties of the test samples was reasonably assumed, hence the exercise ended when the last participant laboratory completed its measurements.

4 RESULTS AND DISCUSSION

Figure 1a, 2a and 3a show the results of the measured relative angular transmittance $\tau_{rel}(\theta)$ as a function of the angle of incidence θ by the six participant laboratories and for Sanyo, REC and Gintung, respectively. The charts highlight in red the outlying measurements; the assigned IAM is indicated by the continuous black lines. All measurements were performed with axis of rotation parallel to the busbars. All laboratories measured at θ ranging at least from −80° to 80°, step 10°.

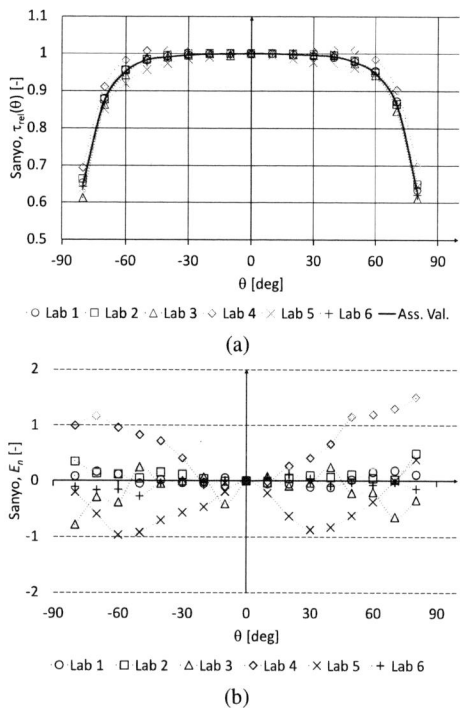

(a)

(b)

Figure 1: Results on Sanyo: (a) Relative angular transmittance (the continuous line indicates the assigned value); (b) E_n scores. Red data indicate outlying measurements. All measurements were performed with axis of rotation parallel to the busbars.

Figure 1a shows that Lab 4 and 5 were outlying on the Sanyo sample, while Lab 1, 2, 3 and 6 showed reproducible measurement results. Lab 4 showed 3-5% higher relative angular transmittance at $|\theta| > 30°$. The outliers of Lab 5 indicate lower relative angular transmittance at $20° < |\theta| < 60°$.

Figure 2a (REC) shows slightly different findings: here Lab 2, 3, 4 and 6 showed reproducible measurement results with slightly higher relative angular transmittance reported by Lab 6 at wide negative incidence angles, which is tight to measurement uncertainty and indicates a minor asymmetry in $\tau_{rel}(\theta)$ that was not confirmed by Lab 2, 3 and 4. Lab 5 was still outlying, showing the same trends already observed for the Sanyo sample.

Figure 3a (Gintung) shows the best reproducibility between all laboratories except for Lab 5 and the measurements at ±80° from Lab 4 only.

Moving the attention to the E_n scores, Figure 1b, 2b and 3b show the comparison between all laboratories for the relative angular transmittance of Sanyo, REC and Gintung samples, respectively. Measurements with $|E_n|>1$ are marked in red: these indicate cases where the measurement procedure caused a bias between the measured and the assigned values, which is not consistent

by the reported measurement uncertainty.

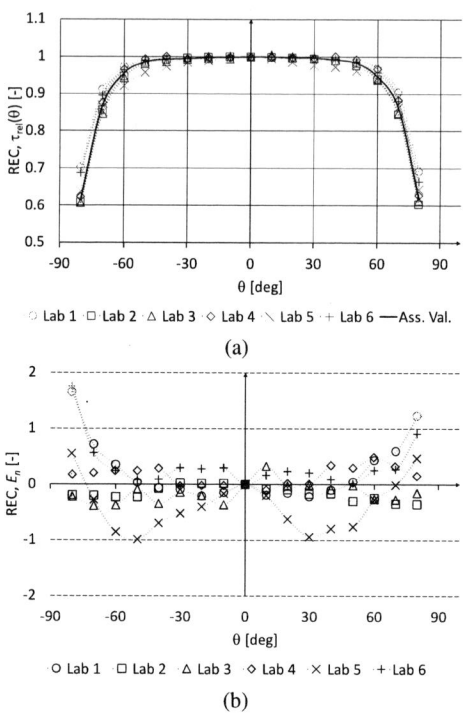

(a)

(b)

Figure 2: Results on REC: (a) Relative angular transmittance; (b) E_n scores.

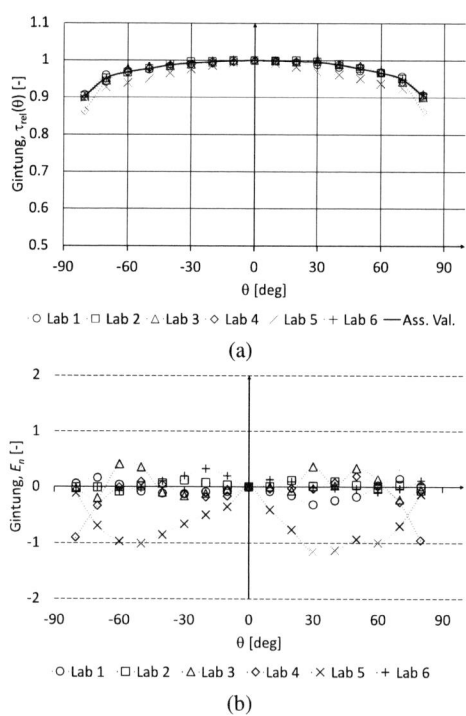

(a)

(b)

Figure 3: Results on Gintung: (a) Relative angular transmittance; (b) E_n scores.

Figure 1b confirms that Lab 4 is outlying, measuring

higher relative angular transmittance on the Sanyo module than Lab 1, 2, 3 and 6.

The Gintung sample (Figure 3b) is where the degree of reproducibility is the highest, with all labs showing very good performance ($|E_n|$<0.5), except for Lab 5 and questionable E_n scores only at ±80° from Lab 4. This is significant since this sample has a very peculiar antiglare treatment that ultimately did not represent an additional challenge for the participants.

Lab 1, 2, 3 and 6 remarkably obtained overall reproducible results using the four different measurement methods (indoor method 2 and 3 and outdoor method 1 and 2). Since outdoor method 1 is the method where the measurement conditions are the closest to the ideal conditions of irradiance (1000 W/m² total irradiance, AM1.5 spectral irradiance, and collimated direct irradiance, with subtraction of the diffuse component), the overall good reproducibility represents the ultimate validation of the new indoor methods 2, 3 and outdoor method 2.

It should also be noted that the level of reproducibility between Lab 1, 2, 3 and 6 is generally superior to the one observed between Lab 2, 4 and 5 that used the same indoor method 2. However, it is also important to notice that, among these three laboratories, Lab 2 was the only one adopting a procedure for the subtraction of the diffuse component of irradiance indoors. The requirement of diffuse irradiance (straylight) subtraction is not present in Ed. 1.0 of IEC 61853-2, while the results of this interlaboratory comparison demonstrate its relevance.

The measurements of relative angular transmittance by Lab 5 showed a peculiar triangular shape at ±40° around normal incidence. Conventional PV modules show cosine response in very good approximation between ± 20°, corresponding to a nearly zero slope in the $\tau_{rel}(\theta)$ curve around $\theta = 0$. As reported by Lab 5, its equipment setting does not allow a complete rotation from −80° to +80°, therefore the measurements were performed in 2 steps: first at positive angles; then the measurements at negative angles were obtained with the same verses of rotation, with the module rotated 180°. The findings of Figure 1a, 2a and 3a are consistent with a systematic misalignment of the normal incidence condition by approximately +2°, resulting in the triangular-shaped $\tau_{rel}(\theta)$. The possible misalignment was confirmed by Lab 5.

5   CONCLUSIONS

The incidence angle effect is showing growing interest among module manufacturers, developers and PV stakeholders as transmittance losses at wide angles may represent an important factor in the energy yield of PV installations. This work aimed at presenting the results of the validation the new methods for measuring this effect that have been presented by various authors in recent years and are currently proposed for publication in the next edition of IEC 61853-2. Method validation was performed via interlaboratory comparison of the measurements of three commercial-size PV samples by six different testing laboratories with renowned expertise in the field.

The results of the exercise showed very good reproducibility on all samples, with some significant outliers or measurement issues only from Lab 4 and 5. These highlighted the importance of two possible causes of measurement bias, e.g. the effect of stray light and angular misalignment. These effects can be corrected with accurate procedures that are going to be included as requirements in edition 2.0 of IEC 61853-2.

ACKNOWLEDGMENTS

The authors would like to acknowledge SERIS for having provided the test samples. SERIS is a research institute at the National University of Singapore (NUS) and is supported by NUS, the National Research Foundation Singapore (NRF), the Energy Market Authority of Singapore (EMA), and the Singapore Economic Development Board (EDB).

REFERENCES

[1] IEC 61853-2, "Photovoltaic (PV) module performance testing and energy rating – Part 2: Spectral responsivity, incidence angle and module operating temperature measurements" (2016).

[2] M. H. Saw, H. L. Soh, A. Ng, K. E. Birgersson, S. E. R. Tay, M. Pravettoni, IEEE J. Photovolt. 13(2) (2023) 267-274.

[3] W. Herrmann, M. Schweiger, L. Rimmelspacher, Proceedings 29th European Photovoltaic Solar Energy Conference and Exhibition (2014) 2403-2406.

[4] W. Herrmann, G. Bardizza, L. Rimmelspacher, M. H. Saw, M. Pravettoni, N. Riedel-Lyngskær, M. Babin, I. Kröger, Proceedings 40th European Photovoltaic Solar Energy Conference and Exhibition (2023).

[5] H. Al Husna Binti Mohd Nasim, PhD thesis, Loughborough University (2018).

[6] M. W. Amdemeskel, G. A. dos Reis Benatto, N. Riedel, B. Iandolo, R. S. Davidsen, O. Hansen, P. B. Poulsen, S. Thorsteinsson, A. Thorseth, C. Dam-Hansen, Proceedings 34th European Photovoltaic Energy Conference and Exhibition (2017) 1723-1726.

[7] F. Plag, I. Kröger, T. Fey, F. Witt, S. Winter, Prog Photovolt Res Appl. 26 (2018) 565–578.

[8] B. H. King, D. Riley, C. D. Robinson and L. Pratt, Proceeding IEEE Photovoltaic Specialist Conference (PVSC), New Orleans, LA, USA, (2015).

[9] J. Coston, C. Robinson, B. King, J. Braid, D. Riley and J. S. Stein, Proceedings 48th IEEE Photovoltaic Specialists Conference (PVSC), (2021) 1499-1503.

[10] S. Riechelmann, D. Friedrich, M. Müller, F. Schmaljohann, H. Sträter, S. Winter, Proceedings of WCPEC-8 (2022) 474-476.

[11] A. van der Heide, A. Tuomiranta, M. Daniels, N. Capiot, M. Daenen, S. Wendlandt, L. Tous, Proceedings of WCPEC-8 (2022) 708-711.

[12] IEC 60904-9, "Photovoltaic devices - Part 9: Classification of solar simulator characteristics" (2020).

[13] ISO 13528, Statistical methods for use in proficiency testing by interlaboratory comparison (2022).

Climate Specific Energy Rating (CSER) analysis of outdoor PV field data

Ismael Medina[1,2], Teodora Lyubenova[1], Ewan Dunlop[1]

[1] Joint Research Center, ESTI, European Comision, Ispra, Italy

[2] Institute for Computer Science, Uni Göttingen, Germany

Climate Specific Energy Rating (CSER)

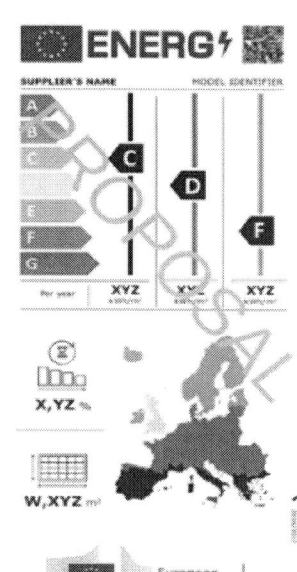

- IEC 61853 standard series: 6 reference climatic profiles.

- Tabulated hourly values of climatic parameters (irradiance, temperature, wind speed, …)

- Allows for more nuanced comparison of performance of a PV system under real operating conditions, beyond comparing power or efficiency at STC.

- Figures of merit: yearly energy yield and CSER.

$$CSER = \frac{E_{year}/P_{STC}}{G_{year}/1000}$$

Climate Specific Energy Rating (CSER)

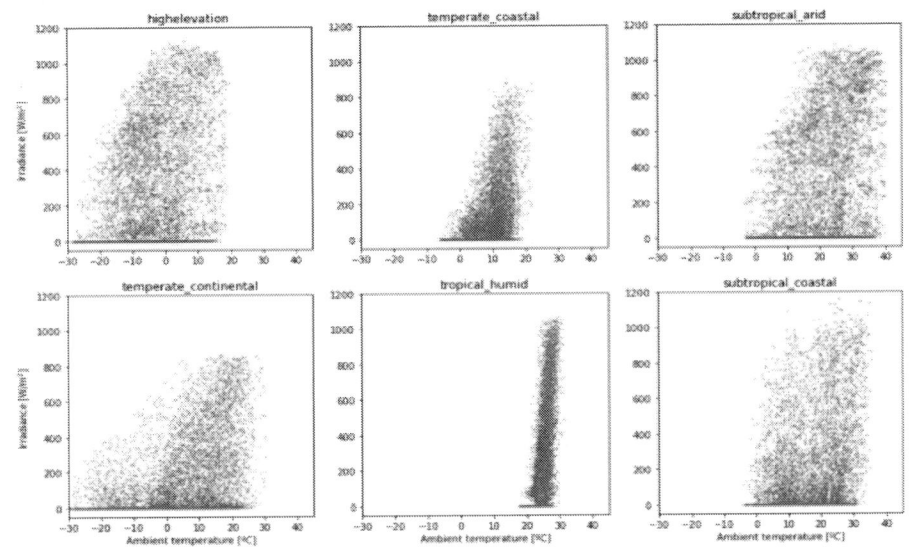

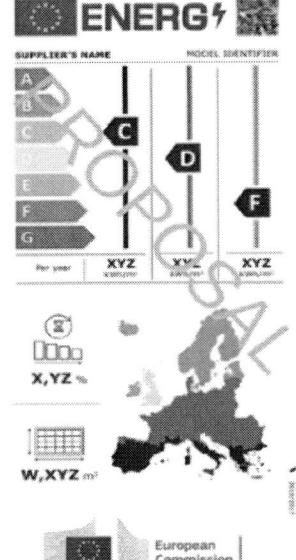

CSER computation

- One of the key inputs for the computation of CSER: G-T power matrices according to IEC 61853-1.

- Issue: measuring power matrices requires expensive equipment (solar simulator, climatized chamber), and specialized workforce.

- Question: are outdoor measurements a viable alternative? Can they provide reliable CSER values?

Irradiance (Wm^{-2})	Module temperature (°C)			
	15	25	50	75
1100	60.79	58.06	51.25	44.44
1000	55.01	52.61	46.61	40.61
800	43.97	42.11	37.44	32.77
600	32.87	31.49	28.03	24.58
400	21.64	20.74	18.5	16.27
200	10.43	9.96	8.79	7.63
100	4.82	4.60	4.06	3.52

Our setting

- Device under study: polycrystalline silicon monofacial PV module (0.5m2, 36 cells, and nominal power 60W).
- Mounting conditions compliant with IEC 61853 outdoors procedure.
- Monitoring of climatic parameters: irradiance, temperature, wind speed.
- IV curves measured every 5 minutes during 36 consecutive months (01/10/2018 to 31/09/2021).
- Unreliable measurements removed according to the standard.

Preliminary analysis

- Not all G-T conditions observed. In particular, no measurements for 75º C.
- We opt for a fit with the linearized King model[1] to populate the power matrices.

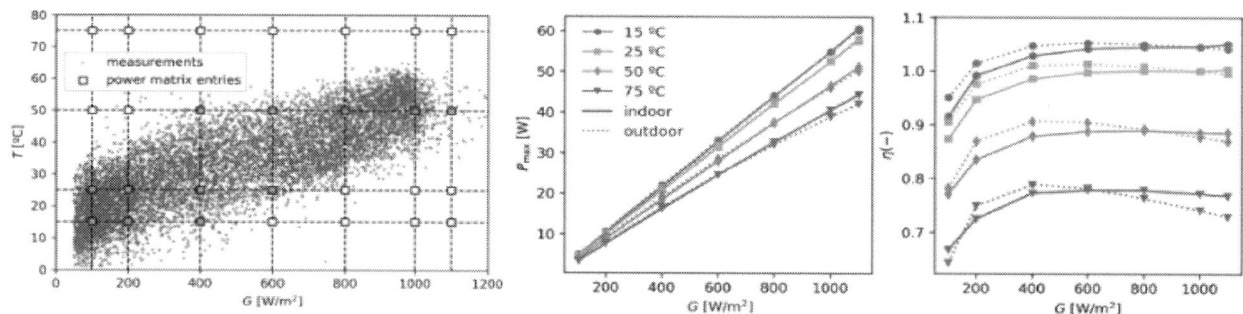

[1] Thomas Huld et al., *A power rating model for crystalline silicon PV modules*, Solar Energy and Solar Cells, 2011.
More details in poster sesion 3AV.3, Teodora Lyubenova, *Power Matrix Validation of Long-Term Outdoor PV Field Data.*

CSER calculation comparison

- We run the CSER calculation on the indoors and outdoors performance matrices.

	High elevation	Subtropical arid	Subtropical coastal	Temperate coastal	Temperate continental	Tropical humid
indoors	0.989	0.895	0.955	0.979	0.975	0.899
outdoors	0.989	0.901	0.962	0.987	0.981	0.910
difference (%)	0.02	0.62	0.74	0.84	0.61	1.18

Influence of measurement period

- How many months are sufficient for reliable CSER?

- Power matrix and CSER computation on subsampled periods of 3, 6, 9, 12, 18 and 24 months.

- Using each subsampled period, we populate power matrix and compute CSER.

Subsampled subtropical coastal CSER

Period length	Initial month	Final month	CSER
3	Oct-18	Dec-18	0.904
3	Nov-18	Jan-19	0.894
3	Dec-18	Feb-19	0.921
3	Jan-19	Mar-19	0.913
...	...	...	...
24	Jul-19	Jun-21	0.892
24	Aug-19	Jul-21	0.891
24	Sep-19	Aug-21	0.889
24	Oct-19	Sep-21	0.888
36	Oct-18	Sep-21	0.891

Period and seasonality analysis

- Each point corresponds to the CSER computed in one subsampled period.

- Warmer colors represent warmer months, cooler colors represent colder months.

Key takeaways

- 12 months results consistently within the uncertainty of indoors CSER.

- Using 6 months *that do not include the full winter* is comparable to using 12 months.

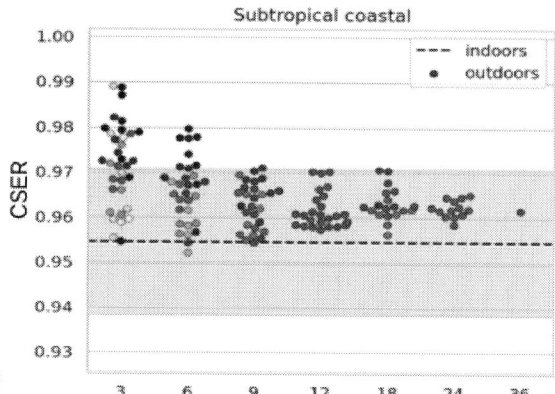

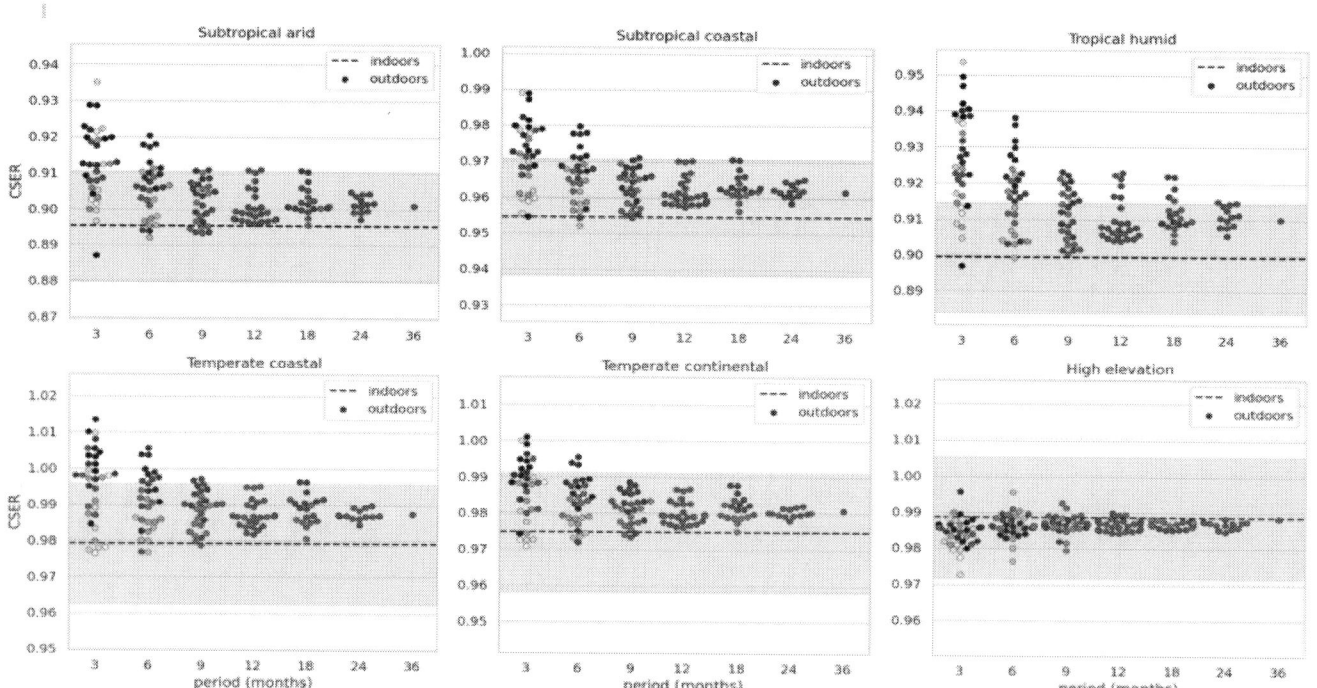

Conclusions

- We investigated the computation of CSER from outdoors data.
- Due to reduced range of meteorological conditions, we populate the performance matrices by fitting the outdoors data.
- Outdoors and indoors CSER show good agreement.
- Using shorter periods of time:
  - 12 months of measurements provide a good compromise between accuracy and time resources.
  - 6 months might be sufficient if enough meteorological variability is recorded.

Thank you

and special thanks to my collaborators Teodora Lyubenova and Ewan Dunlop

© European Union 2024
Authorised under the CC BY 4.0 license.
Slide 1: Polverini, Davide & Gracia Amillo, Ana & Taylor, Nigel & Sample, Tony & Salis, Elena & Dunlop, Ewan. (2021). *Building Criteria for Energy Labeling of Photovoltaic Modules and Small Systems*. Solar RRL. 6. 10.1002/solr.202100518.

41st European Photovoltaic Solar Energy Conference and Exhibition

MODULE PARAMETERS EXTRACTION FOR ASSESSING PHOTOVOLTAIC ENERGY YIELD
A COMPARATIVE APPROACH

A.Hashem[1], H. Sanchez[1], F.Xu[3], SL.Mortazavifar[1], C. Monokroussos[3], R. Gottschalg[1,2]

[1]Hochschule Anhalt, University of Applied Sciences, D-06366 Köthen

[2]Fraunhofer-Center für Silizium-Photovoltaic CSP, D-06120 Halle (Saale)

[3]TÜV Rheinland (Shanghai) Co., Ltd., Shanghai China

ABSTRACT: Parameter extraction techniques play a crucial role in accurately modelling the performance of Photovoltaic (PV) modules. Extracting the modelling parameters from measurements is critical for ensuring accurate performance predictions and energy yield estimates. Various modelling approaches are available, each providing different methods for parameter extraction.. The aim of this paper is to identify suitable combinations of models and extraction methods for system performance modelling. Data for modules is taken from measurements according to the international electrotechnical commission (IEC) energy rating series (IEC61853), specifically the irradiance-temperature matrices. A variety of extraction methods for different modelling approaches including the one diode model, Lambert function are employed and the accuracy of these approaches is determined by re-calculating the energy matrices and evaluating statistical parameters for agreement quality. The impact of the different combinations of modelling and extraction approaches within IEC61853 framework is demonstrated by the variation in modelling results using the standardized environmental datasets from IEC61853-4. The aim is not to determine a 'best' modelling approach, but to quantify the differences that can be expected due to different approaches, enabling an estimation of uncertainties and discussion discussion of which model is sufficient for various applications.

Keywords: System Modelling, Parameter Extraction, Energy Yield

1 INTRODUCTION

Accurate energy yield estimation for PV modules is essential for optimizing performance, achieving efficient energy production, and informing investment decisions in solar energy projects. Additionally, precise energy yield estimations are key for system designers to select the best appropriate PV technology for specific climatic conditions and to optimize system configurations for maximum efficiency. The IEC 61853 standards are fundamental to this process, as they provide a standardized framework for the performance assessment of PV modules in real-world conditions. By providing a globally recognized methodology for performance evaluation, the IEC 61853 standards ensure consistency and comparability across various PV technologies and installations.

Over the years, various methods have been developed to accurately predict the performance of photovoltaic (PV) systems under different environmental conditions. A simplified model for estimating the annual energy production of crystalline PV modules in Germany and Italy utilized a supervised learning approach [1]. This model employed multivariate linear regression with three key inputs: total annual solar radiation, yearly average temperature, and the PV panel's temperature coefficient. The comparison parameter was the ratio of the actual yearly measured energy to the estimated energy. Energy Yield for 15 modules of varying types including crystalline silicon, Cadmium telluride (CdTe) and heterojunction (HJT) was estimated across four different locations Cologne, Ancona, Arizona and Chennai [2]. The specific energy yield showed high differences reaching 12% in Italy, 13% in Germany, 21% in Arizona, and 23% in India. These variations are attributed to factors such as spectral irradiance, temperature effects, and performance stability. A comparison was made between mono- and multi-crystalline silicon PV modules using synthetic hourly meteorological data from five sites through MeteoNorm in PVsyst [3]. The energy rating adhered the IEC 61853-3 draft and was compared with PVsyst results. It was shown that IEC 61853 testing may reveal maximum

differences of up to 5% in the specific yield of PV-module among mainstream silicon- based technologies. The comparison has been expanded to include a wider variety of commercially available c-Si-based PV module types, utilizing newly available data [4]. The characterization of crystalline silicon PV modules including mono and multi c-Si Al-BSF, mono c-Si PERC, and n-type HJT and tunnel oxide passivate contact rechnology (TOPCon) cells, was conducted on samples from 27 PV module manufacturers. High-efficiency PV modules, including PERC, HJT, and TOPCon, demonstrate higher energy yields compared to mono c-Si and mc-Si modules with Al BSF. Specific relative differences of up to 7.34% were observed among crystalline-based PV modules of the same technology.

In this study, three techniques are employed to fit and extract parameters from the single diode model on a modern HJT half-cut module with 132 cells under varying environmental conditions (irradiance and temperature). The power matrices for the actual and predicted I-V curves, as outlined in IEC 61853-1, are computed and compared to evaluate the efficiency of the modeling. Subsequently, the energy yield of the module is calculated for various climatic conditions.

2 MODELLING & PARAMETER EXTRACTION TECHNIQUES

Accurate modelling and parameter extraction are essential for enhancing the performance and efficiency of PV modules. These techniques facilitate the prediction of module behavior under various environmental conditions. In this section the fitting and extraction technqiues used in this study will be discussed.

2.1 Fitting technique

This technique is part of the pvlib library, a Python package used to simulate PV system performance where individual I-V curves are fitted in two phases [5]. First, the single-diode equation is rearranged and approximated for both the linear and exponential components of the I-V curve. The linear region of the curve is fitted using a linear regression model, while the exponential region is fitted by calculating the logarithm of the equation. The resultant fits yield regression coefficients, which are used to compute the key variables of the single-diode model.

2.2 Technique proposed by Celik

The technique adopts an analytical approach based on the methods introduced by Phang [6]. The series and shunt resistances are estimated by analyzing the slopes of the I-V curves. R_{so} and R_{sho} which are the are the reciprocals of the slopes at the open-circuit and short-circuit point, respectively are calculated first. The shunt resistance is assumed to be the R_{sho}. The remaining four parameters are then calculated as follows [7]:

$$R_s = R_{so} - \left[\frac{m\,V_t}{I_o}\right] exp\left(-\frac{V_{oc}}{m\,V_t}\right) \qquad (1)$$

$$I_{ph} = I_{sc}\left(1 + \frac{R_s}{R_{sh}}\right) + I_o\left[\left(exp\,\frac{I_{sc}R_s}{m\,V_T}\right) - 1\right] \qquad (2)$$

$$I_o = \left(I_{sc} - \frac{V_{oc}}{R_{sh}}\right) exp\left(-\frac{V_{oc}}{m\,V_t}\right) \qquad (3)$$

$$n = \frac{V_{mp} + I_{mp}R_{so} - V_{oc}}{N_s V_t\left[\ln\left(I_{sc} - \frac{V_{mp}}{R_{sh}}\right) - \ln\left(I_{sc} - \frac{V_{oc}}{R_{sh}}\right)\right] + \left(\frac{I_{mp}}{I_{sc} - \frac{V_{oc}}{R_{sh}}}\right)} \qquad (4)$$

where where I_{ph} is the photo current, I_o is the reverse saturation current of the p–n diodes, R_s is the series resistance, R_{sh} is the shunt resistance, N_s is the number of the cells in the module, n is the diode ideality factor, m is defined by N_s*n and V_t is the thermal voltage.

2.3 Technique proposed by Villalva

This technique for extracting the parameters of the one-diode model relies on several simplifying assumptions. First, the ideality factor is assumed, and the photocurrent is set equal to the short-circuit current (I_{sc}), an iterative process is used to fine-tune R_s , which is incrementally increased from zero. R_{sh} is subsequently calculated by utilizing the principle that a unique combination of R_s and R_{sh} ensures the calculated maximum power matches the measured maximum power. The detailed equations for the shunt resistance and the reverse saturation current are shown below [8]:

$$I_o = \frac{I_{sc}}{exp\left(\frac{V_{oc}}{a\,V_t}\right)} - 1 \qquad (5)$$

$$R_{sh} = \frac{V_{mp}(V_{mp} + I_{mp}R_s)}{V_{mp}I_{ph} - V_{mp}I_o\,exp\left(\frac{(V_{mp} + I_{mp}R_s)}{N_s a V_t}\right) + V_{mp}I_o - P_{max,m}} \qquad (6)$$

where $P_{max,m}$ is the measured maximum power

3 SIMULATION RESULTS

Simulations are conducted using pvlib python, a community developed toolbox that provides a set of functions and classes for simulating the performance of PV energy systems [9]. The flow chart for the work in this study is shown in Figure 1 where the system input consists of a set of I-V curves for the HJT module at various irradiances and temperature. First, the key parameters of each curve is calculated which will be fed to the fitting and extraction technqiues. Then, the five single diode model parameters are estimated for each curve, which will be used to replot the estimated versus the actual I-V curves. The power matrix is then calculated as specified in standard IEC 61853-1, which is critical for accurate energy yield estimation.

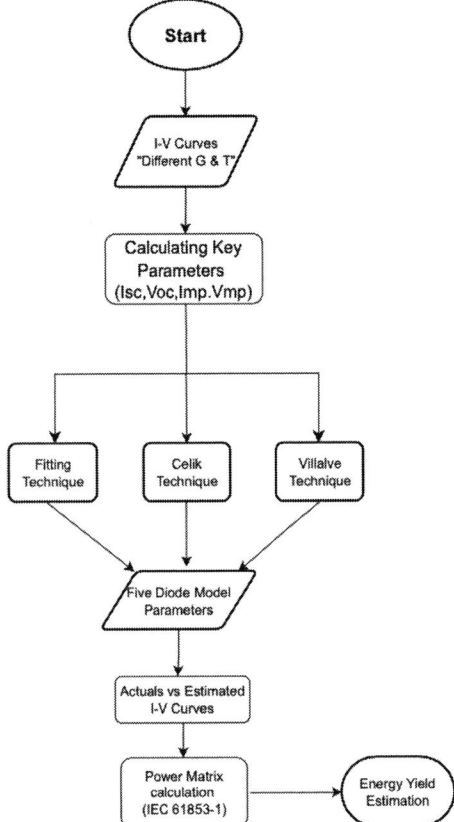

Figure 1: Simulation flow used for energy yield calculation

3.1 Power Matrix

The matrix corresponding to the maximum power is computed for each environmental condition, serving as a critical input for the IEC 61853 standards used in energy yield calculations. Subsequently, the matrix of measured values is compared with the matrix of estimated values to validate the accuracy of the estimation technique.

Due to the assumption required for the ideality factor in the third technique used prior to the estimation of other parameters, trial and error is applied for choosing the optimum value starting from 1.3 ending at 0.8, as shown in Figure 2. Selecting a value outside of this range results in a significantly higher percentage difference in power.

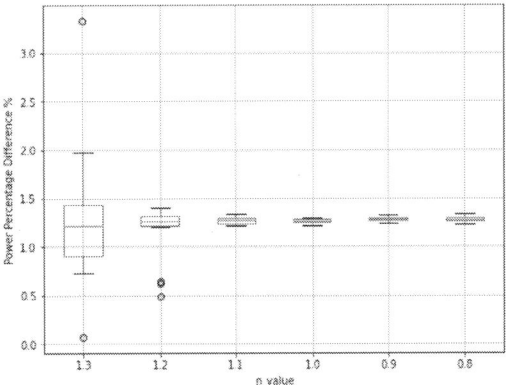

Figure 2: Impact of n value for power difference in Villalva technique

41st European Photovoltaic Solar Energy Conference and Exhibition

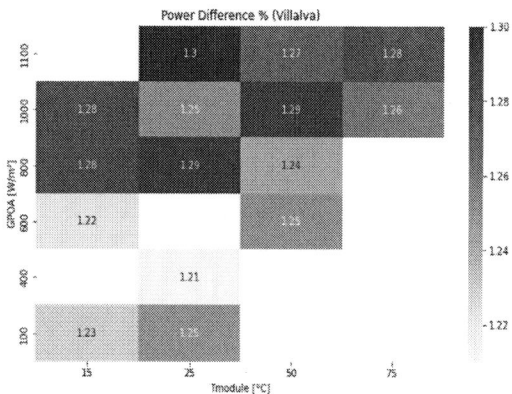

Figure 3: Power difference percentages for Villalva techniqe

It has been observed that starting with an n-value of 1.3 results in an average percentage error of 1.4%, with some outliers peaking at 3%. A reduction in the n-value correspondingly decrease the power difference. An optimum n-value of 1 was selected due to its minimal variance and an average error of 1.26%.

The percentage difference in power for maximum output for the fitting and the celik techniques are shown in Figure 4 and Figure 5 respectively. For the fitting techniuqe it can be seen that a relatively high percentage differences occur at extreme points with maximum values at 25°C,100 W/m², and 75°C,1100 W/m². The Celik technique also exhibits very low differences with only one outlier presenting a 1.72% difference at 15°C and 1000 W/m².

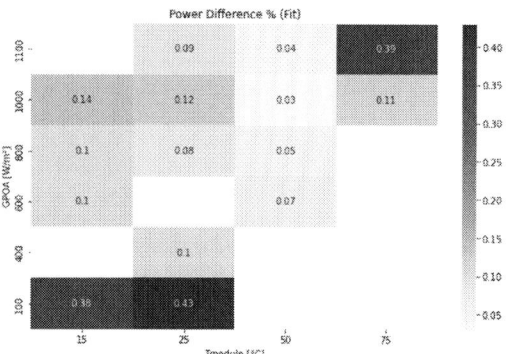

Figure 4: Power difference percentages for Fitting techniqe

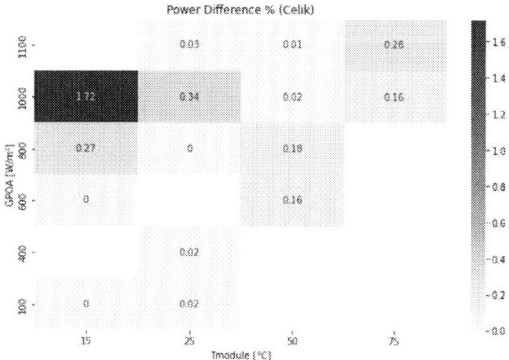

Figure 5: Power difference percentages for Celik Technique

3.2 IEC 61853 Standard

The standard methodology assesses the energy output of a PV module according to a defined climatic profile utilizing annual hourly meteorological data. Initially, the method adjusts the beam and diffuse solar irradiance components based on the angle of incidence (AOI). Following this adjustment, the irradiance values undergo a spectral correction to mitigate spectral losses. The calculation of module temperature is a critical step, as it significantly influences the operational performance of the PV module. The power output of the module is then determined by applying the corrected temperature and irradiance data. Energy production is calculated for each time step, typically one hour, by multiplying the power output by the time step duration. These hourly energy outputs are aggregated to provide the module's total annual energy yield in kilowatt-hours (kWh).

3.3 Energy Calculation

Energy calculations will be performed across various climatic types, including subtropical arid, subtropical coastal, tropical, temperate continental, temperate coastal, and high elevation regions. Subtropical arid climates ~~is~~ are defined by low rainfall and high temperatures, while subtropical coastal climates experience significant rainfall, high humidity, warm to hot summers, and mild winters. Tropical climates maintains warm temperatures throughout the year, accompanied by substantial rainfall and high humidity levels. In contrast, temperate continental climates have relatively hot summers and cold winters, with moderate rainfall. Temperate coastal climates, on the other hand, have milder conditions year-round, with cool summers, mild winters, and high humidity. Finally, high elevation climates, typically found in mountainous regions, are characterized by cooler temperatures and higher precipitation, often in the form of snow.

Figure 6 compares the actual energy values with those estimated by the three previously mentioned techniques, highlighting notable differences in energy levels across various locations. Subtropical arid and high elevation regions exhibit the highest energy values compared to other locations. This is likely due to stronger solar radiation in the subtropical arid climates and the increased solar irradiance at higher altitudes. Despite the cooler temperatures in high elevation areas, the thinner atmosphere results in less scattering and absorption of sunlight, which enhances solar energy levels. ~~When~~ Among temperate climates, temperate continental regions show slightly higher energy levels due to greater irradiance. In contrast, temperate coastal areas experience milder year-round conditions, characterized by moderate temperature fluctuations, mild winters, cool summers, and high humidity.

The close alignment between actual and estimated values demonstrates the accuracy and reliability of the prediction methods. Figure 7 shows the percentage error, where both the fitting and Celik's approaches achieve superior accuracy, with errors below 0.2%. Although Villalva's method exhibits ~~a~~ slightly higher error, its maximum error of 1.2% remains better than most studies in the literature.

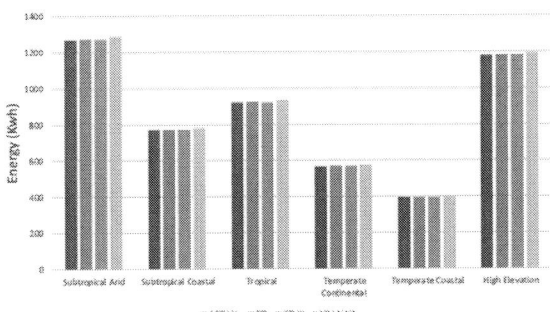

Figure 6: Energy yield estimation in different locations

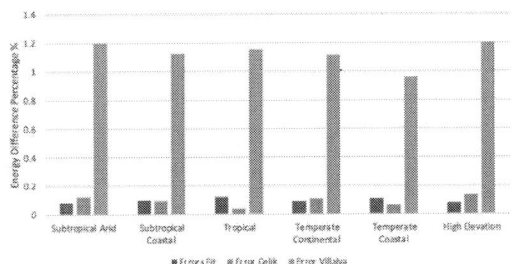

Figure 7: Energy Estimation error for the three techniques

4 CONCLUSION

Accurately energy yield estimation is crucial for optimizing PV system performance and ensuring efficient energy generation. This study has examined three distinct modeling approaches to evaluate the energy yield of PV modules, following the IEC 61853 standard across various climatic conditions. The results demonstrate the accuracy and reliability of these prediction methods, with a maximum deviation of just 1.2% between actual and estimated values. Furthermore, the comparison of energy outputs across different climate types emphasizes the significant impact of environmental factors, such as irradiance, temperature variability, and humidity, on energy yield. This underscores the importance of tailoring PV system designs to specific climatic conditions.

5 REFERENCES

[1] Aste, N., Del Pero, C., Leonforte, F., & Manfren, M. (2013). A simplified model for the estimation of energy production of PV systems. *Energy, 59*, 503-512.

[2] Schweiger, M., & Herrmann, W. Comparison of Energy Yield Data of Fifteen PV Module Technologies Operating in Four Different Climates, 2015. In *42nd IEEE Photovoltaic Specialists Conference, New Orleans, Louisiana.*

[3] Monokroussos, C., Zhang, X. Y., Schweiger, M., Etienne, D., Liu, S., Zhou, A., ... & Zou, C. (2017). Energy rating of c-Si and mc-Si commercial PV-Modules in accordance with IEC 61853–1,-2,-3 and impact on the annual yield. In *Proc. 33rd Eur. Photovolt. Sol. Energy Conf* (pp. 1438-1444).

[4] Monokroussos, C., Zhang, Y., Lee, E. W., Xu, F., Zhou, A., Zhang, Y., & Herrmann, W. (2023). Energy performance of commercial c-Si PV modules in accordance with IEC 61853-1,-2 and impact on the annual specific yield. *EPJ Photovoltaics, 14*, 6.

[5] C. B. Jones, C. W. Hansen, "Single Diode Parameter Extraction from In-Field Photovoltaic I-V Curves on a Single Board Computer", 46th IEEE Photovoltaic Specialist Conference, Chicago, IL, 2019.

[6] Phang, J., Chan, D., Phillips, J., 1984. Accurate analytical method for the extraction of solar cell model parameters. Electron. Lett. 20, 406–408.

[7] Celik, A. N., & Acikgoz, N. (2007). Modelling and experimental verification of the operating current of mono-crystalline photovoltaic modules using four-and five-parameter models. *Applied energy, 84*(1), 1-15.

[8] Villalva, M. G., Gazoli, J. R., & Ruppert Filho, E. (2009). Comprehensive approach to modeling and simulation of photovoltaic arrays. *IEEE Transactions on power electronics, 24*(5), 1198-1208.

[9] Anderson, K., Hansen, C., Holmgren, W., Jensen, A., Mikofski, M., and Driesse, A. "pvlib python: 2023 project update." Journal of Open Source Software, 8(92), 5994, (2023).

Scuola universitaria professionale della Svizzera italiana
Dipartimento ambiente costruzioni e design
Istituto sostenibilità applicata all'ambiente costruito
Laboratorio SUPSI PVLab

SUPSI

PERFORMANCE AND DEGRADATION EVALUATION OF c-Si MODULES UNDER DIFFERENT OPEN-RACK AND RESIDENTIAL MOUNTING CONFIGURATIONS

Gabi Friesen, Ebrar Özkalay, Mauro Caccivio

SUPSI PVLab – University of Applied Sciences and Arts of Southern Switzerland, Mendrisio, Switzerland

EUPVSEC, 23-27 September 2024

Vienna, Austria

26 settembre 2024

SUPSI   41st European Photovoltaic Solar Energy Conference, Wien, 2024

History: From the 1st to the 14th Test Cycle at SUPSI

| 1991 | ... | 2007 | ... | 2011 | ... | 2018 | ... | 2021 |

| c-Si silicon modules | ... | BIPV | ... | thin films | ... | bifacial/coloured | ... | **high efficiency** |

Performance benchmarking of latest c-Si module technologies

Technology specific key questions

- How much energy is gained with the new module technologies?

- How stable are the new technologies?

- How reliable are new module concepts?

Methodology specific key questions

- How is benchmarking impacted by the mounting configuration?

- How do stabilization and degradation mechanisms impact benchmarking?

26/09/2024

Trend: share of cell technologies

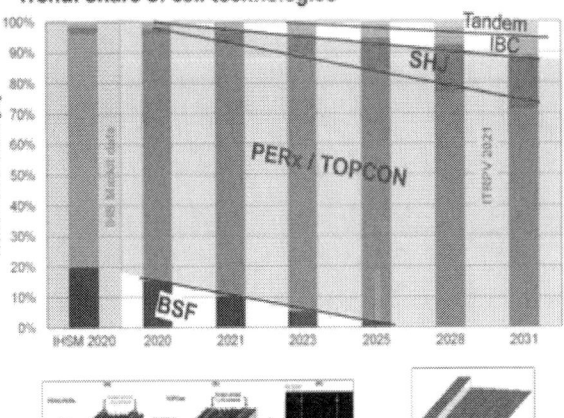

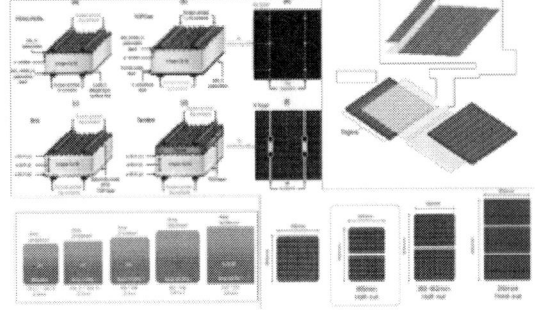

Test Cycle 14 Module Technologies

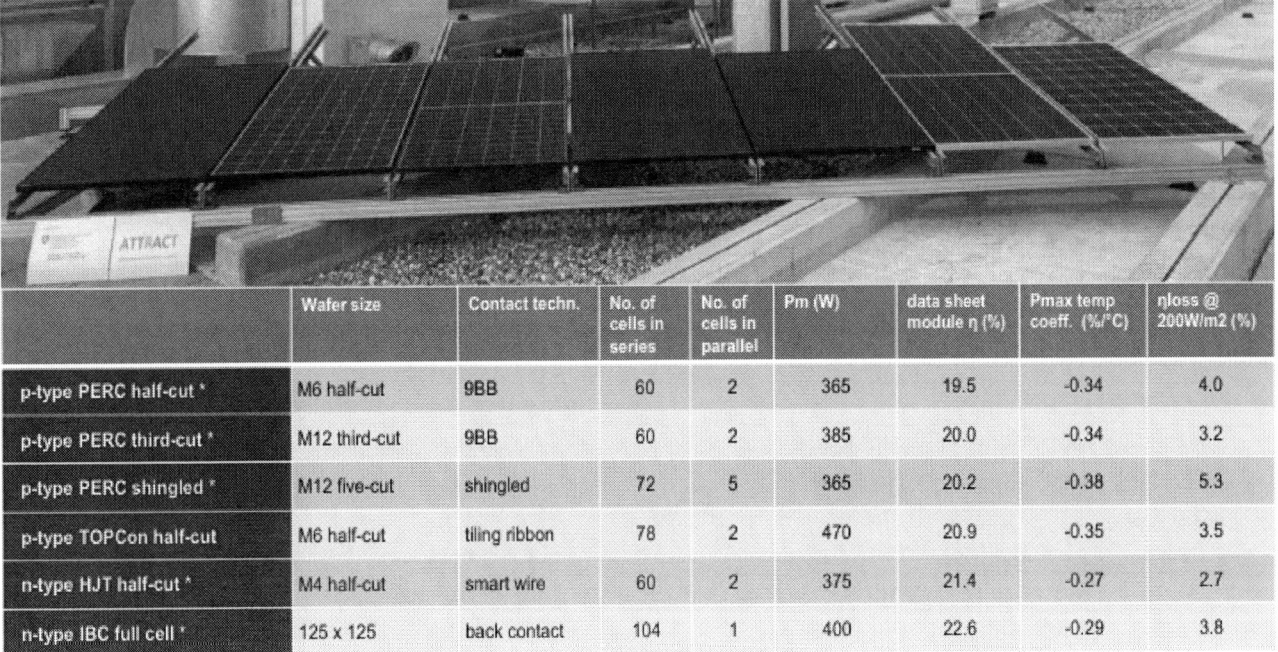

	Wafer size	Contact techn.	No. of cells in series	No. of cells in parallel	Pm (W)	data sheet module η (%)	Pmax temp coeff. (%/°C)	ηloss @ 200W/m2 (%)
p-type PERC half-cut *	M6 half-cut	9BB	60	2	365	19.5	-0.34	4.0
p-type PERC third-cut *	M12 third-cut	9BB	60	2	385	20.0	-0.34	3.2
p-type PERC shingled *	M12 five-cut	shingled	72	5	365	20.2	-0.38	5.3
p-type TOPCon half-cut	M6 half-cut	tiling ribbon	78	2	470	20.9	-0.35	3.5
n-type HJT half-cut *	M4 half-cut	smart wire	60	2	375	21.4	-0.27	2.7
n-type IBC full cell *	125 x 125	back contact	104	1	400	22.6	-0.29	3.8

commercial modules purchased in 2021

Technology Inter-comparison under different mounting configurations

Real world vs. testing environment (Swiss scenarios)

field mounted

flat roof

BIPV tilted roof

30° inclination / open-rack

10° inclination / open-rack /soiling

30° inclination / insulated / shading

Performance testing & monitoring

FIELD MONITORING

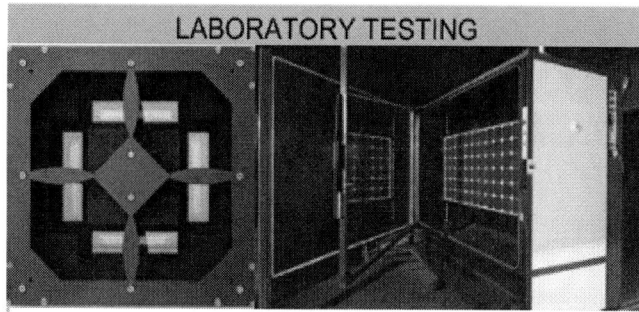

LABORATORY TESTING

- Module level IV curve tracers (Gantner, SUPSI MPPT3000)
- Maximum power point tracker
- Temperature sensors (2 PT100 per module)
- Calibrated pyranometers (2 per inclination)
- Meteo station
- Spectrum radiometer (EKO)

- Pulsed solar simulator (class A+A+A+)
- Thermal control box (15-75 °C)
- Spectral neutral irradiance filter (100-1100 W/m²)
- 29 spectral bandpass filters (311-1200nm)
- Rotational rack (0-80°)

max. measurement uncertainty at STC

$uP_{max}[k=2] = 1.6\%$

SUPSI 41st European Photovoltaic Solar Energy Conference, Wien, 2024

Performance analysis of open-rack modules
Monthly performance ratio over 29 months
Comparison of two modules per technology mounted at 30° tilt

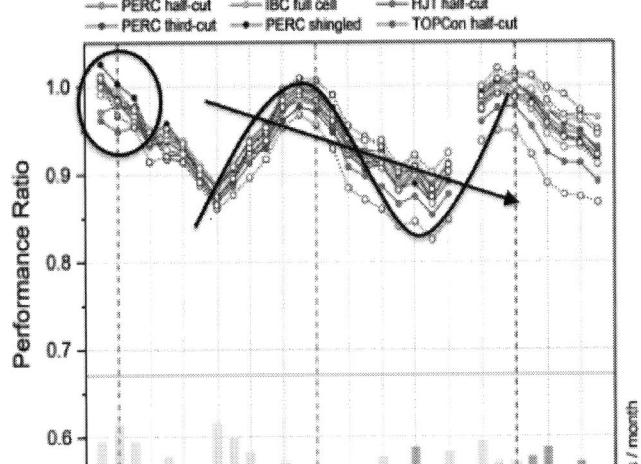

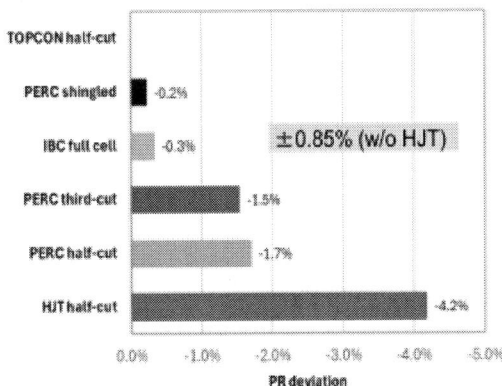

±0.85% (w/o HJT)

SUPSI 41st European Photovoltaic Solar Energy Conference, Wien, 2024

Performance analysis of open-rack modules
Stabilisation of modules before outdoor exposure
Indoor IEC 61215 vs. outdoor stabilization procedures

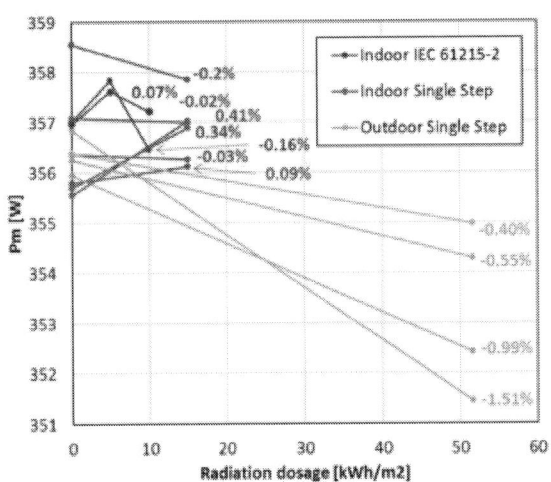

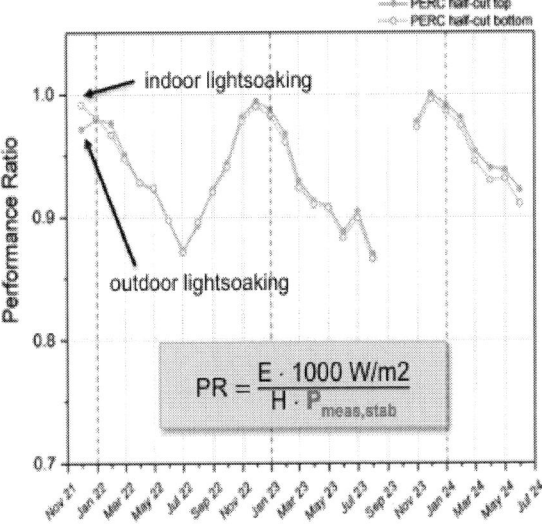

indoor lightsoaking

outdoor lightsoaking

$$PR = \frac{E \cdot 1000\ W/m2}{H \cdot P_{meas,stab}}$$

Performance analysis of open-rack modules
Degradation occurring during field operation
Indoor STC measurements after 29 months of outdoor exposure

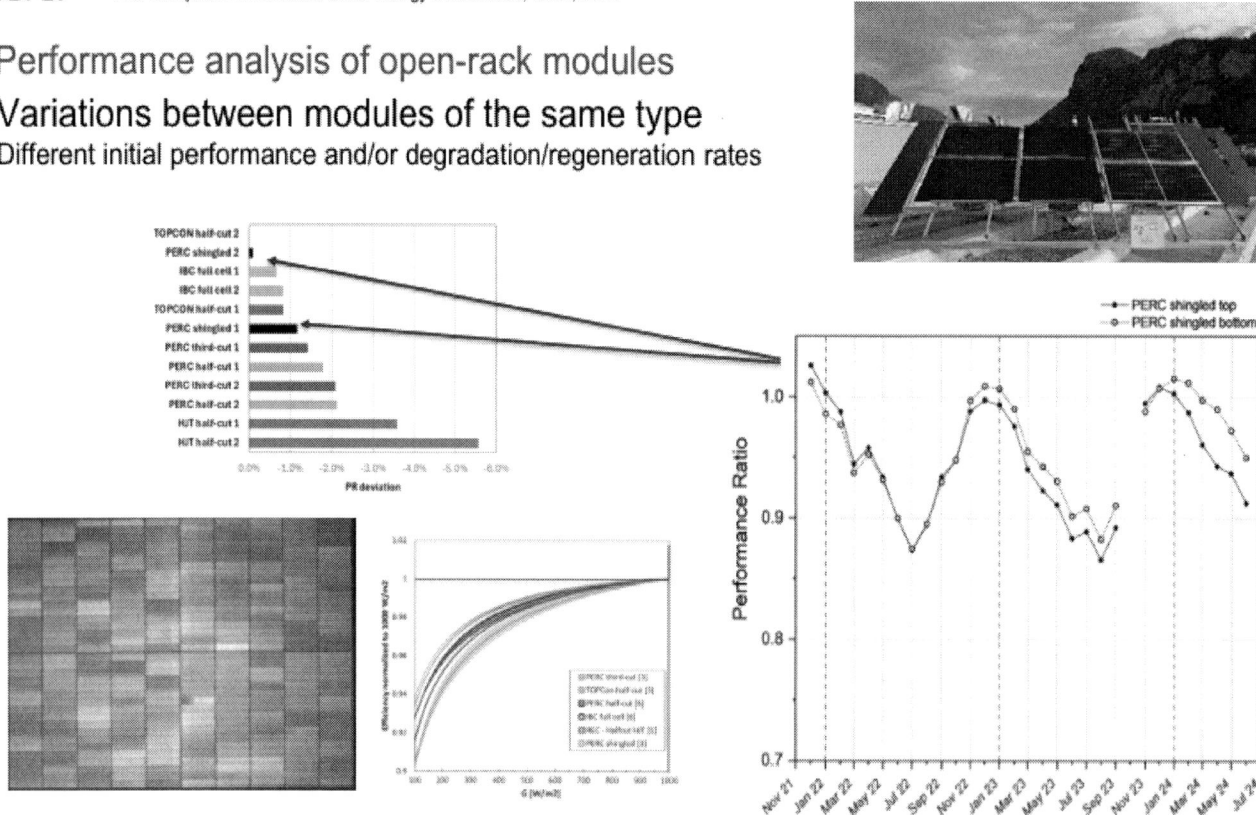

Performance analysis of open-rack modules
Variations between modules of the same type
Different initial performance and/or degradation/regeneration rates

SUPSI 41st European Photovoltaic Solar Energy Conference, Wien, 2024

Performance analysis of low tilt modules

Monthly performance ratio over 29 months
Comparison of modules mounted at 10˚ tilt

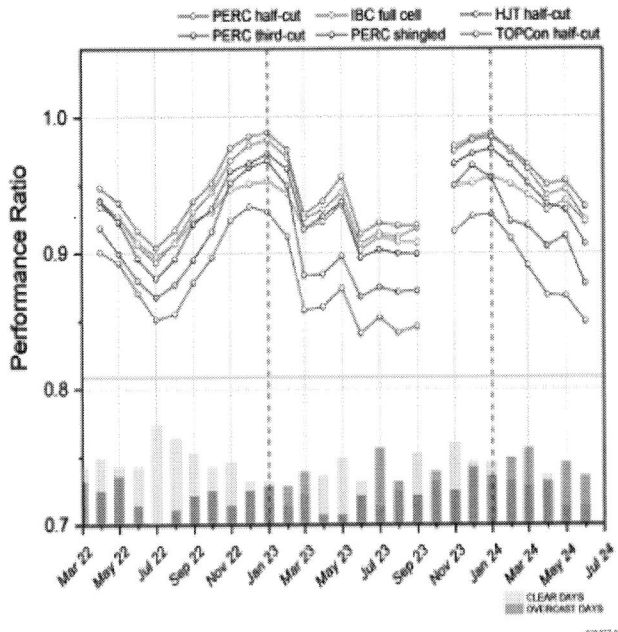

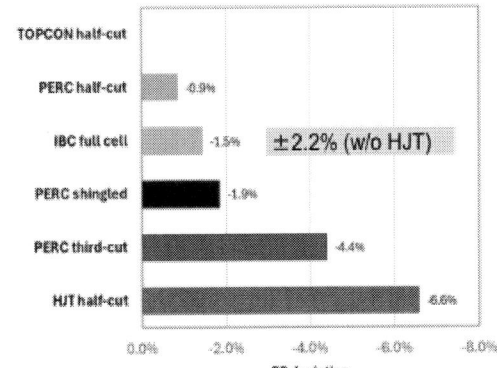

SUPSI 41st European Photovoltaic Solar Energy Conference, Wien, 2024

Performance analysis of low tilt modules

Monthly performance ratio over 29 months
Comparison of modules mounted at 10˚ tilt

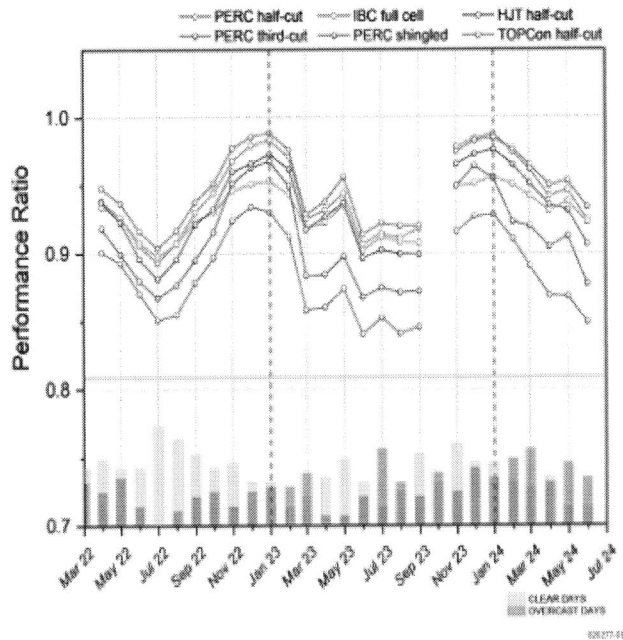

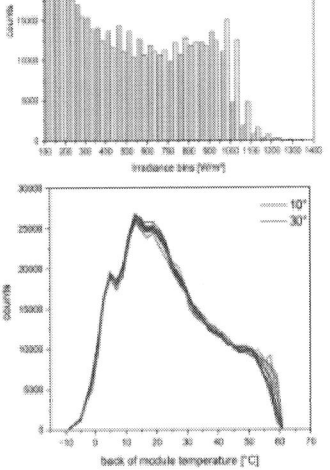

Performance analysis of low tilt modules

Angle of incidence losses
Higher frequency of AOI>50° in winter noon time

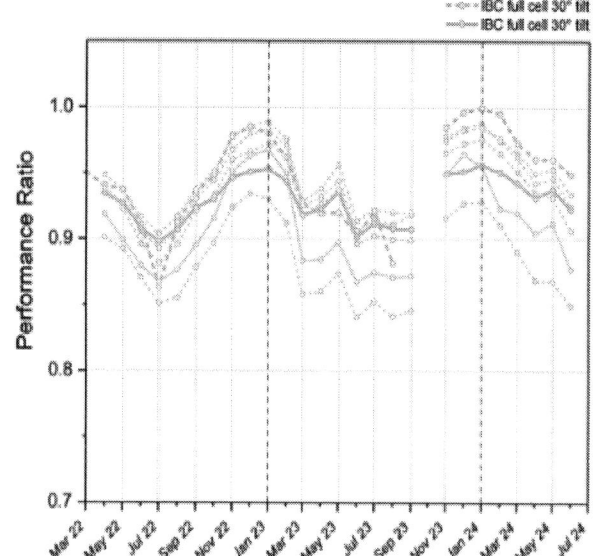

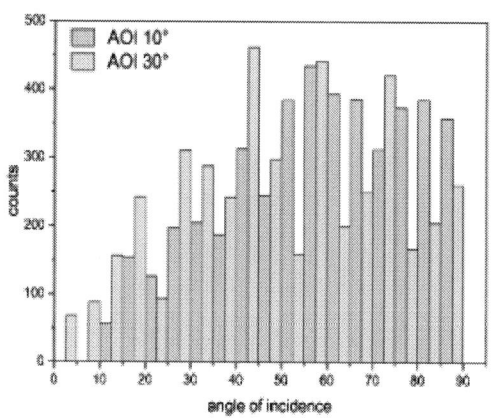

Performance analysis of 10° open-rack modules

Soiling losses
Bottom frame soiling with different cell to frame distance

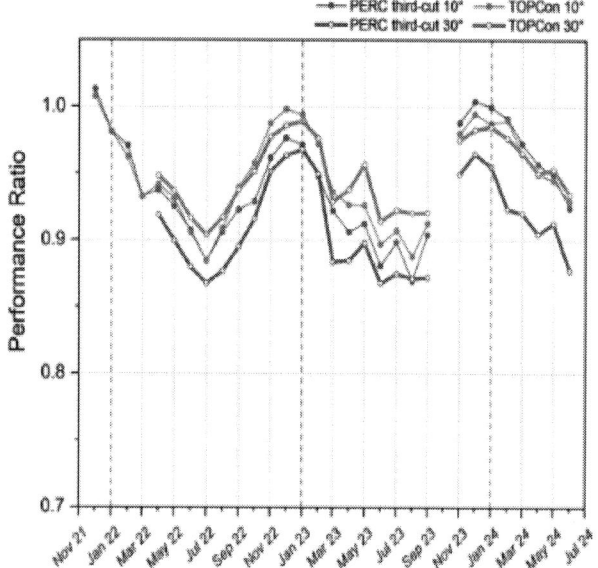

SUPSI   41st European Photovoltaic Solar Energy Conference, Wien, 2024

Performance analysis of 30° BIPV modules
Monthly performance ratio over 26 months
3 module types with rear insulation

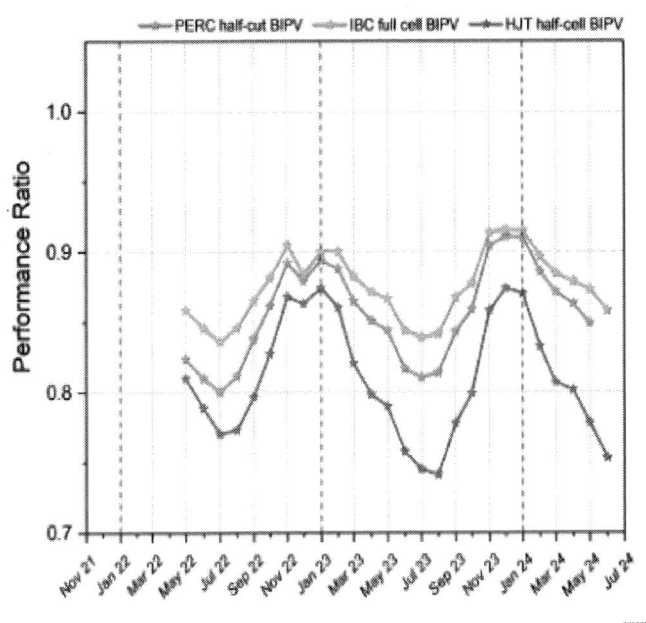

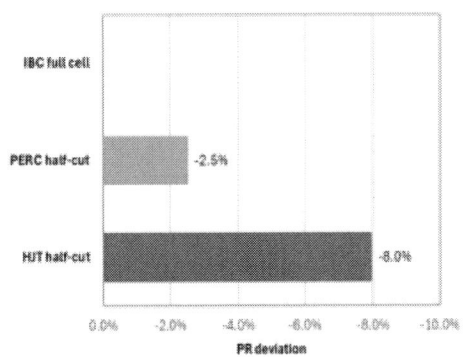

SUPSI   41st European Photovoltaic Solar Energy Conference, Wien, 2024

Performance analysis of 30° BIPV modules
Monthly performance ratio over 26 months
3 module types with rear insulation

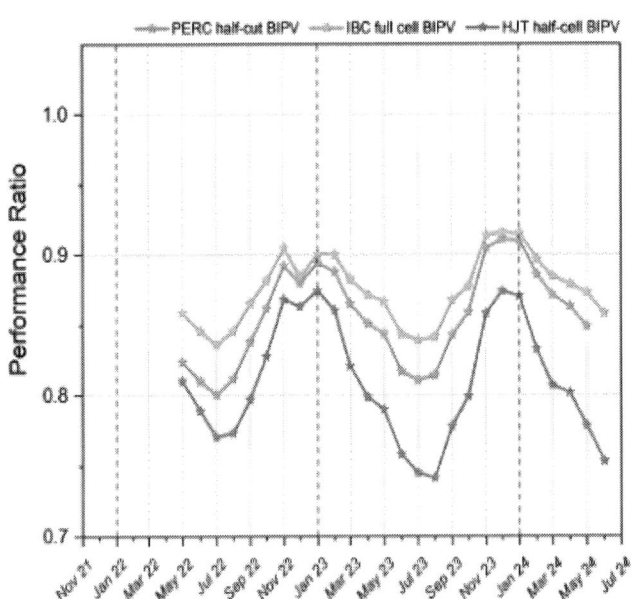

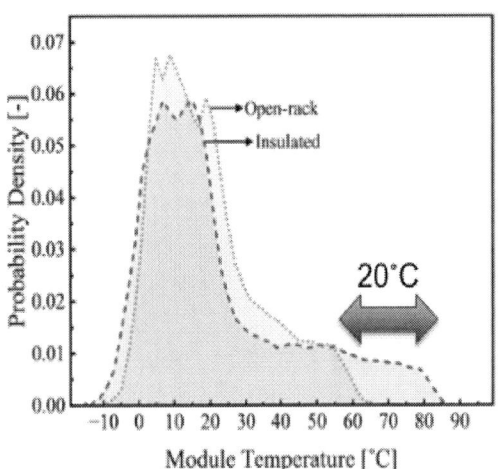

Performance analysis of 30° BIPV modules

Thermal losses
Temperature coefficient related power loss

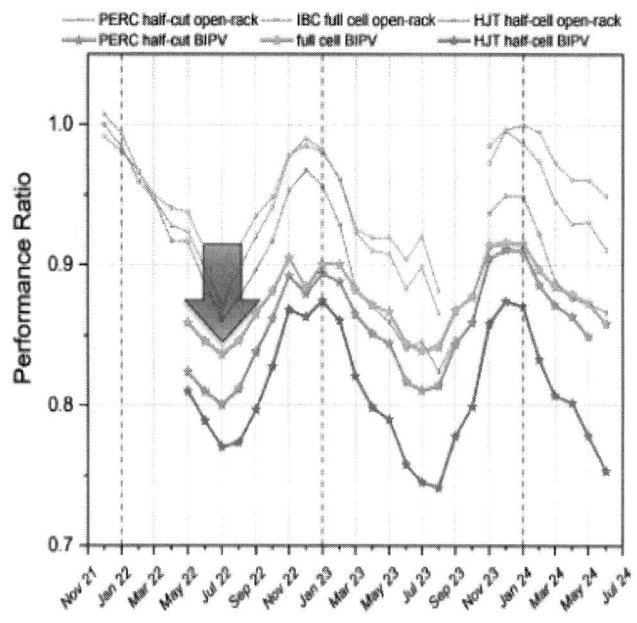

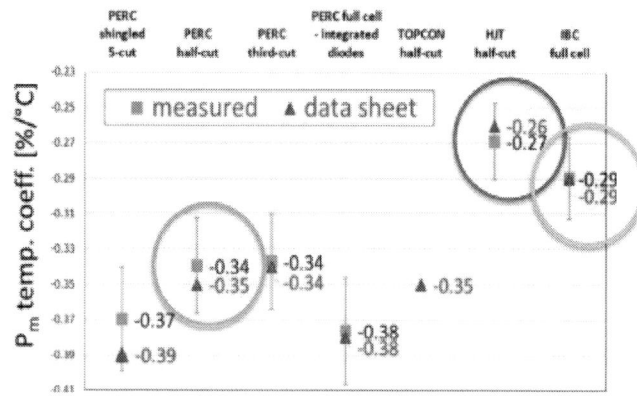

Performance analysis of 30° BIPV modules

Degradation rates under harsh conditions
Open-rack vs. no rear ventilation and constant shading

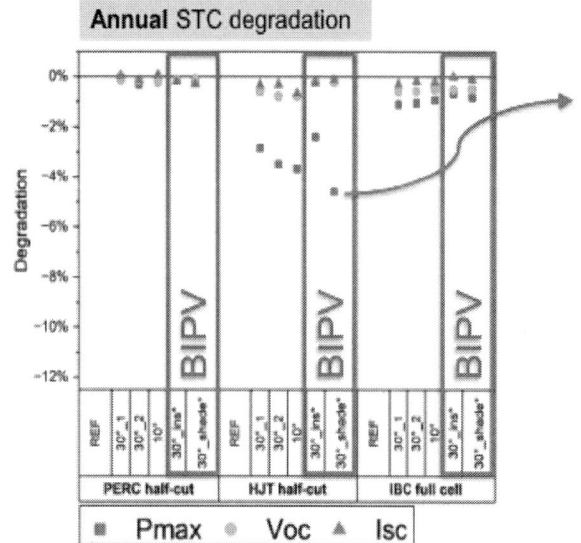

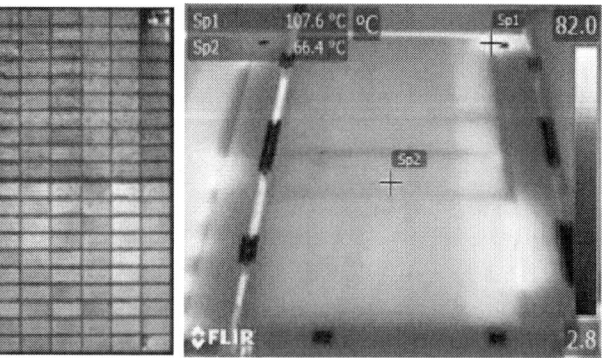

E. Özkalay et al., *EPJ Photovoltaics (2024), "The effect of partial shading on the reliability of photovoltaic modules in the built-environment", doi: 10.1051/epjpv/2024001.*

Performance analysis of 30° BIPV modules
Degradation rates under harsh conditions
Open-rack vers. no rear ventilation and constant shading

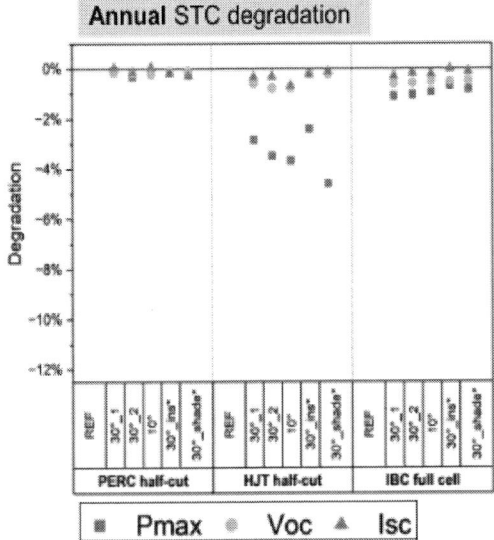

Non-linear degradation is observed due to combination of different failure routes and degradation/regeneration processes

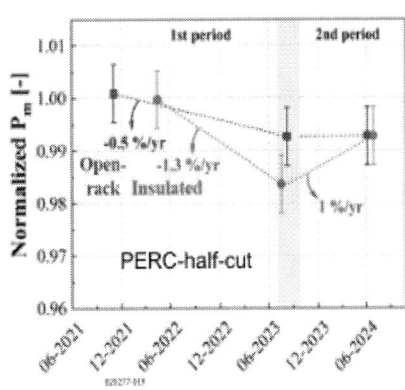

3BO.15.3, Ebrar Özkalay, "Outdoor Accelerated Ageing Test using Additional Thermal and Thermomechanical Stresses"

Summary

→ General

1. PR differences in the range of ±0.85 % for PERC, TOPCON and IBC modules (30° open-rack)

2. Major degradation on HJT modules due to wrong BOM (humidity penetration)

3. Nonlinear degradations are observed with higher rates in the first year

4. Initial stabilization is crucial for accurate technology benchmarkings

5. Degradation / regeneration (e.g LeTid) with significant differences between modules

→ BAPV/BIPV

1. PR differences increases up to ±2.2% for low inclination or insulated modules

2. Operation at higher temperatures does not necessarily increases the annual degradation rates

3. Shading of modules can be critical depending on module & cell technology (e.g. HJT)

Conclusion & Outlook

1. Field performance of new cell technologies are less predictable compared to BSF module.

2. Application specific energy benchmarking/rating is recommended.

3. Degradation rates and in-stabilities needs to be considered in technology benchmarking/rating.

4. Stabilisation procedures has to be reviewed and adapted for the new cell technologies.

5. More long-term data are needed from the field in particular for sub-optimal conditions.

Measurements and analysis will go on with new technologies added the next year

SUPSI

Thank you for your attention!

Acknowledgement:
This work is funded by the **Swiss Federal Office of Energy, BFE.**
SUPSI PVLab and **SUPSI ISAAC Engineering team** researchers and personnel

Schweizerische Eidgenossenschaft
Confédération suisse
Confederazione Svizzera
Confederaziun svizra

Swiss Federal Office of Energy SFOE

Gabi Friesen
gabi.friesen@supsi.ch

SUPSI PVLab
Virtual Tour:

www.pvlab.solar

An Accurate Data-Driven Physical Model
for Bifacial PV Power Estimation

Ali Sohani [a], Marco Pierro [b], David Moser [b], Cristina Cornaro [a]

[a] Department of Enterprise Engineering, University of Rome Tor Vergata, Rome, Italy
[b] Institute for Renewable Energy, Eurac Research, Bolzano, Italy

---

Outline

- Motivation
- Objective
- Bifacial PV power (BFPV) modeling
  - PVlib approach
  - Proposed approach
- Case-studies
  - Fixed-tilt (FT)
  - Single-axis tracking (SAT)
- Accuracy comparison
- Summary of results

eurac research

Motivation

Application of **Bifacial PV** (BFPV) is **increasing** all around the **world**. Several projects have been defined for BFPV development, e.g., European Union Horizon projects TRUST-PV (**power production**) and REGACE projects (**Agri-PV**).

Estimation of power, as the main output of a BFPV system plays an important role in **designing**, **operation and maintenance (O&M)**, **fault detection**, and **performance evaluation**.

eurac research

Objective

Typically, an **effective irradiance** is calculated to obtain power, which **combines both** the **global (front)** tilted irradiance (GTI) and the **rear tilted irradiance** (RTI) into a **single input parameter**.

Here, a **data-driven** model has been developed. The novel proposed data-driven model is able to estimate the **rear side production separately**, which helps to **analyze and diagnose** a BFPV system **better**.

The model could be beneficial in the framework of TRUST-PV (**power production**) and REGACE projects (**Agri-PV**).

BFPV power modeling: PVlib approach

PVlib approach is to **use the monofacial** PV (MFPV) **estimation model** but **inputting** the **effective irradiance** (G_E) [1]:

$$P^{BF(est\ model)_{PVlib}} = P^{MF(est\ model)_{PVlib}}(G_E, T_{amb})$$

Where $G_E = GTI + \varphi\,RTI$ is effective irradiance and φ is the **bifaciality coefficient**. **GTI** and **RTI** are **global tilted irradiance** and **rear tilted irradiance**, respectively.

[1] W.F. Holmgren, C.W. Hansen, M.A. Mikofski. pvlib python: A python package for modeling solar energy systems. Journal of Open Source Software. 3 (2018) 884.

BFPV power modeling: Proposed approach

In **our approach**, BFPV power (P^{BF}) is considered in the form of:

$$P^{BF} = P^{MF} + P_n^{BF,front}\ DBPG$$

P^{MF}: Monofacial PV power
$P_n^{BF,front}$: **Nominal power** of **front side** of bifacial PV

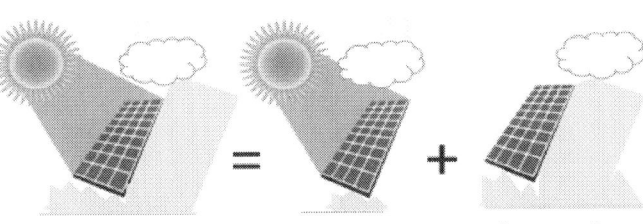

| BFPV power | MFPV power | Dynamic Bifacial power gain (DBPG) |

eurac research

BFPV power modeling: Proposed approach

$$P^{BF} = P^{MF} + P_n^{BF,front} \, DBPG$$

P^{MF} denotes power of **a MFPV** with the **same specifications** at the **same GTI** and **ambient temperature (T_{amb})** as BFPV.

An installed BFPV plant usually **consists of only bifacial modules** and there is **no MFPV** with the same front side specifications there (like the investigated case-studies). Consequently, **simulation** is utilized to **obtain MFPV power**.

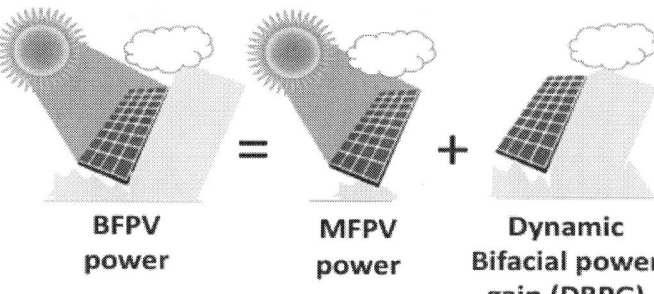

BFPV power **=** **MFPV power** **+** **Dynamic Bifacial power gain (DBPG)**

---

eurac research

BFPV power modeling: Proposed approach

Dynamic Bifacial Power Gain (DBPG):

$$DBPG = \frac{P^{BF} - P^{MF}}{P_n^{BF,front}}$$

In the proposed approach, **DBPG** is **obtained** by **data-driven model** (DDM).

BFPV power **=** **MFPV power** **+** **Dynamic Bifacial power gain (DBPG)**

BFPV power modeling: Proposed approach

Fixed-tilt (FT):

$$DBPG^{(est\ model)}{}_{DDM} = g_{FT}(cos(\boldsymbol{AOI}), \boldsymbol{K_{CS}})$$

Single axis tracking (SAT):

$$DBPG^{(est\ model)}{}_{DDM} = g_{SAT}(cos(\boldsymbol{AOI}), \boldsymbol{K_{CS}}, \boldsymbol{AZ}, \boldsymbol{EL})$$

AOI: Angle of incidence

K_{CS}: Clearness index

AZ: Solar azimuth angle

EL: Solar elevation (altitude) angle

Case-study: FT plant

Technical University of Denmark (**DTU**), Rosklide, **Denmark** [1,2]

Parameter	Value/Condition
Number of PV (rows×PV per row)	4×22
Tilt angle (°)	25
BFPV technology	Passivated emitter and rear contact (PERC)
Nominal front capacity of the used modules ($P_n^{BF,Front}$)	295 W

[1] Riedel N, Berrian D, Alvarez Mira D, Protti AA, Poulsen PB, Libal J, et al. Data used in "Validation of Bifacial Photovoltaic Simulation Software against Monitoring Data from Large-Scale Single-Axis Trackers and Fixed Tilt Systems in Denmark". Technical University of Denmark. Dataset. https://doi.org/10.11583/DTU.13580759.v3. 2021.

[2] Riedel-Lyngskær N, Berrian D, Alvarez Mira D, Aguilar Protti A, Poulsen PB, Libal J, et al. Validation of bifacial photovoltaic simulation software against monitoring data from large-scale single-axis trackers and fixed tilt systems in Denmark. Applied Sciences. 2020;10(23):8487.

eurac research

Case-study: SAT plant

EURAC, Bolzano, **Italy** (Owned and managed by EURAC)

Parameter	Value/Condition
Number of PV (rows×PV per row)	2×12
BFPV technology	Heterojunction technology (HJT)
Nominal front capacity of the used modules ($P_n^{BF,Front}$)	375 W

---

eurac research

Accuracy comparison (FT)

Good accuracy is obtained using **our approach** for FT case:

Normalized root mean square error (**NRMSE**): **1.87%**

Normalized mean bias error (**NMBE**): **0.40%**

Note that since **RTI** data is **not available** for this case for **a large period of time**, effective irradiance could not be obtained in that period. Consequently, **PVlib has not been applied**.

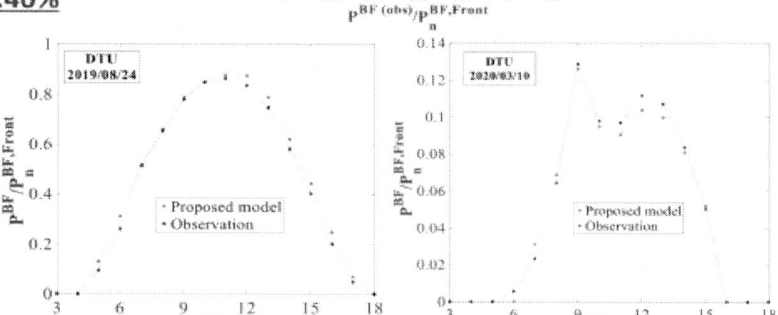

Accuracy comparison (SAT)

For **SAT** case with **black ground**, **high accuracy** is seen like FT plant:

Normalized root mean square error (**NRMSE**):
2.99% ➔ **0.98%**

Normalized mean bias error (**NMBE**):
2.64% ➔ **0.30%**

Considerable improvement has been seen **compared to PVlib** in terms of **both NRMSE** and **NMBE**.

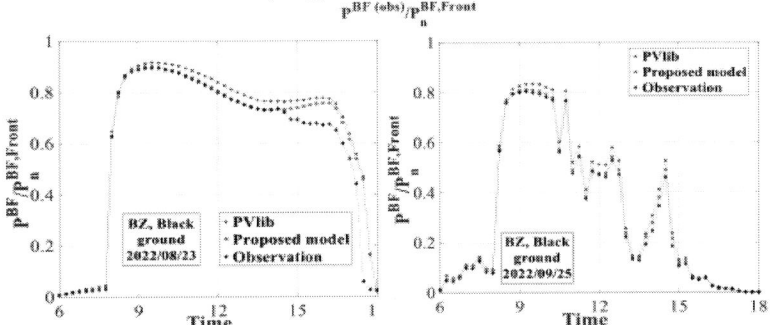

Accuracy comparison (SAT)

The **accuracy** is also high for **SAT** plant with **white ground**:

Normalized root mean square error (**NRMSE**):
4.59% ➔ **1.55%**

Normalized mean bias error (**NMBE**):
3.86% ➔ **0.28%**

The **greater accuracy improvement compared to PVlib** has been achieved for **SAT** plant **white ground than black ground.**

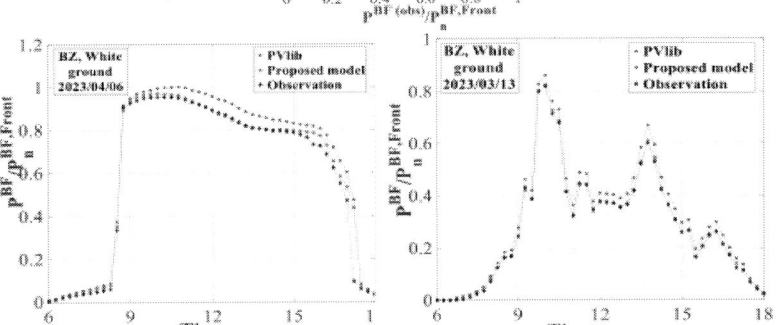

eurac
research

Summary of modeling results

- **Our proposed model** provides **high accuracy** (Good values for both **NRMSE** and **NMBE**).
- **Better accuracy than PVlib** for **SAT** has been observed.
- The **data-driven model does not need albedo data**, which is **not available** in some **real cases**.

Approach	NRMSE (%)			Data requirement	
	FT	SAT (Black ground)	SAT (White ground)	Albedo data	Historical data
PVlib	N.A.	2.99	4.59	✓	X
The proposed data driven model	1.38	0.98	1.55	X	✓

eurac
research

Acknowledgment

Thank you very much for your attention!

Questions/feedback form

ali.sohani@uniroma2.it

41st European Photovoltaic Solar Energy Conference and Exhibition

This presentation was selected by the Sc. Committee of the EU PVSEC 2024 for submission of a full paper to one of the EU PVSEC's collaborating peer-reviewed journals.

COMPARATIVE ANALYSIS OF TEMPERATURE ESTIMATION MODELS IN BIFACIAL PHOTOVOLTAIC MODULES

Aline Kirsten Vidal de Oliveira
Universidade Federal da Santa Catarina
Florianópolis, Brasil

Marília Braga
Universidade Federal da Santa Catarina
Florianópolis, Brasil

Isadora Maciel Queiroz
Universidade Federal da Santa Catarina
Florianópolis, Brasil

Helena Naspolini
Universidade Federal da Santa Catarina
Florianópolis, Brasil

Ricardo Rüther
Universidade Federal da Santa Catarina
Florianópolis, Brasil

ABSTRACT: This study presents a comparative analysis of temperature estimation models for bifacial photovoltaic (PV) modules, focusing on three widely used models: Ross, Faiman, and PVsyst. The research evaluates the performance of these models across different temporal resolutions and assesses the effect of incorporating reflected irradiance, a key consideration for bifacial systems. The findings indicate that the Faiman and PVsyst models, particularly when recalculated using field data, outperform the Ross model in terms of accuracy, especially for bifacial modules. Additionally, the inclusion of reflected irradiance improves temperature prediction for Faiman and PVsyst models, while it negatively impacts the Ross model, which is more suited for monofacial modules. The results also show that PVsyst's default coefficients (Uc=29 W/m²K, Uv=0 W/m³K) lead to significant overestimation of module temperature, suggesting the need for recalculated or alternative coefficients (Uc=40 W/m²K). This study underscores the importance of updating thermal parameters and integrating reflected irradiance to enhance the accuracy of temperature predictions, particularly as bifacial PV technology becomes more prevalent in the solar energy sector.
Keywords: photovoltaics (PV); bifacial PV modules, temperature models, performance simulation.

1 INTRODUCTION

As photovoltaic (PV) generation capacity continues to grow rapidly worldwide, attention has increasingly shifted toward ensuring technological reliability and enhancing the monitoring and forecasting of system performance. Despite the advancements, there is still a limited understanding of how PV systems perform over extended periods in diverse environmental conditions. It's widely acknowledged, however, that factors such as dirt accumulation, light spectral composition, temperature, and humidity levels can have a significant impact on PV performance, with these variables differing greatly depending on the region and the specific PV module technology in use. Moreover, the quality, calibration, and maintenance of instruments used to gather data for performance predictions and levelized cost of energy (LCOE) assessments can significantly affect the accuracy of those projections [1]. A more refined grasp of environmental factors and improvements in data collection protocols could lead to more accurate PV output predictions, minimizing uncertainties. This is especially critical when calculating energy costs, optimizing grid energy dispatch, and enhancing PV reliability.

Among the key determinants of PV system efficiency, cell operating temperature plays a critical role, coming second only to irradiance. Therefore, making accurate estimates of module operating temperatures is essential for projecting energy output. While module temperature is a well-researched topic, the accuracy of various methods used to predict module temperatures—particularly when rear-side temperature measurements are unavailable—has yet to be fully evaluated, especially for newer PV technologies such as bifacial modules.

Several temperature models for PV modules are well-documented in the literature [2], [3], [4]. However, a gap remains in understanding how these models perform with bifacial PV modules, which now make up over 30% of the global PV market and are expected to continue expanding [5]. Thus, investigating the accuracy and dependability of temperature prediction models for bifacial modules is crucial.

This study seeks to assess the precision of three commonly used models for estimating module temperature in bifacial crystalline silicon glass-glass modules. Besides evaluating the models, the research also explores how the temporal resolution of the data and the parameterization of these models impact their accuracy. The study compares measured module temperature data with temperature estimates derived from environmental factors and model coefficients specific to the test location. The aim is to enhance the current models for forecasting energy production and system performance in PV systems with bifacial modules, particularly as this technology sees wider adoption in large-scale solar farms globally.

2 EXPERIMENTAL SETUP

Fig. 1 shows the pilot plant analyzed in this work with a highlight to the subsystem which the results are based on. The study site has a humid subtropical climate according to the Köppen classification (Peel, Finlayson and McMahon, 2007). The PERC type monocrystalline silicon glass-to-glass bifacial modules are from a Tier 1 manufacturer with 645 Wp of individual power. The modules are divided into subsystems installed on single-axis trackers and 26 modules divided into two strings, totaling 16.77 kWp.

Figure 1: Aerial image of the pilot photovoltaic plant located at the headquarters of the Fotovoltaica/UFSC laboratory in Florianópolis-SC. North is pointed approximately to the lower left corner of the image.

Two PT1000 class A thermistors with 4-wire measurement (labeled as T1 and T2 in Fig. 1) were used to record the temperature on the back of the modules. The sensors have been attached to the surface of the PV module with a double-sided adhesive tape provided by the sensor manufacturer and are covered by an aluminized adhesive tape (Fig. 2a). Measures have been taken in order to reduce shading on the back of the module as much as possible, in accordance with the technical standard IEC 61724:2021 (International Electrotechnical Commission, 2017). Thermal imaging was performed to ensure that no hot spots were formed due to the installation of the sensor.

An EE08-SS sensor monitored ambient temperature, and a Gill WindSonic anemometer recorded wind speed, (measured at 2 m and PV systems at 1.5 m from the ground). Sensors installed along the trackers measure the incident front and back irradiance in the plane of the modules using reference cells (IMT Si) and class A pyranometers (EKO MS-80S), shown in Fig. 2b and Fig. 2c. Data were sampled every second and averaged over one minute. Data collection took place from October 2022 to October 2023.

The temperature models (Ross, 1981; Faiman, 2008; PVsyst, 2023) were implemented using the PVLib library in Python (Holmgren, Hansen, and Mikofski, 2018). It is worth noting that these models were not originally developed for bifacial modules. To assess their compatibility with the new technology, the calculations were performed in two different scenarios, for comparison purposes:

1. $G = G_{POA}$: the irradiance data used (G) are those measured in the plane of the module (G_{POA}) (Fig. 2c);
2. $G = G_{POA} + b * G_{Ref}$: the irradiance data used (G) are those measured in the module plane plus the reflected irradiance (G_{ref}) (Fig. 2b) multiplied by the module bifaciality factor (b), provided by the manufacturer in its *datasheet* as being 70 %.

3 TEMPERATURE MODELS

Below is a detailed description of the PV module temperature estimation models used.

3.1 Ross Model
The Ross Model [2], which represents a more traditional approach, was applied using the following Eq. (1).

$$T_R = T_a + \left[\frac{NOCT - 20}{80}\right] \times G \qquad (1)$$

In the equation:
- T_c is the cell temperature in °C;
- T_a is the ambient temperature in °C;
- NOCT refers to the Nominal Operating Cell Temperature in °C (in this study, a temperature of 37.6°C was used);
- G represents the irradiance incident on the module plane, measured in mW/cm².

3.2 Faiman Model
The **Faiman Model** is based on an empirical heat loss factor model [3] and is adopted in IEC performance standards. The corresponding Eq. (2) is outlined below.

$$T_F = T_a + \left[\frac{G}{U_0 + U_1 v}\right] \qquad (2)$$

Where:
- T_F is the module temperature in °C;
- T_a is the ambient temperature in °C;
- v is the wind velocity in m/s;
- G represents the irradiance incident on the module plane, measured in W/m²;
- U_0 is the coefficient for heat dissipation from the module, measured in W/m²K;
- U_1 accounts for the effect of wind, expressed in Ws/m³K.

U_0 and U_1 are coefficients that depend on the type of system installation and are determined through linear regression using measured data on irradiance, ambient temperature, and module temperature. This procedure is followed in this study, with the coefficients being calculated for each variation in the time interval.

3.3 PVSyst Model
The cell temperature model used in the PVsyst software [6] has a structure similar to the Faiman model, but it includes additional parameters for efficiency and absorption, as defined in Eq. (3).

$$T_P = T_a + \left[\frac{\alpha G(1 - \eta_m)}{U_c + U_v v}\right] \qquad (3)$$

Where:
- T_P is the module temperature in °C;
- T_a is the ambient temperature in °C;
- v is the wind velocity in m/s;
- G represents the irradiance incident on the module plane, measured in W/m²;
- U_c is the constant coefficient for module heat loss in W/m²K;
- U_v accounts for the impact of wind on heat loss, measured in Ws/m³K;
- α is the absorption coefficient for solar irradiance, with a standard value of 0.9;
- η_m is the photovoltaic module efficiency. In this study, the efficiency used is the one provided by the manufacturer on the datasheet for STC, without bifacial gain (standard), which is 20.8%.

Just like in the case of Faiman, the coefficients U_c and U_v can be determined through linear regression using measured data on irradiance, ambient temperature, and module temperature. By default, the software sets Uv to zero due to the extremely limited availability of high-frequency, high-quality wind speed data. Moreover, this parameter is typically measured at higher altitudes and in open environments, which are not representative of the operating conditions for a photovoltaic system. In this study, various coefficients were evaluated for comparison:

- U_c and U_v were calculated using linear regression from the measured data. In this instance, the temperatures obtained are the same as those in the Faiman model [7];
- U_v=0 and U_c calculated through linear regression of the measured data to assess the effect of wind speed on the software's temperature estimation;
- U_c=29 and U_v=0, values recommended by PVSyst for systems mounted on ground-based

structures;

- U_c=40 and U_v=0, values suggested by manufacturers for simulating glass-glass bifacial modules.

4 RESULTS

To assess module temperature estimation models across various data granularities, the methods were tested with different temporal resolutions, as reflected in Table 1's Mean Absolute Error (MAE) results. Faiman and PVsyst models had identical temperature estimates, obtained through linear regression, and are thus presented together. Another comparison presented refers to the effect of the inclusion of reflected irradiance as input in the temperature prediction models, performed through the bifaciality coefficient.

Table 1: MAE results for the temperature models for 15, 30 and 60-minute data, evaluating the effect of including irradiance reflected by the soil as input to the models. The best results of each model are highlighted in bold.

MAE (°C)	Ross			Faiman/PVsyst		
	15 min	30 min	60 min	15 min	30 min	60 min
$G = G_{POA}$	2,06	2,00	**1,94**	2,02	1,96	**1,91**
$G = G_{POA} + b * G_{Ref}$	2,10	2,03	**1,97**	1,96	1,90	**1,85**

The analysis shows that the models perform better for lower temporal resolutions, which is expected due to their steady-state nature. However, the increase in the temporal resolution did not significantly impact the errors obtained, especially for the method used in simulations of PV systems. This comparison, however, was performed by calculating the heat transfer coefficients through linear regression for the same temporal resolution and considering the wind speed, which are not the standard conditions used in PVsyst. The Faiman model showed the smallest errors among the models, for all resolutions.

In the case of the Ross model, the use of irradiance reflected by the soil had a negative impact on the results, mainly because parameters used in the calculation were determined for old, monofacial modules. In contrast, the Faiman and PVsyst models benefit from recalculating heat transfer coefficients using total irradiance measurements, adapting to new datasets. Including reflected irradiance in thermal models, especially in PVsyst software, enhances performance projections for photovoltaic systems. It's important to note that field-measured data were used for reflected irradiance, while software typically estimates these values. The quality of the irradiance data modeled by PVsyst is not the object of this study, but is directly related to the accuracy of the temperature estimation method using this variable in the *software*.

The predictions of the analyzed models were calculated using different scenarios of coefficients, as well as standard values commonly used in PVsyst analyses. The results are shown in Tab. 2. Results indicate that while wind speed influences temperature estimation, the resulting errors are negligible for practical purposes. Also, the high-quality data used in the analysis is uncommon in PV projects. However, employing standard PVSyst coefficients (U_c=29 and U_v=0) leads to significant errors in temperature estimates for bifacial modules, affecting energy simulation accuracy.

Table 2: Results of MAE and RMSE for the Faiman and PVsyst temperature models comparing different strategies for the calculation of temperature coefficients.

MAE(°C)	Faiman/ PVsyst - Regression		PVSyst default values	
	Calculated U0 and U1	Calculated U0 and U1=0	Uc=29 Uv=0	Uc=40 Uv=0
$G = G_{POA}$	1,91	1,94	2,12	2,08
$G = G_{POA} + b * G_{Ref}$	1,85	1,88	2,30	1,92

Therefore, it is not recommended to use these coefficients for bifacial module simulation. A better approximation is achieved with U_c=40 and U_v=0, eliminating the need for linear regression with measured data. Another important conclusion is the recommendation to account for reflected irradiance for temperature estimation, which offered better results in almost all alternatives evaluated.

Fig. 2 shows a scatter plot of the data measured (in black) and estimated by the models evaluated (colored lines) considering total irradiance data at the input of the models, including the irradiance reflected by the soil. The use of standard PVsyst coefficients significantly overestimates module temperature (cyan line), while using alternative coefficients (U_c=40 W/m²K and U_v=0 W/m³K) results in lower, more accurate temperatures. The Ross and Faiman/PVsyst models, with coefficients derived from measured data, yield results close to the desired outcome, with Ross tending to slightly overestimate and Faiman/PVsyst tending to underestimate. Despite the Ross model's small overestimation, it proves more practical, not requiring wind speed data.

Fig. 3 reinforces these conclusions, comparing temperatures of measured and modeled modules with different input parameters from 08/26/2023. The high values of temperature estimated for the PVsyst model with U_c=29 W/m²K are justified by the disregard of the effect of wind on the reduction of temperatures and by the small value for the parameter Uc, which is responsible for accounting for the heat loss of the module. Thus, a small value will generate less heat loss and therefore an overestimation of the temperature. The Faiman/PVsyst model, when excluding wind speed impact (green), results in higher temperatures. The thermal parameter calculated for PVsyst without considering wind (U_c=38.47) closely aligns with U_c=40 in the graph, demonstrating their similarity.

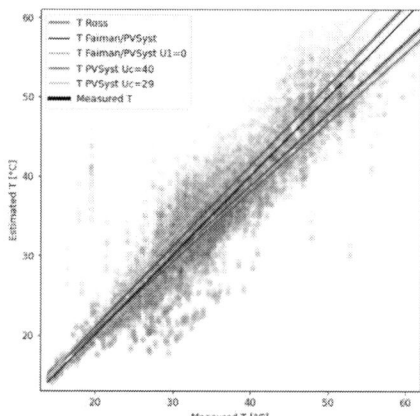

Figure 2: Scatter plot and regression lines for comparison between measured and estimated data.

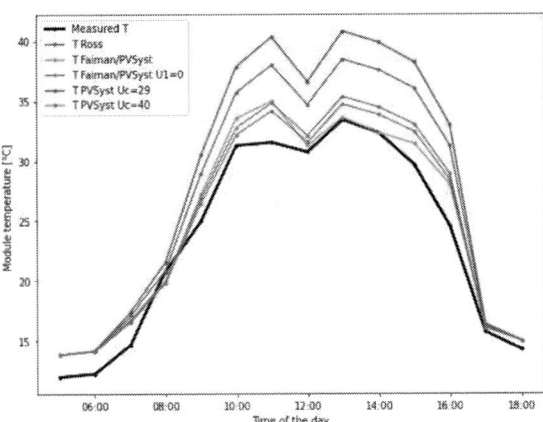

Figure 3: Comparison between measured and estimated data through the models and coefficients analyzed for 08/26/2023.

5 CONCLUSIONS

The comparative analysis of module temperature estimation models reveals that lower temporal resolution enhances accuracy across the models, as expected due to their steady-state nature. However, the impact of increasing temporal resolution on error was minimal. The Faiman/PVsyst models consistently produced the lowest Mean Absolute Error (MAE), especially when recalculating heat transfer coefficients based on field data, demonstrating better adaptability to current bifacial module datasets compared to the Ross model.

The inclusion of irradiance reflected by the soil improved model performance for the Faiman/PVsyst models, highlighting its relevance in simulations, particularly for bifacial modules. Meanwhile, the Ross model was negatively affected by reflected irradiance, likely due to outdated parameters suited to monofacial modules. This reinforces the necessity of updating thermal parameters for newer bifacial technologies.

A critical finding is that using PVsyst's default thermal coefficients (Uc=29 W/m²K, Uv=0 W/m³K) leads to significant temperature overestimation, especially for bifacial modules, underscoring the importance of recalculating these coefficients or employing alternative values such as Uc=40 W/m²K. Moreover, the negligible influence of wind speed on practical temperature estimation accuracy simplifies modeling efforts, particularly for large-scale simulations.

In summary, incorporating reflected irradiance and recalculating heat transfer coefficients for bifacial modules improve temperature prediction accuracy. Standard PVsyst coefficients should be reconsidered for bifacial modules, as more tailored approaches yield significantly better results. These findings support the need for accurate thermal modeling to enhance energy yield projections in modern photovoltaic systems.

6 AKNOWLEDGEMENTS

The authors would like to thank CTG Brasil for the financial support to this study through the PD-10381-0620/2020 project, carried out in partnership with SENAI-RN, ISI-ER and UNESP-Ilha Solteira, within the scope of the Research and Development (R&D) program of the Brazilian National Electric Energy Agency (ANEEL). The authors also thank the colleagues from the Fotovoltaica-UFSC laboratory who did not participate directly in this research but helped in the installation and maintenance of the systems and sensors used in the study, and for the fruitful discussions on the topics addressed in this article. In particular, the authors would also like to thank the colleagues Thamires Alves da Silva and Lessandro Formagini, who assisted in the collection and processing of the data used. Marília Braga is also grateful for the support received from the Coordination for the Improvement of Higher Education Personnel – Brazil (CAPES) through her doctoral scholarship. This work was conducted during a scholarship supported by the International Cooperation Program PROBRAL at the University of Santa Catarina. It was financed by Capes – Brazilian Federal Agency for Support and Evaluation of Graduate Education within the Ministry of Education of Brazil and supported by German Academic Exchange Service (DAAD).

7 REFERENCES

[1] J. S. Stein e B. H. King, "Modelling for PV plant optimization", *Photovoltaics International*, vol. 19th, nº February, p. 101–109, 2013.

[2] Ross, R. G. Jr., "Design Techniques for Flat-Plate Photovoltaic Arrays", em *15th IEEE Photovoltaic Specialist Conference*, Orlando, FL., 1981.

[3] D. Faiman, "Assessing the outdoor operating temperature of photovoltaic modules", *Progress in Photovoltaics: Research and Applications*, vol. 16, nº 4, p. 307–315, jun. 2008, doi: 10.1002/pip.813.

[4] P. Mora Segado, J. Carretero, e M. Sidrach-de-Cardona, "Models to predict the operating temperature of different photovoltaic modules in outdoor conditions", *Progress in Photovoltaics*, vol. 23, nº 10, p. 1267–1282, out. 2015, doi: 10.1002/pip.2549.

[5] VDMA, "International Technology Roadmap for Photovoltaics (ITRPV) - 2023 Results", 15. Edition, May 2024, 15. Edition, maio 2024. Acesso em: 3 de julho de 2024. [Online]. Disponível em: https://itrpv.vdma.org/en/

[6] PVsyst, "Array thermal losses". Acesso em: 30 de novembro de 2023. [Online]. Disponível em: https://www.pvsyst.com/help/thermal_loss.htm

[7] K. Anderson, J. Kemnitz, e M. Boyd, "Evaluating cell temperature models and the effect of wind speed in PV system capacity testing", em *2021 IEEE 48th Photovoltaic Specialists Conference (PVSC)*, Fort Lauderdale, FL, USA: IEEE, jun. 2021, p. 1663–1669. doi: 10.1109/PVSC43889.2021.9519077.

41st European Photovoltaic Solar Energy Conference and Exhibition

This presentation was selected by the Sc. Committee of the EU PVSEC 2024 for submission of a full paper to one of the EU PVSEC's collaborating peer-reviewed journals.

APPARENT INTENSITY DEPENDENCE OF SHUNTS IN PV MODULES
REVISION OF THE SHUNT PARAMETERIZATION IN THE DE SOTO MODEL AND PVSYST

Nils-Peter Harder and José Cano Garcia
TotalEnergies
7-9 Boulevard Thomas Gobert, 91120 Palaiseau, France - Nils.Harder@TotalEnergies.com

ABSTRACT: It is common practice in PV system simulation to use the De Soto model, which describes how to use the 1-diode equivalent circuit model for modules. De Soto's model scales the shunt with irradiance, making it disappear towards zero W/m². Also, the commercial software PVsyst uses a parameterization that reduces the shunt effect when the irradiance goes down. However, the solar cells that make up a module typically do not have an illumination dependent shunt. We therefore investigate the origin of the intensity dependent apparent shunt in modules. We show that this apparent shunt (derived from the slope of the quasi-linear region from I_{SC} ònwards) is a misinterpretation for module I-V curves and has little to do with a shunt conductance, although this slope method serves well for determining the shunt conductance of individual cells. Instead, the module I-V curve slope of the quasi-linear region from I_{SC} onwards is strongly influenced by even small I_{SC} mismatches between the cells. Such mismatch can occur from small illumination inhomogeneity even for A+ solar simulators in the laboratory, or from:cell production variation. Abandoning the practice of using the I-V curve slope to determine the shunt value for equivalent circuit models of modules (and the corresponding shunt scaling in the De Soto model or PVsyst), contributes to physically more meaningful I-V curve parameterizations and bears the opportunity for further improved accuracy of PV system energy yield prediction.
Keywords: Module performance, I-V curve parameterization, Shunt characterization.

1 INTRODUCTION

For single solar cells, the slope at I_{SC} [or rather: the slope between I_{SC} and some mid-sized voltage well before the maximum power point (MPP) where the recombination current is still very small] represents the shunt conductance $\Delta I / \Delta V = 1 / R_{Sh}$. For single cells, this shunt value can be used for equivalent circuit model representations of such single cells, for example in a 1-diode or 2-diode model. We call this quantity $R_{Sh.Slope}$, as it is derived from the I-V curve slope between I_{SC} and some mid-sized voltages before the exponentially increasing recombination current becomes notable. By applying this "slope analysis" to modules instead of single cells, De Soto et al. [1] and Mermoud et al. [2], for example, observed that module $I\neg V$ curves have an intensity-dependent $R_{Sh.Slope}$. De Soto et al. and Mermoud et al. also provided parameterizations of this slope as a function of illumination intensity, and interpreted this slope as the shunt conductance, to be used for their 1-diode model parameterization of modules. De Soto parameterizes the shunt conductance $1 / R_{Sh}$ as being directly proportional to the irradiance. Mermoud's shunt parameterization, which is used in the commercial PV system simulation software PVsyst, describes the module shunt resistance R_{Sh} as decreasing exponentially with irradiance, starting from a finite value at zero irradiance and approaching asymptotically the shunt of the light I-V curve measured at the reference standard test conditions (STC, irradiance = 1000 W/m²). However, an intensity dependent shunt in modules is surprising, or possibly even implausible, because on cell level there is typically no indication that an equivalent-circuit model would need an intensity-dependent shunt for describing the maximum power point (MPP) of a solar cell, not even for describing the MPP for a very wide range of irradiance values as reported by Bunea *et al.* [3] for monocrystalline silicon cells. Grunow *et al.* reported multicrystalline cell efficiency results for different irradiances, measured between 100 W/m² and 1000 W/m² [4], and also Grunow *et al.* described the cell performance with simulations models that do not have an

intensity dependence of the shunt.

(Note that Robinson [5] discovered an often-ignored effect, later explained by Breitenstein [6], that is fundamental for *pn*-junction solar cells. It looks like an intensity-dependent shunt conductance in the light I-V curve that we call "photoshunt". Although academically interesting, it is a very small effect that vanishes well before the MPP. It affects in no conditions known to us the MPP. It is therefore of no relevance for an equivalent circuit model that seeks to describe the power production of a cell or module, which is probably also why the photoshunt is largely ignored in literature. It is nevertheless fair to say that to some small degree Robinson's photoshunt will have some contribution to the appearance of an illumination intensity dependence of the module I-V curve slope in the quasi-linear region from I_{SC} onwards. However, we deem this contribution to be marginal in practice for PV modules.)

We show in this paper the origin of the intensity dependence of $R_{Sh.Slope}$. For modules, this slope-derived $R_{Sh.Slope}$ is not related to an actual shunt conductance, and instead it is the result of I_{SC}-mismatches between the cells, combined with their reverse-bias characteristics. Note that such I_{SC} mismatch might stem from an illumination inhomogeneity in a module measurement [7], from other optical effects such as different white spaces around the cells (e.g. module edge versus center), from variations in cell manufacturing, or a combination of those effects.

Importantly, and as has been discussed already in our recent module characterization study [8], using $R_{Sh.Slope}$ as shunt resistance in an equivalent circuit model for the I-V curve of modules leads to physical inconsistencies: $R_{Sh.Slope}$ typically has a rather low value, but (as this present paper will show in detail) does not represent an actual shunt conductance. Using this value $R_{Sh.Slope}$ nevertheless for the shunt in equivalent circuit models (e.g. a 1-diode model) and trying to reproduce V_{OC} and MPP values, leads to needing to use a too low ideality factor value for the module I-V curve. In our recent characterization study of a TOPCon module [8] we found exactly such phenomenon:

1033

PAN file specified low shunt values that correspond to the measured $R_{Sh.slope}$, in combination with an implausibly low ideality factor in the PAN file. Note that a wrong (too low) ideality factor in turn, affects the temperature-dependency of the modeled module performance. As a consequence, the simulation software PVsyst introduces without physical basis a temperature dependency for the ideality factor (by a temperature coefficient "muGamma" [9]).

It shall be pointed out that a simple model, that neither uses the complication of an intensity-dependent shunt nor an unphysical feature such as a temperature dependent ideality factor, is not only advantageous by virtue of its simplicity and better physical interpretability. As is shown here in Fig. 1, taken from reference [8], such simpler model also leads to accurate (and also more accurate than modeling with intensity-dependent shunt) module performance parameterization over a wide range of illumination intensities and temperatures. (Flash tester certified illumination inhomogeneity of +/- 0.4%.)

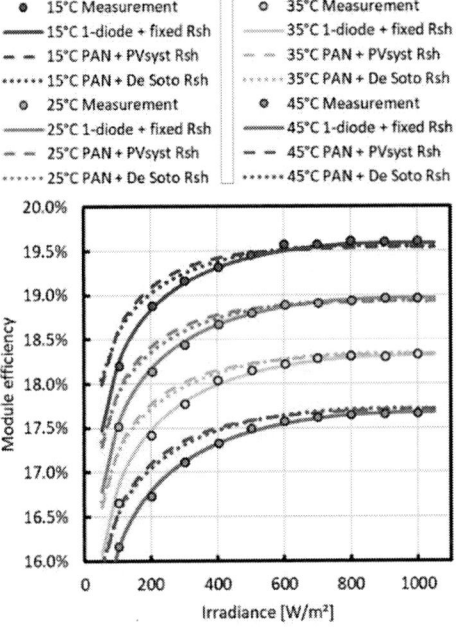

Figure 1: Measured efficiencies of a TOPCon module (symbols) and calculated efficiencies by the 1-diode model (solid lines), with a fixed shunt R_{Sh}, and using a simple series resistance R_S, independent of temperature and intensity, from reference [8]. The dashed and dotted lines are calculated efficiencies based on PVsyst's [2] (dashed lines), or De Soto's [1] (dotted lines) approach, which both use an intensity dependent shunt.

We set out in this paper to explore the origin of the intensity dependent *I-V* curve slope from I_{SC} onwards (i.e. the intensity dependent $R_{Sh.Slope}$) even when measuring new modules in the laboratory. We pay attention to these conditions as these are typically the case when extracting *I-V* curve parameters for the generation of PAN files that define the module properties of PV systems and therefore can have notable economic impact. (In fielded modules, be it from partial shading or degradation effects, much more severe distortions of the *I-V* curve will happen than what we describe here in this paper.)

For the purpose of this paper, that is describing the origin of an intensity-dependent apparent shunt $R_{Sh.slope}$ in

modules, we use a simple 1-diode model for the individual cells, connected in series to a 60-cell module. Note that the 1-Diode model is a highly simplifying cell model that does not claim to represent all aspects of reality accurately. For example, it is known that an analysis of solar cells in detail shows that shunt leakage currents generally are not linear, that defects in the material can cause effects requiring a voltage-dependent ideality factor, and also metal corrosion effects can cause *I-V* curves that are only very crudely approximated by the 1-Diode model. A non-exhaustive list of papers on these topics are references [10, 11, 12, 13, 14]. Our use of the simplified 1-Diode model does not negate these more complex effects. Instead, the 1-Diode model is a simple proxy for the general property that also those more detailed solar cell descriptions share with the 1-Diode model: The light *I-V* curve of the solar cell can be represented with very good approximation as the I_{SC}-shifted dark *I-V* curve (plus series resistance effects), which implies that the *I-V* curve slopes and hence $R_{Sh.Slope}$ are independent of illumination intensity. While the paper has been written with particular attention to Si technologies, the result that cell I_{SC} mismatch can create and increase an intensity dependent slope in the quasi-linear region of the module *I-V* curve $R_{Sh.Slope}$) seem rather general and applicable for many PV technologies.

In this paper, we will refer to $R_{Sh.Slope}$ as "apparent shunt", if this slope is not due a shunt in the literal and physical meaning of the word: A conductance path for electrical current in parallel to the (rest of the) device. The simple 1-diode model obviously (and in our approach: deliberately) does not contain the effects of the apparent shunt discovered by Robinson [5] and explained by Breitenstein [6]. By focusing on a simple model (such as the 1-Diode model), we can highlight in a very pointed manner the main origin of the intensity-dependent $R_{Sh.Slope}$.

2 MODEL DESCRIPTION

We represent individual solar cells in the module circuitry by the "1-Diode model" circuit model in Fig. 2. The equation of its *I-V* curve is given in equation 1:

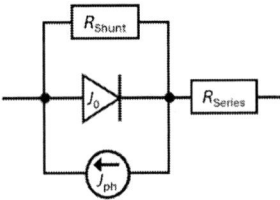

Figure 2: 1-Diode model used in this paper for representing one individual cell in the module circuit.

$$
\begin{aligned}
I[V] = \quad & I_{PH}[\Phi] \\
& - I_0 \times \left(Exp\left[\frac{q(V + I[V] \times R_S)}{n \times kT} \right] - 1 \right) \\
& - \frac{V + I[V] \times R_S}{R_{Sh}}
\end{aligned}
\tag{1}
$$

$I_{PH}(\phi)$ is the photogeneration of current in the solar cell, which is directly proportional to the illumination intensity ϕ. In the typical parameter range of well-working silicon solar cells, $I_{PH}(\phi)$ is numerically almost identical to the short-circuit current I_{SC} at illumination intensity ϕ. I_0 is the saturation current density, R_S is the series resistance, R_{Sh} the shunt resistance, V is the voltage at the terminals of the solar cell, q the elemental charge, kT is the product of the Boltzmann constant k and temperature T. The ideality factor is the variable n. Except for $I_{PH}(\phi)$, all variables are independent of the illumination intensity ϕ. The illumination intensity (or irradiance) ϕ is typically given in W/m², where standard test condition (STC) corresponds to 1000 W/m². We will refer to this standard intensity in this paper as "1 Sun", and 0.5 Sun or 0.2 Sun refer therefore to 500 W/m² and 200 W/m², respectively.

Unless otherwise stated, the standard parameter set for the area-specific values of the parameters in equation 1 are those listed in Table I.

Table I: Standard set of parameter values of the 1-Diode parameterization used in this study

$I_{PH}[\phi]$	R_S	R_{Sh}	I_0	n
40 mA/cm² per Sun	0.5 Ωcm²	10 kΩcm²	58.7 fA/cm²	1

Figure 3 shows a schematic of the module structure in this study. The bypass diodes are modeled with an ideality factor $n = 1$, letting our standard cell I_{SC} pass at 0.2 V.

For our study of the effect of I_{SC} mismatches between the individual cells of the module, we make use of an illumination inhomogeneity map published by Ramspeck [7], describing the cetisPV-Moduletest4 system of halm elektronik GmbH. The 270 cm x 160 cm illumination area of this module tester has an inhomogeneity of +/- 0.4%, defined by (Max − Min)/(Max + Min)). Fig. 4 shows a center section of 6 x 10 cell positions of the illumination intensity map published by Ramspeck. In case of completely identical cells in the module, the illumination intensity map in Fig. 4 translates directly into an identical I_{PH} photogeneration map. It therefore describes the best expectable I_{SC} inhomogeneity distribution in the lab, limited only by the illumination inhomogeneity of a state-of-the-art module flash tester.

We will also explore the effect of larger inhomogeneities, by using I_{PH} maps that are scaled from the +/- 0.4% inhomogeneous map of Fig. 4 to inhomogeneities of +/- 1.0% and +/- 2.0%. These larger inhomogeneities can be understood as either representing measurements with a less ideal module flash tester, or representing I_{PH} inhomogeneity distributions originating from a combination of cell production quality variation in combination with an illumination area inhomogeneity. Note that according to iTeh standard IEC 60904-9, the solar simulator classification "A+" allows for an illumination inhomogeneity of up to +/- 1%, and classification "A" allows up to +/- 2%.

Note that our inhomogeneity map consists of only seven different values. Our cases of increased inhomogeneity only spread these seven values over a larger minimum-to-maximum range. Fig. 4 shows that the average values, and also the minimum and maximum values of the three strings, are very similar. It can therefore be expected that the bypass diodes will have only a minor effect on the I-V curves, which is also what we found in our simulations.

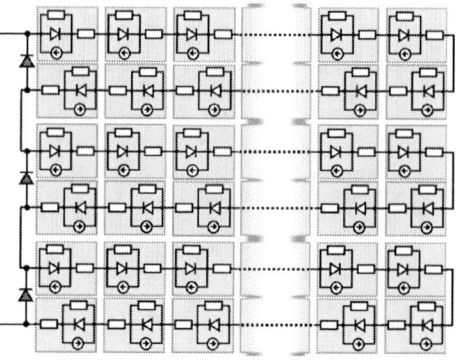

Figure 3: Equivalent circuit for representing a 60-cell module. Each individual cell in this circuit is represented by the 1-Diode model shown in Fig. 2. The diode symbols in grey on the left represent bypass diodes.

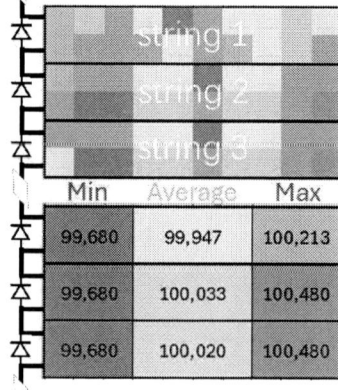

Figure 4: Inhomogeneity map [in "% of average"] of the photogeneration J_{PH} used for the simulation of modules with I_{SC}-mismatched cells. The 6 x 10 distribution in the upper schematic is taken from a subset of the intensity data map of the paper of Ramspeck [7]. The lower schematic lists minimum, average and maximum photogeneration in the cells of the three different cell strings in the module, and it additionally serves as color scale for upper schematic with the +/- 0.4% inhomogeneity case shown here.

However, in an inhomogeneity scenario, individual cells will be forced into reverse bias along the lower voltage range of the I-V curves, also in the presence of bypass diodes. We therefore have to consider the reverse-bias characteristics of the individual cells when exploring the effects of I_{PH} inhomogeneity.

Clement *et al.* published reverse-bias characteristics for different solar cell technologies [15] and identified fundamentally different types. Their IBC solar cell (homojunction) has the most conductive reverse characteristics. The red curve in Fig. 5 shows its reverse characteristics (case 1, labelled "IBC homojunction"), a digitized version of the curve published by Clement. Case 3, the black curve in Fig. 5, can be regarded as the simplest case, as it is using for reverse characteristics the same R_{Shunt} resistance as used in the forward direction. Hence case 3 corresponds to using equation 1 in forward and reverse direction. The yellow curve in Fig. 5, i.e. case 4, uses in reverse direction a three-fold higher shunt resistance than in forward direction.

41st European Photovoltaic Solar Energy Conference and Exhibition

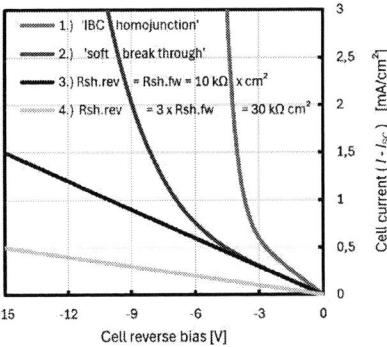

Figure 5: Reverse-characteristics of cell I-V curves used in the simulations in this paper for modules with I_{SC}-mismatched cells.

Similar to case 3, the blue curve, case 2, also uses the same R_{Shunt} in forward and reverse characteristics, but combined with an added reverse bias break-through that is "softened" compared to the "IBC homojunction" of case 1. Note that neither cases 3 nor 4 (black and yellow curve in Fig. 5) feature a reverse bias break-through and have only a linear reverse-bias characteristics. While real solar cells do have a break-through at some reverse-bias voltage, these purely linear reverse bias cases can be regarded as representing cells with a very high break-through voltage.

3 RESULTS

In this section we explore how an intensity dependence of the apparent shunt can arise in the I-V curve of a module that is constructed out of cells, where each individual cell does not feature an illumination intensity dependent shunt. The "intensity dependent shunt" shall be defined here as an illumination intensity dependence of the slope of the I-V curve between I_{SC} and some mid-sized voltage This slope represents for single cell appropriately the shunt conductance, but we will show that for modules this slope is affected by other mechanisms. As an experimental example, Fig. 6 shows the measured intensity dependence of the (apparent) shunt $R_{Sh.Slope}$ of a commercial silicon TOPCon cell module, taken from reference [8]. Additionally, Fig. 6 shows the R_{Sh} parameterizations according to De Soto and Mermoud (PVsyst) as solid and dashed lines, respectively.

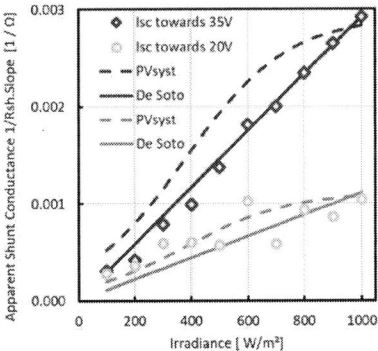

Figure 6: $\Delta I / \Delta V$ Slopes ($1/R_{Sh}$) of I-V curves (taken from Ref. [7]) measured at $T_{module} = 25°C$, compared to parameterizations by De Soto [1] and Mermoud [2]., using PVsyst's recommended parameters for silicon modules.

3.1 I_{SC}-matched cells

In this section we analyze whether I_{SC}-matched cells can lead to an illumination intensity-dependent shunt behavior of the I-V curve. For this scenario, we only have to consider the parameters parameters I_0, n, R_{Sh} and R_S, because I_{PH} is virtually identical to I_{SC} for all practical scenarios that we need to consider here. Variations of the series resistance R_S amongst the cells can be lumped together into one big effective resistance, which does not depend on the distribution of individual cell R_S.

It is therefore of bigger interest to examine what effect have distributions of I_0 values, ideality factor values n, and distributions of shunt values R_{Sh}. In a voltage range where the exponential part of equation 1 dominates the current for all cells, we can approximate equation 1 as:

$$V = n \frac{kT}{q} \times Ln\left[\frac{I(V) - I_{PH}}{I_0}\right] - I(V) \times R_S \qquad (4)$$

$I(V)$ is identical for all cells, and so is I_{PH}, because we considered only the case of I_{SC}-matched cells. It therefore follows for the module voltage V_M:

$$V_M =$$
$$\frac{kT}{q}\left(Ln[I(V) - I_{PH}]\sum_{i=1}^{N} n_i - \sum_{i=1}^{N} n_i \, Ln[I_{0,i}]\right) \qquad (5)$$
$$-I(V) \times R_L$$

…where N is the number of cells in series, R_L is the sum of all cell series resistances $R_{S,i}$, and the index "i" in equation 5 refers to the i-th cell. With the definition of n_{av} as the arithmetic average of all ideality factors n_i, we find:

$$q\frac{V_M + I(V) \times R_L}{kT \; n_{av} \; N} = Ln\left[\frac{I(V) - I_{PH}}{\prod_{i=1}^{N} I_{0,i}^{\frac{n_i}{n_{av} N}}}\right] \qquad (6)$$

…which is essentially the 1-Diode equation for a module without shunt, as can be seen by rewriting it as:

$$I(V) =$$
$$I_{PH} - I_{0,M}\left(Exp\left[\frac{q(V_M + I(V) \times R_L)}{N \times n_{av} \times kT}\right] - 1\right) \qquad (7)$$

…where $I_{0,M} = \prod_{i=1}^{N} I_{0,i}^{\frac{n_i}{n_{av} N}}$, and we added the "-1" of equation 1 that we dropped in equations 4 to 6. We conclude that no mix of ideality factors n_i and saturation currents $I_{0,i}$ will create the emergence of a term that appears like a shunt, nor as an intensity dependent shunt.

Similarly one can show for the voltage range where the exponential part of equation 1 does not dominate for any of the cells (and thus only their linear shunt terms dominate the current) that the I-V curve is described by a shunt term that stems from the sum of all individual $R_{Sh,i}$:

$$I(V) = I_{PH} - \frac{V_M + I(V) \times R_L}{\sum_{i=1}^{N} R_{Sh,i}} \qquad (8)$$

For I_{SC} matched cells it remains to consider cases were the with voltage ranges where the exponential of some cells <u>and</u> the linear shunt term of other cells of equation 1 are relevant for the module current. To verify whether such combination can lead to I-V curve shapes that could be interpreted as intensity dependent shunts, we plot in Fig. 7 the I-V curves of modules with different shunt value distributions across the cells in the module.

41st European Photovoltaic Solar Energy Conference and Exhibition

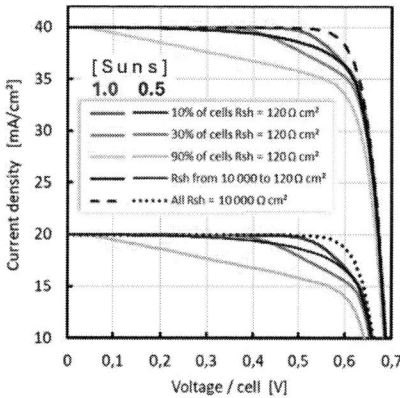

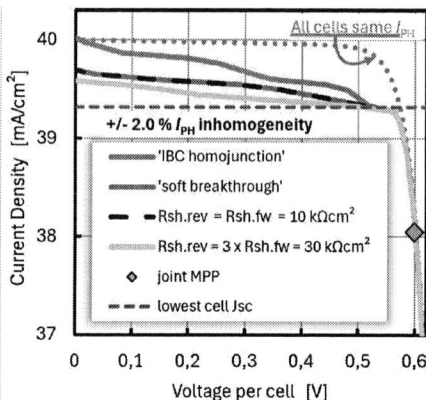

Figure 7: Module *I-V* curves with I_{SC}-matched cells at 1 Sun (upper curves) and 0.5 Sun (lower curves), assuming different R_{Shunt} distribution amongst the cells.

The colored curves in Fig. 7 represent module circuits where different amounts of the cells in the module have a notably lower shunt resistance of 120 Ωcm² than the rest of the cells with a 10 kΩcm² shunt of the standard set of 1-Diode cell model parameters in Table I. The dashed lines represent modules with identical cells, each of them described by the parameter set of Table I. The solid black lines represent a linear variation of the of shunt conductance $1/R_{Sh}$ amongst the cells across the modules.

Fig. 7 shows that inhomogeneous shunt distributions across the cells in the module circuit all produce *I-V* curve shapes that clearly cannot be described by a simple shunt model: No single value of a voltage-independent effective shunt can produce such *I-V* curve shapes. However, apart from a minor series resistance effect, the *slopes* of the curves are identical when comparing 1 Sun to 0.5 Sun curves. Note that *I-V* curve slopes that do not change between 0.5 and 1 Sun underline that in these curves, e.g. unlike in the De Soto parameterization, there is no illumination intensity-dependent shunt. We conclude, also shunt *distributions* cannot create *I-V* curve shapes that could be interpreted as an illumination dependent shunt.

3.2 Modules with I_{SC} mismatched cells

In this section we show simulation results for modules where each individual cell is described by the standard model parameters of Table I, except for a variation of the photogeneration current I_{PH}. The photogeneration current distribution amongst the cells is according to the inhomogeneity map shown in Fig. 4. We scale the inhomogeneity distribution of this map to three different levels of maximum-to-minimum variations: +/- 0.4%, +/- 1.0%, and +/- 2.0%. Since photogeneration current inhomogeneities amongst the cells will drive some cells into reverse bias, we explore the effect of the four different reverse bias characteristics shown in Fig. 5.

Fig. 8 provides a closer look at simulated module *I-V* curves for the case of +/- 2.0% inhomogeneity of the photogeneration I_{PH} of the cells in the module circuit. Note that all cell definitions have completely identical forward-bias characteristics. Nevertheless, the forward characteristics of the module *I-V* curves are different for a wide range of forward voltages up to close to the maximum power voltage V_{MP} at the MPP. Very unlike a real shunt, this effect ends in a sharp kink as all *I-V* curves merge rather abruptly into the grey dotted *I-V* curve, representing a module with homogeneous I_{PH} for the cells.

Figure 8: Zoom into upper current range of 1-Sun *I-V* curves with +/- 2.0% photogeneration inhomogeneity, plotted for the four cases of different cell reverse bias characteristics in Fig. 6. (Also shown for comparison: *I-V* curve with homogeneous I_{PH} distribution, grey dotted line.) Despite different $R_{Sh.Slope}$ values, all *I-V* curves share the same MPP.

Note that despite clearly visible *I-V* curve differences in Fig. 8, the four curves agree perfectly well with each other in the voltage range of the MPP. Thus, despite different $R_{Sh.Slope}$ values, the maximum power production of these four different modules is identical and unaffected by different values of the apparent shunt $R_{Sh.Slope}$.

Fig. 8 also shows that the black dashed line *I-V* curve [i.e. the module of case 3 in Fig. 6] is indistinguishable from the blue line *I-V* curve [case 2 in Fig. 6]. Despite their differences for larger reverse bias voltages, both reverse bias characteristics are nevertheless virtually identical for the first 4 volts, from zero to -4 V. Since the moderate current inhomogeneities considered here do not cause cells in these two module types to exceed reverse bias voltages of -4V, their *I-V* curves in Fig. 9 do not differ.

Fig. 9 shows four plots, each representing a module where all cells in the module have one of the four different reverse bias characteristics of Fig. 5. In each plot there are different curves that correspond to different degrees of inhomgeneities of the photogeneration across the cells in the module circuit: +/- 0.4%, +/- 1.0%, and +/- 2.0%. The quantity plotted in these graphs is the slope of the simulated *I-V* curves between I_{SC} and 450 mV/cell. (Very similar results are obtained for other choices, such as the slope between I_{SC} and 300 mV/cell.) The inverse of these slopes is the apparent shunt resistance $R_{Sh.Slope}$. The horizontal axis of the graphs in Fig. 9 is the illumination intensity expressed in Suns (1 Sun = 1000 W/m²).

The grey dotted line in each of the for graphs of Fig. 9 is the R_{Sh} that all of the cells have and therefore also the shunt of the dark *I-V* curve of the modules. Hence its intersection point with the left vertical axis is the point that all three cures in each of the four graphs converge towards. All three curves in each of the four graphs would collapse to the grey dotted line when inhomogeneity of the cell I_{SCS} would be reduced to zero.

The lower right graph in Fig. 9 features an additional fourth curve with a dotted red line, showing for the case of +/- 2% I_{SC} inhomogeneity the curve from a module circuit without bypass diodes, while the solid line represents the situation with bypass diodes. In the other cases of this study, the difference due to the bypass diodes is negligibly small and therefore not shown explicitly here in Fig. 9.

1037

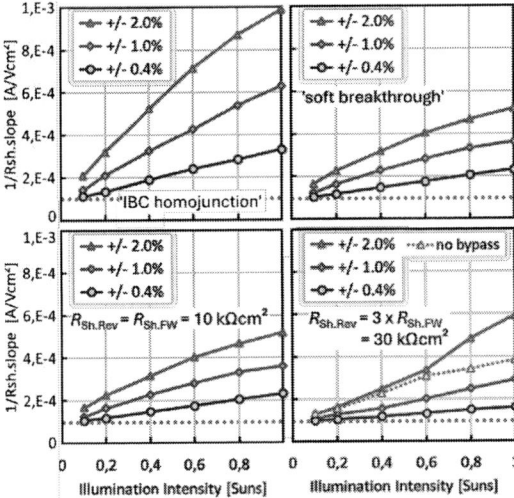

Figure 9: Inverse apparent shunt resistances $R_{Sh.Slope}$ derived from the I-V-curve slopes between I_{SC} and 450 mV/cell, plotted as a function of illumination intensity in Suns. Each graph represents one type of reverse bias characteristics (see Fig. 6), and each curve refers to one photogeneration inhomogeneity distribution +/- 0.4%, +/- 1.0%, and +/- 2.0%. The dotted grey line is the R_{Sh} value of the cells.

We can clearly see in Fig. 9 that cell photogeneration inhomogeneities of the cells in the module circuit cause the emergence of an intensity dependent apparent shunt $R_{Sh.Slope}$, even though all individual cells have an identical and intensity independent shunt. The magnitude of the apparent shunt $R_{Sh.Slope}$ does not only depend on the illumination intensity and inhomogeneity, but also on the cells' reverse bias characteristics. Generally, the more efficiently the cells are able to conduct current in reverse bias direction, the more pronounced is the illumination intensity dependence of the apparent shunt conductance $1/R_{Sh.Slope}$. Note that towards zero illumination intensity the value of $1/R_{Sh.Slope}$ converges towards the value of 1×10^{-4} $1/(\Omega \text{cm}^2)$. This value corresponds to the forward-bias shunt value of 10 kΩcm^2 that all cells in these simulations share (see also Table I).

We consider again Fig. 8, which shows I-V curves of four modules, where the average photo generation in the cells is for all four modules identical. Nevertheless, we can observe that the I-V curves in this figure produce different short-circuit currents I_{SC}, depending on the reverse bias characteristics. We therefore explore in Fig. 10 how the module I_{SC} varies as a function of illumination inhomogeneity and average illumination intensity for modules with different cell reverse characteristics.

We can see from the variation of the ratio I_{SC}/Suns in Fig. 10 that the I_{SC} of modules is not linear with illumination intensity if the photogeneration I_{PH} of the individual cells is not homogeneous across the module. This is an interesting observation and potentially of importance in situations where one may be tempted to use the module I_{SC} as a measure for the average photogeneration, such as when analyzing soiling of modules in the field. In such cases it may be better to observe the maximum power current, as is suggested by Fig. 8, where all curves shared the same MPP, which is also shared by the grey dotted I-V curve that that represents the case of homogeneous cell I_{SC} distributions across the

module, which is equal to the average cell I_{SC} value of the inhomogeneous distributions of the other curves.

The non-linearity of I_{SC} with (average) intensity of the illumination, as explored in Fig. 10, does not mean a model for power production of PV systems would necessarily need to take into account such non-linearity. Instead, Fig. 8 has shown that the MPP, which lies for reasonable non-degenerate cell photogeneration distributions well outside the voltage range affected by cell I_{SC} current mismatch, can be well described with using the average photogeneration. The average photogeneration I_{PH} scales well with the illumination intensity. Hence, the results presented here rather underline that an equivalent circuit model (e.g. 1-Diode model) for PV system power production simulation can and should scale the photogeneration I_{PH} linearly with illumination intensity. The challenge when measuring and characterizing a module lies in the determination of a suitable 1-Diode parameterization of the module I-V curve, including the determination of the average cell I_{PH} ($\approx I_{SC}$) for use in the 1-Diode model parameterization of the module.

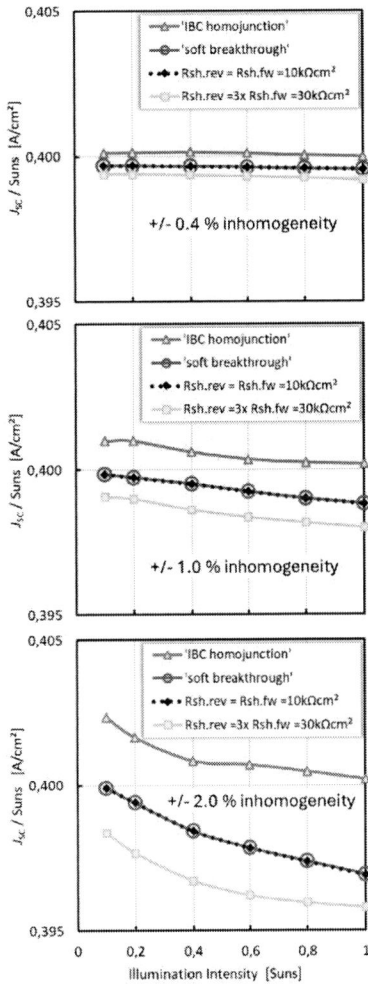

Figure 10: Normalized short-circuit current density I_{SC} of modules, as a function of illumination intensity. The quality of proportionality between I_{SC} and illumination intensity reduces with increasing illumination inhomogeneity, while the I_{SC} of every single cell in the module is near-perfectly proportional to intensity.

4 DISCUSSION OF PRACTICAL CONSEQUENCES

Figure 8 shows that there is a voltage range between I_{SC} and MPP that is distorted from the combined effect of cell short-circuit current mismatches and the cells' reverse bias characteristics. For typical levels of I_{SC} mismatch in new modules under laboratory test conditions (be the mismatches from actual differences in the cells or from illumination inhomogeneity in the module measurement set-up), this distortion vanishes before the MPP, i.e. before the maximum power voltage V_{MP}. Beyond this point, the module I-V curve can be well described by a 1-Diode (or 2-Diode) model (see also reference [8]).

Consequently, it seems presently the best-known-method for fitting equivalent (1- or 2-) diode model parameters for a PV module to proceed as follows: Ignore the distorted part in the lower voltage range of the module light I-V curve and fitting the remaining main part of the I-V curve with a diode model that has a shunt R_{Sh} without intensity independence. Reference [8] used exactly this approach, resulting in a 1-Diode model that provides for a wide range of illumination intensity and different temperatures an excellent fit to the measured data, as was shown in Fig. 1.

5 CONCLUSIONS

We have shown how an apparent shunt emerges in a module I-V curve when combining cells with inhomogeneous photogeneration in the module circuit. Since the difference of photogeneration scales with illumination intensity, this apparent shunt in the module I-V curve scales with illumination intensity. However, this effect is not at all an actual shunt conductance and such intensity-dependent shunt appears only when the shunt conductance is estimated by using the slope of the module I-V curve in the quasi-linear range from I_{SC} towards larger voltages. While this "slope method" is appropriate for cells, it is not a good indication for a shunt conductance in modules. When the cells in the module have different photogeneration, which can be caused by illumination inhomogeneities or by cell production variations, this slope is strongly influenced by current mismatch effects.

We show that this mismatch induced slope in the quasi-linear range of the I-V curve has for typical laboratory settings of measuring new modules no influence on the I-V curve near MPP. However, interpreting this slope as a shunt (or rather: "pseudo shunt") and using this value as shunt conductance in an equivalent circuit model for the module, such as the 1-Diode model, does affect the MPP. Consequently, the slope-derived pseudo shunt value should not be used as shunt conductance in equivalent circuit models for representing the module power production performance.

5 REFERENCES

[1] W. De Soto, S.A. Klein and W.A. Beckman, "Improvement and validation of a model for photovoltaic array performance", Solar Energy 80 (2006), p. 78-88.

[2] A. Mermoud and T. Lejeune, "Performance assessment of a simulation model for PV modules of any available technology", 25th EU-PVSEC (2010).

[3] G.E. Bunea, K.E. Wilson, Y. Meydbray, M.P. Campbell and D.M. De Ceuster, "Low light performance of mono-crystalline silicon solar cells", Proc. IEEE 4th WCPEC, 2006, pp. 1312-1314.

[4] P. Grunow, S. Lust, D. Sauter, V. Hoffmann, C.Beneking, B.Litzenburger, and L. Podlowski, "Weak light performance and annual yields of PV modules and systems as a result of the basic parameter set of industrial solar cells", Proc. 19th EU-PVSEC, 2004, pp. 2190-3.

[5] S. J. Robinson, A. G. Aberle, and M.A. Green, "Departures from the principle of superposition in silicon solar cells", J Appl Phys 76, (1994), p. 7920.

[6] O. Breitenstein, "An alternative one-diode model for illuminated solar cells", Energy Procedia 55 (2014) p. 30 – 37.

[7] K. Ramspeck, C. Böhmer, and M. Meixner, "Measurement Uncertainty Analysis for Large Area High-Efficiency Modules", 40th European Photovoltaic Solar Energy Conference, (2023), p. 020213-001 - 020213-005.

[8] N.-P. Harder, and J. Cano Carcia, "TOPCon module characterization at different temperatures and intensities: Revision of shunt parameterization by De Soto and PVsyst", 52nd IEEE Photovoltaics Specialists Conference (2024), proceedings in print.

[9] https://www.pvsyst.com/help/pvmodule_model _parameters.htm

[10] S. Dongaonkar, J.D. Servaites, G.M. Ford, S. Loser, J. Moore, R.M. Gelfand, H. Mohseni, H.W. Hillhouse, R. Agrawal, M.A. Ratner, T.J. Marks, M.S. Lundstrom; M.A. Alam, "Universality of Non Ohmic Shunt Leakage in Thin Film Solar Cells", J. Appl. Phys. 108, 124509 (2010).

[11] O. Breitenstein, P.P. Altermatt, K. Ramspeck, and A. Schenk, "The Origin of Ideality factors n > 2 of Shunts and Surfaces in the Dark I-V Curves of Si Solar Cells", Proc. 21st EU-PVSEC (2006), p. 625.

[12] S. Steingrube, O. Breitenstein, K. Ramspeck, S. Glunz, A. Schenk, and P.P. Altermatt, "Explanation of commonly observed shunt currents in c−Si solar cells by means of recombination statistics beyond the Shockley-Read-Hall approximation", J. Appl. Phys. 110(1):014515, (2011).

[13] R. Asadpour, , Xingshu Sun, and M.A. Alam, "Electrical Signatures of Corrosion and Solder Bond Failure in c-Si Solar Cells and Modules", IEEE Journal of Photovoltaics Vol. 9(3), 2019, p. 759.

[14] S. Dongaonkar, C. Deline, and M.A. Alam, "Performance and Reliability Implications of Two Dimensional Shading in Monolithic Thin Film Photovoltaic Modules", IEEE Journal of Photovoltaics 3(4), 2013, p. 1367-1375.

[15] C.E. Clement, J. Prakash Singh, E. Birgersson, Y. Wang, and Y. Sheng Khoo, "Illumination Dependence of Reverse Leakage Current in Silicon Solar Cells", IEEE Journal of Photovoltaics, Vol. 11(5), 2021, p. 1285.

41st European Photovoltaic Solar Energy Conference and Exhibition

This presentation was selected by the Sc. Committee of the EU PVSEC 2024 for submission of a full paper to one of the EU PVSEC's collaborating peer-reviewed journals.

CORRELATING FIELD EXPERIMENTATION AND IMAGE ANALYSIS FOR THE ASSESSMENT OF INDUCED LOSSES FROM THIN OBJECT SHADING ON PHOTOVOLTAIC SOURCES

Matthew Axisa[1], Luciano Mule'Stagno[1], Marija Demicoli[1]
[1]University of Malta, Institute of Sustainable Energy, Marsaxlokk, Malta

ABSTRACT:

This research investigates the correlation between the size and intensity of both umbra and penumbra shadow formation and the resulting power loss, based on field experimention using various thin objects to cast shadows on a photovoltaic (PV) module. The results from the Spearman correlation matrix clearly demonstrate that the significance of both the umbra and penumbra size increases as the thickness of the shading object increases. Additionally, it has been noted that penumbra intensity plays a crucial role in power loss for thin objects ranging between 2.8 and 12mm in thickness. This study identifies the range of distances at which thin objects, ranging from 2.8 mm to 12 mm in thickness, produce zero power loss, indicating no observable effect on the performance of the PV module. By understanding these key aspects, other researchers and PV system owners can better assess the impact of shadow factors on their respective systems. Consequently, the findings from this study are also valuable for determining the impact of certain thin objects typically found on rooftops, helping to optimize the design of PV systems without compromising the functionality of the building.

Keywords: solar, photovoltaics, shading, thin objects, umbra, penumbra, image, analysis

1 Introduction

Renewable energy technologies are increasingly becoming critical in addressing global energy demands and mitigating environmental impact, with solar photovoltaic (PV) systems playing a pivotal role in the transition to sustainable energy. Despite their theoretical peak power ratings, PV systems often underperform due to various challenges, with shading being one of the most significant. Even minimal shading can disrupt sunlight uniformity on PV cells, leading to substantial power losses [1]. This underscores the need to understand and mitigate shading to ensure optimal solar performance. The growing use of image analysis techniques has proven effective in identifying and analyzing shading issues on PV systems, enabling quicker and more accurate assessments to support mitigation efforts. This approach is crucial for achieving the United Nations' Sustainable Development Goal 7 (SDG 7), which aims to provide affordable, reliable, and sustainable energy for all. Optimizing solar performance in real-world conditions is essential to meet this objective.

Kour et al. [2] define image processing as the process of altering an image to enhance its visual information for human interpretation or autonomous machine perception. With advancements in technology, images have become sharper and more detailed, making it possible to extract partial shading data. This data can be used to detect faults, cell hotspots, defects, and dust, among other issues [3]. For instance, partial shading on a PV module was detected using an optical camera, and this method was further advanced by using a drone-mounted camera to inspect large PV plants [4].

In a study using only a standard single-lens reflex (SLR) camera, Ryad et al. [5] developed an algorithm to track the Global Maximum Power Point (GMPP) by detecting partial shading on a PV module. The authors used an open-source tool to convert a standard Red, Green, Blue (RGB) image into Hue Saturation Value (HSV) [6]. The Canny method, originally developed by John F. Canny in 1986 and later refined by McIlhagga [7], is the most well-known HSV conversion technique. Similarly, Patel et al. employed HSV for edge detection in PV modules using the Kirsch operator [8]. However, HSV conversion accuracy has been surpassed by newer methods, particularly those that analyze anomalies at solar cell level, taking advantage of higher-resolution images. For example, S. Sural et al. [9] demonstrated improved results using greyscale images and thresholding analysis, which assigns a numerical value from 0 (black) to 255 (white) and represents it as a matrix. Gutiérrez et al. further developed this concept by identifying the percentage of shadow on each cell within a PV module but did not differentiate between light and dark shadows or quantify shadow intensity [10]. Rehman et al., however, distinguished between light and dark shadows by classifying them based on solar radiation levels— 500 W/m² for light shading and 100 W/m² for dark shading [11]. Despite these advancements, none of the studies have adequately categorized shadows into the optical formations of umbra and penumbra and have determined the effectiveness of light shadows in isolation to darker shadows.

Researchers with a strong theoretical understanding of shadow formation have achieved better outcomes in image analysis applications. This is demonstrated in Sinha et al.'s study [12], where comprehensive shadow principles, including the classification of umbra and penumbra, were combined with the Otsu Method [12], [13] for thresholding to effectively remove cast shadows from images. Sinha also explores various shadow detection techniques, such as intensity and texture analysis, segmentation, mask construction, and edge-based approaches, concluding that Otsu's method is more effective for shadow removal rather than for extraction. Dimitros expanded on Sinha's work by examining shadow fuzziness and border issues through the lens of quantum mechanics, introducing the concept of fuzzy sets to describe broad penumbra shadows [14].

Huang et al. [15] further demonstrated fuzzy shadows on white paper, showing how cropping regions of interest and plotting pixel intensity can reveal a sigmoidal curve function. This approach is based on the observation that umbra shadows maintain consistent intensity, while penumbra shadows vary as seen in Figure 1. By identifying where umbra intensity diminishes and penumbra begins, the boundary between the penumbra and the background can be accurately determined. This principle will be foundational in the current study's approach to analyzing shadow boundaries.

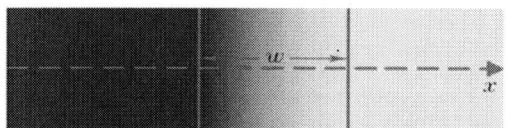

Figure 1. Dynamic shadow boundaries with w representing the penumbra region while x depicting the distance of pixels.

2 Research goal

Despite extensive research in the field of PVs, two significant gaps remain in the application of image analysis for shading assessment. Firstly, as seen in previous studies, the penumbra shadow gets lighter as the shading object moves away from the surface. The gap remains at which point the penumbra shadow diminishes to such an extent that it no longer has any effect on the PV module for each given thin object thickness. Secondly, there is limited understanding of how the penumbra shadow correlates with the umbra as variables, like shadow thickness and intensity, change. Additionally, with the quantification of shadow size and intensity, extrapolation of the relationship between distance and power loss beyond the field experimentation limitations could determine at which point a penumbra only shadow produces zero power loss, thus rendering no impact on the performance of the PV module.

This study aims to address these gaps by conducting field experimentation, applying regression models and employing innovative image analysis algorithms to quantify both umbra and penumbra shadows and to investigate their correlation with power loss on a PV module. In doing so, image analysis is introduced as a method to better understand shadow dynamics and their impact on PV performance.

3 Methodology

3.1 Field experimentation methodology

To accurately determine the point at which the shadow cast by a thin object becomes negligible in terms of its impact on the power output of a PV module, a detailed polynomial regression analysis was conducted. This method enabled the modeling and extrapolation of power loss data as a function of the distance between the shadow-casting object

and the PV module. By fitting polynomial functions to the observed data, it became possible to predict power loss at a wider spectrum of distances and identify the critical point where the shadow effect on the PV module is effectively zero.

The process began with the collection of data, specifically measuring power loss percentages at varying distances from a 20W experimental PV module, with distances ranging from 25 cm to 400 cm, with the latter being the limitational distance of the field experimentation. To analyze the relationship between distance and power loss, the best model fit was determined. In the majority of cases, the best fit was attained by a second-degree polynomial regression model, with the exception of two instances where a third-degree polynomial regression model was deemed a better fit (better R^2) to capture the non-linear behavior of power loss as a function of distance. Using the polynomial feature from the sklearn library in Python v3.12.2, the data was modeled and fitted to derive the polynomial equations that best represent the relationship between distance and power loss, as given below.

Second-degree Polynomial model equivalent to:

$$\text{Power loss} = ax^2 + bx + c$$

while the Third-degree Polynomial model equivalent to:

$$\text{Power loss} = ax^3 + bx^2 + cx + d.$$

Where x represents the distance at each increment, and a, b, c, and d, are the coefficients obtained from the regression model.

Using this methodology, and through field experimentation, it was observed that at a maximum distance of 400 cm, objects with thicknesses up to 12 mm created penumbra-only shadows. Consequently, the power loss attributed solely to the penumbra could be extrapolated to the point where the power loss reaches zero percent.

3.2 Image analysis methodology

The goal of this image analysis approach is to accurately measure the percentage of umbra and penumbra in shadows cast by thin objects. The following shadow quantification process was specifically designed to assess varying shadow sizes on a white background. This method utilizes the dynamic range of pixel values, ranging from 0 (black) to 255 (white), with intermediate values representing 256 shades of grey. This range allows for detailed greyscale representation, simplifying the analysis by highlighting light and dark variations.

The image analysis process starts with the user loading an image into the tool, where it is automatically converted from RGB to greyscale, allowing for a focus solely on intensity values. From the greyscale image, the average pixel intensity for each column is calculated, and the darkest pixel in this column is identified as the reference for normalization. During normalization, the darkest pixel is assigned a value of 1, and the lightest is assigned a value of 0, rescaling the intensity values accordingly. The normalized data is then used to plot a dual mirrored sigmoidal curve indicated by a dashed red line, as shown in Figure 2. This graph visually represents the umbra and penumbra regions, with a sharp peak indicating a smaller umbra and a broader peak indicating a larger one. Two sigmoidal curves are fitted to this data, as illustrated by the blue plot, in this way ascertaining that a good fit of the sigmoidal curve is present since the dashed red line and the solid blue line almost overlap.

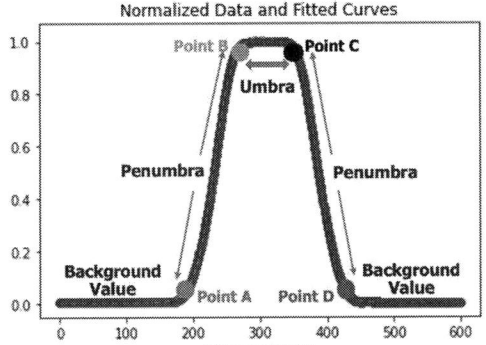

Figure 2. Sigmoidal curve fitting and noise point identification.

To determine the start and end points of the umbra and penumbra, this study utilizes the method proposed by Mirylenka et al. [16], where changes in the curvature of sigmoidal curves indicate the active noise range's impact on performance as represented by Z^* and Z_* in Figure 3. In our study, the noise level corresponds to the penumbra, as it is the only shadow formation with dynamic intensity changes, whereas the umbra has a constant intensity since no light passes through the opaque object. According to Mirylenka et al., the top and bottom 0.05 normalized distance values from the darkest intensity are used as cut-off points for the penumbra, as shown by Points A (red) and B (green) on the left-side sigmoidal curve, and Points C (black) and D (purple) on the right-side curve in Figure 2. These points will hereon be the indicative measure of where the penumbra begins and ends, both above and below the midpoint.

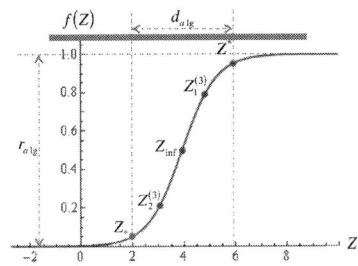

Figure 3. Sigmoidal function and points of interest [16].

The analysis examines two scenarios: one where both umbra and penumbra are present, and another where only the penumbra exists. The boundaries for these scenarios are determined using denormalized values, with example values provided for illustration. Pixel counts are then calculated for each category (above and below the identified points), and the total pixel count for each category is computed. The percentage of pixels in each category is obtained by dividing the pixel count by the total number of pixels in the averaged column.

In some instances, the distance increases to the point at which no umbra value is present, thus leaving only penumbra shadow which in nature is known to increase in size but gets lighter in intensity. For the developed image analysis tool, a scenario is classified as 'Only Penumbra' if the difference between the darkest pixel and either Point B or Point C is 3 pixels or less, indicating minimal or no umbra presence. This threshold was established after analyzing 96 images (each combination of thickness and distance instances) to identify cases with only penumbra shadows. As previously mentioned, umbra shadows have a constant intensity, and if only penumbra is present, no umbra should appear. The image analysis tool detects this situation by initially attempting to quantify the umbra. If the top of the sigmoidal curve forms a sharp peak without a plateau, the 3-pixel value difference between the darkest pixel and the noise level triggers the tool to categorize the instance as having a penumbra-only shadow. The tool then determines whether both umbra and penumbra are present, or only penumbra, based on this analysis.

Establishing the methodology for the quantification of umbra and penumbra size and intensity aims to facilitate the correlation metric data for each individual object thickness. Furthermore, considering that the experimental PV module is composed of 36 series-connected quarter cut solar cells, it was established that the maximum thickness of any thin object placed at a distance of 400 cm would not exceed the width of the quarter cut solar cell, which is set at an object thickness limit of 12 mm. This limitation is designed to contain the penumbra shadow within the horizontal quarter cut cells, preventing any spillover of penumbra shadow onto the upper and lower cells.

4 Results and discussion

4.1 Determining the distance at which a penumbra-only shadow produces zero power loss for each object thickness

Based on the methodology explained in Section 3.1, in most instances, it was sufficient to use a second-degree polynomial curve to model the relationship between distance and power loss. However, other cases required the more complex third-degree polynomial curve so as to obtain a better fit to account for the non-linearities observed. Moreover, for the thinnest objects with 2.8 mm and 3.2 mm thickness, negligible power loss was already observed throughout the entire range of distances studied, up to 400 cm. As a result, no polynomial regression was necessary for these cases, as the power loss was consistently minimal, indicating that these thin objects are not expected to have a significant impact on the PV module's output when placed at any distance away from the PV module.

For the other thin objects, polynomial regression was utilized to derive the equations that best represented the relationship between distance and power loss. These equations allowed for the prediction of power loss at new, untested distances beyond 400 cm. Additionally, by solving these polynomial equations for the point where power loss equals zero, the critical distance at which the shadow of the thin object no longer significantly affects the PV module's power output was derived, as visualized in Figure 4.

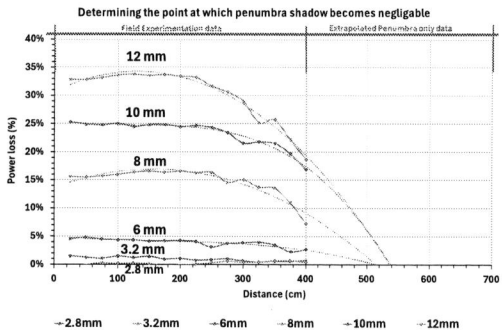

Figure 4. Extrapolating the point at which penumbra shadow produces zero power loss.

The scatter plots shown in Figure 4 show the actual data points represented by a solid line trend as displayed on the left portion of the graph, together with the extrapolated faint line plots on the right side that represent the polynomial regressions. The intersection points, where the power loss drops to zero, are clearly marked on the plots, illustrating the distances at which the shadows of the various thin objects become negligible. The specific distances at which each thin object's shadow renders the power loss negligible are summarized in Table 1.

Table 1. The resultant distance range at which power loss becomes zero for each studied thin object thickness.

Thin object rod thickness (mm)	Distance range at which power loss becomes zero (cm)
2.8 - 3.2	<400
6 - 8	500-525
10-12	550-600

This section has identified the critical distances at which shadows cast by thin objects, ranging from 2.8 mm to 12 mm thickness, become negligible in their impact on PV module power output. Through polynomial regression analysis, a clear methodology has been identified for predicting power loss based on distance. However, it is essential to note that the identified distances should not be universally applied without considering that this experiment observed the raw effects of shading without accounting for Maximum Power Point Tracking (MPPT) and bypass diodes, which are commonly used to mitigate shading effects. Moreover, since the study was conducted under peak weather conditions, the power loss observed represents the maximum possible loss. In real-world scenarios, actual power loss is unlikely to exceed these values. Homeowners and PV installers can now use this dataset as a reliable benchmark, confirming that a thin object at the identified distances produces negligible power loss. This ensures they can retain such objects without negatively impacting the performance of their PV systems.

4.2 Correlating the various factors affecting a shadow

The algorithm detailed in the methodology section of this study was implemented as open-source code using Python v3.12.2, with a graphical user interface (GUI) to facilitate the display of graphs and figures, as shown in Figure 5. A total of 96 images, representing 6 different object thicknesses at 16 distances ranging from 25 to 400 cm, were analyzed, and the resulting umbra and penumbra values were recorded. Figure 6 presents a sample image dataset for thin objects analyzed at various distances within this range. The contrast in shadow intensity and formation across the various incremental distances is significantly noticeable, where the dark portion of the shadow dominated in presence at the first 200cm while instance between 225 and 400cm the penumbra shadow is distinctively the majority portion of the shadow.

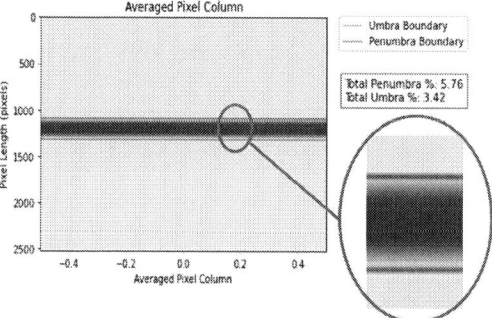

Figure 5. The image analysis tool deriving the percentage umbra and penumbra.

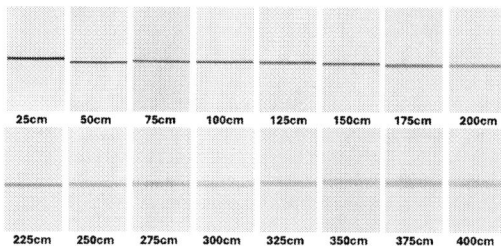

Figure 6. Sample dataset illustrating image analysis performed on a medium thin object (8 mm).

To examine the relationships between power loss and the various influencing factors, this research employs the Spearman correlation matrix. Spearman correlation, a non-parametric measure, evaluates the strength and direction of monotonic relationships between variables. It was selected over Pearson correlation due to its robustness against outliers and non-normal distributions, making it well-suited for analyzing complex data interactions. Power loss is the primary variable compared with factors such as the size and intensity of both umbra and penumbra, as well as distance, as shown in
Figure 7. Moreover, this study uses the correlation coefficient metric from Hair et al. [17, p. 358] to assess the relationships between the identified factors.

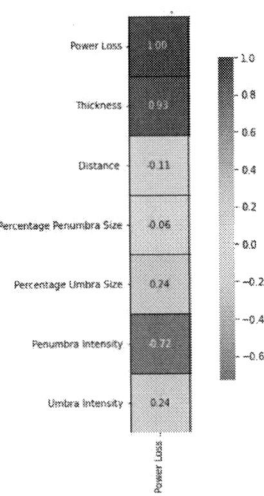

Figure 7. Spearman correlation matrix representing thin objects ranging from 2.8 – 12 mm.

The correlation between power loss and distance shows a generally weak relationship across the object thickness, indicating other factors are more correlated to power loss. The negative correlation coefficient for distance indicates that as the distance between the thin object and the PV source increases, power loss decreases. This trend is illustrated in Figure 6, where the shadow shifts from a high-intensity umbra to a more favorable penumbra at greater distances, leading to reduced overall shadow intensity and power loss.

The correlation between power loss and penumbra size is weak oval while the negative correlation indicates that as the size of the penumbra shadow increases, power loss decreases, likely because a larger penumbra occurs at greater distances where the more intense umbra diminishes. While the impact of penumbra size on power loss is minimal, penumbra intensity shows a much stronger negative correlation, highlighting the significant role of shadow intensity in power loss. This is attributed to the fact that these thin objects depend more on penumbra intensity rather than the umbra shadow which is seen in thicker objects above 12mm. Future research should explore the interaction between penumbra size and intensity in more detail to better understand their combined effect on power loss.

The correlation between power loss and both the size and intensity of the umbra is consistent, suggesting a weak relationship. As previously mentioned, this can be attributed to the abrupt transition to penumbra shadow with increasing distance. As the thickness of the object increases and it becomes darker, power loss also rises. This indicates that objects thicker than 12 mm are likely to demonstrate a stronger correlation with umbra shadow intensity and size, rather than penumbra shadow. Further investigation will be conducted to validate this hypothesis.

5 Conclusions

This study demonstrated the ability of image analysis in quantifying the size and intensity of umbra and penumbra shadows and their impact on PV module power loss. Moreover, the results showed that object thicknesses equal to or under 12 mm in thickness will all produce zero power loss on a PV module, given that the thin object is situated more than 600 cm away. Subsequently, from the Spearman correlation analysis, it is evident that the intensity of the penumbra plays a critical role in thin shading objects, contrasting the correlation with both umbra size and intensity as a weak correlation to power loss is observed. This research highlights the potential of image analysis as a valuable tool which, with further development, could consist of analyzing the shadow in relation to commercially available PV module designs as well as predicting models of power loss to obtain a usable tool in real-life scenarios.

6 Future Works

Future research will focus on validating the methodology of the image analysis tool by comparing its results for umbra/penumbra size and intensity with real-life shading scenarios, such as those created by horizontal overhead wires, as well as by assessing the points at which the penumbra shadow becomes negligible through tests on the PV module using varying object thicknesses and distances.

7 Funding

This research, along with the presentation at the EU PVSEC 2024 Conference, was funded by the University of Malta's RIDT Doctoral Student Overseas Conferences Grant 2024.

8 References

[1] M. A. A. Mamun, M. Hasanuzzaman, and J. Selvaraj, "Experimental investigation of the effect of partial shading on photovoltaic performance," *IET Renewable Power Generation*, vol. 11, no. 7, pp. 912–921, Jun. 2017, doi: 10.1049/iet-rpg.2016.0902.

[2] A. Kour, V. Yv, V. Maheshwari, and D. Prashar, "A Review on Image Processing," *International Journal of Electronics Communication and Computer Engineering*, vol. 4, pp. 2278–4209, Jan. 2012.

[3] A. Salazar and E. Q. Macabebe, "Hotspots Detection in Photovoltaic Modules Using Infrared Thermography," *MATEC Web of Conferences*, vol. 70, p. 10015, Jan. 2016, doi: 10.1051/matecconf/20167010015.

[4] C. Henry, S. Poudel, S.-W. Lee, and H. Jeong, "Automatic Detection System of Deteriorated PV Modules Using Drone with Thermal Camera," *Applied Sciences*, vol. 10, no. 11, 2020, doi: 10.3390/app10113802.

[5] A. K. Ryad, A. M. Atallah, and A. Zekry, "An accurate partial shading detection and global maximum power point tracking technique based on image processing," *Eng. rev. (Online)*, vol. 42, no. 1, 2022, doi: 10.30765/er.1636.

[6] J.-P. Villegas-Ceballos, M. Rico-Garcia, and C. A. Ramos-Paja, "Dataset for Detecting the Electrical Behavior of Photovoltaic Panels from RGB Images," *Data*, vol. 7, no. 6, p. 82, Jun. 2022, doi: 10.3390/data7060082.

[7] W. McIlhagga, "The Canny Edge Detector Revisited," *International Journal of Computer Vision*, vol. 91, Feb. 2011, doi: 10.1007/s11263-010-0392-0.

[8] A. V. Patel, L. McLauchlan, and M. Mehrubeoglu, "Defect Detection in PV Arrays Using Image Processing," in *2020 International Conference on Computational Science and Computational Intelligence (CSCI)*, Las Vegas, NV, USA: IEEE, Dec. 2020, pp. 1653–1657. doi: 10.1109/CSCI51800.2020.00304.

[9] S. Sural, Gang Qian, and S. Pramanik, "Segmentation and histogram generation using the HSV color space for image retrieval," in *Proceedings. International Conference on Image Processing*, Sep. 2002, p. II–II. doi: 10.1109/ICIP.2002.1040019.

[10] A. Gutiérrez Galeano, M. Bressan, F. Jiménez Vargas, and C. Alonso, "Shading Ratio Impact on Photovoltaic Modules and Correlation with Shading Patterns," *Energies*, vol. 11, no. 4, p. 852, Apr. 2018, doi: 10.3390/en11040852.

[11] H. Rehman *et al.*, "Neighboring-Pixel-Based Maximum Power Point Tracking Algorithm for Partially Shaded Photovoltaic (PV) Systems," *Electronics*, vol. 11, no. 3, p. 359, Jan. 2022, doi: 10.3390/electronics11030359.

[12] A. Sinha, "COMPARISION OF TECHNIQUES USED IN SHADOW DETECTION IN AN IMAGE".

[13] J. Yousefi, *Image Binarization using Otsu Thresholding Algorithm*. 2015. doi: 10.13140/RG.2.1.4758.9284.

[14] D. Dendrinos, "ON THE FUZZY NATURE OF SHADOWS," *academia.edu*, Mar. 2017.

[15] Xiang Huang, Gang Hua, J. Tumblin, and L. Williams, "What characterizes a shadow boundary under the sun and sky?," in *2011 International Conference on Computer Vision*, Barcelona, Spain: IEEE, Nov. 2011, pp. 898–905. doi: 10.1109/ICCV.2011.6126331.

[16] K. Mirylenka, G. Giannakopoulos, and T. Palpanas, "SRF: A Framework for the Study of Classifier Behavior under Training Set Mislabeling Noise," in *Advances in Knowledge Discovery and Data Mining*, vol. 7301, P.-N. Tan, S. Chawla, C. K. Ho, and J. Bailey, Eds., in Lecture Notes in Computer Science, vol. 7301. , Berlin, Heidelberg: Springer Berlin Heidelberg, 2012, pp. 109–121. doi: 10.1007/978-3-642-30217-6_10.

[17] J. F. Hair, *Research Methods for Business*. John Wiley & Sons, 2007. [Online]. Available: https://archive.org/details/researchmethodsf0000u nse_j4k9/page/358/mode/2up?q=moderate

41st European Photovoltaic Solar Energy Conference and Exhibition

LABORATORY INTERCOMPARISON ON A SHADING RESISTANCE CLASSIFICATION OF PV MODULES

Stefan Riechelmann[1*], Hendrik Sträter[1], Laura Stenzig[1], Giorgio Bardizza[2], Werner Hermann[2], Ebrar Özkalay[3], Gabi Friesen[3] Özcan Bazkir[4], Alexandra Schmid[5] and Stefan Winter[1]
*Corresponding author: stefan.riechelmann@ptb.de

[1]*Physikalisch-Technische Bundesanstalt* (PTB), Braunschweig, Germany
[2]TÜV-Rheinland, Cologne, Germany
[3]Scuola Universitaria Professionale della Svizzera Italiana (SUPSI), Istituto di Sostenibilità Ambientale Applicata
[4]TÜBITAK, Turkey
[5]Fraunhofer ISE, Freiburg, Germany

ABSTRACT: Shading of PV modules can significantly reduce the performance of PV installations, particularly in rooftop installations. We have developed a classification scheme based on a shading metric to rate PV modules according to their shading tolerance. The results of a laboratory intercomparison, where four participating labs applied the metric to four PV modules with different shading mitigation strategies, are presented. Indoor and outdoor measurements were conducted using a strict procedure developed for reproducibility and future standardization of the scheme. A lab also provided simulations of the PV modules for comparison with the measured results. Consistent ratings were obtained by all participating labs for all examined PV modules under two of the three proposed shading conditions.
Keywords: Performance, PV Module, Shading

1 INTRODUCTION

As awareness of the problems caused by shading of PV modules increases, various mitigation strategies have been developed to enhance the shading resistance of PV systems. The shading resistance of a PV system can be improved during the planning phase and on the inverter level, or by using individual power optimizers directly connected to individual PV modules. Certain technologies have been developed on the PV module level, such as the installation of bypass diodes [1], shingled solar cells [2], or the division of the module into parallel connected submodules [3]. It is essential to remember that shading resistance always involves a trade-off between resistance, power output, and cost.

Innovative module designs are often promoted for their enhanced shading resistance. However, comparing different technologies and designs becomes intricate when shading resistance lacks a well-defined definition and there are no standardized methods to test PV modules for their shading resistance, in comparison to common technologies.

At a previous EUPVSEC in 2022 we presented a shading classification scheme that improves the comparison of shading resistance in typical PV modules [4]. This work shows the results of a laboratory intercomparison where four different labs executed well-defined indoor and outdoor measurement procedures. The results demonstrate the scheme's effectiveness on commercially available PV modules of varying designs.

2 METHODOLOGY

2.1 Metric definition

We assessed a classification scheme that evaluates the relative power reduction of a PV module under various shading patterns, where the expected power reduction solely relies on the shaded portion of the PV module. An idealized response to shading would correspond to a power reduction equivalent to the shaded region's size. For instance, if 20 % of the module region is shaded, a PV module with ideal shading resistance would only endure a

20 % reduction in its power output. Any further loss in addition to this minimal power loss is primarily a result of the solar cells' behavior and circuit design of the PV module. This additional loss (AL) is defined as:

$$AL = \left(1 - \frac{P_{\text{shaded}}}{P_{\text{unshaded}}} - \frac{A_{\text{shaded}}}{A_{\text{total}}}\right) \cdot 100\ \% \qquad (1)$$

We define AL $\geq$ 0 %, e.g. due to non-linear effects, are not considered. The optimal scenario is when AL equals 0 %, indicating that shading does not decrease power generation of the unshaded section of the module. On the other hand, the worst possible outcome is when $AL = \left(1 - \frac{A_{\text{shaded}}}{A_{\text{total}}}\right) \cdot$ 100 %, indicating that the shaded area of the module leads to full power loss of the unshaded portion.

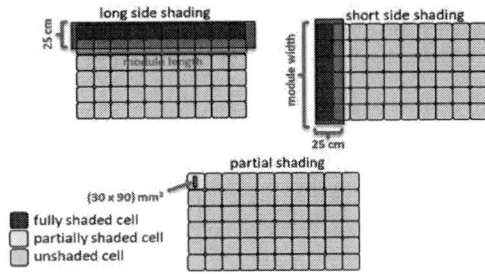

Figure 1: The shading patterns, namely S_{long}, S_{short} and S_{spot}, showcased on a typical 60-cell PV module. Dark red cells are completely shaded, light red areas are partially shaded by the applied shading material.

2.2 Shade patterns

We defined three fixed shading patterns to ensure standardization and high reproducibility:
- shading S_{long} along the long side of the PV module with $A = L_{\text{Module}} \times 25$ cm
- shading S_{short} along the short side of the PV module with $A = W_{\text{Module}} \times 25$ cm
- spot shading centered on one solar cell with $A = 9$ cm $\times 3$ cm,

as shown in Figure 1. The width of S_{long} and S_{short} measures 25 cm and guarantees complete shading for at least one cell row even for the new generation of M12 solar cells. The S_{spot}, with dimensions of 9 cm x 3 cm, is designed to position the pattern in such a way that only a single solar cell is influenced, even if it is a third-cut or shingle cell, thus maximizing the shading effect. We propose that shading be achieved using a material with 0% transmittance for the sake of simplicity. The shading material must be positioned flush with the module edge during the long and short side test to avoid any discrepancies between framed and frameless modules in the rating.

2.3 Classification

Table 1 lists the classification of PV modules based on their additional loss AL for various shading patterns. Four classes are defined, ranging from A (best) to D (worst). The class limits for long and short side shadings are equivalent as they show similarity. However, spot shading produces considerably smaller AL values than the other patterns, leading to the selection of distinct class limits.

Table 1: Classification scheme for the shading resistance depending on the additional loss AL.

Class	long side shading / %	short side shading / %	spot shading / %
A	$AL < 5$	$AL < 5$	$AL < 2$
B	$5 \le AL < 25$	$5 \le AL < 25$	$2 \le AL < 5$
C	$25 \le AL < 50$	$25 \le AL < 50$	$5 \le AL < 10$
D	$AL \ge 50$	$AL \ge 50$	$AL \ge 10$

2.4 Shading material

For measurements conducted at PTB, an opaque material made of black PVC was used. S_{long} and S_{short} patches measuring 200 cm x 25 cm and 120 cm x 25 cm respectively, were cut for the shading experiments. To prevent distortion of PVC due to significant heating of the solar cell under constant light, a polished aluminum stripe measuring 9 cm x 3 cm was used for the spot shading. The shading material shall be fixed precisely at the edges of the PV module, flush with its outer boundary. This shading material is readily available and can be altered by substituting it with simple cardboard, provided it is opaque.

2.5 Indoor measurement procedure

Measurements shall be performed near to STC at irradiance $E = (1000 \pm 100)$ W/m² with module temperature $T = (25 \pm 5)$ °C. A spectral mismatch correction is not mandatory, provided that the spectral fluctuations within a series of relative measurements obtained from the solar simulator being utilized are minor. The irradiance drift during the procedure shall be monitored with a reference cell. In the event of a drift greater than 0.5 % for E or 1 °C for T between all conducted measurements, an IV curve correction must be carried out according to IEC 60891. Note that there is no need to calibrate neither the reference cell nor the PV module, as the values are being used comparatively. A total of 9 individual IV curve measurements shall be performed with a device under test (DUT) according to the following protocol:

1. Unshaded DUT measurement
2. S_{long} shade applied to a long side of the module (material flush on the edge of the DUT).
3. Repeating S_{long} measurement by applying the shading to the opposite long side of the DUT.
4. S_{short} shade applied to a short side of the module.
5. Repeating S_{short} measurement by applying the shading to the opposite short side of the DUT.
6. Performing S_{spot} measurement at Position 1 specified in Figure 2 centered on a cell.
7. Performing S_{spot} measurement at Position 2 specified in Figure 2 centered on a cell.
8. Performing S_{spot} measurement at Position 3 specified in Figure 2 centered on a cell.
9. Performing S_{spot} measurement at Position 4 specified in Figure 2 centered on a cell.

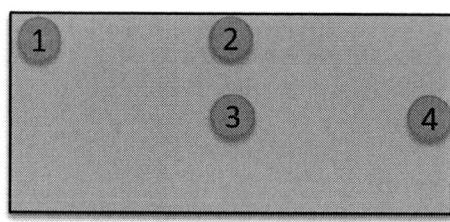

Figure 2: Approximate position of the spot measurement on the PV module surface.

2.6 Outdoor measurement procedure

Measuring shading tolerance outdoors presents additional challenges as a continuous light source can be unstable and the PV module is strongly affected by temperature variations. To mitigate these challenges, it is recommended to use a second module of the same type as a reference device and to conduct unshaded measurements more frequently in between shaded measurements. A total of 11 individual IV curve measurements shall be performed with a device under test (DUT) according to the following protocol:

1. Unshaded DUT measurement 1
2. S_{long} shade applied to a long side of the module (material flush on the edge of the DUT).
3. Repeating S_{long} measurement by applying the shading to the opposite long side of the DUT.
4. Unshaded DUT measurement 2
5. S_{short} shade applied to a short side of the module.
6. Repeating S_{short} measurement by applying the shading to the opposite short side of the DUT.
7. Unshaded DUT measurement 3
8. Performing S_{spot} measurement at Position 1 specified in Figure 2 centered on a cell.
9. Performing S_{spot} measurement at Position 2 specified in Figure 2 centered on a cell.
10. Performing S_{spot} measurement at Position 3 specified in Figure 2 centered on a cell.
11. Performing S_{spot} measurement at Position 4 specified in Figure 2 centered on a cell.

It is again possible to conduct an IV curve correction analogous to the indoor procedure according to IEC 60891 by using a reference cell. A simpler correction method can

41st European Photovoltaic Solar Energy Conference and Exhibition

Table 2: Results of the shading metric intercomparison performed in A3.3.1. While results of long side and short side shading are robust, spot shading resistance measurements tend to scatter substantially between labs.

	module type	L1 Indoor		L1 Outdoor		L2 Indoor		L3 Indoor		L4 Outdoor		L5 Simulation		Mean Lab 1-4	
long-side shading	full-cell	9.93	B	10.73	B	9.92	B	10.09	B	9.45	B	10.47	B	10.02	B
	half-cell	11.00	B	11.37	B	11.00	B	11.26	B	10.90	B	11.49	B	11.11	B
	shingle	78.04	D	78.03	D	78.00	D	78.07	D	-	-	-	-	78.03	D
	diodes	9.55	B	9.58	B	9.51	B	9.49	B	10.25	B	32.63	C	9.68	B
short-side shading	full-cell	85.03	D	84.96	D	84.96	D	85.03	D	84.95	D	84.83	D	84.99	D
	half-cell	35.80	C	35.48	C	35.75	C	35.82	C	36.05	C	35.88	C	35.78	C
	shingle	8.42	B	8.23	B	8.97	B	9.30	B	-	-	-	-	8.73	B
	diodes	22.85	B	20.52	B	23.44	B	23.72	B	21.95	B	27.47	C	22.49	B
spot shading	full-cell	2.35	B	1.14	A	2.13	B	3.16	B	2.23	B	1.57	A	2.20	B
	half-cell	4.27	B	3.75	B	4.28	B	1.47	A	5.23	C	3.81	B	3.80	B
	shingle	4.46	B	5.39	C	6.11	C	4.93	B	-	-	-	-	5.22	C
	diodes	1.93	A	1.46	A	2.03	B	1.91	A	3.78	B	1.59	A	2.22	B

be conducted if a second identical PV module is available and used as a reference device. A simplified drift correction can be conducted as follows:

$$P_{DUT,corr}(S) = P_{DUT}(S) \cdot \frac{P_{ref}}{P_{ref,S}} \quad (2)$$

where $P_{DUT,corr}(S) = P_{DUT}$ is the corrected power of the DUT,

$P_{DUT}(S)$ the uncorrected power of the DUT derived from the IV curves with applied shade S, P_{ref} the power of the reference module measured during the unshaded DUT measurements and $P_{ref,S}$ the power of the reference module during the DUT measurement with applied shade S. The corrected power values are then used for calculating AL values and the corresponding classification of the DUT for each shade condition.

3 RESULTS AND DISCUSSION

Table **2** provides a summary of the measurement outcomes carried out by four participating laboratories on four different modules:

- **Full-cell module**: Heckert Solar 330 W
- **Half-cell module**: JA Solar 365 W
- **Shingled cell module**: Hyundai 400 W
- **Integrated diodes**: AE Solar 320 W

Laboratory 1 conducted measurements both indoors and outdoors, whilst Laboratory 4 investigated a different type of shingle module containing more bypass diodes, and hence results on this device are not displayed. The intercomparison measurements show good agreement under long side and short side shading scenarios – all laboratories rated the modules under test identically. Significant scattering was observed during the spot shading resistance test. In-depth parameter studies were conducted to investigate the impact of the spot shading position on indoor simulators. Our findings revealed a strong correlation with light field non-uniformity, resulting in fluctuations of up to ± 1.5 % for additional losses in these measurements. Considering the scale of these variations, it is impossible to establish a clear classification for the proposed spot-shading. Laboratory 5 performed simulations with a self-developed simulation tool, that calculates PV module performance based on cell-level shading and the electrical design of the individual modules. A good agreement between simulated shading and measured shading is reached especially for full-cell and half-cell modules. The losses of the single diode

module were overestimated, probably due to lack of knowledge of the exact characteristics of the integrated diodes on cell-level.

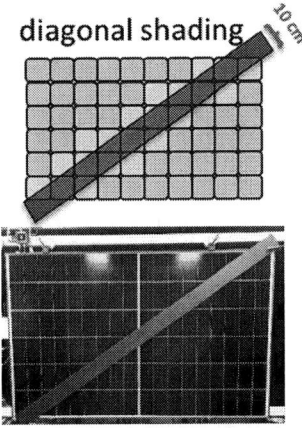

Figure 3: Top: Diagonal Shading S_{diag} introduced as a substitution for spot shading S_{spot}. Bottom: Photo of the shading S_{diag} applied to a half-cell module.

3.1 Adjustment of the test procedure

Due to the inconclusive results of S_{spot}, Laboratory 3 suggested substituting the spot shading with a well-defined diagonal shading, S_{diag}, as shown in Figure 3. This type of shading is typically cast by cables or poles mounted near PV systems. Measurements with the original three shading conditions, along with the additional S_{diag}, were repeated on all four PV modules using an overhauled LED-based solar simulator and under outdoor conditions. The results of this second test series are shown in Table **3**. Repeated measurements conducted by PTB, both indoors and outdoors, showed very good agreement for S_{short} and S_{long}. However, S_{spot} again demonstrated poor reproducibility in both environments. Notably, the shingled module was classified differently at both test sites compared to the earlier measurement series. S_{diag} showed less than 3 % variation when comparing PTB's indoor and outdoor results. Compared to Lab 3, which also tested all modules with S_{diag}, we observed differences of up to 5 % in comparison to PTB's measurements. We assume this discrepancy is due to the high sensitivity to shade placements - small variations can lead to significant differences in module power. It is much easier to place shading material flush against a specific side of a module than to position it diagonally. A precise definition of the position and thorough placement of the shading material is crucial in this case. Nevertheless, in terms of classification, the results with S_{diag} are more stable compared to S_{spot}, though less reproducible than those obtained with S_{short} and S_{long}.

4   CONCLUSIONS

A qualitative metric defining the shading tolerance of PV modules has been developed and tested. The metric comprises three distinct, clearly defined shading conditions that can be readily reproduced by various laboratories. Two of the three proposed shading scenarios have been demonstrated to be robust in a laboratory intercomparison. However, the spot shading condition was found to be insufficiently reproducible, and an alternative diagonal shade condition has been proposed. The results obtained in this study can contribute to future discussions within the IEC TC82 WG2 standardization working group, aimed at developing a standard for the shade tolerance classification of PV modules.

5   ACKNOWLEDGMENTS

The project 19ENG01 (Metro-PV) leading to this publication has received funding from the EMPIR programme co-financed by the Participating States and from the European Union's Horizon 2020 research and innovation programme.

6   REFERENCES

[1] R. G. Vieira, F. M. U. de Araújo, M. Dhimish, M. I. S. Guerra: *A Comprehensive Review on Bypass Diode Application on Photovoltaic Modules*, Energies, MDPI, vol. 13(10), pages 1-21, 2020

[2] H. Hanifi, J. Schneider, J. Bagdahn: *Reduced Shading Effects on Half-Cell Modules – Measurement and Simulation*, 31st EU PVSEC, 2015

[3] N. Klasen, D. Weisser, T. Rößler, D. H. Neuhaus, A. Kraft: *Performance of shingled solar modules under partial shading*, Prog. Photovolt. Res. Appl. **30**, 2022

[4] H. Sträter, S. Riechelmann, S. Winter: *An Approach for a Shading Resistance Classification of PV Modules*, WCPEC-8, 2022

Table 3: Results of repeated measurements of all shade situations, including the additionally defined diagonal shading.

	module type	PTB Old LED 23		PTB New LED 24		PTB Outdoor 23		PTB Outdoor 24		L3 Indoor	
long-side shading	full-cell	9,93	B	10,16	B	10,73	B	10,54	B		
	half-cell	11,00	B	11,20	B	11,37	B	11,73	B		
	shingle	78,04	D	77,78	D	78,03	D	77,74	D		
	diodes	9,55	B	9,65	B	9,58	B	9,71	B		
short-side shading	full-cell	85,03	D	84,73	D	84,96	D	84,44	D		
	half-cell	35,80	C	35,70	C	35,48	C	34,94	C		
	shingle	8,42	B	8,76	B	8,23	B	7,04	B		
	diodes	22,85	B	23,04	B	20,52	B	21,77	B		
spot shading	full-cell	2,35	B	2,08	B	1,14	A	1,86	A		
	half-cell	4,27	B	3,47	B	3,75	B	2,91	B		
	shingle	4,46	B	5,82	C	5,39	C	4,41	B		
	diodes	1,93	A	2,08	B	1,46	A	1,48	A		
diagonal shading	full-cell	-	-	48,82	C	-	-	46,09	C	50,79	D
	half-cell	-	-	47,68	C	-	-	47,75	C	46,86	C
	shingle	-	-	76,76	D	-	-	79,03	D	80,43	D
	diodes	-	-	28,18	C	-	-	28,47	C	32,17	C

41st European Photovoltaic Solar Energy Conference and Exhibition

This presentation was selected by the Sc. Committee of the EU PVSEC 2024 for submission of a full paper to one of the EU PVSEC's collaborating peer-reviewed journals.

EXPLORING DUST PARTICLE PROPERTIES AND PV SOILING MAPPING: A CASE STUDY IN THE ARID LANDSCAPE OF A DESERT ENVIRONMENT

Brahim Aïssa[1], Atef Zekri[2], Mosab I.A. Kareem Subeh[1]

[1]Qatar Environment & Energy Research Institute (QEERI), Hamad bin Khalifa University (HBKU), P.O. Box 34110, Doha, Qatar
[2]HBKU Core Laboratories, Hamad bin Khalifa University (HBKU), P.O. Box 34110, Doha, Qatar
*Corresponding author: baissa@hbku.edu.qa

ABSTRACT: During the settlement of dust particles onto PV module glass surfaces, the incident light is either reflected or absorbed, leading to reduced light transmission to the solar cell. Soiling-induced PV energy-yield losses can be substantial, exceeding 1% per day in dusty environments [1]. Although regular cleaning is essential, it raises O&M increasing costs. Soiling is increasingly pertinent due to expected growth in PV installations, particularly in regions highly impacted by soiling, such as parts of China, India, the USA, and the Middle East and North Africa (MENA) [2]. Factors like airborne dust concentration and dry period length aid in predicting average soiling rates at specific locations. Short-term predictions require additional parameters like wind speed, humidity, and temperatures [3,4]. In desert environments, dew formation can occur, enhancing particle adhesion through processes like cementation, particle caking, and capillary aging. This study builds on prior investigations into soiling microstructures on glass surfaces in Doha, Qatar. Qatar's climate, characterized as a subtropical dry, hot desert climate, influences particle cementation, mainly attributed to the clay mineral palygorskite, challenging the common perception of salt-induced cementation. This research addresses open questions by performing outdoor tests, exposing glass specimens, and employing various characterization methods. Meteorological parameters and dust collected from PV modules were analyzed, emphasizing their influence on soiling formation and optical losses. Results are compared across multiple characterization methods, providing insights into dust particle physical properties perspectives, soiling rates, and relevance of using commercial Dust IQ sensors [4,5].
Keywords: Soiling, Arid region, Dust particle, Particle size distribution.

1 INTRODUCTION

The deployment of photovoltaic (PV) systems in desert environments presents significant challenges due to the pervasive issue of soiling, primarily caused by the accumulation of dust particles on solar panels. Soiling is known to impede the efficiency and overall performance of PV systems by reducing the amount of solar irradiance reaching the active layers of solar modules, thereby limiting their energy output. In arid regions, where dust storms and dry climatic conditions are frequent, this issue becomes even more pronounced, leading to substantial energy losses over time. According to recent studies, soiling can result in power losses ranging from 1% to 7% per day, depending on the geographical location, dust properties, and weather conditions [6-8].

Understanding the physical and chemical properties of dust particles is critical to assessing their impact on PV soiling. Dust characteristics such as particle size distribution, chemical composition, and surface adhesion properties vary across different desert environments and influence the rate of soiling accumulation on PV modules. For instance, in regions with fine particulate matter, dust tends to adhere more firmly to solar panels, making cleaning operations more challenging and frequent. Additionally, environmental factors such as wind speed, relative humidity, and air temperature also play crucial roles in the deposition and adhesion of dust particles.

Several mitigation strategies have been proposed to counteract the effects of soiling in desert regions, ranging from mechanical cleaning systems to anti-soiling coatings and advanced sensor-based monitoring networks. Among these, the strategic selection of site locations for PV installations and the development of accurate soiling-risk maps have emerged as primary methods for minimizing energy losses due to soiling. These maps incorporate both geographical and seasonal data on dust accumulation, allowing stakeholders to make informed decisions regarding cleaning schedules and technology deployment. [9-11]

In recent years, advances in sensor technologies have enabled real-time monitoring of soiling effects, particularly through the use of specialized dust sensors like the DustIQ. These sensors, integrated into national monitoring networks, provide high-resolution data on dust deposition rates and related meteorological parameters, such as wind direction, air pressure, and particulate matter concentrations. For example, the Qatar Environment and Energy Research Institute (QEERI) has successfully implemented a National Soiling and Environmental Parameters Monitoring Network that tracks soiling rates, albedo variations, and particulate matter concentrations across multiple sites. The data gathered from these monitoring stations contribute to the creation of a soiling-risk atlas, which serves as a predictive tool for optimizing the positioning and maintenance of PV systems [12-15].

This study focuses on the exploration of dust particle properties and their impact on PV soiling in a desert environment, using Qatar as a case study. The research aims to analyze the characteristics of dust particles, assess their influence on PV performance, and develop a detailed soiling-risk map based on real-world data collected from QEERI's monitoring network. By combining empirical data with advanced modeling techniques, this study seeks to provide a framework for mitigating soiling losses in PV plants located in arid regions, ultimately contributing to the improved efficiency and long-term sustainability of solar energy systems.

2 METHODOLOGY

The efficient operation of photovoltaic (PV) systems is paramount for ensuring sustainable energy production, particularly as the world increasingly shifts towards renewable energy sources. One critical factor that significantly impacts the performance of PV systems is the accumulation of dust and particulate matter, a phenomenon commonly referred to as soiling. This research presents a detailed investigation into the effects of PV soiling within a desert environment, specifically using a case study conducted in the State of Qatar. This study adopts a multi-faceted approach, encompassing the analysis of the morphological, structural, and compositional properties of the dust particles, along with a thorough examination of the soiling effects.

To quantify the impact of soiling, various methods are employed, including hemispherical transmittance measurements at different wavelengths and multiple soiling quantification techniques. These techniques include assessments of particle mass, surface coverage, transmission loss, and the cleanliness index (ΔCI). By comparing the results obtained from these different methodologies, the study reveals significant insights into the complex nature of soiling and its implications for PV efficiency. Notably, the results indicate an exceptionally low mismatch of approximately 0.02% between the measurements taken by the Dust IQ sensors and the ΔCI values, thus validating the efficacy of these advanced sensors for accurate soiling assessment [16-17].

The research incorporates outdoor exposure experiments utilizing borosilicate glass coupons conducted at the Outdoor Test Facility (OTF) in Doha, Qatar, situated about 10 km from the coast. Borosilicate glass was deliberately selected for these experiments due to its hydrolytic stability, which is essential for maintaining the integrity of the samples during prolonged exposure to environmental conditions. The experiments were structured based on prior investigations and involved varying test periods to assess the accumulation of dust effectively. A total of five glass coupons, each measuring 5×5 cm² and 1.6 mm thick, were exposed at the OTF for durations ranging from 1 to 30 days. These samples were installed around noon on July 2, 2023 ("Day #0"), and subsequently removed after intervals of 1, 4, 8, 15, and 30 days.

Upon removal, the rear sides of the samples were meticulously cleaned with a dry cloth, and they were then stored in a temperature-controlled environment to prevent further contamination. Each sample was precisely weighed using a microbalance before and after exposure, allowing for accurate determination of the mass of soil accumulated. The meteorological conditions during the 30-day period were characterized by sunny weather, with global horizontal irradiation levels consistently exceeding 750 W/m², alongside rain-free conditions and light winds. To ensure comprehensive data collection, meteorological parameters at the site were recorded every minute throughout the experiments.

In addition to the soiling assessments, accumulated dust analyses were compared with soiling values obtained from Kipp & Zonen Dust IQ sensors. To further understand the mineral and chemical composition of the collected dust, analysis was performed using a Rigaku - SmartLab system. The quantitative mineralogical composition was determined through Rietveld analysis, employing High Score Plus software for accurate interpretation. The size distribution of dust particles was assessed using a Quantachrome CILAS 930 particle size analyzer, allowing for detailed insights into the particulate composition. Moreover, microstructural investigations and morphological characterizations of the dust samples were carried out using scanning electron microscopy (SEM), with images obtained from a FEI-Quanta

650/Talos system. Finally, UV-Visible spectroscopy was performed using a Perkin Elmer-Lambda 1050 instrument, utilizing 4 nm steps to capture a detailed spectrum of the dust samples.

Through this comprehensive methodology, the research aims to deepen the understanding of how dust contamination impacts PV performance in arid environments, ultimately contributing to the development of effective mitigation strategies that can enhance the efficiency of solar energy systems in such challenging conditions.

3   RESULTS AND DISCUSSION

In addition to the composition of dust, the distribution of particle sizes plays a crucial role in the soiling of photovoltaic (PV) surfaces, particularly in terms of optical losses and the forces that govern particle adhesion. The size of the dust particles significantly influences how light is scattered when it interacts with the PV modules. Research has demonstrated that the optical losses incurred from a specific mass of dust deposited per unit area tend to be greater when smaller particles are present [18].

This phenomenon can be attributed to the fact that smaller particles are more effective at scattering light, thereby reducing the amount of sunlight that reaches the PV cells and ultimately decreasing their efficiency.

Moreover, the adhesion and removal forces acting on dust particles are highly dependent on their size. Forces such as Van der Waals attraction, capillary action, and wind drag vary considerably with particle dimensions. Smaller particles may exhibit stronger adhesive properties due to their larger surface area relative to their volume, which can enhance the impact of adhesion forces like Van der Waals forces. Conversely, larger particles might be more susceptible to removal by wind drag due to their greater mass, but they can also present challenges in terms of adhesion in specific environmental conditions.

Understanding these dynamics is essential for developing effective cleaning strategies and optimizing the performance of PV systems in dusty environments. By studying both the composition and size distribution of dust particles, researchers can better predict the effects of soiling on energy yield and devise targeted interventions to mitigate these losses. Consequently, thorough analysis of dust characteristics should be a key component of any investigation into the soiling phenomenon, particularly in arid and semi-arid regions where dust accumulation is prevalent.

Figures 1 (a-d) illustrate the scanning electron microscopy (SEM) images of dust particles along with the corresponding image processing results, which were obtained using a modified in-house developed code based on the Kolmogorov function. This analysis allows for accurate determination of surface coverage and the quantification of the number of particles present on the surfaces examined. The SEM images reveal a diverse range of dust particle sizes and irregular distribution patterns, providing a clear visual representation of the complex nature of the dust (as exemplified in Fig. 1e).

Additionally, Figure 1f presents the mapping of particle size distribution across the State of Qatar, highlighting the spatial variation of dust characteristics throughout the region. To calculating surface coverage, the dust particles have been modeled as spherical entities, with their cross-sectional area corresponding to the shape of their associated circular profile defined by a specific radius. The analysis indicates that the deposited particles exhibit a size range extending from 0.6 μm to 50 μm, with a notable proportion of the particles—approximately 20 μm in diameter—contributing significantly to the soiling effects on photovoltaic surfaces. This detailed understanding of the particle size distribution is essential for evaluating the impact of dust accumulation on energy generation efficiency in photovoltaic systems, particularly in arid environments like Qatar.

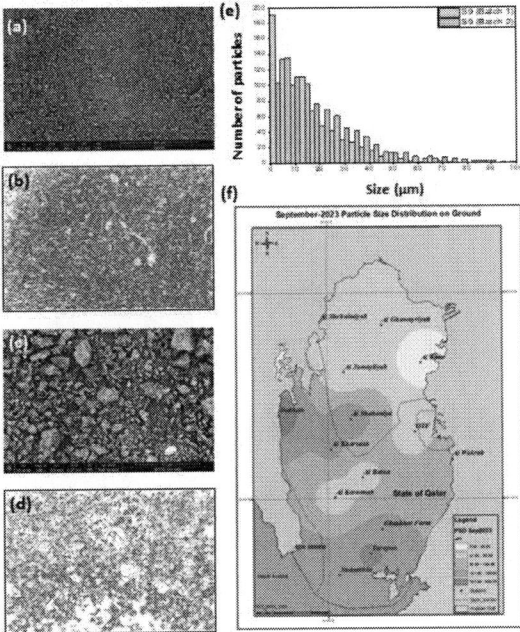

Figure 1: Representative SEM (a & c) and associated images of particle size recognition (b & d) of the dust particles deposited onto glass coupons. (e) Histogram showing a typical example of the size particle distribution. (f) Representative GIS mapping of the particle size distribution over the state of Qatar.

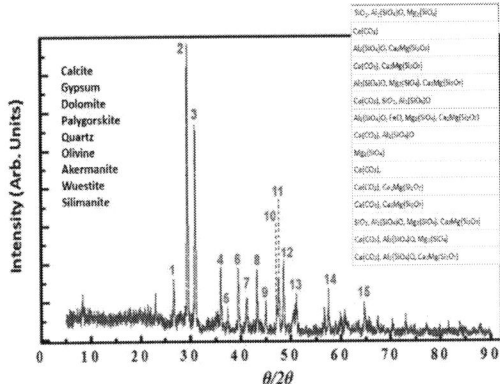

Figure 2: X-ray diffraction (XRD) diagrams of dust samples collected from the Outdoor Test Facility (OTF).

Figure 2 presents the results of the X-ray diffraction (XRD) measurements conducted on dust samples collected from the Outdoor Test Facility (OTF) during July 2023. The diagrams prominently feature peaks corresponding to the most abundant minerals identified within the dust, including calcite, dolomite, quartz, gypsum, olivine, akermanite, wüstite, sillimanite, and

palygorskite, which are clearly marked by numbered indicators. The analysis reveals that all samples collected from Qatar predominantly exhibit strong peaks for calcite and dolomite, with only minor variations observed for other minerals, such as gypsum.

Subsequent to the mineralogical analysis, we examined the optical properties of the exposed glass coupons and performed a detailed analysis of the collected dust particles. Figure 3a illustrates the results of the light transmittance measurements for these glass coupons. Notably, the results indicate a significant reduction in light transmission through the glass as the duration of outdoor exposure increases. The graphs for samples experiencing high soiling losses display transmittance values that are "shifted" towards lower values, which also suggests a slight wavelength dependency. Specifically, increased transmission losses are observed for shorter wavelengths, particularly below 600 nm (data not shown).

To quantify the corresponding transmission losses for all samples, we employed the following Equation (1), as demonstrated in Figures 3 (a-d). The hemispherical total transmittance of the soiled glass samples was measured utilizing a Perkin Elmer Lambda 1050 spectrometer, configured in conjunction with a 150 mm integrating sphere, over a wavelength range of $\lambda = 320–1300$ nm. Both clean and soiled glass samples were analyzed to ascertain the loss in transmission, denoted as TLoss. This loss is calculated as the weighted difference between the integral transmittance of the clean samples T0 and the soiled samples Tn . This thorough analysis not only highlights the impact of dust accumulation on the optical performance of PV surfaces but also underscores the necessity for effective soiling mitigation strategies in desert environments.

$$T_{Loss}[\%] = \int_{320 \ nm}^{1300 \ nm} (T_0 - T_n)d\lambda \Big/ \int_{320 \ nm}^{1300 \ nm} T_0 d\lambda$$

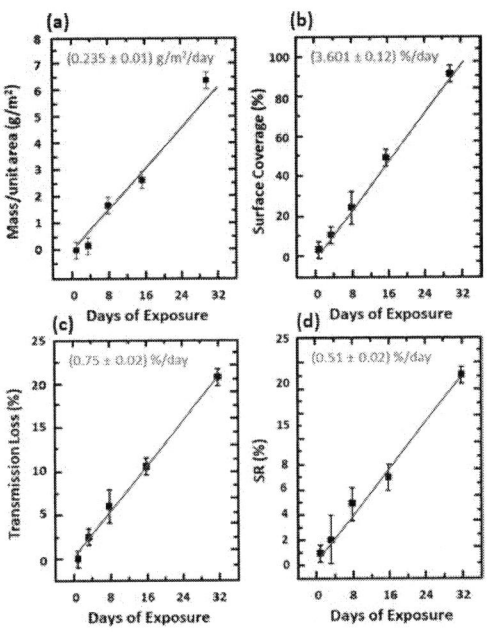

Figure 3: Hemispherical transmittance of soiled glass samples depending on wavelength after 1, 4, 8, 15 and 30 days of outdoor exposure. Comparison of different soiling quantification methods, including (a) particle mass per

unit area, (b) surface coverage, (c) transmission loss, and (d) cleanliness index (SR%). The data was linearly fitted, the corresponding values for the slopes are displayed in red.

Figures 3 (a-d) illustrate the comparative results of soiling quantification through various metrics, specifically mass per unit area, surface coverage, transmission loss, and the change in cleanness index (referred to here as SR%), as a function of exposure time. Across all measurement methods, a consistent linear increase in soiling is observed with prolonged exposure time, demonstrating the impact of dust accumulation on the surfaces.

To determine the soiling rate for each quantification method, linear fitting was applied to the collected data, with the intersection point on the y-axis standardized to zero. The soiling rate calculated from mass per unit area is expressed in grams per square meter per day (g/(m²/day)), while the other three methods are represented in percentage per day (%/day). Although all methods exhibit a similar linear trend in soiling as exposure time increases, they yield distinct values for the soiling rate. This discrepancy can be directly attributed to the differences in the measurement techniques employed.

The linear relationships observed in the data underscore the reliability of these quantification methods in tracking soiling progression over time. However, the variations in the calculated soiling rates highlight the necessity for careful consideration when selecting a method for assessing soiling impacts. Understanding these differences is crucial for accurately evaluating the performance of photovoltaic systems and devising effective strategies for soiling mitigation, particularly in arid and semi-arid environments where dust accumulation can significantly affect energy yield.

Figure 4 presents the spatial mapping of the Soiling Rate percentage (SR%) across the state of Qatar for July 2023, as derived from data collected by the soiling DustIQ sensors. In total, 15 strategically selected geographical sites throughout Qatar have been equipped with Kipp & Zonen DustIQ sensors, which provide precise measurements of the soiling effect on surfaces. These sensors are installed at a tilt angle of 22°, matching the orientation of the solar photovoltaic (PV) modules and glass coupons utilized in the Outdoor Test Facility (OTF).

The soiling rate measured at the OTF was approximately 0.49%, which closely aligns with the rate calculated through the Change in Cleanness Index (ΔCI) loss, recorded at 0.51%, also represented here as SR%. The remarkably low discrepancy of less than 0.02% between these two measurements underscores the accuracy and reliability of the DustIQ sensors in quantifying the soiling effects on PV systems.

This mapping provides valuable insights into the spatial variability of soiling across different regions of Qatar, highlighting areas with potentially higher soiling rates and enabling targeted maintenance and cleaning strategies for PV installations. By understanding these variations in soiling rates, stakeholders can make informed decisions regarding the deployment and upkeep of solar energy systems, ultimately optimizing energy production in arid environments where dust accumulation is a critical concern.

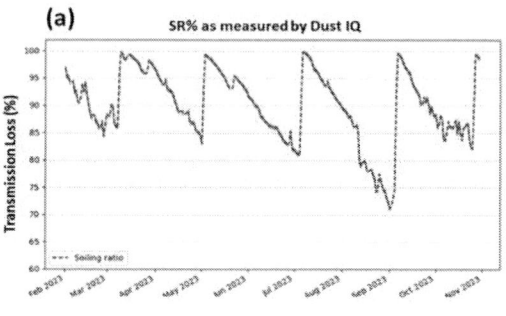

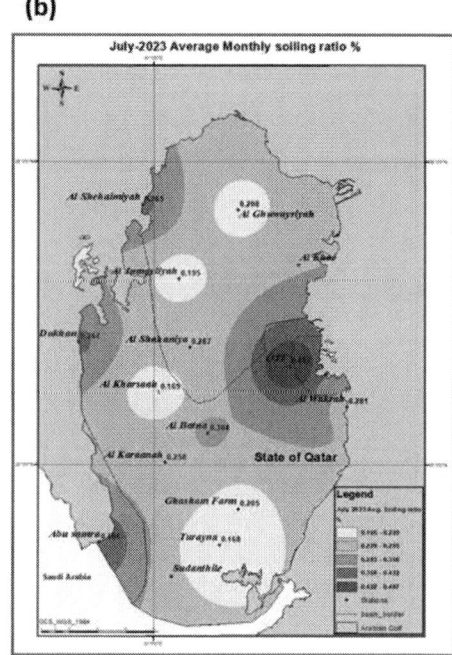

Figure 4: (a) Transmittance Loss measured by Dust IQ sensor located in OTF, for the period of Feb.- Nov. 2023.(b) GIS mapping of the SR% over the state of Qatar for the month of July 2023.

4 SUMMARY AND CONCLUSIONS

The contamination of photovoltaic (PV) modules due to the accumulation of dust and particulate matter is a significant challenge that results in substantial energy yield losses. This issue is becoming increasingly critical as the deployment of solar power expands rapidly in arid and semi-arid regions around the globe. These environmental conditions, marked by frequent dust storms and prolonged dry weather, exacerbate the soiling phenomenon, creating serious obstacles to the efficient operation of PV systems. In response to this pressing concern, a comprehensive series of soiling experiments were conducted at the Qatar Environment and Energy Research Institute's (QEERI) Outdoor Test Facility (OTF). This research specifically focused on the exposure of glass specimens to various dust compositions and conditions that are typical of the region. The experimental findings from these investigations indicated that energy yield losses attributed to dust soiling averaged more than 0.5% per day. This highlights the severity of the soiling issue, particularly in the context of energy production, where every fraction of efficiency is critical for maximizing output. The dust prevalent in Qatar plays a pivotal role in these soiling losses and is composed of a diverse array of minerals. Among these, significant amounts of carbonates such as calcite and dolomite are present, alongside other minerals like quartz, palygorskite, and gypsum. This mineral composition is reflective of the broader Middle East and North Africa (MENA) region, where similar conditions prevail.

Importantly, both dolomite and calcite contribute extensively to the soiling effect, as their unique physical properties enhance their ability to adhere to the surfaces of PV modules. The dust particles that accumulate on these surfaces exhibit a wide range of sizes, spanning from as small as 0.6 μm to as large as 50 μm. Notably, there is a particularly high volume of particles clustering around an average diameter of approximately 20 μm. This size distribution is critical, as it suggests that the soiling material can effectively obscure the surface of the solar cells, significantly hindering their ability to convert sunlight into electricity and reducing overall energy yield. To accurately quantify the soiling rate, various methodologies were employed, including assessing surface coverage, measuring light transmission losses, and evaluating the mass per unit area of the deposited dust. These quantification techniques were rigorously cross-referenced with values recorded by advanced monitoring systems, such as DustIQ sensors, revealing a robust correlation between the different methods. Specifically, the method based on the Change in Cleanness Index (ΔCI%) exhibited a minimal mismatch of only 0.02%, indicating a high degree of accuracy in the characterization of dust soiling.

The insights gained from these investigations into soiling characterization methods not only illuminate the specific challenges faced in desert environments but also possess broader applicability. We believe that the findings from this research can serve as a valuable reference for PV deployment in diverse climatic conditions. This understanding allows for enhanced strategies to manage soiling losses, ultimately improving the efficiency and reliability of solar energy systems worldwide. As the global community increasingly seeks to rely on renewable energy sources, it is critical to optimize the performance of PV technologies across varied geographical and environmental contexts, ensuring that the transition to sustainable energy is both effective and resilient.

References

[1] A. Bernecker et al., Renewable Energy 150 (2020) 725–735.

[2] B. Guo, W. Javed, B.W. Figgis, T. Mirza, "Proceedings of the First Workshop on Smart Grid and Renewable Energy (SGRE)," IEEE, 2015, pp. 1–6.

[3] B.W. Figgis, B. Guo, W. Javed, K. Ilse, S. Ahzi, Y. Rémond, Proceedings of the 5th IEEE International Renewable and Sustainable Energy Conference (IRSEC'17), Tangier, 2018.

[4] D. Bello et al., Renewable and Sustainable Energy Reviews 112 (2019) 805–815.

[5] I. Javed, M.B. Mansoor, M. Siddiqui, H.S. Arif, Renewable Energy 160 (2020) 211–222.

[6] K. Ilse, J. Rabanal, L. Schonleber, M.Z. Khan, V. Naumann, C. Hagendorf, J. Bagdahn, IEEE J. Photovolt. 8(1) (2018) 203–209.

[7] K. Ilse et al., Joule 3(10) (2019) 2303–2321

[8] K. Midtdal, B.P. Jelle, Sol. Energy Mater. Sol. Cells" 109 (2013) 126–141.

[9] T. Mekhilef et al., Solar Energy Materials and Solar Cells 220 (2021) 110907.

[10] S.A. Kalogirou, Renewable Energy Reviews 119 (2020) 503–511.

[11] S. Rajput, T.N. Singh, Energy Conversion and Management 173 (2018) 705–715.

[12] W. Javed, B. Guo, B.W. Figgis, Sol. Energy 157 (2017) 397–407.

[13] L. Micheli, M. Muller, Prog. Photo.: Res. Appl. 25(4) (2017) 291–307.

[14] H. Shiraiwa et al., Energy Procedia 147 (2020) 73–82.

[15] G. Hassan, G. Pozza, F. Lettner, M. Kober, Energy Procedia 155 (2018) 292–299.

[16] K. Ilse, M. Werner, V. Naumann, B.W. Figgis, C. Hagendorf, J. Bagdahn, Phys. Status Solidi RRL, 10(7) (2016) 525–529.

[17] N.M.H. Gavi, B.D. Ngom, A.C. Beye, A.M. Strydom, B. Aissa, V.V. Srinivasu, M. Chaker, Journal of Magnetism and Magnetic Materials 324(6) (2012) 1172–1176.

[18] M.A. El Khakani, V. Le Borgne, B. Aïssa, F. Rosei, C. Scilletta, E. Speiser, et al., Applied Physics Letters 95(8) (2009).

[19] M.A. Habib, M. Barkat, B. Aissa, T. Denidni, Progress In Electromagnetics Research 88 (2018) 135–148.

3DO.20.6

PV MODULE CLEANING UNDER HOT DESERT CONDITIONS

Gerhard Mathiak, Afra Seentakath, Nithin Sha, Shashank Suvarn
Prashanth Gabbadi, Arumugham Muthusamy, Nabeel Ibrahim
DEWA R&D Center, Dubai, UAE

 # Agenda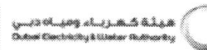

➢ **Introduction and Motivation**
- Need for Efficient Cleaning in MENA Region (Dust and Dry)
  - 25-Year Cleaning can be almost expensive than the Purchase of the PV Modules [Ilse 2019]
- Cleaning Test Approach
➢ **Environment at Site at Dubai**
- Soiling Rate, Weather and Condensation
- **Cleaning Test Field Layout**
  - MBR Solar Park in Small
  - PV Modules
  - Fixed Mounting and Single-Axis Tracker
  - Data Monitoring
➢ **Testing and Evaluation**
- Evaluation of Cleaning Systems
  - Cleaning Efficiency, Availability of Robots
  - Reliability of Modules (ARC, IV, EL)
➢ **Summary and Outlook**

> **Cleaning Test Approach**
> - Realistic conditions
>   - No test samples, real products
>   - Various modules, and mounting configurations
> - Relevance for MENA region
>   - Fully automated (dry) cleaning
>   - Methods of Solar Parks in MENA region
> - First Test Campaign of CTF
>   - Daily cleaning for one year
> - Cleaning Time and Conditions
>   - Consultation with robot manufacturers

Soiling at CTF Dubai

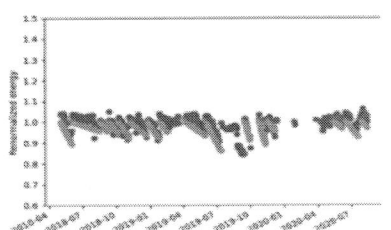 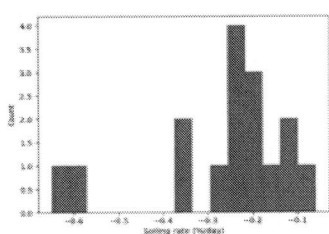

Sand and Dust
- ➢ Quartz Sand on Dunes
- ➢ Dust Devils and Shamal Dust Winds
  - ▪ Sticky Dust (High Calcium Content)

Soiling at OTF Dubai
- ➢ Single Module Analysis at OTF
  - ▪ Crystalline Module, 25° Tilt Angle
  - ▪ Fit (orange) to Renormalized Energy
    - ▪ More than 10% maximum
  - ▪ Histogram (Mean at 0.25%/day)

Soiling at CTF
- ➢ Fixed Structure
  - ▪ Soiling Band at the Bottom Frame
  - ▪ Complete Soiling with Drop Tracks
- ➢ Single-Axis Tracker
  - ▪ Dust Pattern when Horizontal

Soiling on Modules and Bristles

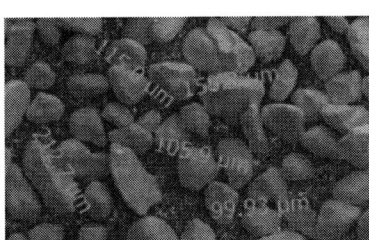

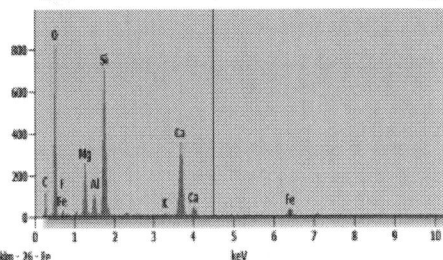

Dust collectected from Module (EDX with Ca, Fe, Mg, Al)

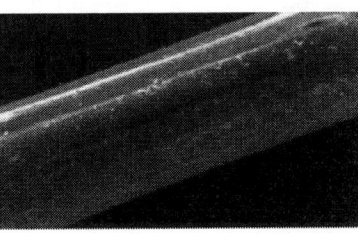

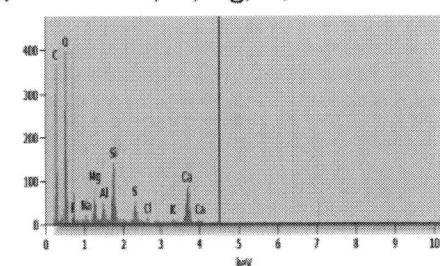

Dust collectected in sand trap after sand storm
- Same as from ground
- SiO2 (100-200 μm)

Dust on Bristle (EDX with Ca, NaCl, S, and C and N=Polyamide)

CTF Site Layout Sketch

Docking Station

➤ Different Mounting Configurations [Mathiak 2023]
  ○ Four Fixed Tables facing South
  ○ Eight Single-Axis Tracking Tables moving from the East to the West
➤ Each Table has two Strings with 16 modules (12 modules on fixed tables 1 and 2)
➤ Docking stations for the Cleaning Robots (2 m) and Space between two Strings (1.5 m)
  ○ Docking Station Construction in Cooperation with Robot Manufacturers

PV Array Layout (167 kWp)

MBR Solar Park in Small

- Framed Modules, Frameless Modules were impossible to Purchase

Table with 2 strings	Height	Module	Facial	Inverter
Fix 1	2p	Trina	monofacial	Inverter 1
Fix 2	2p	Trina	monofacial	Inverter 1
Fix 3	2p	Longi	monofacial	Inverter 2
Fix 4	2p	Longi	monofacial	Inverter 2
Tracker 1	2p	Longi	monofacial	Inverter 3
Tracker 2	2p	Longi	monofacial	Inverter 3
Tracker 3	2p	Longi	bifacial	Inverter 4
Tracker 4	2p	Longi	bifacial	Inverter 4
Tracker 5	1p	Longi	bifacial	Inverter 5
Tracker 6	1p	Longi	bifacial	Inverter 5
Tracker 7	1p	Longi	bifacial	Inverter 6
Tracker 8	1p	Longi	bifacial	Inverter 6

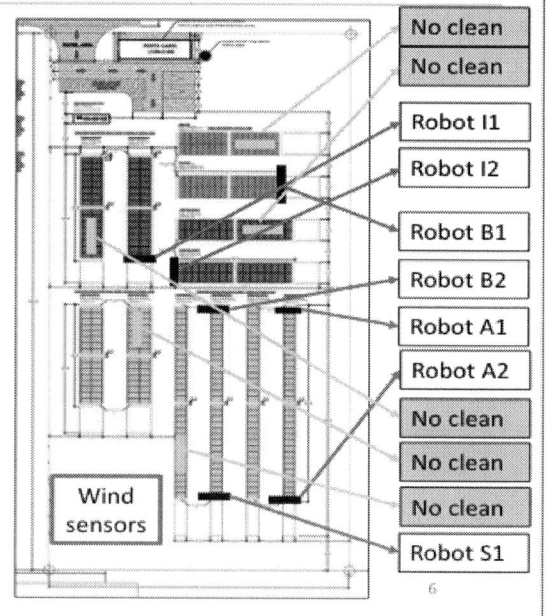

  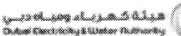

Data Monitoring

- **Inverters (five-minute intervals)**
  - Export Energy
  - Smart Inverter
    - IV Curve of each String, AC and DC
    - Performance Ratio of the PV System,
    - Frequency of Grid,
    - Grid Uptime, Fault Recording
  - Measurement of Current for soiling detection
- **Weather stations (one-minute intervals)**
  - Ambient Temperature, Module Temperature, Wind Speed, Solar Irradiance, Albedometer
- **Tracker systems (one-minute intervals)**
  - Tilt Angle, Wind speed
    - Gale Protection 0° horizontal
    - Night Return Angle changed to 20° East
    - Wet Cleaning Angle (weekly) 50°

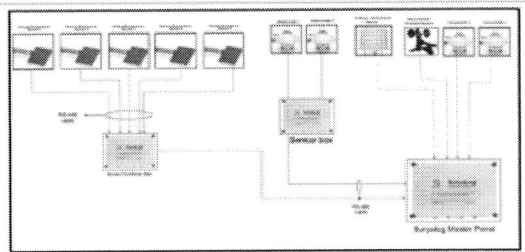

Layout of Weather Station Connection

Huawei Smart String Inverter

Impp on Fixed Structure

Impp - Comparison of 4 Strings of Inverter 1

- PV 4 with the Robot
- PV 6 Shading in the Evening

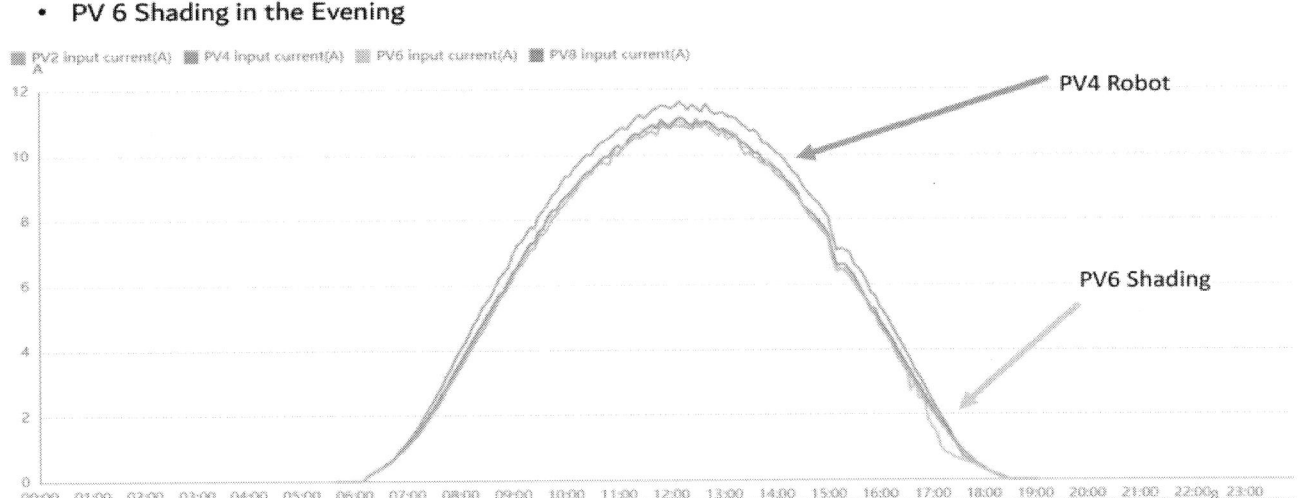

  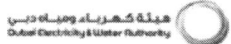

Impp on Single-Axis Tracker

Impp – Comparison of 4 Strings of Inverter 5: PV 1 with the Robot
- In the Morning similar Shading of all 4 Strings,
- In the Evening first Shading on PV 7 and no Shading on PV 5

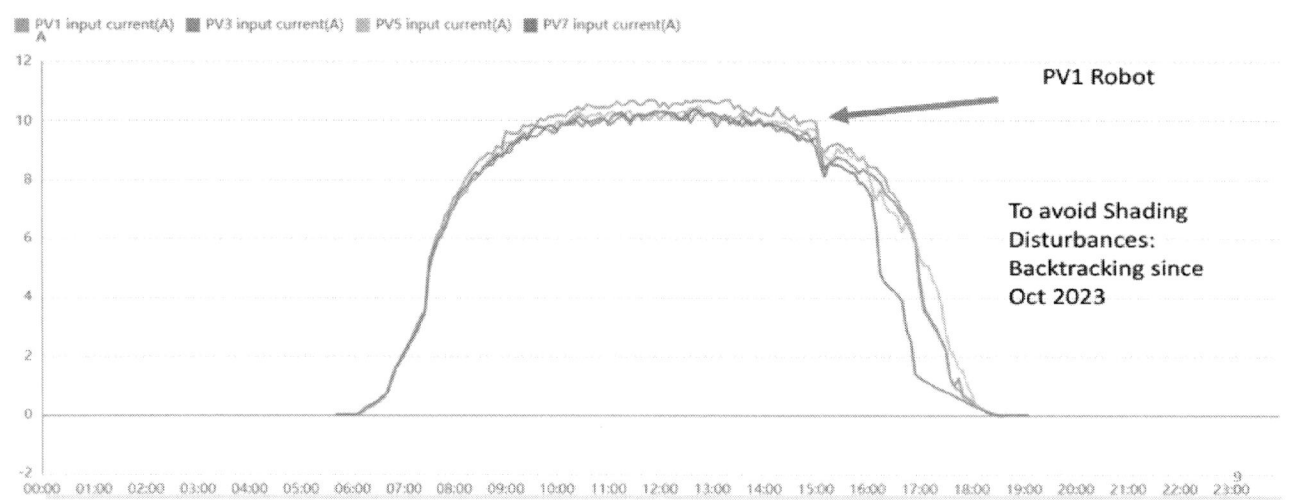

PV1 Robot

To avoid Shading
Disturbances:
Backtracking since
Oct 2023

I-V on Single-Axis Tracker

Isc Measurements at one Smart Inverter
- String I-V Curve Measurements of two Cleaned with Robots (blue and red) and two Non-cleaned (green and violet)
- Relative Isc Values over one Clear-sky Day. The Isc Value is normalized to the Average of the non-cleaned strings.
- Blue has higher Isc than Red.

Comparison of Relative Isc and Yield
- Relative Yield Gain over a Period of about one hour of two Module Strings cleaned with Robots (blue and red) and two Non-cleaned Strings (green and violet) on the same Clear-sky Day. The Yield Gain is normalized to the Average of the Non-cleaned Strings.
- The Yield Gain of blue is 2.5% higher than of red.
- The Measured Cleaning Effect of the Robots (Blue and red) is higher in the Morning and in the Afternoon than at Noon.

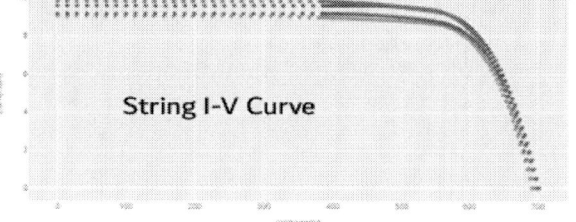

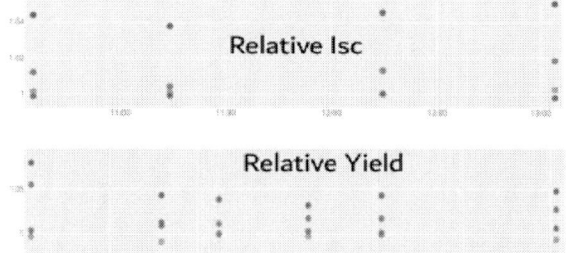

One Layer Interference

One Layer Porous SiO2 Coating

- Roller Coating (Sol-Gel-Process)
- Typically Closed Pores
- Color Simulation by [Karin 2022]
- Higher Porosity, lower Refractive Index n

TEM cross-section Image of ARC coating by [Law 2023]

White → Brown → Dark blue → Blue → White → Orange → Brown → Dark Blue to Blue → Turquoise → Yellow → Orange → Violet → Blue → Green

| 50 | 90 | 130 | 180 | 210 | 250 | 290 | 360 | 400 | 420 | 450 | 520 | 560 nm |

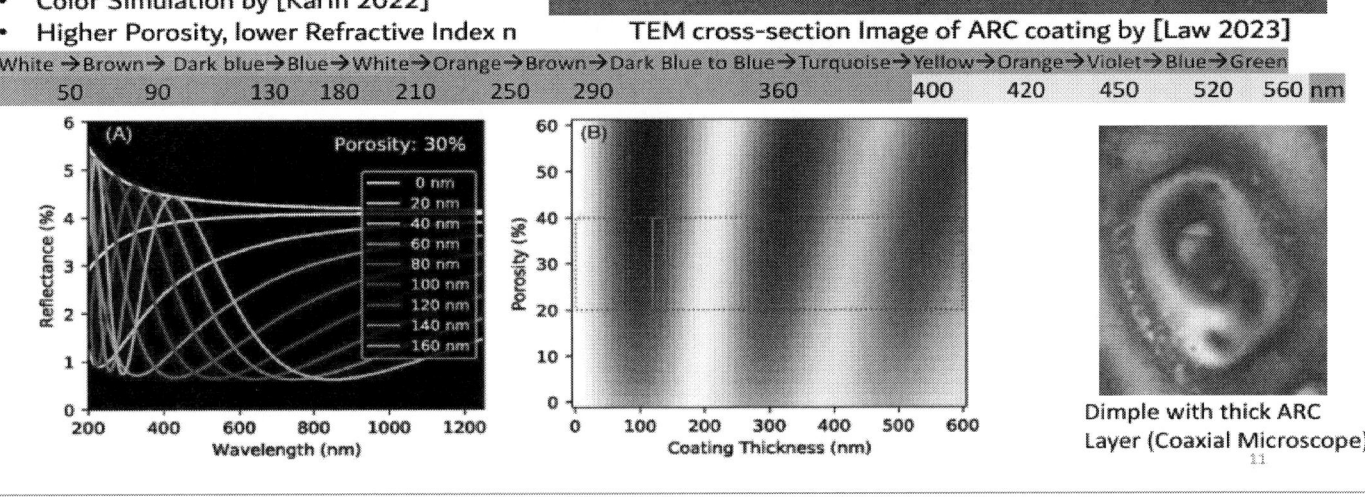

Dimple with thick ARC Layer (Coaxial Microscope)

Initial Reflectance and Microscopy

CTF Control Modules Reflectance Curves_2024

— Trina — Longi Monofacial — Longi Bifacial — Longi_bifacial_R

Longi

Trina

Longi Bifacial-445W

Front   Rear

Trina   Longi mono

ARC of both modules are different

- Longi is blue (50 to 150 nm) with brown hillocks
  - Monofacial and bifacial are slightly different
  - Longi Backglass is without ARC
- Trina has deeper dimples (yellow, orange)

Profilometer and Microscopy

White→	Brown→	Dark blue→	Blue→	White→	Orange→	Brown→	Dark Blue to Blue→	Turquoise→	Yellow→	Orange→	Violet→	Blue→	Green
50	90	130	180	210	250	290	360	400	420	450	520	560 nm	

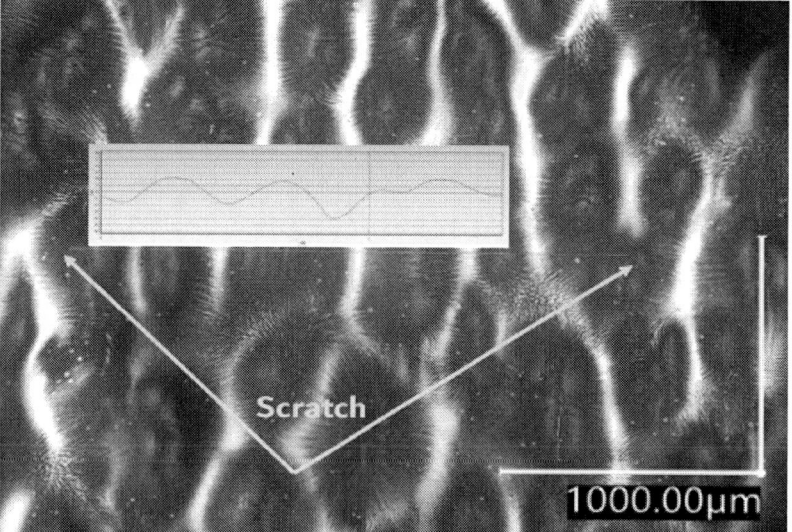

Due to Roller Coating Process
- Thin ARC (Brown) is on Hill
- Thick ARC (Bluewhite) is in Valley

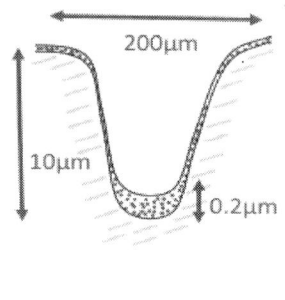

1-Year Degradation in CTF

  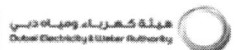

All installed Cleaning Robots were commissioned successfully
- Manual Control, Time-triggered, Internet-triggered successful test runs

Robots survived Dust, Strong Storms and Strong Rain
- But Exchange of Battery Controller Board was necessary for one Robot

Comparison Bristles (e.g. Polyamide) with Microfiber Cloth
- Cleaning Efficiency higher, but higher Degradation of ARC Coating
- Tracks of Dark Bristles on modules visible (also Dark Lines in Microscope)
- Tracks of Wheels on Modules visible
- Bristles get brittle and dusty, Cloth gets dirty too

CTF Modules, Tracker, Equipment
- PERC Modules are LETID Degrading
- Trackers are Retrofitted
- Inverters are better suited for Quality Checks
- Coaxial Microscopy is highly valuable

 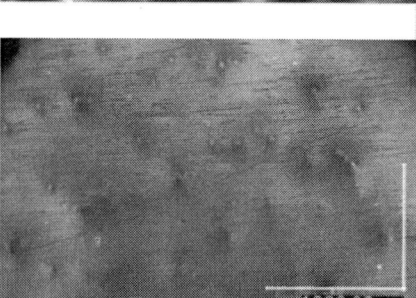

1-Year Degradation in CTF

Laboratory I-V curve Determination of Seven modules

- **Degradation of Isc, Voc, Pmpp and FF from Initial Lab Measurement**
- Degradation of Voc between 2.88% (no clean) and 3.99% (no clean)
  - Degradation of PERC modules can be LETID
    - Patchwork Pattern in EL (initial and 1-year)
- Degradation of Isc between 1.07% (no clean) and 2.31% (Cloth)
  - Ranking Longi (Cloth, Bristles, No Clean, Bristles, Cloth)
  - Ranking Trina (No Clean, Bristles)

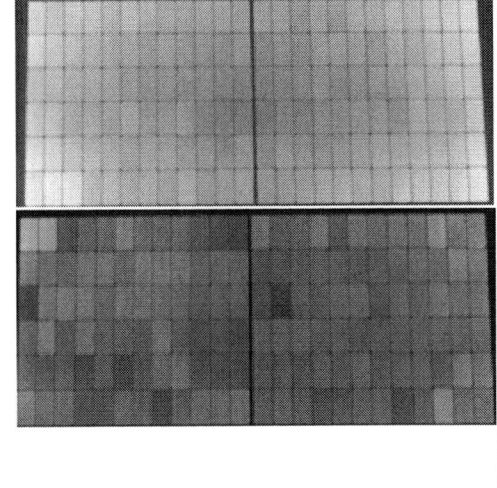

Module	Cleaning	I_{sc} (A)	V_{oc} (V)	P (W)
Longi	No Clean	1.51%	3.99%	5.62%
Longi	Cloth	1.42%	3.16%	4.54%
Longi	Bristles	1.43%	2.99%	4.30%
Longi	Cloth	2.31%	3.72%	5.60%
Longi	Bristles	1.69%	3.51%	5.39%
Trina	No Clean	1.07%	2.88%	3.91%
Trina	Bristles	2.13%	2.98%	4.51%

Field Degradation

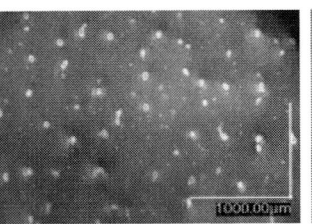 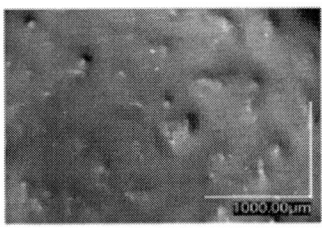

3-Year Field Degradation with Robotic Cleaning

- Color is Fading Out

Reflectance Simulation by [Khan 2020]

- Coating Thinning
- Coating Removal

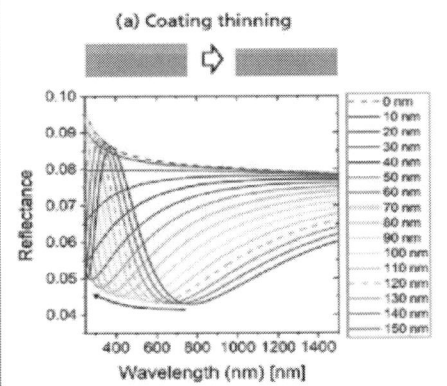 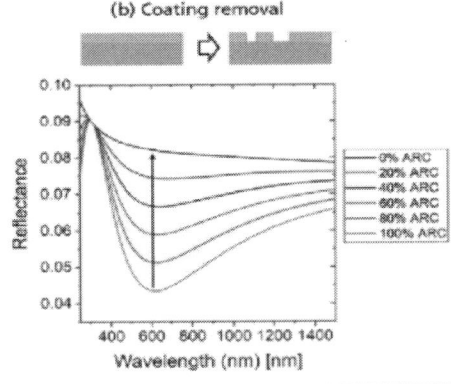 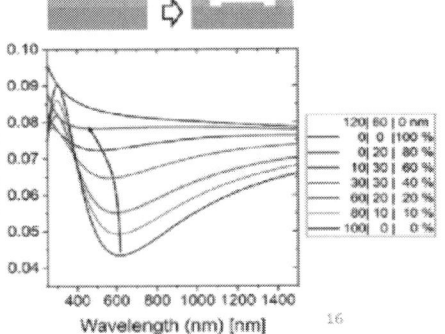

1-Year Degradation in CTF

Reflectance

- **CONTRADICTION between Measurement and Simulation (Coating Thinning)**
- **Thickening or Selective Removal ?**

Initial and current reflectance measurement results

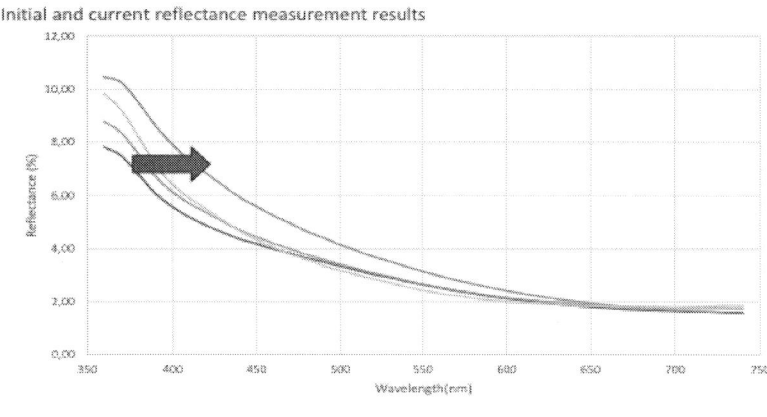

(a) Coating thinning

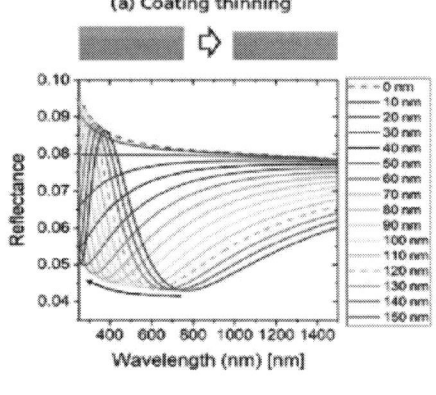

| ▬▬ Microfiber Cloth | ▬▬ Initial (Reference Module) |
| ▬▬ Bristle Brush | ▬▬ Non-cleaned Module |

Summary and Outlook

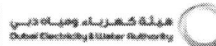

➢ **Optimize PV Cleaning systems**
  -  Standardization of Cleaning Procedures and Cleaning Tests (IEC): Trilateral (Soiling – Modules – Robots)
  - Optimization of Cleaning Schedule (daily/weekly)
  - Looking for alternative Soiling Mitigation Concepts, e.g. Anti-Soiling Coatings

➢ **Compare Outdoor and Indoor testing [IEC 62788-7-3]**
  - Blowing Sand Test
    - Typical Glass Chipping was not found in the Field
  - Indoor Robotic Cleaning
    - 10000 Cycles = 25-Year daily Cleaning
    - With Artificial Soiling [Ferretti 2019a]

References

[Ferretti 2019a] N. Ferretti, A. El-Issa, and L. Podlowski, "Standardized test procedure for PV module cleaning equipment", *Proc. 36 EUPVSEC*, 2019.

[Ferretti 2019b] N. Ferretti, "PV Module Cleaning Market Overview and Basics", *White paper PI Berlin*, 2019

[IEC] IEC 62788-7-3:2022, "Measurement procedures for materials used in photovoltaic modules - Part 7-3: Accelerated stress tests - Methods of abrasion of PV module external surfaces".

[Ilse 2019] Klemens Ilse et al.: Techno-Economic Assessment of Soiling Losses and Mitigation Strategies for Solar Power Generation, Joule 3/10, 2019, 2303-2321

[Karin 2022] Todd Karin, Mason Reed, Jim Rand, Robert Flottemesch, Anubhav Jain: Photovoltaic module antireflection coating degradation survey using color microscopy and spectral reflectance, Progress in Photovoltaics, 30/11, 2022, 1270-1288.

[Khan 2020] Khan, M.Z., Pfau, C., Schak, M., Miclea, P.-T., Naumann, V., Debess, A., Hagendorf, C., Ilse, K.: Resilience of industrial pv module glass coatings to cleaning processes. J. Renewable Sustainable Energy 12 (5), 2020, 053504.

[Law 2023] Adam M. Law, Luke O. Jones, John M. Walls: The performance and durability of Anti-reflection coatings for solar module cover glass — a review. Solar Energy 261 (2023) 85-95

[Mathiak 2023] G. Mathiak, A. Seentakath, A. Muthusamy, J. J. John, "PV Module Cleaning Under Hot Desert Conditions: Creating Test Standards for PV Module Cleaning", 2023 Middle East and North Africa Solar Conference (MENA-SC), pp.1-6, 2023.

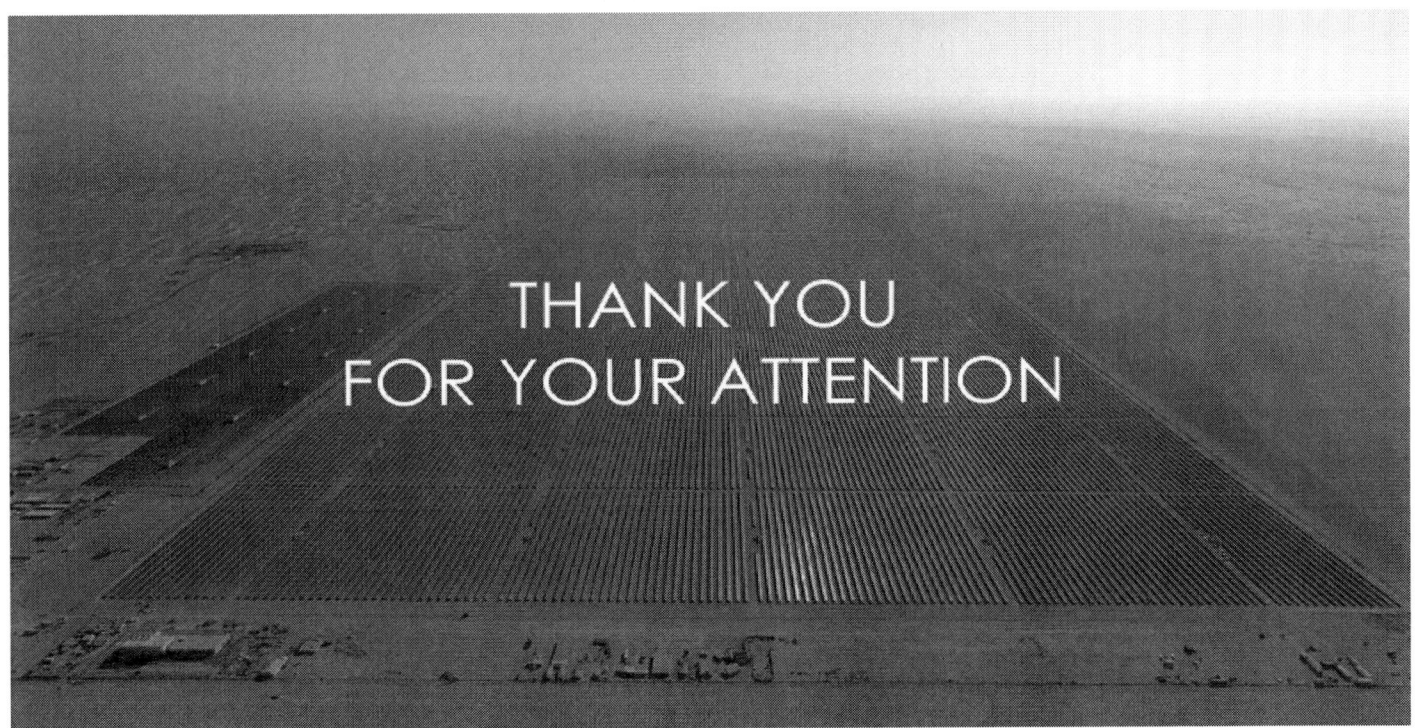

From Fab to Field: Quality control with a mobile PV laboratory

Vienna EU PVSEC, September 27th 2024

In Field Characterisation of PV Modules | BoS Components in Operation

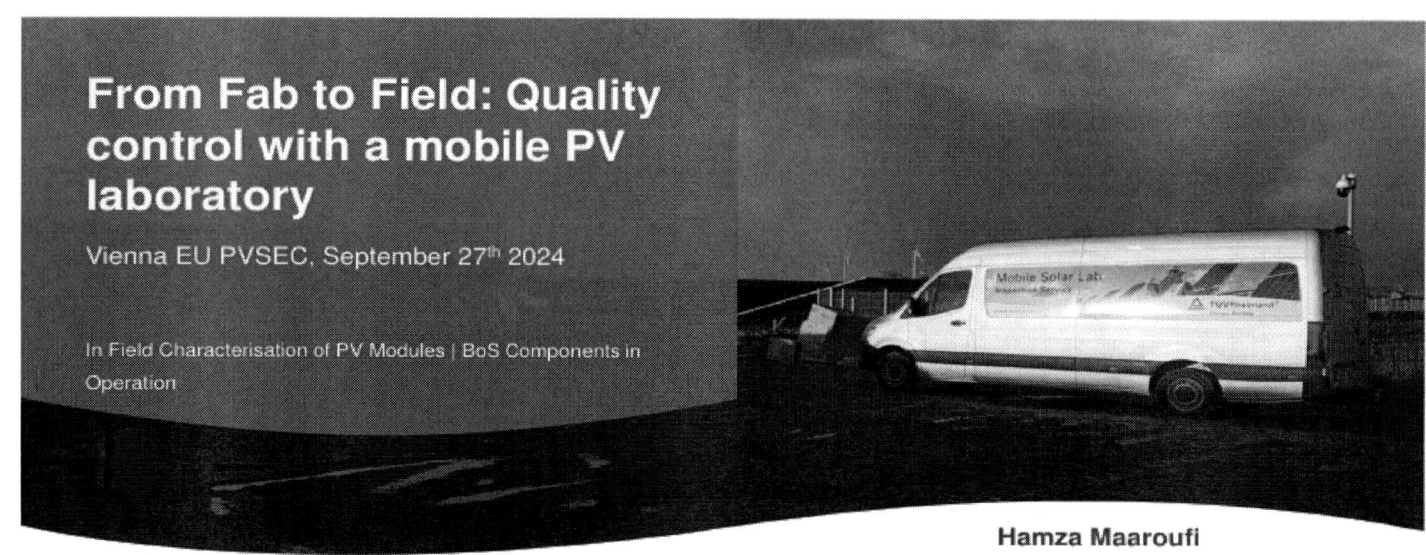

Hamza Maaroufi

Team coordinator PV Power Plants

TÜV Rheinland, Germany

www.tuv.com   TÜVRheinland® Precisely Right.

From Fab to Field: Quality control with a mobile PV laboratory

TÜV Rheinland

40 years of experience in PV

20,000 employees

250+ Solar Experts

6 PV Laboratories

> 20 GW inspected PV projects

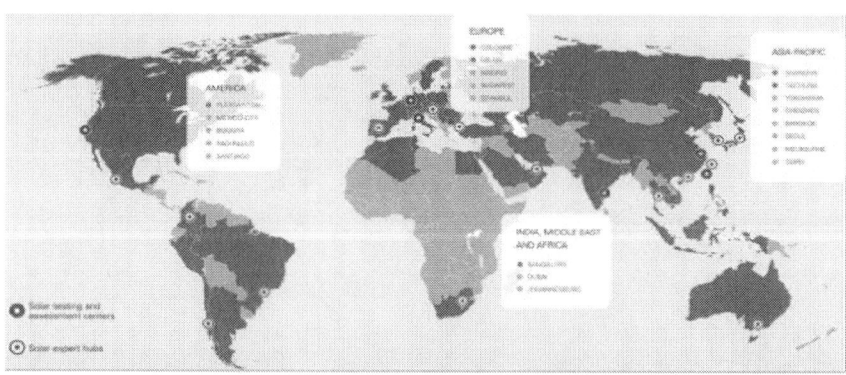

| Development | Production | Planning | Procurement | Construction | Commissioning | Operation |

Motivation for Quality Control

Market evolution cell technologies

Trend: share of cell technologies

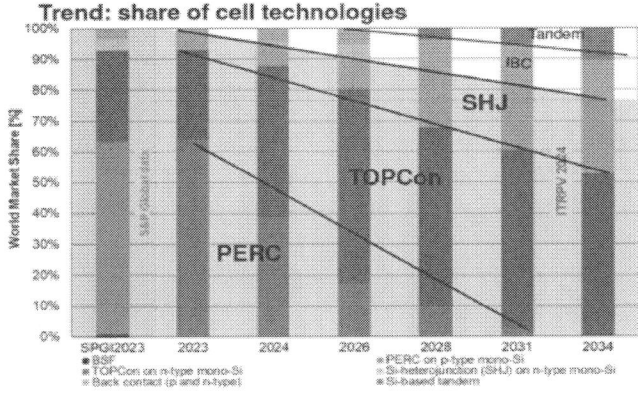

Source: ITRPV 2024

- AI-BSF dominated in the past, was phased out in 2023.

- PERC on p-type mainstream in 2023, < 40% in 2024.

- TOPCon on n-type, strong growth, 25% in 2023, 49% share in 2024.

- HJT and IBC still a niche market, but steadily growing

- Si-Perovskite tandem, expected after 2026 onwards

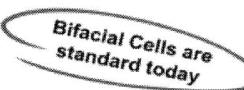

Bifacial Cells are standard today

TÜVRheinland®
Precisely Right.

Motivation for Quality Control

Market evolution PV-Modules size

				2019	2021	2023
Wafer Thickness	500 µm	300 µm	180-200 µm	170-180 µm	150-170 µm	130-150 µm
Solar Cell Size	100 mm	125 mm	156.75/158.75 mm	166 mm	182/210 mm	230R/182R mm
Model Dimensions and Power Rating	30-80 cells, 0.5 – 1.0 m² 60-100W	72 cells, 1.2 – 1.5 m² 180-200W	60/72cells, 1.6 – 2.1 m² 350-420W	60/72cells, 1.7 – 2.1 m² 380-460W	60/72cells, 2.5 – 3.0 m² 510-670W	72cells, 2.6 – 3.1 m² 630-740W

Market share based on module sizes

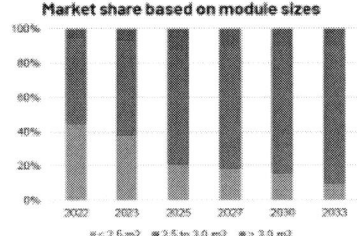

- Modules with a size between 2.5 m² and 3.0 m² will be the mainstream.

- M10 (182 x 182 mm²) / G12 (210 x 210 mm²) modules dominating the market

Trend: G2G will dominate in module technology

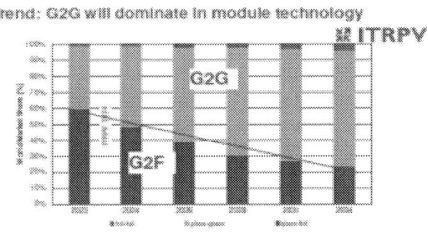

Source: PVInfoLink and ITRPV 2024

TÜVRheinland®
Precisely Right.

Motivation for Quality Control

The potential risks for PV-Modules

STATE-OF-THE ART PV-MODULES	POTENTIAL RISKS.
▪ PERC on p-type ▪ TOPcon on n-type ▪ Large PV modules up 2.6m² ▪ Thinner glass / Thinner cells ▪ Glass-Glass-PV modules / Transparent Backsheet ▪ Encapsulants : EVA / POE	▪ Overestimation of performance ▪ LID and Light and elevated Temperature Induced Degradation (LETID) ▪ Potential Induced Degradation (PID) ▪ Ultraviolet Induced Degradation (UVID) ▪ Glass breakage and cracks ▪ Impact of mounting systems on mechanical durability

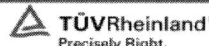

Failure and risk databases for Quality Control

IEA PVPS PV Failure Fact Sheets    TRUST-PV Risk Matrix

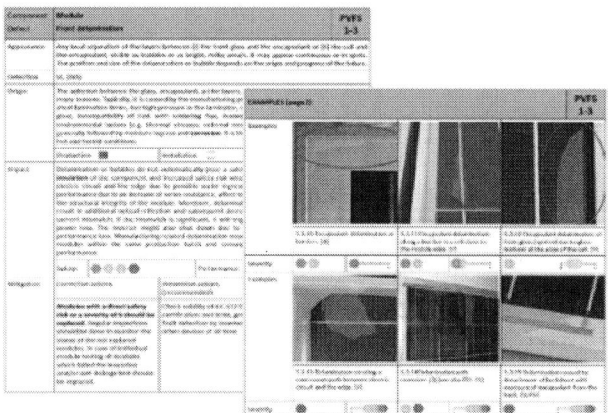

Herz, M. Friesen, G. Jahn, U. Koentges, M, Lindig, S. Moser, D. Identify, analyse and mitigate—Quantification of technical risks in PV power systems. Prog Photovolt Res Appl. 2022; 1 - 14. doi:10.1008/pip.3633

https://iea-pvps.org/research-tasks/performance-operation-and-reliability-of-photovoltaic- systems/documents/

Lindig, S. Gallmetzer, S. Herz, M, et al. Towards the development of an optimized Decision Support System for the PV industry: A comprehensive statistical and economical assessment of over 35,000 O&M tickets. Prog Photovolt Res Appl. 2022; 1 - 20. doi:10.1002/pip.3637

https://trust-pv.eu/reports/risk-matrix/

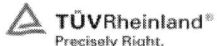

Quality Control during Construction

WHAT IS MOBILE LAB?

* It consists of a LED solar simulator class A+A+A+ acc. to IEC 60904-3 and IEC 60904-9 installed in the truck, an I-V curve analyzer and an EL inspection system / Low measurement uncertainty $(P_{max}) \leq \pm 2.3\%$ (k = 2) / high throughput: Measurement of approx. 100-150 PV modules per day

QUALITY ASSURANCE:

Post shipment inspection:

* Out-of-box measurement of PV modules prior to installation to check product quality
* Focus on transportation damage

WARRANTY CHECK

* Indicative measurements to verify warranty commitments or to support technical due diligence prior to asset sale
* Performance assessment

FAILURE ANALYSIS

* Diagnostic measurements of fielded PV modules (dismounting required)
* Degradation Investigation
* Bypass diode
* Rs

Measurement functions

Output power measurement of PV module

Electroluminescence record

MOBILE LAB: MEASUREMENT FUNCTIONS

Bypass diode functionality test

Low Irradiance measurements / Determination of PV module I-V correction parameters B1, B2, Rs

Quality Control during Construction

Post shipment / Pre-installation

* **4000 PV Modules:** pre-Installation Analysis 2023
* Cell Technology: **PERC** and **TOPcon**
* Measurements device: **Mobile lab** - MBJ LED Solar Simulator
* Tests:
  - **Visual inspection**
  - **Power measurement** @ STC
  - **Electroluminescence Images**

Source: TÜV Rheinland

Quality Control during Construction

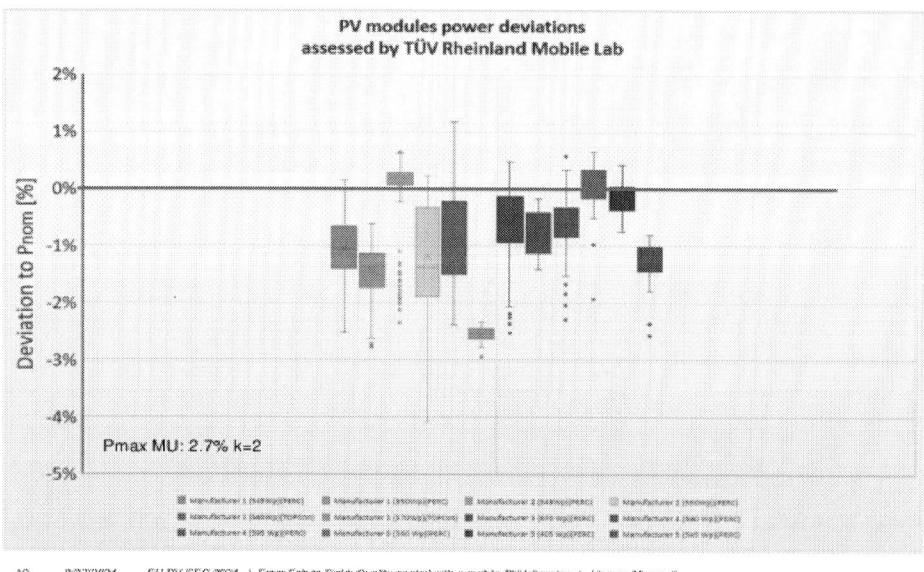

* Deviations from nominal power vary by manufacturer.

* Overestimation of performances

* Same module types - Large differences - **must be taken into account during sampling**

Quality Control during Construction

Post shipment / Pre-installation Inspection

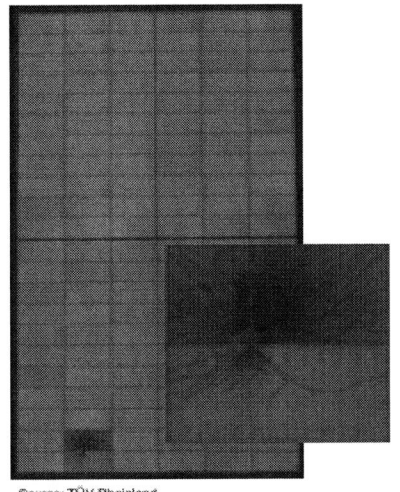 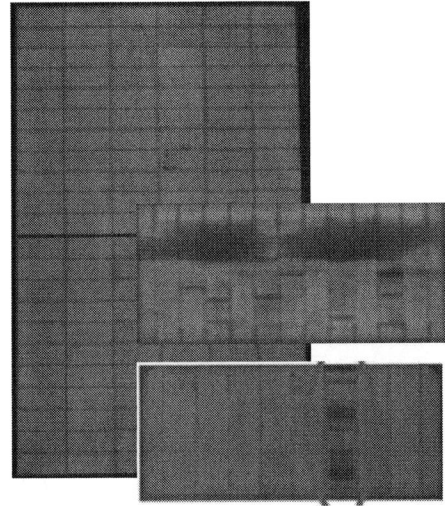

Source: TÜV Rheinland

- Post-shipment test campaign prioritizes transport damage assessment.

- Cracks detected are due to transportation.

- Manufacturing defects frequently identified instead.

- **Defects to undergo laboratory testing for potential future impact evaluation.**

Quality Control during Operation

EL Inspection after weather events

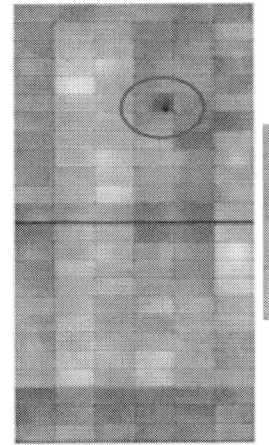 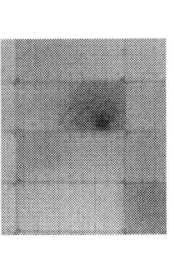 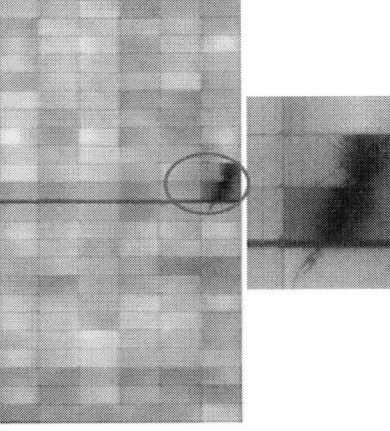 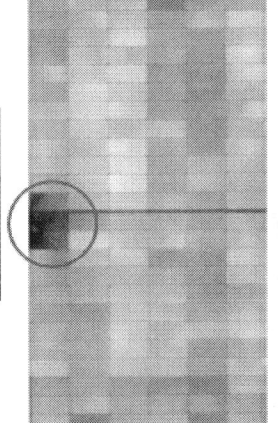 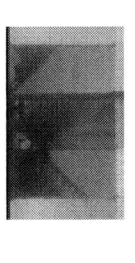

Conclusions

PV MODULES TECHNOLOGIES ARE DEVELOPING RAPIDLY

DEVELOPMENT AND RETESTING

POST SHIPMENT INSPECTION

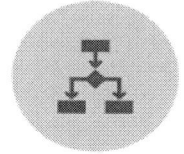

INSPECTION DURING OPERATION

Testing and field **experience** is required to assure **quality** and **long-term reliability**

41st European Photovoltaic Solar Energy Conference and Exhibition

REDUCTION OF UNCERTAINTY OF OUTDOOR PV MODULE CHARACTERIZATION: TEST FIELD EXPERIENCES

Mariella Rivera, Christian Reise
Fraunhofer Institute for Solar Energy Systems ISE
Heidenhofstraße 2, 79110 Freiburg, Germany
mariella.josefina.rivera.aguilar@ise.fraunhofer.de

This research provides insights from outdoor monitoring and characterization of PV modules. We compare results from two test sites: a previously used rooftop installation and the newly established test field of Fraunhofer ISE in Merdingen, Germany. The primary focus is on analyzing uncertainties in outdoor current-voltage (IV) curve measurements caused by effective irradiance and module temperature. The study addresses factors affecting irradiance measurements, including device calibration, low irradiance levels, high angles of incidence, and site mounting configurations that caused partial shading. Additionally, it examines uncertainties related to module temperature, such as measurements non-uniformity at array level and differences from cell to PV module's backsheet. We bring practical insights from Fraunhofer ISE's test field experiences and by assessing the impact of installation conditions on PV module measurement accuracy, we hope to contribute on the reduction of these uncertainties. The findings are intended to guide researchers in optimizing monitoring setups and highlight the importance of accurate irradiance and module temperature measurements.

Keywords: uncertainty, outdoor monitoring, test field

1 INTRODUCTION

Characterizing and monitoring the performance of photovoltaic (PV) modules under outdoor conditions are crucial both for PV research & development and for a sustainable business in the PV industry. This process relies on accurate measurements of solar irradiance, module temperature and module IV curves. Previous research has explored the impact of environmental factors on the accuracy of IV curve measurements and yield prediction of PV systems [1, 2]. Numerous studies have highlighted the importance of precise irradiance measurements, identifying key sources of uncertainty [3, 4] such as spectral mismatch for photodiode pyranometers; directional response for silicon reference cells; and deviations related to calibration, spectral mismatch, angular response and thermal offset for thermopile pyranometers.

Similarly, temperature measurement is another critical influencing factor on PV module performance evaluation. Studies, such as Ref. [5], have demonstrated that inaccuracies in temperature measurements – particularly those caused by non-uniform temperature distribution across PV modules, the number and position of sensors, and mounting configuration– cause significant challenges.

This publication aims to contribute with practical knowledge and experience in outdoor monitoring and PV module characterization by comparing a rooftop-installed monitoring station under not completely ideal conditions with an open-rack system at a newly established test field at Fraunhofer ISE in Germany. The focus is on analyzing the uncertainty of outdoor IV curve measurements due to two main sources: effective irradiance and module temperature measurements.

For the irradiance measurements, this study considers the influences of the device calibration (e.g., sensitivity, linearity, stability), temperature dependence, angular and spectral response, as well as the impact of using different types of irradiance sensors (reference cells and thermopile pyranometers), and the effects of partial shading conditions. Regarding module temperature, the study addresses uncertainties arising from sensor deviations and temperature non-uniformity at both the module and array levels, considering the differences in mounting configurations (**Figure 1**).

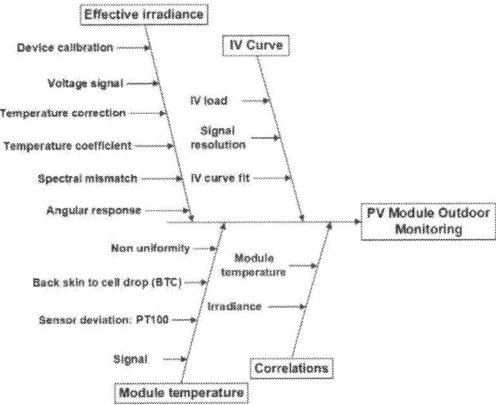

Figure 1: Main sources of uncertainty considered for outdoor PV module characterization by IV curve measurements.

This comparison not only provides insights into the impact of installation conditions on the accuracy of PV module measurements but also contributes to optimizing monitoring setups. Understanding irradiance and temperature variations is crucial for accurately predicting and optimizing PV module performance. The practical focus of this study, incorporating real-world experiences from Fraunhofer ISE's new test field, adds valuable relevance, bridging the gap between theoretical considerations and practical challenges.

2 METHODOLOGY

2.1 Test field measurements

The uncertainty analysis was first conducted on the former rooftop-mounted monitoring station of Fraunhofer ISE in Hochdorf, Germany (**Figure 2**). This station has provided over eight years of monitoring data, with PV modules mounted on a close-back rack with an azimuth angle of 157° and a tilt of 15°. The setup includes irradiance measurements in POA using silicon reference cells from Fraunhofer ISE and a thermopile pyranometer, alongside weather sensors monitoring ambient temperature, relative humidity, and wind speed.

41st European Photovoltaic Solar Energy Conference and Exhibition

Further, the study considered PV module measurements on the new test site in Merdingen, Germany, where modules are mounted on open racks facing south (azimuth 180°) and a tilt angle of 30°. The site is equipped with complete characterization of the climate conditions through irradiance sensors, spectroradiometers, a sun tracker, and weather sensors. The site's open-racks and unobstructed horizon allows for better thermal homogeneity between modules due to improved airflow and reduces shading effects from obstructions or interrow shading.

The monitoring data was collected from four samples of monofacial monocrystalline silicon PV modules with a nominal power of 235 W and a glass/EVA/backsheet (white) structure. These modules were initially installed at the Hochdorf test station and were later moved to the Merdingen test site. In this work, we use data from the full year 2020 for Hochdorf and from the full year of 2023 for Merdingen.

Figure 2: (Top) Former rooftop-mounted monitoring station (Hochdorf, Germany), (Bottom) new test field in Merdingen, Germany.

3 UNCERTAINTY BUDGET

The uncertainty calculations presented here are based on the "Guide to the Expression of Uncertainty in Measurement" (GUM). [6] First, we consider the relationship between the measured quantities and the dependent variables to estimate the sources of uncertainty, which may arise from environmental conditions, instruments accuracy, and measurement conditions. For each source of uncertainty, a value representing its standard uncertainty (x_i) contribution has been estimated. Since the measurements are conducted outdoors, it is not possible to obtain repeated observations under identical conditions; therefore, most of our uncertainty sources are classified as Type B, which are uncertainties based on calibration certificates and deviations or scattering in

measurement data. We also assumed a rectangular distribution for the standard uncertainties, meaning all measurements are equally likely to fall within the specified range. The standard uncertainties are calculated as $u(x_i) = \text{std.dev}/\sqrt{3}$.

The overall uncertainties from a measurement parameter (y) are calculated using the law of propagation. The combined standard uncertainty is defined as the square root of the sum of the squared sensitivity coefficients (partial derivatives of the measurement model with respect to each input parameter) multiplied by the squared standard uncertainties. The total expanded uncertainty, corresponding to a 95% confidence interval in the measured quantity, is calculated by multiplying the combined uncertainty by a coverage factor k of approximately 2. [7]

$$U_c(y) = \sqrt{\sum_{i=1}^{N} \left(\frac{\partial y}{\partial x_i}\right)^2 u(x_i)^2} \tag{1}$$

3.1 Effective Irradiance

3.1.1 Device deviations

The irradiance measurement of the reference cell (Equation 2) depends on the measured voltage (V), the calibration factor (K) and the temperature of the reference cell (T) for the corresponding temperature correction.

$$Geff = \frac{V\,K}{1 + \alpha(Tcell - 25°C)} \tag{2}$$

The overall uncertainty is defined by these sources of uncertainty. The combined uncertainty for the effective irradiance is expressed in Equation 3 as a function of the standard uncertainties (u_i) and sensitivity coefficients (c_i) according to the law of propagation.

$$U_c(Geff) = \sqrt{ \begin{aligned} &\left(\frac{\partial Geff}{\partial K}u_K\right)^2 + \left(\frac{\partial Geff}{\partial V}u_{V-signal}\right)^2 + \\ &\left(\frac{\partial Geff}{\partial Tcell}u_{PT100}\right)^2 + u_{Temp.corr} \end{aligned}} \tag{3}$$

In Hochdorf, the effective irradiance was measured using Fraunhofer ISE's outdoor reference cells. These cells were calibrated at Fraunhofer ISE's calibration laboratory (CalLab), on a solar simulator against a c-Si primary reference cell, which was calibrated at PTB, with a limit deviation of 2.5% and an stability between calibrations of 1.9% [8]. In Merdingen, a reference cell from Mencke&Tegtmeyer with original manufacturer calibration values was used, with a calibration deviation of 1.2% (as class A acc. to IEC 61724-1) and a calibration stability of 0.75%.

For over a year of measurements the median uncertainty (k=1) due to device measurements in Hochdorf is around 1.84%, while in Merdingen, it is around 0.91%, with the highest contribution of uncertainty due to the calibration factor of the reference cells (**Table 1**).

3.1.2 Spectral mismatch

Spectral mismatch between the reference cell and the PV module is a significant source of uncertainty when estimating the effective irradiance. To correct for this, a

1073

Spectral Mismatch Factor (SMM) is applied. The uncertainty sources in the SMM, we estimated using a Monte Carlo simulation, which accounts for the dependencies between all the parameter deviations. These sources of deviation have been thoroughly analyzed by CalLab [8, 9]. The analysis considers deviations from the measurements of the spectral response (SR) of the PV module and the reference irradiance sensor, as well as the measured spectra in the field which are obtained using a EKO MS-710/712 spectroradiometer.

Figure *3* shows the level of uncertainty in the SMM when using a pyranometer and reference cell as irradiance sensors. The higher mismatch between the pyranometer and the PV module results in median uncertainties of 0.9% and values reaching up to 1.2% at low irradiances levels. In contrast, when using a silicon reference cell, the SMM uncertainty remains at a consistently low value of 0.023% across all levels of irradiance.

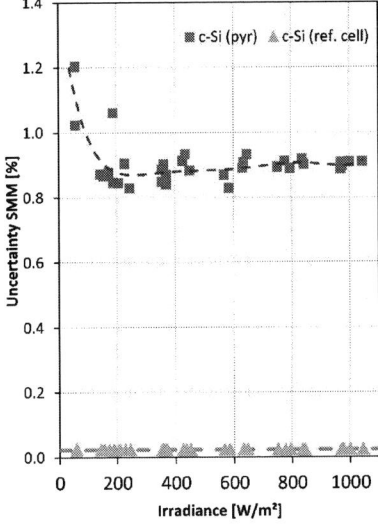

Figure 3: Spectral Mismatch uncertainty (k=1) based on measurements with a pyranometer (blue) and a silicon reference cell, relative to a c-Si PV module.

3.1.3   Angular response
To estimate the deviation in the measurement of effective irradiance between the PV module and the reference cell based on the AOI, the effective irradiance on the PV module was calculated using a Self Reference Algorithm (SRA). This algorithm calculates the effective irradiance and temperature as a function of the module's short-circuit current (Isc) and open-circuit voltage (Voc), and their respective temperature coefficients (alpha and beta). [10] The effective irradiance on the PV module was compared to that measured by the reference cell on clear sky days at both monitoring stations.

In Hochdorf, deviations reached up to 80% (approximately 172 W/m²) in the morning when the AOI were greater than 60°. The asymmetry in deviation between morning (~ 80%) and afternoon (~ 20%) values suggests the presence of partial shading, particularly in the early morning, mainly happening in summer.

The more ideal conditions in Merdingen lead to symmetrical deviations between the PV module and the reference cell, reaching up to 25% (approximately 17 W/m²) during the early morning and late afternoon when the AOI exceeded 80°. These deviation values (**Figure** *4*) were used to calculate the uncertainty in irradiance

measurements, resulting in a median AOI-related contribution to uncertainty of 2.27 % for Hochdorf and 1.06% for Merdingen over a year of measurements (**Table 1**).

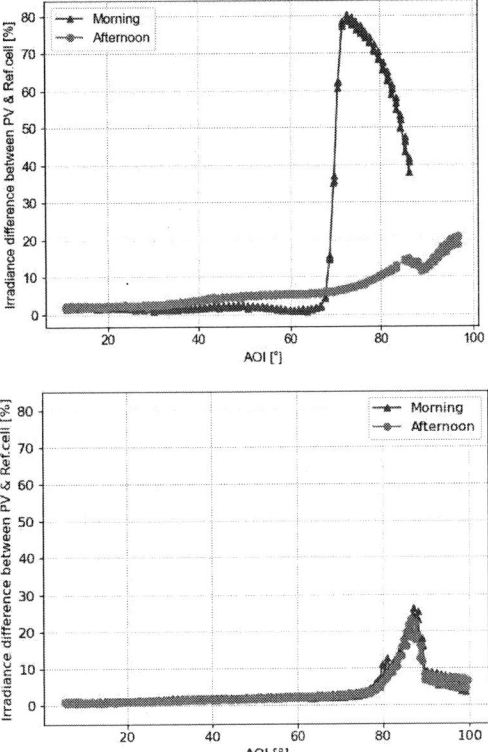

Figure 4: Percentage difference between the effective irradiance measured by the PV module and the silicon reference cell during morning (blue) and afternoon (red) hours. The top graph shows the results for Hochdorf, lower graph for Merdingen.

Table 1: Sources of uncertainty in the measurement of effective irradiance over a year of outdoor exposure in Hochdorf and Merdingen. Median uncertainty value (k=1).

Uncertainty (k=1)	Source of uncertainty	Hochdorf	Merdingen
	Calibration	4.14 W/m²	1.87 W/m²
	Voltage signal	0.07 W/m²	0.04 W/m²
	Temp. meas. (PT100)	0.01 W/m²	0.01 W/m²
	Temp. correction	0.58 W/m²	0.58 W/m²
u-Irrad	Linearity	0.53 W/m²	0.13 W/m²
	Stability	N/A	0.66 W/m²
	Device	4.21 W/m²	2.07 W/m²
	Device	1.84 %	0.91 %

SMM	0.023 %	0.023 %
AOI	2.27 %	1.06 %
u-Irrad [%]	3.01 %	1.44 %
u-Irrad [%](k=2)	**6.03 %**	**2.88%**

3.2 Broadband Irradiance

At Fraunhofer ISE's monitoring stations, the performance evaluation of modules is based on the measurement of broadband irradiance using a thermopile pyranometer, such as Kipp & Zonen CMP21. To accurately measured the usable solar radiation [4] with this sensor, it is necessary to account for the deviations due to spectral mismatch (as discussed in section 3.1.2) and the angular response of the sensor relative to the PV module. Previous calculations, referenced in Appendix B of Ref. [11], were used to analyze these deviations. When comparing the performance of the same pyranometer at both monitoring sites, the primary difference was due to misalignment. Assuming a maximum difference in angle between the mounted PV module and the POA of the pyranometer of 0.5° [11], significant improvement were observed with the installation in Merdingen. This setup reduced the effect of high AOI and the overall uncertainty by 1.32% (k=2).

Table 2: Sources of uncertainty in the measurement of broadband irradiance over a year of outdoor exposure in Hochdorf and Merdingen. Median uncertainty value (k=1).

Uncertainty (k=1)	Source of uncertainty	Hochdorf [%]	Merdingen [%]
u-Broadband irradiance	Signal processing	0.0003	
	Calibration	0.72	
	Misalignment	0.14	0.099
	Temperature dependence	1.0	
	Linearity	0.3	
	Stability	0.5	
	Zero offset (<200 W/m²)	6.53	6.77
	SMM	0.99	
	u-Irrad-pyrano [%]	2.10	1.44
	u-Irrad-pyrano [%](k=2)	**4.21 %**	**2.89%**

3.3. Module temperature

The sources of uncertainty considered for the module temperature measurements include the calibration of the PT100 sensors, the difference between the backsheet and cell temperature (Backsheet to Cell drop or BTC), and deviations from voltage signal processing, as expressed in Equation 4.

$$U_{c-Tmod} = \sqrt{u_{PT100}^2 + u_{Non-unif}^2 + u_{BTC}^2 + u_{V-signal}^2} \quad (4)$$

3.3.1 Non-uniformity

To estimate temperature measurement variability, four PV module samples of the same technology and manufacturer, initially installed in Hochdorf and later moved to Merdingen, were selected. Non uniformity was calculated as the mean temperature difference between the maximum and minimum measurements from the samples for the whole year of exposure (**Figure 5**). The new open raw structure in Merdingen proved to reduce the module temperature measurement differences between samples to less than 2 °C, compared to up to 3 °C in Hochdorf.

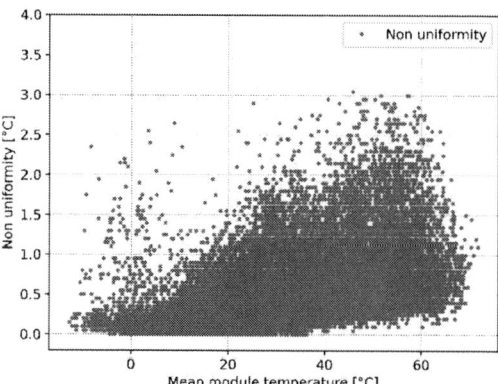

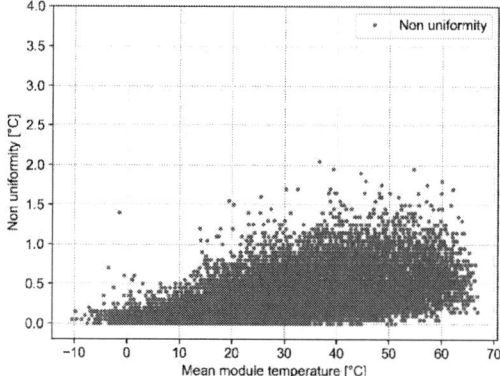

Figure 5: Temperature non-uniformity at the row level between four PV module samples installed in Hochdorf (top) and Merdingen (bottom).

3.3.2 Backsheet to cell drop

BTC is the temperature difference between the back of the module and the cell, estimated by calculating the module's effective temperature (cell temperature) using the SRA method, with an estimated uncertainty of ±0.63 K [10]. The uncertainty in BTC values was calculated across irradiance intervals of 100 W/m², where the values showed a normal distribution. The standard deviation was assigned as the uncertainty source for each irradiance interval, which is defined as Type A uncertainty by $u(x_i) = std. dev/\sqrt{number\ of\ measurements}$. On clear sky days, filtering data with irradiance fluctuations below 2 W/m², the largest deviations occurred at low irradiance levels (<100 W/m²), with BTC reaching up to

7.3 °C in Hochdorf and 5.7 °C in Merdingen. However, the overall BTC uncertainty for the entire year was slightly higher for Merdingen (0.41 °C) compared to Hochdorf (0.38 °C) (k=1) (**Table** 3).

Table 3: Sources of uncertainty in module temperature measurements over a year of outdoor exposure in Hochdorf and Merdingen. Median uncertainty value (k=1).

Uncertainty (k=1)	Source	Hochdorf	Merdingen
	Voltage signal	0.006 °C	0.006 °C
	PT100	0.008 °C	0.008 °C
	Non-uniformity	0.17 °C	0.12 °C
u-Tmod	Back-sheet to Cell Drop (BTC)	0.38 °C	0.41 °C
	u-Tmod	0.54 °C	0.49 °C
	u-Tmod	0.18%	0.16%
	u-Tmod [%] (k=2)	**0.36%**	**0.32%**

3.4. IV curve measurements

The IV curve measurements are taken using electronic loads which perform a voltage sweep while recording voltage and current values. For the IV curve measurements, the primary sources of uncertainty considered were the resolution and calibration of the IV tracer, the extrapolation uncertainties for Voc and Isc values, and the fit variations to the MPP values. The same IV measurement devices were used in both Hochdorf and Merdingen, resulting in uncertainties of 0.5% for Voc, 0.6% for Isc and 0.94-0.95% for Pmpp.

3.5 Overall uncertainty

Since the IV curve depends primary on irradiance and module temperature, the combined uncertainty must consider the correlation between these parameters. According to [6], the correlation coefficients are calculated using Equation 5-7, where σ_{ij} is the covariance matrix between the parameters, ρ_{ij} is the correlation coefficient, $s_i s_j$ are the standard deviations of x_i and x_j and their covariance s_{ij}. This equation, applied to Pmpp, requires the correlations shown in Equation 8. The results for the main IV curve parameters (Voc, Isc and Pmpp) are presented in **Table 4**. Since the correlation to irradiance measurements is taken from the silicon reference solar cells, and due to the large deviations found in Hochdorf due to partial shading at high AOI, the combined uncertainty for the IV curve measurements is filter for conditions below 60°. For higher AOI we find the propagation of uncertainty reach up to 150% (k=2) for Pmpp, which would be erroneous to include in the median uncertainty result. Nonetheless, for Merdingen, with measurements on a complete range of AOI, the median uncertainty is of Pmpp is 3.32% (k=2) and up to 50% for AOI larger than 80°.

$$Uc(y)^2 = \sum_{i}^{N\,sources} \left(\frac{\partial y}{\partial x_i} u(x_i)\right)^2 + 2\sum_{i=1}^{N-1}\sum_{j}^{N} \left(\frac{\partial y}{\partial x_i}\frac{\partial y}{\partial x_j}\sigma_{ij}\right) \quad (5)$$

$$\sigma_{ij} = u_{xi}u_{xj}\rho_{ij} \quad (6)$$

$$\rho_{ij} = \frac{s_{ij}}{s_i s_j} \quad (7)$$

$$\begin{aligned}
&Uc^2(Pmpp)\\
&= \left(\frac{\partial Pmpp}{\partial G}u_G\right)^2 + \left(\frac{\partial Pmpp}{\partial Tmod}u_{Tmod}\right)^2\\
&+ \left(\frac{\partial Pmpp}{\partial Isc}u_{Isc}\right)^2 + \left(\frac{\partial Pmpp}{\partial Voc}u_{voc}\right)^2\\
&+ 2\left[\frac{\partial Pmpp}{\partial G}\frac{\partial Pmpp}{\partial Tmod}\sigma_{G-Tmod}\right.\\
&+ \frac{\partial Pmpp}{\partial Isc}\frac{\partial Pmpp}{\partial G}\sigma_{Isc-G}\\
&+ \frac{\partial Pmpp}{\partial Voc}\frac{\partial Pmpp}{\partial G}\sigma_{Voc-G}\\
&+ \frac{\partial Pmpp}{\partial Isc}\frac{\partial Pmpp}{\partial Tmod}\sigma_{Isc-Tmod}\\
&+ \left.\frac{\partial Pmpp}{\partial Voc}\frac{\partial Pmpp}{\partial Tmod}\sigma_{Voc-Tmod}\right]
\end{aligned} \quad (8)$$

Table 4: Combined uncertainty (k=2) for the measurements of Pmpp, Voc and Isc in Hochdorf and Merdingen, with AOI below 60°.

Uncertainty (k=2)	Hochdorf [%]	Merdingen [%]
uc-Pmpp	5.72	3.32
uc-Voc	1.13	1.09
uc-Isc	5.37	3.21

4 DISCUSSION OF RESULTS

The analysis of a year-long monitoring period identified the primary sources of uncertainty in effective irradiance measurements, in the calibration of the reference cell and in the angular response (**Table 1**). Spectral mismatch effects were minimal due to the SR alignment between the reference cells and the monitored PV modules. When examining the effects of AOI, it became apparent that minor factors, such as partial shading during early mornings in Hochdorf when AOI exceeded 60°, can lead to significant discrepancies. These discrepancies could cause differences up to 172 W/m² between the available irradiance on the PV module and the actual measured value from the reference cell. In contrast, the Merdingen site, without horizon obstructions or partial shading, showed much smaller discrepancies (below 17 W/m²) under high AOI. Note that the irradiance differences should be also partly attributed to the inherent uncertainties in the self-reference method used for calculating PV module irradiance measurement, which has a tolerance of ±28.7 W/m² according to Ref. [10].

For broadband irradiance measured with a thermopile pyranometer, the main sources of uncertainty were the sensor's zero offset at irradiances below 200 W/m², spectral mismatch relative to the monitored modules, and temperature dependence. The main difference between the monitoring sites is the effect of misalignment, where we assumed a maximum misalignment of 0.5° between the sensor and the POA (**Table 2**). [4]

Therefore, we emphasize the importance of improving the measuring conditions at low irradiances by using mounting racks that minimize shading effects and applying measurement corrections for high AOI and spectral mismatches. These improvements are particularly beneficial when monitoring the available irradiance for benchmarking of different cell technologies.

The uncertainty analysis of module temperature measurements revealed as main contribution the difference between backsheet and cell temperature (BTC). This was followed by non-uniformity in temperature measurements across an array of PV modules, with variations of 2–3°C between samples, particularly at high temperature levels (**Table 3**). At the Merdingen test site, BTC values were higher and non-uniformity between the samples was lower, likely due to the open rack structure, which allowed for better airflow, equalizing conditions across all modules. However, this structure also affected the intra-module temperature distribution, causing temperature measurements to deviate from actual cell temperatures depending on the sensor's position. Therefore, we recommend using multiple temperature sensors per PV module to accurately characterize the cell temperature.

When comparing the overall uncertainty of IV curve parameters, from both monitoring stations, we notice higher uncertainties at low irradiance levels. Although, these have a lower impact on the overall energy production and performance calculation. The uncertainty in power measurements at MPP (**Figure 6 and 7**), when considering correlations with irradiance (with AOI below 60°) and temperature levels, showed a significant reduction at the Merdingen site, with less scattering at the different irradiance levels throughout the monitoring year.

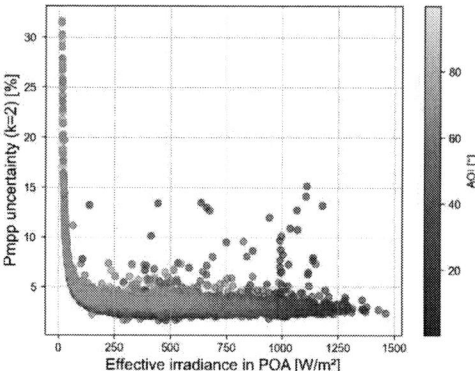

Figure 7: Combined uncertainty in Pmpp (k=2) for Merdingen, as a function of effective irradiance level [W/m²], with data points colored by AOI [°] up to 60°.

To assess the sensitivity of the performance ratio (PR) to deviations in irradiance and module temperature, we calculated the temperature corrected DC PR [12] using both measured values of effective irradiance (*Geff*) and module temperature (*Tmod*), as well as self-referenced values obtained from PV module current and voltage measurements (G_{SRA} and $Tmod_{SRA}$) (Equation 9). In Merdingen, the PR calculated with measured input values closely matched the self-referencing ideal measurement of PR, with mean deviations of 1.9% (**Figure 8**). In contrast, Hochdorf exhibited a mean PR difference of 4.7% between measured and self-referenced values, with higher deviations in winter up to 7.5% (**Figure 9**).

$$PR_{corrected-Measured}$$
$$= \frac{G_{STC} \sum P_{mpp}}{\sum (P_{STC} \times Geff \times (1 - \frac{\delta}{100}(T_{avg} - Tmod)))} \quad (9)$$

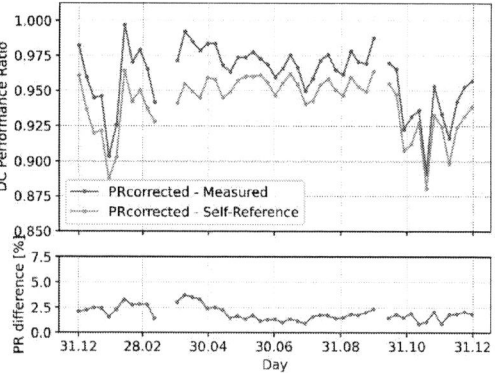

Figure 8: Weekly temperature corrected DC Performance Ratio based on measured input values (blue) and self-referenced input (orange), for one year of monitored data in Merdingen (bottom).

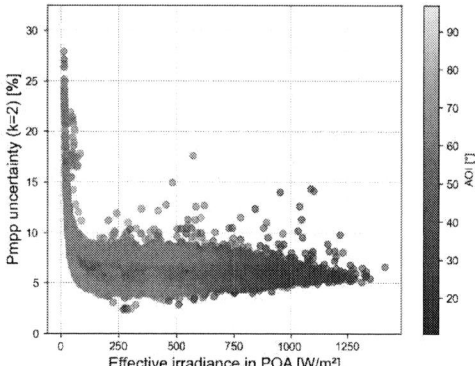

Figure 6: Combined uncertainty in Pmpp (k=2) for Hochdorf, as a function of the effective irradiance level [W/m²], with data points colored by AOI [°] up to 60°.

41st European Photovoltaic Solar Energy Conference and Exhibition

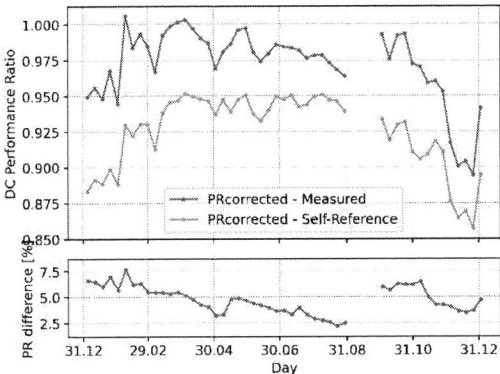

Figure 9: Weekly temperature corrected DC Performance Ratio based on measured input values (blue) and self-referenced input (orange), for one year of monitored data in Hochdorf.

5 CONCLUSIONS

This study highlights the factors that influence the accuracy of outdoor PV module monitoring, with a focus on irradiance and temperature measurements. The calibration of reference cells and the impact of high AOI and partial shading showed to be the primary sources of uncertainty in effective irradiance measurements. The larger discrepancies in Hochdorf, due to partial shading obstructions, and the lower uncertainties in Merdingen, underscore the importance of design of experiment and site-specific conditions on measurement accuracy. These findings suggest that minimizing shading effects and applying corrections for high AOI and spectral mismatches, is essential for obtaining reliable data, especially for studies focused on low irradiance performance characterizations.

Furthermore, the study identified significant uncertainties in module temperature measurements, particularly related to the difference between backsheet and cell temperatures and the non-uniformity across PV module samples. The open rack structure at Merdingen reduced non-uniformity but also highlighted the need for multiple temperature sensors per module to accurately characterize cell temperatures. When comparing the overall uncertainties in IV curve parameters, Merdingen demonstrated more stable measurement conditions, with reduced uncertainty and less scattering across different irradiance levels. Additionally, the analysis of DC PR deviations highlights the importance of considering site-specific conditions and temporal factors (such as weather conditions) and highlights the advantages of summer monitoring, where uncertainties are expected to be lower. The use of self-referencing methods for baseline evaluation of the available effective irradiance and module temperature is also recommended for improving the accuracy of performance assessments.

6 ACKNOWLEDGEMENTS

M. R. is thankful for the financial support of Nagelschneider Foundation (Munich, Germany). The project work was partially funded by the German Federal Ministry for Economic Affairs and Climate Action (BMWK) (funding number 03EE1149A).

7 REFERENCES

[1] D. Thevenard and S. Pelland, "Estimating the uncertainty in long-term photovoltaic yield predictions," *Solar Energy*, vol. 91, pp. 432–445, 2013, doi: 10.1016/j.solener.2011.05.006.

[2] D. Dirnberger, J. Bartke, A. Steinhüser, K. Kiefer, and F. Neuberger, Eds., *Uncertainty of Field I-V-Curve Measurements in Large Scale PV-Systems*, 2010, doi: 10.4229/25THEUPVSEC2010-4BV.1.62.

[3] A. Driesse, Ed., *Uncertainty of Tilted Irradiance Measurements using Photodiodes and Reference Cells*, 2021, doi: 10.4229/EUPVSEC20212021-5BO.6.5.

[4] J. Meydbray, E. Riley, L. Dunn, and K. Emery and S. Kurtz, "Pyranometers and Reference Cells: Part 2: What Makes the Most Sense for PV Power Plants?," 2012, Art. no. NREL/JA-5200-56718.

[5] C. H. Rossa, F. Martinez-Moreno, and E. Lorenzo, Eds., *Reducing Uncertainty in Outdoors PV Module Characterisation*, 2018, doi: 10.4229/35THEUPVSEC20182018-5CV.1.23.

[6] Joint Committee for Guides in Metrology, *Evaluation of measurement data - Guide to the expression of uncertainty in measurement*, 2008.

[7] J. M. Carrillo, F. Martínez-Moreno, C. Lorenzo, and E. Lorenzo, "Uncertainties on the outdoor characterization of PV modules and the calibration of reference modules," *Solar Energy*, vol. 155, pp. 880–892, 2017, doi: 10.1016/j.solener.2017.07.028.

[8] D. Dirnberger and U. Kraling, "Uncertainty in PV Module Measurement—Part I: Calibration of Crystalline and Thin-Film Modules," *IEEE J. Photovoltaics*, vol. 3, no. 3, pp. 1016–1026, 2013, doi: 10.1109/JPHOTOV.2013.2260595.

[9] D. Dirnberger, B. Müller, and C. Reise, "On the uncertainty of energetic impact on the yield of different PV technologies due to varying spectral irradiance," *Solar Energy*, vol. 111, pp. 82–96, 2015, doi: 10.1016/j.solener.2014.10.033.

[10] B. Hüttl, L. Gottschalk, S. Schneider, D. Pflaum, and A. Schulze, "Accurate performance rating of photovoltaic modules under outdoor test conditions," *Solar Energy*, vol. 177, pp. 737–745, 2019, doi: 10.1016/j.solener.2018.12.002.

[11] W. Marion *et al.,* "User's Manual for Data for Validating Models for PV Module Performance," 2014.

[12] T. Dierauf, A. Growitz, S. Kurtz, J. L. B. Cruz, E. Riley, and C. Hansen, "Weather-Corrected Performance Ratio," 2013, doi: 10.2172/1078057.

41st European Photovoltaic Solar Energy Conference and Exhibition

MPP TRACKING LOSSES OF MODULE LEVEL POWER ELECTRONICS AT PARTIAL MODULE SHADING

Franz Baumgartner, Markus Klenk, Adrian Widler, Linus Baumann
ZHAW, Zurich University of Applied Sciences, School of Engineering, IEFE
www.zhaw.ch/~bauf, Technikumstr. 9, CH-8401 Winterthur, Switzerland; Email: bauf@zhaw.ch

ABSTRACT: Power optimisers aim to reduce shading effects when operating a PV system and promise a massive increase in yield. In recent years, the real performance benefits of such Module Level Power Electronic (MLPE) components have been -tested in a few isolated trials. The results support the statement that PV systems with power optimisers deliver an increased yield compared to string inverter systems in certain cases with complex and pronounced shading situations. On the contrary, their effect may even be detrimental in low complexity scenarios for several reasons.
This paper focuses on a specific topic, the analysis of losses associated with Maximum Power Point (MPP) tracking. The MPP adjustiment is analyzed in an indoor test setup and the cases correspond to typical outdoor partial shading using a commercially available MLPE. It is shown that the adaptation to changing conditions works well with small changes in the associated MPP voltage, but fails with larger voltage jumps, such as those that occur with increased shading and activated bypass diodes. The observed effects are also verified in a real system. The preference for a higher voltage and lower current can be advantageous with regard to hot spots but reduces the overall yield.
Keywords: Optimiser, module level power electronics, shading, yield, MPP tracking

1 INTRODUCTION AND MOTIVATION

Power optimisers can improve the performance of photovoltaic (PV) systems with modules under inhomogeneous irradiation conditions, as caused by partial shading or differing module orientations. The demand for optimisers has remained high for years, particularly because manufacturers promise high additional yields. Shading losses and their occurrence are generally known and understandable even to laypersons. The widespread use of optimisers and the corresponding advertising by manufacturers therefore leave customers in no doubt that their use should have a positive effect on the yield of a PV system by suppressing shading effects [5].

On the other hand, commercial PV simulation tools have not yet been able to adequately map the effects of shading on a PV system as far as the power electronics are concerned and must rely on the manufacturer's efficiency specifications for yield calculations [1] [10]. The stated maximum efficiencies are rarely achieved in annual operation [2] [3] [4] [5]. As a result, the benefits of optimisers are usually overestimated by both laypeople and experts, as a realistic estimate or even calculation is not possible and depends heavily on the specific system conditions including their real MLPE losses.

In principle, the results of our investigations support the statement that the use of power optimisers is advantageous in complex shading situations. However, their use in scenarios with low shading complexity can even result in losses, compared to standard systems with string inverters (SINV) and becomes more inefficient if there is no shading. [5] [6] [7].

Besides thermal losses due to the additional electronics, there are other effects that have a negative impact on both the predicted and the achieved system efficiency. As already mentioned, the efficiency of MLPE optimisers is overestimated because the efficiency under optimum conditions is simply assumed to be a fixed value at all operating points. Another issue in this context is setting the operating points. The Maximum Power Points (MPP) can have local and global maxima at module and system level. Problems in determining and correctly setting them are obvious sources of error.

In the IEC standards the typical MPP losses are included in the total efficiency which adds typical a tenth of a percent due to tracking to the static efficiency [8].

In this work experimental investigations on an exemplary chosen MLPE (SE 500) are presented. By varying the module IV-characteristics (corresponding to partial shading) local maxima are deliberately generated and the behavior of the optimiser is analyzed.

2 MEASUREMENTS

Experimental data is presented based on the device investigation at indoor laboratory level. The observed effects are then additionally validated with measurement data from a roof installation.

2.1 Indoor Laboratory Measurements

The measurement of PV systems at outdoor conditions is inherently prone to fluctuations. As an alternative approach a complex indoor test-field was set up at the ZHAW to imitate the output of real modules by an array of solar module simulators [2] [3] [4]. In the test setup described a string with eight solar modules connected in series is imitated (see Fig. 1). Each of the eight modules is emulated by a solar array simulator powering the optimiser. The optimiser input is measured by a power meter PPA 1530.

One of the solar array simulators (IT6018C) is capable to generate an IV-curve of a PV module with three substrings at different irradiation levels separated by bypass-diodes. Accordingly, it is possible to generate the output of a partially shaded module (see Fig. 2, 3 and 11). The bypass-diode is activated only if the module current exceeds the limiting module substring current due to shading [4]. This causes the typical humps in the IV-curve which leads to local and absolute maxima in the Power-Voltage characteristics (Fig.2 and Fig. 3).

As in a real PV system, the output of all eight module optimizers is connected in series. Thus, a common DC current is set via a control loop of the PV DC/AC inverter, which is fed by all MLPE DC/DC optimizers.

The MPP tracking performance of the MLPE optimiser was analysed based on the response to changing conditions according to a schedule of varying imitated irradiation leading to different MPPs over a nine-minute

1079

41st European Photovoltaic Solar Energy Conference and Exhibition

period with changes after every minute (Tab. 1).

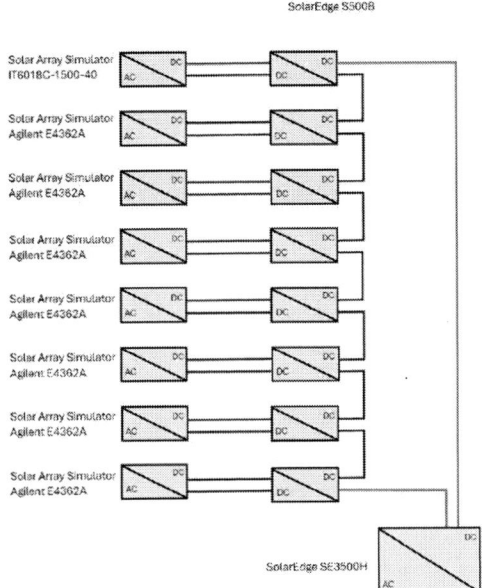

Figure 1: ZHAW indoor laboratory setup of eight optimisers in series connection powered by solar array simulators. The IT6018C output, imitating a partial shaded PV module with three bypass diodes is measured by a power meter and is shown in Fig. 2 [9].

Table I: Irradiance settings of the three substrings as emulated by the IT6018C.

Period of each 1 minute	Sub-string 1 $[Wm^{-2}]$	Sub-string 2 $[Wm2]$	Sub-string 3 $[Wm2]$
1	2	1000	1000
2	750	1000	1000
3	500	750	1000
4	250	500	750
5	500	250	500
6	750	500	250
7	1000	750	500
8	1000	1000	750
9	2	1000	1000

The output of the systems measured with the PPA 1530 power meter is recorded and compared to the given power-voltage curves resulting from the adjusted conditions according to Tab. 1. In this way, the changes in the simulated shading conditions can be directly linked to the respective MPP settings. The progression over time can be seen from the position of the measuring points by number on the power-voltage curves shown in Fig. 2. Correctly identified MPPs are marked by green squares, while wrongly adjusted operating points are indicated by red squares starting at open circuit condition at point 0 (black square).

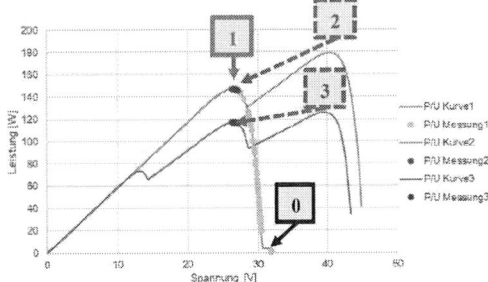

Figure 2: Changes of the measured S500B optimizer MPP input according to the setup of Fig. 1 as response to the changing shading conditions shown in Tab. 1 until step 3.

In the first step, (Tab. 1, line 1) the current of one PV module substring is almost completely blocked by shading, while the other two strings are unshaded. The first MPP (green "1") was correctly found by the optimiser system at the PV modules global MPP at 27 V and 146 W after 45 seconds from starting at the Voc.

After changing the setting (Tab. 1, step 2), by reducing the shading of the first PV module substring while the other two substrings remain unshaded at the same solar irradiance as before. Due to the changed irradiation conditions of the first substring a global maximum at 41 V and 180 W should be accessed by the MPP algorithm. However, the optimiser does not adapt to the higher voltage level but continues to work with the same operating state (red "2", dotted arrow), which is not the global MPP. Sticking to this operation point instead of adjusting to the global MPP causes a power loss of 34 W. According to the almost fixed position of the measuring points, there is no indication of an MPP search at a different voltage level.

In the next setting (Tab. 1, step 3), the irradiance is set differently for all three strings, resulting in three maxima, with the global maximum remaining at the high voltage level of 40 V (red "3", dashed arrow in Fig. 2). Again, there is no indication for a search at different voltage level.

The reduction in irradiance is continued for all PV module substrings according to step 4 of Tab. 1. For better clarity, the results are shown in a new depiction (Fig. 3).

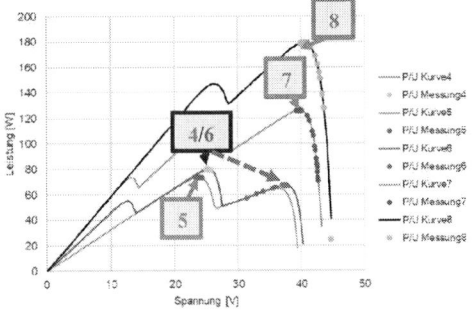

Figure 3: Continued visualization of the results from Fig. 2 and shading settings according to Tab. 1, starting with step 4.

In this constellation, the optimiser succeeds in finding the global MPP (green "4") and in readjusting the voltage. The necessary voltage step of 1V to 26 V to reach the new maxima is small.

Lowering the intensity again for step 5 leads to another

slightly shifted maximum (green "5"), which is also correctly found and adjusted by the optimiser.

With the next step 6 the same level of shading of the three substrings are reached as in step 4, resulting in the same power-voltage curve. However, the optimiser has failed to stay at global MPP and has found only the local MPP at 63 W and 38 V (red "6", dashed arrow), resulting in 17 W losses. In this case, a voltage with a significantly different, higher value is set.

The subsequent changes to the partial shading in steps 7 and 8 lead to the successful setting of the respective global MPP.

The demonstrated non-adjustment of the MPPs obviously leads to a reduced efficiency. Figure 4 shows the course of the settings according to Table 1 and the respective efficiencies achieved with the optimiser.

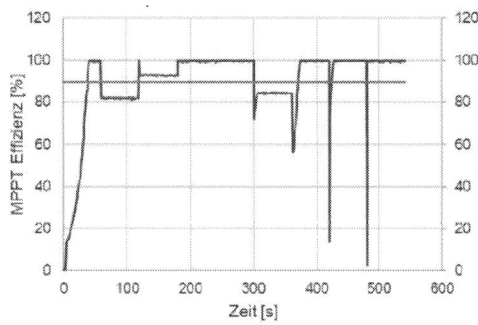

Figure 4: Progression of settings according to Table 1 and efficiency achieved with the optimiser.

Time is apparently not the decisive factor for readjusting the operating point as indicated by results shown in Figure 5.

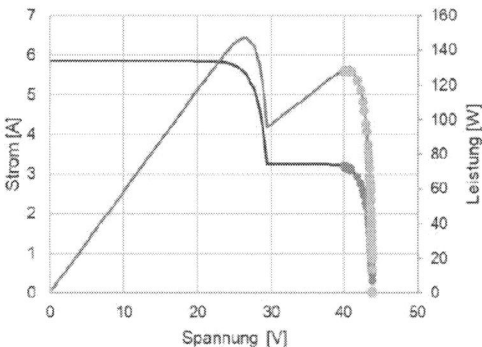

Figure 5: Course of the operation point adjustment after starting the MPP tracking (SolarEdge S500B) [9]

Figure 5 shows the progression of the operating point setting after the start of MPP tracking for a SolarEdge S500B optimiser. First, it is noticeable that it is not the global MPP but a local maximum at a higher voltage that is identified as the MPP. Then there are no measuring points at a lower voltage, which would indicate the search for optional MPPs by an algorithm. The time course of the measurement is shown in Fig. 6. A stable state at the local maximum and at a lower efficiency than for the MPP is achieved after 40 seconds. A similar starting settling time was already shown for the previously described experiment (Fig. 4). Even after an hour of constant shading condition, it was not possible to detect any indication that

an optimizer algorithm was searching for the global maximum at 27 V and therefore remained in the local MPP with the associated power losses.

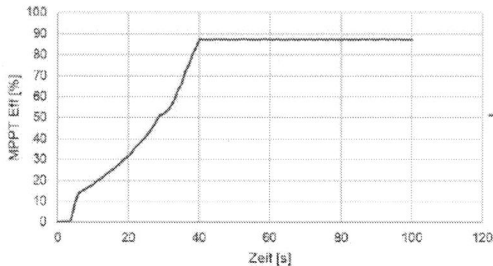

Figure 6: Time and efficiency according to the measurements in Figure 5.

ZHAW has also tested other commercial MLPE products, such as the Huawei 450W-P, using the same methodology with very similar results to the previously described tests, as shown in Figure 7 [5].

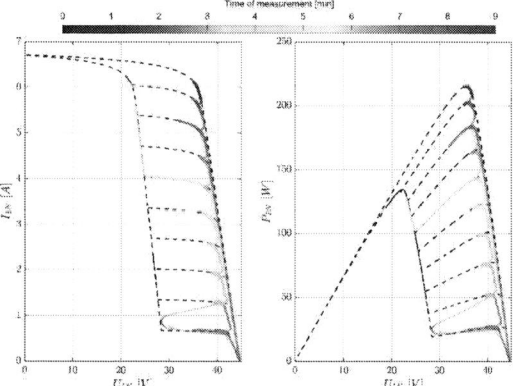

Figure 7: Progress of MPP tracking performance in one minute steps of Huawei 450W-P MLPE under indoor test at ZHAW also failing to track the global maximum power of the shades module first half of the test [5].

2.2 Outdoor measurements

The laboratory measurements showed that the optimiser could not find a global MPP if the associated voltage was lower than that of a previously determined local maximum. Such a readjustment towards a lower voltage and higher current only worked for very small shifts within a local maximum in the range of volts. In these cases, the small voltage shift did not trigger a search for a new global maximum for this MLPE product. On the other hand, changing irradiation conditions leading to MPP shifts towards higher voltages triggered successful readjustments.

A similar behavior, the preferred shift towards higher voltages, is shown for an example of outdoor measurements on a PV rooftop system (Fig. 8) in Constance, Germany. The roof has a rather complex shading situation, which is favorable for the successful use of power optimisers. On a north-facing roof, 14 PV modules in butterfly half-cell configuration (JAM60S21 370/MR) are mounted on dormers (7 modules facing east and 7 facing west) as schematically shown in Fig. 8, each equipped with a Solaredge S440 optimiser and connected to an SE5000H inverter.

41st European Photovoltaic Solar Energy Conference and Exhibition

Figure 8: Outdoor measurements are carried out on a roof with complex shading situation. The module indicated by the red arrow ("module 9") is particularly heavy shaded.

One module ("No. 9"), indicated by the red arrow in Figure 8, is particularly heavy shaded in the morning hours compared to the other modules with the same or similar orientation.

Figure 9 shows PV module currents for a specific day and assigns them to some of the individual modules. The morning hours can be seen in Fig. 10 in more detail. Particularly the current of module 9 remains at a low level in the morning until unshaded conditions are established. Over the morning hours, the shading of module 9 is gradually reduced until the module is fully illuminated at around 10:00 am. The changing shading on this specific module is schematically depicted by the red line in the lower picture in Fig. 10. The arrow indicates the progression of the illuminated PV module area with time.

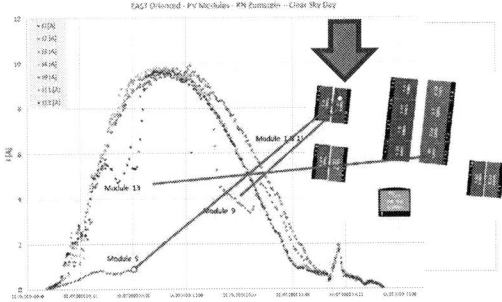

Figure 9: PV module currents for a specific day and assignment to the individual modules. Again module 9 is indicated by a red arrow.

The persistent low current level which represents the current of the diffuse illumination in the affected module substring is surprising because it does not reflect the increasing irradiation during the morning hours. For this module type with three bypass-diodes and the observed shading the I_{Mpp} should be at the same level as for the unshaded cells, but at lower voltage at the global MPP. This is schematically shown for the PV module characteristics in detail in Fig. 11 for similar shading conditions as depicted in the lower photograph of Fig. 10.

Shading of the lower section of the module leads to current-voltage and power-voltage graphs similar to the ones shown in Fig. 11. The shaded case is respectively given by the full lines, while the unshaded condition is represented by the dotted graphs.

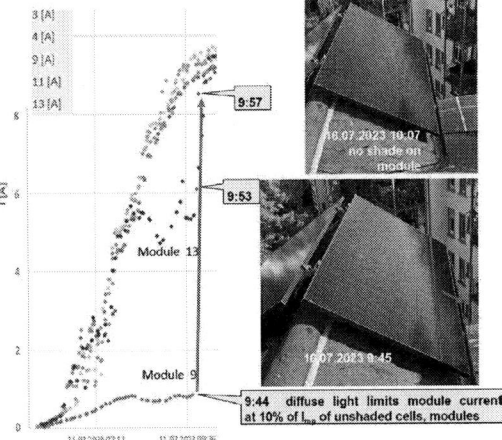

Figure 10: The current from module 9 remains at a low level until shadow-free conditions prevail at around 10:00.

The global MPP for the shaded case (black X) is at a lower voltage but with the same I_{MPP} delivering 2/3 of power of the unshaded case (black square). At this condition one bypass diode is activated. With more pronounced shading earlier in the morning there is another condition with two activated bypass diodes which leads to a global MPP at again lower voltage but the same I_{MPP} (not depicted in Fig. 11) which was not found by the MLPE.

Instead of adjusting to the respective MPPs during the course of the morning hours the persistent low voltage in Fig. 10 indicates that the optimiser adjusts to the local power maxima at higher voltage and lower current (black dot). Obviously, the power out at this local maximum is lower than at the MPP. Thus, a similar behavior to the indoor measurements (see Fig. 5) is observed, in which a sustained low current associated with operating points at high voltage prevails.

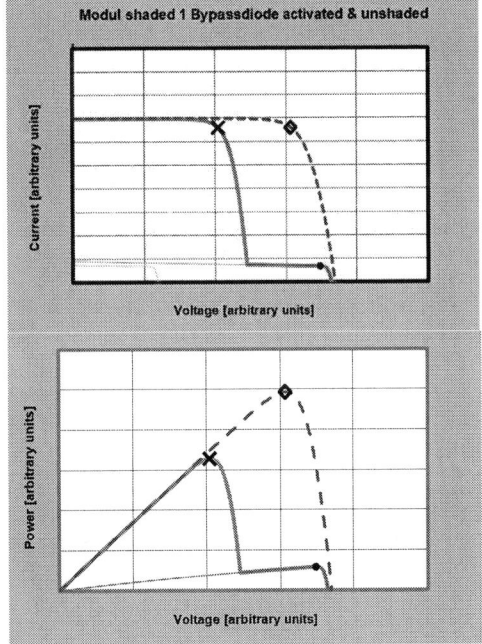

Figure 11: Schematic depiction of the current-voltage (above) and power-voltage (below) graphs due to shading conditions similar to the ones in Fig. 10.

1082

In Figure 12 the currents during the afternoon hours are shown in more detail as in Fig. 9 together with an indication of the progressing shading at module 9 starting at about 12:47. At this point, the shading object is the dormer on the south side, which is why the shadow moves across the narrow side of the module, unlike in Figure 10. At these conditions the optimiser succeeds in adjusting to intermediate currents levels.

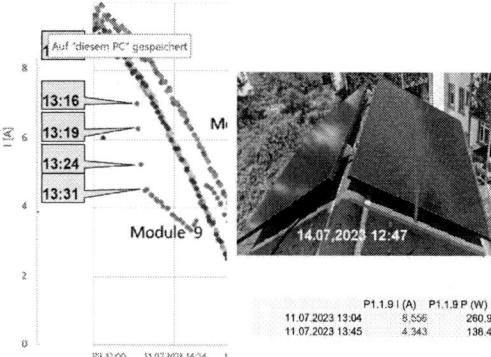

Figure 12: Currents during the afternoon hours are shown in more detail as in Fig. 9 together with an indication of the progressing shading at module 9 [10].

There is currently no explanation or information available from the manufacturer as to the cause of the algorithm failure resulting to reduced power output for the customer, even if it concerns small amounts of energy in this example.

3 SUMMARY AND DISCUSSION

Measurements at the output of a power optimiser under different shading conditions showed a preference for operating points with high voltage and low current. The optimiser could not find a global MPP if the associated voltage was lower than that of a previously determined local maximum at a higher voltage. Such a readjustment towards a lower voltage caused by partial shading events only worked for very small shifts within a local maximum in the volt range. In these cases, the small voltage shift did not trigger a search for a new global maximum by the MLPE. On the other hand, changing irradiation conditions that led to MPP shifts towards higher voltages resulted in successful readjustments.

Why the known methods for finding the global MPP are not used here is not explained by the manufacturers. In some ways, it is beneficial to have such a mechanism, as lower currents reduce the risk of localized heating (hot spots) due to shunt paths in the solar cell but is associated with performance losses. However, this is not mentioned anywhere that the use of this power optimiser always reduces the hot spot risk and one can assume that this would be mentioned accordingly or that there would be an option that can be chosen. On the other hand, the name optimizer of the device suggests that its purpose is to maximize performance.

Perhaps the manufacturers will provide an explanation, or the customer or the PV plant designer is free to choose between different sub-variants of the MLPE products with different methods of MPP tracking.

ACKNOWLEDGEMENT

The ZHAW is involved in the analysis of shading scenarios, to identify characteristic, benchmark shading situations, as a part of the collaboration by the International Energy Agency (IEA) PVPS Task 13 ST2.5 [5]. The IEA T13 Report is supported by the German Federal Ministry for Economic Affairs and Climate Action (BMWK) under contract no. 03EE1120B. The MLPE research of the ZHAW is funded by the Swiss Federal Office of Energy, with Project Number: SI/502247-01 and SH/81000380-02-01-46 [7].

REFERENCES

[1] F. P. Baumgartner, R. Vogt, C. Allenspach, und F. Carigiet, „Performance Analysis of Shaded PV Module Power Electronic Systems", *38th European Photovoltaic Solar Energy Conference and Exhibition; 650-654*, S. 5 pages, 20673 kb, 2021, doi: 10.4229/EUPVSEC20212021-4CO.3.1.

[2] C. Allenspach, F. Carigiet, und F. Baumgartner, „Lab Measurements of Power Optimizer Efficiency and Performance Simulation of Partially Optimized Systems Affected by Shading", *40th European Photovoltaic Solar Energy Conference and Exhibition*, S. 020412-001-020412–005, 2023, doi: 10.4229/EUPVSEC2023/4DO.1.1.

[3] C. Allenspach, F. Carigiet, A. Bänziger, A. Schneider, und F. P. Baumgartner, „Performance Analysis of Power Optimizers by Indoor Lab Testing and Shading Simulation", *8th World Conference on Photovoltaic Energy Conversion; 851-858*, S. 8 pages, 27078 kb, 2022, doi: 10.4229/WCPEC-82022-3DV.1.40.

[4] C. Allenspach, F. Carigiet, A. Bänziger, A. Schneider, und F. Baumgartner, „Power Conditioner Efficiencies and Annual Performance Analyses with Partially Shaded Photovoltaic Generators Using Indoor Measurements and Shading Simulations", *Solar RRL*, Bd. 7, Nr. 8, S. 2200596, Apr. 2023, doi: 10.1002/solr.202200596.

[5] F. Baumgartner, C. Allenspach et. al., *Shaded PV Generators Operated by Optimized Power Electronics*, Report IEA-PVPS T13-27:2024; ISBN 978-3-907281642, https://iea-pvps.org/publications/ [Online: variable Oct.2024].

[6] C. Allenspach, „Master Thesis, Module Level Power Electronics Dynamic and Static Performance in Partial Shaded Photovoltaic Systems". Zürcher Hochschule für angewandte Wissenschaft, Januar 2023. [Online]. Verfügbar unter: https://digitalcollection.zhaw.ch/bitstream/11475/27 358/3/2023_Allenspach_Cyril_MSc_SoE.pdf

[7] F. Baumgartner et. al., final project report of "EFFPVSHADE" public funded by the Swiss Federal Bureau of Energy (SFOE) under the contract Nr. SI/502247-01, Dec 2023. https://www.aramis.admin.ch/ ID=71181; SFOE follow-up project started in Dec 2023, WebPVShade; see public website https://srv-lab-t-579.zhaw.ch

[8] N. Pearsall, *The performance of photovoltaic (PV) systems: modelling, measurement and assessment, Chapter 5: F. P. Baumgartner et. al, Photovoltaic (PV) balance of system components, ISBN 978-1-78242-336-2.* in Woodhead Publishing series in

energy, no. number 105. Amsterdam ; Boston: Woodhead Publishing, 2017.

[9] A. Widler und L. Baumann, „Bachelor Thesis ‚Optimales Photovoltaik-Systemdesign mit Optimizer oder String-Inverter'". Zürcher Hochschule für angewandte Wissenschaft, 7. Juni 2024. [Online]. To be published at https://digitalcollection.zhaw.ch/

[10] F. Baumgartner, M. Golgroodbari, C. Bucher, B. Matthew, F. Valencia, U. Jahn, "Performance of partial shaded PV generators operated by optimized power electronics review - an IEA PVPS T13 activity", in *41st European Photovoltaic Solar Energy Conference and Exhibition (EUPVSEC)*, Vienna, 2024

41st European Photovoltaic Solar Energy Conference and Exhibition

IMPROVEMENT OF TRACKING ALGORITHMS USING DEEP LEARNING

Sarra Ben Brahim[2], Kai Saegebarth[2], Martin Dennenmoser[3], and Alsayed Algergawy[1]

[1] University Of Passau , Passau, Germany
[2] BayWa r.e. Solar Projects GmbH, Arabellastr. 4, 81925 Munich [3] BayWa r.e. Solar Projects GmbH, Kaiser-Joseph-Str. 263, 79098 Freiburg im Breisgau
*Phone: +491726420193 ; Email: sarra.benbrahim@baywa-re.com

ABSTRACT: This paper explores innovative solutions to enhance the efficiency of photovoltaic (PV) tracking systems using deep learning based tracking algorithms. Conventional solar trackers perform well on sunny days but struggle with optimal orientation during cloudy weather and the dynamic path of the sun. We propose dynamic solutions that adapt to real-time weather conditions using on-site measured and satellite data. By leveraging deep learning algorithms and historical time series data, our approach forecasts the optimal angle for the upcoming hour, maximizing the energy output of single-axis solar trackers.
Keywords: Artificial intelligence, machine learning, deep learning, PV system, PV Yield, Solar Tracker, time series forecasting.

1 INTRODUCTION

Fossil fuels have played a key role in driving economic growth, but they pose significant environmental risks. Consequently, there is an increasing shift toward renewable energy to safeguard the environment. Technological advancements in capturing and converting solar energy are advancing swiftly [1]. Photovoltaic (PV) panels, solar collectors, and concentrating solar power systems have become the most efficient methods for transforming solar radiation into usable energy across various applications. However, the efficiency of these solar energy systems is often limited by low energy conversion rates and weather conditions [2]. One crucial factor in optimizing energy output is adjusting the placement angle of the systems to maximize their exposure to solar radiation. Therefore, by equipping these solar devices with a solar tracking system to follow the sun's movement throughout the day, the energy output can be significantly increased [3]. A solar tracking system adjusts the alignment of PV panels, solar collectors, or other solar devices to follow the sun's direct rays, ensuring maximum sunlight exposure and enhancing energy efficiency [4], [5]. However, their effectiveness depends on factors such as geographic location, weather conditions, and the design of the tracking mechanism. Single-axis and dual-axis solar trackers are widely utilized to account for the seasonal changes in the sun's position [6]. The primary difference between the two lies in their motion capabilities: single-axis trackers rotate along one axis, offering a single degree of freedom, while dual-axis trackers provide two degrees of freedom, allowing for more flexible adjustment [7]. In this study we focus only on single axis trackers. Various solar tracking systems have utilized both machine learning (ML) and deep learning (DL) techniques. While conventional ML models are effective, they struggle with processing large datasets and demand a deep understanding of data representation, including feature selection, extraction, and reduction [8]. In contrast, advancements in big data technologies have highlighted the potential of DL models, which are now being applied to a range of tasks. For instance, DL is used to estimate PV system energy yield [9], forecast solar power and temperature [10], and predict wind and solar irradiance [11]. Researchers examined two types of solar tracking algorithms: standard and DL-based. This study explores their impact on PV system performance and inverter output, focusing on single-axis trackers, which are crucial for utility-scale systems.

- **Standard Tracking Algorithm:**
1) Astronomical Tracking Algorithms:
The astronomical tracking algorithm is the simplest algorithm used to calculate the sun's relative position, specifically the incidence angle of solar rays. It uses longitude, latitude, time, and other relevant data based on the solar-terrestrial relationship. The guiding principle is that minimizing the angle between the module's normal vector and incident solar rays results in higher irradiance received by the module. It dynamically adjusts the tracking angle throughout the day based on the sun's changing position. The algorithm assumes clear sky conditions, which may not always hold true due to cloud cover, pollution, or atmospheric disturbances. In such cases, the accuracy of tracking may decrease.
2) Backtracking Algorithms: The backtracking strategy orients the modules on the boundary of row-to-row shading, allowing the modules to face the sun as much as possible while remaining unshaded. Unlike the astronomical tracking calculation, calculating the backtracking algorithms consider the row distances and often also the topographical terrain of the PV site.

- **Deep Learning based Tracking Algorithm:**
Deep learning tracking algorithms adapt the target angle based on learnt historical data. They outperform standard solar trackers, especially in changing weather conditions like cloudy or partially cloudy days. Thus, we propose in this paper to study deep learning algorithm for single axis solar tracker using data collected from our on-site data (pyranometer, temperature sensor, wind vanes, anemometers) combined with satellite-based data and geo-location of the site. The data is collected to form time-series data which is then used to train deep learning-based algorithms. The proposed workflow will be then explained further in the paper. Three different deep learning algorithms will be used to forecast the solar tracking position (Angle in Degree). The remainder of this paper is structured as follows. Section 2 presents the solar tracking forecasting methodology. The solar tracking forecasting methodology in this paper is described in Section 3. This section emphasizes a comprehensive overview of data collection, data preparation, DL models, optimization methods and evaluation metrics used to develop intelligent

solar tracking algorithms. The conclusion is presented in Section 4.

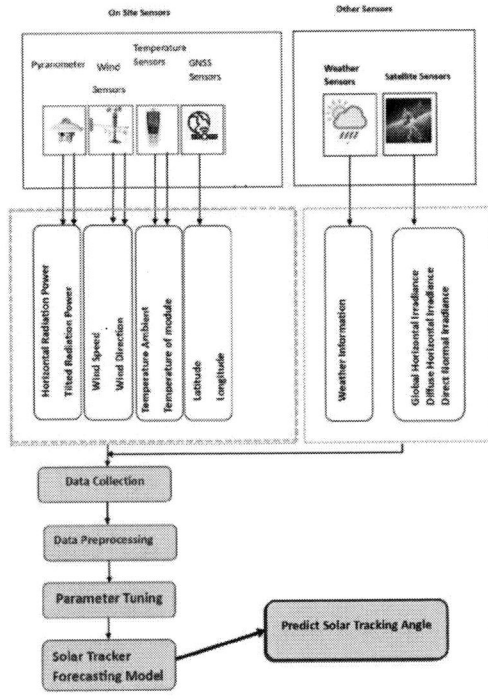

Figure 1: Proposed workflow of the deep learning based solar tracking forecasting model.

2 SOLAR TRACKING FORECASTING METHOD

The proposed method consists of four phases: i) data collection, ii) data pre-processing, iii) parameter tuning, and iv) deep learning model design, as illustrated in Figure. 1. Sensor fusion is therefore required. Collected signals can be detected at unequal intervals so resampling them is an essential step, which will be investigated in the pre-processing phase of the framework. After processing the data, we acquire a time series of data points that are resampled every 15 minutes. Before feeding them into the forecasting model, we do the parameter tuning to determine the optimal combination of hyperparameters for the deep learning model that results in the most accurate predictions. Finally, once we have the model trained, we forecast the optimal angle for the next hour using the last 2 hours of time series data as input.

In this research, we compare multiple deep learning models that are best suited for forecasting solar tracking angle under various weather situations. For the output of the forecasting solar tracking model, we will use the angle of our solar trackers that are already built in different locations to transfer knowledge from historical time series data and build a more robust model. We will compare the deep learning and standard tracking systems using some plots, under partially cloudy and completely cloudy conditions with focus on the potential gain of deep learning tracking algorithms compared to standard tracking algorithms. For completely sunny days, we noticed that deep-learning solar tracking algorithms and standard tracking algorithms show the same behavior. The target angle differences between deep learning algorithms and standard tracking ones result in impacting the output

of the inverters' AC active power, which is observed in partially cloudy conditions as shown in Figure. 3 or completely cloudy conditions, as shown in Figure. 2 .

The gain of the AC active power of inverter is calculated as follows :

gain = (DL_norm_AC power − STD_norm_AC power)
where :

- DL_norm_AC power is the normalized AC active power value for the deep learning algorithm
- STD_norm_AC power is the normalized AC active power value for the standard algorithm.

The diagram in Figure. 4, highlights the gain in Active Power (specific) [kW/kWp] between deep learning-based tracking and standard tracking for a single cloudy day.

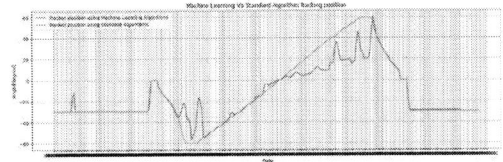

Figure 2: : Comparison Position of tracker's position using DL-based tracking and standard algorithm for tracking for partially cloudy day

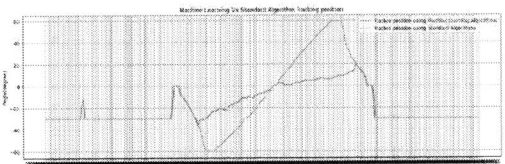

Figure 3: Comparison Position of tracker's position using DL-based tracking and standard algorithm for tracking for cloudy day

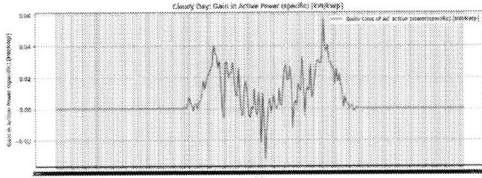

Figure 4: Cloudy Day: Gain in Active Power (specific) [kW/kWp]

3 SOLAR TRACKING FORECASTING MODEL PHASES

3.1 Data collection

When working with time series data, it is important to choose the right temporal resolution. This involves balancing the level of detail with the amount of data collected. Several factors should be considered when selecting the temporal resolution, such as the frequency of the phenomenon being measured, the duration of the study, and the desired level of accuracy. For solar tracking algorithms, it is important to consider mechanical requirements such as the speed at which the tracker can turn and the frequency of change. Therefore, we aim to sample time series data every 15 minutes. After the data is collected, our focus would be on the features. The features are discussed here in detail and are divided into 2 main categories:

- On-site data
- Satellite based weather data

On-site data is collected from pyranometers, wind sensors, temperature sensors. For more accurate results we divide the irradiation measured in module plane into its direct and diffuse component.

3.2 Data Pre-processing

We propose to pre-process the collected data before initiating the model training.

1) Preparation step: To ensure that all captured signals from sensors are available at the same frequency, resampling the time series data is essential. This involves changing the frequency of our time series observations so that they are at the same frequency which is 15 minutes.

2) Standardization: To ensure consistent scales across features, we standardize each feature. Standardization (also known as z-score normalization) [12] transforms each feature to assure that they have a standard deviation of 1 and a mean of 0. Standardized features prevent any single feature from dominating the model during training.

3.3 Parameter Tuning

Hyperparameter tuning is a process that customizes forecasting models to generate precise outputs. It involves controlling the learning process of a deep learning model, requiring different constraints, weights, or learning rates for different data models. In this study, we will focus on recurrent neural networks thus parameters like learning rate, number of units, regularization, and dropout are tuned, as described in Table I.

3.4 Deep Learning Models

In this study, we investigate three different forecasting models to forecast the solar tracking position:

 i. LSTM (Long Short-Term Memory)
 ii. GRU (Gated Recurrent Unit)
 iii. SimpleRNN (Recurrent Neural Network)

All of the mentioned models are all types of recurrent neural networks (RNNs). RNNs are a type of artificial neural network that can process sequential data, such as time series data by maintaining an internal state or memory of previous inputs. They are used in different types of applications of forecasting in energy domain [13], [4]. Simple RNNs are the most basic form of RNNs, which use a single hidden layer to process sequential data. While they are easier to implement, they often struggle with long-term dependencies due to the vanishing gradient problem. LSTM networks are a type of RNN that can learn long-term dependencies, making them effective for time series forecasting. They use special units called memory cells to store information over long periods, which helps in capturing patterns in sequential data. GRUs are a simplified version of LSTMs that use fewer parameters, making them faster to train while still maintaining the ability to capture long-term dependencies. They achieve this through two gates: the reset gate and the update gate. RNNs, including LSTM, GRU, and Simple RNN, are widely used in various forecasting applications within the energy domain. They help in predicting energy consumption, load forecasting, and optimizing renewable energy sources [14], [15].

This study proposes a comparative analysis of these RNN architectures (LSTM, GRU, Simple RNN) to achieve effective multi-step univariate time series forecasting performance for solar tracking algorithms. By evaluating their performance, we aim to identify the most efficient model for enhancing the accuracy of solar tracking systems, thereby improving the efficiency and output of photovoltaic (PV) systems [16].

$$MSE = \frac{1}{n} * \sum_{i=1}^{n}(y_i - \hat{y}_i)^2 \tag{1}$$

$$RMSE = \sqrt{\frac{1}{n} * \sum_{i=1}^{n}(y_i - \hat{y}_i)^2} \tag{2}$$

Where:

- n total number of data points.
- y_i is the observed angle (actual target) for the ith data point.
- $\hat{y}_i$ is the predicted angle (output of the forescating model) for the ith data point.

The mean squared error (MSE) measures the quality of the forecasting. Since the data (input and target) is normalized so the range value of MSE is typically falls in the range of 0 and 1. A lower MSE indicates better model performance, as it reflects smaller prediction errors. We also incorporated Root Mean Squared Error as a metric to evaluate the performance of the deep leaning based solar tracking algorithm. The RMSE is derived from the MSE and provides a more interpretable metric. It calculates the square root of the average squared difference between predicted and actual values, as described in Eq. 2.

Based on the comparison between the three different RNN algorithms described in Table I, and after splitting our dataset into 70% of training and 30% of testing, we find that GRU was the most promising model in terms of minimizing the error between the forecasting and the actual value of the solar tracking resulting in MSE of 4.7×10^{-2} .

Table I: Results of the hyperparameter optimization

Training Parameters	LSTM	GRU	SimpleRNN
Loss Function	MSE	MSE	MSE
Best Parameters	Learning rate= 0.01, Units=64 Dropout=0.1	Learning rate= 0.001, Units=64 Dropout=0.01	Learning rate= 0.001, Units=64 Dropout=0.1

4 CONCLUSION

In this paper, we have investigated the solar tracking forecasting algorithms using on-site sensors and satellite-based data sampled by 15 minutes. After collection of our time series data under different weather conditions, we fed our data into our AI-agent to forecast the solar tracking position. Results have shown the ability of deep learning models to achieve a MSE of 4.7×10^{-2} .This framework has shown promising results and can be further extended across Europe to dynamically forecast the solar tracking position in nearly real time manner.

5 REFERENCES

[1] M. Asif and T. Muneer, "Energy supply, its demand and security issues for developed and emerging economies," Renew. Sustain. Energy Rev., vol. 11, pp. 1388–1413, 2007. [Online]. Available: https://doi.org/10.1016/j.rser.2005.12.004

[2] D. Huynh, T. Nguyen, M. Dunnigan, and M. Mueller, "Comparison between open- and closed-loop trackers of a solar photovoltaic system," in 2013 IEEE Conference on

Clean Energy and Technology (CEAT). Langkawi, Malaysia: IEEE, 2013, pp. 128–133. [Online]. Available: https://doi.org/10.1109/CEAT.2013.6775613

[3] A. Verma and S. Singhal, "Solar pv performance parameter and recom- mendation for optimization of performance in large scale grid connected solar pv plant — case study," 2015.

[4] R. Banerjee, "Solar tracking system," 2015.

[5] A. Hafez, A. Yousef, and N. Harag, "Solar tracking systems: Technologies and trackers drive types – a review," Renew. Sustain. Energy Rev., vol. 91, pp. 754–782, 2018. [Online]. Available: https://doi.org/10.1016/j.rser.2018.03.094

[6] S. Rustemli, F. Dincadam, and M. Demirtas, "Performance comparison of the sun tracking system and fixed system in the application of heating and lighting," 2010.

[7] S. Racharla and K. Rajan, "Solar tracking system – a review," Int. J. Sustain. Eng., vol. 10, pp. 72–81, 2017. [Online]. Available: https://doi.org/10.1080/19397038.2016.1267816

[8] N. Chauhan and K. Singh, "A review on conventional machine learning vs deep learning," in 2018 International Conference on Computing, Power and Communication Technologies (GUCON). Greater Noida, Uttar Pradesh, India: IEEE, 2018, pp. 347–352. [Online]. Available: https://doi.org/10.1109/GUCON.2018.8675097

[9] A. Catalina, A. Torres-Barra´n, C. Ala´ız, and J. Dorronsoro, "Machine learning nowcasting of pv energy using satellite data," Neural Process. Lett., vol. 52, pp. 97–115,2020.[Online].Available:https://doi.org/10.1007/s11063-018-09969-1

[10] V. Gundu and S. Simon, "Short term solar power and temperature forecast using recurrent neural networks," Neural Process. Lett., vol. 53, pp. 4407–4418, 2021. [Online]. Available: https://doi.org/10.1007/s11063-021-10606-7

[11] N. AL-Rousan, N. Mat Isa, and M. Mat Desa, "Efficient single and dual axis solar tracking system controllers based on adaptive neural fuzzy inference system," J. King Saud Univ. - Eng. Sci., vol. 32, no. 7, pp. 459–469, 2020.

[12] K. Cabello-Solorzano, I. Ortigosa de Araujo, M. Pen˜a, L. Correia, and A. J. Tallo´n-Ballesteros, "The impact of data normalization on the accuracy of machine learning algorithms: a comparative analysis," in International Conference on Soft Computing Models in Industrial and Environmental Applications. Springer, 2023, pp. 344–353.

[13] A. Tokgo¨z and G. U¨ nal, "A rnn based time series approach for fore- casting turkish electricity load," in 2018 26th Signal processing and communications applications conference (SIU). IEEE, 2018, pp. 1–4.

[14] H. Habbak, M. Mahmoud, K. Metwally, M. M. Fouda, and M. I. Ibrahem, "Load forecasting techniques and their applications in smart grids," Energies, vol. 16, no. 3, p. 1480, 2023.

[15] M. S. Chowdhury, N. Nabi, M. N. U. Rana, M. Shaima, H. Esa, A. Mitra, M. A. S. Mozumder, I. A. Liza, M. M. R. Sweet, and R. Naznin, "Deep learning models for stock market forecasting: A comprehensive comparative analysis," Journal of Business and Management Studies, vol. 6, no. 2, pp. 95–99, 2024.

[16] Y. Kong, Z. Wang, Y. Nie, T. Zhou, S. Zohren, Y. Liang, P. Sun, and Q. Wen, "Unlocking the power of lstm for long term time series forecasting," arXiv preprint arXiv:2408.10006, 2024.

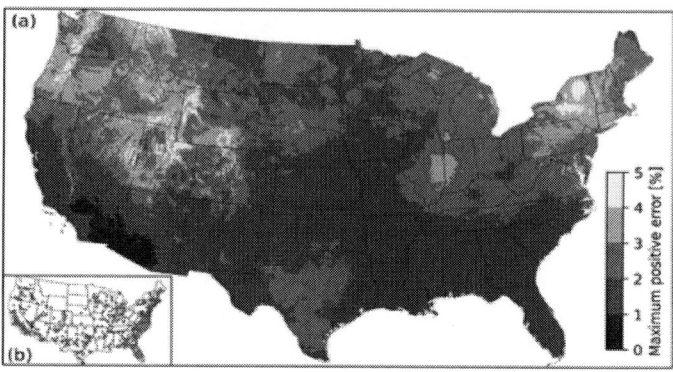

Best Practice Guidelines for the Use of PV System KPIs
Marios Theristis, Sandia National Laboratories

EUPVSEC 2024, 23.09.2024

SAND2024-12535C

Technology Collaboration Programme
by iea

Introduction

- IEA PVPS Task 13: "provides a **common platform to summarize and report** on technical aspects affecting the **reliability, quality, and performance of PV systems** in a wide variety of environments and applications. Combine and integrate knowledge into valuable **summaries of best practices** and methods for ensuring PV systems perform at their optimum."

- Subtask 3.4: "Mapping economic and reliability KPIs" focuses on gathering KPIs and examining the differences in their usage, definitions and methods of calculation. It also analyzes their sensitivity to data quality and climatic variations, offering best practices to reduce uncertainty.

What are KPIs and why are they important?

Key Performance Indicators (KPIs) are metrics used to assess various aspects of a PV system, including technical and financial performance

Technical KPIs allow:

Performance optimization

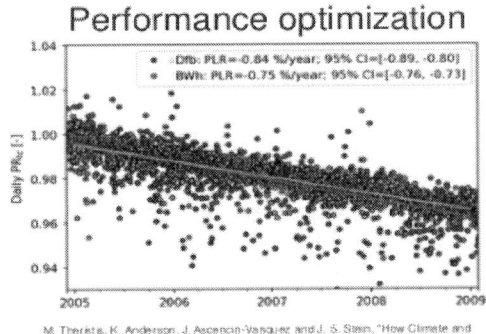

M. Theristis, K. Anderson, J. Ascencio-Vasquez and J. S. Stein, "How Climate and Data Quality Impact Photovoltaic Performance Loss Rate Estimations," *Solar RRL*, vol. 8, no. 2, 2023.

They can also be contractual so $$$ are on stake

Asset owner — O&M provider

Are KPIs not standardized?

Many standards exist

1. **IEC 61724-1:2021:** Guidelines for monitoring PV system performance, including instrumentation, data acquisition, KPIs (e.g., PR, energy yield), data quality, and reporting.

2. **IEC 61724-2:2016:** Methods for evaluating PV system capacity, focusing on capacity-related KPIs (e.g., capacity factor, specific yield), data requirements, and evaluation procedures.

3. **IEC 61724-3:2016:** Energy performance evaluation methods for PV systems, covering energy yield, system efficiency, PR, data requirements, and evaluation procedures.

4. **IEC TS 63019:2019:** Information model for assessing PV system availability, defining availability KPIs (e.g., operational availability, downtime), data requirements, and reporting.

5. **IEC TR 63292:2020:** Roadmap for robust PV system reliability, including reliability KPIs (e.g., failure rates, degradation rates), best practices, risk management, lifecycle management, and case studies.

Limitations of current standards (examples)

- Different approaches for different metrics
- Recommended approaches might be vague and subject to Reader interpretation

For example:

- High level of freedom with proposed data quality routines
- Biases can affect contractual agreements resulting in incorrect financial outcomes

Are KPIs consistent across regions, markets, contract types?

Industry Insights: Feedback on Technical KPIs

Direct input through bilateral contacts

Workshop at Solar Quality Summit Europe 2024

LinkedIn survey

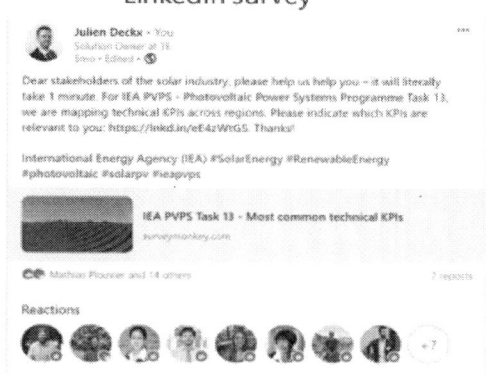

Industry Insights: Feedback on Technical KPIs

High-level conclusions

- Regional variations: while there are trends per region, a globalized world and market means that there are no strict differences to be seen.

- Contractual KPIs: no consistency in which KPI variations are mostly used contractually. E.g. PR vs temperature-corrected PR, time-based vs energy-based Availability, etc.

Industry Insights: Feedback on Technical KPIs

Detailed description

	Energy Performance Index (EPI) IEC TS 61724-3:2016	ASTM Capacity Test ASTM-E2848-13
Description	Calculated as: $$EPI = \frac{Y_{meas}}{Y_{exp}}$$ with measured yield Y_{meas} and expected yield Y_{exp}, calculated based on a pre-agreed performance model.	Power at "reporting conditions" (RC): $$P_{RC} = G_{RC}(a_1 + a_2 \cdot G_{RC} + a_3 \cdot T_{RC} + a_4 \cdot v_{RC})$$ with reference irradiance G_{RC}; temperature T_{RC}; wind speed v_{RC}; regression coefficients a_i, based on measured data.
Application	(Mostly) EU: replacement or complement to performance ratio (PR).	(Mostly) US: acceptance testing of PV plant
Advantages	Less affected by system and weather conditions than PR	More accessible calculation method than EPI
Challenges	Complex and hard-to-reproduce calculation method	Still more complicated and less transparent than PR. Linearity assumption and exclusion of all non-linear events

Example: Challenges due to data quality and climatic variability

PR calculated every 4 x 4 km in CONUS and compared against corrupted time-series

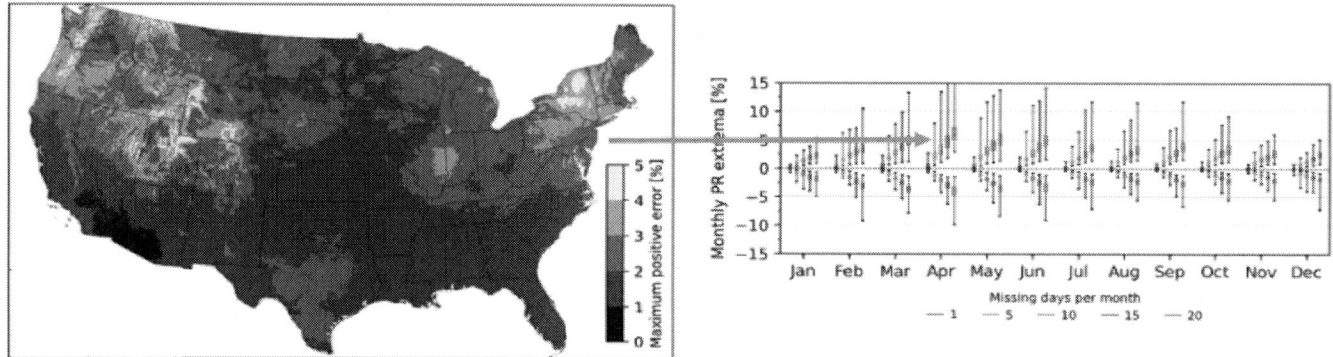

Unified data quality approaches and best practices are needed to minimize bias

S. Lindig, M. Herz, J. Ascencio-Vásquez, M. Theristis, B. Herteleer, J. Deckx and K. Anderson, "A comprehensive review of technical photovoltaic key performance indicators and the role of data integrity," under review, 2024.

Conclusions & Outlook

- High variability in KPIs, not clearly segmented by region, market or contract type

- Lack of transparency not only in definition but also method of calculation → standards fall short

- KPI estimation accuracy depends on data quality and climatic variations → very important because KPIs can make or break contracts with $$$$ consequences

We need to:

- Harmonize definitions and methods of KPI calculations

- Provide best practices to minimize KPI uncertainty

www.iea-pvps.org

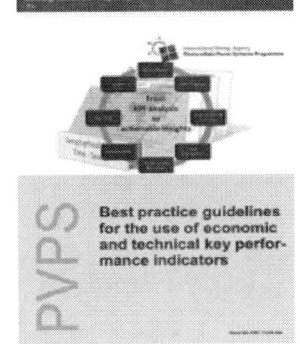

Thank you!

The IEA Task 13 ST3.4 Team
S. Lindig, M. Herz, J. Acencio-Vásquez, M. Theristis, B. Herteleer, J. Deckx, K. Anderson

mtheris@sandia.gov

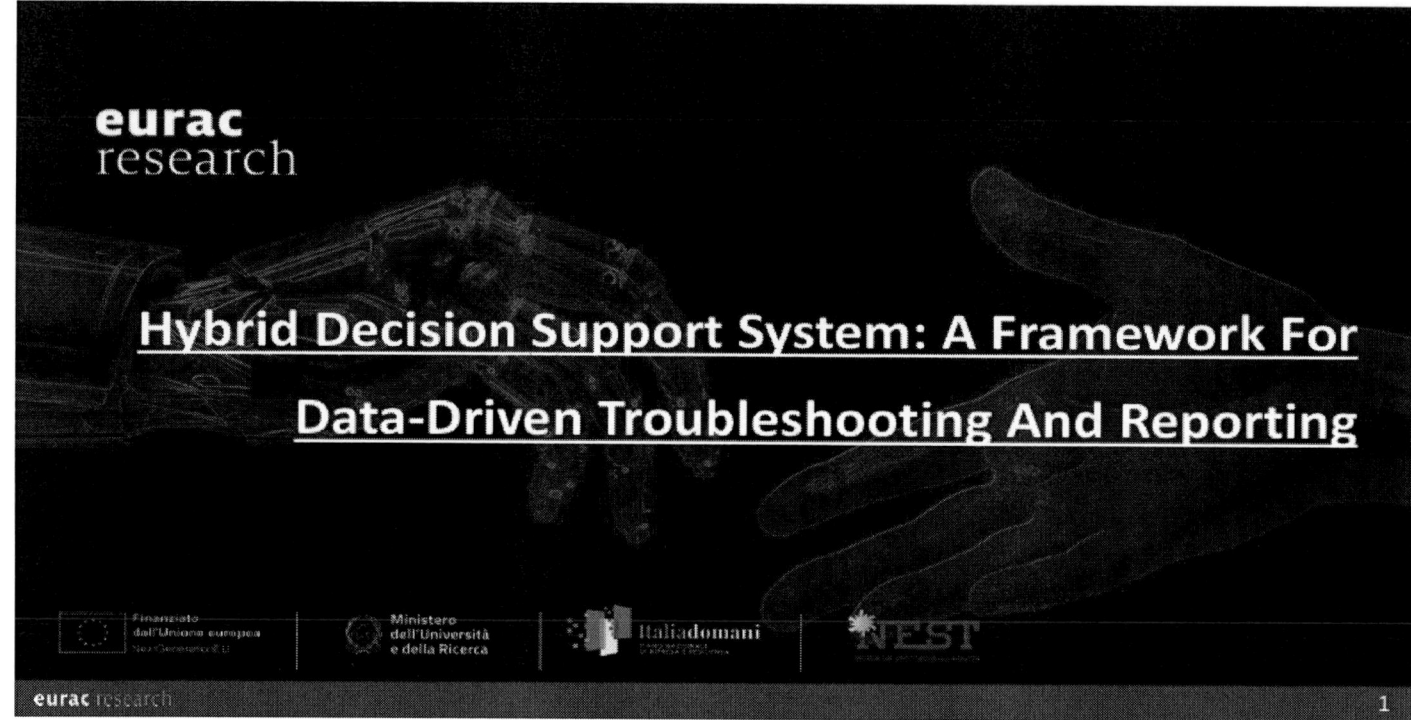

Outline

- **Introduction**
  - Digitalization Evolution
  - Human operator and Machine
- **Proposed Framework**
  - Data-driven Pipeline
  - Human-in-the loop and active learning
- **Toward Unbiased Machine**
  - Operator profiling
  - Inspection Validation
  - Agents in the loop

Digitalization of O&M in PV Systems
Shifting from Human Operators to AI-Driven Automation

Manual Monitoring and Reactive Maintenance

Basic SCADA & Offline Analysis

Centralized Real-Time Monitoring and Remote Diagnostics

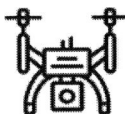

Automation with Drones, Robotics, and Cloud-Based Platforms

Main goal:

- **Increased efficiency and uptime.**
- **Reduced costs and downtime.**
- **Scalable management with minimal human intervention.**

Human Operator & Machine

	Great at	Poor at
	• dealing with uncertainty • weighing context • adapting to new domains	• precise and repetitive tasks • perfect recall
	• pattern recognition • identifying correlations	• identifying context • dealing with uncertainty • incorporating knowledge

Introducing the machine

eurac research

---

1 PV Failure Detection & Reporting Framework

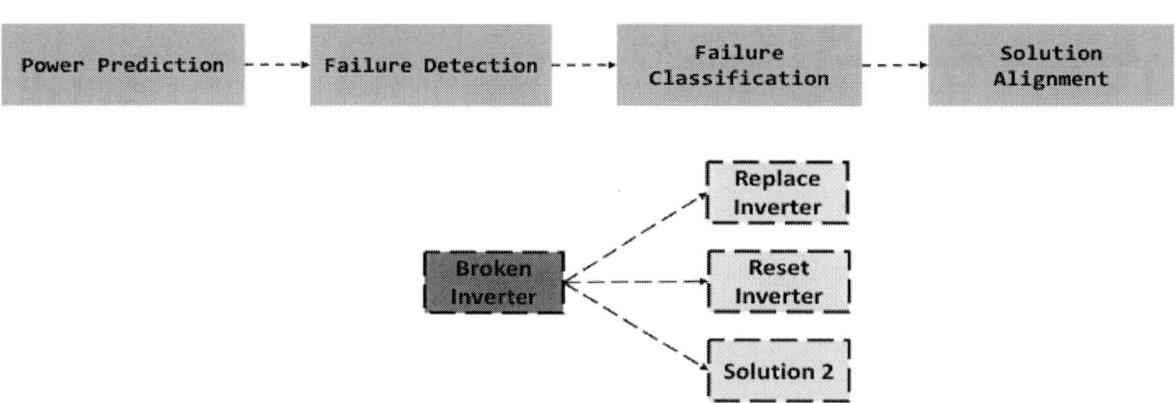

Towards the development of an optimized Decision Support System for the PV industry: A comprehensive statistical and economical assessment of over 35,000 O&M tickets - Lindig - 2023 - Progress in Photovoltaics: Research and Applications - Wiley Online Library

eurac research

To what extent we can trust the machine?

PV Failure Detection & Reporting Framework
Human-in-the loop

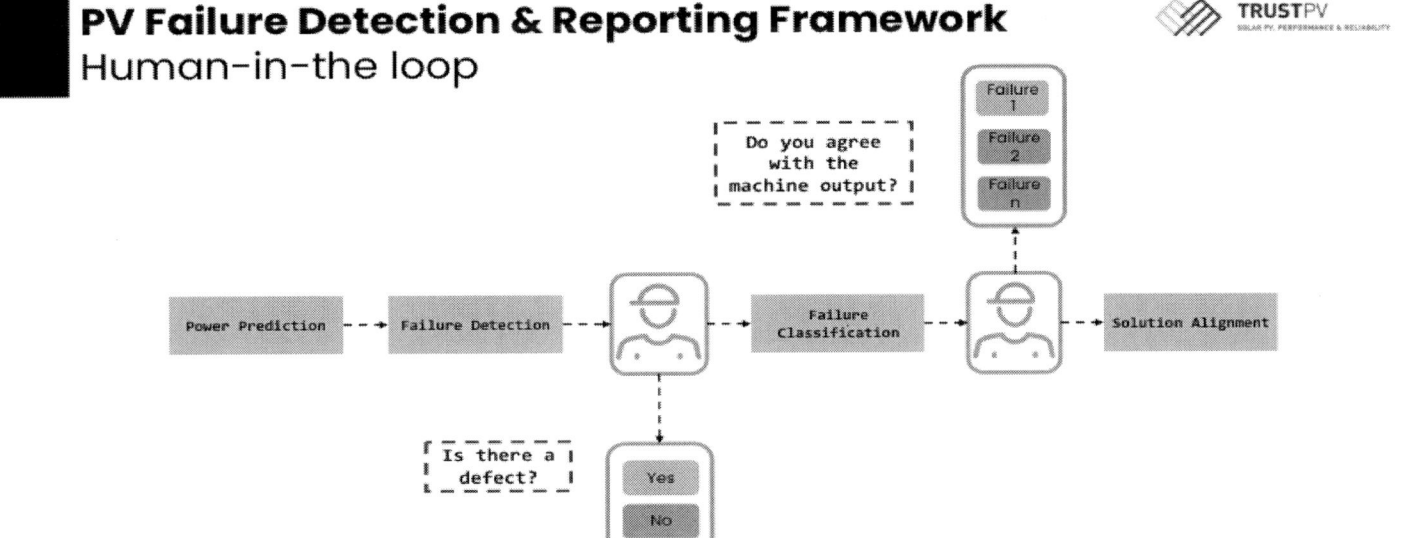

Active learning

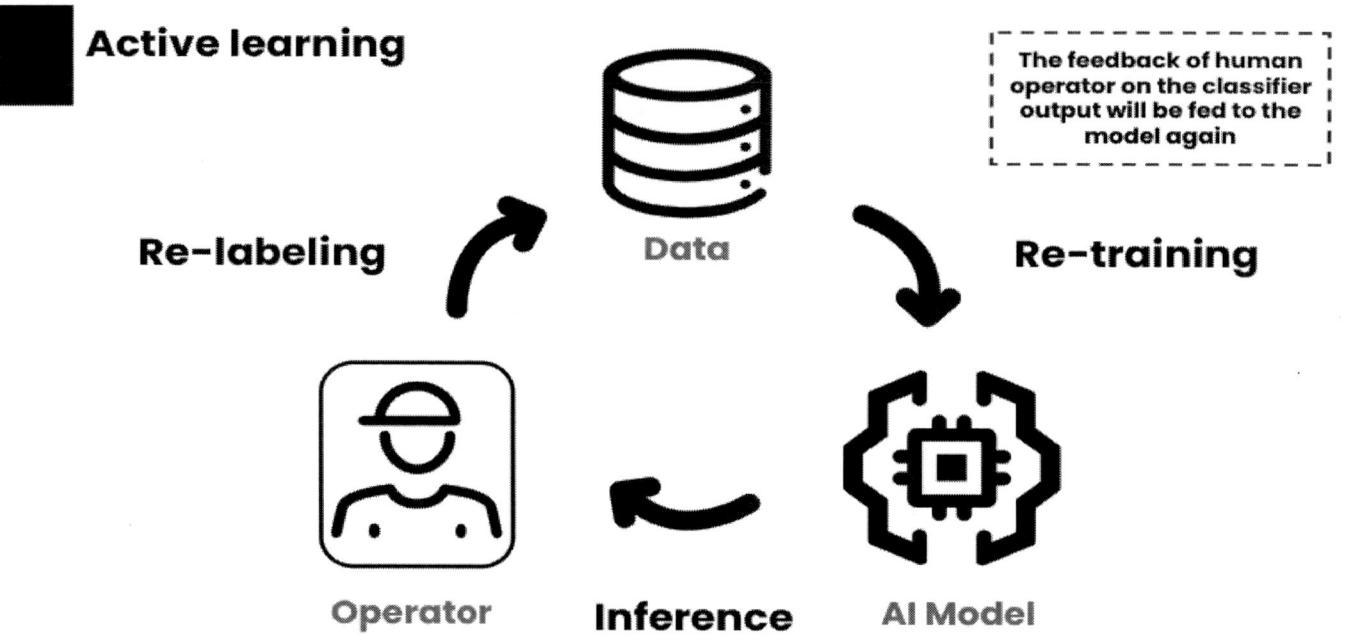

Re-labeling · **Data** · **Re-training**

The feedback of human operator on the classifier output will be fed to the model again

Operator · **Inference** · **AI Model**

How do we know we are not transmitting the human bias to the machine?

1+2 Operators Profiling & Inspection Validation
Experience & Validation

Experienced
- Can improve the machine
- Follows the taxonomy

In-experienced
- Can improve the machine
- Follows the taxonomy

Inspection Validation

Operator inspection can be proved or negated by checking and analyzing the available data regarding the reported failure

---

3 Reporting Agent

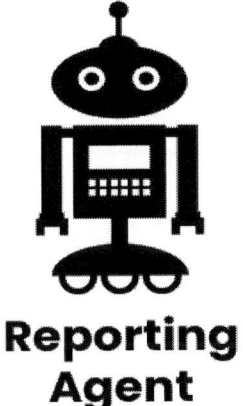

- An artificial intelligence (AI) agent refers to a system or program that is capable of autonomously performing tasks on behalf of a user or another system by designing its workflow and utilizing available tools.

Reporting Agent

Mitigating human bias in the machine

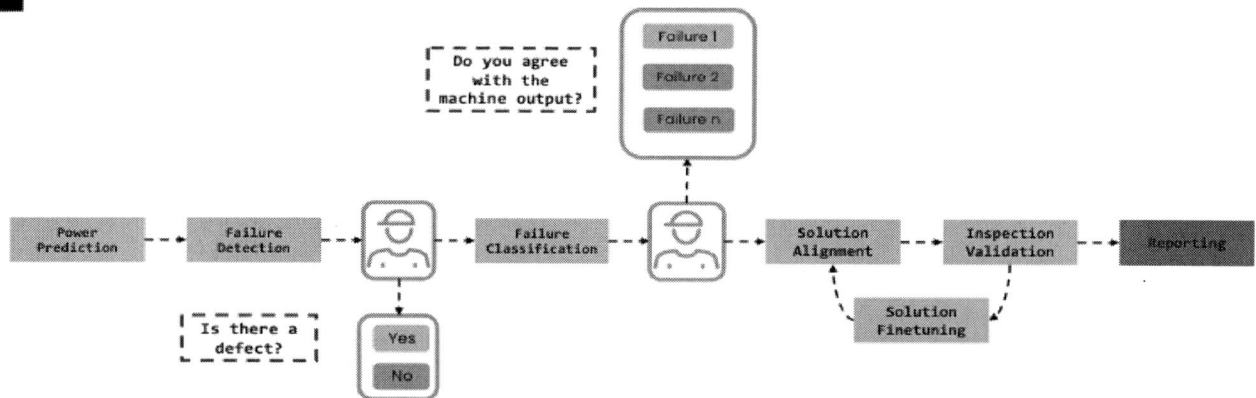

Human oversight over algorithm

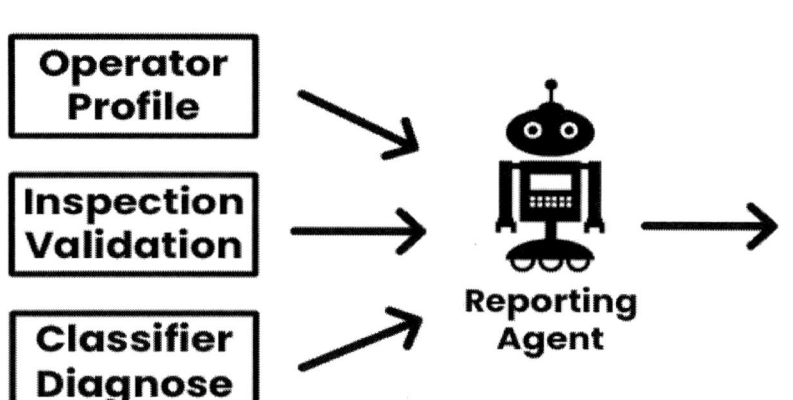

O&M Report

New digitalization paradigm: Human & Machines

Classifier

Reporting Agent

Improves

Operator

Teaches

Key take-aways
Hybrid DSS

- **Unlocking the full potential of generative AI, predictive AI, and human-in-the-loop systems requires the introduction of new digitalization paradigms**

- **Data-driven pipelines are already in place, but if human involvement is necessary, we must fully leverage and maximize the value of their presence.**

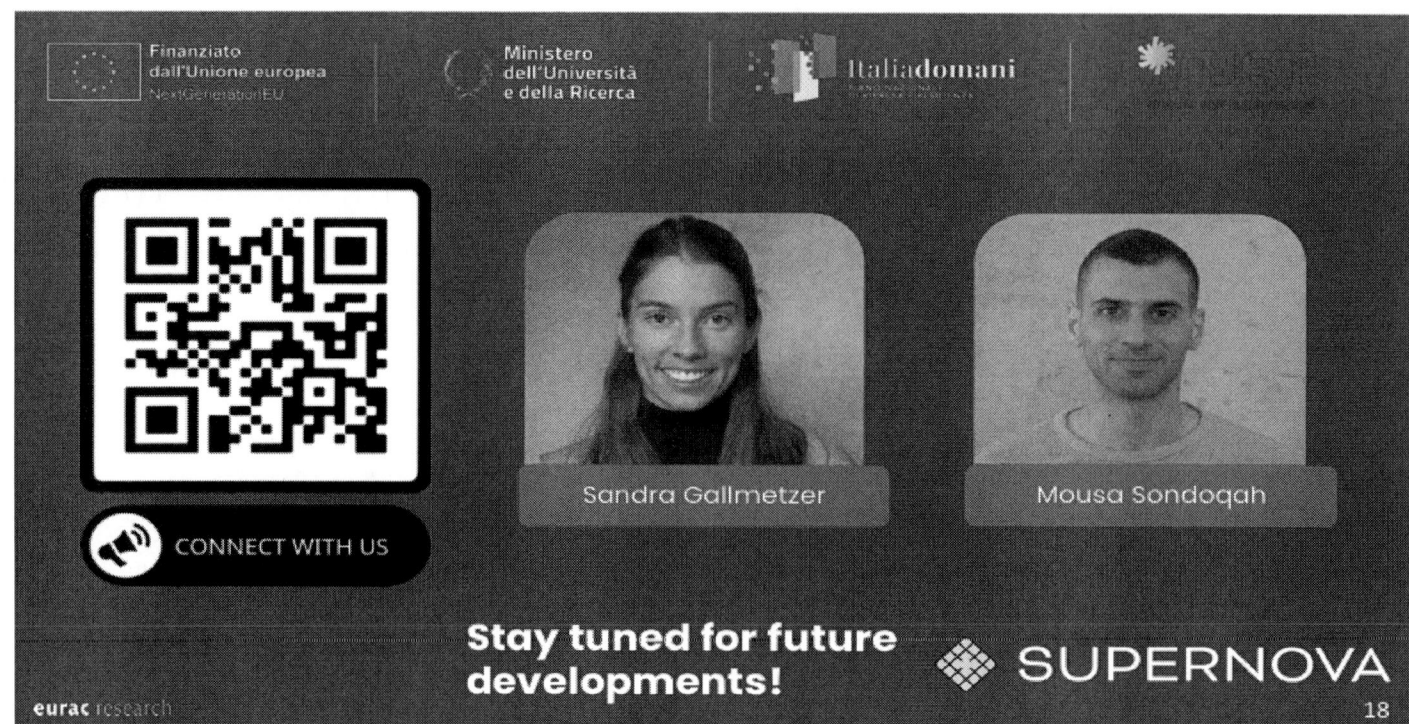

41st European Photovoltaic Solar Energy Conference and Exhibition

IDENTIFYING DISTINCT PERFORMANCE PATTERNS IN UTILITY-SCALE PHOTOVOLTAIC PLANTS USING AN UNSUPERVISED MACHINE LEARNING MODEL

Ali Shakiba[1*], Brendan Wright[1], Ziv Hameiri[1]
[1]University of New South Wales, Sydney, Australia
*a.shakiba@unsw.edu.au

Photovoltaic (PV) systems have achieved an impressive cumulative capacity of 1.2 TW in 2022, emerging as a pivotal component in energy production. However, this significant adoption also highlights the crucial need for accurate estimation of PV system power generation, for gauging electricity availability, assessing investment feasibility, and determining profit margins. Although a large amount of data is collected from operational PV plants, it is challenging to determine their performance, as electrical parameters need to be normalised to account for environmental conditions — a procedure that is often associated with significant uncertainty.

The concept of a digital twin, a virtual replica of the PV plant, is a powerful method to determine the expected behaviour, and therefore, can be used to detect deviations in performance or identify faults. However, the reliance on digital twins based upon physical models poses challenges, as they require many input parameters (some with significant associated uncertainties) and certain simplifications or assumptions (such as linear degradation or soiling), thereby, imposing limitations on achievable accuracy and predictive capability.

This study proposes a novel method to overcome these limitations by leveraging machine learning (ML) algorithms to create learnt digital twins of PV plants, using only the collected raw daily operational data. Hence, the creation of these digital twins does not require the imposition of any physical models with underlying assumptions or biases. The developed data-driven approach is demonstrated using operational data from a utility-scale bifacial PV plant. We show that the developed method automatically learns to represent latent factors within the raw operational data, enabling the identification of different behavioural patterns, such as operation under clear-sky or cloudy conditions. It also correctly infers correlations between electrical and meteorological features, and uncovers implicit temporal characteristics in daily performance, without being provided with this explicit knowledge or information.

This novel ML-based digital twin methodology marks a critical step toward completely automated data-driven Operations and Maintenance (O&M) systems that are based on real-world operational behaviour, with the ability to continuously update and improve over time. This will broadly enhance the reliability of utility-scale PV, supporting the much-needed transition to renewable energy.

Keywords: Learnt digital twin, Performance patterns, Operations and maintenance

1 AIM AND APPROACH

Photovoltaics (PV) has emerged as a dominant contributor to electricity production, reaching a staggering cumulative installed capacity of 1.2 TW in 2022, showcasing a remarkable 24% year-on-year growth in new installations [1]. However, this rapid expansion brings forth a critical challenge, the precise estimation of energy generation by a PV power plant. An accurate assessment of energy production is a vital requirement for optimising the plant's performance. This precision in estimation is equally essential during the design phase as it dictates the payback time and expected profit, both of which are critical factors for the bankability of the project.

The ongoing monitoring of PV plant operations leads to the accumulation of substantial amounts of electrical and meteorological data. However, evaluating their performance based on this collected operational data has proven to be a challenging task. Firstly, the assessment of the electrical performance of the PV plant is significantly influenced by weather and environmental conditions, necessitating accurate correction and normalisation, primarily to temperature and irradiance. Secondly, the collected data often contains noise, inaccuracies, and missing values. Collectively, these transformations can introduce considerable uncertainty that profoundly affects the reliability of performance analysis.

Digital twins are employed as virtual replicas of PV plants to forecast the anticipated energy generation and subsequently identify performance deviations. Typically, these digital twins rely on physical models to simulate the plant's performance. However, the reliance on physical models can introduce significant limitations as these

models often oversimplify real-world behaviour, such as assuming a linear degradation [2]. Additionally, these models frequently require a large number of input parameters, some of which are inherently challenging to measure with high precision or fundamentally subject to large measurement uncertainties. Furthermore, certain essential inputs are not directly measured but instead estimated, introducing additional uncertainties into the model. For instance, the PV module operating temperature is only rarely measured in the field, and instead is estimated using other parameters such as plane-of-array (POA) irradiance, ambient temperature, and wind speed [3].

To address these challenges, this study presents a novel, machine learning (ML)-driven method for constructing digital twins of PV plants, **solely relying on raw daily operational data** (which includes power output, POA irradiance, and ambient temperature). This eliminates the need for physical models and their associated limitations or bias, resulting in more robust and adaptable digital twins that effectively capture the complex dynamics of PV plants. Furthermore, we use the latent feature embedding learnt by the developed ML-based digital twin model to cluster operational days of performance by similarity.

The proposed method employs three distinguished component ML algorithms: (1) β-Variational Autoencoders (β-VAE) [4], (2) t-distributed stochastic neighbour embedding (t-SNE) [5], and (3) hierarchical density-based spatial clustering of applications with noise (HDBSCAN) [6]. The β-VAE automatically extracts and learns a disentangled normal distribution of independent latent features from the raw daily operational data. The t-

SNE algorithm performs a non-linear embedding to reduce the dimensionality of the learnt latent space while preserving pairwise similarity between samples. Finally, the HDBSCAN algorithm robustly identifies clusters of similar operational days, even with varying relative shapes and densities.

As a pre-processing step, each feature in the raw daily operational data is independently min-max normalised, projected into the range [0,1], followed by substituting any missing data with a constant [zero]. The block diagram of the developed pipeline is illustrated in Figure 1.

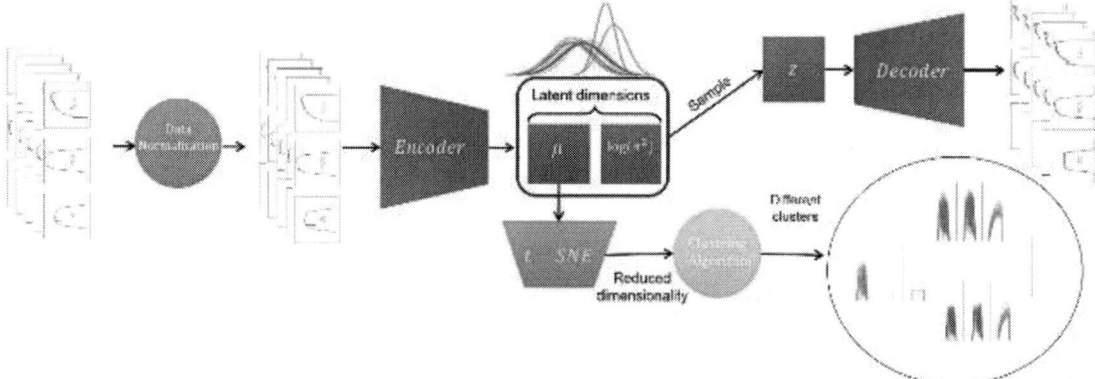

Figure 1: Block diagram of the developed approach to identifying different performance patterns in PV plants: (green) pre-processing of the electrical and meteorological data to extract and normalise the daily operational data, followed by (blue) training a β-VAE to learn a set of disentangled latent features (named 'learnt latent space'), (orange) using the t-SNE algorithm to compress the learnt latent features, and finally (yellow) clustering with the HDBSCAN density-based clustering algorithm. Each cluster maps to a performance pattern of the PV plant.

2 SCIENTIFIC INNOVATION AND RELEVANCE

Traditional digital twins for PV modules often face limitations due to their dependence on physical models and their inherent assumptions, mainly driven by their oversimplification. They are also limited by the large uncertainty of many of the input variables. To address these challenges, we propose a novel ML-based method that operates as a digital twin. This approach extracts distinct performance patterns of PV plants *directly* from their raw daily operational data, avoiding reliance on any physical models and their associated limitations. This data-driven method marks a critical step toward a fully automated, data-driven O&M approach for utility-scale PV plants based on *continuously evolving* digital twins which can update and improve over time. Furthermore, we showcase an application of this ML-based digital twin in clustering operational days based on similarity of performance and behaviour.

3 RESULTS AND CONCLUSIONS

To evaluate the developed method, we utilised the daily operational data from a single array of 28 bifacial PV modules installed within a utility-scale PV plant in Australia. The dataset spans 33 consecutive months, recorded at 5-minute intervals, and includes the array inverter output power (generated power), POA irradiation, and ambient temperature. The ML-based digital twin was trained to automatically learn latent features from the raw daily operational data, enabling it to reconstruct the

original information. To evaluate the effectiveness of the trained ML-based digital twin model in capturing the temporal patterns and the correlations between the different input daily operational data, the t-SNE algorithm was employed to embed the latent features into two dimensions considering all the days within the training dataset.

Figure 2 presents these two dimensions graphs; each point in these graphs corresponds to an individual day while the colour indicates the: (a) daily aggregated generated power, (b) daily aggregated POA irradiance, (c) maximum daily ambient temperature, and (d) day of the year (as a proxy for the seasonal patterns). Note that the distance between points represents their similarity. Importantly, the model captures a clear relation between the learnt latent features and the specified colour-coded variables. Figure 2(a) and (b) show general regions based on generated power and POA irradiance, respectively. While temperature extremes form separated regions in Figure 2(c), overlap occurs for days within the range of 5-25°C. This potential ambiguity might be due to missing information about the albedo (these are bifacial PV modules), or modules' operational temperature. Finally, Figure 2(d) highlights the model's ability to infer temporal patterns without explicitly providing any date information to the model. The day-of-year colouring aligns with the cluster positioning, demonstrating the model's capacity to learn temporal dependencies from only the raw daily operational data.

41st European Photovoltaic Solar Energy Conference and Exhibition

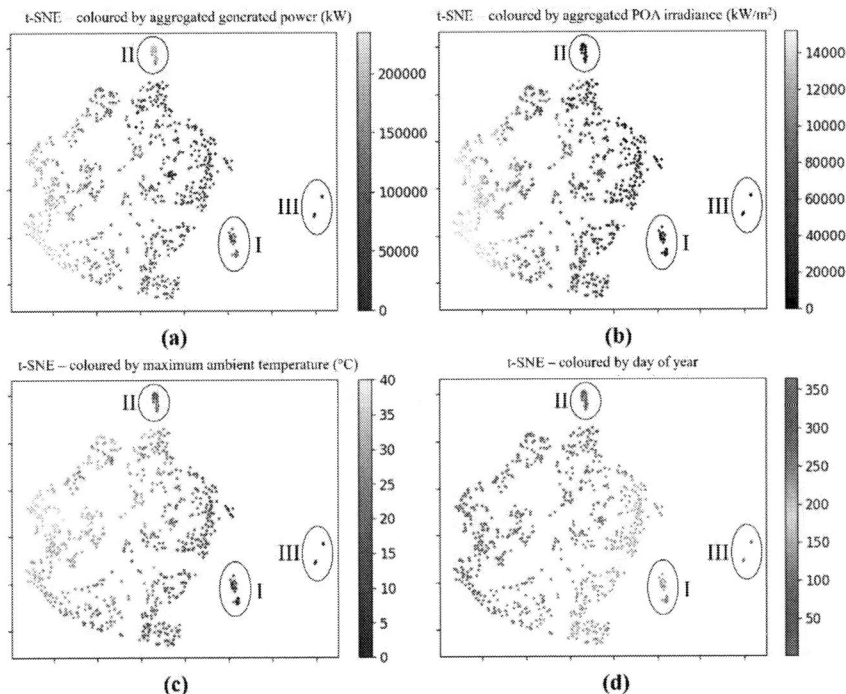

Figure 2: The two-dimensional embedding of daily operational data of a PV array consisting of 28 bifacial modules installed in a utility-scale plant in Australia, where each point corresponds to an operational day, coloured by the (a) daily aggregated generated power, (b) daily aggregated POA irradiance, (c) maximum daily ambient temperature, and (d) day of the year. Three isolated clusters are labelled as (I), (II), and (III) correspond to outliers in the operational data.

Some interesting clusters worthy of discussion include those labelled in Figure 2 [(I), (II), and (III)] which represent days with zero values for either the aggregated daily generated power, POA irradiation, or maximum ambient temperature. For example, points labelled as (I) and (II) are those with daily aggregated POA irradiation of zero where non-zero power is generated; this is, of course, impossible, and the model accurately identified these cases. Note the difference between the generated power among the points labelled as (I) and (II), visually evident in Figure 2(a) through darker blueish colours signifying lower generated power on days in the cluster (I) in comparison to brighter yellowish colours denoting higher generated power on days in the cluster (II). Finally, days labelled as (III) are those with communication blackouts, i.e. no data is collected at all.

Figure 3 presents a representative selection of three cluster regions, each comprising operational days of high similarity, colour-coded by the daily aggregated generated power. Plotting the original daily operational data of all days within each cluster reveals further validation of the model's capability to separate behavioural patterns based

on similarity, even though their daily operational data is not identical. While all the three clusters in Figure 3(a) to (c) map to cloudy days, a deeper investigation reveals the model's effectiveness in learning latent features capable of differentiating behavioural patterns of PV plants on cloudy days. Both clusters (a) and (b) consist of days in spring/summer, while the cluster (c) consists of days in autumn/winter. The days in the cluster (a) are characterised by higher generated power (orangish colours, with an average generated power of 188 MW and a standard deviation of 13 MW) in comparison to days in the cluster (b) (with an average generated power of 110 MW and a standard deviation of 31 MW). On the other hand, days in cluster (a) have clouds mostly in the afternoon, whereas days in clusters (b) and (c) are completely cloudy days. The average generated power in the cluster (c) is 31 MW with a standard deviation of 15 MW. Therefore, this model provides a novel capability to classify the operational days solely through raw daily operational data, without relying on explicit performance computations.

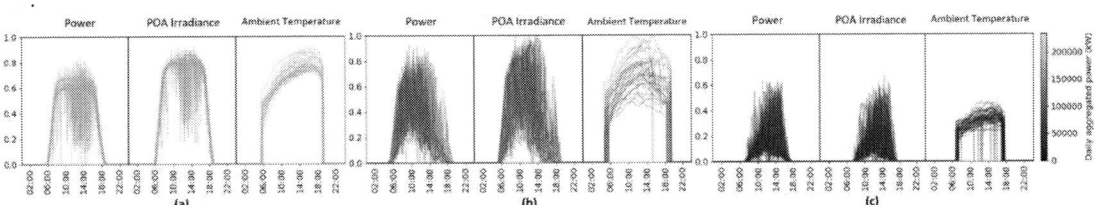

Figure 3: Representative examples of clusters coloured by the daily aggregated generated power, depicting (a) days in spring/summer with cloudy afternoons (28 days), (b) completely cloudy days in spring/summer (28 days), and (c) completely cloudy days in autumn/winter (53 days).

In conclusion, this paper introduced a novel ML-based method that acts as a digital twin of PV plants, based only on their raw operational data, avoiding reliance on any physical models and their associated limitations. The effectiveness of this data-driven method is demonstrated with an array of 28 bifacial PV modules installed in a utility-scale PV plant. A thorough analysis revealed that it could automatically extract latent features from the raw operational data and effectively use them to identify different performance patterns (e.g., operating under clear-sky or cloudy conditions), and infer correlations and hidden temporal characteristics in the daily operational data without any access to any explicit knowledge. Furthermore, the extracted features are utilised to cluster the PV plant's operational days based on performance. This ML-based, continuously evolving, digital twin marks a critical step towards fully automated, data-driven O&M systems. This advancement is integral to enhancing the reliability of utility-scale PV, contributing significantly to the critical transition to renewable energy.

4 REFERENCES

[1] 'Snapshot of global PV markets 2023', International Energy Agency (IEA), IEA-PVPS T1-44:2023, 2023. [Online]. Available: https://iea-pvps.org/wp-content/uploads/2023/04/IEA_PVPS _Snapshot_2023.pdf

[2] D. C. Jordan, S. R. Kurtz, K. VanSant, and J. Newmiller, 'Compendium of photovoltaic degradation rates', Prog. Photovolt. Res. Appl., vol. 24, no. 7, pp. 978–989, 2016, doi: 10.1002/pip.2744.

[3] J. Kratochvil, W. Boyson, and D. King, 'Photovoltaic array performance model', SAND2004-3535, 919131, 2004. doi: 10.2172/919131.

[4] I. Higgins et al., 'beta-VAE: Learning basic visual concepts with a constrained variational framework', in International Conference on Learning Representations, 2016.

[5] L. van der Maaten and G. Hinton, 'Visualizing data using t-SNE', J. Mach. Learn. Res., vol. 9, no. 86, pp. 2579–2605, 2008.

[6] R. J. G. B. Campello, D. Moulavi, and J. Sander, 'Density-based clustering based on hierarchical density estimates', in Advances in Knowledge Discovery and Data Mining, vol. 7819, Springer, Berlin, 2013, pp. 160–172.

Enhancing Fault Diagnosis in Photovoltaic Plants: A Comprehensive Approach to Simultaneous Failures

Giosué Maugeri, Salvatore Guastella, Andrea Rossetti

23/09/2024

Paper Submitted in EPJ Photovoltaics, Sep. 2024

Agenda

- General introduction to the Fault Diagnosis* problem
- How to manage the Simultaneity of Faults#
- Result on Simultaneity of Bypass Diodes, Series Resistance Increases, and Partial Shading Effects Fault
- Conclusions

---

*Diagnosis: in this study, Diagnose a fault means recognize the fault nature by analyzing the monitored parameter
#Simultaneity of Fault: when two or more faults occur simultaneously, overlapping their effects

Introduction

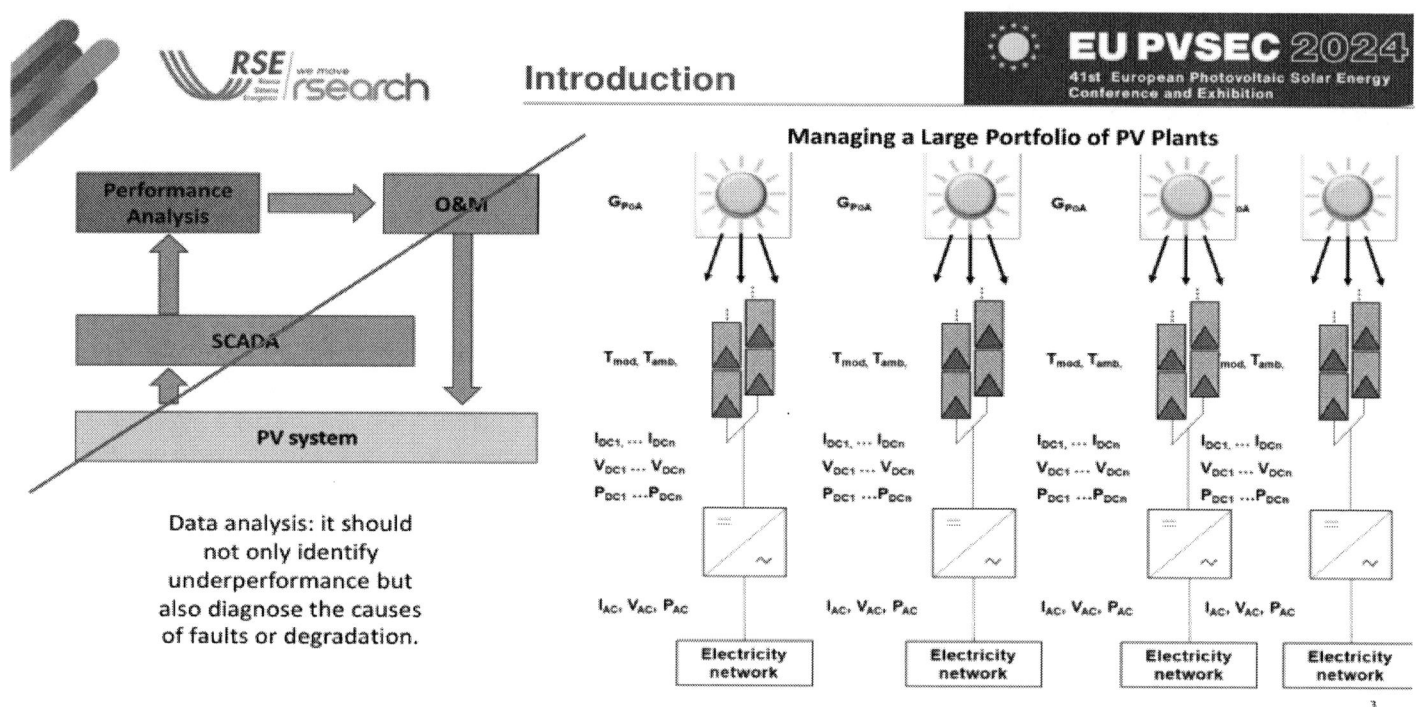

Data analysis: it should not only identify underperformance but also diagnose the causes of faults or degradation.

Introduction

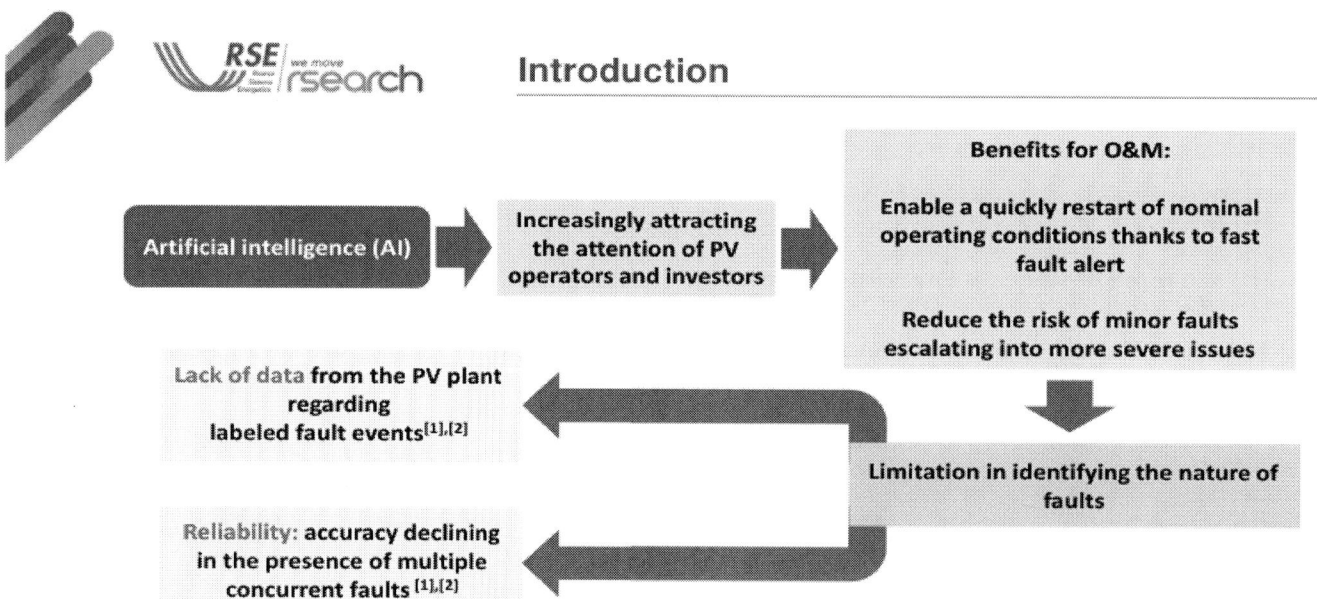

[1] Ihsan Ullah Khalil; Azhar Ul-Haq et al. NUST College of Electrical and Mechanical Engineering - "Comparative Analysis of Photovoltaic Faults and Performance Evaluation of its Detection Techniques," IEEE Access, vol. 8, pp. 1-1. 10.1109, 2020.
[2] Renewables 2023 - Analysis and forecast to 2028, IEA - International Energy Agency Website: www.iea.org, 2024.

RSE's PV Fault Facility in Milan, Italy[3]

Generated dataset of over 1.6 million fault-labeled records

An experimental setup to generate PV datasets with appropriately labeled failure events for training and validation of ML algorithms.

BENEFITS
- Support the development of ML algorithms in **identifying the most significant features** for the training phase.
- **Offer large data sets of labeled PV** failures to the PV Researchers
- **ML algorithms can be trained** taking into account the problems that can arise in the real field
- **Demonstrate the applicability of the techniques** developed to real photovoltaic systems

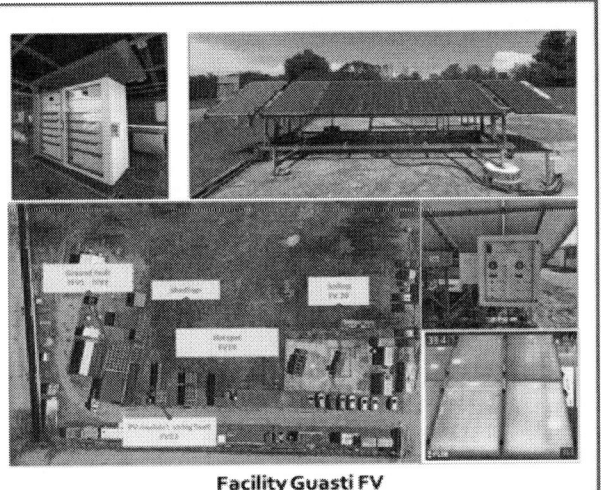

Facility Guasti FV

PV DataLake

PV fault characterizzation

PV fault detection and diagnosis models

[3] G. Maugeri, et. al, "A Facility Test to Generate Data from Real PV Systems Affected by Faults," in WCPEC-8, the 8th World Conference on Photovoltaic, Milan, Italy, 2022.

Methodology adopted in the RSE's PV Fault Facility

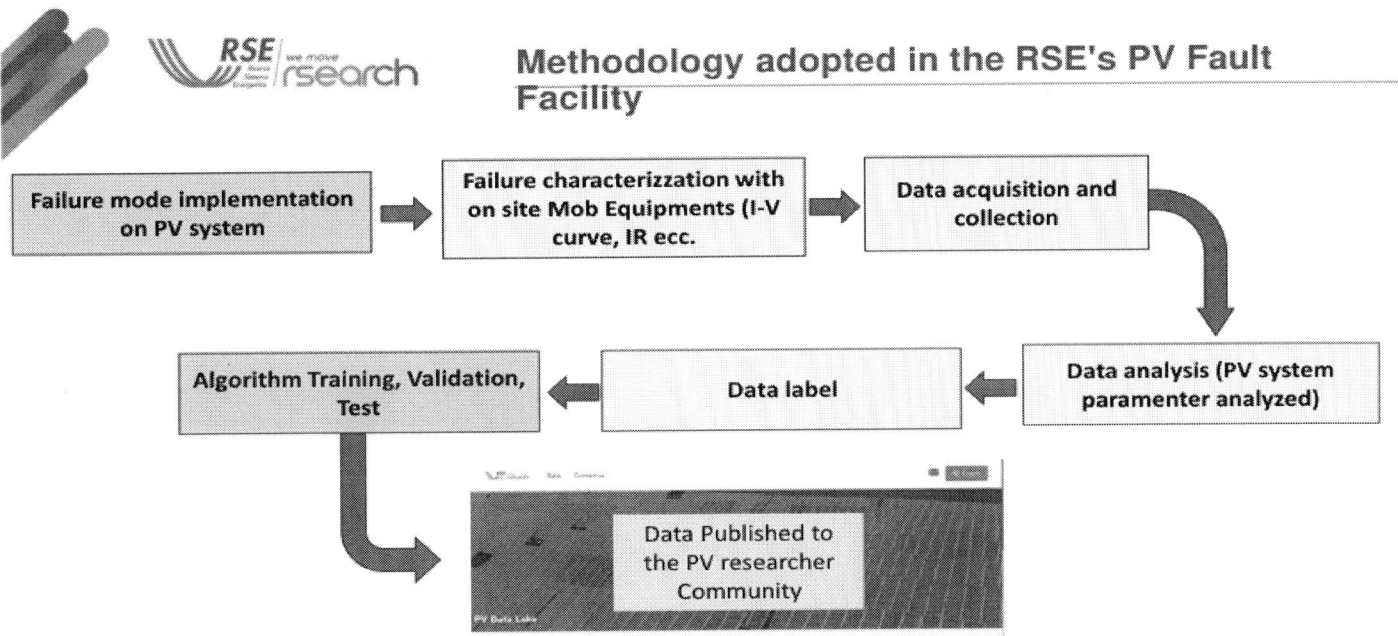

Failure mode implementation: scheduling

Three specific Fault type focused in this work: Diode Fault; Increased Series Resistance and Partial shading

☐ **No faults**   ▨ **Diode fault**   ▨ **Increased Series Resistance**   ▨ **Partial shading**

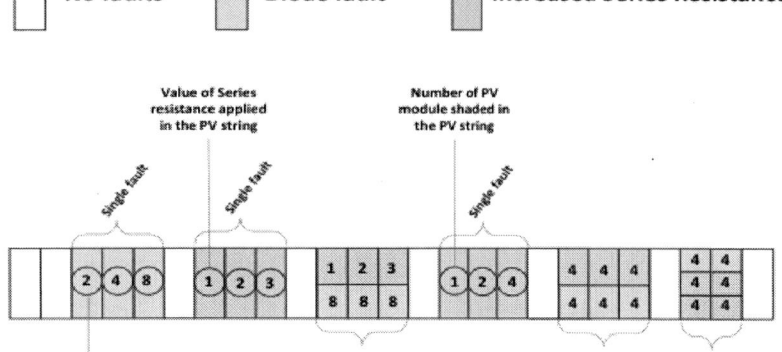

Fault datasets were generated according to a specific testing program.

- periods **free of faults**, with single faults and combined faults.
- The duration of each fault test was determined by the meteorological conditions during the **monitoring period** and typically lasted **no less than one week**.
- For each implemented failure mode, field measurements were **conducted using a current-voltage (I-V) curve tracer**

An interlude about the three-fault selected

String Voltage and Current monitored by Scada system and compared with predicred value based on free fault period; orange line is the predicted value, blue line is the measured value. Both failure are visible with a decrease on String Voltage

A day in which **shorted diode** fault were artificially replicated

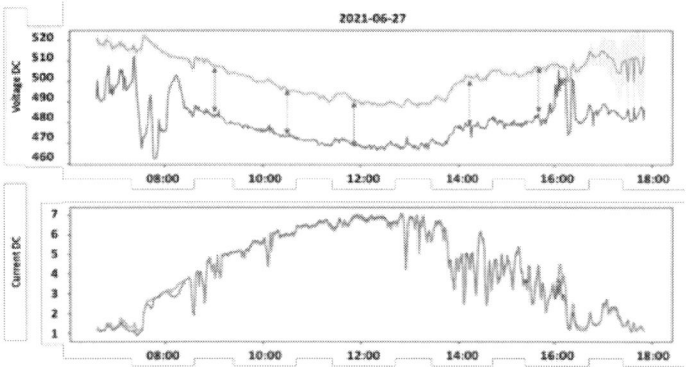

A day in which **increased string series resistance** fault is artificially replicated.

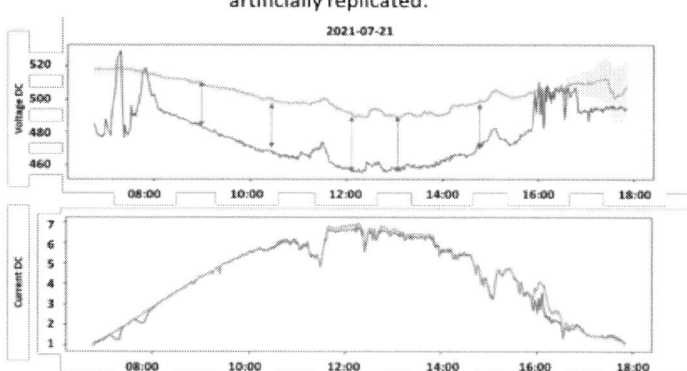

Experimental Setup and Failure characterizzation

Series Resistance and Diodes fault

Fault Type	Effect	Test case
Progressive increase in series resistance;	Gradual system degradation	Add series resistance 5Ω
Short circuited Diodes.	Voc Degradation	Short circuit of 4 Diode

Impact of Series Resistance and Diode Faults on I-V Curve

The Voc value is affected by diode faults.

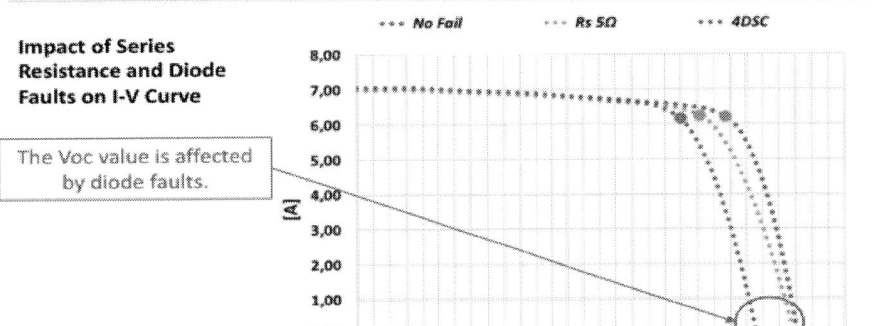

Experimental Setup and Failure characterizzation

Partial shading, combined with series resistance and Diode fault

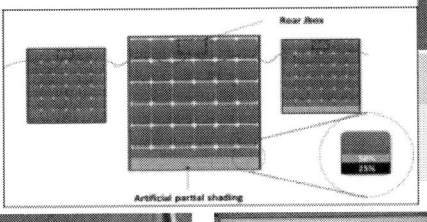

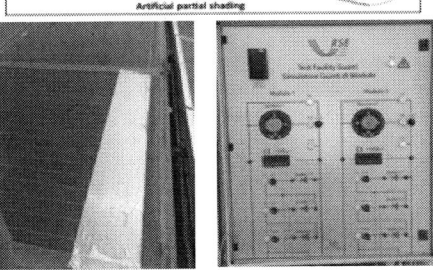

Fault Type	Effect	Test case
Partial shading	String Voltage\ Current reduction	Partial shading in 4 of 24 PV string modules
Combined fault Partial shading; series resistance; 4 Diode short-circuited	Overlapping effects in string voltage reduction	Partial shading in 4 of 24 PV string modules Short circuit of 4 Diode Add series resistance 5Ω

Impact of Partial shading and Series Resistance and Diode Faults on I-V Curve

The Voc value is affected by diode faults.

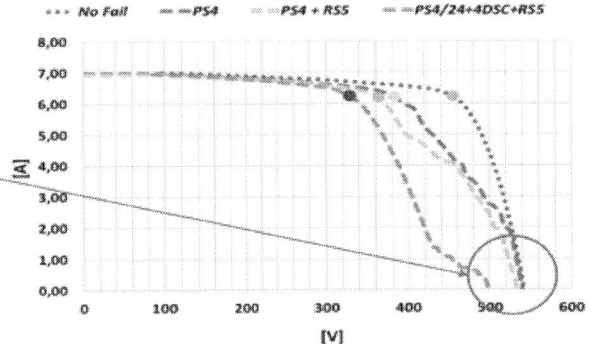

Data acquisition and collection

PV system analyzed: TFV1 (P. 7,44 kW); 2 PV Strings of 24 PV Modules each; c-Si Technology

List of implemented and analyzed failure in *TFV1* and number of records analyzed for each failure mode

Failure type	Acronym	Number of records labeled with faults	Period of monitoring
No Failure in the PV String	No_Fail	427.078	From:
n. 2 Diode Fault	2DSC	35.265	01/07/2023
n. 4 Diode Fault	4DSC	31.470	
n. 8 Diode Fault	8DSC	39.604	To:
Series Resistance 1Ω	Rs1	23.804	
Series Resistance 2Ω	Rs2	19.502	31/07/2024
Series Resistance 3Ω	Rs3	10.980	
n. 2 Modules in Partial shading	PS2/24	7.086	Total days:
n. 4 Modules in Partial shading	PS4/24	7.319	395
TOTALS		**602.108**	

XDSC – n. X Diodes faulted in short circuit

RsX – Series Resistance @ X Ω

PSX/24 – Partial shading applied on X modules of 24 string modules

Data acquisition and collection

PV system analyzed: TFV3 (P. 3,44 kW); 1 PV Strings of 24 PV Modules; c-Si Technology

List of implemented and analyzed failure in TFV3 and number of records analyzed for each failure mode

Failure type	Acronym	Number of records labeled with faults	Period of monitoring
No Failure in the PV String	No_Fail	9.180	From:
n. 4 Diode Fault	4DSC	7.875	13/06/2024
Increased Series Resistance @ 5Ω	Rs5	10.441	
n. 4 Modules in Partial shading	PS4/24	11.522	To:
n. 4 Diode Fault + n. 4 Modules in Partial shading	4DSC+PS4/24	14.634	02/09/2024
n. 4 Diode Fault + Series Resistance 5Ω	4DSC+Rs5	11.932	Total days:
Series Resistance 5Ω + n. 4 Modules in Partial shading	Rs5+PS4/24	6.240	81
TOTALS		**71.824**	

4DSC – 4 Diodes faulted in short circuit

Rs5 – Series Resistance @ 5 Ω

PS4/24 – Partial shading applied on 4 modules of 24 string modules

4DSC + PS4/24 – Combined fault

4DSC+Rs5 – Combined fault

Rs5 + PS4/24 – Combined fault

Data analysis PV data monitoring

Analysis of string voltage at the maximum power point as the number of short-circuited diodes and the applied series resistance on the string increase

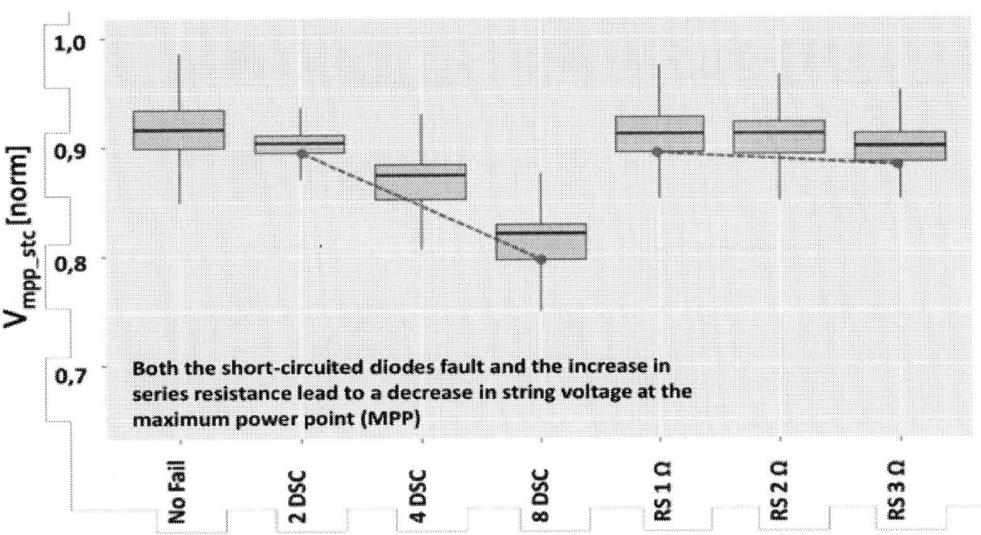

Both the short-circuited diodes fault and the increase in series resistance lead to a decrease in string voltage at the maximum power point (MPP)

Question mark?

How to separate the effects of these faults?

How to recognize these faults through PV parameter monitored by SCADA system?

Daily String Voltage (Vstart) analysis

Defining $V_{start,}$ daily measured by SCADA

- low string DC current values ($I_{dc} << 0.5$ A)
- solar irradiance values between [10÷100] W/m²
- data acquired between 05:00 and 10:00 each day
- temperature-corrected voltage (V_{MPP_STC}) values to compensate for thermal effects on the string voltage

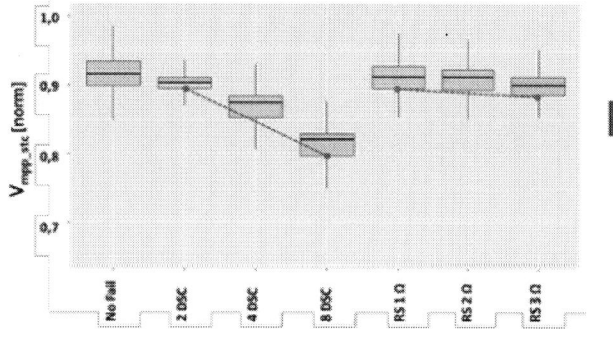

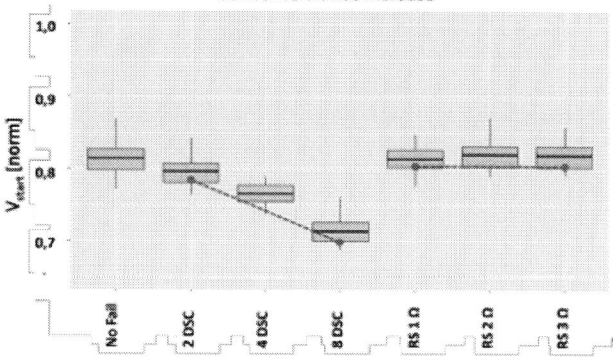

Daily String Voltage (Vstart) analysis

Comparison of V_{start} in the presence of bypass diode failure, string series resistance increase, and partial shading

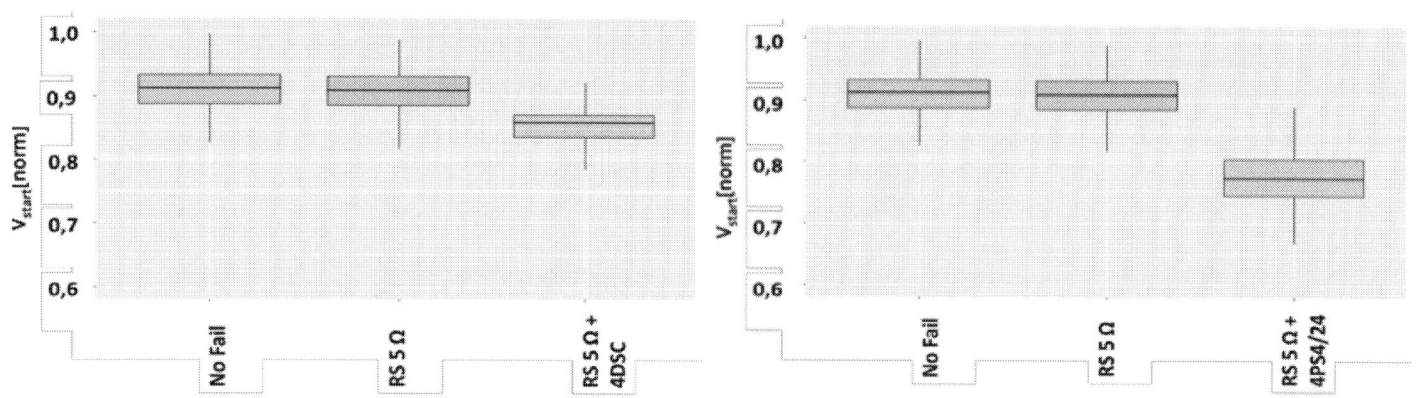

Degradation phenomena have little to no impact on the string voltage V_{start}

ML Fault detection and diagnosis - Flowchart

Each day starts processing the PV string voltage V_{start}
- Low string DC current values (Idc << 0,5 A).
- Solar irradiance between [10–100] W/m².
- Data collected between 05:00 and 08:00 each day.

V_{start} values are corrected in temperature to account thermal effects

V_{start} values are analysed to identify any signs of short-circuited bypass diode faults

If short-circuited diode condition is detected, the V_{start} is used to estimate the number of diodes or PV modules affected by the fault.

The final fault diagnosis of the flowchart, which can be implemented by Machine Learning techniques, will integrate the information gathered from these routines

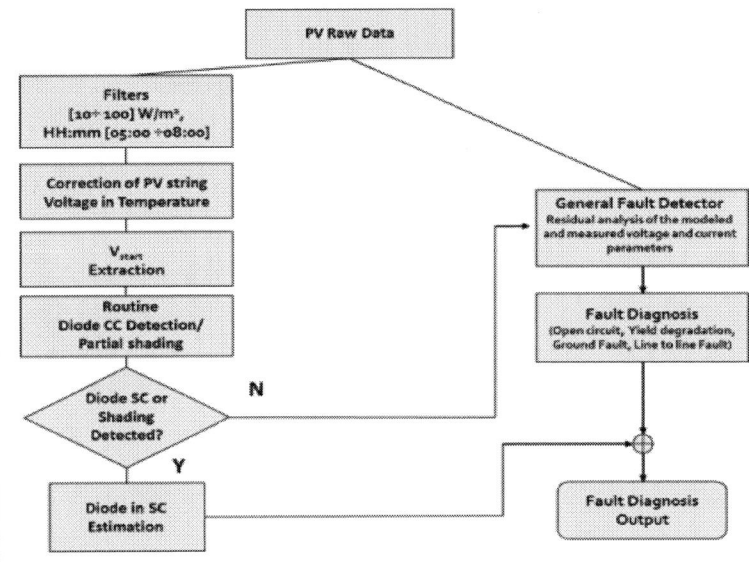

Highlight & Conclusions

- The RSE's PV Fault Facility generated **dataset of over 1.6 million** fault-labeled records enabling a detailed analysis of PV failure mode

- **Three fault conditions have been deeply investigated** in this study, including the **overlapping of their effect, due to their similar effect on String Voltage**: increased series resistance, short-circuited bypass diodes, and partial shading

- The **Vstart analysis helps to recognize these faults effectively reducing misdiagnosis risks,** even with overlapping malfunctions.

- Future work will **expand this study to include additional PV failure modes and validate the methodology on a large-scale PV plant**. This routine will be integrated into AI-based algorithms for enhanced fault detection.

Giosuè Maugeri

 giosue.maugeri@rse-web.it

#wemoversearch

 www.rse-web.it

 @Ricerca sul Sistema Energetico - RSE SpA

 @RSEnergetico

 RSE SpA - Ricerca sul Sistema Energetico

This work has been financed by the Research Fund for the Italian Electrical System under the Three-Year Research Plan 2022-2024 (DM MITE n. 337, 15.09.2022), in compliance with the Decree of April 16th, 2018".

41st European Photovoltaic Solar Energy Conference and Exhibition

DESIGN AND APPLICATION OF INTELLIGENT SCALABLE AUTOMATIC FAULT DETECTOR FOR COMMERCIAL PHOTOVOLTAIC SYSTEMS

Mücahid Candan[1,2], David Melgar[1], Christian Schill[1], Mete Çubukçu[2],
Eduardo Sarquis Filho[3], Björn Müller[3], Duarte Kazacos[4]
[1]Fraunhofer Institute for Solar Energy, [2]Solar Energy Institute of Ege University, [3]Enmova GmbH, [4]Mondas GmbH
[1]Freiburg/Germany, [2]İzmir/Türkiye, [3]Freiburg/Germany
mucahidcandan@gmail.com, david.melgar@ise.fraunhofer.de, christian.schill@ise.fraunhofer.de,
eduardo.sarquis@enmova.de, bjoern.mueller@enmova.de , duarte.kazacos@mondas-gmbh.de

ABSTRACT: Fault detection (FD) in photovoltaic power systems (PVPS) is critical for maintaining system efficiency and reliability. This study leverages real-time time series monitoring data from over 80 rooftop PVPS, collecting measurements and analyzing maintenance logs. These datasets undergo re-organization, filtering, and quality assurance processes to create comprehensive training, testing, and validation sets, considering variations in the number of inverters. Various dimensional extraction methods are applied to evaluate the accuracy of existing dimension reduction techniques, alongside the proposed feature extraction method. Comparative analysis demonstrates that the proposed method outperforms current approaches in accuracy and efficiency. Additionally, the intelligent fault detection process, including diagnosis and post-processing, enables precise localization and characterization of fault properties, providing deeper insights for fault management in PVPS.
Keywords: Photovoltaic fault detection, intelligent fault detection, fault detection per sample, dimensionless fault detection, ensemble fault detection

1 INTRODUCTION

Photovoltaics (PV) play a critical role in addressing the growing global energy demand, especially as fossil fuel reserves diminish, and greenhouse gas (GHG) emissions exceed acceptable thresholds. Beyond the installation of PV systems, ensuring their maintenance and operational sustainability is essential to preserve energy yield and to contribute effectively to reducing GHG emissions. Given the necessity of sustaining the performance of established systems, automatic fault detection (FD) is becoming increasingly significant, particularly with the global expansion of commercial photovoltaic power systems (PVPS). Automatic FD mitigates energy losses and prevents damage caused by system faults.

While previous studies using artificial intelligence have predominantly focused on detecting anomalies at the module level, in strings, or in inverters within photovoltaic power systems (PVPS) of constant inverter numbers, the detection of anomalies in PVPS with varying numbers of inverters necessitates more generalized fault detection methods. Existing literature on fault detection often employs methods such as current-voltage (IV) curves, power-voltage (PV) curves, yield prediction (Y), and performance ratio (PR) to detect anomalies, with alarms typically triggered over aggregated time periods, such as on a daily basis [1–16]. Artificial intelligence (AI) methods, however, have primarily been applied in model-based approaches, often relying on limited data samples, non-commercial laboratory measurements or synthetic data. Many AI-based techniques are also focused on generalized approaches rather than addressing the core causes of system faults [17–31]. The common approach of training neural networks through feature extraction, guided by engineering expertise, often suffers from a limited feature set, resulting in the omission of critical factors that occur in real-world systems[32].

The dataset utilized in this study was collected over a span of more than six years from commercial rooftop systems installed in Germany. It includes corresponding periodic maintenance reports and logs from the company responsible for the operation of these systems. Notably, the data collection was conducted prior to and independently of the present study.

The methodology in this study is based on the development of a neural network (NN)-based fault detection (FD) system. The approach begins with feature extraction, leveraging the practical knowledge and experiences of the monitoring and maintenance teams. Multiple automated feature extraction methods, including dimensionality reduction techniques [33,34], as well as hybrid and ensemble approaches combining auto-encoders and dimensionality reduction methods, are explored. The fault types are redefined, extracted from maintenance reports, and classified into pre- and on-duty fault categories. These classifications are then subjected to comparative analysis. Ultimately, an online fault detection system, capable of running on each measurement of real-time monitoring data, is achieved.

This work has been partially funded by the German Federal Ministry of Economic Affairs and Climate Action for this research under Grant Number 03EE1188C (Project IPVPro), and TÜBİTAK (The Scientific and Technological Research Council of Türkiye) project.

2 SYSTEM DATASET

The dataset, covering January 2016 to October 2020, was created from maintenance tickets and reports, and PV monitoring data. It contains inverter measurements including DC current and voltage, AC power, and AC voltage, as well as ambient temperature readings for some systems. In addition, irradiance in plane of array measurements were collected, varying between single or multiple sensors depending on the tilt and orientation of the arrays. Among these, the most commonly available data are DC current, DC voltage, AC power from inverter measurements, and irradiance from plane of array sensors. The measurement data in this dataset were collected at intervals ranging from 5 to 15 minutes, depending on the PVPS from which the measurements were taken.

Daily monitoring tickets include reports detailing fault explanations, fault types, sub-fault types, occurrence dates, and the status of faults (ongoing or resolved). The fault types are defined primarily by the monitoring team, with specific sub-fault types categorized accordingly, as presented in Table I.

Table I: Fault types in ticket files

Fault Category	Sub-Fault Category
Events	inverter changed
External influence	snow
Installation	misaligned sensor
	none
	system failed
	breakdown
Inverter	switched off
	disconnected
	delayed switches
	insulation
	none
	monitoring cabinet
	router problems
Monitoring	sensor defect
	sensor damaged
	sensor shaded
	wrong values
Unclassified	system failure
Other	none
	none
	module defect
Solar Generator	hotspots
	insulation
	string failure
OK Status	none

Among the main fault categories presented in Table I, *Events, external influence, unclassified* and *ok status* types do not present the faulty conditions. Additionally, for each fault class, explanation reports are provided by the monitoring team.

3 METHODOLOGIES

This section covers dataset handling and filtering, feature extraction methods, and the design of the artificial intelligence system, each presented in separate subsections. The primary goal of the artificial intelligence part is to facilitate the classification of faults, along with their diagnosis. To achieve this, the collected data, including fault logs, are thoroughly examined in conjunction with current photovoltaic literature, IEA Task reports, and expertise from Fraunhofer ISE. [9,21,32,35–39]. Figure 1 illustrates the overall structure of the system.

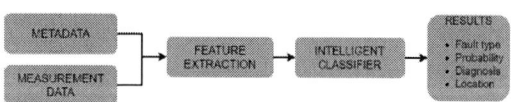

Figure 1. General structure of the system

3.1 Dataset Handling and Filtering

Filtering fault tickets is of utmost importance. Among the primary fault categories listed in Table I, those classified as Events, External Influence, Unclassified, and OK Status are excluded from further analysis, as they do not reflect actual fault conditions. Additionally, the None sub-faults require further clarification, which is achieved through the examination of fault reports.

To refine the dataset, two steps are taken: First, non-informative sub-fault categories or those with a minimal number of samples, such as System Failure in the Inverter

category and Insulation in the Solar Generator category, are eliminated. Second, the *None* subclasses are defined through a text-filtering operation applied to the technical descriptions of the corresponding sub-fault types. This process yields new main categories, such as Soiling and Shading, along with new sub-fault categories including Soiling, Shading, Solar Generator, Inverter, String Failure, and Monitoring. Subsequently, the fault categories are renamed, reorganized, and reclassified, as illustrated in Table II. In Table II, main fault categories begin with a capital letter, whereas sub-fault categories are presented in lowercase.

Table II. Rebuilt Ticket Dataset

Fault Class	New Fault Category	Ex Sub-Fault Categories
F1	*Fault in PV Module/s*	module defect
F2	*Fault in PV String/s*	solar generator
		string failure
F3	*Inverter Faults*	inverter
		system Failed
		breakdown
		disconnected
F4	*Shading*	shading
N	*Normal Condition*	normal condition
F5	*Sensor/s Alignment Fault*	misaligned sensor
F6	*Sensor/s Fault*	monitoring
		sensor defect
		sensor damaged
		sensor shaded
		wrong values
F7	*Communication Fault*	monitoring cabinet
		router Problems
		no Measurements
F8	*Inverter/s Off*	switched off
F9	*Inverter/s Late Wake-up*	delayed switches

Fault classification is more feasible following the reclassification presented in Table II. The categories highlighted in turquoise in Table II are directly detected by the neural network ensemble (F1-F4, N), while the orange rows (F4-F9) are identified through diagnostic algorithms employed in the artificial intelligence pre- and post-processing phases.

For the measurement data, the newly reclassified fault categories (as shown in Table II) are imported as a new time series, rescaled for each measurement sample. This process results in the assignment of a new fault type for each measurement sample. However, faults are recorded on a daily basis, and during the review of fault ticket reports, only a limited number of faults are predicted instead of conducting site inspections. To eliminate non-faulty measurements that have been incorrectly classified as faulty samples, a fault type verification filter is implemented.

Before the feature extraction process, a windowed box filter is applied to remove outliers from the dataset[39]. Time series data with fault labels are processed, eliminating invalid numbers, null measurements, and readings with very low irradiance (<50 W/m²) [40]. This also removes samples taken at night. To determine normal conditions without meteorological data, we propose analyzing the Performance Ratio (PR) [41] and the irradiance ratio (G/$G_{clearsky}$: Measured Irradiance to clear sky plane of array irradiance (Perez Model)) together. This approach introduces a limitation, as it may discard the incoming data during cloudy conditions. However, despite this limitation, there is still a substantial amount of data available for training. While PR can be noisy when evaluated per sample, adding the irradiance ratio addresses this issue without requiring meteorological data.

3.2 Feature extraction process

Feature vectors are numerical representation of an object, image, system, data that captures its essential characteristics of "features" for analysis in data processing or artificial intelligence tasks. Each element in the feature vector represent a specific observable aspect of the object.

Feature extraction is a particularly important process to obtain better accuracies in fault classification. In this sub-section, at first auto-feature extraction processes and then the extended feature extraction process are investigated. Auto feature extraction techniques are employed by the dimension reduction techniques and auto-encoders.

In addition to automated feature extraction, the manual feature extraction process draws on the technical expertise developed through years of fault detection and analysis. This approach can achieve higher accuracy than automated feature extraction, as the faults and their properties have been extensively studied and understood by professionals on the monitoring team over the past 30 years.

Furthermore, in addition to the experiments conducted with automated feature extraction, manual feature extraction processes are also investigated. These processes utilize measurements, characteristics, variables, and their statistical calculations related to the photovoltaic (PV) system, as illustrated in Figure 3.

Table III. Key elements of feature extraction process

Statistics	Analytics	Characteristics
Mean	DC/AC Yields	Performance Ratio
Variance	DC/AC Powers	Fill Factor
Standard Deviation	Irradiance	Yields
Maximum	DC Current	Inverter Efficiency
Minimum	DC Voltages	DC Efficiency
Summation	Calculated Irradiance	MPP Normalizations

By using combinations of multiple elements of the Feature extraction table shown in Table III, more than 105 features are generated and then a couple of them are eliminated afterwards by using the mutual information theory[42].

3.3 Design of the intelligent fault classifier

The design of artificial intelligence (AI) has progressed significantly, focusing on large language models and image classification. Despite increasing interest in time-series applications,

This study develops an intelligent fault classifier for photovoltaic (PV) systems by leveraging deep neural networks and feature extraction techniques to enhance fault detection accuracy. The dataset comprises an original imbalanced set and a balanced set created through under-sampling [43], divided into training (60%), validation (20%), and testing (20%) portions.

Artificial neural networks (ANN), one-dimensional convolutional neural networks (CNN), and random forests (RF) are employed. Normalization parameters and results are recorded for each training iteration. An ensemble

algorithm merges predictions using macro averages from confusion matrices to optimize accuracy. The weights of each predictor are based on their macro average accuracies, enabling a weighted prediction merging.

This approach calculates a weighted average of predictions, avoiding gradient boosting as none of the learners are weak.

4 RESULTS

The results of the automatic feature extraction methods, along with the proposed manual feature extraction method, are presented in Table IV. The confusion matrix for the ensemble classifier test is illustrated in Figure 3. Each cell in the matrix indicates the percentage of instances from a true class (represented by the rows) that were predicted as belonging to another class (represented by the columns). For instance, the value located in the F1 (Fault in PV Modules) row and F2 (Fault in PV Strings) column reflects the percentage of F1 instances that were misclassified as F2. Diagonal cells (e.g., F1 predicted as F1) signify correct classifications, while off-diagonal cells indicate instances of misclassification.

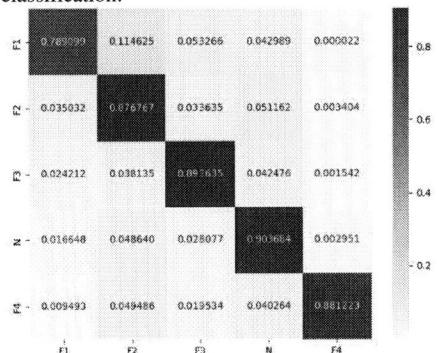

Figure 2. Confusion matrix of ensemble classifier

Table IV. Feature extraction techniques and accuracies

Feature Extraction	Training	Accuracy
Proposed	1-D CNN	88.1%
Proposed	RF	83.9%
Proposed	ANN	83.4%
FS	1-D CNN	82.9%
RFA	1-D CNN	81.4%
SVD	1-D CNN	80.6%
Sparse PCA	1-D CNN	79.9%
Sparse PCA	RF	79.7%
SVD	RF	78.9%
PCA	1-D CNN	78.7%
RP	1-D CNN	78.5%
RP	RF	78.4%
ICA	1-D CNN	77.1%
RFA	RF	76.9%
RFA	ANN	75.2%
FS	ANN	75.1%
RP	ANN	74.7%
PCA	RF	74.1%
FA	1-D CNN	73.7%
ICA	RF	72.9%
FA	RF	71.9%
Sparse PCA	ANN	71.5%
LLE	RF	71.2%
FS	RF	70.3%
PCA	ANN	70.0%
SVD, FA, ICA, LLE	ANN, 1-D CNN	Below 70%

In Table IV, Feature Selection using k-best (**FS**), Random Forest Autoencoder (**RFA**), Singular Value Decomposition (**SVD**), Principal Component Analysis (**PCA**), Random Projection (**RP**), Independent Component Analysis (ICA), Factor Analysis (**FA**) and Locally Linear Embedding (**LLE**) are investigated with the training structures Random Forest (RF), 1-Dimensional Convolutional Neural Network (1-D CNN) and Artificial Neural Network (ANN).

For the test dataset, which comprises 20% of the entire dataset, the counts of correct classifications and misclassifications are presented in Table V.

Table V. Classification Results

Fault Type	Total	Correct	False
Fault in PV Module/s	33639	26544	7095
Fault in PV String/s	148112	129860	18252
Inverter Faults	93493	83549	9944
Shading	9182	8091	1091
Normal Condition	208692	188592	20100
Sensor/s Alignment Fault	340	229	111
Sensor/s Fault	2	2	0
Communication Fault	4600	4600	0
Inverter/s Off	48	48	0
Inverter/s Late Wake-up	10301	9233	1068
TOTAL	508409	450747	57662
		88.65%	11.45%

In conjunction with Table VI and Figure 3, the performance of the ensemble model, which evaluates fault detection on a sample-by-sample basis, demonstrates that the AI classifies the state of the photovoltaic power system (PVPS) with an accuracy of 88.65%.

The DEFNE system provides four different output representations, with additional formats currently in development. In Table VI, the AI's evaluation is presented per measurement. The **Time** column provides temporal information, while the second column (**F Type**) lists the fault type (e.g., Fault in PV Module/s). The third column having **%** symbol as a column name indicates the probability of the fault expressed as a percentage, and the comment column outlines the cause of the fault. The final column (**L**) specifies the locations of the faults. For instance, Inverter measurement 2 indicates a Higher Impedance, as does Inverter 1. The probability figures represent the predictions generated directly by the AI.

Table VI. One Example of DEFNE Output

Time	F Type	%	COMMENT	L
20.03.2018 16:35	F1	54.76	Higher Impedance, Higher Impedance!	2 1
20.03.2018 16:40	F1	55.10	Higher Impedance, Higher Impedance!	2 1
20.03.2018 16:45	F1	55.61	Higher Impedance, Higher Impedance!	2 1

5 CONCLUSIONS

This study introduces an intelligent and scalable automatic fault detection (FD) system for commercial photovoltaic power systems (PVPS). Leveraging real-time time series monitoring data from over 80 rooftop PV systems, the developed system uses a combination of feature extraction techniques, both automated and manual, to enhance the accuracy and efficiency of fault detection per monitoring measurement. The incorporation of dimensionality reduction and ensemble methods—such as one-dimensional convolutional neural networks (1-D CNN), artificial neural networks (ANN), and random forests (RF)—has demonstrated the superiority of the proposed approach over existing methods.

The methodology allows for the detection and classification of faults across a range of system components, including photovoltaic modules, strings, inverters, and sensors, with an overall classification accuracy of 88.65%. This higher level of performance confirms the system's efficacy in managing the complexity and variability of commercial PVPS, especially those with different numbers of inverters. The reorganization and reclassification of fault categories, coupled with advanced AI techniques, offer a robust solution for real-time fault diagnosis, providing deeper insights into the nature and causes of system failures.

The proposed approach addresses key limitations of traditional fault detection methods, such as their reliance on daily data aggregation and their limited ability to identify core fault causes. By employing a sample-by-sample detection strategy (per measurement), the system enhances the timeliness and precision of fault identification, enabling faster maintenance and operational decisions.

6 ACKNOWLEDGEMENTS

Candan conceptualized the system, developed the software, conducted the research, and authored the paper as part of his PhD thesis. Melgar contributed PV expertise, conducted tests, assisted with the writing, and supervised the work. Schill and Mete provided supervision and critical review. Filho and Müller contributed to the dataset and provided valuable insights on data handling and maintenance logbooks.

7 REFERENCES

[1] Abd el-Ghany HA, ELGebaly AE, Taha IB. A new monitoring technique for fault detection and classification in PV systems based on rate of change of voltage-current trajectory. International Journal of Electrical Power & Energy Systems. 2021;133:107248.

[2] Ammiche M, Kouadri A, Halabi LM, Guichi A, Mekhilef S. Fault detection in a grid-connected photovoltaic system using adaptive thresholding method. Solar Energy. 2018;174:762–9.

[3] Bressan M, El Basri Y, Galeano AG, Alonso C. A shadow fault detection method based on the standard error analysis of IV curves. Renewable energy. 2016;99:1181–90.

[4] Dhimish M, Holmes V. Fault detection algorithm for grid-connected photovoltaic plants. Solar Energy. 2016;137:236–45.

[5] Dhimish M, Holmes V, Dales M. Parallel fault detection algorithm for grid-connected photovoltaic plants. Renewable energy. 2017;113:94–111.

[6] Dhoke A, Sharma R, Saha TK. A technique for fault detection, identification and location in solar photovoltaic systems. Solar Energy. 2020;206:864–74.

[7] Fazai R, Mansouri M, Abodayeh K, Trabelsi M, Nounou H, Nounou M. Machine Learning-Based Statistical Hypothesis Testing for Fault Detection. In: 2019 4th Conference on Control and Fault Tolerant Systems (SysTol) [Internet]. Casablanca, Morocco: IEEE; 2019 [cited 2024 Apr 26]. p. 38–43. Available from: https://ieeexplore.ieee.org/document/8864776/

[8] Harrou F, Sun Y, Taghezouit B, Saidi A, Hamlati ME. Reliable fault detection and diagnosis of photovoltaic systems based on statistical monitoring approaches. Renewable energy. 2018;116:22–37.

[9] Hocine L, Samira KM, Tarek M, Salah N, Samia K. Automatic detection of faults in a photovoltaic power plant based on the observation of degradation indicators. Renewable Energy. 2021;164:603–17.

[10] Iqbal MS, Niazi YAK, Khan UA, Lee BW. Real-time fault detection system for large scale grid integrated solar photovoltaic power plants. International Journal of Electrical Power & Energy Systems. 2021;130:106902.

[11] Liu Y, Ding K, Zhang J, Li Y, Yang Z, Zheng W, et al. Fault diagnosis approach for photovoltaic array based on the stacked auto-encoder and clustering with IV curves. Energy Conversion and Management. 2021;245:114603.

[12] Madeti SR, Singh SN. Online fault detection and the economic analysis of grid-connected photovoltaic systems. Energy. 2017;134:121–35.

[13] Mallor F, León T, De Boeck L, Van Gulck S, Meulders M, Van der Meerssche B. A method for detecting malfunctions in PV solar panels based on electricity production monitoring. Solar Energy. 2017;153:51–63.

[14] Mansouri M, Al-Khazraji A, Hajji M, Harkat MF, Nounou H, Nounou M. Wavelet optimized EWMA for fault detection and application to photovoltaic systems. Solar Energy. 2018;167:125–36.

[15] Taghezouit B, Harrou F, Sun Y, Merrouche W. Model-based fault detection in photovoltaic systems: A comprehensive review and avenues for enhancement. Results in Engineering. 2024 Mar 1;21:101835.

[16] Yurtseven K, Karatepe E, Deniz E. Sensorless fault detection method for photovoltaic systems through mapping the inherent characteristics of PV plant site: Simple and practical. Solar Energy. 2021;216:96–110.

[17] Amiri AF, Kichou S, Oudira H, Chouder A, Silvestre S. Fault Detection and Diagnosis of a Photovoltaic System Based on Deep Learning Using the Combination of a Convolutional Neural Network (CNN) and Bidirectional Gated Recurrent Unit (Bi-GRU). Sustainability. 2024 Jan;16[3]:1012.

[18] Chen Z, Wu L, Cheng S, Lin P, Wu Y, Lin W. Intelligent fault diagnosis of photovoltaic arrays based on optimized kernel extreme learning machine and I-V characteristics. Applied Energy. 2017 Oct 15;204:912–31.

[19] Dhimish M, Holmes V, Mehrdadi B, Dales M, Mather P. Photovoltaic fault detection algorithm based on theoretical curves modelling and fuzzy classification system. Energy. 2017 Dec 1;140:276–90.

[20] Eskandari A, Milimonfared J, Aghaei M. Line-line fault detection and classification for photovoltaic systems using ensemble learning model based on I-V characteristics. Solar Energy. 2020 Nov 15;211:354–65.

[21] Garoudja E, Harrou F, Sun Y, Kara K, Chouder A, Silvestre S. Statistical fault detection in photovoltaic systems. Solar Energy. 2017;150:485–99.

[22] Hajji M, Yahyaoui Z, Mansouri M, Nounou H, Nounou M. Fault detection and diagnosis in grid-connected PV systems under irradiance variations. Energy Reports. 2023;9:4005–17.

[23] Harrou F, Dairi A, Taghezouit B, Khaldi B, Sun Y. Automatic fault detection in grid-connected photovoltaic systems via variational autoencoder-based monitoring. Energy Conversion and Management. 2024 Aug 15;314:118665.

[24] Hussain M, Dhimish M, Titarenko S, Mather P. Artificial neural network based photovoltaic fault detection algorithm integrating two bi-directional input parameters. Renewable Energy. 2020 Aug 1;155:1272–92.

[25] Kara Mostefa Khelil C, Amrouche B, Benyoucef A soufiane, Kara K, Chouder A. New Intelligent Fault Diagnosis (IFD) approach for grid-connected photovoltaic systems. Energy. 2020 Nov 15;211:118591.

[26] Khalil IU, Haq AU, ul Islam N. A deep learning-based transformer model for photovoltaic fault forecasting and classification. Electric Power Systems Research. 2024;228:110063.

[27] Saravanan S, Kumar RS, Balakumar P. Binary firefly algorithm based reconfiguration for maximum power extraction under partial shading and machine learning approach for fault detection in solar PV arrays. Applied Soft Computing. 2024;154:111318.

[28] Sarquis Filho EA, Kiefer K, Holland N, Kollosch B, Müller B, Branco PC. Analysis of automatic fault detection methods for commercially operated pv power plants. Ratio. 2021;7:18.

[29] Seghiour A, Abbas HA, Chouder A, Rabhi A. Deep learning method based on autoencoder neural network applied to faults detection and diagnosis of photovoltaic system. Simulation Modelling Practice and Theory. 2023;123:102704.

[30] Vieira RG, Dhimish M, de Araújo FMU, da Silva Guerra MI. Comparing multilayer perceptron and probabilistic neural network for PV systems fault detection. Expert systems with applications. 2022;201:117248.

[31] Zhong Y, Zhang B, Ji X, Wu J. Fault Diagnosis of PV Array Based on Time Series and Support Vector Machine. Coccia G, editor. International Journal of Photoenergy. 2024 Jan 20;2024:1–10.

[32] Ghosh R, Das S, Kumar Panizrahi C. Classification of Different Types of Faults in a Photovoltaic System. In: 2018 International Conference on Computation of Power, Energy, Information and Communication (ICCPEIC) [Internet]. 2018 [cited 2024 Jun 3]. p. 121–8. Available from: https://ieeexplore.ieee.org/document/8525170

[33] Ayesha S, Hanif MK, Talib R. Overview and comparative study of dimensionality reduction techniques for high dimensional data. Information Fusion. 2020;59:44–58.

[34] Todo W, Laurent B, Loubes JM, Selmani M. Dimension Reduction for time series with Variational AutoEncoders [Internet]. arXiv; 2022 [cited 2024 Aug 28]. Available from: http://arxiv.org/abs/2204.11060

[35] Köntges M. Performance and reliability of photovoltaic systems, subtask 3.2: review of failures of photovoltaic modules. IEA PVPS Task 13 External Final Report, IEA-PVPS [Internet]. 2014 [cited 2024 Apr 26]; Available from: https://cir.nii.ac.jp/crid/1372541648439450880

[36] Chen Z, Wu L, Cheng S, Lin P, Wu Y, Lin W. Intelligent fault diagnosis of photovoltaic arrays based on optimized kernel extreme learning machine and IV characteristics. Applied energy. 2017;204:912–31.

[37] Mellit A, Tina GM, Kalogirou SA. Fault detection and diagnosis methods for photovoltaic systems: A review. Renewable and Sustainable Energy Reviews. 2018 Aug 1;91:1–17.

[38] Loutzenhiser PG, Manz H, Felsmann C, Strachan PA, Frank T, Maxwell GM. Empirical validation of models to compute solar irradiance on inclined surfaces for building energy simulation. Solar Energy. 2007 Feb 1;81[2]:254–67.

[39] Spiliotis E. Time Series Forecasting with Statistical, Machine Learning, and Deep Learning Methods: Past, Present, and Future. In: Hamoudia M, Makridakis S, Spiliotis E, editors. Forecasting with Artificial Intelligence: Theory and Applications [Internet]. Cham: Springer Nature Switzerland; 2023 [cited 2024 Sep 2]. p. 49–75. Available from: https://doi.org/10.1007/978-3-031-35879-1_3

[40] Little RJ, Rubin DB. Statistical analysis with missing data [Internet]. Vol. 793. John Wiley & Sons; 2019 [cited 2024 Jan 17]. Available from: https://books.google.com/books?hl=tr&lr=&id=BemMDwAAQBAJ&oi=fnd&pg=PR11&dq=Statistical+Analysis+with+missing+data&ots=FCyT7WCVXZ&sig=zYzK_Q_funMluO1WZPWOKhCTv9I

[41] IEC TS 61836:2016 Solar photovoltaic energy systems - Terms, definitions and symbols. International Electrotechnical Committee; 2016.

[42] Carrara N, Ernst J. On the Estimation of Mutual Information [Internet]. arXiv; 2019 [cited 2024 Jun 20]. Available from: http://arxiv.org/abs/1910.00365

[43] Spelmen VS, Porkodi R. A review on handling imbalanced data. In: 2018 international conference on current trends towards converging technologies (ICCTCT) [Internet]. IEEE; 2018 [cited 2024 Jan 19]. p. 1–11. Available from: https://ieeexplore.ieee.org/abstract/document/8551020/?casa_token=J-1_C0cu9k0AAAAA:D65WnOIG8rH1_w149oc_vMI293-y7Q7cdeJJB4BY38w5h7T5ZDoOX7-_a46zxM7otkl4HP7dNg

Uncertainty-Aware Estimation of Inverter Field Efficiency Using Bayesian Neural Networks in Solar Photovoltaic Plants

Gerardo Guerra[1], Pau Mercade-Ruiz[1], Gaetana Anamiati[1], Lars Landberg[2]
[1]GreenPowerMonitor a DNV Company
[2]DNV

23 September 2024

Content

About us

Objectives

Definitions

Model

Optimisation

Fleet modelling

European efficiency calculation

Case Study

Conclusions

About us

DNV

The purpose of DNV is to safeguard life, property, and the environment. DNV offer world-renowned testing, certification and technical advisory services to the energy value chain including renewables, oil and gas, and energy management.

For more than 20 years the company has carried out R&D activities within all their business areas.

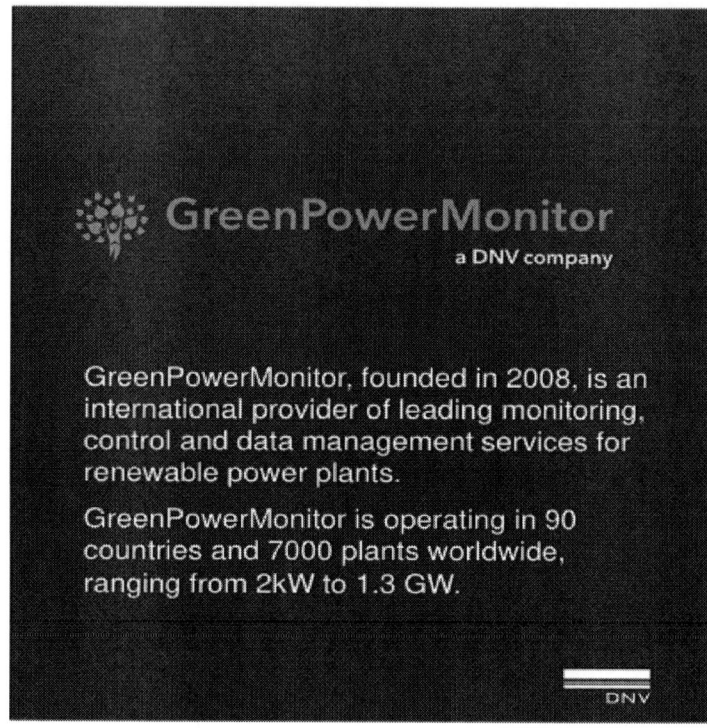

GreenPowerMonitor, founded in 2008, is an international provider of leading monitoring, control and data management services for renewable power plants.

GreenPowerMonitor is operating in 90 countries and 7000 plants worldwide, ranging from 2kW to 1.3 GW.

Objectives

Data-driven inverter efficiency

Re-create European efficiency

Develop ML model for calculations

Definitions

Efficiency

$$\eta = \frac{P_{AC}}{P_{DC}} \cdot 100$$

Tolerance

$$\eta_{tol} = \eta - 0.2 \cdot \left(1 - \frac{\eta}{100}\right) \cdot \eta$$

European efficiency

$$
\begin{aligned}
\eta_{Euro} = \quad & 0.03 \cdot \eta_{5\%} \\
+& 0.06 \cdot \eta_{10\%} \\
+& 0.13 \cdot \eta_{20\%} \\
+& 0.10 \cdot \eta_{30\%} \\
+& 0.48 \cdot \eta_{50\%} \\
+& 0.20 \cdot \eta_{100\%}
\end{aligned}
$$

Percentage of rated power output

Laboratory conditions

T_{AMB}	25ºC
V_{AC}	Nominal
Q_{AC}	0 VAR
Simulator/Power supply	PV array simulator

Model

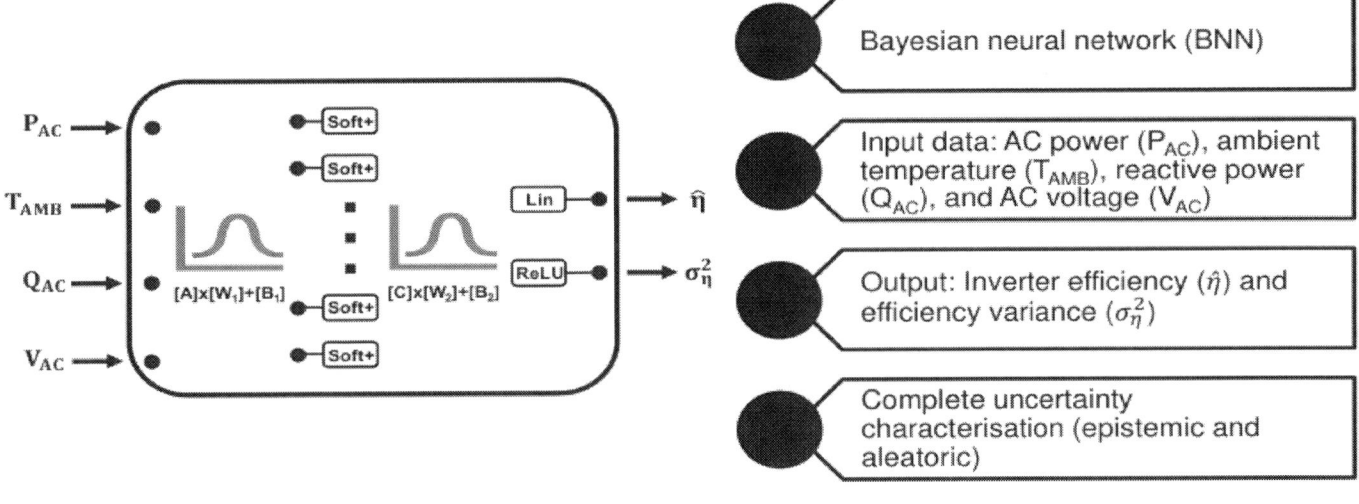

- Bayesian neural network (BNN)
- Input data: AC power (P_{AC}), ambient temperature (T_{AMB}), reactive power (Q_{AC}), and AC voltage (V_{AC})
- Output: Inverter efficiency ($\hat{\eta}$) and efficiency variance (σ_η^2)
- Complete uncertainty characterisation (epistemic and aleatoric)

Optimisation

Parameters	neurons in hidden layer (H), sigma prior (σ_p), regularisation term (lr)
Optuna package	Tree-structured Parzen Estimator algorithm
Loss function (BNN)	$-\log p(\eta\|\hat{\eta}, \sigma_\eta^2) + \text{lr} \cdot \text{KLD}$
Objective function (Optuna)	$\text{mse}(\eta, \hat{\eta}) + \text{mebe}(\eta, \hat{\eta}) + \text{skw}(\eta, \hat{\eta}) - \sigma_{ep} + \text{loss}_{\text{BNN}} + \frac{H}{2000}$

- KLD: Kullback-Leibler divergence
- mse: mean square error
- mebe: median bias error
- skw: skewness
- σ_{ep}: epistemic uncertainty

Fleet modelling

Collect fleet data

Choose base inverter

Optimise model for base inverter

Train fleet model
Train single-device models

European efficiency calculation

Define power levels (P_{lvl}) and weights (W) for evaluation

Define T_{AMB}, Q_{AC}, and V_{AC} for laboratory conditions

Define number of evaluations (N) for inference

For power in P_{lvl} **do**

    For n in N **do**

$$\hat{\eta}_n, \sigma^2_{\hat{\eta}_n} = BNN(power, T_{AMB}, Q_{AC}, V_{AC})$$

Calculate $\hat{\eta}_{power} = \frac{1}{N}\sum_{n=1}^{N}\hat{\eta}_n$ $\sigma^2_{ep_{power}} = \frac{1}{N-1}\sum_{n=1}^{N}\left(\hat{\eta}_n - \hat{\eta}_{power}\right)^2$ $\sigma^2_{al_{power}} = \frac{1}{N}\sum_{n=1}^{N}\sigma^2_{\hat{\eta}_n}$ $\sigma_{power} = \sqrt{\sigma^2_{ep_{power}} + \sigma^2_{al_{power}}}$

For n in N **do**

$$\eta_n = \sum_{power}^{P_{lvl}} W_{power} \cdot \left(\hat{\eta}_{power} + \mathcal{N}(0,1)\cdot\sigma_{power}\right)$$

Calculate $\hat{\eta}_{Euro} = \sum_{power}^{P_{lvl}} W_{power}\cdot\hat{\eta}_{power}$ $\eta_{nv} = \frac{1}{N}\sum_{n=1}^{N}\eta_n$ $\sigma_{Euro} = \sqrt{\frac{1}{N-1}\sum_{n=1}^{N}(\eta_n - \eta_{nv})^2}$

Case Study

Test site		Inverter	
Number of inverters	43	Rated power output	2750 kW
Sampling frequency	5 minutes	Nominal AC voltage	600 V
Start date	2022/10/31	European efficiency	98.5%
End date	2023/10/31	Efficiency tolerance	98.2%

Case Study

Optimisation results	
H	20
σ_p	0.9
lr	0.0026

Fleet model training [%]	
RMSE	0.49
MBE	6e-4
σ_{ep}	0.012
σ_{al}	0.48

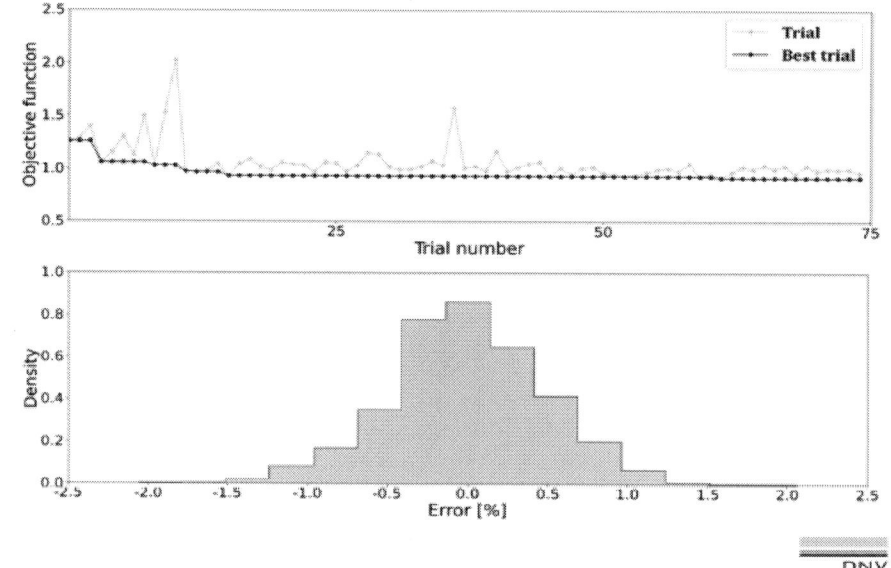

Case Study

Fleet efficiency [%]		
P_{lvl}	$\hat{\eta}_{power}$	σ_{power}
5	95.5	0.74
10	97.3	0.55
20	97.9	0.48
30	98.0	0.47
50	97.8	0.45
100	96.9	0.46

$\hat{\eta}_{Euro}$	σ_{Euro}
97.6	0.24

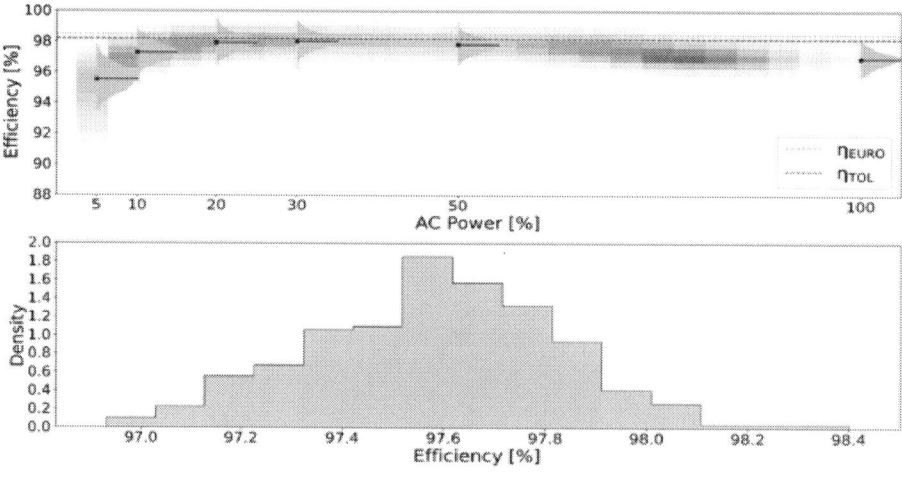

Case Study

	Device efficiency [%]	
	Mean	**Std Dev**
RMSE	0.21	0.03
MBE	-2e-3	3e-3
σ_{ep}	6e-3	5e-4
σ_{al}	0.19	0.03
$\hat{\eta}_{Euro}$	97.6	0.41
σ_{Euro}	0.11	0.02

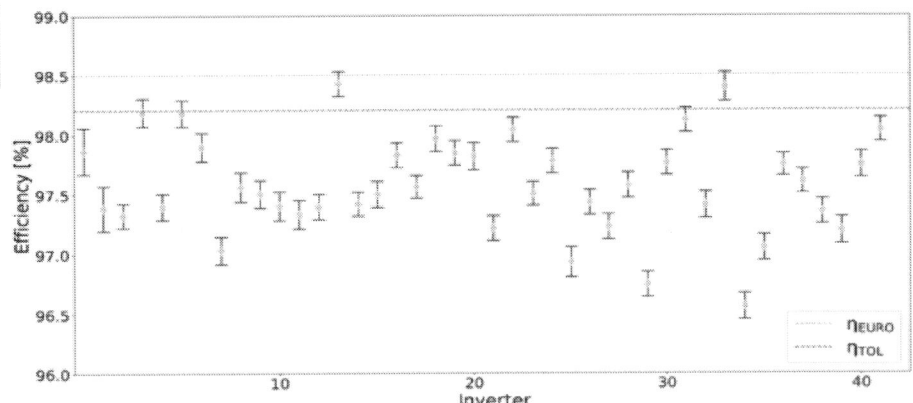

Conclusions

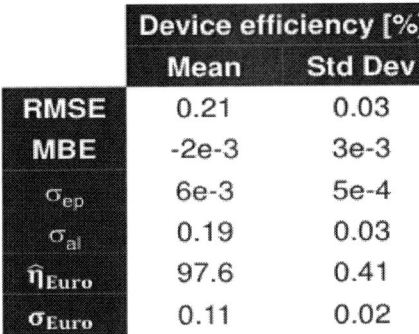

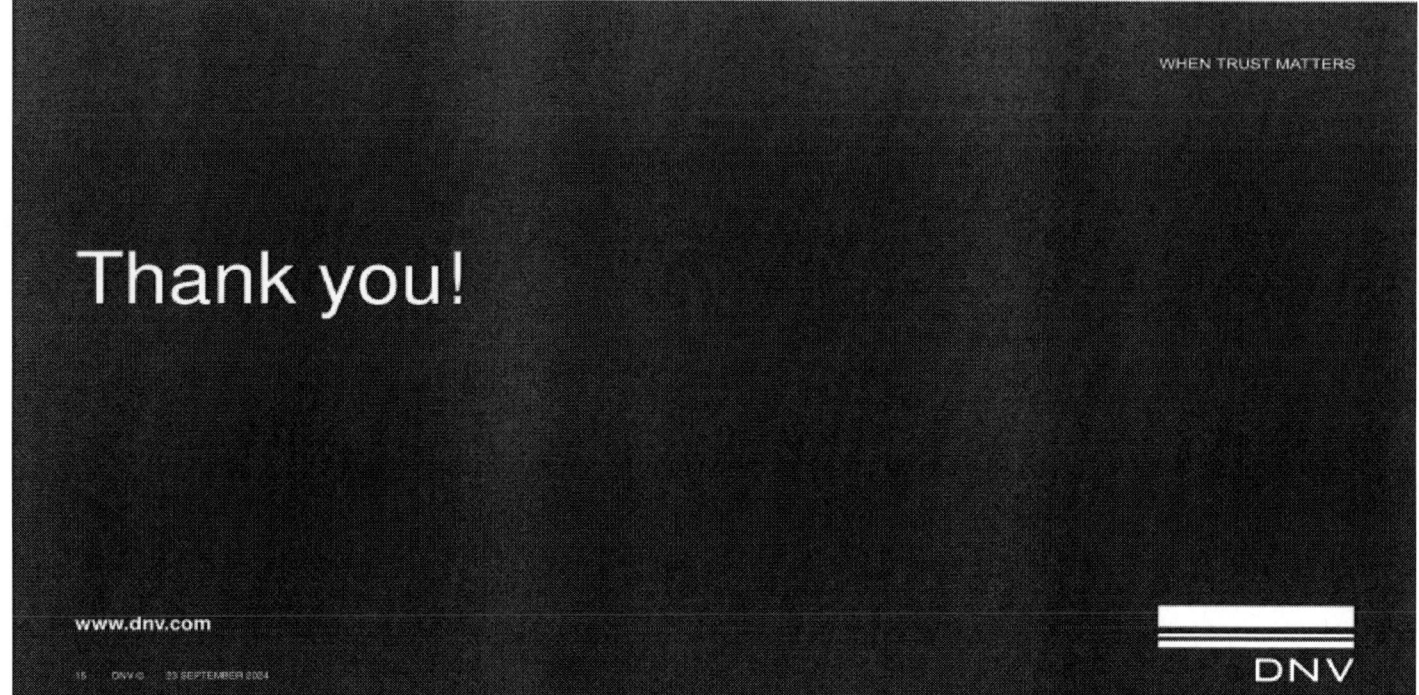

41st EU PVSEC 2024, 4AO.8.2, 23.09.2024

Analysis of Fault Detection and Defect Categorization in Photovoltaic Inverters for Enhanced Reliability and Efficiency in Large-scale Solar Energy Systems

Stephanie Malik[1], David Daßler[1], Dharm Patel[1], Carola Klute[1], Robert Klengel[1], Andreas Dietrich[2], Kai Kaufmann[3], Carsten Hennig[4], Danny Wehnert[5], Matthias Ebert[1]

[1]Fraunhofer IMWS, [2]DiSUN - Deutsche Solarservice GmbH, [3]DENKweit GmbH, [4]saferay holding GmbH, [5]Leipziger Energiegesellschaft mbH & Co. KG

Motivation

Experience from operators:
Daily messages / information from the monitoring systems in general

> **1**
>
> Portfolio size: 300 MWp (various system sizes with string and central inverters)
>
> "**1 to 15 error messages per day**, sometimes significantly more, depending on the weather conditions and external influences such as grid fluctuations"
>
> → This results in **3-5 tickets** per day, which require a service call with an **on-site diagnosis** and, if necessary, a repair.

> **2**
>
> Portfolio size: 130 MWp (distributed over 12 systems, consisting of central inverters)
>
> „**30 to 60 error messages per day**"

⇨ There is a great need in O&M in large system portfolios to detect a deviating behavior at an early stage.

⇨ Objective: Detect faulty inverter behavior, limit the causes of defects and identify similarities in patterns.

Outline

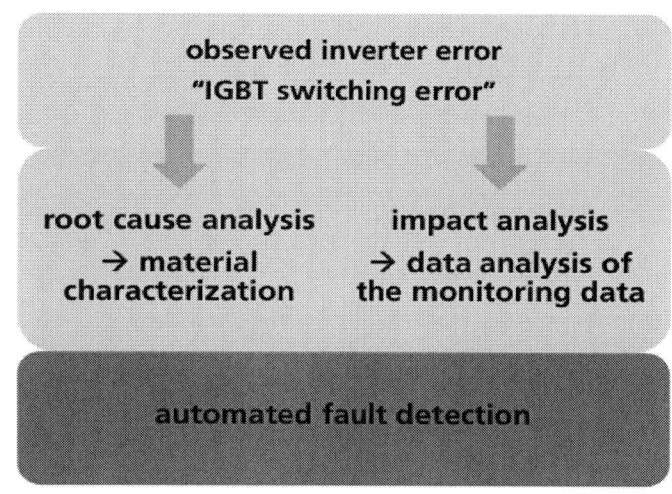

observed inverter error
"IGBT switching error"

root cause analysis
→ **material characterization**

impact analysis
→ **data analysis of the monitoring data**

automated fault detection

Material characterization

Material characterization
Non-destructive analysis

—

* **a** Impairments in plane of the wire bond connections

* **b** The emitter metallization of the IGBT components is completely degraded and the associated wire bond connections to the DCB substrate are also destroyed.

* **c** Re-melting of the solder connections of the chips and the pins of the connection terminals is visible.

Power module after field ageing

Power module after failure

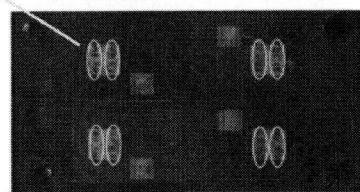

Characterization of the failure mechanisms in inverter components,
above: Power modules without housing and potting
bottom: non-destructive scanning acoustic microscopy (SAM) analysis (wire bond level) before removal of housing and potting

---

Material characterization
Non-destructive analysis

—

* **a** Impairments in plane of the wire bond connections

* **b** The emitter metallization of the IGBT components is completely degraded and the associated wire bond connections to the DCB substrate are also destroyed.

* **c** Re-melting of the solder connections of the chips and the pins of the connection terminals is visible.

Power module after field ageing

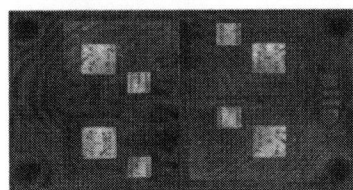

Power module after failure

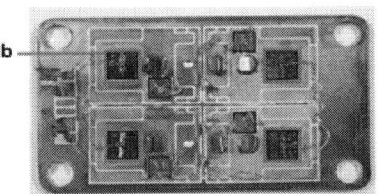
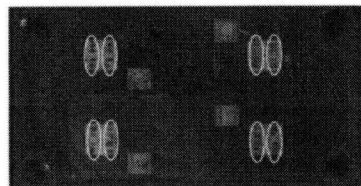

Characterization of the failure mechanisms in inverter components,
above: Power modules without housing and potting
bottom: non-destructive scanning acoustic microscopy (SAM) analysis (wire bond level) before removal of housing and potting

Material characterization
Non-destructive analysis

---

* **a** Impairments in plane of the wire bond connections

* **b** The emitter metallization of the IGBT components is completely degraded and the associated wire bond connections to the DCB substrate are also destroyed.

* **c** Re-melting of the solder connections of the chips and the pins of the connection terminals is visible.

Power module after field ageing

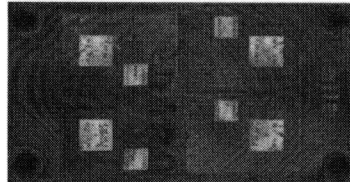

Power module after failure

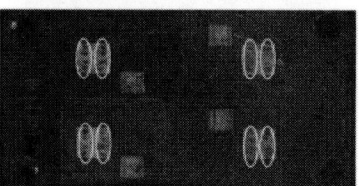

Characterization of the failure mechanisms in inverter components,
above: Power modules without housing and potting
bottom: non-destructive scanning acoustic microscopy (SAM) analysis (wire bond level) before removal of housing and potting

---

Material characterization
Cross-sectional view – light microscope

---

* wire bond contacts are destroyed and not visible

* partially missing chip and solder area up to the DCB substrate

* The module was exposed to enormous heat, causing melting the chip and solder materials locally.

Conclusion

1. all four IGBT components are equally affected
2. the associated diodes show no damage
3. the peripheral extent of damage is less drastic overall

* a longer lasting overtemperature load has led to degradation of the semiconductors

* In the end, a breakthrough occured and the final damage was caused by the short-term release of massive heat.

Power module after field ageing

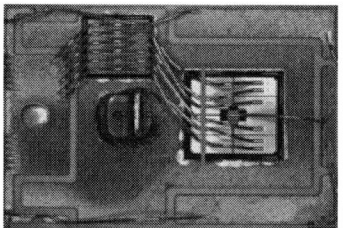

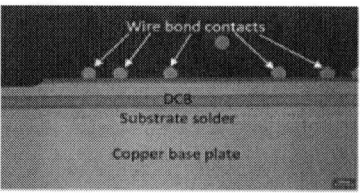

Power module after failure

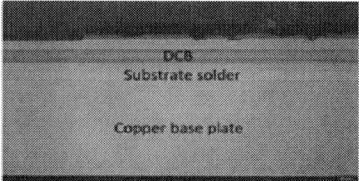

Cross sectional view of the bonding wires and solder contact using light microscope imaging of the IGBTs

Take away message:

- root cause analysis done

- failure mechanism understood

→ overtemperature load over a longer period → degradation of the semiconductors

→ Final damage: breakthrough happened and a release of massive heat

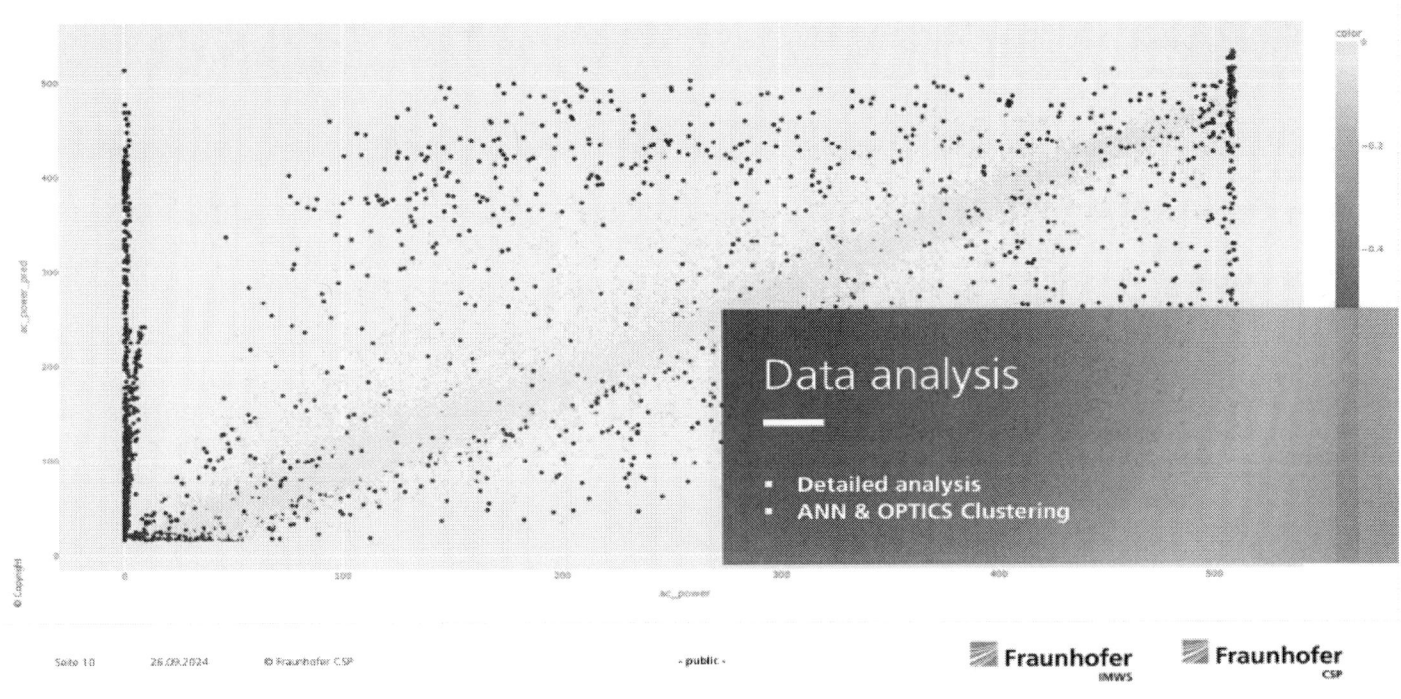

Data analysis

- Detailed analysis
- ANN & OPTICS Clustering

Data analysis
Detailed analysis

Detailed investigation of individual parameters of the inverters

- Deviating inverter behavior recognizable during a grid-related curtailment of inverter 1

- Different states are approached alternately ("underload" - "wake-up" - "underload", "night") in contrast to inverter 2

- Occurrence of the error message "IGBT switching error" ½ h after curtailment ended → IGBT module defective

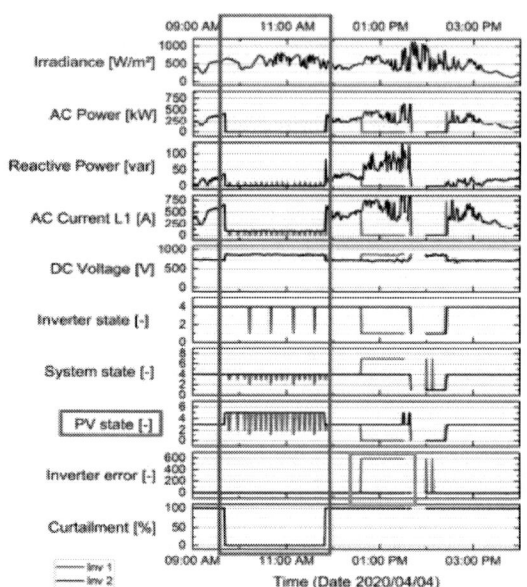

Comparison of measured electrical variables and status information of Inv 1 and a comparable Inv 2 in the same system

---

Data analysis
Application of System Portfolio

How often does curtailment occur?

Is the pattern also visible on other inverters?

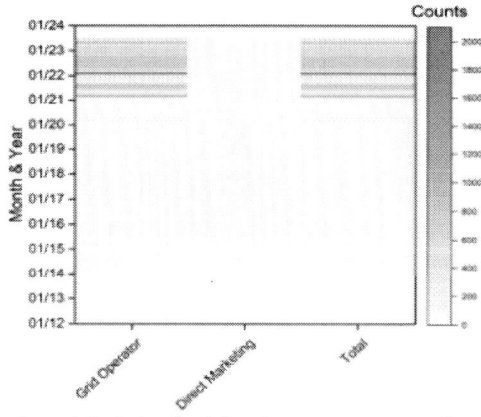

Amount of minutes of curtailment occurrence, aggregated per month and year, split between grid operator and direct marketing

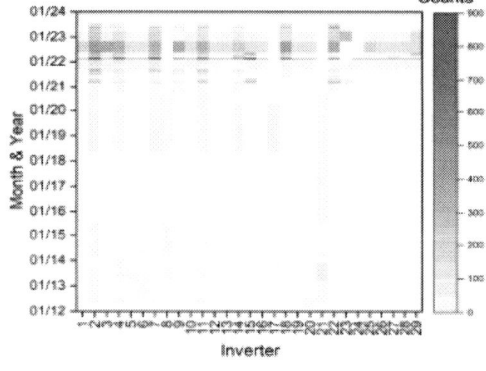

Amount of faulty inverter behavior during curtailment; determined by the state information and it's change from one minute to the next

Take away message:

- impact analysis done
- pattern found and possible affected systems identified

→ But is it possible to detect it automatically?

Data analysis
ANN & OPTICS Clustering – Results

ANN Modelling → First OPTICS Clustering → Second OPTICS Clustering

Determination of reference behavior

- input
    - irradiance, ambient temp
    - sun position
- output
    - DC & AC power

- 1 year (10 min interval) – year 2021

- RMSE
    Training 15.5 kW
    Test 15.8 kW

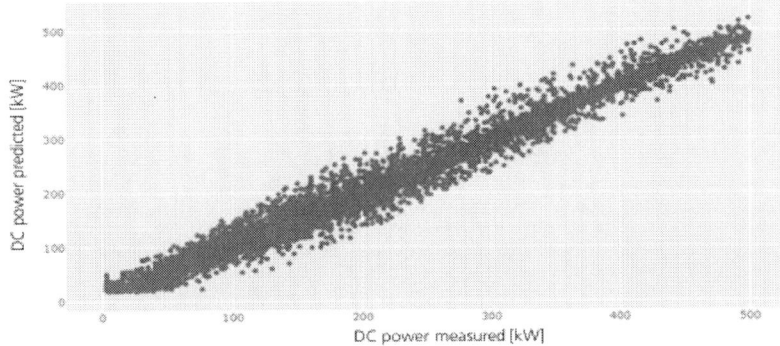

Data analysis
ANN & OPTICS Clustering – Results

ANN Modelling   **First OPTICS Clustering**   Second OPTICS Clustering

legend cluster
0
-1

Distinguishing between "normal" and "irregular" behavior

parameters for clustering:
* electrical inverter parameters
* pred. DC & AC power from ANN
* irradiance, ambient temp
* sun position

* 1 year (10 min interval) – year 2022
* exclusion of night-time values

* metric: Mahalanobis
* minPts=3000   ξ=0.001

cluster 0 → 89% from data set   cluster -1 → 11% from data set

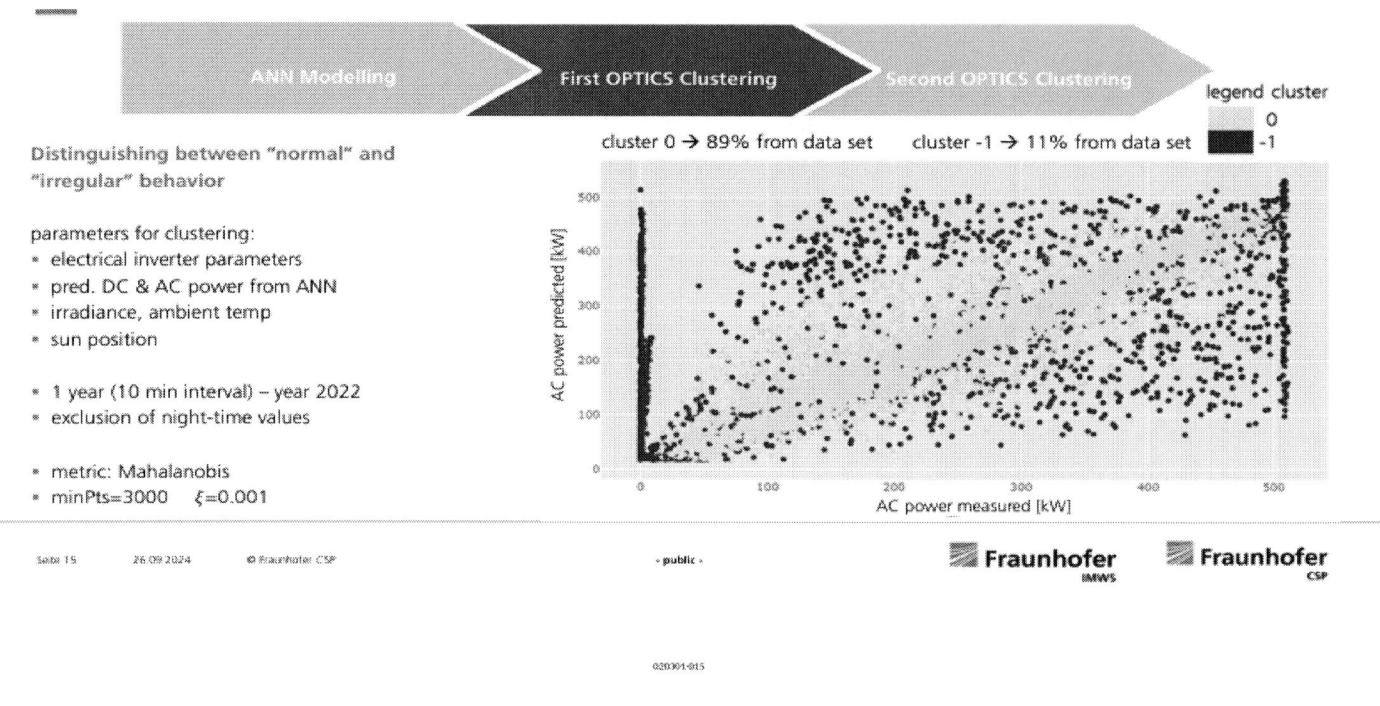

Data analysis
ANN & OPTICS Clustering – Results

ANN Modelling   First OPTICS Clustering   **Second OPTICS Clustering**

A closer look at irregular behavior

Parameters for clustering:
* "outliers" of first clustering (year 2022)
* electrical inverter parameters

* metric: euclidean
* minPts=100   ξ=0.001

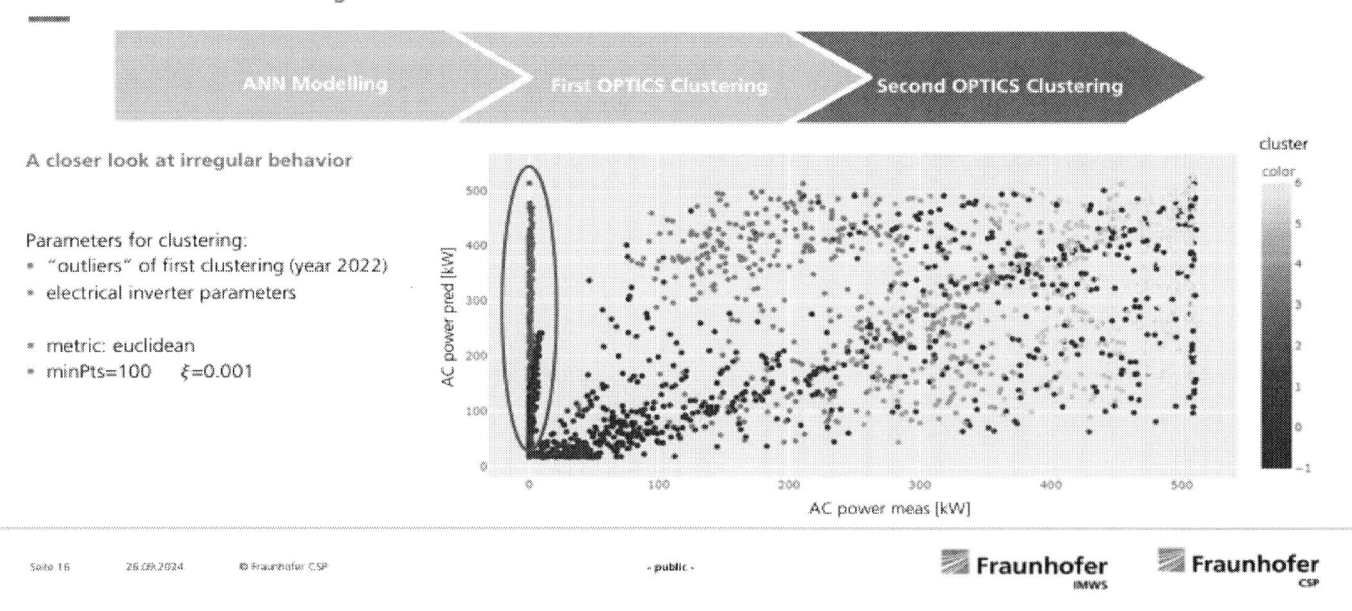

Conclusion

observed inverter error
"IGBT switching error"

failure mechanism understood

pattern found and impact of the entire system identified

automatically detected in the field

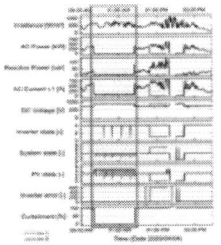

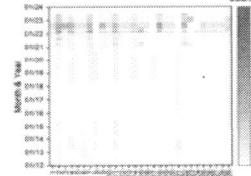
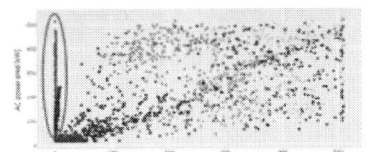

Thank you for your attention

Contact

Stephanie Malik
PV Systems and Integration
Phone: +49 (0) 345 5589 5212
Stephanie.Malik@csp.fraunhofer.de

Fraunhofer Center for Silicon Photovoltaics CSP
Otto-Eißfeldt-Straße 12
06120 Halle (Saale)
www.csp.fraunhofer.de

This work was funded by

Supported by:

 Federal Ministry
for Economic Affairs
and Climate Action

robStROM (FKZ: 03EE1163B)

on the basis of a decision
by the German Bundestag

ANOMALY DETECTION IN SIMILARLY BEHAVING SOLAR INVERTERS

Pau Mercade Ruiz[1], Gerardo Guerra[1], Gaetana Anamiati[1], Lars Landberg[2]

[1]GreenPowerMonitor a DNV company, Gran Via de les Corts Catalanes, 130, Barcelona, Spain; Email:
pau.mercade@dnv.com
[2]DNV Denmark, Tuborg Parkvej 8, Hellerup, Denmark; Email: lars.landberg@dnv.com

ABSTRACT: Inverters are the primary cause of ticket creation and energy loss in PV plants. Automatic detection of inverter malfunction can help PV operators optimize maintenance, thereby reducing overall downtime and maintenance costs. To this end, anomaly detection is used on performance data generated by a group of similarly behaving inverters. At any point in time, the first and third quartiles of the data for all the inverters serve as baseline for the range of values that would be generated by a normally operating inverter. Two measures of abnormality based on this principle are thereby devised. One tracking the differences between the distributions of the data generated by the inverter and baseline, while the other tracking the cumulated differences between inverter and baseline data. Finally, the two abnormality measures are combined into a single anomaly score ranging from 0 to 1 for increasing degree of abnormality. A threshold is then set on the anomaly score above which an alarm goes off warning of inverter malfunction. The methodology has been tested on a real PV plant for the specific task of detecting inverter faults yielding 80 and 25 percent detection and false alarm rates, respectively.
Keywords: inverter, anomaly detection, fault, malfunction, maintenance, data

1 INTRODUCTION

Automatic detection of solar photovoltaic (PV) plant component malfunction is a cost-effective tool that allows for better maintenance planning and overall downtime reduction. Nowadays, PV plants are typically equipped with a monitoring infrastructure that continuously records its operational status, which has enabled the development of various anomaly detection methodologies [1–7]. Nonetheless, there are some aspects of these that can be improved on.

Using only the most relevant signals (parameters) in terms of defining operational status and performance can make the methodology more robust and help ease interpretation and explainability. Better interpretation and explainability of the underlying causes that trigger the anomaly detection methodology may also help other complementary tasks make use of it, such as when it comes to the development and implementation of predictive maintenance [3, 8].

Avoiding the use of idealised physical assumptions in the development of the anomaly detection methodology can make the detection process more robust since they may not hold for noisy measurements under real operation conditions. Instead, using a complete data-driven methodology has been proven effective in dealing with real-world conditions.

Another limitation in most advanced anomaly detection methodologies is the need of periodic retraining or fine-tuning of the model used as baseline for normal operation. Since this model typically relies on a training dataset consisting of data points generated under all possible operational modes, any change in operation, e.g., due to aging, maintenance intervention or environmental condition, would require gathering data for the new operational modes and retraining.

The present work aims at automatically identifying instantaneous inverter malfunction using anomaly detection and improving on some of the aspects mentioned above.

The focus is set on the inverter because it is generally responsible for most of the plant's energy losses and the main cause of ticket creation [9, 10], and so the PV component that offers the highest return for better maintenance.

The methodology to detect inverter malfunction requires only AC power, DC voltage and DC current data for a group of similarly behaving inverters. These signals are chosen because they are typically measured in most PV plants nowadays and they can capture most of the AC/DC operational status of the inverter. The methodology does not make any assumption on the relationship between the inverter signals, nor it requires a baseline model to be periodically retrained. Instead, inverter malfunction is here assessed comparing data points of one inverter with those of other similarly behaving inverters. At any point in time, the first and third quartiles of the data for all the inverters are used as the baseline for the normal range of values that an inverter operating normally would measure. Finally, significant deviations from baseline are quantified by an anomaly score for AC power, DC voltage and DC current separately. A high anomaly score is then considered as an indication of inverter malfunction.

Two assumptions underpin the present methodology. The first being that the inverters of a PV plant that have the same technical specifications and operate nearby would measure similar values of AC power, DC voltage and DC current. The second being that a faulty inverter would consistently report significantly different values of AC power, DC voltage and/or DC current as those measured by most of the group of similarly behaving inverters.

The present work is structured in 3 sections: section 2 explains the anomaly detection methodology, section 3 presents and discusses the results of testing the methodology in a real PV plant, and section 4 provides relevant conclusions to the present work.

2 ANOMALY DETECTION

The purpose of the anomaly detection methodology is to detect inverter malfunction.

The methodology requires a group of inverters having similar technical specifications, operating under similar environmental conditions and recording AC power (PAC), DC voltage (VDC) and DC current (IDC) every 5 minutes for at least one year.

The group of inverters is used as the baseline for normal operation and deviations from it are considered a proxy for inverter malfunction. Normal operation here refers to the operation of a healthy inverter.

The electrical signals PAC, VDC, and IDC are chosen because they can provide a comprehensive status of the inverter's operation and are typically measured in most PV plants nowadays. Nonetheless, the methodology is flexible enough to take on any other inverter signal that would be deemed relevant for defining the operational status of the inverter (see section 2.3, additional signals).

The 5-minute (or shorter) sampling period is needed to be able to spot changes in the operation of the inverter that can suggest inverter malfunction.

The one-year (or longer) historical dataset (HD) is needed to both preprocess the data and set the appropriate thresholds for anomaly detection.

Both the 5-minute sampling period and the one-year HD requirements are normally attainable by most PV plants nowadays.

Overall, the input of the methodology are time series of PAC, VDC and IDC data generated by a group of inverters. Nighttime is removed from the time series though since it does not bring any relevant information about the operational status of the inverter.

2.1 Preprocessing

The data preprocessing step aims to make the data generated by the group of inverters comparable to one another. Specifically, for every inverter and electrical signal, the data is transformed so that the data become uniformly distributed over the range of values that the normally operating inverter could measure.

Firstly, for every inverter, the HD is arranged as a table, where every row (data point) consists of 3 columns which are populated by the instantaneous measurements of PAC, VDC and IDC. The number of rows is hence determined by the sampling period and the length of the historical record.

Secondly, data points are removed from HD when any of the following conditions is met by either PAC, VDC or IDC:

- Out-of-range values, i.e., values below/above the first/third quartile minus/plus 1.5 times the interquartile range.
- Periods where consecutive values remain constant.

Thirdly, based on the highly correlated PAC and irradiance measurements, data points associated with extreme inverter underperformance or corrupted PAC readings are removed from HD, which would be outliers in the 2D irradiance-PAC dataset. The local outlier factor method [11] is used to find and remove them from HD; however, any other outlier detection method could be used instead provided that only the most extreme outliers are removed. That is, this cleaning step is meant to remove only the most extreme outliers. If no irradiance measurements are available, they may be replaced by the median of PAC values of all the inverters at any point in time.

Finally, for every electrical signal, a data point is transformed into the proportion of data points in the cleaned HD that have a value less or equal to that of itself.

In the remaining of the text, any data point that is mentioned, also those that make up HD, is assumed to have been transformed accordingly.

2.2 Anomaly score

The anomaly score is meant to rate data points according to the degree of abnormality of the inverter that generated them.

Two measures of abnormality are used for anomaly score calculations, the one-sided two-sample Kolmogorov-Smirnov test statistic (KS) and a bias metric (BM). KS is meant to track differences between the distributions of inverter and baseline recent data, while BM is meant to track differences between inverter and baseline recent data. Both KS and BM calculations use weights to weigh recent data points higher as these are more representative of the current operational status of the inverter.

The first (Q1) and third (Q3) quartiles of PAC, VDC and IDC values measured by all the inverters at any point in time are used as the baseline for the range of values that the normally operating inverter would measure.

KS and BM definitions, as measures of abnormality, assume that at any point in time and for every electrical signal separately, the distribution of the data of all the inverters combined is unimodal. Therefore, presumably any inverter operating under normal conditions would tend to measure values close to the interval (Q1, Q3). A normally operating inverter could also still measure values outside the interval (Q1, Q3); nonetheless, doing so either in a consistent manner or with a value far from (Q1, Q3) would suggest some sort of inverter malfunction.

KS calculation: At any point in time, for every electrical signal and inverter, the weighted empirical distribution functions of the two samples consisting of the last N inverter (F), Q1 (G1) and Q3 (G3) data points are calculated. Then, KS1 is calculated as the minimum difference between F and G1. Similarly, KS3 is calculated as the maximum difference between F and G3. An inverter operating under normal conditions would be expected to yield $F \leq G1$ and $F \geq G3$, and so $KS1 \leq 0$ and $KS3 \geq 0$. Normal data points are then disregarded by setting KS1 and KS3 to zero whenever $KS1 \leq 0$ and $KS3 \geq 0$, respectively.

BM calculation: At any point in time, for every electrical signal and inverter, BM1 and BM3 are calculated as the weighted sum of the last M differences between Q1 and the inverter data points and between Q3 and the inverter data points, respectively. An inverter operating under normal conditions would be expected to yield $BM1 \leq 0$ and $BM3 \geq 0$. Normal data points are then disregarded by setting BM1 and BM3 to zero when $BM1 \leq 0$ and $BM3 \geq 0$, respectively.

Weights calculation: For a sequence of i = 1,...,N consecutive data points, the weight at position i is equal to the sine of argument $i\pi$ divided by 2N radians. Then, every weight is normalized by the sum of all weights, so that the sum of all normalized weights is equal to 1. Then, for every data point in the sequence, the i-th data point is assigned the i-th normalized weight.

Parameter settings: N and M must be set for KS and BM calculations, respectively. By default, N and M are recommended to be set to 125 and 25, respectively. This setting translates into processing windows of length 10 (approximately one day of operation) and 2 hours for KS and BM calculations, respectively, which agree with the design purpose of KS and BM as anomaly indicators and have been seen to lead to stable results, i.e., less susceptible to noise. Smaller N and M values could lead to bad estimates of F and G, and overestimate inverter underperformance and malfunction, respectively. Larger

N and M values could mask relevant differences between F and G, and underestimate inverter underperformance and malfunction, respectively. Further optimization of N and M would be desirable after the anomaly detection has been run for a year or more in real operation conditions and data of the performance of the methodology has been gathered.

Finally, the anomaly score (AS) is calculated as the geometric mean of KS1 and BM1 for PAC and as the maximum between the geometric mean of KS1 and BM1 and the geometric mean of KS3 and BM3 for VDC and IDC. The resulting anomaly score is hence a value ranging from 0 to 1, where higher values indicate increasingly anomalous operation.

2.3 Alarm system

An alarm system is devised to warn of inverter malfunction. The alarm system is based on an upper bound (threshold) to the anomaly score above which an alarm is triggered warning of inverter malfunction.

The cleaned HD is used to set the threshold of the anomaly score for every inverter and electrical signal separately. After the anomaly score is calculated for every data point in the cleaned HD, the threshold is calculated as the exponential of the third quartile plus 1.5 times the interquartile range of the logarithm of all the anomaly scores that are greater than 0. If no data points in the cleaned HD have an anomaly score greater than 0, the threshold is set to 0.

Since the distribution of the anomaly scores is in general positively skewed, the log transformation in the threshold calculation is meant to help operate over a rather more symmetrical distribution and obtain a better estimate of the acceptable range of anomaly scores.

Finally, an inverter is warned to be in a state of malfunction and an alarm goes off, when any of the following conditions is met:

- The anomaly score for PAC is greater than its threshold.
- The anomaly score for PAC is greater than 0 and either the anomaly score for VDC or IDC is greater than their thresholds.

Additional signals: If other inverter signals in addition to PAC, VDC and IDC were deemed relevant for defining the operational status of the inverter and wanted to be integrated into the anomaly detection methodology, they would play the same role as VDC and IDC do. That is, the anomaly score and threshold would be calculated for every additional signal, and alarms would then be triggered if any of the following conditions were met:

- The anomaly score for PAC is greater than its threshold.
- The anomaly score for PAC is greater than 0 and any of the anomaly scores for VDC, IDC and the extra signals are greater than their thresholds.

3 CASE STUDY

The methodology has been tested in a PV plant consisting of 42 inverters for one year of operation, during which a register of inverter faults and maintenance actions performed on the inverters (maintenance log) was made available. The time series of PAC, VDC and IDC were made available for the entire PV plant life.

The anomaly detection system (N=125 and M=25)

was run for the year of operation and both anomaly scores and alarms were gathered for testing.

The maintenance log consists of a list of dates when either an inverter fault or a maintenance action happened, together with some brief description of the event.

Although the maintenance log provides valuable insights into the operational status of the inverters, it still lacks further information to allow for a more thorough test of the anomaly scores and alarms. An ideal test would use daily assessment of the actual operational status of the inverters for comparison with the assessment derived from the anomaly scores and alarms.

Instead, the test presented here takes the dates in the maintenance log associated with only inverter faults (fault record) and compares them with the dates associated with any alarm being triggered (alarm record). The dates in the maintenance log associated with maintenance actions (maintenance record) are excluded from the test. Hence, the comparison leads to 4 outcomes: the number of coinciding dates in both records (271), the number of dates that appear only in the fault record (67), the number of dates that appear only in the alarm record (89), and the number of remaining dates for all the 42 inverters throughout the one-year testing period except dates in the maintenance record (13692). Despite the huge class imbalance inherent in the fault detection problem, the results may still reveal what performance can be expected from the presented anomaly detection methodology.

In terms of fault detection performance, the faults are mostly detected by the alarm system, 271 out of 338 dates with faults are successfully detected. The 67 undetected faults go unnoticed because the fault either does not yield any clear drop in inverter PAC in comparison with the PAC of the other inverters (Fig. 1) or if it does, the drop is too brief for the alarm system to catch (Fig. 2).

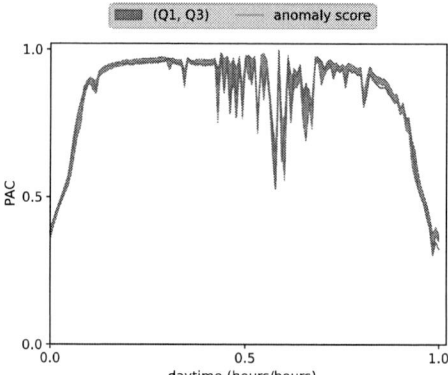

Figure 1: Undetected inverter fault. No apparent inverter underperformance can be seen from PAC data (blue line) when compared to the normal range (grey area) and therefore the anomaly score (orange line) remains 0.

41st European Photovoltaic Solar Energy Conference and Exhibition

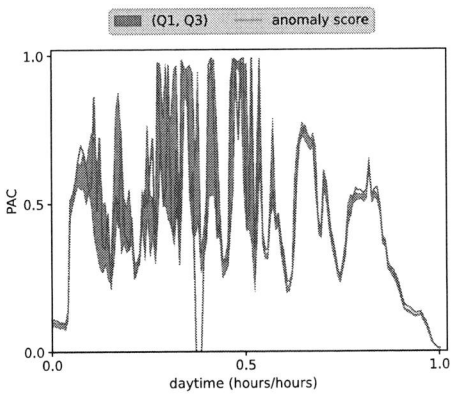

Figure 2: Undetected inverter fault. The drop in performance seen on PAC data (blue line) is too brief in time for the anomaly score to react to it and therefore the anomaly score (orange line) remains 0.

In terms of false alarms, while 89 dates out of the total 360 dates with alarms are not successfully associated with any fault in the maintenance log, they may still warn of inverter underperformance to some degree (Fig. 3). Unless PAC readings were corrupted, inverter underperformance could be caused by either some sort of inverter malfunction (Fig. 3) or another PV component malfunction, such as e.g. a problem upstream from the inverter (Figs. 4 and 5). If it was caused by inverter malfunction, the problem would have had to either resolve by itself, without human intervention, or both the maintenance action that set the inverter back to normal operation and the fault be omitted in the maintenance log.

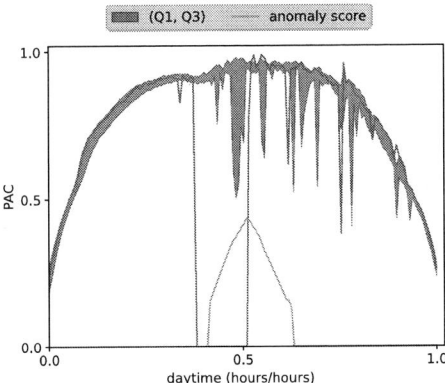

Figure 3: False alarm. The anomaly score (orange line) reacts to the sudden drop in performance seen on PAC data (blue line); however, no inverter fault was registered in the maintenance log.

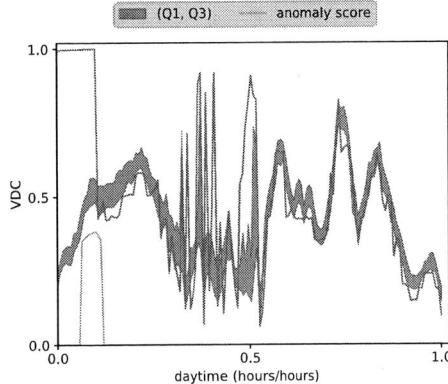

Figure 4: False alarm. The anomaly score (orange line) reacts to the sudden increase of VDC (blue line); however, no inverter fault was registered in the maintenance log. The data could have been corrupted by a problem upstream from the inverter, which led VDC to diverge from normal behaviour (grey area).

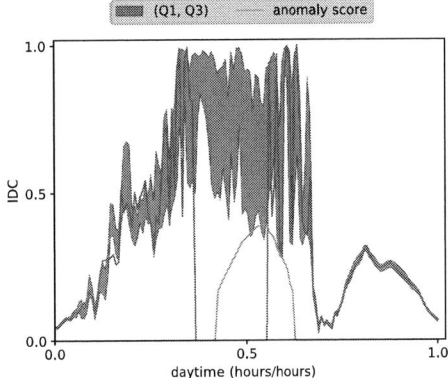

Figure 5: False alarm. The anomaly score (orange line) reacts to the sudden drop in IDC (blue line); however, no inverter fault was registered in the maintenance log. A problem caused upstream from the inverter could have caused IDC to drop.

4 CONCLUSIONS

The anomaly detection methodology developed in this work can be applied to most PV plants nowadays. It is data-driven, which has been proven effective when it comes to dealing with real-world noisy conditions. It uses the data generated by a group of similarly behaving inverters to detect anomalies on the individual inverters. This approach has the advantage of having a baseline for normal operation that remains up to date as long as the inverters remain similar in behaviour. The methodology is also easily interpretable and explainable, so that the detected anomalies and the underlying mechanisms that lead to the detection can be easily interpret and debugged. The methodology has been tested in a real PV plant for the problem of fault detection and it has shown positive results. The test outlines the challenges associated with detecting inverter malfunction that does not yield a decrease in inverter performance, and that alarms may be susceptible to be triggered by disfunction of any other PV component upstream from the inverter. That is, any problem in the PV plant that can make the performance of the inverter drop will inevitably make the alarm system react to it. Finally, the test results also reveal the challenges associated with keeping track of all the maintenance activities and faults that may happen at inverter level, which the present anomaly detection methodology could help overcome, with the added advantage of being data-driven and hence transparent.

5 REFERENCES

[1] Benninger, Moritz, Martina Hofmann, and Marcus Liebschner. "Anomaly detection by comparing photovoltaic systems with machine learning methods." In NEIS 2020; Conference on Sustainable Energy Supply and Energy Storage Systems, pp. 1-6. VDE, 2020.

[2] Benninger, Moritz, Martina Hofmann, and Marcus Liebschner. "Online monitoring system for photovoltaic systems using anomaly detection with machine learning." In NEIS 2019; Conference on Sustainable Energy Supply and Energy Storage Systems, pp. 1-6. VDE, 2019.

[3] De Benedetti, Massimiliano, Fabio Leonardi, Fabrizio Messina, Corrado Santoro, and Athanasios Vasilakos. "Anomaly detection and predictive maintenance for photovoltaic systems." Neurocomputing 310 (2018): 59-68.

[4] Zhao, Yingying, Qi Liu, Dongsheng Li, Dahai Kang, Qin Lv, and Li Shang. "Hierarchical anomaly detection and multimodal classification in large-scale photovoltaic systems." IEEE Transactions on Sustainable Energy 10, no. 3 (2018): 1351-1361.

[5] Mekki, Hamza, Adel Mellit, and Hassen Salhi. "Artificial neural network-based modelling and fault detection of partial shaded photovoltaic modules." Simulation Modelling Practice and Theory 67 (2016): 1-13.

[6] Chine, W., A. Mellit, Vanni Lughi, A. Malek, Giorgio Sulligoi, and A. Massi Pavan. "A novel fault diagnosis technique for photovoltaic systems based on artificial neural networks." Renewable Energy 90 (2016): 501-512.

[7] Zhao, Ye, Brad Lehman, Roy Ball, Jerry Mosesian, and Jean-François de Palma. "Outlier detection rules for fault detection in solar photovoltaic arrays." In 2013 Twenty-Eighth Annual IEEE Applied Power Electronics Conference and Exposition (APEC), pp. 2913-2920. IEEE, 2013.

[8] Guerra, Gerardo, Pau Mercade Ruiz, and Lars Landberg. "A Machine Learning-Based Predictive Maintenance System for Solar Inverters." In 36th European Photovoltaic Solar Energy Conference and Exhibition (EU PVSEC), 2019.

[9] Gunda, Thushara. Inverter Faults & Failures: Common modes & patterns. No. SAND2020-2323C. Sandia National Lab. (SNL-NM), Albuquerque, NM (United States), 2020.

[10] Golnas, Anastasios. "PV system reliability: An operator's perspective." In 2012 IEEE 38th Photovoltaic Specialists Conference (PVSC) PART 2, pp. 1-6. IEEE, 2012.

[11] Kriegel, Hans-Peter, et al. "LoOP: local outlier probabilities." Proceedings of the 18th ACM conference on Information and knowledge management. 2009.

Towards Higher Efficiency: Data Analysis and Optimization of PV String Wiring in a Long-Running Solar Power Plant

Žiga Miklič, Janez Krč, Marko Topič

University of Ljubljana, Faculty of Electrical Engineering, LPVO

EUPVSEC 2024

Introduction of 6,9 kW roof-integrated PV power plant

(launched in 2012)

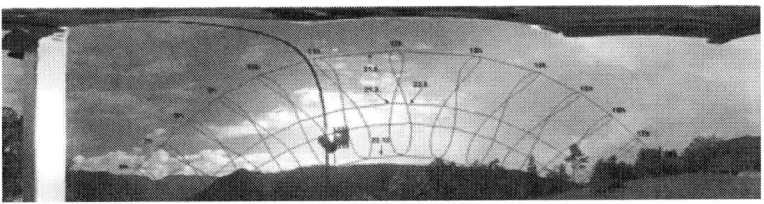

~ **7 %** annual loss due to shading (hills, mauntains, trees...)

~ **3.5 %** annual loss due to non ideal orientation

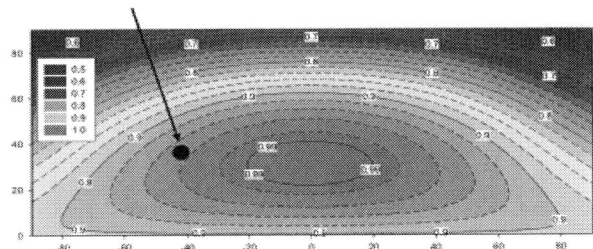

Orientation of a "southern" part of the roof
(**-43°** from S to E, **36°** tilt)

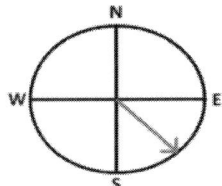

PV Components

- **120** PV modules (CiS – Thin film)
- PV generator power = **6,9 kWp**
- Roof integrated (SolRif)
- PV module power = **57,5 Wp**
- PV generator area = **100 m2**
- Inverters (3 x 1ph) – SMA Sunny Boy 2500HF
- Inverter efficiency = **96 %**
- In operation since **2012**

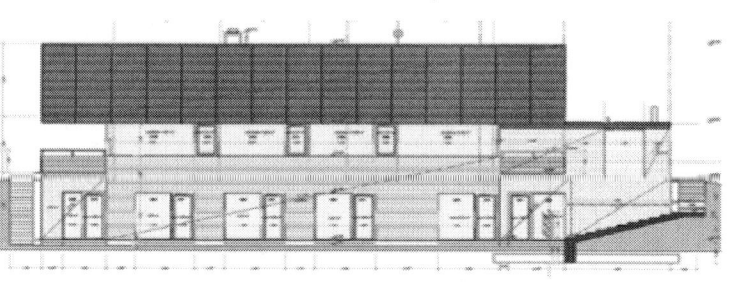

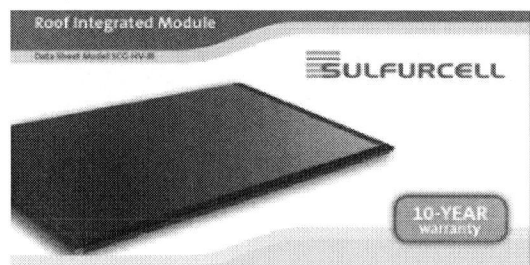

Electrical Characteristics at 1000 W/m², 25°C, AM 1,5				
Modul	SCG52-HV-RI	SCG55-HV-RI	SCG57-HV-RI	SCG60-HV-RI
Rated Power**	52,5 W	55 W	57,5 W	60 W
Tolerance	+8/-2%	+8/-2%	+8/-2%	+8/-2%
Module efficiency	6,4%	6,7%	7,0%	7,3%
Voltage at Vmpp	37,8V	38,8V	39,7V	40,3V
Current at Impp	1,39A	1,42A	1,45A	1,49A
Open-Circuit Voltage (Voc)	50,2V	50,9V	51,4V	52,1V
Short-Circuit Current (Isc)	1,67A	1,69A	1,71A	1,74A
max. system Voltage	1000V	1000V	1000V	1000V
Reverse Current Load	5A	5A	5A	5A

Wiring of PV plant

CiS PV modules

| 10 | 10 | 10 | 10 | | 10 | 10 | 10 | 10 | | 10 | 10 | 10 | 10 |

4 parallel strings   4 parallel strings   4 parallel strings

1 phase inverter   1 phase inverter   1 phase inverter

GRID

Measurement instruments

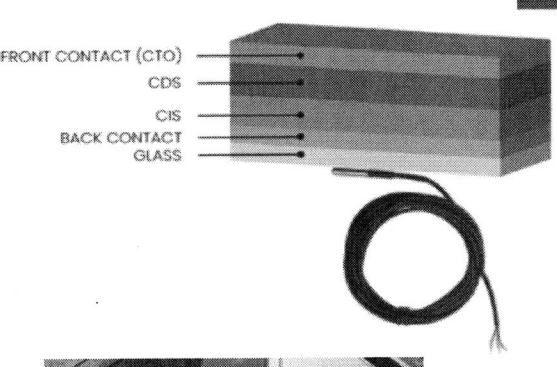

- **Metrel MI 3108 EurotestPV** [string IV curve]
- **Metrel A 1378 EutrotestPV Remote** [module temperature and solar irradiance]
- **SMA Sensorbox (SN: 21009)** [module temperature and solar irradiance]

MI 3108 EurotestPV

A 1378 EurotestPV Remote

Data analysis

- Inverter data analysis (output power, solar irradiance, module temperature).
- Solar irradiance data filter **(>800 W/m2)**
- **Temperature-compensated** calculated PR factor
- Measurement data analysis (Metrel instruments data)
- IV curve plot for each and paralel PV string

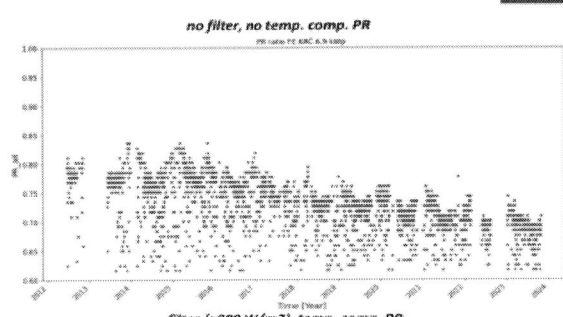

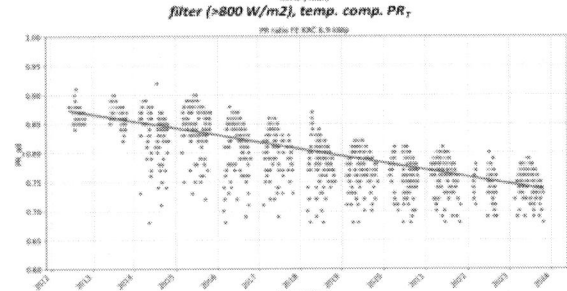

Actual PV string wiring

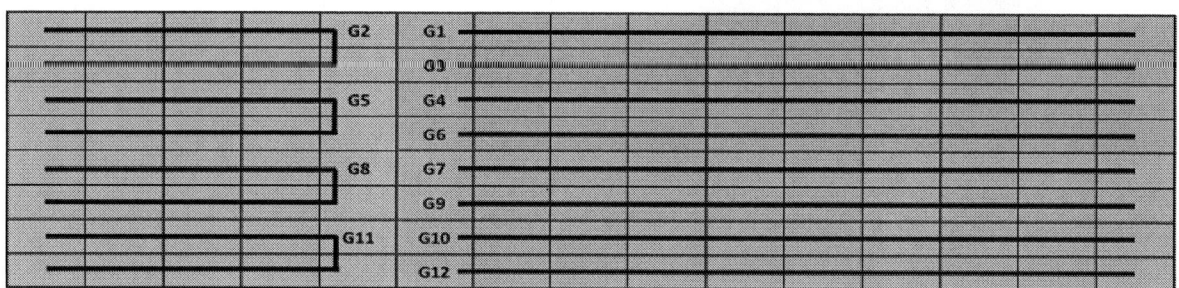

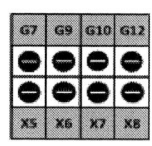 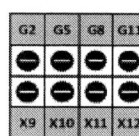 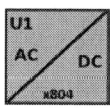 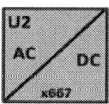 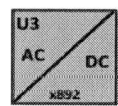

PV string measurements (18.12.2023)

Parallel strings:

- G1, G3, G4, G6
- G7, G9, G10, G12
- G2, G5, G8, G11

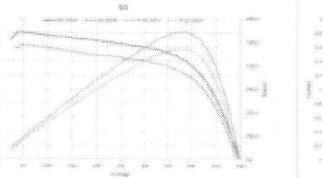 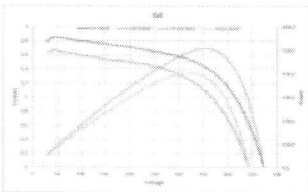 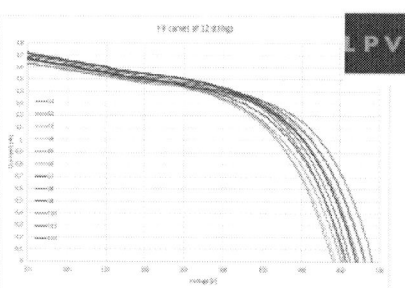

String	Voc [V]	deg. Voc	Isc [A]	deg. Isc	FF	deg. FF	Vmpp [V]	deg. Vmpp	Impp [A]	deg. Impp	Pmpp [W]	deg. P	Eff. [%]	deg. Eff
Nom.	514	0%	1,71	0%	0,65	0%	397	0%	1,45	0%	575	0%	6,46%	0%
G1	490	-4,7%	1,57	-8,2%	0,61	-6,5%	380	-4,3%	1,24	-14,5%	472	-17,9%	5,30%	-18%
G3	482	-6,2%	1,58	-7,6%	0,59	-9,7%	369	-7,1%	1,22	-15,9%	451	-21,6%	5,06%	-22%
G4	442	-14,0%	1,64	-4,1%	0,55	-15,5%	329	-17,1%	1,22	-15,9%	402	-30,1%	4,51%	-30%
G6	439	-14,6%	1,65	-3,5%	0,57	-13,4%	326	-17,9%	1,26	-13,1%	409	-28,9%	4,59%	-29%
G7	457	-11,1%	1,64	-4,1%	0,56	-13,9%	341	-14,1%	1,24	-14,5%	424	-26,3%	4,76%	-26%
G9	480	-6,6%	1,64	-4,1%	0,59	-10,3%	364	-8,3%	1,27	-12,4%	463	-19,5%	5,20%	-19%
G10	454	-11,7%	1,67	-2,3%	0,58	-10,9%	335	-15,6%	1,32	-9,0%	442	-23,1%	4,96%	-23%
G12	462	-10,1%	1,68	-1,8%	0,58	-12,1%	349	-12,1%	1,28	-11,7%	448	-22,1%	5,03%	-22%
G2	472	-8,2%	1,64	-4,1%	0,58	-11,6%	353	-11,1%	1,27	-12,4%	448	-22,1%	5,03%	-22%
G5	451	-12,3%	1,63	-4,7%	0,56	-15,1%	335	-15,6%	1,22	-15,9%	409	-28,9%	4,59%	-29%
G8	470	-8,6%	1,64	-4,1%	0,58	-10,9%	354	-10,8%	1,27	-12,4%	451	-21,6%	5,06%	-22%
G11	469	-8,8%	1,67	-2,3%	0,58	-11,2%	356	-10,3%	1,28	-11,7%	455	-20,9%	5,11%	-21%

Visual representation of degradation

Degradation Voc [%]

Degradation Vmpp [%]

Degradation Isc [%]

Degradation Impp [%]

Degradation FF [%]

Degradation Pmpp [%]

PR$_T$ factor of PVS [2012-2023]

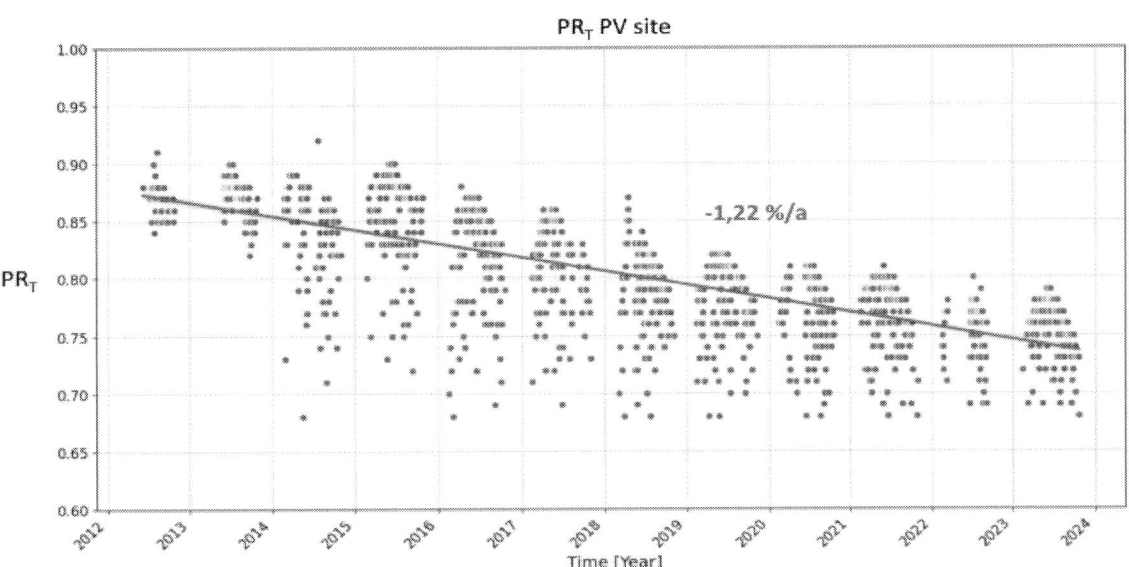

PR$_T$ PV site

-1,22 %/a

PR$_T$ factor

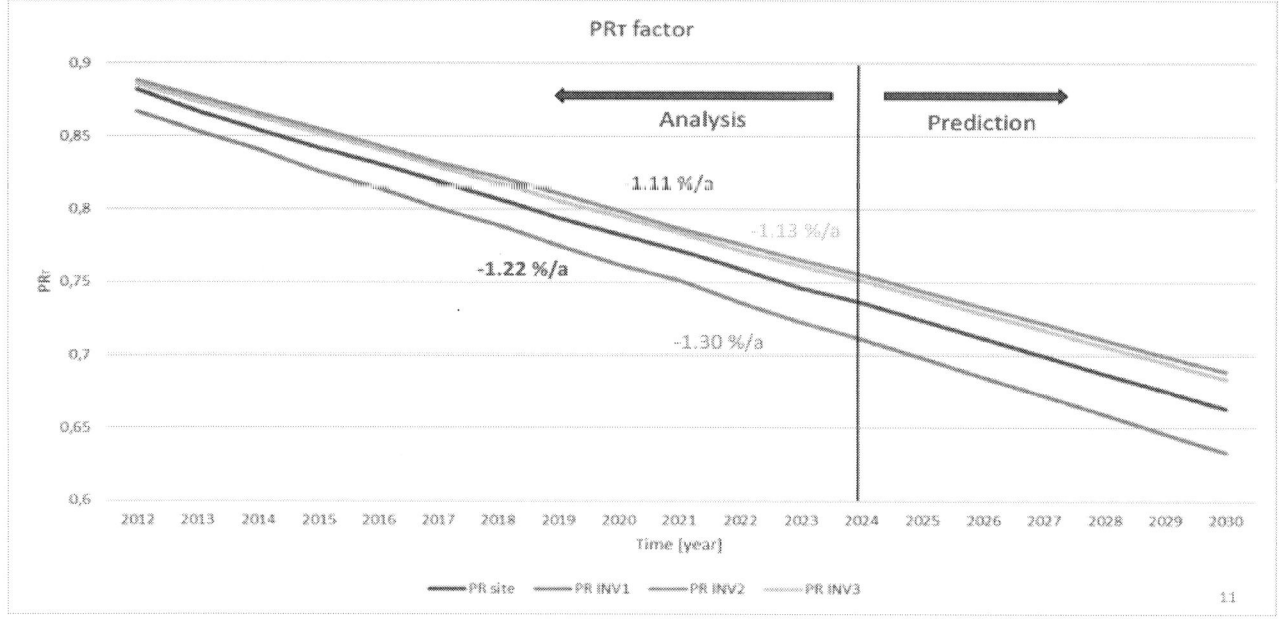

Optimal re-wiring of PV gen. („Vmpp")

String	Voc [V]	deg. Voc	Isc [A]	deg. Isc	FF	deg. FF	Vmpp [V]	deg. Vmpp	Impp [A]	deg. Impp	Pmpp [W]	deg. P	Eff. [%]	deg. Eff
Nom.	514	0%	1,71	0%	0,65	0%	397	0%	1,45	0%	575	0%	6,46%	0%
G1	490	-4,7%	1,57	-8,2%	0,61	-6,5%	380	-4,3%	1,24	-14,5%	472	-17,9%	5,30%	-18%
G3	482	-6,2%	1,58	-7,6%	0,59	-9,7%	369	-7,1%	1,22	-15,9%	451	-21,6%	5,06%	-22%
G9	480	-6,6%	1,64	-4,1%	0,59	-10,3%	364	-8,3%	1,27	-12,4%	463	-19,5%	5,20%	-19%
G11	469	-8,8%	1,67	-2,3%	0,58	-11,2%	356	-10,3%	1,28	-11,7%	455	-20,9%	5,11%	-21%
G8	470	-8,6%	1,64	-4,1%	0,58	-10,9%	354	-10,8%	1,27	-12,4%	451	-21,6%	5,06%	-22%
G2	472	-8,2%	1,64	-4,1%	0,58	-11,6%	353	-11,1%	1,27	-12,4%	448	-22,1%	5,03%	-22%
G12	462	-10,1%	1,68	-1,8%	0,58	-12,1%	349	-12,1%	1,28	-11,7%	448	-22,1%	5,03%	-22%
G7	457	-11,1%	1,64	-4,1%	0,56	-13,9%	341	-14,1%	1,24	-14,5%	424	-26,3%	4,76%	-26%
G10	454	-11,7%	1,67	-2,3%	0,58	-10,9%	335	-15,6%	1,32	-9,0%	442	-23,1%	4,96%	-23%
G5	451	-12,3%	1,63	-4,7%	0,56	-15,1%	335	-15,6%	1,22	-15,9%	409	-28,9%	4,59%	-29%
G4	442	-14,0%	1,64	-4,1%	0,55	-15,5%	329	-17,1%	1,22	-15,9%	402	-30,1%	4,51%	-30%
G6	439	-14,6%	1,65	-3,5%	0,57	-13,4%	326	-17,9%	1,26	-13,1%	409	-28,9%	4,59%	-29%

Actual PV string wiring

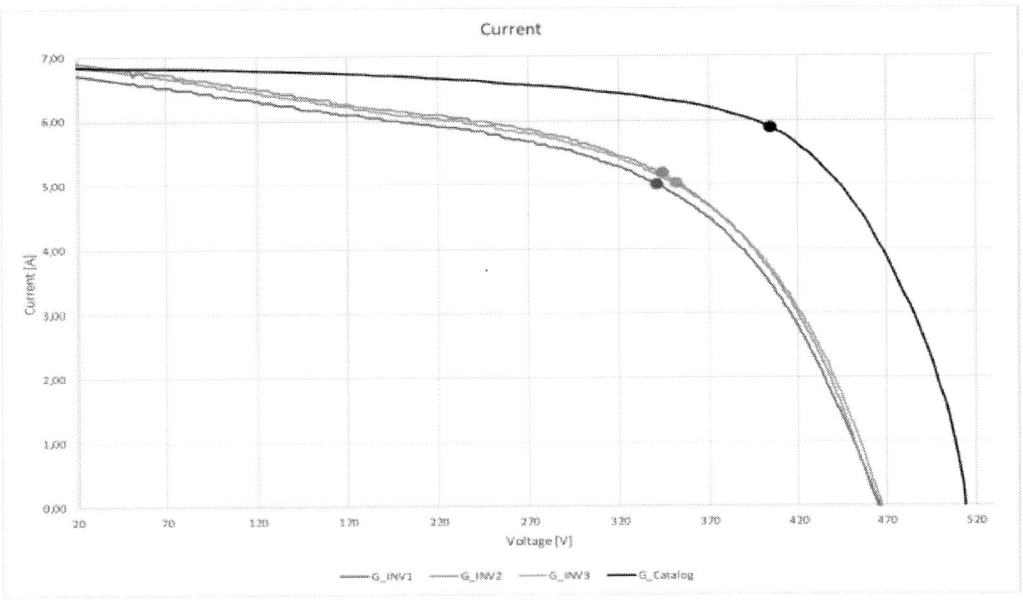

Optimal PV string wiring

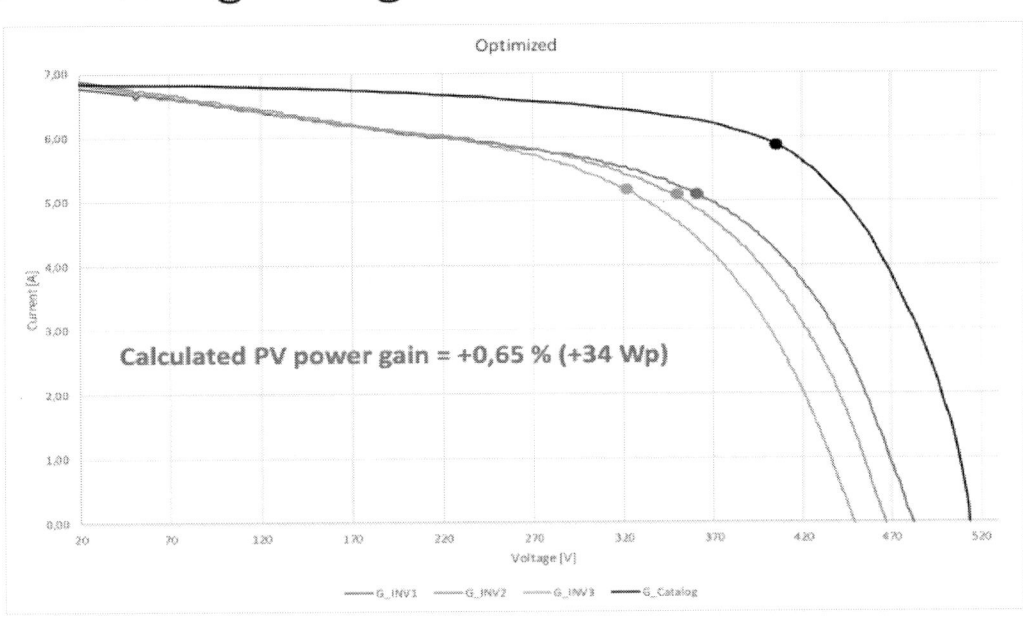

PR$_T$ factor after optimization

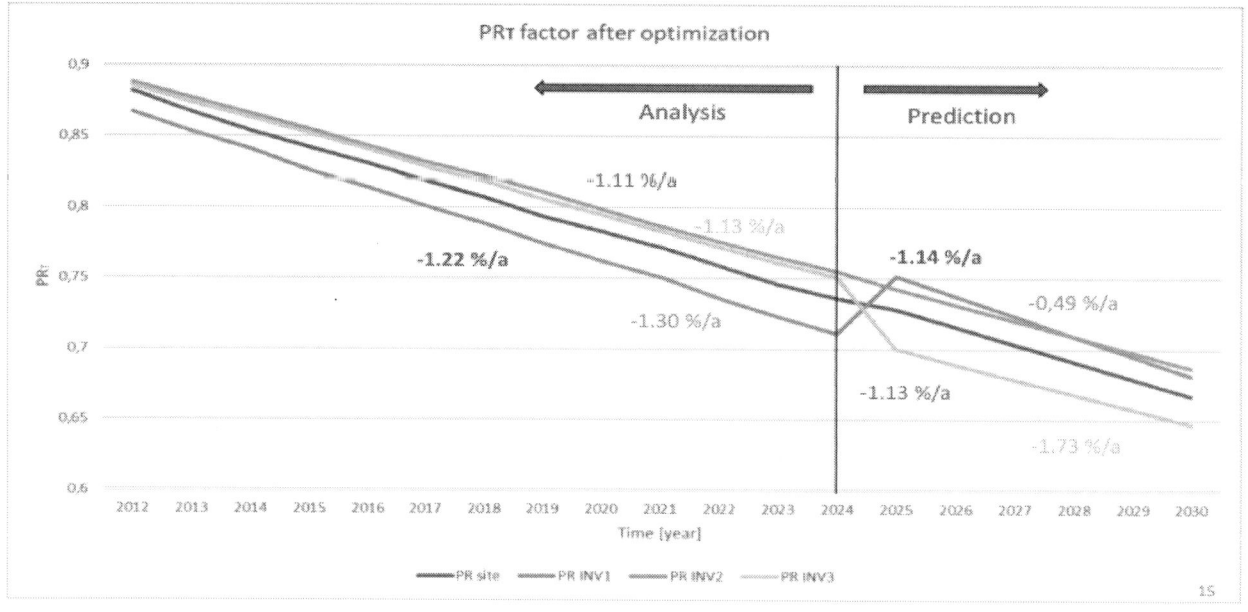

Soiling measurements of PV modules

- PVS **4.6 kWp** 2011-2024
- 20 Si monocrystalline PV modules
- Nominal module power: **230 Wp**
- In operation for over **14 years**
- Without module lever power optimizers
- Traditional 3-phase string inverter

Soiling measurements

- Triple-flash measurement (1, 0.6, 0.2 sun)
- First measurements of soiled PV modules
- Second measurements of cleaned PV modules
- Electroluminescence images of selected PV modules
- IR pictures of the soiling on surface

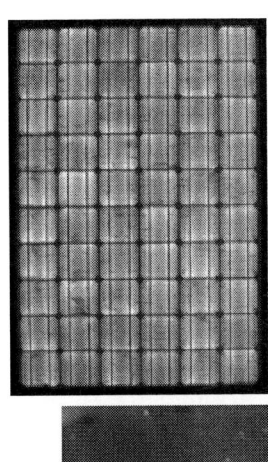

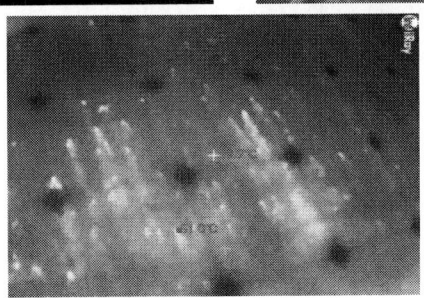

Soiling measurements results

- Pre-Cleaning degradation: **6,23%** (0,45 %/a)
- Post-cleaning degradation: **3,49%** (0,25 %/a)
- Cleaning Pmpp gain: **2,94%**

* Relative to nominal Pmpp (230 Wp) @STC!
* Initial Pmpp power of PV module was not measured!

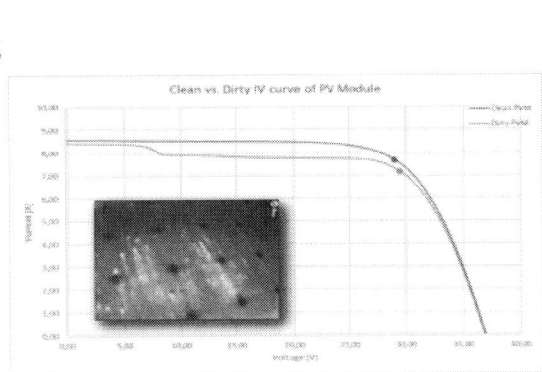
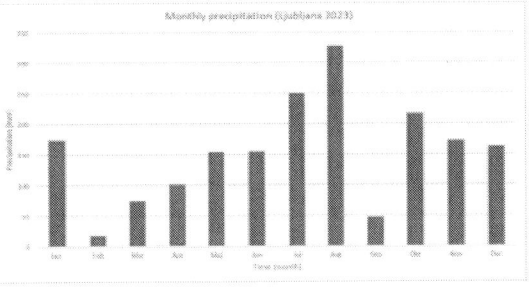
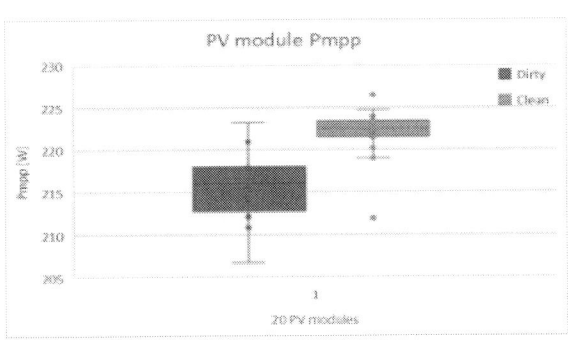

Conclusion

  ChatGPT

- Addressed efficiency losses due to suboptimal orientation and soiling
- Optimized PV string wiring led to a **0.65%** power gain
- Soiling mitigation resulted in a **2.94%** improvement in degraded modules
- Minor adjustments can enhance long-term solar plant performance
- This contributes to sustainable and efficient energy production.

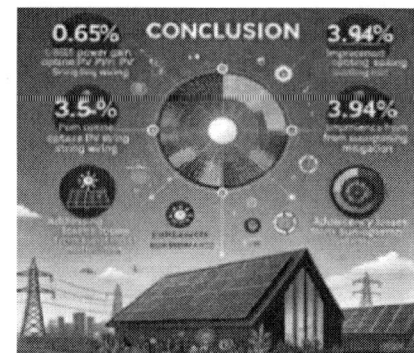

 Thank you!

The Research was funded by the Slovenian Research and Innovation Agency (ARIS), program P2-0415. Ž.M. thanks the Slovene research agency ARIS for funding his PhD program.

41st European Photovoltaic Solar Energy Conference and Exhibition

This presentation was selected by the Sc. Committee of the EU PVSEC 2024 for submission of a full paper to one of the EU PVSEC's collaborating peer-reviewed journals.

QATAR DUST ATLAS PROJECT: DEPLOYMENT OF A NATIONAL FIELD SOILING AND ENVIRONMENTALPARAMETERS MONITORING NETWORK

Brahim Aïssa[1*], Mohamed Abdelrahim[2], Mosab Subeh[1], Amir A. Abdallah[1], Benjamin W. Figgis[1], Juan Lopez-Garcia[1], Veronica Bermudez Benito[1]

[1]Qatar Environment & Energy Research Institute (QEERI), Hamad bin Khalifa University (HBKU), P.O. Box 34110, Doha, Qatar
[2]Bin Omran Trading & Telecommunications, P.O. Box 288, Doha, Qatar
*Corresponding author: baissa@hbku.edu.qa

ABSTRACT: The impact of soiling on PV power production exhibits significant spatial and temporal variability on both large and small scales, even within the radius of a solar power plant [1]. Variations in soiling are prominent across different locations in a country like Qatar, influenced by specific environmental and meteorological conditions. To enhance the accuracy, cost-effectiveness, and reliability of soiling monitoring, a new class of sensors has emerged, leveraging the optical characteristics of dust particles on PV glass. These stand-alone sensors, exemplified by Atonometrics' Mars Soiling Sensor and Kipp & Zonen's Dust IQ, require minimal maintenance and have no moving parts. Since 2013, Qatar environment and energy research institute (QEERI) has become a regional leader in PV soiling research, with focus put on desert environments, exploring both experimental (characterization, innovative measurement methods, mapping) and modeling aspects (sandstorm prediction, dust deposition) [2]. The Qatar Dust Atlas project, carried out at QEERI's Energy Center, is dedicated to evaluating the spatio-temporal mapping of soiling across Qatar and forecasting it using Machine Learning and WRF-CHEM modeling. Key elements addressed in the Dust Atlas encompass a thorough understanding of soiling in Qatar through real world data, incorporating atmospheric aerosol (primarily PM10), field-soiling data, and Albedo [3]. This effort leads to the creation of a spatio-temporal map of dust particles, alongside the development of a land suitability index and a soiling-risk map for prospective PV installations. This paper provides an overview of recent developments and common outcomes from the Dust Atlas project, with a focus on the initiation and establishment of a "National Soiling and Environmental Parameters Monitoring Network". This network integrates monitoring stations and advanced sensors to directly assess soiling and other influential factors, playing a pivotal role in crafting AI-based solutions for forecasting soiling on PV solar panels at diverse spatial and temporal resolutions throughout Qatar.
Keywords: Soiling-ratio, Photovoltaic, Environmental parameters, Space-time variation, Particulate matter

1 INTRODUCTION

Soiling is a multifaceted and ever-changing issue influenced by numerous variables. These include site-specific characteristics, prevailing weather patterns, the tilt angle and orientation of photovoltaic (PV) panels, the material used for panel surfaces, as well as the physical and chemical properties of dust particles [4-7]. Addressing soiling effectively necessitates consistent monitoring since mitigation strategies must be customized to the distinct environmental and operational conditions of each PV installation. A one-size-fits-all approach is impractical in this context. Though reactive solutions like cleaning PV modules are widely available, significant challenges remain in determining the optimal timing, frequency, and methodology for these cleaning interventions.

Alternatively, a more proactive strategy involves gaining a detailed understanding of local dust composition and behavior. This knowledge plays a crucial role in enhancing decision-making processes for selecting ideal PV plant locations and determining the most appropriate PV technologies to implement. To this end, the Qatar Environment and Energy Research Institute (QEERI) is spearheading the development of a Dust Atlas mapping project specifically for Qatar. This initiative aims to comprehensively assess the country's solar energy potential by understanding the impact of dust and soiling.

In 2021, QEERI embarked on establishing a nationwide network of specialized sensors designed to measure soiling effects and all associated environmental factors that influence it. This network covers the entire geographical expanse of Qatar and is intended to provide real-time, high-precision data on dust accumulation and related phenomena. By February 2023, the full installation

of this national sensor network was completed.

How does soiling impact PV efficiency?

Soiling, the accumulation of dust, dirt, or other particles on the surface of photovoltaic (PV) panels, has a significant impact on the efficiency and overall performance of solar energy systems. The main ways in which soiling reduces PV efficiency include:

☐ Reduction in sunlight reaching the PV cells: Dust and other particles block sunlight from fully reaching the photovoltaic cells, which directly reduces the amount of solar energy converted into electricity. Even a thin layer of dust can significantly affect light transmission, lowering energy output.

☐ Scattering and reflection of sunlight: Dust particles on the surface of the PV panels scatter and reflect incoming sunlight away from the cells, further reducing the efficiency of the energy capture process. This effect is particularly pronounced in areas with high dust accumulation or during periods of heavy dust storms.

☐ Non-uniform shading and hotspot formation: Soiling can cause uneven shading on different parts of a PV panel. This can lead to localized heating, known as "hotspots," which can damage the cells and further decrease the system's efficiency. Hotspots also reduce the overall lifespan of the PV modules.

☐ Decreased performance under specific environmental conditions: Soiling can interact with other environmental factors, such as humidity or rain, causing dust to harden or adhere more strongly to the surface of the panels. This can make cleaning more difficult and further impact long-term efficiency. In regions with minimal rainfall, such as arid and desert climates, soiling can become a persistent issue as natural cleaning by rain is rare.

☐ Cumulative energy losses over time: Even a small efficiency loss due to soiling can accumulate over time, leading to significant reductions in overall energy production. The longer a PV system goes without proper cleaning or maintenance, the more pronounced the efficiency drop will be.

In heavily polluted or dusty areas, such as desert regions or industrial zones, the impact of soiling on efficiency can be substantial, leading to energy output reductions of 10% to 40% or more, depending on the severity of the soiling. Regular monitoring and appropriate cleaning strategies are essential to mitigate these losses and ensure the PV system operates at optimal efficiency.

This paper provides an in-depth review of the key findings and results from this project, with particular emphasis on data gathered at QEERI's Outdoor Test Facility (OTF) site. The OTF serves as a focal point for monitoring and analyzing soiling effects in controlled and real-world conditions, offering valuable insights into the broader implications for solar energy generation in Qatar.

2 METHODOLOGY

For the Qatar Dust Atlas project, high-quality monitoring stations have been deployed at 15 strategically selected locations across the country to provide comprehensive data on dust and environmental conditions. Each station is equipped with a combination of 15 Dust IQ sensors and 15 meteorological stations that continuously measure a wide range of parameters, including air temperature, relative humidity, wind speed and direction, and atmospheric pressure. In addition to these environmental metrics, the stations also monitor particulate matter (PM) concentrations at three levels: PM 1, PM 2.5, and PM 10. These stations are capable of transmitting minute-resolution data averages to QEERI servers at intervals of every ten minutes. Importantly, all equipment complies with the IEC 61724-1 standard for photovoltaic (PV) system performance monitoring [8], ensuring high data accuracy and reliability.

Figures 1a and 1b showcase the QEERI Outdoor Test Facility (OTF), which is dedicated to testing and analyzing PV performance in real-world conditions. Figure 1c maps out the 15 sites that make up QEERI's National Dust and PV Soiling Monitoring Network, offering a visual representation of the station locations throughout Qatar. These sites were carefully selected to provide full coverage of the country, ensuring that the distance between any two stations remains below 30 km, thus forming a dense and cohesive network capable of capturing localized variations in dust and soiling.

Each monitoring station consists of a standalone, self-powered system housed on a weather monitoring pole. These stations feature an embedded solution manufactured by Libelium, which integrates efficient sensors for both meteorological and particulate matter monitoring. At the heart of each setup is the Kipp & Zonen DustIQ sensor, mounted at a 22-degree tilt angle to match the inclination of the solar PV modules at the OTF. This configuration allows the DustIQ to deliver precise measurements of the soiling effect on solar panels. The sensor is powered by an adjacent photovoltaic panel and supported by a deep cycle battery, ensuring continuous data collection, with one sample recorded every minute. The collected data is automatically transmitted to QEERI's server infrastructure, located at the HBKU data center, where it undergoes further analysis.

To manage and store this influx of data, QEERI employs a robust server architecture that utilizes a local PostgreSQL database to store the logger data. This database is then replicated to another QEERI server for redundancy and data security. Additionally, QEERI is developing a web application on the main server to facilitate real-time visualization and analysis of the soiling data. This platform will display information in a variety of formats, including maps, graphs, and tables, enabling users to easily interpret the data for different applications.

Furthermore, an additional set of 15 albedometers is being installed throughout the country, with 5 stations already operational. When fully deployed, these albedometers will work in conjunction with the DustIQ and meteorological sensors to provide a ground-calibrated, highly accurate map of solar energy resources in near real-time. This map will offer high spatial resolution, giving detailed insights into the soiling impact and solar resource distribution across Qatar [9-12].

All monitoring stations are designed to measure daily soiling ratios along with meteorological parameters such as temperature, humidity, pressure, wind speed and direction, and particulate matter concentrations (PM 1, 2.5, and 10). While most of the stations were commissioned in February 2023, the OTF site has been operational since 2013, making it an invaluable long-term source of soiling data and trends in the region. The combination of data from these stations will allow for a detailed and real-time assessment of soiling effects, which will be critical for optimizing the performance of PV systems in Qatar.

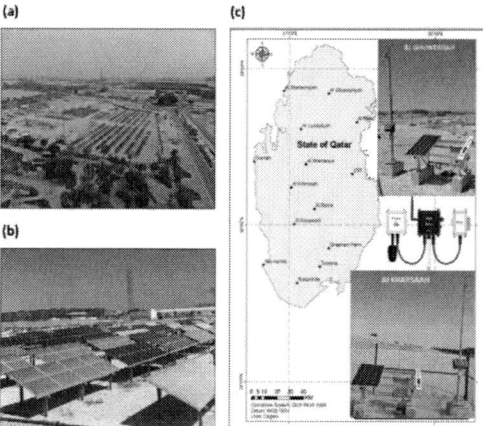

Figure 1: QEERI Outdoor Test Facility's PV testing area (a & b). (c) The 15 sites of QEERI's National Dust and PV soiling monitoring Network. Inset in (c) are typical photos of the Dust IQ sensor along with meteorological and PM sensors.

3 RESULTS AND DISCUSSION

At the QEERI Outdoor Test Facility (OTF) site, Figure 2a illustrates the temporal variations in atmospheric aerosol concentrations (specifically PM10) recorded from July 2021 through January 2024. During this two-and-a-half-year period, the daily mean PM10 concentration was calculated to be $100.36 \pm 22.13 \, \mu g/m^3$. Notably, the data shows that PM10 levels were higher during the winter and fall seasons compared to the summer months, indicating a seasonal variation in dust accumulation, which could be attributed to changes in weather patterns, wind direction, and humidity levels [13-16].

In Figure 2b, the annual average albedo measurements are depicted, providing valuable insights into surface reflectivity at the OTF site. The albedo, a measure of how much solar radiation is reflected by a surface, exhibited a consistent yearly average value of 0.45 over multiple years. This stability suggests minimal variation in surface properties that could impact the overall energy balance at the test facility. A complementary histogram (Figure 2c) further emphasizes the distribution of albedo values, with the data being tightly centered around this average, reinforcing the consistent nature of the site's reflective characteristics.

Figure 2d focuses on the mean soiling rate observed over a 24-month period, indicating an average soiling rate of 0.69 ± 0.03%. This value represents the reduction in PV panel efficiency due to dust accumulation. Over the two years, the soiling rate remained relatively stable, though Figure 2e, which presents a histogram of the daily soiling rates over three consecutive years (2021, 2022, and 2023), reveals a noticeable upward trend. The data shows a gradual increase in the mean daily soiling rates, rising from 0.5% in 2021 to 0.68% in 2022, and reaching 0.69% in 2023. This progressive increase may reflect either a change in environmental conditions or a shift in dust accumulation patterns over time.

These findings are critical for understanding the long-term impact of dust on solar energy production at the OTF site and for developing targeted mitigation strategies to maintain PV panel performance in Qatar's challenging environmental conditions.

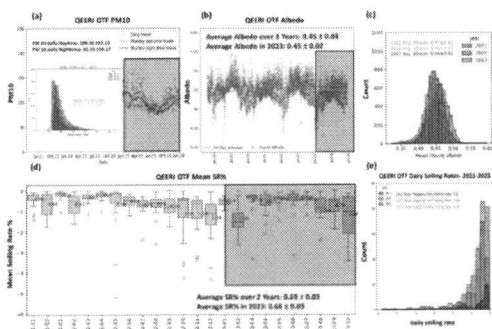

Figure 2: OTF site: (a) variation of the PM 10, and (b) Albedo over time. (c) Distribution histogram of the Average Albedo for the year 2021, 2022 and 2023. (d) Variation of the mean soiling rate % from November 2021 to November 2023. (e) Distribution histogram of the mean SR% for the year 2021, 2022 and 2023.

Figure 3 presents detailed insights into the spatial and temporal variations of soiling and albedo across Qatar, using data from the QEERI National Dust and PV Soiling Monitoring Network.

Fig. 3a displays a Geographic Information System (GIS) map showing the daily soiling ratio percentage across Qatar, derived from field measurements collected by the DustIQ network throughout 2023. This assessment spans from the commissioning of the National Monitoring Network in February 2023 to the end of December 2024. The map highlights variations in soiling rates across different regions, underscoring the site-specific nature of soiling losses. Notably, the OTF site, situated in an area marked by urban development and high traffic, recorded an extreme soiling rate of 0.69% per day. This high rate of soiling can be attributed to the increased dust and

particulate matter generated by human activity in the vicinity. The map emphasizes that soiling rates are highly dependent on geographic location and local environmental factors, reinforcing the need for tailored mitigation strategies based on regional conditions [17-21].

Fig. 3b presents the yearly evolution of albedo values, collected from several monitoring stations equipped with albedometers. Albedo, which measures the reflectivity of a surface, plays a crucial role in determining the amount of solar radiation absorbed by PV panels. The albedo measurements reveal a consistent trend across all monitoring stations: albedo values are generally higher in the second half of the year (June to December) compared to the first half (January to June). This seasonal variation suggests that environmental factors such as dust accumulation and changes in surface properties may be influencing the albedo levels, possibly due to increased dust deposition during the drier months.

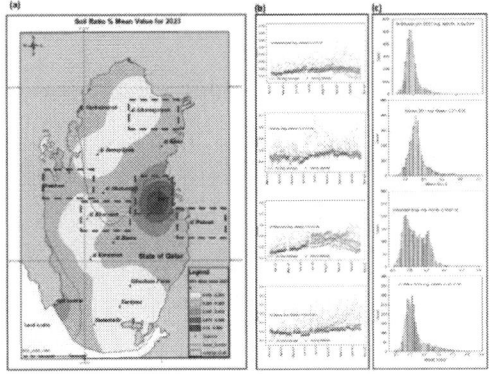

Figure 3: (a) Field measurement of the daily soiling ratio percentage in Qatar in 2023. (b) Variation of the Average Albedo and (c) the associated distribution histogram.

The yearly average albedo values for different locations vary, with the lowest recorded value of 0.37% in Dukhan, located on Qatar's west coast, and the highest value of 0.45% at the OTF site. This range indicates regional differences in surface reflectivity, which could be influenced by factors such as land use, vegetation cover, and dust deposition patterns. The consistently higher albedo values at the OTF site may also be linked to the urban nature of the area, which could result in higher reflectivity due to buildings and paved surfaces [22].

Fig. 3c provides a distribution histogram of albedo values over the year, showing the spread and frequency of albedo measurements at the various monitoring stations. This histogram confirms the stability of albedo patterns across the monitoring period, with values clustering around site-specific averages. These data sets will soon be cross-referenced with particulate matter (PMx) data, which is currently being processed and is expected to offer further insights into the relationship between PM concentrations and albedo variations. The processed PMx data will be available by September 2024 for presentation at an upcoming conference. This correlation will likely shed light on how particulate matter contributes to both soiling and changes in surface reflectivity, offering a more comprehensive understanding of the environmental factors affecting PV performance in Qatar.

Through these detailed assessments, Figure 3 underscores the importance of understanding localized environmental conditions in managing soiling and optimizing the performance of solar energy systems. By continuously monitoring both soiling and albedo, QEERI is working towards developing more effective strategies for mitigating losses and enhancing the efficiency of PV installations across Qatar.

4 CONCLUSIONS

For upcoming PV plant projects in desert environments, we emphasize that the key to effective soiling mitigation lies in the strategic selection of the PV site location. Backed by the recent deployment of the "National Soiling and Environmental Parameters Monitoring Network," QEERI is committed to providing continuous, dynamic geographical and seasonal mapping of dust across Qatar. This extensive analysis not only monitors PV soiling but also includes factors such as albedo, airborne particulate matter (PM), and various meteorological elements, all of which are critical to understanding their collective impact on PV performance.

The ultimate goal of this effort is to develop a land suitability index, and a soiling-risk map tailored for future PV installations. These tools will help identify optimal locations for PV plants, considering site-specific environmental conditions and dust accumulation patterns. By utilizing the vast experimental data gathered through this ongoing initiative, QEERI aims to provide reliable input for forecasting models that will help optimize cleaning schedules for PV modules.

This data-driven approach offers valuable insights to PV deployment companies, addressing critical concerns such as maintenance planning and site selection. Historical data are being rigorously analyzed, and forecasting tools are under active development to produce a dynamic rendition of the Qatar Soiling Atlas. This evolving product will guide the improved positioning of PV plants in relation to soiling risks, ensuring that installations are both efficient and cost-effective over time. The advanced mapping and forecasting capabilities offered by QEERI will allow stakeholders to make informed decisions, enhancing the long-term performance and sustainability of solar energy systems in desert environments.

References

[1] K. Ilse et al., Joule, 3, 2303, 2019.
[2] "IEC 61724-1 Ed. 1.0 en:2017 - Photovoltaic system performance - Part 1: Monitoring," International Electrotechnical Commission (IEC), 2017.
[3] Y. Cui, J. Xiao, J. Xiang, J. Sun, and E. M. Vitucci, Frontiers in Energy Research, 9, 1, 2021.
[4] M. Gostein, B. Littmann, J. R. Caron, and L. Dunn, in Conference Record of the IEEE Photovoltaic Specialists Conference (PVSC), pp. 3004–3009, 2013.
[5] M. Gostein, T. Duster, and C. Thuman, in IEEE 42nd Photovoltaic Specialists Conference (PVSC), 2015.
[6] S. Kagan et al., in 2018 IEEE 7th World Conference on Photovoltaic Energy Conversion (WCPEC), 3432–3435, 2018.
[7] A. Marquis, M. Gostein, and B. H. King, in 2022 IEEE 49th Photovoltaics Specialists Conference (PVSC), 2022, pp. 0291–0294.
[8] "Photovoltaic System Performance – Part 1: Monitoring," IEC Standard 61724-1, 2017.
[9] G. Hassan, G. Pozza, F. Lettner, and M. Kober, Energy Procedia, vol. 155, 292, 2018.
[10] D. Bello et al., Renewable and Sustainable Energy Reviews, 112, 805, 2019.
[11] A. Kazemian, H. C. Hottel, and R. S. Dani, Solar Energy, 203, 74, 2020.
[12] C. Drews, C. Chokani, and R. S. Abhari, IEEE Journal of Photovoltaics, 8, 1778, 2018.
[13] I. Javed, M. B. Mansoor, M. Siddiqui, and H. S. Arif, Renewable Energy, 160, 211, 2020.
[14] S. Rajput, T. N. Singh, Energy Conversion and Management, 173, 705, 2018.
[15] T. Mekhilef et al., Solar Energy Materials and Solar Cells, 220, no. 110907, 2021.
[16] S. A. Kalogirou, Renewable Energy Reviews, 119, 503, 2020.
[17] E. Abad, J. Muñoz, A. Louroiro, and R. Streckiene, Energy, 172, 898, 2019,
[18] A. J. Toor, R. Sharma, and M. Shaikh, Environmental Research Letters, 15, 105, 2020.
[19] H. Shiraiwa et al., Energy Procedia, 147, 73, 2020.
[20] A. Bernecker et al., Renewable Energy, 150, 725, 2020.
[21] Gavi, N. M. H. et al., Journal of Magnetism and Magnetic Materials, 324(6), 1172, 2012.
[21] El Khakani, M. A., Le Borgne, V., Aïssa, B., Rosei, et al. Applied Physics Letters, 95(8), 2009.
[22] Habib, M. A., Barkat, M., Aissa, B., and Denidni, T. Progress In Electromagnetics Research, 88, 135, 2018.

41st European Photovoltaic Solar Energy Conference and Exhibition

This presentation was selected by the Sc. Committee of the EU PVSEC 2024 for submission of a full paper to one of the EU PVSEC's collaborating peer-reviewed journals.

QUALITY ASSURANCE FROM LABORATORY TO FIELD: NOVEL TEST SOLUTIONS FOR SOILING-PRONE PV SYSTEMS

John (Ioannis) A. Tsanakas[1], Rodrigo Moretón[2], Eric Pilat[1], Jorge Solórzano[3], Kévin Garcia[5]

[1] CEA, Liten, Univ. Grenoble Alpes Campus INES, 73375 Le Bourget du Lac, France
[2] Qualifying Photovoltaics (QPV), 28031 Madrid, Spain
[3] Entec Solar, 28033 Madrid, Spain
[4] Compagnie Nationale du Rhône (CNR), 69004 Lyon, France

ABSTRACT: Site-specific quality assurance, from laboratory to the field, is crucial for enhancing the performance and reliability of photovoltaic (PV) plants, especially in challenging environments with stress/loss factors like soiling, ultraviolet (UV) radiation, humidity or moisture ingress. The impact of soiling on power losses in PV installations, particularly in arid/desert areas, marine environments, and areas near mines or intense agriculture, underscores the necessity for proactive monitoring and quality control for such PV plants. Existing commercial soiling test solutions are not standalone devices while their output data (e.g. soiling ratio measurements), do not directly correlate with the overall impact of soiling on PV plant's power output. In this paper, we present the development, prototyping, validation and first field results of two innovative solutions, "E-Dust" and "SoilRatio" developed in the framework of H2020 SERENDI-PV, leaping beyond the state-of-the-art in soiling-specific PV quality control. The advanced in-field assessments of soiling provided by these two novel tools are further complemented by a lab-based setup and protocol, developed by CEA, for (controlled) soiling testing, including PV design and cleaning optimization.

Keywords: PV modules; PV systems; PV quality assurance; soiling; PV operations and maintenance; soiling losses.

1 RATIONALE and AIM

Site-specific quality assurance plays a pivotal role in improving performance and reliability of PV plants, while optimizing their O&M, particularly for installations susceptible to challenging environmental conditions and stressors such as soiling buildup, ultraviolet (UV) exposure, humidity or moisture ingress [1-3].

Soiling is an issue commonly confronted in the O&M context, especially for PV installations in certain regions and (micro)climatic profiles, i.e. prone to dust or air pollution, humidity, water/salt mists, organic matter and seasonal effects such as pollen and airborne sand. These conditions are experienced notably in arid/desert areas but also in marine/saltwater environments (e.g. floating PV) or sites in the proximity of mines or intense agricultural activity (as in the case of agrivoltaics) [4]. Return-of-experience (REX) from such soiling prone PV plants is highly suggestive of significant associated power losses, emphasizing the need for proactive – if not predictive – monitoring of soiling and quality control in the field; to be leveraged, in turn, for optimizing soiling mitigation plans. In addition, PV systems at the proximity of water surfaces can be subject to relatively more aggressive combined levels of humidity/moisture, salt mist spraying and UV radiation (due to reflectance).

In this context, advances in site-specific qualification protocols and testing solutions are in the R&D spotlight, being indispensable for establishing rigorous quality assurance and environmental stress-resilience of PV plants, from material/design (in-lab testing) level to system operations (field-testing). In the EU-funded H2020 project SERENDI-PV [5], we have been striving to address exactly such needs of the PV industry for quality control and assurance in emerging PV technologies (such as bifacial or

floating PV) and site-specific conditions.

In this paper, we present the development, prototyping, validation and (upcoming) demonstration results of two innovative solutions, beyond the state-of-the-art in soiling-specific PV quality control:

1. "E-Dust", a novel in-field soiling test kit for PV plants, developed (and soon to be commercialized) by QPV;
2. A soiling testing framework by CEA, encompassing an in-lab (controlled) soiling testing protocol and "SoilRatio", a novel setup for in-field PV soiling monitoring and evaluation, based on patented features [6].

2 METHODOLOGY

2.1 Concept and novelty of "E-Dust" soiling test kit

The most common methods for estimating electrical energy losses due to soiling are to use optical sensors or to compare the short-circuit current of two solar devices (clean vs dusted). These approaches may be sufficient to calculate soiling irradiance losses, but are not designed to address the real effective power losses that soiling causes in PV modules. In fact, as soiling deposition is dominated by prevailing winds, dust patterns are usually non homogeneous (Fig. 1, left and middle). This kind of dust distributions create "steps" in the I-V curve of the modules, thus affecting the maximum power point, PM, more than the ISC (which can be understood as the irradiance affection). As an example, Fig. 1 (right) presents the I-V curves of the PV modules, before and after being cleaned. Power output (P_{out}) loss is 30% higher than short-circuit current (I_{sc}) loss.

41st European Photovoltaic Solar Energy Conference and Exhibition

Figure 1: Left and Middle: The most usual soiling deposition in a PV plant, dominated by prevailing winds. Right: This type of soiling causes "steps" in the I-V curve of the modules (when comparing the two I-V curves before and after cleaning), increasing the soiling impact, as showing in this example case.

To address such limitation and quantify the real (effective) PV power losses due to soiling, QPV has worked on a novel soiling sensing approach, implemented today under the name "E-Dust". E-Dust is a soiling sensor based on the use of two PV modules (clean and dusted) from the same batches, installed in the same structures (PV arrays) and exposed to the same wind and tracking conditions as the rest of PV modules that make up the PV plant. Again, the full I-V curve of both modules is simultaneously and periodically measured. As a comparative measure is carried out, no specific calibration is needed, as the system automatically calibrates itself when both modules are cleaned.

2.2 Concept and novelty of "SoilRatio" soiling test kit

CEA-INES has been working, over almost a decade, on evolving a concept for autonomous in-field monitoring and precise quantification of the soiling rate, including soiling losses evaluation, in operating PV plants. As of today, most commercial soiling test kits employ two devices or "sensors", one being a (standard) PV module and the other, so-called reference, typically being either a single-cell PV module or a PV mini-module. Other commercialized concepts involve a photodiode optical sensor that has to be calibrated with the target type of soiling. These sensors are preferably mounted on the same frame, with the (pre)requisite that the reference device/module is regularly cleaned. None of the existing (commercial) soiling test kits can be considered as stand-alone device for soiling measurements and assessments in PV plants. Besides the output data (soiling ratio measurements) generated by such solutions are are not directly associated with the overall (potential) impact of soiling on the power output (i.e. soiling losses) at PV plant level; hence, requiring post-treatment and interpretation.

CEA's soiling testing concept proposes a "intervention-free" approach, i.e. a stand-alone setup of two devices/sensors, of whice one is constantly kept under a protective cover, throughout the measurement intervals, thus remaining perfectly clean before each soiling measurement. This way, the sensor does not require any cleaning, which guarantees optimal accuracy, with barely any need for interevention, thus minimal operating or maintainance costs for the whole setup.

Following its proof-of-concept (PoC) and preliminary demonstration stages, an important milestone was recently reached, in the framework of SERENDI-PV project, with the manufacturing and validation of the final version of the industrial prototype, called "SoilRatio" (Fig. 2). In its hardware and firmware, the current SoilRatio prototype incorporates three features beyond the state-of-the-art:

1. Fully autonomous (i.e. stand-alone, self-powered and self-calibrated) monitoring of soiling ratio in the field;
2. Algorithm for post-treatment (analysis) of soiling data;
3. Precise qualification of the soiling type and quantification of associated power losses.

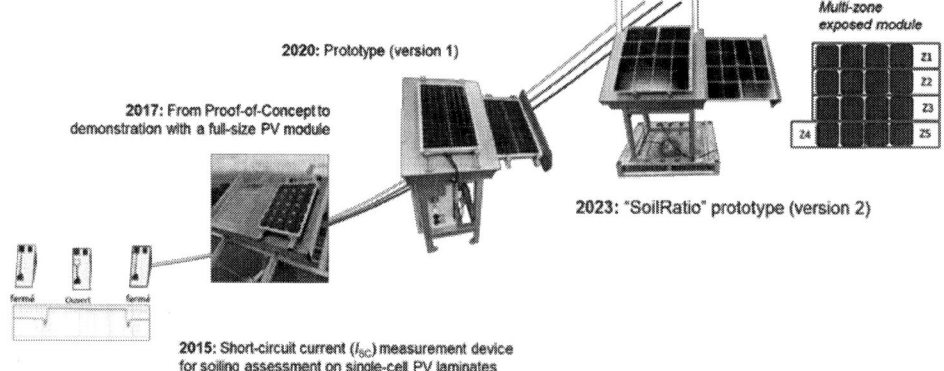

Figure 2: The development of SoilRatio, from PoC to today's prototype towards demonstration.

The SoilRatio setup integrates a customized multi-zone PV module (overview in Fig. 2, upper right; detail in Fig. 3), each zone being an independent sensor, a configuration that allows precise qualification of the soiling type (e.g. uniform/localized). The soiling data analysis is based on the post-treatment of the multi-zone

1161

measurements and their extrapolation to PV array, up to PV plant level. For the latter, SoilRatio's algorithm takes into account the electrical architecture of the PV installation, to then calculate the PV plant's power output losses associated to soiling. Both the multizone PV module (and its electrical configuration) concept and the self-calibration procedure are features patented by CEA.

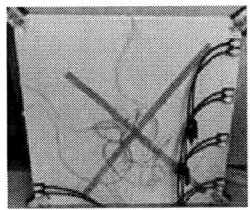  

Figure 3: Implementation and schematic of the "multizone module" electrical architecture, employed in the current version of SoilRatio prototype.

2.2.1 Quantification of the impact of module fouling/soiling with the SoilRatio measurement kit

The methodology consists of comparing the short-circuit current of a PV module-zone exposed to the ambient (thus soiling contamination) conditions, with that of a PV module-zone that is covered insider a drawer (thus protected from soiling) for the time interval between two soiling test measurements. The protected PV module is exposed to soiling only for the minimum time necessary for a measurement. This comparison of short-circuit currents is expressed by the parameter called "Soiling Ratio" defined by the following formula:

$$SR = \frac{Isc_{corr\_exp}}{Isc_{corr\_prot}}$$

The current measurements are normalized to obtain STC (Standard Test Conditions) corrected values. The correction uses PV module temperature values measured (also within the SoilRatio kit) for each PV module-zone. The PV module-zones integrated in the protected panel are used as references. We therefore consider that the irradiance ratio is equivalent to the current ratio on each of these protected PV module-zones.

$$Isc_{corr\_prot} = (1 + \alpha(Tmod_{prot} - 25)) \times Isc_{stc\_prot}$$

$$Isc_{corr-exp} = Isc_{mes\_exp}(1 + \alpha(Tmod_{exp} - 25)) \times \frac{Isc_{stc\_prot}}{Isc_{mes-prot}}$$

We thus obtain the simplified formula for the calculation of the Soiling Ratio:

$$SR = \frac{Isc_{mes\_exp}(1 + \alpha(Tmod_{exp} - 25))}{Isc_{mes\_prot}(1 + \alpha(Tmod_{prot} - 25))}$$

2.2.2 Overview of the auto-calibration process

SoilRatio's self-calibration process, defined and patented by CEA research team, is based on the observation that two identical PV modules in the same environment do not generate the same short-circuit current (I_{sc}) and that, most importantly, this difference varies linearly with the angle of incidence (AOI) of the light beam. Hence an additional correction to be made to the short-circuit current of the exposed module, as follows:

$$Isc_{Tot\_Corr\_exp} = Isc_{corr\_exp} - (a\theta + b) \times {Isc_{prot}}/{100}$$

This affine function model is verified for absolute AOI values below 35°. The procedure consists in carrying out

the measurements during several hours in order to cover the widest range of angles and that during a clear-sky day (zero cloud coverage).
Three conditions ("filters") are applied, as follows:
1. Temperature difference between exposed and protected modules < 75%.
2. Current measured on each PV module zone > 4A
3. Absolute value of the angle of incidence < 35°.

Indicatively, Table 1 shows the result of the calibration performed at INES outdoor PV testing premises (N 45.6419 E 5.8753 ; Le Bourget-du-Lac, France), on the 15th February 2023. The linear regression model is confirmed by sufficiently high correlation coefficients.
The regression coefficients are low, which shows that the difference generated by the AOIs is negligible. Taking into account the "offsets" allows correcting significantly the soiling ratios between 0.3% and 2%. On that period of the year, the sun (solar altitude) is still considerably low, hence the AOI in relation to the surface of the sensors inclined at 30° is at least 28°. It is therefore strongly recommended to repeat a calibration closer to the summer solstice (i.e. between late May and late June, for the northern hemisphere), when days are longest, in order to benefit from a wider range of AOIs and therefore a better accuracy on the regression coefficients and offsets.

Table 1: Calibration results for the SoilRatio, carried out at INES premises, on the 15/02/2023.

	Zone 1	Zone 2	Zone 3	Zone 4	Zone 5
Coef. Reg.	0.006	0.007	0.006	0.013	0.014
Offset	-1.178	-0.868	0.355	1.983	1.685
R²	0.851	0.899	0.895	0.911	0.903
SR before	0.989	0.996	1.009	1.022	1.018
SR after	1.001	1.005	1.006	1.002	1.001

2.2.3 Improving measurement accuracy by identifying the smallest AOI

As mentioned earlier, the angle of incidence can cause a deviation in the measurement. Even if this is corrected by the calibration process, it is still preferable to minimize it as much as possible by carrying out the measurements at the time when the AOI is the lowest, for the specific location and day of the year. To identify these moments, a calculation is performed by integrating the parameters entered at the time of installation, namely Global Positioning System (GPS) coordinates, tilt, orientation and altitude. The results are shown in Fig. 4. Once calculated, the optimal opening times were made available to the SoilRatio kit which took them into account when it was set at "automatic measurement" mode on 28/02/2023.

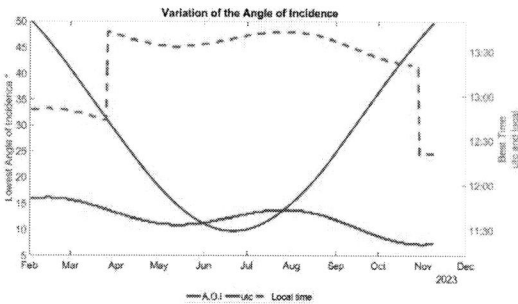

Figure 1: Minimum incidence angle time for each day, for the period from February 2023 to February 2024.

2.3 In-lab setup for controlled soiling and cleaning tests

In parallel to SoilRatio prototype's delvelopment, we have also designed a complete in-lab protocol for controlled replication of soiling and cleaning tests, for different bill of materials (BOM) of PV modules, as well as soiling type characterization/analysis.

Two specific test benches developed and operating at CEA are employed in the designed protocol (Fig. 5): a soiling chamber (Fig. 5, up) and a cleaning chamber (Fig. 5, down). For all designed tests, the impact of soiling is quantified by means of electrical (I-V) characterization measurements of the studied PV modules/laminates, at STC, carried out in a Class A+A+A+ PASAN solar simulator.

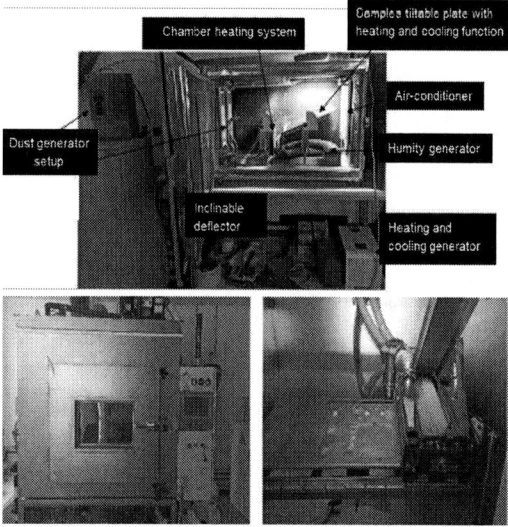

Figure 5 : The lab setup at CEA-INES developed and employed for controlled replication of soiling, cleaning tests and dust characterization. Up: the soiling chamber and its different components. Down: External and inside view of the cleaning chamber.

The soiling chamber, with its dust generator, allows to carry out a homogeneous and repeatable soiling of the modules. A precise mass of dust previously dehydrated is placed in one piston. This piston is part of the dust generator, which allows a good control of the injected dust volume flow. Carried by dry air under controlled pressure, this dust arrives on a deflector thus generating a dust cloud inside the chamber. The suspended dust will then settle on the modules installed on the sample holder plate, which can also be tilted between 0° and 90°. As such, a complete

and controlled soiling operation (or "injection") is concluded. Different devices, also components of the soiling chamber, allow adjusting key parameters, such as the temperature and the humidity for each test. Specifically, the interior temperature can be adjusted between 15°C and 50°C, that of the sample holder plate between 10°C and 50°C, while relative humidity can be set up to 90%.

The studied samples, also fabricated at CEA-INES facilities, comprise of "mini PV modules" (i.e. single-cell PV laminates), using solar cells of silicon heterojunction (SHJ) technology, in different bill of materials (BOM) combinations, i.e. different layouts of front cover coatings and encapsulants. More details on the tested scenarios are given in the results, Section 3.3. Indicatively, Fig. 6 shows one of the tested single-cell PV laminates, before (clean state) and after a soiling injection (soiled state).

The dust (soiling) type and sample used for the study originates from actual soiling collected in the field, in April 2023, in a utility-scale PV plant site located in southern France. This site has been identified by CNR (SERENDI-PV partner and PV plant owner) as susceptible to soiling due to its proximity to a quarry as well as to significant agricultural activity, both generating potentially important levels of soiling, in a region with generally little rainfall ("hot dry-summer" climate, classified as *Csa*, per the Köppen climate classification). A total 12 kg of dust/soiling was collected and sieved with a 1.6mm mesh, to be then used for the experiment.

Figure 6: Example of one the studied PV laminates-samples, before and after being subjected to a soiling injection cycle.

We have adjusted parameters taking into account the onsite real conditions and the limits of the artificial soiling equipment:

- Scenario 1: the deposit of dust occurs when the panel is wet due to high relative humidity (RH>70%) and because the panel is cooled (Panel T° below 17°C and chamber T° above 22°C).

- Scenario 2: the deposit occurs when the PV module and chamber are at 30 °C and humidity is below 30%.

The entire experimental sequence applied in this study comprises four main steps:

1. Measurement (I_{sc} and P_{max}) of a clean PV laminate-sample, which serves as reference,
2. Soiling injection,
3. Measurement,
4. Cleaning by hand; return to a completely clean module.

Through the designed testing protocol, overall aim of the specific study was to assess the impact of: i) different materials selection (BOMs) and ii) different tilt angles during injection, on their proneness of PV modules to soiling buildup (and therefore soiling losses).

3 FIRST RESULTS and DISCUSSION

Figure 7 shows the installation of an early E-Dust prototype in a PV plant, together with the clean and dirty PV modules to which it is connected.

Figure 7 : (a) The installed E-Dust prototype, (b) the PV array with a cleaned and dirty modules to which E-Dust is connected.

Figure 8 shows the evolution of the clean and dirty modules' power along a day, as measured by two E-Dust prototypes installed in the corresponding PV plants in Portugal and Mexico. Beyond the performance anomalies (as the tracking deviation observed in Figure 8-top) significant differences in the soiling ratio are observed among different hours of the day (thus, among different incidence angles). A key conclusion in this area is the evolution of power losses due to soiling over the course of the day. Several analyses have shown that hourly losses increase at lower incidence angles, therefore, this effect is more pronounced at the beginning and end of the day. As Figure 8-bottom shows, during the first two hours of the day as irradiance increases losses also increase, reaching maximum values well over 10%. At the point of most direct impact, losses are at their minimum. Once more in the evening, the losses escalate. These results make the case for continuous measurements, in order to obtain a more accurate soiling loss estimation.

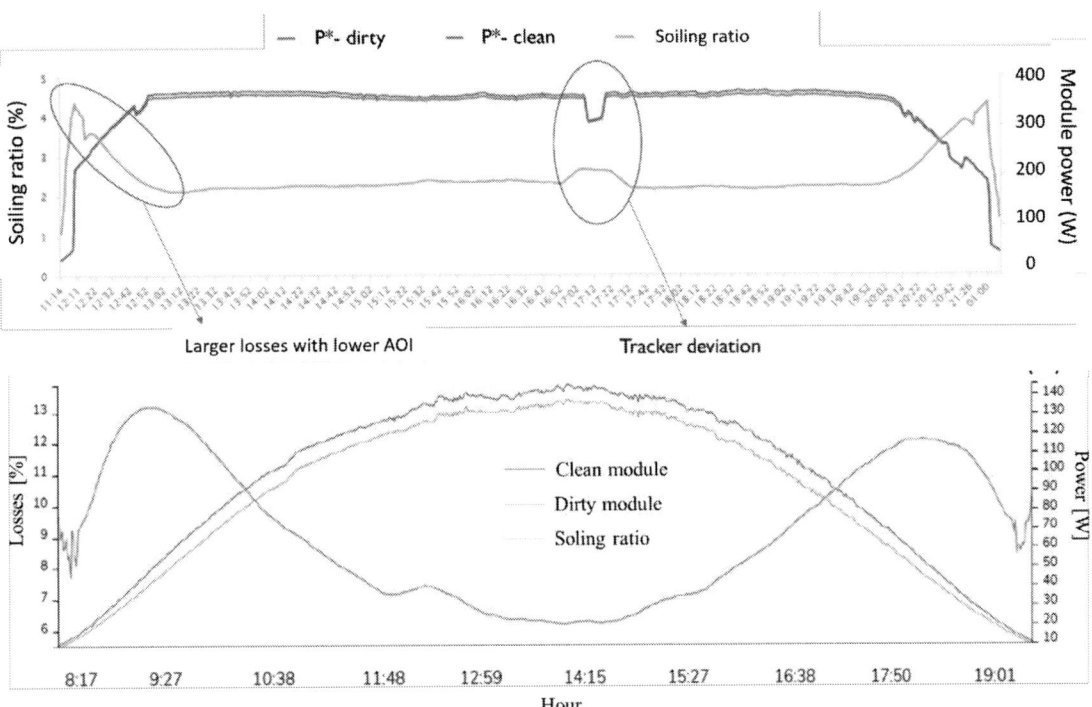

Figure 8 : Day-long evolution of clean/dirty modules' power and soiling ratio as measured by E-Dust prototypes installed in Portugal (top) and Mexico (bottom).

Furthermore, Fig. 9 and Fig. 10 present the daily evolution of the soiling ratio and the energy losses due to soiling as measured by two E-Dust prototypes installed in two PV plants in Portugal and Chile, respectively. Together with the soiling ratio, real energy losses are calculated based on the PV plant production. Expected interannual effects can be observed, as the impact of the rainy season in the cleanliness of the PV modules (Portugal), together with other more local effects, as the self-cleaning effect of the windy season (Atacama desert, Chile). The obtention of

this information paves the way to optimize the cleaning campaigns (frequency, cleaning method, etc) and, therefore, the final performance of the PV plant. For example, it does not make sense to clean the PV modules at the end of the summer (Portugal case) given the cleaning effect of the rainy season coming right afterwards. Instead of that, accomplishing a cleaning campaign early August would have reduced the overall energy losses due to soiling by between 20% and 30%.

41st European Photovoltaic Solar Energy Conference and Exhibition

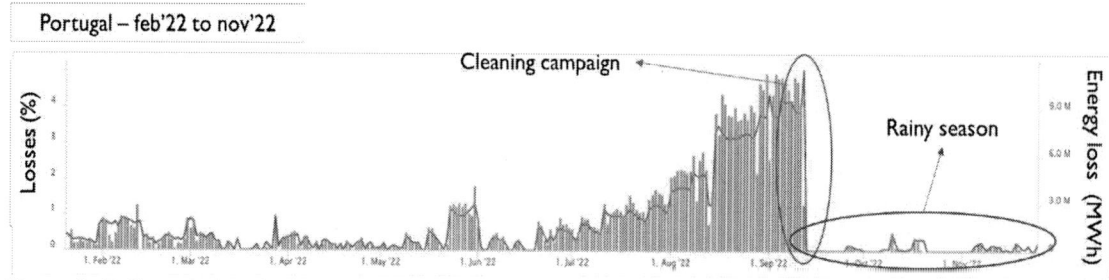

Figure 9 : Daily evolution of soiling ratio and soiling losses as measured by an E-Dust prototype installed in a Mediterranean area (Portugal).

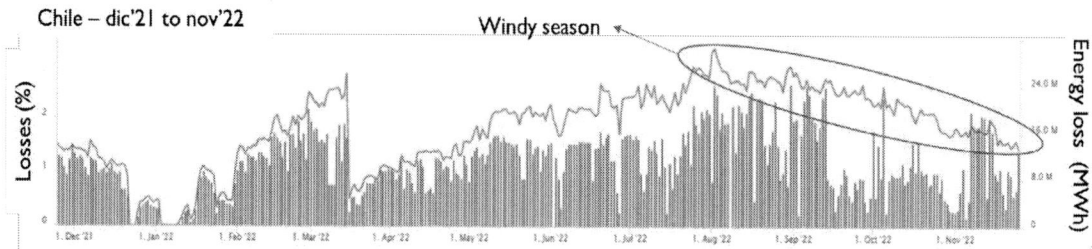

Figure 10 : Daily evolution of soiling ratio and soiling losses as measured by an E-Dust prototypes installed in the Atacama desert (Chile).

3.2 SoilRatio validation / test campaign

A preliminary validation/test campaign of SoilRatio's post-treatment functionalities (especially regarding the post-treatment and soiling loss calculation) has been carried out in the Q4 of 2023 at the outdoor test PV facilities of INES campus (Le Bourget-du-Lac, France). For the calculation of daily soiling ratios, the developed algorithm uses three different parameters/formulas: i) the average of the measured current ratios, ii) the median of the measured current ratios and iii) the linear regression coefficient. In order to facilitate the soiling type diagnosis, the algorithm have been developed so as to treat multiple indicators e.g. all measurement points of I_{sc}, temperature, etc. Then, similarly to the case of daily ratios, for the measurement and calculation of periodic (multiple-day) soiling ratios the developed algorithm makes use of: i) the average, ii) the median and iii) the irradiance-weighted average of daily ratios. Fig. 11 presents example (preliminary) results of the test campaign, for the period from 30/09/2023 to 03/01/2024. Figure 11 left, particularly, shows the results of the daily soiling ratio measurements per module zone and their evolution over the assessment period. Observing towards the end of the test period, SoilRatio's zone-specific measurements are already indicative of a moderate inhomogeneity of the soiling propagation on the test module's surface, witnessed as slight deviations between the I_{sc} (hence the soiling ratio) in the different zones. Further characterization measurements (in-field and in-lab) are following up, in order to validate such observations regarding both the quantity the and the type/quality of soiling. and the calculated soiling ratios per zone, for the exposed (uncleaned) module. Figure 11 right, presents the evolution of the (relative) power loss attributed to soiling, at PV module level, for the same tested sample (exposed, uncleaned module). In particular, we provide here a comparison between the power losses (i.e. soiling losses) calculated real-time by the built-in analysis algorithm of SoilRatio and the the power losses measured by an IV tracing bench, at the same measurement intervals and the same testing period as of Figure 11, left. A first judgement of such comparison indicates that SoilRatio's calculations present very promising accuracy, deviating in most cases by less than 0.5% in terms of relative value, from the actual measured power loss, generally showing a slight overestimation of the soiling losses. Further results and assessment on that aspect are ongoing and their presentation will follow in future work.

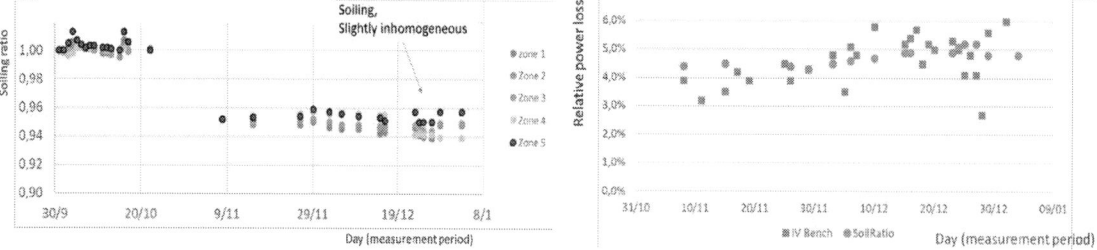

Figure 12: Left: daily soiling ratio calculated for all five sub-module zones of the exposed (uncleaned) PV module. Right: comparison of power losses calculated by SoilRatio vs measured by an IV tracing bench, for the same test period.

1165

3.2 In-lab controlled soiling tests: Results

We have carried out a sequence of controlled lab-based soiling testing, characterization and loss analyses for seven different BOM scenarios applied to corresponding PV laminates/samples (Table 2), with a focus on assessing the impact of front cover type/coating and the encapsulant on soiling buildup.

Table 2: The different PV laminates-samples and BOM scenarios subjected to the indoor soiling testing protocol

Sample module	Front cover	Encapsulant	
S1	plain PV glass	TPO1[1]	*AR: Anti-Reflection coating
S2	plain PV glass	POE[2]	**AS: Anti-Soiling coating
S3	AR*+AS**	TPO	***HPB: Hydrophobic
S4	AR+AS	POE	[1]TPO: Thermoplastic Polyolefin
S5	AR	TPO	[2]POE: Polyolefin Elastomer
S6	HPB***	TPO	
S7	White glass	TPO	

Two key metrics were defined and calculated for the study.

• The short circuit current loss rate (after soiling):

$$Icc_{loss}\% = \frac{Icc_{dirty} - Ic_{ref}}{Icc_{ref}} \times 100$$

• The short circuit gain (after cleaning):

$$Icc_{rel.gain}\% = \frac{Icc_{cleaned} - Icc_{dirty}}{Icc_{ref} - Icc_{dirty}} \times 100$$

Results with regard to the BOM assessment study (Fig. 12) show that samples with white glass as front cover present the higher soiling losses, whereas anti-soiling and hydrophobic coatings seem to better "resist" against soiling buildup.

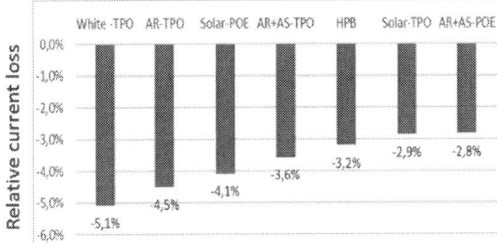

Figure 12: Comparative results on the relative current losses due to the applied soiling, for the different BOMs.

With regard to the soiling study in function of environmental conditions/scenarios (decribed in Section 2.3), results (Figure 13) indicate that the impact of fouling (soiling) on the I_{sc} losses is significantly higher under Scenario 1 (RH>80%, T=17°C/22°C) vs Scenario 2 (RH<30%, T=25°C/30°C) test conditions, due to dew effect favored under the former scenario, which, in turn, favors the soiling buildup.

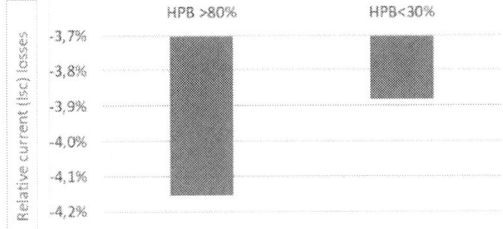

Figure 13: Comparative results on the relative current losses due to the applied soiling, for Scenario 1 vs 2.

Further to the BOM- and environmental- specific soiling tests, we have also experimented on the impact of different tilt angles on the soiling loses. The studied tilt angles were 0°, 20°, 40°, 60°. We concentrated our tests on three sample types, i.e. white glass, AR+AS-POE and Solar-TPO, only using the dry condition injection process. This dry process generates typically fewer soiling losses, as mentioned earlier, but it is easier to control and as consequence offers excellent repeatability.

Results shown in Figure 14 indicate that soiling losses at 20° and 0° are similar, around 4.6%, whereas they are slightly lower at 40°. On the other hand, we see that the losses are more than twice lower at 60°. By extrapolating the trend curve, it is reasonable to consider that from 70° the losses would become negligible.

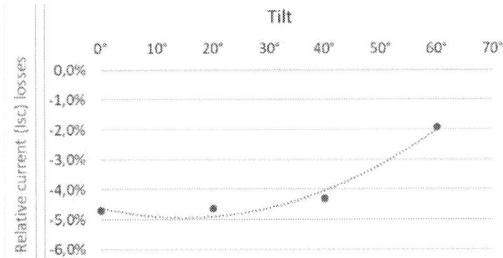

Figure 14: Comparative results on the impact of different tilt angles on the relative current (soiling) losses.

4 CONCLUDING REMARKS

We presented the development and first validation results of two novel soiling testing prototypes, addressing known limitations of commercial solutions, beyond state-of-the-art, especially with regard to the quantification and qualification of different soiling types. Both tools offer refined estimations of actual power losses, as well as (for the case of SoilRatio) calibration-free, intervention-free standalone operability. First results provide already better insights and precision in soiling measurements, especially in the case of inhomogeneous soiling, validated at test and real-scale PV sites.

In addition, the presented indoor lab-based protocol for controlled soiling tests has been proven to provide extremely valuable insights into soiling mechanisms and their interdependence with PV-, site- and environmental-specific features and conditions.

Currently both E-Dust SoilRatio are prepared to be installed and demonstrated in two utility-scale PV plants operated by SERENDI-PV partners and PV asset owners CNR and Galp. The specific PV installations, located in southern France and central-north Spain respectively, are expectedly prone to soiling buildup and (potential) losses, due to their location to an arid dusty area (for the case of Galp site) and their proximity to agricultural activities and an active quarrying site (for the case of CNR site). The demonstration period is scheduled for (at least) a full-year and will also include a benchmarking study with already installed commercial tools, along with a cleaning schedule optimization for CNR's and Galp O&M teams, as well as a validation of soiling models developed by SERENDI-PV partners.

In the near future, we also aim to establish an extensive "soiling library" for indoor analyses, as well as to explore exploitation and commercialization routes for both tools and the indoor testing protocol.

ACKNOWLEDGMENTS

Part of this work has been carried out in the framework of the H2020 SERENDI-PV project. SERENDI-PV project has received funding from the European Union's Horizon 2020 research and innovation programme under grant agreement No. 953016. For CEA team, part of this work was also supported by the French National Program "Programme d'Investissements d'Avenir - INES.2S" under Grant Agreement ANR ANR-10-IEED-0014 0014-01.

REFERENCES

[1] SolarPower Europe (2021): "Operation & Maintenance: Best practice guidelines Version 5.0". ISBN: 9789464444247.

[2] U. Jahn, et al. Report IEA-PVPS T13-25: 2022

[3] J.A. Tsanakas et al. (2023). Towards Climate-Specific O&M for PV Plants: Guidelines and Best Practices. In: Proceedings EU PVSEC 2023.

[4] C. Schill et al. Report IEA-PVPS T13-21: 2022.

[5] SERENDI-PV (Smooth, reliable and dispatchable integration of PV in EU Grids) [https://serendipv.eu/]. Grant agreement ID: 953016

[6] S. Arbaretaz, E. Pilat, et al. (2022). Quantifying the Energy Impact of Soiling Thanks to the Tool SoilRatio. In: Proceedings WCPEC-8.

Degradation root-cause numerical analysis of around 100 PV modules installed in hot and arid desert environment

Shahzada Pamir Aly*, Kaushal Chapaneri, Baloji Adothu, Jim Joseph John, Gerhard Mathiak, and Vivian Alberts

DEWA Research & Development Center, P.O.Box: 564 Dubai, United Arab Emirates
* corresponding author: phone +971 507094595 | shahzada.pamir.aly@gmail.com

Abstract—The voluminous amount of data produced by a modern solar farm suggests opportunities for smart farm management, just-in-time cleaning and maintenance, predictive logistics, etc. In this regard, a physics-based statistical approach called the Suns-Vmp method has been used to determine the pseudo I-V characteristics of a PV system based exclusively on measured time-series data for DC power output. These pseudo I-V characteristics have then been used to decode the degradation pathways within the photovoltaic (PV) system. In this work, we compare real-time field-measured I-V characteristics obtained through a high-speed data logger with pseudo I-V characteristics derived from the Suns-Vmp method. The objective is to evaluate the potential and limitations of both approaches in interpreting field data and to assess the frequency of calibration required for the Suns-Vmp method to yield robust insights into solar farm performance. The methodology was rigorously tested on 96 PV modules of different technologies and configurations installed in the outdoor test facility of Dubai Electricity & Water Authority (DEWA) R&D, Dubai, UAE. The power degradation results from the 96 PV modules indicated that the median degradations of the mono- and poly-crystalline technologies were estimated to be around 0.21%/year and 0.49%/year. This research sheds light on the feasibility and reliability of the Suns-Vmp method for assessing the performance and reliability of solar farms, offering valuable insights for effective solar farm management and maintenance.

Keywords—PV Reliability; Field Measured Data; Physic-based Degradation; PV Health; Maximum Power Point (MPP); I-V Measurements.

I. Introduction

A steady decline in the levelized cost of electricity (LCOE) of photovoltaic (PV) systems has resulted from considerable research and development (R&D), which has driven a tremendous growth in the deployment of PV systems among other renewable energy technologies. As of now, the LCOE is calculated using the assumption that the PV system deteriorates linearly over time, as indicated by the degradation rate. However, it is now widely known that the kinetics of degradation in PV modules are inherently non-linear [1].

There are two primary types of methodology used to assess the health of PV modules: data-dependent online techniques and classic offline/qualification testing. In order to provide a thorough and accurate characterization, offline methods study temporal deterioration by periodically and momentarily disconnecting PV modules. This allows them to pinpoint the main degradation mechanisms of the deployed PV module. Conversely, the online methods use the continuous data collected from PV modules and a variety of statistical and numerical methods as well as machine learning (ML) approaches to measure the modules' temporal degradation.

The main aim of this work is to estimate not only the overall degradation of the photovoltaic (PV) systems using collected field data but also to carry out the root-cause analysis for identifying the underlying degradation pathways. This information helps plan preventive and corrective maintenance, as well as provide information for developing location-agnostic PV module designs. The approach used to achieve the above aim is by implementing a physics-augmented statistical technique to extract the PV parameters of a PV module using limited field dataset. In this regard, the Suns-Vmp method was developed to combine the capabilities of both online and offline techniques [2,3].

II. Methodology

In the Suns-Vmp method, to recreate pseudo I-V curves of the PV modules at various operating conditions, this method makes use of only the field measured data compromising of maximum power point (MPP) current (Imp) and voltage (Vmp), plane of array (POA) irradiance and module temperature. It fits this data via a two-diode electrical model to extract the 5 PV parameters and then use these parameters to recreate the I-V curves. Since each PV parameter has physical significance, one can also identify the root causes of the overall degradation. The following version of the two-diode model has been adopted in this work:

$$I = I_{ph} - I_{o1}\left[exp\left(\frac{V + IR_s}{V_{th}}\right) - 1\right]$$
$$-I_{o2}\left[exp\left(\frac{V+IR_s}{2V_{th}}\right) - 1\right] - \frac{V+IR_s}{R_{sh}} \quad (1)$$

where the 5 parameters are light current (I_{ph}), diode-1 saturation current (I_{o1}), diode-2 saturation current (I_{o2}), shunt resistance (R_{sh}) and series resistance (R_s). These 5 parameters can be solved using different equivalent electrical models [4–7]. Similarly, for the two-diode model in Eq. (1), the module temperature can be estimated using various thermal models [8–13].

III. RESULTS AND DISCUSSIONS

To test the Suns-Vmp model's capabilities, it was applied to 120 PV modules of different technologies and configurations installed in the outdoor test facility of Dubai Electricity & Water Authority (DEWA) R&D, Dubai, UAE (Figure 1). All these modules were installed in March 2015. For each module there is a dedicated temperature sensor, POA irradiance sensor, an electronic load for measuring MPP data as well as tracing complete I-V curves every 10 minutes. Each row's tilt angle and cleaning frequency is also displayed in Figure 1. The exact same PV module is repeated within the same column, to study the effect of tilt angle and soiling on the PV performance and degradation. Out of 120 modules, only 96 modules mono-facial crystalline modules were selected for this analysis. So basically, there are 24 different modules analyzed but repeated in four different rows (96 in total). The data analyzed was from March 2015 to October 2022 (approximately 7 years).

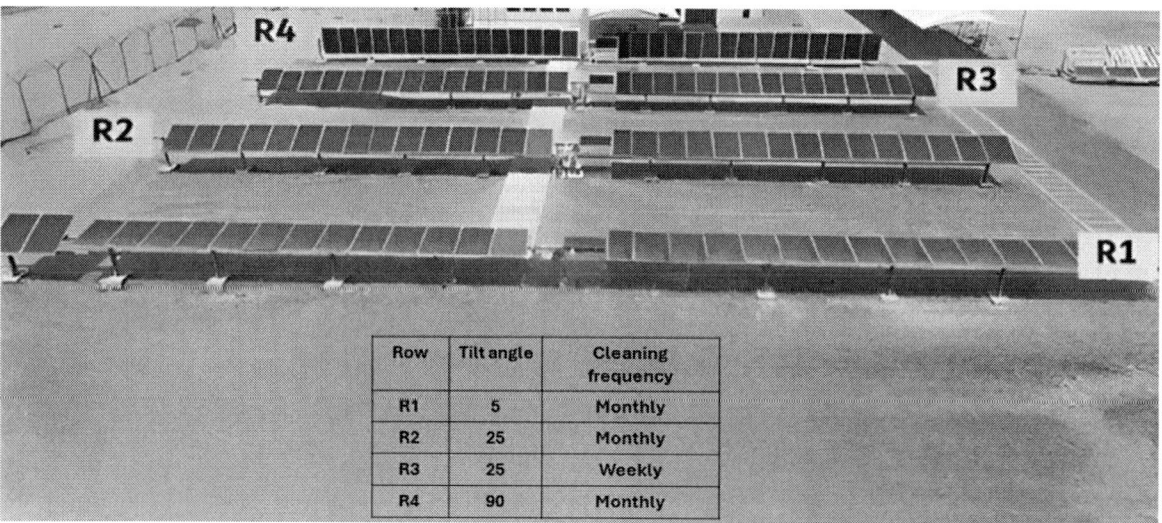

Row	Tilt angle	Cleaning frequency
R1	5	Monthly
R2	25	Monthly
R3	25	Weekly
R4	90	Monthly

Figure 1   Outdoor test facility (OTF) of DEWA R&D (Dubai, UAE) with over 120 PV modules installed, of different technologies and configurations, in a stand-alone configuration, with dedicated electronic loads and temperature sensors.

For demonstration purposes, only one example module's detailed I-V fitting and degradation analysis is displayed in Figure 2, Figure 3 and Figure 4. Figure 2 shows the recreated pseudo I-V curves along with the measured I-V curves for the chosen example module from the OTF, for different timestamps having varying levels of incident irradiance and module temperature. This comparison ascertains how accurate the Suns-Vmp method is in mimicking the actual I-V curves while using limited field data. Figure 3 shows the temporal evolution of the 5 PV circuit parameters for the chosen example module. Using these 5 parameters, we can also do the overall degradation analysis over time, along with identifying the contribution of each of these 5 parameters, as further elucidated in Figure 4.

Thus, using the Suns-Vmp method, we can estimate both the overall degradation (non-linear with time) as well as the underlying pathways contributing to this degradation. Where the underlying pathways are identified using the rate of change of each PV parameter, as shown in Figure 3.

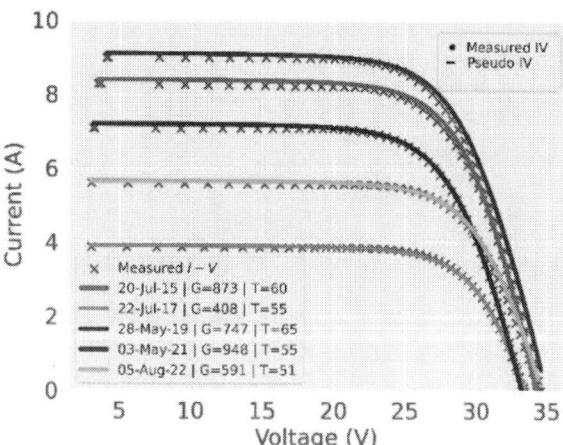

Figure 2   Recreated pseudo I-V vs actual I-V at various environmental conditions (G = incident irradiance and T = cell temperature) and different periods of time, for the chosen poly-crystalline module.

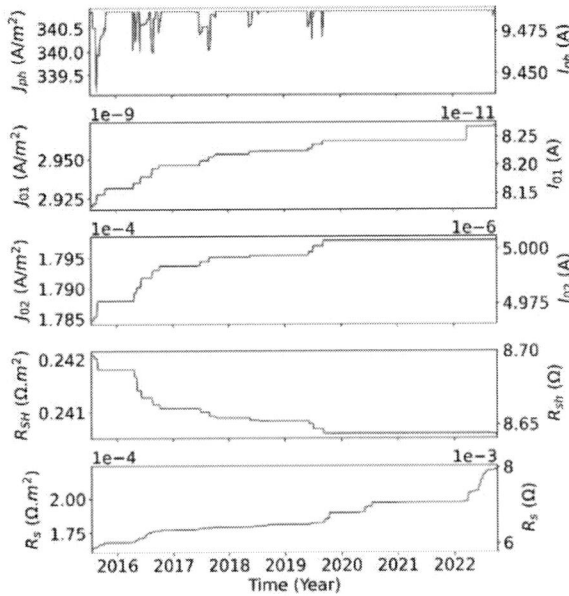

Figure 3   The extracted PV parameters at STC via the Suns-Vmp model.

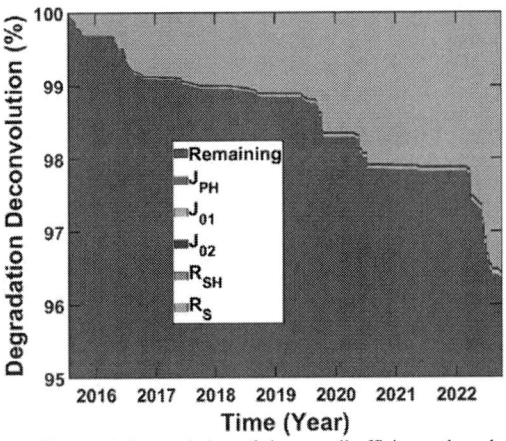

Figure 4   Temporal deconvolution of the overall efficiency degradation to highlight the least and most prominent degradation pathways the chosen example module.

Figure 5 shows the module-wise degradation of all the 96 chosen PV modules from the outdoor test facility (OTF-A). The color contours are independent for each row. Red shades show higher degradation while green shades show lower degradation. The same information is further displayed in Figure 6, where it can be seen that as expected Row1 (90 degrees tilt) shows the minimum degradation, while Row 3 and Row 4 with optimal tilt angles of 25 degrees show maximum degradation. Furthermore, within these chosen mono-facial crystalline modules, median degradations of the mono- and poly-crystalline technologies were estimated to be around 0.21%/year and 0.49%/year.

	Row1		Row2		Row3		Row4		#
POLY	129	-0.36	229	-0.77	329	-0.37	429	-0.02	24
MONO	128	-0.66	228	-0.74	328	-0.55	428	-0.02	23
MONO	127	-0.30	227	-0.35	327	-0.36	427	0.00	22
MONO	126	-0.50	226	-2.81	326	-0.70	426	0.00	21
POLY	125	-0.36	225	-0.20	325	-0.36	425	0.00	20
MONO	123	-0.11	223	-0.16	323	-0.07	423	-0.01	19
POLY	122	-0.54	222	-0.32	322	-0.25	422	-0.13	18
MONO	121	-0.24	221	-0.37	321	-0.41	421	-0.20	17
POLY	120	-0.37	220	-0.43	320	-0.35	420	-0.05	16
POLY	119	-0.58	219	-0.66	319	-0.54	419	-0.79	15
MONO	118	-0.15	218	-0.11	318	-0.08	418	0.00	14
POLY	117	-1.16	217	-0.52	317	-0.04	417	-0.88	13
POLY	114	-0.18	214	-0.08	314	-0.18	414	-0.01	12
POLY	112	-0.31	212	-0.39	312	-0.75	412	-0.20	11
POLY	111	-0.43	211	-0.48	311	-0.69	411	-0.33	10
POLY	110	-0.26	210	-0.33	310	-0.38	410	-0.01	9
POLY	109	-0.22	209	-0.32	309	-0.80	409	-0.18	8
MONO	108	-0.08	208	-0.26	308	-0.33	408	-0.19	7
POLY	107	-0.50	207	-0.24	307	-0.73	407	-0.97	6
POLY	106	-0.39	206	-0.32	306	-0.35	406	-0.69	5
POLY	105	-1.82	205	-1.46	305	-0.58	405	-0.41	4
POLY	104	-0.57	204	-0.25	304	-0.32	404	-0.17	3
POLY	103	-0.36	203	-0.37	303	-0.46	403	-0.49	2
POLY	102	-0.40	202	-0.76	302	-0.65	402	-0.56	1
	-0.45		-0.53		-0.43		-0.26		Avg Degrad

Figure 5   Module-wise STC power degradation of all the 96 chosen PV modules from the OTF-A.

IV. CONCLUSIONS

In order to assess the pseudo, I-V characteristics of PV modules using restricted MPP (Imp and Vmp) and environmental (irradiance and module temperature) data using a two-diode model, we critically examined the Suns-Vmp approach in this work. The real field measured I-V characteristics of 96 PV modules of various technologies installed over a period of 7 years and 9 months at the outdoor test facility of Dubai Electricity and Water Authority (DEWA)

R&D (Dubai, UAE) within four rows of varying inclination angles and cleaning frequencies were then compared to these pseudo I-V characteristics. The following is a summary of the study's main findings:

- The series resistance (R_s) was determined to be the largest contributor to the overall degradation for all 96 PV modules [14–16].

- The power degradations of the mono-crystalline, poly-crystalline, and thin-film technologies are projected to be between 0.21% and 0.49% annually, with respect to technology-specific power deterioration.

- Row 4 (90-degree tilt angle and monthly cleaning) modules displayed the least amount of power degradation among the various tilt angles and cleaning frequencies, while Row 2 (25-degree tilt angle and monthly cleaning) modules displayed the most.

- The inability of the Suns-Vmp method to replicate I-V curves for partially shaded modules, which feature I-V curves with steps or kinks, is one of its main shortcomings. This is so because the PV parameters that the Suns-Vmp technique extracts are indicative of the module.

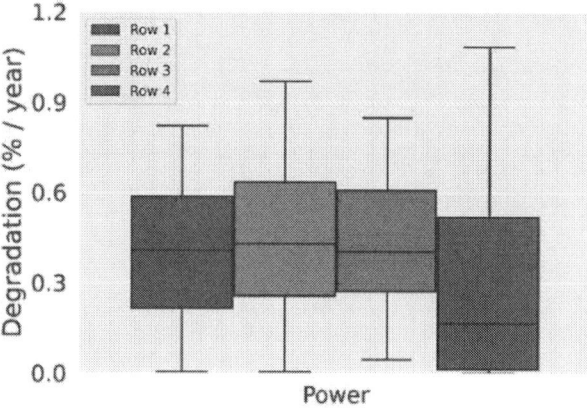

Figure 6   Row wise I-V characteristics mean degradation (%/year) of 96 PV modules at DEWA R&D OTF, Dubai, UAE, with tilt angles and cleaning frequencies: Row 1 @ 5° (monthly), Row 2 @ 25° (monthly), Row 3 @ 25° (weekly) and Row 4 @ 90° (monthly).

REFERENCES

[1]   Jordan DC, Silverman TJ, Sekulic B, Kurtz SR. PV degradation curves: non-linearities and failure modes. Progress in Photovoltaics: Research and Applications 2017;25:583–91. https://doi.org/10.1002/PIP.2835.

[2]   Aly SP, Chapaneri K, John JJ, Mathiak G, Alberts V, Alam MA. Suns-Vmp method for health monitoring of 110 PV modules. Renewable and Sustainable Energy Reviews 2024;202:114653. https://doi.org/10.1016/j.rser.2024.114653.

[3]   Sun X, Chavali RVK, Alam MA. Real-time monitoring and diagnosis of photovoltaic system degradation only using maximum power point—the Suns-Vmp method. Progress in Photovoltaics: Research and Applications 2019;27:55–66. https://doi.org/10.1002/PIP.3043.

[4]   Aly SP. Numerical models development for simulating optical, thermal and electrical performance, as well as structural degradation of PV modules. PhD Thesis. Hamad Bin Khalifa University (HBKU), 2019.

[5]   Aly SP, Ahzi S, Barth N. An adaptive modelling technique for parameters extraction of photovoltaic devices under varying sunlight and temperature conditions. Appl Energy 2019;236:728–42. https://doi.org/10.1016/j.apenergy.2018.12.036.

[6]   Ahzi S, Aly SP, Barth N. New Developments in the Modeling and Simulations of the Thermal Behavior and Electrical Yield of Photovoltaics Panels With the Consideration of Desert Environmental Conditions. 2017 International Renewable and Sustainable Energy Conference (IRSEC), IEEE; 2017, p. 1–6. https://doi.org/10.1109/IRSEC.2017.8477380.

[7]   Aly SP, Chapaneri K, John JJ, Mathiak G, Alam MA. Retrofitting the Translation Equations of the One-Diode Model for Photovoltaic Modules. 2023 Middle East and North Africa Solar Conference (MENA-SC), IEEE; 2023, p. 1–4. https://doi.org/10.1109/MENA-SC54044.2023.10374515.

[8]   Aly SP, Barth N, Ahzi S. A Three-Dimensional Finite Element Based Dynamic Thermal Model of PV Modules with an Improved Thermal Network. 2017 International Renewable and Sustainable Energy Conference (IRSEC), IEEE; 2017, p. 1–6. https://doi.org/10.1109/IRSEC.2017.8477421.

[9]   Aly SP, Ahzi S, Barth N, Figgis BW. Two-dimensional finite difference-based model for coupled irradiation and heat transfer in photovoltaic modules. Solar Energy Materials and Solar Cells 2018;180:289–302. https://doi.org/10.1016/j.solmat.2017.06.055.

[10]   Aly SP, John JJ, Mathiak G, Albadwawi O, Pomares L, Alberts V. A thermal model for bifacial PV panels. 2022 IEEE 49th Photovoltaics Specialists Conference (PVSC), IEEE; 2022, p. 457–9. https://doi.org/10.1109/PVSC48317.2022.9938549.

[11]   Chapaneri K, Aly SP, John JJ, Mathiak G, Alberts V, Alam MA. Self-Thermometry of PV Panels. 2023 IEEE 50th Photovoltaic Specialists Conference (PVSC), Institute of Electrical and Electronics Engineers (IEEE); 2023, p. 1–3. https://doi.org/10.1109/PVSC48320.2023.10359913.

[12]   Aly SP, Ahzi S, Barth N, Abdallah A. Using energy balance method to study the thermal behavior of PV panels under time-varying field conditions. Energy Convers Manag 2018;175:246–62. https://doi.org/10.1016/j.enconman.2018.09.007.

[13]   Aly SP, Barth N, Figgis BW, Ahzi S. A fully transient novel thermal model for in-field photovoltaic modules using developed explicit and implicit finite difference schemes. J Comput Sci 2018. https://doi.org/10.1016/j.jocs.2017.12.013.

[14]   Aly SP, Ahzi S, Barth N. Effect of physical and environmental factors on the performance of a photovoltaic panel. Solar Energy Materials and Solar Cells 2019;200:109948. https://doi.org/10.1016/j.solmat.2019.109948.

[15]   Ahzi S, Aly SP, Barth N, Abdallah A. Modeling and Simulation of the Longtime Behavior and Fatigue Failure of Photovoltaic Modules under Desert Environment. 2019 7th International Renewable and Sustainable Energy Conference (IRSEC), 2019, p. 1–5. https://doi.org/10.1109/IRSEC48032.2019.9078198.

[16]   Aly SP, Ahzi S, Barth N, Abdallah A. Numerical analysis of the reliability of photovoltaic modules based on the fatigue life of the copper interconnects 2019. https://doi.org/10.1016/j.solener.2020.10.021.

41st European Photovoltaic Solar Energy Conference and Exhibition

A SYSTEMATIC APPROACH FOR THE INTEGRATION OF BIPV PLANNING INTO THE CONSTRUCTION PLANNING PROCESS

Frank Ensslen[1], Mona Mühlich[1], Jan-Bleicke Eggers[1], Tilmann E. Kuhn[1], Bruno Bueno[1]
[1]Fraunhofer Institute for Solar Energy Systems ISE, Heidenhofstraße 2, 79110 Freiburg/Breisgau, Germany
Corresponding author: Frank Ensslen | Phone: +49 (0)761 4588 5650 | E-mail: Frank.Ensslen@ise.fraunhofer.de

ABSTRACT: The integration of PV into buildings is an essential cornerstone for the energy transformation. It also allows power systems in an urban context to be aesthetically appealing. Despite their potential, as economically feasible and space-saving renewable energy converters, building-integrated photovoltaic (BIPV) installations are not yet well integrated into the construction planning process. The lack of a systematic approach for BIPV planning is one of the factors preventing a far-reaching deployment of this technology. Therefore, substantial knowledge of relevant BIPV planning phases is essential. In the optimum case, BIPV is considered in the planning of the building from the very beginning, because it has a direct impact on its architectural design, on the building energy and sustainability concepts, as well as its economic viability. Preliminary technological options, as well as product availability and requirements from building regulations, must be examined early in the plan of works. This study describes an ideal planning flow for BIPV as a general approach, detailing sub-processes with hereby generated and to be exchanged information, and stakeholders' responsibilities. The BIPV sub-processes have higher or lower priority in the different planning stages which is documented in detail in the paper.
Keywords: Building-integrated photovoltaic (BIPV), BIPV planning process, BIPV stakeholders

1 AIM AND APPROACH

The lack of knowledge or experience in the construction sector regarding planning, engineering, and implementation of PV technologies in facades and roofs of buildings has been identified as one of the main barriers for building-integrated photovoltaic (BIPV). The construction sector only adapts new technologies slowly as it consists of numerous small and medium-sized stakeholders, which often lack efficient communication and common goals. This often hinders the realization of BIPV projects. Even if BIPV solutions are considered at all, they are often implemented only at advanced phases of the construction planning process, resulting in missed opportunities, poor realization concepts and unnecessarily high costs and delays, mostly leading finally to cancelling envisaged BIPV projects. Moreover, looking at the interfaces between participating stakeholders, liability for BIPV installation and its components is not yet clearly defined.

This study aims at a better and holistic understanding of BIPV planning within the construction planning process to lower one of the barriers for widely realizing BIPV. Seamlessly integrating the planning for it into the conventional planning phases leads to better and significantly more cost-effective BIPV solutions. This is realized through a systematic description of the BIPV planning stages in combination with a "Stakeholder-Sub-Process Model", which starts already at the very beginning of the construction project. Accordingly, this study introduces a methodology addressing the interactions between the different BIPV stakeholders in their changing roles throughout the planning process, associated with the documentation needed. Further, a concept structuring the whole BIPV planning process usefully is described.

2 SCIENTIFIC INNOVATION AND RELEVANCE

Although successfully demonstrated many times, the planning and installation of BIPV systems both still face significant barriers (e.g., cooperation between different trades, intersections of responsibilities and liabilities) and are therefore challenging. Especially the planning process is highly complex, as the integration of PV requires its planning to be included within the complete planning process for a building.

To the best of our knowledge, recent research on BIPV planning has been focusing on the digitalisation and the design phase of the planning process. Jakica et al. [1] described the different phases of the whole BIPV planning process and further investigated the design and performance of modelling tools for BIPV. Building up on this, Yang et al. [2] investigated especially the conceptual design phase, arguing that the PV integration must be introduced early in the planning process.

In earlier research work by Hemmerle [3], the BIPV planning flow is described according to the German building planning structure, explaining what needs to be done in each single phase and accordingly, which actions should be taken for the BIPV planning. Similarly, in a wider context, the research project "BIPV-Initiative Baden-Württemberg" (Baden-Württemberg is a German State) provides information on the BIPV planning process in the form of a specially designed web-based solution [4]. Here, many topics are addressed that need to be considered when planning BIPV and, thus sorted according to the sequential planning and building phases of the overall building planning process. The examples given provide a profound overview of the BIPV planning phases. Again, the importance of considering BIPV early in the overall planning process is emphasized.

3 PLANNING AND CONSTRUCTION PROCESS

3.1 Comparison of planning and construction processes in different countries.

An understanding of the complete planning and construction process is essential to achieve an effective integration of BIPV systems. In Germany, for instance, construction projects are divided into nine work phases (with legal significance). This set of phases is known by the German acronym HOAI ("fee structure for architects and engineers") [5]. The stages are: 1) "Basic research", 2)

Preliminary planning", 3) Design Planning", 4) "Approval planning", 5) "Implementation planning", 6) "Preparation of Contract Award", 7) "Participation in Contract Award", 8) "Construction Supervision" and 9) "Project Monitoring". In some cases, the planning process starts with a preliminary work phase zero: "Demand planning" or "Feasibility study".

By comparing the plan of works in England (RIBA Plan of Work) [6], Germany (HOAI) [5], France (Missions de base) [7] and Italy (DM 560/2017) [8], it becomes evident that single stages are divided into more detail in some countries and less in others. Nevertheless, the procedure and planning stages are very similar in the mentioned countries. The RIBA Plan of Work [6] represents the workflow in a compact manner and is therefore used as a role model to describe the BIPV planning and construction process along these stages. The actual planning takes place in stages 0 to 4. Stage 5 deals mainly with realization of works and stage 6 with the handover of the building. Finally, stage 7 covers the ongoing use of the building.

3.2 The core tasks of the RIBA Plan of Work

All core tasks of the RIBA Plan of Work [6] are structured into seven stages containing different sub-processes (Table I).

Table I: RIBA stages with core tasks [6]

RIBA stages	Core Tasks
0: Strategic Definition	Client Requirements Project Risks and Project Budget Site Appraisals

1: Preparation and Brief	Project brief with
	  ▫ Project Outcomes
	  ▫ Sustainability Outcomes
	  ▫ Quality Aspirations
	  ▫ Spatial Requirements
	Feasibility Studies
	Project Budget
	Site Information with Site Surveys
	Project Programme
	Project Execution plan

2: Concept Design	Architectural Concept including
	  ▫ Strategic Engineering
	  ▫ Cost Plan
	  ▫ Project Strategies
	  ▫ Outline Specification
	Project Brief Derogations
	Design Reviews with Project Stakeholders
	Design Programme
	(Building Regulations)

3: Spatial Coordination	Architectural Concept including
	  ▫ Design Studies
	  ▫ Engineering Analysis
	  ▫ Cost Exercises
	Spatially Coordinated Design to Updated

	▫ Cost Plan
	  ▫ Project Strategies
	  ▫ Outline Specification
	Change Control Procedures
	Design Programme
	Building Regulations

4: Technical Design	Architectural and Engineering Technical Design Building Systems Design Programme Building Regulations Application Construction phase plan
5: Manufacturing and Construction	Site Logistics Building Systems Construction Programme Construction Quality Site Queries Commissioning Building Manual
6: Handover	Plan for Use Strategy Project Performance Commissioning Aftercare Post Occupancy Evaluation
7: Use	Facilities Management Asset Management Post Occupancy Evaluation Verification of Project Outcome Sustainability Outcomes

In the following sections, relating to the BIPV planning process, only stages 0 to 4 are comprehensively addressed.

4 BIPV PLANNING PROCESS

4.1 Novel "Stakeholder-Sub-Process Model"

The "Stakeholder Sub-Process Model" was developed by Fraunhofer ISE based on the Swiss Model for Building Documentation in Structural Engineering [9] to describe the interdependence of stakeholders, sub-processes and documentation supporting an ideal BIPV planning and organisational process (Figure 1).

In this model the whole BIPV planning process is first structured into meaningful sub-processes, each in turn with its own tasks or sub-tasks. Thereby, each sub-process with its (sub)-tasks gets a certain priority for individual actions according to the planning stage. Sub-processes get input from other sources or generate information by themselves, either for their own use or other stakeholders' purposes.

Second, every sub-process/task must be assigned to a certain stakeholder, the so-called "process owner". Stakeholders taking a specific role are predominantly responsible for implementing (multiple) tasks in (multiple) sub-processes. However, the same player can be responsible for different required sub-processes. As the works and planning stages progress, stakeholders can also pass on responsibility to other protagonists. Within the scope of their role, stakeholders are responsible for the existence of proper documentation and getting input from the sub-processes.

The documentation is a specific collection of information satisfying the information demand of each stakeholder as a "process owner". A certain format of the documentation serves for thoroughly organizing and handling relevant information between sub-processes.

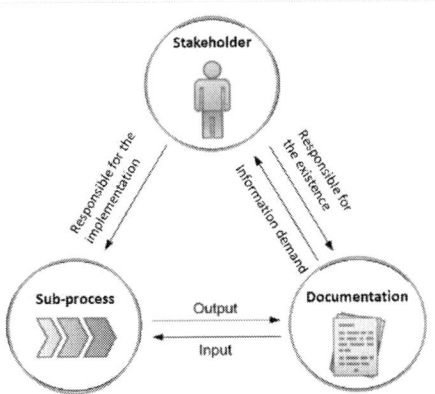

Figure 1: General interdependence of stakeholder, sub-process, and documentation based on [9].

4.2 BIPV sub-processes, (sub-)tasks, and documentation

The following paragraphs describe relevant BIPV sub-processes within RIBA planning stages 0 to 4 [6].

BIPV Design

For the successful integration of BIPV into building design, it is important to consider it from the outset. This means BIPV should already be represented in the conceptual phase, where the first ideas and impressions, among other things, emerge through visualizations. Several topics should be considered in the overall design of the building for a successful PV integration. The orientation determines the suitable and potential building envelope surfaces, the surroundings can cause shading, and the PV modules can cause glare, depending on their design (see Figure 2).

A glare risk assessment is therefore recommendable for identify hotspots in the surrounding urban environment of BIPV installations. Locations that can be particularly affected by the specular reflection from BIPV modules can be identified visually, or by applying raytracing simulation tools [10]. To reduce reflectance and to increase transmittance of the front glass, many PV modules are equipped with textured or anti-reflective-coated front glasses. These measures may lead to a mitigation of the glare risk, but the scattering effect may indeed also increase the glare source.

Figure 2: Example of a conventionally glazed facade causing glare in the centre of Freiburg, Germany. © Fraunhofer ISE

The dimensioning of BIPV surfaces depends on both the building design and the overall energy concept, as well as on the related yield targets, and the project's cost framework. Initial ideas about the choice of technology and colour of the PV modules must be formulated at an early stage, embedding the BIPV into the overall design (see Figure 3). For the concrete design of the roof or facade surfaces, available products on the market should be researched and common dimensions considered in the design.

Figure 3: Sports hall of the school "im Rot" with a heritage protection character in Eppingen (Germany). The colour of the PV modules was selected to ensure low visual contrast to the old roof. © Fraunhofer ISE

Consideration of legal requirements already in the early design phase is recommended. The local building code can include design regulations relevant for the integration of PV.

Specific tasks that arise in BIPV planning include the visualization of solarized surfaces and the design of roof or facade surfaces, ideally considering common dimensions and component availability. At a later stage the focus lies on the specific BIPV module design or the selection of BIPV modules on the market, as well as the research on and consideration of building regulations applicable to BIPV.

BIPV Structural Engineering

The structural requirements for the integration of BIPV must be considered and aligned with building regulations in the overall planning. This means paying attention to the (preliminary) static dimensioning of the BIPV modules already in the early planning stages. If prefabricated BIPV systems are used, the corresponding mounting systems need to be integrated into the building construction. Additionally, the building science properties must be considered when integrating PV, as these will take on the functions of e.g. wall or roof cladding. When using standard BIPV modules which are commercially offered as such by their manufacturers, the mounting system must be included in the structural planning. For the development of customised BIPV modules, tasks must be executed like aligning the dimensioning with the module's statics, designing the configuration of the module itself, and developing an appropriate substructure.

Furthermore, technical approvals must be checked. In many cases, the structural integration of BIPV modules requires project-specific customization of the modules, e.g., as shown in [11]. Difficulties arise from the fact that product characteristics may change and thus certificates and approvals may become invalid. One possible solution is to certify several sub-categories for a product and to characterize each of the sub-categories with a limited number of parameters. Installation typologies can be found in the two-part European BIPV standard EN 50583 and the two-part international BIPV standard IEC 63092.

BIPV Fire Safety

Fire safety regulations vary by country or even region. For a complete (goal-oriented) fire protection concept, the requirements in the building context must always be observed. It is advisable to familiarize oneself with the applicable regulations early on and align them with the structure and mounting system of the BIPV module.

The fire safety behaviour of BIPV modules must be proven if the building regulation codes require a certain classification of the fire behaviour (e.g., non-combustible, or difficult to ignite respectively flame retarding with lower propagation), depending on the building usage building class. Further information on fire safety of BIPV can be found in the IEA PVPS related report from 2023 [12].

BIPV Economic Efficiency and Sustainability

Economic viability plays a crucial role in the implementation of integrated PV. Available subsidies should be considered and examined, for example, in terms of a building's sustainability concept.

Sustainability certification or life cycle assessment (LCA) for PV products and technologies is becoming increasingly relevant the construction sector. Information about the resources consumed and waste generated during the complete lifecycle of construction products can be aggregated into, for example, Environmental Product Declarations (EPD) that serve as the basis for complete building environmental LCA calculations. As well, aspects of energy pay-back time (EPBT) and global warming potential (GWP) of the BIPV modules used play a role that is worth knowing.

Public funding opportunities depend on the country and region, and time of installation of the BIPV system. To determine economic viability, investment and operating costs must be compared with savings and revenues, in combination with the energy utilization concept.

BIPV Energy Concept

The decision on how to use the PV electricity depends on the energy concept of the entire building. For this, the energy demand must be determined, and the local possibilities of feeding electricity into the grid and self-consumption of the electricity within or near the building must be checked. Depending in these conditions, different operator models for the PV system are suitable, i.e. [13].

BIPV Yield Determination

The determination of the expected yield occurs in several steps. First, the solar potential of the building's surfaces must be assessed. Based on this, the anticipated performance and yield are calculated, considering the planned BIPV module configurations. Possible shading (see Figure 4), the orientation of the modules in the azimuthal (compass) direction and the tilt of the BIPV modules are decisive for the annual yield and the temporal distribution of solar power generation. A BIPV system can be planned with differently oriented sub-systems. Then a more uniform electricity generation profile can be achieved over the day, which can help to minimize the load on the electricity grid and lead to higher self-consumption of the generated electricity.

BIPV System Design

In the electrical planning, the various components of the BIPV system must be coordinated and external factors, like (frequent) partial shading, considered. The detection of partial shading requires a detailed analysis of the solar irradiance on the surface which is to be equipped with PV. Because shading can occur on different spots at different times, the analysis must be highly resolved temporally as well as spatially. To account for effects such as frequent partial shading, a detailed calculation that considers the individual cell interconnections is necessary for a meaningful result. Alternatively, the innovative matrix interconnection also increases the yield of solar modules in those cases.

The electrotechnical details are important for the safety of the system; the inverters should be placed strategically, and if storage systems are used, they must appropriately be sized. Additionally, the wiring should be planned as efficiently as possible, based on the yield estimation. The electronics of the BIPV system must also be geometrically and spatially integrated into the roof or facade.

Figure 4: Partial shading caused by trees on a building facade. The interconnection between BIPV modules can be optimized or, for instance, the matrix interconnection can be used in BIPV modules to increase the resilience of the installation against partial shading. © Fraunhofer ISE

4.3 BIPV-related information exchange

As in the general planning process, different information is required and generated in the various sub-processes of the BIPV planning phases. In the following, the necessary inputs, and generated outputs of each BIPV sub-process are listed (Table II).

Table II: BIPV sub-processes with input and output.

BIPV Sub-process	Input	Output
Design	Building Design, Location, Budget, Solar Potential from Yield Determination, Energy Concept	Facade/Roof Design, Positioning of BIPV Modules
Structural Engineering	Structural Design, BIPV Module Type, Fire Protection Requirements	Facade/Roof Structural Concept and BIPV Module Mounting
Fire Safety	BIPV Module Type, Facade/Roof Structural Concept, Overall Protection Goal-oriented Fire Protection Concept	Fire Behaviour Classification of BIPV Modules and Mounting System, Concept of BIPV fire safety (possibly of Facade/Roof)
Economic Efficiency and Sustainability	Budget, Results from Yield Determination Analyses, Information on possible Subsidies, Feed-in Remuneration Requirements for Building's Sustainability, LCA of BIPV Modules	Investment Concept
Energy Concept	Building Energy Concept, Results from Yield Determination Analyses, Feed-in Options, Self-Consumption	BIPV Roll into the Energy Concept
Yield Determination	Building Envelope Design, Location, Orientation of Relevant Surfaces, BIPV Module Specifications	Expected Yield Determination, Specific Performance of BIPV Modules
System Design	BIPV Design, Building Energy Concept, Results from Yield Determination	BIPV System Design

4.4 BIPV sub-process priorities in different RIBA planning stages

The BIPV sub-processes have higher or lower priority in the different stages of the overall RIBA planning process [6]. The priority of a sub-process in a stage is outlined in Table III, on a scale from zero to three, taking zero as no consideration and three as the most important topic to be treated. Here, only planning stages 0 to 4 are considered.

Table III: BIPV sub-process priorities in RIBA planning stages 0 to 4 [6].

RIBA Planning stage → BIPV sub-process ↓	0	1	2	3	4
Design	1	0	3	3	2
Structural Engineering	0	0	2	3	3
Fire Safety	0	0	2	3	3
Economic Efficiency and Sustainability	1	2	3	2	3
Energy Concept	1	2	3	0	0
Yield Determination	2	1	1	3	3
System Design	0	0	1	2	3

0-3: Indication of the priority (zero none, three high) of the sub-process in the RIBA planning stage [6].

4.5 BIPV stakeholders

The number and expertise or background of the stakeholders involved in BIPV projects differs according to the features of the construction, i.e., is it a refurbished or new building envelope? Is it a large, medium-size, or small building? Is it a public or commercial building? Is it a residential, office or industrial building?

However, at an early stage, it is useful already to identify the BIPV stakeholders who will assume the different role-related responsibilities for the required planning tasks, and the associated interfaces between different trades and professions. Here, first interactions between different stakeholders take place and possible conflicting goals are identified during the evaluation of the design, construction, and electrical options.

Some of the players in charge of the different sub-processes are the following (not exhaustive):

- Client (C)
- Building User (BU)
- Building Operator (BO)
- General Contractor (GC)
- Architect or Planner (AP)
- Engineering: Structural Concept (ES)
- Engineering: Facade/Roof (EF)
- Engineering: Fire Protection (EFP)
- Engineering: PV-System (EPV)
- Engineering: Electrical Concept (EE)
- Engineering: Energy Concept (EEC)

Table IV provides an exemplary overview of when and which stakeholders are involved in the various planning BIPV sub-processes.

Table IV: Stakeholder's assignment to BIPV sub-processes dependent on RIBA planning stage [6].

RIBA Planning stage → BIPV sub-process↓	0	1	2	3	4
Design	AP		AP, EPV, EF		
Structural Engineering			AP, EPV, ES, EF		
Fire Safety			AP, EPV, EF, EFP		
Economic efficiency and Sustainability		AP, EPV			
Energy concept		AP, EEC, EPV			
Yield Determination			AP, EPV, EE		
System Design			AP, EPV, EF, EE		

Legend: see list above

5 CONCLUSIONS

This study aims at a better and holistic understanding of BIPV planning within the construction planning process to lower one of the barriers for widely realizing BIPV.

Therefore, a systematic approach for structuring the BIPV planning (here considered stages 0 to 4 according to the RIBA Work of Plan [6]) by applying a "Stakeholder-Sub-Process Model" is introduced. The assignment of stakeholders to each BIPV sub-process and in relation to the RIBA planning stage is shown in an exemplary overview. Relevant BIPV sub-processes with their individual required inputs and generated outputs, which are both accompanied by important documentation, are described, and prioritized according to each planning stage. It is important to note that ideally some sub-processes are already considered in the earliest planning stage.

6 ACKNOWLEDMENT

The authors would like to thank the European Union's Horizon 2020 Research and Innovation programme for funding the PROBONO project under Grant Agreement No. 101037075.

7 REFERENCES

[1] N. Jakica, et al. "BIPV Design and Performance Modelling: Tools and Methods," 2019. Available online: https://iea-pvps.org/wp-content/uploads/2020/01/IEA-PVPS_15_R09_BIPV_Design_Tools_report.pdf

[2] R.J. Yang et al. (2023), "Digitalizing building integrated photovoltaic (BIPV) conceptual design: A framework and an example platform," Building and Environment. Vol. 243, p. 110675. DOI: https://doi.org/10.1016/j.buildenv.2023.110675.

[3] C. Hemmerle (2015), Photovoltaik in der Gebäudehülle: Wertung bautechnischer Anforderungen. Photovoltaics in Building Envelopes: Evaluation of Structural Requirements (Doctoral thesis). Retrieved from https://nbn-resolving.org/urn:nbn:de:bsz:14-qucosa-204836.

[4] Initiative für Bauwerkintegrierte PV-Anlagen Baden-Württemberg (2022), Leitfaden zur Planung, Bau und Betrieb von bauwerkintegrierten Photovoltaikanlagen. Available online: https://bipv-bw.de/.

[5] Verordnung über die Honorare für Architekten- und Ingenieurleistungen (Honorarordnung für Architekten und Ingenieure - HOAI, Engl. „Fee Structure for Architects and Engineers"), 2021. Available online: https://www.hoai.de/hoai/volltext/hoai-2021/

[6] Royal Institute of British Architects (2020), RIBA Plan of Work. Available online: https://www.architecture.com/knowledge-and-resources/resources-landing-page/riba-plan-of-work

[7] Ministère de la Transition écologique et solidaire. (2018) Missions de base de la maîtrise d'œuvre. Available online: https://www.legifrance.gouv.fr/codes/section_lc/LEGITEXT000037701019/LEGISCTA000037726513

[8] Decreto Ministeriale n. 560 del 1 dicembre (2017), sull'introduzione di metodi e strumenti digitali nelle opere pubbliche (DM 560/2017). Available online: https://www.mit.gov.it/normativa/decreto-ministeriale-numero-560-del-01122017

[9] KBOB/IPB (2016), Bauwerksdokumentation im Hochbau, Dokumentationsmodell BWD – Engl. "Building Documentation in Structural Engineering, BWD Documentation Model", Switzerland. Available online: https://www.kbob.admin.ch/dam/kbob/de/dokumente/Publikationen/Bauwerksdokumentation%20Hochbau/KBOB_IPB_Modellbeschrieb_BWD_2016.pdf.download.pdf/KBOB_IPB_Modellbeschrieb_BWD_2016.pdf.

[10] R. Schregle et al. (2019), A Variable Resolution Approach for Zonal Glare Assessment. DOI: 10.13140/RG.2.2.17955.81442

[11] T.E. Kuhn et al. (2021), Review of technological design options for building integrated photovoltaics (BIPV), Energy and Buildings, Volume 231, Elsevier. DOI: https://doi.org/10.1016/j.enbuild.2020.110381

[12] S. Boddaert et al. (2023), Fire safety of BIPV. International mapping of accredited and R&D facilities in the context of code and standards. Report of International Energy Agency, Photovoltaic Power Systems Programme.

[13] H. Gholami, et al. (2020), Economic analysis of BIPV systems as a building envelope material for building skins in Europe Energy, Volume 204, Elsevier. DOI: https://doi.org/10.1016/j.energy.2020.117931

41st European Photovoltaic Solar Energy Conference and Exhibition

This presentation was selected by the Sc. Committee of the EU PVSEC 2024 for submission of a full paper to one of the EU PVSEC's collaborating peer-reviewed journals.

SEMITRANSPARENT BIFACIAL PV WINDOWS WITH INTEGRATED BLINDS: EXPERIMENTAL AND MODELLING RESULTS

Simona Villa*[1], Martin Hurtado Ellmann[1,2], Roland Valckenborg[1]
[1]TNO Energy and Materials Transition, High Tech Campus 21, 5656AE, Eindhoven, The Netherlands
[2]Photovoltaic Materials and Devices group, Delft University of Technology, 2628CD Delft,
The Netherlands
*simona.villa@tno.nl

ABSTRACT: The stricter requirements for the energy performance of buildings are creating a market for several building integrated photovoltaic (BIPV) technologies, including PV windows. Herein, we present an innovative multifunctional PV window concept designed to enhance energy generation while providing over-heating protection for better indoor thermal and visual comfort. This concept utilizes bifacial c-Si solar cell strips combined with Venetian blinds, all embedded in a unique insulating glazing unit. The bifacial technology increases the energy yield by using the blinds as reflectors, directing more irradiance to the cells' rear side. Twelve demonstrators were installed and monitored for a full year. Several measurement campaigns were conducted to assess the impact of different blind types and tilt angles on the total yield. It was found that the highest energy boosts occurs when the blinds are fully closed at a 75° angle with their convex side facing outwards. Blinds with the highest specular reflectance achieve a maximum performance increase of 25% on sunny days and a daily average increase of 13% compared to the case of no blinds. Simulations were conducted to better understand the complexity of the system and to assess the potential of this bifacial PV window concept.
Keywords: PV windows, bifacial, Venetian blinds, outdoor performance, modelling

1 INTRODUCTION

High electrical efficiency and low cost are keys for the success and wide deployment of building integrated photovoltaic (BIPV) windows. Without sufficient electricity production, photovoltaic (PV) windows can hardly compete with alternative types of smart and advanced window technologies which aim at reducing the building's energy consumption in alternative ways, for instance with low-e coatings or vacuum double glazing [1]. Therefore, efforts should be directed towards enhancing the electricity output of such BIPV products. On top of that, it is essential not to overlook the additional criteria that a PV window must fulfill, which include: maintaining a satisfactory level of transparency and, ideally, being color-neutral or aesthetically pleasing, in such a way that visual and thermal comfort are assured.

In this context, the research institute TNO has teamed up with the glass company Pilkington and other partners to develop an innovative and multifunctional PV window concept, which aims to boost the energy generation of semitransparent windows while, at the same time, providing over-heating protection and visual comfort [2]. The concept that we propose is based on the combination of bifacial striped c-Si solar cells with Venetian blinds [3]. The use of the bifacial technology allows to boost the energy yield thanks to the embedded blinds, which act as reflectors, increasing the amount of irradiance reaching the rear side of the cells. To the best of our knowledge, no bifacial PV windows exist on the market yet. The embedded Venetian blinds, besides acting as reflectors for the solar cells' rear side, will be an active shading element that will limit the cooling requirement and risk of overheating in summer, that is one of the biggest challenges to be faced in the coming years.

Understanding the electrical performance of this PV window concept is not trivial because of the several parameters that come into play, such as: electrical design, cell layout, Venetian blind material, slats tilt angle, sun position, weather conditions, etc. To this aim, both experimental outdoor measurements and simulations have been performed.

The results of the yearly outdoor testing have been extensively reported in a separate publication [4], and the main conclusions are summarized here in Section 3. The simulations results, which allow to assess the potential of the studied PV window in a more general context, are then presented in Section 4.

2 DESCRIPTION OF THE PV WINDOW CONCEPT

2.1 Concept and operating principle
Figure 1a illustrates the structure of the PV window, consisting of a unique insulating glazing unit (IGU) in which Venetian blinds have been integrated in the spacer between the two glass panes. The outer glass pane consists of a PV laminate, in which bifacial c-Si solar cells have been incorporated, so that their front side directly faces the sky, while the rear-side faces the Venetian blinds, located towards the internal part of the window. Figure 1b illustrates also the operating principle of the multifunctional PV window: in the 'no boost' mode, we have an ordinary solar window (with c-Si cell strips) because the blinds are not deployed; in the 'partial boost' mode, the blinds are deployed but kept open (i.e. horizontally), so they can block the heat to some extent but there is still normal visibility; in the 'maximum boost' position, the blinds are completely closed (i.e. vertically), the output from the PV cell strips is maximized and sunlight towards the inside of the room is blocked. The blinds tilt can vary to follow the sun position or any other manual/automated command depending on the user's requirements and the desired output to be optimized (PV yield, thermal or visual comfort).

2.2 Outdoor demonstrators
In September 2023, 12 different small-scale PV window demonstrators, with size 60 x 45 cm, have been installed on a south-facing test façade at the SolarBEAT [5] outdoor test facility (Eindhoven, the Netherlands), as shown in Figure 2.

41st European Photovoltaic Solar Energy Conference and Exhibition

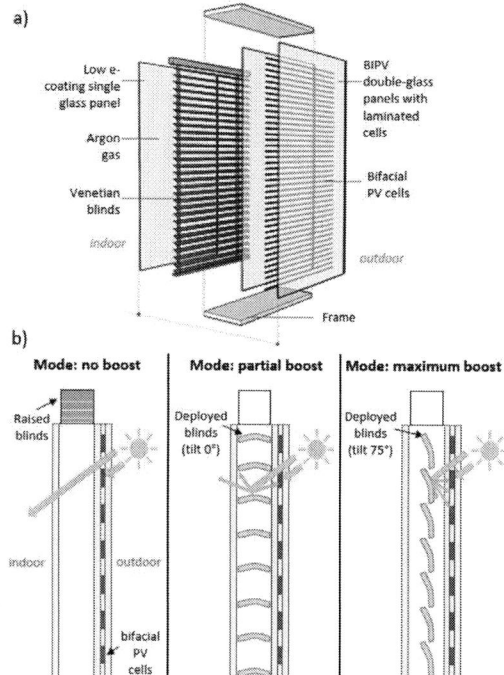

Figure 1: a) Exploded view of the PV window. b) Schematic representation of its operating principle. Three modes are shown: raised blinds (left), partial boost mode with open blinds (middle), and boosted reflection mode with closed blinds (right).

The 12 different PV window design variations are based on the following:

- 2x bifacial c-Si PV technologies
  - interdigited back contact (IBC) cells (by TNO)
  - passivated emitted and rear contact (PERC) cells (by S'Tile [6])
- 2x Coverage Ratio (CR)
  - CR=50% (vertical cell distance = 10 mm)
  - CR=60% (vertical cell distance = 7 mm)
- 3x types of Venetian blinds (by Pellini [7])
  - S157 (grey, most sold commercially)
  - S102 (white, high diffuse reflection)
  - V95 (grey, high specular reflection)

Figure 2: Photographs (with labeling) of the 12 PV windows installed at the SolarBEAT facility.

Each window is connected to an IV-tracer to monitor the electrical parameters and gather full IV curves. Additionally, we measure: in-plane irradiation, airgap temperature inside the window, blinds control (to set tilt and height), weather data (GHI, DHI, T_{amb}, wind, etc.) and shade detection via a webcam.

The venetian blinds embedded in the PV windows are slightly curved aluminum-based slats. The three variations tested in this work are: the S157, which has a silver coating and is the most commercially sold blind type, thus considered our "reference"; the S102, which has a white coating and is highly diffusive; and the V95, which has a special low emissivity and highly reflective coating with silver color appearance. Figure 3 shows the total reflectance of the three blinds and their decomposition in diffuse and specular components (throughout the visible spectrum range).

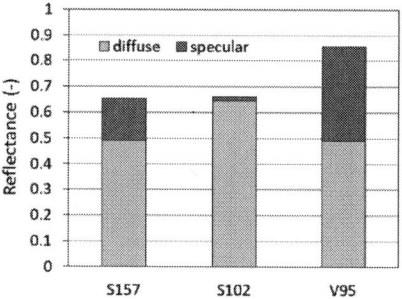

Figure 3. Total reflectance of the three different venetian blind types, distinguishing between diffuse and specular component (in the visible spectrum range).

One full year of data has been recorded, during which several measurement campaigns have been conducted. Four main configurations were tested: 1. no blinds (reference case); 2. blinds deployed and open horizontally (tilt=0°); 3. blinds deployed and closed with convex side facing outwards/upwards (tilt=75°); and 4. blinds deployed and closed with convex side facing inwards/downwards (tilt=-75°).

3 OUTDOOR ANALYSIS RESULTS

3.1 Overall daily performance comparison

Figure 4 shows an overview of the daily temperature-corrected performance ratio (PR$_{T\text{-corr}}$) over the full year, in which it is possible to clearly distinguish between the various measurement campaigns. For better visual clarity, only the 6 PERC windows are shown in the plot. It should be noted that in the PR calculation we used the standard equation for mono-facial PR (since there is no irradiance sensor on the rear side of the cells). Even when the blinds are not deployed, the average PR is around 100-105%; this high value reflects a first boosting effect given simply by the use of bifacial cells, rather than mono-facial. When looking at the periods where the blinds are fully deployed and tilted at t=75°, we clearly see an increase in PR, with the V95 blind (i.e. high specular reflection) outperforming the other types.

The boxplots in Figure 5 summarize the overall performance results, by grouping all days at which the blinds were kept at the same position throughout the whole year. Only one window per type is shown for clarity, but the trend is identical for the others. Here, it can be clearly observed that the blinds position giving the highest boost is at tilt t=75° (convex side facing out/upwards), while the PR at the other positions is either comparable or slightly lower than the case of no blinds.

1179

41st European Photovoltaic Solar Energy Conference and Exhibition

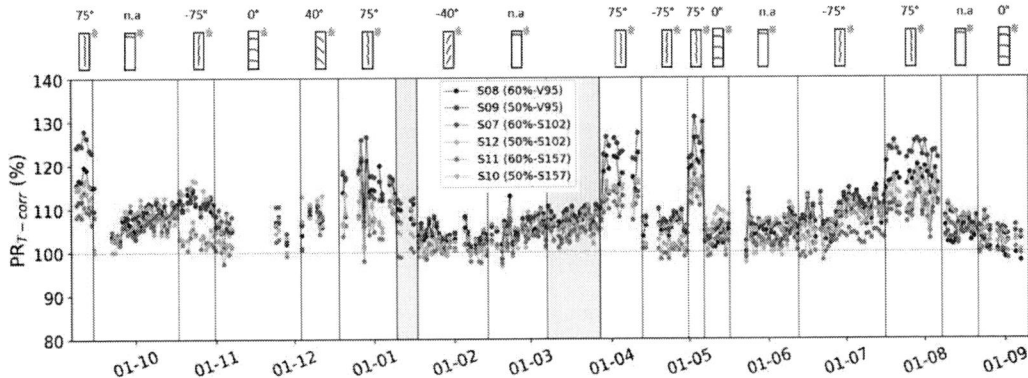

Figure 4. Overview of daily performance ratio for the 6 PERC windows, distinguishing between the different measurement campaigns at the various blinds positions.

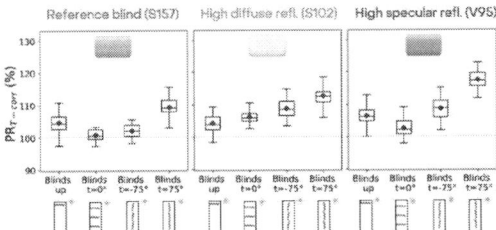

Figure 5. Boxplots showing daily $PR_{T\text{-corr}}$ of each window type (S157, S102, V95) for each blind configuration, including data gathered over the full year.

If we calculate the average of the performance of the various windows with the same blind type, we obtain the results shown in Table I. The table presents the average relative performance increase, or decrease, of the various configurations relative to the reference case, where the blinds are fully raised. It can be concluded that the optimal blind configuration (giving the highest energy boost) is the tilt of 75°. In this position, the most reflective blind V95 leads to a daily boost of 13%. For the white coated blind with high diffuse reflection properties, S102, a significant

boost of 9% is also recorded, and even the most common gray-painted blind, S157, showed a PR increase of 5%.

Table I: Effect on the PR of the various blinds types at the different tilt configurations (calculated with respect to case of blinds raised up).

Blind type	Tilt angle t = 0°	Tilt angle t = -75°	Tilt angle t = 75°
S157	-4%	-3%	+5%
S102	+1%	+4%	+9%
V95	-2%	+2%	+13%

3.1 Performance boost on sunny days

The values presented above are based on daily averages; on sunny days, we saw that the instantaneous energy boost given by the use of the blinds can reach significantly higher values. This can be clearly observed in Figure 6, in which the energy boost of blinds V95 and S102 are tested in two sunny days and compared (on the same day) with the other blinds in which the blinds were raised up. The instantaneous energy yield boost on those days were up 25% for the V95 and 10% for the S102.

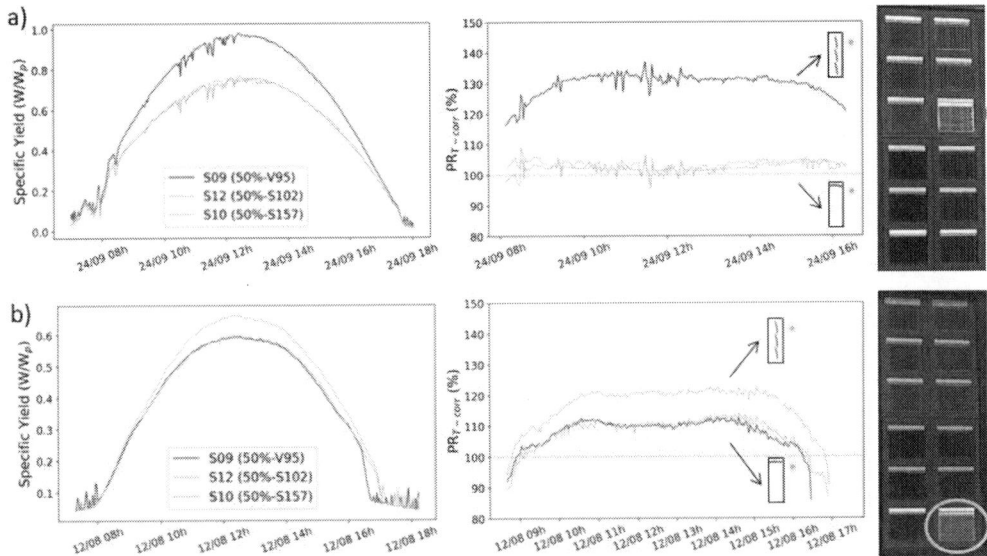

Figure 6. Specific yield and T-corrected PR of the PV windows during a sunny day in which only one blind type was deployed (at tilt 75°) while all the remaining windows had no blinds in use. The tested blinds type were the V95 (a) and S102 (b). On the right, photographs of the setups during these two days.

1180

4 MODELLING RESULTS

4.1 Modelling approach

To better understand the system from an optical and electrical point of view, a simulation model was created in BIGEYE, which is a simulation tool for bifacial PV modelling developed by TNO [8]. The model can provide additional information on the window performance under different operation scenarios, such as with different blind types and configurations or in different weather conditions, so that their boosting potential can be fully assessed.

Figure 7 shows the geometrical model of a PV window created in BIGEYE. As can be seen, the bifacial c-Si PV cell strips were modelled as an homogeneous bifacial PV module for which number of cells, cell area, interconnection pattern, bifaciality factor and transparency values were defined, along with other optical, electrical and thermal parameters describing the properties of the physical windows. A series of reflective slats were modeled at the back side of this bifacial PV module to simulate the Venetian blinds, for which geometrical and optical properties were defined. The rest of the test façade where every window is embedded was added to finish the geometrical model.

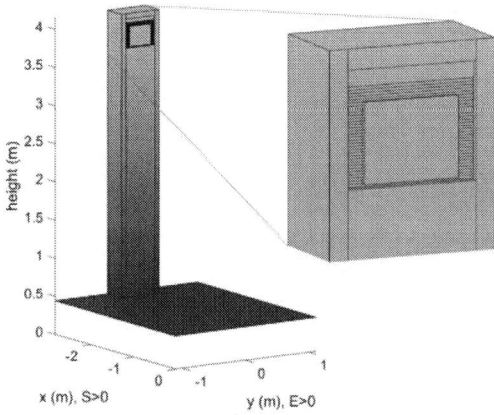

Figure 7. Geometrical model of the PV window.

4.2 Model validation

To evaluate the accuracy and reliability of the model, several validations were conducted comparing the results between simulated performance and experimental measurements. These validations were made for various blind positions experimentally tested as part of the measurement campaigns, as well as for different seasons and weather conditions. The comparison of the results aimed to ensure that the simulated window performance accurately described the behavior of the windows in real outdoor conditions across different scenarios.

Figure 8a, 8c and 8d show a comparison between the measured and the simulated Pmpp, Vmpp and Impp, respectively, for a fully sunny day. As can be noted from this example, the model can accurately predict the power output of the window, and this is true for both cases, with retracted and deployed blinds. The most significant differences in the measured versus simulated results are highlighted in yellow, and can be explained by the fact that the experimental setup experienced some shading from its surroundings in the early morning and late afternoon hours. This undesired effect was not accounted for in the simulation model.

4.2 Modelling results

At first, the simulations were used to gather more detailed information on the rear-side irradiance gain due to the blinds deployment, which could not be directly measured from the experimental setup. The modelled rear side irradiance on a sunny day hitting on a window with the blinds deployed and tilted at $75°$ is shown in Figure 9. All three blinds types are plotted in order to compare their different contribution to the reflected irradiance. For comparison purposes, the front-side total irradiance (both measured and simulated) is also depicted. Thanks to the blinds reflections, the rear side of the cells receive an extra 25% irradiance at noon when the V95 blinds are used, and approximately 16% and 20% extra when the S102 and S157 are used, respectively.

Additionally, the simulation model was used to estimate the potential of the Venetian blinds under test to boost the energy yield of the windows. To this aim, the window performance in a typical meteorological year (TMY) was simulated for every blind type and configuration (for the same location: Eindhoven, NL).

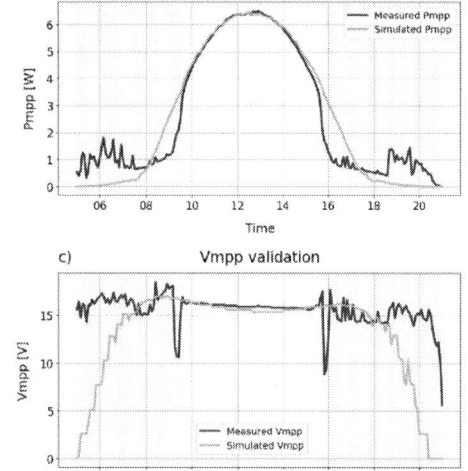

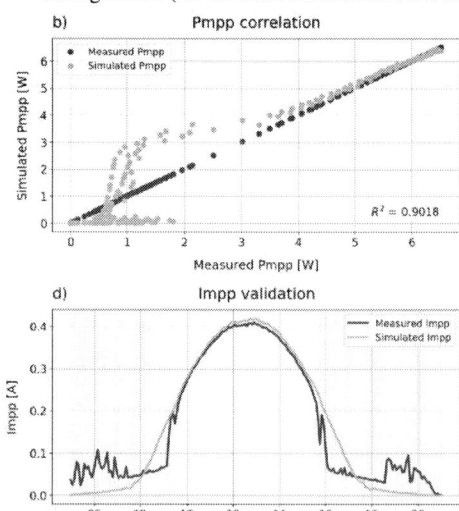

Figure 8. An example of the electrical performance validation for a fully sunny day (29/07/2024).

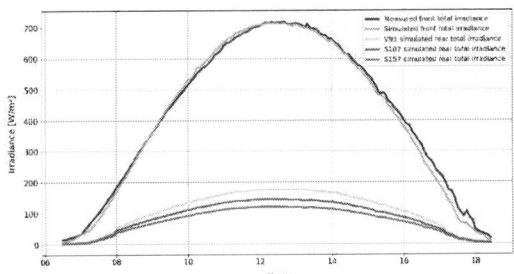

Figure 9: Comparison between measured and simulated irradiance on the window plane, including rear-side contribution of the three blind types, for a fully sunny day (08/09/2023).

Figure 10 shows the results of these simulations. As can be seen from the figure, the V95 blind type has the higher potential to boost the energy yield of this window regardless of its tilt angle, followed by the S102 and the S157 types.

It should be noticed that these simulations are assuming that the blinds are kept at the respective positions for the whole year, which is an extreme and not realistic assumption. A more realistic assumption could be that in the summer months the blinds are mostly deployed in closed position (75°), while in winter they are mostly raised up, with intemediate positions in the spring and autumn. Naturally, in reality, the user could adjust the blinds every day depending on the sun position and irradiance level, in order to ensure thermal and visual confort inside the room. Thus, altought the exact boost given by each blind depends on the users' preferences and the particular sky conditions, the number would lie between 6-19% for the V95 hihgly specular refletive blind, between 4-15% for the S102 hihgly diffusive, and between 3-13% for the standard S157 blind.

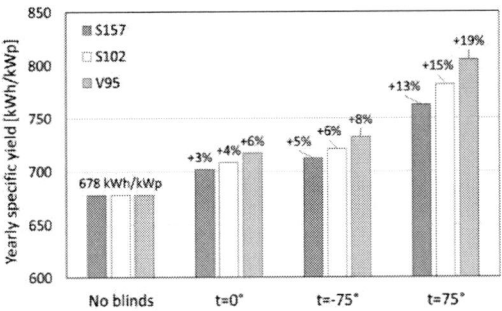

Figure 10: Blinds performance boosting potential under different blinds configurations for a typical meteorological year.

5 CONCLUSIONS

A multifunctional PV window concept aimed at maximizing the energy yield of semitransparent PV windows was presented. The proposed concept utilizes bifacial c-Si solar cell strips combined with Venetian blinds, all embedded in a unique insulating glazing unit. The bifacial technology increases the energy yield by using the blinds as reflectors, directing more irradiance to the cells' rear side. This was demonstrated by a full-year outdoor testing research, and backed-up by simulations. Three boosting effects could be identified. The first is simply given by the use of bifacial cells instead of monofacial. This already makes this PV window a very competitive BIPV product, with a power density of 100 Wp/m² (when coverage ratio is 60%). On top of this, the second boosting effect is given by the deployment of the venetian blinds, especially when used in "closed" position. And the third one is given by the choice of highly reflective slats, in particular the tested V95 blind with high specular reflective component. This type of blind allowed to achieve a maximum performance increase of 25% on sunny days and a daily average increase of 13% compared to the case of no blinds. Simulations also helped to understand further the internal reflection mechanisms and evaluate the potential of this PV window concept. As an outlook for future research, the PV window potential assessment could be performed for different locations worldwide.

REFERENCES

[1] G. Yu, H. Yang, D. Luo, X. Cheng, M. K. Ansah, Renew. Sustain. Energy Rev. 2021, 149, 111355.
[2] TopSector Energie, ZIEZO Public Summary, https://projecten.topsectorenergie.nl/projecten/ziezo-36497 (accessed: May 2024).
[3] M. Ribberink (Pilkington Group Ltd.), US2020332593A1, 2020.
[4] S. Villa, D. Out, N. Guillevin, M. Hurtado Ellmann, M. Ribberink, R. Valckenborg, Solar RRL, 2024, 2400515.
[5] TNO, SolarBEAT, the outdoor test facility for BIPV, www.tno.nl/solarbeat (accessed: May 2024).
[6] S'Tile, S'Tile, https://silicontile.fr/en/ (accessed: May 2024).
[7] Pellini, Venetian blind, https://www.pellini.net/en/tipologie/venetian-blind (accessed: May 2024).
[8] G.J.M. Janssen, B.B. Van Aken, A.J. Carr, A.A. Mewe, Energy Procedia, 2015, 77, 364.

41st European Photovoltaic Solar Energy Conference and Exhibition

This presentation was selected by the Sc. Committee of the EU PVSEC 2024 for submission of a full paper to one of the EU PVSEC's collaborating peer-reviewed journals.

COST-EFFECTIVE ENERGY TRANSITION: ROOFTOP PV IN EUROPEAN UNION BUILDINGS

Carmen Maduta[1]
Delia D'Agostino[1]
Sofia Tsemekidi-Tzeiranaki[2]
Luca Castellazzi[1]

[1]European Commission, Joint Research Centre (JRC), Ispra, Italy
[2]Network Research Belgium S.A. (NRB) EEIG, Herstal, Belgium

carmen.maduta@ec.europa.eu

ABSTRACT

The building sector, recognized as the largest energy consumer within the European Union (EU), plays a pivotal role in achieving the climate ambition of a decarbonized society by 2050. Crucial aspects of the clean transition involve reducing energy demand and increasing reliance on renewable sources, particularly solar energy.

This paper explores the significance of rooftop photovoltaic (PV) systems in reaching cost-optimal levels of primary energy demand in buildings by analyzing recent data from EU Member States. The findings reveal that integrating rooftop PV can yield, on average, energy savings of around 30-35% for new buildings and around 20% for renovated ones. However, the results also show that global costs can both increase and decrease with the installation of PV systems, depending on the level of energy savings achieved. In scenarios where energy savings exceed 30%, PV variants are generally cost-effective, whereas for savings below 10%, PV installations may not be the most economically viable option. These aspects are particularly important since such calculations and their results are used to benchmark the national level of energy performance of buildings.

The findings highlight rooftop PV as a crucial technology for meeting the goal of Zero-Emission Buildings as of 2030, where on-site carbon emissions will be prohibited.

Keywords: cost-optimal energy performance, zero-emission buildings, rooftop photovoltaics

1 INTRODUCTION

The building sector is a significant contributor to energy consumption and greenhouse gas emissions in the European Union (EU), accounting for approximately 40% of the region's total energy use and around 36% of related CO_2 emissions [1]. Addressing these challenges is crucial to meeting the EU's climate goals of achieving a decarbonized society by 2050. Key strategies in this transition involve reducing energy demand and increasing the reliance on renewable energy sources [2].

Among renewable technologies, rooftop photovoltaic (PV) systems are increasingly acknowledged for their ability to improve building energy performance by decreasing reliance on non-renewable energy sources and harnessing renewable energy [3]. The EU policy framework supports the integration of PV systems through various directives and regulations aimed at improving energy efficiency and reducing carbon emissions [4]. The Energy Performance of Buildings Directive (EPBD) plays a central role in this context. The 2012 and 2018 recasts of the EPBD set ambitious targets for improving building energy performance, with the 2024 recast further strengthening these goals by introducing solar-ready requirements for new and existing buildings [5]. The concept of cost-optimal energy performance, as defined by the EPBD, is crucial for evaluating the effectiveness of energy-saving measures, including PV installations [6].

Despite the recognized benefits, challenges remain in fully integrating PV systems into building performance benchmarks [7]. Variations in energy savings, costs, and policy implementation across different regions and building types complicate the evaluation of PV's role in achieving cost-optimal energy performance. This paper addresses these issues by exploring the role of rooftop PV systems in the context of EU energy performance policies and cost-optimal methodologies. It aims to provide insights into the integration of PV systems within the broader framework of building energy performance and policy objectives, contributing to a more comprehensive understanding of their potential impact on achieving the EU's climate and energy goals.

2 METHODOLOGY

The study aims to show the importance of rooftop PV contribution in setting cost-optimal levels of primary energy demand of buildings. The analysis draws upon the most recent data sourced from reports submitted by EU Member States, employing the cost-optimal methodology to compute performance benchmarks for both new and existing residential and non-residential structures.

2.1. Cost-optimal methodology

Introduced in the 2012 recast EPBD, the cost-optimal level is defined as "the energy performance level which leads to the lowest cost during the estimated economic lifecycle". The cost-optimal methodology is instrumental in determining and defining the most cost-effective levels of minimum energy performance requirements for building. These levels undergo periodic revisions, as countries adapt to new technologies and related costs, aligning with evolving energy and climate targets within the building sector. The first round of cost-optimal calculations was submitted by Member States in 2013, the second in 2018, and the third in 2023 [8].

The calculation approach can be summarized in the following steps:

- Establishment of reference buildings. Real or virtual buildings representing the building stock must be selected. A reference building is defined as a building that is representative of their functionality and geographic location, accounting for indoor and outdoor conditions [9]. Reference buildings are defined for four sub-categories by each country: single-family houses (SFH) and multi-family house (MFH) for residential buildings and offices, and at least one another sub-category (e.g., schools, hospital, hotels, etc.) for non-residential buildings.
- Identification of energy efficiency and renewable measures. These must be implemented in new or existing buildings, including different packages of measures or measures of different levels (e.g., from lower to higher insulation levels), which must respect the EU and national legislation. The impact of applying packages of measures on reference buildings should be estimated in the cost-optimal calculation in terms of energy and financial performance.
- Calculation of the (net) primary energy consumption. The energy performance calculation must be based on the current national or CEN standard methodologies for each selected building variant. Framework conditions for the calculations must be defined in terms of climate data, performance of energy systems, primary energy factors, and indoor air quality.
- Calculation of the global cost. The net present value (NPV) should be used at each step, based on 30 years for residential and 20 years for non-residential buildings. The included cost categories are: initial investment costs, running costs (i.e., energy, operational, maintenance, replacement costs), disposal costs, final value, and the cost associated with CO_2 emissions (only for the macroeconomic perspective). For the assessment of the financial performance (global costs) of the chosen combinations of packages, the European Standards EN 15459 is suggested as reference.
- Identification of cost-optimal levels. This should be expressed in primary energy consumption (in kWh/m^2 per year) for each reference building. The cost-optimal configuration presents the lowest costs maintaining a high performance. It can be identified in the lower part of the curve that reports global costs (Eur/m^2) and energy consumption (kWh/m^{2}y).
- Evaluation of the shortfall against current minimum energy performance requirements. If the difference is higher than 15%, Member States must justify the gap or define a plan to reduce it. The national benchmark related to the final outcome of the cost-optimal calculations can be calculated for a financial or a macroeconomic perspective.

2.2 Research aims

Although the cost-optimal methodology has been widely implemented and discussed in the literature, a study focusing on the rooftop PV impact on the benchmarked energy performance of EU buildings is missing. This study addresses this research gap by providing an overview on how rooftop PV is considered by countries in benchmarking their building standards of minimum energy performance requirements. It focuses on a comparative analysis of energy savings resulting from energy efficiency upgrades, examining scenarios both with and without rooftop PV. This comparative exploration spans diverse climatic zones, building sub-

categories, floor areas, and installed peak power, aiming to reveal dominant trends across EU. It is important to note that, as of June 2024, not all countries have submitted updated calculations. Among those that have, not all reports provided a clear consideration of PV systems. Consequently, this analysis focuses on the countries where a clear characterization of PV systems across various building types was available: Belgium, Denmark, Finland, Estonia, Italy, Cyprus, Czechia, Croatia, and Hungary.

3 RESULTS

The predominant trend observed across countries that submitted the cost-optimal calculation updates is the reliance on the contribution of PV systems when calculating and determining the cost-optimal primary energy demand for all four sub-categories (SFH, MFH, office and other), both in new and existing buildings. Most Member States explored PV as viable renewable energy technology. PV emerges as an attractive choice not only for enhancing self-consumption and therefore increasing energy independence but also for its capacity to offset non-renewable primary energy demand through exported energy, as outlined in the methodology framework. This results in significantly lower energy performance levels. However, the extent of this contribution is influenced by climatic conditions and building categories, including both new and existing residential and non-residential buildings.

Therefore, the following sections addresses the PV contribution in reducing the non-renewable energy demand by building type, climatic zone. It also explores the global costs of variants with and without PVs.

3.1 Energy savings by building type

The results show that, on average, in new buildings—whether residential or non-residential—the cost-optimal package with rooftop PVs can reduce, on average, the non-renewable primary energy demand by 30-35%, with a maximum reduction of 63% and a minimum of 7% compared to the no-PV package (Figure 1). Renovated buildings show a slightly lower reduction, around 20%, with a maximum of 55% and a minimum of 2%.

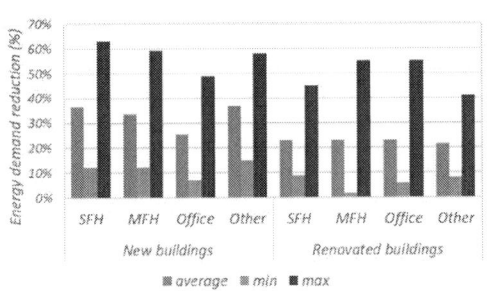

Figure 1: Non-renewable energy demand reduction (%), average, minimum and maximum observed across the Member States, by building type

Examining the various building types, no significant differences are noted among the categories. However, a key observation is that non-residential buildings, particularly those other than offices, are less represented in the data compared to residential buildings, such as SFH and MFH. Consequently, there is less data available to make observations about non-residential buildings.

1184

3.2 Energy savings by climate zones

Climate zones are a crucial factor in determining the effectiveness of PV systems in reducing non-renewable energy consumption. For this analysis, the climate zones defined in the European Commission guidelines on NZEB deployment are used [10]. The following categorisation into climate zones was applied: (1) Mediterranean: Cyprus, Croatia, and Italy (2) Oceanic: Belgium, and Denmark; (3) Continental: Czechia and Hungary and (4) Nordic: Estonia and Finland [11].

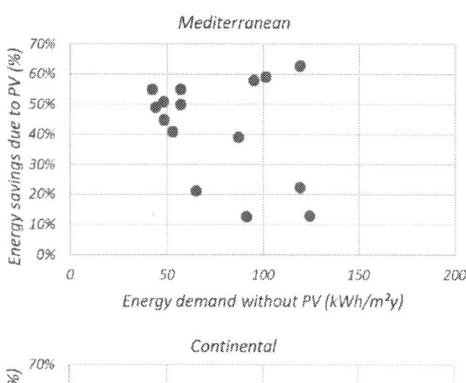

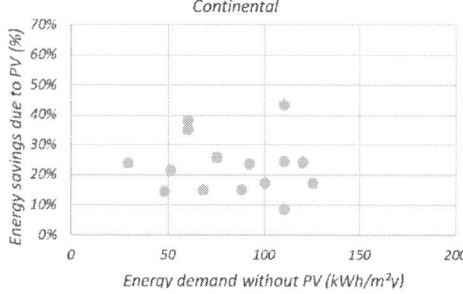

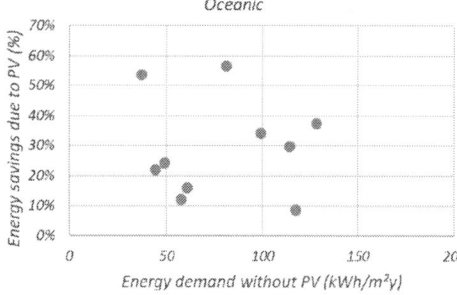

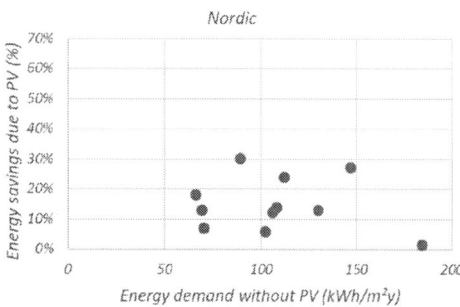

Figure 2: Energy demand without PV (kWh/m²y) and the corresponding energy saving (%) of non-renewable primary energy due to rooftop PV across the building types in Member States, by climate zone.

In Mediterranean countries, the cost-optimal integration of PV results in an average reduction of non-renewable primary energy demand of around 45%, with savings ranging from 13% to 63% depending on the building type (Figure 2). In Nordic countries, the average reduction is about 15%, varying between 2% and 30% across different building types in the cost-optimal configuration. In Continental and Oceanic countries, the average reduction in non-renewable primary energy demand is between 20% and 30%, with a maximum of 57% and a minimum of 8%.

3.3 Global costs comparison

Another important consideration is how the costs change in the variants with and without PV systems, in the cost-optimal configurations. To depict potential trends in global cost changes, Figure 3 shows the difference (in %) between global costs without PV and with PV in relation with the reduction in non-renewable primary energy due to PV, by climate zone. A positive value on the y-axis indicates a lower global cost with PV compared to without PV, while a negative value indicates a higher global cost with PV. It is important to note that such an analysis is only possible for a few countries (Belgium, Denmark, Finland, Czechia, and Hungary), as in most cases, only the cost category corresponding to the cost-optimal scenario—generally with PV—is provided.

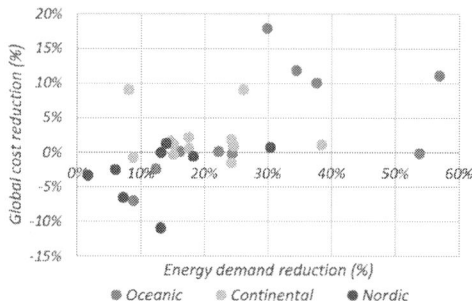

Figure 3: Global cost chance between without and with PV variants and the corresponding reduction of non-renewable primary energy demand due to rooftop PV, by climate zone

A first observation is that in all cases where the projected energy savings due to PVs are higher (e.g., above 30%), the global cost is reduced compared to the non-PV variants. Conversely, for lower energy savings (e.g., below 10%), the PV variant does not appear to be a cost-effective measure.

When examining the different climate zones, the Nordic zone generally shows higher global costs for the PV variant compared to the no-PV variant, except when energy savings exceed 10%, at which point the costs tend to be similar. The Continental and Oceanic zones show diverse results: the Continental zone exhibits both positive and negative global cost changes, even for similar energy savings. These variations may be influenced by country-specific parameters such as the evolution of energy prices and technologies cost, primary energy factors, and discount rates.

4 CONCLUSIONS

The paper presented the significant role of rooftop PV systems in determining the cost-optimal primary energy demand for buildings across the EU. By leveraging recent data submitted by EU Member States, the study provides a comprehensive overview of how rooftop PV installations influence energy performance benchmarks in both residential and non-residential buildings.

Across all building types—both new and existing—rooftop PV systems contribute substantially to reducing non-renewable primary energy demand. The average reduction is between 30-35% for new buildings and around 20% for renovated ones. This aspect is particularly important since such calculations and their results are used to benchmark the national level of energy performance of building.

The impact of PV systems varies significantly by building type and climate zone. Result reveal that global costs can both increase and decrease with the installation of PV systems, depending on the level of energy savings achieved. In variants where energy savings exceed 30%, PV variants are generally cost-effective, whereas for savings below 10%, PV installations may not be the most economically viable option. This is particularly evident in Nordic countries, where the PV variant is often not selected as the benchmarking level. In such cases, countries consider alternative renewable sources. Variations in cost trends across different climate zones and building types highlight the importance of context-specific analysis to inform energy policy and investment decisions.

The study's analysis is constrained by the availability and completeness of data from Member States. While the focus is on countries that provide clear considerations of PV systems, future research could benefit from a broader dataset that includes more countries. Additionally, exploring the impact of dynamic factors such as future changes in energy prices, technological advancements, and evolving policy frameworks will further refine the understanding of PV systems' cost-effectiveness in achieving energy performance goals of buildings.

5 ACKNOWLEDGEMENTS

The views expressed here are purely those of the authors and may not, under any circumstances, be regarded as an official position of the European Commission. The authors wish to thank Christian Thiel and Maurizio Bavetta (JRC) for supporting this research.

6 REFERENCES

[1] European Commission, "Key facts on energy and EU buildings," Energy. Energy Performance of Buildings Directive. [Online]. Available: https://energy.ec.europa.eu/topics/energy-efficiency/energy-efficient-buildings/energy-performance-buildings-directive_en

[2] European Commission, "Communication from the Commission to the European Parliament, the Council, the European Economic and Social Committee and the Committee of the Regions: The European Green Deal," Brussels, COM(2019) 640, Nov. 2019.

[3] D. D'Agostino, D. Parker, P. Melià, and G. Dotelli, "Optimizing photovoltaic electric generation and roof insulation in existing residential buildings," *Energy Build.*, vol. 255, p. 111652, Jan. 2022, doi: 10.1016/j.enbuild.2021.111652.

[4] European Commission, "Solar energy, available at: https://energy.ec.europa.eu/topics/renewable-energy/solar-energy_en." 2024.

[5] European Commission, "Energy performance of buildings (recast). European Parliament legislative resolution of 12 March 2024 on the proposal for a directive of the European Parliament and of the Council on the energy performance of buildings (recast). Texts adopted." 2024.

[6] P. Moran, J. Goggins, and M. Hajdukiewicz, "Super-insulate or use renewable technology? Life cycle cost, energy and global warming potential analysis of nearly zero energy buildings (NZEB) in a temperate oceanic climate," *Energy Build.*, vol. 139, pp. 590–607, Mar. 2017, doi: 10.1016/j.enbuild.2017.01.029.

[7] V. Kapsalis *et al.*, "Bottom-up energy transition through rooftop PV upscaling: Remaining issues and emerging upgrades towards NZEBs at different climatic conditions," *Renew. Sustain. Energy Transit.*, vol. 5, p. 100083, Aug. 2024, doi: 10.1016/j.rset.2024.100083.

[8] P. Zangheri, D. D'Agostino, R. Armani, and P. Bertoldi, "Review of the Cost-Optimal Methodology Implementation in Member States in Compliance with the Energy Performance of Buildings Directive," *Buildings*, vol. 12, no. 9, p. 1482, Sep. 2022, doi: 10.3390/buildings12091482.

[9] I. Ballarini, S. P. Corgnati, and V. Corrado, "Use of reference buildings to assess the energy saving potentials of the residential building stock: The experience of TABULA project," *Energy Policy*, vol. 68, pp. 273–284, May 2014, doi: 10.1016/j.enpol.2014.01.027.

[10] European Commission, "Commission Recommendation 2016/1318 of 29 July 2016 on Guidelines for the promotion of nearly zero-energy buildings and best practices to ensure that, by 2020, all new buildings are nearly zero-energy buildings," 2016.

[11] C. Maduta, D. D'Agostino, S. Tsemekidi-Tzeiranaki, L. Castellazzi, G. Melica, and P. Bertoldi, "Towards climate neutrality within the European Union: Assessment of the Energy Performance of Buildings Directive implementation in Member States," *Energy Build.*, vol. 301, p. 113716, Dec. 2023, doi: 10.1016/j.enbuild.2023.113716.

41st European Photovoltaic Solar Energy Conference and Exhibition

This presentation was selected by the Sc. Committee of the EU PVSEC 2024 for submission of a full paper to one of the EU PVSEC's collaborating peer-reviewed journals.

DYNAMIC BIPV SHADING SYSTEMS: PERFORMANCE ANALYSIS FOR HIGH TRL VALIDATION AND MARKET TRANSFER

Tian Shen Liang[a], Paolo Corti[a], Pierluigi Bonomo[a], Francesco Frontini[a]

[a] SUPSI – University of Applied Sciences and Arts of Southern Switzerland, Institute of Applied Sustainability to the Built Environment (ISAAC), Via Flora Ruchat-Roncati 15, CH 6850, Mendrisio, Switzerland

ABSTRACT: This paper evaluates the performance of an innovative, dynamic, and vertically-oriented building-integrated photovoltaic (BIPV) shading device. The research is part of a Swiss Pilot and Demonstration project, advancing the technology from Technology Readiness Level (TRL) 5 to 7, with the goal of validating its operational consistency, replicability, and cost-effectiveness for market deployment. The shading slats consist of two-string PV modules, featuring laminated glazing with an outer layer of white satin glass. Two designs are tested using a small-scale mock-up to assess their effectiveness in temperature reduction and energy production: 1) an optimised version with each string connected to a bypass diode, and 2) a standard version where two strings share a single bypass diode. Results show that the optimised design consistently achieves lower module temperatures and higher energy output, with more than a 20% increase during spring and summer. No additional risks are identified, as the pilot installation does not experience extreme temperature or humidity conditions. The first-floor system's specific energy yield is lower due to partial shading, but this impact would have been greater by using standard PV modules, underscoring the need for BIPV-specific expertise for successful implementation.

Keywords: building-integrated photovoltaics (BIPVs), photovoltaic shading device (PVSD), technology readiness level (TRL)

1 INTRODUCTION

Building-integrated photovoltaic (BIPV) plays a critical role in transforming the building envelope to generate renewable energy, and it can be integrated as roofs, façades, and external structures [1,2]. Photovoltaic shading device (PVSD) falls into the last category, with examples like vertical and horizontal PV louvres, PV window blinds, PV fins, and PV egg crates [3]. These devices help reduce energy consumption by shading interiors from direct sunlight, maintaining thermal comfort, and generating renewable energy on-site [4,5].

However, PVSD implementations are still relatively rare, with limited studies or products available. As noted in our review paper, this scarcity is due to two main challenges [3]. First, integrating PV technology with building structures is technically complex, requiring a combination of skills in PV systems, architecture, and construction, along with unconventional project management strategies. Second, the high initial costs of traditional BIPV shading solutions deter investors, especially those accustomed to large-scale PV power plants, as they often overlook the benefits of integrating these systems with architectural and functional elements.

To bridge this gap, SUPSI, in collaboration with Sunage SA and Aziende Industriali di Lugano (AIL) SA, initiated a Pilot and Demonstration project. The project demonstrates the techno-economic potential of an innovative, aesthetically pleasing, and pre-fabricated dynamic PVSD system installed on the façade of the new auditorium at Franklin University Switzerland (FUS) in Sorengo (Figure 1, Top). It aims to advance the technology from Technology Readiness Level (TRL) 5 to 7 and facilitate its market introduction. The project also sought to develop expertise in manufacturing and installation while raising awareness of PVSD technology among building users and the public.

The research includes the validation of a small-scale PVSD mock-up (Figure 1, Bottom) with an optimised design in a relevant environment to achieve TRL 6, focusing on improving temperature and energy performance compared to standard PV solutions. Then it demonstrates the system's readiness in full-scale buildings

to achieve TRL 7, by ensuring its reliability and replicability.

This paper presents the findings from a year-long monitoring of the mock-up and pilot installation, which involved analysing datasets on temperature, humidity, and electrical performance of the PVSD.

Figure 1: Top) pilot installation at FUS, and bottom) mock-up at SUPSI Mendrisio campus rooftop.

2 METHODOLOGY

2.1 Technical Details of SUPSI Mock-up

The quantitative performance analysis and validation of the mock-up involved comparing optimised and standard PV module designs (Figure 2), with two key

1187

performance indicators (KPIs) guiding the assessment: 1) the optimised module should have a lower temperature than the standard module, and 2) the energy yield outdoors should be at least 20% higher than that of the standard module. These KPIs were established based on parametric analysis [6], simulations [7,8], and preliminary studies aimed at minimising the effects of shading mismatches on PV slats [9,10,11,12].

The optimised PV module was designed by dividing its internal circuit vertically into two electrically independent strings, each containing six cells and connected to a bypass diode. In contrast, the standard PV module does not feature such a division, resulting in a single circuit. Both designs utilised 12 crystalline silicon cells, each measuring 158.75 x 158.75 mm (Table I).

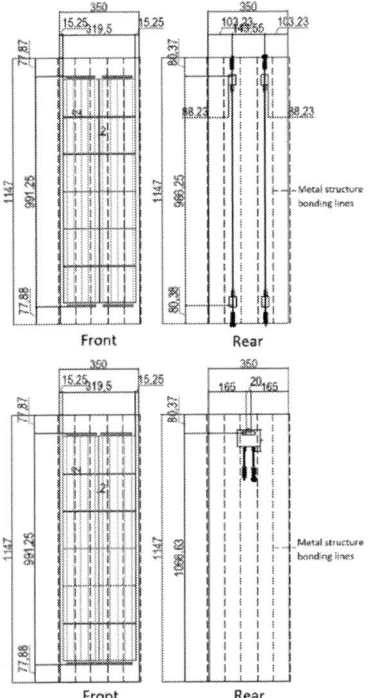

Figure 2: Schematic diagram of top) optimised, and bottom) standard PV modules.

Of the six PV modules (refer to Figure 1, bottom), four are active: the left strings of slats L2 and L3 were connected to one maximum power point tracker (MPPT), the right strings to a second MPPT, and L4 and L5 to a third MPPT. The PV modules were bonded to an aluminium structure using industrial adhesive (thus known as PV slats). PV slats rotate around a vertical axis to track the Sun's movement based on the site's geographic coordinates. The rotation limits, along with start and end times, are controlled remotely via software developed by the project partner, with an option for manual adjustment, bypassing solar tracking.

Temperature measurements were taken on the diodes, the rear-side glass of the left and right strings (approximately at the centre of the cells), and the air gap between the module's rear side and the aluminium enclosure for slats L2 and L5. The MPPT recorded the maximum voltage and current. Additionally, three EKO ML-02 silicon pyranometers were installed on the front glass of slat L2 to measure the plane-of-array (POA) irradiance on the left string, right string, and at the vertical

centerline of the slat. A meteorological station on the SUPSI rooftop provided data on irradiance, ambient temperature, and humidity.

Table I: Details of PV Module

Item	Description
Front glass	Float satin glass with 4 mm thickness; uniform Suncol colour "Bianco Traffico"
Encapsulant	Polyolefin Elastomer (POE)
Cell type	Crystalline silicon cells (wafer size: G1); black ribbon
Back glass	Float clear with a 4 mm thickness
Weight	23.75 kg/m^2
Module total thickness	9.5 mm (-0.5 mm / +1.5 mm)
Nominal power	67.7 Wp per module

2.2 Techincal Details of Pilot Installation

The PVSDs on the façade of the demonstration building are separated into two systems, one for the first floor and another for the second floor (Figure 3). The auditorium's curved façade faces multiple directions. Each moving group consists of three PV slats, creating a total of 21 active groups and 63 PV slats: seven groups facing west, eight facing south, and six facing east. The position of the three slats within each group is controlled simultaneously by a motor and speed reducer, with the motor system managed by a programmable logic controller (PLC). The PLC is integrated with the building's home automation system, allowing users to manually adjust the PV slats. This capability enables complete opening or closing of all slats to control the light intensity inside the auditorium or to position the slats for maintenance and cleaning.

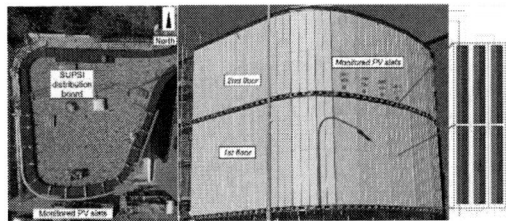

Figure 3: The PVSDs installed on the façade of the demonstration building are grouped into first and second-floor systems.

The optimised PV module design was also used in the pilot installation, with the main difference from mock-up being the nominal power and dimensions: 158 W$_p$ (2310 x 350 mm^2) for the first floor and 124 W$_p$ (1830 x 350 mm^2) for the second floor. The variation in electrical output and height is due to the difference in height between the two floors, resulting in nominal power outputs of 10.3 kW$_p$ for the first floor and 8.1 kW$_p$ for the second floor. Each floor's PV slats consist of two PV modules. As illustrated in Figure 3, the left (red) and right (blue) strings of the PV modules in each group of three slats are connected in series to mitigate the shading effects at extreme angles of opening. Each string is equipped with a power optimiser, and the PV slats on each floor are connected to a SolarEdge inverter. Energy production data for the first and second-floor systems, along with temperature of diodes, air gaps, and PV modules and humidity for selected PV slats on the second floor's south side (as shown in Figure 3), were collected. Data was analysed following the procedures outlined in Figure 4. Data availability was calculated for each month, and sky conditions were categorised for each day to accurately analyse energy yield, as it is directly proportional to the irradiance received by the PV modules.

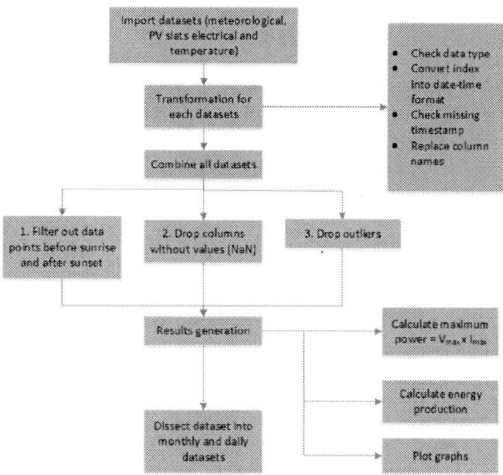

Figure 4: Steps for data preparation and analysis.

3 RESULTS AND DISCUSSION

3.1 SUPSI Mock-up
Temperature Evaluation

Evaluating the maximum temperature is particularly important because it indicates the effectiveness of optimised module in reducing self-shading mismatch. During indoor temperature tests, the optimised module showed a lower maximum temperature under two shading conditions: 2.0°C and 3.0°C cooler in moving shadow and hot spot tests, respectively. Figure 5 provides descriptive statistics of outdoor temperature measurements for PV modules, diodes, and air gaps, showing similar results to the indoor tests. Both strings of the L2 (optimised) module consistently recorded lower maximum and average temperatures compared to the L5 (standard) module, successfully meeting the target for KPI 1.

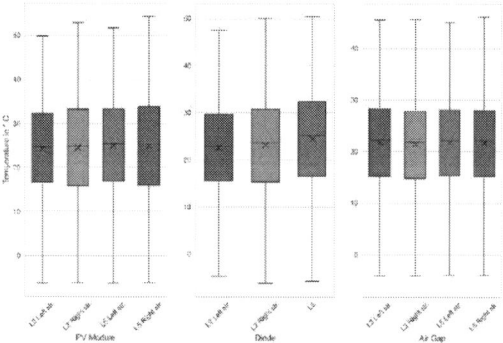

Figure 5: Comparison of left) module, centre) diode, and right) air gap temperature between optimised and standard PV modules. The box's horizontal line and "X" mark denote median and mean values, respectively.

In the L5 module, partial shading on one string caused the current generated by the unshaded string to be limited to the level of the shaded string, resulting in lost power being dissipated as heat in the shaded string and increasing the module temperature [13]. Additionally, the right string of L2 exhibited higher mean and maximum temperatures than the left string, a trend that can be explained in the next section with daily temperature profiles.

No temperature anomalies were detected in the diodes, with maximum temperatures remaining lower than those of the PV modules. The diodes used in the pilot installation are the same as those in the mock-up slats, with a specified maximum junction temperature of 200 °C, a maximum reverse bias voltage of 40 V, and a short-circuit current of 7.5 A. The electric current generated by the mock-up slats was relatively low, ranging between 3 A and 4 A (with a voltage of less than 14 V). The active diode only heated to a relatively low temperature when the lower current from the unshaded strings passed through it, a condition that occurred only in the L5 module and potentially increasing the risk of early failure.

The maximum diode temperatures for both L2 and L5 are significantly below 140 °C, which is the maximum temperature recorded in indoor thermal test conducted in accordance with IEC 61215-2:2021 (MQT 18.1). This confirms the thermal safety and long-term reliability of the bypass diode in outdoor conditions.

Similarly, the air gap temperatures show no abnormalities and are consistently lower than the PV module temperatures. The high thermal conductivity of thermo-lacquered aluminium improves conductive heat transfer, and the support structures within the air gap (as shown by the structure bonding lines in Figure 2) increases the heat dissipation area [14,15].

Figure 6 and Figure 7 present the temperature time series, showing trends on clear-sky and cloudy days. The effectiveness of the L2 module in managing shading and mismatch is evident, as it consistently maintained lower module and diode temperatures throughout the day compared to the L5. This temperature control translates into higher energy performance, which is discussed later.

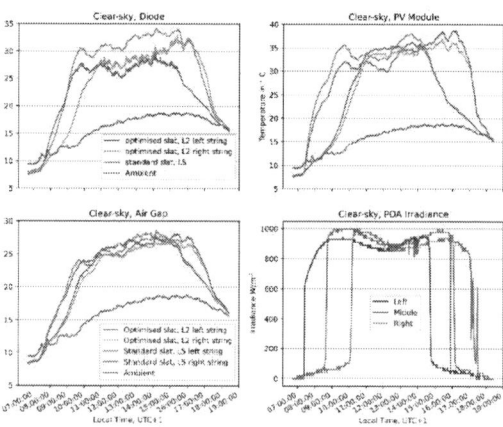

Figure 6: Temperature profile during a clear-sky day (12th March), POA irradiance (measured by EKO ML-02) plot at the bottom right.

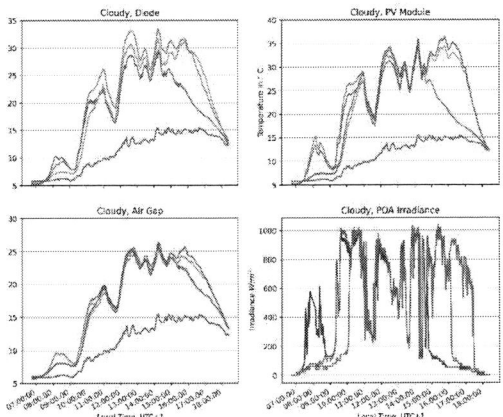

Figure 7: Temperature profile during a cloudy day (11th March). For legends, refer to the previous figure.

All temperature components of L2's right string show peak higher than those of the left string, due to rising ambient temperatures in the afternoon, which generally peak between 3 p.m. and 5 p.m. The right strings of L2 and L5 slats were more exposed to irradiance, as indicated by the red POA irradiance curve in Figure 6, while the left strings had lower temperatures due to shading from adjacent slats (as shown by the blue POA irradiance curve). Similar temperature trends were observed on cloudy days.

Energy Evaluation
During the monitoring period, as shown in Figure 8, the total energy yield for the optimised PV slats (L2 and L3) and the standard PV slats (L4 and L5) are 40.8 kWh and 36.1 kWh, respectively. The energy yield for L2 and L3 was calculated by summing the yields from the left and right strings. Due to the lower solar altitude in winter, vertical PVSDs produced more energy than in summer. However, energy yields in July and December are relatively lower: the data availability are 55% and 45%, respectively, and all recorded days in July are cloudy.

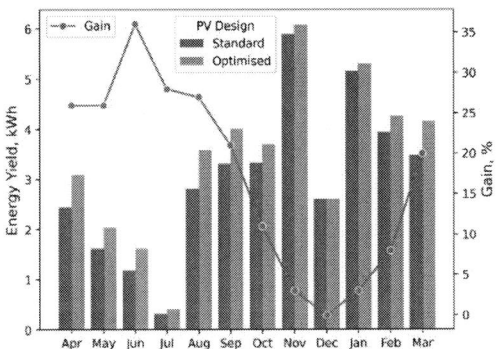

Figure 8: Monthly energy yield for the mock-up.

Nevertheless, the optimised PV slats achieve over 20% energy yield gain for seven months, meeting the KPI 2. The overall annual yield gain is 13%, with gains between 3% and 11% from October to February, except in December, where no gain was observed. The current from the unshaded strings of L4 and L5 is constrained by the partially shaded strings, resulting in lower power output and energy yield [10,16]. In contrast, the optimised design consistently outperformed the standard design in both

temperature and energy performance, as highlighted in the yellow area of Figure 9. The orange area indicates that standard PVs produced higher energy at noon, which is due to the lower peak power and fill factor of the optimised PVs. Although this was not confirmed, it suggests that the optimised PV's energy gain could have been even greater. By splitting the two strings and adding a bypass diode to each, the optimised design reduced the temperature impact of shading mismatch and achieved significant energy yield gains, especially during the summer when the solar altitude is higher.

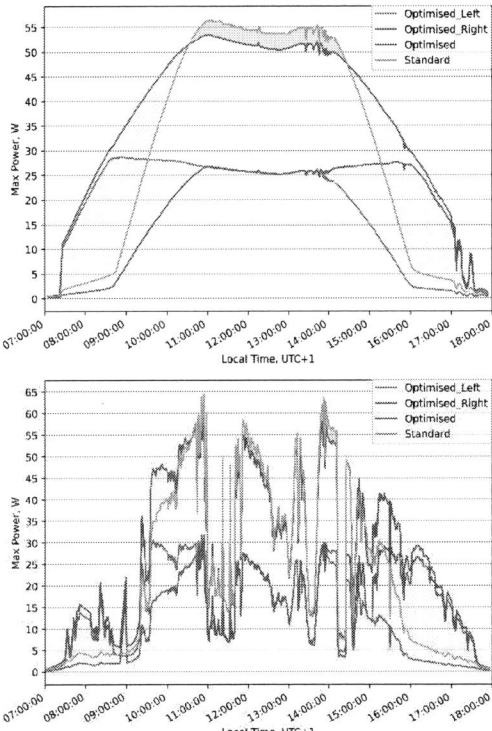

Figure 9: Time series of irradiance and power yield for top) clear sky day (12th March) and bottom) cloudy day (11th March).

3.2 Pilot Demonstration
Temperature Evaluation
Figure 10, left, displays the descriptive statistics for temperature measurements of slats 200-104, as referenced in Figure 3. Group 1 includes the temperatures of Air_1, Module_1, and Diode_1, which correspond to the air cavity, the rear side of the PV module, and the diode temperature of the slat's left string; group 2 refers to the corresponding measurements on the right string. Figure 10, right, shows the descriptive statistics for relative humidity.

Only the measurements of slat 104 are shown here because the temperature data across all monitored slats are fairly consistent. For the other slats (107, 110, and 113), the mean and maximum PV temperature ranges differ by no more than 0.4 °C and 0.2 °C, respectively. Similarly, for diode and air gap temperatures, the mean and maximum variations are within 0.8 °C and 2.8 °C, and 0.4 °C and 1.1 °C, respectively. The minimal temperature variation among slats may suggest consistent manufacturing quality, with differences primarily driven by slat rotation and irradiance exposure at different times

1190

of the day.

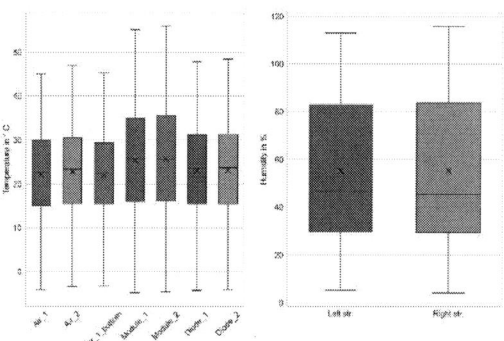

Figure 10: Descriptive statistics of left) temperature components and right) relative humidity measurements for slat 104.

The figure also highlights statistics for Air_1 and Air_1_bottom in slat 104, where the maximum temperature difference is only 0.2 °C. However, for the other three slats, this difference ranges between 1.0 °C and 2.3 °C. A notable temperature difference was observed on August 22nd, a clear summer day, where the top sections of all slats recorded higher temperatures than the bottom sections. The mean and maximum relative humidity values for the other slats show minimal differences compared to slat 104, with variations of around 0.7% and 2%, respectively. Overall, the data are consistent with the mock-up results, showing that the maximum temperatures are the highest for the module, followed by the diode and air gap temperatures. Similarly, the right string generally shows higher maximum temperatures than the left string.

Humidity levels are responsive to daily weather conditions with noticeable differences for both clear and cloudy days (as illustrated in Figure 11): air becomes drier when temperature increases [17]. The diode temperatures closely follow the patterns observed in the PV modules. Consistent with the mock-up data, the diode temperatures do not show any anomalies and remain below the values recorded during indoor thermal testing.

The mean and maximum temperatures of the monitored PV slats are comparable to those of the optimised and standard PV slats of the mock-up, indicating that the temperature performance of the pilot installation aligns well with the mock-up system. Additionally, the daily temperature profiles during clear and cloudy days in the pilot installation mirror those observed in the mock-up, with the group 2 temperature components being higher than those in group 1, largely due to the increased ambient temperatures in the afternoon.

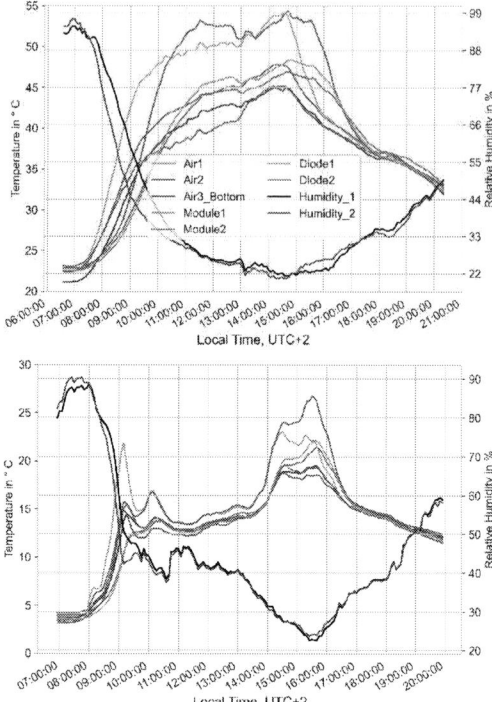

Figure 11: Time series of temperature and relative humidity measurements for left) clear-sky (22nd August) and right) cloudy day (7th April).

Energy Evaluation

Figure 12 presents the monthly energy and specific energy yield from April 2023 to March 2024. The PVSD system generated a total energy yield of 4.85 MWh on the first floor and 4.19 MWh on the second floor. In February, the system was non-operational due to maintenance. Although the first-floor system consistently produced more monthly energy than the second floor, its specific energy yield was lower. This can be attributed to differences in irradiance, as shown in the irradiance simulation in Figure 13. The figure demonstrates the irradiance received by the selected slats at various façade orientations on June 21st. All slats are oriented southwest, controlled by the same tracking algorithm. During June and summer months, the higher solar altitude means the upper sections of the second-floor slats receive more direct irradiance, leading to a higher specific energy yield.

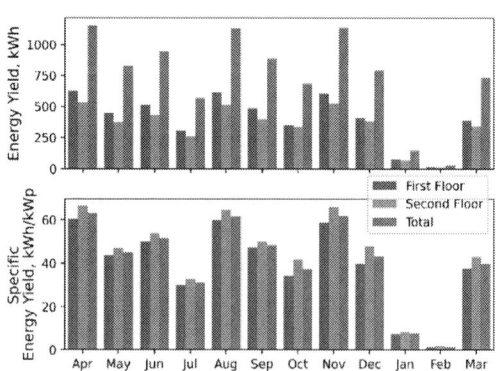

Figure 12: Monthly energy and specific energy yield for pilot installation.

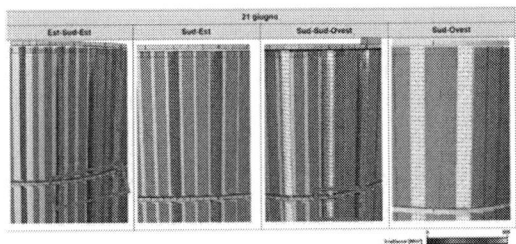

Figure 13: The top parts of the first floor's slats are shaded, resulting in lower exposure to irradiance.

In contrast, the top parts of the first-floor slats are shaded and receive less irradiance, particularly in the east-southeast and south-southwest orientations. Additionally, it was observed that occupants often manually adjusted the slats, leaving them in suboptimal positions for energy generation. This suggests a need to enhance the control algorithm, for example, by programming the slats to automatically return to their optimal tracking positions when the auditorium is unoccupied, which could be detected using motion sensors or computer vision techniques.

4 CONCLUSIONS

This paper evaluates the optimised performance of PVSDs in terms of temperature reduction and increased energy yield. It includes a technical validation and demonstration in an operational environment, highlighting the benefits of PV technologies in multifunctional building façades without sacrificing aesthetics or energy efficiency.

The study addresses complex scenarios such as varying orientations and partial shading, showing that optimising the electrical layout by splitting strings and incorporating diodes mitigates mismatch effects between shaded and unshaded areas. The customised design leads to improved temperature control and energy output, meeting the KPIs. Specifically, the optimised L2 module of the mock-up maintains lower temperatures compared to the standard L5 module, regardless of sky conditions. In scenarios with partial or complete shading, standard modules (L4 and L5) exhibit lower current output due to their connected strings. In contrast, the two-diode design significantly enhances energy yield, particularly during spring and summer when solar angles are higher.

No extreme temperature or humidity levels were detected in the monitored slats of the pilot installation, indicating no added risks. Descriptive statistics of slat 104 was consistent with that from other slats, and overall trends aligned with those observed in the mock-up system.

The first floor's system showed lower specific energy yield compared to the second floor due to shading, especially when the sun is higher in the sky. However, the mock-up results suggest that the impact of shading and mismatch would be more pronounced by using the standard PV design, which emphasise the importance of specialised BIPV consultation since the project's inception.

Our research also suggests that integrating a digital twin could further enhance PVSD system control. This would enable real-time monitoring and diagnostics by comparing actual energy yields with predictions based on time series data and local meteorological inputs. Additionally, indoor sensors could be used to optimise energy performance while maintaining occupant comfort.

ACKNOWLEDGEMENT

The project is supported by the Pilot and Demonstration programme of the Swiss Federal Office of Energy (SFOE). The authors acknowledge the SFOE for the support in the development of the presented work. The authors thank the support provided by project partners: Sunage SA and Aziende Industriali di Lugano (AIL) SA. The authors also thank AIL for providing access to the power and energy data for the pilot installation at Franklin University Switzerland. Open access funding was provided by SUPSI.

5 REFERENCES

[1] P. Bonomo and F. Frontini, "Building Integrated Photovoltaics (BIPV): Analysis of the Technological Transfer Process and Innovation Dynamics in the Swiss Building Sector," Buildings, vol. 14, no. 6, p. 1510, May 2024, doi: 10.3390/buildings14061510.

[2] P. Bonono et al., "Categorization of BIPV applications," 2021.

[3] P. Corti, P. Bonomo, and F. Frontini, "Paper Review of External Integrated Systems as Photovoltaic Shading Devices," Energies (Basel), vol. 16, no. 14, p. 5542, Jul. 2023, doi: 10.3390/en16145542.

[4] W. Bahr, "A comprehensive assessment methodology of the building integrated photovoltaic blind system," Energy Build, vol. 82, pp. 703–708, Oct. 2014, doi: 10.1016/j.enbuild.2014.07.065.

[5] G. Martinopoulos, A. Serasidou, P. Antoniadou, and A. M. Papadopoulos, "Building Integrated Shading and Building Applied Photovoltaic System Assessment in the Energy Performance and Thermal Comfort of Office Buildings," Sustainability, vol. 10, no. 12, p. 4670, Dec. 2018, doi: 10.3390/su10124670.

[6] J. Hofer, A. Groenewolt, P. Jayathissa, Z. Nagy, and A. Schlueter, "Parametric analysis and systems design of dynamic photovoltaic shading modules," Energy Sci Eng, vol. 4, no. 2, pp. 134–152, Mar. 2016, doi: 10.1002/ese3.115.

[7] L. Karlsen, P. Heiselberg, and I. Bryn, "Occupant satisfaction with two blind control strategies: Slats closed and slats in cut-off position," Solar Energy, vol. 115, pp. 166–179, May 2015, doi: 10.1016/j.solener.2015.02.031.

[8] E. Taveres-Cachat and F. Goia, "Investigating the performance of a hybrid PV integrated shading device using multi-objective optimization," J Phys Conf Ser, vol. 1343, no. 1, p. 012086, Nov. 2019, doi: 10.1088/1742-6596/1343/1/012086.

[9] B. B. Pannebakker, A. C. de Waal, and W. G. J. H. M. van Sark, "Photovoltaics in the shade: one bypass diode per solar cell revisited," Progress in Photovoltaics: Research and Applications, vol. 25, no. 10, pp. 836–849, Oct. 2017, doi: 10.1002/pip.2898.

[10] J. C. Teo, R. H. G. Tan, V. H. Mok, V. K. Ramachandaramurthy, and C. Tan, "Impact of Partial Shading on the P-V Characteristics and the Maximum Power of a Photovoltaic String,"

Energies (Basel), vol. 11, no. 7, p. 1860, Jul. 2018, doi: 10.3390/en11071860.

[11] H. P. Ikkurti and S. Saha, "A comprehensive techno-economic review of microinverters for Building Integrated Photovoltaics (BIPV)," Renewable and Sustainable Energy Reviews, vol. 47, pp. 997–1006, Jul. 2015, doi: 10.1016/j.rser.2015.03.081.

[12] S. Lee, J.-B. Eggers, J. Eisenlohr, and K. Lee, "Learning From Tetris: A New Approach for the Automated Configuration of the Interconnection Layout of BIPV Modules for Large-Scale Application," IEEE J Photovolt, vol. 11, no. 3, pp. 770–778, May 2021, doi: 10.1109/JPHOTOV.2021.3064585.

[13] A. A. Aminou Moussavou, A. K. Raji, and M. Adonis, "IMPACT STUDY OF PARTIAL SHADING PHENOMENON ON SOLAR PV MODULE PERFORMANCE," in 2018 International Conference on the Industrial and Commercial Use of Energy (ICUE), 2018, pp. 1–7.

[14] E. Bellini, "Multi-level fin heat sinks for solar module cooling," https://www.pv-magazine.com/2022/02/01/multi-level-fin-heat-sinks-for-solar-module-cooling/.

[15] S. Khan, A. Waqas, N. Ahmad, M. Mahmood, N. Shahzad, and M. B. Sajid, "Thermal management of solar PV module by using hollow rectangular aluminum fins," Journal of Renewable and Sustainable Energy, vol. 12, no. 6, Nov. 2020, doi: 10.1063/5.0020129.

[16] A. DJALAB, N. BESSOUS, M. M. REZAOUI, and I. MERZOUK, "Study of the Effects of Partial Shading on PV Array," in 2018 International Conference on Communications and Electrical Engineering (ICCEE), IEEE, Dec. 2018, pp. 1–5. doi: 10.1109/CCEE.2018.8634512.

[17] A. K. Shrestha, A. Thapa, and H. Gautam, "Solar Radiation, Air Temperature, Relative Humidity, and Dew Point Study: Damak, Jhapa, Nepal," International Journal of Photoenergy, vol. 2019, pp. 1–7, Dec. 2019, doi: 10.1155/2019/8369231.

41st European Photovoltaic Solar Energy Conference and Exhibition

STUDY ON IMPROVEMENT OF POWER GENERATION FOR A WINDOW BY SOLAR RADIATION REFLECTED FROM THE LOW-E COATING OF A SEMI-TRANSPARENT PHOTOVOLTAIC MODULE THAT IS EQUALLY ARRANGED LINEAR DOUBLE-SIDED SOLAR CELLS

Kazuhiko Umeda[1*], Nobusato Kobayashi[1*], Akira Yamaguchi[1*]
Akihiko Nakajima[2*], Kengo Maeda[2*], Akihiro Kuraoka[2*], Naoki Kadota[2*]
[1*] TAISEI CORPORATION, Japan [2*] KANEKA CORPORATION, Japan
[1*] 344-1, Nase-cho, Totsuka-ku, Yokohama, Japan, [2*] 1-12-32, Akasaka, Minato-ku, Tokyo, Japan

ABSTRACT: TAISEI CORPORATION and KANEKA CORPORATION developed see-through PV (Photovoltaic module) for windows. This PV is the building-integrated photovoltaic module, which has high quality design for the purpose of installing exterior building windows. Especially it keeps the basic functions, which are views(daylighting), heat insulation and heat shielding, as well as solar generation.

This research reports the results of outdoor verification and simulation implemented from the perspective of paying attention to the improvement of solar generation by reflecting the near infrared rays on the low-E coating. The main results are as follows.

In the case of aperture ratio 50%, as a result of outdoor verification, it was confirmed that annual power generation by Low-E coating increased by approximately 9%. Also, as results of simulation, the ratio of incident radiation on the indoor side of liner solar cells was larger than that of traditional square solar cells. The results imply the liner shape is advantageous against square shape in the case of solar generation on the indoor side by the incident radiation reflected from the low-E coating. Furthermore, some periodic fluctuations on the PV using linear bifacial solar cells were confirmed through the simulation.

Keywords: double-side power generation, mono-crystal silicon solar cells, low-emissivity glass, simulation, outdoor verification

1 INTRODUCTION

Application of photovoltaic power generation to buildings is important for popularization and expansion of ZEB (Net Zero Energy Building). When a photovoltaic module (PV) is installed on a wall, the installation place is rarely limited by equipment, etc., as on a rooftop, and the demand for PV is expected to increase in the future. On the other hand, a wall has a feature that it is easier to see from the surroundings than a rooftop, and the demand for PV that harmonizes with the building design is expected to increase. Especially, PV installed on a window requires not only high design and power generation performance, but also the basic functions of a window such as view, lighting, heat shielding and heat insulation.

Based on the above situation, the authors examined a new daylight-type PV(hereafter, STPV : Semi-transparent photovoltaic module) by double-sided power generation. In previous reports [1][2][3], we reported the results of outdoor verification of power generation improvement effect by Low-E reflected solar radiation. In this report, we report the revised results of previous reports [4][5], which confirmed the incident characteristics of Low-E reflected solar radiation to the indoor side of solar cells by simulation.

2 PV CONSIDERED FOR WINDOWS (STPV)

2.1 Architectural issues in conventional see-through PV

There are two types of conventional photovoltaic PV: Type A, in which the gap between adjacent crystalline Si solar cells is transparent, and Type B, in which many fine holes are formed in the power generation layer of amorphous Si solar cells formed on a glass substrate. Architectural issues of the conventional technology are shown in Fig. 1, and the issues are summarized.

① Obstruction of View by Solar Cells (Type A)

Square (about 5 inch square) crystalline Si solar cells dispersed in the PV obstruct the view, causing a sense of oppression.

② Conspicuous Wiring (Type A)

Wiring connecting the crystalline Si solar cells is laid in the gap between the adjacent cells, spoiling the appearance.

③ Low Power Generation Performance (Type B)

In the case of amorphous Si PV, since the power generation layer is formed by coating on a glass substrate, the arrangement of Si atoms becomes irregular and amorphous, resulting in low power generation efficiency per the same power generation area as that of crystalline Si solar cells.

④ Opposite Relationship between Power Generation and View (Lighting)

When the power generation area is expanded, the opening area that can be viewed shrinks and the view is obstructed. On the other hand, when the view is improved, the opening area expands and the power generation area shrinks.

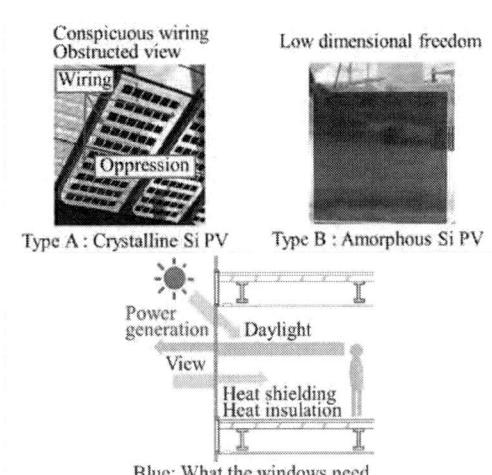

Figure 1: Architectural issues in installing PV to windows

⑤ Heat Shielding and Insulation Required

PV itself does not have the heat shielding and insulation functions required for windows, so it is necessary to devise additional functions.

2.2 STPV Solving Architectural Problems

A linear solar cell is fabricated by connecting strip-shaped single-crystal Si solar cells capable of generating double-sided power with high power generation performance in the longitudinal direction, and a daylight-type PV arranged at equal intervals above and below is incorporated as the outdoor side glass of the low-E double glazing of the solar radiation acquisition type[6]. This makes it possible to simultaneously achieve views, daylighting, heat shielding, heat insulation, and power generation.

Solar cells capable of generating double-sided power using single-crystal Si with high power generation efficiency [7] are intended to improve power generation performance, and the near-infrared radiation mainly reflected by the low-E coating [8] can increase power generation on the back of the solar cells.

The concept of a daylight-type PV that embodies the above is shown in Fig. 2.

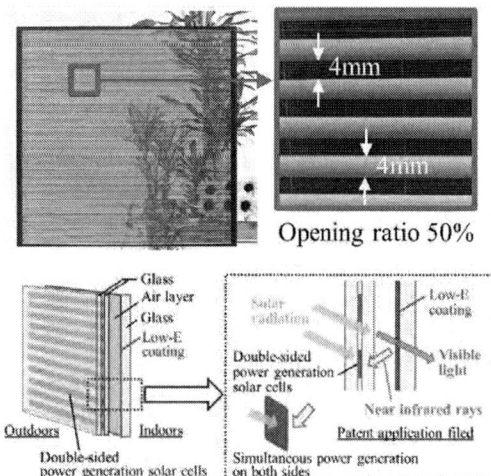

Figure 2: Features of STPV

3 OUTDOOR VERIFICATION

3.1 Summary of verification

In order to compare the difference of power generation by Low-E coating of STPV, the test specimens with the same specifications except for the presence of Low-E coating were exposed outdoors. The place was on the roof of the laboratory building of Taisei Advanced Center of Technology (in Yokohama). The verification period was one year from January 2019 to December 2019.

Fig. 3 shows the cross-sectional configuration of the test specimens. The dimensions of the light-receiving surface of the test specimen are approximately 1.2 m wide and 0.9 m high, and the module aperture ratio is 50%. The solar cells fabricated in a line shape with a height of 4 mm are uniformly arranged with a gap of 4 mm vertically. The PV is integrated as the outdoor side glass of the double-glazed glass in test specimen A, and as the outdoor side glass of the low-E double-glazed glass of the solar radiation acquisition type in test specimen B. A Low-E

glass uses a reflective coating with a reflectance of approximately 80% in the near-infrared region.

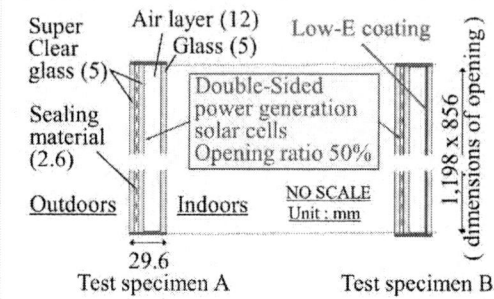

Figure 3: Material structure of test specimens

The thermal performance is shown in Table 1. The solar heat gain coefficient is based on the test standard JSTM K 6101 [9], and the heat transfer coefficient is based on the test standard JIS A 4710 [10]. The values of these thermal performances are the performances including solar cells when the power is not generated.

Table I: Thermal performance of specimens

Thermal index	Specimen A	Specimen B
Solar Heat Gain Coefficient [-]	0.45	0.28
Heat Transfer Coefficient [W/(m²·K)]	2.89	1.85

The test specimen was set up vertically with the light-receiving surface facing south, and the test specimen was placed in a position where there was no installation restriction and the shade of the building adjacent to the southeast side was avoided as much as possible. The schematic arrangement of the test specimen is shown in Fig. 4a. As for the shade of the building adjacent to the southeast side, during the period of low solar altitude from November to February, the time period when the shadow on the test specimen becomes longer and the shadow on the test specimen mainly in the morning. On the other hand, no shadow on the test specimen in June and July, when the sun is high. Since the shade of the test specimen affects the power generation, the measured data were compared only during the evaluation period when both test specimens are generating power without any troubles.

For preventing solar radiation other than that transmitted through the test specimen from entering the test specimen, the test apparatus shown in Fig. 4b was made with a space of 20 cm depth surrounded by wooden boards behind the test specimen.

In addition, a contrivance was made to suppress the reflection of solar radiation transmitted through the test specimen inside the test specimen by making all the surfaces inside the test specimen black matted. Mechanical ventilation by outside air was always carried out in order to suppress the temperature rise inside the test equipment by solar radiation.

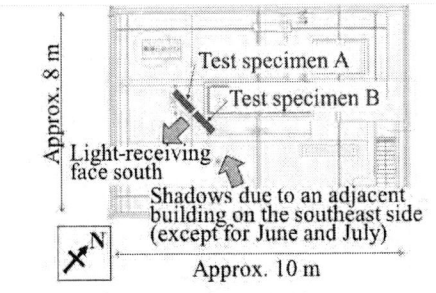

Approx. 8 m

Test specimen A

Test specimen B

Light-receiving face south

Shadows due to an adjacent building on the southeast side (except for June and July)

N

Approx. 10 m

a) Arrangement of specimen on the rooftop

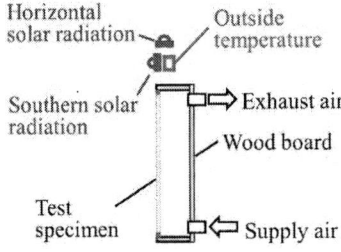

Horizontal solar radiation

Outside temperature

Southern solar radiation

Exhaust air

Wood board

Test specimen

Supply air

b) Cross-section of experimental apparatus

Figure 4: Outline of the test specimens

The measurement items were horizontal solar radiation, southern solar radiation, air temperature, and power generation output. The outline of the measurement position is shown in Fig. 4b, and the main specifications of the measurement equipment are shown in Table 2. The measurement interval is one minute.

Table 2: Measuring devices

Items	Interval	Devices		
		Name	Primary specifications	
Solar Radiation	1 minite	Pyranometer	Range	:285~3000 nm
			Resolution	:<0.5 W/m²
Temperature		Thermometer	Range	:-50~60 °C
			Accuracy	:0.2 °C(-20~50 °C)
Power output		IV curve tracer	Range	:3~300 V, 0.03~10 A
			Rated output	:300 W

3.2 Results of Outdoor verification
3.2.1 Weather conditions

Figure 5 shows the changes in solar radiation and outside temperature during the verification period. Here, solar radiation is the daily accumulated average value and outside temperature is the daily average value. However, in order to exclude the influence of the shadow of the buildings adjacent to the southeast side, the accumulated value of solar radiation and the average value of outside temperature were calculated in the time zones shown in Figure 5. The annual average values of solar radiation were 9.7 MJ/(m²·day) for horizontal solar radiation and 5.3 MJ/(m²·day) for southern solar radiation. Horizontal solar radiation was higher than southern solar radiation. As for the annual transition, horizontal solar radiation was higher in summer when there were no or few excluded time zones and lower in winter when there were many excluded time zones. This annual transition showed a general trend. On the other hand, southern solar radiation generally increased in winter when the solar altitude was low and tended to decrease in summer when the solar altitude was high. Due to the effect of many excluded time zones, southern solar radiation decreased in winter and slightly increased in the intermediate season. As for the outdoor temperature, the annual transition was normal.

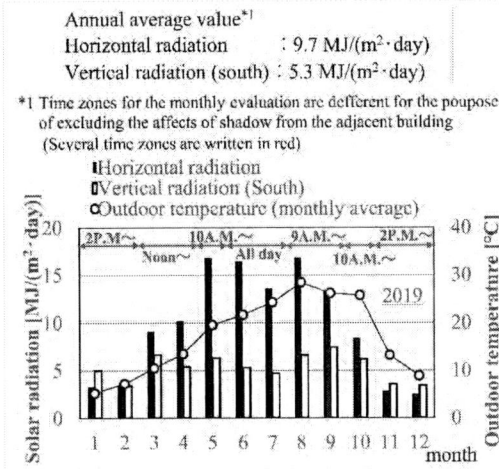

Annual average value[*1]
Horizontal radiation : 9.7 MJ/(m²·day)
Vertical radiation (south) : 5.3 MJ/(m²·day)

[*1] Time zones for the monthly evaluation are defferent for the poupose of excluding the affects of shadow from the adjacent building (Several time zones are written in red)

Figure 5: Radiation and Outdoor temperature

3.2.2 Power generation amount

The transition of power generation is shown in Fig. 6. In Fig. 6, the daily average of power generation of the test specimen was calculated in the same time period as in Fig. 5. Since power generation is proportional to incident solar radiation, power generation shows the same transition as that of southern solar radiation shown in Fig. 5. As shown in Fig. 6, the power generation of Test specimen B with Low-E coating was higher each month than that of Test specimen A without Low-E coating. The increase rate of power generation of Test specimen B relative to Test specimen A was about 10%, and the annual transition tended to be relatively stable. The annual average of power generation was 87.2 Wh/(m²·day) for Test specimen A and 95.7Wh/(m²·day) for Test specimen B. Test specimen B generated about 9% more power than Test specimen A. This suggests that near-infrared radiation reflected by the Low-E coating was incident on the back surface of the solar cell, and the power generation increased.

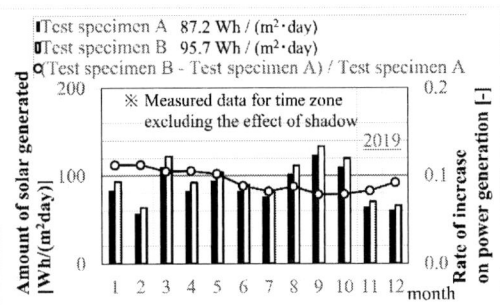

Test specimen A 87.2 Wh / (m²·day)
Test specimen B 95.7 Wh / (m²·day)
(Test specimen B - Test specimen A) / Test specimen A

※ Measured data for time zone excluding the effect of shadow

Figure 6: Power generation amount

4   SIMULATION

4.1 Calculation model

The main specifications of the calculation model are shown in Table 3. In the calculation, only the shape of the solar cell was different, and the radiation analysis [11] was performed by a method without using the results of the outdoor verification. The calculation region is a space of 250 mm square and 22 mm depth, and it is divided by the structure lattice.

In the calculation, in order to reduce the calculation load, the smallest model (hereafter, cell model) including the basic components of the solar cells and the gap between adjacent solar cells was adopted. In addition, the thickness of the sealing material of the solar cell was simplified without using it.

Table 3: Primary specifications of the calculation model

Calculation model	Model 1					Model 2				
See-through PV	STPV					This PV, equally arranged sequre solar cells, is integrated as the outdoor side glass of the Low-E double glazing.				
Measurement	250 mm (W) × 250 mm (H) × 22 mm (D)									
Parts	Outer frame	cell (front)	cell (back)	Super Clear glass	Low-E glass	Outer frame	cell (front)	cell (back)	Super Clear glass	Low-E glass
Polygon number	8	6250	6250	25000	12500	8	5780	5780	25000	12500
Total	50008					49068				

Fig. 7 shows the arrangement and cross-sectional configuration of the PV of the calculation model. In Model 1, the linear solar cell of 4 mm height is evenly arranged with a gap of 4 mm vertically. On the other hand, the solar cell of Model 2 is 127 mm square and evenly arranged with a gap of 40 mm vertically and horizontally.

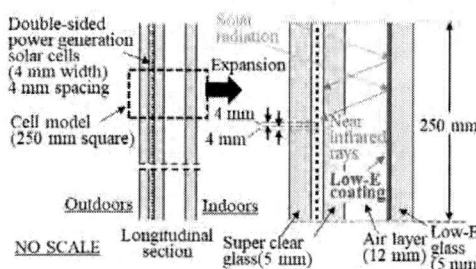

a) Model 1

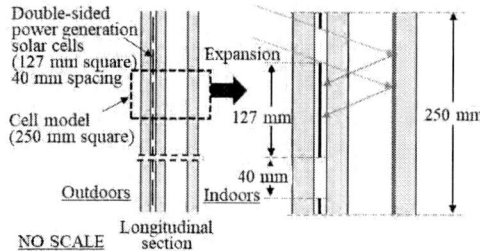

b) Model 2

Figure 7: Calculation model

4.2 Calculation case

The calculation case is shown in Table 4. In order to reduce the calculation load, the calculation is limited to two days of summer solstice and winter solstice instead of annual simulation. The calculation target time is set to two directions of south and west considering symmetry in the time zone when direct solar radiation is incident. In addition, the weather is assumed to be clear because direct solar radiation seems to have a large effect on the reflected solar radiation to the back of the solar cells. The calculation target point is Tokyo.

Table 4: Calculation case

Calculation date	Weather	Calculation target point	Direction	Time Slot
Summer solstice	Clear sky	Tokyo	West	1 P.M.～5 P.M.
Winter solstice			South	8 A.M.～Noon

4.3 Results of simulation

Figure 8 shows the ratio of the amount of solar radiation reflected by the Low-E coating to the amount of solar radiation incident on the outdoor side of the solar cell at 10 minute intervals. The summer solstice is on the west side where the amount of solar radiation incident on the indoor side of the solar cells is large, and the winter solstice is on the south side where the amount of solar radiation incident on the indoor side is large.

A large fluctuation phenomenon is observed in the transition of Model 1. This suggests that the solar radiation reflected by the Low-E coating is not incident on the indoor side of the solar cells but passes through the opening between adjacent solar cells of Model 1 to the outdoor side, and is incident on the indoor side of the solar cells alternately.

The ratio of incident on the indoor side of the solar cells to the outdoor side of the solar cells is 13.7% for Model 1 and 10.4% for Model 2 in the time zone average of both compared cases. It is found that the amount of solar radiation incident on the indoor side of the solar cells is larger for Model 1 than Model 2. It is suggested that the generation performance of Model 1 is higher than that of Model 2 because the amount of power generated increases when the amount of incident solar radiation is larger.

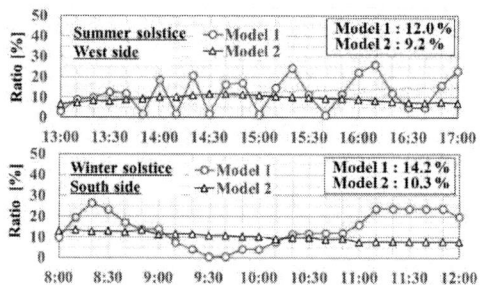

Figure 8: Incident radiation on the indoor side of solar cells

5 CONCLUSION

In this study, as a solution to the architectural problems of conventional daylight-type PV, we proposed a daylight-type PV in which double-sided photovoltaic cells processed in a linear shape are arranged equally in the top and bottom with an aperture ratio of 50%. Regarding a model (Model 1) incorporated as the outdoor side glass of the solar radiation acquisition type Low-E double glazing, we investigated on improvement of power generation. The main findings obtained are as follows.
1) Outdoor verification

The increase of the power generation by the solar reflection of the Low-E coating was about 9% on an annual average.
2) Simulation

The ratio of the solar radiation incident on the indoor side to the solar radiation incident on the outdoor side of the solar cells on a clear day of summer solstice and winter solstice was compared with a model (Model 2). Model 2 is a daylight-type PV with an aperture ratio of 50% using a conventional square double-sided power generation crystalline Si solar cell was incorporated as the outdoor side glass of a solar radiation acquisition type Low-E double glazing. The main results are as follows.

① The average of two calculation cases was 13.7% for Model 1 and 10.4% for Model 2. The amount of solar radiation incident on the indoor side of the solar cells was larger for Model 1 than for Model 2, indicating the power generation superiority of Model 1.

② A large fluctuation phenomenon was observed in the transition of the amount of solar radiation incident on the indoor side of the solar cells in Model 1. This suggests that the solar radiation reflected by the Low-E coating passes through the opening between adjacent solar cells to the outdoor side and enters the indoor side of the solar cell alternately.

6 ACKNOWLEDGMENT

The authors express our sincere gratitude to stakeholders related to the research of STPV.

7 REFERENCES

[1] K. Umeda : Outdoor exposure verification of next-generation see-through photovoltaic module, Summaries of Technical Papers of Annual Meeting, Architectural Institute of Japan, Environmental Engineering, pp.1975-1976, 2020 (in Japanese).

[2] K. Umeda, A. Yamaguchi, N. Kobayashi : Outdoor Exposure Verification of Next-generation See-through Photovoltaic Modules, Effects of Solar Generation and Solar Shading, Taisei Advanced Center of Technology Technical Report No. 53, pp.47-1 - 47-4, 2020 (in Japanese)

[3] K. Umeda, A. Yamaguchi, S. Miyajima, A. Nakajima : Outdoor exposure verification of next-generation see-through photovoltaic module Part 2 Comparison on the solar generation by the direction of linear bifacial solar cells, Summaries of Technical Papers of Annual Meeting, Architectural Institute of Japan, Environmental Engineering, pp.1805-1806, 2024 (in Japanese)

[4] K. Umeda, A. Yamaguchi : Study on improvement of power generation for a window by solar radiation reflected from the low-E coat of a see-through photovoltaic module that is equally arranged liner double-sided solar cells Evaluation on improvement of power generation by using simulation analysis, Summaries of Technical Papers of Annual Meeting, Architectural Institute of Japan, Environmental Engineering, pp.1909-1910, 2023 (in Japanese)

[5] K. Umeda, A. Yamaguchi : Study on improvement of power generation for a window by solar radiation reflected from the low-E coat of a semi transparent photovoltaic module that is equally arranged linear double-sided solar cells, J. Environ. Eng. AIJ. Vol. 89, No.821, pp.399-409, Jul. 2024 (in Japanese)

[6] N. Kadota, K. Maeda, A. Kuraoka, T. Makino, A. Nakajima, K. Umeda, N. Kobayashi, A. Yamaguchi: Development of building façade integrated photovoltaic module for realization of ZEB. Part 1 Characterization of see-through photovoltaic module, Proceedings of JSES conference, pp.145-146, 2020 (in Japanese)

[7] New Energy and Industrial Technology Development Organization (Joint Research Sites) Kaneka Corporation: (FY 2015 – FY 2019 Results Report, Development of technologies to reduce generation costs for high performance, highly reliable photovoltaic power generation / Development of advanced composite technology silicon solar cells and high-performance CIS solar cells / Development of Composite Solar Cell Module Based on Crystalline Si Solar Cell), Report Control Number 20200000000580 (in Japanese)

[8] T. Otsuki, S. Oumi and H. Nakajima: Application of Low-Emissivity Coating to Energy Efficient Window, Journal of the Japan Society of Infrared Science and Technology Vol.7 No.2, 1997 (in Japanese)

[9] Japan testing center for construction materials: JSTM K 6101(2013), Test method for the determination of solar heat gain coefficient and shading coefficient of window's shading devices using solar simulator, 2013

[10] Japanese Industrial Standards: JIS A 4710(2015), Windows and door sets-Thermal resistance test,2015

[11] M. Oguro, Y. Morikawa and H. Motohashi: Development of Heat Island Analysis and Assessment System for Buildings and Building Blocks, Outline of the Program and Analysis Example of Future Eco-city, Taisei Advanced Center of Technology Technical Report No. 42, pp.49-1 - 49-8, 2009 (in Japanese)

Experimental investigation of the temperature distribution in a BIPV facade

Nanna Lysgaard Andersen, Markus Babin, Sune Thorsteinsson
DTU Electro, Technical University of Denmark, Roskilde, Denmark

---

 Agenda

- Introduction & Motivation
- Part 1: Effect of module coloration and mounting configuration on the thermal behaviour of PV modules
  - Methodology
  - Results
- Part 2: Temperature distribution of a vertical BIPV container facade including impact of mounting configuration.
  - Methodology
  - Results
- Conclusions

Introduction & Motivation

- Building-integrated PV vs conventional PV
- Difference in mounting configuration and potential module coloration
- Absorption in colorants impacting temperature
- Operating temperature affects PV Performance
- Temperature distribution of building facades
- Previous studies including smaller scale setups

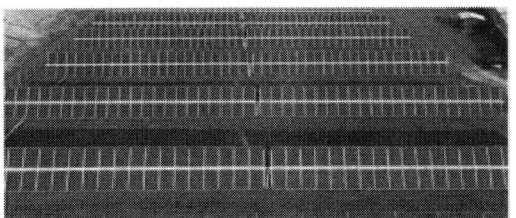

Impact on thermal behaviour?

[1] Babin M. et al., DOI: 10.1051/epjpv/2023028, [2] Ozkalay, E. et al., DOI: 10.1109/JPHOTOV.2021.3114988

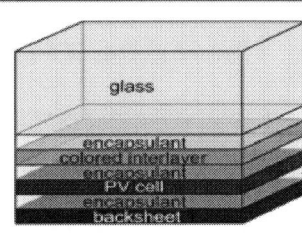

PART 1:

What is the effect of **mounting configuration** and **coloring interlayers** (including additional layers of encapsulant) on the **thermal behaviour** of PV modules?

Can we still apply **well-established temperature models**, such as the **Sandia Model** to describe the relationship between **cell and back-of-module temperature**?

Methodology: Mini-Module Setup

- Located at DTU Risø, Denmark
- Coloured modules, including additional layers of encapsulant
- Temperature sensors
- Insulated vs ventilated configuration
- Operating at Isc

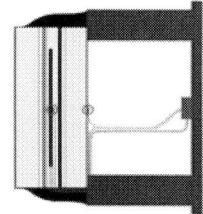

[1] Babin M., Jóhannsson I.H., Jakobsen M.L., Thorsteinsson, *Experimental evaluation of the impact of pigment-based colored interlayers on the temperature of BIPV modules* (2023), 14, art. no. 34, DOI: 10.1051/epjpv/2023028

Methodology: Data Filtering and Modeling

- Clear-sky days
- Data collected at different times between summer 2022 and winter 2023
- Irradiance > 20 W/m^2

Setup	Winter	Spring	Summer	Autumn	Total
Mini-Module, Insulated	5	2	0	1	8
Mini-Module, Ventilated	0	16	12	3	31

- Sandia Model

$$T_{cell} = T_{bom} + \frac{G_{POA}}{G_0} \cdot \Delta T$$

Module Type	Mount	ΔT (°C)
Glass/cell/glass	Open rack	3
Glass/cell/glass	Close roof mount	1
Glass/cell/polymer sheet	Open rack	3
Glass/cell/polymer sheet	Insulated back	0

[2] J. Kratochvil, W. Boyson, D. King, Tech. Rep. SAND2004-3535, 919131 (2004), *Photovoltaic Array Performance Model*, https://www.osti.gov/servlets/purl/919131/

Results: Ventilated Mini-Modules

- Difference in prediction for high and low irradiance data
- ΔT coefficient has largest effect at high irradiance

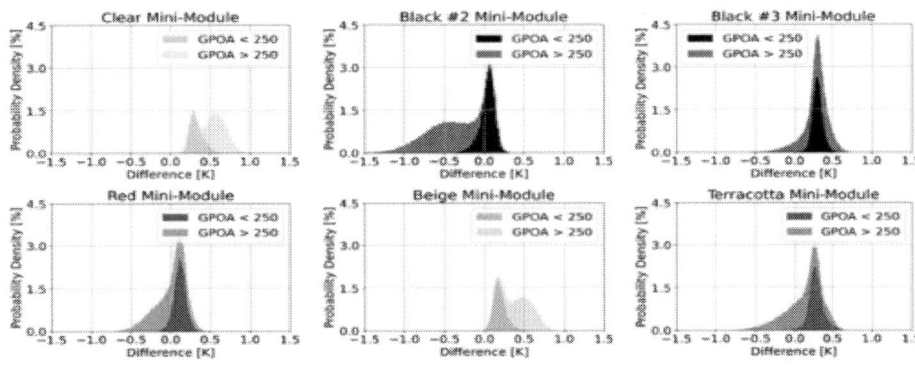

Results: ΔT Coefficients

- Calculating individual ΔT coefficients
- Clear linear dependency for ventilated case

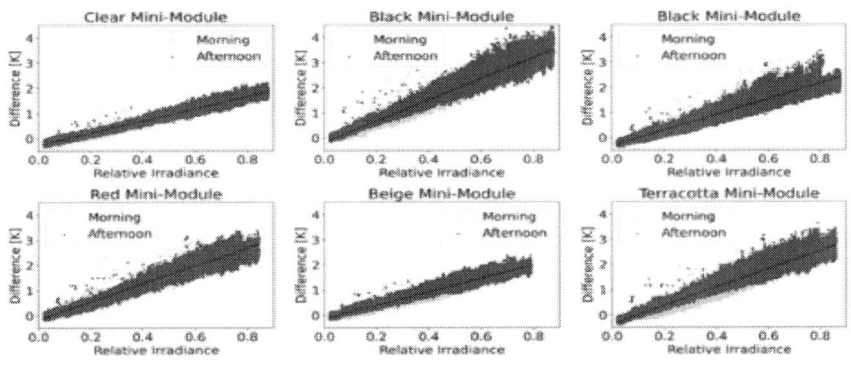

Ventilated configuration

Module	Ventilated ΔT
#1 Clear	2.42
#2 Black	4.21
#3 Black	3.1
#4 Red	3.55
#5 Beige	2.65
#6 Terracotta	3.63
Average	3.26
Sandia Report	3

Results: ΔT Coefficients

- Calculating individual ΔT coefficients
- Hysteresis at insulated case

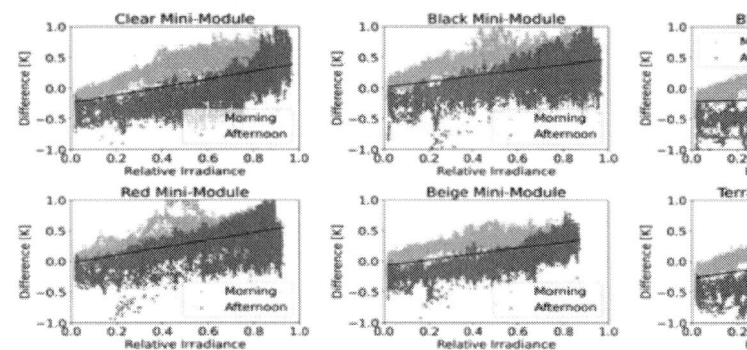

Insulated configuration

Module	Insulated ΔT
#1 Clear	0.65
#2 Black	0.46
#3 Black	0.12*
#4 Red	0.62
#5 Beige	0.48*
#6 Terracotta	0.58
Average	0.49
Sandia Report	0

Results: Customized ΔT Coefficients

- Individual customized coefficients
- Narrower distributions but slight overestimation

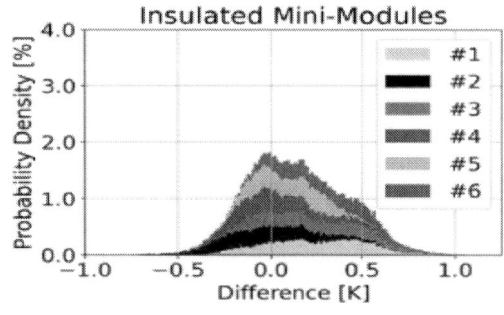

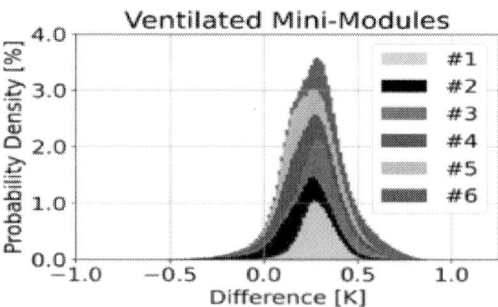

PART 2:

How does the **mounting configuration** affect the **temperature distribution of a vertical BIPV facade**?

Methodology: BIPV Container Setup

- South façade: 16 glass-glass modules
- Rear-ventilated or insulated
- Operating at mpp
- Temperature sensors
- Faulty sensors

- ● Pt100 sensor
- ○ Digital sensor

Results: Facade Temperature Distribution

- Temperature distribution of average maximum temperature
- Insulated vs ventilated
- Distance to insulation, vertical temperature gradient

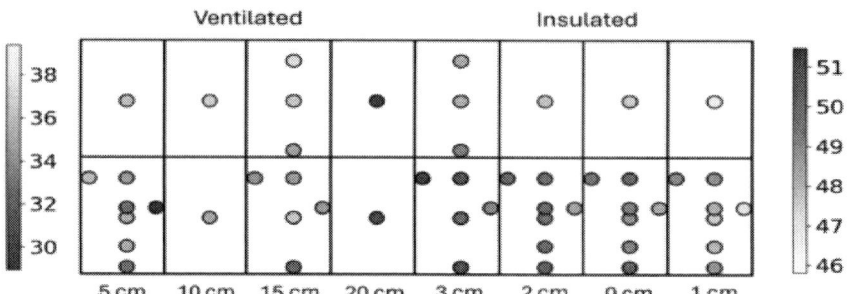

Results: Facade Temperature Distribution

- Temperature distribution of average maximum temperature
- Insulated vs ventilated
- **Distance to insulation**, vertical temperature gradient

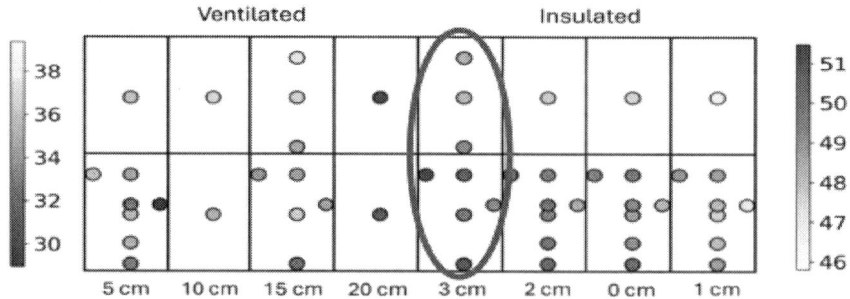

Results: Facade Temperature Distribution

- Temperature distribution of average maximum temperature
- Insulated vs ventilated
- **Distance to insulation**, vertical temperature gradient

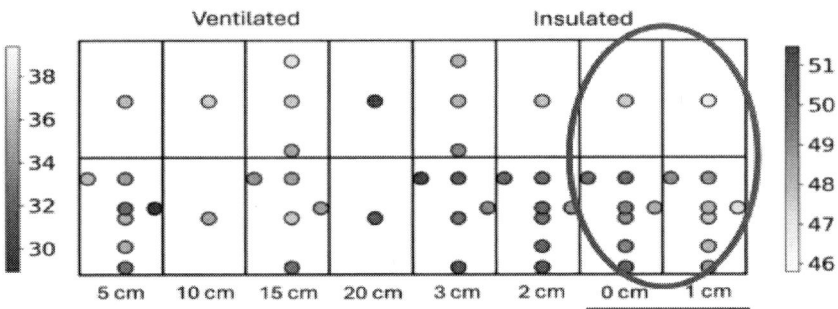

Results: Facade Temperature Distribution

- Temperature distribution of average maximum temperature
- Insulated vs ventilated
- **Distance to insulation**, vertical temperature gradient

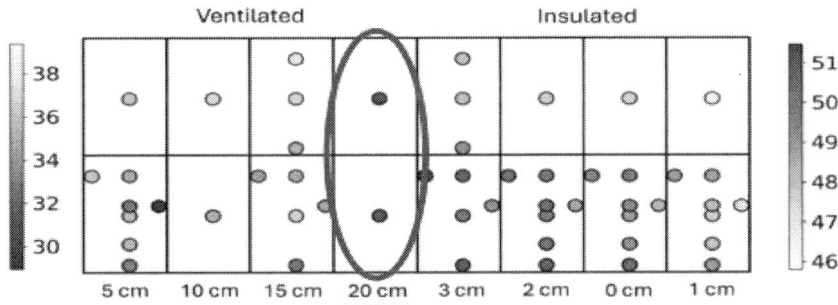

Results: Facade Temperature Distribution

- Temperature distribution of average maximum temperature
- Insulated vs ventilated
- Distance to insulation, **vertical temperature gradient**

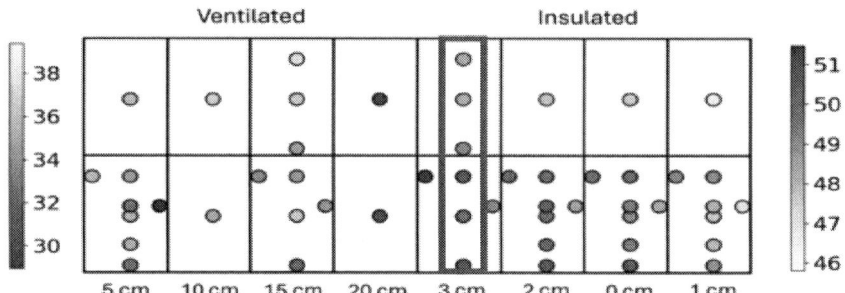

Results: Facade Temperature Distribution

- Temperature distribution of average maximum temperature
- Insulated vs ventilated
- Distance to insulation, **vertical temperature gradient**

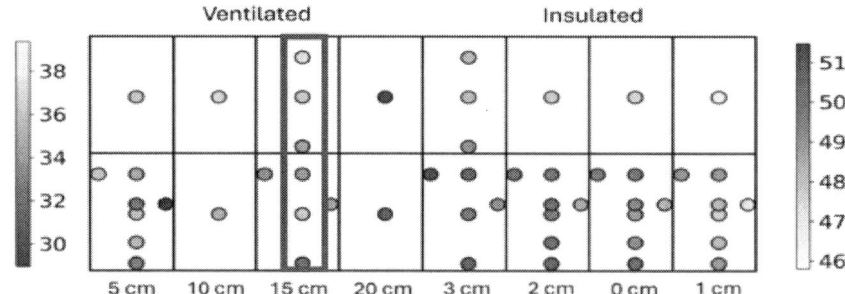

Conclusions

- Colored interlayers impact relationship between cell and back temperature
- The heat transfer – construction and configuration
- Sandia model predicting within +/- 1 °C
- New ΔT coefficients were calculated
- Thermal hysteresis effect

- Differences in max temperatures up to 20 °C
- Temperature variations with height
- Further analysis and modelling is required

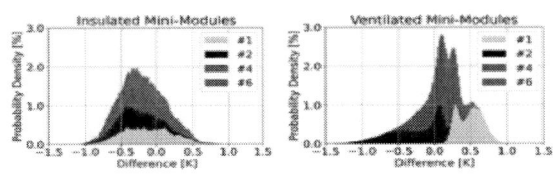

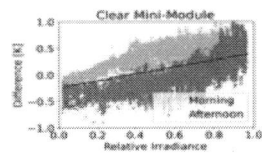

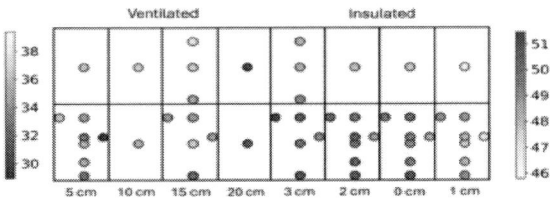

Thank you for your attention!

Supported by:

The Energy Technology Development and Demonstration Programme

64021-1079 UnitSun is supported by the Energy Technology development and Demonstration Program (EUDP)

Thank you!

Markus Babin
Sune Thorsteinsson
SPVS Group
UnitSun project partners

Contact Information:
Nanna Lysgaard Andersen
nalan@dtu.dk
DTU Electro
Frederiksborgvej 399
4000 Roskilde
Denmark

41st European Photovoltaic Solar Energy Conference and Exhibition

This presentation was selected by the Sc. Committee of the EU PVSEC 2024 for submission of a full paper to one of the EU PVSEC's collaborating peer-reviewed journals.

TREND-BASED PREDICTIVE MAINTENANCE AND FAULT DETECTION ANALYTICS FOR PHOTOVOLTAIC POWER PLANTS

Demetris Marangis[1,2], Andreas Livera[1,2], George Makrides[1,2] and George E. Georghiou[1,2]
[1] PV Technology Laboratory, FOSS Research Centre for Sustainable Energy,
Department of Electrical and Computer Engineering, University of Cyprus, 1678 Nicosia, Cyprus
[2] PHAETHON Centre of Excellence for Intelligent, Efficient and Sustainable Energy Solutions, 2109 Nicosia, Cyprus

ABSTRACT: The generation of underperformance alerts is essential for ensuring the long-term performance and cost-effective operation and maintenance (O&M) of photovoltaic (PV) systems. Although various failure detection methods have been developed, a literature survey revealed a notable lack of strategies for effective predictive maintenance, that can accurately anticipate underperformance conditions. This study aims to develop an innovative workflow for fault detection and predictive maintenance in PV systems. The proposed workflow integrates a predictive model based on the eXtreme Gradient Boosting (XGBoost) algorithm to simulate the PV performance, the One-Class Support Vector Machine (SVM) algorithm for fault detection, and the Facebook Prophet algorithm to forecast performance trends and subsequently generate predictive maintenance alerts. The data-driven methodology was validated using a dataset from a 1.8 MWp PV power plant in Greece. The results showed the effectiveness of the workflow for fault detection and prediction. The workflow achieved a sensitivity of 96.9% for fault detection while when generating predictive maintenance alerts up to 7 days in advance, the workflow yielded a sensitivity of 92.9%.
Keywords: predictive maintenance, failure detection, photovoltaic, machine learning.

1 INTRODUCTION

The path to a sustainable future involves the transition from fossil fuels to renewable energy production, minimizing greenhouse gas emissions. The EU's agenda has set ambitious climate and energy goals for 2030, targeting at least 45% renewable energy by 2030 [1].

Solar photovoltaic (PV) systems have emerged as a potential renewable energy option for producing sustainable energy due to their low cost and ease of installation/operation [2]. Global solar power entered the terawatt age in 2022, and it is expected to double again (hitting the 2 TW mark) in three years [3]. Additionally, the levelized cost of electricity is projected to decrease by 1.4% by 2030, making PV systems more affordable and accessible for all [4].

The challenge in maximizing solar power production remains the vulnerability of PV systems to technical and performance issues. An annual loss in revenues of about $14.5 billion by 2024 is estimated in the global PV industry due to inefficient operation and maintenance (O&M) activities [5]. By implementing an optimized O&M strategy, an average of 5.27% annual energy could be recovered (yearly potential recoverable extra income of $10,000 for each MW). The PV industry and researchers alike should thus focus on maximizing the system's annual energy yield by diagnosing, resolving, and predicting underperformance conditions.

Common O&M strategies can be executed to ensure optimal PV performance. These O&M strategies are mainly split into 3 categories: preventive, predictive (or condition-based), and corrective maintenance. Preventive maintenance consists of planned visual and physical inspections, functional testing, measurements, and verification activities to ensure adherence to operating manuals and warranty requirements [6]. Corrective maintenance involves actions to repair failures, malfunctions, and damages identified through remote monitoring or plant inspections [7]. Condition-based maintenance involves utilizing real-time data and forecasts to plan and optimize maintenance activities, such as cleaning and snow removal, while also detecting potential failures at early stages or before they occur.

Within this context, the generation of accurate maintenance alerts could help technicians and operators to detect faults at early stages and anticipate fault issues, thereby minimizing downtime periods and maximizing the PV energy yield. Anomaly detection techniques could also offer valuable information about the fault type, while also providing insights about the health status of the PV system [8], [9]. The use of automated data-driven models could enable technicians to effectively schedule the O&M actions, thereby improving the reliability and performance of PV systems [10], [11], [12], [13], [14].

Machine learning (ML) and artificial intelligence (AI) algorithms have proven to be highly effective for developing data-driven fault diagnostics and analytics due to their exceptional ability to recognize patterns and simulate complex behaviors, such as the nonlinear dynamics of PV systems [15], [16], [17], [18], [19]. These techniques can process vast amounts of data, uncover subtle anomalies, and accurately predict system faults, making them ideal for optimizing real-time monitoring and maintenance. Most of the reported fault detection techniques employ ML principles to simulate the PV performance, and through a comparison with actual performance, enable the real-time identification of malfunctions in PV systems. In this field, the implemented AI-driven techniques have demonstrated high accuracies (over 95%) in diagnosing underperformance conditions (such as partial shading [20], soiling [21], and inverter-related failures [22]).

Current research efforts focus on predictive maintenance techniques to predict potential failures, thus enabling proactive scheduling of the maintenance tasks. A literature survey revealed only a limited number of predictive maintenance algorithms [23], [24], [25]. A promising predictive maintenance model was developed by Betti et al. [23]. The proposed algorithm was based on a self-organizing map and could predict the occurrence of faults over a forthcoming 7-day period. The algorithm was capable of generating fault warning levels up to 10 days before the fault occurrence.

The gap in current methodologies lies in the lack of a robust system for generating real-time fault diagnostics and predictive maintenance alerts for utility-scale PV systems. This study thus aims to develop an innovative workflow for fault detection and predictive maintenance in PV systems. The proposed workflow targets to fill in the literature gap by generating fault detection and predictive maintenance alerts for underperforming conditions.

2 METHODOLOGY

A sequence of steps is followed to develop the proposed predictive maintenance and failure detection models. The workflow steps included: (a) data acquisition of meteorological and electrical data, (b) implementation of data quality routines, (c) creation of a predictive model (leveraging the eXtreme Gradient Boosting algorithm), (d) development of failure detection algorithms and (e) predictive maintenance alerts and warning levels generation.

2.1 Data acquisition

The first step involved the acquisition of historical data from a utility-scale PV power plant in Greece. The system has a nominal power of 1.8 MW$_p$. It comprises of 7824 polycrystalline silicon PV modules, each with a nominal power capacity of 230 W$_p$. The modules are organized into 326 strings, with each string comprising 24 series-connected modules. The entire array is connected to four grid-connected inverters. The performance of the system, along with the weather conditions, is monitored and recorded following the International Electrotechnical Commission (IEC) 61724-1 [26]. The measurements are acquired at 1-minute intervals and recorded at 15-minute resolution.

The quantities measured include the in-plane irradiance (G_I), module temperature (T_{mod}), array DC current (I_A), voltage (V_A) and power (P_A). In this work, a dataset covering a 3-year period was used. For the test PV system, a maintenance log was kept. The log file recorded the different fault types (that occurred during its operation) and the performed O&M actions.

2.2 Data quality routines

The second step involved the application of data quality routines to the acquired measurements to ensure high data fidelity. The data quality stage involved data filtering and imputation techniques, that were applied to filter out and/or fill in missing and erroneous measurements [27], [28]. Low irradiation data $G_I \geq 20W/m^2$ were discarded from the analysis due to the presence of significant errors, as reported in study [29]. The filtered and imputed data were then normalized (by dividing each value by its corresponding nominal value). Finally, the data were aggregated into hourly and daily intervals.

2.3 Predictive model

The eXtreme Gradient Boosting (XGBoost) algorithm was utilized to predict the electrical characteristics (DC power, DC voltage, DC current) of the PV system based on the given meteorological conditions. The XGBoost algorithm was utilized due to its superior performance in regression and classification tasks [30]. The meteorological conditions used as inputs to the predictive model included the T_{mod} and G_I. The first year of data was excluded from the analysis due to the presence of multiple faulty periods (>10%) that could potentially bias the analysis. Thus, the train and test sets each consisted of one year of data.

Additionally, faulty periods were discarded from the train set (that included only normal operation data points) using the maintenance log of the PV system.

2.4 Fault detection algorithm

This step included the application of the One-Class Support Vector Machines (SVM) to detect failures. One-Class SVM algorithm is a semi-supervised anomaly detection algorithm that employs kernel functions to establish a decision boundary that can distinguish normal and anomalous data points [31], [32]. The features used for classifying the normal and anomalous data points included the DC voltage and DC current indicators [33]. The voltage and current indicators were defined as the ratio between the average daily measured DC voltage/current and the average daily simulated DC voltage/current. The simulated values were estimated by the XGBoost predictive model.

The SVM boundary line was thus generated by the algorithm to classify the data points into normal and anomalous. Data anomalies correspond to PV fault conditions. Finally, the data were labeled with a fault detection alert (FD) value of 1 for data anomalies, while normal operating values were given a 0 value.

2.5 Predictive maintenance alerts and warning levels generation

The last step in the developed workflow included the generation of predictive maintenance alerts and warning levels. This was done by applying the seasonal decomposition method to the actual DC power indicator (to extract the performance trend) and forecasting the performance trend of the system under normal operating conditions with predefined threshold limits. Actual trend data points that fall outside the upper and lower limits were marked as predictive maintenance alerts. To reduce false alarms, the gradient of the historical trend was analyzed to identify instances where the PV system's health status is deteriorating.

The DC power indicator was defined as the ratio between the average daily measured DC power and the average daily simulated DC power. The simulated values were estimated by the XGBoost predictive model.

The seasonal decomposition is a statistical method used to separate a time series into its underlying components (trend, seasonal and residuals) to better comprehend its structure and to facilitate further analysis such as forecasting [34]. It was applied to both the train set (that comprised of only normal operating conditions) and the test set to extract the performance trend.

To forecast the performance trend, the Facebook Prophet (FBP) algorithm was used. FBP is a forecasting tool developed for forecasting time series data. It excels with data that has strong seasonal patterns and extensive historical records. It effectively manages missing data, trend shifts, and outliers [35].

In this study, the FBP algorithm was trained using the performance trend data from the train set and then forecasted the expected performance trend for the test set. In addition, the FBP algorithm estimated the 95% confidence interval of the forecasted trend. These values were marked as the upper bound (UB) and lower bound (LB) of normal operating behavior. Comparison of the determined limits (UB and LB) with the actual trend of the test set allowed the generation of predictive maintenance alerts (Pdm). More details about the generation of predictive maintenance alerts can be found in the extended journal publication [36].

PdM alerts were used to notify operators of potential performance issues in advance, enabling them to take preventive actions before failures occur. On the other hand, FD alerts were used to inform operators when a fault has already occurred within the system, requiring immediate action to restore the PV system to normal operation. Together, these values were used to generate the warning levels as shown in Table I.

Table I. Generated warning levels based on the Pdm and FD values.

Warning level	Pdm value	FD value	Color
1	1	0	Yellow
2	2	0	Orange
3	Any	1	Red

2.6 Performance evaluation

The developed workflow was evaluated for both its ability to model the normal PV operation and the effectiveness of generating warning levels. More specifically, the predictive model's accuracy was evaluated using the normalized Root Mean Square Error (nRMSE) [37], while for the generated warning levels, the precision metric was utilized [38].

3 RESULTS

3.1 Predictive model

The predictive model was initially deployed to simulate the electrical characteristics (DC current, voltage, and power) of the PV system under study. The XGBoost predictive model demonstrated exceptional accuracy in simulating the PV electrical measurements. Fig. 1 presents the measured and predicted values of the PV electrical parameters. The predicted DC current and power curves exhibit a strong correlation with the irradiance profile, indicating the model's robustness in accurately capturing the system's electrical characteristics under both clear sky and cloudy conditions. The models for DC power, current and voltage achieved mean nRMSE values of 1.78%, 1.75%, and 1.88%, respectively. The consistently low predictive errors across all electrical parameters confirm the model's effectiveness in accurately simulating the characteristics of the PV system.

3.2 Fault detection

The developed One-Class SVM algorithm was applied to the test set and the results are illustrated in Fig. 2. According to the PV plant O&M record, there were 33 faulty days within the test set. By employing the SVM boundary line, 32 out of 33 faulty days were successfully identified (represented by red dots), demonstrating the algorithm's effectiveness in detecting anomalies – fault conditions. The algorithm achieved a sensitivity of 96.9% over the test set period.

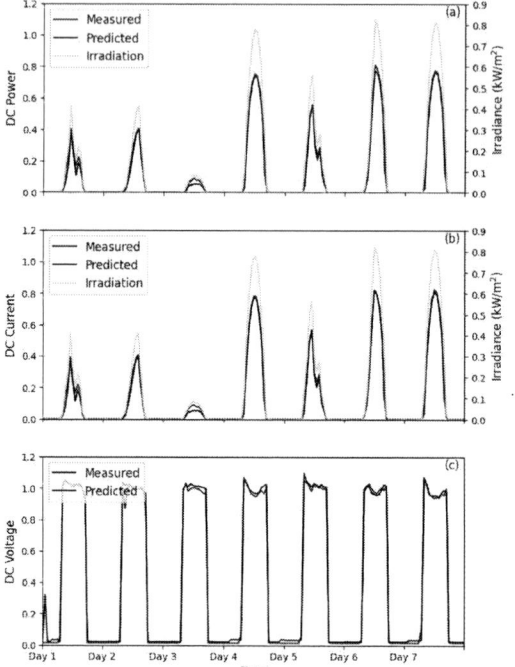

Figure 1: Plot of the measured (black line) and predicted (blue line) values for (a) DC power, (b) DC Current and (c) DC voltage over a weekly period for the PV system under study. The irradiance measurements are also depicted (yellow line). All plotted quantities were normalized to their nominal values.

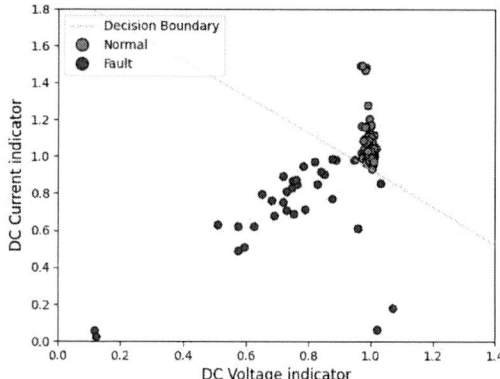

Figure 2: Scatter plot of the DC current and DC voltage indicators. The faulty values are marked in red, whereas the normal operating values are marked in blue. The SVM decision boundary line is plotted using a green dashed line.

3.3 Predictive maintenance

The actual and the forecasted trend is depicted in Fig. 3. A comparison of the forecasted trend's bounds (gray highlighted regions) with the actual trend enables the identification of faulty periods. The comparison highlights various faulty periods throughout the year where the actual trend falls outside the lower bound of the forecasted trend (points outside the 95% confidence interval). This is particularly evident in the first five months, where multiple faulty periods are detected.

41st European Photovoltaic Solar Energy Conference and Exhibition

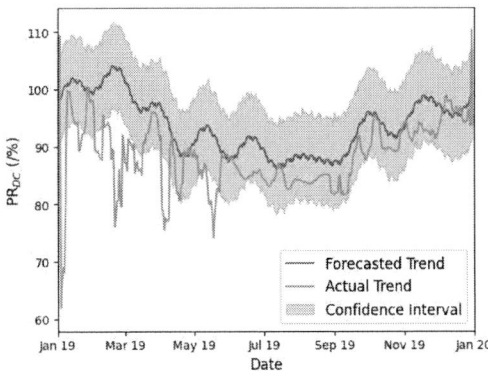

Figure 3: Plot of actual (orange line) and forecasted (blue line) trend by the FBP algorithm over a yearly period. The confidence interval thresholds, that mark the normal operation limits, are gray highlighted.

The generated warning levels (highlighted in yellow, orange, and red) and the corresponding hourly DC power curve for a monthly period in the test set are presented in Fig. 4. The proposed method generated different warning levels, while also predicting several PV underperformance issues. For example, the faults that occurred on the 6th, 14th, 20th, 23rd and 28th were all detected (highlighted in red) by the proposed workflow. Only the fault that occurred on the 13th of February was not detected as it occurred during low irradiation conditions.

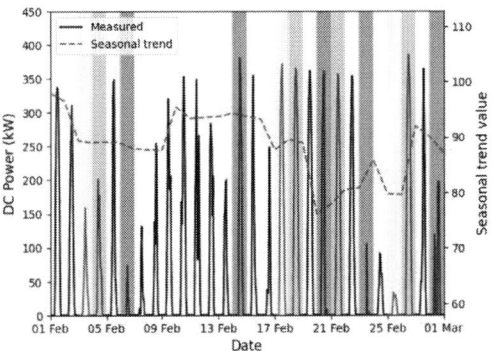

Figure 4: Plot of measured DC power (blue solid line), generated warning levels (colored in yellow, orange, and red) and historical seasonal trend (green dashed line).

The proposed workflow effectively generated predictive maintenance alerts up to 7 days in advance, achieving a sensitivity of 92.9%. Overall, the combination of the Pdm and FD values demonstrated high accuracy in generating warning levels. The results demonstrate the workflow's capabilities for fault detection and prediction.

4 CONCLUSION

Predicting future PV underperformance conditions is essential for maximizing PV energy yield and for minimizing downtime. In this paper, a data-driven workflow was developed to mimic the electrical behaviour of the PV system, to generate fault detection and predictive maintenance alerts. The workflow consists of a predictive model based on XGBoost for simulating the PV performance, the One-Class SVM algorithm for fault

detection and the FBP algorithm for forecasting the PV performance trend and for generating predictive maintenance alerts and warning levels.

The proposed workflow was benchmarked using historical data from a PV power plant in Greece. The workflow was used to detect and predict underperformance conditions using PV inverter-level measurements. The results demonstrated that the workflow's capability to simulate the PV system behavior (presenting nRMSE values < 2%), identify fault conditions (achieving a sensitivity of 96.9%) and generate predictive maintenance alerts up to 7 days in advance (yielding a sensitivity of 92.9%).

Future work will focus on further benchmarking the developed workflow on PV power plants installed at different climatic conditions. Additionally, the workflow in combination with other data analytics will be integrated into a Software-as-a-Service (SaaS) O&M monitoring platform. This platform will be capable of providing informative notifications and recommendations to PV plants and operators about detected faults and predicted future PV underperformance conditions.

5 REFERENCES

[1] European Commission, "2030 Climate Target Plan." Accessed: Jul. 26, 2024. [Online]. Available: https://ec.europa.eu/info/law/better-regulation/have-your-say/initiatives/12265-2030-Climate-Target-Plan_en

[2] A. Khalid, N. Ullah, A. Ahmad Riaz, M. Zeeshan Zahir, and Z. Ali Khan, "Detection and Prediction of Faults in Photovoltaic Solar Panel Using Regression Analysis," *Journal of Mechanical Engineering Research and Developments*, vol. 44, no. 11, pp. 34–49, 2021.

[3] SolarPower Europe, "Global Market Outlook for Solar Power 2022–2026.," 2022. Accessed: Jul. 24, 2024. [Online]. Available: https://www.solarpowereurope.org/insights/market-outlooks/global-market-outlook-for-solar-power-2022

[4] K. Keisang, T. Bader, and R. Samikannu, "Review of Operation and Maintenance Methodologies for Solar Photovoltaic Microgrids," Nov. 05, 2021, *Frontiers Media S.A.* doi: 10.3389/fenrg.2021.730230.

[5] A. Livera, M. Theristis, L. Micheli, E. F. Fernandez, J. S. Stein, and G. E. Georghiou, "Operation and Maintenance Decision Support System for Photovoltaic Systems," *IEEE Access*, vol. 10, pp. 42481–42496, 2022, doi: 10.1109/ACCESS.2022.3168140.

[6] K. Deli and D. Noel, "On-Field Operation and Maintenance of Photovoltaic Systems in Cameroon," *Maintenance Manag*, 2020.

[7] R. Andrews *et al.*, "Operation & Maintenance Best Practice Guidelines/Version 4.0," in *SolarPower Europe: Operation & Maintenance Best Practice Guidelines/Version 4.0*, 2019.

[8] A. Dhoke, R. Sharma, and T. K. Saha, "An approach for fault detection and location in solar PV systems," *Solar Energy*, vol. 194, pp. 197–208, Dec. 2019, doi: 10.1016/j.solener.2019.10.052.

[9] A. Michail *et al.*, "A comprehensive review of unmanned aerial vehicle-based approaches to support photovoltaic plant diagnosis," Jan. 15, 2024, *Elsevier Ltd.* doi: 10.1016/j.heliyon.2024.e23983.

[10] P. Branco, F. Gonçalves, and A. C. Costa, "Tailored algorithms for anomaly detection in photovoltaic systems," *Energies (Basel)*, vol. 13, no. 1, p. 225, 2020.

[11] M. Dhimish and V. Holmes, "Fault detection algorithm for grid-connected photovoltaic plants," *Solar Energy*, vol. 137, pp. 236–245, 2016.

[12] A. Drews *et al.*, "Monitoring and remote failure detection of grid-connected PV systems based on satellite observations," *Solar energy*, vol. 81, no. 4, pp. 548–564, 2007.

[13] A. Livera, M. Theristis, G. Makrides, J. Sutterlueti, and G. E. Georghiou, "Advanced diagnostic approach of failures for grid-connected photovoltaic (PV) systems," in *Proceedings of the 35th European Photovoltaic Solar Energy Conference (EU PVSEC), Brussels, Belgium*, 2018, pp. 24–28.

[14] Y. Zhao, B. Lehman, R. Ball, J. Mosesian, and J.-F. de Palma, "Outlier detection rules for fault detection in solar photovoltaic arrays," in *2013 twenty-eighth annual IEEE applied power electronics conference and exposition (APEC)*, IEEE, 2013, pp. 2913–2920.

[15] A. A. Al-Katheri, E. A. Al-Ammar, M. A. Alotaibi, W. Ko, S. Park, and H. J. Choi, "Application of Artificial Intelligence in PV Fault Detection," *Sustainability (Switzerland)*, vol. 14, no. 21, Nov. 2022, doi: 10.3390/su142113815.

[16] N. Kellil, A. Aissat, and A. Mellit, "Fault diagnosis of photovoltaic modules using deep neural networks and infrared images under Algerian climatic conditions," *Energy*, vol. 263, p. 125902, 2023.

[17] S. Voutsinas, D. Karolidis, I. Voyiatzis, and M. Samarakou, "Development of a machine-learning-based method for early fault detection in photovoltaic systems," *Journal of Engineering and Applied Science*, vol. 70, no. 1, p. 27, 2023.

[18] Z. Yi and A. H. Etemadi, "Fault detection for photovoltaic systems based on multi-resolution signal decomposition and fuzzy inference systems," *IEEE Trans Smart Grid*, vol. 8, no. 3, pp. 1274–1283, 2016.

[19] H. Wang, J. Zhao, Q. Sun, and H. Zhu, "Probability modeling for PV array output interval and its application in fault diagnosis," *Energy*, vol. 189, Dec. 2019, doi: 10.1016/j.energy.2019.116248.

[20] A. Mellit and S. Kalogirou, "Assessment of machine learning and ensemble methods for fault diagnosis of photovoltaic systems," *Renew Energy*, vol. 184, pp. 1074–1090, 2022.

[21] B. Meyers, "Estimation of soiling losses in unlabeled PV data," in *2022 IEEE 49th Photovoltaics Specialists Conference (PVSC)*, IEEE, 2022, pp. 930–936.

[22] F. Pereira and C. Silva, "Machine learning for monitoring and classification in inverters from solar photovoltaic energy plants," *Solar Compass*, vol. 9, p. 100066, 2024.

[23] A. Betti, M. Tucci, E. Crisostomi, A. Piazzi, S. Barmada, and D. Thomopulos, "Fault prediction and early-detection in large pv power plants based on self-organizing maps," *Sensors*, vol. 21, no. 5, pp. 1–16, Mar. 2021, doi: 10.3390/s21051687.

[24] M. De Benedetti, F. Leonardi, F. Messina, C. Santoro, and A. Vasilakos, "Anomaly detection and predictive maintenance for photovoltaic systems," *Neurocomputing*, vol. 310, pp. 59–68, 2018.

[25] T. Huuhtanen and A. Jung, "Predictive Maintenance of Photovoltaic Panels via Deep Learning," in *2018 IEEE Data Science Workshop, DSW 2018 -*

Proceedings, Institute of Electrical and Electronics Engineers Inc., Aug. 2018, pp. 66–70. doi: 10.1109/DSW.2018.8439898.

[26] "Photovoltaic system performance-Part 1: Monitoring (IEC 61724-1:2017)," 2017.

[27] A. Livera, M. Theristis, E. Koumpli, G. Makrides, J. Stein, and G. E. Georghiou, "Guidelines for ensuring data quality for photovoltaic system performance assessment and monitoring.," In Proceedings of the 37th European Photovoltaic Solar Energy Conference (EU PVSEC), 2020, pp. 1352–1356.

[28] A. Livera *et al.*, "Data processing and quality verification for improved photovoltaic performance and reliability analytics," *Prog Photovolt*, vol. 29, no. 2, pp. 143–158, 2021.

[29] B. H. Hamadani and M. B. Campanelli, "Photovoltaic characterization under artificial low irradiance conditions using reference solar cells," *IEEE J Photovolt*, vol. 10, no. 4, pp. 1119–1125, 2020.

[30] T. Chen and C. Guestrin, "Xgboost: A scalable tree boosting system," in *Proceedings of the 22nd acm sigkdd international conference on knowledge discovery and data mining*, 2016, pp. 785–794.

[31] M. Amer, M. Goldstein, and S. Abdennadher, "Enhancing one-class support vector machines for unsupervised anomaly detection," in *Proceedings of the ACM SIGKDD workshop on outlier detection and description*, 2013, pp. 8–15.

[32] B. Schölkopf, J. C. Platt, J. Shawe-Taylor, A. J. Smola, and R. C. Williamson, "Estimating the support of a high-dimensional distribution," *Neural Comput*, vol. 13, no. 7, pp. 1443–1471, 2001.

[33] S. Silvestre, M. A. da Silva, A. Chouder, D. Guasch, and E. Karatepe, "New procedure for fault detection in grid connected PV systems based on the evaluation of current and voltage indicators," *Energy Convers Manag*, vol. 86, pp. 241–249, 2014.

[34] R. G. Vieira, M. A. Leone Filho, and R. Semolini, "An Enhanced Seasonal-Hybrid ESD technique for robust anomaly detection on time series," in *Anais do XXXVI Simpósio Brasileiro de Redes de Computadores e Sistemas Distribuídos*, SBC, 2018, pp. 281–294.

[35] S. J. Taylor and B. Letham, "Forecasting at scale," *Am Stat*, vol. 72, no. 1, pp. 37–45, 2018.

[36] D. Marangis, A. Livera, G. Tziolis, G. Makrides, A. Kyprianou, and G. E. Georghiou, "Trend-Based Predictive Maintenance and Fault Detection Analytics for Photovoltaic Power Plants," *Solar RRL [Submitted]*, 2024.

[37] S. Theocharides, G. Makrides, G. E. Georghiou, and A. Kyprianou, "Machine learning algorithms for photovoltaic system power output prediction," in *2018 IEEE International Energy Conference (ENERGYCON)*, 2018, pp. 1–6. doi: 10.1109/ENERGYCON.2018.8398737.

[38] A. Dong, Y. Zhao, X. Liu, L. Shang, Q. Liu, and D. Kang, "Fault diagnosis and classification in photovoltaic systems using scada data," in *2017 International Conference on Sensing, Diagnostics, Prognostics, and Control (SDPC)*, IEEE, 2017, pp. 117–122.

41st European Photovoltaic Solar Energy Conference and Exhibition

This presentation was selected by the Sc. Committee of the EU PVSEC 2024 for submission of a full paper to one of the EU PVSEC's collaborating peer-reviewed journals.

PV MODULE OPERATING TEMPERATURE:
RELIABLE EXTRACTION OF MODEL PARAMETERS FROM DYNAMIC FIELD DATA

Anton Driesse[1], Jesus Polo[2]

[1] PV Performance Labs, Emmy-Noether-Str. 2, Freiburg, Germany,
anton.driesse@pvperformancelabs.com

[2] Photovoltaic Solar Energy Unit, Renewable Energy Division, CIEMAT, Spain

Module temperature is widely perceived to be a secondary and much less significant factor than irradiance for determining PV system yield. This is understandable to the extent that irradiance is the fuel and temperature merely affects the conversion efficiency of that fuel; however, the *uncertainty* in module temperature and *uncertainty* in irradiance can make similar contributions to *uncertainty* in yield. Thus, accurate models for PV module operating temperature *and* accurate parameters for those models are essential for accurate yield assessments.

In this work we develop and demonstrate an effective strategy to account for the effects of thermal mass in both parameter extraction and system simulation processes. This strategy can be used with a variety of empirical steady-state equations, but the best results are achieved when the steady-state equation accounts for radiative losses to the sky. We therefore propose a new model to accompany the fitting strategy—one that sacrifices a little bit of the simplicity of industry-standard models to achieve substantial gains in modelling accuracy.

Keywords: PV module, PV system, operating temperature, cell temperature, system simulation, system performance, yield assessment, uncertainty.

1 INTRODUCTION

Module temperature is widely perceived to be a secondary and much less significant factor than irradiance for determining PV system yield because irradiance is the fuel and temperature merely affects the conversion efficiency of that fuel. But how do these two factors affect *uncertainty* in yield? An uncertainty of ±2% in irradiance translates almost directly into a yield uncertainty of ±2% (disregarding other sources of uncertainty). The same level of yield uncertainty can result from an uncertainty of ±4 °C in operating temperature combined with a temperature coefficient of -0.5%/°C. While these numbers are only indicative, they demonstrate that the *uncertainty* in module temperature and *uncertainty* in irradiance can both make similar contributions to the overall uncertainty in yield.

Uncertainty doesn't affect the fuel or efficiency directly, but it does affect the cost of financing. Hence, improvements in operating temperature modelling are also a way to bring the cost of PV generation down. While the uncertainty argument applies to all systems, it is even more relevant for novel system designs and configurations such as agrivoltaics and floating systems, where there are more variables, less collective experience and less empirical performance data available.

To model PV module temperature accurately, suitable models are needed, but also accurate *parameters* for those models. Most of the commonly used temperature models describe steady-state conditions only, and most common parameter extraction methods reject dynamic conditions in order to focus on perceived stable conditions. It is easy to argue that dynamic models are more suitable and should be used instead, but steady-state models are embedded in standards and software and will not disappear so quickly. Furthermore, there is no single obvious choice for a dynamic replacement, let alone one that is coupled with a suitable parameter extraction method.

This work establishes a very useful link between the steady-state and dynamic worlds. It demonstrates that the effect of thermal mass can be modelled precisely by applying a first-order low-pass filter to either the inputs or the output of the simplest steady-state model. The same strategy also works for models that include wind speed and/or radiative losses, although it is no longer mathematically exact. Extracting model parameters from dynamic data is now simplified because it can be done using low-pass filtered inputs, whereas subsequent PV system simulations can be carried out by applying a low-pass filter to the steady-state model output. We also propose new model equation that accounts for radiative losses to the sky in a simplified but effective manner, and demonstrate that this new model, when coupled with the low-pass filter processing strategy, produces substantially more accurate PV module operating temperature predictions.

2 METHOD

Before discussing module temperature, we must clarify what this term refers to. PV system yield simulations require internal cell temperature, whereas experimental measurements provide a sensor temperature, which gives a more or less accurate indication of the (back-of) module temperature. The thermal resistance of a typical foil or glass module back is low, therefore cell and (back-of) module temperature should differ by less than ~1°C even at high irradiance levels, and a well-attached temperature sensor of suitable design should be able capture this quite accurately. We therefore simply refer to module temperature throughout this work without regard for small differences from cell or sensor temperature. If such differences are large in a particular installation, then site-specific corrections could be implemented.

2.1 Single parameter models

There are a variety of module temperature models that have different names and parameters but are algebraically identical to the fundamental equation:

$$T_m = T_a + \frac{G}{U} \qquad (1)$$

where T_m is the module temperature, T_a is the ambient temperature, G is the plane-of-array irradiance, and U is a heat transfer coefficient that represents the combined average effect of all heat loss mechanisms. Given a suitable set of measurements, a value of U can be found by linear regression.

Equation (1) is rearranged as an energy balance in order to incorporate a thermal mass or capacitance, C:

$$G - (T_m - T_a)\, U - C\frac{dT_m}{dt} = 0 \qquad (2)$$

Model inputs T_a and G are time-varying quantities in PV systems, and solving equation (2) for time-varying module temperature leads to the exact solution:

$$T_m(t) = (f * T_a)(t) + \frac{(f * G)(t)}{U} + c_o e^{-\frac{U}{C}t} \qquad (3)$$

$$f(t) = \frac{U}{C} e^{-\frac{U}{C}t}, \; t \geq 0 \qquad (4)$$

where $f(t)$ is the impulse response of a first-order low-pass filter and * is the convolution operator, which effectively applies the low-pass filter to each of the two time-varying model inputs. The ratio C/U is the time constant of the low-pass filter, which is replaced by τ (tau) for clarity, as shown in equation (5).

$$f(t) = \tau^{-1} e^{-t/\tau}, \; t \geq 0 \qquad (5)$$

The trailing c_0 term in equation (2) represents a possible start-up transient. This can be ignored because it is very small when a time series starts at low irradiance, as is usual. In in any case, it typically becomes insignificantly small after 30-60 minutes.

Given equations (3) and (5) together with a suitable set of time series measurements, the two parameters U and *tau* can be found by using a non-linear optimization algorithm, which can be thought of as a smart trial-and-error method. However, trying a new *tau* value requires reapplying the low-pass filter to the entire input time series, which is much more computationally costly than trying a new U value. For this reason, we propose a two-stage optimization: an outer loop varies the *tau* value; and for each candidate *tau*, a much quicker inner loop finds the corresponding optimal U.

Once the optimal U and *tau* parameters have been found, module temperature can be predicted for other time series in two simple steps: first, apply the low-pass filter to the inputs to account for the effect of thermal mass; then, use the steady-state equation to calculate the final module temperature from the filtered inputs. However, for this simple model it is mathematically *identical* to reverse these steps: first, calculate steady-state module temperatures using the steady-state equation; then, apply the low-pass filter to obtain the time-varying response accounting for thermal mass.

2.2 Simplifying the low-pass filter

In some software environments it may be difficult, expensive, or impossible to implement an exact first-order low-pass filter. In such situations, it is possible to

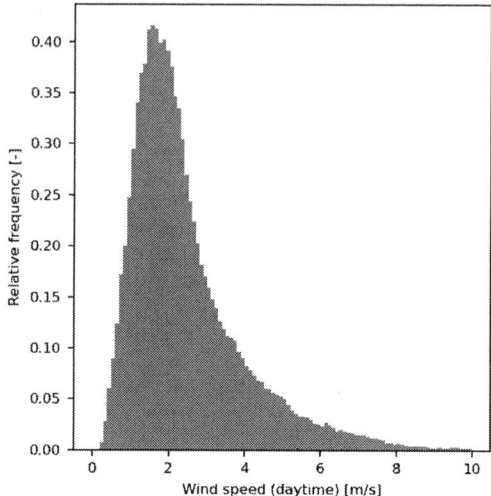

Figure 1 Day-time distribution of wind speed with a narrow spread around the most frequent value in the data set used. (See section 3.) Similar profiles are seen in many other locations, although not always as sharply defined as this.

replace it with a moving average low-pass filter to achieve nearly the same effect. In the two-stage parameter extraction process, the width or *span* of the moving average window is varied instead of *tau*. Experimentally, we have observed a ratio of *tau:span* of approximately 3:5.

2.3 Adding wind speed

Wind promotes PV module cooling. This means that the heat transfer coefficient U increases with wind speed and this is often modelled as a linear increase. Multiple model variations found in the literature contain this linear dependency and, unsurprisingly, the performance of these models is usually found to be identical (or nearly so). Here we work with the model equation attributed to Faiman [1] as representative of the industry standard:

$$T_m = T_a + \frac{G}{u_0 + u_1 \cdot V_w} \qquad (6)$$

where *u0* and *u1* define the relationship between U and wind speed.

Unfortunately, this equation is no longer a linear one in the mathematical sense. Nevertheless, it can be rewritten as an energy balance to incorporate thermal mass as before:

$$G - (T_m - T_a)(u_0 + u_1 \cdot V_w) - C\frac{dT_m}{dt} = 0 \qquad (7)$$

The exact analytical solution to equation (7) is substantially more complicated that the solution to equation (2), and from the fact that the U value varies with wind, it can be seen that the time constant C/U or *tau* is no longer constant. However, when we consider histograms of day-time wind speed, we note that they frequently have a tall and narrow peak. (See **Figure 1**.) This suggests that we may view the effect of wind speed on temperature as being fairly constant with some perturbations, and that we may treat the non-linear energy balance equation as a linear one with some parameter perturbations. Following the pattern of the exact solution in equation (2), we therefore propose the following approximate solution for equation (7):

$$T_m(t) = (f * T_a)(t) + \frac{(f * G)(t)}{u_0 + u_1 \cdot (f * V_w)(t)} \quad (8)$$

$$f(t) = \tau^{-1} e^{-t/\tau}, \ t \geq 0 \quad (9)$$

In this solution the same low-pass filter that is applied to the ambient temperature and irradiance is also applied to the wind speed.

As with the single-parameter model, the parameters *u0*, *u1* and *tau* can be extracted from observations using a two-stage optimization and the low-pass filter can be applied either to the inputs or to the output of the steady-state equation for simulation. Furthermore, the moving average low-pass filter can be used to simplify calculations here as well.

2.4 Subtracting radiative losses

A substantial fraction of the solar energy absorbed by PV modules is radiated out to the sky. Although this energy flow varies by climate and weather conditions, it has been demonstrated previously that PV module temperatures can be predicted more accurately when this is effect is included explicitly, for example [2], [3], [4].

Including the radiative loss in the thermal balance using the Stefan–Boltzmann law leads to a non-linear and circular dependency on T_m. Although the resulting equation can still be numerically solved with a few iterations, a small simplification can deliver most of the benefit without that complexity as explained in [2]. The radiative loss is divided into two parts: module → ambient and ambient → sky. The former is approximated to be linear and lumped with U, whereas the latter does not depend on T_m and can therefore be calculated exactly without iteration.

We proposed modifications to several well-known models to include this radiative loss term [2]. One example is the modified Faiman model:

$$T_m = T_a + \frac{G - F \cdot \epsilon \cdot (\sigma T_a^4 - L_{down})}{u_0 + u_1 \cdot V_w} \quad (10)$$

where L_{down} is the down-welling long-wave radiation, σ is the Stefan–Boltzmann constant, ϵ is the long-wave emissivity of the module front surface, and F is the effective view factor of the sky (dependent on PV module tilt angle). This model is available in pvlib python as the function faiman_rad().[5]

Both F and ϵ can be estimated based on prior knowledge, but they cannot both be extracted from time-series measurements; only their product can be extracted. Therefore, for model fitting one of them must be kept constant, or they can be replaced with a single empirical parameter Fe.

2.5 Putting it all together

A complete PV module operating temperature modelling method can now be assembled. It requires a steady-state equation with four inputs: ambient temperature, short-wave solar radiation, long-wave radiation and wind speed, but what is the best way to formulate this?

First, we note that not all short-wave radiation is converted to heat. This can be conveniently captured in an auxiliary equation because thermal mass does not affect optical and electrical energy flows:

$$G_{net} = G \cdot (1 - \rho - \eta - \tau_o) \quad (11)$$

G_{net} is the amount of short-wave radiation that is thermalized in the module; therefore, in equation (11) we reduce the incoming short-wave radiation G by the fractions that are reflected, ρ, converted to electricity, η, and transmitted τ_o. These fractions can be nominal values or more sophisticated time-varying quantities depending on simulation needs, but strictly speaking those details are outside the boundary of the thermal model.

The net long-wave radiation calculation is also moved to an auxiliary equation because slightly different equations may be required depending on the available data source: pyrgeometer readings, sky temperature, T_s, approximations or a numerical weather model variables such as down-welling long-wave radiation L_{down}. Two possible equations are:

$$L_{net} = \sigma T_a^4 - \sigma T_s^4 \quad (12)$$

$$L_{net} = \sigma T_a^4 - L_{down} \quad (13)$$

L_{net} is the net long-wave radiation exchange between the sky and a black body at ambient temperature.

With the help of these auxiliary equations, the core steady-state model equation simplifies to:

$$T_m = T_a + \frac{G_{net} - \text{Fe} \cdot L_{net}}{u_c + u_w \cdot V_w} \quad (14)$$

One could ask whether the parameter Fe should be included in the auxiliary equation rather than in the main equation. Indeed this would be logical, but in this work the value of Fe is purely empirical, therefore it is practical to position it with the other empirical parameters in the core steady-state equation.

Thermal mass does not affect the auxiliary equations; thus, to obtain a dynamic model it is sufficient to rewrite the core equation as an energy balance with added capacitance:

$$G_{net} - \text{Fe} \cdot L_{net} - (T_m - T_a)(u_c + u_v \cdot V_w) - C\frac{dT_m}{dt} = 0 \quad (15)$$

The two approximate solutions to equation (15) are shown in equations (16) and (17), and equation (18) repeats the specification of the low-pass filter for completeness.

$$T_m(t) = (f * T_a)(t) + \frac{(f * G_{net})(t) - \text{Fe}(f * L_{net})(t)}{u_c + u_w \cdot (f * V_w)(t)} \quad (16)$$

$$T_m(t) = \left(f * \left(T_a(t) + \frac{G_{net}(t) - \text{Fe} L_{net}(t)}{u_c + u_w \cdot V_w(t)} \right) \right)(t) \quad (17)$$

$$f(t) = \tau^{-1} e^{-t/\tau}, \ t \geq 0 \quad (18)$$

To summarize, we propose a core steady-state model in equation (14), which turns into differential equation (15) when thermal mass is added. We then propose two approximate analytical solutions to this differential equation: the first, equation (16), is needed in the two-stage parameter extraction process; the second, equation (17), is an optional alternative to (16) for system simulations using known parameters. Both solutions apply the same first-order low-pass filter, whose impulses response is shown in equation (18).

41st European Photovoltaic Solar Energy Conference and Exhibition

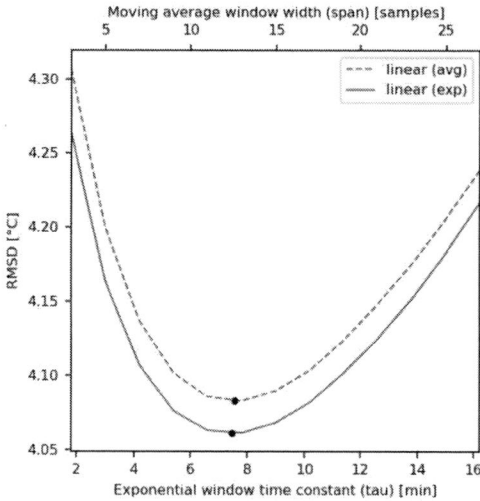

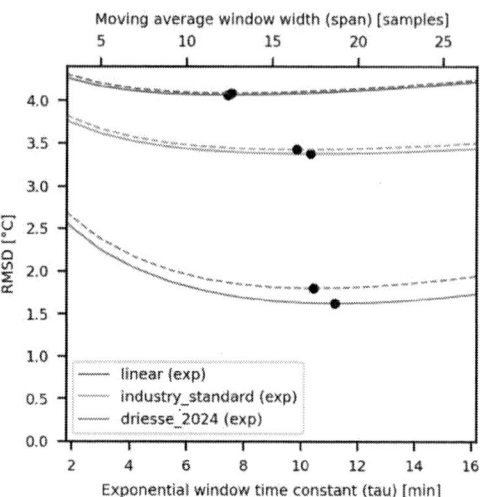

Figure 3 Variation of RMSD with low-pass filter parameter *tau* (first-order filter, solid line) and *span* (moving average filter, dashed line) for the single-parameter linear model.

Figure 2 Variation of RMSD with low-pass filter parameter *tau* (first-order filter, solid line) and *span* (moving average filter, dashed line) for three temperature models.

3   RESULTS

In this section we demonstrate the extraction of model parameters and evaluate the accuracy of the resulting model predictions using an open data set [6]. The data set contains one year of relevant measurements at Sandia National Labs in Albuquerque, NM, USA at one-minute resolution and is fully documented in the paper that accompanies it. Local measurements are augmented by down-welling long-wave radiation obtained from the ERA5 reanalysis data set [7].

From the data set we select the period from February to December because some maintenance activity interrupted monitoring activity in January. Furthermore, we only use data records where irradiance exceeds 5 W/m^2 so that the reported performance metrics represent the more relevant day-time model performance. By applying these two constraints it should be possible to replicate the results shown in this work.

3.1 Two-stage parameter extraction

The objective for optimal parameter extraction is to minimize the mean-squared deviation between measured and predicted module temperatures. This is the most common optimization objective, although others are possible.

Figure 3 illustrates the search for the optimal value of *tau* in the case of the single-parameter linear model. For each value of *tau*, an optimal U value is determined by linear regression, and the resulting RMSD values are plotted. The final optimal combination of *tau* and U is the one having the overall lowest RMSD—the lowest point on the graph. It is important to note that such a unique point actually exists, as this confirms the validity of extracting *tau* from the measurements.

The dashed line in **Figure 3** illustrates the optimization using a moving average low-pass filter. The optimal U values found are nearly identical: 37.71 using the first-order filter and 37.79 using the moving average. Naturally, it is not necessary to carry out the calculations for the full range of *tau* (or *span*) values in order to find the optimum one. A golden-section search, for example, would be a more efficient way to find the optimal *tau* in practice.

3.2 Comparisons between models

We now use the two-stage fitting procedure with each of the three main models discussed in Section 2 in order to compare them: the single-parameter linear model (equation (1)), a representative of the two-parameter industry-standard model (equation (6)), and our new three-parameter model (equation (14)). Scatter-plots of the individual results with and without thermal mass are shown in the appendix while the most important results are summarized in **Figure 2**. Here, we observe that incorporating wind speed brings a relatively small improvement, which can be explained by the narrow wind speed profile (**Figure 1**). Taking into account long-wave radiation clearly provides the greatest incremental improvement: the new model reduces RMSD by half compared to the industry standard.

3.3 Low-pass filtering options

As discussed in section 2, it is possible to use an exact first-order low-pass filter or a moving average approximation; and the chosen low-pass filter can be applied to the inputs of the steady-state model or to the outputs. **Figure 2** shows that as model accuracy improves sensitivity to both the *type* of low-pass filter and the filter *parameter* (*tau* or *span*) increases. Nevertheless, even in the worst case (best model), the steady-state model parameters that were found using the moving average filter differed by less than 4% from those found using the first-order filter.

1217

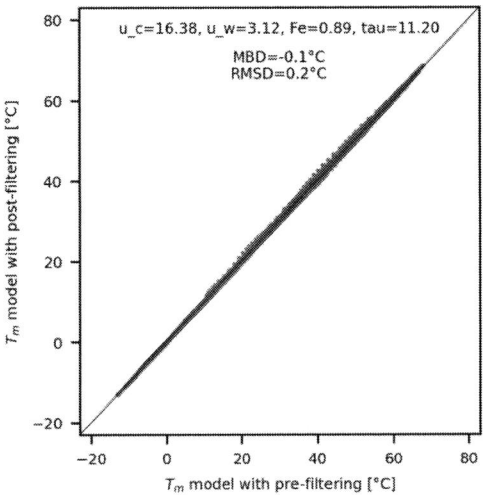

Figure 4 Comparison between low-pass filtering of output vs. inputs for the new model shows negligible differences.

For the single-parameter linear model, applying the low-pass filter to its inputs is identical to applying the same low-pass filter to its output. As discussed, this is no longer exactly true for the other models, but in practice the difference appears to be negligible (**Figure 4**).

4 DISCUSSION AND CONCLUSIONS

In this paper we have developed and explained the basis for using low-pass filters to account for thermal mass in the context of PV module temperature modeling. We had used this technique previously in [8] and several other authors have also recently applied low-pass filters to steady-state model inputs [3], [9] or output [10]. However, the rationale for these filters remained somewhat vague in those publications.

Low-pass filtering is by no means the only way to accommodate thermal mass in temperature simulations. The differential model equations (2), (7) and (15) can be solved numerically by iterative methods, which avoid the approximations we discussed. However, extracting model parameters becomes much more computationally expensive compared to the two-stage optimization using low-pass filters. Furthermore, given the low RMSD values we achieved, there is not so much potential for further improvement using exact numerical solutions.

While using low-pass filters reduces the RMSD of all models, by far the greatest reduction in RMSD is achieved with the new model that incorporates a radiative loss term. In [8] we needed to use a low-pass filter to illustrate the benefit of that radiative loss term, and in this work we need the radiative loss term to fully demonstrate the benefit of low-pass filters. The final results of these efforts are a new temperature model and an accompanying parameter extraction method that stand on a firm foundation. With minor approximations we achieve major gains in accuracy and usability while avoiding unnecessary complexity.

5 AUTHOR CONTRIBUTIONS

Anton Driesse: Conceptualization, Formal analysis, Methodology, Software, Validation, Visualization, Writing–original draft, Writing–review & editing. **Jesus Polo**: Validation, Writing–review & editing.

6 REFERENCES

[1] D. Faiman, 'Assessing the outdoor operating temperature of photovoltaic modules', *Prog. Photovolt: Res. Appl.*, vol. 16, no. 4, pp. 307–315, Jun. 2008, doi: 10.1002/pip.813.

[2] A. Driesse, J. Stein, and M. Theristis, 'Improving Common PV Module Temperature Models by Incorporating Radiative Losses to the Sky', SAND2022-11604, 1884890, 709196, Aug. 2022. doi: 10.2172/1884890.

[3] J. Barry *et al.*, 'Dynamic model of photovoltaic module temperature as a function of atmospheric conditions', *Adv. Sci. Res.*, vol. 17, pp. 165–173, Jul. 2020, doi: 10.5194/asr-17-165-2020.

[4] K. R. McIntosh *et al.*, 'The influence of wind and module tilt on the operating temperature of single-axis trackers', in *2022 IEEE 49th Photovoltaics Specialists Conference (PVSC)*, Philadelphia, PA, USA: IEEE, Jun. 2022, pp. 1033–1036. doi: 10.1109/PVSC48317.2022.9938577.

[5] K. S. Anderson, C. W. Hansen, W. F. Holmgren, A. R. Jensen, M. A. Mikofski, and A. Driesse, 'pvlib python: 2023 project update', *JOSS*, vol. 8, no. 92, p. 5994, Dec. 2023, doi: 10.21105/joss.05994.

[6] A. Driesse, M. Theristis, and J. Stein, 'PV Module Operating Temperature - Data and Resources'. EMN-DURMAT (EMN-DuraMAT); Sandia National Laboratories (SNL-NM), Albuquerque, NM (United States), 2023. doi: 10.21948/2204673.

[7] A. Inness *et al.*, 'The CAMS reanalysis of atmospheric composition', *Atmos. Chem. Phys.*, vol. 19, no. 6, pp. 3515–3556, Mar. 2019, doi: 10.5194/acp-19-3515-2019.

[8] A. Driesse, *Module operating temperature model parameter determination*. (Dec. 30, 2022). Sandia National Labs. doi: 10.5281/ZENODO.10003736.

[9] B. Herteleer, A. Kladas, G. Chowdhury, F. Catthoor, and J. Cappelle, 'Investigating methods to improve photovoltaic thermal models at second-to-minute timescales', *Solar Energy*, vol. 263, p. 111889, Oct. 2023, doi: 10.1016/j.solener.2023.111889.

[10] M. Prilliman, J. S. Stein, D. Riley, and G. Tamizhmani, 'Transient Weighted Moving-Average Model of Photovoltaic Module Back-Surface Temperature', *IEEE J. Photovoltaics*, vol. 10, no. 4, pp. 1053–1060, Jul. 2020, doi: 10.1109/JPHOTOV.2020.2992351.

41st European Photovoltaic Solar Energy Conference and Exhibition

APPENDIX

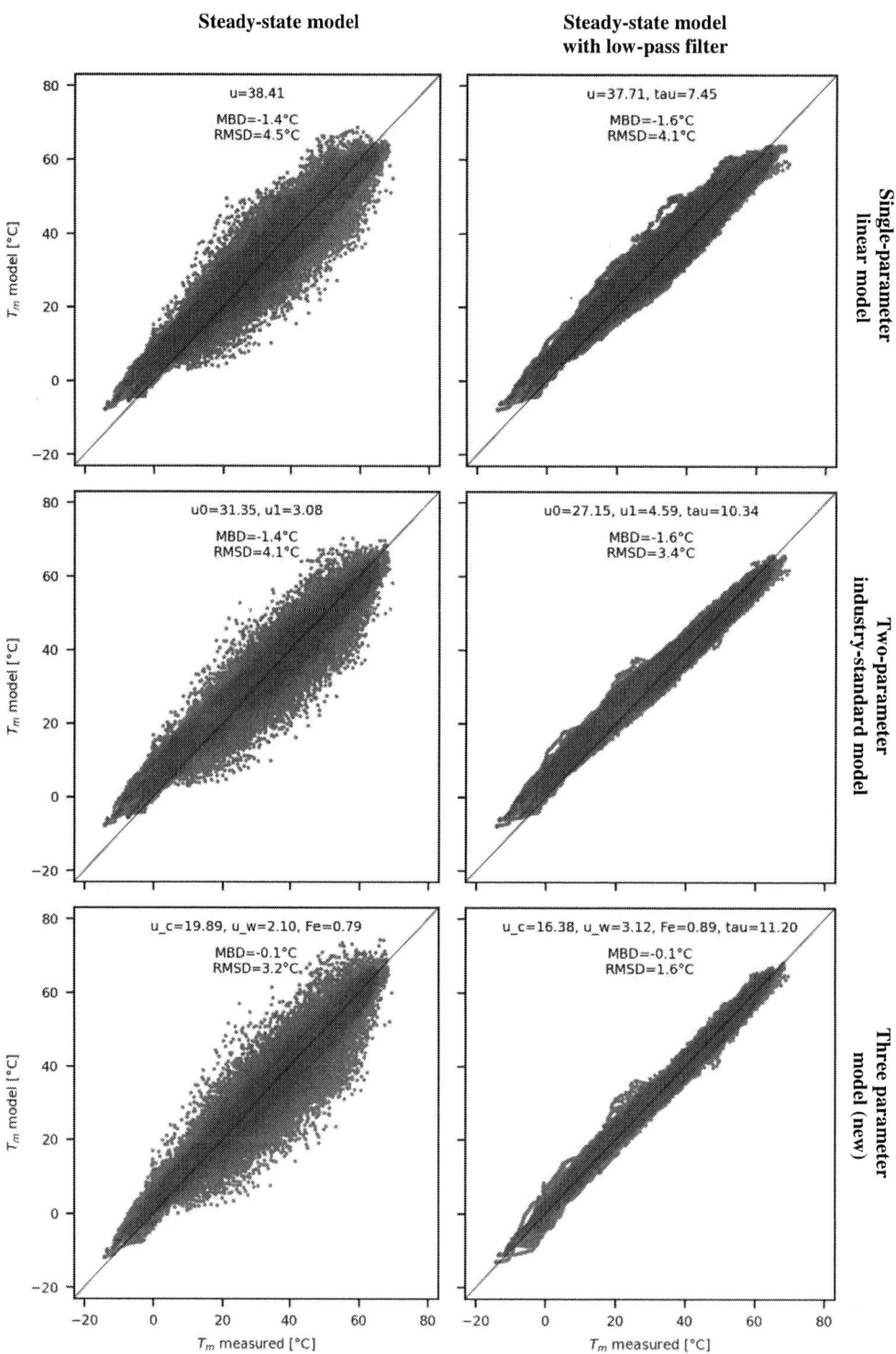

Figure 5 Accuracy of three temperature models using the same data set. Incremental improvements are seen when adding additional explanatory variables, but by far the greatest improvement is achieved when both thermal mass *and* radiative losses are accounted for (three-parameter model with low-pass filters applied).

41st European Photovoltaic Solar Energy Conference and Exhibition

This presentation was selected by the Sc. Committee of the EU PVSEC 2024 for submission of a full paper to one of the EU PVSEC's collaborating peer-reviewed journals.

FROM PIXELS TO INSIGHTS:
A SOFTWARE PROTOTYPE FOR AI-DRIVEN COMPLETE DIAGNOSTICS OF PV PLANTS

John (Ioannis) A. Tsanakas[1], Murielle Stepec[1], Philippe Marechal[1], Duy-Long Ha[2]
[1] CEA, Liten, Univ. Grenoble Alpes Campus INES, 73375 Le Bourget du Lac, France
[2] Entech smart energies, 29000 Quimper, France

ABSTRACT: Although aerial infrared (aIRT) imagery-based solutions for diagnostics of PV plants demonstrate impressive time-efficiency and spatial resolution, they also suffer from considerable drawbacks: limited automation (hence, expert dependence) and insufficient quantitative insights. In this paper, we introduce a software prototype, evolved from an innovative diagnostics framework researched and developed by CEA-INES over the last years, which integrates aIRT imagery with deep learning-based algorithms and physical/electrical modeling. With such approach, unlike conventional ones, we worked on reaching both qualitative fault detection and quantitative (power loss) insights, with a focus on various spatial granularity levels within PV systems. Leveraging advanced deep learning techniques, first results show that we can achieve automated PV module detection and fault identification/classification, with associated power loss analysis at PV system, string/inverter or module level. Further real-life validation efforts are underway, in utility-scale PV plants. Future developments aim to enhance further our PV diagnostic framework, through data fusion with SCADA outputs and integration with maintenance and end-of-life (EoL) management tools.

Keywords: PV systems; PV fault diagnostics; infrared imagery; machine learning.

1 RATIONALE and AIM

As of today, monitoring-based detection and assessment of underperforming photovoltaic (PV) plants are typically executed in a semi-manual top-down approach, analyzing low performing components or failures (e.g. in PV modules), by drilling down from substations, inverters to strings and junction boxes [1]. Particularly for utility-scale PV systems, which can exceed hundreds of thousands of PV modules, monitoring-based diagnostic needs can be significantly complex and demanding.

Relying solely on PV monitoring data, has two significant intrinsic limitations: i) expert-dependence (e.g. misconfigured reference performance data or misinterpreted deviations) and ii) insufficient spatial granularity. As a result, for several underperformance issues and failures – especially on PV module and submodule level – they may either remain undetected, trigger "false alarms" or their root-cause stays unidentified. On this basis, the need for advanced diagnostics, i.e. at higher spatial granularity, is typically addressed today by additional data streams, i.e. from advanced PV inspections, notably aerial infrared thermographic (aIRT) imagery. Although aIRT imagery data analytics demonstrate impressive time-efficiency and spatial resolution (inspection rates of several MW/hour; fault detection and classification down to submodule/cell level), they also suffer from considerable limitations. On one hand, image-based fault detection features (still) lack today of sufficient automation, thus remaining highly expert-dependent and prone to false negatives and/or positives. On the other hand, they remain decoupled from PV monitoring, barely correlated with e.g. real-time performance data ; thus only providing a rather qualitative instantaneous assessment, with practically inexistent temporal granularity, than a complete diagnosis with quantitative root-cause analysis [1].

Over the last 7 years, CEA-INES has built an innovative (and continuously evolving) diagnostics framework for PV plants, based on a patented methodology that leverages aIRT imagery, coupled with physical/electrical modelling [2-4]. What differentiates today such methodology from the state-of-the-art, is the fact that it is made to provide not only qualitative fault diagnosis (i.e. detection, classification), but also primary quantitative insights (statistical analysis,

severity, power loss estimations) for different PV failures, at different spatial granularity levels, i.e. from PV system and inverter/string level, down to module and submodule level. Although several high-impact publications and ongoing research, development and innovation programs demonstrate the scientific and industry relevance of such approach, there is still much room for further development and improvement:

1. Need for more precise qualification and root-cause analysis features for PV faults, thus minimizing false negatives/positives, e.g. by correlation with SCADA (supervisory control and data acquisition) outputs or employment of multi-stream data fusion.
2. Need for more accurate power loss analysis (beyond today's ~80-90% demonstrated accuracy) and complete diagnosis, again by leveraging data fusion from multiple streams (e.g. monitoring, imagery).
3. Need for higher time-efficiency, thus automation in recognition/analysis and classification of PV module and failure signatures, in diverse PV plant configurations.

In the context of the EU-funded H2020 project SERENDI-PV [5], we have evolved the aforementioned diagnostic framework in the form of "ASPIRE", a software prototype. Considering the above three needs, our efforts – at this stage – have focused on implementing a deep learning-based approach for fully automated recognition of PV modules and their failure signatures, along with enhanced precision for the qualification and analysis of PV failures. A last (aimed) feature, i.e. the correlation/fusion with SCADA data, is under exploration. In this paper, we present the main elements of our approach, the software prototype functionalities and example results from one our first validation/case studies. Further real-life demonstrations are underway, in targeted utility-scale PV plants.

2 METHODOLOGY

2.1 Deep learning approaches used and training database

The automated detection/recognition feature of ASPIRE has been based on coupling two known deep learning (DL) approaches (Fig. 1): i) the real-time object detection pipeline YOLO (You Only Look Once) and ii) the Mask Region-based Convolutional Neural Network (Mask R-CNN). The former approach yields higher sensitivity to detection, while the latter is more informative about the shape/pattern (of the aIRT signatures, in our case).

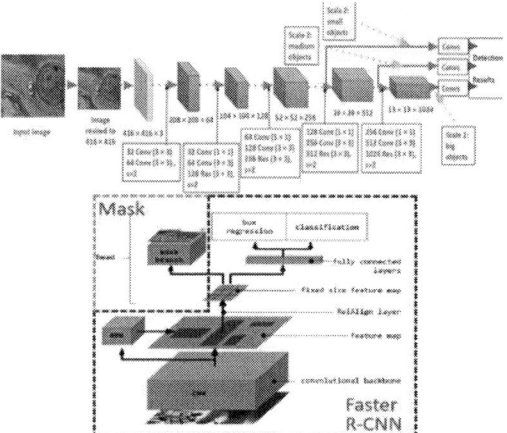

Figure 1: Example architecture of the real-time object detection pipeline YOLO (on the left) and Mask R-CNN (on the right) algorithms [6-8].

For the training of the coupled algorithms, we have implemented and employed a database of anonymized real-case aIRT images of PV installations (Fig. 2). Such images were intentionally chosen to include diverse PV layouts and conditions (e.g. in terms of size, tilt, brightness) and interfering objects (e.g. roofs, cars, substations, trees/vegetation, chimneys, etc), in order to maximize the extraction and optimize the detection of PV module features.

Figure 2: Example aIRT images of diverse features, employed in the database for auto-detection training.

The variation of the characteristics and the hyperparameters of the algorithm (i.e. the convolution and pooling mask sizes for the CNN, or the size of the ROI (region of interest) Align for the Mask R-CNN) allows to train and optimize its performance. To avoid the appearance of parasitic phenomena (i.e. false features due to overlapping or poor up-convolution) and to increase the tuning possibilities, we have chosen to set the size of each image dimension to 2^9 pixels, eventually employing images of infrared (pseudocolor) type and of 512x512x3 size in pixels. To optimize the calculations and increase the possibilities of training by batch, mini-batch, and for

parallel training on GPU, we have created a training database (train set and validation set) of size 2^{11} pixels, to which 256 images are added for the test phase. In general, the size of the test data set is 20% of the size of the training set, the ratio 256/2048 is slightly lower, this can be an opportunity to complete such test sets with use-case specific data.

2.2 Implementation steps – Training/test parameters and initial outputs

The overall auto-detection function in ASPIRE, has been implemented in four main steps: i) the algorithm implementation and pre-training, ii) the transfer learning, iii) the training/test database and parametrization and iv) the detection test outputs (validation) and loop-learning.

The Mask R-CNN algorithm's backbone is a residual network (ResNet deep network), with 50 convolution layers, which skips some layers during training to avoid the gradient vanishing phenomenon and to optimize the learning. For the training of our neural network we were inspired by a study of Bommes et al. [9] and we used a Mask R-CNN pre-trained on a COCO (Common Objects in Context) database. Following the algorithm implementation and its pre-training, we have applied transfer learning, for which – in our case study – we have two clases: the PV modules and the background. Therefore, to train our network in recognizing these two classes, our transfer learning is applied on the last two ResNet layers; thus, in turn, the training adapts the weights to our database on its last layers.

Next to the transfer learning, we have employed the database described in Subsection 2.1, to train and further optimize/tune the auto-detection algorithm for ASPIRE. The Mask R-CNN was trained with 2048 images, whereas a test set consists of 256 images. The training is carried out with a batch of size 2 (i.e. the update of the coefficients of the network is done every 2 images). We perform 30 epochs (i.e. the data set is seen 30 times by the network) and a learning rate adjusted after each epoch at initial LearnRate equal to 0.001 (i.e. no advancement in the gradient).

Finally, the optimization method applied in our case is the SGDM (stochastic gradient descent with momentum). From the outputs of the initial detection tests that we performed, we observe that, while the detection of PV module objects is correctly learned by the Mask R-CNN and false detections are avoided, the PV module masks lack precision as they do not always follow the exact PV module edges and outline (Fig. 3).

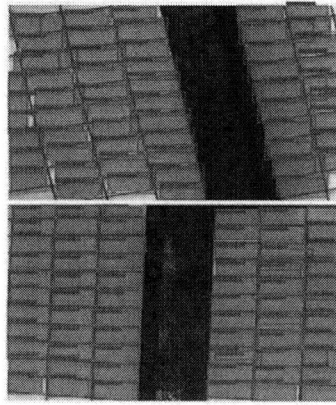

Figure 3: Example outputs from initial auto-detection tests on aIRT images.

Further training, optimization and coupling with the YOLO algorithm allowed us to improve the precision of our auto-detection feature, eventually included in the ASPIRE.

3 FIRST RESULTS and DISCUSSION

The developed version of ASPIRE prototype has three functional blocks: i) Auto-detection of PV modules and anomalies (i.e. potential failures), ii) Classification and estimation of power losses and iii) complete analysis dashboard, with primary qualitative and quantitative insights. In the following sub-sections we give an overview of representative (example) results for each function.

3.1 Results on the auto-detection feature

At its base, the software's algorithm auto-detects PV modules and anomalies, from image and/or radiometric data, using segmentation, based on the determination of the PV module edges. In real-world complex cases of overlapping PV arrays or complex scenes, this method alone cannot auto-detect all PV modules separately; that is why we have recently developed and applied instance segmentation based on the coupled DL approach presented in Section 2. In Fig. 4, we present an example of the resulting workflow in ASPIRE, starting from raw aerial infrared and visual (red-green-blue, RGB) images (Fig. 4A). The aIRT and RGB images are combined by image fusion. The fused image is post-treated to improve contrast and reduce parasitic elements, to then undergo automatic adaptive thresholding (CEA patent pending), for auto-segmentation, to automatically detect the PV modules (Fig. 4B). From the latter, outlier features are extracted for anomalies (potential faults) detection (Fig. 4C). Fault detection in ASPIRE is implemented on two levels: PV module and PV plant. The sub-Section that follows discusses further on the faults' analysis.

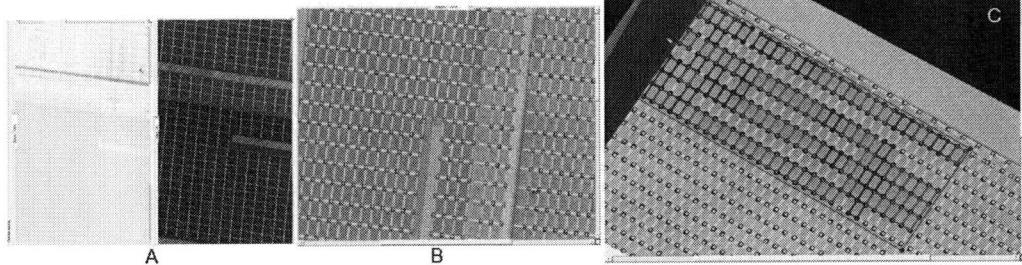

Figure 4: (A) Example raw aIRT and aerial RGB images imported in ASPIRE; (B) Detection of PV modules, from a fused aIRT/RGB image, with auto-segmentation; (C) Auto-detection of anomalies in a fused aIRT/RGB image (example case of open-circuited modules/strings).

3.2 Results on the classification and loss analysis feature

At PV module level, ASPIRE looks for faults such as hot spots, failed (e.g. shunted) bypass diodes, junction box failures, potential induced degradation (PID), as well short- or open- circuited PV modules and submodules. We have created signature (detection) masks for each fault that we apply to the pixels that correspond to the PV module's IR signature. Using expert-set thresholds, the algorithm predicts whether there is a fault (or faults) and specifies its type (classification). At PV plant level, the diagnosis takes into account the electrical architecture of the PV plant, i.e. the physical and electrical layout of inverters and strings. When, for instance, open-circuited PV modules are detected on the same string, automatically all modules of the string are displayed as disconnected as well.

Next to the detection/classification steps, the power loss analysis feature is based on the application of the pixel-level temperature (radiometric) data of the PV modules (in particular, the delta-temperature (ΔT) between normal and fault pixels) to CEA's physics model-based methodology [3]. For the visualization of the power loss analysis, ASPIRE creates hollow matrices (Fig. 5) of "healthy" (fault-free) and faulty PV modules, using a smoothing algorithm. For each faulty PV module, a color delta (hence, ΔT), i.e. between the PV module's healthy and faulty state, is then associated to a calculated power loss (in %) at PV module, string or system level.

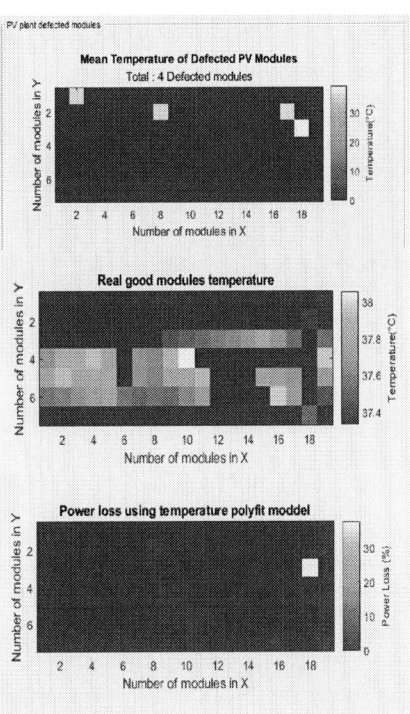

Fig. 5. Top: Initial (raw) hollow matrix of PV modules from an aIRT image, in ASPIRE. Middle: Hollow matrix with corrected PV modules in fault-free mode. Bottom: Final hollow matrix for power loss calculation/analysis in ASPIRE

41st European Photovoltaic Solar Energy Conference and Exhibition

3.3 Results on complete analysis dashboard

The last functional block of ASPIRE is a dashboard for overview and analysis of a PV plant's condition, that encompasses insights from the auto-detection, fault classification and power loss analysis features. Figure 6-left presents the results from an aIRT inspection case study on a test PV site (rooftop PV system, consisting of 26 strings of 21 PV modules/string). The complete analysis dashboard shown in this case includes the mapping of PV modules and failures (power loss/hollow matrix), the fault-specific percentage of power output losses for the different diagnosed/classified failure modes ("defect classes") in the inspected PV plant, as well as power production and loss data at string level, when information on the electrical layout is available. In the presented example, the majority of diagnosed faults are associated as disconnected strings, whereas few cases of hot spots, bypass diode or junction box failures are also identified. The loss analysis at string production level is suggestive of such faults, with losses ranging from 1-1.5% for failures of low/moderate severity (e.g. individual hot spots) up to 20-30%, for open-circuited strings or array level failures. Further, it is also possible to carry out a focused diagnostic study at PV module level, down to submodule and cell level. Figure 6-right shows an example of such dashboard for targeted analysis, at PV module level, for the case of a diagnosed PID (flagged as "multi-hotspot"). In another example of targeted analysis at PV module level, Fig. 7 left, shows the function of ASPIRE's "defect checker" (detection mask) feature and how the latter is used to confirm a bypass diode failure: i) comparing with and excluding thermal patterns of other PV failure modes and ii) implementing a sub-module level analysis to pinpoint the location of the failed component, i.e. the bypass diode 1 (left one) in this case. Then, Fig. 7 right, similarly to Fig. 6 right, displays the overall analysis dashboard for the PV module impacted from the bypass diode failure.

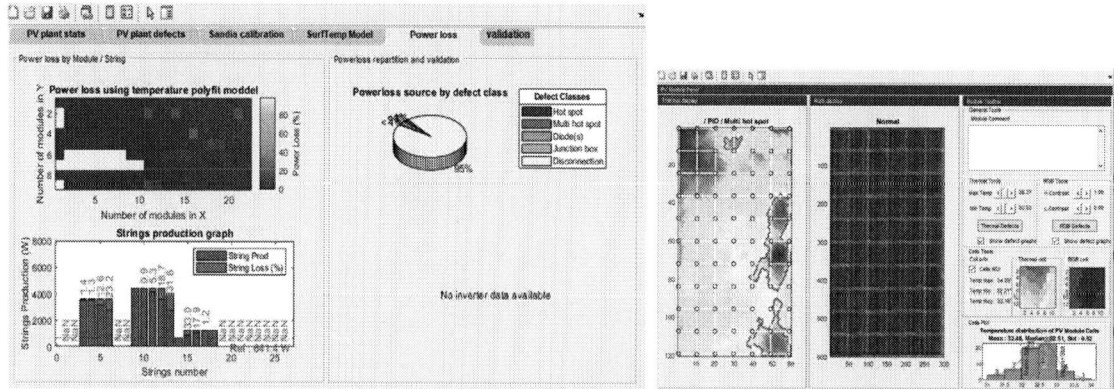

Figure 6: Left: ASPIRE's insights and analysis dashboard for an example aIRT inspection case in a test PV site. Right: PV module-level dashboard; example of a diagnostic analysis for the case of a PID.

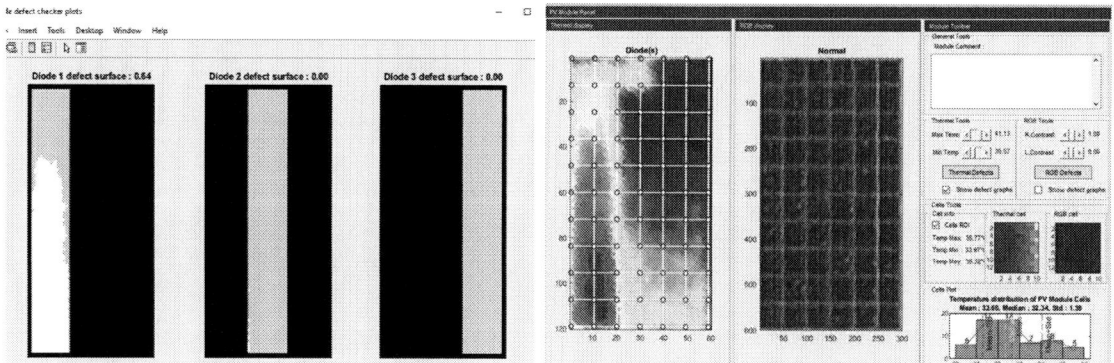

Figure 7: Left: ASPIRE's "defect checker" (detection mask) feature at PV module level, for the case of a bypass diode failure (Diode 1, left). Right: Output generated in the PV module-level dashboard, for the full analysis of the identified bypass diode failure.

Overall evaluation of ASPIRE's acquired results, against actual PV data, indicates that ASPIRE's outputs are generally coherent (losses between 0-100%) and consistent with deterministic models of (known) losses (e.g. open-circuits or bypass diode failures). Results from global power loss validation show an approximate 1% to 3% mean overestimation of the actual power production compared to the inverter and I-V data for the studied test site. In terms of diagnostic accuracy, for the test PV site, ASPIRE has (so far) correctly diagnosed ~89% of the anomalies, whereas ~9.7% of diagnosed cases were false negatives and ~1.7% false positives.

4 CONCLUDING REMARKS

In this paper, we introduce ASPIRE, a software prototype for complete fault diagnostics in operating PV plants, evolved from an innovative diagnostics framework researched and developed by CEA-INES over the last years, which integrates aIRT imagery with deep learning-based algorithms and physical/electrical modeling. The developed approach, unlike conventional ones, enables both qualitative fault detection and quantitative (power loss) insights, at various levels of spatial granularity within PV systems. First results, mainly on a PV test site (with controlled or known fault scenarios), confirm ASPIRE's capacity to achieve automated PV module detection and fault identification/classification, with associated power loss analysis at PV system, string/inverter or module level.

All things considered, although results are considerably promising, for both initially set objectives, i.e. higher time-efficiency (through the DL-driven automated detection) and higher accuracy in fault classification/loss analysis, we are currently working on further validating, evolving and improving this software prototype, on three future goals:

- Correlation and fusion with other data streams, to minimize false negatives and (most importantly) maximize the spatiotemporal granularity and accuracy of the PV faults diagnosis.
- Interoperability with on-site automated maintenance tools, to streamline (data-driven) corrective maintenance interventions, such as robot-assisted cleaning (soiling mitigation) and vegetation management.
- PV qualification feature, to be introduced for prepare-for-repair/reuse operations.

ACKNOWLEDGMENTS

Part of this work has been carried out in the framework of the H2020 SERENDI-PV project. SERENDI-PV project has received funding from the European Union's Horizon 2020 research and innovation programme under grant agreement No. 953016. Part of this work was also supported by the French National Program "Programme d'Investissements d'Avenir - INES.2S" under Grant Agreement ANR ANR-10-IEED-0014 0014-01.

REFERENCES

[1] SolarPower Europe (2021): "Operation & Maintenance: Best practice guidelines Version 5.0". ISBN: 9789464444247.

[2] J.A. Tsanakas, L.-D. Ha, F. Al Shakarchi (2017). Renewable Energy 102, 224-233.

[3] D.-L. Ha, J.A. Tsanakas, Patent No: WO2016189052/A1 (International), FR3036900 (France), CEA.

[4] U. Jahn, M. Herz, M. Köntges, et al. Report IEA-PVPS T13-10:2018.

[5] SERENDI-PV (Smooth, reliable and dispatchable integration of PV in EU Grids) [https://serendipv.eu/]. Grant agreement ID: 953016

[6] Figure available via license: Creative Commons Attribution 4.0 International

[7] A. Ammar, A. Koubaa, M. Ahmed, et al. (2021). Vehicle Detection from Aerial Images Using Deep Learning: A Comparative Study. Electronics 10, 820.

[8] T. -Y. Lin, P. Dollár, R. Girshick, et al. (2017). Feature Pyramid Networks for Object Detection. In: Proc. 2017 IEEE Conference on Computer Vision and Pattern Recognition (CVPR), Honolulu, HI, USA, pp. 936-944.

[9] E. Shelhamer, J. Long and T. Darrell. (2017). Fully Convolutional Networks for Semantic Segmentation. IEEE Transactions on Pattern Analysis and Machine Intelligence 39(4), 640-651, doi: 10.1109/TPAMI.2016.2572683.

[10] L. Bommes, T. Pickel, C. Buerhop-Lutz, et al. (2021). Computer vision tool for detection, mapping, and fault classification of photovoltaics modules in aerial IR videos. Prog Photovolt Res Appl. 29(11).

[11] M. Mattei, G. Notton, C. Cristofari, et al. (2006). Calculation of the polycrystalline PV module temperature using a simple method of energy balance, Renewable Energy 31(4), 553-567.

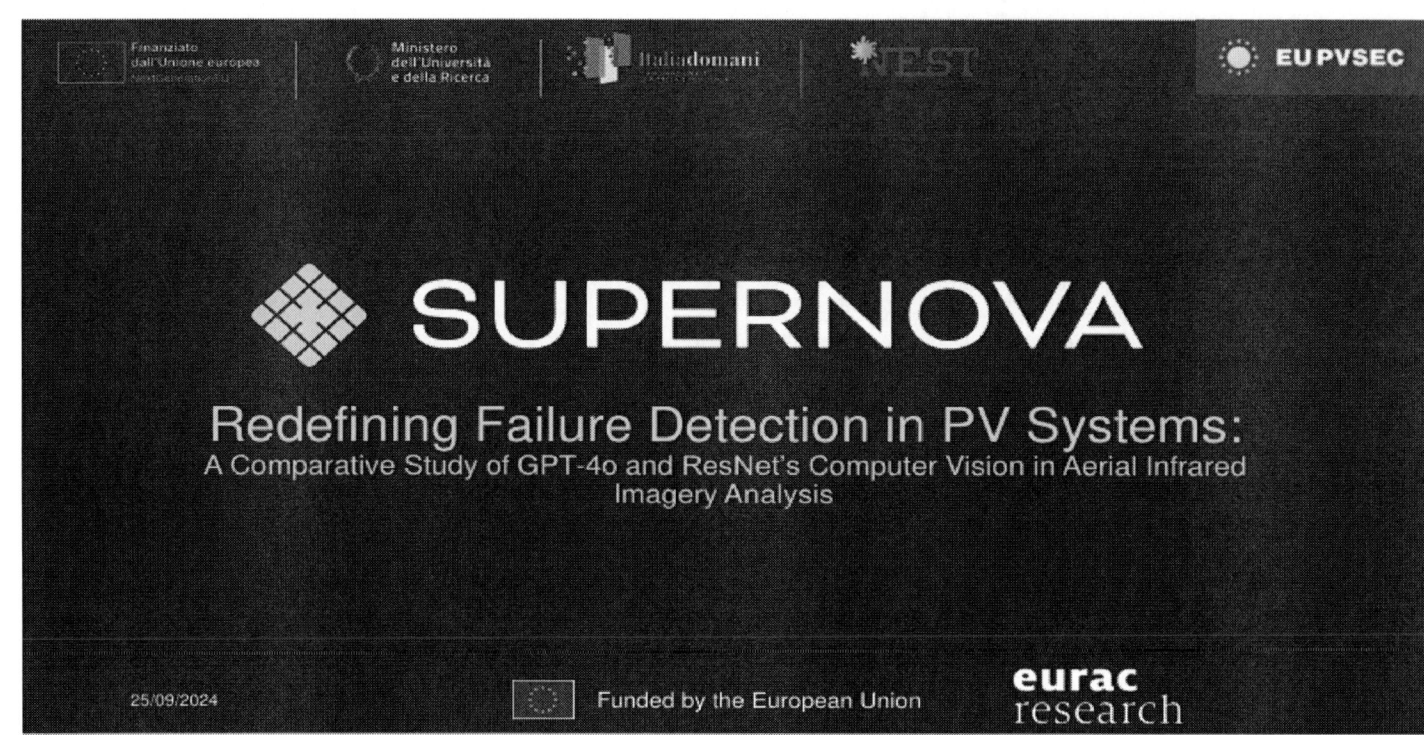

OUTLINE

SUPERNOVA

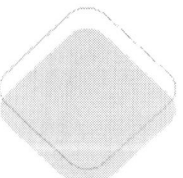

MODELS

DATASET

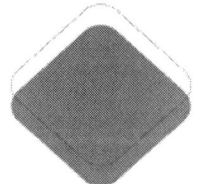

RESULTS

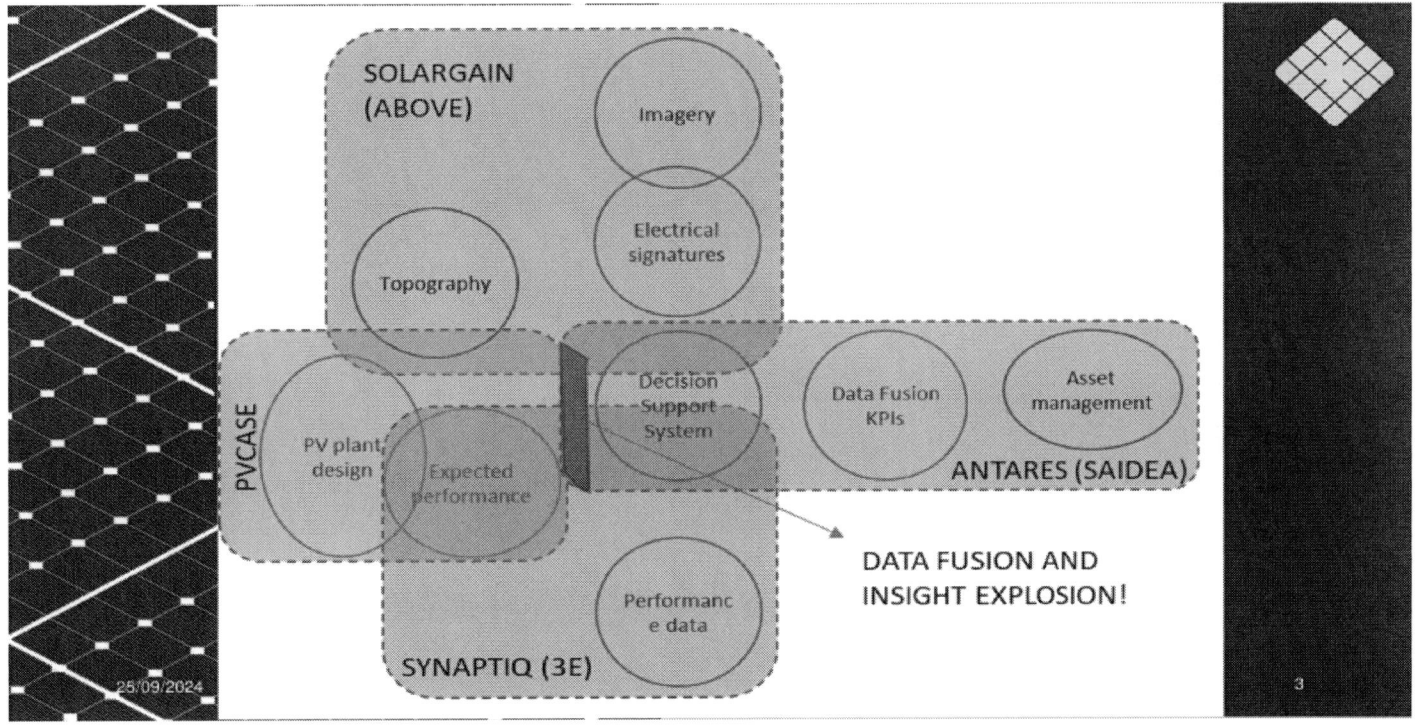

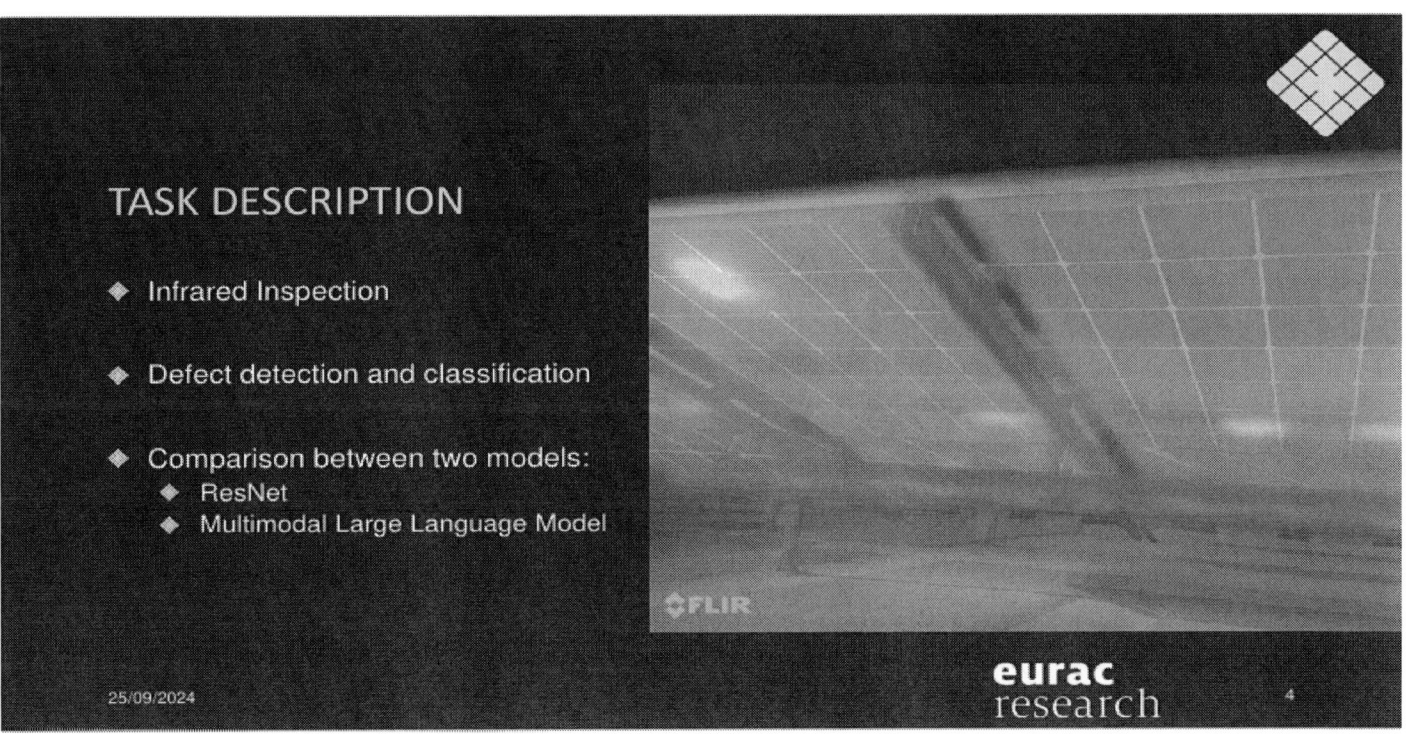

RESNET

◆ A deep neural network introduced by Kaiming He in 2015.

◆ Residual learning: introduces skip connections that bypass one or more layers.

◆ State-of-the-art performance in many image classification tasks, object detection and more.

25/09/2024

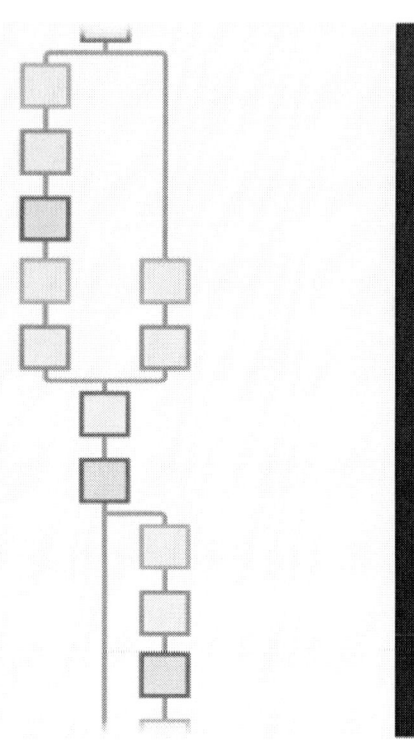

MULTIMODAL LARGE LANGUAGE MODEL

◆ Advanced AI models that can process and integrate multiple types of input:
  ◆ Text
  ◆ Images
  ◆ Audio

◆ Combines language understanding with visual perception

◆ Generates text at human level

◆ For image captioning, visual question answering, and contextual reasoning...

25/09/2024

eurac
research

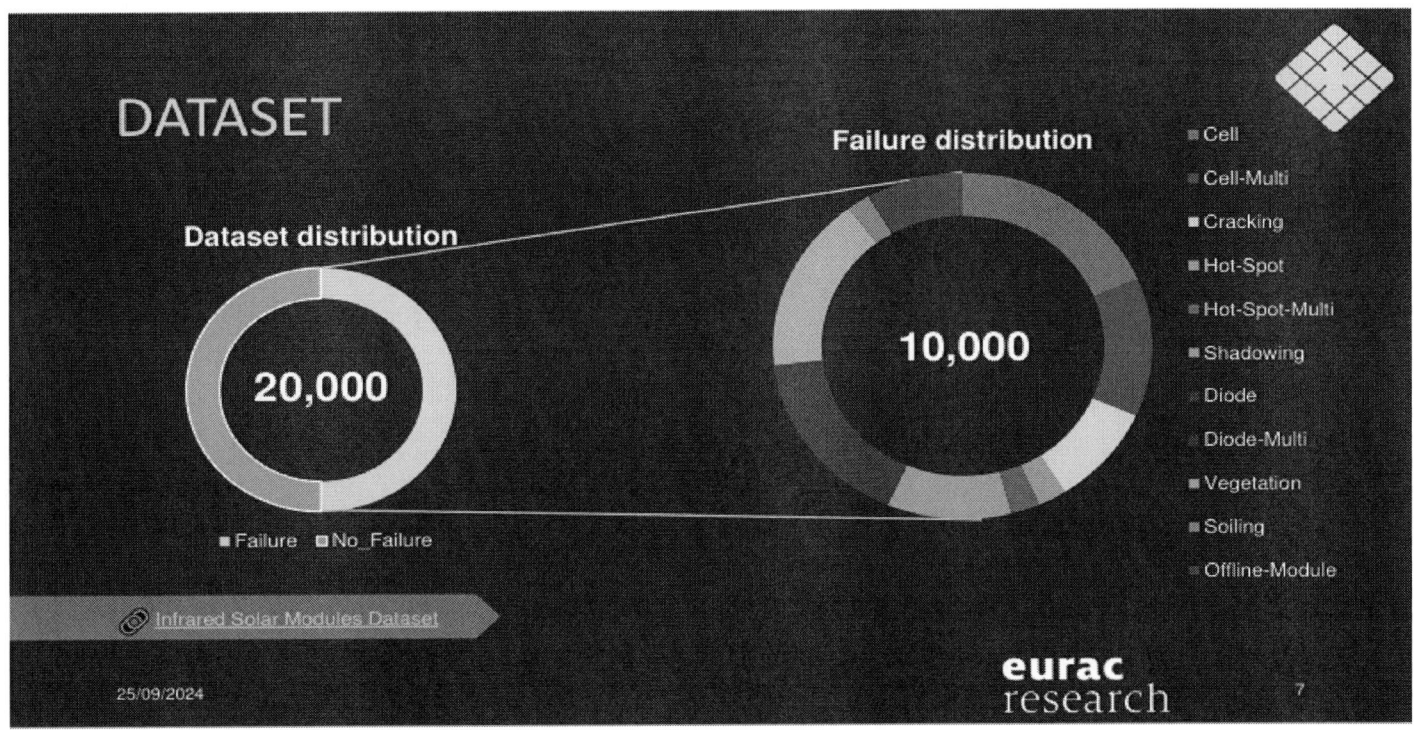

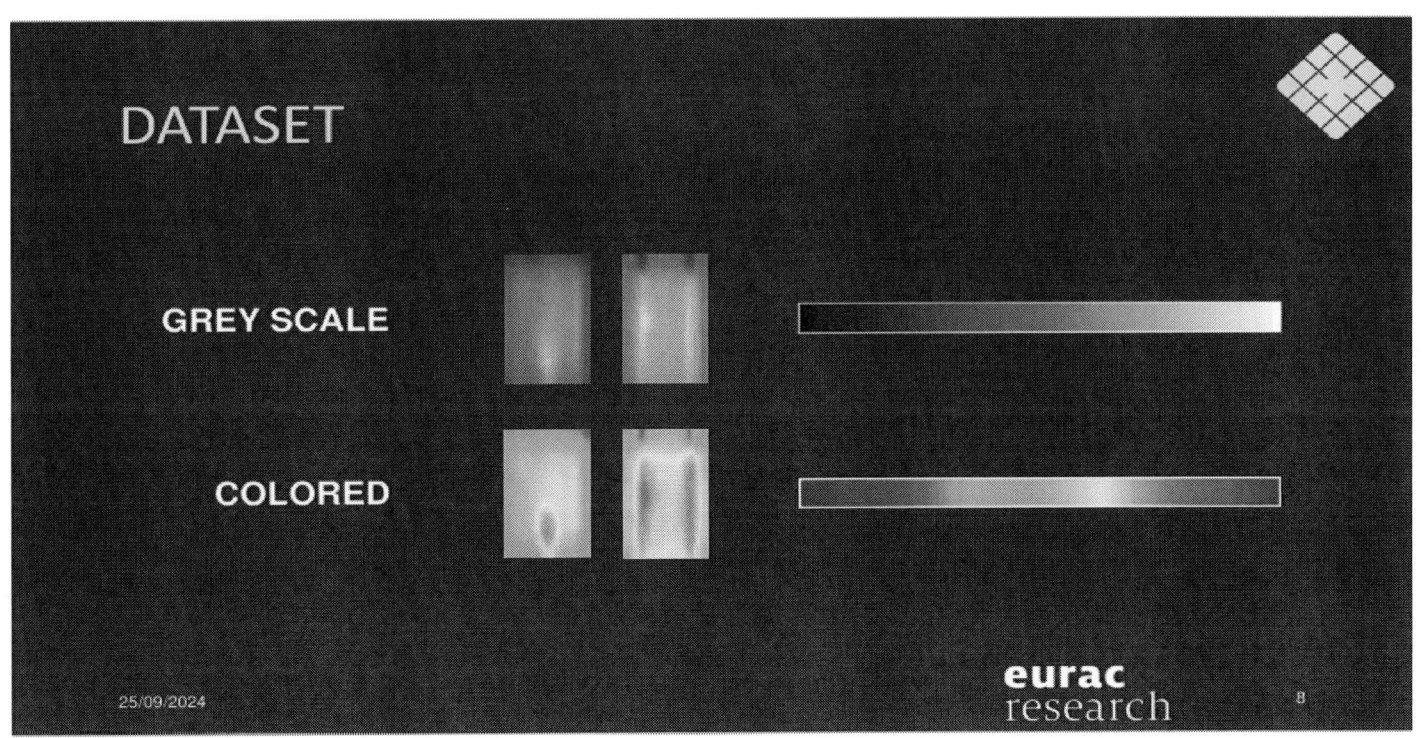

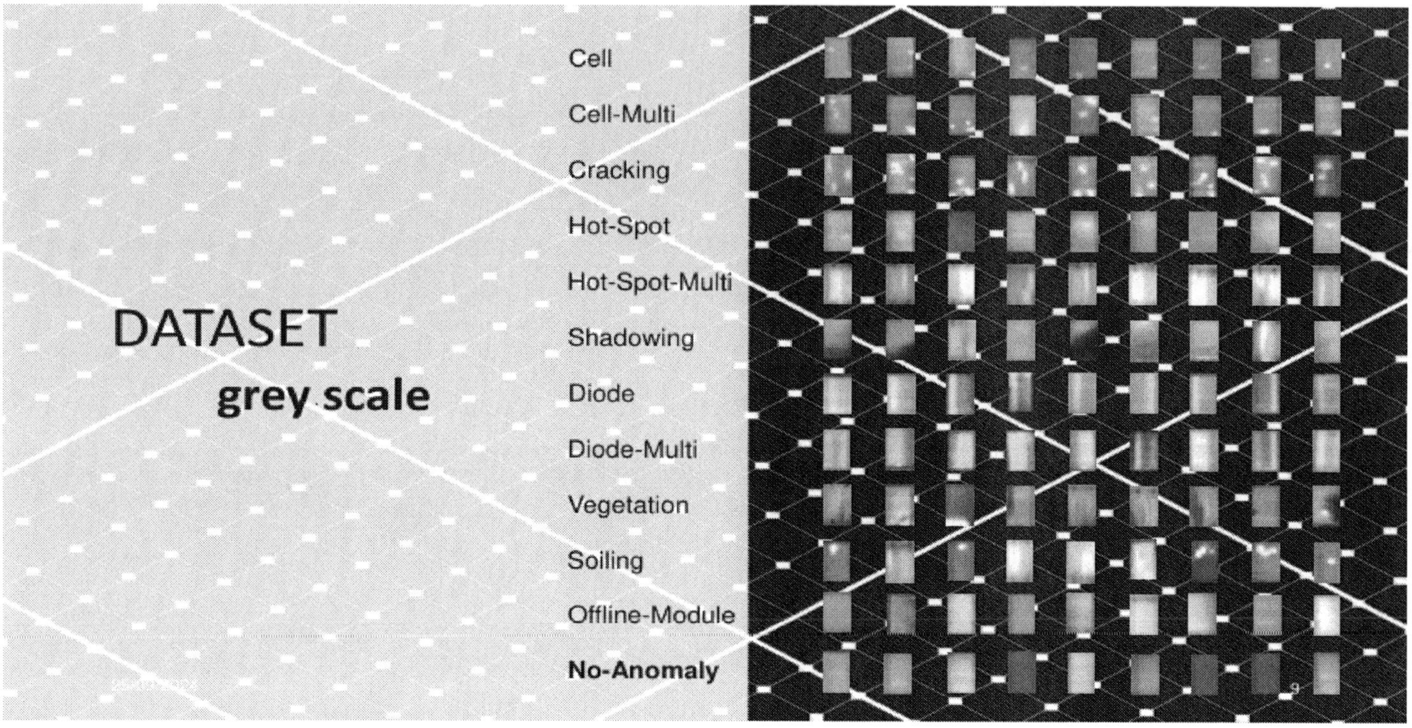

020319-009

41st European Photovoltaic Solar Energy Conference and Exhibition

020319-010

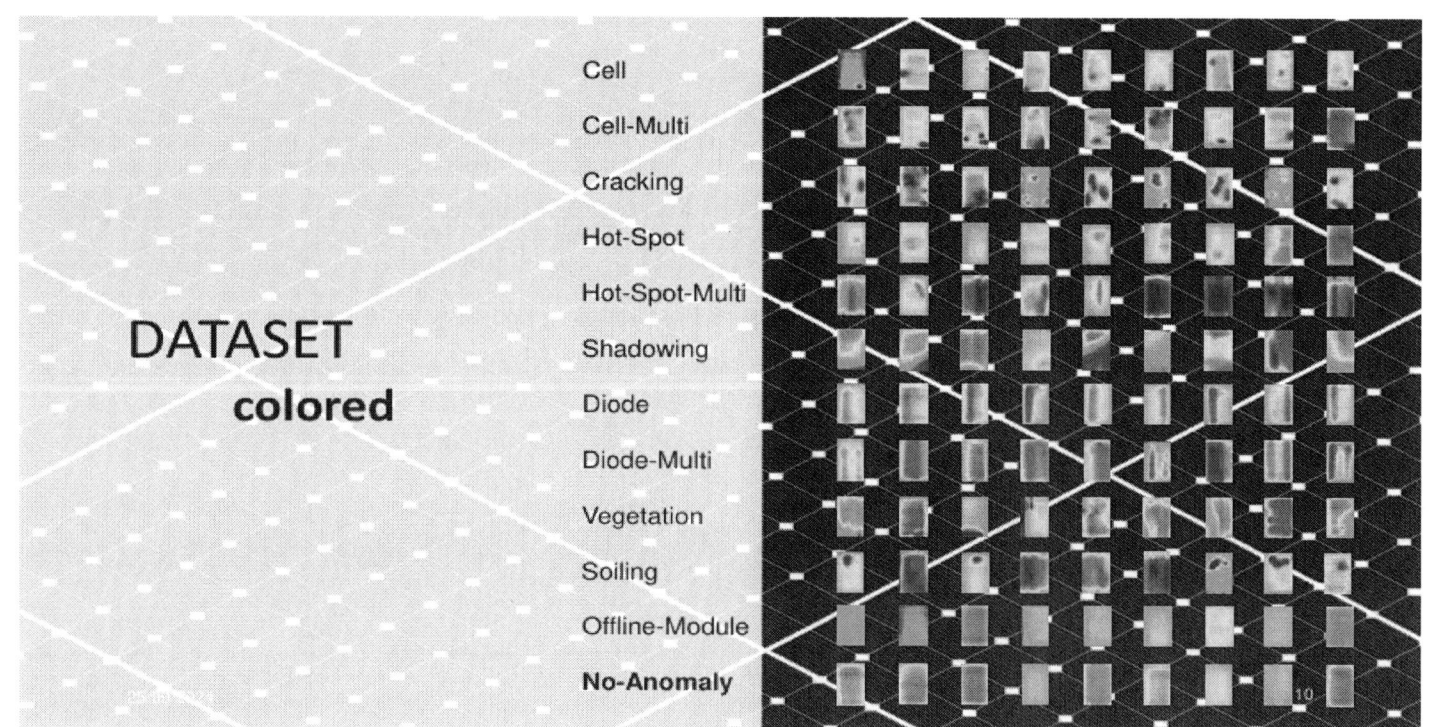

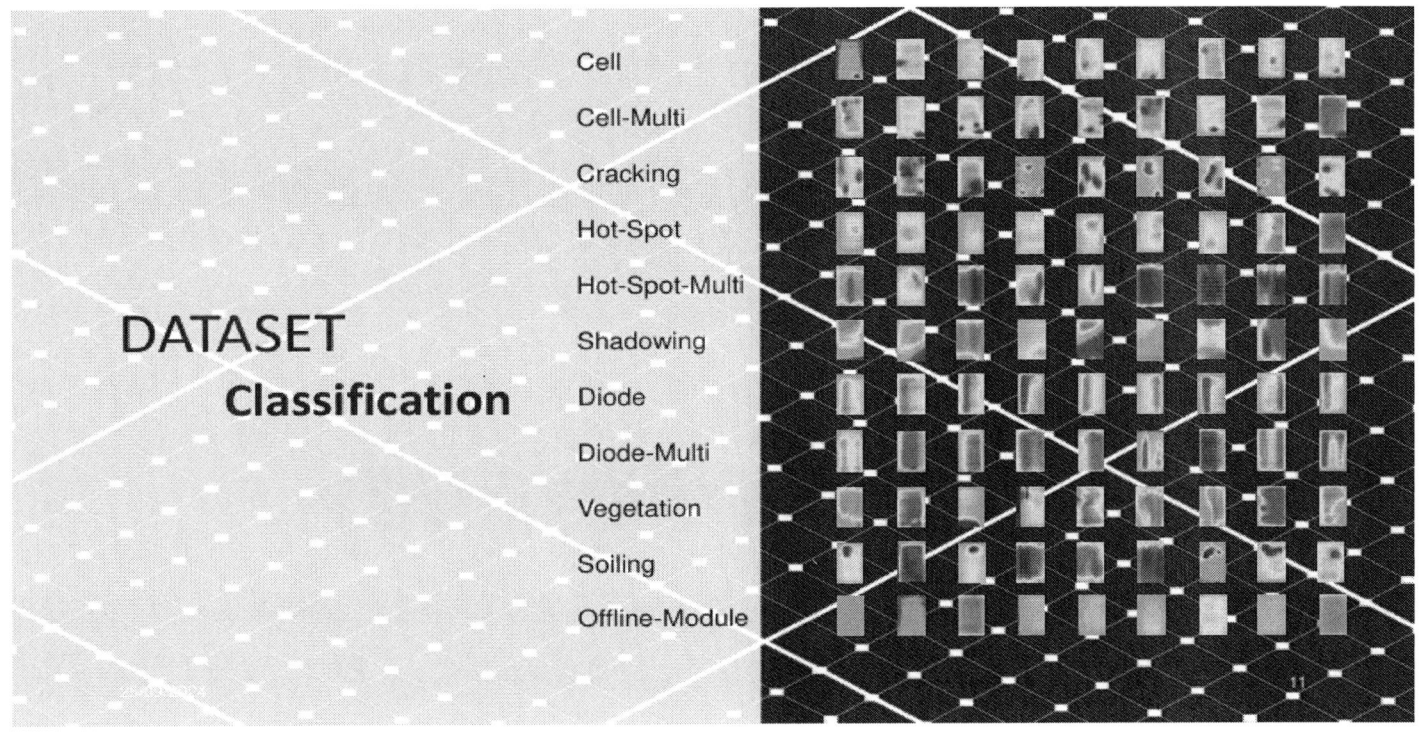

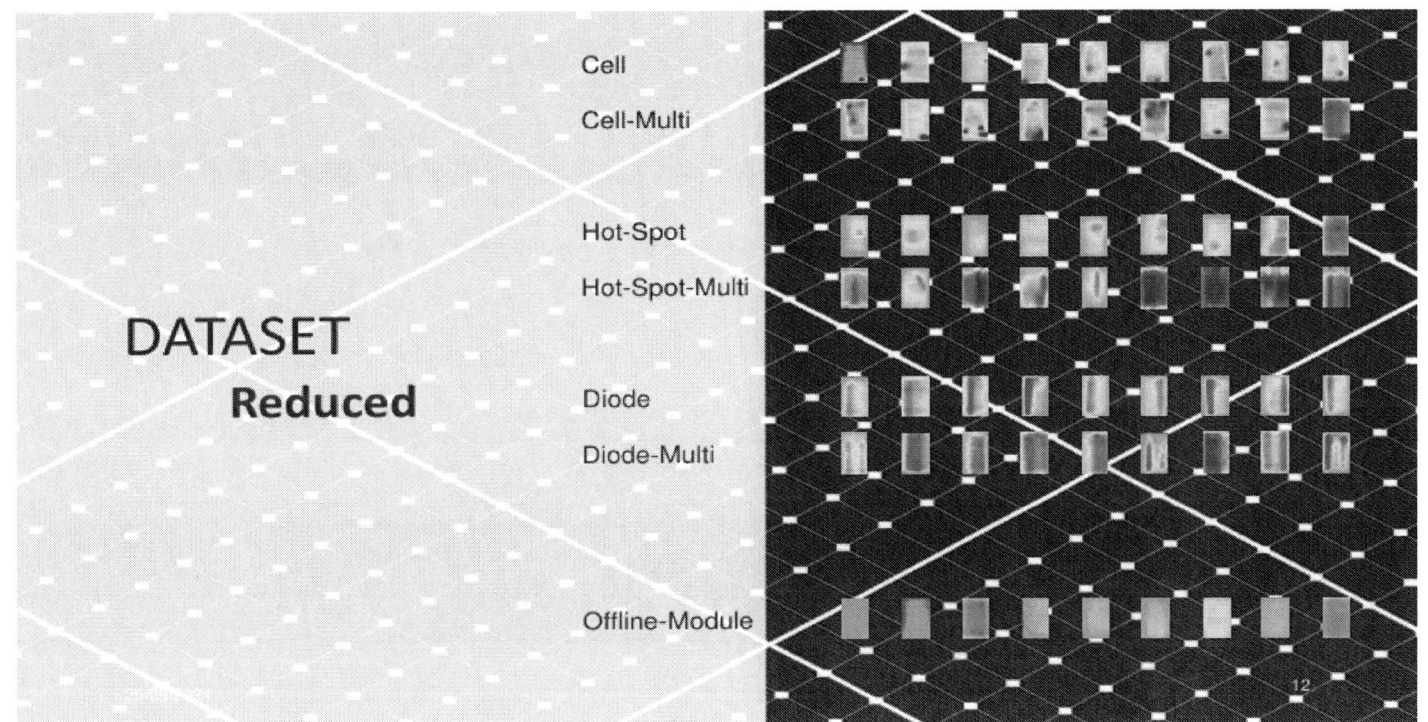

RESULTS

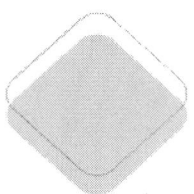

QUANTITATIVE ANALYSIS
ResNet versus GPT-4o

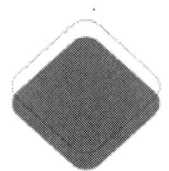

QUALITATIVE ANALYSIS
GPT-4o capabilities

RESULTS
Quantitative Analysis

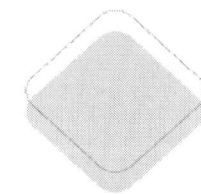

		ACCURACY			
		Detection	Classification Total	Classification Reduction	5 Failure Classes
ResNet50	Grey Scale	0.79	0.61	0.75	
	JetColor	0.84	0.66	0.79	
GPT-4o Basic Prompt	Grey Scale	0.72	0.10	0.15	
	JetColor	0.73	0.12	0.15	
GPT-4o Description	Grey Scale	0.72	0.16	0.36	0.63
	JetColor	0.69	0.24	0.28	0.71

RESULTS
Qualitative Analysis

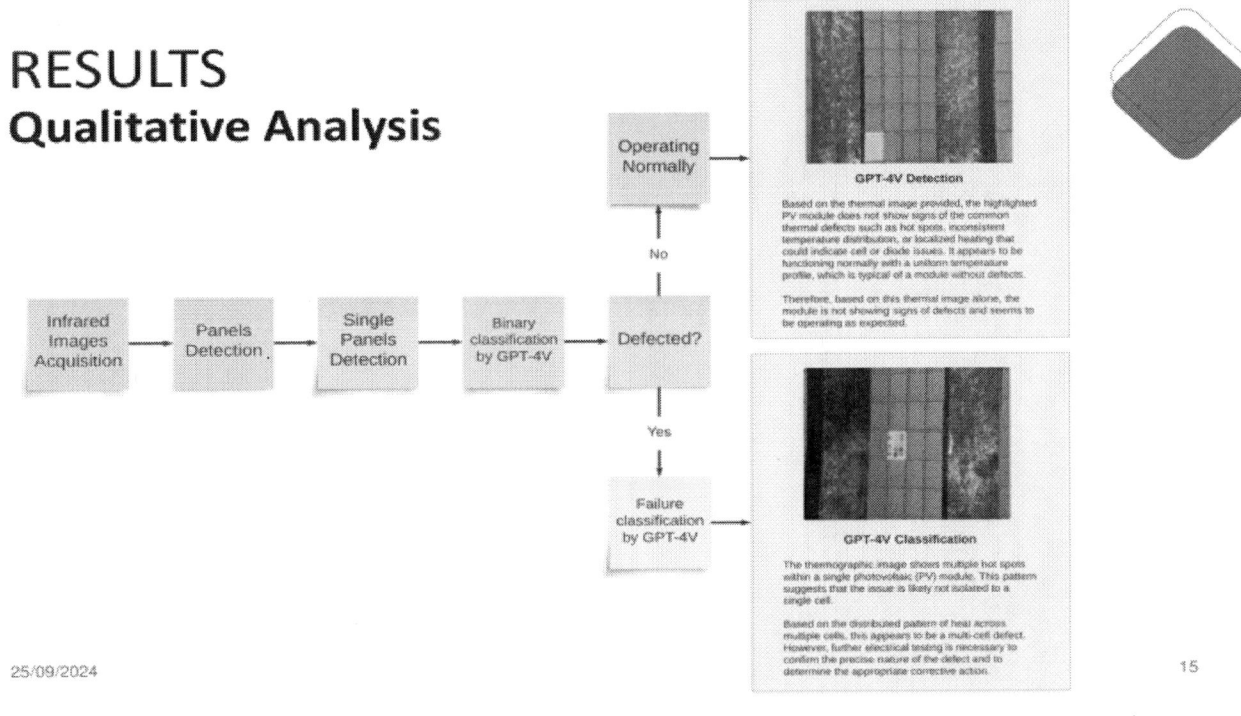

CONCLUSIONS

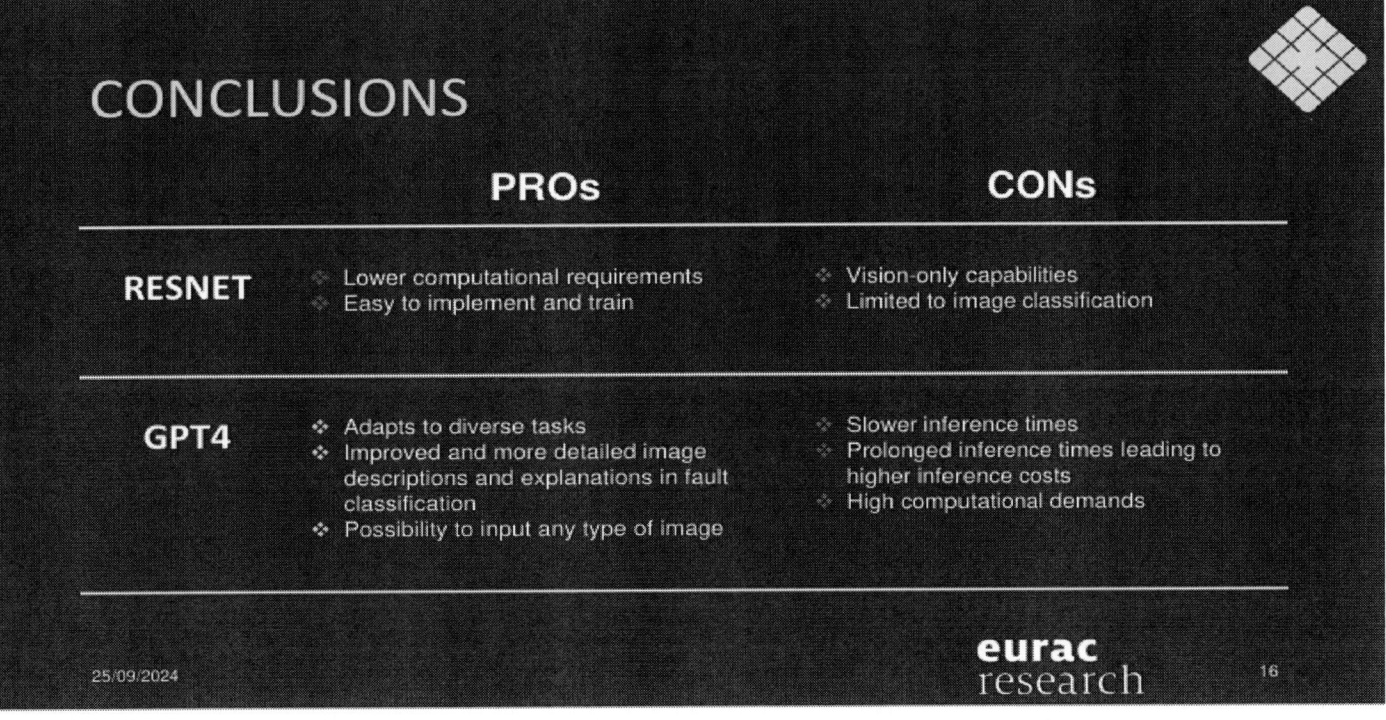

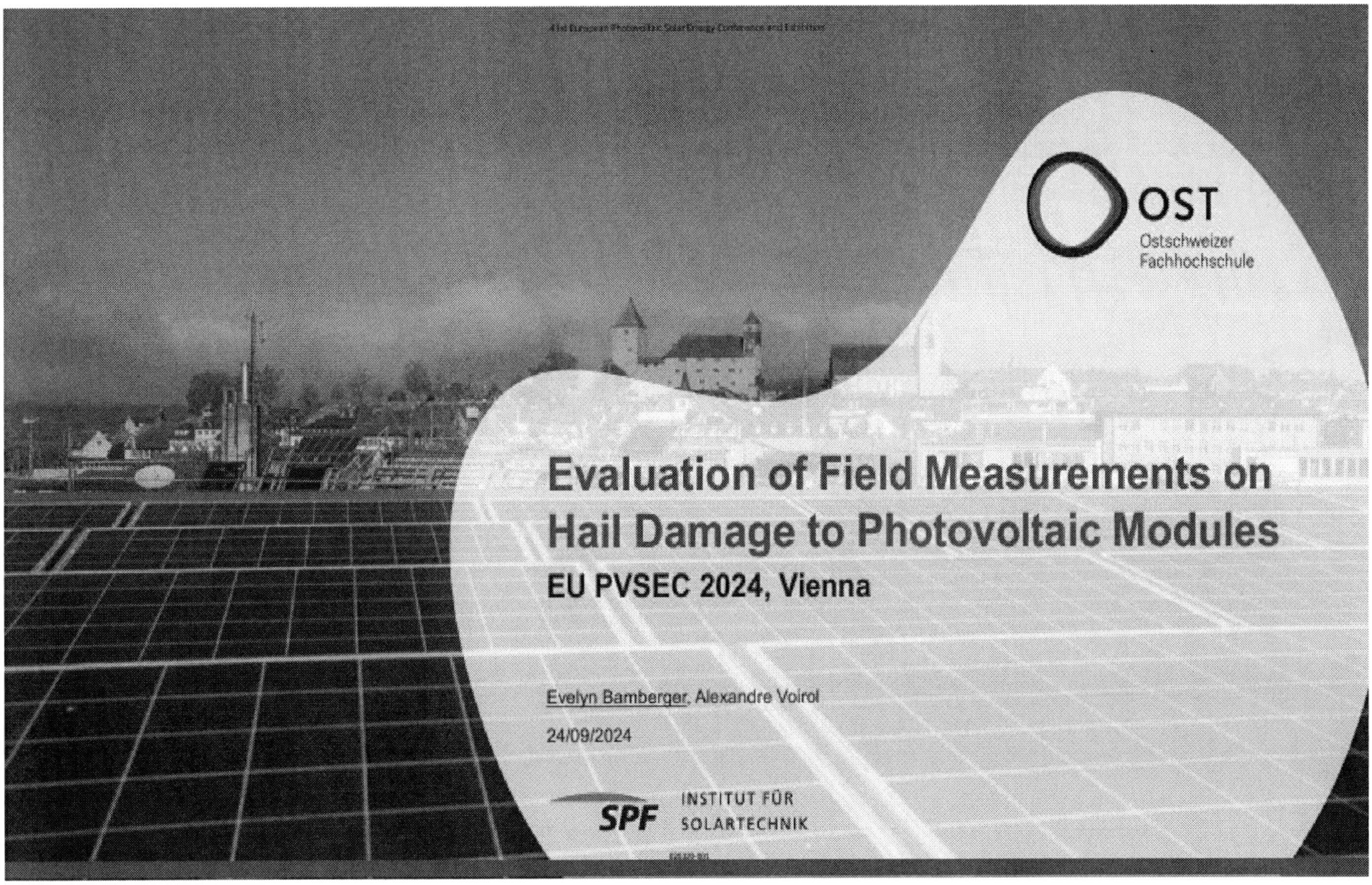

Evaluation of Field Measurements on Hail Damage to Photovoltaic Modules

EU PVSEC 2024, Vienna

Evelyn Bamberger, Alexandre Voirol

24/09/2024

Background

Project ACHILLES

Advanced CHaracterIsation of PhotovoLtaics for HaiL RESistance

- Increasing hail events and PV installations = increasing damage to PV modules due to hail

- **Which** PV modules should be replaced after a hail event and **when**, because safety risks or yield losses can occur?

  - Analysing measurement data from the mobile PV laboratory

  - Development of a methodology for accelerated ageing to estimate the long-term development of cell damage

  - Carrying out tests on various PV modules and validation with data from the field

  - Creation of a methodology for the assessment of hail damage

Funded by

Project partners

SUPSI — University of Applied Sciences and Arts of Southern Switzerland

SWISSOLAR

Background

Switzerland: Big Hailstorms in June and July 2021

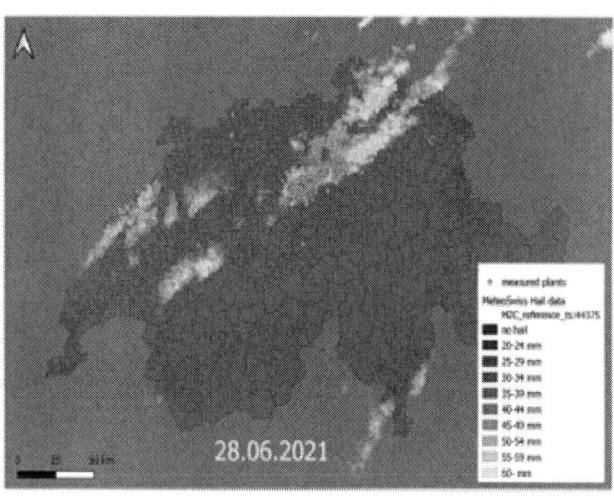

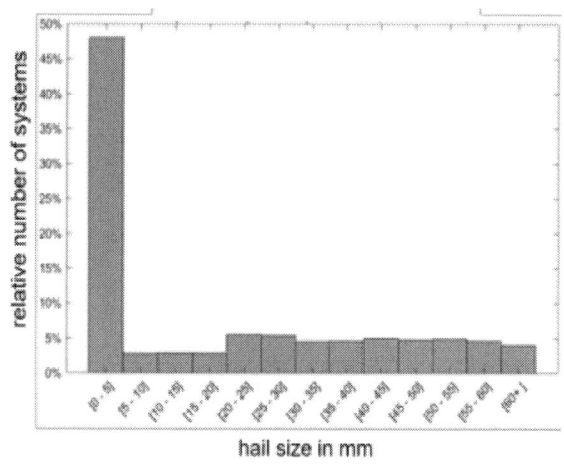

Around **15%** of all PV systems were in an area that was hit by hailstones larger than **50 mm** in June or July 2021.

3 | Hail Damage to PV Modules   24/09/2024

Procedure

Module Damages: Glass Breakage and Microcracks

Image source: Energie Netzwerk

- Glass breakage
  - Collection of data from installers

- Microcracks
  - Electroluminescence (EL) imaging with mobile PV lab

4 | Hail Damage to PV Modules   24/09/2024

Procedure

EL Images and Microcracks: Cell and Module Judgement

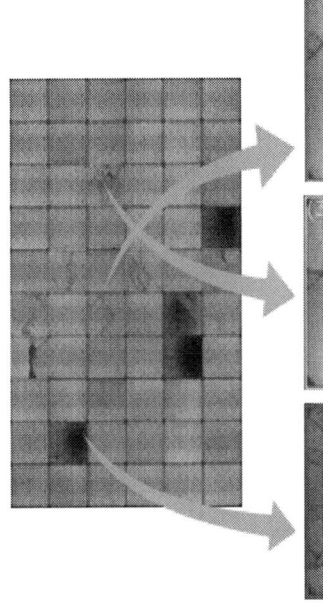

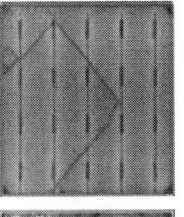

- non-critical cracks that will lead to no or only minimal cell area disconnection (max. 1%)

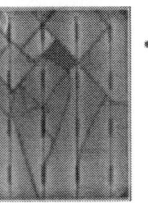

- critical cracks that can lead to cell area disconnections of up to 20% in the future

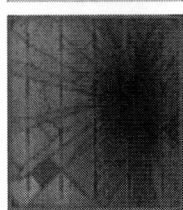

- very critical cracks that can lead to cell area disconnections of more than 20% in the future

- **Module Judgement**

class	green	yellow	red	total
A	<10% and	0% and	0% and	<10%
B	<20% and	<10% and	0% and	<20%
C	≥20% or	≥10% and	<10% and	<30%
D			≥10% or	≥30%

Acc. to MBJ Solar Module Judgment Criteria Rev. 3.4, 2019

5 | Hail Damage to PV Modules

Procedure

Available Data and Data Sources

Total power of the measured systems	34.96 MW
Number of systems	344
Number of modules installed	126'124
Number of systems with glass breakage	180
Number of modules with glass breakage	4'283
Number of modules measured	5'209
Number of modules in class C or D	3'037 (58%)
Number of modules in class C	1'725 (33%)
Number of modules in class D	1'312 (25%)

Data from mobile PV lab and installers

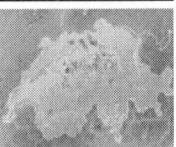

MeteoSwiss

Schröer et al. Hagelklima Schweiz. Daten, Ergebnisse und Dokumentation: Fachbericht MeteoSchweiz No. 283. MeteoSchweiz; 2023. https://doi.org/10.18751/PMCH/TR/283.HAGELKLIMASCHWEIZ/1.0

Sonnendach.ch
Klauser and Schlegel, Dokumentation Geodatenmodell Solarenergie: Eignung Dächer (Sonnendach.ch) Solarenergie: Eignung Fassaden (Sonnenfassade.ch). Bern: BFE; 2016.

Pronovo
Rohrbach, Dokumentation «minimales Geodatenmodell» Elektrizitätsproduktionsanlagen. BFE; 2022

6 | Hail Damage to PV Modules

Evaluation
Hail Size and Microcracks / Glass Breakage

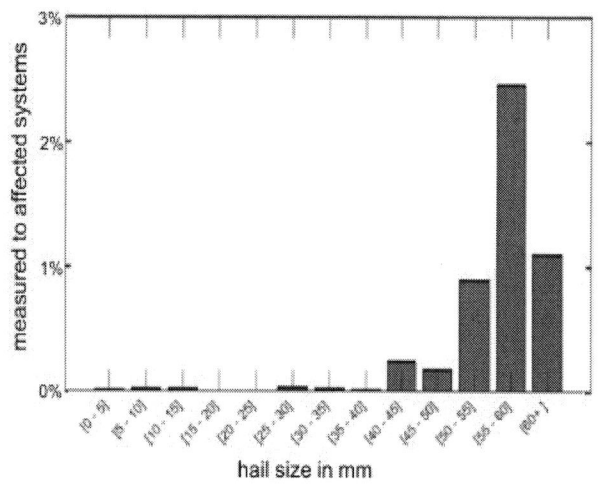

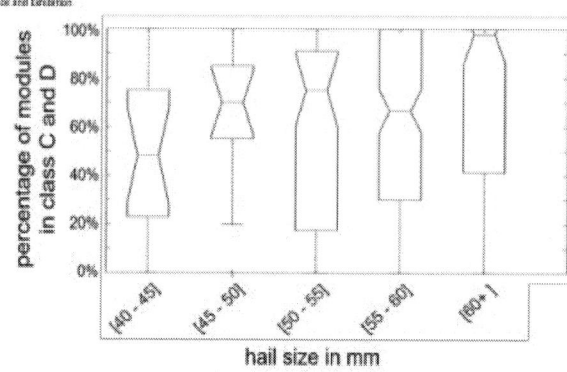

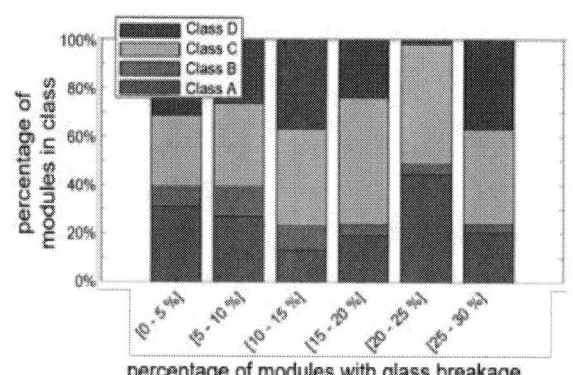

Hail Damage to PV Modules   24/09/2024

Evaluation

Microcracks and Inclination/Orientation

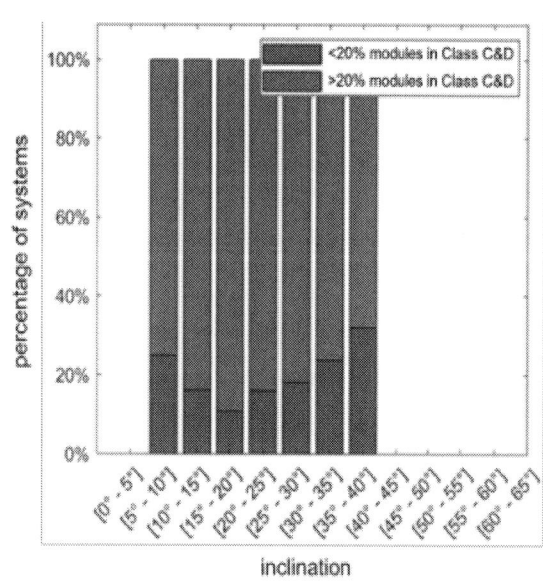

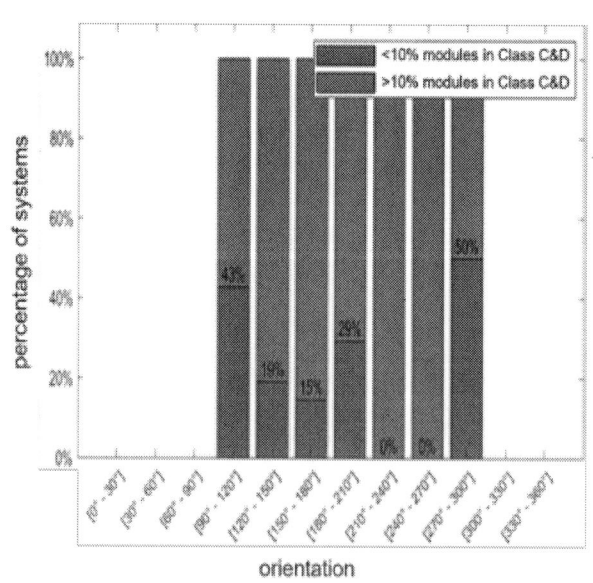

Hail Damage to PV Modules   24/09/2024

Evaluation

Inclination and Hail Size

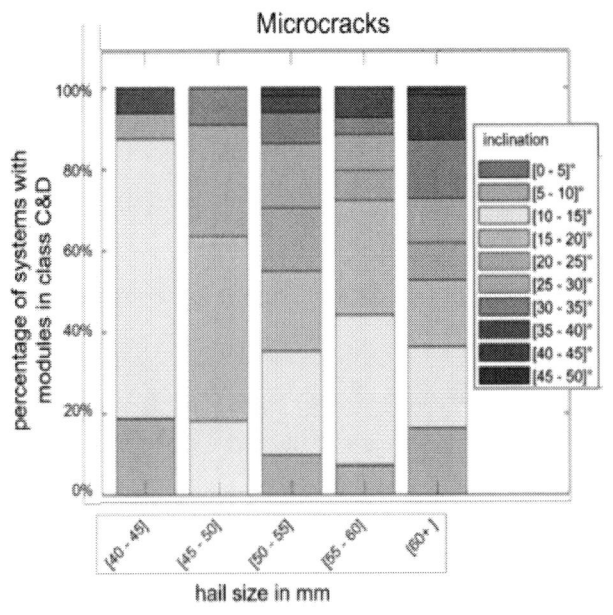

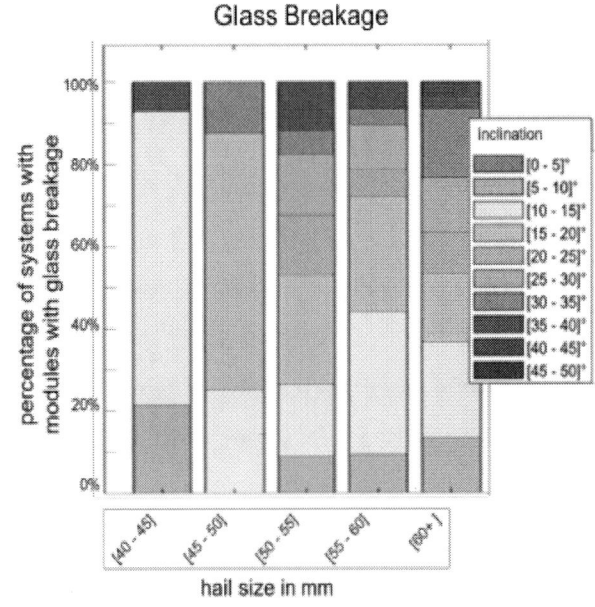

Hail Damage to PV Modules

24/09/2024

Evaluation

Year of Construction and Microcracks / Glass Breakage

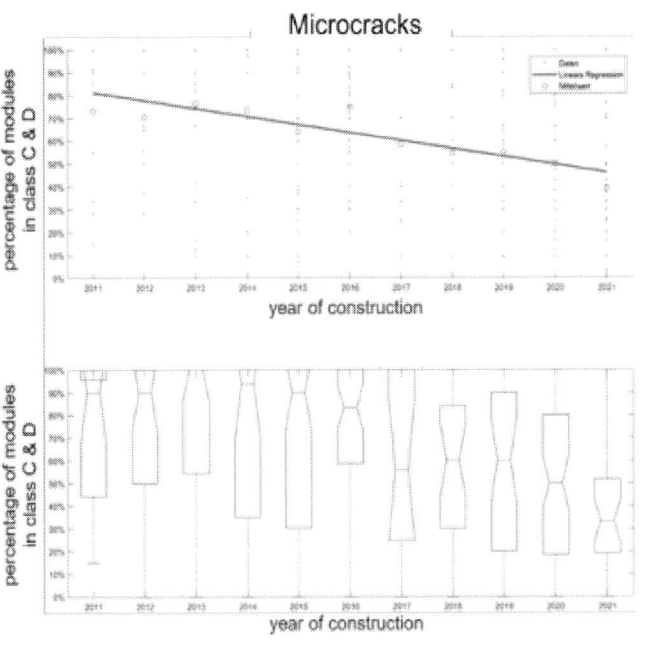

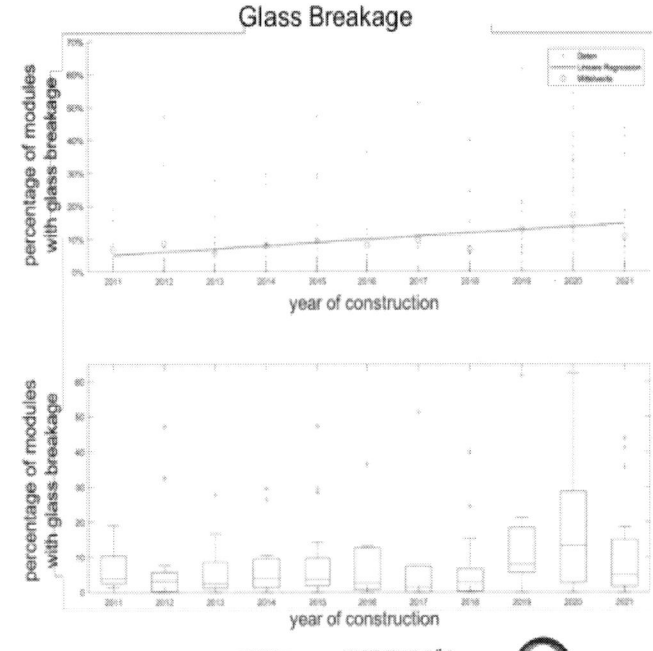

Hail Damage to PV Modules

24/09/2024

Conclusions

- Large hail events in June and July 2021 in Switzerland: PV systems affected on a large scale for the first time

  - 344 systems inspected, EL measurements of around 5,200 modules

  - 58% of modules with critical cell damage (module class C and D)

  - Damage can typically occur from hailstone size of 4 cm, from 5 cm it becomes probable

  - Cell cracking becomes more likely as hailstone size increases, but there is no significant correlation between cell cracking and glass breakage

  - Inclination and orientation are more relevant for smaller hailstones

  - Cell cracking is less common in newer modules, while glass breakage is slightly more common

  - Large spread of data

Limitations and Outlook

- Study limitations

  - Measurements only on request, no random selection

  - Accuracy of hail data

  - Data collection from installers is prone to error (glass breakage data)

- Outlook

  - Indoor and outdoor ageing of modules with cell damage due to hailstorms

  - Measurement of field systems with cell damage due to hailstorms after 3 and 5 years of operation

Many Thanks for your Attention!

- Evelyn Bamberger
  evelyn.bamberger@ost.ch
  www.spf.ch
  www.ost.ch

Project funded by

13 | Hail Damage to PV Modules   24/09/2024

41st European Photovoltaic Solar Energy Conference and Exhibition

This presentation was selected by the Sc. Committee of the EU PVSEC 2024 for submission of a full paper to one of the EU PVSEC's collaborating peer-reviewed journals.

EVALUATION OF DAYLIGHT FILTERS FOR ELECTROLUMINESCENCE IMAGING INSPECTIONS OF C-SI PV MODULES

Gisele Alves dos Reis Benatto*[1], Thøger Kari[1], Rodrigo Del Prado Santamaría[1], Aysha Mahmood[1],
Liviu Stoicescu[2], Sergiu Viorel Spataru[1]
[1]DTU Electro, Technical University of Denmark (DTU), Frederiksborgvej 399, 4000 Roskilde, Denmark.
[2]Solarzentrum Stuttgart GmbH, Rotebühlstr. 145, 70197 Stuttgart, Germany.

ABSTRACT: Outdoor daylight electroluminescence and photoluminescence of PV modules for defect and fault detection are of high interest in Operations & Maintenance industry to expand available inspection time to daylight hours and increase safety – compared with more established nighttime imaging – while getting the PV diagnosis accuracy only provided by luminescence imaging. Literature reports that filtering away sunlight is mandatory, although it also overlaps with the solar cell luminescence emission range. Furthermore, image processing is required for images taken in bright daylight to increase the signal-to-noise ratio and achieve the best possible image quality under any solar conditions. In this work, we evaluate the performance of six optical filter configurations indoors, using artificial lighting as a controlled noise source, and outdoors in overcast and sunny conditions. The results indicate that filters with 50 to 300 nm of transmission bandwidth over the electroluminescence spectrum peak provide proper balance of signal-to-noise ratio regarding the electroluminescence image quality, given that they block efficiently the wavelengths detected by the camera without PV luminescence signal. In a high noise scenario such as bright daylight conditions, a filter with narrower bandwidths that allow higher iris aperture, provides the best electroluminescence images.
Keywords: Daylight electroluminescence, Field inspections, PV modules, PERC cells

1 INTRODUCTION

The integrity of PV modules and solar cells can directly affect the power production yield from a PV power plant. In recent years, infrared thermography inspection of PV plants has been conducted to a larger and larger extent and frequency by inspection companies, even with its somewhat limited capability of detecting and distinguishing between PV module faults in early and even late stages [1], [2]; but the Operations & Maintenance companies are slowly becoming aware of the need for accurate inspection methods, especially preventive maintenance.

Luminescence-based imaging such as Electroluminescence (EL) and photoluminescence (PL) have the potential for localizing, identifying, and indicating the severity of a wide range of PV-related defects and are now routinely used in PV research and manufacturing; however, still faces challenges for wide application on field inspections mainly due to the negative impact of sunlight on the luminescence image acquisition.

One of the first reports on daylight luminescence imaging in the literature [3], describes that outdoor EL and PL imaging in the daytime require a "daylight filter" to avoid most of the sunlight being captured by the camera simultaneously. However, the daylight filter parameters are not strict or optimized. Most of the daylight EL and PL setups reported [4]–[7] use an Indium Gallium Arsenide (InGaAs) camera (normally with a detection range from 900-1700 nm) and bandpass (BP) filters with center wavelengths ranging from 1140 to 1160 nm, bandpass width of 25 to 30 nm, optical density $\geq$ 4, and bandwidth transmission $\geq$ 90%. Other reports [8], [9] use a bandwidth of 150 nm or a 1000 nm long-pass (LP) filter only, thereby allowing a full detection range of 1000 to 1700 nm of the sunlight spectrum; indicating that a broad range of filters can be successfully used to acquire EL and PL images in broad daylight as long as other – not always reported – parameters and camera settings are adjusted, such as lens iris aperture, camera exposure time, among others. However, which filter bandwidth range will minimize the number of images required to reach a desirable signal-to-

noise ratio (SNR) in different sunlight irradiance conditions and ultimately maximize the robustness of the method is still a question to be answered. The importance of answering this question increases when additional noise is present due to camera motion (in the case of aerial drone inspections) and sun variations (environmental conditions), restricting the imaging acquisition to very short exposure times.

Most recently, the comparison between optical filters for sunny weather revealed that daylight EL imaging can be very sensitive to even small changes in weather conditions, allowing only indications of the filter comparison instead of accurate results [10]. To better understand how the modulated EL signal (required for daylight EL) is affected by light noise, in this work, we use an indoor setup with a controlled halogen light source, which provides a similar intensity of light noise during modulated EL imaging with an InGaAs camera, as experienced during EL imaging in a low light overcast day. This study aims to compare optical filter performance indoors with controlled ambient light noise and outdoors under sunny conditions with uncontrolled natural ambient light noise; understand noise/signal balance for optical filtering in terms of EL image quality; and define the main parameters for more reliable daylight EL image acquisition.

2 METHODOLOGY

The indoors EL imaging setup consisted of a set of halogen bulbs that were positioned 3.4 meters from the PV module. The irradiance on the module surface, measured with a reference cell at the plane of the array (POA), was 23.7 W/m^2. From the InGaAs camera point of view, this irradiance was enough to saturate the camera sensor when no filter was in use, requiring smaller lens apertures and therefore equivalent to a bright cloudy day, yet with more direct light component.

The current is modulated through the PV panel as described in [11], and the acquired image sequences are processed via Fast Fourier Transform (FFT) using the methodology from [12].

In previous works, many efforts were made to define an SNR marker that correlates well with the actual EL image quality [5], [12]. Among the methods available, SNR_{Kari} demonstrated to be the most accurate marker available to date [11], [13]. The calculated SNR_{Kari} value for minimal satisfactory EL image quality is in the range of $4 > SNR_{Kari} > 1$, which should allow to identify macroscopic cell defects such as C-cracks and ribbons damage [14]. When $SNR_{Kari} \geq 4$, the EL image is assigned to have excellent quality, being comparable to an EL image taken indoors in a darkroom.

It is expected that fluctuations on calculated SNR_{Kari} occur when the global, random noise is very high compared to EL signal, or when the noise has compatible frequences to the framerate, as that will directly affect the calculations made through FFT, or indeed any modulation-based algorithm.

3 RESULTS AND DISCUSSION

3.1 Filter characterization

Fig. 1a shows the spectrally resolved EL emission using a PERC c-Si cell on the PV module studied. Table I shows the datasheet specifications of the studied filter configurations, and the measured filter specs in parenthesis calculated from the spectra in Fig. 1a.

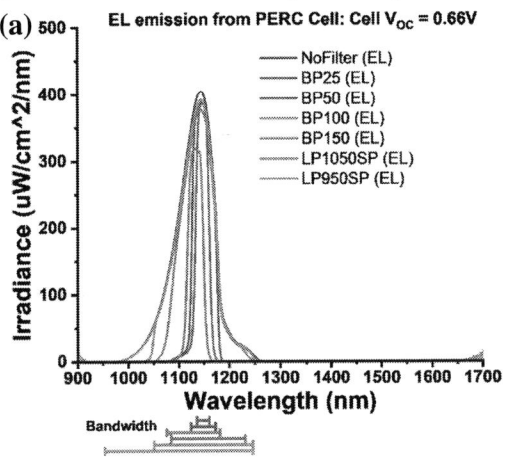

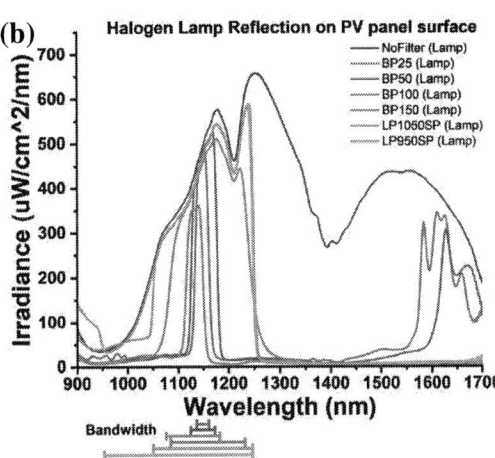

Figure 1: Response of the different filters to the EL emission (a) and halogen lamp light reflected (b) on the PERC PV module.

Note that especially for the bandwidth measured at the full width at half maximum (FWHM) for BP100, which is the most relevant factor for the transmission of the EL emission, the magnitude changes significantly from the one specified in the datasheet, which describes the bandwidth at a lower transmission region than 70% or FWHM. Transmission for the LP+SP filter sets block only sections of the tails of the c-Si EL spectrum, and thereby admit almost all the EL signal, while eliminating the noise from ambient light (i.e. sunlight or halogen lamp). These filter sets with 300 and 200 nm transmission bandwidth over the EL spectrum range can be considered the logical next step in bandwidth from the *No filter* situation.

Finally, Fig. 1b shows the filter performance in blocking the light noise from the halogen lamp reflected by the PERC module surface. This measurement revealed that BP25 and BP50 do not block the light at 1520-1680 nm. This was originally unknown since the manufacturer only provides the transmission and blocking properties from 200-1500 nm.

Table I: Datasheet and measured details of the studied optical filter configurations.

Filter configuration	Central wavelength	Bandwidth (Meas. FWHM)	Transmission (Meas. at the peak)	Optical density
BP25	1150 nm	25 nm (26.5 nm)	> 90% (97%)	>4
BP50	1150 nm	50 nm (49.2 nm)	> 90% (97%)	>4
BP100	1130 nm	106 nm (32.5 nm)	> 70% (76%)	4
BP150	1160 nm	150 nm (75.7 nm)	> 88% (92%)	4
LP1050+SP1250	1150 nm	200 nm (80.8 nm)	> 97% (97%)	>5 + >4
LP950+SP1250	1100 nm	300 nm (80.8 nm)	> 97% (97%)	>5 + >4

3.2 Filter performance at EL imaging

Fig. 2 shows the results of SNR_{Kari} performance from modulated EL at 100% I_{SC} current bias at indoor conditions with halogen light noise. These values are extracted during EL image reconstruction via FFT. EL images are acquired with a range of exposure times (1-3 ms) to evaluate the correlation with possible contributing factors for the filter performance. The narrow exposure time range was chosen because short exposure times are preferred and often required for sequences acquired in motion (UAV) but also for minimal use of the lens iris, as it will be an additional variable to the filter comparison. Each measurement acquires two repeated image sequences at three different exposure times. The iris opening for each measurement is indicated on the SNR_{Kari} results. The exposure time bars have indicative colors for each filter, also denoted in Table I.

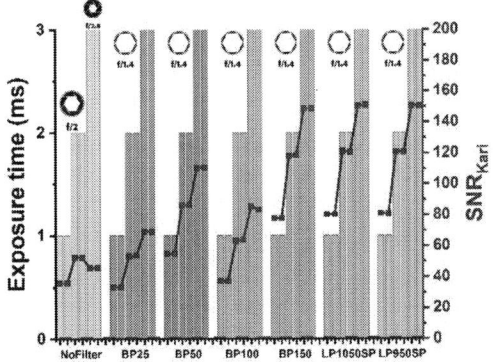

Figure 2: SNR_{Kari} from modulated EL at 100% I_{SC} current bias indoors with halogen lamp light noise of 23.7 W/m² at the PV module POA.

1242

Proportionally, the lower the blocking ratio towards the EL signal, the higher the values for both markers. With the added halogen light noise, the iris aperture was f/2 for 1 ms and f/2.8 for 2 and 3 ms for the *No filter* scenario. For this reason, the SNR_{Kari} results were overall lower than the other filter configurations, which had the iris fully open, and indicating that iris opening highly relevant for image quality.

Fig. 3 shows the results of SNR_{Kari} performance from modulated EL at 100% I_{SC} current bias at outdoors sunny condition with 905 ± 8 W/m^2 of sun irradiance in the POA. For such ambient light conditions, the fluctuations are significant from one measurement to the other, even when they are just subsequent repetitions. In this high noise scenario, the filter performance is far from what could be expected observing the filter characterization alone. The iris aperture must be closed to its minimum (f/16) when *No filter* was used and mostly closed (f/8) for the other filters.

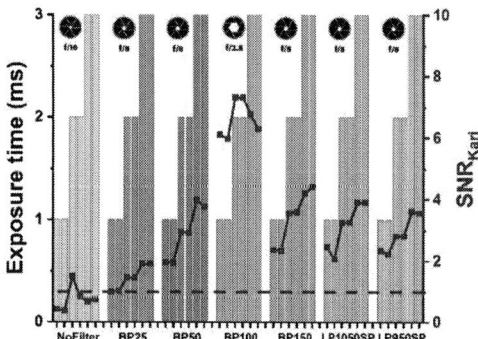

Figure 3: SNR_{Kari} from modulated EL at 100% I_{SC} current bias at outdoor sunny conditions with sun irradiance of 905 ± 8 W/m^2 at the PV module POA.

For BP100, due to its high blocking range and low transmission compared to the other filters, the iris could be almost fully open (f/2.8), justifying the distinguished good performance of the SNR_{Kari} image quality marker. *No filter* scenario here didn't satisfy the minimum satisfactory quality threshold for SNR_{Kari}, even at 3 ms exposure time. At the instances when $SNR_{Kari} > 4$, the EL image reconstruction can be considered be robust with less images than used for the analysis shown in Fig. 2 and 3 (128 frames) and even using the least performing filter (BP25), image quality can be improved with significant higher number of frames. Note that BP25 has similar EL spectrum transmission characteristics compared with BP100 (see Fig. 1a), however the higher transmission at the bandwidth and the unwanted transmission in the 1520-1680 nm range contribute to the need of a lower iris aperture to avoid sensor saturation, and this may also be the case for BP50.

Fig. 4 shows the EL image of four central cells in the PV module for processed image quality reference with examples of the worst, medium, and highest SNR_{Kari} values, confirming the correspondence of the calculated SNR with the actual EL image quality. In the examples, excellent quality (Fig. 4 right, with $SNR_{Kari} > 4$) can be observed with an image comparable to laboratory/indoors EL imaging, while the satisfactory image quality example (Fig. 4 center, with $4 > SNR_{Kari} > 1$) shows an EL image with visible noise but distinguable features that would enable the identification of most of macroscopic PV cell faults.

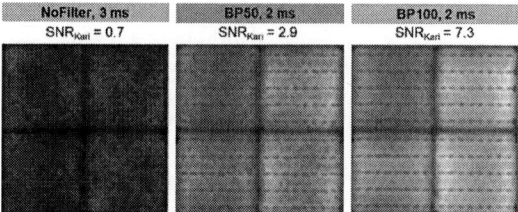

Figure 4: Examples of processed 100% I_{SC} EL images of four central cells in the PV module taken at outdoors sunny conditions. Left: Below satisfactory quality; Center: Satisfactory quality; Right: Excellent quality.

4 CONCLUSIONS

The study performed in this paper attempts a fair comparison between daylight optical filters under controlled test conditions for outdoors EL imaging under sunlight. The results made clear that lowering iris aperture has an overall higher influence in decreasing the EL signal and image quality than the bandwidth and transmission properties of the daylight optical filters. Regarding the filter properties, successful daylight EL imaging was performed with filters bandwidth within the main peak of the EL emission spectrum, with high transmission in the range of 50-300 nm bandwidth, blocking most of the remaining wavelengths from external light noise.

The following recommendation can be made, with the obvious caveat that cameras are all different and may have different base-noise and saturation levels: If you can only opt for one filter, everything points to a moderately narrow filter between 25-50 nm, without transmission out of the bandwidth and low transmission ratio, as this is likely to always exceed $SNR_{Kari} \gg 4$ in low light and allows an increased aperture and integration time in broad sunlight.

5 ACKNOLEDGEMENTS

This research has been carried in the Eurostars project no. 115687, Automated Daylight Electroluminescence Inspection of Large Photovoltaic Systems (ADELI).

6 REFERENCES

[1] IEA-PVPS T13-10, *Review on Infrared (IR) and Electroluminescence (EL) Imaging for Photovoltaic Field Applications*, no. March. 2018.

[2] S. A. Rahaman, T. Urmee, and D. A. Parlevliet, "PV system defects identification using Remotely Piloted Aircraft (RPA) based infrared (IR) imaging: A review," *Sol. Energy*, vol. 206, no. June, pp. 579–595, 2020, doi: 10.1016/j.solener.2020.06.014.

[3] L. Stoicescu, M. Reuter, and J. . Werner, "Daysy: Luminescence Imaging of PV Modules in Daylight," *29th Eur. Photovoltaics Sol. Energy Conf. Exhib. Amsterdam, Netherlands*, pp. 2553–2554, 2014, doi: 10.4229/EUPVSEC20142014-5DO.16.2.

[4] R. Bhoopathy, O. Kunz, M. Juhl, T. Trupke, and Z. Hameiri, "Outdoor photoluminescence imaging of photovoltaic modules with sunlight excitation," *Prog. Photovoltaics Res. Appl.*, vol. 26, no. 1, pp. 69–73, Jan. 2018, doi: 10.1002/pip.2946.

[5] G. Alves Dos Reis Benatto *et al.*, "Drone-Based Daylight Electroluminescence Imaging of PV

Modules," *IEEE J. Photovoltaics*, vol. 10, no. 3, pp. 1–6, 2020, doi: 10.1109/jphotov.2020.2978068.

[6] Y. Augarten, A. Wrigley, A. Gerber, B. Pieters, and U. Rau, "Quantitative Outdoor Imaging: Analysis of Solar Modules in Daylight," in *32nd European Photovoltaic Solar Energy Conference and Exhibition*, 2016, pp. 1846–1848. doi: 10.4229/EUPVSEC20162016-5BV.1.17.

[7] M. Vuković, I. Eriksdatter Høiaas, M. Jakovljević, A. Svarstad Flø, E. Olsen, and I. Burud, "Photoluminescence imaging of silicon modules in a string," *Prog. Photovoltaics Res. Appl.*, vol. 30, no. 4, pp. 436–446, 2022, doi: 10.1002/pip.3525.

[8] M. Guada *et al.*, "Daylight luminescence system for silicon solar panels based on a bias switching method," *Energy Sci. Eng.*, no. June, pp. 1–15, 2020, doi: 10.1002/ese3.781.

[9] T. J. Silverman, M. G. Deceglie, K. VanSant, S. Johnston, and I. Repins, "Illuminated Outdoor Luminescence Imaging of Photovoltaic Modules," in *2017 IEEE 44th Photovoltaic Specialist Conference (PVSC)*, Jun. 2017, pp. 3452–3455. doi: 10.1109/PVSC.2017.8366288.

[10] G. A dos Reis Benatto, A. A. Mayordomo, T. Kari, R. Del Prado Santamaría, P. B. Poulsen, and S. V Spataru, "Characterizing the Performance of Daylight Filters for Electroluminescence Imaging of Crystalline Silicon PV Modules," in *40th European Photovoltaic Solar Energy Conference and Exhibition*, 2023, pp. 020372-001-020372–005. doi: 10.4229/EUPVSEC2023/4CV.1.6.

[11] G. A. dos Reis Benatto *et al.*, "Daylight Electroluminescence Imaging Methodology Comparison," in *40th European Photovoltaic Solar Energy Conference and Exhibition*, 2023, no. 1, pp. 020374-001-020374–005. doi: 10.4229/EUPVSEC2023/4CV.1.7.

[12] G. A. dos Reis Benatto *et al.*, "Daylight Electroluminescence of PV Modules in Field Installations: When Electrical Signal Modulation is Required?," in *8th World Conference on Photovoltaic Energy Conversion*, 2022, pp. 735–739. doi: 10.4229/WCPEC-82022-3BV.3.44.

[13] M. Dhimish and A. M. Tyrrell, "Optical Filter Design for Daylight Outdoor Electroluminescence Imaging of PV Modules," *Photonics*, vol. 11, no. 1, 2024, doi: 10.3390/photonics11010063.

[14] M. Köntges *et al.*, "Review of Failures of Photovoltaic Modules," *IEA-Photovoltaic Power Syst. Program.*, pp. 1–140, 2014, doi: 978-3-906042-16-9.

41st European Photovoltaic Solar Energy Conference and Exhibition

IN-SITU MAINTENANCE-FREE MEASUREMENT OF
SOILING-INDUCED POWER LOSSES IN PV ARRAYS

Michael Gostein[1], Damien Cosme[2], Quentin Berthet-Rayne[2], Julien Chapon[3], Lluvia Ochoa[3], Bill Stueve[1],
Dhanup Somasekharan Pillai[4], Brahim Aïssa[4], Benjamin W. Figgis[4], Juan Lopez-Garcia[4], Veronica Bermudez Benito[4]
[1]Atonometrics, Austin, Texas, USA; [2]TotalEnergies, Doha, Qatar; [3]TotalEnergies, Paris, France;
[4]Qatar Environment & Energy Research Institute (QEERI), Hamad Bin Khalifa University (HBKU), Education City,
P.O. Box: 34110, Qatar Foundation, Doha, Qatar
michael.gostein@atonometrics.com

ABSTRACT: The accumulation of dust, dirt, and other contaminants, referred to as soiling, is a major contributor to energy loss in photovoltaic (PV) systems, particularly in desert environments. Soiling is primarily quantified by the soiling ratio, representing the ratio of actual PV power output to that expected under clean conditions. Although different types of soiling sensors are available, accurate soiling measurements often require comparing the maximum power output of soiled PV modules to the expected output for clean modules, particularly when soiling accumulates non-uniformly on module surfaces. In-situ module-level I-V measurement units now provide a direct means to measure the maximum power output of soiled modules within a PV array, eliminating the necessity for standalone reference modules. However, determining the anticipated power output in a clean state for modules typically involves a comparison with regularly cleaned reference modules, posing challenges for operations and maintenance. In this study, we assess an instrumentation system designed to measure actual power losses due to soiling without relying on the periodic cleaning of reference devices. In-situ I-V measurements determine the actual power output, while soiling-compensated irradiance measurements are used to determine the expected clean-state power without the need for regular maintenance. Testing is ongoing in the desert environment of Qatar. Results from the first four months of testing reveal good performance in soiling measurement compared to a control system comprising a soiled module and a regularly cleaned module whose power outputs are directly compared.
Keywords: soiling, bifacial PV, rear irradiance, maintenance

1 INTRODUCTION

Soiling, the accumulation of dust, dirt, and other contaminants, is one of the principal reasons for energy loss in PV systems [1][2] and therefore must be quantified accurately to assess system performance. Soiling measurement is used for various purposes, including pre-construction resource assessment, identifying losses during commissioning and capacity tests, assessing performance against guarantees, and of course, optimizing cleaning schedules for operations and maintenance (O&M).

Soiling is quantified by the soiling ratio (SR), the ratio of actual PV power output to expected output under clean conditions [3]. Numerous types of soiling sensors have been developed to measure soiling ratio. However, many of these sensors only provide an approximation of the true soiling ratio, for example by measuring the short-circuit current (I_{sc}) of reference devices as a proxy for module power or by utilizing small-area devices such as reference cells or optical dust sensors. The most accurate measurements of soiling losses require comparing the maximum power (P_{max}) of measured soiled PV modules to the expected maximum power output for a clean module.

Figure 1: Outdoor Test Facility (OTF) of the Qatar Environment and Energy Institute (QEERI), showing extensive soiling.

41st European Photovoltaic Solar Energy Conference and Exhibition

Figure 2: No-maintenance soiling measurement instrumentation at QEERI's OTF. Left: in-situ I-V unit (RDE300i) mounted beneath module racking, which measures periodic I-V sweeps on one of the modules without disconnection from the array. Center: rear-side reference cells (RC22), which are designed to mount compactly within the module frame area in multiple positions without shading the rear side of the PV cells. Right: front-side RC22 reference cell coupled with Mars™ optical soiling sensor, to provide soiling-compensated irradiance values without cell cleaning.

This is especially true when soiling accumulation is non-uniform across the modules, with concentrated bands of soiling at module edges [4][5][6][7], which causes a divergence between true soiling ratio and approximate measurements from many typical sensors.

In-situ module-level I-V measurement units [7] now enable the direct measurement of the maximum power output of soiled PV modules, the numerator of *SR*, while they are still connected to a PV array, eliminating the need for standalone reference modules.

How is the expected power, the denominator of *SR*, to be determined? A simple approach to *SR* measurement is to compare the P_{max} of a soiled module to that of a regularly cleaned module. In this approach ("module-module") the clean module power output provides the expected power for the denominator of *SR*. However, this requires regularly cleaning the clean reference module, which may impose a burden on busy operations and maintenance (O&M) personnel.

Here we evaluate a new instrumentation approach that allows measuring actual power losses due to soiling to measure true *SR*, but without the need for any cleaning operations by O&M. The approach uses in-situ I-V measurements to determine the power output of a soiled module within a PV array while incorporating measurements of soiling-compensated irradiance, achieved by coupling a PV reference cell with an optical soiling sensor, to determine the expected clean-state power of the soiled module. In this appraoch ("module-cell-optical") measurements are performed without requiring regular maintenance for cleaning reference devices.

2 EXPERIMENT

Our study is being conducted in the harsh and dusty desert environment at the Outdoor Test Facility (OTF) of the Qatar Environment and Energy Research Institute (QEERI), shown in Figure 1. The soiling rate at this location is very high, typically ~0.3-0.5% per day [2][8], making it an excellent location for testing soiling measurement.

Equipment for evaluating the module-cell-optical soiling measurement approach was installed in early 2024. The system is installed on a string with 440 W bifacial crystalline silicon modules and is comprised of several

instruments, shown in Figure 2. An in-situ I-V measurement unit (RDE300i) periodically measures I-V curves of its associated PV module and determines the module's P_{max} while the module contributes to normal power production in between I-V sweeps. Two rear-side reference cells (RC22) are installed to the east and west sides of the measured module to measure rear-side irradiance contributions. An additional reference cell (RC22) is installed on the front-side of the array together with an optical soiling sensor (Mars™) used for correcting the front-side irradiance reading, eliminating the need for sensor cleaning. The Mars sensor has been previously validated at this site and found to measure soiling rates in agreement with other methods [2][9].

Soiling ratio measurements from the system under evaluation are compared to a control system which comprises a pair of soiled/clean reference modules which are continuouslly held at Pmax by an max-power-point tracking electronic load system. The clean module in this pair is cleaned weekly.

The test string and the control system are normally completely cleaned every two month providing a periodic reset to the experiments.

Tests began in early 2024 and are ongoing, expected to continue through mid 2025.

3 EQUATIONS

We calculate soiling ratio *SR* as the ratio of the measured power of the soiled module to its expected power when clean, using the formula

$$SR = \frac{P_{max}^{soiled}}{P_{max,0}^{soiled} \cdot \left(1 + \gamma \cdot \left(T^{soiled} - T_0\right)\right) \cdot \left(G_{tot}/G_0\right)} \qquad (1)$$

where P_{max}^{soiled} is the maximum power of the soiled module determined from its I-V curve, $P_{max,0}^{soiled}$ is the maximum power of the soiled module when it is completely clean and at reference conditions of irradiance and temperature, γ is the module's maximum power temperature coefficient, T^{soiled} is the module's measured temperature, T_0 is the reference condition temperature (e.g. 25 °C), G_{tot} is the total effective front plus rear POA irradiance, and G_0 is the reference condition irradiance (e.g. 1000 W/m²). We

41st European Photovoltaic Solar Energy Conference and Exhibition

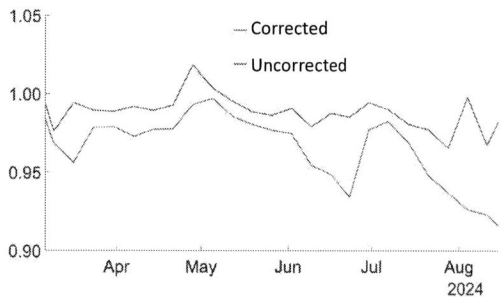

Figure 3: Normalized ratio of uncorrected (soiled) and corrected (soiling-compensated) front-side RC22 reference cell irradiance readings to readings from a regularly-cleaned reference cell.

use the two rear-side and one front-side reference cells to estimate the average total effective irradiance illuminating the soiled module from front and rear, according to the formula

$$G_{\text{tot}} = G_{\text{front}} + \varphi \cdot (G_{\text{rear,E}} + G_{\text{rear,W}})/2 \qquad (2)$$

where G_{front} is the irradiance measured by the front reference cell after correction for soiling losses by the reading of the Mars sensor, $G_{\text{rear,E}}$ and $G_{\text{rear,W}}$ are the irradiances measured by the rear east and west cells, and φ is the PV module bifaciality factor. Averaging the two rear-side cells allows for estimating the average irradiance across the module's rear plane. The bifaciality factor φ appears in eq. (2) to account for the lower responsivity of the module to rear-side light relative to front side light.

The front-side irradiance G_{front} is calculated by

$$G_{\text{front}} = G_{\text{front}}^{\text{uncorr}} \cdot (1 + SL) \qquad (3)$$

where $G_{\text{front}}^{\text{uncorr}}$ is the uncorrected irradiance read from the non-cleaned front-side RC22 reference cell and SL is the fractional soiling loss recorded by the Mars soiling sensor to which the front-side RC22 is mounted (see Fig. 2).

4 RESULTS

The main principle behind the approach evaluated here is that the non-cleaned front-side RC22 reference cell can be compensated for soiling losses using the Mars optical soiling sensor. Fig. 3 presents a validation of this principle. The plot shows the normalized ratio of the weekly average uncorrected (soiled) and corrected (soiling-compensated, using eq. (3)) front-side RC22 irradiance to a irradiance readings from a separate regularly cleaned front-side reference cell. The ratio of the uncorrected readings declines as soiling builds up on the RC22 and the Mars sensor, reaching losses of approximately 6% and 8% in late June and late August, respectively, and increases when the system is cleaned, for example at the beginning of July, while the ratio of the corrected readings stays near 1.0, demonstrating the soiling compensation.

Fig. 4 shows soiling ratios measured by multiple methods during the study to date. The black line shows soiling ratio measured by the Mars soiling sensor. The pink trace shows the P_{max}-based soiling ratio of the PV module in the test system determined using the module-cell-optical combination of the RDE300i, Mars, and RC22 devices, calculated using eq. (1), (2), and (3). The blue trace shows a similar I_{sc}-based soiling ratio, calculated using a modified form of eq. (1) in which $P_{\text{max}}^{\text{soiled}}$ is replaced by $I_{\text{sc}}^{\text{soiled}}$, the I_{sc} of the soiled module measured by the RDE300i, and $P_{\text{max,0}}^{\text{soiled}}$ is replaced by $I_{\text{sc,0}}^{\text{soiled}}$, the clean-state value of I_{sc} at reference conditions. The dashed brown trace shows soiling ratio of the control system, calculated by the ratio of P_{max} between the soiled and clean modules in the control module pair.

At the beginning of May, July, and September the test string with the module-cell-optical system (RDE300i, Mars, and RC22 devices) was cleaned to reset the experiment, while the Control system was cleaned at the beginning of May and September only.

Fig. 4 shows that the soiling ratios measured by the module-cell-optical system are similar to those measured by the control system in the two-month period May to July, with soiling reaching approximately 6% loss at the end of June. The data also appear to show a lower soiling ratio for the P_{max}-based SR than for the I_{sc}-based SR, which is typical of conditions where soiling accumulates non-uniformly. Fig. 5 shows a photograph of soiling on the test

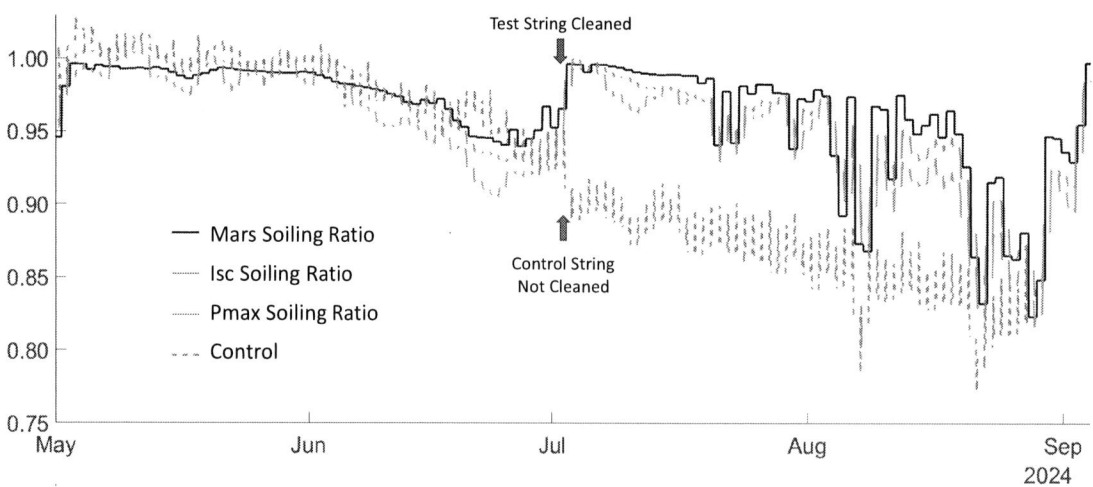

Figure 4: Soiling ratios calculated using multiple methods over a 4-month period.

1247

module taken at the end of June, prior to cleaning, which indeed appears to show soiling accumulation near the bottom edge of the module, consistent with the lower Pmax soiling ratio. The ability to measure this phenomenon is the benefit of using I-V data in soiling measurements.

During July and August the module-cell-optical test system results cannot be directly compared to the Control system, which was not cleaned in July. However, qualitatively, the data in Fig. 4 show both systems measuring approximately 6% increase in soiling loss between the beginning of July and end of August.

The rear-side reference cells remained clean during the study period, as verified by visual observation.

Several technical points of the data are under review. We wish to evaluate how well the rear-side irradiance sensors capture the average rear-side irradiance contribution sensed by the soiled PV module. A previous study on a single-axis tracking system [10] demonstrated good agreement between the average rear-side reference cell irradiance and the module performance; however, the current test system is fixed tilt and may therefore behave differently. Collecting data for a longer period will help us evaluate these points by observing any seasonal variations.

Note that while the study was performed without cleaning the front-side Mars/RC22 combination, except during the bi-monthly experiment resets, the measurement approach would still be valid if the Mars/RC22 combination were occasionally cleaned. The approach eliminates the requirement of regular cleaning – easing the burden on O&M staff – while still allowing for irregular cleaning if desired to improve measurement accuracy.

CONCLUSIONS

Our ongoing study evaluates an approach to maintenance-free measurement of power losses due to soiling in a PV array using a combination of in-situ I-V measurements, front and rear reference cells, and an optical soiling sensor. The combination of instruments allows measuring true soiling ratio, including the effect of non-uniform soiling on module power losses, without requiring periodic cleaning of reference devices. Results from the first four months of testing validate the approach. Additional data will be collected at this site through mid-2025.

5   ACKNOLWEGEMENTS

This material is partially based upon work supported by the U.S. Department of Energy's Solar Energy Technologies Office under Award Number DE-SC0020831.

6   REFERENCES

[1] K. Ilse *et al.*, "Techno-Economic Assessment of Soiling Losses and Mitigation Strategies for Solar Power Generation," *Joule*, vol. 3, no. 10, pp. 2303–2321, Oct. 2019, doi: 10.1016/J.JOULE.2019.08.019.

[2] B. Aïssa *et al.*, "A Comprehensive Review of a Decade of Field PV Soiling Assessment in QEERI's Outdoor Test Facility in Qatar: Learned Lessons and Recommendations," *Energies*, vol. 16, no. 13, 2023, doi: 10.3390/en16135224.

[3] "IEC 61724-1 Ed. 1.0 en:2017 - Photovoltaic system

Figure 5: Soiled module at the end of June, showing non-uniform soiling accumulation near the module edge.

performance - Part 1: Monitoring."

[4] Y. Cui, J. Xiao, J. Xiang, J. Sun, and E. M. Vitucci, "Characterization of Soiling Bands on the Bottom Edges of PV Modules," *Front. Energy Res.*, vol. 9, no. April, pp. 1–7, 2021, doi: 10.3389/fenrg.2021.665411.

[5] M. Gostein, B. Littmann, J. R. Caron, and L. Dunn, "Comparing PV power plant soiling measurements extracted from PV module irradiance and power measurements," *Conf. Rec. IEEE Photovolt. Spec. Conf.*, pp. 3004–3009, 2013, doi: 10.1109/PVSC.2013.6745094.

[6] M. Gostein, T. Duster, and C. Thuman, "Accurately measuring PV soiling losses with soiling station employing module power measurements," *2015 IEEE 42nd Photovolt. Spec. Conf. PVSC 2015*, Dec. 2015, doi: 10.1109/PVSC.2015.7355993.

[7] S. Kagan *et al.*, "Impact of Non-Uniform Soiling on PV System Performance and Soiling Measurement," *2018 IEEE 7th World Conf. Photovolt. Energy Conversion, WCPEC 2018 - A Jt. Conf. 45th IEEE PVSC, 28th PVSEC 34th EU PVSEC*, pp. 3432–3435, Nov. 2018, doi: 10.1109/PVSC.2018.8547728.

[8] E. Kam-Lum, B. E. Meyers, D. Cosme, B. Aissa, and G. Scabbia, "Soiling Rate Determination from Referenced Systems in Desert Climate using PVInsight Soiling Algorithm," in *Conference Record of the IEEE Photovoltaic Specialists Conference*, 2021, no. March, pp. 2552–2554, doi: 10.1109/PVSC43889.2021.9518459.

[9] B. Aïssa, G. Scabbia, B. W. Figgis, J. Garcia Lopez, and V. Bermudez Benito, "PV-soiling field-assessment of Mars™ optical sensor operating in the harsh desert environment of the state of Qatar," *Sol. Energy*, vol. 239, no. December 2021, pp. 139–146, 2022, doi: 10.1016/j.solener.2022.04.064.

[10] M. Gostein, A. Marquis, R. Campbell, P. Wolffersdorff, and M. Bila, "Soiling And Irradiance Measurements In Bifacial Systems Using In-Situ Reference Modules And Rear-Side Reference Cells," *52nd IEEE Photovoltaics Spec. Conf.*, 2024.

IMPROVING PERFORMANCE RATIO CALCULATIONS THROUGH OPTIMIZING FRONT POA IRRADIANCE SENSOR POSITIONING

Marc A.N. Korevaar[1], Damon Nitzel[1], Shuo Wang[2], Nate Solofra[3]
[1]OTT Hydromet B.V., Delft, the Netherlands
[2]New Energy Research Group, Turku University of Applied Sciences, Joukahaisenkatu 7, 20520 Turku, Finland
[3]Merit Controls, 46 East Main Street, Suite 302, Somerville, NJ 08876, New Jersey, United States of America

ABSTRACT: Underperformance of PV plants against forecast can happen for many reasons. Here we investigate the placement locations of the irradiance sensors. With well-placed accurate irradiance sensors, the performance ratio can be more accurately determined. And this reduction in performance ratio variability, due to sensor locations, reduces the risks of the solar park investment.
Kipp & Zonen produces pyranometers that accurately measure front and rear side plane of array (POA) irradiance as well as surface albedo. The primary site investigated is a bifacial, single-axis tracker system. For the front side, we evaluate the mounting position of the pyranometer within the plant and across the PV module. We have performed simulations using a ray-tracing toolkit, Bifacial Radiance, from NREL and in addition a 7 month validation in a utility scale PV plant with multiple pyranometers is underway.
Within the plant both simulations as well as measurements show that edge mounted pyranometers measures about 2% more. Across the module for cumulative data, the variation in the simulation is larger than in the measurements. For the intra-day variation there is a good match between simulation and measurements.
To help improve the performance ratio calculation a guidance for optimal placements of irradiance sensors is given.
Keywords: Pyranometer, Irradiance measurement, Single Axis Tracker, Performance ratio

1 INTRODUCTION

Performance ratio (PR) calculations are a vital part of PV plant evaluation, influencing various aspects such as contract payments, bonuses as well as the PV plant valuation. Therefore, accurate calculations are of utmost importance. For the PR calculation the meteorological input of largest influence is the irradiance measurement. Irradiance measurements with pyranometers can provide uncertainties, in economic relevant hours of the day, as low as 1.3% - 1.7%.

In the past there has been work done on the influence of rear POA mounting position on the irradiance measurement accuracy for bifacial plants [1,2,3]. Only recently the placement of the front POA sensor has been investigated in a single-axis tracked site [4,5].

Here we build upon the previous research of front POA sensor placement with simulations as well as a longer validation measurements period of now 7 months.

The simulations and field data presented show that performance modeling uncertainty can be reduced with well-placed POA pyranometers both within the array and across the module.

2 METHOD

2.1 Simulations

The simulations were performed with the ray-tracing toolkit developed by NREL: Bifacial Radiance[6]. The simulations were run in solar time instead of local time, to overcome a small asymmetrical effect, which occurs when the simulated site has solar noon significantly before or after 12:00 local time. The primary site investigated, near San Angelo, Texas, was selected as a representative sample of common utility-scale PV sites using bifacial modules and single-axis trackers. The parameters used to perform the Bifacial Radiance simulations are shown in Table I, below. The simulations

Table I: Simulation parameters used in bifacial radiance.

Parameter	Value	Parameter	Value
Hub Height	1.3 m	Module rows	6
Tilt	Backtracking on	Modules per row	80
Azimuth	N-S axis	Number of sensors	20
Ground Coverage Ratio (GCR)	0.48	Location	San Angelo, TX USA
Albedo	0.35	Weather data	Energyplus.net
Module orientation	Portrait	Module dimensions	1.98 m x 0.98 m
Nr. Modules above each other	1	Torque tube	Yes
Type of plant	Tracked	Type of simulation	Cumulative sky yearly

Software versions: Python 3.11.5, Radiance 5.3 (2020-09-03), Bifacial radiance V0.4.2 (2023-03-11)²⁵, PVLIB V0.10.2

calculate the expected irradiance at regular points across the array, representing 80 modules per row and 20 'sensor' segments across each module; each irradiance value corresponds to a 20 by 25 cm area. Simulations were run for 1-year and for a 7-month window to match the available field-measurement data.

Primarily, the simulations were run for a module size area in different locations of the PV plant, as shown in Figure 1.

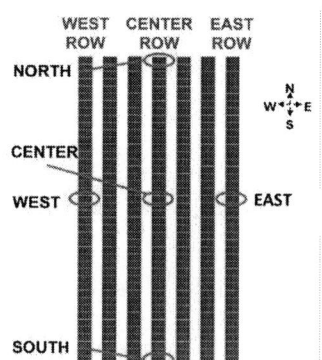

Figure 1: Simulated module locations *within the array.*

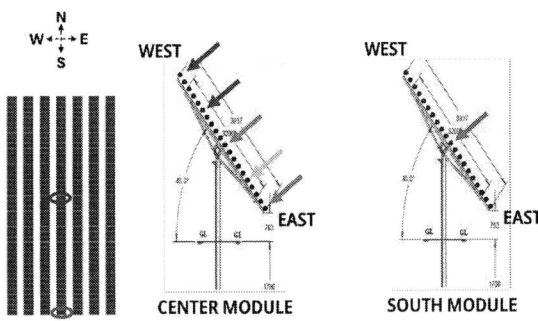

Figure 2: Simulated module locations *across the module.*

Secondly, the simulations were also run for distinct measurement positions across the module, as shown in Figure2.
Finally, full simulations of the modelled plant were run allowing the creation of heat maps showing irradiance differences with the average of the plant.

2.2 Measurements

We are actively validating the simulation results with an arrangement of pyranometers at this site in San Angelo.
The installation locations of the POA pyranometers within the array are shown in Figure 3. The field data collection campaign will last at least one year, here we present data from the first 7 months of data up to in August 2024 compared to the simulation results.

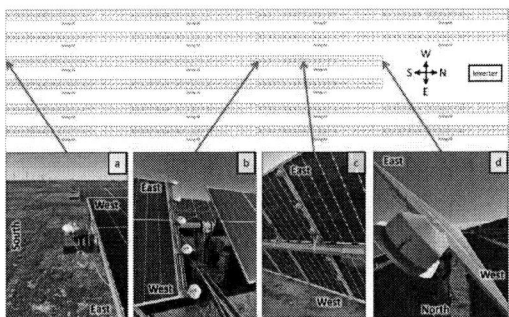

Figure 3: Field-test site installation setup. (a) South-center mounted SMP12. (b)East-center, East-mid-center, Center-center, West-mid-center, West-center mounted SMP12s. (c)Rear-facing pyranometers not discussed here. (d)North-center mounted SMP10-V housed in a CVF4 heater ventilator.

To minimize sensor-to-sensor variation, the Kipp & Zonen SMP12 pyranometers installed were from a single production batch and were calibrated indoors against the same reference sensor. Furthermore, the sensor data was normalized by averaging 3600 two-second samples taken near solar noon (level sensors) on a clear-sky day (July 10).

3 RESULTS

3.1 Simulations

Comparison of irradiance simulation data that represents the area of a solar module at different locations *within the*

array (see Figure 1) shows that edge-located modules received up to 2.3% more irradiance than center mounted modules, see Table II, below.

Table II: Yearly edge-located module irradiance differences from center-located module irradiance.

	North	South	West	East
Irradiance diff. [%]	+2.0	+2.2	+2.3	+2.1

The irradiance simulation results over a year, in the form of a heatmap showing irradiance difference from the average of the plant are shown in Figure 4. The differences of the annual simulations at each 'sensor' (20 by 25 cm segment) from the average of the entire plant range from 4.3% higher to -3.3% lower than average. The maximum irradiance difference, 7.6%, was observed between the 'sensor' position at the eastern edge of the central module and the western edge of a southernmost module in the western row.

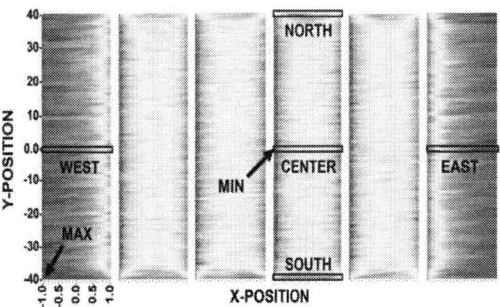

Figure 4: Heatmap of yearly irradiance difference of position within the array to the average of array.

Simulating the yearly irradiance values shows variation *within the array* and *across the module* (see Figure 5).
The eastern edge of the eastern row (yellow) receives the most irradiance because it 'sees' more diffuse sky throughout the day, and vice versa for the western edge of the western row module (red). The edges of the north (green), center (black) and south (blue) center-row modules receive less irradiance due to diffuse fraction shading from adjacent rows at higher zenith angles.

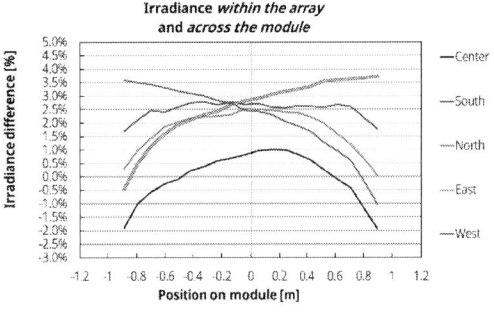

Figure 5: Simulated yearly irradiance differences of each position across the module compared to the average, relative to the center module within the array.

41st European Photovoltaic Solar Energy Conference and Exhibition

3.2 Validation

For the center-position of the center-row (Figure 5, black line), the field measurements in San Angelo show a difference of 0.8%. Using Typical Meteorological Year (TMY) weather files, simulated irradiance values show a difference of 2.6% (Figure 6). One factor in the difference between simulation and field data is caused by an increased number of clear skies in 2024 compared to the average TMY data. The center position reports the irradiance relative to the average over the module (+0.23%), the -0.5 m and 0.5 m positions have a difference of 0.13% and 0.25% respectively. The edge positions under-report by up to -0.52% and -0.09% (Figure 6).

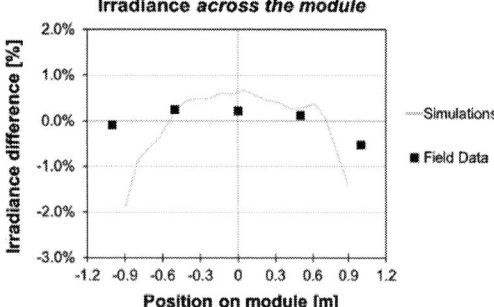

Figure 6: Irradiance differences across the module for center module of center row; 7 months field data.

Simulated and on-site intraday data under clear sky (Figure 7) show center position with near zero percent difference from the average through the day. The edge and middle positions both over and underreport irradiance depending on tilt angle by up to ±3% in simulations and ±5% in field data.

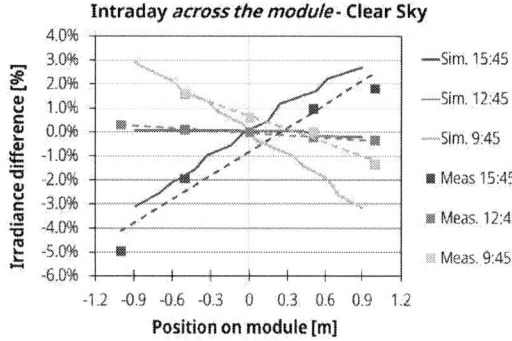

Figure 7: Intraday irradiance for the center module positions in the morning, at noon, and in the afternoon on a clear day.

Also, on a cloudy day the center position shows close to zero percent difference from the daily average, the other positions show differences of approximately ± 10% in simulations and ± 10% in field data as shown in Figure 8.

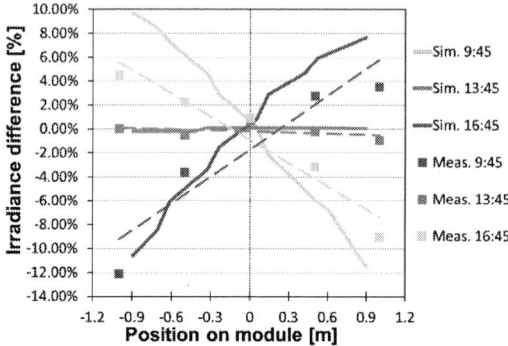

Figure 8: Intraday irradiance for the center module positions in the morning, at noon, and in the afternoon on a cloudy day.

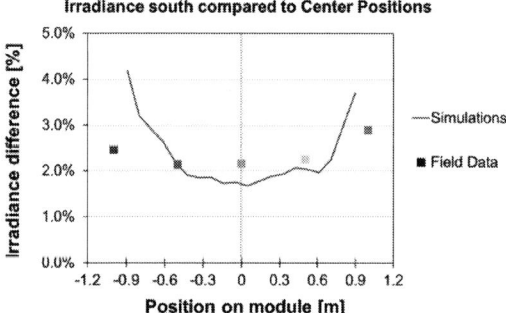

Figure 9: South-center irradiance minus Center-center positions; 7 month field data

The simulation difference between the south-center and centered irradiance locations is 2.2% (Table II). The observed field measurement differences between the south pyranometer (Figure 3 a) and the center pyranometers (Figure 3 b) is 2.2%. Plotting the other center-positions against the south-center value shows reasonable correlation between the simulations and field measurements (Figure 9). The north-center pyranometer (Figure 3 d) field measurements measured 0.38 % more than the center pyranometer (Figure 3 b).Based on these measurement results, siting recommendations in the green area, within the plant are given in Figure 10.

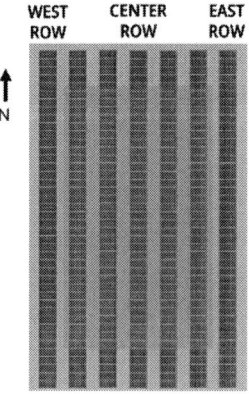

Figure 10: Considerations for siting POA sensors. Yellow: 2% of modules,Green: 98% of modules.

1251

3 CONCLUSIONS

Sensors located on edge modules within the array over-report irradiance for simulated and on-site data in San Angelo. High irradiance inflates the expected energy in performance metrics, resulting in apparent underperformance. To avoid artificially high irradiance values, the data confirms that POA pyranometers should be interior to the array.

Placing a POA pyranometer in the center of the module is optimal for intra-day POA irradiance, both for clear-sky and cloudy-sky conditions.

For long-term averaging of POA irradiance, positions between -0.5 and 0.5 m are within 0.25% of the average over the module. Therefore, mounting POA pyranometers in centered positions is the ideal for single sensor deployments.

The Bifacial Radiance software has a good correlation with the on-site irradiance measurements collected, allowing for qualitative understanding of the irradiance at different locations within a simulated PV park.

4 ACKNOWLEDGEMENTS

We extend great appreciation to Nate Solofra and Chad Paden, among others at Merit for their help enabling the on-site measurement opportunity and to the knowledgeable staff at NREL for the creation of the Bifacial Radiance software.

5 REFERENCES

[1] "Simulation and Validation of Bifacial Irradiance Sensor Mounting Position", Korevaar et al. EU PVSEC 2020

[2] "Strategies for Rear Irradiance Monitoring", Riedel et al., PVPMC 2023

[3] "Bifacial photovoltaic Technology: recent advancements, simulation and Performance measurement", M. Aghaei et al., 2022, ISBN:978-1-83968-858-4

[4] "Simulation of POA front irradiance sensor mounting position", Korevaar et al., PVPMC 2023, Mendrisio.

[5] "Simulation of POA front irradiance sensor mounting position", Korevaar et al., PVPMC 2023, Mendrisio.

[6] Ayala Pelaez and Deline, (2020). bifacial_radiance: a python package for modeling bifacial solar photovoltaic systems. Journal of Open Source Software, 5(50), 1865, https://doi.org/10.21105/joss.01865

41st European Photovoltaic Solar Energy Conference and Exhibition

A METHOD FOR DETECTING PV MODULE'S DEGRADATION DUE TO INCREASED LOCAL RESISTANCE IN POWER PLANT

Tohru Kohno*, Jun Tsunoda*
Hitachi, Ltd. Research & Development Group
1-280 Higashi-Koigakubo, Kokubunji, Tokyo 185-8601, Japan
E-mail: toru.kono.cw@hitachi.com

ABSTRACT: To create a reuse standard, we devised and demonstrated a method to detect the deterioration of PV (Photovoltaic) modules caused by an increase of local resistance from inverter operation data (voltage, current, irradiation). After that, a method for determining the quality ranking of PV modules focused on the local resistance was devised.The PV system equipped with a control unit performs the proposed detection method using one time zone of MPPT (maximum power point tracking) and two time zones of clipping of the output to detect local resistance increases. We obtained the operating data from the 53.8 kW PV power generation system installed 11 years ago. By applying the proposed method, we succeeded in detecting an increase in local resistance for a module that has deteriorated by about 7% compared to its rating. Over the past four years, the degradation rate is equal to or slightly higher than the guaranteed performance of the module of 0.7% / year. In addition, we investigated the PV modules in this system and found many parts of corrosion. This showed that the proposed method is useful.
Keywords: degradation, local resistance, MPPT, clipping, corrosion

1 INTRODUCTION

Since the introduction of the FIT system in 2012, the number of PV modules installed in Japan has increased significantly. As of 2023, it has been further introduced in the future due to the spread of self-consignment, PPA (Power Purchase Agreement) and the Tokyo Metropolitan Government's mandatory installation of housing. On the other hand, according to the Ministry of the Environment, the amount of PV module waste is expected to increase sharply, from 3,000 tons in 2025 to 20,000 tons in 2030 and 200,000 tons in 2035. 20,000 tons is equivalent to 15 million PV modules.

In Japan, there is a disposal cost reserve system that started in July 2023 and guidelines for the recycling and reuse of PV power generation facilities, but as of January in 2024, there is no clear legislation. Since the final disposal sites for PV modules are limited, it is important to promote the reuse and recycling of PV modules to build a circular economy society.

The reason why the market for reuse and recycling PV modules has not been launched is that they are not eligible for subsidies unlike new modules, and there are no clear reuse standards. For example, it is necessary to decide whether to reuse, recycle, or dispose of PV modules in power plants that have been affected by a disaster. However, due to the lack of clear reuse standards and differences in quality depending on the type of PV module, in many cases, it is disposed of without consideration. In the operation and maintenance of tha large-scale PV power plants, there is a need to avoid inspection by drone or inspection of PV modules at exposure sites because of the cost, and to establish reuse standards based only on actual operation data.

Due to the rapid increase in the use of inexpensive PV modules in 2011~14 due to the introduction of FIT, we developed our own accelerated test as a method for assessing the reliability of PV modules [1-2]. In this study, we focused on the peeling of the solder of the interconnector, which is responsible for the wiring of PV modules, and positioned the degradation factor of PV modules as an increase in the overall resistance component (increase in series resistance), and compared the degree of degradation for each PV module type.

However, due to the frequent performance degradation due to PID (potential-induced degradation) at each power plant, and the replacement of butyl as a sealant for PV modules with inexpensive silicon, there were concerns about performance degradation due to moisture intrusion. Since 2017, the main cause of degradation of PV modules has been shifted from an increase in series resistance to a decrease in shunt resistance, and a method for distinguishing between PID and moisture intrusion and a method for detecting and quantifying shunt resistance with high accuracy have been developed [3-4].

At the same time, when the series resistance component of the PV module increased locally in 2014~2015, the series resistance increased in the I-V measurement in the darkroom, and the shunt resistance decreased in the I-V measurement under exposure (light irradiation) [1]. This phenomenon later came to be known as a fake shunt.[5] Ten years have passed since the implementation of the FIT, and when we investigated the PV modules, it became clear that the series resistance component of the PVmodules was increasing locally. In this case, it is difficult to estimate performance degradation from I-V measurements by exposure operation data alone, since the I-V characteristics are such that the shunt resistance decreases under light irradiation. Therefore, in this study, we developed a diagnostic technique that can estimate the local increase in the series resistance of a PV module from the data obtained during operation.

2 DETECTING METHOD OF LOCAL RESISTANCE

2.1 I-V characteristics when the local resistance increases

Fig. 1 (a) shows the front and back views of the PV cell, and Fig. 1 (b) shows an equivalent circuit diagram of a PV cell. A finger can be represented as a series resistance that connects multiple PVs between busbars. An increase of local resistance is defined as a partial increase in busbar and finger's resistance. As the local resistance increases, the I-V curve of the STC (Standard condition) becomes characterized as if the shunt resistance has decreased, as shown in Fig. 1(c). In the dark I-V curve, this phenomenon is not seen, as shown in Fig. 1(d) [1]. Due to the above characteristics, when the local resistance increases, the loss appears to be larger near MPP (Maximum Power Point) than in the high-voltage region of the I-V curve where the clipping of output. Therefore, a feature of the proposed method is to detect an increase of local resistance by using clipping of output and MPPT control in combination.

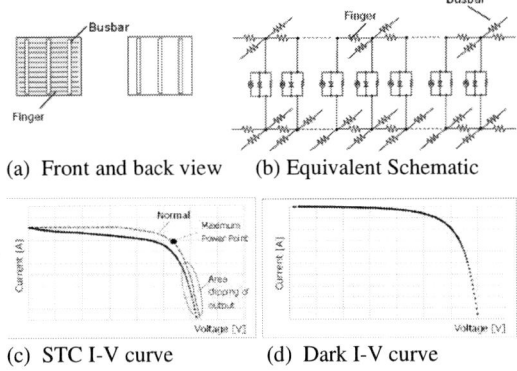

(a) Front and back view   (b) Equivalent Schematic

(c) STC I-V curve   (d) Dark I-V curve

Figure 1: Three-dimensional equivalent circuit of a PV cell and I-V curve

2.2 Algorithm for detecting an increase in local resistance from operational data

Fig. 2 (a) shows the configuration diagram of the PV system. The PV system consists of a power conditioner driving a plurality of PV modules arrayed. The control device equipped with the deterioration factor estimation function controls the output power of PV (clipping) through controlling the PV inverter.

Fig. 2 (b) shows an example of the change in output power over time in a PV system. As a general rule, the output power follows the maximum power point tracking (MPPT) as the amount of solar radiation increases, but the output power may be intentionally suppressed. Fig. 2 (b) illustrates the clipping of output power when such output controls are performed. In this example, the output controls are performed twice, and the amount of irradiation is larger in the second time.

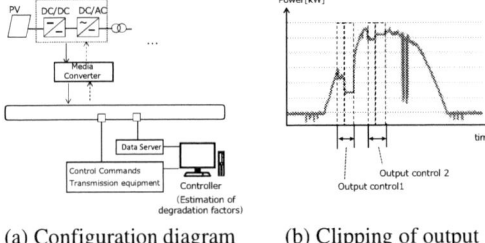

(a) Configuration diagram   (b) Clipping of output

Figure 2: Configuration diagram and control of the PV system

Fig. 3 shows the I-V and P-V curves of PV arrays. When clipping the output, PV system is usually operated in region of ($\partial P / \partial V$ < predetermined value 3) instead of the region of ($\partial P/\partial V$> predetermined value 2) in the P-V curve. This condition is ($\partial I / \partial V$ < predetermined value 1) in the I-V curve. On the other hand, the maximum power point corresponds to the voltage region in which the value of differentiating the power in the P-V curve by the voltage is nearly 0.

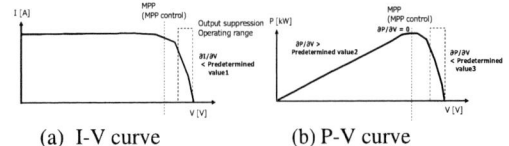

(a) I-V curve   (b) P-V curve

Figure 3: Operating area when clipping of output

Fig. 4 is a flowchart for diagnosing an increase of local resistance. At steps 1, 2, and 4, measurements of voltage, current, and irradiation in each time zone are taken to calculate losses and resistances (R_1, R_2, R_3). If 2nd loss is less than or equal to 1st loss (step 3 : No), it is diagnosed that the degradation factor of PV is not due to the increase in series resistance (step 5). If 2nd loss is greater than 1st loss, and if R_3 is larger than R_1 and R_2 (step 6 : Yes), it is diagnosed that the degradation factor is a local increase in the series resistance of PV (step 7). Otherwise, it is estimated that the overall uniform increase in the series resistance of PV is the degradation factor (step 8).

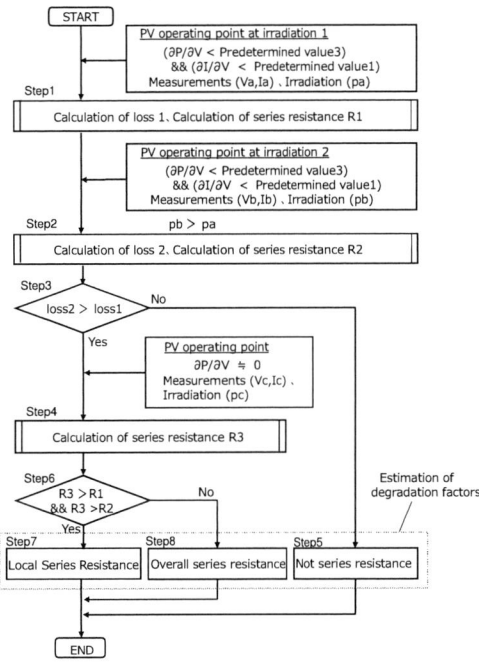

Figure 4: Flowchart for diagnosing PV degradation

Fig. 5 shows the flowchart of calculating R_1 and the first loss. The calculation of R_2, second loss, and R_3 are similar procedures. The measured values are acquired from inverter at step 0. The short-circuit current I_{sc} corresponding to irradiation p_a is calculated at step 1. At step 2, the ratio j of the operating current relative to I_{sc} is calculated. At step 3, the operating voltage V_p is calculated using kT/q = 0.026 at 298 K. At step 4, R_1 is the series resistance per cell and the voltage drop V'_p is calculated. At step 5, the PV's estimated temperature T is calculated. At step 6, the temperature characteristics of the short-circuit current α [%/°C] and T are used to convert the short-circuit current corresponding to T. If step 2 ~ step 6 is repeated more than the prescribed number of times (for example, 3 times), skip to step 8. At step 8, the measured

1254

current I_a is converted to the STC current I_{p0}. After that, the delta_I is calculated by comparing it with the current at j of STC I-V curve extracted from the database. If the delta_I is small enough, skip to step 12, otherwise increment R_1 and return to step 2. Finally, the first loss is calculated at step 12. The derivation of the formula for each step is described in the paper [6].

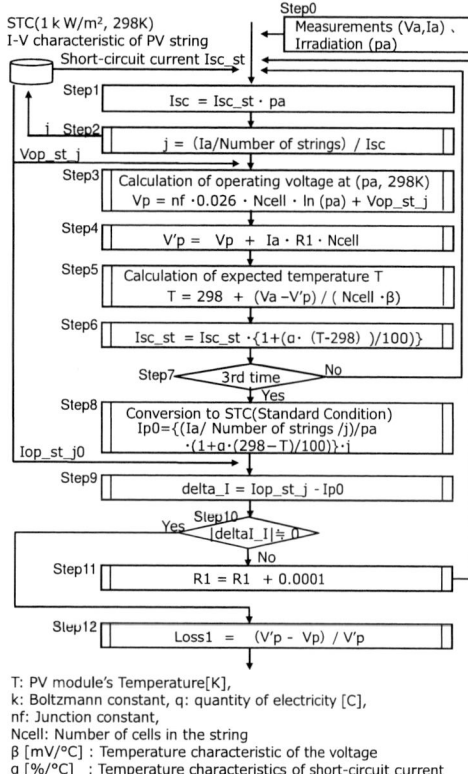

T: PV module's Temperature[K],
k: Boltzmann constant, q: quantity of electricity [C],
nf: Junction constant,
Ncell: Number of cells in the string
β [mV/°C] : Temperature characteristic of the voltage
α [%/°C] : Temperature characteristics of short-circuit current

Figure 5: Flowchart for calculating series resistance and loss at step 1, 2, and 4 in Fig. 4

Fig. 6 shows the flow of ranking the quality of PV systems based on the diagnosis results. If Rs_1 is below the first threshold and Rs_3/Rs_1 is below the second threshold, it is judged that it is normal or uniformly deteriorated (Quality Ranking 1). If Rs_1 is less than or equal to the first threshold and Rs_3/Rs_1 is above the second threshold, it is judged that the series resistance is slightly increased locally (Quality Ranking 2). If Rs_1 exceeds the first threshold and Rs_3/Rs_1 is equal to or greater than the third threshold, it is judged that the series resistance is increasing locally (Quality Ranking 3). If Rs_1 exceeds the first threshold and Rs_3/Rs_1 is less than the third threshold, recycling is recommended (Quality Ranking 4).

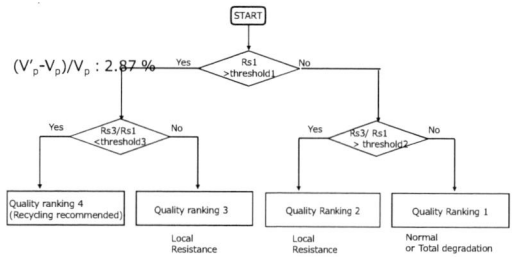

Figure 6: Flow for ranking the quality of a PV system

3   EXPERIMENTAL RESULTS

Fig. 7 shows the measurement data of the 53.8 kW PV system. In Time zone 1 (Clipping), the average value of 14 pieces of per minute data from 10:30 to 10:43 was applied to the algorithm for calculating the R_1 in Fig. 4. Similarly, in time zone 2 (Clipping), the average value of 14 pieces of per minute data from 11:05 to 11:18 was applied to the algorithm for calculating the R_2. In time zone 3 (MPPT), the average value of per minute data from 9:55 to 10:08 was used for calculating the R_3.

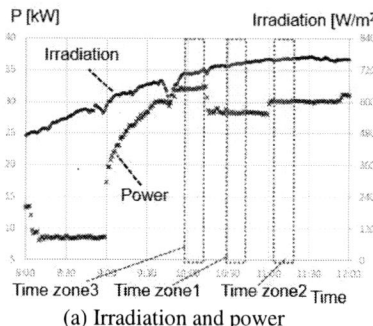

(a) Irradiation and power

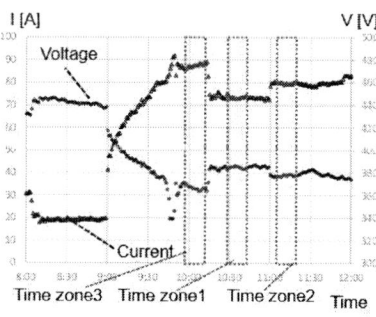

(b) Voltage and current

Figure 7: Measurement data of 53.8kW PV system

Table 1~3 shows the results in each time zone. These results (R_1 and R_2: 0.0035Ω, loss 1:2.87%, loss 2: 3.15%, and R3:0.0095Ω) indicate detecting an increase of local resistance. Table 4 shows the STC outputs of the 3 PV modules extracted from the 53.8 kW PV system. These PV modules are a 7% drop to the rating, an average decline of 0.72% per year was observed.

Table 1: Calculation result of the series resistance R_1 by the proposed algorithm in time zone 1 (Clipping)

Series Resistance [Ω]	J	Vp	V'p	PV temperature [K]	Ip0 [A]	delta_I [A]
0	0.7017	456.99	456.99	340.93	8.0760	0.0010
0.001	0.7010	456.99	460.84	343.22	8.0757	0.0007
0.002	0.7004	456.99	464.70	345.52	8.0755	0.0005
0.003	0.6998	456.99	468.55	347.81	8.0752	0.0002
0.0035	0.6994	456.99	470.48	348.96	8.0750	0.0000

Table 2: Calculation result of the series resistance R_2 by the proposed algorithm in time zone 2 (Clipping)

Series Resistance [Ω]	J	Vp	V'p	PV temperature [K]	Ip0 [A]	delta_I [A]
0	0.7441	450.66	450.66	341.51	8.0760	0.0010
0.001	0.7434	450.66	454.85	344.00	8.0757	0.0007
0.002	0.7427	450.66	459.04	346.50	8.0754	0.0004
0.003	0.7419	450.66	463.23	348.99	8.0751	0.0001
0.0035	0.7416	450.66	465.32	350.24	8.0750	0.0000

$(V'_p - V_p)/V_p$: 3.15 %

Table 3: Calculation result of the series resistance R₃ by the proposed algorithm in time zone 3 (MPPT)

Series Resistance [Ω]	J	Vp	V'p	PV temperature [K]	Ip0 [A]	delta_I [A]
0	0.8831	408.66	408.66	322.60	8.0776	0.0026
0.001	0.8822	408.66	413.25	325.33	8.0774	0.0024
0.002	0.8812	408.66	417.83	328.05	8.0772	0.0022
0.003	0.8803	408.66	422.41	330.78	8.0770	0.0020
0.004	0.8793	408.66	426.98	333.50	8.0767	0.0017
0.005	0.8784	408.66	431.56	336.23	8.0765	0.0015
0.006	0.8774	408.66	436.14	338.96	8.0762	0.0012
0.007	0.8765	408.66	440.72	341.68	8.0756	0.0006
0.008	0.8755	408.66	445.30	344.41	8.0756	0.0006
0.009	0.8746	408.66	449.88	347.13	8.0752	0.0002
0.0095	0.8741	408.66	452.16	348.49	8.0750	0.0000

$(V'_p-V_p)/V_p$: 9.62 %

Table 4: Measurement results of output power of PV modules extracted from 53.8 kW PV system

Identification	Power [W] 2019	Power [W] 2023	Rate of change
PV module 1	235.9	230.5	-2.29 %
PV module 2	234.5	225.7	-3.75 %
PV module 3	233.7	227.6	-2.61 %
Average	234.7	227.9	-2.88 %

● 7% reduction relative to rating 245 W (227.9 W / 245 W)
● 0.74% decline per year (2.88 W / 4)

Considering that the manufacturer's warranty for PV modules is 0.7% degradation per year, and that the rate of degradation has become equivalent, it is reasonable to conclude that the quality ranking in Figure 6 has shifted from 1 to 2. Fig. 8 shows the I-V curve in the STC of "PV module 2", which had the lowest output. Fig. 8 (a) shows the I-V curve when swept from a current of 0 A (open circuit voltage) to a short-circuit current (around 8.6 A), and Fig. 8 (b) shows an enlargement of the STC I-V characteristics of Fig. 8 (a) from a current of 8 A to a short-circuit current (near 8.6 A).

Similarly, Fig. 9 shows the I-V curve (dark I-V curve) of "PV module 2" in a dark room. Fig. 9 (a) shows the I-V curve when the current is swept from 0 A (open circuit voltage) to -7 A, and Fig. 9 (b) shows the dark I-V characteristics of Fig. 9 (a) from 0 A to -1 A. In Fig. 8(b), when the voltage is 20 V or less, the current slope is about 0.08 A, but in Fig. 9(b), it is found that almost no current is generated. As a result, it was determined that the cause of the deterioration was not the shunt resistance, but the performance degradation due to the increase in the local series resistance.

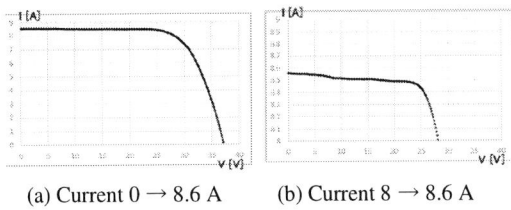

(a) Current 0 → 8.6 A   (b) Current 8 → 8.6 A

Figure 8: STC I-V curve of PV module 2

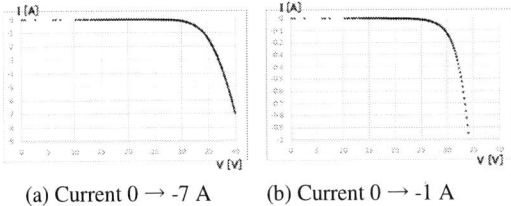

(a) Current 0 → -7 A   (b) Current 0 → -1 A

Figure 9: Dark I-V curve of PV module 2

Fig. 10 shows an EL image of "PV modul 2". A large number of non-luminescent cells were observed in the lower part of the electrode and fingers of some PV cells. This suggests that the degradation factor of this PV module is likely to be the increase in local resistance.

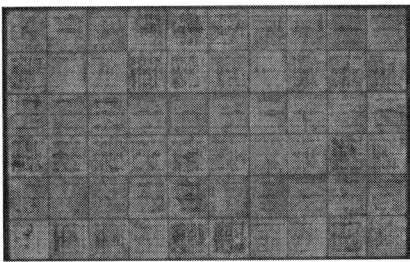

Figure 10: EL image of PV module 2

Fig. 11 shows an observation photograph of one part (4 cells) of "PV module 2". It was confirmed that the PV module had a lot of blackening on the busbar. This phenomenon is called "corrosion", and it is expected that the performance of PV modules that have been confirmed to be corroded will decline sharply from a certain period. In other words, it was suggested that the quality ranking in Fig. 6 may change at an early stage, from 2→3→4 to 4. This corrosion needs to be investigated in the future, targeting PV modules introduced after FIT in Japan, such as the correlation between the rate of increase in local resistance and corrosion.

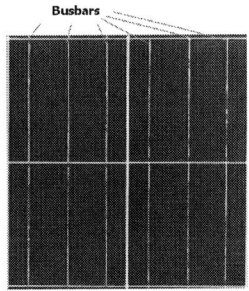

Figure 11: Observation photo of PV module 2
(Corrosions are noticeable on the busbar)

5 CONCLUSION

We developed and demonstrated a method to detect the deterioration of PV modules caused by an increase of local resistance from PV inverter's operation data (voltage, current, irradiation). Based on the results of detection, a method for determining the quality ranking of PV (Photovoltaic) modules focused on the local resistance was devised. By applying the proposed method for the 53.8 kW PV power generation system installed 11 years ago, we succeeded in detecting an increase in local resistance for a module that has deteriorated by about 7% compared to its rating. Over the past four years, the degradation rate is equal to or slightly higher than the guaranteed performance of the module of 0.7% / year. In addition, we investigated the PV modules in this system and found many parts of corrosion. This showed that the proposed method is useful.

5   ACKNOWLEDGEMENT

The authors would like to thank Mr. K. Morita, and the members of PVSQ management, LLC for the evaluation of PV module and their valuable suggestions.

6   REFERENCES

[1] K. Morita et al., Proceedings European Photovoltaic Solar Energy Conference 2015 (2015) pp. 2515 - 2520.

[2] M. Fujimori et al., Proceedings European Photovoltaic Solar Energy Conference 2015 (2015) pp. 1911 - 1914.

[3] T. Kohno et al., Proceedings European Photovoltaic Solar Energy Conference 2020 (2020) pp. 1072 - 1075.

[4] T. Kohno et al., 47th IEEE Photovoltaic Specialists Conference (2020) pp. 359 - 364.

[5] E. A. Gaulding et al., IEEE Journal of Photovoltaics, vol. 12, Issue. 3 (2022) pp. 690–695

[6] T. Kohno et al., IEEE Journal of Photovoltaics, vol. 9, Issue. 3 (2019) pp. 780–789.

41st European Photovoltaic Solar Energy Conference and Exhibition

DEPENDENCE OF SERIES RESISTANCE ON IDEALITY FACTOR AND SHUNT RESISTANCE IN ONLINE PHOTOVOLTAIC MODULE PARAMETRIC IDENTIFICATION

Heidi Kalliojärvi, Kari Lappalainen
Tampere University, Electrical Engineering Unit, P. O. Box 692, FI-33101 Tampere, Finland
Email: heidi.kalliojarvi@tuni.fi, kari.lappalainen@tuni.fi

ABSTRACT: Ageing of photovoltaic (PV) modules can be comprehensively detected by investigating measured current–voltage curves. The curves can be subjected to fitting the single-diode model describing the operation of a PV cell. The model parameters provide numerical information of the condition of the PV cells. Especially, series resistance is a specific indicator for ageing, whence it should be reliably identified. As the parameters depend on the selected identification procedure and the operating conditions, those effects should be distinguished. Therein, a problem is the strong interconnection of the parameters: variation in one parameter affects the others too. Particularly, values of the ideality factor and the series and shunt resistances depend on each other. This must be taken into account before drawing diagnostical conclusions. At practical PV applications, only such identification procedures that can reproduce the operating conditions are of interest. However, the effects of variations of ideality factor and shunt resistance on the series resistance when using such procedures are unclear. This article addresses this issue by investigating the effects of these parameters on the series resistance by using two identification procedures that identify also the operating conditions. The results provide guidance for online condition monitoring of PV systems.
Keywords: condition monitoring, current–voltage curve, series resistance, shunt resistance, ideality factor

1 INTRODUCTION

Manufacturers of PV modules usually guarantee a performance warranty of 25–30 years for their modules. This means that a PV module should maintain at least 80% of its rated power production capability when reaching the end of the warranty period [1]. However, this condition is not always satisfied in practice. Indeed, field-exposed PV modules suffer from different ageing and degradation phenomena during their lifespan [2]. This results from the surrounding environmental conditions that are often harsh. Ageing of PV modules causes harm to the installation and its users. Firstly, aged PV modules produce less power than non-aged modules. Secondly, the ageing effects tend to be cumulative, i.e., aged PV modules stress also the other parts of the PV installation, making them more vulnerable to degradation [3]. In this light, the detection of ageing in time is crucial when monitoring the condition of the PV system. Then, appropriate maintenance procedures for the PV system could be optimally timed.

A comprehensive way to monitor the condition of PV systems is to analyze the current–voltage (I–U) curves measured from the terminals of a PV module or a larger configuration of PV modules [4]. Indeed, the curves can be subjected to single-diode model fitting, which can be done also online. The single-diode model analysis provides quantitative information of the condition of the PV module. The obtained numerical results are of high value when assessing the severity level of the possible ageing and degradation. It is well-known that ageing of PV cells typically manifests as increments of series resistance parameter [5]. Hence, reliable identification of this parameter is very important.

Unfortunately, reliable analysis of ageing based on series resistance identification is exposed to certain practical challenges. Firstly, it is not only the possible ageing but also the operating conditions that affect the identified parameters [6]. However, the lack of irradiance and/or temperature sensors at practical PV sites is a rule rather than an exception [7]. In order to overcome this problem, the used single-diode model parameter identification procedure must be selected so that it can identify the PV module operating irradiance and

temperature jointly with the actual parameters. A few such identification procedures with online applicability do exist [8, 9, 10]. Secondly, it is known that a large number of identified parameters increases the computational cost, which is a real problem in practical applications. It also increases the risk of the fitting algorithm to get trapped at a local solution producing parameter values with no physical meaning [11]. Moreover, the number of identified parameters might affect their stability. A straightforward solution for these issues is to restrict the number of identified parameters to minimum extent so that only those parameters necessary for diagnosis are identified. However, changes in the number of identified parameters cause changes in other parameters; the root mean square error (RMSE) of the fitted I–U curve might be improved if the number of the identified parameters are reduced [12]. In ageing detection, the changes in series resistance originating from variations in ideality factor and shunt resistance are of major interest, as these three parameters are particularly sensitive to variations in each other. For instance, the author of [13] observed that series resistance is a linearly decreasing function of ideality factor.

The present study contributes this discussion by investigating the mutual effect of identified parameters. The parameters are identified by applying two online applicable single-diode model identification procedures with limited numbers of identified parameters. The effect of ideality factor and shunt resistance on the series resistance is analyzed by statistical means using these identification procedures. The findings of the present study will serve as guidelines for ageing detection of real-case PV systems.

The remainder of the paper is organized as follows. Section 2 is dedicated for describing the single-diode model of a PV cell jointly with the used identification procedures and the measurement data. In Section 3, the experimental results are presented and discussed. Finally, Section 4 closes the paper.

2 METHODS AND DATA

2.1 Single-diode model

The operation of a PV cell, module or larger unit can be described with the widely used single-diode model [14]. It expresses the PV current as an implicit function of the current and the voltage as

$$I = I_{\mathrm{ph}} - I_o \left(e^{\frac{U+IR_s}{AU_T}} - 1 \right) - \frac{U + IR_s}{R_{\mathrm{h}}}. \tag{1}$$

Therein, I and U are the PV output current and voltage. I_{ph} is the photocurrent accounting for the electron–hole pairs that make the current flow through the PV cell, I_o is the dark saturation current, A is the ideality factor, and R_s and R_{h} are the parasitic series and shunt resistances that describe the loss mechanisms occurring within the PV cells. $U_T = N_s k_B T/q$ is the thermal voltage, where N_s is the number of series-connected PV cells in the PV module, k_B is the Boltzmann constant, T is the PV module operating temperature, and q is the electron charge.

2.2 Parameter identification procedures

In this paper, two alternative versions of the identification procedure proposed in [9] are used. The four-parameter version produces the parameter set $\{I_{\mathrm{ph}}, T, R_s, R_{\mathrm{h}}\}$ as its output. The three-parameter version produces the set $\{I_{\mathrm{ph}}, T, R_s\}$ as its output; the shunt resistance is set constant. Otherwise, the used equations are the same in both versions. The dark saturation current is derived directly from Eq. (1) as

$$I_o = \frac{I_{\mathrm{ph}} - \dfrac{U_{\mathrm{OC}}}{R_{\mathrm{h}}}}{e^{\frac{U_{\mathrm{OC}}}{AU_T}} - 1}, \tag{2}$$

where the open-circuit (OC) voltage U_{OC} is calculated as in [15] via

$$U_{\mathrm{OC}} = U_{\mathrm{OC, STC}} + \alpha_U(T - T_{\mathrm{STC}}) + AU_T \ln(G_{\mathrm{eff}}). \tag{3}$$

Therein, $U_{\mathrm{OC, STC}}$ is the OC voltage in standard test conditions (STC), α_U its temperature coefficient, T_{STC} the PV module operating temperature in STC, and the effective irradiance G_{eff} is obtained by

$$G_{\mathrm{eff}} = \frac{I_{\mathrm{SC}}}{I_{\mathrm{SC, STC}} + \alpha_I(T - T_{\mathrm{STC}})}. \tag{4}$$

In Eq. (4), $I_{\mathrm{SC, STC}}$ is the short-circuit (SC) current in STC and α_I is its temperature coefficient. As a slight modification to the method of [9], the SC current I_{SC} is calculated via

$$I_{\mathrm{SC}} = \frac{I_{\mathrm{ph}}}{1 + \dfrac{R_s}{R_{\mathrm{h}}}}. \tag{5}$$

The actual fitting is performed with the fit.m algorithm of Matlab. In order to speed up the calculation, the iterative procedure of [9] has been replaced by explicit equations providing I as function of U as in

$$I = \frac{R_{\mathrm{h}}(I_{\mathrm{ph}} + I_o) - U}{R_{\mathrm{h}} + R_s} - \frac{AU_T}{R_s} W(\theta_I) \tag{6}$$

with

$$\theta_I = \frac{R_s R_{\mathrm{h}} I_o e^{\frac{R_s R_{\mathrm{h}}(I_{\mathrm{ph}} + I_o) + R_{\mathrm{h}} U}{AU_T(R_s + R_{\mathrm{h}})}}}{AU_T(R_s + R_{\mathrm{h}})}, \tag{7}$$

where $W(\cdot)$ denotes the Lambert W function [16].

It is remarkable that the identification procedure produces the PV module operating temperature as one of the direct output parameters. In turn, the irradiance received by the PV module is calculated after the actual fitting via

$$G = \frac{G_{\mathrm{STC}} I_{\mathrm{SC}}}{I_{\mathrm{SC, STC}} + \alpha_I(T - T_{\mathrm{STC}})}. \tag{8}$$

Throughout the most part of the present study, the value of ideality factor is varied. However, its STC value A_{STC} is required for investigating the dependence of series resistance on shunt resistance so that the effect of ideality factor is excluded. A_{STC} is calculated as in [16] via

$$A_{\mathrm{STC}} = \frac{\alpha_U - \dfrac{U_{\mathrm{OC, STC}}}{T_{\mathrm{STC}}}}{U_{T, \mathrm{STC}}\left(\dfrac{\alpha_I}{I_{\mathrm{ph, STC}}} - \dfrac{3}{T_{\mathrm{STC}}} - \dfrac{E_{\mathrm{g, STC}}}{k_B T_{\mathrm{STC}}^2}\right)} \tag{9}$$

The initial guesses for the identified parameters are given as follows. For the first I–U curve in the dataset, the initial guess is $\{I_{\mathrm{MPP}}/\gamma_{I, \mathrm{STC}}, T_{\mathrm{STC}}, R_{s, \mathrm{STC}}, R_{\mathrm{h, STC}}\}$ for the four-parameter version of the identification procedure; the three-parameter version utilizes similar initial guess with the shunt resistance excluded. Therein, I_{MPP} is the current at the maximum power point (MPP) retrieved from the measured I–U curve, and $\gamma_{I, \mathrm{STC}} = I_{\mathrm{MPP, STC}}/I_{\mathrm{SC, STC}}$, where $I_{\mathrm{MPP, STC}}$ and $I_{\mathrm{SC, STC}}$ are the MPP and SC currents in STC. The parasitic resistances $R_{s, \mathrm{STC}}$ and $R_{\mathrm{h, STC}}$ in STC are obtained by the procedure of [18] which takes as its inputs the set $\{I_{\mathrm{SC, STC}}, I_{\mathrm{MPP, STC}}, U_{\mathrm{OC, STC}}, U_{\mathrm{MPP, STC}}, A\}$, where $U_{\mathrm{MPP, STC}}$ is the MPP voltage in STC. For this particular purpose of the calculation of the initial guess for the parasitic resistances for the first I–U curve, a constant value of $A = 1.1$ for the ideality factor has been used in accord with the original choices made in [9, 19]. For the remainder of the I–U curves, the initial guess for I_{ph} is again provided by $I_{\mathrm{MPP}}/\gamma_I$, where the other parameters take the identified parameters of the previous I–U curve in the dataset as their initial guesses. The upper and lower limits for the identified parameters are reported in Table I.

Table I: Upper and lower limits for the identified parameters

Parameter	Lower limit	Upper limit
I_{ph}	0 A	11 A
T	273.15 K	343.15 K
R_s	0.1 Ω	5 Ω
R_{h}	50 Ω	1000 Ω

2.3 Measurement data

The measurement data used in this study was gathered from an individual PV module from the solar PV power research plant located on the rooftop of a campus building of Tampere University, Tampere, Finland [17]. The PV power plant consists of 69 PV modules (NAPS NP190GK) fabricated from multi-crystalline silicon. The operating temperature of the PV module and the irradiance received by it were registered with a Pt100 thermocouple and an

SPLite2 photodiode, respectively. The I–U curves were measured by using an I–U curve tracer utilizing IGBTs as an electronic load with the sampling frequency of 1 Hz.

Every measured I–U curve consists of 4000 measurement points that are evenly spaced in terms of time Such an operational principle makes the measurement points spread unevenly along the resulting I–U curve. This renders the need to preprocess the measured I–U curves before the actual fitting. This has been done as follows. Firstly, the clearly abnormal measurement points have been removed by using a statistical interquartile range method as in [20]. Secondly, the remaining measurement points have been redistributed by replacing the points with redundant voltage values by averaging their currents as in [21]. The latter procedure mitigates the uneven weighting of the measurement points.

It was observed in [9] that the electrical characteristics of the investigated PV module in STC slightly differ from their values reported in the manufacturer's datasheet. Indeed, the PV panel suffers from mild ageing. Hence, the values have been redetermined for 2023, and they are reported in Table II jointly with the other relevant characteristic of the PV module.

Table II: Electrical characteristics of the PV module in STC

Parameter	Value
$I_{SC, STC}$	8.86 A
$U_{OC, STC}$	32.8 V
$I_{MPP, STC}$	8.06 A
$U_{MPP, STC}$	22.5 V
N_s	54
α_I	0.0047 A/K
α_U	-0.124 V/K
$R_{s, STC}$	0.8162 Ω
$R_{h, STC}$	741.6 Ω
A_{STC}	1.0654

In the present paper, a set of 100 I–U curves measured under high irradiance conditions (at least 800 W/m^2) on 24 August 2023 between 11:27:44 and 11:36:59 (UTC+2) was investigated. The set of curves consisted of four blocks of consecutive curves; the medium- and low irradiance curves lying between these blocks were removed. However, one of the remaining 100 high-irradiance curves was subject to minor deformation due to slight partial shading, and it was hence discarded. Fig. 1 shows the measured operating irradiance and temperature of the PV module as functions of the number of the measured I–U curve. The vertical lines indicate the points where the measured I–U curves are not successive.

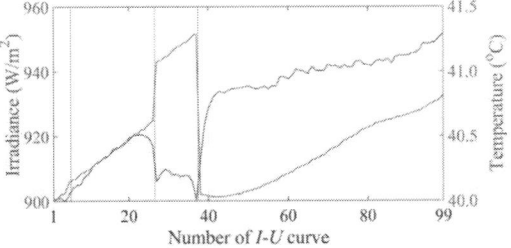

Figure 1: Measured irradiance and temperature of the investigated PV module as functions of the number of the measured I–U curve.

3 RESULTS AND DISCUSSION

The main purpose of the study is to investigate the dependence of R_s and R_h on A and each other. The ideality factor was varied within the range of 1.0–1.5 with steps of 0.01, and its effect on the parasitic resistances was investigated.

Fig. 2 shows the identified series resistances of the 99 I–U curves jointly with their mean values as a function of ideality factor for the four- and three-parameter versions of the identification procedure. It is observed that series resistance is a linear and decreasing function of the ideality factor. In particular, the mean value of the identified series resistances of consecutive I–U curves for a fixed constant ideality factor decreased with increasing ideality factor. These findings are in accord with the results presented in [13]. The dependence of the series resistance on the choice of the ideality factor is something that must be taken into account when evaluating the series resistance in practical applications.

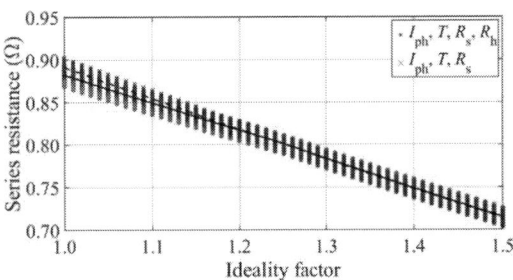

Figure 2: Identified series resistance (red and blue markers) jointly with their averages (solid and dashed black lines) as a function of the ideality factor for 99 high-irradiance I–U curves by using the four- and three-parameter versions of the identification procedure.

In the light of Fig. 2, the identified series resistances for any fixed selected ideality factor seemed to exhibit rather similar deviations. This result was confirmed by Fig. 3 representing the standard deviation of the identified series resistance values of the 99 I–U curves as a function of the ideality factor. It was observed that when the ideality factor was 1.27 or lower, the four-parameter version of the identification procedure produced higher standard deviation in R_s than the three-parameter version. The largest percentual difference, when compared to the three-parameter version, was slightly less than 10% and occurred at the ideality factor value around 1.2. However, the difference between the standard deviations of the series resistances produced by the four- and three-parameter versions started to decrease when the ideality factor was further increased. From the value $A = 1.28$, the standard deviations practically coincided. Most importantly, the standard deviation of R_s was for both versions of the identification procedure a nonlinear and nonmonotonic function of the ideality factor. These findings indicate that the stability of identified series resistance values is dependent on the choice of the ideality factor as well as the used identification procedure. However, the differences were small.

41st European Photovoltaic Solar Energy Conference and Exhibition

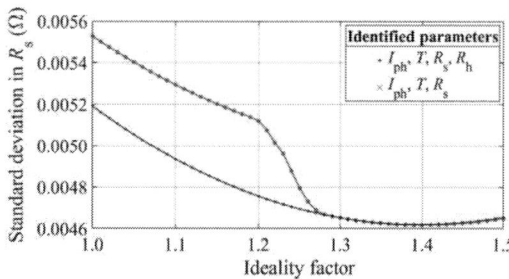

Figure 3: Standard deviation of the identified series resistance as a function of the ideality factor for 99 high-irradiance I–U curves by using the four- and three-parameter versions (red and blue markers, respectively) of the identification procedure.

Fig. 4 illustrates the behavior of the identified shunt resistance values jointly with their mean obtained by using the four-parameter identification procedure as a function of the ideality factor. The shunt resistance values increased in a nonlinear fashion with the increasing ideality factor. Deviation in identified shunt resistance was the smallest when the ideality factor was the lowest. The deviation in R_h gradually increased with increasing ideality factor. When the ideality factor was 1.2 or higher, some of the identified R_h values reached their upper bound. From the value $A = 1.28$ onwards, all the shunt resistance values were at the upper bound. This explains the coinciding standard deviations of R_s in Fig. 3 for high values of ideality factor: the three-parameter and four-parameter begin to resemble each other as both have constant high R_h values. Indeed, the $R_{h,STC}$ value reported in Table II is high as well. However, it should be noted that the identified parameter values depend on the used parameter identification procedure.

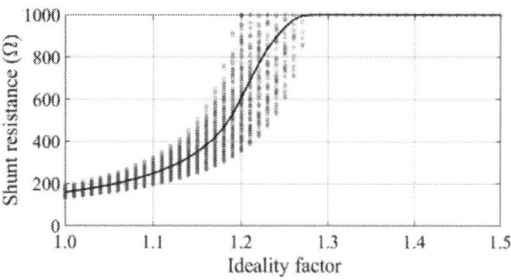

Figure 4: Identified shunt resistance (red circles) jointly with its average (black solid curve) as a function of the ideality factor for 99 high-irradiance I–U curves by using the four-parameter version of the identification procedure.

In Fig. 5, the identified shunt resistance is shown as a function of the identified series resistance by applying the four-parameter version of the identification procedure to the 99 I–U curves. It is observed that the lowest series resistance values were produced in the cases where R_h was at the upper bound. When the series resistance values kept on increasing, all the R_h values were not anymore fixed at the upper bound but began to decrease in a nonlinear fashion. For those R_s values that produce the steeply decreasing part of the average R_h curve, the deviation in identified R_h is large; the range of variation is several hundreds of Ω. This is a normal finding since single-diode model is rather insensitive with respect to the shunt

resistance [22]. When the series resistance increased even more, the deviation in R_h decreased, indicating the increasing stability of the identified shunt resistance as a function of the series resistance. Diagnostically, this is an interesting observation since traditionally large deviation in identified R_h values has constituted a limitation for reliable analysis of the shunt resistance [23].

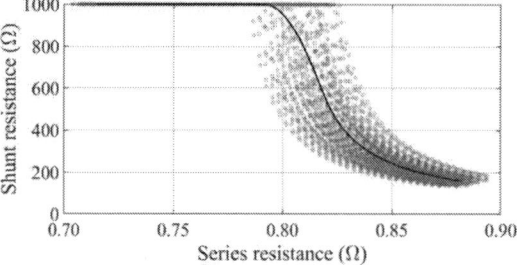

Figure 5: Identified shunt resistance (red circles) jointly with its average (black solid curve) as a function of the identified series resistance for 99 high-irradiance I–U curves by using the four-parameter version of the identification procedure.

Fig. 6 provides the identified series resistance values and their average as a function of the shunt resistance obtained by using the three-parameter version of the identification procedure when the shunt resistance was fixed to certain values. The ideality factor was set to its STC value. It is observed that the identified average series resistance is a nonlinearly increasing function of the shunt resistance; the average R_s values began to stabilize towards a certain level when the shunt resistance was sufficiently high. Such a finding indicates that when ideality factor is kept constant, a wide range of high R_h values produces rather similar patterns of R_s values.

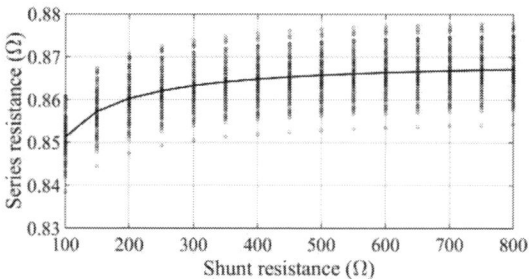

Figure 6: Identified series resistance (blue circles) jointly with its average (black solid curve) as a function of the shunt resistance for 99 high-irradiance I–U curves by using the three-parameter version of the identification procedure and a constant ideality factor.

According to Fig. 6, the deviation of identified R_s values remains approximately similar throughout the variation of shunt resistance. A closer look at this issue is provided by Fig. 7 representing the standard deviation of the identified series resistance as function of the shunt resistance. The standard deviation of R_s was a monotonically increasing nonlinear function of the shunt resistance. In particular, the smallest standard deviation was produced by the lowest shunt resistance values. This finding is in line with [23].

41st European Photovoltaic Solar Energy Conference and Exhibition

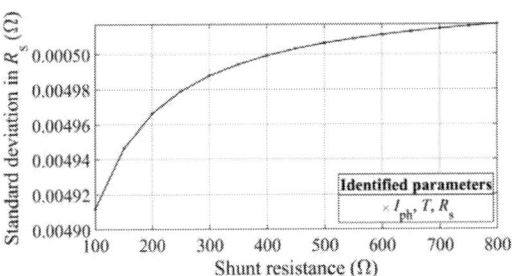

Figure 7: Standard deviation of identified series resistance as a function of the shunt resistance for 99 high-irradiance *I–U* curves by using the three-parameter version of the identification procedure and a constant ideality factor.

When performing diagnostical single-diode model analysis, the information of the operating conditions plays a central role. Fig. 8 shows the percentage difference between the average calculated and measured irradiances and temperatures as a function of ideality factor for the four-parameter and three-parameter versions of the identification procedure. The differences between the calculated and measured values were rather small for both quantities. The irradiance difference was nearly a linear increasing function of the ideality factor for the three-parameter model. In contrast, the corresponding behavior for the four-parameter version was nonlinear and nonmonotonic. The difference between calculated and measured temperature was a monotonically increasing function of the ideality factor for both versions of the identification procedure. The temperature difference was higher for the four-parameter version than for the three-parameter version when the ideality factor was low. However, the temperature differences of the four-parameter and three-parameter versions of the identification procedure were almost the same for higher ideality factor values.

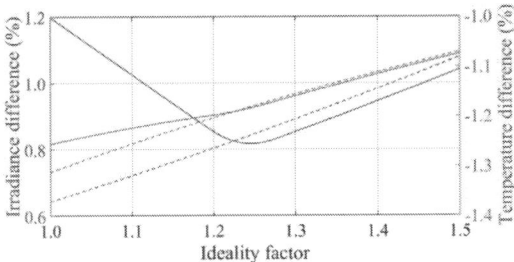

Figure 8: Average percentage differences between calculated and measured irradiance and temperature as a function of the ideality factor for 99 high-irradiance *I–U* curves by using the four-parameter (solid lines) and three-parameter (dashed lines) versions of the identification procedure.

4 CONCLUSIONS

This paper focused on investigating the dependencies of the series and shunt resistances and the ideality factor of the PV cell single-diode model. The model parameters were obtained by using an identification procedure that is suitable for online condition monitoring at practical PV sites due to its capability to reproduce the PV module operating irradiance and temperature jointly with the actual single-diode model parameters. Two versions of the identification procedure were compared, one of which produces the shunt resistance as an identified parameter (the four-parameter version) while the other keeps it fixed at its STC value (the three-parameter version).

Firstly, the effect of variable ideality factor on the identified series resistance values was analyzed. It was observed that series resistance is a linearly decreasing function of the ideality factor for both versions of the used identification procedure. However, the series resistance values obtained with the three-parameter version exhibited slightly better stability than those obtained with the four-parameter version when expressed as a function of the ideality factor.

Secondly, the effects related to the shunt resistance identified by four-parameter version via the variation of ideality factor were investigated. It was observed that low ideality factor values produce lower and stabler shunt resistance values. In addition, it was shown that these shunt resistance values were tied to rather high series resistance values.

Thirdly, the behavior of the identified series resistance as a function of variable shunt resistance using the three-parameter version of the identification procedure was studied. Then, the ideality factor was fixed at its STC value. It was shown that lower shunt resistance values corresponded to lower series resistance values. However, the average series resistance values began to stabilize asymptotically towards an upper limit when the shunt resistance was increased.

Finally, the effects of variable ideality factor on the capability of the identification procedure to reproduce the PV module operating conditions were analyzed. It was shown that the differences between the average calculated and measured irradiances and temperatures were very small, but that such differences depend on the value of the ideality factor as well as the used version of the identification procedure.

In conclusion, the variation of the ideality factor has obvious effects on the other parameters obtained in the single-diode model fitting. This fact should be considered when performing the diagnosis based on single-diode model fitting to measured current–voltage curves.

ACKNOWLEDGMENTS

The authors acknowledge the financial support from KAUTE Foundation, Business Finland (grant number 1191/31/2022), the Research Council of Finland (funding decision 348701) and Otto A. Malm Foundation for the research reported in this paper.

REFERENCES

[1] E. Karatas, R. Gottschalg, IEEE Journal of Photovoltaics 13(6) (2023) 945.

[2] J. Kim, M. Rabelo, S. P. Padi, H. Yousuf, E. Cho, J. Yi, Energies 14(14) (2021) 4278.

[3] P. Manganiello, M. Balato, M. Vitelli, IEEE Transactions on Industrial Electronics 62(11) (2015) 7276.

[4] S. Gallardo–Saavedra, L. Hernándes–Gallejo, M. del Carmen Alonso–García, J. D. Santos, J. I. Morales–Aragonés, V. Alonso–Gómez, Ã. Moretón–

Fernández, M. Ã. Gonzáles–Rebollo, O. Martínez–Sacristán, Energy 205 (2020) 117930.

[5] D. Sera, R. Teodorescu, P. Rodriguez, 34th Annual Conference of IEEE Industrial Electronics, Orlando, FL, USA (2008) 2195.

[6] H. Ibrahim, N. Anani, Energy Procedia 134 (2017) 276.

[7] J. D. Bastidas–Rodriguez, E. Franco, G. Petrone, C. A. Ramos-Paja, G. Spagnuolo, Mathematics and Computers in Simulation 131 (2017) 101.

[8] K. Lappalainen, P. Manganiello, M. Piliougine, G. Spagnuolo, S. Valkealahti, IEEE Journal of Photovoltaics 10(3) (2020) 852.

[9] H. Kalliojärvi–Viljakainen, K. Lappalainen, S. Valkealahti, Energy Reports 8 (2022) 4633.

[10] K. Lappalainen, M. Piliougine, G. Spagnuolo, Energy Conversion and Management 258 (2022) 115526.

[11] A. Laudani, F. R. Fulginei, A. Salvini, Solar Energy 103 (2014) 316.

[12] S. Wang, C. Wang, Y. Ge, S. Liu, J. Xu, R. A. Amer, Energy 291 (2024) 130345.

[13] A. Zaatri, European Journal of Sustainable Development Research 8(1) (2024) em0244.

[14] G. Petrone, C. A. Ramos–Paja, G. Spagnuolo, John Wiley & Sons (2017).

[15] J. A. Kratochvil, W. E. Boyson, D. L. King, Sandia National Laboratories, Albuquerque, NM, and Livermore, CA, USA (2003).

[16] J. Accarino, G. Petrone, C. A Ramos–Paja, G. Spagnuolo, Proceedings International Conference on Clean Electrical Power (2013) 62.

[17] D. Torres Lobera, A. Mäki, J. Huusari, K. Lappalainen, T. Suntio, S. Valkealahti, International Journal of Photoenergy 2013(1) (2013) 837310.

[18] M. G. Villalva, J. R. Gazoli, E. R. Filho, IEEE Transactions on Power Electronics 24(5) (2009) 1198.

[19] V. Stornelli, M. Muttillo, T. de Rubeis, I. Nardi, Energies 12(22) (2019) 4271.

[20] H. Kalliojärvi–Viljakainen, K. Lappalainen, S. Valkealahti, Proceedings 47th IEEE Photovoltaic Specialists Conference (2020) 0117.

[21] K. Lappalainen, S. Valkealahti, Applied Energy 301 (2021) 117436.

[22] M. R. Rashel, A. Albino, T. Goncalves, M. Tlemcani, Proceedings 10th International Conference on Software, Knowledge, Information Management and Applications (2016) 333.

[23] H. Kalliojärvi, K. Lappalainen, S. Valkealahti, Energies 15(23) (2022) 9079.

41st European Photovoltaic Solar Energy Conference and Exhibition

PREDICTIVE MAINTENANCE AND ANOMALY DETECTION ANALYTICS FOR UTILITY-SCALE PHOTOVOLTAIC PLANTS

Jesus Montes-Romero[1,2], Demetris Marangis[1,3], Andreas Livera[1,3],
George Makrides[1,3], Juergen Sutterlueti[4], Steve Ransome[5] and George E. Georghiou[1,3]
[1] PV Technology Laboratory, FOSS Research Centre for Sustainable Energy,
Department of Electrical and Computer Engineering, University of Cyprus, 1678 Nicosia, Cyprus
[2] Advances in Photovoltaic Technology (AdPVTech), CEACTEMA, University of Jaén, 23071 Jaén, Spain
[3] PHAETHON Centre of Excellence for Intelligent, Efficient and Sustainable Energy Solutions, 2109 Nicosia, Cyprus
[4] Gantner Instruments GmbH, Montafonerstraße 4, 6780 Schruns, Austria
[5] Steve Ransome Consulting Ltd, KT2 6AF #99 Kingston upon Thames, United Kingdom

ABSTRACT: Ensuring optimal lifetime performance and reduced levelized cost of electricity (LCoE) of photovoltaic (PV) systems necessitates the anticipation of underperforming conditions and cost-effective maintenance interventions. Even though, many data-driven preventive maintenance techniques have been proposed, a main challenge remains the lack of a standardised solution for exploiting data stream signals to predict optimum operation and identify possible problems for utility-scale PV power plants. The scope of this work is to address this challenge by presenting a novel trend-based performance loss routine for predicting potential underperforming conditions at utility-scale PV power plants. The proposed predictive maintenance and anomaly detection workflow operates on acquired performance data streams and evaluates performance trend deviations by analysing performance deviations. The results demonstrated that the constructed workflow proved robust in predicting the performance (at both AC and DC sides) since absolute performance ratio (PR) deviations <0.3% were exhibited at different weather type categories. Furthermore, the trend-based evaluation routines were capable to predict potential induced degradation (PID) loss decreases once the difference between the predicted and forecasted daily average voltage slopes deviate >2% at utility-scale. Similarly, the predictive model for monitoring the insulation resistance (R_{iso}) demonstrated effectiveness to forecast low insulation incidences by analysing the daily R_{iso} patterns. Finally, useful information is provided for the replication of predictive maintenance routines that form integrated components of utility-scale PV plant monitoring systems.

Keywords: data analytics, failures, machine learning, performance, photovoltaic

1 INTRODUCTION

Transitioning into a decarbonized era necessitates a profound shift in energy and digital paradigms, facilitated by intelligent and highly adaptable solutions [1]. Solar energy assumes a paramount role in this transformation, given its capacity to curtail carbon emissions swiftly and extensively through innovative energy generation, grid support capabilities and revenue generation. Within this framework, optimizing the performance and lifetime durability of photovoltaic (PV) systems through advanced data-driven predictive operations becomes imperative [2]. Moreover, the early detection and classification of commonly exhibited faults (e.g., open- and short-circuit faults, inverter faults, etc.) and underperforming conditions (e.g., shading, clipping, curtailment, etc.), is an industrial requirement towards improved performance and improved lifetime reliability [1].

Over the past years, many academic institutions and industrial organisations have been particularly active in the development of failure diagnostic algorithms for PV systems [3]–[5]. Recently, the implementation of predictive operation and maintenance (O&M) algorithms for anticipating faults and optimizing field operations at utility-scale PV power plants has attracted a lot of interest [6]–[8]. In this field, the largest share of research efforts focused on data-driven statistical and machine learning (ML) algorithms aiming to accurately diagnose anomalies, derive long-term performance loss trends and generate predictive maintenance [6], [8], [9].

Even though, failure diagnosis and predictive modelling has been extensively investigated at site-specific small-scale environments, the fact that the developed algorithms are yet not fully verified at actual utility-scale plants is the main reason why a standardised method does not yet exist. Moreover, there are many models capable to detect different common failures (e.g., inverter shutdown, soiling, shading, etc.) [10], however, there is yet no standardized methodology to classify each fault based on the exhibited patterns and failure signature conditions. To this end, the challenge of achieving accurately performing predictive maintenance and anomaly detection routines for faults such as potential induced degradation (PID) and insulation fault conditions (which have not been studied before) at utility-scale remains a fundamental industrial need. Also, classifying PV system faults by considering minimum input features is another important factor towards the achievement of cost-effective O&M processes.

The scope of this work is to present an accurate, transferable, and location-independent predictive maintenance routine for forecasting performance losses at utility-scale PV power plants. The proposed routine targets to fill-in the gap of knowledge to accurately predict PID conditions and insulation resistance (R_{iso}) faults at different system scales and locations. The proposed routine further involves an anomaly detection stage applied to univariate datasets for the early diagnosis of anomalous conditions in PV power plants. The application of the proposed architecture to high-resolution data streams enables the effective evaluation of PV systems. Overall, the paper contributes with new knowledge in the following domains:

- Robust predictive maintenance routine for predicting PID and insulation fault conditions that are transferable and applicable at MW-scale.

- Classification of predictive performance based on novel weather type categories that allow fast and comparable limits.
- Benchmarking the predictive performance of data-driven analytics to forecast underperforming conditions for utility-scale plants.
- Evaluating the performance of statistical techniques for the early detection of extreme conditions.

Ultimately, useful information is provided on the workflow's robustness when applied to different site locations and topologies. The outcomes of this study are relevant to a large stakeholder target group ranging from policy makers and utilities, plant operators, engineering procurement construction contractors and investors, that are ushering in a new era of digitally enhanced and effective data-driven monitoring.

2 EXPERIMENTAL APPARATUS

The performance of the proposed workflow was validated using 1-year historical data streams from a test-bench PV system installed at the University of Cyprus in Nicosia, Cyprus. The test-bench PV system of 1 kW$_p$ nominal capacity, comprises of 5 crystalline silicon PV modules. The PV modules are connected in series to form a string at the input of a grid-connected smart inverter. They are installed in an open field arrangement at a tilt angle of 30°.

The actual demonstration of the workflow was verified at data streams (inverter level) from a utility-scale PV power plant (>1MW$_p$) administered by Gantner Instruments (over a 1-year period).

At both sites the main PV operational parameters along with the prevailing meteorological conditions are recorded following the requirements set by the IEC 61724-1 [11]. The meteorological measurements include the in-plane global irradiance (G_I), wind direction (W_a), wind speed (W_s) and ambient temperature (T_{amb}). The PV system's operational measurements include the module temperature (T_{mod}), array DC current (I_{dc}), voltage (V_{dc}) and power (P_{DC}) and AC power (P_{AC}). All the field measurements are acquired from a high-performance edge controller (Gantner Instruments Q.station XT) at 1-second resolution and stored at high capacity, distributed streaming platforms.

3 METHODOLOGY

The approach followed to implement the proposed workflow comprised of a sequence of steps (see Fig. 1) that involved the: (a) naming convention and normalisation, (b) data sanity and enrichment, (c) development of digital twin (DT) models and statistics, (d) universal fault signatures, (e) artificial intelligence (AI)-driven fault diagnosis and predictive analytics and (f) performance evaluation and demonstration. The detailed workflow methodology is provided in [1].

3.1 Naming convention and normalisation

Initially, a post-processing step was conducted on all parameters of the acquired data streams to provide unique naming for all parameters. A common convention for all system PV power plant topologies was followed. Subsequently, all PV system electrical parameters were normalized (e.g., nP_{DC} stands for the normalized array DC

power) based on nominal and rated values. This step ensures model transferability regardless of location and system topology.

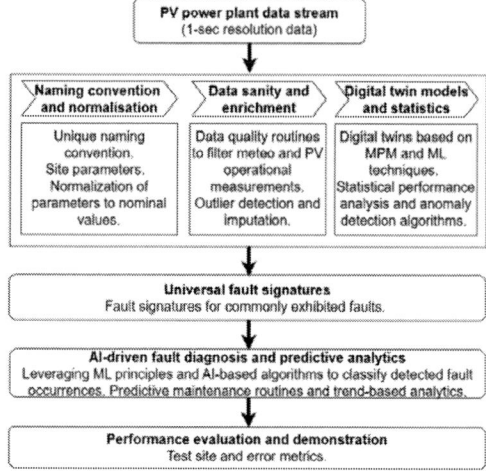

Figure 1: Flowchart of the steps for the proposed data-driven predictive maintenance and anomaly detection workflow for PV systems.

3.2 Data sanity and enrichment

Filtering routines were applied to all data streams based on irradiance and temperature levels. The resulting normalized and filtered data against set irradiance, temperature, angle of incidence (AOI) and clearness index (k_T) conditions, form the basis for the development of the data-driven models. Moreover, outlier detection and data imputation techniques were applied to the available irradiance data streams to detect and treat outlying data points [12]. High-quality datasets were thus constructed.

3.3 Digital twin models and statistics

This step involved the development of DT models using both the Mechanistic Performance Model (MPM) [1] the eXtreme Gradient Boosting (XGBoost) [1].

The MPM is a mechanistic model, applied to the normalized performance data, that yields meaningful coefficients for the measured weather parameters of G_I, T_{mod} and W_s [8]. The MPM was developed using the best features of existing models and it had 5 coefficients C_1 to C_5 defined by (1):

$$PR_{DC} = C_1 + C_2 T_{mod} + C_3 \text{Log}_{10}(G_I) + C_4 G_I + C_5 W_s \quad (1)$$

where PR_{DC} is the performance ratio at the DC side. The same equation can be used to predict the performance ratio at the AC side (PR_{AC}). The MPM was fitted using a month of filtered data to extract the 5 coefficients and it was then used to predict the PV performance over the yearly evaluation period.

The XGBoost is an ensemble algorithm that combines several decision trees using the boosting method to generate the desired output prediction [1]. The core of XGBoost is to optimize the objective function's value by using gradient descent to create new trees based on the residual errors of previously trees [1]. Once the decision trees were trained, ensemble modelling by weighted averaging was performed to pool the results from multiple trees and averaged using weights based on accuracy to minimize the error [1].

The XGBoost model was trained using a supervised regime (random 70:30% train and test set approach).

Statistical performance analysis (i.e., residual analysis, hypothesis testing) was then performed to improve the performance of the devised DT models and to detect anomalous conditions. The utilisation of density plots and the application of the Seasonal Hybrid Extreme Studentized Deviates (S-H-ESD) [1] algorithm enabled the detection of anomalies (i.e., data points that deviate significantly from the expected or normal behaviour). In parallel, fault alarms were also generated in case of repetitive daily errors of >2% between the predicted and actual power values.

The above statistical procedure formed the first fault detection step of the workflow.

3.4 Universal fault signatures

To gain insights into the fault conditions and diagnose the root-cause, the daily profiles (with detected faults) obtained from PV system data streams were compared against labelled daily profiles of commonly exhibited faults (shading, clipping, open- and short-circuit module faults, inverter faults and PID). The labelled fault profiles were constructed by emulating fault conditions/patterns over a daily period. The emulated failure signature profiles utilized normalized parameters to compile universal fault signature timeseries, that are suitable for training supervised fault classifiers. Finally, each failure signature was associated with a specific label to assist the fault classification process. More details about the fault emulation procedure can be found in [1].

3.5 AI-driven fault diagnosis and predictive analytics

This stage involved the utilisation of ML principles and AI-based algorithms for fault detection and classification. Different AI-based classifiers were developed and trained using the labelled daily fault profiles (i.e., the universal fault signature timeseries).

Apart from fault classification, this step also involved the development of predictive analytics. More specifically, trend-based performance loss routines were developed for predicting potential underperforming conditions in PV systems.

In this work, focus is shed on predicting PID conditions and insulation resistance faults at different system scales and locations. The predictive maintenance algorithm for PID and R_{iso} conditions included an initial daily average voltage and R_{iso} calculation step and a systematic trend analysis of the weekly slope deviations. PID conditions are detected once the difference between the predicted and actual daily average voltage slopes deviates >2%. For the prediction of low R_{iso} conditions, an autoregressive model is applied to the daily resistance measurements to forecast the duration where the limiting value of either 700 kΩ required by DIN VDE 0126-1-1 or 100 kΩ set by most inverter manufacturers, is reached [13].

3.6 Performance evaluation

The performance of the constructed DT models was evaluated using the normalized root mean squared error (nRMSE), calculated by (2):

$$\text{nRMSE} = \frac{1}{P_{nominal}} \sqrt{\frac{\sum_{1}^{n}(y_{predicted,i} - y_{measured,i})^2}{n}} \quad (2)$$

where n is the amount of data points, $y_{measured,i}$, $y_{predicted,i}$ is the measured and predicted power, respectively and $P_{nominal}$ is the nominal capacity of the investigated PV system.

4 RESULTS

In this section, results on the digital twin's performance, effectiveness of statistical performance analysis for outliers' detection and prediction of PID conditions and insulation resistance faults are be presented. A full analysis of different underperforming conditions along with fault and predictive analytic results can be found in [1].

4.1 Digital twin models' predictions

The developed DT models were trained and tested using historical field measurements (over a 1-year period) from the investigated PV systems. The MPM was fitted using only a 1-month of data, while the XGBoost model was trained using a 70% portion of the entire dataset. The results obtained when applying the MPM and XGBoost models to the test-bench PV system, demonstrated that both DT models were capable to follow the daily exhibited DC power production profiles, as shown in Fig. 2a. The performance of the DT models remained consistent regardless of the PV system size, as high accuracies were also achieved when applied to the inverter level data of the PV power plant (see Fig. 2b). More specifically, the MPM yielded power predictive errors (as given by the average RMSE normalized to the nominal capacity of the system) <2% over the yearly evaluation period. Similar results (error <2%) were also obtained for the XGBoost DT model. The performance of the XGBoost model was finally validated on low duration data streams. When trained with a month of data, the ML model yielded a nRMSE <4%.

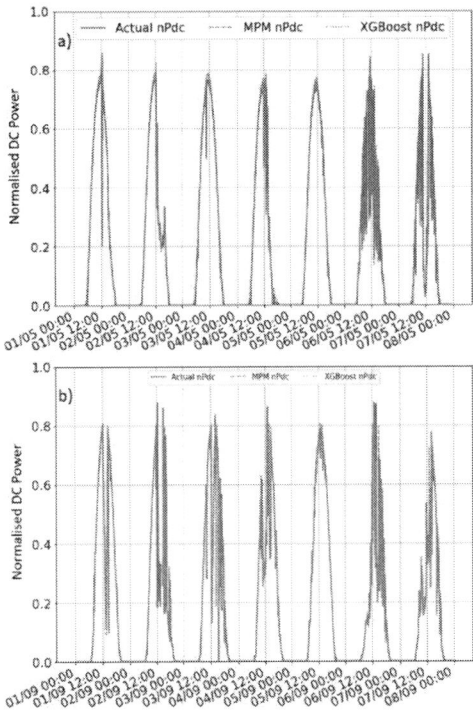

Figure 2: Plots of actual and predicted nP_{DC} by the MPM and XGBoost model for the: (a) small-scale test-bench PV system of 1 kW$_p$ in Cyprus and (b) utility-scale PV power plant of 3.5 MW$_p$ in Germany (inverter level data).

It worth noting here that the models analyse the data in real-time and require only the measured irradiance and ambient temperature data.

Moreover, the performance of the devised MPM was validated at varying irradiance conditions and using the inverter level data (see Fig. 3). Four different weather types (i.e., clear morning, clear noon, clear evening, diffuse sky and other) [14] were thus used to analyse the MPM performance. It is obvious from Fig. 3 that low PR_{DC} deviations of <0.3% were exhibited between the MPM predictions and actual measurements for the utility-scale PV power plant over a month evaluation period. The performance of the model was thus not affected by irradiance conditions since high accuracies were obtained for both clear and overcast days, signifying the robustness of the DT model to varying weather conditions.

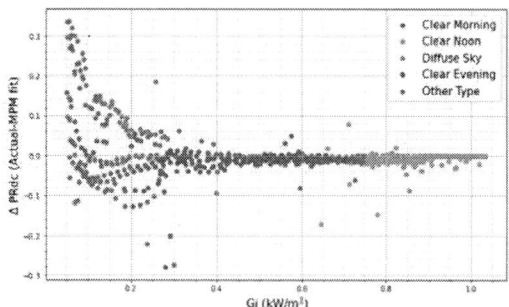

Figure 3: Plot of PR_{DC} deviations at different weather type categories over a monthly period for the utility-scale PV power plant of 3.5 MW$_p$ in Germany (inverter level data).

The DT models were finally used to predict the AC power production profiles over the yearly period. The heatmap of the daily PR_{AC} deviations between the MPM predictions and actual measurements for the utility-scale PV plant is depicted in Fig. 4. The results underscore the model's resilience to diverse weather conditions, as evidenced by the consistently low hourly absolute PR_{AC} deviations (<0.2), throughout the diurnal cycle.

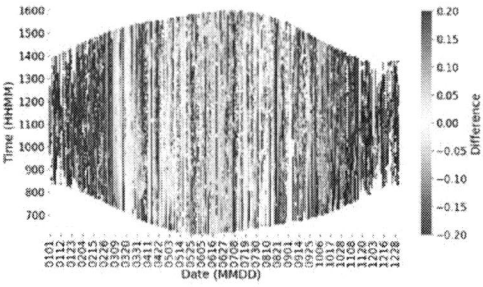

Figure 4: Average hourly PR_{AC} deviations heatmap between MPM predictions and actual measurements for the utility-scale PV power plant over a yearly period.

4.2 Statistical performance analysis

The application of hypothesis testing statistical techniques (density plots and two-Sample T-test) to the PR_{DC} inverter time series of the utility-scale PV plant proved capable to detect anomalies in the dataset that did not follow the normal distribution (see Fig. 5). Additionally, the data points detected as falling outside the normal distribution fit of the predictive MPM were filtered for further investigation. This method provides a first level

verification stage for detected fault conditions. Outliers were also statistically detected by filtering the incidents falling below the 95% lower bound confidence interval.

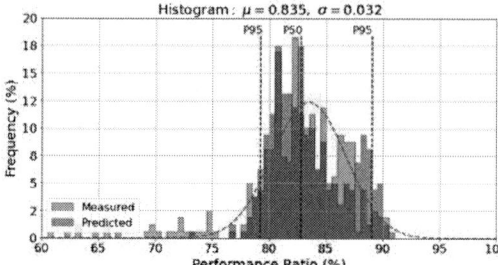

Figure 5: Histogram of exhibited and predicted PR_{DC} (MPM fit) of the utility-scale PV power plant and distribution overlay for a winter month.

4.3 Predictive analytics

As regards the diagnosis process for PID conditions, the predictive maintenance routine proved capable in predicting the imputed PID conditions by systematically analysing the weekly deviations of the daily average voltage slopes. Fig. 6 presents the plot of the daily voltages and imputed PID profiles. PID conditions are detected once the difference between the predicted and actual daily average voltage slope deviates >2%.

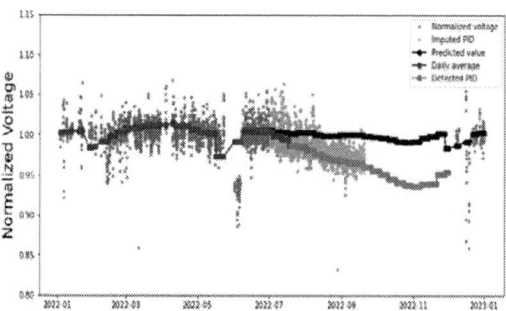

Figure 6: Utility-scale PV power plant plot of normalized voltage (blue dots) and imputed PID (orange dots) [1].

Similarly, the boxplot of Fig. 7 shows the actual daily R_{iso} captured pattern and the applied trend used to forecast the degradation of the imputed R_{iso} deterioration. Specifically, the trend analysis of the daily insulation level conditions demonstrated that the model effectively forecasted the duration for the insulation to degrade to the set limiting value of 100 kΩ.

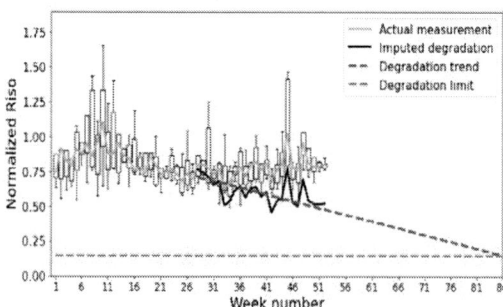

Figure 7: Boxplot of normalised R_{iso} (values were normalized to 700 kΩ) and trend forecast for the utility-scale PV power plant.

4 CONCLUSIONS

Accurate and reliable failure diagnosis and prognosis of PV power plants is amongst the main industrial challenges to uphold improved lifetime performance and reliability. This work presented the method followed to develop a predictive maintenance and an anomaly detection workflow for PV systems. The proposed workflow operates on acquired PV performance data streams and evaluates performance trend deviations by analysing performance deviations.

The developed routines were benchmarked at a multitude of scale PV systems and across different locations. The results demonstrated that the robustness and reliability of the workflow for predicting the PV performance, PID loss conditions (when the forecasted daily average voltage slope deviated >2%) and insulation resistance faults in PV systems.

Finally, useful information is provided for the replication of predictive maintenance routines that form integrated components of utility-scale PV plant monitoring systems. Integrating such advanced fault and predictive analytics into monitoring systems can help optimise field O&M activities (e.g., cleaning activities, replacement of faulty equipment, etc.).

6 REFERENCES

[1] J. Romero *et al.*, "Novel data-driven health-state architecture for photovoltaic system failure diagnosis," *Sol. Energy*, vol. 279, no. July, p. 112820, 2024, doi: 10.1016/j.solener.2024.112820.

[2] A. Livera, G. Paphitis, M. Theristis, J. Lopez-Lorente, G. Makrides, and E. George, "Photovoltaic system health-state architecture for data-driven failure detection," *Solar*, vol. 2, no. 1, pp. 81–89, 2022, doi: https://doi.org/10.3390/solar2010006.

[3] N. V. Sridharan, S. Vaithiyanathan, and M. Aghaei, "Voting based ensemble for detecting visual faults in photovoltaic modules using AlexNet features," *Energy Reports*, vol. 11, no. March, pp. 3889–3901, 2024, doi: 10.1016/j.egyr.2024.03.044.

[4] B. Ren *et al.*, "Machine learning applications in health monitoring of renewable energy systems," *Renew. Sustain. Energy Rev.*, vol. 189, p. 114039, 2024, doi: 10.1016/j.rser.2023.114039.

[5] S. Idrissi Kaitouni *et al.*, "Implementing a Digital Twin-based fault detection and diagnosis approach for optimal operation and maintenance of urban distributed solar photovoltaics," *Renew. Energy Focus*, vol. 48, no. November 2023, p. 100530, 2024, doi: 10.1016/j.ref.2023.100530.

[6] I. A. Zulfauzi, N. Y. Dahlan, H. Sintuya, and W. Setthapun, "Anomaly detection using K-Means and long-short term memory for predictive maintenance of large-scale solar (LSS) photovoltaic plant," *Energy Reports*, vol. 9, pp. 154–158, 2023, doi: 10.1016/j.egyr.2023.09.159.

[7] A. Livera, D. Marangis, G. Tziolis, V. Paraskeva, G. Makrides, and G. E. Georghiou, "Predictive analytics for maximizing the photovoltaic system performance," in *52nd IEEE Photovoltaic Specialist Conference (PVSC)*, 2024, pp. 1–6.

[8] M. De Benedetti, F. Leonardi, F. Messina, C. Santoro, and A. Vasilakos, "Anomaly detection and predictive maintenance for photovoltaic systems," *Neurocomputing*, vol. 310. pp. 59–68, 2018, doi: 10.1016/j.neucom.2018.05.017.

[9] J. Ramirez-Vergara, L. B. Bosman, E. Wollega, and W. D. Leon-Salas, "Review of forecasting methods to support photovoltaic predictive maintenance," *Clean. Eng. Technol.*, vol. 8, no. January 2021, p. 100460, 2022, doi: 10.1016/j.clet.2022.100460.

[10] A. Livera, M. Theristis, G. Makrides, and G. E. Georghiou, "Recent advances in failure diagnosis techniques based on performance data analysis for grid-connected photovoltaic systems," *Renew. Energy*, vol. 133, pp. 126–143, 2019, doi: 10.1016/j.renene.2018.09.101.

[11] IEC, "IEC 61724-1:2021: Photovoltaic system performance - Part 1: Monitoring," 2021.

[12] A. Livera *et al.*, "Data processing and quality verification for improved photovoltaic performance and reliability analytics," *Prog. Photovoltaics Res. Appl.*, vol. 29, pp. 143– 158, 2021, doi: 10.1002/pip.3349.

[13] DIN, "DIN VDE V 0126-1-1: Automatic disconnection device between a generator and the public low-voltage grid," 2013.

[14] J. Sutterlueti, S. Ransome, R. Kravets, and L. Schreier, "Characterising PV modules under outdoor conditions: what's most important for energy yield," in *26th European Photovoltaic Solar Energy Conference and Exhibition*, 2011, pp. 3608–3614.

41st European Photovoltaic Solar Energy Conference and Exhibition

SAFETY ANALYSIS OF PV SYSTEMS FOR SOUNDPROOF TUNNEL
BASED ON VOLTAGE AND CURRENT MISMATCH

Juhee Jang, Chongmin Kim, Sujeong Oh
Electrical Safety Research Institute Korea Electrical Safety Corportation
111, Anjeon-ro, Iseo-nyeon, Wanju-gun, Jeollabuk-do, 55365, Korea

ABSTRACT: Nowadays, the installation of photovoltaic(PV) systems integrated into structure is increased. Traditional PV systems are installed with an optimal tilt and azimuth angle. However, suitable PV systems are difficult to install with an optimal installation angle due to shape and aesthetics of building. Sometimes, theses are configured with different angles in a series as like tunnel PV systems. Various angles of PV systems are caused by current and voltage mismatch conditions. Due to mismatch of systems, by-pass diode fault can occur and it can be developed into fire accidents. Therefore, the safety issues depending on current and voltage mismatch have to consider for the PV systems which are consisted of difference angle in a string. In this paper, the voltage and current mismatch of tunnel PV system are confirmed with monitoring data. The mismatch of module in the string of tunnel PV systems are analyzed using optimizers input and output data.

Keywords: PV systems, mismatch, optimizer

1 INTRODUCTION

As growth of PV systems market trend, various types of PV systems such as Building Integrated PV(BIPV) systems and Road Integrated PV(RIPV) are increased.[1]-[2] The traditional PV systems are usually installed with same angle. The power mismatch of traditional system is confirmed 5-10%.[3] On the other hands, integrated PV systems are difficult to install with an optimal angle, sometimes these types of PV systems have to install with multiple angles[4] These are reduced the generated energy and caused of power mismatch in a string. A Module level power electronic (MLPE) is largely divided into optimizer and microinverter. The major advantage of MLPE is maximized the power through regulating the output of a PV module. Therefore, to reduce the losses energy depending on the mismatch, installation of the MLPE is increased.[5]

To mitigate the losses energy of PV systems have been analyzed using MLPE. However, current and voltage mismatch between modules are not confirmed. Mismatch in a string can be accelerated degradation of system and leading by-pass diode fault. Also, it is developed hot spot of PV systems. Therefore, output of MLPE have to analyze not only point of reducing the power but also the voltage and current mismatch between module in a string. In this study, the output power is compared with between two PV system which are configured with MLPE or not. Also, PV module power mismatch is calculated based on monitoring data.

2 METHODOLOGY

2.1 Test bed set up

Figure 1 is a soundproof tunnel PV system which is analyzed. These are represented arch shape and consisted of two systems (System 1, System 2). Depending on arch shape, these systems have voltage and current mismatch between modules in a string. Table I is detail installation conditions of System 1 and System 2. These are divided depending on installation of optimizer. Also, each system is configured with 3 strings which are consisted of different installation angles. Figure 2 presents module configuration of each string. The orientation of string 2 is

320°(North-west direction) and string 2 is 140°(South-East direction). The azimuth angles of string 3 are composed 320° and 140° to check the highest mismatch conditions in a string.

Figure 1 : (Test bed) Tunnel PV system

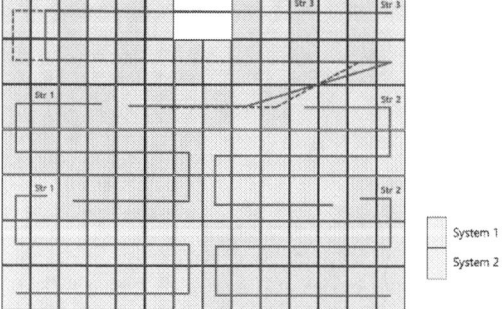

Figure 2 Module configuration of each string

Table I : Installation conditions of test bed

	System 1	System 2
System power	15.84kWp	
Configuration	3 parallel × 16 series	
Tilt angle	0°~5°	
Azimuth angle	140°(South East), 320°(North West)	
Power optimizer	Voltage control	-

2.2 Analysis method

In this paper, the module voltage mismatch (V_{mis}) and module current mismatch (I_{mis}) are analyzed using mismatch rate. These are calculated by bellowing equation (1), (2). Each value is acquired by optimizers.

(1) $$I_{mis} = 2 \times \frac{I_{max} - I_{min}}{I_{max} + I_{min}} \times 100\%$$

(2) $$V_{mis} = 2 \times \frac{V_{max} - V_{min}}{V_{max} + V_{min}} \times 100\%$$

I_{min} : minimum current in a string
I_{max} : maximum current in a string
V_{min} : minimum voltage in a string
V_{max} : maximum voltage in a string

The available data used for analysis mismatch. The monitoring period on this system is listed below.

Monitoring period: 2024.05.20. ~2024.07.31.

3 RESULTS

In this paper, monitoring data analysis is divided into two. First, to confirm the power compensated based on optimizer, daily energy is compared between each system in the testbed. Also, voltage and current mismatch between each module is compared.

3.1 Energy comparison between systems

Table II is total energy yield of each system. Total energy is cumulated energy of each PV system during monitoring periods. As Table II, generated energy of system1 which is configured with optimizer is confirmed 3.70% higher than system2.

Table II : Total energy yield of each system

	System 1	System 2	Deviation ratio
Energy yield [kWh/kWp]	4.32	4.16	3.70%

Figure 3 is daily energy yield comparison result between each system. As Figure 3, energy of system1 is confirm higher than system 2 when the system generated high energy. Energy yield of system 2 is analyzed higher than system 1 reversely.

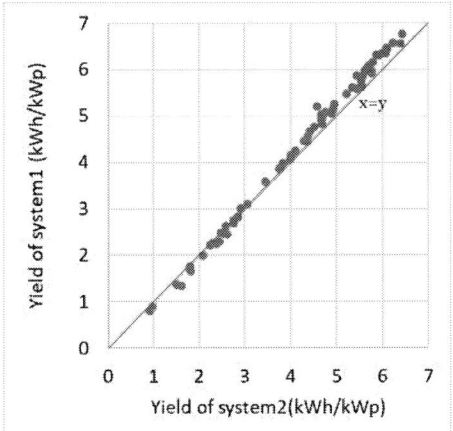

Figure 3 : Comparison between module yield

3.2 Mismatch rate of module in string

Module voltage mismatch rate is confirmed in Table III and Figure 4.

Table III : Module voltage mismatch rate

	String 1	String 2	String 3
Min	2.17%	2.46%	2.50%
Max	10.54%	25.09%	48.24%
Average	5.28%	6.57%	8.37%
Median	5.30%	6.31%	5.84%

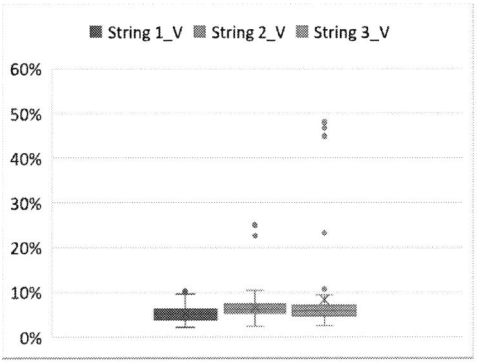

Figure 4 Module voltage mismatch rate

Also, Module current mismatch rate is confirmed in Table IV and Figure 5.

Table IV : Module current mismatch rate

	String 1	String 2	String 3
Min	5.30%	4.73%	7.57%
Max	63.00%	59.52%	52.91%
Average	20.15%	21.60%	24.22%
Median	17.58%	19.60%	22.18%

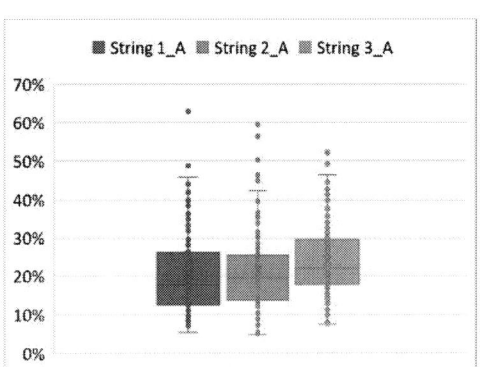

Figure 5 Module current mismatch rate

As you can see in above figures and tables, the average voltage mismatch is confirmed 5.28% to 8.37% and current mismatch is analyzed 20.15% to 24.22%. The highest voltage and current mismatch are shown at the string 3 which is consisted of various azimuth and tilt angles. The current mismatch is calculated higher than voltage mismatch.

The module number which is checked highest and lowest current of each string are confirmed as Table V. The lowest current of each string is shown at center of the module. Figure 6 is IR(infrared) image of a PV tunnel system. Hot spots are observed at the middle of the PV system. These modules which are located at the middle of the system have 0° tilt angle and shown soiling(see in Figure 7).

Table V : Module numbers which are shown highest and lowest current in a string

	Lowest	Highest
String 1	8	5
String 2	6	9
String 3	7	4

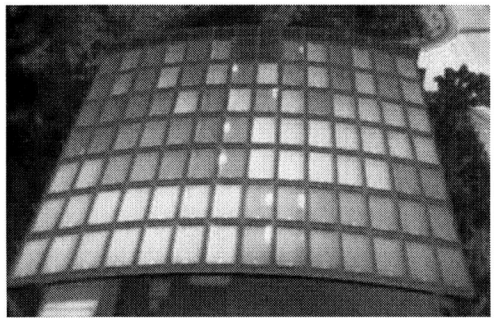

Figure 6 IR image of a PV tunnel system

Figure 7 Soiling of PV system

The voltage control type optimizers are kept inverter input voltage to set point. Therefore, depending on the configuration the string current can be lower or higher than module current. In this system, the string current depending on optimizer is 17% lower than module current. The module current after the optimizer is calculated through the module power and optimizer voltage. The current mismatch is seen in following Table VI and Figure 8. The average mismatch of optimizer output current is lower than module current mismatch. It confirmed largest decreased in the string 3 which is configured with various installation angles.

Table VI : Optimizer output current mismatch rate

	String 1	String 2	String 3
Min	2.81%	2.16%	2.01%
Max	54.85%	54.87%	48.16%
Average	14.17%	14.50%	13.24%
Median	10.35%	10.74%	11.08%

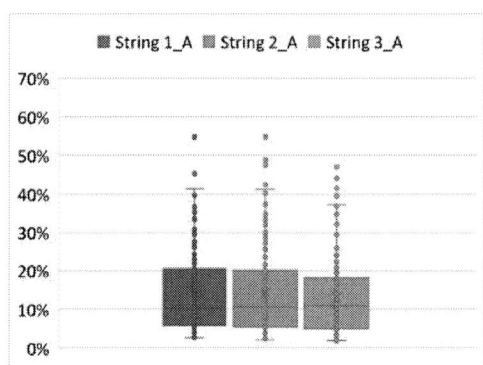

Figure 8 Optimizer output current mismatch rate

4 CONCLUSIONS

In this paper, generated energy and mismatch rate of a soundproof PV system is confirmed.

Optimizers elevated the total energy of PV system by 3.7%. However, daily energy when PV system generated low energy is decreased by optimizer. It is confirmed that consuming energy by optimizer is higher than improved energy.

The average voltage mismatch is confirmed 5.28% to 8.3%. The average current mismatch is shown 20.15% to 24.22% depending on tilt and azimuth. The module which is located at center of system is confirmed the lowest current because of soiling. The soiling of module is confirmed based on tilt angle. To reduce the soiling issue and current mismatch, the system should have to install with tilt over 0° or clean periodically for safety.

The soundproof tunnel PV system can be installed with multiple angles. In this case, generated power can be improved and mismatch effects can be reduced through the optimizer. The optimizer output current is mitigated through the optimizer. Also, the current mismatch rate is decreased 22% to 14%. Especially, current mismatch decreased 11% by optimizers in the string 3. However, the difference is confirmed moer than 10% even installed with optimizer. It represents that mismatch is more influenced by installation angles (tilt, azimuth) than power improvement by optimizers.

The current mismatch can lead the degradation of PV module. Therefore, to reduce the current mismatch, it recommended that configurated with similar tilt and azimuth angle.

5 REFERENCES

[1] C. Peng, et al. Building-integrated photovoltaics (BIPV) in architectural design in China., Energy and buildings 43.12 (2011): 3592-3598.

[2] J. D. Huyeng, et al. Technical aspects for road integrated photovoltaics towards a more sustainable mobility sector, WCPEC-8 8th conference,26-30 Sep. 2022.

[3] H. Bayat, et al. A power mismatch elimination strategy for an MMC-based photovoltaic system. IEEE Transactions on Energy Conversion 33.3 (2018): 1519-1528.

[4] S. Yaday, et al. Optimum azimuth and inclination angle of BIPV panel owing to different factors

influencing the shadow of adjacent building. Renewable Energy 162 (2020): 381-396.

[5] K. A. K Niazi, et al. Performance assessment of mismatch mitigation methodologies using field data in solar photovoltaic systems., Electronics 11.13 (2022), 1938.

Acknowledgement

This research was supported by New & Renewable Energy of the Korea Institute of Energy Technology Evaluation and Planning (KETEP) grant fund by the Korea government Ministry of Knowledge Economy (No. 20213030010280)

41st European Photovoltaic Solar Energy Conference and Exhibition

IMPROVED MODELLING OF PV SYSTEMS WITH SNOW SOILING
FOR OPTIMIZED LOCAL ENERGY SHARING

Ida Fuchs, Ole-Morten Midtgård
Norwegian University of Science and Technology / Department of Electric Energy
ida.fuchs@ntnu.no, ole-morten.midtgard@ntnu.no

ABSTRACT: Snow soiling on photovoltaic (PV) systems can significantly impact their performance, particularly in regions with heavy snowfall. This study investigates the effects of snow soiling on energy sharing efficiency within local energy communities, focusing on Norwegian cabin fields. We analyze the occurrence of snow soiling across different panel mounting configurations and evaluate its impact on energy sharing models using both simulated and measured PV data. The study employs a detailed experimental setup in Trondheim, Norway, where five PV panel orientations with varying tilt angles are monitored over three years (2021-2023). The results indicate that snow soiling frequently occurs, with significant variability across panel configurations. Panels with lower tilt angles and specific mounting designs experience prolonged snow coverage, while those with higher tilt angles and configurations allowing snow to slide off show reduced snow coverage. In terms of energy sharing, the impact of snow soiling is found to be minimal. Although snow soiling leads to a worse initial condition, the improvement potential for energy sharing remains consistent across different scenarios. The study reveals that variations in generation profiles due to snow soiling are too small to significantly influence the benefits of energy sharing, suggesting that other factors, such as load profiles, may play a more dominant role. Future research should explore the effects of snow soiling at higher altitudes, include local shadow effects in simulation models, and analyze specific periods such as spring holidays to refine energy sharing strategies. This research provides valuable insights into optimizing PV system performance in snowy environments and supports the development of effective energy sharing models in local communities.
Keywords: PV system modelling, Snow soiling, Solar home systems, Energy sharing, Decentralized energy systems

1 INTRODUCTION

Snow soiling on photovoltaic (PV) systems refers to the accumulation of snow on the surface of solar panels, which can significantly impact their performance and efficiency [1], [2], [3]. Solar panels convert sunlight into electricity, and any obstruction, including snow, reduces the amount of light reaching the photovoltaic cells, thereby decreasing the electrical output.

In regions that experience heavy snowfall, snow soiling becomes a critical issue for the operation and maintenance of PV systems [4], [5]. Snow not only blocks sunlight but can also adhere to the panels, forming a layer that persists even after other snow has melted, especially on panels with sub-optimal tilt angles or flat installations. This persistence can lead to prolonged periods where the panels generate little to no power, affecting the overall energy yield and efficiency of the solar power system.

Several studies have analyzed the effects of snow soiling and the resulting losses in PV energy output. Many of these studies focus on determining average energy yield losses for specific regions and months of the year [4]. Such average values can be used in simulation software to predict energy yield for new systems. However, these averages are sometimes unsuitable for specific applications, such as energy sharing. Energy sharing between prosumers in local energy communities benefits when different systems produce varying outputs at the same time. Therefore, applying average loss values uniformly across all days and hours of a month results in a less detailed understanding of the actual energy sharing potential.

To better understand the impact of snow soiling, we conducted a detailed observation of a real-world PV system, specifically targeting snow accumulation. This study analyzed production data spanning three years, from 2021 to 2023, and included photographic documentation of snow coverage, carefully cross-referenced with the collected data. Featuring panels with various azimuth and tilt angles, the studied PV system provides comprehensive insights into snow soiling behavior at the location.

The results of the snow soiling analysis were then integrated into an energy sharing model for PV and battery systems in Norwegian cabin fields and compared with simulated PV data without the snow soiling effect. This comparison highlights the impact of snow soiling on the energy sharing model and suggests further steps to improve the modeling process.

Based on these considerations, the following research questions are posed:

- How often does snow soiling occur throughout the winter season on different panel mounting configurations?
- What is the extent of snow soiling's impact on energy sharing efficiency in local energy sharing models for Norwegian cabin fields?

The remainder of this paper is structured as follows: Section 2 outlines the methodology used in the study, Section 3 presents the results of our analysis, and Section 4 discusses the implications and limitations of our findings for future research and practice. The paper concludes in Section 5.

2 METHODS

This section outlines the methodology used to address the research questions. The first part of the study involves an experimental setup to observe snow accumulation and soiling effects on PV panels with different orientations. The second part compares these results with simulated data from the experimental system to determine the number of days the system is covered by snow. This step provides valuable insights into improving snow soiling modeling for energy sharing systems. In the third part, both measured and simulated data are used to evaluate the impact of snow-covered days on the performance of the energy sharing model.

2.1 Experimental setup

The test system is located in Trondheim, Norway, and is mounted on the roof of the Faculty of Information Technology and Electrical Engineering at the Norwegian University of Science and Technology (NTNU). The PV system is part of the National Smart Grid Laboratory, co-owned by NTNU and SINTEF. Table I provides key specifications for the total installed system as well as details on the subsystem used for the experiment.

Table I: Key data of total PV system and experimental subsystem

	Total system	Subsystem
Number of panels	62	5
Number of inverters	1	1
Number of optimizers	62	5
Peak power per panel [W_P]	325	325
Number of orientations	5	5

Figure 1 shows the layout of the entire system, with the five panels selected for this experiment marked by a yellow star. One panel from each available orientation was chosen. Table II provides details on the various orientations, including their azimuth and tilt angles. Figure 2 presents an image of the PV system site, highlighting the different panel groups. Group 1 represents the optimal configuration for maximum yield in this region, with a 180-degree azimuth (south orientation) and a 45-degree tilt. Group 2 faces towards southwest with a tilt of 48 degrees. Groups 4 and 5 consist of low-tilt mounted panels, located in front of Group 3, which has panels mounted at a high tilt of 55 degrees (standing panels).

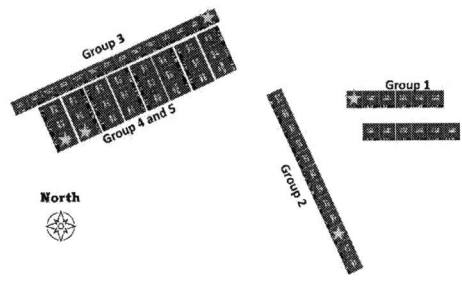

Figure 1: PV system layout. Experimental subsystem highlighted with stars.

Table II: Overview of PV panel orientations

Group	Orientation	Azimuth	Tilt
1	South (S-45)	180	45
2	Southwest (SW-48)	246	48
3	Southeast (SE-55)	157	55
4	Southwest (SW-15)	247	15
5	Northeast (NE-15)	67	15

Data is collected through optimizers attached to each panel, which measure several parameters for each individual panel. The optimizers transmit the data to the inverter platform, where it can be accessed and downloaded by the user at different time resolutions. For this experiment, the power output of the selected panels was downloaded at 15-minute intervals. The data spans the years 2021 to 2023.

Figure 2: PV system site with groups.

2.2 Evaluation of snow soiling

To assess the impact of snow soiling, the measured PV power output data is compared to simulated data for each panel. The simulations are performed using Pvsyst, where each panel's specific tilt and azimuth are modeled. The simulation is based on typical meteorological year (TMY) data, with the horizon included in the model. However, local shading is not accounted for, which will be discussed as a source of potential mismatch. In this step, snow soiling is intentionally excluded from the simulation, as the objective is to compare simulated data without snow soiling to the measured data that includes it, in order to identify the effect of snow soiling.

The measured PV power output is then compared with the simulated output. For any day where the measured power output is close to zero throughout the entire day, while the simulated data indicates significant power generation, that day is marked as fully snow soiled. The number of such days is counted for each month. Additionally, plots illustrating when these days occur are included in the results presentation.

2.3 Snow soiling in energy sharing models

To evaluate how the inclusion of snow soiling affects the outcome of energy sharing, both simulated and measured PV data are used to model a local energy community consisting of four PV systems for four different mountain cabins. The orientations of the PV systems match those in the experimental setup, allowing the measured data to be used for modeling the PV system output of the cabins. Figure 3 illustrates the simulated energy community.

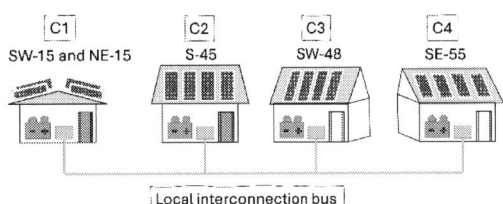

Figure 3: Energy community of four mountain cabins and their PV system orientations.

The first cabin, C1, is modeled with two panels oriented SW and two panels oriented NE. Each cabin has four panels with a capacity of 325 Wp, consistent with the setup of our PV system. Cabin C2 is oriented southward, with all four panels positioned at a 45-degree angle. The orientations for each cabin are summarized in Table III.

Table III: PV panel orientations per modelled cabin

Cabin	Orientation	Azimuth	Tilt
C 1	SW and NE	247, 67	15
C 2	S	180	45
C 3	SW	246	48
C 4	SE	157	55

Table IV presents the key simulation data for the off-grid PV and battery system used by each modeled cabin. These specifications fall within the range of what is currently commercially available for off-grid systems in Norwegian cabins.

Table IV: Key simulation data for off-grid system

Name	Number
PV panel size	325 W (mono crystalline)
Number of PV panels	4
Battery capacity	20 kWh (lithium)
Inverter max power	10 kW
Battery max power	8 kW

The demand for each cabin is simulated using the same principles outlined in [6]. Different cabin categories are selected, ranging from low consumption in C1 and C2 to higher consumption in C3 and C4. Various usage patterns are modeled, including every second weekend, every third weekend, individual weekends, and holidays only. The demand is plotted at a 15-minute time resolution for the entire year in Figure 4, illustrating both power consumption limits and usage patterns.

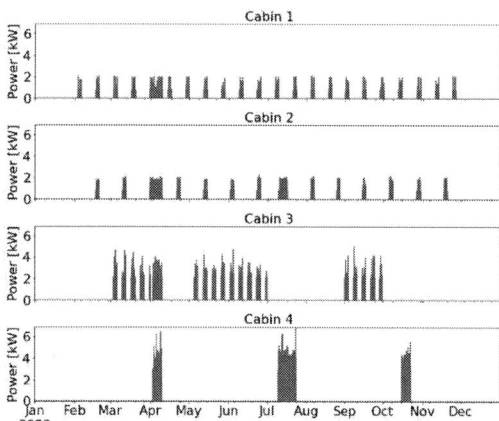

Figure 4: Demand patterns for the four simulated cabins.

Energy sharing between the cabins is simulated using a rule-based approach, implemented in the controllers connected to the local sharing grid. The rule-based energy sharing follows the principles outlined in [7].

To evaluate the results of the energy sharing, the improvement potential is assessed in terms of the lost load. The lost load is the percentage of demand that cannot be met by the off-grid system or local interconnected network. The improvement potential is calculated as the difference between the lost load without energy sharing and the lost load with energy sharing.

To examine the impact of including snow coverage in the energy sharing model, this improvement potential is compared between simulated PV data (without snow) and measured PV data from several years with snow-covered panels. Variations in panel orientations may lead to different snow coverage behaviors, and such increased diversity in PV generation could result in enhanced energy sharing and greater potential benefits.

In addition to the direct comparison between simulated data without snow and measured data with snow, the measurements are used to create a snow mask that can be applied to the simulated data. This process is done for each of the measured years. Finally, simulated PV data without snow is compared to simulated PV data with the snow mask when used in the energy sharing model. This final comparison provides a realistic assessment of the sole impact of snow soiling on energy sharing models.

3 RESULTS

3.1 PV power output

Figures 5a-c show the daily PV yield measured for the years 2021-2023 for one panel from each of the five studied groups, representing five different PV panel orientations. The data is presented alongside the simulated data for comparison.

In 2023, two weeks of data (weeks 28 and 29) were missing in mid-July. To fill this gap, week 28 was replaced with data from week 27, and week 29 was replaced with data from week 30 of the same year.

For the S-45, SW-48, SW-15, and NE-15 orientations, the measured data closely matches the simulated data. However, for the SW-55 orientation, the simulated data consistently shows higher power peaks compared to the measured data, a pattern observed across all three years.

This discrepancy is due to a mounting issue, where Group 3 (SW-55) is positioned in a way that Groups 4 and 5 cast shadows on it throughout the year.

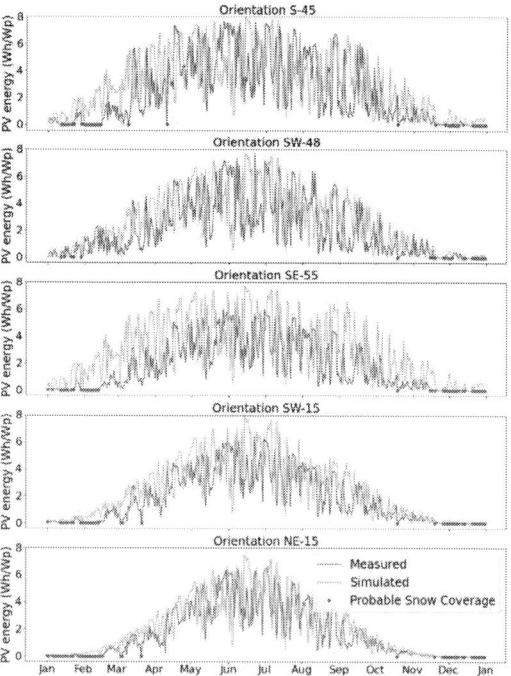

Figure 5a: Comparison of measured PV daily yield data from **2021** and simulated daily yield for each PV panel orientation. Identified snow-coverage days marked in red.

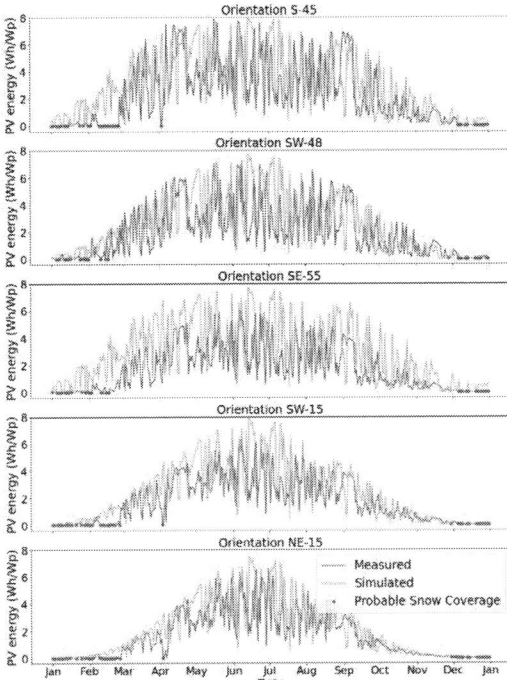

Figure 5b: Comparison of measured PV daily yield data from **2022** and simulated daily yield for each PV panel orientation. Identified snow-coverage days marked in red.

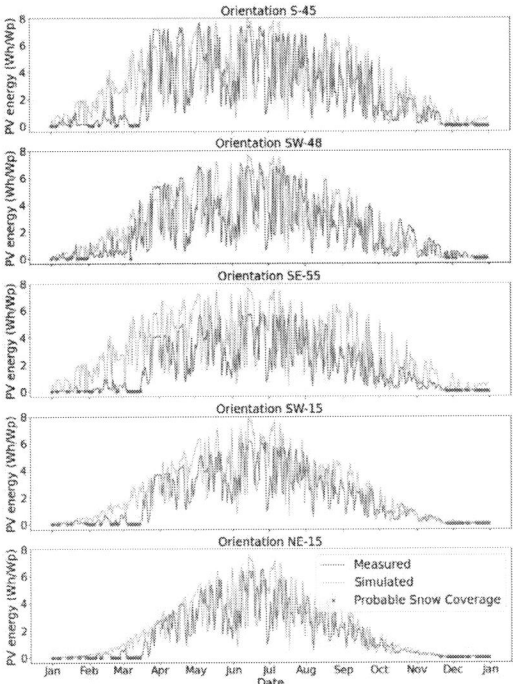

Figure 5c: Comparison of measured PV daily yield data from **2023** and simulated daily yield for each PV panel orientation. Identified snow-coverage days marked in red.

Additionally, days with full snow soiling typically occur during periods of low potential generation, but snow coverage reduces the actual generation to near zero. These snow-soiled days often occur in clusters, with several consecutive days being affected.

3.2 Days with snow soiling

The results for the number of snow-soiling days over the studied years are shown in Figure 6. The snow-soiling days are displayed for the five different panel orientations corresponding to the five groups. The results are visualized as a color map, where white indicates a high number of snow-soiling days during a given month, and blue represents fewer or no snow-soiling days.

It can be observed that December consistently shows that panels are snow-covered for almost two-thirds of the month, which coincides with the lowest solar yield due to limited solar irradiation. November, however, shows fewer snow-covered days compared to January and February. March presents an interesting case: while there were no snow-covered days in 2022 across any orientations, and only a few in 2021, there were significant numbers in 2023. Given that March has considerable solar irradiation, even a single snow-covered day could lead to a substantial loss of energy yield.

Number of days with full snow soiling

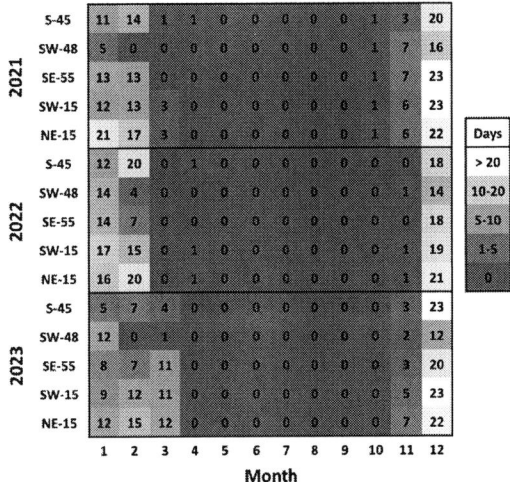

		1	2	3	4	5	6	7	8	9	10	11	12		Days
	S-45	11	14	1	1	0	0	0	0	0	1	3	20		> 20
	SW-48	5	0	0	0	0	0	0	0	0	1	7	16		10-20
2021	SE-55	13	13	0	0	0	0	0	0	0	1	7	23		5-10
	SW-15	12	13	3	0	0	0	0	0	0	1	6	23		1-5
	NE-15	21	17	3	0	0	0	0	0	0	1	6	22		0
	S-45	12	20	0	1	0	0	0	0	0	0	0	18		
	SW-48	14	4	0	0	0	0	0	0	0	0	1	14		
2022	SE-55	14	7	0	0	0	0	0	0	0	0	0	18		
	SW-15	17	15	0	1	0	0	0	0	0	0	1	19		
	NE-15	16	20	0	1	0	0	0	0	0	0	1	21		
	S-45	5	7	4	0	0	0	0	0	0	0	3	23		
	SW-48	12	0	1	0	0	0	0	0	0	0	2	12		
2023	SE-55	8	7	11	0	0	0	0	0	0	0	3	20		
	SW-15	9	12	11	0	0	0	0	0	0	0	5	23		
	NE-15	12	15	12	0	0	0	0	0	0	0	7	22		

Month

Figure 6: Color map of days with full snow soiling for each panel orientation plotted over the months of the three studied year 2021-2023.

It is evident that SW-48 experiences the fewest snow-covered days in all years. Despite SE-55 having a higher tilt, it performs no better than SW-48, and in 2023, it is even less effective than S-45, which has an even lower tilt than SW-48. The increased number of snow-covered days for S-45 and SE-55 can be attributed to their mounting configuration. These panels are installed in such a way that snow accumulates on the roof and cannot slide off, while SW-48 benefits from an open space beneath the panels that allows snow to slide off. This indicates that snow sliding off is influenced not only by a higher tilt but also by the presence of space for snow to slide.

Figure 7 visually illustrates this effect. The left column of pictures shows February 5, 2024, where new snow has accumulated on the panels in addition to existing snow. A week later, the snow has melted and slid off from Group 1 and Group 2, as shown in the right column of pictures in Figure 7. However, Group 3 still retains partial snow coverage due to the lack of space for the snow to slide off. Groups 4 and 5 experience longer periods of snow coverage due to their low tilt angles.

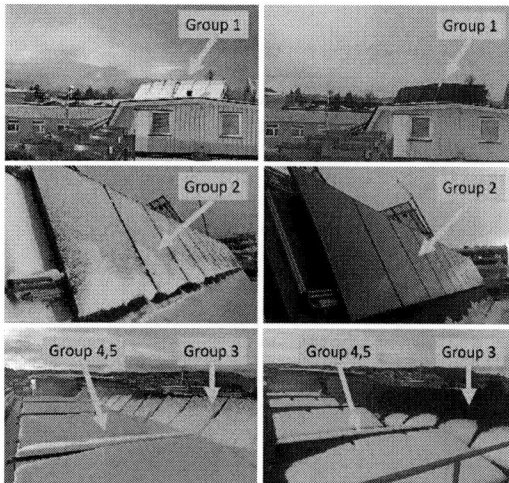

Figure 7: Two days of snow soiling for the studied PV system with marked groups of different orientations. Left: 5ᵗʰ February 2024. Right: 13ᵗʰ February 2024.

3.3 Snow soiling in energy sharing models

The impact of snow soiling was assessed using an energy sharing model for a Norwegian mountain cabin field. This model includes four mountain cabins with the specifications outlined in Section 2.3.

Table V presents the results, comparing the simulated PV data (Sim PV) to the measured PV data (M PV) as inputs for the energy sharing model. The table shows the percentage of remaining lost load for both the off-grid (OG) and the energy sharing (ES) case. The reduction in the percentage of lost load represents the improvement potential of energy sharing. We analyze whether this improvement potential varies depending on the PV data used and whether snow soiling is included in the model.

Table V: Lost load in percentage for each cabin (C1-C4) and the total local grid comprised of the four cabins modeled with simulated PV data (Sim PV) compared with measured PV data (M PV) for 2021-2023 for the off-grid case (OG) and energy sharing case (ES)

Cabin	Sim PV OG/ES [%]	M PV 21 OG/ES [%]	M PV 22 OG/ES [%]	M PV 23 OG/ES [%]
C1	17 / 2	27 / 6	25 / 3	25 / 8
C2	5 / 2	15 / 7	5 / 1	15 / 6
C3	48 / 10	57 / 18	52 / 16	56 / 14
C4	81 / 40	87 / 52	83 / 38	87 / 51
Total	47 / 17	56 / 25	51 / 19	55 / 24

The values for lost load in the off-grid case presented in Table V clearly indicate that the measured data for all three years results in a worse starting point for the cabins compared to what the simulated data predicts. The lost load for all cabins is almost always higher in the off-grid case with the measured PV data than with the simulated data. Consequently, when energy sharing is applied, the final lost load is generally worse with the measured data compared to the simulated data, except for two instances in 2022 for cabins C2 and C4, where measured data shows a lower lost load than the simulated data, suggesting a significant improvement for those cabins that year.

Figure 8 presents the energy sharing improvement potential for both the simulated PV data and measured PV

data, for each household and the total community. It is evident that the improvement potential remains within a similar range for each household, with some deviations but no consistent trend. For the total community, the improvement potential appears to be more consistent, indicating a balancing effect across the houses. This suggests that, despite variations in individual households over the years, the overall potential for energy sharing remains relatively stable when considering the entire community, independent of different PV input data.

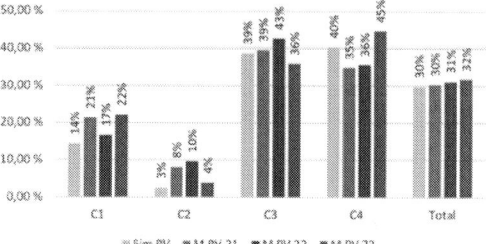

Figure 8: Energy sharing improvement potential with simulated (Sim PV) and measured (M PV) PV data for 2021-2023.

Table VI and Figure 9 present the final issue studied: the sole influence of snow soiling. For this analysis, the simulated PV data was used with a snow mask applied according to Figure 6, where all days with snow soiling were set to zero in the simulated data. This procedure was repeated for all three years, followed by another energy sharing analysis.

Table VI: Lost load in percentage for each cabin (C1-C4) and the total local grid comprised of the four cabins modeled with simulated PV data (Sim PV) compared with simulated PV data with a snow soiling mask (SiS PV) for 2021-2023 for the off-grid case (OG) and energy sharing case (ES)

Cabin	Sim PV OG/ES [%]	SiS PV 21 OG/ES [%]	SiS PV 22 OG/ES [%]	SiS PV 23 OG/ES [%]
C1	17 / 2	19 / 3	21 / 3	22 / 3
C2	5 / 2	5 / 2	7 / 4	5 / 2
C3	48 / 10	48 / 10	48 / 10	49 / 12
C4	81 / 40	81 / 42	81 / 40	81 / 40
Total	47 / 17	48 / 18	48 / 18	48 / 18

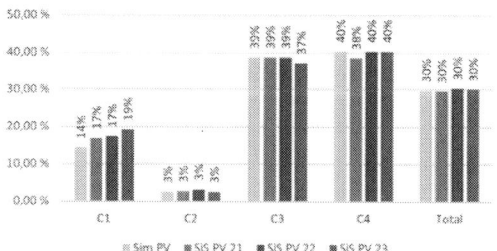

Figure 9: Energy sharing improvement potential with Sim PV data with a snow mask (SiS PV) for 2021-2023.

Table VI shows the lost load for this adjusted simulated data, revealing a similar effect of a worse initial situation, though the difference is minimal. Cabin C1, which has the flat tilt panel and is therefore most affected by snow coverage, shows only slight deterioration across all years. Cabin C2 experiences a worse starting situation only in 2022, and Cabin C3 in 2023, with differences of just 1 or 2 percentage points. For the total community, the change in both the off-grid and energy sharing cases is only 1 percentage point across all simulations with snow.

Figure 9 illustrates that the improvement potential remains very similar, indicating that modeling snow coverage for energy sharing has a minor impact on the outcome. The changes are minimal, highlighting that snow soiling only causes a small variation in the results.

4 DISCUSSIONS

This study presents valuable insights into the impact of snow soiling on energy sharing within a specific type of energy community, namely a swarmgrid for a cottage area. The analysis demonstrates that snow soiling affects the energy yield and sharing potential, but the overall impact on improvement potential appears minimal. This finding aligns with previous research suggesting that the improvement potential in energy sharing is predominantly influenced by variations in load profiles rather than differences in generation profiles [6].

One notable aspect of this study is the experimental setup's location in Trondheim, Norway, which contrasts with the higher altitudes typical of many cabin locations. The elevation could significantly affect the amount of snow and the duration of snow soiling. Future research should consider conducting similar studies at higher altitudes to assess the impact of extended snow coverage.

The mounting discrepancy observed in Group 3, compared to Groups 2 and 1, highlights an important factor in snow soiling. The mounting configuration influences snow accumulation and sliding, emphasizing that snow soiling is not solely determined by panel tilt but also by the presence of space for snow to slide off. This finding underscores the need for detailed attention to mounting configurations in future studies to optimize panel performance in snowy conditions.

Local shadows were not incorporated into the simulation models, which could affect the accuracy of the simulated data. Moreover, this study focused on days with total snow coverage and did not account for partial coverage or coverage only for part of the day. Future research should explore these factors to provide a more nuanced view of snow soiling effects.

The study's annual perspective provides a broad view of energy sharing potential throughout the year. However, spring holidays, particularly in Norway, are a crucial period where cabins are frequently used. Analyzing the impact of energy sharing specifically during the spring months (March and April) could offer additional insights and help households make more informed decisions about participating in such energy communities.

5 CONCLUSIONS

This study addresses the impact of snow soiling on PV systems and its influence on energy sharing efficiency, specifically within Norwegian cabin fields. By investigating snow soiling across different panel mounting configurations and assessing its effect on energy sharing models, this research provides valuable insights into optimizing PV system performance in snowy conditions.

Our analysis reveals that snow soiling occurs with significant variations across different panel mounting configurations. Panels with lower tilt angles and specific mounting setups, such as those in Groups 3, 4 and 5, are more prone to prolonged snow coverage. Conversely, panels with higher tilt angles and configurations that allow snow to slide off, such as those in Group 2, experience fewer snow-covered days. This variability highlights the importance of not only panel orientations but also mounting design in mitigating the effects of snow soiling.

The impact of snow soiling on energy sharing efficiency was assessed by comparing models using both simulated and measured PV data. The results indicate that snow soiling has a minimal effect on the overall improvement potential for energy sharing. Although snow soiling reduces the energy yield, leading to a worse initial condition, the relative benefit of energy sharing remains consistent. The study indicates that variations in generation profiles due to snow soiling are too small to significantly influence the benefits of energy sharing.

Future research should consider conducting studies at higher altitudes, where snow coverage may be more prolonged. Additionally, incorporating partial snow coverage could offer a more detailed understanding of snow soiling impacts. A focused analysis of energy sharing efficiency during critical periods, such as spring holidays, could also provide valuable insights for optimizing energy sharing strategies in seasonal contexts.

In summary, this research underscores the importance of panel design and mounting configuration in managing snow soiling effects and highlights that the overall potential for energy sharing remains robust despite variations in snow coverage. The findings contribute to optimizing PV system performance in snowy environments and offer guidance for future studies aiming to enhance energy sharing models in diverse settings.

6 REFERENCES

[1] B. Marion, R. Schaefer, H. Caine and G. Sanchez, "Measured and modeled photovoltaic system energy losses from snow for Colorado and Wisconsin locations", Sol. Energy, 2013.

[2] L. Powers, J. Newmiller and W. Townsend, "Measuring and modeling the effect of snow on photovoltaic system performance", Proc. Photovolt. Specialists Conf., 2010.

[3] A. Skoczek, O. Osvald, B. Schnierer, L. Helienek, T. Harcinikova, "Assessing PV Energy Loss due to Snow with Meteorological Models", Proceedings of the EU PVSEC 2023.

[4] M. Øgaard, H. Nygard Riise, J. Helene Selj, "Estimation of Snow Loss for Photovoltaic Plants in Norway, Proceedings of the EU PVSEC 2021.

[5] H. Nygard Riise, M. Øgaard, T. Uberg Nærland, Soiling and Snow Impact on a PV Plant at a Farm in Norway, Proceedings of the EU PVSEC 2021.

[6] I. Fuchs, H. O. Tørrisplass, S. Völler, Energy sharing in solar and battery off-grid systems with advanced PV generation modelling: A case study of Norwegian cabin fields, Proceedings of the EU PVSEC 2023.

[7] I. Fuchs, C. S. Sanchez, S. Balderrama, G. Valkenburg, Swarm electrification for Raqaypampa: Impact of different battery control setpoints on energy sharing in interconnected solar homes systems, *Preprint at SSRN: http://dx.doi.org/10.2139/ssrn.4872057*

41st European Photovoltaic Solar Energy Conference and Exhibition

ENSURING PHOTOVOLTAIC MODULE INTEGRITY THROUGH ELECTROLUMINESCENCE IMAGING AND MACHINE LEARNING SOLUTIONS

Daniel J. Castillo Patton [1]*, Lucas Viani [1], Fernando García [2], Vicente Parra [1], Sofía Rodríguez-Conde [1], Jesús Cuaresma [1]

[1] Enertis Applus+, Parque Empresarial Las Mercedes, C/ de Campezo, 1, 28022 Madrid, Spain

[2] Carlos III University of Madrid, Av. de la Universidad, 30, 28911 Leganés, Madrid, Spain

*e-mail: daniel.castillo.p@enertisapplus.com

ABSTRACT: This paper presents a tool for detecting and preventing defects in photovoltaic (PV) modules using Electroluminescence (EL) imaging and machine learning. The tool employs adaptive defect detection models and convolutional neural networks (CNN) to improve detection accuracy and efficiency. Our dataset includes diverse images from different inspections across different regions, enhancing the robustness of the training process. The tool's phases include synchronized defect detectors, cell segmentation, and automated criticality assessment, allowing for precise defect identification and prioritization of maintenance actions. Testing demonstrates significant improvements in defect detection and criticality evaluation, reducing the need for manual inspections and optimizing PV system maintenance.

1 INTRODUCTION

Photovoltaic (PV) systems are crucial for sustainable energy production, converting sunlight into electricity without harmful emissions. Ensuring the integrity of PV modules is vital for maximizing energy output and extending their lifespan. A major challenge in PV manufacturing is the detection and mitigation of defects that can impair module performance.

Electroluminescence (EL) imaging is a non-destructive technique widely used to detect defects such as cracks and shunts in PV cells, which are invisible to the naked eye. However, manual inspection of EL images is labor-intensive and prone to human error, limiting its effectiveness [1]. Recent advancements in machine learning, particularly convolutional neural networks (CNNs), offer promising solutions for automating defect detection in PV modules. CNNs can accurately identify and classify defects in EL images, reducing the need for manual inspections and lowering associated labor costs [2].

This paper introduces a comprehensive tool that integrates EL imaging with machine learning to enhance defect detection accuracy and efficiency in PV modules. The tool uses a diverse dataset of EL images from inspections conducted across various regions, ensuring robust model training and evaluation. It operates in phases, including synchronized defect detection, cell segmentation, and automated criticality assessment, allowing for detailed analysis and targeted interventions based on defect severity.

The paper's objectives are to describe the tool's development, validate its performance through rigorous testing, and explore its implications for PV system maintenance and operational efficiency. By automating defect detection, the tool aims to improve the reliability and longevity of PV installations, thereby supporting more sustainable solar energy production.

2 RELATED WORKS

The field of defect detection in photovoltaic (PV) cells has advanced significantly, especially with the integration of machine learning and computer vision techniques. Several studies have explored various methods to improve the accuracy and efficiency of defect detection in PV modules.

Electroluminescence (EL) imaging is a widely used non-destructive technique for detecting defects in PV cells. This method captures emitted light from PV modules under forward bias, revealing defects such as cracks and shunts that are not visible to the naked eye. Traditional image processing methods often require extensive manual intervention and are limited by human error and fatigue. Some works provide a comprehensive review of EL-based defect detection, highlighting the transition from manual inspection to automated approaches using computer vision and machine learning. This review underscores the need for efficient and accurate defect detection methods in the PV industry. [1]

Deep learning, particularly convolutional neural networks (CNNs), has revolutionized defect detection across various fields, including PV systems. CNNs can learn complex patterns in image data, making them well-suited for identifying and classifying defects in developing a deep-learning-based method for detecting PV cell defects, achieving significant improvements in detection accuracy through data enhancement and feature fusion techniques [3].

Other works explored the use of deep learning for early defect diagnosis in PV modules using thermal images. This approach integrates thermal image analysis with machine learning algorithms to detect defects, offering an efficient and robust solution for PV module inspection [4].

Advancing the use of Convolutional Neural Networks (CNNs) by integrating multiple architectures [5] enhances the accuracy and robustness of defect detection in photovoltaic cells. Their approach utilizes the strengths of various models, such as ResNet152, Xception, and others, by combining them to form a more reliable system. This method addresses the limitations of individual models, leading to improved performance in detecting defects in electroluminescence images. The integration of diverse architectures helps mitigate issues related to data imbalance and enhances feature extraction capabilities through techniques like coordinate attention mechanisms.

In addition to CNNs, other machine learning techniques like support vector machines (SVMs) and random forests have been explored for defect detection in PV cells. For example, in [6] a proposed hybrid model combining SVMs and CNNs, which demonstrated improved accuracy in detecting micro-cracks and other subtle defects.

Overall, the integration of EL imaging with advanced machine learning techniques has significantly improved the defect detection process in PV systems. These advancements reduce the reliance on manual inspections, enhance detection accuracy, and provide valuable insights into the criticality of detected defects, thereby contributing to the reliability and efficiency of PV systems.

3 METHODOLOGY

This section details the methodology employed to develop the tool for detecting and preventing defects in photovoltaic (PV) modules using Electroluminescence (EL) imaging and machine learning techniques. The methodology is divided into several key phases: data collection, model development, system implementation, and performance evaluation. Each phase is described in detail below.

3.1 Data Collection

The dataset comprises a series of images obtained during various inspections conducted by Enertis Applus+ at factories and photovoltaic panel processing centers. Given that hundreds of thousands of panels are manufactured, a sample from each batch is selected to assess their condition. Additionally, many batches are reviewed by external agents before being shipped to the installation site to perform a comprehensive analysis of the modules' states. Although some defects are not considered critical, these analyses help identify potential structural failures in the future.

The images obtained are electroluminescence images. Electroluminescence is an imaging technique used to analyze and evaluate the quality of photovoltaic materials and devices, such as solar panels. This technique is based on the emission of light from a semiconductor material when an electric current is applied. The resulting images

provide detailed information on defects, homogeneity, and efficiency of the devices [7].

From these sets, several images have been selected, and all the major defects and anomalies considered important for this analysis have been identified:

Table I: Dataset.

Data	Images	Defects
Cross-crack	638	2003
Soldering	764	2478
Scratch	35754	57900
Black-spot	35500	49325

Each of these groups has been used in various combinations for the tool presented here, aiming to find the best combination among the groups to develop robust models. The dataset used in this study comprises EL images collected from various inspections conducted by Enertis Applus+ across multiple geographical regions. The diversity of the dataset, which includes different environmental contexts, soil types, panel variations, and vegetation, enhances the robustness of the training process.

3.2 Model Development

The model development phase focuses on creating and training convolutional neural networks (CNNs) to detect and classify defects in EL images.

3.2.1 Defect Detection Models - Fast RCNN

Adaptive defect detection models were developed to tackle varying defect complexities, all utilizing the same CNN architecture due to its effectiveness. The selection of the Faster R-CNN (Region-based Convolutional Neural Network) architecture:

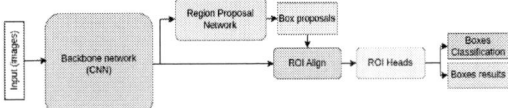

Figure 1: Basic architecture for a Faster RCNN model, showing the different steps performed to realize a detection.

This model was implemented in the Detectron2 library, and was crucial for defect detection in solar panels from electroluminescence images. This choice is justified by Faster R-CNN's proven empirical performance [8], ability to detect objects of various dimensions, adaptability, and strong support from comprehensive documentation and an active scientific community [10]. These factors validate Faster R-CNN as a robust solution for defect detection in solar panels.

3.2.2 Panel Detection Model - Mask RCNN

In this work, a CNN model was developed to identify individual cells in photovoltaic panels, enabling comprehensive panel analysis. Mask R-CNN was chosen for this task due to its advanced capabilities compared to Faster R-CNN.

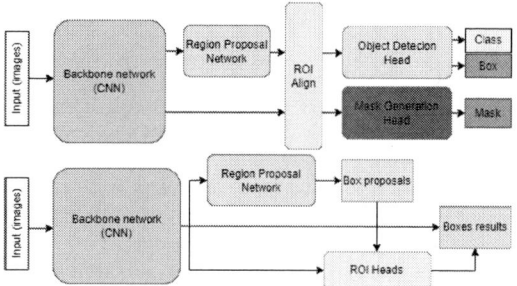

Figure 2: Basic architecture for a Mask RCNN model, showing the different steps performed to generate a segmentation process.

To implement Mask R-CNN, the Detectron2 framework was used for its flexibility and efficiency. This framework allows for fine-tuning specific to defect detection tasks and supports scalable, real-time processing, making it well-suited for industrial applications [13].

4 MODELS TRAINING AND EVALUATION

4.1 Training

In this section, we will review the training features used for the different models, as well as the parameters that have been adjusted and why.

4.1.1 Defect Detection - Faster RCNN

First, we will examine the training processes for the models that use Faster RCNN. All detectors are based on this model because it provides the best results.

For each type of defect, a set of parameters that are crucial for the detection of small defects is established.

Table II: Training selection for small defects.

	Parameters settings
Backbone	ResNet FPN
ResNet out features	4
Anchor Generator	[32, 64, 128, 256]
ROI Heads score	0.2
Learning Rate	0.0002
Regularization	L2 (0.0001)
Learning Schedule	(200000,250000)
Input size	(220, 440)

To optimize small defect detection in photovoltaic cells, we selected key parameters to enhance model performance. We utilized a multi-layer backbone to

capture features at different scales, enabling detection of both large and small defects, and a low score threshold of 0.2 was set to increase sensitivity by accepting lower confidence detections. The model was fine-tuned with a base learning rate of 0.0002, crucial for capturing small details.

For the rest of defect we have adjust the parameters to a more standard ones:

Table III: Training selection.

	Parameters settings
Learning Rate	0.0001
Iterations	60000
Batch Size	1
Loss Function	SmoothL1Loss + CrossEntropyLoss
Optimizer	SGD
NMS threshold	0.3
Regularization	L2 (0.0001)
Learning Schedule	(48000,54000)
Activation Function	ReLU

The loss function is essential for evaluating the difference between the model's predictions and actual training data. We use SmoothL1Loss to minimize outlier impact by accurately estimating bounding box locations, and CrossEntropyLoss to align predicted probabilities with true categories, ensuring correct label assignment. To prevent duplicate detections, a Non-Maxima Suppression threshold of 30% is applied.

Hyperparameters are carefully chosen for optimal model performance, balancing agility and stability. A moderate learning rate ensures a gradual descent to the loss function's global minimum, reducing divergence risks. The evaluation strategy verifies model accuracy and consistency, ensuring reliable results.

4.1.2 Cell Detection - Mask RCNN

The model used for cell detection in solar panels is a Mask R-CNN, configured with key parameters to optimize its performance for this specific task. Below, the most important parameters are justified:

Table IV: Training selection.

	Parameters settings
Backbone	ResNet FNP
Model Weights	X-101-32x8d
Batch Size	1
Learning Rate	0.001
Regularization	L2 (0.0001)
Learning Schedule	(24000,30000)
Activation Function	ReLU

These parameters are selected to maximize the model's accuracy and efficiency in the specific task of cell detection and segmentation in solar panels, providing a balance between computational complexity and predictive performance.

The Mask R-CNN model is used for the detection and segmentation of cells in solar panels. The selected parameters are designed to maximize the model's accuracy and efficiency. These include using ResNet FPN as the backbone to capture features at multiple scales, a batch size of 1 due to memory limitations, and a learning rate of 0.001 to ensure stable convergence. L2 regularization prevents overfitting, and a tailored learning schedule allows for precise model tuning. The ReLU activation function enhances training efficiency. These adjustments ensure a balance between computational complexity and the model's predictive performance.

4.2 Evaluation Metrics

For the evaluation of the models, we have chosen the most common metrics to measure the accuracy and quality of the trained models. These metrics are as follows: Precision (1), Recall (2), and F1-Score (3). Precision measures the accuracy of positive predictions, while recall measures the ability to identify all relevant instances.

$$(1) \; Precision = \frac{TP}{TP+FP} \quad (2) \; Recall = \frac{TP}{TP+FN}$$

In the domain of defect defect, recall in our case is the metric we consider most important of all, as it determines the total number of defects found in the image as a whole. For our case, it is paramount to find all defects during an inspection, and although it is important not to make a mistake with the defect class, it is something that is easily correctable afterwards, while failure to find a defect and miss it is a major problem in the long run.

$$(3) \; F1 \; Score = 2 \cdot \frac{Precision \cdot Recall}{Precision+Recall}$$

The F1-score, is a key metric in the evaluation of classification models, provides a balanced measure of accuracy and recall. Its contribution lies in considering both false positives and false negatives, providing a complete picture of the model's performance in class identification, especially in unbalanced scenarios. In summary, the F1-score helps to understand how the model balances accuracy and completeness in its classification capability.

4.3 Evaluation Results

For the evaluation of the models, a separate evaluation set (previously separated during the training process) that has never been visible to the model was chosen. This ensures that the metrics presented later are reliable under new circumstances.

Presented below are the performance metrics for each class for the detection models (see Table V):

Table V: Defect Detector Type Model.

Class	Precision	Recall	F1-Score
black-spot	0.836	0.885	0.856
scratch	0.844	0.883	0.863
cross-crack	0.859	0.945	0.9
soldering	0.906	0.93	0.918

Average total	0.861	0.919	0.889

And for the model of cell detection (see Table VI):

Table VI: Cell Detector Model Type.

Class	Precision	Recall	F1-Score
cell	0.99	0.99	0.99

Firstly, for the detection models, we can observe that the results are quite high, especially for the classes cross-crack and soldering, which align with the training using the module set. This occurs because the nature of these defects causes them to span multiple cells of the panel, which greatly aids the deep learning model in easily identifying these defects. In contrast, the defects such as scratch and black-spot are found in more homogeneous environments. Additionally, the number of these latter two defects is vastly greater, which typically lowers the precision due to the increased possibility of errors with so many instances of the same class.

Let's review some cases to see how the model performs (Figures 5-7):

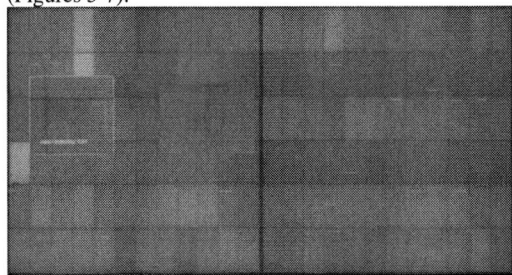

Figure 3: Panel with defects of type cross-crack and soldering annotated.

As we can see, all instances of soldering and cross-crack defects in this panel have been found. Despite cross-crack defects not being easily noticeable to the naked eye, the model successfully detected them.

Now, lets review a case for the cell segmentation model:

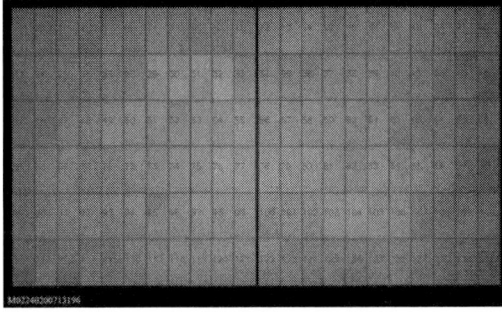

Figure 4: Panel with all the cells segmented and counted.

In this figure, we can see that all the cells are correctly annotated, even with external elements such as the black border and the image name in the bottom left corner. Additionally, during the detection phase of the modules, we organize the cells in a readable manner to archive this information for future reference accurately.

5 SYSTEM IMPLEMENTATION

The developed models were integrated into a comprehensive system that includes several interconnected phases for defect detection, cell segmentation, and automated criticality assessment.

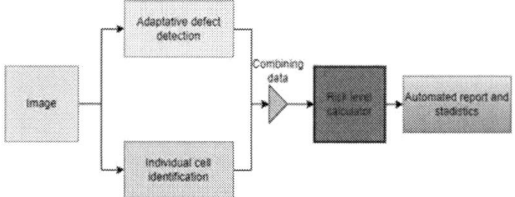

Figure 5: Diagram simplification of the tool proposed.

This diagram illustrates the system's operation. Next, we will explain each of the steps separately.

5.1 Cell Segmentation Phase

Individual cells within the PV panels were segmented and analyzed, allowing for the creation of a detailed history for each cell. This segmentation enables the identification of recurrent issues and patterns.

This section focuses on the complexity during the training phase, where the model identifies all the photovoltaic cells in a panel and stores the information in the following format: the starting pixel on the horizontal axis of the cell (x), the starting pixel on the vertical axis (y), the height in pixels, the width in pixels, and the cell number, which is assigned sequentially from 1 to n, with n being the total number of cells in the panel.

5.2 Defect Detection Phase

Synchronized detectors were employed to identify defects, selectively using different methods based on the defect type. This phase significantly enhances detection accuracy and minimizes false positives.

This phase is designed with two types of detectors:

- **Cell Detector:** These detectors operate exclusively on a single cell. They require information from the previous step to identify the cell and then perform the analysis on it. This step is repeated *n* times, where *n* is the total number of cells in the panel.
- **Panel Detector:** These detectors operate on the entire panel image, thus not requiring any prior image preprocessing.

The rationale for using these two distinct types of detectors stems from a meticulous trial-and-error study for each defect type. Some defects are better identified when analyzed in isolation, focusing only on the cell

region where a detector can be applied based on specific image features, such as texture and color. This approach allows the algorithm to identify subtle variations within a confined region, optimizing the precision in detecting defects intrinsic to a single cell. Conversely, other defects require more context, as they occur at the cell edges and/or span across multiple cells, necessitating an analysis that considers the interactions between adjacent cells and the surrounding environment for more effective detection [14].

5.3 Criticality Assessment Phase

Automated assessment of defect criticality was implemented, considering both the type and specific impact of defects on cell areas. This assessment is pivotal for prioritizing maintenance actions.

This phase encompasses the two previous steps, as it requires data from the cells as well as the location of the defects. With this information, a predefined risk threshold is established for each type of defect. This threshold corresponds to a percentage of the cell, meaning that if the defect comprises a greater percentage of the cell than the established threshold, it is classified as critical.

This step serves as a safety measure, providing an effective way to quickly identify the most severe issues a panel might have, without the need for human intervention to spot them.

5.4 Automated report and statistics

Once all the steps are completed, we can generate a report for the identified image, including:

- Identification of all cells
- Detected defects
- Locations of defects
- Critical defects

By aggregating all the images from an inspection, we can perform a comprehensive analysis of the entire plant.

6 RESULTS AND PRELIMINARY REPORT

In this section, we will examine the final results produced by the system. Referring to Figure 6, from the moment we input an image and execute all the steps, the resulting image appears as follows:

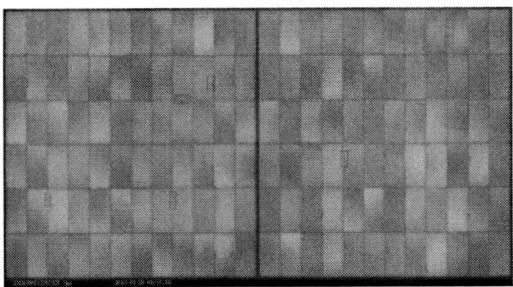

Figure 6: Photovoltaic panel with affected cells (green) and corresponding defects (blue).

As we can see in the final image, the system returns the image with all detected defects and the affected cells. All this information is also stored in a data table with the following details:

- Defect location
- Number of affected cells
- Cell location
- Percentage of affected cell
- Type of defect

With all this information combined from all images in an inspection, we can extract more advanced data to assess the inspection status through statistical analysis (Figure 7):

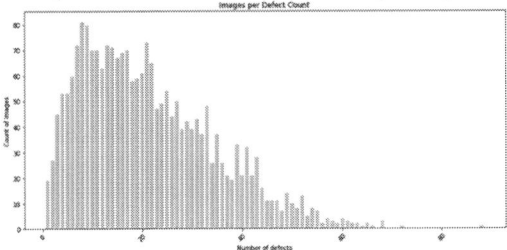

Figure 7: Graph of Number of images versus Number of defects.

7   USE CASE

In this final section, we will present a use case in collaboration with one of the teams from Enertis Applus+ during the factory inspection of a set of defects.

Table VII: Performance in a real-case scenario.

Class	Accuracy Rate
black-spot	99.99 %
scratch	95.05 %
cross-crack	96.08 %
soldering	99.99 %

These results differ from the test outcomes due to the dataset's extensive variability, which includes a wide range of examples not typically encountered in real inspections for certain panel types. This difference highlights the robustness of the model in handling diverse scenarios while ensuring reliability in real-world conditions. Due to the sensitivity of the data and confidentiality agreements, further details about the specific inspection cannot be disclosed.

7   CONCLUSIONS AND FUTURE WORKS

In this work, we developed a runtime environment for electroluminescence images in factories, capable of identifying a wide range of defects and segmenting individual cells in photovoltaic panels. This approach allows for detailed analysis, creating a historical record for each cell and assessing defect criticality to determine potential risks that could render a panel unusable. The aggregated data enables comprehensive reporting for each module, enhancing its value throughout its lifespan by detecting critical failures early and facilitating continuous monitoring.

Additionally, this work aims to significantly reduce inspection processing times, streamline results, and improve upon human labor limitations.

For future work, it is essential to expand the dataset to include more defect types, creating a more comprehensive ecosystem that addresses both existing and emerging defects. Furthermore, the adaptability of new models should be considered to enhance results.

8   REFERENCES

[1] Hussain, T., Hussain, M., Al-Aqrabi, H., Alsboui, T., & Hill, R. (2023). A review on defect detection of electroluminescence-based photovoltaic cell surface images using computer vision. Energies, 16(10), 4012. https://doi.org/10.3390/en16104012

[2] Wang, J., Bi, L., Sun, P., Jiao, X., Ma, X., Lei, X., & Luo, Y. (2023). Deep-learning-based automatic detection of photovoltaic cell defects in electroluminescence images. *Sensors, 23*(1), 297. https://doi.org/10.3390/s23010297

[3] Deitsch, S., Christlein, V., Berger, S., Buerhop-Lutz, C., Maier, A., Gallwitz, F., & Riess, C. (2019). Automatic classification of defective photovoltaic module cells in electroluminescence images. *Solar Energy, 185*, 455-468. https://doi.org/10.1016/j.solener.2019.02.067

[4] Pierdicca, R., Paolanti, M., Felicetti, A., Piccinini, F., & Zingaretti, P. (2020). Automatic faults detection of photovoltaic farms: solAIr, a deep learning-based system for thermal images. *Energies, 13*(24), 6496. https://doi.org/10.3390/en13246496

[5] Wang, J., Bi, L., Sun, P., Jiao, X., Ma, X., Lei, X., & Luo, Y. (2023). Deep-learning-based automatic detection of photovoltaic cell defects in electroluminescence images. *Sensors, 23*(1), 297. https://doi.org/10.3390/s23010297

[6] Sowthily, C., Senthil Kumar, S., & Brindha, M. (2021). Detection and classification of faults in photovoltaic system using Random Forest algorithm. In Bhateja, V., Peng, S.L., Satapathy, S.C., & Zhang, Y.D.

(Eds.), *Evolution in Computational Intelligence. Advances in Intelligent Systems and Computing* (Vol. 1176). Springer, Singapore. https://doi.org/10.1007/978-981-15-5788-0_72

[7] Bertoldo, S., Caico, E. M., Corti, F., Coletti, G., & Acciarri, M. (2019). Passive electroluminescence and photoluminescence imaging acquisition of photovoltaic modules. *Sensors, 19*(11), 2588. https://doi.org/10.3390/s19112588

[8] Ren, S., He, K., Girshick, R., & Sun, J. (2017). Faster R-CNN: Towards real-time object detection with region proposal networks. *IEEE Transactions on Pattern Analysis and Machine Intelligence, 39*(6), 1137–1149. https://doi.org/10.1109/tpami.2016.2577031

[9] Girshick, R. (2015). Fast R-CNN. *2015 IEEE International Conference on Computer Vision (ICCV).* https://doi.org/10.1109/iccv.2015.169

[10] Wu, Y., & Kirillov, A. (2019). Detectron2. GitHub Repository.

[12] Bertoldo, S., Caico, E. M., Corti, F., Coletti, G., & Acciarri, M. (2019). Passive electroluminescence and photoluminescence imaging acquisition of photovoltaic modules. *Sensors, 19*(11), 2588. https://doi.org/10.3390/s19112588

[13] Li, H., Qi, X., Li, X., & Qi, X. (2023). Dilated heterogeneous convolution for cell detection and segmentation based on Mask R-CNN. *Sensors, 23*(5), 2543. https://doi.org/10.3390/s23052543

[14] Tang, W., Yang, Q., Xiong, K., & Yan, W. (2020). Deep learning based automatic defect identification of photovoltaic module using electroluminescence images. *Solar Energy, 201*, 453-460. https://doi.org/10.1016/j.solener.2020.03.049

41st European Photovoltaic Solar Energy Conference and Exhibition

University of Applied Sciences and Arts
of Southern Switzerland

SUPSI

Institute for Applied Sustainability to the Built Environment

Schweizerische Eidgenossenschaft
Confédération suisse
Confederazione Svizzera
Confederaziun svizra

Swiss Federal Office of Energy SFOE

RACONT2050
Reliability And COmparison of New PV Technologies

D. Chianese, M. Caccivio, G. Friesen

SUPSI-PVLab, Via Flora Ruchat-Roncati 15, 6850 Mendrisio, Switzerland, email: domenico.chianese@supsi.ch

INTRODUCTION

Fast and simultaneously innovations introduced in recent years by the PV manufacturers are manifold but too many changes at once have to be balanced with quality, reliability, and lifetime and, ultimately, return on investment (ROI). In this project we analyse and verify the reliability and long time performance of selected new technologies in real systems with the highest allowable voltage conditions (with Voc @(-10°C) voltages between 1400 and 1500V) and combined polarisation of the photovoltaic field toward ground. Seven PV technologies, four of which are bifacial, for a total of 132 kWp, are installed on the two green roofs of the university campus of USI-SUPSI in Viganello/Lugano and monitored with diagnostic tools. Reliability and performance analysis in the laboratory are correlated with field measurements.

A second installation with five strings (six modules per technology) was built on the roof of the Campus Trevano, adjacent to the remaining 1982 old TISO installation, as low-voltage string references and comparison.

		PV TECHNOLOGIES and manufacturer data						Warranty	
	Cell technology	Ptot [kW]	Pnom [W]	Bifi	Module efficiency [%]	Pm γ coeff. [%/°C]	First year Pm degradation [%]	Year 2-x annual Pm degradation [%]	
Tech #1	PERC mono-facial half-cut	23.22	540		21.1	-0.35	2	0.55	
Tech #2	PERC bifacial shingled	18.48	660	70±5%	21.2	-0.34	2	0.45	
Tech #3	TOPCon n-type mono-facial half-cut	14	560		21.7	-0.3	1	0.4	
Tech #4	TOPCon n-type bifacial half-cut	14.55	560	80±5%	21.7	-0.3	1	0.4	
Tech #5	HJT bifacial half-cut	27.75	375	90±2%	20.9	-0.26	1	0.2	
Tech #6	PERC mono-facial half-cut	18.995	655		21.1	-0.34	2	0.55	
Tech #7	TOPCon n-type bifacial half-cut	15.12	560	80±5%?	21.7	-0.3	1	0.4	

Table of module technologies and picture of PV system at SUPSI Campus in Lugano-Viganello (Switzerland)

INITIAL INDOOR MEASUREMENTS

For a batch of modules (reference modules and modules on the low voltage strings in Trevano) indoor measurement were performed:

	TESTED MODULES				MEASURED VALUES				
	N.meas. modules	cell technology	Pnom [W]	Avg initial Pmax	Avg initial Pmax deviation	Avg LID after 10kWh/m²	Max LID after 10kWh/m²	Avg deviation after indoor 10kWh/m²	Avg deviation after outdoor exposure
Tech #1	11	PERC mono-facial	540	536.71	-0.61%	-0.20%	-0.39%	-0.37%	-1.57%
Tech #2	10	PERC bifacial shingled	660	636.76	-3.52%	-0.19%	-0.31%	-3.62%	-4.53%
Tech #3	16	TOPCon n-type mono-facial	560	551.47	-1.52%	-0.26%	-0.43%	-1.78%	-2.13%
Tech #4	10	TOPCon n-type bifacial	560	547.63	-2.21%	-0.26%	-0.40%	-2.20%	-2.82%
Tech #5	12	HJT bifacial	375	364.93	-2.69%	-0.06%	-0.15%	-2.4%	-4.57%
Tech #6	13	PERC mono-facial	655	653.92	-0.17%	-0.09%	-0.32%	-0.29%	-1.82%
Tech #7	4	TOPCon n-type bifacial	560	548.05	-2.13%	-0.08%	-0.35%	-2.22%	-2.50%

- **Average initial Pm deviation** compared to the factory nominal values differ from -0.17% to -3.52% (Table 2), while St.Dev varies from 0.12 to 0.35%.
- On average, the manufacturers have a measured power output that is almost -2% lower than the nominal values and are therefore below the declared initial values with a production distribution of -0% to +3%.
- The **bifaciality** measurement results, according to IEC TS 60904-1-2:2019, show relative values consistent with what is declared on the datasheets (i.e. 66.7% for PERC; 89.3% for HJT; 79.7% and 77.5% for TOPCons).

INDOOR AND OUTDOOR LID DETERMINATION

In the indoor stabilisation four reference modules of each of the 7 technologies were exposed to > 10kWh/m2 of irradiation in our laboratory's Steady State Sun Simulator (CCB class) prior to be installed into the fields.

- The difference between initial Pm values and after 10kWh/m^2 is small and on average does not exceed -0.4%.
- The deviation was -0.35% up to -0.43% in TOPCon modules; from -0.11% up to -0.39% in PERC modules and -0.15% in HJT modules.

After approx. 1220 kWh/m2 in the field, four reference modules per string were replaced.

- The deviations of the power (Pm) from the values provided by the manufacturers (Pnom) are consistent with the deviation data from the Pmax values after indoor LID and within the confidence interval of the measurements. These deviations are important in HJT (- 4.57%) and in PERC bifi Shingled (-4.53%).

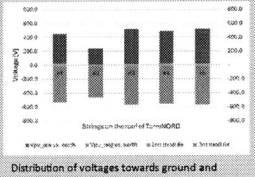

Distribution of voltages towards ground and position of the 4 reference modules on the string.

MEASURED IRRADIANCE DEPENDENCY

As one might expect, the normalised efficiency at 1000W/m^2 shows decreasing values with decreasing irradiance for each technology. Tech#2 technology shows a better performance at low irradiances, probably due to a low shunt resistance and high losses at STC condition.

Normalised efficiency at various irradiances

FIRST FAULTS DETECTION

- In this first year of operation, some traces of degradation could be observed in one type of module and two other modules broke for unknown reasons.

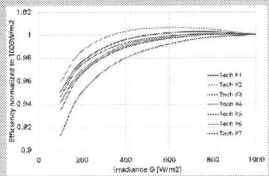

- Thermographic analysis of the module arrays in both locations showed no hot spots in the modules apart in the broken two modules.

Low-voltage strings on Campus Trevano, for references and comparison and the remaining 42 years old TISO installation

CONCLUSIONS

- Initial indoor analysis show deviation in Pm, on average is almost -2% lower than the nominal values ,
- LID after 10 kWh/m^2 in Steady State Simulator show small differences and on average does not exceed -0.4%.
- LID after 1220 kWh/m^2 in the field show consistent differences
- Two broken modules was found in the system
- After one year of exposure, no particular problems were observed in the performance of the strings.
- A new equipment was developed for "Long-Term Monitoring of Degradation and Defect in High-Voltage Strings Through Dark I-V Measurements": see poster 4BV.3.35 .

ACKNOWLEDGEMENTS: The here presented results are part of the project RACONT funded by the Swiss Federal Office of Energy under the contract SI/502264-01.

Comparative analysis of string IV measurement methods for fault detection in photovoltaic systems

Martin Bartholomäus, Peter B. Poulsen, Sergiu V. Spataru

Technical University of Denmark, Department of Photonics Engineering, Roskilde, Sjælland, 4000, Denmark

Abstract—**This work assesses the extent to which three string current-voltage (IV) measurement methods - a portable IV tracer, an inline IV tracer and an inverter - can be utilized to detect faults in photovoltaic (PV) strings in the field. The analyzed faults include potential induced degradation, cell cracks, ribbon damage, short-circuited and open-circuited bypass diodes and shading. We analyse the imprint of fault penetration in the strings with simple diagnostic markers such as the series resistance and shunt resistance, estimated from the slopes of the IV characteristic curve, as well as the I_{MP}-I_{SC} ratio and V_{MP}-V_{OC} ratio, to evaluate their applicability to detect faults in PV string under various conditions outdoors. Results show all three devices measuring very similar IV curves under the described faults and on healthy strings. We demonstrate, that the shunt resistance is suitable to detect early PID at a power loss smaller than 0.5%. Shading proved to be detectable from the series resistance and V_{MP}-V_{OC} ratio from measurements with all three devices.**

Index Terms—**IV-curve, IV tracer, inverter, fault detection**

I. INTRODUCTION

The current-voltage (IV) characteristic curve of a photovoltaic (PV) device contains diagnostic information, such as series and shunt resistance, which allow to detect and identify degradation and faults in PV panels and strings [1]. This has led to the use of module or PV string IV measurements for condition monitoring and fault detection purposes [2, 3, 4]. However, the widespread application of string IV monitoring for diagnosis is limited due to the availability of inverters and commercial PV monitoring equipment capable of continuously monitoring the IV characteristic of PV strings. At the same time, researchers are investigating the additional benefits of using IV curve data compared to maximum power point (MPP) monitoring data typically acquired by inverters. Popular approaches to this involve machine learning [5, 6].

In practice, several methods for measuring string IV characteristics are available commercially. First, portable IV tracers are highly accurate but require a trained person to be physically present at the PV string of interest, connect the device, and take measurements. This method is time-intensive, especially if multiple strings are tested; therefore, the number of measurements is limited.

Second, some PV inverters are capable of taking string IV measurements [7] [8] [9]. This method has the advantage that no additional hardware needs to be installed. However, we have shown in previous work that string IV curves measured by an inverter show large differences at low irradiance and under the influence of shade compared to measurements with an IV tracer [10]. This could lead to difficulties detecting

specific faults that are irradiance-dependent and require IV measurements at different irradiance levels for confirmation, such as shunting-type potential induced degradation, which is more visible on the IV curve at lower irradiance [11]. The applicability of inverter measured string IV curves to fault detection methods might therefore be compromised because of weak or misleading diagnostic marker signals.

Third, inline devices can be deployed, which are dedicated fixed measurements devices installed at the PV string. They require additional hardware but spare manual labor after installation. They can also be used to retrofit older plants that do not have inverters with IV measurement capability, as measurements can be scheduled or triggered remotely. For example, Joglekar and Hedge utilized an online IV tracer that can be retrofitted at junction box level [12]. ThamizhMani et al. presented a non-contact IV-tracing solution that is capable to measure IV curves at a sub-string level using multiple voltage probes [13].

This work compares string IV measurement methods and investigates which method is most suitable for fault detection under various operating conditions. For this purpose, we compare marker signals, such as the current or voltage ratios extracted from the IV measurements of each device. We test strings with planted faults such as potential induced degradation (PID), cell cracks, ribbon damage, short-circuited or open-circuited bypass diodes and analyze the IV curve imprints detected by each method compared to healthy strings.

II. EXPERIMENTAL SETUP

A. Measurement equipment and procedure

Our experiments were conducted at a PV plant in Denmark (lat. 55.7, long. 12.1), which is part of the Risø Campus at the Technical University of Denmark (DTU). The PV module substructures are south-facing fixed-tilt arrays with a tilt angle of 25°. We conducted the tests on four PV strings with 22 monofacial modules of type Trina TSM-305DD05A.08 with 305 Wp each. Figure 1 shows a single line diagram of the strings. IV curve measurements were taken on each string individually with each IV tracer device.

Prior to our experiments, the strings had been deployed for about five years. As the deployment time and conditions were the same for each string, we assume a relatively even degradation between all four strings.

String level IV curves were measured with a Huawei Sun2000 185-KTL inverter, an inline string IV tracer by Pordis (a first-of-its-kind prototype for in-situ IV measurement of

41st European Photovoltaic Solar Energy Conference and Exhibition

TABLE I: Specifications of IV tracer devices

Device	Solmetric PVA 1500V	Pordis 231 series	Huawei Sun2000 185KTL
Sweep time [s]	~0.2	~0.1	~1.2
No of IV points	500	~750	64
Sweep technology	charging a load capacitor	charging a load capacitor	DC/DC converter
Accuracy (I,V)	± 0.5% (0.25V, 40mA) [14]	unknown	± 0.5% [7]

multiple 1500V strings), and Solmetrics' PV Analyzer 1500 (PVA). Table I gives an overview of the specifications for each tracing device.

The first two devices allowed for the recording of a large amount of traces, as they can be triggered remotely. The Pordis acquired measurements every five minutes and we considered a total of 7313 curves taken between March 25 and July 19 (3835 after filtering). The Huawei measured 228 curves between July 19 and September 3, 2024. A Kipp and Zonen SMP10-V Pyranometer (spectrally flat class A) is used to record the irradiance in the plane of array (G_{POA}) at every minute and the data is matched to the IV curves with the nearest matching timestamp.

To incorporate measurements from the PVA portable tracer, we conducted experiments on two days in July and August 2024, measuring with all three devices in close timely proximity. Here, we first triggered measurements with the inverter and inline tracer device. For measurements with the PVA 1500V IV-tracer, the strings were then isolated and connected to the device, where the measurement was triggered from the Solmetric software.

To ensure similar measurement conditions, the time difference should be small, so that sun's position, irradiance and temperature conditions are similar. For comparisons, corresponding measurements were corrected to common temperature and irradiance conditions following procedure 1 of IEC 60891.

During measurements with the PVA, we measured both G_{POA} and temperature at the back of the module with Solmetrics' Solsensor. Its irradiance sensor was used as a single central sensor for plane of array irradiance, and a type K thermocouple served as the temperature sensor, though only one module on String 7.4 could be measured. The temperature reading is assumed to be representative for all four strings.

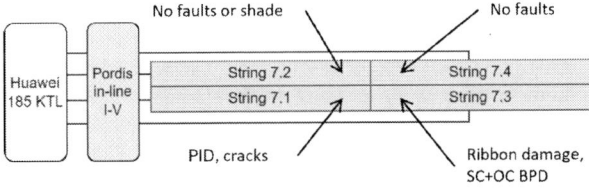

Fig. 1: Array layout of tested strings from DTU Risø campus. The strings each have individual inputs at the inverter and inline tracer and can be measured separately by each device. Each string consists of 22 modules, all of the same make.

TABLE II: Faults introduced in the PV strings. Power loss is measured at standard testing conditions (STC) and given relative to total string power (6.71 kWp).

String	Faults	Power loss at STC
String 7.1	6x PID	-27.1 W (-0.40 %)
	8x cell cracks	-51.9 W (-0.77 %)
String 7.2	No faults or shade	
String 7.3	8x ribbon damage	-98.7 W (-1.47 %)
	2x short-circuited BPD	-302.4 W (-4.51 %)
	2x open-circuited BPD	-0 W (-0 %)
String 7.4	No faults	

B. Introduction of faults

We tested the devices by introducing temporary and permanent faults within the measured PV strings. First, two of the four strings were equipped with modules with known, artificially created faults. Table II gives an overview of the type of faults and amount of modules that were introduced into the PV strings as well as the power loss connected to the faults. To quantify the power loss, indoor flash tests were performed before and after exposing the modules to stress. The solar simulator used was an Endeas Quicksun 540XLi large area flasher (accuracy class AAA per IEC 60904-9 Ed.1).

For String 7.1, we created six artificially degraded modules with PID through voltage stress with the foil method following IEC TS 62804-1. The modules were put under at negative 1500V bias for more than three weeks. Additionally, cell cracks were introduced to eight modules with mechanical force by walking on the modules. Electroluminescence images were used to confirm the formation of the cell cracks.

On String 7.3, eight modules were damaged by cutting ribbons through the backside sheet. The openings from cutting the ribbons were re-sealed with plastic sealant. Further, two modules were modified with open-circuited bypass diodes (BPD) by cutting the diodes, and two modules with short-circuited bypass diodes (by soldering a wire across the diodes in the module junction box). Of the short-circuited modules, one had one of three BPD short-circuited, and the other two out of three.

The introduction of the described faults is explained in greater detail in a masters thesis by Yu [15].

In addition, we conducted experiments with artificial shade, using cardboard applied to the front side of the solar modules, covering and shading full cells. We increased the severity of the shade in four stages, by introducing an increasing number of shaded cells/modules into the PV string, taking

41st European Photovoltaic Solar Energy Conference and Exhibition

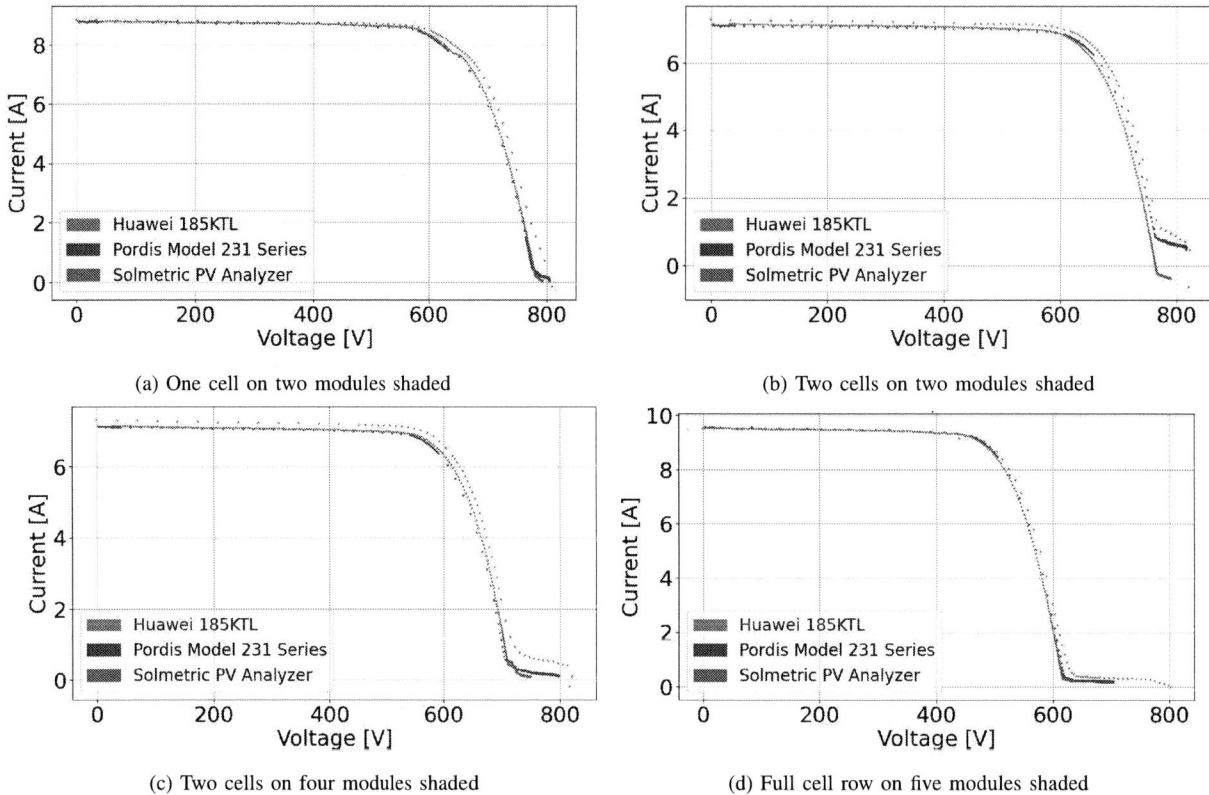

(a) One cell on two modules shaded

(b) Two cells on two modules shaded

(c) Two cells on four modules shaded

(d) Full cell row on five modules shaded

Fig. 2: Example IV curves under increasing severity of shade on the string.

IV measurements at each step and at the initial, non-shaded state. The stages were: shading one cell on two modules (in the 22-module string), shading two cells on two modules (the two cells are on two different cell strings within the modules), shading two cells on four modules, and shading an entire row of cells on five modules (modules are in portrait mode).

C. Calculation of diagnostic markers

Simple methods are employed to detect faults through the IV measurements obtained with each device. Here we use the I_{MP}-I_{SC} or V_{MP}-V_{OC} ratio, as well as series (R_S) and shunt resistance (R_{SH}) as markers and detect the strongest signal among the different IV measuring methods.

The I_{MP}-I_{SC} ratio is calculated as I_{MP}/I_{SC} and the V_{MP}-V_{OC} ratio as V_{MP}/V_{OC}. The shunt and series resistance are calculated with the slope method. We are selecting one third of the range from I_{SC} to I_{MP} (for R_S) and V_{OC} to V_{MP} (for R_{SH}) respectively and fit a linear function to the IV curve in that range. The inverse of the slope of the linear function gives the resistance of the PV device. Note that these resistances are calculated as indicators and do not necessarily have a physical meaning, for example under the influence of shade.

III. RESULTS

A. Comparison of IV tracing devices

We first show example IV curves measured with all three measurement devices. We cover non-faulty strings, a string with PID and cracks, a string with ribbon damage, open-circuited and short-circuited bypass diodes and the influence of shade of increasing severity on a healthy string (Table II in Section II-B).

Figure 2 shows selected IV curves from measurements under the influence of artificial shade on a healthy PV string. Measurements were done with all three measurement devices and were corrected to common temperature and irradiance conditions. For all measurements, G_{POA} was high, ranging from 670 to 970 W/m^2. Under low shade (one cell shaded on two modules), the inverter displays a slightly different curve, with the two other tracer showing a more distinct deformation round the maximum power point. In the second shade scenario, the agreement is quite well and differences in the position of the inflection points are mostly caused by the correction procedure, correcting for relatively high irradiance mismatch between PVA and the others (around 90 W/m^2). In the third scenario, there is a deformation around the inflection point, which the inverter does not record, whereas the other two

1289

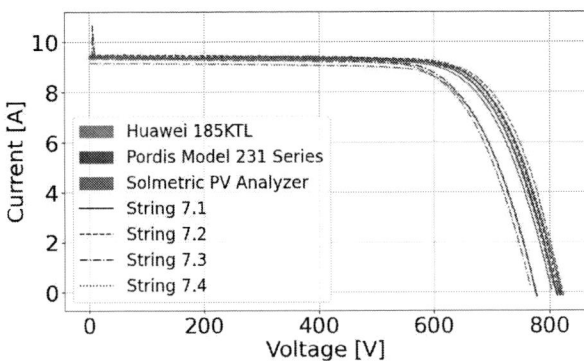

Fig. 3: IV curves of the 4 strings without shade

tracers do. In the most severe shade scenario, the three methods agree well again.

Figure 3 shows IV curves on all four strings without shade measured with the three different devices. All curves are corrected to common temperature and irradiance conditions. It is clearly visible that String 7.3 has a lower open-circuit voltage, which can be attributed to the short-circuited bypass diodes, which exclude the cells on the short-circuited cell strings from adding to the string voltage. All three devices capture this effect and show similar IV curves. The string with PID (String 7.1) shows slightly less power, but the effect is marginal. In general, the agreement between the devices is very high, considering that uncertainty around measuring the ambient conditions contributed to the differences between them. Overall, the three IV measurement devices showed very comparable measurements on healthy, faulty and shaded strings in relatively high irradiance conditions.

B. Application to fault detection

The following section displays simple markers that are utilized to detect the faults introduced to the PV strings. It shall be noted, that the physical meaning of the calculated resistances may be limited, but is not important in this context.

1) Shunt resistance diagnostic marker: Figure 4a shows the shunt resistance estimated with the slope at I_{sc} method as a marker for fault detection from 3835 measurements with the Pordis inline tracer. Different strings (and therefore faults) are marked with colors and a regression line of an inverse function is added. It shows that there is significant clustering between the non-faulty and faulty strings, especially visible at low irradiance. It is therefore possible to detect faults such as PID using the shunt resistance as a marker, despite of the relatively low power loss caused by the faults. The lower shunt resistance in the PID and cracks string is likely cause by the PID, which is known to create shunt paths and therefore lower the shunt resistance. The string with ribbon damage and bypass diode failures also presents with significantly lower shunt resistance in low light. Shade does not alter the shunt resistance marker to a relevant extend.

For comparison, Figure 4b shows the 228 measurements with the Huawei inverter. The strings cluster, but the regression fitting is poor due to low amount of data points especially in the low irradiance range. The results still present the PID-infused string with the lowest shunt resistance across all irradiance levels including low light. For two other two strings, the inverter data is inconclusive, unlike the larger data set recorded with the inline tracer. To reliably detect the faults, a wide range of irradiance conditions needs to be covered, focusing on lower irradiance. In-situ devices are more suitable for such a task compared to a portable IV tracer.

2) Series resistance diagnostic marker: Figure 5 shows the estimated series resistance as a function of G_{POA} for both the inline tracer and inverter. Regression lines (for inverse functions) are calculated for each fault type except shade and displayed in the graphs.

In the inverter data, there is a significant difference between the non-faulty and faulty strings. However, the inline tracer data does not show the same effect; instead, all data clusters overlap and regression lines are very close together. The difference between results from both devices is likely due to the limited number of data points in the Huawei data, especially in the low irradiance range. Series resistance effects are expected to be most visible in the high irradiance range, but here, differences between the strings are minimal. We therefore cannot use the series resistance calculated from the slope of the IV near V_{OC} as a fault marker for the presented faults.

However, the series resistance marker changes significantly (up to more than ten-fold) under the influence of shade, seen in both devices (inline tracer: series resistance of up to 300 Ω, not shown, PVA not shown). This is because partial shading primarily affects the higher voltage part of the string IV, where the series resistance marker is calculated. It should be noted that the data on the influence of shade is limited to the high irradiance range ($> 600~W/m^2$). Under these conditions, the results show that this marker can be utilized as a fault detection marker for shade.

3) I_{MP}-I_{SC} ratio diagnostic marker: The I_{MP}-I_{SC} ratio is depicted as a function of G_{POA} in Figure 6. No significant clustering between the strings with and without faulty modules can be observed. Even under the influence of shade, the I_{MP}-I_{SC} ratio remains stable and is not useful to differentiate shaded from unshaded strings.

4) V_{MP}-V_{OC} ratio diagnostic marker: The V_{MP}-V_{OC} ratio as a function of G_{POA} is shown in Figure 7. No significant clustering between the strings with and without faulty modules can be observed. However, under the influence of shade, the V_{MP}-V_{OC} ratio drops significantly, as partial shading impacts the shape of the string IV curve quite significantly, shifting the maximum power points due to the presence of the bypass diodes, and making it a suitable marker to detect shade in string IV.

41st European Photovoltaic Solar Energy Conference and Exhibition

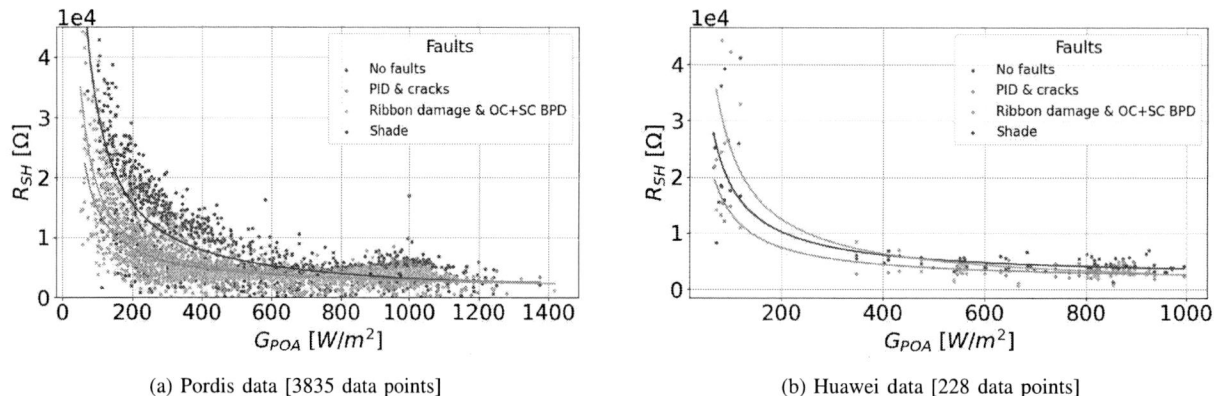

(a) Pordis data [3835 data points]

(b) Huawei data [228 data points]

Fig. 4: Shunt resistance as a fault detector as a function of GPOA with the inline tracer and inverter.

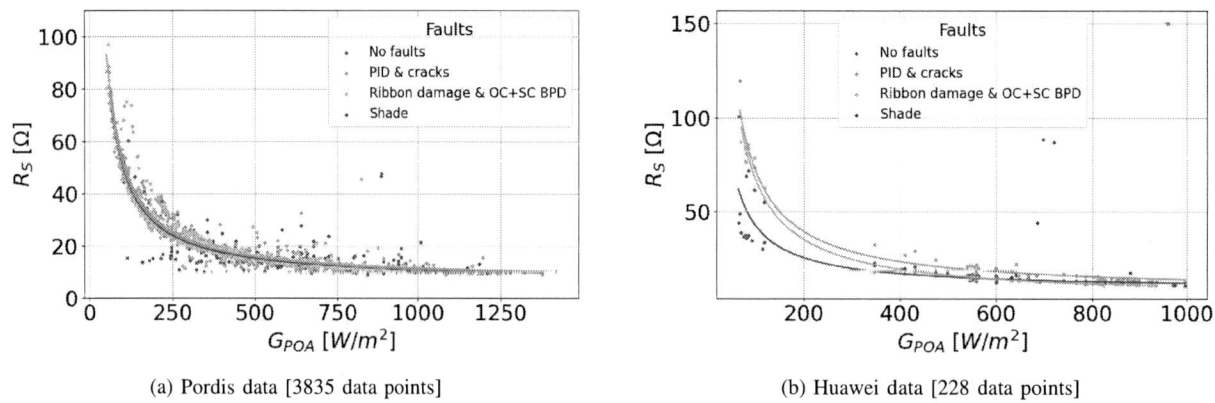

(a) Pordis data [3835 data points]

(b) Huawei data [228 data points]

Fig. 5: Series resistance as a fault detector as a function of GPOA with the inline tracer and inverter.

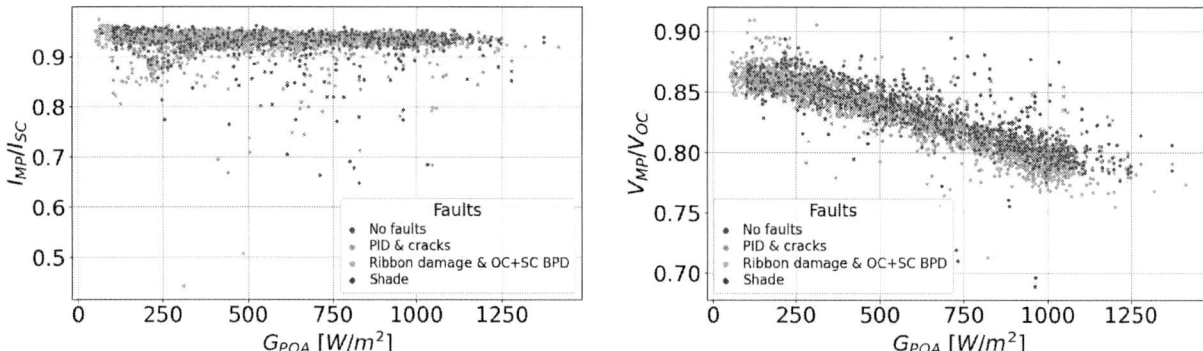

Fig. 6: I_{MP}-I_{SC} ratio as a fault detector with the Pordis inline string IV tracer

Fig. 7: V_{MP}-V_{OC} ratio as a fault detector with the Pordis inline string IV tracer

IV. SUMMARY AND CONCLUSION

In this work, three IV tracing devices, a portable IV tracer, a string inline IV tracer and an inverter, were utilized to record string IV curves on strings with faulty modules and under the influence of shade. The measurements were compared and simple fault detection markers were tested for their ability to reliably detect the faults.

We found, that all three devices produce IV curves of similar quality on healthy and faulty strings as well as under the influence of shade. The comparison of the IV tracing devices is limited by the fact that IV curves could not be acquired simultaneously with all devices on the same string, making IV curve corrections to common temperature and irradiance conditions necessary.

From the fault detection markers, the shunt resistance derived from the slope of the IV curve near I_{sc} can be used to detect degradation in PV strings due to PID and cell cracks, even at low fault penetration (power loss: -0.4% and -0.77%, respectively). Further, the V_{MP}-V_{OC} ratio and series resistance can be used to detect shade.

Additional testing should be done to confirm faults and keep different fault types apart, which will be further investigated in future work. For example, the results could be improved by separating the faults in different strings, increasing their severity and conducting additional experiments, such as partial shading on modules with open-circuited bypass diodes.

Future work should also investigate the tested markers as a function of module temperature, which was not attempted here due to missing data. In addition, more advanced fault detection markers and strategies should be tested and bench marked against fault detection using simple maximum power point monitoring data.

V. ACKNOWLEDGEMENT

This research received support from the TwInSolar project funded by the European Union's Horizon Europe research and innovation program, grant number 101076447.

REFERENCES

[1] Y. Zhu and W. Xiao. "A comprehensive review of topologies for photovoltaic I–V curve tracer". en. In: *Solar Energy* 196 (Jan. 2020), pp. 346–357. ISSN: 0038092X. DOI: 10.1016/j.solener.2019.12.020. URL: https://linkinghub.elsevier.com/retrieve/pii/S0038092X19312344 (visited on 02/01/2023).

[2] Sergiu Spataru et al. "Monitoring and Fault Detection in Photovoltaic Systems Based On Inverter Measured String I-V Curves". en. In: 2015, p. 8.

[3] Mohamed Hassan Ali et al. "Real Time Fault Detection in Photovoltaic Systems". In: *Energy Procedia* 111 (2017), pp. 914–923. ISSN: 1876-6102. DOI: https://doi.org/10.1016/j.egypro.2017.03.254.

[4] Luís Guilherme Monteiro et al. "Field IV curve measurements methodology at string level to monitor failures and the degradation process: A case study of a 1.42 MWp PV power plant". In: *IEEE Access* 8 (2020), pp. 226845–226865.

[5] Haohui Liu et al. "Automatic IV Curve Diagnosis with Deep Learning". en. In: *2021 IEEE 48th Photovoltaic Specialists Conference (PVSC)*. Fort Lauderdale, FL, USA: IEEE, June 2021, pp. 2242–2246. ISBN: 978-1-66541-922-2. DOI: 10.1109/PVSC43889.2021.9519033. URL: https://ieeexplore.ieee.org/document/9519033/ (visited on 01/18/2023).

[6] Michael W. Hopwood et al. "Neural Network-Based Classification of String-Level IV Curves From Physically-Induced Failures of Photovoltaic Modules". en. In: *IEEE Access* 8 (2020), pp. 161480–161487. ISSN: 2169-3536. DOI: 10.1109/ACCESS.2020.3021577. URL: https://ieeexplore.ieee.org/document/9186596/ (visited on 02/01/2023).

[7] Ltd. Huawei Technologies Co. *Smart I-V Curve Diagnosis Technical White Paper*. en. 2020.

[8] Ltd Sungrow Power Supply Co. *Sungrow New String Inverters for C&I PV Applications — SG33/40/50/110C*. en. 2020.

[9] SB Kjær, O Oprea, and U Borup. "Adaptive sweep for PV applications". In: *26th European Photovoltaic Solar Energy Conference and Exhibition, Hamburg, Germany*. 2011, pp. 3708–3710.

[10] Martin Bartholomäus et al. *Evaluating the Accuracy of Inverter Based String IV Measurements*. en. 2023. DOI: 10.4229/EUPVSEC2023/4CV.1.4. URL: https://userarea.eupvsec.org/proceedings/EU-PVSEC-2023/4cv.1.4.

[11] Wei Luo et al. "Elucidating potential-induced degradation in bifacial PERC silicon photovoltaic modules". en. In: *Progress in Photovoltaics: Research and Applications* 26.10 (2018), pp. 859–867. ISSN: 1062-7995, 1099-159X. DOI: 10.1002/pip.3028. URL: https://onlinelibrary.wiley.com/doi/10.1002/pip.3028.

[12] Ashish V Joglekar and Balachandra Hegde. "Online IV Tracer for per string monitoring and maintenance of PV panels". In: *IECON 2018-44th Annual Conference of the IEEE Industrial Electronics Society*. IEEE. 2018, pp. 1890–1894.

[13] G TamizhMani et al. "Simultaneous non-contact IV (NCIV) measurements of photovoltaic substrings and modules in a string". In: *2021 IEEE 48th Photovoltaic Specialists Conference (PVSC)*. IEEE. 2021, pp. 1792–1794.

[14] a Fluke Company Solmetric. *PV Analyzer I-V Curve Tracers*. 2024. URL: https://www.solmetric.com/pv-analyzer-i-v-curve-tracers/ (visited on 10/09/2024).

[15] Congrui Yu. "Comprehensive Evaluation of Inspection Techniques for Fault Detection in Ground-Mounted Photovoltaic Systems". MA thesis. Technical University of Denmark, 2024.

41st European Photovoltaic Solar Energy Conference and Exhibition

AI-SAFEPV: AN AI-BASED FAULT DETECTION SOFTWARE PACKAGE TO PROVIDE SAFETY IN PHOTOVOLTAIC ARRAYS

A. Eskandari[1], J. Milimonfared[2], A. Nedaei[2], P. Parvin[3], M. Braga[4], and M. Aghaei[5,6*]

[1]Department of Electrical Engineering, Iran University of Science and Technology, Tehran, Iran
[2]Department of Electrical Engineering, Amirkabir University of Technology, Tehran, Iran
[3]Department of Physics and Energy Engineering, Amirkabir University of Technology, Tehran, Iran
[4]Universidade Federal de Santa Catarina – UFSC, Florianópolis, SC 88040-900 Brazil
[5]Department of Ocean Operations and Civil Engineering, Norwegian University of Science and Technology, Ålesund, Norway
[6]Department of Sustainable Systems Engineering, University of Freiburg, Freiburg, Germany

*mohammadreza.aghaei@ntnu.no

ABSTRACT: Faults diagnosis in photovoltaic (PV) systems may not always be possible using conventional protection devices due to the nonlinear behavior of PV characteristics, its dependency on the operating environment, and operation of Maximum Power Point Tracking (MPPT) algorithms. To date, numerous studies have been carried out to overcome this challenge through Artificial Intelligence (AI) techniques. However, most of the AI-based techniques require a large dataset and also suffer overfitting problem. In this study, we propose AI-SafePV which is an intelligent and automatic fault diagnosis software package using a small dataset for the training process through feature extraction and selection processes, as well as using an ensemble learning algorithm to classify open-circuit (OC) and line-line (LL) faults in PV systems. For this purpose, firstly, the AI-SafePV software package extracts the key features of the operating current and voltage of the PV arrays. Secondly, the Lasso penalty is applied as a feature selection technique to determine the best subset of features. Thirdly, an ensemble learning algorithm consisting of three individual learning algorithms namely Logistic Regression (LR), Support Vector Machine (SVM), and k-Nearest Neighbors (KNN) is used in the classification stage to predict conditions of PV systems based on a weighted voting approach. Moreover, we apply a genetic algorithm (GA) to optimize the weights assigned to the algorithms in order to detect electrical faults in PV systems with higher accuracy. The experimental output of AI-SafePV software package demonstrate high efficiency and reliability to diagnose open-circuit (OC) and line-line (LL) faults in PV arrays under challenging conditions with an outstanding accuracy of 99%.

Keywords: Photovoltaic (PV) system; Fault diagnosis; Autonomous monitoring; Ensemble learning algorithm; Genetic algorithm (GA)

1 INTRODUCTION

At the end of 2023, photovoltaics (PV) installation exceeded 1.5 TW [1]. PV systems may experience numerous failures during their lifetime due to environmental conditions, human errors, and degradation [2]. Therefore, fault diagnosis and comprehensive monitoring play a pivotal role for improving the service life of PV systems. However, conventional protection devices such as Ground Fault Protection Devices (GFPDs) and Over Current Protection Devices (OCPDs) are seemingly suffering from several negative points. To overcome the challenges, various PV monitoring techniques were deployed for detection and classification of a wide range of electrical faults in PV systems [3], [4]. These techniques are generally classified into: a) difference measurement, b) signal processing, and c) artificial intelligence (AI) methods. Among AI algorithms, the ensemble learning algorithms have received much attention in recent years because they can obtain a better prediction accuracy compared to the single classifiers and also reduce the overfitting chance in complex models. Ensemble learning algorithms combine the predictions of multiple learning algorithms [5]. To date, the ensemble learning algorithms have been successfully used in renewable energy systems and PV systems [6].

In this study, an intelligent and autonomous software package named AI-SafePV is proposed to diagnose the faults in PV arrays using a weighted ensemble learning algorithm based on a genetic algorithm (GA). The AI-SafePV software package includes two steps. The first step is related to creating the dataset. In this step, the main feature is extracted via analyzing the PV array current and voltage under normal and faulty conditions. These features are based on statistical parameters in the time and frequency domain. Then, the least absolute shrinkage and selection operator (Lasso) penalty algorithm as a feature selection technique was used for finding the best feature and obtaining a higher performance accuracy in the classification process. In the second step, the weighted ensemble learning algorithm is applied for fault diagnosis. The proposed ensemble learning algorithm comprises three individual learning algorithms namely Logistic Regression (LR), Support Vector Machine (SVM), and k-Nearest Neighbors (kNN). The final result of the classification is obtained from the weights combination of the result of each of the individual learning algorithms. Here, we have selected a GA to optimize the weights of these learning algorithms in order to achieve a higher accuracy in fault diagnosis process. The experimental output of AI-SafePV confirms the reliability and accuracy of this software package in diagnosing the faults with a smaller number of datasets, under PV systems challenging operation conditions which is high fault impedance and low mismatch levels.

2 THE AI-SAFEPV SOFTWARE PACKAGE

In this section, the AI-SafePV which an intelligent and autonomous diagnosis software package for electrical faults detection in PV systems is introduced. This software package is based on statistical features which are extracted from the current and voltage of PV arrays as well as an optimal weighted ensemble learning model using a GA. Figure 1 sketches a clear schematic diagram of how the developed AI-SafePV software package is structured. This package includes two main stages which are 1) the dataset collection which is followed by a preprocessing stage and 2) the fault diagnosis stage based on a developed ensemble learning algorithm. The mentioned preprocessing stage is comprised of three main sub-stages which are (i) feature extraction, (ii) data normalization, and (iii) feature

selection. After the preprocessing stage is fulfiled, the developed ensemble learning model is implemented for the faults diagnosis. In this stage, the input dataset is divided into two subsets which are training and validation datasets, such that training dataset is applied to train the model and then the validation dataset is used to validate the performance of the AI-SafePV software package.

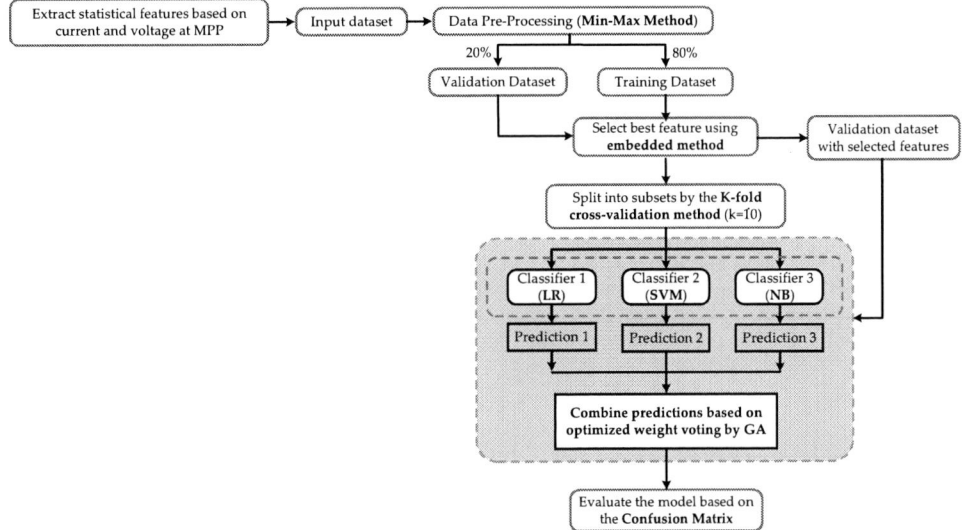

Figure 1. The intelligent framework of AI-SafePV software package for PV array fault diagnosis.

2.1. Statistical Features Extraction

In this step, the statistical features in the frequency and time domains [7] are extracted from the current and voltage of the PV array at MPP condition. In this study, 12 features in the time domain and 3 features in the frequency domain were extracted from the current and voltage of PV arrays to create the datasets. Therefore, the total number of features is equal to 30 for the current and voltage. The frequency domain and time domain features are listed in Table I, respectively.

Table I
Frequency domain and time domain features.

Extracted Features		
$f_1 = \dfrac{1}{N}\sum\limits_{i=1}^{N} x_i$	$f_2 = \sqrt{\dfrac{1}{N}\sum\limits_{i=1}^{N}(x_i\text{-}f_1)^2}$	$f_3 = \sqrt{\dfrac{1}{N}\sum\limits_{i=1}^{N} x_i^2}$
$f_4 = \dfrac{1}{N}\sum\limits_{i=1}^{N}(x_i\text{-}f_1)^2$	$f_5 = \dfrac{1}{N}\sum\limits_{i=1}^{N}\left(\dfrac{x_i\text{-}f_1}{f_2}\right)^4$	$f_6 = \dfrac{1}{N}\sum\limits_{i=1}^{N}\left(\dfrac{x_i\text{-}f_1}{f_2}\right)^3$
$f_7 = \left(\dfrac{1}{N}\sum\limits_{i=1}^{N}\sqrt{|x_i|}\right)^2$	$f_8 = \max\dfrac{|x_i|}{f_3}$	$f_9 = \dfrac{\max|x_i|}{1/_N \Sigma_{i=1}^{N}|x_i|}$
$f_{10} = \max\dfrac{|x_i|}{f_7}$	$f_{11} = \dfrac{f_3}{f_1}$	$f_{12} = \max(x_i)\text{- min}(x_i)$
$f_{13} = \dfrac{1}{N}\sum\limits_{i=1}^{N} f_i$	$f_{14} = \sqrt{\dfrac{1}{N}\sum\limits_{i=1}^{N}(f_i\text{-}f_1)^2}$	$f_{15} = \sqrt{\dfrac{1}{N}\sum\limits_{i=1}^{N} f_i^2}$

2.2. Data Normalization

In this study, the Min-Max Normalization was used because it maintains the relation in the original dataset and performs a linear transformation on the original dataset. This method can transform the data into a pre-defined boundary. the data were normalized in the range of 0 and1 by Eq. (1):

$$x' = \frac{x - \min_Y}{\max_Y - \min_Y} \qquad (1)$$

where x is the original value of a feature Y, x' is its normalized value, and $\max_Y$ and $\min_Y$ are the maximum and minimum values of feature Y.

2.3. Lasso Algorithm

In this study, we have used the embedded method to select the best features because it has an interaction with the learning algorithm construction, and has a less computational complexity than wrapper methods. Among embedded methods, the Lasso algorithm has received much attention in recent years because it can produce exact zero estimated coefficients for irrelevant features. Hence, these features can be removed from the original dataset. The Lasso algorithm is a regularization method that is used with logistic regression for classification problems [8], expressed through Eq. (2):

$$\beta_{Lasso} =$$
$$\arg\min_{\beta}\left\{\frac{1}{2}\sum_{i=1}^{N}\left(y_i - \beta_0 - \sum_{j=1}^{p} x_{ij}\beta_j\right)^2 + \lambda\sum_{j=1}^{p}|\beta_j|\right\} \qquad (2)$$

where $y_i \in\{0,1,2\}$ represents class labels, $x_i\in R^p$ denotes feature vector with p-dimensional, β is classification weight vector, and λ is a tuning parameter that controls estimated coefficients.

2.4. Ensemble Learning Algorithm

In this study, to obtain a higher accuracy and a better performance in the process of faults diagnosis, we have proposed a weighted ensemble learning algorithm which is a combination of three individual predictive algorithms namely Logistic Regression (LR), Support Vector

Machine (SVM), and k-Nearest Neighbors (kNN). To combine the final predictions, here, a weighted voting technique is used in which the final result is based on the weights (real numbers between 0 and 1) assigned to the output of each learning algorithm. Here, we have employed a Genetic Algorithm (GA) to obtain the optimized weights. The prediction output of the ensemble model based on weighted voting is computed by Eq. (3) [9]:

$$\acute{y} = \arg\max_j \sum_{i=1}^{m} w_i p_{ji}(x) \qquad (3)$$

where $\acute{y}$ denotes the predicted output of ensemble learning model, j represents the number of class labels which in our study $j = 3$, m is the number of learning algorithms which is also equal to three in this paper, p_{ji} shows the predicted output of i_{th} learning algorithm based on probability occurrence, w_i is the optimal weight which is assigned via a Genetic Algorithm (which is discussed in detail throughout the next section) to the i_{th} learning algorithm, i.e., $w = (w_1, w_2, w_3)$, $w_i \in [0,1]$.

3 RESULTS AND DISCUSSION

3.1. Experimental Setup and Dataset

In this experimental study, a small-scale PV system with capacity of 180 W has been implemented in order to prepare the datasets of the normal and fault experiments under various conditions, as well as to evaluate the performance of the proposed model. Figure 2 shows the configuration of the PV array which is composed of 3 PV strings in parallel and each string has 6 modules in series.

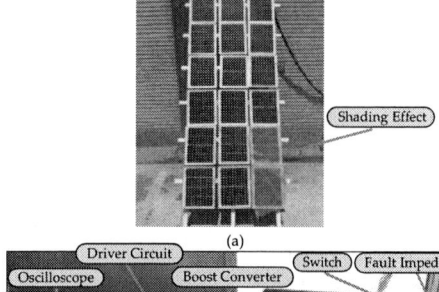

(a)

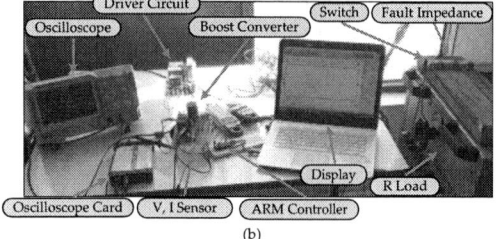

(b)

Figure 2. The experimental prototype: (a) the PV array, and (b) the boost converter and data collection station.

As mentioned previously, we have emulated several OC and LL faults in the PV system to build up the datasets. The LL faults which are based on mismatch levels and fault impedances include the LL faults with 16%, 33% and 50% mismatching. Also, the impedances ranging from 0 Ω to 25 Ω with the increment step of 5 Ω for these LL faults. In addition, as the fault impedance range is increased, the instantaneous changes in current (or reverse current) in the LL fault is become smaller resulting in even more difficult to be detected. Moreover, the OC faults have been created by simply disconnecting the electrical cables on the second string in the PV array. As discussed

before, all the faults were emulated under various amounts of irradiance and temperature. As discussed, all these faults were emulated under various irradiance and temperature. In this experimental study, 1445 data samples have been recorded under three conditions, including 433 normal cases, 580 LL fault events, 432 OC fault cases.

3.2. Experimental Results

In this section, the experimental results are presented and discussed. As mentioned in the subsection 4.1, the datasets were recorded under normal and faulty events through developed data acquisition system. Then, according to the Table I, statistical features related to the collected datasets were extracted. In total, 1445 data instances have been collected to train and validate the AI-SafePV software package. 1156 (80%) and 289 (20%) data samples were randomly selected for training and validation datasets, respectively. In the training dataset, there are 339 normal events, 350 OC faults, and 467 LL faults. Besides, the validation dataset includes 94 normal condition, 82 OC faults, and 113 LL faults.

Here, we have pre-processed datasets and normalize the data as well as the best features were selected from training dataset using Lasso penalty algorithm. This dataset includes the 30 features from current and voltage characteristics. Table II lists the selected features based on their importance in terms of the estimated coefficients. As listed in Table II, sixteen features have been selected and other features have an estimated coefficient of zero.

Table II
Estimated coefficients and elected features

Selected features	Lasso Coefficients	Selected features	Lasso Coefficients
f_{12} (V)	26.4	f_5 (V)	1.39
f_5 (I)	18.43	f_{14} (V)	1.38
f_8 (V)	5.78	f_1 (V)	1.18
f_6 (I)	5.12	f_3 (V)	0.77
f_6 (V)	3.89	f_7 (I)	0.23
f_{11} (I)	2.89	f_7 (V)	0.16
f_1 (I)	2.38	f_{15} (V)	0.12
f_8 (I)	2.12	f_{15} (I)	0.013

V=related to voltage   I=related to current

After creating fully the desireble dataset, the fault diagnosis process enters the second stage where the proposed weighted ensemble learning model is exploited to solve a multi-class classification problem i.e. to diagnose the faults and classify them into three pre-defined classes namely: the normal condition, the OC faults, and the LL faults which are labeled as the first, the second, and the third class, respectively.

The following performance criteria are used to evaluate the performance of the AI-SafePV software package: (1) The training accuracy of the ensemble models and the individual learning algorithms is measured by the 10-fold cross-validation method. Therefore, mean accuracy and standard deviation are used to evaluate training results. (2) The classification accuracy, precision, recall, and false positive rate (FPR) metrics have been used to evaluate the ensemble models and the individual learning algorithms by the validation datasets as Eq. (4)-(7):

$$Accuracy = \frac{TP + TN}{TP + TN + FN + FP} \qquad (4)$$

$$\text{Recall} = \frac{TP}{TP+FN} \tag{5}$$

$$\text{Precision} = \frac{TP}{TP+FP} \tag{6}$$

$$\text{FPR} = \frac{FP}{FP+TN} \tag{7}$$

where, TP represents the number of the first class samples the have been correctly predicted as first class, FN is the number of the first class samples that have been incorrectly classified as the second class, FP represents the number of the second class samples that have been incorrectly predicted as the first class, TN is the number of the second class samples that have been correctly classified as the second class.

First, the optimal weights of the individual learning algorithms were obtained through a Genetic Algorithm (GA) using the original dataset. In the first step, we have trained the model via training dataset which was acquired based on the selected features. The training results including mean accuracy and standard deviation based on 10-fold cross-validation have been presented in Table III.

Table III
Training results of the learning algorithms

Learning Algorithm	Train Acc.	STD
Ensemble model	98.96 %	1.14%

In the final stage, the validation dataset was used to evaluate the performance of the AI-SafePV software package in order to show its effectiveness. Table IV summarizes the results of our learning algorithm using the validation dataset, which uses Recall and Precision criteria to accurately evaluate the performance of the learning algorithm.

Table IV
Validation results of the learning algorithms

Algorithm	Accuracy	Precision	Recall	FPR
Ensemble model	98.61%	P_1= 99% P_2= 100% P_3= 97%	R_1= 97% R_2= 100% R_3= 99%	FPR_1=0.52% FPR_2=0% FPR_3=1.67%

Partial shading is another challenge in fault detection schemes. Therefore, it is important to detect accurately the faults under partial shading conditions. For the validation of the AI-SafePV software package, a thin plastic sheeting has covered in PV array (see Figure 2) in order to implement partial shading conditions. In this regard, 30 cases of fault conditions under partial shading conditions and 30 cases partial shading disturbances without fault conditions have been measured. The AI-SafePV software package has correctly detected all 60 events. The experimental results demostrated that the AI-SafePV software package is able to recognize faults from a partial shading situation. This is due to the selected features from voltage and current of PV array at MPP condition.

For further validation, the performance of the AI-SafePV software package has been tested under LL faults with a blocking diode in each string. To do this, 60 events of LL faults has been created under different conditions, including fault impedance (0,10, and 20), mismatch percentage (16%, and 33%), and irradiance. The AI-SafePV software package has correctly detected 58 out of 60 LL fault events in the presence of string blocking diodes.

4 CONCLUSIONS

In this paper, we have proposed a an intelligent AI-SafePV software package to diagnose the LL and OC faults in PV arrays. This package only used the current and voltage of the PV array to extract the features of the faults. In addition, the Lasso penalty algorithm was used to extract the best features and use them in the training process to achieve a better performance in diagnosing the faults. In this AI-SafePV software package, ensemble learning algorithm was applied including three individual learning algorithms, namely k-Nearest Neighbors (KNN), Support Vector Machine (SVM), Logistic Regression (LR). The final result of the classification was obtained from the weighted combination of the results of these three algorithms. Therefore, to achieve a higher accuracy in the classification stage, we have utilized the optimal weights of each algorithm in the ensemble learning algorithm using a GA. The output results of the AI-SafePV software package showed that this package was able to diagnose the faults with an average accuracy of 99%.

REFERENCES

[1] "https://www.pv-magazine.com/2022/03/15/humans-have-installed-1-terawatt-of-solar-capacity/".

[2] M. Aghaei et al., "Review of degradation and failure phenomena in photovoltaic modules," *Renew. Sustain. Energy Rev.*, vol. 159, p. 112160, May 2022, doi: 10.1016/J.RSER.2022.112160.

[3] T. Berghout, M. Benbouzid, T. Bentrcia, X. Ma, S. Djurović, and L. H. Mouss, "Machine Learning-Based Condition Monitoring for PV Systems: State of the Art and Future Prospects," *Energies 2021, Vol. 14, Page 6316*, vol. 14, no. 19, p. 6316, Oct. 2021, doi: 10.3390/EN14196316.

[4] S. R. Madeti and S. N. Singh, "Monitoring system for photovoltaic plants: A review," *Renew. Sustain. Energy Rev.*, vol. 67, pp. 1180–1207, Jan. 2017, doi: 10.1016/J.RSER.2016.09.088.

[5] L. Rokach, "Ensemble-based classifiers," *Artif. Intell. Rev.*, vol. 33, no. 1–2, pp. 1–39, Feb. 2010, doi: 10.1007/S10462-009-9124-7/METRICS.

[6] A. Eskandari, M. Aghaei, J. Milimonfared, and A. Nedaei, "A weighted ensemble learning-based autonomous fault diagnosis method for photovoltaic systems using genetic algorithm," *Int. J. Electr. Power Energy Syst.*, vol. 144, 2023, doi: 10.1016/j.ijepes.2022.108591.

[7] W. Caesarendra and T. Tjahjowidodo, "A Review of Feature Extraction Methods in Vibration-Based Condition Monitoring and Its Application for Degradation Trend Estimation of Low-Speed Slew Bearing," *Mach. 2017, Vol. 5, Page 21*, vol. 5, no. 4, p. 21, Sep. 2017, doi: 10.3390/MACHINES5040021.

[8] I. Kamkar, S. K. Gupta, D. Phung, and S. Venkatesh, "Stable feature selection for clinical prediction: Exploiting ICD tree structure using Tree-Lasso," *J. Biomed. Inform.*, vol. 53, pp. 277–290, Feb. 2015, doi: 10.1016/J.JBI.2014.11.013.

[9] A. Mellit and S. Kalogirou, "Assessment of machine learning and ensemble methods for fault diagnosis of photovoltaic systems," *Renew. Energy*, vol. 184, pp. 1074–1090, Jan. 2022, doi: 10.1016/J.RENENE.2021.11.125.

41st European Photovoltaic Solar Energy Conference and Exhibition

DETECTIVEPV: A DETECTION PACKAGE FOR ELECTRICAL FAULTS IN PHOTOVOLTAIC ARRAYS BASED ON MACHINE LEARNING

A. Eskandari[1], J. Milimonfared[2], A. Nedaei[2], P. Parvin[3], M. Braga[4], and M. Aghaei[5,6*]

[1]Department of Electrical Engineering, Iran University of Science and Technology, Tehran, Iran
[2]Department of Electrical Engineering, Amirkabir University of Technology, Tehran, Iran
[3]Department of Physics and Energy Engineering, Amirkabir University of Technology, Tehran, Iran
[4]Universidade Federal de Santa Catarina – UFSC, Florianópolis, SC 88040-900 Brazil
[5]Department of Ocean Operations and Civil Engineering, Norwegian University of Science and Technology, Ålesund, Norway
[6]Department of Sustainable Systems Engineering, University of Freiburg, Freiburg, Germany

*mohammadreza.aghaei@ntnu.no

ABSTRACT: Conventional protection devices in PV arrays may not be able to detect line-line (LL) and line-ground (LG) faults due to these faults are not detectable under high fault impedance and low mismatch level. In recent years, many efforts have been devoted to overcome these challenges using intelligent methods. However, these methods could not classify the type of faults and diagnose their severity. This paper proposes DetectivePV which is a novel and intelligent fault monitoring package to detect and classify LL and LG faults at the DC side of PV systems. In this package, firstly, the main features of PV array current-voltage (I-V) curves under different fault events and normal conditions are extracted. Then the faults are classified using a machine learning (ML)-based Hierarchical Classification (HC) platform. DetectivePV package aims to reduce the amount of dataset which is required for the ML algorithms training process and still obtain a high accuracy in detecting and classifying the fault events especially at low mismatch levels and high fault impedance. The experimental output results of DetectivePV verify that the package precisely detects and classifies LL and LG faults in PV systems under different conditions with the accuracy of 96.66% and 91.66%, respectively.

Keywords: Fault detection and classification, Hierarchical classification, Line-Line and Line-Ground faults, Machine learning, Photovoltaic monitoring

1 INTRODUCTION

Photovoltaic (PV) systems experience various faults due to environmental conditions, human errors, and equipment failure during their service life. Line-ground (LG) and line-line (LL) faults are among the frequent faults and cannot be detected without appropriate protection devices, such as the Ground Fault Protection Device (GFPD) and the Over Current Protection Device (OCPD). However, it is difficult to detect these faults in PV systems by conventional methods due to low current value of the faults, which caused by high impedance in series or in low mismatch level, and they remain hidden in PV systems. To date, many efforts have been devoted to developing reliable and robust methods using Machine Learning (ML) techniques to overcome these challenges. In these studies, various machine learning methods like probabilistic neural network [1], random forest learning [2], the stage-wise additive modeling using multi-class exponential loss function based on the classification and regression tree [3], and conventional neural network [4] were used for faults diagnosis in PV systems. Despite, these studies could not propose a precise model to recognize the patterns of the faults, because these methods have not paid enough attention to LL and LG faults under low mismatch or with high impedance.

However, some studies have attempted to consider these challenging issues using various ML methods such as decision tree-based method [5], Kernel-based Extreme Learning Machine (KELM) algorithm [6], Graph-Based Semi-Supervised Learning (GBSSL) algorithm [7], fuzzy inference system [8], and two-stage support vector machine [9]. Nonetheless, these models for the faults detection have several disadvantages and drawbacks namely, 1) a big dataset required for the learning process, 2) low detection accuracy for the faults with low-

percentage mismatch or high impedance, 3) these fault detection methods were not reliable enough due to only one classifier was used in these methods. Moreover, these methods have not used any comparative process to choose the optimal classifier and most of them were not able to classify the type of faults and diagnose their severity.

To overcome the discussed problems and challenges, here we propose DetectivePV, which is an intelligent PV monitoring package to detect and classify LL and LG faults especially under critical conditions with different severity levels. For this purpose, first, various normal and faulty states have been elicited from the PV array I-V curve to create the initial dataset. Then, the package aims to constuct several useful features from the initial dataset. Third, an ML-based Hierarchical Classification (HC) method which is composed of four models is developed to improve the accuracy of the classification procedure [10]. The first and second models aim to detect the fault condition and classify the faults type. The third and fourth models are developed to classify the severity of the faults. The fourth step is to use the ML techniques to opt for the best classifier and features in order to obtain a higher performance for faults detection and classification. DetectivePV aims to reduce the number of datasets required for the training process by designing the HC platform and using the feature selection algorithm, and improve the accuracy of faults detection and classification compared to previous models and packages.

2 FAULT ANALYSIS AND FEATURE EXTRACTION

The most important impact of LG and LL faults is the voltage drop in the faulty strings, which leads to drawing a reverse-current from the normal strings. Normally, OCPDs and GFPDs are applied to interrupt this reverse-current. It should be noted that detection of LG faults with high mismatch level and LL faults with a low mismatch level is a challenge due to the amount of reverse- current is drastically reduced in these situations and it is not possible to cut it off with conventional protection devices.

In this study, we have firstly investigated the effect of LG and LL faults occurred on PV systems by I-V curves analysis. For this purpose, a 10×3 PV array with an output power of 4.5 kW has been designed, and the faults have been simulated under different conditions. Five points of the I-V curve, namely (A) short circuit current (I_{SC}), (E) open-circuit voltage (V_{OC}), (C) Maximum Power Point (MPP) (V_{MPP}, I_{MPP}), (B) half short-circuit current ($I_{SC}/2$), and (D) half open-circuit voltage ($V_{OC}/2$) are considered to evaluate the performance of PV arrays under normal and faulty conditions. Whenever LL faults occur and the Maximum Power Point Tracking (MPPT) is active, the MPPT decreases the faults' impact on the PV array through shifting its operating point to the new MPP.

This package is able to extract sixteen features, which have been defined based on five specified points in the PV array I-V curves. Thses points are: I_{SC}, V_{OC}, $0.5I_{SC}$, $0.5V_{OC}$, V_{MPP} (with their corresponding current/voltage values). These features are normalized by the parameters' value (at STC) of the PV arrays in order to make the scheme scalable. These features are listed in Table I.

TABLE I
FEATURES EXTRACTION PROCEDURE

Extracted features		
$f_1 = I_{SC}/I_{SC(STC)}$	$f_2 = V_{OC}/V_{OC(STC)}$	$f_3 = V_{MPP}/V_{MPP(STC)}$
$f_4 = I_{MPP}/I_{MPP(ST}$	$f_5 = \frac{I_{V_{OC}}}{2}/I_{SC(STC)}$	$f_6 = \frac{V_{I_{SC}}}{2}/V_{OC(STC)}$
$f_7 = f_4/f_3$	$f_8 = f_3/f_2$	$f_9 = f_4/f_1$
$f_{10} = (I_{MPP}\text{-}I_{SC})/(V_{MPP})$		$f_{11} = (\text{-}I_{MPP})/(V_{OC}\text{-}V_{MPP})$
$f_{12} = (I_{V_{OC}}\text{-}I_{SC})/(\frac{V_{OC}}{2})$		$f_{13} = (I_{MPP}\text{-}I_{\frac{V_{OC}}{2}})/(V_{MPP}\text{-}\frac{V_{OC}}{2})$
$f_{14} = (\frac{I_{SC}}{2}\text{-}I_{MPP})/(V_{\frac{I_{SC}}{2}}\text{-}V_{MPP})$		$f_{15} = (\text{-}\frac{I_{SC}}{2})/(V_{OC}\text{-}V_{\frac{I_{SC}}{2}})$
$f_{16} = \frac{FF}{FF_{(STC)}} \rightarrow FF = \frac{V_{MPP} \times I_{MPP}}{V_{OC} \times I_{SC}}, FF_{(STC)} = \frac{V_{MPP(STC)} \times I_{MPP(STC)}}{V_{OC(STC)} \times I_{SC(STC)}}$		

3 DETECTIVEPV PACKAGE: THE PROPOSED FAULT DETECTION MODEL

In this paper, an ML-based HC platform has been developed to improve the accuracy of the classification procedure and reduce the overall training dataset during the training process (see Figure 1).

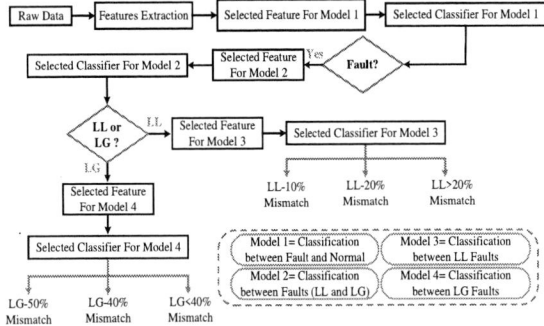

Figure 1. Flowchart of proposed HC for precise faults classification.

For this purpose, a multi-classification algorithm is divided into binary classification algorithms, which leads to a systematic and effective classification in each category. This method consists of four models. The first and second ones aim to detect the fault condition and classify the faults type through a binary classification. The third and fourth models include the multiple classes, which classify LL and LG faults severity based on mismatch percentage. This package proposes a comprehensive ML approach to detect and classify the faults on the I-V characteristics of PV arrays. It should be noticed that three different classifiers are applied here, namely Support Vector Machine (SVM) [11], Naive Bayes (NB) [12], and Logistic Regression (LR) [13] in order to select the best classifier in each model to increase the performance and reliability of the models. The proposed ML approach consists of two main steps, namely feature selection algorithm in order to choose the best features for each classifier and the classifier selection, which aims to choose the best classifier based on the performance metrics. This approach has been applied to each model developed by the HC platform. Figure 2 illustrates the structure of the proposed ML approach.

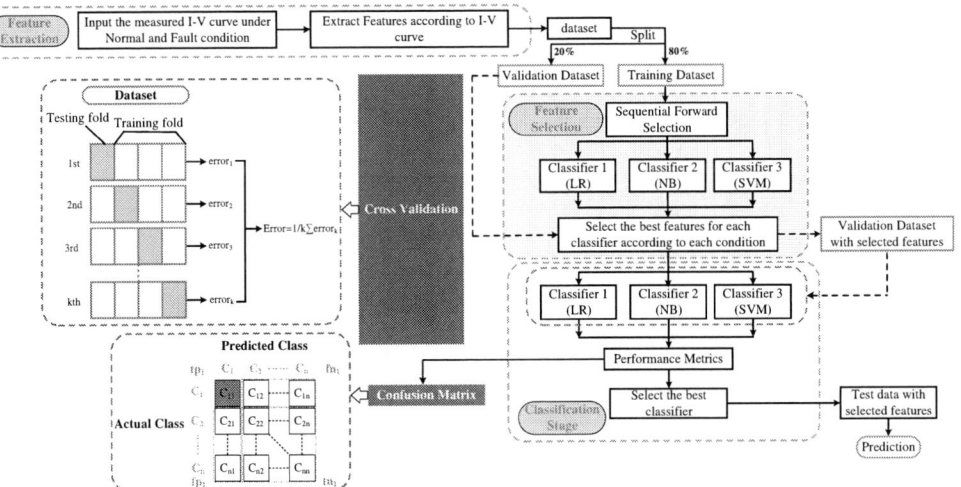

Figure 2. The intelligence structure for DetectivePV package in fault detection and classification with a feature selection approach.

1298

1) Feature selection

In this study, the SFS algorithm is used for each model created by the HC platform, and each model uses three classifiers. This means that the SFS algorithm is used for each of these classifiers. In this proposed monitoring package, the SFS algorithm aims to improve the performance of classifiers because irrelevant features are eliminated in the classifier and also, this algorithm ranks the importance of the features so that it provides a better understanding which features efficiently contribute to each model.

2) Performance metrics

In this study, various performance metrics are used like accuracy, recall, precision.

$$Accuracy = \frac{\sum tp_i}{the\ total\ number\ of\ datasets} \quad (1)$$

$$R_i = (tp_i)/(tp_i + fn_i) \quad (2)$$

$$P_i = (tp_i)/(tp_i + fp_i) \quad (3)$$

Where tp_i is the case that a sample data in the *ith* class correctly identified as the *ith* class, tn_i represents the case that a sample data in the *jth* class is correctly recognized as the *jth* class, fp_i indicates the case which a sample data belonging to the *jth* class which is incorrectly identified as the *ith* class, and fn_i is the case that a sample data belonging to the *ith* class is incorrectly assigned to the *jth* class.

4 EXPERIMENTAL VERIFICATION

A PV system has been designed and constructed to validate DetectivePV package. For experimentation, a 6 (series)×3 (parallel) PV array has been built using Yingli YL010D-18b (see Figure 3(a)). Moreover, to extract maximum power from the PV array, a DC-DC boost converter was designed in order to implement the MPPT algorithm using Perturbation and observation (P&O) method as shown in Figure 3(b). To execute the MPPT algorithm, the P&O algorithm has been coded on an STM32f103C8T6 ARM controller. The PV system has been connected to a resistance load of 200 Ω. In addition, voltages and currents have been recorded at a sampling frequency of 1KHz per second using an InstruStar-ISDS205A oscilloscope card and a personal computer.

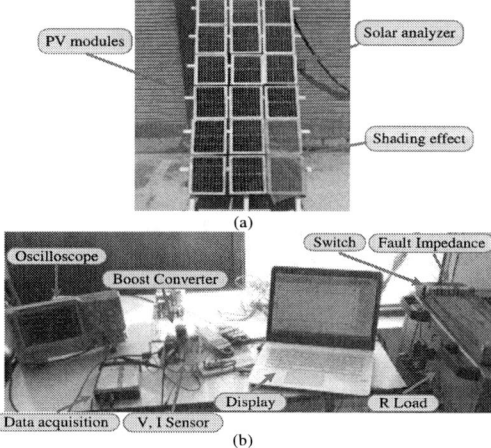

Figure 3. Design of the PV system and experimental procedure: (a) PV array, and (b) dc-dc converter and data acquisition system.

In this experimental study, 360 events of LL faults, 360 events of LG faults, and 300 normal conditions have been measured on the PV array under various operating conditions, as summarized in Table II.

TABLE II
EXPERIMENTAL DATASET UNDER VARIOUS CONDITIONS

Fault	Variable	
	Mismatch level (%)	fault impedances (Ω)
LL fault	1/6 (16.67%), 2/6 (33.3%)	0, 5, 10, 15, 20
LG fault	4/6 (66.67%), 3/6 (50%)	0, 5, 10, 15, 20

Among these cases, 60 LL fault events and 60 LG fault events are used to test DetectivePV package, while the rest has been used to train the classifiers. To evaluate the performance of DetectivePV package in the PV system, firstly, the features are extracted for each model. According to the HC method (see Figure 1), the fault events have been tested for each model. The models have been evaluated with the second scenario. Table III presents the experimental results of DetectivePV package, which have been tested on the PV system. According to the proposed detection procedure, the first step was to detect the faults through model 1. According to the proposed detection procedure, the first step was to detect the faults through model 1. The experimental results show that the accuracy is decreased when the impedance of faults increased. However, two samples of the faults were not detected by model 1 under the impedance of 20 Ω, which includes two LL faults. There are 58 cases of LL faults and 60 cases of LG faults for the validation of model 2. Model 2 could successfully classify LL and LG faults with an accuracy of 100% and 95%, respectively. Thus, the next models should be applied to classify LL and LG faults severity based on the mismatch percentage through model 3 and model 4, respectively. According to experimental results, the models are fully capable of classifying these faults. Nonetheless, the model 4 could not classify two LG fault cases with 66.67 % mismatching and under impedance of 20 Ω. There were 60 LL and 60 LG fault events, which 58 cases related to LL faults and 55 cases related to LG faults which detected by DetectivePV package precisely.

TABLE III
EXPERIMENTAL RESULTS OF FAULTS CLASSIFICATION

Model Type	Class	Fault impedance			Average Accuracy
		0	10	20	
1	Fault	40/40	40/40	38/40	98.33%
2	LL	20/20	20/20	18/18	100%
	LG	20/20	19/20	18/20	95%
3	LL-16.67%	10/10	10/10	8/8	100%
	LL-33.3%	10/10	10/10	10/10	100%
4	LG-66.67%	10/10	10/10	7/9	93.1%
	LG- 50%	10/10	9/9	9/9	100%
Final Success	LL	58/60			96.66%
	LG	55/60			91.66%

5 CONCLUSION

In this study, a novel intelligent monitoring package named DetectivePV was proposed to detect and classify LG and LL faults in PV systems. In this package, an ML-based hierarchical platform was developed to organize the input data into four models in order to obtain a higher accuracy in fault classification. First, sixteen features were extracted from the PV array I-V curves under normal and fault conditions. Then, according to the models created by the hierarchical platform, the best features have been selected using the SFS method for each classifier in each model. The classifiers were evaluated by the best selected features in order to select the best classifier in each model. DetctivePV package aimed to reduce the amount of dataset required for the training process and also obtained a high accuracy in detecting and classifying the LG and LL fault events at low mismatch levels and high fault impedances. The experimental results of DetectivePV package showed that it precisely detected and classified LL and LG faults in PV systems under different critical conditions with the accuracy of 96.66% and 91.66%, respectively.

6 REFERENCES

[1] M. N. Akram and S. Lotfifard, "Modeling and Health Monitoring of DC Side of Photovoltaic Array," *IEEE Trans. Sustain. Energy*, vol. 6, no. 4, pp. 1245–1253, Oct. 2015, doi: 10.1109/TSTE.2015.2425791.

[2] Z. Chen *et al.*, "Random forest based intelligent fault diagnosis for PV arrays using array voltage and string currents," *Energy Convers. Manag.*, vol. 178, pp. 250–264, Dec. 2018, doi: 10.1016/j.enconman.2018.10.040.

[3] J. M. Huang, R. J. Wai, and W. Gao, "Newly-designed fault diagnostic method for solar photovoltaic generation system based on IV-Curve measurement," *IEEE Access*, vol. 7, pp. 70919–70932, 2019, doi: 10.1109/ACCESS.2019.2919337.

[4] X. Lu *et al.*, "Fault diagnosis for photovoltaic array based on convolutional neural network and electrical time series graph," *Energy Convers. Manag.*, vol. 196, pp. 950–965, Sep. 2019, doi: 10.1016/J.ENCONMAN.2019.06.062.

[5] Y. Zhao, L. Yang, B. Lehman, J. F. De Palma, J.

Mosesian, and R. Lyons, "Decision tree-based fault detection and classification in solar photovoltaic arrays," *Conf. Proc. - IEEE Appl. Power Electron. Conf. Expo. - APEC*, pp. 93–99, 2012, doi: 10.1109/APEC.2012.6165803.

[6] Z. Chen, L. Wu, S. Cheng, P. Lin, Y. Wu, and W. Lin, "Intelligent fault diagnosis of photovoltaic arrays based on optimized kernel extreme learning machine and I-V characteristics," *Appl. Energy*, vol. 204, pp. 912–931, Oct. 2017, doi: 10.1016/J.APENERGY.2017.05.034.

[7] Y. Zhao, R. Ball, J. Mosesian, J. F. De Palma, and B. Lehman, "Graph-based semi-supervised learning for fault detection and classification in solar photovoltaic arrays," *IEEE Trans. Power Electron.*, vol. 30, no. 5, pp. 2848–2858, May 2015, doi: 10.1109/TPEL.2014.2364203.

[8] Z. Yi and A. H. Etemadi, "Fault detection for photovoltaic systems based on multi-resolution signal decomposition and fuzzy inference systems," *IEEE Trans. Smart Grid*, vol. 8, no. 3, pp. 1274–1283, May 2017, doi: 10.1109/TSG.2016.2587244.

[9] Z. Yi and A. H. Etemadi, "Line-to-line fault detection for photovoltaic arrays based on multi-resolution signal decomposition and two-stage support vector machine," *IEEE Trans. Ind. Electron.*, vol. 64, no. 11, 2017, doi: 10.1109/TIE.2017.2703681.

[10] A. Eskandari, J. Milimonfared, and M. Aghaei, "Fault Detection and Classification for Photovoltaic Systems Based on Hierarchical Classification and Machine Learning Technique," *IEEE Trans. Ind. Electron.*, vol. 68, no. 12, pp. 12750–12759, Dec. 2021, doi: 10.1109/TIE.2020.3047066.

[11] R. Gholami and N. Fakhari, "Support vector machine: principles, parameters, and applications," in *Handbook of Neural Computation*, Elsevier, 2017, pp. 515–535.

[12] G. I. Webb, "Naïve Bayes BT - Encyclopedia of Machine Learning," C. Sammut and G. I. Webb, Eds., Boston, MA: Springer US, 2010, pp. 713–714. doi: 10.1007/978-0-387-30164-8_576.

[13] D. Kuonen, *Book Review: Regression modeling strategies: with applications to linear models, logistic regression, and survival analysis*, vol. 13, no. 5. 2004. doi: 10.1177/096228020401300512.

41st European Photovoltaic Solar Energy Conference and Exhibition

WET LEAKAGE AND INSULATION TEST ON STRING LEVEL THROUGH IEC 61215

*Mario Martínez[1], *Sergio Suarez[1], Jose Cantisano[1], Jonathan Vilela[1], Jose Maria Alvarez[1], Jose Manuel Rivas[1], Sofía Rodríguez-Conde[1]
[1]Enertis Applus Parque Empresarial Las Mercedes, C/ de Campezo, 1, 28022 Madrid, Spain
www.enertisapplus.com
*Phone: +34 91 651 70 21, *sergio.suarez@enertisapplus.com, *Mario.martinez@enertisapplus.com*

ABSTRACT: This study explores the adaptation of the IEC 61215 wet leakage and insulation resistance test to the string level in photovoltaic (PV) systems. Traditionally, insulation tests are conducted at the module level, but scaling up to large PV installations with interconnected strings presents new challenges. A mathematical model was developed to calculate the equivalent resistance of strings, considering the parallel connection of modules. Simulations involving 5 million strings of 30 modules were conducted, and new pass/fail criteria were defined based on failure thresholds and confidence limits. These criteria enable early detection of insulation failures, with three distinct zones—pass, uncertainty, and failure—established to provide a robust framework for evaluating insulation at the string level. Results show that string-level testing offers a more comprehensive assessment of insulation integrity, improving reliability and reducing testing time in large installations. Future work will involve validating the model through further field trials and experimental measurements under various conditions.
Keywords: Photovoltaic systems. Insulation resistance. Wet leakage. IEC 61215.

1 INTRODUCTION

The IEC 61215 standard for photovoltaic (PV) modules includes various tests designed to ensure the reliability and durability of PV systems, one of which is the wet leakage and insulation resistance test. These tests are traditionally conducted at the module level to assess insulation integrity by measuring the resistance between the conductive parts of the module and the ground under specific conditions. While this method is effective at detecting insulation issues at the individual module level, it becomes more complex when scaling up to large PV installations, where hundreds or thousands of modules are interconnected in strings. In these cases, module-level testing can be time-consuming, and insulation failures that occur at the string level may be more difficult to detect through individual module assessments.

This study explores the adaptation of the IEC 61215 wet leakage and insulation test to the string level, where multiple modules are connected in parallel. Testing at the string level has the potential to provide a more comprehensive assessment of the insulation resistance across an entire PV system, while also reducing the time required for testing large installations. However, conducting the test at this level presents new challenges, as the insulation resistance value measured by the instrument represents the equivalent resistance of the string, which is a parallel combination of the insulation resistances of the individual modules.

In parallel circuits, the module with the lowest insulation resistance has the greatest impact on the overall equivalent resistance of the string. Therefore, a string with one or more faulty modules exhibiting low insulation resistance will result in a reduced equivalent resistance. Given this behaviour, it is necessary to develop new pass/fail criteria specifically for string-level testing, which can account for the impact of low-resistance modules on the overall measurement.

2 METHODOLOGY AND MODEL DEVELOPMENT

This section presents the mathematical model for calculating equivalent resistance in module strings and outlines the methodology for defining pass/fail criteria for string-level insulation testing.

2.1 Mathematical Model for Equivalent Resistance

When testing at the string level, the insulation resistance of the string is not simply the sum of individual module resistances. Instead, the modules are connected in parallel, meaning the equivalent resistance (R_{eq}) of the string is the result of the combined effect of each module's insulation resistance ($R_1, R_2, ..., R_N$). The equivalent resistance is determined by equation (1)

$$R_{eq} = \frac{1}{R_{total}} = \sum_{i=1}^{N} \frac{1}{R_i} + R_{Ground} \qquad (1)$$

In this configuration, the module with the lowest insulation resistance dominates the overall resistance of the string, significantly lowering R_{eq} if any module experiences insulation failure. Therefore, it is crucial to account for the parallel connection of modules when testing at the string level.

Additionally, we define the average resistance ($R_{average}$) a parameter that helps compare and study the model, described in equation (2).

$$R_{average} = \frac{1}{N} \sum_{i=1}^{N} R_i + R_{Ground} \qquad (2)$$

To further simplify the calculations, the ground resistance (R_{Ground}) is considered negligible in this model, focusing only on the insulation resistances of the modules themselves.

2.2 Simulation and data generation

To define appropriate thresholds for passing or failing a string of modules based on insulation resistance, a large-scale simulation was conducted. The simulation involved generating insulation resistance values for 5 million strings, each composed of 30 modules. Each module was assigned a random insulation resistance value between a defined maximum of 5 GΩ and a minimum of 5 kΩ.

1301

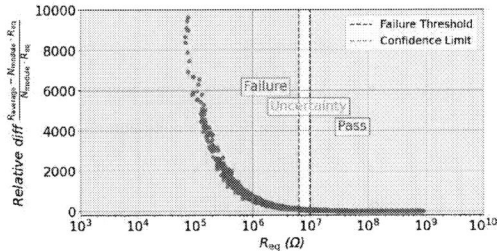

Figure 1. Plot showing the relative difference between the $R_{average}$ and the R_{eq} versus the equivalent resistance R_{eq}. The graph illustrates the three zones: Pass (green), where the equivalent resistance exceeds the confidence limit, Failure (red), where the equivalent resistance falls below the failure threshold, and Uncertainty (yellow), indicating borderline results that require further investigation. The dashed lines represent the failure threshold and the confidence limit, respectively.

The resulting data was used to calculate the equivalent resistance for each string, providing a distribution of values that served as the basis for developing pass/fail criteria. These criteria are designed to account for the fact that a single module with low resistance can drastically reduce the equivalent resistance of the entire string.

3   PASS/FAIL CRITERIA

3.1 Criteria for individual modules

For individual photovoltaic modules, the pass/fail threshold is based on the insulation resistance ($R_{individual}$) of the module. The limit for passing or failing a module is set as follows:

$$R_{limit\ PASS/FAIL} = 40\ \text{M}\Omega \cdot \text{m}^2$$

If the measured insulation resistance of the module (R_{eq}) meets or exceeds this threshold, the module passes the insulation test. For a single module, the equivalent resistance is simply equal to the individual insulation resistance.

3.2 Criteria for strings of modules

At the string level, the R_{eq} is affected by the parallel connection of multiple modules. In a parallel circuit, the module with the lowest insulation resistance has the greatest influence on the total equivalent resistance, meaning that even one module with low insulation resistance can drastically reduce the equivalent resistance of the entire string.

To address this, the pass/fail criteria for strings of modules are based on two main thresholds: the failure threshold and the confidence Limit. The pass/fail limit is defined as:

$$R_{limit\ PASS/FAIL} = R_{Failure\ Threshold}$$

The criteria can be summarized as follows:

- Pass: If the equivalent resistance of the string is greater than the failure threshold ($R_{eq} > R_{Failure\ Threshold}$), the string is considered to pass the test. In this case, the equivalent resistance is approximately equal to the average resistance of all the modules in the string. This is represented by:

$$N \cdot R_{eq} \approx R_{average}$$

- Fail: If the equivalent resistance of the string falls below the confidence limit ($R_{eq} < R_{Confidence\ Limit}$), it indicates the presence of at least one module with a very low insulation resistance, leading to an overall failure. In this case, the equivalent resistance is dominated by the lowest resistance module R_{min}:

$$R_{eq} \approx R_{min}$$

- Uncertainty zone: There exists a small range between the failure threshold and the confidence limit, known as the uncertainty zone. In this zone, the equivalent resistance falls between the two thresholds, and it is unclear whether a significant insulation failure exists. Further investigation at the module level may be necessary to confirm the status of the string.

The Figure 1 illustrates these zones, plotting the relative difference between the average resistance and the equivalent resistance against the equivalent resistance. The pass/fail threshold and the confidence limit are marked, with regions of failure, uncertainty, and passing clearly distinguished.

4   RESULTS

In this section, we analyse the behaviour of R_{eq} of strings composed of 30 modules, based on the previously defined model and the pass/fail criteria. The results provide insight into how the relative differences in resistance values behave depending on the individual module resistances and the equivalent resistance of the string.

4.1 Analysis of relative difference for R_{min}

In Figure 2, we observe the relationship between the R_{min} and the R_{eq} for the entire string. The graph is segmented into three key zones: pass, uncertainty, and failure, with thresholds set by the failure threshold (10 MΩ) and the confidence limit (4.8 MΩ), marked by dashed lines.

In the failure zone (highlighted in red), the data reveals a pronounced relative difference, which suggests that the performance of the string is significantly impacted by the

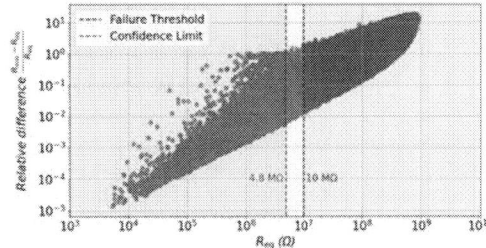

Figure 2. Relative difference between the R_{min} and the R_{eq} of a string of 30 modules. The graph is divided into three zones: pass (green), uncertainty (yellow), and failure (red), determined by the failure threshold (10 MΩ, purple dashed line) and the confidence limit (4.8 MΩ, green dashed line). The higher the relative difference in the failure zone, the more the string's performance is influenced by modules with lower insulation resistance.

module with the lowest insulation resistance. As the equivalent resistance increases and approaches the pass zone (green), this influence diminishes, and the relative difference shrinks. This transition reflects a more uniform distribution of insulation resistance across the modules, indicating that no module is disproportionately affecting the string's overall performance.

This analysis demonstrates that, by focusing on the minimum insulation resistance, we can effectively detect when a single faulty module begins to dominate the string's behavior, providing early warning signs of potential insulation issues.

4.2 Analysis of relative difference $R_{average}$

In Figure 3, we shift the focus to the relative difference between the $R_{average}$ and the R_{eq}. Like the previous analysis, the graph is divided into the pass, uncertainty, and failure zones, with the same critical thresholds.

In the failure zone, the relative difference is notably larger, emphasizing that the string's insulation integrity is compromised by a small number of modules with lower-than-average resistance values. As we move towards the pass zone, the relative difference steadily decreases, illustrating a much more consistent performance across the entire string. This indicates that, in the pass zone, the insulation resistance of the individual modules is more evenly distributed, contributing to a higher overall equivalent resistance.

The analysis of $R_{average}$ offers a broader perspective on the string's health. While R_{min} highlights the worst-performing modules, $R_{average}$ provides insight into how consistently the insulation resistance is distributed across all modules. A smaller relative difference in this metric correlates with better overall string performance, making it a valuable parameter for ensuring long-term reliability.

5 CONCLUSIONS

Enertis Applus's trials show that adapting the IEC 61215 test for measuring insulation in a string of photovoltaic modules is effective. This method, validated through field measurements and simulations, enhances the accuracy and efficiency of insulation assessment, improving the reliability of photovoltaic systems.

The study developed a model for determining the equivalent resistance of strings, considering the parallel connection of modules. The defined pass/fail criteria, based on failure thresholds and confidence limits, effectively identify insulation failures and ensure the overall integrity of photovoltaic installations. By establishing distinct zones for passing, failing, and uncertainty, this approach provides a robust framework for evaluating insulation at the string level, addressing the limitations of module-by-module testing.

The analysis of relative differences between the equivalent resistance, minimum resistance, and average resistance highlights how the string's performance is influenced by the presence of low-resistance modules. The results demonstrate that string-level testing allows for early detection of insulation failures, which can help prevent energy losses and maintain the long-term performance of the system.

Future work will focus on the further validation of this model through additional field trials and experimental

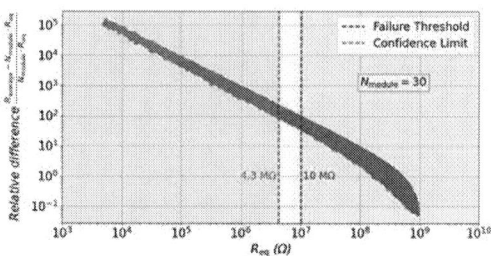

Figure 3. Relative difference between the $R_{average}$ and the R_{eq} for a string of 30 modules. The graph is divided into three zones: pass (green), uncertainty (yellow), and failure (red), determined by the failure threshold (10 MΩ, purple dashed line) and the confidence limit (4.3 MΩ, green dashed line). The relative difference decreases as R_{eq} increases, reflecting more uniform insulation resistance across the string in the pass zone.

measurements. This will include testing under various environmental conditions and for different types of photovoltaic modules, ensuring that the methodology is robust and adaptable to different real-world scenarios.

6 REFERENCES

[1] N. Sitthiphol, C. Sirisamphanwong, N. Ketjoy, and K. Sriprapha, "Insulation Resistance and Leakage Current in PV Modules and Strings with Different Grounding Configurations," *Applied Mechanics and Materials*, vol. 839, pp. 114–118, Jun. 2016, doi: 10.4028/www.scientific.net/amm.839.114.
[2] S. Pingel, O. Frank, M. Winkler, S. Oaryan, T. Geipel, H. Hoehne and J. Berghold, Potential Induced Degradation of solar cells and panels, 35 th, IEEE Photovoltaic Specialists Conference (PVSC), (2010), 2817–2822.
[3] International Electrotechnical Commission, Standard IEC 61215. Crystalline silicon terrestrial photovoltaic (PV) modules, Design qualification and type approval, Edition 2.0 (2005)
[4] D. Alberto, L. Sonia, M. Giampaolo and O. Emanuele, Investigation on performance decay on photovoltaic modules, snail trails and cell microcracks, IEEE journal of photovoltaics, 4 (2014) 1204–1211.
[5] J.C. Hernandez, P.G. Vidal and A. Medina, Characterization of the insulation and leakage currents of PV generators Relevance for human safety, Renewable Energy, 35 (2009) 593-610.
[6] S. Nopphadol, S. Chatchai, K. Nipon and S. Kobsak, The Study of Decrement in Insulation Resistance of PV String and its Effects on PV System Degradation, Key Engineering Materials, Vols. 675-676, (2016), 734-738

41st European Photovoltaic Solar Energy Conference and Exhibition

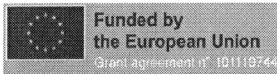

Funded by the European Union
Grant agreement n° 101119744

 TALOS

roboTics and Artificial intelligence Living labs improving Operations in PV Scenarios

Context

Photovoltaic (PV) systems are predicted to produce 23% of the world energy demand by 2050. However, PV systems have a low power density, causing utility-scale projects to increasingly face competing uses of land. These conflicts can be mitigated if PV systems are successfully integrated in other sectors, such as floating on reservoir lakes or sheltering crops.

Land-based PV, Acampo Arpal, Spain

General challenges in PV

- High space requirements
  - Conflicts with other sectors
  - Scaling operation & maintenance
- Maintenance
  - PV health monitoring
  - Cleaning
  - Vegetation management (shade, fire hazard)
- If autonomous systems are used for operation & maintenance:
  - Collaboration
  - Diagnosis and decision-making
  - Navigation

Importance of automation

Deployment of PV systems can benefit from the advancement of robotics and AI for their monitoring, as well as operation and maintenance. As PV sites become more complex, this could not only improve their performance but also reduce their costs and the need for human intervention in dangerous, dull and dirty jobs.

Understanding the cost reduction possibilities in PV plants is necessary for the energy transition.

If just 1% of arable land was dedicated to producing solar energy, it would be able to meet the world's energy demand.

Solar PV Power Potential is Greatest Over Croplands, Adeh et al., 2019

Problem statement

Current methods for operation and maintenance in PV plants rely heavily on human intervention. Robotic solutions are not fully autonomous

- Potentially dull, dirty or dangerous tasks
- Poor scalability
- High cost (up to 25% of Levelised Cost of Energy)
  - Limited frequency of maintenance activities (fixed calendar or as required)
  - Problems not spotted early (defects, soiling, shading from vegetation)
  - Deterioration and economic losses
  - Trade-off between operation & maintenance costs and plant performance

Pros
- Space-saving
- No shade
- Evaporation reduction
  - Limit water shortage

To study
- Environmental impact
- Durability
- Site access
- Soiling factors of the PV
- Regulations for operations

Floating PV on reservoir Alqueva dam, Portugal

Objectives

Deploy an autonomous AI-powered robotic fleet for unmanned monitoring, inspection, operation and maintenance in 3 PV scenarios: land, floating and agriculture

- Inspection, diagnosis
- Cleaning (smart water consumption)
- Vegetation management (cutting, collection)
- Collaborative (robot-robot, human-robot)
- Digital twins for prediction
- Inspection possible 24/7

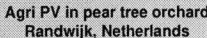

Agri PV in pear tree orchard Randwijk, Netherlands

Pros
- Space-saving
- Protection from weather
  - Crops
  - Grazing animals for mowing

To study
- Balance crop yield and energy production

Expected results

Preventive actions will improve plant performance and save resources:

- ↘ greenhouse gas emissions (>450 tons/year)
- ↘ water consumption (up to 35%)
- ↘ operation and maintenance costs (up to 5%)
- ↘ Maintenance time (>70%)
- ↘ Inspection time (20-50% depending on scenario)
- ↘ Risk exposure of workers (>90%)
- ↗ PV cost-effectiveness: 10%
- ↗ PV lifecycle: up to 3 years

Contact

🌐 talosproject.eu

in TALOS EU PROJECT

X @Talos_EUproject

Thanos Balafoutis
a.balafoutis@certh.gr
https://ibo.certh.gr

CERTH – CENTRE FOR RESEARCH & TECHNOLOGY HELLAS

iBO – Institute for Bio-Economy and Agri-Technology

 edp Ingeteam DTA INESCTEC ICONS alsys

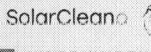 isotrol Eden Library SolarCleano CERTH

 ΓΕΩΠΟΝΙΚΟ ΠΑΝΕΠΙΣΤΗΜΙΟ ΑΘΗΝΩΝ AGRICULTURAL UNIVERSITY OF ATHENS WAGENINGEN UNIVERSITY & RESEARCH iBO

41st European Photovoltaic Solar Energy Conference and Exhibition

HARMONISING MULTI-SITES MEASUREMENT OF PHOTOVOLTAIC SYSTEMS
COMPREHENSIVE FRAMEWORK FOR REAL-LIFE TEST CONDITIONS IN A MALTESE ENVIRONMENT

Brian Bartolo[1, 2], Brian Azzopardi[1, 2, 3, 4] *, Alexandre Mignonac[5], Marcus Rennhofer[6], Bernhard Kubicek[6], Rita Ebner[6],
Carlos Meza[7], Melodie de l'Epine[8], Eugenia Zugasti[9], Steve Zerafa[2, 3, 10,] Kenneth Scerri[3, 1]

[1] The Foundation for Innovation and Research – Malta (FiR.mt)
[2] Malta College of Arts, Science and Technology (MCAST)
[3] The University of Malta
[4] Azzopardi & Associates, Malta – Lithuania
[5] Commissariat à l'Energie Atomique et aux Energies Alternatives (CEA) France
[6] AIT Austrian Institute of Technology GMBH, Austria
[7] Anhalt University of Applied Sciences, Germany
[8] Becquerel Institute, Belgium
[9] Fundación CENER—National Renewable Energy Centre, Spain
[10] PIXAM Ltd., Malta

* Brian.Azzopardi@FiR.mt

ABSTRACT: The growing demand for renewable energy has accelerated the installation of photovoltaic (PV) systems. However, optimising the performance of existing systems is critical, especially in regions with challenging environmental conditions like Malta. This paper presents a comprehensive framework for harmonising multi-site PV measurements tailored to the Mediterranean climate. By utilising advanced sensor technology, real-time data acquisition, and open-source platforms, the framework ensures accurate monitoring of PV system performance across multiple sites. The sensors capture electrical, meteorological, and environmental data, which are transmitted and analysed to identify factors affecting system efficiency, such as temperature, humidity, and soiling. This study is driven by the PROMISE project, which introduces innovative solutions to enhance PV monitoring and reliability. The framework also integrates single-module monitoring for detailed technological insights, allowing comparisons across diverse sites in Europe. Preliminary results highlight the framework's capacity to enhance PV system efficiency, scalability, and real-time performance analysis, addressing both plant-level and module-specific challenges. The use of open, interoperable systems enables future expansion and adaptation. This paper outlines the objectives, methodology, and initial findings, positioning this framework as a key tool for advancing PV performance monitoring in regions facing similar environmental conditions.

Keywords: Photovoltaic System Monitoring, Multi-Site Measurements, Renewable Energy Optimisation, Mediterranean Climate PV Performance

1 INTRODUCTION

The European Green Deal, along with the "Fit for 55" package, mandates that EU member states reduce their carbon emissions by at least 55% by 2030, with the ultimate goal of achieving net-zero emissions by 2050. Solar energy, particularly photovoltaic (PV) systems, is pivotal in achieving this transition due to its potential to provide clean, affordable, and scalable energy solutions. Among the various renewable energy sources, PV systems are gaining traction as one of the fastest-growing technologies in the energy sector.

In regions with long sunny periods, like Malta, the large-scale adoption of PV systems offers a practical approach to meeting renewable energy targets. However, simply increasing the number of installations is not enough. It is equally critical to enhance the performance of existing systems to ensure optimal energy output, reliability, and longevity. The optimisation of current systems can provide valuable insights into the future design and development of PV technology. This need has driven the push towards real-time monitoring and performance evaluation of installed systems.

Malta's unique Mediterranean climate introduces specific challenges to PV performance, including high temperatures, humidity, and atmospheric salinity, all of which can impact the long-term reliability and efficiency

of PV installations. Additionally, internal factors such as electrical faults, system degradation, and component failures further contribute to the reduction in system performance. Therefore, the development of a robust, comprehensive monitoring system is essential to address these challenges and ensure sustainable energy generation in such environments. The increasing adoption of solar photovoltaic (PV) systems as a sustainable energy solution highlights the critical need to optimize their efficiency and reliability, prompting extensive research in this area [1],[2],[3],[4].

1.1 Aim and Scope

This paper proposes a comprehensive framework for enhancing the efficiency and reliability of PV systems in Malta through real-time monitoring. The primary aim of this study is to optimize PV system performance by leveraging data from multiple experimental PV installations across the country. This is achieved using advanced data acquisition technologies, including the Modbus RTU protocol, high-precision sensors, and API-based data management systems. By utilising open-source platforms, the proposed framework ensures flexibility and scalability for future expansion while minimizing reliance on proprietary software.

The key objectives of this framework are:
    i.   To analyse the environmental challenges

impacting PV systems in Malta,

ii. To develop a centralised data analytics platform for real-time performance monitoring,

iii. To evaluate system faults and maintenance requirements,

iv. To address grid-related issues associated with PV system integration,

v. To validate manufacturer claims regarding system performance post-installation, and

vi. To extrapolate data for predictive analysis of future PV performance trends.

This framework is designed with scalability in mind and can be applied to regions facing similar environmental challenges. The centralised data analytics platform developed in this study provides a flexible, adaptable tool for performance evaluation, enabling better understanding of PV systems under challenging conditions.

1.2 PROMISE PV Living Laboratories

The framework described in this paper is part of the PROMISE project, a three-year research initiative funded by the European Commission. The project aims to establish a platform for studying the reliability of both existing and emerging PV module technologies in the Mediterranean climate. PROMISE supports the development of digitalisation techniques, predictive algorithms, and optimisation tools to aid the energy transition. In addition to technological advancements, the project also includes capacity-building activities, such as workshops, training programs, and knowledge transfer, to create a robust research ecosystem in Malta.

The PROMISE project introduces two main frameworks:

i. A research platform to study the reliability of PV systems and develop innovative solutions for optimisation.

ii. A knowledge transfer platform that supports training, internships, and workshops to enhance the research and engineering capabilities within Malta's PV sector.

1.3 Data Monitoring Constraints

One of the main challenges in PV system monitoring is differentiating between genuine system faults and data anomalies caused by noise. Various studies highlight the difficulties of filtering noise from performance data without losing critical fault signals. Data anomalies, such as those caused by shading or environmental noise, can often be mistaken for system faults, leading to inaccurate performance assessments [6],[7].

To improve fault detection accuracy, effective noise filtering techniques must be employed [8]. For example, setting appropriate threshold levels, such as disregarding data with solar irradiance values below 20 W/m² [6], can help eliminate noise while retaining relevant data. Further, selecting peak sunlight hours (typically between 10 AM and 4 PM) reduces uncertainties related to shading that occur during early morning or late afternoon periods [9].

While filtering strategies are important, it is more efficient to minimize noise at the source by using high-quality sensors designed to operate in challenging climatic conditions. These sensors should adhere to Original Equipment Manufacturer (OEM) standards to ensure precision and reliability in the data acquisition process [6]. Moreover, granular data acquisition, which captures real-time uncertainties, can significantly enhance the detection of transient anomalies.

This paper is structured as follows: Section 2 provides an overview of the ten PV laboratories in Malta, detailing system specifications and sensor selections, Section 3 describes the data acquisition and processing framework, discussing how data is filtered and prepared for analysis, Section 4 presents the preliminary results from the multi-site monitoring framework, and finally Section 5 concludes the study, summarizing the findings and identifying areas for future research.

2. SYSTEM OVERVIEW AND MONITORING SITES

This section provides an overview of the ten PV living laboratories in Malta focusing on system specifications and sensor selections. The sites are strategically distributed across the island to capture variability in solar irradiance, weather conditions, and environmental factors. This diverse site selection ensures that the performance of photovoltaic (PV) systems can be analyzed across different geographic and environmental conditions. The data collected from these sites offers valuable insights into the factors affecting PV system efficiency, enabling the identification of hidden influences such as environmental matrices.

2.1 Site Representation

Each site in the Malta PV Living Laboratories is equipped with a range of sensors designed to monitor both electrical performance and environmental conditions. These sensors are deployed across a variety of locations, including urban, rural, and coastal areas, to ensure comprehensive data collection. Figure 1 shows the geographic distribution of the sites.

2.2 Sensor and Device Selection

The monitoring system relies on high-precision electrical and meteorological sensors, which capture data at 10-second intervals. The data acquisition is based on the Modbus RTU protocol using RS485 communication, ensuring cost-effective and efficient data transmission. Critical parameters such as DC current, voltage, and weather conditions are monitored with certified sensors

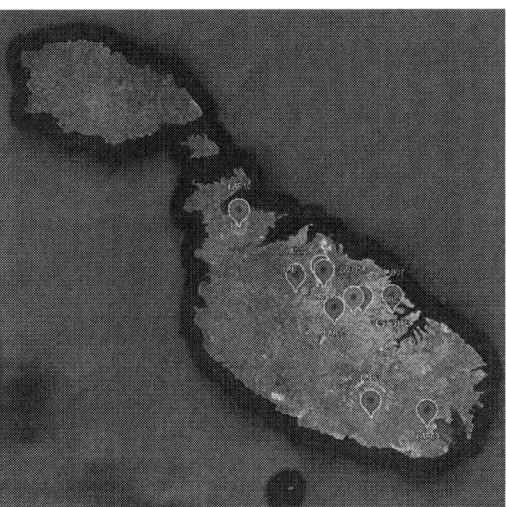

Figure 1: Malta PV Living Laboratories

compliant with ISO/IEC 17025 and ISO 9001 standards. These sensors ensure high accuracy and reliability in performance monitoring, enabling real-time analysis and early fault detection. Figure 2 shows the in-house local data monitoring panel.

3. DATA PROCESSING

This section outlines the sensor integration and data processing framework used for PV system monitoring. The sensors are controlled by a Raspberry Pi, which acts as the master device in the Operational Technology (OT) system, bridging the gap between OT and Information Technology (IT). Each sensor is assigned a unique ID, and data is collected using the Modbus RTU protocol over an RS485 communication system. This setup ensures robustness against noise and transmission errors, adhering to strict specifications like baud rates and terminating resistors.

The Raspberry Pi runs on a Linux-based system for stability, using Python libraries for data acquisition. This open-source approach facilitates customisation without the added cost of proprietary software. Data is transmitted to API platforms for live monitoring and storage. Additionally, remote access is enabled via Real VNC, allowing for system adjustments after setup. Figure 3 shows the comprehensive framework.

3.1 Data Treatment

Data is initially captured at 10-second intervals to ensure granularity, then averaged into 30-second intervals to reduce the dataset size while preserving essential resolution. Preliminary data cleaning involves removing irrelevant entries, such as non-numeric values (NaNs) and data with solar irradiance below 20 W/m², which eliminates noise during night hours. Further processing includes consolidating data from different channels into a unified table for each site, and excluding periods of low sunlight, such as early morning or late evening, to minimize the impact of shadows.

Figure 2 In-house local data monitoring panel.

4. PRELIMINARY RESULTS

The multi-site monitoring framework has been successfully implemented across six PV living laboratories so far. Electrical and meteorological sensors, along with the central processing unit, are housed in IP67-rated enclosures, ensuring protection against harsh environmental conditions. Data from electrical and weather sensors is transmitted to a shared data bus, enabling real-time monitoring. Figure 4 shows the public dashboard of LAB01. Initial results demonstrate the system's robustness in tracking performance across diverse environmental conditions.

5. CONCLUSION

This study introduces a comprehensive framework for monitoring and optimising PV systems in Malta, addressing climate challenges through real-time data

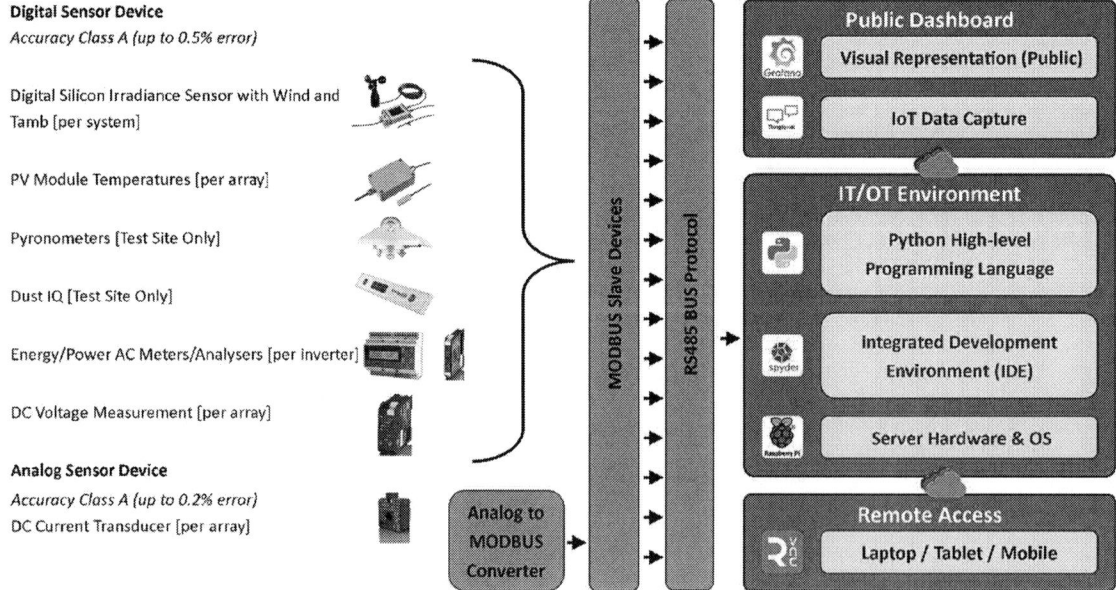

Figure 3: The Comprehensive Framework

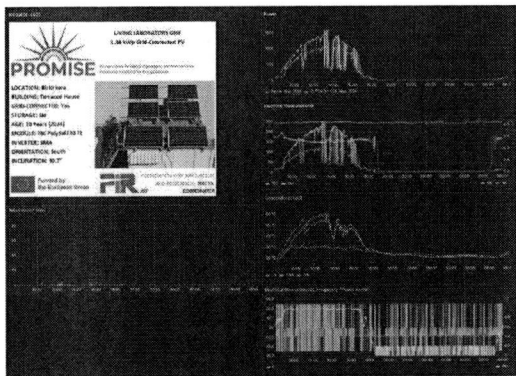

Figure 4: LAB01 Public Dashboard

acquisition and advanced sensor integration. The PROMISE project highlights the value of harmonized multi-site measurements in improving system reliability and performance. Key findings include the identification of performance indicators like energy yields and fault detection. Future work will focus on integrating air quality sensors to assess pollution impacts, refining the data analytics platform, and expanding the system's scalability to larger networks, further supporting renewable energy efficiency and sustainability efforts.

REFERENCES

[1] M. Green, E. Brill, B. Jones, and J. Dore, Improving efficiency of PV systems using statistical performance monitoring: International Energy Agency Photovoltaic Power Systems Programme: IEA PVPS Task 13, Subtask 2: report IEA-PVPS T13-07:2017. Paris: International Energy Agency, 2017.

[2] D. L. King, 'More "efficient" Method for Specifying and Monitoring PV system performance', presented at the , 37th IEEE Photovoltaic Specialists Conference, pp. 219-224, 2011.,

[3] 'Technical Specification for on-site energy efficiency testing and assessment of photovoltaic (PV) power plants, 2020.', Chin. Soc. Electr. Eng. CSEE TCSEE 0160.

[4] Matthias Littwin,Franz P. Baumgartne,Mike GreenWilfried van Sark,Ulrike Jahn, 'Performance of New Photovoltaic System Designs', International Energy Agency IEA, IEA-PVPS T13-15:2021, Apr. 2021. [Online]. Available: https://iea-pvps.org/wp-content/uploads/2021/03/IEA-PVPS_Task-13_R15-Performance-of-New-PV-system-designs-report.pdf

[5] PROMISE: Photovoltaics Reliability Operations and Maintenance Innovative Solutions for Energy Alliance, funded by the European Union's Horizon Europe program, Grant Agreement No. 101079469. PV-PROMISE.EU

[6] Shimshon Rapaport,Mike Green,Ulrike Jahn, 'The Use of Advanced algorithms in PV failure monitoring', International Energy Agency IEA, t IEA-PVPS T13-19:2021, Sep. 2021. [Online]. Available: https://iea-pvps.org/wp-content/uploads/2021/10/Final-Report-IEA-PVPS-T13-19_2021_PV-Failure-Monitoring.pdf

[7] S. Vergura, 'A Statistical Tool to Detect and Locate Abnormal Operating Conditions in Photovoltaic Systems', Sustainability, vol. 10, p. 608, Feb. 2018, doi: 10.3390/su10030608.

[8] Å. Skomedal, M. Øgaard, J. Selj, H. Haug, and E. Marstein, 'General, Robust and Scalable Methods for String Level Monitoring in Utility Scale PV Systems', Oct. 2019. doi: 10.4229/EUPVSEC20192019-5BO.5.4.

[9] M. Hazza, H. Attia, and K. Hossin, 'Solar Photovoltaic Power Prediction Using Statistical Approach-Based Analysis of Variance', Sol. Energy Sustain. Dev. J., vol. 13, pp. 45–61, Jun. 2024, doi: 10.51646/jsesd.v13i2.181.

[10] 'IEC TS 61724-2 - Photovoltaic system performance - Part 2: Capacity evaluation method | GlobalSpec'. Accessed: Aug. 08, 2024. [Online]. Available: https://standards.globalspec.com/std/10047074/iec-ts-61724-2

[11] 'ISO/IEC 17025:2017(en), General requirements for the competence of testing and calibration laboratories'. Accessed: Sep. 20, 2024. [Online]. Available: https://www.iso.org/obp/ui/en/#iso:std:iso-iec:17025:ed-3:v1:en

[12] 'ISO 9001:2015(en), Quality management systems — Requirements'. Accessed: Sep. 20, 2024. [Online]. Available: https://www.iso.org/obp/ui/en/#iso:std:iso:9001:ed-5:v1:en

ACKNOWLESGEMENTS

Partly funded by the European Union under Grant 101079469 PROMISE "Photovoltaics Reliability Operations and Maintenance Innovative Solutions for Energy Alliance" project, under Grant 101075747 and UK Research and Innovation (UKRI) TRANSIT "TRANSITion to sustainable future through training and education" project, European Union, Xjenza Malta under Grant REP-2023-061 RoOFPEVs "Robust Optimization Framework for PVs and EVs Integration at Low Voltage Network" project.

Views and opinions expressed are, however, those of the author(s) only and do not necessarily reflect those of the granting authorities/agencies nor that the granting authorities/agencies can be held responsible for them.

Furthermore, the authors would like to thank the owners of the PV systems deployed as FiR.mt Living Laboratories in Malta.

AUTHORS CONTRIBUTIONS

Conceptualisation (BB, BA, AM, MR, BK), Data curation (BB), Formal analysis (BB), Hardware/Software (BB, BA), Funding acquisition (BA), Investigation (BB, BA), Methodology (BB, BA), Project administration (BA), Resources (BA), Supervision (BA, KS), Validation (BA), Visualisation (BB, BA), Writing – original draft (BB, BA), Writing – review and editing (RE, CM, ME, EZ, SZ,).

41st European Photovoltaic Solar Energy Conference and Exhibition

MEDITERRANEAN CLIMATE IMPACT ON PHOTOVOLTAIC SYSTEMS
INSIGHTS FROM MALTA AND IMPLICATIONS FOR FUTURE EUROPEAN INTEGRATION

Brian Bartolo[1, 2], Brian Azzopardi[1, 2, 3, 4 *], Alexandre Mignonac[5], Marcus Rennhofer[6], Bernhard Kubicek[6], Rita Ebner[6],
Carlos Meza[7], Melodie de l'Epine[8], Eugenia Zugasti[9], Steve Zerafa[2, 3, 10,] Kenneth Scerri[3, 1]

[1] The Foundation for Innovation and Research – Malta (FiR.mt)
[2] Malta College of Arts, Science and Technology (MCAST)
[3] The University of Malta
[4] Azzopardi & Associates, Malta – Lithuania
[5] Commissariat à l'Energie Atomique et aux Energies Alternatives (CEA) France
[6] AIT Austrian Institute of Technology GMBH, Austria
[7] Anhalt University of Applied Sciences, Germany
[8] Becquerel Institute, Belgium
[9] Fundación CENER—National Renewable Energy Centre, Spain
[10] PIXAM Ltd., Malta

* Brian.Azzopardi@FiR.mt

ABSTRACT: As the demand for clean energy grows, optimising photovoltaic (PV) systems in climate-specific environments is crucial to achieving energy transition goals. This paper investigates the impact of the Mediterranean climate on PV systems, focusing on insights gained from Malta through the PROMISE project. By leveraging real-time data from five PV living laboratories, the study examines key environmental challenges such as high temperatures, humidity, atmospheric salinity, and soiling, which significantly affect system performance. Seasonal variations in temperature and irradiance are analyzed to assess their impact on PV efficiency and energy yield. These findings are further extrapolated to address broader European challenges in integrating PV systems across diverse climates. The research offers critical recommendations for enhancing PV technology in Mediterranean regions and explores the role of PV installations in urban heat mitigation. The developed data-driven framework provides scalable solutions for improving system resilience and forecasting performance, contributing to future PV integration efforts across Europe.

Keywords: Mediterranean Climate, Photovoltaic System Performance, Environmental Factors, Real-time Data Monitoring, PV Degradation and Soiling

1 INTRODUCTION

The European Green Deal [1] and the Fit for 55 package [2] set ambitious targets for reducing carbon emissions by at least 55% by 2030 and achieving net-zero emissions by 2050. Solar energy, particularly photovoltaics (PV), is pivotal in this transition, offering clean, scalable energy solutions [3]. Mediterranean countries, like Malta, with their abundant sunshine, are well-positioned to contribute significantly to these energy transition goals. However, while the expansion of PV installations has been supported by various government incentives, enhancing the performance and reliability of existing PV systems remains a crucial challenge.

PV systems in Mediterranean climates face unique operational challenges due to environmental factors such as high temperatures, humidity, atmospheric salinity, and soiling, which degrade system performance. Existing studies [4], [5], [6], [7] have highlighted the need for focused research in these climates to understand these externalities. Internal issues such as electrical faults and system degradation further complicate the optimisation of PV systems, emphasising the need for advanced monitoring and fault detection strategies to sustain long-term performance [8], [9], [10].

The aim of this paper is to assess the challenges posed by the Mediterranean climate on PV systems and explore strategies to enhance system efficiency and reliability. Through real-data capture from PV living laboratories in Malta, this study investigates how environmental factors affect PV performance and provides recommendations for broader European PV integration. Additionally, this research evaluates the potential role of PV installations in mitigating urban heat, offering dual benefits in energy generation and urban climate adaptation.

The assessment presented in this paper is part of the broader PROMISE project, a three-year research initiative funded by the European Commission. PROMISE aims to establish a comprehensive platform for studying the reliability of both current and emerging photovoltaic (PV) technologies, specifically within the context of the Mediterranean climate. The project focuses on advancing digitalisation techniques, predictive algorithms, and optimization tools to support the energy transition. Beyond technological innovations, PROMISE also fosters capacity building through workshops, training programs, and knowledge transfer, thus strengthening Malta's research and engineering ecosystem.

PROMISE operates through two core frameworks: a research platform dedicated to the study and optimization of PV system reliability, and a knowledge transfer platform designed to enhance skills through training, internships, and workshops, further building the capacity of Malta's PV sector.

This paper is structured as follows: Section 2 outlines the geographic locations and technical specifications of the PV sites under investigation. Section 3 presents the results of the study, including the impact of seasonal variations on PV efficiency. Finally, Section 4 provides conclusions and recommendations for future research.

2. GEO-LOCATIONS AND TECHNICAL SPECS

This study focuses on five PV sites in Malta, labeled LAB01 to LAB05, as part of the Malta PV Living Laboratories Platform, Table 1. These sites are strategically located across the island to capture variability in irradiance and environmental conditions such as air quality, clearness index, and meteorological factors. Each site is equipped with high-precision sensors that monitor critical parameters like irradiance, temperature, and humidity, capturing data at 10-second intervals for detailed analysis.

The data acquisition system employs the Modbus RTU protocol over an RS485 interface, ensuring cost-effective and reliable data transmission. Sensors used for string DC current measurements and environmental monitoring are certified according to ISO/IEC 17025 and ISO 9001 standards to guarantee accuracy and minimize interference. Key performance metrics, such as Performance Ratios (PR), energy yields, and kilowatt-hours per kilowatt peak (kWh/kWp), are calculated to assess the operational reliability and efficiency of each site.

For the selected days of analysis, May 10, 2024, showed an average temperature of 19.08°C with 48% humidity, while August 4, 2024, experienced a higher average temperature of 29.15°C and 62% humidity. These contrasting periods provided a comprehensive assessment of how seasonal variations, heat, and humidity levels influence PV system performance under distinct Mediterranean climatic conditions.

The distributed setup across varied environments enables comprehensive analysis of how local climatic factors influence PV performance, providing insights into optimizing systems for both urban and rural settings under Mediterranean climate conditions.

3. IMPACTS OF SEASONAL VARIATIONS

This section presents the results of the study, Figure 1, focusing on the impact of seasonal variations and environmental factors, including temperature, humidity, soiling, and shadowing, on the performance of the PV systems. Data were collected from several finalized systems equipped with electrical and meteorological sensors, providing detailed insights into key parameters such as irradiance, temperature, power generation, and the effect of shadows.

Table 1: Malta PV Living Five Laboratories

	LAB 01	LAB 02	LAB 03	LAB04	LAB05
Geo-location	Central	Central	Central	Central	Central
Building Type	Terraced House	Residential (Public)	Semi-Detached Villa	Semi-Detached Villa	Semi-Detached Villa
System	Rooftop	Rooftop	Rooftop	Rooftop	Rooftop
Inverter	Central	Central	Central	Central	Small (x3)
Grid-Connected	Yes	Yes	Yes	Yes	Yes
Single/Thre	Single	Three	Single	Three	Three
Inverter	SMA	KOSTAL	SMA	GROW	
Monitoring	No	No	Yes	Yes	No
Orientation	21° SW	26° SE	7° SW	22° SW	24° SW
Inclination	31°	15°	13°	9°	16°
Age (Years)	12	5	9	7	7
Number of	6	35	12	20	14
Rated	1.38	11.55	3.12	5.4	3.64
Recording	Jan-24	Mar-24	Mar-24	Apr-24	Apr-24

The results from May 10, 2024 (spring), and August 4, 2024 (summer), highlight significant differences in PV system efficiency due to seasonal changes. In May, the moderate average temperature of 19.08°C and 48% humidity contributed to optimal PV performance, with relatively high power output. However, in August, the average temperature rose to 29.15°C with 62% humidity, which negatively affected the efficiency of the PV systems. This reduction in efficiency was attributed to several factors, including the rise in ambient and module temperatures, which is a well-known contributor to performance loss in PV panels. The elevated temperatures led to an increase in both cell and back sheet temperatures, particularly at LAB02, which was enclosed and had restricted air circulation. On the other hand, LAB04 and LAB05, located in open areas with better ventilation, experienced somewhat lower temperature impacts but still showed performance losses due to other environmental factors.

A notable factor during the summer months was the reduction in Global Plane of Array (GPOA) levels and power generation due to air quality issues. The higher humidity levels and absence of rainfall during August increased atmospheric water content and airborne particles, lowering irradiance levels. This effect was more pronounced at LAB04 and LAB05, which were located near dust-prone areas such as a quarry and construction site, exacerbating the soiling problem. Soiling, compounded by the lack of rain, led to significant losses in power output, as indicated by the general decrease in GPOA and overall energy generation across all sites.

Shadowing also played a role in affecting system performance, particularly during the early morning and late evening periods when sunlight was obstructed by nearby structures. In urban locations such as LAB01 and LAB02, partial shadowing was observed during certain times of the day, further contributing to performance reductions. To mitigate the effects of shadowing on data accuracy, the study excluded periods of heavy shading from the analysis. This ensured that the data reflected true performance during peak solar hours when shadowing was minimal.

Overall, the results clearly show that PV systems in the Mediterranean climate are highly sensitive to seasonal variations and external factors such as soiling, temperature, humidity, and shadowing. The combined effects of these factors led to a significant reduction in efficiency during the summer months, emphasizing the need for tailored mitigation strategies to maintain system performance throughout the year.

4. CONCLUSION

This study provides critical insights into the impact of Mediterranean climate conditions on photovoltaic (PV) system performance, with a focus on real-time data collected from living laboratories in Malta. The results clearly demonstrate that elevated temperatures, high humidity, and environmental soiling significantly reduce PV performance, particularly during the summer months when these conditions are most severe. Variations in ambient temperatures due to local environmental factors, such as proximity to construction sites or areas with heavy dust and pollution, further exacerbate the reduction in energy generation.

41st European Photovoltaic Solar Energy Conference and Exhibition

REAL-TIME DATA INSIGHTS

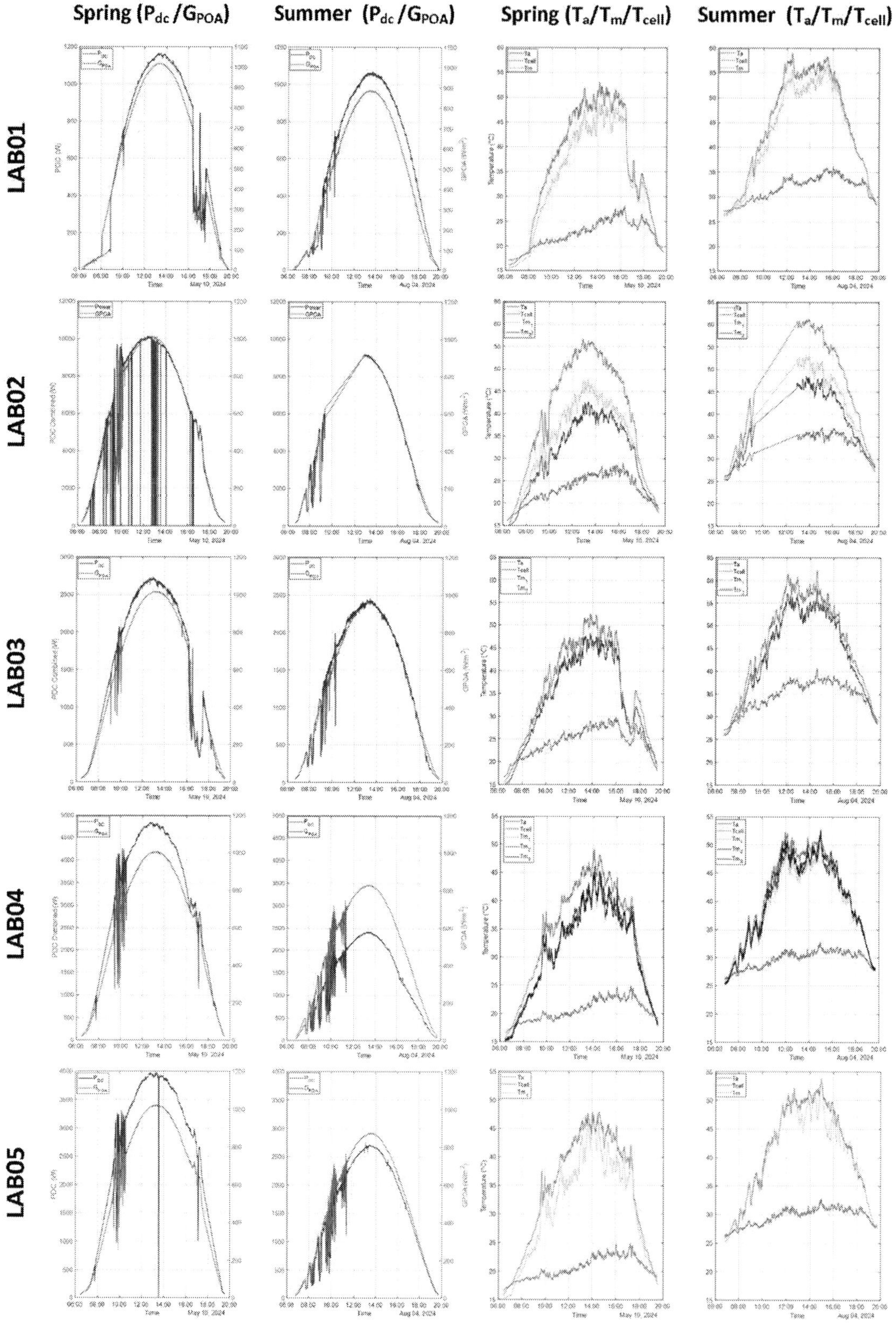

Figure 1: Real-Time Data Insights from collected data

The findings emphasise the importance of accounting for specific environmental challenges when optimising PV systems in Mediterranean regions. The results also underline the need for tailored mitigation strategies, such as improved cleaning methods to address soiling and better thermal management to reduce the impact of elevated temperatures on PV modules. The differences in system performance across the monitored sites highlight the influence of micro-environments, such as urban versus open areas, and suggest that localised solutions are critical for maximizing efficiency.

Future research should focus on expanding the study framework to other Mediterranean regions to gather more comprehensive data and explore regional variations in PV performance. Additionally, the integration of AI-driven tools for real-time performance monitoring and predictive analysis, particularly in Performance Ratio (PR) forecasting, will enhance the accuracy of system assessments across different climatic conditions. This approach will not only optimise PV generation in Malta but also provide scalable solutions for other regions with similar environmental challenges.

To further advance the field, future studies could incorporate the role of shadowing more extensively and develop algorithms to predict and mitigate the effects of partial shading, particularly in urban areas. Additionally, future work should explore advanced materials and technologies that are more resilient to high-temperature and high-humidity environments, as well as develop cleaning and maintenance protocols tailored to areas prone to soiling.

By refining the existing framework and integrating advanced monitoring techniques, this research offers a pathway to enhancing the resilience and efficiency of PV systems across Mediterranean and other climates experiencing similar environmental stressors.

REFERENCES

[1] 'The European Green Deal - European Commission'. Accessed: Sep. 25, 2024. [Online]. Available: https://commission.europa.eu/strategy-and-policy/priorities-2019-2024/european-green-deal_en

[2] 'Fit for 55 - The EU's plan for a green transition', Consilium. Accessed: Sep. 25, 2024. [Online]. Available: https://www.consilium.europa.eu/en/policies/green-deal/fit-for-55/

[3] 'Solar', IEA. Accessed: Sep. 25, 2024. [Online]. Available: https://www.iea.org/energy-system/renewables/solar-pv

[4] M. Green, E. Brill, B. Jones, and J. Dore, *Improving efficiency of PV systems using statistical performance monitoring: International Energy Agency Photovoltaic Power Systems Programme: IEA PVPS Task 13, Subtask 2: report IEA-PVPS T13-07:2017.* Paris: International Energy Agency, 2017.

[5] D. L. King, 'More "efficient" Method for Specifying and Monitoring PV system performance', presented at the , 37th IEEE Photovoltaic Specialists Conference, pp. 219-224, 2011.,

[6] 'Technical Specification for on-site energy efficiency testing and assessment of photovoltaic (PV) power plants, 2020.', *Chin. Soc. Electr. Eng. CSEE TCSEE 0160.*

[7] Matthias Littwin,Franz P. Baumgartne,Mike GreenWilfried van Sark,Ulrike Jahn, 'Performance of New Photovoltaic System Designs', International Energy

Agency IEA, IEA-PVPS T13-15:2021, Apr. 2021. [Online]. Available: https://iea-pvps.org/wp-content/uploads/2021/03/IEA-PVPS_Task-13_R15-Performance-of-New-PV-system-designs-report.pdf

[8] H. A. Kazem and M. Chaichan, 'Effect of humidity on photovoltaic performance based on experimental study', vol. 10, pp. 43572–43577, Dec. 2015.

[9] G. Makrides, B. Zinsser, M. Norton, and G. Georghiou, 'Performance of Photovoltaics Under Actual Operating Conditions', 2012. doi: 10.5772/27386.

[10] T. Tsoutsos, N. Frantzeskaki, and V. Gekas, 'Environmental impacts from the solar energy technologies', *Energy Policy*, vol. 33, no. 3, pp. 289–296, Feb. 2005, doi: 10.1016/S0301-4215(03)00241-6.

[11] 'Electrical Isolation Design and Electrochemical Corrosion Available: https://www.iso.org/obp/ui/en/#iso:std:iso-iec:17025:ed-3:v1:en

ACKNOWLESGEMENTS

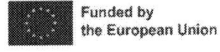

Partly funded by the European Union under Grant 101079469 PROMISE "Photovoltaics Reliability Operations and Maintenance Innovative Solutions for Energy Alliance" project, under Grant 101075747 and UK Research and Innovation (UKRI) TRANSIT "TRANSITion to sustainable future through training and education" project, European Union, Xjenza Malta under Grant REP-2023-061 RoOFPEVs "Robust Optimization Framework for PVs and EVs Integration at Low Voltage Network" project.

Views and opinions expressed are, however, those of the author(s) only and do not necessarily reflect those of the granting authorities/agencies nor that the granting authorities/agencies can be held responsible for them.

Furthermore, the authors would like to thank the owners of the PV systems deployed as FiR.mt Living Laboratories in Malta.

AUTHORS CONTRIBUTIONS

Conceptualisation (BB, BA, AM, MR, BK), Data curation (BB), Formal analysis (BB), Hardware/Software (BB, BA), Funding acquisition (BA), Investigation (BB, BA), Methodology (BB, BA), Project administration (BA), Resources (BA), Supervision (BA, KS), Validation (BA), Visualisation (BB, BA), Writing – original draft (BB, BA), Writing – review and editing (RE, CM, ME, EZ, SZ,).

41st European Photovoltaic Solar Energy Conference and Exhibition

Zentrum für Sonnenenergie- und Wasserstoff-
Forschung Baden-Württemberg

Comparison of physical, Machine Learning and hybrid models of monofacial and bifacial PV systems

Jonas Petzschmann, Dirk Stellbogen and Manuel Heim
Zentrum für Sonnenenergie- und Wasserstoffforschung Baden-Württemberg (ZSW), Meitnerstrasse 1, 70563 Stuttgart, Germany
jonas.petzschmann@zsw-bw.de, dirk.stellbogen@zsw-bw.de

Situation

Data-driven models for PV systems can represent complex operation conditions. Physics based models for PV systems require less information. Hybrid models may combine the respective advantages.

Objective

The study benchmarks the three modelling methods by applying them to monofacial and bifacial PV systems, which have been operated and monitored in the same environmental conditions at the ZSW outdoor test facilities in Southern Germany. Different configurations of physical, data-based and hybrid models and input data sets have been evaluated.

Experimental Basis

Test setup with two PV module strings operated at independent inverter MPPTs:

- 11 bifacial PV modules in one string
- 11 'monofacial' PV modules in the second string, from the same module type but covered on the backside.

The geometrical layout is optimized to prevent bias from view angle differences.

Figure 1: top: backside covered bifacial PV modules; right: PV test system with module string scheme (light blue: bifacial modules, dark blue: modules made monofacial; grey: not connected)

Modelling

The **physical modelling** uses single-diode based PV models and a transposition models for deriving plane-of-array (POA) irradiance. **Machine Learning** models are based on Artificial Neural Networks (ANN). The **hybrid models** use ANNs to relate the physical model output to the observations.

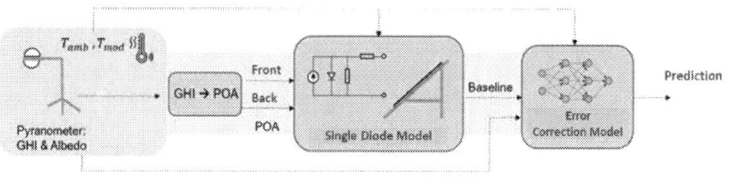

Figure 2: Functional diagram of the physical and Machine Learning hybrid modelling.

Results

The **physical, data-driven Neural Network and hybrid modelling** schemes are tested with different sets of input information.

The **physical model** shows good results for high irradiances related to clear sky situations but higher errors for other conditions. The **data related models** have considerably smaller errors and are able to cope with anomalies like tree shading by learning from the recorded observations.

More enhanced data sets lead to better modelling accuracies, but the **data-based and hybrid models** have always lower error margins. The **ML model** can even exceed the **physical model** when applied to a more scarce data set.

The **hybrid model** achieves error metrics similar to the level of the **fully data driven ANN** model. It even performs slightly better for the monofacial case.

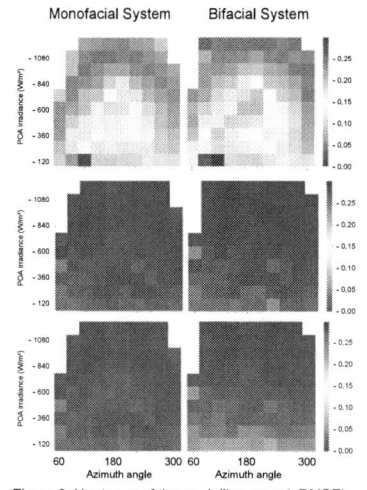

Figure 3: Heat map of the modelling error (nRMSE) vs. irradiance and sun azimuth angle; top: physical model, middle: ML model, bottom: hybrid model.

Variant	Irradiation Data				Temperatures	
	GHI	Albedo	POA$_{front}$	POA$_{back}$	T_{amb}	T_{mod}
meteo	✓	-	-	-	✓	-
poa	✓	-	✓	-	✓	-
full	✓	✓	✓	✓	✓	✓

Table: Input data set variations for modelling tests

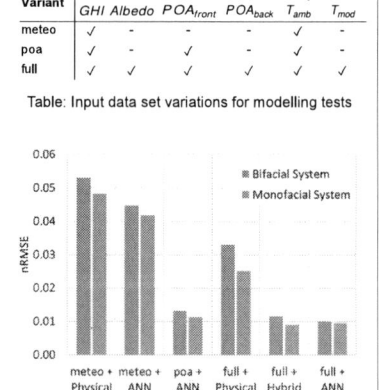

Figure 4: Total error margins (nRMSE) for various sets of input data (see Table) and modelling schemes (Physical model, data driven ANN model and Hybrid model)

Conclusion

The study showed, that **physical models** are being outperformed by purely data-based **Machine Learning** (ML) models as well as by **hybrid models** in all test cases. Thereby, the **data-driven models** are able to reduce the gap in modelling accuracy between bifacial and monofacial PV systems. The error margins of the **hybrid models** arrive at similar levels as the pure **ML models**.

QUANTITATIVE SHADE DETECTION FOR PV SYSTEMS BASED ON CLEARSKY DATA

Achim Schulze[1], Markus Panhuysen[2], Darwin Daume[3],
Maximilian Schönau[2,3]

[1]Rosenheim Technical University of Applied Sciences, Germany, achim.schulze@th-rosenheim.de
[2]Smartblue AG, Kistlerhofstr. 75, Munich, Germany
[3]Coburg University of Applied Sciences, Coburg, Germany

ABSTRACT: In modern monitoring applications, losses from shading should be quantitatively separated from other loss mechanisms such as soiling, degradation or total failures. Based on clearsky reference data, we introduce a shading matrix which enables accurate hourly description of shading losses of PV strings over the year. This matrix can be used to improve failure detections and reference yield data in PV systems. Furthermore, a shading matrix with respect to sun position is introduced to improve the spatial understanding of shading caused by surrounding objects.

Keywords: PV Systems, Failure detection, Shading, Clearsky

1 INTRODUCTION

Ideally, losses from shading can be quantitatively separated from other loss mechanisms such as soiling, degradation or total failures.

Due to its periodic behaviour and although often resulting in large power losses, partial shading normally is not regarded as a serious fault - in contrast to soiling and degradation. In order to be able to analyze the latter precisely, a precise temporal and quantitative analysis of shading is required.

Partial shading is often detected by directly analyzing IV-curves, see for example [1, 2, 3], but for most PV systems, complete IV-curves cannot be obtained directly.

Our approach is based only on MPP power data provided by DC/AC power inverters which in most cases is the only available data for monitoring applications.

We analyze the shading behaviour of PV systems by comparing the actual power output with the theoretical power output from physical models at points in time with clear sky.

2 CLEARSKY DETECTION AND POWER MODEL

For a PV system, all clearsky events during a year (or several years) are detected - either directly from weather data using weather codes or cloudcover parameters or by analyzing irradiation data, e.g. GHI (global horizontal irradiation) and/or BNI (beam normal irradiation), cf. [4].

After obtaining suitable atmospherical parameters, clearsky irradiations E_{bni} (clearsky beam normal irradiation) and E_{dif} (clearsky diffuse horizontal irradiation) are calculated for each point in time with clear sky by a clearsky model, e.g. [5, 6, 7, 8, 9]. This irradiation data then is converted to irradiation data in the inclined surface (POA, plane of array) given

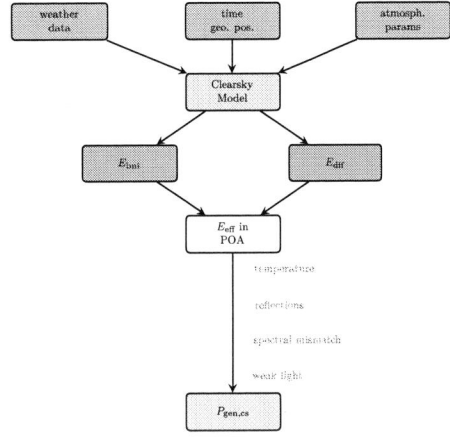

Figure 1: Clearsky irradiation and power model

by the metadata of the system.

Finally, the theoretical electrical power output $P_{\mathrm{gen,cs}}$ (at clearsky) of the PV system is predicted by a physical model, where temperature, weak light and optical losses are taken into account. The complete procedure is depicted in Figure 1. The ideal clearsky power $P_{\mathrm{gen,cs}}$ is then used as a reference for shadow detection at clearsky times.

3 SHADOW DETECTION AND RESULTS

The observed (measured) output power of the PV array at clearsky times is denoted by P_{meas}.

In order to create a time-resolved shading matrix, time-averaged bins are introduced: The mean value of P_{meas} for each pair (i,j), where i is the month of the year, i.e. $i = 1, 2, \ldots, 12$, and j is the hour in local time, i.e. $j = 6, 7, \ldots, 21$, is denoted by $P^{i,j}_{\mathrm{meas}}$. The same method of averaging is applied to $P_{\mathrm{gen,cs}}$

which yields the mean theoretical clearsky power $P_{\mathrm{gen,cs}}^{i,j}$.

Taking the ratio of the mean actual power output to the mean theoretical value at each clearsky hour of each month, the time-resolved clearsky shading matrix $\eta^{i,j}$ is calculated:

$$\eta^{i,j} = \frac{P_{\mathrm{meas}}^{i,j}}{P_{\mathrm{gen,cs}}^{i,j}}, \quad i = 1,2,\dots 12, \quad j = 6,7,\dots,21. \quad (1)$$

Small values of the denominator, i.e. specific power values $P_{\mathrm{gen,cs,spec}}^{i,j} < 100$ W/kWp are dropped to avoid bins with very low direct irradiation and large uncertainty.

We apply this method for a PV rooftop power plant in Switzerland near Lake Constance. The orientation of the PV modules is south-east ($120°$) with a slope of $10°$. We get the clearsky shading matrix shown in Figure 2.

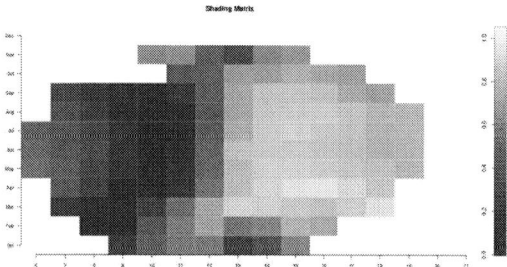

Figure 2: Shading Matrix for year 2022, shading is visible in the morning hours

In Fig.2, the x-axis depicts the daily hour (of each month) and white color visualizes missing data (i.e. no clearsky event was found). The shown data is the ratio of mean actual power of the PV system to the mean theoretically calculated clearsky power.

For the winter months, a snow model is taken into account in order to drop the days when pv-modules are covered with snow.

In Fig. 2, shading of the pv modules is detected in the morning hours and no shading is oberserved in the afternoon (except in the winter months). A satellite picture of the site reveals that the shading is caused by a large tower block, see Fig 3.

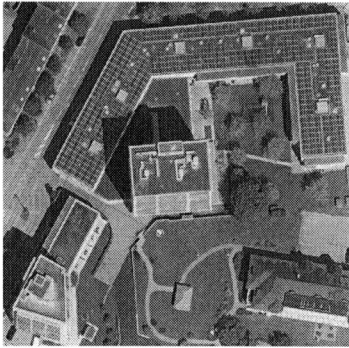

Figure 3: Satellite picture of the site, provided by Google Maps, Airbus Maxar Technologies ©2024

Using the matrix from Eq.(1), shading effects can be separated from other failures such as string out-

ages and unintended failure warnings due to periodic shading can be avoided.

4 SHADING MATRIX WITH RESPECT TO SUN POSITION

To improve understanding of the spatial distribution of shading, the shading matrix from Eq.(1) can be redefined with respect to sun position. For this purpose, spatial bins are introduced for each pair (i,j) of sun position angles, where $i = 0,1,2,\dots,90$ denotes the sun elevation angle in degrees and $j = 0,1,2,\dots,359$ denotes the sun azimuth angle in degrees.

Similar to section 3, the averaged values

$$P_{\mathrm{meas}}^{i,j}, \quad P_{\mathrm{gen,cs}}^{i,j}$$

and the shading matrix

$$\eta^{i,j} = \frac{P_{\mathrm{meas}}^{i,j}}{P_{\mathrm{gen,cs}}^{i,j}}, \, i = 0,1,\dots 90, \, j = 0,1,2\dots,359 \quad (2)$$

are established with respect to sun position. The shading matrix from Fig. 2 for the site in Switzerland then is converted to the matrix with respect to sun position as shown in Fig. 4.

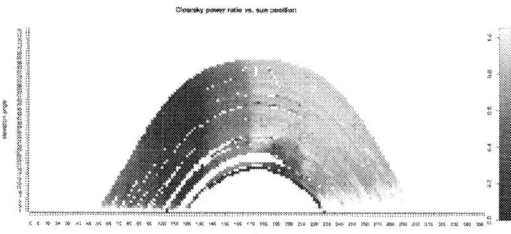

Figure 4: Clearsky shading matrix with respect to sun position, a shade caused by a tower block can be clearly seen

The sharp edges of the tower block can be identified in Fig. 4 and also a smaller shading problem can be seen in the early afternoon for low sun elevations due to another tower block in the south-west, see Fig. 3. With the modified shading matrix, identification and quantification of shades caused by surrounding objects is improved.

5 FURTHER TOPICS

Firstly, using the shading matrix (1), a shadow-corrected reference power model based on sensor or satellite irradiation data can be developed. Since the beam separation is known from the clearsky model, a good estimate for direct shading is obtained.

Furthermore, if only a small amount of data is available and/or only very few clearsky times are identified, gaps in the clearsky shading matrix can be filled using machine learning methods - this is subject of our current research.

Finally, comparing the shading matrices from several years can reveal gradual degradation (or gradual development of other failures).

ACKNOWLEDGEMENT

The authors would like to thank the Bayerische Forschungsstiftung for the financial support of the project KICK-PV (KI-basierte Charakterisierung und Klassifizierung von PV-Anlagen zur prädiktiven Wartung), grant number AZ-1564-22.

REFERENCES

[1] S. Sugumar, D. Winston, and M. Pravin, "A novel on-time partial shading detection technique for electrical reconfiguration in solar PV system," *Solar Energy*, vol. 225, pp. 1009–1025, 2021.

[2] S. Fadhel *et al.*, "PV shading fault detection and classification based on IV curve using principal component analysis: Application to isolated PV system," *Solar Energy*, vol. 179, pp. 1–10, 2019.

[3] M. Schönau, D. Daume, B. Hüttl, and D. Landes, "Improving IV Curve Classification by Machine Learning Methods Using Deep Autoencoders," *40th EU PVSEC*, 2023.

[4] M. Schönau, Daume, M. Panhuysen, A. Schulze, B. Hüttl, and D. Landes, "Verbesserte Clear-Sky-Erkennung durch hybrides Maschinelles Lernen," *7th RETCon Nordhausen*, 2024.

[5] P. Ineichen and R. Perez, "A New airmass independent formulation for the Linke turbidity coefficient," *Solar Energy*, vol. 73, pp. 151–157, 2002.

[6] R. Perez *et al.*, "A New Operational Model for Satellite-Derived Irradiances: Description and Validation," *Solar Energy*, vol. 73, pp. 307–317, 2002.

[7] M. Lefèvre *et al.*, "McClear: a new model estimating downwelling solar radiation at ground level in clear-sky conditions," *Atmos. Meas. Tech.*, vol. 6, pp. 2403–2418, 2013.

[8] B. Gschwind *et al.*, "Improving the McClear model estimating the downwelling solar radiation at ground level in cloud free conditions," *Meteorol. Z./Contrib. Atm. Sci.*, vol. 28, pp. 147–163, 2019.

[9] C. Gueymard, "REST2: High-Performance solar radiation model for cloudless-sky irradiance, illuminance, and photosynthetically active radiation - Validation with a benchmark dataset," *Solar Energy*, vol. 82, pp. 272–285, 2008.

41st European Photovoltaic Solar Energy Conference and Exhibition

ASSESSING ELECTROLUMINESCENCE IMAGE QUALITY WITH MACHINE-LEARNING AND GREY-LEVEL CO-OCCURRENCE MATRIX TEXTURE DESCRIPTORS

Thøger Kari, Aysha Mahmood, Rodrigo del Prado Santamaria, Gisele Alves dos Reis Benatto,
Peter Behrensdorff Poulsen and Sergiu V. Spataru.
Department of Electrical and Photonics Engineering, Technical University of Denmark.
Frederiksborgvej 399, 4000, Roskilde, Denmark.

ABSTRACT: Electroluminescence (EL) based fault detection and power prediction is highly depending on image quality. When it comes to daylight imaging and moving acquisition, however, existing Signal-to-Noise Ratio (SNR) metrics are unreliable; a substantial obstacle for plant-scale automation of postprocessing. This work involves using Grey-Level Co-occurrence Matrix (GLCM) texture features for estimating noise and blur severity in EL images of Photovoltaic (PV) modules. The aim is to directly gauge, rather than predict, image quality. Given the absence of a reliable baseline metric, blur and noise were added artificially to lab quality images, generating exactly known labels for training and validation. Candidate models were kept simple and suitable for linear regression. Hyper-parameters have a great impact on GLCM features and were optimized concurrently with the regression models. To stress-test the model, they were validated against synthesized datasets from completely different and intentionally damaged modules. The proposed method shows a lot of promise, being capable of linearly tracking the noise in the synthesized datasets. For blur, however, it is less accurate. More work is required, including diversifying the applied training models and texture features, and finding a solution to validating them on daylight-captured datasets.
Keywords: Daylight Electroluminescence, Texture Analysis, Image Quality, Machine Learning, PV plant monitoring.

1 INTRODUCTION

Electroluminescence imaging is an increasingly popular technique for inspecting the state of solar cells in PV modules, with a much higher range of failure detection modes than infrared imaging. In daylight, since sunlight exceeds the EL emission from a PV panel by several orders of magnitude, it is usually necessary to modulate the current, with images recorded by a short-wave infrared (SWIR) camera at a suitable framerate and a relatively short integration time – typically below 10 ms. This allows the modulated signal to be extracted by a lock-in method or frequency analysis [1] [2]. Up until now, the quality of the available methods for automatically measuring or predicting the quality of the resulting EL images are based on SNR measures, calculated from the entire stack of recorded images, not gauged from the actual reconstructed EL image [1], [3], [4]. These methods, however, have proven unreliable at predicting the actual image quality in daylight conditions or when recording on a moving platform [5], [6].

Figure 1 shows four examples of daylight EL imaging of PV modules, cropped to six cells. **A** and **B** show the reconstructed EL images of two recordings (SNR$_{AVG}$ values of 32.7 and 20.9 respectively) made almost immediately after each other, with the camera in the exact same position. For comparison, **C** shows a much higher quality, low-light example with SNR$_{AVG}$=34.0. All values are much higher than the threshold (SNR$_{AVG}$>5) suggested from literature to be adequate for outdoor imaging, while (SNR$_{AVG}$>20) should generally result in excellent quality. This is clearly not always the case. This unreliability in image quality, in turn leads to unreliability in further post-processing. For almost any kind of post-processing, such as fault or crack detection, or power prediction, the garbage in – garbage out principle applies, and poor-quality images should be either reliably pre-processed or discarded. To do this, however, reliable detection of poor-quality images is required.

This research aims to create robust, predictable, and understandable metrics, that can be used to not only bridge the gap between EL image reconstruction from the field

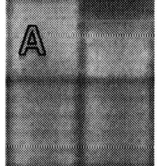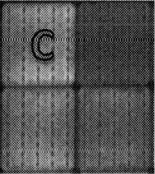

Figure 1. Noise in daylight EL imaging. **A** and **B**: Two sequences taken less than a minute apart. SNR$_{AVG}$ values 32.7 and 20.9 respectively. **C**: High quality low-daylight recording. SNR$_{AVG}$ =34.0.

and the digital-twin post-processing, but also to optimize the EL image reconstruction itself.

The means to achieve this will be grey-level texture analysis, specifically Grey-Level Co-occurrence Matrices (GLCMs), which are the basis for the most common texture features in computer vision and image classification. They are directional, and features derived from a GLCM can describe a variety of information about a region of an image, such as the orderliness, variation, contrast, sharpness, and entropy [3]. Given how solar cells and PV modules have well defined texture-like patterns, such as the busbars, fingers, cell spacing, and crystalline cell-characteristics, it follows intuitively, that texture descriptors could be used to gauge how well these patterns are imaged.

The use of texture descriptors for this purpose is novel and PV modules are uniquely suited for this type of processing. In a PV plant, and generally during inspection, it is well known what type of PV module the camera is capturing. This allows for a lowering of generalizability and increase of specificity. Noise-detection models can be tailored to particular module-types with either full specificity or categorized by common physical features ("textures") and be hot-swapped on demand.

Given the absence of a reliable metric to serve as a benchmark to train the model, the work in this paper employs synthesized datasets by adding noise and blur to high-quality image-series captured in lab-conditions.

41st European Photovoltaic Solar Energy Conference and Exhibition

1	1	0	0	0
0	2	3	1	0
0	2	4	2	0
0	0	0	1	1
0	0	0	0	2

+

1	0	0	0	0
1	2	2	0	0
0	3	4	0	0
0	1	2	1	0
0	0	0	1	2

=

2	1	0	0	0
1	4	5	1	0
0	5	8	2	0
0	1	2	2	1
0	0	0	1	4

Figure 3. Left: 5x5 image patch with 5 grey levels. **Right:** Corresponding GLCM for a distance of 1 to the right, plus its transpose (middle), which corresponds to adding the GLCM for the opposite direction, creating a symmetric matrix. The size of the GLCM is determined by the number of grey-levels not the size of the analyzed region.

2   INTRODUCTION TO GLCMs

Gray-level-matrix texture analysis in general aims to replicate some of the qualities of human visual perception, by quantifying certain characteristics of the image using so-called texture features (see Fig. 2). These include graininess, orderliness, disorder and the appearance of gradients. As the name suggests, images are first converted to greyscale before processing. For meaningful measures, the number of grey levels must usually be reduced via binning and any region being analyzed by a single texture feature is typically much smaller than the size of the full image. GLCMs and their derived texture features, are the earliest variety of this. For each grey level, a GLCM counts the number of times another given grey level is found between pixels at a particular distance and angle, and the number is recorded in a matrix, row-column style. That is to say, one GLCM covers all n grey levels in an $n \times n$ matrix, for one particular set of distance and angle.

An example of a hypothetical image patch with 5 grey levels and its GLCM for both one step to the right plus one step to the left can be seen in Figure 3. Including the step in the opposite direction is identical to adding the transpose of the matrix to itself, creating a symmetrical matrix. This makes sense to do for most features both from a visual standpoint, as well as for computational purposes. Multiple texture features can be calculated from these matrices, popularly known as Haralick features, including correlation, contrast, homogeneity, coarseness, and entropy, the definitions of which can be found in [7].

2   METHODOLOGY

The work comprised four stages: 1) Recording reference sequences from the selected PV module type. 2) Synthesizing varied datasets from the benchmark sequences. 3) Preliminary analysis for feature selection. 4) Hyper-parameter exploration and individual model optimizations for noise and blur. 5) Validation.

2.1 Reference Measurements

Lab-quality EL images of the same module type were recorded to serve as the baselines for texture metrics. The images were recorded indoors with an InGaS camera, in

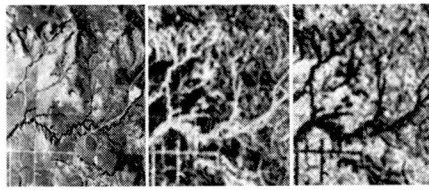

Original   Entropy   Homogeneity

Figure 2. From a GLCM, many different metrics can be calculated for either a full image using a moving window approach – like above – or at select locations, and in turn be used to gauge and classify the textures in the image.

the dark, and a modulated current was injected into the panels, as is customary for outdoor, daylight imaging. Two healthy modules were recorded for the model training itself, along with two cracked modules, two ribbon-damaged modules, and a module affected by Potential Induced Degradation (PID). Cropped sections of three damaged modules are shown in Fig. 4. The damaged modules were selected intentionally for the validation phase, to make sure that the method is robust also on non-healthy modules.

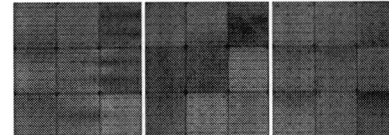

Figure 4. Three damaged reference modules. **Left:** Ribbon Damage. **Middle:** Cracked. **Right:** PID affected.

2.2 Data Synthesis

Noise and blur were added synthetically to each of the modulated image series, both from the reference modules and the validation modules, with noise assumed to be gaussian and the amount of blur controlled by changing the radius of a gaussian blur filter. This approach generates a dataset where the amount of noise and blur is exactly known. Adding noise directly to the modulated dataset simulates the random, ambient noise on a sunny day, while the effect of blurring the final image or the entire dataset should be roughly equivalent. However, to calculate matching SNR values for comparison, it was necessary to apply blur to all the images in an acquisition.

If successful, the data-synthesis approach will facilitate training a robust noise and blur model on different PV module types using at most a handful of initial images.

The design of experiment method Optimal Latin Hypercube Design (OLHD) [8] was used to create unbiased datasets for each module, with varying levels of noise and blur. Combinations of the extremes of the noise and blur, from none to maximum, applied to a healthy module acquisition can be seen in Fig. 5.

2.3 Feature Selection

The available number of GLCM texture-features in the chosen software package was 20, but it is known that a number of them are relatively redundant. To reduce the

Sharp	Blur	Noise	Noise & blur

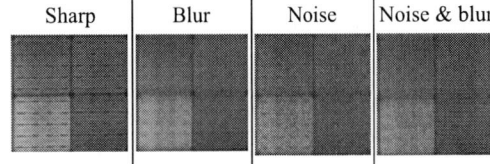

Figure 5. Extrema of the data-synthesis output, with minimum and maximum levels of noise and blur.

optimization problem, a preliminary analysis was therefore conducted using simple Principal Component Analysis (PCA) on the training dataset.

Since the goal was to create 2 separate models, one for blur and one for noise, the 20 features were analyzed along with those two parameters individually. Another goal for this fairly new research was to reduce the number of features in each model to one dependent (blur and noise respectively) and two explanatory features. Therefore, features that did not contribute positively to the first two principal components were successively removed, leaving a total of 9 candidate features. These could be different for different module types and will therefore not be listed here.

2.4 Optimization and Hyper-Parameter Exploration

GLCM features are sensitive to hyper-parameters such as the direction and distance of co-occurrence, the number of grey-levels *(bit-depth)* into which the image is binned, the area of the probe region, and the scaling and outlier-management of the dataset. These hyper-parameters had to be optimized at the same time as the model itself.

Artificial hyper-parameters created for the GLCM calculations relates to the recentering of the pixel intensities, as well as the selection of regions of interest (ROI) in the image. EL images typically consist of mostly bright pixels, some dark, and very little information in between. Because a GLCM scales quadratically with the number of grey-levels, and the most important information about EL image quality resides in the brightest areas, it is necessary to isolate and (see Fig. 6). This was achieved by calculating the standard deviation around the histogram mean of the brightest pixels, with the histogram based on a trust-region. This results in two hyper-parameters: The size of the trust region, and the size of the inclusion region, expressed as a factor, (*d* in Fig. 6), of the standard deviation around the histogram mean.

The pixels in the inclusion region are binned into N grey levels and divided into evenly spaced ROIs of $m \times m$ pixels, separated by a stride s. The top P percentage patches in terms of highest median intensity and lowest standard deviation were then picked, as it was assumed that these spots would be less likely to contain faults and therefore have more consistent texture descriptors. GLCM and Haralick Features were then calculated for all ROIs.

Due to the highly directional features of the modules, the horizontal and vertical directions were chosen for GLCM calculations, at 2 pixels distance rather than 1, as this tends to capture higher contrasts. All texture features were averaged over these two directions. Settling on two fixed directions and a single distance reduced the number of hyper-parameters to optimize, leaving the following:

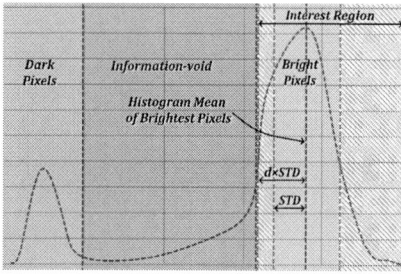

Figure 6. To avoid unnecessarily large GLC matrices with many non-informative bins, images should be culled and centered around the brightest pixels.

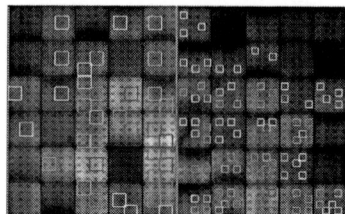

Figure 7. Two examples of GLCM hyper-parameters, incl. probe area, coverage area, stride, bit-depth, and inclusion factor. The larger inclusion-factor to the right is seen as an increase in dark areas and general contrast.

1. Trust-region size for the histogram mean.
2. Scaling factor for the inclusion region around the histogram mean.
3. Number of grey levels in the recentered image.
4. Square ROI side-length in pixels.
5. Stride between neighboring ROIs.
6. Area of all the ROIs combined as a percentage of total image area.

Via code- and data-optimization, the processing speed of the evaluation was fast enough to elect for exploration of the hyper-parameters using brute force in a dense solution space. The procedure is as follows:

1. An OLHD is used to create an unbiased GLCM hyper-parameter space. An example of ROI selection for two different points in this space are shown in Fig. 7.
2. For each point in the GLCM hyper-parameter space, the 9 candidate texture features must be calculated for each image in the training set.
3. Once the texture features are calculated for a given set of hyper-parameters, two regression models are found for noise and blur respectively by testing each pairwise combination of the 9 candidate features as input variables to a regression.
4. The combination of texture features that most often are found to produce the best regression models during hyper-parameter exploration, can be considered the most robust. Therefore, the final two models (comprising both regression coefficients and hyper-parameters) are chosen among models found in this group.

2.5 Validation

To investigate if the method would work on unhealthy modules, validation was performed on the datasets from the damaged modules. This was simply a matter of calculating the GLCM features of the final models on the datasets, using the GLCM hyper-parameters representing the best model. From this, a weighted, normalized, root-mean-square deviation (WNRMSD) was calculated for each validation dataset, along with a coefficient of determination (R^2).

3 RESULT AND DISCUSSION

3.1 Model Optimization

A total of 200 different, semi-stochastic noise and blur combinations were generated for the two healthy modules, and SNR_{Kari} and SNR_{AVG} values were calculated for each synthesized image series and stored for later comparison.

Similarly, 200 sets of candidate hyper-parameter

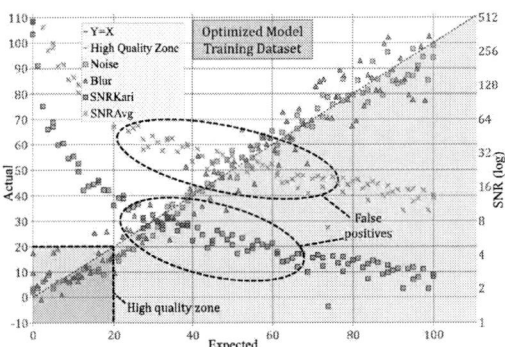

Figure 8. Expected vs modeled noise and blur levels (normalized to 100) from training the linear regression models. SNR_{Kari} and SNR_{AVG} values for a given dataset are shown on the right axis.

settings were generated using OLHD, and for each of them, the 9 texture features were calculated for each of ROIs in the 200 images. For each image, the median and the standard deviation of the texture features in these ROIs were calculated, resulting in two input values for each texture feature instead of one.

For each point in hyper-parameter space, regression models were trained on the corresponding $200 \times 9 \times 2$ texture-feature values, using linear least squares with the 12-coefficient model shown in Eq 1:

Eq. 1.:
$$f(c, x_i) = c_0 + c_1 x_{i,0} + c_2 x_{i,1} + c_3 x_{i,0} x_{i,1}$$
$$+ c_4 x_{i,2} + c_5 x_{i,3} + c_6 x_{i,2} x_{i,3} + c_7 x_{i,0} x_{i,2} + c_8 x_{i,0}^2$$
$$+ c_9 x_{i,2}^2 + c_{10} x_{i,0}^2 x_{i,1} + c_{11} x_{i,2}^2 x_{i,3}$$

Here, $x_{i,0}$ and $x_{i,2}$ are the medians of the 1st and 2nd texture features for a given feature-combination, calculated for an image i in a dataset. $x_{i,1}$ and $x_{i,3}$ are their standard deviations. Given 9 texture features, there are 36 combination for blur and noise each, resulting in 72 candidate models per point in hyper-parameter space. For each regression model, 5 models with highly random starting points were optimized, and one final optimization was then performed starting from the average of the coefficients of the best 3 models. This gives a total of 86.400 regressions to perform on the training set, which can be computed in a few minutes on an ordinary laptop. **Note:** Images in the high-quality region below 20% noise and blur were deemed more important and given 5x higher weight in the regressions.

After the entire set of hyper-parameters was explored, the most often superior combination of texture-features was found, which is shown in Table 1. This particular combination was the best in 64 of the 200 points in hyper-parameter space, indicating that these features are the most descriptive for blur and noise out of the 9 selected candidates. The best two final models were therefore picked from these.

Table 1. Index of the combination of texture features that most often resulted in the best regression models during hyper-parameter exploration (64 out of 200 times).

Target	1st Texture Feature	2nd Texture Feature
Noise	Difference Entropy	Texture Homogeneity
Blur	Difference Entropy	Entropy

The resulting fit of the training dataset can be seen in

Figure 8. Here, it can be seen that the model is able to fit the noise and blur very linearly, in particular the noise. The graph also shows how the SNR metrics overpredicts in a large range, well above the point where image quality starts to degrade significantly. Indeed, both metrics consistently show values that are supposed to be at least acceptable for the entire range, but as seen in Fig. 5 it is far from the case that the images in the extremes are acceptable. It is worth noting, that the SNR-metrics showed such high non-linearity to the noise in the images, that it was necessary to graph them on a log-scale. Especially the SNR_{AVG} is not very responsive to even large fluctuations in noise.

3.2 Validation

For each of the validation modules, a distinct, semi-random set of blur and noise combinations were generated, like for the training set. 160 combinations were created for each module, and the selected texture features were calculated and the two models (blur and noise) were tested on each of the resulting datasets. Figures are not shown here, as they are quite similar to Fig. 8, save for more unpredictable and sometimes highly biased values for blur.

Table 2. shows an overview of the coefficient of determination and NWRMSD of the validation on the three most individually different modules. The R^2 gives an estimate of goodness of fit, and shows how well the model explains the variance in the dataset, while the NWRMSD gives a measure of the overall error, normalized in a way that corresponds roughly to percentages. In general, the model was stable and robust when it came to noise for all validation sets, but not when it came to blur. The values were almost completely independent of re-randomization of the blur and noise combinations, indicating that both the stability of the noise model and the unreliability of the blur model is systemic.

Table 2. An overview of selected R^2 and NWRMSD values from the validation modules. For noise, the values were overall stable and promising, while blur showed both high error and poor fit for some of the modules.

Validation Module	Target	R^2	NWRMSD
Ribbon Damaged	Noise	0.902	8.0
	Blur	0.133	21.54
Cracked	Noise	0.974	4.25
	Blur	0.734	12.06
PID Affected	Noise	0.928	7.32
	Blur	0.870	9.15

4 CONCLUSIONS AND OUTLOOK

Clearly, using the texture descriptors, even the simple, linear regression is capable of generalizing noise quite well after hyper-parameter optimization, even on unknown and highly different modules, and in a much more linear and predictable fashion than the SNR metrics. Blur, however, is not quite as well generalized.

The proposed method therefore shows great promise for estimating image noise in a linear and predictable fashion, but more work is required if the goal remains to find suitable grey-level features for blur.

The optimization process to find a model for a particular PV module type requires a couple of healthy modules, the synthesizing of a dataset with noise, and fitting regression models to this dataset. The synthesization process is the lengthiest of these steps, taking a couple of hours of computation with the current approach. However, this is likely to be overkill, as it is quite plausible that both blur and a well-simulated noise can simply be applied to the final EL image and not necessarily the entire dataset. More research is needed to answer this question.

Once a dataset has been created, the model-optimization itself only takes a few minutes, and as for evaluation, it is currently possible to evaluate the noise in roughly 50 images per second, making it a very computationally inexpensive approach compared to the SNR metrics that each take around 1 second per image.

One downside of the method is, that it is specific to particular PV module-types, but training requires only a few lab-quality images, and models can be hot-swapped. It is quite likely, that this type of model, with relatively stable and statistically based features, will be more robust and easy to train on e.g. multi-crystalline modules, where other methods can struggle because of the random orientations and shades of the crystals in the cells.

Polynomial, linear regression might not be the most suitable model for this work, as clearly, this type of model should be stable around zero, and linear polynomial models rarely are. Ensemble modeling or non-linear models could be investigated.

Using the GLCM features for individual directions instead of averaging them, could provide further improvements.

As for the lack of reliability for blur, there are, a number of practical ways to deal with this before acquisition, so it is not as big of an issue as the noise. However, choosing different or additional texture-feature candidates – or a different model – could possibly improve blur predictions as well.

It should be said, naturally, that even though an effort was made to make it difficult for the model, by validating on unhealthy modules, the image acquisitions were done under controlled conditions, and the noise model applied assumed a Gaussian distribution, which might not be realistic. Eventually, it will become necessary to find a way to validate the model under daylight field conditions or with a more realistic noise synthesis. This will be a challenge, given the lack of a benchmark metric, but it is a problem that should be solved to give confidence in the method. Motion-based capture is another area where further research is needed, along with investigating the effect of different resolutions as well as keystoning effects caused by tilt of the PV modules or obscure camera angles.

Overall, the proposed method shows promise for potentially solving the garbage-in – garbage-out problem posed with daylight captured EL-images, and could pave the way for plant-scale automation of analyses for fault-detection, prognosis and maintenance purposes.

5   ACKNOWLEDGEMENTS

This research has been carried in the Eurostars project: Automated Daylight Electroluminescence Inspection of Large Photovoltaic Systems (ADELI).

6   REFERENCES

[1] G. A. dos R. Benatto *et al.*, "Daylight Electroluminescence of PV Modules in Field Installations: When Electrical Signal Modulation is Required?," 2022, *EU PVSEC*. doi: 10.4229/WCPEC-82022-3BV.3.44.

[2] T. Kari, S. Spataru, A. A. Santamaria Lancia, H. Parikh, P. B. Poulsen, and G. A. Dos Reis Benatto, "Computer Vision Method for Extracting an Induced Electroluminescence Signal from Photovoltaic Modules in Daylight Conditions Using Drone-Captured Images," in *EU PVSEC*, 2020, pp. 1573–1579.

[3] "IEC Standard TS 60904-13," 2018.

[4] C. Mantel *et al.*, "SNR Study of Outdoor Electroluminescence Images under High Sun Irradiation," in *2018 IEEE 7th World Conference on Photovoltaic Energy Conversion (WCPEC) (A Joint Conference of 45th IEEE PVSC, 28th PVSEC & 34th EU PVSEC)*, 2018, pp. 3285–3289. doi: 10.1109/PVSC.2018.8548264.

[5] T. Kari, R. Del Prado Santamaria, G. A. dos Reis Benatto, P. Koelblin, L. Stoicescu, and S. V. Spataru, "Evaluation of Motion-Induced Noise and Pixel-Bleeding in Electroluminescence Field Inspection of PV Modules," in *50th PVSC*, Institute of Electrical and Electronics Engineers (IEEE), Dec. 2023, pp. 1–5. doi: 10.1109/pvsc48320.2023.10359690.

[6] G. A. Benatto Dos Reis, A. A. Mayordomo, T. Kari, R. Santamaria Del Prado, P. B. Poulsen, and S. V Spataru, "Characterizing the Performance of Daylight Filters for Electroluminescence Imaging of Crystalline Silicon PV Modules," in *40th EUPVSEC Proceedings*, 2023. doi: 10.4229/EUPVSEC2023/4CV.1.6.

[7] R. M. Haralick, "Statistical and structural approaches to texture," *Proceedings of the IEEE*, vol. 67, no. 5, pp. 786–804, 1979, doi: 10.1109/PROC.1979.11328.

[8] D. Akinlana and L. Lu, "Multiple Objective Latin Hypercube Designs for Computer Experiments," Joint Statistical Meetings Proceedings, 2021, pp. 457–474.

41st European Photovoltaic Solar Energy Conference and Exhibition

FORECASTING THE LIFETIME OF PHOTOVOLTAIC MODULES THROUGH COUPLING A PHYSICS-BASED DEGRADATION MODEL WITH 3D HEAT TRANSFER SIMULATIONS

Timofey Golubev
ThermoAnalytics, Inc.
23440 Airpark Blvd,
Calumet, Michigan, 49913, USA
tg@thermoanalytics.com

ABSTRACT: This study introduces a novel methodology for forecasting the lifetime of photovoltaic (PV) modules with a focus on accurately accounting for module temperatures, a critical factor influencing PV degradation. Traditional degradation models, often empirical and based on constant rates over time, fail to relate observed degradation rates with environmental stressors, leading to inaccuracies, especially when applied to PV systems in new locations or configurations. To address this limitation, we integrate a physics-based degradation model with 3D heat transfer simulation in TAITherm™, which provides accurate module temperatures. We model Siemens M55 mono-crystalline silicon modules in free-standing and integrated PV applications, including building rooftops and vehicles, to compare their degradation rates in a hot and temperate climate. In both climates, integrated applications are predicted to exhibit higher degradation rates due to increased module temperatures resulting from less efficient heat dissipation. Compared to free-standing modules, rooftop and vehicle-integrated modules experience lifetime reduction of 2 and 4 years, respectively. Finally, we explore the use of phase change materials (PCMs) for lifetime extension. Our simulations predict that integration of 4 cm and 10 cm slabs of calcium chloride hexahydrate PCM into free-standing modules in Phoenix, can extend their lifetime by 2 and 10 years, respectively.
Keywords: degradation, lifetime, thermal-electrical, heat transfer, phase change material, TAITherm

1 INTRODUCTION

Forecasting photovoltaic degradation is essential for long-term financial planning and reliability assessment of PV systems. PV degradation is often modeled using data-driven models [1]. However, since these data-driven models do not relate the fundamental factors such as temperature, UV exposure, and humidity to module degradation, they can fail to accurately predict the degradation of systems in new locations or in different configurations, such as integrated into buildings or vehicles. In contrast, physics-based degradation models, aim to relate stressors to a module's degradation rate. While still needing empirical module-specific data for model calibration, physics-based models allow to forecast the impact of climate factors and PV system configuration on its lifetime.

A literature review of PV degradation models through 2018 is presented in Ref. [1]. Physics-based models describe loss in PV performance over time with consideration of factors such as temperature, relative humidity (RH), and UV irradiation. As described in [2], most physics-based degradation models relate climatic stress factors to module electrical parameters. Calibration of such models requires current-voltage curve measurements as the modules degrade, which is not readily available outside of a research lab environment. Two state-of-the-art models have been proposed for degradation rate prediction directly from climatic stress factors: Bala Subramaniyan et al. [3] and Kaaya et al. [4, 5] (hereafter referred to as the Bala and Kaaya model, respectively).

The Bala model describes the PV degradation rate as the product of temperature, RH, and UV terms

$$Rate(T, \Delta T, UV, RH)$$
$$= \beta_0 \exp(\frac{-\beta_1}{k\,T_{max}})(\Delta T_{daily})^{\beta_2}(UV_{daily})^{\beta_3}(RH_{daily})^{\beta_4}$$

where the terms with subscript "daily" refer to daily averages.

The Kaaya model describes PV degradation in terms of rates of hydrolysis (R_{Dh}), photo-degradation (R_{Dp}), and thermomechanical degradation (R_{Dt}) using Arrhenius law relationships as

$$R_{Dh} = A_h rh_{eff}^n \exp\left(\frac{-E_{ah}}{k_B T_m}\right)$$
$$R_{Dp} = A_p UV^x (1 + rh_{eff}^n) \exp\left(\frac{-E_{ap}}{k_B T_m}\right)$$
$$R_{Dt} = A_t C_n (\Delta T_m)^\theta \exp\left(\frac{-E_{at}}{k_B T_m}\right)$$

where rh_{eff}^n is the average effective humidity inside the module, UV is the total UV irradiance received by the module, T_m is the average module temperature, and ΔT_m is the difference between the module minimum and maximum temperatures experienced in a typical meteorological year (with min and max temperatures defined as the 5th and 95th percentiles), C_n is the number of effective temperature cycles calculated using a Rainflow counting algorithm, and k_B is the Boltzmann constant. $A_h, A_p, A_t, E_{ah}, E_{ap}, E_{at}, n, x,$ and θ are model fitting parameters. Note that some references describing the Kaaya model have $(273 + \Delta T_m)^\theta$ instead of $(\Delta T_m)^\theta$ in the R_{Dt} equation, which the author believes is a typographical error because a temperature difference (ΔT_m) whether in Kelvin or in Celsius does not need a unit conversion.

The overall PV power degradation is

$$\frac{P(t)}{P_{initial}} = 1 - exp\left(-\left(\frac{\Gamma}{R_{DT} \cdot t}\right)^\mu\right)$$

where

$$R_{DT} = (1 + R_{Dh})(1 + R_{Dp})(1 + R_{Dt}) - 1$$

and Γ and μ are module-dependent empirical parameters.

Physics-based PV lifetime models require an estimate for module temperatures, which are typically calculated using empirical equations [6, 7, 8]. While empirical models have been shown to be able to accurately predict module temperatures for free-standing modules, they are not necessarily applicable for estimating PV temperatures in integrated applications. Additionally, even for free-standing modules, these empirical models need to be calibrated for each location and PV technology [9, 10]. In this study, we use 3D heat transfer simulation to generate accurate module temperature predictions without needing calibration to each location and PV module type. The 3D heat transfer software, TAITherm™, is coupled to the Kaaya degradation model for PV lifetime prediction.

High PV module temperatures reduce both the power conversion efficiency and overall lifespan of the panels. Therefore, strategies to reduce solar cell temperatures are beneficial. Among the various approaches explored, the incorporation of phase change materials (PCMs) has emerged as a particularly promising solution. PCMs possess the ability to absorb and store excess thermal energy by converting it into latent heat. In the PV application, PCMs can store excess thermal energy during periods of high solar irradiance, which reduces module operating temperatures during the day. As ambient temperatures drop at night, the PCM restores back to its lower-energy phase, releasing the stored heat when the PVs are not in operation. A recent review of PCMs for PVs highlighted calcium chloride hexahydrate ($CaCl_2$-6 H_2O) as one of the most effective pure PCM cooling systems [14]. To demonstrate the applicability of our modeling methodology for the design of PV thermal management solutions, we simulate the impact of calcium chloride hexahydrate integration on the lifetime of a freestanding PV module.

2 METHODS

2.1 Degradation Model

In order to choose the degradation model, we compared predictions from both the Bala and Kaaya models described above for a wide range of climatic conditions. We found the Bala model to be very sensitive to impacts of the climatic variables where, for example, the PV lifetime of the same module is predicted to be 20 years in one location and 100 years in another. Very similar results were also reported in another study that tested the model [2]. Another related issue with the model formulation is that as relative humidity (RH) or UV irradiance approaches 0%, the degradation rate goes to zero, and the panel lifetime goes to infinity. We find that the Kaaya model generally predicts PV lifetimes that are closer to those reported in the literature than the predictions from the Bala model. Additionally, the model behaves better in the limits of low RH and UV. In our study, the model parameters of the individual rate models are taken from Table 4.3 of Ref. 5.

2.2 Thermal-Electrical Modeling

To calculate the PV module temperatures, we use the commercial heat transfer software TAITherm™, which uses a numerical, finite volume method based on first principles physics to solve for transient heat transfer due to conduction, convection, and radiation. In TAITherm™, a surface or volume mesh provided by user is used to describe the system geometry. From this mesh, a thermal nodal network for calculating lateral and vertical heat conduction is automatically generated. The mesh is also used to calculate view factors for modeling radiation exchange. Each component in the system is assigned thermal material properties, such as thermal conductivity, density, and specific heat, to accurately simulate transient thermal responses. Optical surface properties, including solar absorptivity and thermal emissivity, are specified for each surface to control radiation exchange. Convection can be modeled with varying degrees of fidelity. Options include the use of bulk fluid nodes or fluid streams with standard fluid flow correlations, mapping imported computational fluid dynamics (CFD) results onto surfaces, or employing the built-in RapidFlow™ 3D flow solver.

Module temperatures depend on the PV power production since that portion of incident solar energy is converted to electricity instead of heat. TAITherm™'s extension for thermal-electrical modeling of PVs is based on the author's previous work [10, 11]. PV power production can be modeled with either an equivalent circuit model or using a nominal efficiency at standard operating conditions (STC) and a temperature coefficient. Since this study is not focused on detailed electrical behavior of the modules, we chose the simpler nominal efficiency and temperature coefficient method where the efficiency of a perfectly clean module under normal irradiance is calculated as

$$\eta(T) = \eta_{nom} + \beta_\eta \left(T - T_{ref} \right)$$

TAITherm™ also allows one to consider angular losses in terms of an incidence angle modifier (IAM). The IAM can be calculated using either the Martin-Ruiz empirical model [12] or a user-defined polynomial. For this study, we use the Martin-Ruiz model option, which defines the IAM for direct radiation as

$$K_\alpha^{dir} = 1 - \left[\frac{\exp\left(-\frac{\cos(\alpha)}{\alpha_r} \right) - \exp(-\frac{1}{\alpha_r})}{1 - \exp(-\frac{1}{\alpha_r})} \right]$$

where α is the angle of incidence (AOI) and α_r is an empirical coefficient. We use α_r=0.16 in this study, which is a typical value for clean silicon panels [12]. The IAM for diffuse radiation is estimated by averaging the IAMs for direct radiation for all AOIs (denoted by $\overline{K_\alpha^{dir}}$). The overall IAM is then calculated by

$$K_\alpha = \frac{E_{dir}^{POA} K_\alpha^{dir} + (E_{dif}^{POA} + E_{alb}^{POA})\overline{K_\alpha^{dir}}}{E_{total}^{POA}}$$

where E_{dir}^{POA}, E_{dif}^{POA}, E_{alb}^{POA} are the plane-of-array (POA) direct, diffuse, and albedo irradiance, respectively.

While TAITherm™ also allows including user-specified soiling losses, we neglect soiling losses in this work to simplify the comparisons of degradation rates.

2.3 Coupling Degradation and Thermal-Electrical Models

As PV modules age, their temperatures will increase with decreasing PV efficiency due to a lower portion of absorbed solar energy being converted to electricity. Therefore, the temperature-dependent degradation model needs to be coupled to the thermal-electrical model. Our coupling methodology is as follows. First, we use TAITherm™ to simulate the module temperatures during a typical meteorological year (TMY) at the location of interest when the module is new (i.e. beginning of life, BOL) and at its end-of-life (EOL), which is defined as the time when power output equals 80% of the new module's power output. To estimate module temperatures during the

PV's lifetime, we interpolate between TAITherm™'s results for BOL and EOL. Unlike the Kaaya et al. approach, which calculates the degradation rate parameters from TMY weather only once, we re-calculate the degradation parameter (R_{DT}) for each year of the module's life, using the module temperature estimates based on our heat transfer simulations. We then compute the power degradation using the differential form of the Kaaya power equation

$$P_{n+1} = P_n + \Delta t \left.\frac{dP(t)}{dt}\right|_{t=n+1/2}$$
$$= P_n$$
$$+ \Delta t \cdot P_{initial} \left[\frac{-1}{t} exp\left(-\left(\frac{\Gamma}{R_{DT} \cdot t}\right)^{\mu}\right) \mu \left(\frac{\Gamma}{R_{DT} \cdot t}\right)^{\mu}\right]\Bigg|_{t=n+1/2}$$

where n is the time-step (number of years).

2.4 Case Study Models

In this study, we model and compare the degradation of a free-standing Siemens M55 module (nominal efficiency 12.88%, temperature coefficient −0.4334 %/°C) with the degradation that would be experienced if these modules are integrated into a house rooftop and vehicle roof and hood. Figure 1 shows the 3D heat transfer model geometries.

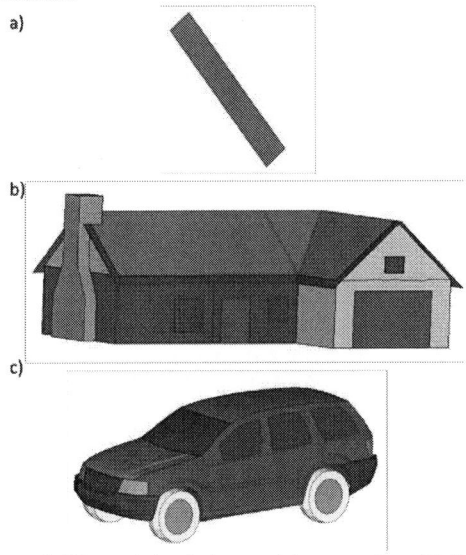

Figure 1: Thermal simulation model geometries with PVs shown in red. a) Free-standing PV module, b) residential rooftop PV, c) vehicle-integrated PV.

The PV module thermal properties are taken from literature [13] and are shown in Table I. Note that the front glass, front EVA encapsulant, and silicon active layer were combined into a single "PV active layer" layer (with mass-weighted averaged properties) in order to enable using a standard opaque multi-layer part type for the thermal-electrical PV model. While it is possible to define a part as being a transparent material, this generally results in more computationally expensive simulations. Our testing has shown that the thermal and electrical results for the active layer from equivalently setup models using the standard opaque part type and transparent part type are very similar. Therefore, we recommend using the simpler and less computationally expensive opaque part types for PV modeling, when detailed analysis of individual PV layer

temperatures and optical properties is not necessary.

Table I: Thermophysical properties of PV module

Layer	Thick. mm	Dens. kg/m³	Conduc. W/m-K	Sp. Heat J/kg-K
Glass/EVA/Si	3.75	2329	0.76	783
EVA	0.4	960	2090	0.26
Backsheet	0.3	1200	1250	0.26

We verified our selection of PV module material properties by confirming that a steady-state simulation of the free-standing module under normal operating test conditions (NOCT) predicted the manufacturer-reported value of 45 °C. To develop a baseline model to compare against, we fitted the module-dependent Γ and μ parameters of the Kaaya model to obtain the measured degradation rate of 1.45%/year for M55 modules in Phoenix [3]. We found Γ=53.7 and μ =0.35. Then, using the same model parameters, we predict the degradation in all other scenarios. The analysis is performed for a hot and a temperate location, Phoenix, Arizona, and Westchester, New York, respectively. The free-standing modules are assumed to have tilt angles equal to the latitude and azimuth angle of 180°. In the house model, the roof has a pitch of 41°.

To predict the impact of calcium chloride hexahydrate PCM integration on PV lifetime, we calculate module temperatures with and without the PCM. TAITherm™ supports accurately modeling PCMs by using the Phase Change Material part type and specifying key thermophysical properties. The PCM must be modeled as an additional model part that is connected to the rest of the geometry through a "Thermal Link" to allow for heat exchange. Material properties for calcium chloride hexahydrate were taken from literature [15] and are shown in Table II. We assume the PCM is contained inside a slab attached to the back of the module with the same area as the module.

Table II: Thermophysical properties of CaCl₂·6H₂O [15]

Melting Temperature	29 °C
Density (solid/liquid)	1706/1538 kg/m³
Conductivity (solid/liquid)	1.09/0.546 W/m-K
Specific Heat (solid/liquid)	2060/2230 J/kg-K
Heat of Fusion	170 kJ/kg

3   RESULTS AND DISCUSSION

Figure 2 compares the predicted degradation of free-standing, rooftop, and vehicle-integrated Siemens M55 modules in Phoenix, Arizona and Westchester, New York. For lifetime prediction, the end of life (shown with dashed lines) is defined when the power production reduces to 80% of the power at beginning of life. For the rooftop and vehicle-integrated scenarios, which have multiple PV modules, we report the average of the degradation values. The free-standing modules are predicted to have the longest lifetimes of about 13 years in Phoenix and 19 years in Westchester. The rooftop modules experience an approximately 2 year reduction in lifetime in both locations. The vehicle-integrated modules have a 4 year reduction in lifetime. The shorter lifetimes predicted in integrated applications is due to generally higher module temperatures than for the free-standing module due to lack of convection on their backsides (Fig. 2 c and f). Note that our rooftop model simulates panels that are directly

41st European Photovoltaic Solar Energy Conference and Exhibition

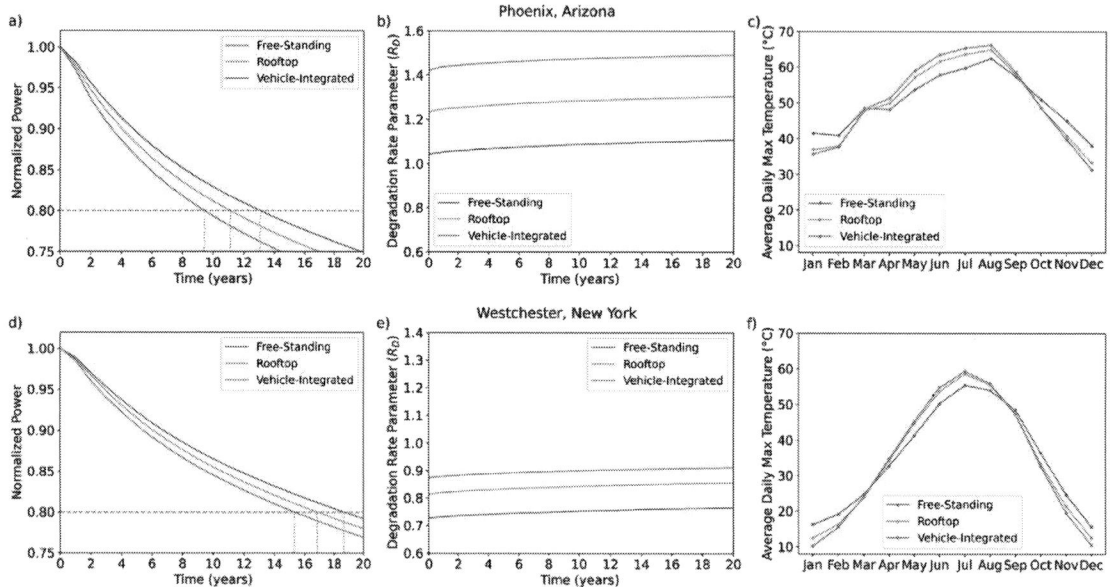

Figure 2: Comparison of degradation and average daily max temperatures for free-standing, rooftop, and vehicle-integrated Siemens M55 modules in Phoenix, Arizona and Westchester, New York. Module temperatures in panels c and f are plotted at beginning of life.

integrated into the roof without an air gap (i.e., solar tiles/shingles). The degradation parameters increase over time (Fig. 2b and e) due to increasing module temperatures as the power conversion efficiency decreases, further accelerating degradation processes. The free-standing module temperatures are higher than for the integrated cases in lower irradiance months due to more optimal solar angle of incidence. However, according to the Kaaya et al. degradation model, most of the degradation damage occurs during the higher irradiance months when the high temperatures accelerate the degradation processes. Our results suggest that for a given location, relative panel temperatures of different PV installations during high irradiance months are correlated with relative degradation rates.

Figure 3 shows a comparison of the impacts of module placement and orientation on the temperature and degradation of PV modules installed on a building rooftop. Note that the North/South and East/West labels refer to the orientation of the main house. As seen in Fig. 1b., the modules on the garage roof always have a perpendicular orientation to those on the main roof, (e.g., a North/South facing house will have the Main House PVs facing north and south and the Garage PVs facing east and west. While the temperature profiles reveal a clear seasonal variation, the temperature trends between the different modules are consistent for almost all months. Variations in solar exposure and incident angles cause differences in PV temperatures, leading to a predicted variability in lifetimes of 1-2 years.

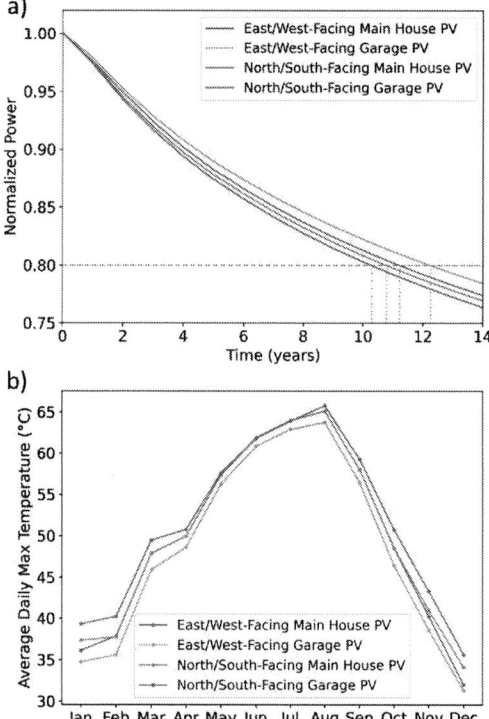

Figure 3: Power degradation and temperature (at beginning of life) predictions of modules integrated into a house rooftop in Phoenix, Arizona.

1325

Figure 4 shows a comparison of the impacts of module placement and orientation for the vehicle-integrated panels. The PV cells on the hood of the vehicle receive less solar irradiance than those on the roof due to intermittent shading from the vehicle's cabin. This causes the hood-mounted PV cells to be consistently several degrees cooler than the roof's PV cells, thereby extending their lifetime by an estimated 2 years. This analysis underscores the role of module placement and installation specifics on their lifetime.

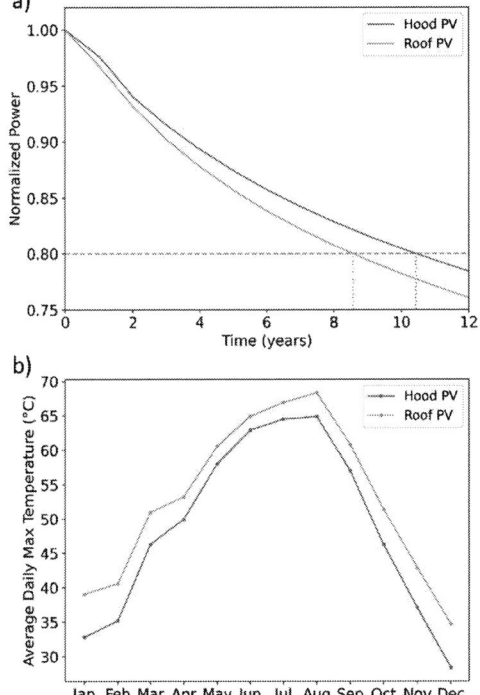

Figure 4: Power degradation and temperature (at beginning of life) predictions of modules integrated into a vehicle body in Phoenix, Arizona.

Finally, to demonstrate the applicability of our modeling methodology for the design of PV thermal management solutions, we simulate the impact of integrating calcium chloride hexahydrate phase change material (PCM) into free-standing Siemens M55 modules in Phoenix. The panels were assumed to have a tilt angle of 33.4° (equal to the latitude of Phoenix) and azimuth angle of 180°.

Figure 5 shows TAITherm™ temperature predictions for free-standing modules without PCM compared to those with 4 cm and 10 cm thick layers of PCM attached to their backs for three representative days in April and July in Phoenix. In April, the PCM absorbs heat from the system, reducing panel temperatures. Periods of time when the temperature curves are nearly horizontal are the phase transition periods where almost all heat energy entering and leaving the system is in the form of latent heat. We can see in this example that while a 4 cm thick layer of PCM significantly reduces the panel temperatures during the initial daily temperature increases, all of the PCM material changes phase within a few hours, after which the panel temperatures spike to levels that are close to those in the no-PCM case. A 10 cm PCM layer is much more effective at reducing peak PV temperatures.

In the summer (Fig. 5b), 4 cm of PCM completely changes phase during the first few hours of hot days and is therefore insufficient to offset panel heating during the entirety of the day. Additionally, the PCM layer prevents cooling of the panels at night due to the phase transition back to solid state, which occurs when the panel temperatures decrease to the melting point of 29 °C. Therefore, the panels start the next day at a warmer temperature, leading to higher max temperatures than if there was no PCM. While reducing nighttime cooling is detrimental for daytime panel temperatures, it can have the benefit of preventing dew formation, which has been reported to cause soiling cementation [16].

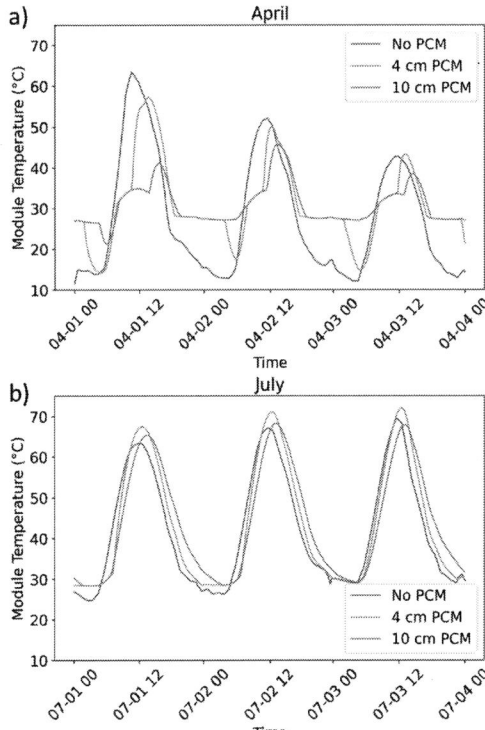

Figure 5: PV module temperatures for free-standing modules without PCM, with 4 cm PCM, and 10 cm PCM during four representative summer days in Phoenix, Arizona.

Figure 6 compares the power degradation and average daily maximum temperatures for free-standing modules in Phoenix with varying amounts of PCM attached to their backs. The modules with 4 cm of PCM have only a modest improvement in lifetime of about 2 years, while the modules with 10 cm of PCM benefit from a lifetime extension from 13 to 23 years.

In Figure 6b, we can see that while the 4 cm PCM layer is effective at reducing max temperatures during the cooler months, it results in higher module temperatures in June through September since it prevents panel cooling at night, while being insufficient to offset the solar heat load during the day. Nevertheless, the decrease in maximum temperatures during the cooler months offset the counterproductive effect of the PCM during the summer and the net result of the 4 cm PCM layer is an increase in lifetime. For a more optimal use of a 4 cm PCM layer, one could remove the PCM during the summer months to avoid the counterproductive effect.

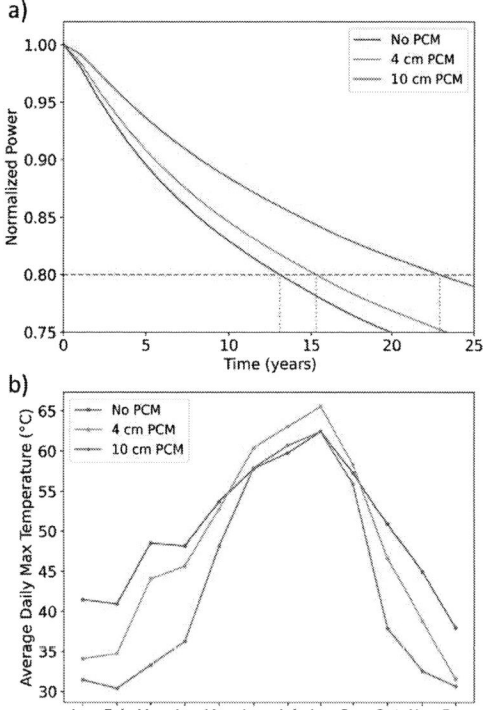

Figure 6: Power degradation and temperatures (at beginning of life) of modules with and without integrated PCM in Phoenix, Arizona.

With a 10 cm PCM layer, the modules experience lower average daily maximum temperatures for all months except July. In cooler months, the PCM layer results in a 10-15 °C decrease in daily maximum module temperatures. Because the 10 cm PCM layer is very effective at reducing panel temperatures, it is predicted to extend the PV lifetime by 10 years.

In addition to extending lifetime, integration of PCMs can also improve PV power yields due to increased PV efficiencies at lower operating temperatures. However, in this case study, we found the effect to be small with the 10 cm PCM layer increasing PV energy yield by only 3.4% compared to the no-PCM baseline.

4 CONCLUSIONS

This study demonstrates the application of a novel methodology that integrates a validated physics-based PV degradation model with 3D heat transfer simulations to forecast the lifetime of PV modules under varying environmental conditions. Our results highlight the significant impact of module temperatures on degradation rates and reveal the limitations of traditional empirical models, particularly in scenarios involving integrated PV applications. By offering a more precise estimation of module temperatures and subsequently, degradation rates, this methodology enables more reliable predictions of PV system performance over time. We demonstrate how our modeling methodology can be used to help engineer techniques for PV thermal management and lifetime extension by simulating PV modules with an integrated calcium chloride hexahydrate phase change material. In future work, this approach can be combined with battery lifetime models for comprehensive analysis of PV-battery systems.

5 REFERENCES

[1] S. Lindig, I. Kaaya, K. Weib, D. Moser, M. Topic, "Review of Statistical and Analytical Degradation Models for Photovoltaic Modules and Systems as Well as Related Improvements," IEEE J. Photovolt., vol. 8, no. 6, 2018.

[2] I. Kaaya, J. Ascencio-Vasquez, K. Weiss, M. Topic, "Assessment of uncertainties and variations in PV modules degradation rates and lifetime predictions using physical models," Solar Energy, vol. 218, 2021.

[3] A.B. Subramaniyan, R. Pan, J. Kuitche, G. TamizhManni, "Quantification of Environmental Effects of PV Module Degradation: A Physics-Based Data-Driven Modeling Method," IEEE Journal of Photovoltaics, vol. 8, no. 5, 2018.

[4] I. Kaaya, M. Koehl, A. Mehilli, S. Mariano, K. Weiss, "Modeling Outdoor Service Lifetime Prediction of PV Modules: Effects of Combined Climatic Stressors on PV Module Power Degradation," IEEE Journal of Photovoltaics, vol. 9, no. 4, 2019.

[5] I. Kaaya, "Photovoltaic Lifetime Forecast: Models for long-term photovoltaic degradation prediction and forecast," Ph.D. thesis, University of Malaga, 2020.

[6] David Faiman, "Assessing the outdoor operating temperature of photovoltaic modules. Progress in Photovoltaics: Research and Applications," 16(4):307–315, 2008.

[7] John Duffie. Solar engineering of thermal processes. Wiley, Hoboken, 2013.

[8] Modeling Guide. PV Performance Modeling Collaborative (PVPMC). Sandia National Laboratories. https://pvpmc.sandia.gov/modeling-guide/2-dc-module-iv/cell-temperature/

[9] M. Koehl, M. Heck, S. Wiesmeier, J. Wirth, "Modeling of the nominal operating cell temperature based on outdoor weathering," Solar Energy Materials & Solar Cells, vol. 95, 2011.

[10] T. Golubev, "Multi-physics modeling and simulation of photovoltaic devices and systems," Ph.D thesis, Michigan State University, 2020.

[11] T. Golubev, R.R. Lunt, "Evaluating the Electricity Production of Electric Vehicle-Integrated Photovoltaics via a Coupled Modeling Approach," in 2021 48th IEEE Int. Photovoltaics Specialists Conference (PVSC), 2021.

[12] N. Martín and J. M. Ruiz, "A new model for PV modules angular losses under field conditions," Intl. Journal of Solar Energy, vol. 22, no. 1, 2002.

[13] T. Silverman et al., "Reducing Operating Temperature in Photovoltaic Modules," IEEE Journal of Photovoltaics, 2018.

[14] R. Kassar, A. Takash, J. Faraj, M. Khaled, H. Ramadan, "Phase change materials for enhanced photovoltaic panels performance: A comprehensive review and critical analysis," Energy and Built Environment, 2024.

[15] C. Pan et al, "Experimental, numerical and analytic study of unconstrained melting in a vertical cylinder with a focus on mushy region effects," Intl. Journal of Heat and Mass Transfer, vol. 124, 2018.

[16] K. Ilse et al., "Techno-Economic Assessment of Soiling Losses and Mitigation Strategies for Solar Power Generation," Joule, vol. 3, no. 10, 2019.

41st European Photovoltaic Solar Energy Conference and Exhibition

DEVELOPMENT OF A MODEL
TO ENSURE THE SAFETY OF PV SYSTEMS USING FMEA

Sujeong OH, Chongmin KIM, Juhee JANG
Korea Electrical Safety Corporation / Electrical Safety Research Institute
111, Anjeon-ro, Iseo-myeon, Wanju_gun, Jeonbuk-do, Korea

ABSTRACT: To date, research aimed at ensuring the safety of photovoltaic (PV) systems has primarily focused on improving power generation efficiency and identifying failure types at the component level. In most cases, when a PV accident occurs, the faulty equipment is simply replaced, and the system is resumed. Additionally, an analysis of Korea fire statistics and investigation data for PV systems over the past five years reveals that causes include tracking, poor contact, and insulation degradation, with unidentified short circuits accounting for 27% of the total accidents. This paper aims to ensure the safety of PV system operations by identifying major failure modes through the examination of Korea PV accident cases and defining new failure modes centered on these failure phenomena. Furthermore, using Failure Modes and Effects Analysis (FMEA), the study evaluates potential failure causes in terms of severity, occurrence, and detection, assigns risk priorities, and proposes risk impact assessments and improvement measures for each failure mode. The failure modes are categorized according to the major components of PV systems, such as modules, inverters, junction boxes, connectors, and cable are classified by the possible failures associated with each piece of equipment. Notably, considering the significant proportion of PV systems installed in mountainous area within Korea, the study presents failure modes that comprehensively account for structural factors.
Keywords: FMEA, RPN, PV systems, Failure Modes

1 INTRODUCTION

The installation of photovoltaic (PV) systems is increasing globally, and with this rise, the incidence of related accidents is also on the rise. Recent fire accidents in PV systems in Korea [1], as shown in Figure 1, have revealed that the causes include tracking, poor contact, and insulation degradation, with unidentified short circuits accounting for 27% of all cases.

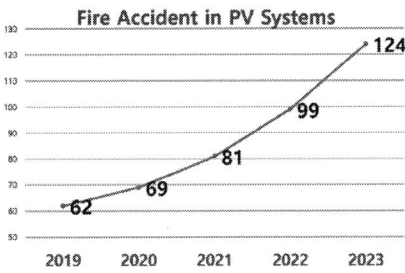

Figure 1: PV Fire Accident in Korea

Consequently, there is growing concern in Korea that, rather than focusing on further construction, there is a pressing need for safety measures to ensure the safe operation of existing installations. Additionally, due to the geographical characteristics of Korea, a significant number of PV systems are installed in mountainous areas. Approximately 20% of all PV power plants require special management due to factors such as landslide history and other accident risks. For these reasons, the analysis of potential failures for safe operation has become increasingly important. However, research on PV failure analysis has so far been more extensively conducted from the perspective of improving power generation efficiency [2]. Regarding the response to PV system accidents, the current approach often involves replacing the faulty equipment or introducing new technologies, such as Module-Level Power Electronics (MLPE), to ensure safety, rather than addressing the root cause of the problems.

This paper develops failure modes to secure the safety of PV systems through Failure Modes and Effects Analysis (FMEA). The failure modes are structured based on the components of the PV system, with a focus on failure phenomena (e.g., inverter shutdowns, fires). A total of 29 failure modes are identified, and for each mode, the risk priority analysis, the impact on the system, and factors that accelerate failure are defined.

2 FMEA

The analysis of failures in PV systems can be conducted using various methods such as Fault Tree Analysis (FTA) [3,4], Failure Modes and Effects Analysis (FMEA) [5,6], and machine learning [7,8]. However, this paper employs FMEA, a widely used method for prioritizing potential defects, to propose criteria for enhancing system safety. Through FMEA, system issues can be identified, and the causes and impacts of each failure mode can be clearly distinguished. Additionally, FMEA allows for the identification of critical issues that need to be addressed on-site to ensure the safety of the system.

The FMEA in this study was conducted through the following steps:

1. Analysis of system components
2. Classification of potential failure modes and their causes
3. Evaluation of Risk Priority Numbers (RPN)
4. Identification of failure acceleration factors
5. Derivation of the FMEA table

To define the failure modes, the components of the PV system were first categorized. The major components of the PV system were divided into five categories: arrays, junction boxes, inverters, connectors, and cables. These were further subdivided into a total of ten subcomponents. The components and subcomponents used to classify each failure mode are listed in Table 1.

Table I: PV component and sub-component

Component	Sub-component
Array	Cell/Module
	Junction Box
Combiner Box	Fuse
	Blocking Diode
	Circuit Breaker
Inverter	Inverter
	Inverter Component
	FAN
Connector	Connector
Cable	Cable

3 CLASSIFICATION OF FAILURE MODES

Previous studies on the classification of failure modes have primarily focused on module defects and inverter system failures. However, these studies often lack clear classification criteria, as failure types and causes are not distinctly separated and are frequently used inconsistently. Therefore, it is deemed necessary to establish a clear classification of failure modes to ensure the safety of photovoltaic (PV) systems.

In this paper, the failure modes of PV systems are classified by first defining the failure types for each major component, such as inverters, junction boxes, and arrays, and considering the impact of each failure on surrounding components. This approach allows for a clear distinction between failure modes and causes, as well as an understanding of the effects of failures on the overall system.

Each failure mode is defined based on the type of damage caused by the failure (e.g., fire, overheating, insulation failure, inverter shutdown). This method improves upon previous classifications where failure causes and modes were often conflated. A total of 29 failure modes have been defined across 10 subcomponents.

4 RISK PRIORITY NUMBER(RPN) EVALUATION

The defined failure modes were evaluated in terms of severity, occurrence, and detection to assess their risk. Severity refers to the extent of the impact a failure has on the system, occurrence measures how frequently the cause of the failure occurs, and detection assesses the ability to detect the failure. Each of these aspects was evaluated using a 7-point Likert scale to derive the Risk Priority Number (RPN). The Likert scale is widely used in FMEA for both quantitative and qualitative assessments. The RPNs were determined through a survey of professionals in the photovoltaic (PV) industry, including designers, operations and maintenance (O&M) personnel, and safety managers, who rated each aspect using the 7-point Likert scale. The criteria for each evaluation item on the Likert scale are presented in the table below.

Table 2: 7-point Likert scale criteria

	1 Point	7 Point
Severity	Causes a reduction in power generation	Results in fire, electric shock, or other human injuries
Occurrence	Occurs less than once per year	Occurs more than once per month
Detection	Can be detected visually	Requires precise diagnostics

The survey received a total of 53 responses, and the reliability of these responses was analyzed, with unreliable responses being excluded. In Likert scale surveys, the reliability of responses is typically assessed to ensure that participants have answered consistently and sincerely. While various methods such as repeating similar questions or analyzing response times can be used, this survey focused on volatility analysis and identifying disinterested responses. Volatility analysis checks whether a respondent gave the same score for all items, which could indicate a lack of engagement. Standard deviations were calculated, and responses from three participants with low standard deviations were deemed unreliable and excluded. Disinterested responses were identified by checking if participants frequently chose the middle value (4), which might indicate neutral or thoughtless answers. However, no responses were excluded for this reason, as there were no cases of excessive neutral responses.

After analyzing the reliability of the survey responses, scores for severity, occurrence, and detection were defined based on 50 valid responses. Typically, in Likert scale surveys, central tendencies such as the mean, median, and mode are used to represent the data. Given the distribution of responses, which included several extreme values, the median was selected as the representative value. The median, which represents the middle value when the scores are ordered, is particularly useful for reflecting the central tendency of the data. The RPN for each failure mode was then calculated by multiplying the severity, occurrence, and detection scores. The RPNs derived for all failure modes are presented in the table 3.

Table 3: RPN tables

Component	Sub-component	Failure Mode	RPN
Array	Cell/Module	Burn marks	12
		PID	42
		Delamination	18
		Discoloration	18
		Corrosion	12
		Breakage	36
		Module open	36
		Module shading	12.5
		Soiling	15
	Junction box	Bypass diode Open/short	40
		Bypass diode connection failure	60
		Enclosure damage	24
		Enclosure corrosion	12
Combiner box	Fuse	Open	42
		Not operation	27
	Blocking diode	Diode open	32
	Circuit Breaker	Open	30
		Not operation	15

Table 3: (continued)

Component	Sub-component	Failure Mode	RPN
Inverter	Inverter	Damaged	14
	Inverter component	Switching control failure	42
		Switching element damage	48
		DC link capacitor damage	32
	FAN	FAN failure	36
Connector		Defect	40
		Mismatch	36
		Corrosion	36
Cable		Breakage	40
		Corrosion	60
		Enclosure Grounding corrosion	54

As shown in Table 3, in terms of safety aspects of photovoltaic (PV) systems, failures related to junction boxes, cables, connectors, and bypass diodes are considered to be of higher risk priority compared to module failures. Although module failures occur more frequently, they are assessed as having a lower priority due to their lower potential for causing serious incidents such as fires or personal injury. Failures with higher risk priority are typically characterized by higher occurrence rates and detection rates.

5 FAILURE ACCELERATION FACTORS

In Korea, a significant proportion of photovoltaic (PV) systems are installed in mountainous areas, necessitating the consideration of complex failure modes that account for both structural and environmental factors. In the 2023 special safety inspection of PV sites in mountainous area within Korea, findings revealed that out of a total of 1,408 sites, 78.5.% exhibited sediment accumulation in drainage channels, 68.8% showed erosion due to runoff, and 45.8% experienced soil loss within the site. To ensure system safety, this paper introduces additional elements that can accelerate failures based on the proposed failure modes. First, failure modes that can be accelerated by structural deformation include module damage, connector damage, poor connector contact, and cable breakage. Soil erosion, while an environmental factor, affects failure modes similar to those influenced by structural deformation. Environmental factors such as heavy rain, heavy snow, and typhoons impact failure modes related to cells/modules, as well as junction box corrosion and connector/cable issues.

6 CONCLUSIONS

With the increase in the installation of photovoltaic systems and their operating lifespans, diverse research has been conducted on risk assessment for safe operation. This paper introduces new failure modes and determines their risk priorities using Failure Mode and Effects Analysis (FMEA), one of the types of failure analysis. Failure modes were categorized based on the types of damage they could cause (such as overheating/fire, insulation failure, and inverter shutdown). This categorization clarified the mixed-use failure modes and defined 29 specific failure modes. A survey was conducted with experts in the photovoltaic field to determine the risk priorities for these failure modes. Additionally, the paper considers structural and environmental factors that can accelerate failures.

In the future, the defined failure modes and risk priorities will be used to conduct experimental validations of the impact of each failure mode on the system. Moreover, a safety index will be developed for system-level risk assessment, which will be utilized in a safety grading system.

[1] National Fire Data System(Korea), (https://www.nfa.go.kr), Fire Statistics, 2024-09-03

[2] Wang M-H, et al. "Two-stage fault diagnosis method based on the extension theory for PV power systems", Int J Photoenergy 2012, (2012)

[3] Pramod R. Sonawane, et al., "Reliability and Criticality Analysis of a Large-Scale Solar Photovoltaic System Using Fault Tree Analysis Approach", Sustainability, 15 (2023)

[4] Ong, N. A. F. M. N. et al., "Fault tree analysis of fires on rooftops with photovoltaic system", Journal of Building Engineering, (2022)

[5] Alessandra Colli, "Failure mode and effect analysis for photovoltaic systems", Renewable and Sustainable Energy Reviews, 50, 804-809 (2015)

[6] Jing Li, et al., "Risk Evaluation of photovoltaic power systems: An improved failure mode and effect analysis under uncertainty", Journal of Cleaner Production 414, (2023)

[7] Basnet B, et al., "An intelligent fault detection model for fault detection photovoltaic systems.", J Sens, (2020)

[8] Aref Eskandari, et al., "Fault Detection and Classification for Photovoltaic Systems Based on Hierarchical Classification and Machine Learning Technique", 68, 12750-12759 (2021)

Acknowledgement

This research was supported by New & Renewable Energy of the Korea Institute of Energy Technology Evaluation and Planning (KETEP) grant fund by the Korea government Ministry of Knowledge Economy (No. 20213030010280) and Korea Electrical Safety Corporation (No.2024-0107).

41st European Photovoltaic Solar Energy Conference and Exhibition

LONG-TERM MONITORING OF DEGRADATION AND DEFECT IN HIGH-VOLTAGE STRINGS THROUGH DARK I-V MEASUREMENTS

Samuele Chiesa*, Gian Carlo Dozio*, Domenico Chianese**
University of Applied Science of Southern Switzerland (SUPSI-ISEA*, SUPSI-ISAAC**)
Via Serafino Balestra 16, CH-6900 Lugano
samuele.chiesa@student.supsi.ch, giancarlo.dozio@supsi.ch, domenico.chianese@supsi.ch

ABSTRACT: For on-site monitoring of photovoltaic (PV) strings, most traditional techniques, such as daylight I-V measurements, thermography, and electroluminescence, require stable weather conditions and are often time-consuming. Dark I-V measurements, conducted at night under consistent environmental conditions, provide a reliable alternative. To continuously monitor the degradation trend and detect potential faults in PV systems, a dark I-V measurement system (DIVMS) was developed. During measurements, the PV string is disconnected from the inverter and connected to a high-voltage source, where a controlled reverse current is injected to record the I-V curve. However, in 1500V systems, a single string may require up to 30 kW of instantaneous peak power, making traditional high-voltage power supplies both expensive and impractical for industrial use. As a cost-effective alternative, a local energy storage system charged by a low-power flyback converter was developed.
Keywords: dark I-V, measurements, degradation, high-voltage.

1 INTRODUCTION

The photovoltaic sector has recently seen important and rapid progress, including advancements in cell technologies like BSF, PERC, TOPCON, and SHJ, as well as half-cells, varied cell sizes, Multi Bus Bar (MBB) or shingled cell connections, and bifacial cells. However, these major changes must be balanced with considerations for quality, reliability, lifespan, and return on investment. Additionally, increased system voltage up to 1500V adds further stress on PV modules.

While long-term monitoring of key performance indicators (KPIs) can provide insights into the productivity of a plant, it does not offer a detailed analysis of possible defects in PV strings and modules. Traditional on-site evaluation techniques, such as daylight I-V measurements, thermography, and electroluminescence, are highly weather-dependent and time-consuming.

To address these challenges and continuously track performances and detect possible faults in PV systems, an automatic monitoring system that measures the dark I-V characteristics at night was developed. These in-situ dark I-V measurements allow for the detection of key parameter changes, such as bypass diode failures, potential-induced degradation (PID) at the string level, increased series resistance, and electrical mismatch between strings.

2 THEORETICAL ANALYSIS

To capture the I-V curve, the DIVMS must supply up to 1500V at 20A, resulting in a peak power of 30 kW. Since this power is required for durations of only tens of milliseconds, implementing a traditional power supply for this purpose is impractical. The solution lies in locally storing the energy needed to acquire the I-V curve. The faster the curve is traced, the less energy is required. However, this conflicts with the intrinsic capacitance of the panels, which imposes a minimum time limit for accurately tracing the curve. Therefore, a tradeoff exists between the speed of tracing (which reduces energy consumption) and the accuracy required for proper measurements.

2.1 Simulations

To quantify the acquisition speed limits and, consequently, the energy to be stored, simulations were conducted in PLECS, using the single diode model for PV modules.
The models chosen for the simulation correspond to the types of modules installed at SUPSI's Viganello Campus as part of the RACONT research project (PVLab).
Figure 1 presents the results of these simulations, providing insights into the relationship between measurement speed and accuracy.

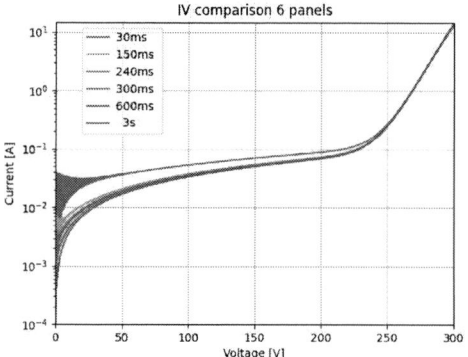

Figure 1: Plot of the I-V curve for the same 6-module string at different measurement speeds.

The relationship between measurement speed and accuracy also varies with the number of modules in the string, making it difficult to present an exact figure for the minimum time required to accurately acquire the I-V curve. However, an estimated time of approximately 300ms is provided for a reasonably accurate measurement. Further studies are needed to better understand this phenomenon. With this given time, the energy required to trace the I-V curve of a 1500V, 20A string is roughly 2.1kJ.

1331

Other simulations were conducted to better visualize the impact of series and shunt resistance changes in PV modules.

Figure 2 and Figure 3 show the results.

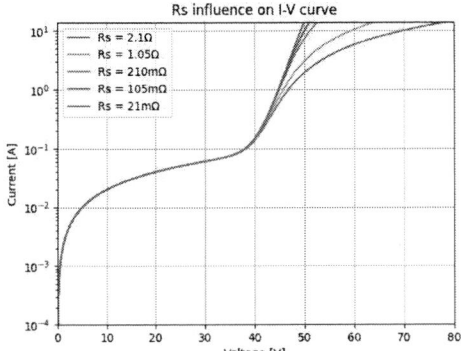

Figure 2: Series resistance influence on I-V curve

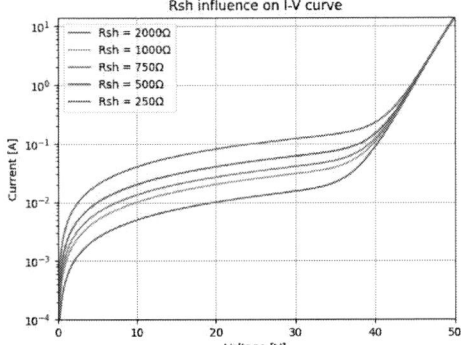

Figure 3: Shunt resistance influence on I-V curve

3   DESIGN

As previously indicated, to trace the I-V curve of a 1500V, 20A PV string in 300ms, approximately 2 kJ of energy is required, a value determined through extensive simulations.

Given the need for rapid energy release during the measurement, capacitors are used for energy storage.

As the I-V curve is traced, the voltage applied to the PV string gradually increases from 0V to 1500V. The energy storage system must be able to keep this voltage until the end of the measurement. However, as the energy is supplied, the capacitors discharge, causing their voltage to decrease. This means the initial stored voltage must exceed 1500V to ensure sufficient voltage at the end of the measurement.

To achieve this within the estimated measurement time, a charging voltage of 2000V and a capacitance of 2400μF (three 800μF capacitors in parallel) is required. A parallel connection between capacitors is preferred over a series connection to avoid the need for capacitor balancing, simplifying operation and enhancing system reliability.

3.1 High voltage generation – flyback converter

High voltage for capacitor charging is generated using an isolated AC/DC converter, employing a current-mode flyback topology that operates in either critical conduction mode (CrCM) or discontinuous conduction mode (DCM), depending on output voltage conditions. Operation in CrCM and DCM ensures constant power delivery. Due to the output capacitance and the lack of strict voltage regulation requirements, no feedback loop from the secondary to the primary side of the circuit is necessary. For this design, the L6565 quasi-resonant zero-voltage-switching (ZVS) controller was selected.

The control loop is driven solely by the primary peak current of the inductor, which, in conjunction with the applied voltage, determines the energy transferred per switching cycle. The complete demagnetization of the transformer marks the end of each cycle, enabling efficient capacitor charging at nearly constant power. As the voltage across the output capacitors increases, the switching frequency also rises. To prevent the switching frequency from exceeding a certain limit, the controller transitions from CrCM to DCM at around 200 kHz.

The output voltage is digitally controlled and compared with a reference value generated by a digital-to-analog converter (DAC). The flyback converter is deactivated once the output voltage reaches the reference level. This voltage can be regulated according to the string voltage, up to a maximum of 2000V, ensuring that only the energy required to trace the I-V curve is stored.

3.2 Voltage and current management – IGBTs

A linear regulator utilizing insulated-gate bipolar transistors (IGBTs) was developed to control the current flowing through the PV string during measurements.

Using a single IGBT for linear regulation of the PV voltage is impractical due to excessive power dissipation, which would require a prohibitively expensive component. The solution involved connecting multiple IGBTs in series, allowing the power dissipation to be distributed across each IGBT, improving performance and significantly reducing overall costs.

The number of transistors, determined to be four, was selected based on the Safe Operating Area (SOA) graph of the chosen IGBT and the maximum dissipated power, which was established through simulations.

During measurements, voltage balancing across the IGBTs is managed via software control. The Collector Emitter voltage across each transistor is monitored, and the gate voltage of the IGBT with the highest value is incrementally adjusted. This ensures even voltage distribution across the transistors during the entire measurement process.

As the capacitors charge, the IGBTs are automatically balanced through the voltage dividers that monitor the voltage across each device. This setup also provides protection in the case of a transistor failure, safeguarding the remaining IGBTs. In the worst-case scenario, when the capacitors are fully charged to 2000V, each IGBT will handle a maximum voltage of 500V.

3.3 High voltage measurements

To achieve accurate voltage measurement across the wide range of 0V to 1500V in the PV string, a differential measurement technique is employed. Direct measurement over such a broad range while ensuring high precision presents significant challenges. To achieve this, a system of selectable voltage dividers is integrated, segmenting the full voltage range into smaller, more manageable intervals. This approach ensures that accuracy is maintained throughout the entire measurement process.

The primary advantage of the selectable voltage divider system lies in its ability to dynamically adjust the division ratio based on the measured voltage. For lower voltage values, a finer division ratio is applied to enhance accuracy. As the voltage increases, the system automatically shifts to a higher division ratio, ensuring measurements remain within the operational limits of the sensing components while preserving precision.

This dynamic adjustment of the voltage divider allows the measurement system to function efficiently and accurately across the entire voltage range, optimizing both measurement precision and overall system performance.

3.4 Current measurements

Similar to the approach used for voltage measurement, achieving high precision in current measurement requires dividing the full current range into smaller, more manageable intervals. This is accomplished by configuring three shunt resistors in series, with each shunt being bypassed as the current exceeds a predefined threshold.

This method allows the system to maintain high measurement precision across the entire current range while simultaneously minimizing power dissipation in the shunt resistors. By dynamically bypassing shunts as needed, the system ensures both efficient power management and accurate current measurement.

3.5 Data acquisition

To acquire the PV voltages and currents, the differential voltage signal is first processed through an instrumentation amplifier, which scales the signal to a range between 0V and 5V.

The conditioned signal is then fed into an analog-to-digital converter (ADC) for simultaneous sampling. This simultaneous acquisition enhances the overall precision of the measurements by minimizing timing discrepancies between voltage and current readings.

The acquired readings are processed by the controller, which interprets the values according to predefined calibration parameters. This approach ensures a high degree of measurement accuracy and precision throughout the system.

The measurement ranges and resolutions are presented in Tables I and II.

Table I: Voltage measurement ranges and resolution

Range Nr.	Measured Range [V]	Resolution
1	0.05 to 8	244μV
2	8 to 45	1.13mV
3	45 to 203	4.82mV
4	203 to 1008	24.55mV
5	1008 to 2000	30.26mV

Table II: Current measurement ranges and resolution

Range Nr.	Current ranges [A]	Resolution
1	0.00015 to 0.04	1.22μA
2	0.04 to 1	29.28μA
3	1 to 20	579.51μA

Every range ensures that the measurement error is always less than 1%.

4 RESULTS

4.1 Setup

To test the developed DIVMS, SUPSI PVlab provided a set of 26 TSM-260PC05A PV modules. These modules, sourced from a solar plant that was decommissioned after a hailstorm, were ideal for testing purposes.

Figure 4 shows the setup used for these measurements.

The DIVMS is directly connected to the modules using a 4-wire configuration. The modules are stacked on top of each other to optimize space and are covered with a black blanket to guarantee night conditions.

Figure 4: Measurement setup at SUPSI Mendrisio campus

4.2 Measurements

Figure 5 presents the results of the initial set of measurements. The goal of this test was to evaluate each module individually for comparative analysis.

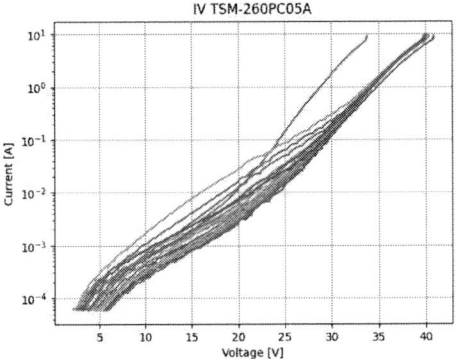

Figure 5: Comparison of the I-V curve for 26 identical modules following a hailstorm.

The damage inflicted on these modules was unequal, resulting in varied characteristics across each unit. The parameter most significantly affected by the damage appeared to be the shunt resistance, although in some instances, the series resistance was also notably impacted.

Subsequent measurements were conducted on a string composed of the 26 damaged modules connected in series. These tests were designed to increase the power and evaluate the DIVMS's performance for its intended purpose of measuring PV strings. Specifically, three tests were performed. The first test involved measuring the I-V characteristic of the string in its existing condition with the damaged modules. The second test introduced a series resistance of 2.5Ω to simulate a fault in the series resistance, using the initial measurement as a baseline. Finally, a third test involved inserting a shunt resistance of 400Ω in parallel with a single module.

Due to the unavailability of data from the string prior to the damage, artificial defects were introduced to simulate potential faults.

Figure 6 presents the results of test performed with an added series resistance, compared to the baseline measurement.

Figure 7 presents the results of the test performed with an added shunt resistance, compared to the baseline measurement.

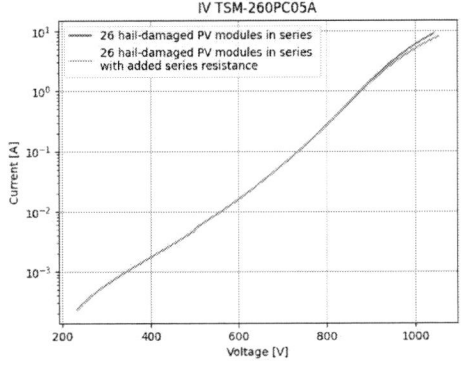

Figure 6: Comparison of the I-V curve of 26 hail-damaged PV modules with and without an added series resistance of 2.5Ω

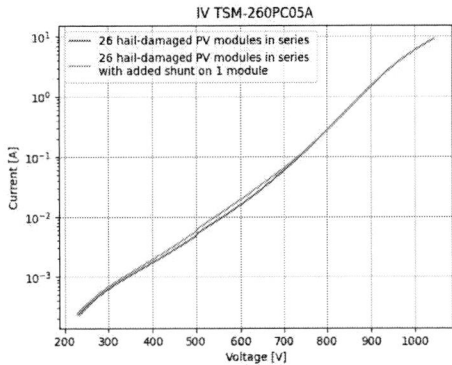

Figure 7: Comparison of the I-V curve of 26 hail-damaged PV modules with and without the added shunt resistance of 400Ω

The results clearly indicate that the designed DIVMS is able of detecting even small variations in individual PV modules within a large string.

5 CONCLUSIONS

This work has developed an innovative, low-cost diagnostic tool for measuring the dark I-V characteristics of PV strings, capable of handling voltages up to 1500V and currents up to 20A. The tool provides a practical solution for monitoring PV string performance, enabling the detection of degradation trends and defects over time.

By measuring dark I-V characteristics, the DIVMS can track degradation indicators such as increased series resistance and shunt faults. Its high precision allows for detection of even minor defects in individual PV modules, ensuring early fault identification and targeted maintenance, which is critical for maintaining the efficiency and longevity of large PV installations.

The DIVMS will be deployed as part of the RACONT research project at the Viganello SUPSI Campus, where it will monitor PV strings nightly. This real-world application will help optimize system performance while serving as a platform for further improvements.

In summary, the developed diagnostic tool offers a scalable, cost-effective solution for PV string monitoring, providing high accuracy and early fault detection to enhance the reliability and performance of photovoltaic systems.

ACKNOWLEDGMENTS

The here presented results are part of DTI-ISEA Master Thesis and of the project RACONT funded by the Swiss Federal Office of Energy under the contract SI/502264-01.

41st European Photovoltaic Solar Energy Conference and Exhibition

MACHINE LEARNING TECHNIQUES FOR THE ASSESSMENT OF OPEN CIRCUIT VOLTAGE LOSSES IN PHOTOVOLTAIC SYSTEMS

Sandra Riaño[1*], Jose D. Santos[1], Miguel Esteras[1], Amaia Abanda[1], Javier Del Ser[1,2]

[1]TECNALIA, Basque Research & Technology Alliance (BRTA), 48160 Derio, Bizkaia, Spain
[2]University of the Basque Country (UPV/EHU), 48013 Bilbo, Bizkaia, Spain
*Corresponding author: sandra.riano@tecnalia.com

ABSTRACT:
This investigation introduces a pioneering approach for remotely monitoring photovoltaic (PV) module open-circuit voltage (Voc) using SCADA data, without incurring energy production losses. By synergistically combining machine learning algorithms for data modeling with fundamental physics-based knowledge, our methodology successfully identifies Voc values from voltage SCADA data and provides a suite of Voc-related performance metrics. These metrics enable the Failure Detection and Diagnosis (FDD) through monitoring deviations in the performance of the module. This FDD strategy allowing for the optimization of the PV operation and intelligent maintenance (O&M) of the asset. The proposed approach is validated using synthetic data, and its robustness is assessed under diverse operational conditions. The findings reported in this study have significant implications in the enhancement of the reliability, efficiency, and overall performance of PV installations, ultimately contributing to the widespread adoption of solar energy as a clean, sustainable, and robust power source.
Keywords: PV systems, O&M, Voc, Machine Learning, Failure Detection and Diagnosis

1 INTRODUCTION

As the green energy transition accelerates sharply over time, Photovoltaic (PV) systems play a crucial role in sustainable power generation. Consequently, PV systems should be reliable, and their performance should be optimized during the entire life span.

From an electronics perspective, the current-voltage (IV) curve of a PV system defines its performance. Among its characteristic parameters, the open-circuit voltage (Voc) stands out as a pivotal parameter, standing for the maximum voltage of the PV system when there is no current flow.

Monitoring deviations of Voc from its nominal (*design*) value can allow for the identification of failure modes, like potential induced degradation (PID) or bypass diode short-circuit, among others. Moreover, recent studies [1] have pointed out that different degradation phenomena in new module technologies, such as Silicon Heterojunction (Si-HJ) and Passivated Emitter Rear Cell (PERC), can be caused by mechanisms that could mainly lead to Voc losses. Thus, remote Voc monitoring of commercial PV systems becomes a potentially useful task for improving the reliability of these assets. For this purpose, Machine learning (ML) techniques are currently being applied to detect failures in PV systems. Commonly used ML techniques include artificial neural networks, k-nearest neighbors (KNN) or fuzzy systems [2][3].

It has been found in the literature algorithms for estimating the open-circuit voltage (Voc), but they are primarily designed for Maximum Power Point Tracker (MPPT) systems [4][5]. These algorithms continuously adjust the current-voltage relationship of solar panels to ensure operation at the maximum power point, thereby maximizing energy generation.

This work builds upon this prior research to propose a new ML method to detect Voc deviations from its design value in PV installations without interrupting power flow. Voc estimation is aimed to optimize the operation and maintenances (O&M) of PV assets. In doing so, the approach leverages SCADA data, including irradiance, voltage, and module temperature. The proposed method relies on different ML techniques, such as clustering and regressions [6], hybridizing them with problem-specific

physics knowledge to estimate the value of Voc without affecting the PV system generation. We verify the detection performance of this hybrid model over synthetic data of a simulated PV system, aiming to inform with empirical evidence the responses to four research questions (RQs):

- *RQ1: Is it possible to identify Voc from maximum power point voltage (Vmp) using PV system SCADA data?*
- *RQ2: Which model describes most accurately Voc dependencies on operation conditions?*
- *RQ3: Is it possible to estimate, from SCADA data, Voc-related metrics such as Voc at standard test conditions (v-oc-stc), Voc temperature coefficient (v-oc-tc), or thermal voltage coefficient (v-th), so that the health status of PV modules can be estimated more reliably?*
- *RQ4: How can real experimental conditions impact on the proposed methodology?*

The rest of the manuscript is organized as follows: Section 2 describes the proposed method. The experimental setup is detailed in Section 3, whereas results are presented and discussed in Section 4. Finally, Section 5 ends the work by summarizing the main findings and future research.

2 PROPOSED METHOD

Figure 1 displays the steps of presented method. Firstly, self-supervised labelling is needed to identify Voc from voltage measurements. Secondly, Voc metrics are estimated using measurements labeled as Voc.

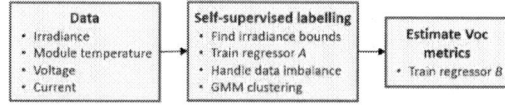

Figure 1: General diagram of method workflow

2.1 Self-supervised labelling

Some background is offered before going through the steps defined in Figure 1 for self-supervised labelling. Although theoretically Voc values should appear with zero current, in real-world assets low current levels appear in practice in voltage levels very close to Voc. This phenomenon complicates the identification of Voc based solely on current levels. Nevertheless, the physics behind photovoltaics is useful for separating Voc and Vmp values into two groups.

Let G be the Global Tilted Irradiance (GTI) measured in W/m², and let T be the module temperature measured in ºC, and V the voltage in V. The Shockley diode equation [7], could be rewritten[8] to express dependencies between V, G and T as

$$V(G,T) = b_0 + b_1 \ln\left(\frac{G}{1000}\right) + b_2\,T, \qquad (1)$$

where $\theta = \{b_0, b_1, b_2\}$ is a set of coefficients that can be learned from data by means of a least square regressor[9].

The dependency of synthetic voltage with the logarithmic transformation of irradiance and the temperature is shown in Figure 2. Higher voltage measurements where $\ln\left(\frac{G}{1000}\right) < -4$, corresponding to 18 W/m², are Voc values. The linear dependence of Voc with $\ln(G)$ is lost at high irradiance, where $\ln\left(\frac{G}{1000}\right) > -2, 135$ W/m². Figure 3 illustrates the same dependencies using experimental data, where the two voltage groups (Voc and Vmp) exhibit less distinct separation.

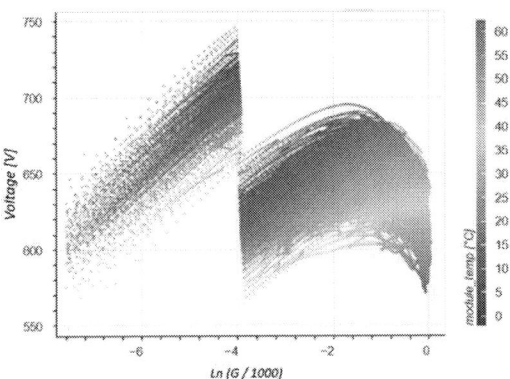

Figure 2: Voltage dependency with transformed irradiance and temperature (synthetic data, April-August 2023).

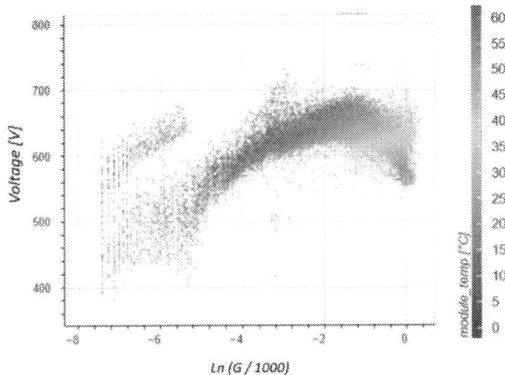

Figure 3: Voltage dependency with transformed irradiance and temperature (experimental data, April-August 2023).

Vmp behaves logarithmically with respect to irradiance at low irradiance levels (where $\ln\left(\frac{G}{1000}\right) > -2$), as Voc does. In addition, let V' be the voltage

measurement adjusted to their corresponding values at 25°C using the v-oc-tc provided by the PV modules manufacturer. Eq. (1) could be rewritten as

$$V'(G) = b_0 + b_1 \ln\left(\frac{G}{1000}\right), \qquad (2)$$

where $\theta = \{b_0, b_1\}$ is a set of coefficients that can be learned from data by means of a least square regressor.

Thus, a regressor, called A, can be trained with SCADA data to find optimal parameters θ in Eq. (2). For a given irradiance level, the Voc should be greater than Vmp due to the voltage drop in the series resistance of the electrical circuit. Therefore, the error in prediction should present two different groups, Voc and Vmp values. A Gaussian Mixture Model (GMM) [10] trained with the prediction error of regressor A as feature can be used to differentiate and label the voltage values into two clusters. However, this hypothesis is only true at certain irradiance levels that must be found before training the regressor A.

2.1.1 Find irradiance bounds

Figures 5 illustrates this process in the results and discussion section of the paper. V appears in orange, while V' values are colored in purple. Voltage measurements at lowest irradiances are not used to train the regressor A since these points belong to Voc. So, a lower irradiance bound to train this regressor must be defined. Then, irradiance values are split in evenly spaced intervals, and for each interval the median of irradiance for V is given to smooth the voltage measurements (blue line). The first drop of this smoothed voltage indicates the irradiance level where the transition from Voc and Vmp to only Vmp measurements. This irradiance value corrected with an additional offset to ensure that Voc values are left out is taken as the lower irradiance bound.

The upper irradiance bound of the regressor A training must leave out voltage values without linear dependency. Figure 5 shows how V' do not drop so clearly as V does. So, in this case, a rolling mean is applied to voltage measures without temperature correction (red line). The lower bound estimated in the previous step is used to filter the values at lower irradiances. Then, this voltage mean is fitted to a third order polynomial (black line). The inflection point to this curve is the irradiance level where the linear relationship does not hold any longer, and it can be taken as the upper bound of irradiance when training the regressor A.

2.1.2 Train regressor A

Once the irradiance bounds are set, the regressor A is trained to fit Eq. (2) and the prediction relative error is estimated.

2.1.3 Handle data imbalance

Voc has few occurrences per day. In Figure 2, Voc values are 5% of the total voltage measurements. Voc is higher than Vmp by physical meaning. Therefore, considering only almost positive error values to train GMM is a first attempt to reduce imbalance between Voc and Vmp clusters. Then, a logarithmic transformation is applied to the frequency axis of error histogram. A kernel density estimation (KDE) [11] function is estimated to have a probabilistic kernel of this error distribution. Finally, random samples of the kernel function are taken, reducing the imbalance between both voltage groups. Figure 6 in results section shows the handling of data imbalance, where the distribution of prediction error higher than -7% are presented.

2.1.4 GMM clustering

Quantiles 5% and 95% of the regression error are introduced to the GMM as the initial means of the clusters (dotted vertical lines in Figure 6). Then, the GMM is trained, and the cluster with lowest irradiances and highest voltages is labeled as Voc. Voltage values not labeled as Voc are labeled as Vmp.

2.2 Estimate Voc metrics

Identified Voc values from SCADA measurements are used to train a regressor model which considers relationships explained at Eq. (1), where V is Voc.

The Voc regressor, called regressor B, underestimates considerably v-oc-tc, as can be seen in Table II. Some experiments using Voc estimation from single diode model [16] for the whole range of the irradiance confirm a dependency between v-oc-tc and irradiance. This dependency is not considered in Eq. (1). In the experiments, Voc estimation is split in evenly spaced irradiance levels. Figure 4 shows the v-oc-tc estimation for each irradiance level presented against the median irradiance of training data. The estimated value converges to design value where irradiance levels are close to Standard Test Condition (STC). Other studies confirm this dependency [12].

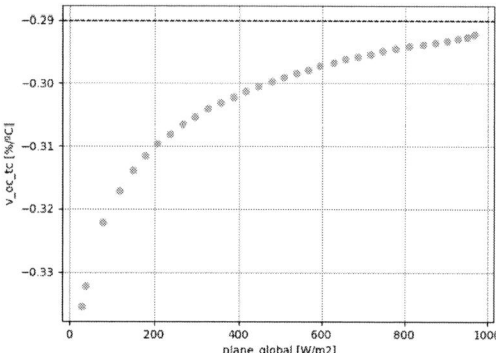

Figure 4: Dependency of v_oc_tc with irradiance

Consequently, a new model is proposed for regressor B following these dependencies:

$$V_{oc}(G,T) = b_0 + b_1 \ln\left(\frac{G}{1000}\right) + b_2 T + b_3 \ln\left(\frac{G}{1000}\right)T, \quad (3)$$

where $\theta = \{b_0, b_1, b_2\}$ is a set of coefficients that can be learnt by an ordinary least square regressor.

As regressor B has a physical meaning [13], its accuracy is manifest in the estimation errors of these metrics:

- v-oc-stc, the estimated voltage by regressor at G=1000 W/m^2 and T=25°C.
- v-oc-tc, the dependency on temperature, as estimated by regressor, obtained as the relative voltage reduction per degree Celsius at constant G=1000 W/m², measured in %/°C.
- v-th, the thermal voltage coefficient, obtained as the sum of coefficient b_1 and b_3 multiplied by T=25°C.

3 EXPERIMENTAL SETUP

3.1 Characteristics of simulated PV system

The presented method has been tested with synthetic data. Data is generated taking a real-world PV system as reference. The objective of the synthetic data is to explore the performance and the limits of the method. The reference PV system is located at Derio, Spain, and it is a string of 16 modules connected in series. They are installed with a tilt of 10° and azimuth of 189°. The electrical parameters of the PV system are described in Table I.

Besides of normal operation data, the simulation includes some disturbances that reflect real experimental conditions. These disturbances or controlled bias are related to SCADA sampling ratio and to the mismatch between irradiance measurements and the plane of modules.

Table I: Electrical parameters of the simulated PV system.

	Module	String
P-max (Wp)	450	7202
V-oc-stc (V)	49.70	795.20
I-sc-stc (A)	11.50	11.50
V-mp (V)	41.30	660.80
I-mp (A)	10.90	10.90
Efficiency (%)	20.40	20.40
Solar cell technology	c-Si	c-Si
N. of cells	144	1152
V-oc-tc (%/°C)	-0.29	-0.29
I sc-tc (%/°C)	0.05	0.05
P-max-tc (%/°C)	-0.37	-0.37

3.2 Normal operation data generation

Components of irradiance Global Horizontal Irradiance (GHI), Diffuse Horizontal Irradiance (DHI), and Direct Normal Irradiance (DNI) are downloaded from CAMS[14]. These measurements have one minute-level frequency and belong to the specified location. The dataset spans the period from January 2021 to May 2024 (not included). Irradiance components are transformed to the plane global irradiance of PV site (tilt=10, azimuth=189). Ambient temperature and wind components for the same time-period are downloaded from Copernicus [15] and they are interpolated to one minute frequency.

The module temperature is estimated using Sandia Array Performance Model (SAMP)[16][17] with CAMS and Copernicus meteo data.

The production variables Voc, Vmp. Imp and short circuit current Isc are generated through pvlib features [16] [18] [19], using the previously explained meteorological data and the module temperature estimation, with PV design values of Table I.

Single diode model estimates Voc and Vmp as separated time-series. The inverter of the real PV asset continuously monitors Voc values of modules until the minimum power threshold required for energy injection into the Alternating Current (AC) side is reached. This threshold is needed to merge Voc and Vmp estimations in a unique voltage time-series emulating data coming from SCADA. The power is calculated using the Vmp and Imp estimations from single diode model. A quantile of the power based on its distribution and the specification of the inverter is used. When the power is under this quantile, maximum power current Imp is set to zero and Vmp is overwritten with single diode model Voc estimated values. The 0.08 quantile of power for all available data is used to merge Voc and Vmp in the experiments carried out, which corresponds to 130 W.

Firstly, the method is validated through 5 months of data (April 2023 to August 2023 (5 months) are used to validate the method. This time span is selected because there are experimental data available in this period.

Secondly, the algorithms are trained with monthly data from all the available period with the aim of detecting potential seasonality.

3.3 Controlled bias data generation

Experimental data can differ from synthetic due to a wide range of circumstances, from measurement errors to degradation of the system. Two types of controlled bias are considered in this analysis: i) sampling averages in SCADA data gathering; and ii) orientation mismatch between irradiance sensor and modules set-up.

In the first case, the one-minute frequency data are sampled with lower frequency rates: 2, 4, 5, 10 and 15 minutes. These values are considered based on the indicated maximum recording interval applied to on-site ground-based [20]. The averaged value for the new period is calculated through three different methods: mean, median and last value. In the second case, components of irradiance are transformed to different tilts and azimuths around the set-up ones. These experiments try to replicate misalignment between the angles where irradiance is measured and the module angles.

4 RESULTS AND DISCUSSION

4.1 Normal operation data

Synthetic data from April 2023 to August 2023 (5 months) are used to validate the method. Figure 5 shows the irradiance bounds detected by the algorithm. Lower bound of $\ln\left(\frac{G}{1000}\right) = -3.72$, corresponding to 24 W/m^2, while upper bound is -1.34, corresponding to 260 W/m^2. Figure 6 shows the prediction error distribution of regressor A and the associated KDE. Performance metrics for GMM are not provided because they cannot be compared with real data where Voc is not labeled.

In response to RQ1, Figure 7 shows clustering results, where voltage colored in orange is labeled as Vmp, and voltage colored in green is labeled as Voc. The algorithm identifies Voc from maximum power point voltage. The comparison between Figure 2 and Figure 3 highlights that the experimental data may present some challenges, but the algorithm can still be applied.

In response to RQ2, Table II holds the Voc estimation metrics obtained using Eq. (1) or Eq. (3) when training regressor B. It shows how regressor B incorporating the irradiance-temperature dependency in Eq. (3) outperforms the model fulfilling Eq. (1).

The clustering and the Voc regressor are applied for synthetic data of each month from 2021 to May 2024 to detect potential seasonality. In response to RQ3, Figure 8 shows this monthly estimation of specified metrics for 40 months. The maximum error in all metrics is 1%. It corresponds to v-oc-tc, which arises as the most sensitive metric. It is never overestimated and an obvious seasonality can be observed. The estimation error is lower in July or August. This fact can be related to the range of temperature of these months when PV modules generation starts or ends (moments when Voc appears).

In a partial response to RQ4, Figure 9 shows a strong dependency between the maximum module temperature

and the error estimating v-oc-tc.

Table II: Voc estimation metrics (April -August 2023)

	Eq. (1)	Eq. (3)
Voc STC (V)	790.25	795.195
Voc STC error (%)	-0.622	-0.0007
Voc TC (%/ °C)	-0.35	-0.2892
Voc TC error (%)	21.465	-0.281
nNsVth (V)	28.640	29.660
nNsVth error (%)	-3.491	-0.056

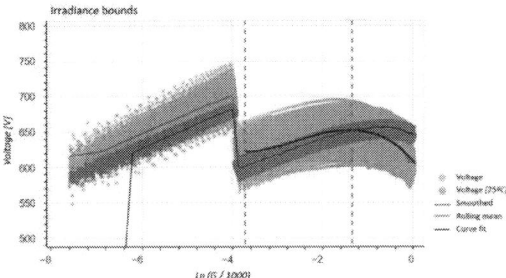

Figure 5: Irradiance bounds to train regressor A (April-August 2023)

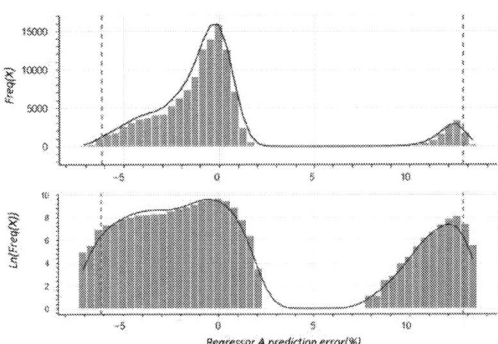

Figure 6: Regressor A error distribution, logarithmic transformation of histogram frequency and KDE (April-August 2023)

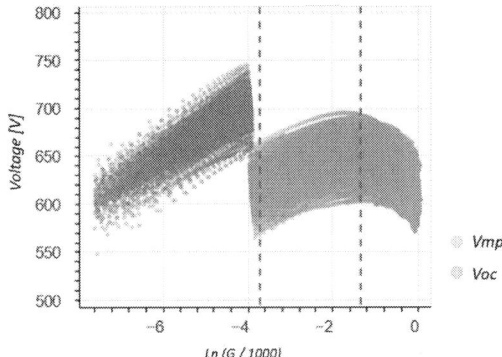

Figure 7: GMM clustering results trained with balanced error of regressor A, with Voc values colored in green and Vmp values in orange. (April-August 2023)

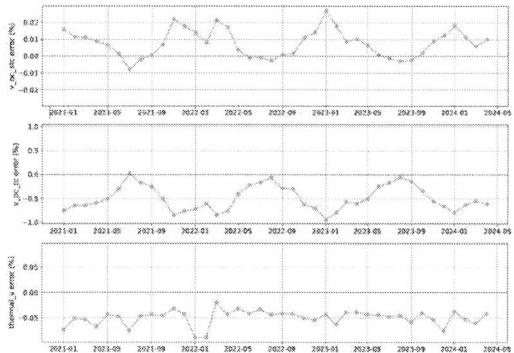

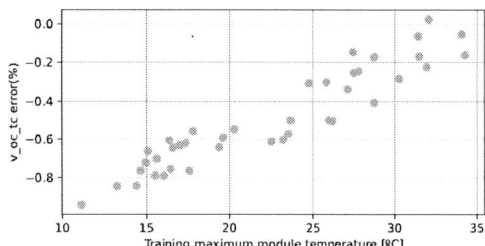

Figure 8: Monthly Voc metrics estimation

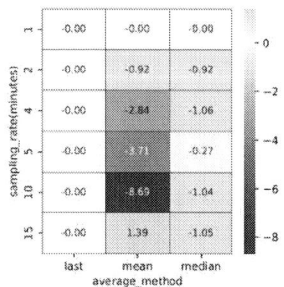

Figure 9: Training max temperature vs v-oc-tc estimation for monthly estimation

4.2 Resampled data

Synthetic data from August 2023, which has lower error rate than other months in the simulated time span (as can be seen in Figure 8), is used to analyze the influence of controlled biases.

This month data is sampled each 2, 4, 5, 10 and 15 minutes to emulate commercial SCADA sampling rates. The last value, the mean and the median of measurements are the three methods used to give an averaged value to the re-sampled period.

In a partial response to RQ4, Figures 10 and 11 depict the estimation error of metrics v-oc-stc and v-oc-tc where resampling is introduced in synthetic data. Results show that using the last value method has the least impact on the algorithm, as it merely removes values from the resampled period without generating new averaged data, which is not the case for the other methods. The mean method can be skewed by extreme values, whereas the median method shows lower errors when sampling rate is reduced. The v-oc-tc is also the most sensitive metric to the studied biases.

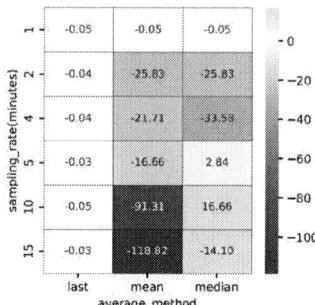

Figure 11: Resampled v-oc-tc estimation error (%) (August 2023)

4.3 Irradiance mismatch

The GTI measured in August 2023 is shifted in tilt ±20° and in azimuth ±30°, making increments every 5 degrees. Only a negative shift of 10° is considered due to the tilt of the modules is 10°.

Expanding the response to RQ4, Figures 12 and 13 show the estimation error of metrics v-oc-stc and v-oc-tc where mismatching in GTI is introduced in synthetic data. The heatmaps show how the errors become larger as the discrepancy between the actual angles and the shifted angles increases. The maximum relative error in metric v-oc-stc remains under ±1.5% in all the experiments carried out. However, v-oc-tc reach a relative error of 44% in one of the worst cases. If tilt is shifted −10°C, the modules are at a tilt of 0°. This fact explains why azimuth shifts do not affect the metrics in the first column of Figures 12 and 13.

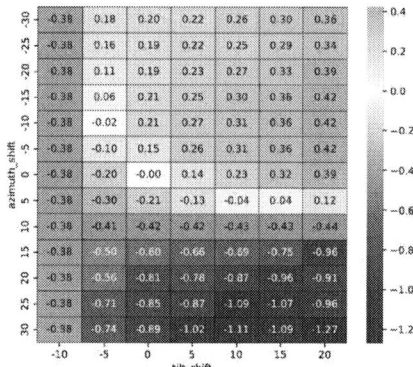

Figure 12: Irradiance mismatch v-oc-stc error (%) (August 2023)

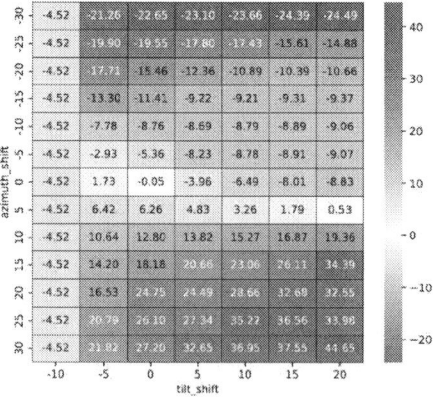

Figure 13: Irradiance mismatch v-oc-tc error (%) (August 2023)

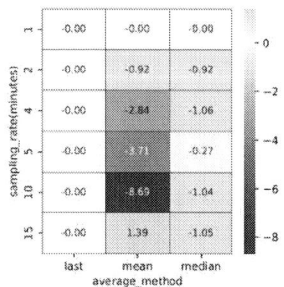

Figure 10: Resampled v-oc-stc estimation error (%) (August 2023)

5 CONCLUSIONS

In conclusion, our proposed method has shown promising performance with synthetic data under normal operating conditions. The clustering approach effectively identifies Voc values from voltage measurements with synthetic data. The performance of regressor B relies on the correct classification of Voc values in self-supervised labelling. Therefore, the performance of the algorithm with experimental data may pose a challenge.

Notably, the polarization of modules from Voc to maximum power point voltage occurs within short time periods of less than 15 minutes, highlighting the potential for inaccuracies in estimated metrics when relying on averaged SCADA data. This issue particularly impacts on v-oc-tc estimation.

Furthermore, our results suggest that mismatches between the plane of array and irradiance measurement planes can significantly impact v-oc-tc estimation, especially when tilt or azimuth deviations exceed 10°.

The v-oc-stc is more robust than the v-oc-tc because it is an estimation from a whole model, where inaccuracies in input data can be compensated among the coefficients.

In contrast, the temperature coefficient emerges as the most sensitive metric to introduced biases, exhibiting strong dependencies on the range of module temperatures. This sensitivity may pose challenges for periodic monitoring of deviations, due to the seasonality of module temperature.

Our research priority in the near future is using real SCADA data to further validate the presented method. Furthermore, we envision other research directions to be tackled in the future, such as including the estimation of regressor B uncertainties.

6 ACKNOWLEGMENTS

This study has received funding from the project IA4TES (*Inteligencia Artificial para la Transición Energética Sostenible*) funded by Ministry of Economic Affairs and Digital Transformation (MIA.2021.M04.0008) and from the project SerendiPV funded by the European Union's Horizon 2020 research and innovation programme under grant agreement No 953016.

7 REFERENCES

[1] A. Sinha, J. Qian, S.L. Moffitt, et al. UV-induced degradation of high-efficiency silicon PV modules with different cell architectures. Progress in Photovoltaics January. 2023; Volume 31(1):36-51. doi: https://doi.org/10.1002/pip.3606

[2] IEA_PVPS. The use of advanced algorithms in PV failure monitoring. URL https://iea-pvps.org/key-topics/theuse-of-advanced-algorithms-in-pv-failure-monitoring/

[3] H. P. -C. Hwang, C. C. -Y. Ku and J. C. -C. Chan, "Detection of Malfunctioning Photovoltaic Modules Based on Machine Learning Algorithms," in IEEE Access, vol. 9, pp. 37210-37219, 2021, doi: 10.1109/ACCESS.2021.3063461.

[4] A. Nadeem, H.A. Sher, et all Online current-sensorless estimator for PV open circuit voltage and short circuit current, Solar Energy, Volume 213, 2021, Pages 198-210, ISSN 0038-092X, https://doi.org/10.1016/j.solener.2020.11.004

[5] A.I. M. Ali, H. R. A. Mohamed, Improved P&O MPPT algorithm with efficient open-circuit voltage estimation for two-stage grid-integrated PV system under realistic solar radiation, International Journal of Electrical Power & Energy Systems, Volume 137, 2022, 107805, ISSN 0142-0615, https://doi.org/10.1016/j.ijepes.2021.107805.

[6] C. M. Bishop, Pattern Recognition and Machine Learning. Springer 2006. ISBN 9781493938438

[7] W. Shockley, "The theory of p-n junctions in semiconductors and p-n junction transistors," in The Bell System Technical Journal, vol. 28, no. 3, pp. 435-489, July 1949, doi: https://doi.org/10.1002/j.1538-7305.1949.tb03645.x

[8] N. Anani, H. Ibrahim. Adjusting the Single-Diode Model Parameters of a Photovoltaic Module with Irradiance and Temperature. *Energies* **2020**, *13*, 3226. https://doi.org/10.3390/en13123226

[9] T. W. Anderson. John B. Taylor. "Strong Consistency of Least Squares Estimates in Normal Linear Regression." Ann. Statist. 4 (4) 788 - 790, 1976. https://doi.org/10.1214/aos/1176343552

[10] Scikit_learn. Gaussian Mixture Models, URL https://scikit-learn.org/stable/modules/mixture.html accessed January 2024

[11] T. Hastie et all, The Elements of Statistical Learning: Data Mining, Inference, and Prediction. New York: Springer. 2001. ISBN 0-387-95284-5. OCLC 46809224

[12] F. P. Gasparin, F. D. Kipper et all, Assessment on the variation of temperature coefficients of photovoltaic modules with solar irradiance, Solar Energy, Volume 244, 2022, Pages 126-133, ISSN 0038-092X, https://doi.org/10.1016/j.solener.2022.08.052.

[13] K.D. Jäger et all, Solar Energy Fundamentals, Technology, and Systems, 2014, Delft University of Technology. ISBN 9781906860738

[14] CAMS solar radiation services https://ads.atmosphere.copernicus.eu/cdsapp#!/dataset/cams-solar-radiation-timeseries?tab=overview, accessed August 2024

[15] Copernicus ERA5 hourly data on single levels https://cds.climate.copernicus.eu/cdsapp#!/dataset/reanalysis-era5-single-levels?tab=overview, accessed August 2024

[16] A. Jensen, K. Anderson, W. Holmgren, M. Mikofski, C. Hansen, L. Boeman, R. Loonen. "pvlib iotools — Open-source Python functions for seamless access to solar irradiance data." Solar Energy, 266, 112092, (2023). DOI: 10.1016/j.solener.2023.112092.

[17] King, D. et al, 2004, "Sandia Photovoltaic Array Performance Model", SAND Report 3535, Sandia National Laboratories, Albuquerque, NM.

[18] A. Jain, A. Kapoor, "Exact analytical solutions of the parameters of real solar cells using Lambert W-function", Solar Energy Materials and Solar Cells, 81 (2004) 269-277.

[19] W. De Soto et al., "Improvement and validation of a model for photovoltaic array performance", Solar Energy, vol 80, pp. 78-88, 2006.

[20] International Electrotechnical Commission. IEC 61724-1:2021 Photovoltaic system performance - Part 1: Monitoring (Ed. 2.0). Ginebra, Suiza.

41st European Photovoltaic Solar Energy Conference and Exhibition

REAL-TIME MONITORING AND DIAGNOSTIC OF ROOFTOP MONOFACIAL PV SYSTEM VALIDATED WITH THERMOGRAPHY

Amr Osama[*,1,2], Giuseppe Marco Tina[1], Antonio Gagliano[1], Gabino Jimenez-Castillo[3], Francisco José Munoz-Rodríguez[4]

[1] University of Catania, Department of Electric, Electronics and Computer Engineering, Viale Andrea Doria 6, Catania, Italy, 95125

[2] Mechanical Power Engineering Department, Faculty of Engineering, Port Said University, Port Fuad, Egypt

[3] Department of Electrical Engineering, Center for Advanced Studies in Earth Sciences, Energy and Environment, University of Jaen, Jaen, Spain

[4] Department of Electronic and Automatic Engineering, Universidad de Jaen, Jaen 23071, Spain

*Corresponding author's E-mail address: amrosama@eng.psu.edu.eg - Tel.: +201099732663

ABSTRACT: Photovoltaics reliability was based on the availability of performance monitoring systems and identifying the causes of deterioration, if any, to ensure the system's efficiency and reliability. The electrical operating conditions have a direct impact on the thermal performance of the module which is rarely considered. Several thermal models and implemented in the simulation software, however, the estimated module temperature's precision for different electrical operating conditions is not defined, which the current work is focused on. A real-time monitoring application has been developed to assess the thermal behavior of three sets of rooftop PV systems performance, at Jean University campus in Spain, operating at different electrical statuses; a short-circuited PV set-1, open-circuited PV set-2, and PV set-3 operate at the maximum power point (MPP). The monitoring revealed that the module temperature of the short-circuited PV system is 8.5% more than the one of the open-circuited PV system. Moreover, the module temperature of the MPP operating PV system is 20.73% and 10.39% less compared to the short-circuited and open circuited one. The NOCT thermal model overestimates the module temperature the PV system operates at MPP by 15.33% while the Faiman model overestimates it by 16.65%.

Keywords: PV monitoring, Thermal model, short circuit, Open circuit, maximum power point

1 INTRODUCTION

Photovoltaics (PV) witnessed a significant installation gross reaching more than 510GW in 2023 achieving a cumulative capacity of about 1.8TW [1]. With the more depending on the PV potential toward renewable and sustainable energy transition, several configurations and installation solutions have been introduced across all segments of the building sector i.e. residential, commercial, and industrial buildings [2]. Moreover, modern PV module technologies with more economical and production benefits added to the market share. With all these new technologies and installation solutions, continuous monitoring and assessment of the performance to this system is required to ensure the reliability and performance optimality of such systems [3].

As the PV system operates in harsh outdoor conditions, faults in the PV system can be stimulated mainly in the PV module as the key component [4]. Recognizing the root causes of failures became increasingly crucial following the realization that 2% of PV modules are anticipated to fail after 11-12 years, which does not meet the manufacturer's guarantee. Thus, the need for monitoring and inspection techniques is crucial to detect the specific fault and its cause [5]. Several studies have tried to tackle the aforementioned issues. The visual inspection can be beneficial in providing the on-site detection, classification, and localization of faults, including snail tracks, bird dropping, slight chalking, degradation, a defective junction box, corrosion of the frame, discoloration, and cell cracking [6]. Yet it needs proper training and expertise, and it may pose an electrical hazard to the inspectors.

The majority of the monitoring techniques were focused on analyzing the electrical performance of the PV system. This can be represented in analyzing the instant IV characteristic curves to detect the type and severity of a fault when compared to a healthy I-V curve [7]. However, this requires the PV system to be out of production during the characteristic curve scanning. An alternative method, power loss analysis (PLA), is to compare the performance of PV plants using artificial neural networks [7]. Yet as the number of PV panels increases, it becomes more complex, making it only applicable to small PV systems.

Real-time monitoring techniques gained a reputation for their dependability and effectiveness. Several approaches have been developed considering the integration of hardware and software for this purpose [8]. Real-time monitoring can provide instant status and evaluation of the overall performance of the PV system that can give such technique an edge compared to other techniques, which prompts further investigation [9]. For fault detection, maximum power point tracking (MPPT) is considered for open circuit (OC) fault, and partial shading conditions based on the inflection points [10]. Monitoring circuits (i.e., voltage and current sensors), is an appealing approach to identifying the PV panel working point in real-time [11]. However, regardless of the sensor cost for even medium- or large-scale plants, such methods neglect thermal faults and can't provide a clear assessment as they depend only on the electrical performance of the modules.

It has been noticed that most of the available monitoring techniques emphasize the PV system productivity and the module health while operating at the maximum power point condition. There was a rare focus on the impact of the electrical operating conditions on the thermal performance of the PV system in addition to the thermal model accuracy for different electrical operating points, which is minimally studied.

Therefore, the present work is directed at real-time investigation of the thermal performance of the PV system for different electrical operating conditions. Open-circuited PV short-circuited PV, and maximum power point operating PV systems are experimentally investigated simultaneously at Jaen University, Spain. For better evaluation, thermal photography has been included in each PV set examined. A comprehensive comparison between the thermal performance measured and the calculated one using thermal models is used for accuracy assessment.

2. EXPERIMENTAL METHODOLOGY

2.1 Experimental Setup

Three sets of rooftop PV systems have been installed on the Faculty of Engineering building at Jaén University campus in Spain. Each PV set has been fixed to a specific electrical operating condition; the first set is short-circuited PV set 1, the second set is open-circuited PV set 2, and the third set is maximum power point operating PV set 3. Each set consists of two identical monofacial PV modules connected in parallel. The six common monofacial PV modules, with the specification mentioned in Table 1, are south-oriented with 1.5 m above the roof surface.

Figure 1: A photograph of the experimental setup

The main climatic variables presented in Ambient temperature (T_a), plane of array irradiance (G_{POA}), and wind velocity (Ws) have been captured with a Calibrated PV cell and Anemometer (model-DATASOL MET) sensors installed on the PV rack, as can be seen in Figure 1. The module temperature (T_m) is measured via the PT100 temperature sensor surface type fixed on the middle rear surface of a module in each set. Moreover, thermal imaging has been included to have a better thermal evaluation according to the captured thermal distribution along the module. As for the electrical performance, the current and voltage of both short-circuited PV set 1 and open-circuited PV set 2 are monitored by data acquisition

devices respectively. While PV set 3 is connected to electronic load EL9000B which has the ability to trace the maximum power point (MPP) periodically to make sure that the PV set 3 is operating at (MPP).

Table 1
Specification of the PV module

Specification	Value
Model name	WS-100/12V
Panel brand	WAAREE
Material	Polycrystalline
Dimensions in mm	$1150 \times 675 \times 35$
Short-circuit current, (I_{sc}) [A]	6.07
Open circuit voltage, (V_{oc}) [V]	21.97
Maximum power, (P_{MP}) [W]	100
Maximum power voltage, (V_{MP})[V]	17.47
Maximum power current, [(I_{MP}) [A]	5.73
Nominal Operating Cell Temperature [°C]	46
Module efficiency at standard test conditions, (η_{STC})[%]	12.88
Open circuit voltage temperature coefficient [%/°C]	-0.2941
Maximum power temperature coefficient [%/°C]	-0.3845

2.2 Measurement approach

The developed monitoring system was based on a developed software simulation on the LabVIEW program that supports real-time measurements and performs analysis accordingly. The environmental variables, thermal, and electrical performance variables were monitored using the developed LabVIEW application. The measurement sampling was 30 seconds, according to IEC 71642. The instant environmental variables are utilized in the thermal models to calculate the module temperature, which is compared with the measured module temperature.

The Nominal Operating Cell Temperature (NOCT) thermal model is one of the most famous thermal models utilized by the simulation software and supported by standard IEC 61215-2:2016 as can be seen in the following formula [12]:

$$T_c = T_a + G_{POA} \cdot \left(\frac{T_{c,NOCT} - 20}{800} \right) \quad \textbf{(1)}$$

Where T_a is the ambient temperature in [°C], G_{POA} is the plane of array irradiance in [W/m^2], $T_{c,NOCT}$ is the nominal operating cell temperature which is reported in the module datasheet in [°C]. This model is proportionated to NOCT conditions (T_a=20°C and G_{POA} =800 W/m^2) as a reference point

Duffie and Beckman's model [13] is considered for the thermal model's accuracy investigation. Duffie and Beckman's model is adopted by SAM simulation software, see equation (2) [14].

$$T_c = T_a + \left(\frac{9.5}{5.7 + 3.8 \cdot Ws} \right) \left(\frac{G_{POA}}{800} \right) \left(T_{c,NOCT} - 20 \right) \left(1 - \frac{\eta(T)}{\tau\alpha} \right)$$
$$(2)$$

Where $\tau\alpha$ are effective transmittance-absorptance product ($\tau\alpha \approx 0.9$), $\eta(T)$ is the electrical efficiency of the PV module as a function of the cell temperature (usually $\eta(T) \approx \eta_{STC}$). In case of no electrical production, it should be considered equal to zero.

The Faiman model [15] is also adopted in the developed software for a comprehensive comparison of thermal performance with other calculated module temperatures. The Faiman model, presented in equation (3), is the main thermal model for PVsyst simulation software [16].

$$T_c = T_a + \frac{G_{POA}\left[\tau\alpha - \eta(T)\right]}{U_0 + U_1 \cdot Ws} \qquad (3)$$

Where U_0: the coefficient describing the effect of the radiation on the module temperature [W/m²°C] and U_1: the coefficient describing the cooling by the wind [W.s/m³.°C]. according to PVsyst for open rack (free standing) installation, U_0 and U_1 are 25 W/m²°C and 1.2 W.s/m³.°C respectively.

It is worth mentioning that all of the stated thermal models estimate the cell temperature T_c. To compare the measured module temperature T_m via temperature sensors with the calculated ones using thermal models, the calculated T_c must be modified to calculate the module temperature $T_{m,c}$ as can be seen in the following formula [17]:

$$T_m = T_c - \Delta T \cdot \frac{G_{POA}}{1000} \qquad (4)$$

Where ΔT is the temperature difference between the PV module's front and back surface in the STC radiation of 1000W/m². It's very sensitive to the module technology and the installation configuration. It can be stated that for a monofacial PV module installed in open rack installation

as in the adopted system, ΔT is equal to 3°C [18].

A thermal comparison between the estimated and measured module temperature is considered for each PV set operating in different electrical operating conditions. The percentage accuracy (PA) of the thermal model relative to the measured T_m can be calculated according to the following formula:

$$PA = \frac{S - M}{M} \times 100 \quad (5)$$

Where S and M represent the simulated (calculated) and measured module temperature, respectively.

3. RESULT AND DISCUSSION

The monitoring of the three rooftop PV sets took place for five days from 15th to 19th of March 2024. The archived measurements were filtered by eliminating the readings for irradiance lower than 50 W/m² and higher than 1200 W/m² according to IEC 61724-2:2016 [19]. The dataset also has been averaged to 5-minute bases to improve the records' precision and stability.

3.1 Thermal Behavior Comparison

The thermal performance presented in measuring the module temperature of the three sets is utilized to evaluate their behavior at different real electrical operating conditions simultaneously during the period of experimentation. The daily trend of the three sets of module temperature is shown in Figure. 2 for the 19th of March as a sample. It can be seen during this day that; the

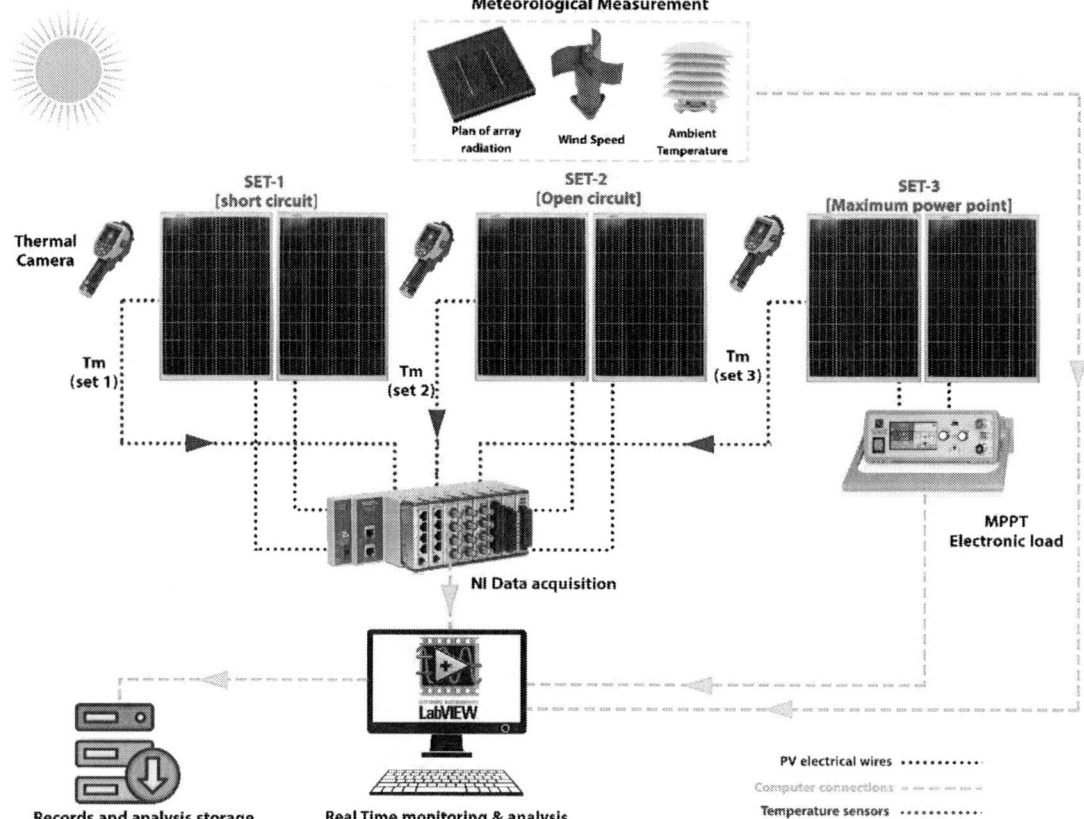

Figure 1: Schematic diagram of the experimental setup

average daily radiation and ambient temperature were 617.2 W/m² and 28.3°C respectively. When the PV set 3 operates at MPP, the T_m achieved a daily average temperature of 40.3°C which is the lowest value compared to other PV sets. When the PV operates at an open circuit (OC) as stimulated in PV set 2, a daily average module temperature of 44.2°C is reached which is 9.7% higher than the system operates at MPP. Moreover, the PV set that operates at short circuit (SC) conditions has a daily average module temperature of 47.5°C. This means that the module temperature of a short-circuited PV system is 17.8% higher than the one the system operates at MPP.

When the PV module operates in outdoor conditions, it receives a continuous rate of solar radiation. According to the heat balance principle, the MPP operation works on the conversion part of the received energy to electrical power that is extracted from the PV module, while the remaining energy elevates the module temperature. Moreover, the open-circuited PV module prevents the extraction of the converted electrical power part, which means that all of the received radiation is converted to thermal energy, elevating the module temperature higher than the ones that work as MPP. Furthermore, for the short-circuited PV set 1, the maximum value of current (Isc) flows through the PV module terminals, facing the resistance of the PV cells themselves which in turn significantly increases the cell temperature. This can explain the elevated module temperature of PV set 1 compared to open circuit ones in PV set 2.

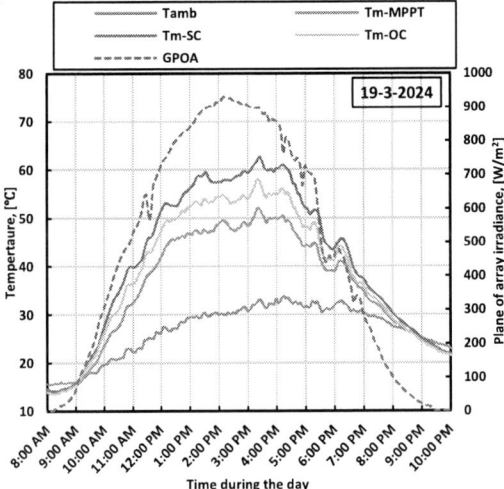

Figure 3: Daily trend of the PV module temperature at different electrical operating points.

An infrared thermal image is taken to the three PV sets simultaneously as can be seen in Figure 4. The temperature distributions between the three sets are clearly noticeable, whereas PV-set 3 has a uniform yellowish color distribution. The reddish colors indicate the higher temperature indicated in Figure 4. The open-circuited PV-set 2 has a uniform reddish temperature distribution, indicating higher temperature. However, short-circuited PV-set 1 has clear signs of multiple hotspots, which means that these specific cells are dissipating more thermal energy rather than the electrical ones [20].

3.2 Thermal Models Evaluation

While the thermal models are the simulation basis for estimating the electrical power, the accuracy of such models needs to be examined. the daily trend of the calculated module temperature using different models relative to the measured one of the PV systems operating at MPP is presented in Figure 5.

It can be seen that the adopted thermal model in this study tends to overestimate the module temperature relative to the measured one. NOCT model estimates the daily average module temperature by 46.59°C, which is higher than the measured one by 15.60%. while the Faiman model estimates it by 47.56°C with 18.01% more than the measured one. Duffie and Beckman's model has the lowest accuracy compared to the other adopted models. It estimates the Tm by 49.17 °C, which is 22% higher than the actual measured T_m of PV set-3 that operates at MPP. Simplifying Duffie and Beckman's model to a formula similar to Faiman models indicated that the wind speed coefficient (U_1=11.61 [W/m²°C]) is significantly higher than the one considered in PVsyst (U_1=1.2 [W/m²°C]). This can explain the high sensitivity to the wind speed, which in turn produced a highly fluctuated estimation, as can be seen in Figure 5.

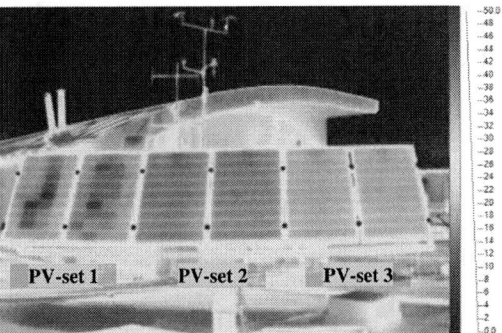

Figure 4: Thermal image of the examined three PV sets at 12:00 pm (G_{POA}=703.5 W/m², T_{amb}=27.3°C).

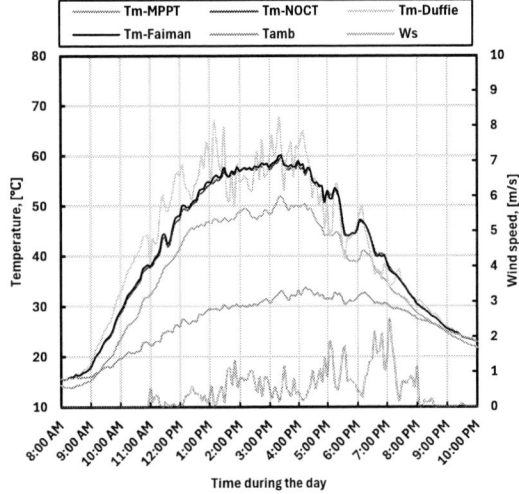

Figure 5: Daily trend of the T_m when operating at MPP compared with the calculated ones via thermal models.

A summary of the thermal performance during the 5 days of monitoring formed in the daily average module temperature measured and calculated via the considered thermal models is shown in Figure 6. It can be deduced that the NOCT thermal model has a better accuracy in estimating the module temperature when the PV module operates in open circuit conditions with an average precision of 4.4% while an average precision of 15.32% for estimating the temperature of the MPP operating PV system. While the Faiman model tends to overestimate the T_m for the PV module that works in MPP conditions by an average of 16.65%, it is found that such a model has a better evaluation capability for estimating the Tm for short circuit conditions with an underestimation of 3.63%. Moreover, during the monitoring period, Duffie and Beckman's model has a low level of accuracy in estimating the MPP operating module temperature of 29.14%.

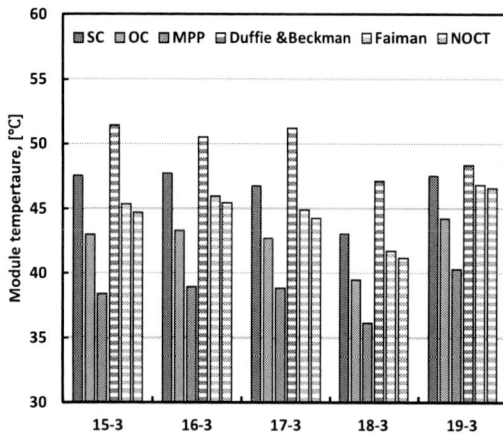

Figure 6: A daily average module temperature of the three sets examined during 5 days of monitoring compared to the calculated one.

4. Conclusion

The evaluation of the electrical operating conditions on the thermal performance of a rooftop mono-facial PV system is considered through a developed real-time monitoring application on LabVIEW software. Three sets were experimentally monitored while operating at short circuit, open circuit, and MPP conditions. Thermal models accuracy is evaluated. In summary, the present work provides developed the following observations:

- The module temperature of the short-circuited PV system is 20.73% higher than the one the system operates at MPP on an average daily basis.
- When the PV module operates at open circuit conditions, the module temperature is elevated by 10.40% compared to the system operating at MPP.
- When the PV module is short-circuited, this stimulates multiple hotspots which pose a significant hazard to the PV module's health.
- The NOCT thermal model tends to overestimate the module temperature of a system operating at MPP by the average value of 15.33%, while the Faiman model overestimates it by 16.65%
- Duffie and Beckman's model has a low level of accuracy in estimating the MPP operating module temperature of 29.14%.

Acknowledgment

This work has been supported by the Italian National PHD in Photovoltaics, CURRICULUM C: Monitoring and Diagnosis has been received. Also, support by MUR, Italy funds in the frame of PRIN 2020 "A Holistic Monitoring and Diagnostic Tool for Photovoltaic Generators (HOTSPHOT)" project (CUP:E63C2001116000).

References

[1] A. Latoui and M. E. H. Daachi, "Real-time monitoring of partial shading in large PV plants using Convolutional Neural Network," *Sol. Energy*, vol. 253, no. March, pp. 428–438, 2023, doi: 10.1016/j.solener.2023.02.041.

[2] K. Chattopadhyay, A. Kies, E. Lorenz, L. von Bremen, and D. Heinemann, "The impact of different PV module configurations on storage and additional balancing needs for a fully renewable European power system," *Renew. Energy*, vol. 113, pp. 176–189, 2017, doi: 10.1016/j.renene.2017.05.069.

[3] I. Høiaas, K. Grujic, A. G. Imenes, I. Burud, E. Olsen, and N. Belbachir, "Inspection and condition monitoring of large-scale photovoltaic power plants: A review of imaging technologies," *Renew. Sustain. Energy Rev.*, vol. 161, no. March 2021, p. 112353, 2022, doi: 10.1016/j.rser.2022.112353.

[4] I. Høiaas, K. Grujic, A. Gerd, I. Burud, and E. Olsen, "Inspection and condition monitoring of large-scale photovoltaic power plants : A review of imaging technologies," *Renew. Sustain. Energy Rev.*, vol. 161, no. March, p. 112353, 2022, doi: 10.1016/j.rser.2022.112353.

[5] S. Ansari, A. Ayob, M. S. Hossain Lipu, M. H. Md Saad, and A. Hussain, "A review of monitoring technologies for solar pv systems using data processing modules and transmission protocols: Progress, challenges and prospects," *Sustain.*, vol. 13, no. 15, 2021, doi: 10.3390/su13158120.

[6] M. Waqar Akram, G. Li, Y. Jin, and X. Chen, "Failures of Photovoltaic modules and their Detection: A Review," *Appl. Energy*, vol. 313, no. February, p. 118822, 2022, doi: 10.1016/j.apenergy.2022.118822.

[7] Y. Y. Hong and R. A. Pula, "Methods of photovoltaic fault detection and classification: A review," *Energy Reports*, vol. 8, pp. 5898–5929, 2022, doi: 10.1016/j.egyr.2022.04.043.

[8] N. Iksan, P. Purwanto, and H. Sutanto, "Real-Time Monitoring of Photovoltaic Systems and Control of Electricity Supply for Smart Micro Grid-PV using IoT," *TEM J.*, vol. 13, no. 1, pp. 514–523, 2024, doi: 10.18421/TEM131-53.

[9] G. Jiménez-Castillo, ; S. Aneli, A. J. Martínez-Calahorro, A. Gagliano, and G. M. Tina, "Monitoring Electrical and Weather Parameters in Photovoltaic Bifacial Systems," *Global Challenges for a Sustainable Society*. pp. 546–553, 2023.

[10] Q. Xu, X. Li, X. Yin, and C. Feng, "A Real-Time Fault Detection Technique Based on MPPE in Photovoltaic Systems," *PEDG 2023 - 2023 IEEE 14th Int. Symp. Power Electron. Distrib. Gener.*

[11] P. Guerriero, V. D'Alessandro, L. Petrazzuoli, G. Vallone, and S. Daliento, "Effective real-time performance monitoring and diagnostics of individual panels in PV plants," *4th Int. Conf. Clean Electr. Power Renew. Energy Resour. Impact, ICCEP 2013*, pp. 14–19, 2013, doi: 10.1109/ICCEP.2013.6586958.

[12] J. H. Bae, D. Y. Kim, J. W. Shin, S. E. Lee, and K. C. Kim, "Analysis on the Features of NOCT and NMOT Tests with Photovoltaic Module," *IEEE Access*, vol. 8, pp. 151546–151554, 2020, doi: 10.1109/ACCESS.2020.3017372.

[13] J. A. Duffie and W. A. Beckman, *Solar Engineering of Thermal Processes*, vol. null. in null, vol. null. 1982.

[14] P. Gilman, A. Dobos, N. DiOrio, J. Freeman, S. Janzou, and D. Ryberg, "System Advisor Model (SAM) Photovoltaic Model Technical Reference Update," *Natl. Renew. Energy Lab.*, no. March, p. 93, 2018, [Online]. Available: https://sam.nrel.gov/%0Ahttps://sam.nrel.gov/images/web_page_files/sam-help-2018-11-11-r4.pdf%0Asam.nrel.gov/content/downloads

[15] D. Faiman, "Assessing the outdoor operating temperature of photovoltaic modules," *Prog. Photovoltaics Res. Appl.*, vol. 16, no. 4, pp. 307–315, Jun. 2008, doi: https://doi.org/10.1002/pip.813.

[16] "https://www.pvsyst.com/help/models.htm."

[17] N. Martín-Chivelet, J. Polo, C. Sanz-Saiz, L. T. Núñez Benítez, M. Alonso-Abella, and J. Cuenca, "Assessment of PV Module Temperature Models for Building-Integrated Photovoltaics (BIPV)," *Sustain.*, vol. 14, no. 3, pp. 1–15, 2022, doi: 10.3390/su14031500.

[18] D. L. King, W. E. Boyson, and J. A. Kratochvil, "Photovoltaic array performance model, SANDIA Report SAND2004-3535," *Sandia Rep. No. 2004-3535*, vol. 8, no. December, pp. 1–19, 2004.

[19] E. Ogliari, A. Dolara, D. Mazzeo, G. Manzolini, and S. Leva, "Bifacial and Monofacial PV Systems Performance Assessment Based on IEC 61724-1 Standard," *IEEE J. Photovoltaics*, vol. 13, no. 5, pp. 756–763, 2023, doi: 10.1109/JPHOTOV.2023.3295869.

[20] M. U. Ali *et al.*, "Early hotspot detection in photovoltaic modules using color image descriptors: An infrared thermography study," *Int. J. Energy Res.*, vol. 46, no. 2, pp. 774–785, 2022, doi: 10.1002/er.7201.

41st European Photovoltaic Solar Energy Conference and Exhibition

Single Image Geospatial Referencing

JÜLICH Forschungszentrum

Evgenii Sovetkin[†], Andreas Gerber[†], Bernhard Kubicek[‡] and Bart E. Pieters[†]

[†] IMD-3 Photovoltaics, Forschungszentrum Jülich, 52425 Jülich, Germany
[‡] AIT Austrian Institute of Technology GmbH, Vienna, Austria

Abstract

We have developed an algorithm to geospatially reference drone images from a single image. Our approach involves leveraging the known physical dimensions of modules and a Pix2Pix neural network to locate module pixel coordinates. By calculating a homography between the image and the ground plane, our method can determine the geographical coordinates of every pixel.

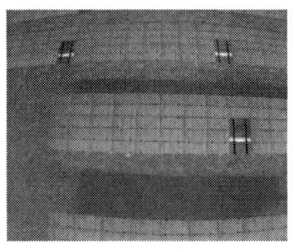 

works with low- and high- resolution images

Pix2Pix-model converts an image to a binary grid [1]

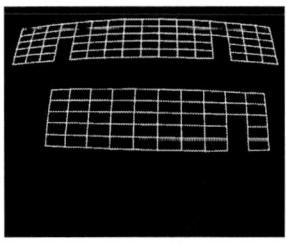 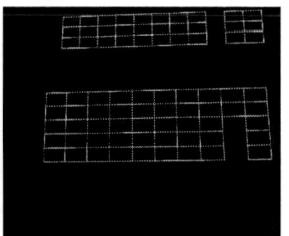

graph-theoretic analysis ensures robustness [1]

compute module corner coordinates

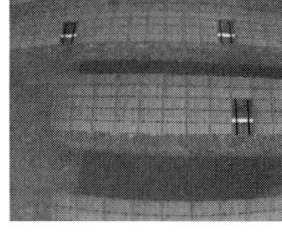 

Motivation

- identify the same modules in drone images across different measurements
- physical dimensions of modules and table tilt are known
- take advantage of regular module arrangement
- use the drone's recorded GPS data

Approach

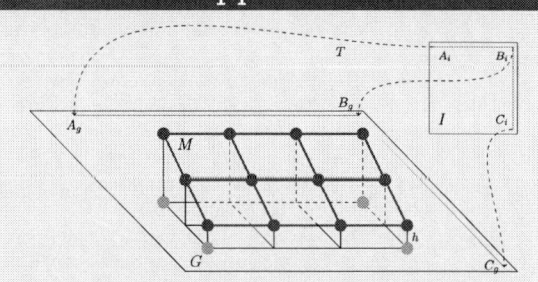

Step 1: estimate camera distortion model (need many images) => pinhole camera model

Step 2: compute camera pose (solvePnP): use the physical dimensions of modules and their identified pixel coordinates => compute pixels of ground points

Step 3: compute the world coordinates of ground projection (wrt to the module)

Step 4: from steps 2 and 3, compute homography, mapping pinhole pixels to geographical coordinates

Conclusions

- associate for every pixel corresponding ground coordinate, taking into account camera distortions and orientation
- method processing time 0.5s due to Pix2Pix and GPU processing

Limitations:

- flat ground assumption: DSM is needed for nonflat surfaces
- same size modules assumption

[1] Evgenii Sovetkin et al. "Fast Cell Detection and Distortion Correction for Outdoor Electroluminescence Images". In: 2023 IEEE 50th Photovoltaic Specialists Conference (PVSC). 2023, pp. 1-6. doi: 10.1109/PVSC48320.2023.10359780.

Acknowledgement: This work is supported by the "ReliaREN-Pro" project (Foerderkennzeichen: 03EI4052)

 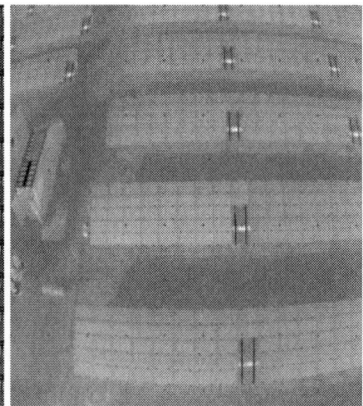

Member of the Helmholtz Association

41st European Photovoltaic Solar Energy Conference and Exhibition

This presentation was selected by the Sc. Committee of the EU PVSEC 2024 for submission of a full paper to one of the EU PVSEC's collaborating peer-reviewed journals.

OUTDOOR EXPOSURE STUDY ON THE PERFORMANCE OF NINE DIFFERENT TYPES OF INDUSTRIAL PV MODULES UNDER 35° AND UNDER 90° TILT

Carolin Ulbrich[1]*, Niklas Albinius[1], Luka Wernke[1, 2], Björn Rau[1], Rutger Schlatmann[1, 3]

[1] Helmholtz-Zentrum Berlin für Materialien und Energie, PVcomB, Schwarzschildstr. 3, 12489 Berlin, Germany
[2] Technische Hochschule Wildau, Hochschulring 1, 15745 Wildau, Germany
[3] Hochschule für Technik und Wirtschaft Berlin, Wilheminenhofstraße 75A, 10313 Berlin, Germany

ABSTRACT: For practical and aesthetic reasons, solar PV modules with black instead of white backsheets or with colored front glass are used in vertical building integrated PV installations. This double use of area is favorable for societal acceptance that is ever more important with the raising number of PV installations. Besides, the variety of installation angles implies a broadening of the electricity generation profile which is indeed advantageous. However, these non-optimized orientations and the use of all-black and of colored modules under non-optimized orientation result in a lower yield compared to optimal PV systems with modules with white backsheet or even bifacial modules. This experimental study quantifies the yield reduction for using an exemplary blue colored versus a reference CIGS module and black instead of white backsheets in a comparison of two CIGS and seven up to date silicon PV modules of various technologies (Silicon heterojunction monofacial black and white backsheet, bifacial, PERC half cells black and white backsheet, shingled with black backsheet, PERC IBC with white backsheet) monitored in parallel under optimum tilt and 90° South orientation. We found that in May vertical installations show a specific yield reduced by 44 % compared to 35° tilted installations on average over all investigated monofacial technologies, and a gain of 4 % in January. Modules with black backsheets installed vertically produced up to 7 % less in May than their counterparts with white backsheets, this loss being up to 4 % for 35° tilted modules. Also in May, the efficiency delta of 1%abs. between the CIGS module with blue-colored front glass and the reference resulted in a reduction of specific yield by 12% on the 90° tilted open rack. All these differences are less significant in winter. These numbers quantify the price to be paid for the advantages of customized non-optimum installations.

Keywords: BIPV, vertical installation, bifacial solar modules, colored solar modules, technology comparison

1 INTRODUCTION

The increasing demand for solar module installations as one renewable energy source has led to significant area need challenging social acceptance which is crucial for a successful energy transition from fossil to renewables. Currently 44% of the global PV capacity is made up of roof installations and 55% are field installations [1]. The expected ramp-up implies area conflicts and acceptance conflicts. Agri-PV and BIPV are growing topics in the PV field, since they are seen to mitigate those conflicts. Also, they could completely cover Europe's electricity demand when deploying those technologies at a large scale [2] [3]. The increasing installations also challenge grid management. Producing energy where it is consumed and under an increased variety of installation orientation, however, broadens the generation profile of these mixed PV installations, especially by the vertical installation in differently oriented facades, and supports morning and evening load requirements [4]. Undoubtedly, there will be more installations under non-optimum angles.

In BIPV applications, PV modules act as both electricity generating solar panels and as building elements. The latter function is the main property in the building skin and typically influences the electrical performance of the PV function. The performance of PV modules is influenced by various factors, mainly by the angles of incidence of sunlight, temperature and shading.

It is known that the non-optimum orientation comes at a cost, e.g. in respect to shallower angles of incidence of the incoming direct light and a resulting lower yield. Also, ventilation for module cooling is not a priority to architects. Sometimes a coloring is chosen which decreases the efficiency. The measurements under STC of a colored and a black module cannot directly be translated to an annual yield difference of a colored BIPV system – with all the contributions from non-perpendicular angles of incidence.

This paper quantifies the reduction in specific yield due to the choice of a black back sheet or a colored front glass and a vertical installation angle. By examining the modules under both optimal and facade orientation, this research seeks to provide insights into their practical application in architectural settings (and Agri-PV). The analysis is grounded in rigorous long-term outdoor exposure data, encompassing also a facade installation for comparison. It presents a comprehensive comparison of commercial PV modules of the below listed technologies oriented southward installed at Berlin, Germany at an optimum angle and under 90° as often the case in BIPV/Agri-PV. The chosen technologies are monofacial Silicon heterojunction modules with white and black backsheet, bifacial silicon heterojunction, PERC half-cells, shingled and IBC with white and black back shields as well as CIGS and CIGS with a blue coloring. Additionally, bifacial modules were oriented in 2024 facing east/west, standing upright (90° angle). The study aims to elucidate the variances in specific energy yield among these modules. A building integrated facade containing the blue colored CIGS modules delivers temperatures in the BIPV context in comparison to those measured in the open rack setup. The resulting reduction of 44% in May 2023 in Berlin by going vertical instead of optimal in a south facing installation is put into context. Installing bifacial modules vertically but east/west instead of south makes up for the difference to the optimal tilt in summer [5]. Black backsheets affect the performance under shallow incidence angles [6] [7] and their specific yield in case of the silicon heterojunction modules was 98% of that of the white backsheet module for that configuration. The blue-colored front glass had a bigger effect. Yet, as area is scarce and PV modules are cheap as well as energy generation in summer is vast, there will be more and more of such installations that do not aim for the highest yield but for the best solution in respect to societal acceptance, area use, etc. [3].

2 MATERIAL AND METHODS

In this contribution we evaluate the data of 9 types of modules installed in duplicate, one row vertically and one under optimum tilt, both rows facing south in Berlin, Germany (52.43°N,13.52°E). Data was acquired over more than one year, starting in Fall 2022.

Figure 1 a) is a picture of the installation. The nine modules installed in duplicate are from left to right:

- Passivated emitter rear cell, PERC, half-cell modules with white and black backsheets monofacial (*PERC HC wbs, PERC HC bbs*)
- Silicon heterojunction, SHJ, half-cell modules with black backsheet monofacial, bifacial, and, with white backsheet monofacial (*SHJ bbs, SHJ bifi, SHJ wbs*)
- Thin film CIGS blue-colored and a reference transparent double glass module monofacial (*CIGS blue, CIGS ref*)
- PERC shingled cell modules with black backsheet (*PERC sh bbs*)
- PERC interdigitated back contact, IBC, full cell module with white backsheet monofacial (*PERC IBC wbs*)

Figure 1 b) shows part of the setup under snowy conditions. The modules are kept in their maximum power point (MPP) individually with IV curves measured additionally every 10s. This data was complimented by continuous acquisition of module rear side temperatures by PT 1000 temperature sensors and irradiance measurements by silicon reference cells (Si-01TC/ IMT Technology) in the plane of array. Global and diffuse irradiance are measured in the horizontal plane by two Kipp and Zonen CMP 11 pyranometers using the shadow ring setup. The ambient temperature as well as wind speed, wind direction, humidity, air pressure and precipitation intensity are acquired by a weather station (from CLIMA SENSOR US/Adolf Thies GmbH & Co.). The latter equipment for the environmental data is connected to a DLx MET datalogger from Adolf Thies GmbH & Co. and records one-minute averaged data. The incident spectrum is recorded every 5 min by two EKO WISER spectroradiometers, tilted to 35°and 90°.

Fig. 1: Picture of the industrial module test setup in the roof laboratory in Berlin, Germany. All modules are MPP-tracked individually while rear side temperatures are measured as well as POA irradiance by monitoring cells. a) sunny day, b) after snow fall.

Figure 2 shows the two bifacial modules that were deinstalled by the end of 2023 and instead are mounted in 2024 in an East-West installation standing upright, one facing East with the optimized frontside and one facing West. Temperature sensors are placed so that they do not shade the active area.

Fig. 2: Picture of the installation of the bifacial modules facing east and west, respectively with the optimized side. The bifacial modules were taken off the open rack and installed as shown here in December 2023.

Figure 3 shows the facade integrated PV system consisting of the same blue CIGS modules. We use this system to compare the operating temperature of the open rack modules to the same modules installed in a facade. Facade installations are expected to result in higher temperatures. The ventilated curtain wall facade has areas of different air gaps between the modules and the underling insolation on the south facing side [9].

Fig. 3: Picture of the CIGS blue modules installed in the facade of the living lab for BIPV owned and monitored by Helmholtz Zentrum Berlin in Berlin, Germany. The modules are installed on the North, West and South facade. The system is equipped with string monitoring and sensors recording rear side temperatures as well as irradiance in the module plane. Weather data is acquired on the building roof.

3 RESULTS

An excerpt of the analysis of the MPP data of the eighteen PV modules are shown here. In 2023 all modules were exposed vertically and 35° tilted and oriented towards south. For a fair comparison we present here the specific yield. Note that we present data of single modules. More statistics would be needed to generalize the observations.

41st European Photovoltaic Solar Energy Conference and Exhibition

3.1 Technology comparison

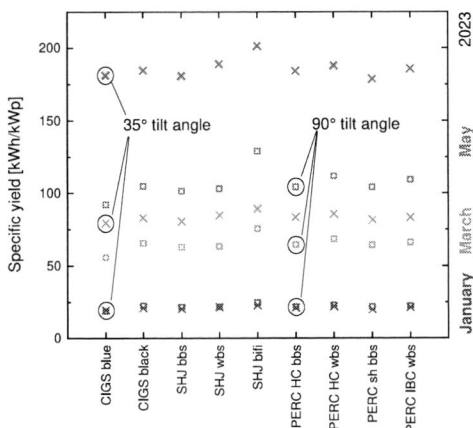

Fig. 4: Specific yield of all nine installed industrial modules as marked on the x-axis. Dark blue marks the specific yield for January 2023, light blue March 2023 and red May 2023. The crosses represent the data of the tilted raw (35°, south facing) while the squares represent the vertical installation (90°, south facing). The bifacial modules profit from the active rear side. Differences in specific yield are highest when there is more irradiance.

Figure 4 shows the measured yield divided by the respective nominal power of each module. For the peak power kWp values, in case of the CIGS modules the module name plate values were used. In all other cases, the STC power measured in specifically commissioned initial indoor characterizations were used.

As expected, the bifacial modules outperform all other modules under 35° and 90° south installation. In the indoor determination of the nominal power the rear side is not illuminated so that the denominator in calculating the specific yield is small. The bifaciality is not considered. Outdoors, the bifacial gain is highest among the shown data in May with more light reaching the rear side despite the suboptimal installation with a broad shadowing holder blocking some light at the rear side. The specific yield for all monofacial modules installed under 35° tilt is very similar with all modules with white backsheets having a

slightly higher specific yield than those with black backsheet or blue front glass.

The monofacial modules installed under 90° tilt have on average 44% less specific yield than those under 35° in May. In January when angles of incoming light are flatter, there is a slight gain of on average 4%. The loss is large and if modules were scarce and area abundant, it would seem unwise to install PV modules in a suboptimal way. Yet, as stated in the introduction, varying installation angles brings about many advantages in respect to the broadening of the generation profile. Installing PV vertically also saves area.

When analyzing Fig. 4, one has to bear in mind that the specific yield as an indicator may lead to an overinterpretation of the performance of different module types. All these modules/module groups have different sizes. The yield summed up over all available data for 2023 per module is highest for PERC IBC and the SHJ modules and CIGS with the smaller module sizes generates less than half as much energy. The yield per m² brings all data closer together. Still CIGS with its lower efficiency performs accordingly.

3.2 Clear sky and seasons

Figure 5 shows exemplarily for the SHJ modules (monofacial bbs squares, wbs circles and bifi as triangles) the quotient of the specific yield over a part of the exposure time. Note that the bifi modules were taken off and installed East/West-facing in 2024, data points end here in December 2023. A color code marks clear days according to a simple calculation of

$$clear\ sky\ index = 1 - \frac{DHI}{GHI}$$

With DHI the diffuse irradiance and GHI being the global irradiance measured by the meteo station on the same roof. Data loss is due to system outage.

The data show that in the winter month on clear sky days, the 90° modules perform better than the 35° oriented modules, the quotient is above 100 %. For diffuse conditions, the 35° module as it sees more of the sky, performs better as is the case also for the direct light under higher angles of incidence.

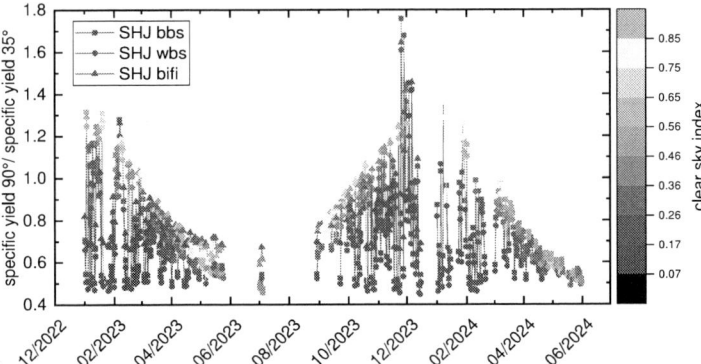

Fig. 4: Quotient of the specific yield of SHJ modules installed under 90° and 35° tilt. Circles mark wbs, squares mark bbs, triangles stand for the bifacial modules. The data is color coded by a simple clear sky index. The data point close to 01/12/2023 marks a snow event where the 35° module was partly covered with snow and the 90° module was free.

1350

3.3 Black and white backsheets

The modules rear temperature for the vertical installations was on average only a few degrees higher for the bbs modules than for the wbs counterpart during the reporting time Jan-May 2023 with no strong change over that period of time.

Fig. 6 shows that for the same time the specific yield of the SHJ bbs is lower than that of the SHJ wbs. In January, when the sun is low and the incidence angles of the light is perpendicular to the 90°-tilted module surface, the quotient of specific yield of the black by the white backsheet is close to 1. For diffuse light and for the direct light in summer the quotient is smaller. Obviously the black backsheet results in losses under shallow angles. (The data under 35° tilt confirm that as the quotient over time has an upward trend in that case – as incidence angles are more fortunate in summer for that tilt.)

3.4 Colored modules and module rear temperature in open rack and in facade installation

The specific yield of CIGS blue (135 Wp) compared to the reference module (140 Wp) of the same series but with transparent front glass is reduced by 12%, 15% and 17% in the month May, March and January under 90°, see Fig. 4. The reduction amounts to 2%, 5% and 8% under 35° tilt. Under STC the efficiency of the CIGS blue module is only 1 % less.

In facade installations next to the less fortunate angle orientation of the modules, higher temperatures occur that do result in further yield reduction depending in the module's temperature coefficient. Our setups allow us to compare the open rack module rear temperature to the temperature in the BIPV facade integration.

Figure 7 shows a histogram of the measured module rear side temperatures of the CIGS blue modules in open rack 90° and in the facade installation. The same color code as above is used to mark January, March and May data for the open rack modules while the facade module temperatures are respectively displayed in grey. Interestingly, the temperature difference is small in the winter month. In May we measure a difference of up to 10 K higher temperatures in the facade than on the open rack for the same type of modules, CIGS blue.

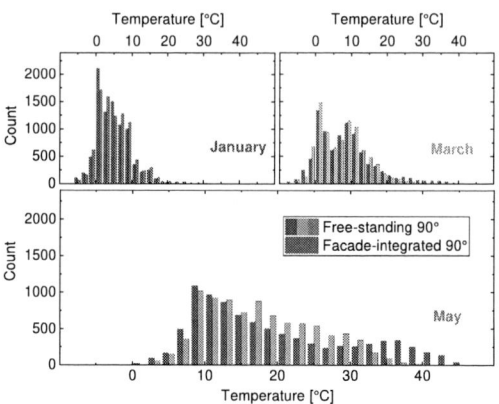

Fig. 7: Histogram of module rear side temperatures for the CIGS blue modules in the 90° open rack (color coded for a) January in blue, b) March in light blue and c) May in red) and the facade installation, respectively in gray for all months. Modules in the facade heat up more than in the open rack in May.

The reduction in specific yield due to the facade integration and color choice has to be put into context. Colored modules are used in the building environment where aesthetics matters more that energy yield [8]. They facilitate the entrance of PV into the market of building skin material that they replace and that would otherwise not produce any energy. The coloring opens opportunities for architects and foster acceptance. Another advantage of vertical modules is the reduced covering by snow, see Fig. 1b). Obviously, tilted open rack installations suffer more from snow shadowing than facades. In a study of a facade installation with the CIGS blue module, see Fig. 3, we found out that despite the lower yield generation, the module installation still amortizes well within the module lifetime [8].

3.5 Bifacial module in east/west configuration

Another typical occasion where vertical orientation is common are the east/west facing rows of bifacial modules in Agri-PV installations.

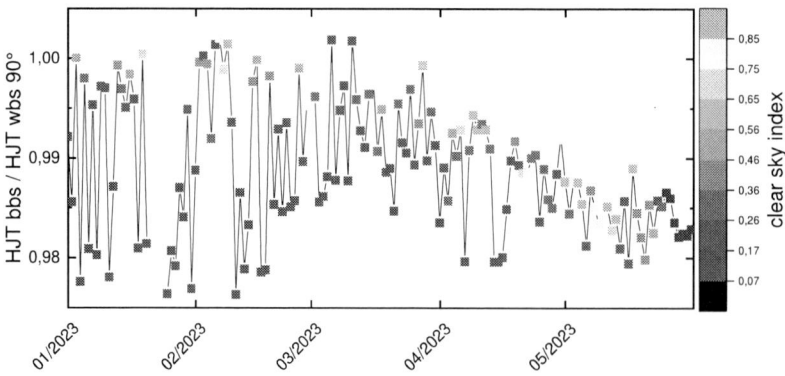

Fig. 6: Quotient of the specific yield of the SHJ bbs and wbs modules installed vertically. The color code is a simple clear sky index.

41st European Photovoltaic Solar Energy Conference and Exhibition

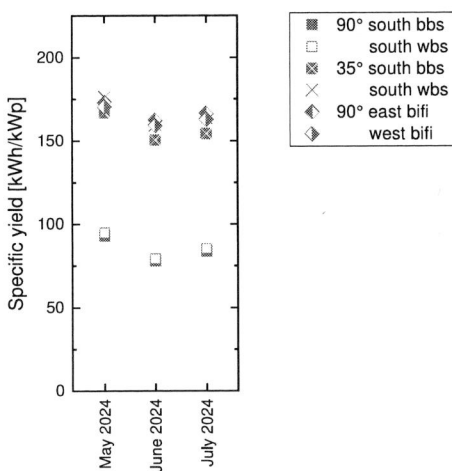

Fig. 5: Specific yield of the SHJ modules under 35° south facing with wbs and bbs compared to their 90° counterparts in May, June, July 2024. The diamonds mark the data of an east/west facing installation of the bifacial modules (optimized site oriented as marked in the legend). The bifaciality makes up for the disadvantage of the 90° tilt.

Figure 8 shows the data for three exemplary summer months in 2024, where the bifacial heterojunction modules were installed facing east/west as shown in the picture Fig. 2. For comparison the data of the vertically installed south-facing SHJ mono bbs (red filled square) and wbs (red open square) are shown. They are as much lower as in the data of 2023. The respective tilted modules (bbs red cross, wbs light red cross) perform as well as observed for May 2023. In June and July, sunshine hours were less than in May with hotter average temperature, the specific yield decreased slightly. The West and East facing modules SHJ bifi are marked by the diamonds with respective color code as given in the legend. It is expected that the specific yield for east facing is slightly higher than for West due to the favorable lower temperatures in the mornings (when sun shines on the east facing side). Additionally, we know that the module installed westwards is shaded in early mornings and late evenings. The data meet this expectation. For all the shown summer months May, June and July, the SHJ bifi modules installed under 90° facing east and the one facing west perform comparable to 35° SHJ bbs modules and SHJ wbs modules. The bifaciality makes up for the tilt loss under 90° in the summer months. This result is not unexpected and underlines the efforts of implementing vertical bifacial modules in east/west configuration in PV-fences and Agri-PV.

4   DISCUSSION

Some of the features shown in our measurements may come as no surprise, such as the outperforming of the bifacial module – even despite the wide bar that casts a shadow on the active rear. Bifacial modules are expected to dominate the market in future and the optimal tilt and

south facing orientation delivers the highest specific yield. An east/west orientation was found to equal out the disadvantages of the 90° tilt and result in specific yields as high as under 35° tilt/south for summer months. PV is abundant in summer at noon and this east-west orientation advantageously broadens generation profiles towards mornings and afternoons. Without the activated rear, the reduction in yield by vertical tilts compared to the optimum tilt is notable. We quantify it to be 44 % in May as an average over all monofacial modules installed, while there is a small gain of 4% in winter. This reduction is significant, the gain is small. Yet, the PV module orientations in building facade integrations are usually vertical and not necessarily south-facing. There is a huge potential in covering facades and the practical advantage of installing PV in facades is that there is no additional land covered while integrated PV even replaces building material. For aesthetics the modules used in the building context are often colored or at least all-black. This change comes at the cost of a lower yield that we figured 12 % in May for the CIGS module that has a blue front glass and 1% less efficiency under STC conditions. Also, the module temperatures are expected to rise in a facade installation compared to the open rack. We measured the temperatures on the module rear side and found them to vary little. Only in summer we found that there was a maximum of 10 K difference in the maximum temperature. The reduction in yield in this installation is significant. However, it is not a showstopper. Seen from the point of view of the builder, such installations amortize well in the lifetime of the solar modules in an appealing building. The low prices make non-optimal installations affordable, the enormous ramp-up of PV in the next years make these installations necessary. Detailed yield simulations for bifacial, colored and all-black modules under non-optimum angles and installation conditions will be a topic of future research and the here presented acquired data serves for validating models.

5   ACKNOWLEDGMENTS

5.1 Funding
Niklas Albinius and Björn Rau acknowledge funding by the German Federal Ministry of Education and Research (BMBF) for the Solar TAP innovation platform under the Helmholtz Innovation Platforms funding line. Maximillian Riedel acknowledges funding by the European partnering project TAPAS (PIE-0015). Carolin Ulbrich and Björn Rau acknowledge funding by the Helmholtz Association under the program "Energy System Design" under grant number ZT-0 0 02.

5.2 Conflict of interest
The authors certify that there are no financial conflicts of interest (e.g., consultancies, stock ownership, equity interest, patent/licensing arrangements, etc.) in connection with this article. Note that six of the 18 modules that we investigated here were provided by one company for free. To comply with the NDA and disguise the origin of the modules, we present here the specific yield.

5.3 Data availability statement
Some of the data presented here was acquired under an NDA. The other data used in this paper will be made available upon request.

1352

5.4 Author contribution statement

Conceptualization, C.U. and M.R.; Methodology, C.U. and M.R.; Software, L.W., N.A. and M.R.; Validation, C.U., M.R., N.A. and B.R.; Formal Analysis, C.U., M.R., N.A. and B.R.; Investigation, C.U., N.A., L.W., M.R., B.R.; Resources, M.R.; Data Curation, L.W.; Writing – Original Draft Preparation, C.U.; Writing – Review & Editing, C.U., N.A., B.R.; Visualization, C.U., M.R., L-W. and N.A.; Supervision, R.S.; Project Administration, C.U., B.R., R.S.; Funding Acquisition, C.U., B.R., R.S.

6 REFERENCES

[1] SolarPower Europe, "Global Market Outlook For Solar Power 2024-2028 - Focus on China," 2024.

[2] K. A. K. Niazi and M. Victoria, "Comparitive analysis of photovoltaic configurations for agrivoltaic systems in Europe," *Progress in Photovoltaics: Research and Applications,* vol. 31, pp. 1101-1113, 2023. https://doi.org/10.1002/pip.3727

[3] BIPV Boost, "BIPV market and stakeholder analysis," 2019.

[4] T. Baumann, H. Nussbaumer, M. Klenk, A. Dreisiebner, F. Carigiet and F. Baumgartner, "Photovoltaic systems with vertically mounted bifacial PV modules in combination with green roofs," *Solar Energy,* 2019. https://doi.org/10.1016/j.solener.2019.08.014

[5] L. Wang, Y. Tang, S. Zhang, F. Wang and J. Wang, "Energy yield analysis of different bifacial PV (photovoltaic) technologies: TOPCon, HJT, PERC in Hainan," *Solar Energy,* no. 238, 2022. https://doi.org/10.1016/j.solener.2022.03.038

[6] H. Lim, S. H. Cho, J. Moon, D. Y. Jun and S. H. Kim, "Effects of Reflectance of Backsheets and Spacing between Cells on Photovoltaic Modules," *Appl. Sci.,* vol. 12, no. 443, 2022. https://doi.org/10.3390/app12010443

[7] G. Makrides, M. Theristis, J. Bratcher, J. Pratt and G. E. Georghiou, "Five-year performance and reliability analysis of monocrystalline photovoltaic modules with different backsheet materials," *Solar Energy,* no. 171, 2018. https://doi.org/10.1016/jsolener.2018.06.110

[8] M. C. López-Escalante, E. Navarrete-Astorga, M. Gabáz Perez and J. R. Ramos-Barrado, "Photovoltaic modules designed for architectural integration without negative performance consequences," *Applied Energy,* no. 279, 2020. https://doi.org/10.1016./j.apenergy.2020.115741

[9] N. Albinius, C. Ulbrich, B. Rau, M. Riedel and R. Schlatmann, "A comprehensive case study of a full-size BIPV facade," *Mauscript submitted,* 2024.

41st European Photovoltaic Solar Energy Conference and Exhibition

PHOTOVOLTAIC OUTPUT POWER MODELING: A HYBRID APPROACH

Santos, Leticia de Oliveira[1]*, Souza, Francisco A. A.[2], AlSkaif, Tarek [3], Carvalho, Paulo C. M.[1]
[1]Federal University of Ceará, Graduate Program of Electrical Engineering, Fortaleza, CE, Brazil
[2]OnePlanet Research Center, imec-NL, Wageningen, The Netherlands
[3]Information Technology Group at Wageningen University, Wageningen, The Netherlands.
*leticia@fisica.ufc.br

ABSTRACT: This study presents a physics-based model aimed at estimating the thermal-electrical performance of photovoltaic (PV) systems. The hybrid model integrates fundamental physical principles with empirical data, thereby improving reliability while reducing data requirements. The physics equations are derived from a thermoelectric dynamic model of the PV module, encompassing reflection losses, module temperature determination, PV module efficiency calculation, and estimation of systematic losses. In addition, Bayesian inference is employed to mitigate uncertainties associated with physical parameters values. Numerical simulations demonstrate the model's ability to accurately describe both module operating temperature and power output, accounting for the temperature effect on PV conversion efficiency. Error analysis underscores its superior accuracy compared to traditional physical modeling approach. This research contributes to advancing PV power modeling by offering a robust and reliable methodology combining physics-based and data-driven techniques.
Keywords: Photovoltaic models, PV power estimation, PV temperature estimation, hybrid modelling.

1 INTRODUCTION

Accurately estimating photovoltaic (PV) power is essential for effective design and operation of PV systems, as inaccuracies can directly impact the electric system, and financial outcomes of PV investments [1]. These analytics methods for PV power estimation generally fall into three categories: physical, data-driven, and hybrid approaches.

For power prediction concerning new PV systems and day-ahead or longer time horizons, physical models are prevalent and widely accepted, being applied even for real-time forecasting and performance analysis [2]. These models require only PV plant information and no historical data. Conversely, data-driven models employ machine learning algorithms to establish relationships between weather variables and the PV power. These models usually excel in effectively capturing intra-hour power fluctuations. Meanwhile, hybrid models integrating both physical principles and data-driven approaches, enhance reliability while reducing data requirements, and minimizing regional biases [3].

In our study, we present a physics-based model for estimating the thermal-electrical performance of PV systems, grounded in the PV system's energy balance. The physical model is optimized based on PV module data, yielding PV module operating temperature (Tm) and output power (Pout). Initially, the model estimates Tm and then predicts Pout, considering the temperature effect during irradiance-to-power conversion [4]. Equations encompasses comprehensive thermoelectric dynamics of the PV module, including reflection losses, cell temperature determination, PV module efficiency calculation, and systematic losses estimation, such as those caused by the inverter. As our model is transient, i.e. tracking PV temperature changes over time, it adeptly responds to short-term variations in irradiance, ambient temperature, and wind speed. Moreover, Bayesian inference is employed to refine the physical model parameters; this data-driven approach captures patterns beyond the physical model's scope due to physical approximations. One of the main advantages of our method is the mitigation of uncertainties related to the PV physical model parameter values, a drawback of physical models.

Our approach represents a novel contribution in PV power modeling, featuring a physics-oriented data-driven framework. This hybrid model integrates detailed electro-thermal modeling accommodating transient effects, with varying environmental parameters (solar irradiance, ambient temperature, and wind speed). In summary, our PV power estimation model, bridging physical insights with data-driven technique, presents robustness, reliability and enhanced accuracy compared to traditional physical modeling approach.

The remainder of this article is structured as follows: Section 2 presents the physical model for PV power estimation, as well as the utilization of Bayesian inference for determining the model's parameters. Section 3 presents the results, comparing the estimated values with the measurements and the benchmark model. Finally, Section 4 presents the conclusions.

2 HYBRID MODEL

The hybrid model described here and applied for thermal-electrical PV performance estimation relies in part on the thermal model described in our previous work in [5]. Here the model is expanded to produce double output Tm and Pout. Additionally, the model is adapted to incorporate systematic losses. The physical assumptions and model description are described in Section 2.1. The parameters optimization approach and model solution are described in Section 2.2.

2.1 Physical model

The proposed model is tailored for poly-Si modules operating under normal conditions, devoid of shading, dust, or any other agents affecting absorptivity. The PV module is treated as a homogeneous block of materials with average thermal characteristics and temperature. We assume that the fraction of solar irradiance absorbed by the PV cell, not converted into electricity, is transferred through radiation and convection heat loss from the PV module to the environment. Conduction in the PV module layers is disregarded, with layer properties factored into the module's heat capacity (Cm). Additionally, we consider the system operating under a time-dependent regime, resulting on the energy balance equation described by Eq. 1 and represented by Fig. 1.

$$C_m \frac{dT_m}{dt} = q_s - q_{conv} - q_{rad} - P_{out} \qquad (1)$$

Here, $\frac{dT_m}{dt}$ denotes the time derivative of the module temperature (Tm). The components of Eq. (1) are delineated as follows:

$$q_s = I_{POA} A_m \alpha_s \qquad (2)$$
$$q_{conv} = h A_m (T_m - T_a) \qquad (3)$$
$$q_{rad} = \sigma A_m (\epsilon_p T_m^4 - F_{t,sky}\epsilon_{sky}T_{sky}^4 - F_{b,roof}\epsilon_r T_a^4) \qquad (4)$$
$$P_{out} = \mu G A_m \eta_{stc}[1 - \beta(T_m - T_{stc})] \qquad (5)$$

Here, q_s symbolizes the solar irradiance absorbed by the module, dependent on the irradiance on the array plane (I_{POA}), module area (A_m), and absorptance (α_s). The components q_{conv} and q_{rad} represent heat loss by convection and radiation, respectively. q_{conv} tends to increase as Tm rises above the ambient temperature (Ta) and incorporates the convective heat loss coefficient (h). We represent h as a linear function of wind speed (Ws) – Eq. (6), where a and b are empirically determined terms. q_{rad}, described according with the Stefan-Boltzmann law, includes the Stefan-Boltzmann constant (σ), PV module emissivity (ϵ_p) to the surrounds, sky and roof emissivity (ϵ_{sky} and ϵ_r) to the PV module, and $F_{t,sky}$ and $F_{b,roof}$, view factors from the top of the module to the sky and from the PV module back-side to the roof. The view factors are represented by Eq. (7) and Eq. (8). $Tsky$ is the sky temperature estimated by Eq. (9). Finally, Pout in Eq. 5 denotes PV module output power, considering temperature effect on the module efficiency (η), which takes into account the temperature coefficient of the PV module (β), and module efficiency and temperature at Standard Test Conditions (STC), Tstc equal to 25°C [6]. The term μ is included to compensate for systematic losses, such as those caused by the inverter.

$$h = a Ws + b \qquad (6)$$
$$F_{t,sky} = \frac{1 + cos\theta}{2} \qquad (7)$$
$$F_{b,roof} = \frac{1 - cos(\pi - \theta)}{2} \qquad (8)$$
$$T_{sky} = 0.0552 T_a^{1.5} \qquad (9)$$

2.2 Parameters optimization and model solution

Certain parameters from Eq. (1) to (5), such as Am, β, ηstc, and the view factors that depend on θ, are well-known, determined by module specifications or installation configuration. Conversely, values for Cm, αs, ϵ_p, ϵ_{sky}, ϵ_r, μ and the terms a and b from h, here referred as unknown parameters, are typically assumed from scientific literature. Consequently, uncertainties may arise due to potential variations in the physical properties of the PV module materials compared to literature values. Hence, our primary objective is to infer optimal values for the unknown parameters.

The parameters fitting is achieved through Bayesian Inference with the No-U-Turn Sampler, an extension of the Markov Chain Monte Carlo (MCMC) algorithm. This technique enables parameter fitting without the need for tuning parameters, sampling from correlated parameter distributions. The approach involves defining a prior probability distribution to represent parameter value

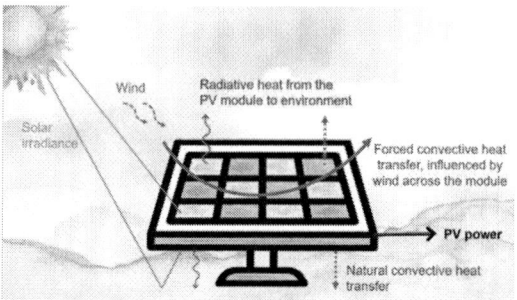

Figure 1: Representation of a free-standing PV system energy balance.

uncertainty. Gaussian, half-Gaussian, and uniform distributions are adopted for priors, utilizing reference values from literature, expert knowledge, and solar panel manufacturer datasheets. The unknown parameters and constant values as well as the distributions adopted for fitting the model are presented in Table I.

Table I: Parameter values of the thermoelectrical model

	Type	Value	Distribution
A_m	Constant	0.94 x 1.9 m²	x
η_{stc}	Constant	17%	x
β	Constant	0.42%/°C	x
θ	Constant	10°	x
σ	Constant	$5.67 \cdot 10^{-8}$ Wm^{-2}K^{-4}	x
C_m	Unknown parameter	$25\,000 \pm 10^4$ J/K	◢
a	Unknown parameter	$a > 0$	◣
b	Unknown parameter	$b > 0$	◣
α_s	Unknown parameter	$0.7 - 0.97$	▥
ϵ_p	Unknown parameter	$0.85 - 0.98$	▥
ϵ_{sky}	Unknown parameter	$0.85 - 1$	▥
ϵ_r	Unknown parameter	$0.6 - 0.9$	▥
μ	Unknown parameter	$0.7 - 0.99$	▥

To numerically solve Eq. (1), incorporating instantaneous values of I_{POA}, Ta, and Ws from the dataset, we employ the Euler scheme to discretize the time derivative. The current state is updated from the previous state based on dynamic equations using the data to account for inertia effect. For discrete time steps $dt \rightarrow \Delta t$, based on Eq. (1) and the set of priors for the unknown parameters, a likelihood is defined as Eq. (10), where τ denotes the scale parameter of the normal distribution.

$$T_{m,n+1} \sim normal(T_{m,n} + C_m^{-1} \Delta t (q_s - q_{conv} - q_{rad} - P_{out}, \tau) \qquad (10)$$

From the likelihood we estimate the unknown parameters based on data from a measurement campaign in Fortaleza, Brazil (3°44'S, 38°34'W) from December 1st, 2022 – June 8th, 2023. The 'best-fit' values are used as parameters of Eq. (1) to generate double output, Tm and Pout.

3   RESULTS

The model is tested using data from June 9th – November 7th, 2023. In Fig. 2 is shown a sample of Tm estimations compared with the reference values obtained from the temperature sensor. Furthermore, Pout was estimated based on Tm by Eq. (5) and compared with measured PV plant power during the same period, as illustrated in Fig. 3.

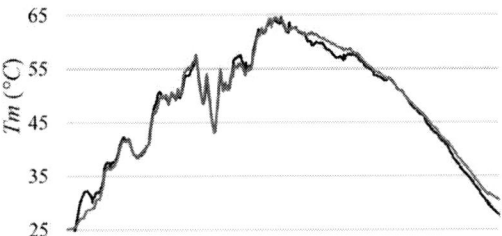

Figure 2: Sample of module temperature measured (black) and estimated (blue) by our hybrid model

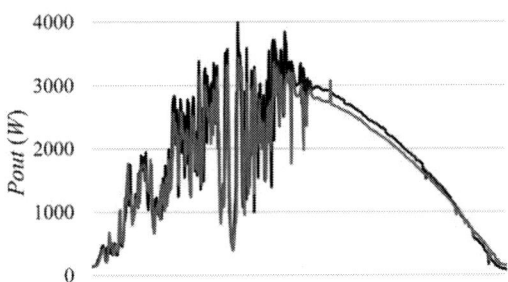

Figure 3: Sample of PV power measured (black) and estimated (blue) by our hybrid model

To quantify the advantages of implementing the data-driven approach, we compare the model with the fitted parameters to the physical model using manufacturer values if available or the most common values found in the literature for the unknown parameters: $C_m = 22{,}222$ J/K, $\alpha_s = 0.97$, $\epsilon_p = 0.9$, $\epsilon_{sky} = 1$, $\epsilon_r = 0.6$, a = 0.01, b = 25, $\mu = 0.98$. The error metrics of both models for Tm and $Pout$ estimation are tabulated in Table II.

Table II: Error metrics for the estimated results

	Tm (°C)		$Pout$ (kW/kWp)[1]	
	MAE	MBE	MAE	MBE
Only Physical	4.31	4.25	0.085	-0.053
Physical Optimized	1.56	0.73	0.059	-0.002

[1]The unit kW/kWp has been used to show the error in relation to the PV plant power.

The proposed model shows superior performance in terms of mean average error (MAE), indicating an advantage in implementing the hybrid modeling approach. The mean bias error (MBE) is positive for Tm estimation, suggesting that the model tends to overestimate the PV module operating temperature. Given that increasing operating temperature leads to reduced efficiency, the model underestimates $Pout$, as evidenced by Table II. However, the model shows good performance in terms of estimating Tm and $Pout$.

4 CONCLUSION

This study introduces a physics-based model for estimating the thermal-electrical performance of PV systems, integrating the robustness of physics-based approaches with the adaptability of data-driven techniques. The hybrid approach incorporates PV module thermoelectric dynamics, and effectively accounts for reflection losses, module temperature, PV module efficiency, and systematic losses. Additionally, empirical data is used to mitigate uncertainties associated with physical parameter values through Bayesian inference. In comparison to purely data-driven or physical models, our approach improves the model's reliability, interpretability and generalizability while reducing data requirements. Numerical simulations confirm the model's capability to accurately describe the module's operating temperature and power output. Error analysis reveals that the hybrid model outperforms the traditional physical model, demonstrating superior accuracy.

5 AKNOWLEGMENTS

This work was supported through research grants by *Fundação Cearense de Apoio ao Desenvolvimento Científico e Tecnológico* (FUNCAP), *Coordenação de Aperfeiçoamento de Pessoal de Nível Superior* (CAPES), and *Conselho Nacional de Desenvolvimento Científico e Tecnológico* (CNPq) - Brazil.

6 REFERENCES

[1] D. Xiao-Jian, et al. "Simultaneous operating temperature and output power prediction method for photovoltaic modules," in Energy 260 (2022): 124909, doi: 10.1016/j.energy.2022.124909.
[2] K. Kamono and Y. Ueda, "Real-Time Estimation of Areal Photovoltaic Power Using Weather and Power Flow Data," in IEEE Transactions on Sustainable Energy, vol. 9, no. 2, pp. 754-762, April 2018, doi: 10.1109/TSTE.2017.2760012.
[3] M. J. Mayer and G. Gróf, "Extensive comparison of physical models for photovoltaic power forecasting," in Applied Energy, Volume 283, 2021, 116239, doi: 10.1016/j.apenergy.2020.116239.
[4] L. d. O. Santos, P. C. M. de Carvalho and C. d. O. C. Filho, "Photovoltaic Cell Operating Temperature Models: A Review of Correlations and Parameters," in IEEE Journal of Photovoltaics, vol. 12, no. 1, pp. 179-190, Jan. 2022, doi: 10.1109/JPHOTOV.2021.3113156.
[5] L. O. Santos, F. A. A. Souza, C. O. Carvalho Filho, P. C. M. Carvalho, T. AlSkaif and R. I. S. Pereira, "Hybrid Modeling for Photovoltaic Module Operating Temperature Estimation," in IEEE Journal of Photovoltaics, vol. 14, no. 3, pp. 488-496, May 2024, doi: 10.1109/JPHOTOV.2024.3372328.
[6] A. Driesse, M. Theristis and J. S. Stein, "A New Photovoltaic Module Efficiency Model for Energy Prediction and Rating," in IEEE Journal of Photovoltaics, vol. 11, no. 2, pp. 527-534, March 2021, doi: 10.1109/JPHOTOV.2020.3045677.

41st European Photovoltaic Solar Energy Conference and Exhibition

ESTIMATION OF ANNUAL POWER LOSS OF A SOLAR PV SYSTEM DUE TO THE RISE IN THE CELL TEMPERATURE: A CASE STUDY FOR INDIAN CLIMATE

Shubham Kumar, P. M. V. Subbarao
Indian Institute of Technology Delhi
Hauz Khas, New Delhi, India - 110016

A rise in the cell temperature may cause a significant decrease in the real field power output of solar PV systems, especially in hot tropical climates such as India. A more accurate forecast of annual output is critical in the planning phase of any solar PV system or plant. The present study implements an experimentally validated 1-D transient thermal model to predict the cell temperature of PV systems. Along with the thermal model, the meteorological data of the last three years, various statistical tools, and an empirical function for PV efficiency are used to develop a framework for forecasting the PV output. When applied to a 2.72 kW poly-crystalline silicon PV array in New Delhi, the annual output is estimated to be 9.65 MWh which is 9.9% less than the forecasted ideal output. This difference is found to be 13.9% in the hot month of April and 2.8% in the colder month of January. Thus, the effect of cell temperature on the output of a PV system is immense and must be incorporated in predicting its real field output. This will enable the stakeholders to make more informed decisions regarding the installation site, PV technology, and requirement of cooling.
Keywords: Temperature coefficient, forecast, cell temperature, power output.

1 INTRODUCTION

The actual efficiency and power of solar photovoltaic (PV) systems often fall below their rated values due to various factors such as dust deposition, low solar irradiation, potential induced degradation, etc. The rise in cell temperature of PV systems is one of such factors. Solar PV systems can only utilize a fraction of the input solar energy to generate electricity. Most of the remaining energy gets converted into heat within the device, leading to a rise in the cell temperature. The cell temperature can be significantly higher than standard testing conditions of 25°C and may exceed even 70°C in hot climates [1,2].

It is well established that an increase in the cell temperature leads to a reduction in the efficiency (η) of PV systems. The relative rate of decrease in PV efficiency with temperature is known as the temperature coefficient (β) and is dependent on the cell material and PV technology. The value of 'β' for typical crystalline silicon (c-Si) PV systems is -0.45%°C^{-1} [3]. This means that if the cell temperature reaches 65°C, the PV system will produce 18% less electricity than its rated value due to the higher cell temperature. Several studies have reported 'β' to be as low as -0.78%°C^{-1} for commercial-grade c-Si panels [4]. Even for advanced PV technology such as silicon heterojunction cells, the value of 'β' is approximately -0.3%°C^{-1} [5].

Installation of a solar PV plant is often preceded by a forecast of annual electrical yield so that the LCOE and possible economic gains can be estimated. To perform that task with greater accuracy, it may be vital to incorporate the effect of cell temperature on the actual output power. This becomes even more critical in hot tropical countries like India where PV plants are being installed at an appreciable rate and the cell temperature is expected to be significantly higher than 25°C for most of the year. This study aims to develop a scientific methodology to estimate the actual annual electrical energy obtainable from a particular PV system and presents an example case study for the geographical climate of New Delhi.

2 METHODOLOGY

2.1 Assumption of input variables

To estimate the actual PV efficiency and output power at any instant, we first need to estimate the instantaneous cell temperature of PV systems. For this, the required input variables are the instantaneous solar intensity, wind speed, and ambient temperature.

Solar irradiation (I) is taken as a sum of direct, diffuse and reflected components, as given in Equation (1) and (2) [6]. Here, I_T is the extra-terrestrial solar intensity.

$$I = k_T \times \left(R_D + \frac{k_D (1 + cos\theta)}{2} - k_D R_D \right. \tag{1}$$
$$\left. + \frac{r(1 - cos\theta)}{2} \right) \times I_T$$

$$I_T = G_{sc} \times \left(1 + 0.033 \cos\left(\frac{2\pi D}{365}\right)\right) \tag{2}$$
$$\times (cos\delta \cos(\phi$$
$$- \theta) cos\omega$$
$$+ sin\delta \sin(\phi - \theta))$$

The sky clearness index (k_T) of the last three years has been taken and their month-wise average values have been used for the forecasting. The diffusion constant (k_D) is estimated using equation (3) [7].

$$k_D = 0.69 - 0.47 \, k_T \tag{3}$$

The ambient temperature of a day has been modelled as a degree 4 polynomial, as given in Equation (4).

$$(T_o - T_{sr})/(T_{max} - T_{sr}) \tag{4}$$
$$= \alpha \left(\frac{t}{t_D}\right) + b \left(\frac{t}{t_D}\right)^2$$
$$+ c \left(\frac{t}{t_D}\right)^3 + d \left(\frac{t}{t_D}\right)^4$$

Where t_D is the duration of sunshine on a day. Maximum temperature of a day (T_{max}) and Temperature at sunrise (T_{sr}) are taken as month-wise averages of the last 3 years' data. Using the regression method, we obtained a=0.6667, b=0.8889, c=2.6667 and d=3.5556. These average statistical calculations should neutralize the minor fluctuations in the weather conditions over a particular month and yield fairly accurate estimates of total annual output.

Instantaneous wind speed (V) is randomized in 1-minute intervals using a Weibull distribution function, as given in Equation (5) [8].

$$f(V) = \frac{k}{s}\left(\frac{v}{s}\right)^{k-1} \exp\left(-\left(\frac{v}{s}\right)^k\right) \qquad (5)$$

The random wind speed distribution will ensure that higher wind speeds are not taken during a particular time interval of the day, and thus, the numerical framework does not produce any bias in the estimation of total output.

2.2 Framework to estimate PV output.

A 1-D transient thermal model has been developed to measure the instantaneous cell temperature and then the instantaneous output of the PV system. Figure 1 depicts the flow chart of this framework.

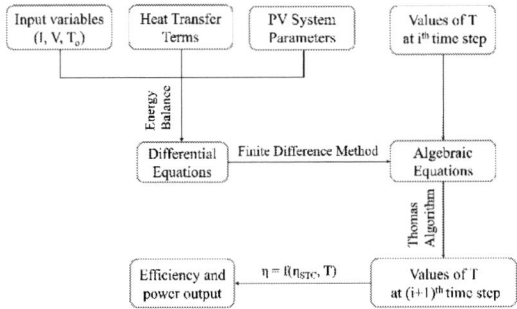

Figure 1: Flow chart depicting the framework to obtain the instantaneous cell temperature and PV output.

The model assumes the PV system to be made up of three domains. The front domain consists of glass and front encapsulant, the cell domain assumes a continuous cell layer and the back domain consists of back encapsulant and back sheet layers. Using the principle of energy conservation, differential equations are written for each domain which involve heat conduction within the PV system and convective and radiative heat transfer between the PV system and surroundings. These differential equations are converted into algebraic equations using the fundamentals of the finite difference method and then solved using the Thomas algorithm to obtain instantaneous cell temperature. The implemented calculation method uses explicit radiation terms and implicit convection terms to linearize the equations and eliminate the need for an iterative process. This numerical model has been validated with the experimental data with a mean error of only 3.9% and 0.95°C [9].

Then the empirical relation between PV efficiency and cell temperature, given in Equation (6), is used to find the instantaneous power output. Adding all the power outputs over a year yields the annual electrical output of the PV system. More elaborate details of this framework can be found in the author's previous work [9]. The overall model can be applied to any PV technology, system size, and geographical location after suitably changing the relevant parameters such as temperature coefficient, standard efficiency, PV panel dimensions, tilt angle, latitude of location, etc.

$$\eta = \eta_{STC}(1 + \beta(T - T_{STC})) \qquad (6)$$

2.3 Details of the studied PV system

The above framework is implemented for a solar PV array consisting of 16 poly-crystalline PV modules (2 m x 1 m) of capacity 340 W each tilted at 23° with horizontal in New Delhi, India. The tilt angle has been chosen to get maximum solar irradiation throughout the year according to the latitude of the geographical location. For Delhi, the average wind speed is 1.54 m/s. Also, k = 1.47, s = 1.62 and the percentage of calm hours is 30% [8].

Some important parameters for the studied solar PV array are reported in Table I. General parameters of the PV system such as the thickness of the PV layers, heat capacity, thermal conductivity, etc. can be found in [9].

Table I: Parameters of the studied PV system

Parameter	Value
Emissivity (ε) of PV surfaces	0.85
Efficiency at STC (η_{STC})	0.17
Temperature coefficient (β)	-0.45%°C^{-1}
Optical loss factor	14%

Month-wise taken values of k_T, T_{sr} and T_{max} for the geographical location of New Delhi are reported in Table II.

Table II: Month-wise values of k_T, T_{sr} and T_{max}

Month	Sky clearness index (k_T)	Temperature at sunrise (T_{sr})	Maximum temperature (T_{max})
January	0.496	10°C	18°C
February	0.576	15°C	26°C
March	0.618	22°C	33°C
April	0.635	26°C	39°C
May	0.583	27°C	40°C
June	0.54	29°C	41°C
July	0.429	31°C	38°C
August	0.433	30°C	36°C
September	0.501	29°C	34°C
October	0.592	24°C	32°C
November	0.521	17°C	27°C
December	0.508	14°C	21°C

3 RESULTS AND DISCUSSIONS

3.1 Forecast of PV Output

When PV efficiency is assumed to be always constant and equal to its rated value at STC, then the annual output of the studied PV array is forecasted to be 10.71 MWh. However, after incorporating the effect of cell temperature on actual yield using Equation (6), the forecast comes down to 9.65 MWh which is 9.9% lower than the ideal output.

Apart from the annual output, the effect of cell temperature on the monthly output should be analysed to comprehend the season-wise effect of the cell temperature on PV output. The month-wise forecasted ideal and actual yield are shown in Figure 2 and Table III. We may observe that the monthly PV output depends strongly on the sky clearness index, whereas the difference between ideal and actual outputs depends maximum on the ambient temperature during a month.

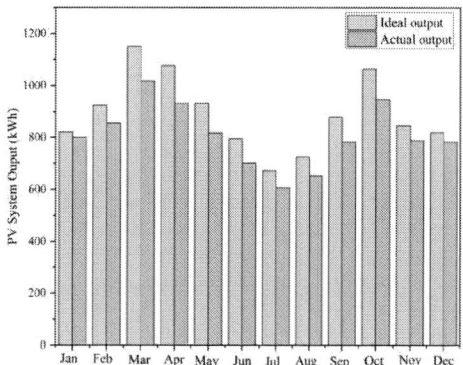

Figure 2: Month-wise forecasted electrical output of the studied PV system.

Table III: Forecast of monthly outputs

Month	Ideal output (kWh)	Actual output (kWh)	% Diff
January	821.6	796.7	3.0%
February	925.1	852.5	7.8%
March	1150.8	1012.9	12.0%
April	1076.9	927.7	13.9%
May	932.9	815.2	12.6%
June	794.7	700.4	11.9%
July	672.7	605.1	10.0%
August	725.8	652.3	10.1%
September	879.3	785.7	11.2%
October	1063.9	943.1	11.4%
November	846.3	785.7	7.2%
December	819	781.2	4.6%

The difference is above 11% in the hot season from March to June. However, the difference is only 3% and 4.6% in the cold months of January and December. This is mainly because of the ambient temperatures in these seasons. Since the cell temperature is usually 15-25 degrees above the ambient temperature, the cell temperature is only a little over STC in the winter season. On the other hand, the cell temperature crosses 60°C on most days in the summer because the ambient temperature is often above 35°-40°C in this season in New Delhi. This means that the effect of temperature on annual output can be ignored with lesser repercussions in cold climates with low ambient temperatures throughout the year. However, incorporating the effect of cell temperature becomes necessary if the annual output of a PV system is being forecasted in hotter conditions such as in India.

3.2 Advantages of the study.

Having a reliable estimate of the power loss a PV system may suffer due to a rise in cell temperature may help in deciding about the preferable PV technology (due to the value of β), choice of site of installation, number of PV modules required, need of cooling methods, etc.

For instance, if we install a PV array of the same standard efficiency but a better temperature coefficient i.e. $\beta = -0.3\%°C^{-1}$, the annual PV output estimated by the numerical model is 10.01 MWh. This is 3.7% higher than the annual output corresponding to $\beta = -0.3\%°C^{-1}$. Thus, replacing the basic poly-crystalline silicon PV systems with more advanced PV technology such as Silicon hetero-junction (SHJ) PV modules will produce a significant enhancement in PV output. The standard PV efficiency at STC and the cost of PV modules are already two key factors in deciding the PV technology to be used in a solar PV system. The above study illustrates the need to incorporate the temperature coefficients of available PV technologies before making the final decision.

Another important decision to be taken may be regarding the need and economic viability of integrating a cooling method with the solar PV system. For instance, if a cooling method claims to double the backside convection coefficient for all wind conditions, the numerical model forecasts the annual output to be 9.76 MWh. This gives a 1.14% enhancement in the annual output. Since the benefits of any cooling method depend significantly on the intensity and wind speed, any improvement claimed only at high solar intensity can be misleading. However, the improvement in annual output is a reliable criterion and can be weighed against its cost to comprehend its viability in the real field.

3.3 Sensitivity analysis

Multiple factors can affect the annual PV output but are often not predictable or controllable with very high precision. Thus, during the planning phase of the installation of the PV system, stakeholders may be interested in knowing how critical the unpredictability of these factors can be. For example, the average wind speed of a particular year can be slightly higher or lower than the previous few years. Similarly, the tilt angle during installation may be 1 or 2 degrees different from the planned value. Table III shows the change in annual output if any of the input parameters/variables changes by a given value.

Table IV: Sensitivity analysis of total annual PV output

Scenario	Change in the forecasted annual PV output
Tilt angle 25° instead of 23°.	Decreases by 0.07% only
Tilt angle 21° instead of 23°.	Decreases by 0.13% only
The average wind speed is 20% more than assumed.	Increases by 0.12% only
The average ambient temperature is 1°C higher than predicted.	Decreases by 0.44%

3.4 Future scope of improvement

The following improvements can be attempted to improve the reliability of these types of numerical studies.

- In the present numerical study, the month-wise wind speed has not been considered. Since winds depend strongly on the season, different average wind speeds for different months will provide more accurate results of the PV output.
- Past data of a greater number of years can be considered, probably with a suitable weighted mean.
- The height and local conditions of solar PV system can be considered to improve the accuracy in the assumed wind speeds. For instance, wind speeds over a multi-storied roof will be very different from those on the ground due to the big height difference.

4 Conclusion

The present study implements an experimentally validated 1-D numerical model to estimate the cell temperature and power output of a solar PV system to forecast the actual annual output of the PV system. The calculations show that ignoring the effect of cell temperature on the PV output may lead to a significant overestimation of up to 10% in hot tropical climates. Such a study can be applied to make important decisions such as the choice of a particular PV module having a given temperature coefficient, viability of cooling methods, location of installation based on local wind speeds, etc. This would eventually lead to better planning before the installation of any big PV system and eventually reduce the LCOE of PV power, thus making renewable energy more competitive and reliable.

References

1. Peng Z, Herfatmanesh MR, Liu Y (2017) Cooled solar pv panels for output energy efficiency optimisation. Energy Conversion and Management 150:949–955.
2. Marini´c-Kragi´c I, Nižeti´c S, Grubiši´c-Cabo F et al (2020) Analysis ˘ and optimization of passive cooling approach for free-standing photovoltaic panel: Introduction of slits. Energy Conversion and Management 204(112):277.
3. Dupré O, Vaillon R, GreenMA (2017) Thermal behavior of photovoltaic devices. Phys Eng 10:978–3
4. Singh P, Ravindra NM (2012) Temperature dependence of solar cell performance - an analysis. Solar Energy Materials and Solar Cells 101:36–45.
5. Abdallah, A., Martinez, D., Figgis, B., & El Daif, O. (2016). Performance of Silicon Heterojunction Photovoltaic modules in Qatar climatic conditions. *Renewable Energy*, *97*, 860-865.
6. Alsadi, S. Y., & Nassar, Y. F. (2017). Estimation of solar irradiance on solar fields: an analytical approach and experimental results. IEEE transactions on sustainable energy, *8*(4), 1601-1608.
7. Jamil, B., & Akhtar, N. (2017). Comparative analysis of diffuse solar radiation models based on sky-clearness index and sunshine period for humid-subtropical climatic region of India: A case study. Renewable and Sustainable Energy Reviews, *78*, 329-355.
8. Sarkar, A., Gugliani, G., & Deep, S. (2017). Weibull model for wind speed data analysis of different locations in India. KSCE Journal of Civil engineering, 21, 2764-2776.
9. Kumar, S., & Subbarao, P. M. V. (2023). An improved numerical model to predict the operating temperature and efficiency of solar photovoltaic systems. Environmental Science and Pollution Research, 1-14.

41st European Photovoltaic Solar Energy Conference and Exhibition

This presentation was selected by the Sc. Committee of the EU PVSEC 2024 for submission of a full paper to one of the EU PVSEC's collaborating peer-reviewed journals.

SNOW LOSSES FOR DIFFERENT PV MODULE DESIGNS: MODELLING AND VALIDATION IN SOUTHERN FINLAND

Shuo Wang[1*], Hugo E. Huerta[1], Sami Jouttijärvi[2], Aleksi Heinonen[1], Juha A. Karhu[3], Anders V. Lindfors[3], Kati Miettunen[2], Samuli Ranta[1]

1. New Energy Research Group, Turku University of Applied Sciences, Joukahaisenkatu 7, 20520 Turku, Finland
2. Department of Mechanical and Materials Engineering, University of Turku, Vesilinnantie 5, 20500 Turku, Finland
3. Finnish Meteorological Institute, Erik Palménin aukio 1, FI-00560 Helsinki, Finland

*Corresponding author: shuo.wang@turkuamk.fi

ABSTRACT: As PV systems are becoming more common in regions with significant snow cover during winter, improving the accuracy of the snow loss prediction and monitoring is crucial. In this work, we propose a novel hybrid snow loss model that differentiates between snow loss discrepancies in full-cell and half-cell modules. Snow coverage on PV modules is evaluated using an image processing algorithm, and the interconnection between PV cells and modules is incorporated through a cell-level electrical model. Using these methods, we performed an analysis of PV snow losses for both module types in a testing system situated in southern Finland during the winter of 2023-2024. The results show that half-cell modules reduce annual snow losses by approximately 26% compared to full-cell modules. In contrast to the Marion model, this mitigating effect of half-cell modules is well reflected in our modeling results. Furthermore, our model outperforms the Marion model in predicting time-dependent snow losses, reducing the RMSE by a factor of around 3. This improvement is essential for accurate performance monitoring and forecasting in snow-affected PV systems.

Keywords: Snow loss, modelling, half-cell module, image processing

1 INTRODUCTION

According to the statistic from Motiva Oy [1] , over 180 of MW-level utility scale PV systems are in plan for the next few years in Finland, with the total expected installation capacity of over 14 GW. As being in Nordics, the annual yield of PV system in Finland can be significantly reduced by up to 10% due to snowfall [2]. In addition to external measures such as adding heaters to the modules, various system design strategies have been proposed to mitigate the power loss due to snow. For instance, half-cell modules exhibit superior shade tolerance compared to full-cell modules and are expected to generate more power when partially covered by snow. Bifacial PV systems can generate power from the backside even when the front side is fully covered by snow. Moreover, backside power generation after a snowfall can be significantly higher than in non-snowy conditions due to the highly reflective snow-covered ground. The use of half-cell or bifacial modules is supposed to reduce the snow loss of PV system. However, to our best knowledge, none of existing snow loss models can accurately evaluate these benefits and provide an accurate estimation of the snow losses of PV systems using half-cell or bifacial modules. In this study, we propose a novel hybrid snow loss model that considers the mitigating effect of bifacial and half-cell modules for improved PV yield estimation in snowy conditions. Using this model and experimental validation, we employed a PV testing system situated in Turku, southern Finland, as a case study to investigate the PV snow losses of full-cell and half-cell modules during the winter season of 2023-2024.

2 SNOW LOSS MODEL

Two main types of models have been proposed in the existing publications to evaluate the snow loss of PV systems, generation loss model and snow cover model. The generation loss model directly predicts the power loss from system configuration parameters and weather conditions. This type of models is mostly realized by data-driven curve fitting approaches [3–6]. They are generally limited to a minimum temporal interval of one day. Recently an approach based on computational intelligent techniques were proposed to improve the modelling to higher temporal resolution [7]. Nevertheless, this approach evaluates two distinct physical processes into one fitting process, which is not favorable for accuracy.

On the other hand, the snow cover models separate the complexity of these processes into two steps of modelling: snow accumulation/removal on the module and PV power generation. As the first step, the system and weather characteristics are used to simulate the snow coverage on the module at a specific time by the first-principle approach [8,9] or the threshold approach [10–12]. The first-principal models use complex physical model to simulate the module temperature and predict when the snow will melt or slide down. It can provide a comprehensive analysis but is constrained by the necessity of making assumptions and the difficulty of ascertaining accurate parameters for different scenarios. The threshold models define empirical threshold limits based on historical data, to judge if the snow start to clear from the module entirely or partly and to estimate the snow coverage. One of the most popular threshold models, the Marion model [12], is implemented in SAM [13] and PVlib [14] and is widely used for snow loss evaluation. Automated image capture and processing were recently employed to directly calculate the snow coverage from onsite photos for more accurate evaluation [15–17]. Using the snow coverage ratio, the snow loss is calculated as the difference between the power output from all the modules and those not covered by snow. However, these models rely on simple approach for PV power calculations, which are not able to account for different module designs, such as half-cell module or bifacial modules.

Here we propose a novel hybrid snow loss model which combines deep learning and analytical methods. The framework of the model is shown in Figure 1. The model consists of three sub-models: snow coverage model, optical model, and electrical model. Instead of the first-principal or threshold methods in the existing models, we employ a deep learning approach in the snow coverage model to address the complexity of snow-related processes. The model is trained using weather conditions, system configurations, and snow coverage ratio that is evaluated from onsite photos through an image processing algorithm. The optical model is realized using either analytical equations for monofacial modules or ray-tracing approach for bifacial modules. In our previous work [18], we have developed a Radiance-based ray-tracing method capable of performing cell-level irradiance simulations on the 3D replication of a specific PV system. The snow coverage ratio is used to calculate the effective irradiance on the PV modules in this step. At last, a cell-level electrical model is implemented to account for the interconnection between cells and modules, enabling precise calculation of the snow losses for various system designs, such as different module orientations and the use of half-cell module or bifacial module. Another advantage of this model is its ability to predict time-dependent snow losses with high temporal resolution, up to one minute in our case.

This paper focuses on investigating the differences in snow losses between full-cell and half-cell modules. Snow loss evaluation is based on snow coverage ratio derived from onsite photos. The development and validation of other components of the model will be discussed elsewhere.

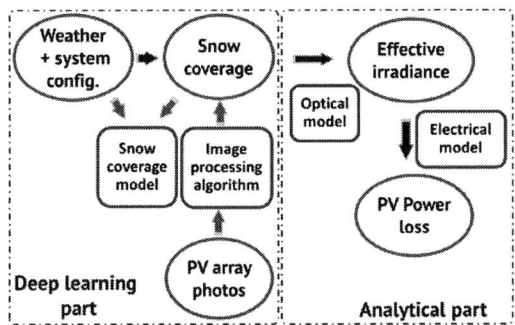

Figure 1: Framework of the snow loss model proposed in this work

3 DATA COLLECTION

Data from a testing system situated in Turku, southern Finland, as shown in Figure 2, was used for snow loss investigation and model validation. The testing system, installed on the rooftop of our university campus building, accommodates eight different PV modules for simultaneous monitoring. The PV array is oriented south with a 45° tilt angle. An IP camera (Reolink RLC-811A) positioned in front of the modules captured images of the modules for snow cover detection. Both the photos and PV performance data were recorded at 1-minute intervals. Onsite meteorological data was collected by a weather station and a solar irradiance station. A pyranometer (Kipp&Zonen SMP10) and two reference cells (Si-mV-85-PT100-4L and Si-mV-85-PT100-4L-B) measured the plane-of-array (POA) irradiance.

The modules under investigation in this study are one full-cell module (SoliTek Standard P.60-S-285, nominal power 285 Wp) and one half-cell module (Trina Honey TSM-DE06M.08, nominal power 335Wp), as indicated in the red box on the left and the green box on the right,

respectively, in Figure 2. The period for analysis is from May 2023 to May 2024. This timeframe allows for the evaluation of the annual performance and losses, covering the entire winter season.

Figure 2: Testing system analyzed in this work. The red box on the left and the green box on the right indicate the full-cell and half-cell module, respectively.

4 RESULTS

4.1 Snow coverage evaluation

The assessment of snow coverage ratio on PV modules employed an image processing algorithm that utilises the k-means clustering method [19] to differentiate between regions with and without snow cover. In comparison to the conventional threshold method, the k-means clustering method enhances the precision of snow coverage estimation under diverse lighting conditions.

Figure 3 shows an example of the snow coverage evaluation process for the full-cell module. The region of the PV module under study is initially extracted from the raw image, then filtered to exclude the gaps between cells and converted into single-color images. Subsequently, the classification is implemented through the k-means clustering method according to the color of the pixels. Then the proportion of the module covered by snow is finally calculated. With this process, the snow coverage ratio of both the full-cell and half-cell modules were calculated for the period under study.

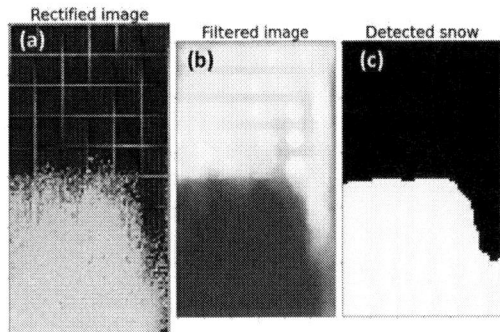

Figure 3. Images obtained from image processing algorithm. (a) the rectified region under study (red box in Figure 2) extracted from the raw image. (b) the filtered image which is converted to single-color image. (c) the image after classification. Note the gaps between cells were excluded during filtering so that (a) is bigger than others.

4.2 Snow loss evaluation

In order to simplify the modeling process, the POA irradiance was directly used to calculate the effective irradiance with or without snow coverage. The irradiance on the part of module covered by snow was lowered by a factor related to the snow depth. Different module schemes of full-cell and half-cell module were realized using PVMismatch tool [20] in the electrical model to calculate the module power with and without snow cover. The power was calibrated in sunny days without snow cover to eliminate the influence from other losses, such as degradation. The snow loss was then calculated as the difference between the modeled power without snow cover and the measured or modeled power with snow cover.

The power outputs of modules in three days from 25 November to 27 November 2023 are shown in Figure 3. The selected dates were chosen because the modules were partially covered by snow and illuminated by good sunshine, allowing for clear observation of the snow losses as the difference between red and black lines in the figure. The calculation shown in Table I illustrates that the cumulative snow loss over the three-day period is reduced from 54.3% for a full-cell module to 37.8% for a half-cell module. The similar mitigating effect of the half-cell module is observed for the annual snow losses from May 2023 to May 2024, as also shown in Table I, although the absolute percentage of snow losses is quite small due to the weather and the high tilt angle of the testing system. The results demonstrate that the use of half-cell module result in a reduction in snow loss by approximately 26%.

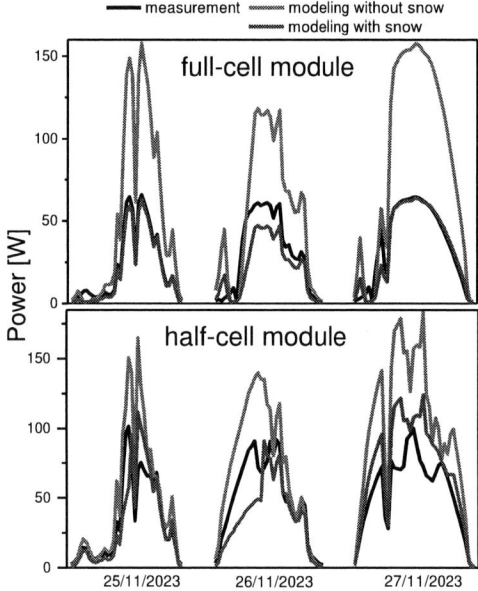

Figure 4: The time dependent power output of the full-cell and half-cell modules in three days from 25 November to 27 November 2023.

The blue lines in Figure 4 represent our model's predicted power outputs with snow cover, showing good agreement with the measured data. Table I also presents the calculated snow losses from our model, which indicate an overestimate of 16-18% for the annual snow losses. Furthermore, the snow losses were evaluated using the Marion model, which yielded similar loss values for both two types modules, indicating it is unable to distinguish snow losses between full-cell and half-cell modules.

The root mean squared error (RMSE) between

measured and modeled time-dependent power output on snowy days was calculated for both our model and the Marion model. Our model achieved RMSE values of 6.5 and 7.6 for full-cell module and half-cell module, respectively, compared to around 20 for both module types from the Marion model. This demonstrates the superior accuracy of our model in predicting time-dependent snow losses, which is crucial for performance monitoring and forecasting in operational PV systems.

Table I: Calculated snow losses from measurement and modelling results

		Full-cell module	Half-cell module
Snow loss 25/11-27/11/2023	Measured	54.3%	37.8%
	Modeled	60.3%	36%
Annual snow loss 5/2023 – 5/2024	Measured	1.03%	0.76%
	Modeled	1.2%	0.9%
	Marion model	1.05%	1.1%
RMSE of power series for snowy period [W]	Modeled	6.5	7.6
	Marion model	20.3	19.3

5 CONCLUSIONS

In this work, we propose a novel hybrid snow loss model that differentiates between snow loss discrepancies in full-cell and half-cell modules. Snow coverage on PV modules is evaluated using an image processing algorithm, and the interconnection between PV cells and modules is incorporated through a cell-level electrical model. Using these methods, we performed an analysis of PV snow losses for both module types in a testing system situated in southwest Finland during the winter of 2023-2024. The results show that half-cell modules reduce annual snow losses by approximately 26% compared to full-cell modules. In contrast to the Marion model, this mitigating effect of half-cell modules is well reflected in our modeling results. Furthermore, our model demonstrates superior performance in forecasting time-dependent snow losses compared to the Marion model, reducing the RMSE by a factor of around 3. This improvement is crucial for accurate performance monitoring and forecasting in snow-affected PV systems.

ACKNOWLEDGEMENTS

The work is funded by the Strategic Research Council (SRC) established within the Research Council of Finland under project RealSolar 359141.

REFERENCES

[1] Motiva Oy, https://aurinkosahkovoimalat.fi/, Accessed on 9th September 2024
[2] H. Böök et al., Solar Energy **204**, 316 (2020).
[3] T. Townsend and L. Powers, in *Conference Record of the IEEE Photovoltaic Specialists Conference* (2011), pp. 003231–003236.

[4] R. W. Andrews and J. M. Pearce, in *38th IEEE Photovoltaic Specialists Conference* (2012), pp. 3386–3391.

[5] M. Zamo *et al.*, Solar Energy **105**, 792 (2014).

[6] A. A. Shishavan, E. C. Foresman, and F. Toor, in *Conference Record of the IEEE Photovoltaic Specialists Conference* (2016), pp. 2625–2630.

[7] B. Hashemi, S. Taheri, and A.-M. Cretu, IET Smart Grid **7**, 221 (2024).

[8] M. M. D. Ross, *Snow and Ice Accumulation on Photovoltaic Arrays: An Assessment of the TN Conseil Passive Melting Technology, Report# EDRL 95-68 (TR)* (1995).

[9] A. Rahmatmand, S. J. Harrison, and P. H. Oosthuizen, in *Conference Proceedings of the ESIM* (2016), pp. 286–296.

[10] E. Lorenz, D. Heinemann, and C. Kurz, Progress in Photovoltaics: Research and Applications **20**, 760 (2012).

[11] L. Bosman and S. Darling, in *IEEE International Conference on Renewable Energy Research and Applications* (2016), pp. 567–571.

[12] B. Marion *et al.*, Solar Energy **97**, 112 (2013).

[13] System Advisor Model, National Renewable Energy Laboratory. Golden, CO. https://https://sam.nrel.gov.

[14] K. S. Anderson *et al.*, J Open Source Softw **8**, 5994 (2023).

[15] J. Ma, A. Khoury, and M. Rodgers, in *IEEE 50th Photovoltaic Specialists Conference (PVSC)* (IEEE, 2023), pp. 1–5.

[16] C. Baldus-Jeursen *et al.*, IEEE J Photovolt **13**, 610 (2023).

[17] X. Zhang and M. T. Araji, Sustainable Energy, Grids and Networks **34**, 101036 (2023).

[18] Hugo E. Huerta *et al.*, in *Proceedings of the 40th European Photovoltaic Solar Energy Conference and Exhibition* (Lisbon, Portugal, 2023), pp. 020297–001.

[19] S. P. Lloyd, IEEE Trans Inf Theory **28**, 129 (1982).

[20] M. Mikofski, B. Meyers, and C. Chaudhari, (2018). "PVMismatch Project: https://github.com/SunPower/PVMismatch". SunPower Corporation, Richmond, CA.

41st European Photovoltaic Solar Energy Conference and Exhibition

DEFECT QUANTIFICATION SYSTEM THROUGH AERIAL INSPECTIONS

*Mario Martínez[1] ,*Sergio Suarez[1], Daniel Jason[1], Daniel Villoslada[1], Jose Rivas[1] , Sofía Rodríguez-Conde[1]
[1]Enertis Applus Parque Empresarial Las Mercedes, C/ de Campezo, 1, 28022 Madrid, Spain

Enertis Solar, S.L. Av. Bruselas 31, 28108 Madrid, Spain; www.enertis.es
Phone: +34 91 651 70 21, *sergio.suarez@enertisapplus.com, mario.martinez@enertisapplus.com,

ABSTRACT: Aerial inspection tests are based on the qualitative interpretation of a series of patterns and defects through some type of photogrammetric test. Enertis Applus improves aerial inspections by implementing a software tool that allows for the quantitative interpretation of defects found in purely qualitative inspections. This tool allows for better identification of some types of defects that are more complicated to identify, such as LetID or PID in strings. The tool manage temperature histogram frequency of each individually module from aerial inspection Metadata, where the data is extrapolated to STC(Standard test conditions) and normalized, can be used to compare two randomly different modules on the PV plant to identify some deviations showing faults or thermal issues from a quantitative way using different statistical analysis tools
Keywords: Analysis, Thermography, Electroluminescence, Dron

1 INTRODUCTION

Aerial inspection tests of photovoltaic (PV) systems have traditionally relied on the qualitative interpretation of patterns and defects through photogrammetric analysis. However, Enertis Applus has advanced these inspections by developing a software tool that enables the quantitative interpretation of defects. This tool significantly enhances the identification of complex issues such as Light and Elevated Temperature-Induced Degradation (LeTID), Potential Induced Degradation (PID), hot spots, activated bypass diodes, and thermal anomalies in module strings.

By analyzing the temperature histogram frequency of each individual module from aerial inspection metadata, the tool allows for a more precise comparison between modules within a PV plant, providing clearer identification of faults and thermal issues that are often challenging to detect through traditional methods.

Enertis Applus also conducts Electroluminescence (EL) and Infrared (IR) inspections using drones. For instance, a photograph of a defect that is difficult to categorize through an IR inspection conducted by the company can be seen below. These challenging-to-categorize defects have inspired the idea of performing this quantitative defect analysis using aerial vehicles

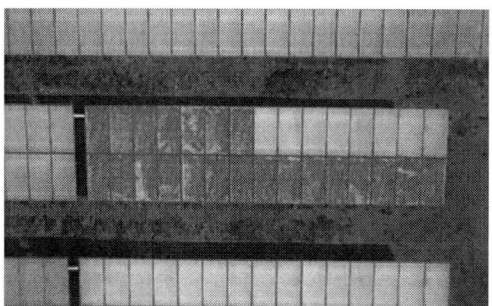

Figure 1: Aerial thermographic inspection of photovoltaic modules conducted by Enertis Applus

Furthermore, the discussion extends to the potential benefits of broadening the IEC 62446 standard to encompass any string of modules, regardless of their characteristics. Currently, this standard offers guidelines for interpreting resistance and leakage values on an individual basis. Extending the standard could provide a more comprehensive framework for evaluating the performance and safety of solar panels, thereby enhancing the effectiveness of the software tool. Standardized measures of insulation resistance for different module sizes could be utilized to better identify and quantify defects, particularly those such as LeTID or PID in strings.

Incorporating analytical tools into aerial inspections can further enhance the interpretation of identified defects. The use of statistical tools allows for a more precise analysis, with normalized parameters determining the type of defect and its correlation with power loss. This approach offers a more comprehensive understanding of the system's performance and highlights potential areas for improvement

2 STATISTICAL RECOGNITIONS

2.1 Mathematical Characterization of Defects

Enertis Applus have developed a software that analyzes thermographic images taken by a drone, extracting the images of individual modules and analyzing their temperature histograms using five key variables. These variables are: Maximum temperature (T_max), Mean temperature ($\bar{T}$), Median temperature (T_median), Standard deviation (σ), and Kurtosis (K).

By analyzing these five variables, it is possible to establish value ranges that can be associated with different types of defects detected in IR images, such as hot spots, short circuits, activated diodes, and Potential Induced Degradation (PID).

$$\bar{T} = \frac{1}{N}\sum_{i=1}^{N} T_i \quad (1)$$

$$\sigma = \sqrt{\frac{1}{N}\sum_{i=1}^{N}(T_i - \bar{T})^2} \quad (2)$$

$$K = \left(\frac{1}{N}\sum_{i=1}^{N}(T_i - \bar{T})^4\right)\sigma^{-4} \quad (3)$$

Complex defects observed during certain aerial thermal inspections are difficult to interpret or correlate with defects found in other types of tests. In this case, defects associated with a mismatch between cells, such as PID or LeTID, can go unnoticed due to low thermal gradients, which worsen with the lack of extrapolation to standard conditions (STC of 1000W/m²).

The analysis of thermal gradients can offer a useful solution when differentiating hot spots produced by failures in the cells of the modules and brightness produced by the tilt of the equipment. It can also help identify solar panels with temperature distributions outside of the norm, even though these may be very difficult to identify with the human eye.

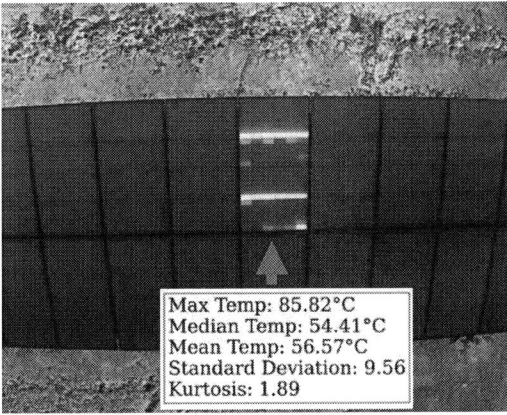

Figure 2: Thermal analysis of a photovoltaic module with a hot spot (red arrow). Statistics shown: max temperature, median, mean, standard deviation, and kurtosis.

The measurement method is based on a model of module segmentation at an individual level through pattern identification. After identifying the module, the pixel intensity and temperature histogram are obtained from the image metadata. The information from the histograms allows us to identify how the temperature is spatially distributed across the module and what the maximum dispersion of the temperatures is.

The differences in pixel images can be visualized using statistical software to understand where the defects on the PV module are and how they affect the Pmax or mismatch. To quantify the existing differences accurately, several histograms are obtained for healthy PV modules and compared with units that present some type of unusual distribution. The most important statistical parameters to define the statistical distribution, such as kurtosis, standard deviation, moment, mean, and median, are obtained.

Figure 3: Software tool workflow showing IR picture, individual module isolation and temperature histogram calculation for shine on a PV module on field

By comparing these statistical parameters between healthy and defective modules, we can precisely identify and characterize the defects. This quantitative approach

allows for a more accurate assessment of the PV system's performance and helps in pinpointing specific issues that may not be easily detectable through traditional inspection methods

2.2 Quantitative Analysis of Defects

In this context, we are identifying defects using quantitative numerical variables rather than relying on qualitative visual indicators in the images. This method leverages statistical analysis of temperature histograms to provide a more objective and precise identification of defects. By focusing on numerical data, we can detect subtle anomalies that might be overlooked in a purely visual inspection, thereby enhancing the reliability and accuracy of our defect detection process.

To illustrate this, we include three images. The first image is an IR test image where almost no differences are noticeable, making it difficult to determine any defects. Among the six modules shown in the IR image, one presents PID while the rest are in good condition. However, this is not easily noticeable at first glance.

Figure 4: Close-up thermography showing a module with PID (bottom right) and the rest of the panels in good condition.

To complement the previously shown IR image, we provide an EL image (Fig. 5) where it is evident that one of the units located in the bottom right corner exhibits advanced PID compared to the neighboring modules.

Figure 5: EL image showing a module with PID, highlighted with a blue square.

Next, we show the histogram of one of the healthy modules located in the upper left corner. The extracted temperature histogram and the EL image corroborate the good health of the module. The distribution is quite normal, centered around the mean, with no asymmetry and low standard deviation.

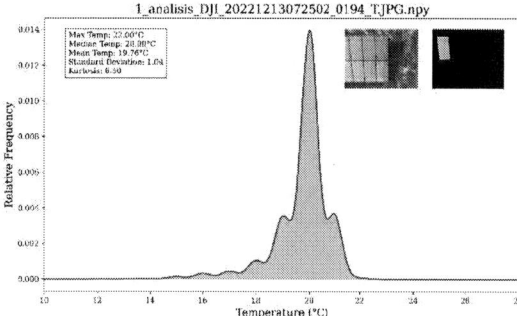

Figure 6: Temperature histogram for a healthy upper module, verified by the accompanying EL image.

On the other hand, we analyze a second image where the module in the lower right corner presents PID, as seen in the EL image. The histogram for this module is not normal, asymmetric, and has a significantly different standard deviation compared to a healthy module.

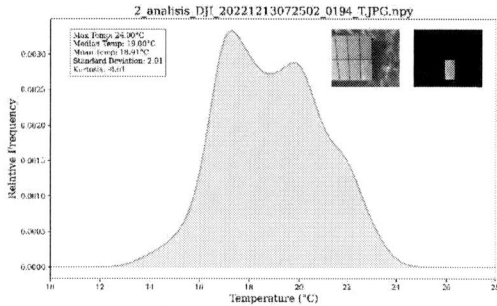

Figure 7: Temperature histogram for a lower module affected by PID, verified by the accompanying EL image.

The figure compares temperature histograms of a defect-free module (top) and one with PID (bottom). The defect-free module shows a homogeneous distribution (high kurtosis, low standard deviation). In contrast, the module with PID has a higher maximum temperature and thermal variability, indicated by a higher standard deviation and negative kurtosis

In this context, we are identifying defects using quantitative numerical variables rather than relying on qualitative visual indicators in the images. This method leverages statistical analysis of temperature histograms to provide a more objective and precise identification of defects. By focusing on numerical data, we can detect subtle anomalies that might be overlooked in a purely visual inspection, thereby enhancing the reliability and accuracy of our defect detection process.

4   CONCLUSIONS

Enertis Applus has developed a quantitative analysis tool that enhances aerial inspections of photovoltaic (PV) modules. By analyzing temperature histograms, the tool supports the detection of defects such as Light and Elevated Temperature-Induced Degradation (LeTID), Potential Induced Degradation (PID), hot spots, activated bypass diodes, and other thermal anomalies. This methodology can also be applied to electroluminescence (EL) images, broadening its use for identifying subtle defects. This approach allows for more accurate module comparisons, aiding in maintenance strategies and overall plant performance.

The integration of advanced software tools for the quantitative analysis of thermographic images represents a significant advancement in the field of PV system inspections. By leveraging statistical parameters such as maximum temperature, mean temperature, median temperature, standard deviation, and kurtosis, a more precise and objective identification of defects is achieved.

This approach surpasses traditional qualitative methods, providing a clearer and more reliable assessment of the PV modules' health. The ability to quantify these defects using numerical data enhances the accuracy of inspections and helps in pinpointing specific issues that could affect the overall performance of the PV system.

The comparison of temperature histograms between healthy and defective modules further underscores the effectiveness of this approach. Healthy modules typically exhibit a homogeneous temperature distribution with high kurtosis and low standard deviation, while defective modules, such as those affected by PID, show higher maximum temperatures and greater thermal variability.

This distinction is crucial for early detection and remediation of defects, ultimately contributing to the longevity and efficiency of PV installations. Moreover, the use of EL imaging in conjunction with thermographic analysis provides an additional layer of verification, ensuring that even subtle anomalies are detected and addressed promptly.

In conclusion, the adoption of quantitative analysis techniques for aerial thermographic inspections marks a significant step forward in the maintenance and optimization of PV systems. By combining advanced software tools with rigorous statistical analysis, a higher level of precision in defect detection is achieved, leading to improved performance and reliability of solar energy installations. This innovative approach not only enhances the effectiveness of inspections but also sets a new standard for the industry, paving the way for more efficient and sustainable energy solutions.

5   REFERENCES

[1] R. Pierdicca, M. Paolanti, A. Felicetti, F. Piccinini, and P. Zingaretti, "Automatic faults detection of photovoltaic farms: Solair, a deep learning-based system for thermal images," Energies (Basel), vol. 13, no. 24, Dec. 2020, doi: 10.3390/en13246496.

[2] Institute of Electrical and Electronics Engineers, Fla. IEEE Photovoltaic Specialist Conference 42 2015.06.14-19 Tampa, Fla. IEEE Photovoltaic Specialists Conference 42 2015.06.14-19 Tampa, and

Fla. PVSC 42 2015.06.14-19 Tampa, 2015 IEEE 42nd Photovoltaic Specialist Conference (PVSC) 14-19 June 2015, New Orleans, LA.

[3] John A, "FEATURE | INFRARED THERMOGRAPHY," 2012.

41st European Photovoltaic Solar Energy Conference and Exhibition

SHAPING EUROPEAN COLLABORATION ON PHOTOVOLTAICS:
A COLLABORATIVE PLATFORM FOR SIMULATION AND MONITORING (COPLASIMON)

Simone Vitale[1], Jonathan Leloux[1], Hervé Colin[2], Eric Pilat[2], Stéphane Mollier[2], Basem Idlbi[3], Rodrigo Moretón[4], Oscar
Anchorena[4], Christophe Salperwyck[5], David Melgar[6], Christian Schill[6]
1) LuciSun, 2) Univ. Grenoble Alpes, CEA, Liten, 3) Smart Grids Research Group, Ulm University of Applied Sciences ,
4) Qualifying Photovoltaics (QPV), 5) MyLight150, 6) Fraunhofer Institute for Solar Energy Systems

ABSTRACT: The Collaborative Platform for Simulation and Monitoring (COPLASIMON) has been developed within
the SERENDI-PV project, funded by the EC Horizon Europe H2020 Programme, to foster collaboration and knowledge
sharing within the photovoltaic (PV) community. COPLASIMON aims to serve the entire PV community by providing
resources and opportunities for collaboration on diverse solar energy topics. In addition, the platform includes valuable
research outputs from the SERENDI-PV project, such as insights on bifacial PV, building-integrated PV (BIPV), and
floating PV technologies, which are available to users seeking information on these emerging areas. The platform
features a range of functionalities, including demo versions of software tools, multiple calls for collaboration on topics
such as bifacial PV, floating PV, BIPV, residential PV, solar resource, quality control, soiling, and grid integration and
flexibility, as well as comprehensive data analytics to support the development and optimisation of PV systems. To
further enhance stakeholder engagement, COPLASIMON offers interactive features, such as a matchmaking page for
PV stakeholders, online workshops for community feedback, and a forum for ongoing dialogue, aiming to support the
wider PV community beyond the specific scope of the SERENDI-PV project.
Keywords: Collaboration, Platform, PV, Demo, Matchmaking, COPLASIMON

1 INTRODUCTION

The Collaborative Platform for Simulation and
Monitoring (COPLASIMON) serves as a central hub for
photovoltaic (PV) research, collaboration, and innovation.
Developed as part of the SERENDI-PV project
(https://serendipv.eu/coplasimon/), it has been
continuously enhanced with new tools, content, and
materials to support a wide range of PV stakeholders.
COPLASIMON provides a comprehensive online
environment designed to facilitate knowledge exchange
and collaboration within the PV community. Key features
include an interactive forum based on the Discourse
framework, which acts as a matchmaking tool for
connecting researchers and industry professionals, a
GitHub repository for sharing source code and software,
and a CKAN-based database for structured data
management. These resources enable a dynamic exchange
of insights and foster cooperation on topics critical to
advancing solar energy technology.

The platform's website is accessible at
https://coplasimon.eu/. Throughout the SERENDI-PV
project, multiple partners have contributed research
outputs, datasets, and simulation tools, many of which are
highlighted in this scientific contribution. COPLASIMON
also hosts collaboration calls to promote research on a
variety of emerging PV topics, including Building
Integrated Photovoltaics (BIPV), floating PV, soiling,
bifacial technologies, and grid integration. The platform
aims to go beyond showcasing the results of SERENDI-
PV by actively supporting engagement and collaboration
across the broader PV community. Through mapping
ongoing research and sharing key findings,
COPLASIMON provides a unique space for stakeholders
to access and contribute to cutting-edge developments in
the solar energy field.

2 THE COPLASIMON PLATFORM

The SERENDI-PV project aimed to address critical
challenges to increase the penetration of photovoltaic (PV)
power in the European energy market by focusing on
advanced modelling, diagnostics, quality control, and grid
integration. The project achieved its objectives by
enhancing the accuracy of modelling for new PV
technologies—such as bifacial PV, floating PV, and
building-integrated PV (BIPV)—innovating in fault
diagnosis, and improving quality control both in the field
and in the lab. This comprehensive approach helped
reduce performance uncertainties, improve the bankability
of new technologies, and explore solutions for managing
higher PV penetration in the grid through enhanced
forecasting and distributed energy resource management.
The Collaborative Platform for Simulation and Monitoring
(COPLASIMON) was created to support these efforts by
providing an accessible space for sharing SERENDI-PV's
content and developments, focusing on modelling, data
analytics, and scenarios for grid-PV interaction. While the
platform showcases SERENDI-PV's results, its primary
purpose is to serve the broader PV community, enabling
knowledge exchange and fostering collaboration both
among project partners and external stakeholders.

Key contributors such as CEA-INES, LuciSun,
MyLight150, THU, and QPV have significantly expanded
the platform's capabilities by adding content and engaging
external stakeholders. This collaborative effort aims to
address common barriers in the PV sector, including the
lack of data management tools and standards, limited
communication channels for PV experts, and the absence
of a centralised repository to track global PV research and
collaborations.

To overcome these challenges, COPLASIMON offers
several key features:

- CKAN data exchange database: The platform
  includes a CKAN-based data exchange database
  (https://ckan.coplasimon.eu/) to facilitate
  structured data sharing. Users can upload
  datasets either publicly or for private data
  exchange. A survey was also conducted to
  gather information on data standards and
  practices within the energy sector to enhance
  interoperability.

- Discourse forum: The Discourse-based forum
  (https://forum.coplasimon.eu/) has been
  structured to promote collaboration and

knowledge exchange. Specific tutorials are available to guide users on how to initiate and manage permanent collaboration calls. Several calls are already open, focusing on topics such as bifacial PV, BIPV, soiling, and floating PV. The platform also includes a dedicated matchmaking tool to connect stakeholders and facilitate partnerships.

- Mapping of PV collaborations: COPLASIMON provides a structured mapping of existing PV collaborations in Europe. Researchers and stakeholders are encouraged to promote their projects, propose new collaborations, and showcase ongoing research initiatives through the platform (https://coplasimon.eu/).

Another major challenge in the PV community is the limited availability of open-access content and demo software tools. To address this, COPLASIMON features a dedicated section for demo software tools, which includes resources for testing, information on tools with free demo versions, and links to open-source software hosted on GitLab or GitHub repositories.

In Section 3, we provide detailed examples of materials and tools contributed by our partners to highlight the benefits gained from participating in COPLASIMON and to demonstrate how the platform supports collaboration and innovation across the PV community.

3 DEMO TOOLS

Several initiatives have been launched within COPLASIMON through the collaboration between SERENDI-PV partners and external stakeholders. These contributions include simulation tools, data sets, and analytical resources to support diverse PV research and development activities. Specifically:

- LuciSun, which led the development of the platform, has provided demo versions of tools focused on PV modelling (LuSim) and the financial planning of residential PV systems within energy communities (Consolectiv).

- CEA-INES has contributed tools such as Trifactors for simulating bifacial PV systems, SQSL for quantifying the impact of soiling, and ASPIRE for fault detection in PV modules using drone imagery.

- MyLight150 has made available a detailed dataset on PV system performance, supporting extensive research and analysis activities.

- THU is involved in simulating PV feed-in and grid interactions at the demo site in Hittistetten, Germany, focusing on voltage and loading impacts. THU has also collaborated with LuciSun to develop an interactive platform for exploring flexibility options in grid management.

- QPV offers a soiling analysis tool and a performance analysis tool that helps optimize PV plant operations by comparing expected and actual production data.

3.1 LuSim (LuciSun)

LuSim is a PV simulation software tool that is particularly suitable for PV systems with complex 3D scenes, complex shading, irregular terrain, bifacial PV,

agrivoltaics, or building-integrated photovoltaics (BIPV) [1], [2], [3].

As part of their contributions, LuciSun has made a demo version of the LuSim software tool available online (https://coplasimon.eu/index.php/lusim-demo-teaser/).

This demo version is designed to provide insights into the different irradiance components at the module cell level for photovoltaic (PV) systems, allowing users to visualize these components in a 3D scenario for a specific timestamp or retrieve a time series for a selected cell. The tool enables users to generate graphs and download the results for further analysis.

The current version of LuSim offers two predefined scenarios where users can modify parameters such as tilt and pitch. However, certain elements, including module type, location, and the number of PV modules or sheds, remain predefined (**Error! Reference source not found.**).

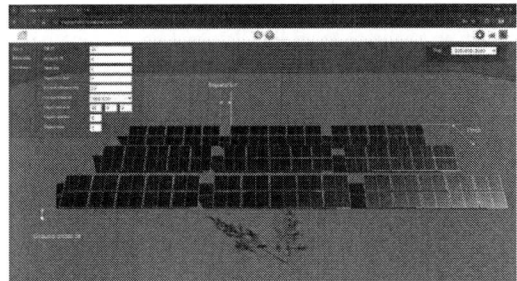

Figure 1: Example of PV plant modelling in the LuSim demo.

3.2 Consolectiv (LuciSun)

LuciSun has published a free online tool, named Consolectiv Europe (for CONsumption, SOLar, and collECTIVe) that can be used through a public website (https://consolectiv.eu/) (**Error! Reference source not found.**).

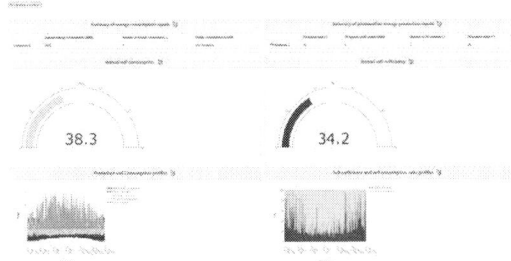

Figure 2: Example of results available from Consolectiv.

The tool includes an interface for calculating the self-consumption ratio, self-sufficiency ratio, and other metrics related to the energy self-consumption of photovoltaic installations within energy communities (ECs) and for prosumers. It is designed to be used by the general public, policymakers, and PV installation professionals. Initially developed with a focus on the Brussels Region in Belgium [4], the tool has since been expanded to cover a broader range of European countries, as shown in **Error! Reference source not found.**.

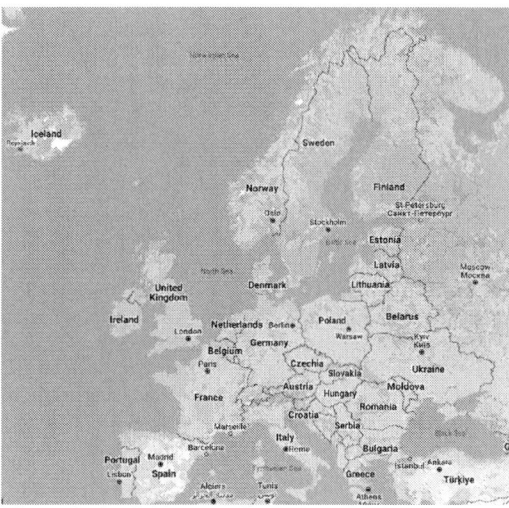

Figure 3: Available countries in Consolectiv Europe.

The tool calculates self-consumption and self-sufficiency ratios for all of Europe at a 10-minute resolution, with subsequent aggregation over intervals of 1 hour, 1 day, 1 month, and 1 year, based on the location and other user-defined parameters. For the consumption data, synthetic profiles in TMY format have been used, corresponding to 17 different types of buildings and their associated consumption patterns. These profiles are generated using a synthetic method developed by the National Renewable Energy Laboratory (NREL) in the USA ([5], [6]) and consider local climate conditions and consumer preferences.

Two main functionalities have been developed:

- The first is aimed at individual prosumers, providing a tool to help them identify the optimal design conditions for their photovoltaic installations, taking into account self-consumption and calculating the corresponding return on investment (ROI).

- The second functionality is designed for energy communities, allowing users to combine different building consumption profiles with various photovoltaic production setups to assess the overall impact on self-consumption and self-sufficiency for an entire neighbourhood of prosumers.

3.3 Trifactors (CEA-INES)

CEA-INES has developed a simulation tool for bifacial PV systems called Trifactors, which uses a 3D view factors approach to calculate the reflected irradiance on the backside of the modules. This method determines the proportion of radiation (in this case, luminous) that leaves a given surface and directly reaches another, accurately accounting for side effects. The resulting irradiance is then fed into electrical and thermal modelling algorithms to calculate the system's electrical output, taking into account the uneven distribution of irradiance on both sides of the modules [7] [8]. Recent updates now enable the tool to consider the impact of nearby obstacles, such as the module support structure.

3.4 ASPIRE (CEA-INES)

CEA-INES has developed the ASPIRE tool to detect PV module faults using aerial images captured by drones.

It consists of three functional blocks:

- PV module detection: Uses a combination of RGB and infrared images to identify PV modules through segmentation, based on the detection of module edges.

- Anomaly detection: Identifies issues such as hot spots, faulty bypass diodes, junction boxes, potential-induced degradation (PID), and disconnected modules by detecting electrical faults or soiling at the module level, and assesses the impact at the PV plant level by considering the plant's electrical architecture.

- Estimation of power losses: Calculates power losses at the module, string, and overall PV plant level.

Figure 4 shows an example of the ASPIRE tool.

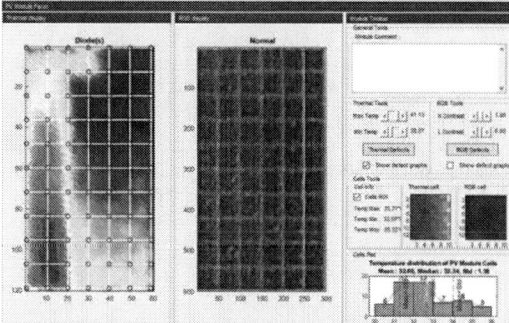

Figure 4: ASPIRE interface showing the detection of a bypass diode fault on a PV module and its analysis.

As ASPIRE manages the electrical architecture of the entire PV plant, it determines the impact of module failures on the affected PV strings. The tool estimates power losses for each module, each PV string, and the entire plant.

The Stochastic Quantifying Soiling Loss (SQSL) method, also developed by CEA-INES, provides a way to measure soiling using standard monitoring data from a PV plant. Unlike other methods, SQSL does not require environmental data or information about artificial cleaning operations. Instead, it calculates the soiling impact directly from the electrical data measured in solar power plants. Unlike conventional approaches that look for change points or rely on identifying clean-only days, SQSL classifies performance metrics into three distinct categories: cleaning days, stable periods, and soiling periods.

The method uses a performance metric (PM), defined as the ratio of relative current (DC current/STC DC current) to relative irradiance (G/STC G). A typical PM profile consists of different periods divided into three types: cleaning days, stable periods, and soiling periods. Once these periods are characterized, the Monte Carlo method is applied to generate multiple random PM profiles using the probability distributions of various parameters such as frequency, duration, and soiling rates. A statistical analysis of these generated profiles allows for the estimation of the most likely average soiling ratio, along with an uncertainty interval.

3.5 MyProSizer (MyLight150)

Within the project, MyLight150 enhanced its residential PV simulation tool, MyProSizer (Figure 5), which enables PV system installers to simulate performance under different renovation scenarios, estimate savings for potential customers, run financial analyses, and generate customizable reports for clients. The MyProSizer simulator integrates real consumption data and appliance usage profiles to accurately model energy projects.

MyLight150 also offers to share a subset of its PV production data in exchange for detailed analysis. The dataset includes 2,037 photovoltaic (PV) systems from 2015 to 2021, primarily located in Belgium and France. The systems, mostly under 12 kWp, have their energy production monitored at 10-minute intervals. The shared PV systems were selected to create a dataset with minimal missing data.

The collaboration calls provided MyLight150 with increased visibility among stakeholders who could benefit from this data. The COPLASIMON platform served as a unified space for connecting multiple actors and businesses, fostering a collaborative environment.

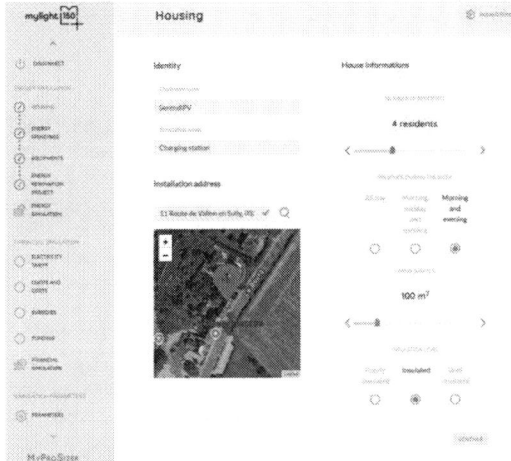

Figure 5: MyProSizer Residential PV simulation tool.

3.6 Flexibility Platform (THU)

Within the project, THU has been involved in simulating PV feed-in from selected households combined with grid simulations, focusing on voltage levels and loading of grid components at the demo site in Hittistetten, Germany. This village, located near the city of Senden in Bavaria, is characterized by a high penetration of PV systems. In Hittistetten, approximately 1.66 MWp of PV capacity is installed, corresponding to an average of around 14 kWp per residential building.

The simulation data is generated using the grid simulation software PowerFactory (https://www.digsilent.de/en/powerfactory.html) and is based on a detailed grid model built from real data provided by the Geographic Information System of the local grid operator, Stadtwerke Ulm/Neu-Ulm Netze GmbH (SWU) (https://www.ulm-netze.de/). The study also explored the available flexibilities of these PV systems, guided by site-specific PV forecasts for the Hittistetten area.

The results of these simulations have been made accessible through the COPLASIMON platform via an interactive PowerBI interface.

The interface is shown in Figure 6. More interactive material can be found in the DSO platform website.

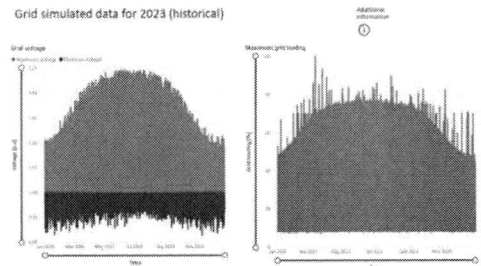

Figure 6: Example of interface of the PowerBI platform.

The simulation results presented on the collaborative platform include key parameters relevant for grid operators, such as the maximum and minimum voltage levels of the grid and the maximum loading of all grid components. These parameters are critical to ensuring that the grid operates within predefined limits under normal conditions, as specified by the European standard DIN EN 50160.

The platform can be accessed at the following link: https://coplasimon.eu/index.php/dso-flexibility-platform/.

3.7 Field lab measurements platform (FHG)

Several innovations related to soiling detection and quantification have been developed by different partners within the project. To gather feedback from relevant industry stakeholders outside the project, several activities were undertaken:

- Online webinar: Key representatives from various sectors, including PV system operators, O&M service providers, and academia, were invited to share their feedback on the soiling-related innovations developed within SERENDI-PV. The general audience also participated, contributing their views on the topic. Highlights from the event have been shared on the COPLASIMON platform.

- Call for PV performance monitoring data to analyze soiling impact: A public call for cooperation was launched on COPLASIMON to gather additional real-world datasets for analysis. Participating companies receive a free analysis of their data and can provide feedback on the proposed solutions and results. Confidentiality has been ensured throughout the process.

- Showcasing soiling innovations: The collected data will be analyzed using various approaches, including those developed within SERENDI-PV and other established methods in the literature. The results will be published on COPLASIMON for the broader community.

3.8 E-Dust (QPV)

The performance and soiling analysis tool enables detailed evaluation of key performance indicators (KPIs), comparison with initial production estimates, and identification of optimization opportunities. By analyzing production and operating data—preferably over periods of one year or more—deviations in the generation and loss chains can be detected and compared to the initial expectations. This allows for targeted recommendations to improve the efficiency of PV systems.

The analysis can also be applied specifically to soiling using a prototype tool called E-Dust, which has been validated through multiple field campaigns. E-Dust has provided extensive insights into real-world soiling patterns, enabling the optimization of cleaning schedules and reducing the cost of cleaning operations based on operating data.

4 DATA SHARING AND COLLABORATION CALLS

The COPLASIMON platform aims to serve as an entry point for continuing to promote connections between stakeholders even after the project's completion. Through the platform, users can access a marketplace to offer or request services and a data-sharing repository where relevant datasets are made available for the community.

Within the SERENDI-PV project, the COPLASIMON platform has been developed to foster data sharing and collaborative research within the photovoltaic (PV) community. Various collaboration calls and data-sharing initiatives have been established to promote participation from external stakeholders and to enhance the usability of the shared data. These calls are aimed at addressing diverse topics such as bifacial PV, floating PV, building-integrated photovoltaics (BIPV), soiling, residential PV, solar resource, quality control, and grid integration. The COPLASIMON platform aims to serve as an entry point for continuing to promote connections between stakeholders even after the project's completion. Through the platform, users can access a marketplace to offer or request services and a data-sharing repository where relevant datasets are made available for the community.

Key data sharing initiatives:

- Public data repositories: The COPLASIMON platform hosts several public data repositories accessible through the CKAN database interface, allowing users to upload and download datasets related to PV performance, solar resource, and grid integration. These repositories are structured to support effective data management and encourage contributions from both SERENDI-PV partners and external stakeholders.

- Confidential data sharing via NDA: For datasets containing sensitive information, sharing is facilitated through non-disclosure agreements (NDAs). For instance, MyLight150, a SERENDI-PV partner, shared operational data from over 2,000 PV systems under specific NDAs, ensuring that data usage and publication are aligned with privacy requirements. These agreements provide a clear framework outlining the use of data, the project context, and approval processes for any published results.

- Public database of solar resource and weather data: A comprehensive database was created to provide high-quality solar resource and meteorological data. It integrates publicly available ground measurements with Solargis satellite model data, allowing users to cross-compare datasets and gain insights into the use and limitations of satellite-based solar resource models. The database interface, built using Google My Maps, is user-friendly and includes links to source data and validation reports. This initiative aims to improve transparency and promote the use of high-quality data in research

and industry(SERENDI-PV-D7.4_Public_...).

- Performance monitoring of non-degraded and degraded PV modules: A detailed performance monitoring system was implemented at Tecnalia facilities in Spain to assess both non-degraded and degraded PV modules, monitored from April to September 2024. This dataset, which includes a mix of artificially degraded units and naturally degraded ones obtained from various PV sites, serves as a benchmark for fault detection and diagnosis (FDD) tools at the PV module level. The data includes operating voltages and currents of PV modules, periodic measurements of key parameters such as short-circuit current (Isc), open-circuit voltage (Voc), maximum power point current (Imp), maximum power point voltage (Vmp), fill factor (FF), series resistance (Rs), and shunt resistance (Rp), along with detailed irradiance and module temperature data. This information supports the analysis of various degradation modes and enables the development of robust diagnostic solutions.

- Collaboration calls for external stakeholders: The collaboration calls within COPLASIMON have allowed partners to engage with a wider audience by providing a public space for showcasing research results and inviting contributions from other organizations. For example, calls for data contributions and feedback have been launched on topics such as soiling and performance monitoring, encouraging a dynamic exchange of knowledge and insights.

- Anonymisation and data quality control: In cases where data sensitivity is a concern, anonymisation techniques have been applied. For example, personal information has been removed and location resolution reduced to maintain confidentiality of shared datasets. A systematic quality control process ensures that shared data meets a high standard of reliability, which is essential for building trust among collaborators.

- Metadata from 18,000 PV systems: The BDPV (Base de Données Photovoltaïque) database is a large French platform dedicated to the collection, analysis, and sharing of data related to PV installations. It provides performance monitoring, data sharing, and benchmarking tools for over 18,000 PV systems. Users can voluntarily share their production data to contribute to a larger dataset that helps track the overall performance of PV systems in France. The metadata include information such as system location, size, and technology, supporting transparency and community engagement in the solar sector.

- Marketplace and service offerings: The COPLASIMON platform also serves as a marketplace where services such as data analysis, simulation tools, and research partnerships can be offered or requested by members of the community. This feature allows participants to leverage shared resources and engage in mutually beneficial collaborations.

By providing these structured mechanisms for data sharing and collaboration, the COPLASIMON platform plays a pivotal role in promoting innovation and cooperation in the PV sector, extending beyond the scope of the SERENDI-PV project and supporting a broad range of solar energy research initiatives. These efforts are complemented by a focus on adhering to the FAIR (findable, accessible, interoperable, and reusable) principles to maximize the impact of shared data and ensure that it is used effectively across the PV community.

5 CONCLUSION

The results from the SERENDI-PV project have demonstrated the value of involving a wide range of external stakeholders from the photovoltaic sector, highlighting the crucial role of collaboration in refining and validating the tools developed throughout the project. Engaging with these stakeholders through workshops and data-sharing initiatives has not only enriched the platform's content but also underscored its potential to serve as a hub for future collaborations in the PV community.

The collaborative environment fostered by COPLASIMON has helped align the interests of various actors, enhanced the understanding of shared research challenges, and informed the development of policies supporting the sustainable growth of PV. The lessons learned have contributed to defining a long-term vision for data sharing and stakeholder engagement, addressing one of the key challenges faced by the PV sector today.

Looking ahead, increasing the number of active users and contributors on the platform will be essential for maintaining its relevance and impact. We invite the scientific community and industry professionals to join us by visiting our contact page and contributing their research and expertise. By expanding the network of contributors, COPLASIMON will continue to evolve as a dynamic and effective resource for promoting innovation and knowledge sharing across the photovoltaic sector.

6 ACKNOWLEDGEMENTS

The development of COPLASIMON was partially funded by European Commission through the Horizon 2020 project SERENDI-PV (https://serendipv.eu/), which belongs to the Research and Innovation Programme, under Grant Agreement 953016.

7 REFERENCES

[1] Robledo J., Leloux J., Lorenzo E., Gueymard C.A., From video games to solar energy: 3D shading simulation for PV using GPU, Solar Energy, https://doi.org/10.1016/j.solener.2019.09.041, 2019.

[2] Robledo J., Leloux J., Sarr B., Gueymard C.A., Driesse A., Dynamic and visual simulation of bifacial energy gain for photovoltaic plants, 38th European Photovoltaic Solar Energy Conference and Exhibition (EUPVSEC38), Online, 2021.

[3] El Boujdaini I., Bruhwyler R., Rajan S.P., Robledo J., Sarr B., Leloux J., Lebeau F., Gueymard C.A., 3D modelling of light-sharing agrivoltaic systems for orchards, vineyards and berries, 40th European Photovoltaic Solar Energy Conference and Exhibition (EUPVSEC40), Lisbon, Portugal, 2023.

[4] Sarr B., Zhao Z., Leloux J., Hendrick P., Robledo J., A free online tool for the simulation of collective self-consumption in Brussels, 38th European Photovoltaic Conference and Exhibition (EUPVSEC38), Online, 2021.

[5] Wilcox S. and Marion W., Users Manual for TMY3 Data Sets, NREL Technical Report TP-581-43156, 2008.

[6] Baechler M.C., Williamson J.L., Gilbride T.L., Cole P.C., Hefty M.G., Love P.M., Building America best practices series: volume 7.1: Guide to determining climate regions by county, Pacific Northwest National Lab.(PNNL), Richland, WA, United States, 2010, https://www.osti.gov/biblio/1068658.

[7] Mollier S., Tsanakas I., Assessing uncertainties from reflected irradiance in bifacial PV simulation through a 3D view factor model and rear sensor measurements, 40th European Photovoltaic Solar Energy Conference and Exhibition (EUPVSEC40), Lisbon, Portugal, 2023.

[8] Melliti D., Tsanakas I.. Ha D.L., Imagery and monitoring data coupling towards complete PV plant diagnostics, 8th World Conference on Photovoltaic Energy Conversion (WCPEC8), Milan, Italy, 2022.

41st European Photovoltaic Solar Energy Conference and Exhibition

PERFORMANCE OF VERTICALLY MOUNTED BIFACIAL PHOTOVOLTAICS ON HIGH-RISE BUILDINGS IN THE NORDIC CONDITIONS

Bergpob Viriyaroj[1]*, Sami Jouttijärvi[2], Matti Jänkälä[1], Kati Miettunen[2]

[1]Department of Architecture, Aalto University, Otakaari 24, 02150 Espoo, Finland
[2]Department of Mechanical and Materials Engineering, University of Turku, Vesilinnantie 5, 20500 Turku, Finland
*e-mail address: Bergpob.viriyaroj@aalto.fi

ABSTRACT: In urban environments, limited land availability creates the demand for PV system installation on buildings. Roof spaces are often used as the installation surface for conventional solar panel systems. In multi-story buildings, which have low roof area compared to building volume, the roof space can be too limited to host sufficiently sized PV systems. Adding PV on building façade to increase the total electricity production is a solution to the limited roof space in multi-story buildings. This study investigates the potential of installing vertically mounted bifacial photovoltaics (VBPV) on multi-story buildings in Helsinki, Finland (60°N). A five-story apartment building with different PV systems, modeled with commercial PVSyst software, is chosen as a case study. The result shows that VBPV on the facade has a lower specific production than rooftop monofacial photovoltaics (MPV), but their values are acceptable. The specific yield of VBPV on the facade is around 700 kWh/(kWpyear), while MPV on the roof is around 1000 kWh/(kWpyear). The percentage of increase in production by integrating VBPV on the facade depends on the amount of PV that can be installed on the roof, ranging from 126% to 7%. The comparison of facade-installed VBPV and MPV demonstrates that their total production is almost equal. MPV appears to be less complicated to install, while VBPV has the benefit of matching peak production with residential electricity consumption and occupying less facade space.

Keywords: Vertical bifacial PV, Built environment, High-latitude conditions, Case studies.

1 INTRODUCTION

Photovoltaics is one of the most widespread sources of renewable energy. The development of photovoltaics has led to a significant reduction in their price as well as the number of photovoltaic products, resulting in the high rate of PV proliferation [1]. The modularity of solar panels has also allowed them to be used in diverse applications, from large-scale solar stations to household usages. Their modularity and scalability provide opportunities to integrate them into urban structures. Even in northern latitude locations, where the amount of solar irradiation is understood to be lower than in locations closer to the equator, the adoption of PV is on the rise.

Although solar panels are becoming more common in northern latitude countries, there are many challenges associated with utilizing them in this environment. Some of the challenges they face in this latitude are the low solar elevation and the prolonged period of low solar exposure during wintertime. These challenges may hinder the proliferation of PV, but they can also promote novel PV products that can overcome these challenges. Bifacial PV is one of the technologies that have been widely studied and shown to have advantages when deployed in Northern latitudes.

A study by Rodríguez-Gallegos et al. shows that simply replacing monofacial PV with bifacial PV with slight adjustments can improve the electricity yield of PV in Northern latitudes [2]. Under specific circumstances, these advantages can be enhanced with a different approach to PV orientations. Vertically installed bifacial photovoltaics (VBPV) appear to be suitable for the low solar angle in Northern latitude locations [3]. When installed facing East and West, the specific yields of this type of solar panels can be competitive with installing south-facing MPV [4]. VBPV has an additional benefit of shifting the peak electricity production from noon to morning and evening. On average, the consumption profile of residential buildings peaks in the evening while at noon the consumption is lower [5]. The shift of peak production away from noon makes VBPV suitable for self-

consumption in residential buildings.

For buildings with limited roof space, there are examples of projects that used their façade skins to install PV. Multi-story buildings generally have a smaller ratio of roof space compared to their floor areas. This means that, in this building type, the electricity being produced by rooftop PV often fails to satisfy the demand for electricity in the building. Installing PV on the façade is one way to address the lack of roof space. Even though the specific yields of PV on the facades are expected to be lower than PV on the rooftop, they appear to be able to increase the total power production of the PV systems on buildings [6]. The losses related to facade installations are strongly correlated with solar elevation angle, and the low angles in Nordic latitudes are expected to yield fewer losses compared to more southern locations.

In a low-rise building, our previous studies have shown that VBPV can yield good performance in northern latitudes if there is no significant shading [7]. Therefore, they can be expected to have good performance on the façade of multi-story buildings. This study explores the possibilities of using VBPV on building facades to supplement electricity production of rooftop MPV on multi-story buildings. The study aims to determine the specific yield and total production of VBPV in the underlying context. A five-story apartment building in Helsinki was chosen as the case study. We will use commercial software, PVSyst, to simulate the PV production in this study.

2 GOALS AND METHODS

2.1 Case study

The study was conducted using a case study site and building located in Helsinki. The case study building is an open-source wooden multistory apartment building design, Make 2.0, developed in 2016 by the Helsinki Housing Service [8]. The architectural design was developed by Harris-Kjisik Oy with the aim of being built on different sites and providing quality affordable housing.

The case study site is located in Helsinki's Oulunkylä

neighborhood, where the first Make 2.0 buildings will be constructed. In the study, PV panels were installed on the roof and the south-facing balcony zone of the building. The performance of the panels was simulated considering the shading caused by the immediate surroundings of the site, such as buildings and vegetation, and by the row-to-row shading of the PV system itself.

2.2 Methodology for PV system simulation

A commercial software, PVSyst [9], was used to simulate the power production of the suggested PV systems. The case study house was implemented in PVSyst using the "Near Shadings" tool, and the PV panels were added individually to suitable locations for each simulation. The location of the house is set to a real site in Helsinki, Finland (60°N).

In the simulation, there are two PV sizes: a standard full-size panel and a smaller panel consisting of one row of PV cells. The dimensions of these two types of panels are shown in Figure 1. For the full-size, a generic "440 Wp Twin 144 half-cell" panel from the PVSyst library was used. The small 5x1 cell panels were implemented in PVSyst manually by downscaling the size and properties of the full-size panels. The inverter(s) were chosen from the PVSyst library, taking the smallest available inverter that avoids forced curtailment of the production. If the smallest inverter in the PVSyst library was too large, a fictional inverter was created by downscaling the properties of the "3 kWac inverter." The exact wiring configuration was optimized to maximize the power output.

PV production simulations were run for one year using the weather data imported from Meteonorm 8.1 (1991-2010) with PVSyst's built-in import tool. The shading from the surroundings and other PV panels was modeled in detail, including the losses related to the electric layout of the PV modules. The total production, specific yield, and the hourly energy production were used as indicators when comparing different PV systems. The VBPV systems were modeled by analyzing the front and rear sides separately as MPV systems.

2.3 Scenarios

To explore multiple possibilities of VBPV in the case study, scenarios were created for the simulation. MPV rooftop installation on top of the building is used as the baseline for comparison to VBPV. There are three scenarios for each PV type. The reason for choosing to create three scenarios for MPV is due to the uncertainties of the roof design in the building. The energy report that has been studied by energy consultants showed several scenarios. Three of those scenarios were chosen to represent three different design solutions, and the details are as follows:

1. 12 panels scenario

In this scenario, the roof space is occupied by other infrastructures (such as air conditioning), significantly limiting the area on which MPV can be installed on the roof. The limitation is drastic, not only because other infrastructure occupies the roof space but also because they create shading on the roof, further limiting the space that can be used to install PV. This scenario represents a building that has not been designed with PV utilization in mind and reflects the original design of the case study in the energy report.

2. 45 panels scenario

In this scenario, the assumption was made that the other building infrastructure systems were installed elsewhere. This provides a significantly larger area to install PV compared to the 12-panel scenario. The scenario represents a building design that is more suitable for PV installation but lacks optimization. The report by the energy consultant presents this scenario as a possible change to the design to make the PV solution more viable.

3. 88 panels scenario

The design of the roof has been modified in this scenario. The result of this modification allowed one side of the roof to be fully covered with solar panels, leaving spaces only for maintenance. This design represents a building that has been highly optimized for installing MPV on the roof.

For VBPV, we chose to install VBPV on the southern side of the case study. This allowed the panels to be exposed to the sun from the east and west. This orientation has the potential to achieve the highest PV efficiency in northern latitudes.

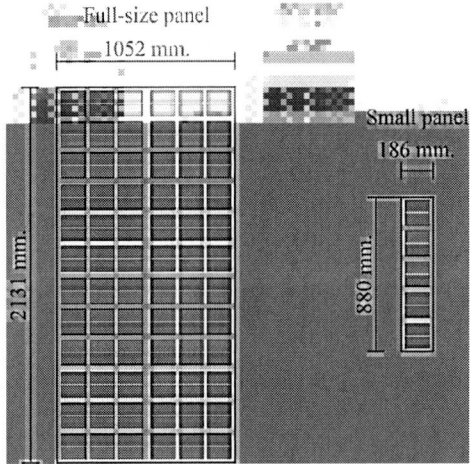

Figure 1. *Size of panels in full-size panels system with 144 half-cell and small panels system with 10 half-cell.*

Figure 2. *Installation scheme of VBPV – facade: full-size panels system The figure was created with PVsyst software (PVsyst, 2024).*

1. Full-size panel scenario

The first scenario of VBPV comprises all full-size panels. They are installed in three arrays, spread evenly on the southern facade of the case study. Each array has seven panels. The distance between each array was calculated to avoid self-shading. Figure 2 shows the installation diagram. A facade MPV scenario, where the number and location of the panels were identical to those in Figure 2, and the only difference was the panel orientation (parallel to the facade), has been provided as a reference.

2. Small panel scenario

This scenario only utilizes small panels as shown in Figure 3. The total number of PV panels in this scenario is 158. The cells were spread throughout the entire facade, in appropriate locations for installation. The surface of balconies has also been used to install these PV.

Figure 3. *Installation scheme of VBPV – facade: small panels system The figure was created with PVsyst software (PVsyst, 2024).*

Figure 4. *Installation scheme of VBPV- facade: hybrid system The figure was created with PVsyst software (PVsyst, 2024).*

3. Hybrid scenario

A combination of full-size panels and small panels was used in this scenario (Figure 4). There are 14 full-size panels and 52 small panels in the hybrid system.

3 RESULTS

The results of the simulation can be seen in Table 1. The annual specific yield of MPV – rooftop in three different scenarios is around 1000 kWh/kWp. This number shows the efficiency of the PV systems and, since the three MPV scenarios are installed in the same location, the number is almost the same. We can see larger differences between the MPV scenarios. The 12-panel, 45-panel, and 88-panel MPV systems have a total annual production of 5320 kWh, 20,020 kWh, and 38,700 kWh, respectively. The 12-panel system produces only 13% of the full roof scenario, while the 45-panel system produces around 50% of the full roof scenario.

Table 2 1. *Total annual production and specific yield of MPV – roof top systems and VBPV - facade systems*

PV scenarios	Total annual production (kWh/year)	Specific yield (kWh/(kWp*year)
MPV - roof top systems		
12 Panels	5320	1010
45 Panels	20,000	1010
88 Panels	38,700	1000
VBPV - facade systems		
Full-size panels	6700	730
Small panels	2810	600
Hybrid	5560	720

For VBPV-façade, the specific yield for the full-size panel system is 730 kWh, and for the hybrid system, it is 720 kWh. These two systems have similar results, while the small panel system has a lower specific production at 600 kWh. The specific yield difference between rooftop MPV and façade VBPV is noticeable, especially for the small panel scenario, but the VBPV performance could still be considered acceptable.

The parameter that sees the largest difference between the results of each scenario is the total annual production. The total electricity production for the full-size panel, small panel, and hybrid systems is 6700 kWh, 2810 kWh, and 5560 kWh, respectively.

4 DISCUSSIONS

Depending on the scenario, VBPV-façade can produce 26% more electricity than MPV – rooftop. This result came from comparing the 12-panel MPV – rooftop system to the full-size VBPV-façade panel system. The worst performance for VBPV-façade is when the small VBPV panels system is compared to the 88-panel MPV – rooftop system. In this comparison, VBPV produced approximately 7% of the electricity produced by the rooftop MPV system. Figure 5 and Table 2 display the comparison between each scenario.

From the results, it appears that the share of VBPV-façade compared to MPV-rooftop depends heavily on the amount of PV on the roof. In a building with limited roof space, the incentives to install PV on the façade are high and should be considered. The case study building is considered a high-rise building in Finland. However, this building may not fit the definition of a high-rise building in other countries. The roof area constraint will be more severe in taller buildings, as they will have a lower ratio of roof area to the total floor area. The larger floor area translates to a larger area for heating and cooling. In taller buildings, the number of occupants also tends to be higher, resulting in higher energy consumption. With limited roof area, achieving a high level of self-sufficiency of electricity from MPV – rooftop will become more difficult, and VBPV-façade will become more profitable.

The small panels provide considerably lower energy production compared to the large ones, and it could be considered if more PV can be installed on the façade. However, the investment to install them may be lower than full-size panels, since small size panels will suffer significantly lower wind load than full-size panels. This system could be worth considering in situations where a project aims to reach a specific amount of on-site electricity production, but the rooftop system is slightly short of this goal.

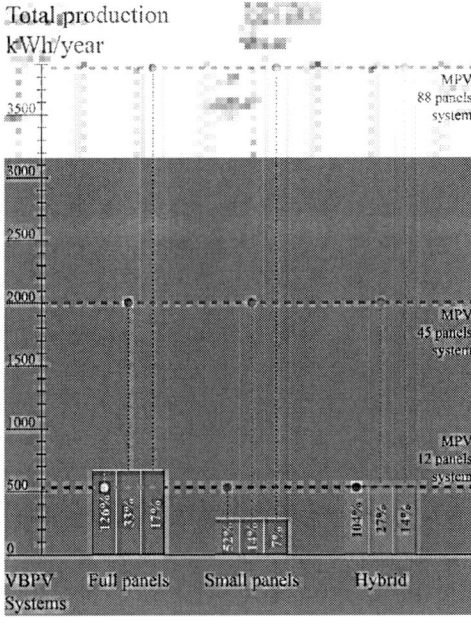

Figure 5. *Comparison between total annual production of MPV – roof top systems and VBPV - facade systems*

4.1 Comparison to MPV - facade

There have been examples of buildings overcoming the constraint of roof space by installing PV on the building's façade without using bifacial PV technology. Since this study focuses on exploring the possibility of VBPV, it is important to also compare the performance of this PV system to the MPV in the same installation location. Another simulation has been created to determine the performance of MPV-façade of the case study. The south façade of the case study is able to host 21 full-size MPV panels, which is the same number of panels as full-size VBPV-façade systems. The results can be seen in Table 3.

Table 2. *Comparison between total annual production of MPV – roof top systems and VBPV - facade systems*

VBPV – façade systems	Full panels	Small panels	Hybrid
Total production (kWh/year)	6700	2810	5560
Compared total annual production			
MPV – roof top: 12 panels system	126%	52%	104%
MPV – roof top: 45 panels system	33%	14%	27%
MPV – roof top: 88 panels system	17%	7%	14%

Table 3. *Total annual production and specific yield of VBPV - facade systems and MPV - facade*

PV systems	Total annual production (kWh/year)	Specific yield (kWh/(kWp*year)
VBPV - facade systems		
Full-size panels	6700	730
Small panels	2810	600
Hybrid	5560	720
MPV - facade systems		
MPV panels	6820	740

Table 4. *The nominal power, annual production and annual specific yield of the analyzed PV systems.*

System	Nominal power (kWp)	Energy production (kWh)	Specific yield (kWh/ kWp)
MPV – roof top: 45 panels system	19.8	20020	1010
MPV - facade	9.24	6820	740
VBPV – Full-size panels	9.24	6700	730
Combination of rooftop PV and facade PV			
MPV – roof top: 45 panels + VBPV – Full-size panels	29.04	26720	920
MPV – roof top: 45 panels + MPV - facade	29.04	26840	930

The installation cost of VBPV full-size panels is likely to be more expensive than MPV-façade, due to them being subjected to higher wind loads. However, VBPV has the additional benefit of an improved temporal match between the PV power production and electricity load, which increases the self-consumption rate and the economic value of the produced electricity [REFS]. When comparing two equally sized (P = 9.24 kWp) façade PV

1378

systems, one with MPV and another with VBPV, the VBPV system has 1.8% lower total annual production than MPV-façade (Table 4), but the production during morning and evening is significantly improved (Figure 6). This highlights the potential of VBPV full-size panels in tailoring the PV production profile. When the façade systems are combined with MPV - rooftop system (Figure 6, Table 4), this shift of production is still observable but significantly smaller: since the MPV - rooftop system has over two times higher nominal power than the façade system, it dominates the overall shape of the energy production. However, changing the façade system from MPV to VBPV lowers and widens the production peak.

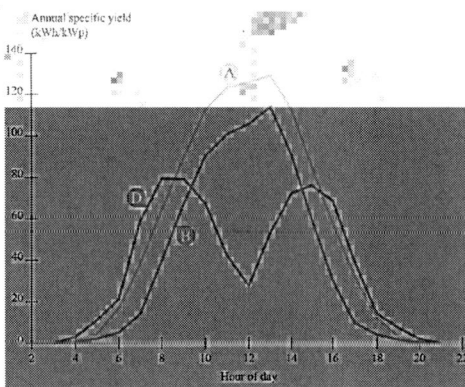

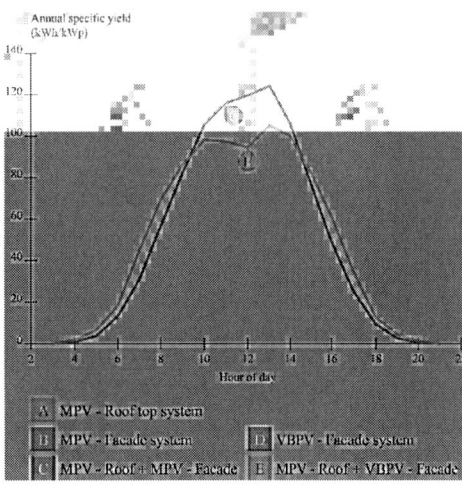

Figure 6. *The diurnal distribution of the specific yield of the MPV – roof top, MPV - facade, VBPV - facade systems and combinations including MPV – roof top and one (MPV or VBPV) facade system.*

Another factor that can make VBPV a better choice than MPV is the availability of area on the building façade. In general, buildings in Northern latitudes tend to be designed in a way that prioritizes access to solar radiation on the southern side. Buildings in the Nordic region, especially residential buildings, are often designed to maximize facade openings on the southern side. This facade design is in conflict with the utilization of PV on the facade, as they compete for the building's facade

surface. Installing more PV panels means that there will be less space for windows or balconies.

VBPV, however, occupies significantly less facade area and could be used to create desired visual obstructions between balconies. The study conducted by [10] demonstrates a VBPV system on roofs that provides space for roof gardens. Using the same principle, VBPV on the facade can provide space for windows and balconies between PV panels.

In the case study, the building has enough facade area to install an equal number of MPV panels and VBPV panels, resulting in comparable performance between the two systems. However, other residential building designs might aim to maximize openings more than the one in the case study. This will result in fewer areas to install MPV panels, making this PV system perform worse than its VBPV system counterparts.

4.2. The influence of shadings

There is one important factor that has not been thoroughly explored in this study, and that is the problem with shading. Our previous study demonstrated that shadings have a significant influence on the performance of VBPV in low-rise buildings [7]. On a multi-story building, however, the results can be different. The site of the case study was selected with minimal shadings; in other neighborhoods, the results can be entirely different. Different countries also have different approaches to the planning of high-rise buildings. The complexity of this subject is high and is beyond the scope of this study. This is one of the topics that should be explored further in other VBPV studies.

5 CONCLUSIONS

In this study, we have demonstrated that VBPV on the facade can be used to supplement the electricity from rooftop MPV on multi-story buildings. The specific yield of VBPV on the facade is slightly lower than that of rooftop MPV, but it remains within an acceptable range.

The utilization of VBPV on the facade can increase the total production of electricity. The degree of this increase can be high or low depending on the amount of electricity produced with MPV on the rooftop.

The performance of facade-installed VBPV panels is similar to that of facade-installed MPV panels considering the total annual production. However, VBPV has an advantage over MPV in residential buildings due to its morning-evening peak electricity production.

Another advantage of using VBPV is that it requires less facade area to install than MPV. The space between PV arrays can still be used to install windows or balconies. In a northern-latitude location, this is a valuable factor, since solar radiation is also used to provide natural light and heat up the living space.

Future studies should explore the effect of shadings on these facade-installed VBPV systems on multi-story buildings. There have been studies that explored this topic, but on much smaller buildings. Changing the height of the buildings affects the degree of shadings they receive. The locations of high-rise buildings also tend to be different from those of low-rise buildings, and different locations in urban environments will result in different types of shading that the buildings receive.

6 REFERENCE

[1] IEA (2023), World Energy Outlook 2023, IEA, Paris https://www.iea.org/reports/world-energy-outlook-2023, Licence: CC BY 4.0 (report); CC BY NC SA 4.0 (Annex A)

[2] Rodríguez-Gallegos, C. D., Bieri, M., Gandhi, O., Singh, J. P., Reindl, T., & Panda, S. K. (2018). Monofacial vs bifacial Si-based PV modules: Which one is more cost-effective?. Solar Energy, 176, 412-438. https://doi.org/10.1016/j.solener.2018.10.012

[3] Jouttijärvi, S., Lobaccaro, G., Kamppinen, A., & Miettunen, K. (2022). Benefits of bifacial solar cells combined with low voltage power grids at high latitudes. *Renewable and Sustainable Energy Reviews, 161*, 112354. https://doi.org/10.1016/j.rser.2022.112354

[4] Jouttijärvi, S., Karttunen, L., Ranta, S., & Miettunen, K. (2024). Techno-economic analysis on optimizing the value of photovoltaic electricity in a high-latitude location. *Applied Energy, 361*, 122924. https://doi.org/10.1016/j.apenergy.2024.122924

[5] Anvari, M., Proedrou, E., Schäfer, B., Beck, C., Kantz, H., & Timme, M. (2022). Data-driven load profiles and the dynamics of residential electricity consumption. Nature communications, 13(1), 4593. https://doi.org/10.1038/s41467-022-31942-9

[6] Jouttijärvi, S., Thorning, J., Manni, M., Huerta, H., Ranta, S., Di Sabatino, M., ... & Miettunen, K. (2023). A comprehensive methodological workflow to maximize solar energy in low-voltage grids: A case study of vertical bifacial panels in Nordic conditions. Solar Energy, 262, 111819. https://doi.org/10.1016/j.solener.2023.111819

[7] Viriyaroj, B., Jouttijärvi, S., Jänkälä, M., & Miettunen, K. (2024). Performance of vertically mounted bifacial photovoltaics under the physical influence of low-rise residential environment in high-latitude locations. Frontiers in Built Environment, 10, 1343036. https://doi.org/10.3389/fbuil.2024.1343036

[8] City of Helsinki. (2022). Make 2.0 Puinen mallikerrostalo. https://puuinfo.fi/suunnittelu/tyyppisuunnitelmat/puinen-mallikerrostalo-make-2-0/

[9] Pvsyst (2024). Logiciel photovoltaïque. Available at: https://www.pvsyst.com/(Accessed October 26, 2024).

[10] Baumann, T., Nussbaumer, H., Klenk, M., Dreisiebner, A., Carigiet, F., & Baumgartner, F. (2019). Photovoltaic systems with vertically mounted bifacial PV modules in combination with green roofs. Solar Energy, 190, 139-146. https://doi.org/10.1016/j.solener.2019.08.014

6 ACKNOWLEDGEMENTS

We acknowledge financial support from Research Council of Finland (project ECOSOL, numbers 34275 and 347277) and from Strategic Research Council within the Research Council of Finland (project RealSolar, number 358542). The content of the publication is the responsibility of the authors.

41st European Photovoltaic Solar Energy Conference and Exhibition

REDUCING THE ANGULAR COLOUR DEPENDENCE OF BUILDING INTEGRATED PHOTOVOLTAIC MODULES BASED ON OPTICAL INTERFERENCE COATINGS

Chang Chuan You [a,*], Ørnulf Nordseth [a], Arne Røyset [b], Tore Kolås [b],
[a] Institute for Energy Technology, NO-2027 Kjeller, Norway
[b] SINTEF Industry, NO-7465 Trondheim, Norway
*Corresponding author: chang.chuan.you@ife.no

ABSTRACT: This work focuses on optical interference coatings applied on the back side of the front glass to achieve the colouration of crystalline silicon-based photovoltaic (PV) modules. Such interference coatings can be designed to obtain desired aesthetic properties combined with a relatively low loss in power conversion efficiency, which makes this colouring technology attractive for building integrated photovoltaics. The visual appearance and power conversion efficiency are influenced by the coating design parameters such as the number of layers, design wavelength, layer thicknesses and refractive indices, and the glass surface's texture. One drawback of using optical interference coatings is that the colour changes with the observation angle. In order to reduce the angular colour dependence, we have fabricated and analyzed PV mini-module prototypes with both different glass textures and different coating design parameters. A simulation model was adopted to investigate relevant coating designs and predict their performance, with the aim of achieving low angular colour dependence and high-power conversion efficiency.
Keywords: BIPV, colour, optical interference coatings, visual appearance

1 INTRODUCTION

When photovoltaic (PV) modules are integrated into the facade of buildings, aesthetic aspects become important, such as colour perception. Optical interference coatings are attractive for adding colour to PV modules since dielectric thin films absorb very few photons. By applying such interference coatings on the front glass, colourful and efficient PV modules can be realized, with losses close to the theoretical minimum limits. In addition, this gives a large design freedom and, in general, a wide range of colours can be obtained. However, one drawback of this technology is the angular colour dependence, i.e., the colour changes with the observation angle, which is often unwanted from an architectural perspective. This work aims to investigate how angular colour dependence is affected by different design parameters, including the surface texture of the front glass and the refractive indices of the thin films. The work is a continuation of a study that was recently published, which involved the analysis of PV modules with interference coatings on planar glass [1].

Different concepts involving interference coatings applied to the front glass of PV modules have been published [2–6]. Here, we investigate how the angular dependence is affected by the various design parameters, including the surface texture. PV mini-module prototypes with both different glass textures and different coating design parameters have been fabricated and analyzed. A simulation model based on 3D optical ray tracing, which takes into account the front glass surface texture, has been established in the present work and validated by comparing simulated reflectance curves to measured reflectance curves. The model will be used for the prediction of the performance of relevant coating designs.

2 METHODOLOGY

2.1 Experimental details

Fig. 1 shows a schematic overview of the experimental sample design for a low-index case, in which the optical interference coating consisted of alternating a-SiOx:H (refractive index n = 1.5) and a-SiNx:H (n = 2.0) thin films prepared by plasma-enhanced chemical vapour deposition

(PECVD). For the high-index case, the optical coating consisted of sputter-deposited thin films of YOx (n = 1.90) and TiOx (n = 2.25). The refractive index (n) and thickness of the dielectric thin film layers were determined using a variable angle spectroscopic ellipsometer (Woollam VASE). The measured ellipsometry data was modelled using the software WVASE32 from Woollam. In the present work, we have chosen the coating design parameters N and M to be N5M1 and N3M3 at a design wavelength $\lambda_D = 630$ nm, where N is the number of layers, and M represents a layer thickness parameter. Here, M is an integer and is the thickness normalised to a quarter-wavelength layer. For M = 1, the thickness is equal to one quarter wave layer, for M = 2, the thickness is equal to 2 quarter wave layers, etc. M is therefore defined as the layer thickness parameter which determines whether the interference peak is the first order (M = 1), second order (M = 3) or third order (M = 5) [1]. The materials and corresponding layer thickness for the different designs analyzed in this study are summarized in Table I.

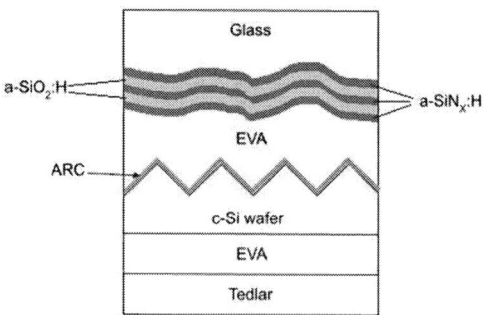

Figure 1: Schematic overview of the experimental sample design. A conventional ~75-nm-thick anti-reflection coating of SiNx:H was deposited on a pyramidal-textured monocrystalline Si wafer, which was laminated to a glass with a textured back surface onto which optical interference coatings were deposited. Note that metal contacts were omitted from the front and rear side of the silicon wafers to simplify the optical reflectance measurements and corresponding analysis.

Table I: Layer specifications for the different designs with a design wavelength $\lambda_D = 630$ nm.

Design	Index case	Layer material	Layer thickness (nm)
N5M1	Low-index	SiO$_2$	105
		SiN$_x$	78.8
N5M1	High-index	YO$_x$	82.9
		TiO$_x$	70
N3M3	Low-index	SiO$_2$	315
		SiN$_x$	236.3
N3M3	High-index	YO$_x$	248.7
		TiO$_x$	210

A detailed experimental sample matrix is presented in Table II. Three types of glass substrates were used: planar glass, frosted microscope slides (Thermo scientific Menzel-Gläsers) and commercial PV module glass. Table III shows the surface topography data for frosted and PV module glass substrates obtained using a NANOVEA ST400 modular optical profilometer.

Table II: Overview of experimental samples.

Sample no.	Texture	Design	Refractive index
1	Planar	N5M1	Low
2	Frosted	N5M1	Low
3	PV module	N5M1	Low
4	Planar	N5M1	High
5	Frosted	N5M1	High
6	PV module	N5M1	High
7	Planar	N3M3	Low
8	Frosted	N3M3	Low
9	PV module	N3M3	Low
10	Planar	N3M3	High
11	Frosted	N3M3	High
12	PV module	N3M3	High

Table III: Surface topography data for frosted and PV module glass substrates. The profile parameter is defined by ISO 4287. R_q is the root-mean-square deviation of the assessed profile. R_{dq} is the root-mean-square slope of the assessed profile. R_{Sm} is the mean spacing of profile elements.

Glass type	R_q [μm]	R_{dq} [°]	R_{Sm} [μm]
Frosted	1.45	11.8	46.9
PV module	1.23	2.82	268

PV mini-module prototypes were made by laminating the coated glass substrates on top of pyramidal-textured mono-crystalline Si wafers, using a tedlar backsheet and EVA at 150 °C in a P.Energy L036A laminator. (100)-oriented, p-type CZ monocrystalline silicon wafers with a thickness of 200 μm were used as substrates. The wafers were surface textured by etching in 30 % KOH at 75 °C for 2 min and subsequently cleaned in 20 % HCl at room temperature for 5 min and rinsed in deionized water. An 80-nm-thick anti-reflection coating of hydrogenated amorphous silicon nitride (a-SiN$_x$:H) was deposited onto the silicon substrate at 350 °C using an Oxford Instruments PlasmaLab System133 PECVD tool with a gas mix of silane, ammonia, and nitrogen [1].

The angle-dependent optical reflectance data of the PV mini-modules was obtained using a monochromatic light source with wavelengths ranging from 350 nm to 1100 nm (with a step size of 10 nm) in a home-built spectral response measurement setup. This setup consists of a Newport Oriel Apex illuminator with a Cornerstone 260 monochromator, a set of collimating and focusing lenses, and a 150 mm integrating sphere (Labsphere model RTC-060-SF) with a centre-mounted sample holder which can be rotated to a desired incident angle. The angle of incidence for the reflectance measurements was varied from 10° to 70°, in steps of 10°.

2.2 Modelling

A simulation model of front glass with optical interference coatings was adopted to investigate selected coating designs and predict their performance. The simulation model also included a simplified method to study the effect of glass surface texture. The simulation model was implemented in the optical ray tracing software TracePro from Lambda Research Corporation. Both low-index cases and high-index cases were modelled for the coating design parameter N3M3.

Optical absorption was added to the EVA layers to ensure that only the first surface of the EVA contributed to reflected light in the simulations. All the other materials simulated were assumed to be non-absorbing. The front surface of the front glass was kept planar, and without any coatings, in the simulation model. The interference coatings described above were applied to the back surface of the glass. In addition, a simplified method to vary the texture of the back surface of the glass was applied, consisting of pits shaped as spherical caps as illustrated in Fig. 2. We have considered three average slope angles for the spherical cap: 0°, 5.4°, and 10.7°. In order to reduce the simulation time, only one such pit was applied to the surface. The rays were contained within a mirror box, providing the same simulation results as would be obtained with a surface with multiple overlapping spherical caps placed in a grid pattern.

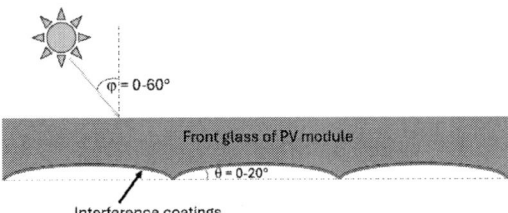

Figure 2: 2D illustration of the spherical cap implemented in the 3D ray tracing model to simulate optical reflection from interference coatings on a textured glass surface.

For both the experimental characterisation as well as the simulation, light was impinging on the surface at selected angles of incidence and the total reflected light resulting from all angles of reflection was measured and quantified. This corresponds to a situation of observation, where diffuse light (e.g., from an overcast sky) is incident on the surface, and the colour of the surface is determined at a given angle of observation. Simulations were carried out with light sources of different wavelengths from 430 nm to 680 nm, with steps of 5 nm, using 10000 initial rays. For each wavelength, two different incidence angles were simulated, 0° (normal incidence) and 60°. For each simulation, the total amount of reflected light was recorded. The spectral peak wavelengths were found, and the wavelength shifts were calculated as the peak wavelength difference between the 0° and 60° cases.

3 RESULTS AND DISCUSSION

Fig. 3(a) shows reflection data curves measured at an incident angle ranging from 10 – 70° for Sample 9 (as an example). As shown, the position of the reflection peak shifted towards shorter wavelengths (i.e., the wavelength shift) when the angle of incidence was increased, indicating a gradual colour shift. This colour shift is visualised by the CIE 1931 xy chromaticity diagram in Fig. 3(b) and by the simulated sRGB colour patches displayed in Fig. 3(c). The colour was simulated using the CIE1931 2° standard observer colour matching functions and CIE standard daylight illuminant D65. The chromaticity diagram was obtained using the free MATLAB colour toolbox [7]. Generally, the colour dependence can be determined by evaluating the wavelength shift at an increasing incidence angle. A small overall wavelength shift indicates better colour stability.

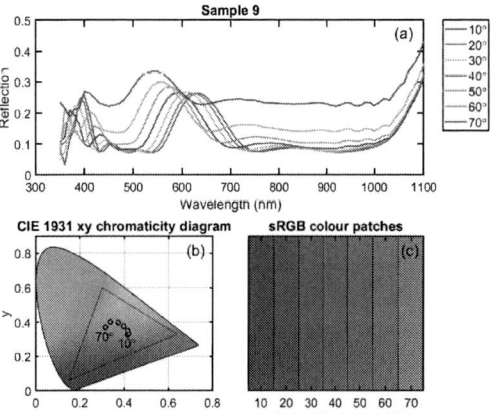

Figure 3: (a) Reflection curves were obtained at different angles of incidence for Sample 9, for which the optical interference coating was deposited on a PV module glass. Note that the position of the reflection peak shifts towards shorter wavelengths when the incident angles are increased. (b) The CIE 1931 xy chromaticity plot for the calculated colours is based on the reflection data acquired at incident angles ranging from 10 – 70°. (c) Simulated sRGB colour patches based on the reflection data for angles of incidence ranging from 10 – 70°.

As shown in Fig. 4(a) and (b) for the low-index cases with the coating design parameter N5M1 and N3M3, respectively, higher colour stability (i.e., low angular dependency) was achieved for the optical coatings prepared on the frosted glass as compared to the planar and PV module glass, as the measured wavelength shift was considerably smaller for the frosted glass. Note that the wavelength shift is the difference between the wavelength corresponding to the reflection peak at the lowest (0° in the model) angle of incidence and the peak wavelength at higher incident angles. Recently, Xu et al. reported a similar suppression of the angular colour dependence when they applied optical interference coatings on sandblasted glass sheets with rough surface texture [8]. For

the N3M3 low-index case, the optical modelling results obtained at an incident angle of 60°, as represented by the green star symbols in Fig. 4(b), show a decrease in the wavelength shift from 83 nm to 72 nm and 54 nm when the average slope angle of the surface texture of spherical caps was increased from 0° (topmost) to 5.4° (middle) and 10.7° (bottom-most), respectively. It should be noted that the simulation results agree relatively well with the experimental data at the incident angle of 60°.

Another way to lower the angular colour dependence is to use dielectric thin films with higher optical refractive indices. This is evident by comparing the wavelength shift as a function of incident angle in Fig. 4(a) and (c) for samples with the N5M1 design parameter, and in Fig. 4(b) and (d) for samples with the N3M3 design parameter. As can be seen, the wavelength shifts are considerably smaller for the high-index case than those for the low-index case, especially for the planar and PV module glass at higher incident angles.

Interestingly, there is almost no apparent effect of the glass surface texture on the colour stability in the high-index cases because the wavelength shift did not change noticeably with rougher glass surface texture, as indicated by Fig. 4(c) and (d). However, for the N3M3 high-index case, the model predicts a considerable improvement in the colour stability at an incident angle of 60° when the average slope angle of the spherical caps was increased from 0 to 10.7°, as shown in Fig. 4(d). This discrepancy between the experimental and simulated results is currently not fully understood and will be subject to further investigation. One possible explanation is that the layers in the thin film stack were not perfectly deposited on the rough surface (i.e., not following the surface contours) of the frosted glass during the film deposition process, leading to incomplete film coverage of the glass substrate.

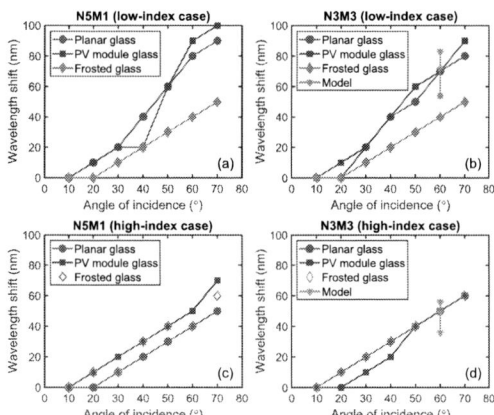

Figure 4: Measured wavelength shift of the reflection peak as a function of angle of incidence for (a) N5M1 low-index case; (b) N3M3 low-index case; (c) N5M1 high-index case; (d) N3M3 high-index case. The wavelength shift is determined as the difference between the wavelength corresponding to the reflection peak at the lowest (0° in the model) angle of incidence and the peak wavelength at higher incident angles. The green star symbols in (b) and (d) represent the simulation results obtained for three average slope angles of the spherical caps: 0° (topmost), 5.4° (middle), and 10.7° (bottom-most). The solid lines are a guide to the eye.

4 CONCLUSION

In this work, we have analyzed the angular colour dependence of coloured building-integrated photovoltaic modules based on crystalline silicon solar cells. The colour was achieved by applying optical interference coatings on the back side of the front glass, and the influence of various coating design parameters, including the surface texture of the glass, on the visual performance was investigated. Our results show that the angular colour dependence can be reduced by increasing the refractive indices of the interference coatings and optimizing the surface texture of the glass. We found that the experimental results obtained for fabricated samples corresponded well with the results obtained using a 3D optical ray tracing model. The simulation model allows for analysis of the performance of relevant coating designs, and thus, can provide useful insights into how such optical interference coatings can be used to tailor-make coloured PV modules that are both visually attractive and provide high efficiency.

5 ACKNOWLEDGEMENTS

The authors gratefully acknowledge the financial support from the Research Council of Norway through The Research Center for Sustainable Solar Cell Technology (RCN project number 257639).

The authors also thank Bent Thomassen and Junjie Zhu at IFE for assisting with the fabrication of optical interference coatings and the lamination of the PV mini-module prototypes.

6 REFERENCES

[1] A. Røyset, T. Kolås, Ø. Nordseth, C. C. You, Optical interference coatings for coloured building integrated photovoltaic modules: Predicting and optimising visual properties and efficiency, Energ. Buildings 298 (2023) 113517.

[2] A. Wessels, A. Callies, B. Bläsi, T. Kroyer, O. Höhn, Modelling the optical properties of Morpho-inspired thin-film interference filters on structured surfaces, Opt. Express 30 (2022) 14586–14599.

[3] Z. Xu, T. Matsui, K. Matsubara, H. Sai, Tunable and angle-insensitive structural coloring of solar cell modules for high performance building-integrated photovoltaic application, Sol. Energy Mater. Sol. Cells 247 (2022) 111952.

[4] B. Bläsi, T. Kroyer, T. Kuhn, O. Höhn, The morphocolor concept for coloured photovoltaic modules, IEEE J. Photovolt. 11 (2021) 1305–1311.

[5] Zhou Z, Jiang Y, Elkin-Daukes N, Keevers M, Green M. Optical and thermal emission benefits of differently textured glass for photovoltaic modules. IEEE J Photovolt. 11 (2021) 131–137.

[6] K. Manwani, M. Lagier, A. Krammer, J. Fleury, A. Schüler, Development of novel orange colored photovoltaic modules with improved angular stability and high energy efficiency, Sol. Energy Mater. Sol. Cells 278 (2024) 113144.

[7] S. Westland, Computational Colour Science using MATLAB 2e, MATLAB Central File Exchange, 2023.

[8] Z. Xu, T. Matsui, K. Matsubara, H. Sai, Effect of multilayer structure and surface texturing on optical and electric properties of structural colored photovoltaic modules for BIPV applications, Appl. Energy, 367 (2024) 123347.

41st European Photovoltaic Solar Energy Conference and Exhibition

DESIGN AND OPTIMIZATION OF STRUCTURAL COLORED INTERLAYERS FOR BUILDING-INTEGRATED PHOTOVOLTAIC APPLICATIONS

Catarina G. Ferreira[1,2], Irina Vyalih[2,3], Jani Lamminaho[2,3], Markus Babin[4], Nanna Lysgaard Andersen[4],
Peter Behrensdorff Poulsen[4], Sune Thorsteinsson[4], Karlis Petersons[5], Joel D. Cox[1,2,6], Morten Madsen[2,3]

[1] POLIMA, University of Southern Denmark, Campusvej 55, 5230 Odense M, Denmark
[2] SDU Climate Cluster, University of Southern Denmark, Odense 5230, Denmark
[3] CAPE, University of Southern Denmark, Mads Clausen Institute, 6400 Sønderborg, Denmark
[4] Technical University of Denmark, Institute of Electrical and Photonics Engineering, 4000 Roskilde, Denmark
[5] Stensborg A/S, 4000 Roskilde, Denmark
[6] Danish Institute for Advanced Study, University of Southern Denmark, Campusvej 55, DK-5230 Odense M, Denmark

ABSTRACT: Building-integrated photovoltaics offer a significant increase in available surface area for photovoltaic (PV) modules, without the need for a new electricity infrastructure, making them a promising solution to fulfill the global energy demand in a green and renewable way. However, in addition to high power conversion efficiencies, PV modules for building integration also benefit from an aesthetically attractive design, which can be achieved by fine-tuning their color according to the architectural features of the building.
This work explores in detail the structural coloration of silicon PV modules arising from the addition of a 1-dimensional thin multilayer structure to the interlayer between the glass substrate and the PV cell. Three different configurations, with varying degrees of periodicity, are considered and compared, to finally demonstrate the outstanding potential of employing optimization algorithms to obtain structural colored PV modules with a broader color gamut.
Keywords: Building-integrated photovoltaics, structural colors, distributed Bragg reflectors, inverse design optimization

1 INTRODUCTION

Photovoltaics (PVs) stand as key enabling technology to fulfill the ever-growing energy demand in a clean and sustainable way. However, despite the continuous increase in global PV installation, only a small share of the total worldwide electricity is currently provided by PV cells and modules [1]. A promising route to significantly expand the surface area coverage of PVs, without the need for a new electricity infrastructure or for additional load on the land resources, relies on integrating PV modules on the roofs and façades of buildings [2]. Building-integrated photovoltaics (BIPV) offer as main advantage on-site production of electricity at attractive costs, by simultaneously acting as energy generators and construction elements of the buildings [3]. Nevertheless, to boost social acceptance of such BIPV technology requires combining high power conversion efficiency (PCE) with an aesthetically attractive design, tuned to the specific architectural needs of the building [4,5].

Such purpose can be achieved through development of colored PV modules, tailored to replicate commonly used raw construction materials (clay, wood, concrete) or to match the surrounding environment of the building [6,7]. For some PV technologies, such as organic, perovskite, and dye-sensitized solar cells, intrinsic coloration can be obtained by selecting active layer materials with specific spectral absorption behavior [8,9]. This methodology, however, is only applicable to a restricted set of materials and allows for a limited selection of colors. An alternative approach consists in adding a colored element (glass, encapsulant, interlayers) in front of the absorbing layer of the PV cell. Pigments or chemical colorants are typically used to achieve such coloration, but they strongly absorb light that should be available to the PV cell, resulting in severe PCE drops [10]. To overcome such hurdles, structural coloration originating from light interference in thin-layer structures composed of non-absorbing dielectric materials have been considered as a promising approach to

obtain PV modules with vivid colors and low optical losses [11,12]. In the present work, an in-depth study will be performed on the properties of such 1-dimensional multilayer structures for structural coloration of silicon PV modules, including their drawbacks, limitations, and future research directions. In addition to the conventional periodic designs, non-trivial configurations obtained from an inverse design approach will be considered to demonstrate the enormous potential of such optimization techniques to achieve structural colors with higher color gamut.

2 METHODOLOGY

2.1 Structural colored PV modules

This work is part of the ColorFoil project, that seeks to achieve homogeneous, low-iridescent, colored silicon PV modules by depositing a planar multilayer stack directly on top of an interlayer (polymer foil incorporating a resin-based diffuser structure), placed in between the glass substrate and the PV cell as shown in Figure 1.

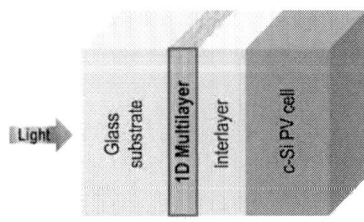

Figure 1: Schematic illustration of the structural colored PV module considered in ColorFoil. Encapsulant layers are omitted, as they are index-matched to the interlayer and glass materials.

This 1-dimensional multilayer structure intercalates two different dielectric materials with contrasting refractive indexes that should be compatible with large-scale deposition techniques. For that reason, TiO_2 is used in this case as high refractive index material while ZnO was chosen as the low refractive index dielectric. The number of thin layers and respective thicknesses were adjusted according to the structural color to be reflected, considering three different approaches, as later described in Section 3.

2.2 Light propagation calculations

Key to achieve structural coloration in ColorFoil modules is the use of a 1-dimensional multilayer structure, for which the reflectance can be tuned by controlling light interference between the different composing thin layers, and which is anticipated to serve as optical filter to the broadband absorbing thick silicon PV cell. For this reason, a simplified configuration (Figure 2) is considered in the present work for the light propagation calculations, which is expected to result in structural colors close to the ones produced by the complete ColorFoil PV module. In this case, the multilayer is placed on the back of a thick layer with a wavelength-independent refractive index of 1.5 (simulating the glass substrate and respective index-matching encapsulant), with light incident from air, and with a semi-infinite final medium with a constant refractive index of 1.5 (simulating the polymer foils and encapsulant used in front of the silicon PV cell).

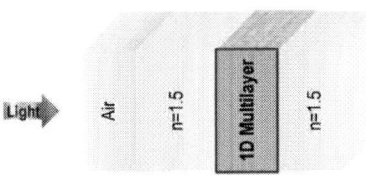

Figure 2: Schematic illustration of the system considered in this work for the light propagation calculations.

Optical simulations of the light propagation in this planar configuration are performed by solving the 1-dimensional Maxwell equations through a generalized version of the transfer matrix method [13,14], in which light is considered to propagate coherently inside the thin multilayer structure, to correctly account for the inherent interference effects derived from having layer thicknesses close to the incident wavelength, and to lose such coherency inside the much larger (several orders of magnitude thicker than the wavelength of the incident light) front substrate. In these simulations, light is considered to be unpolarized (as it happens with sunlight), such that equal amounts of parallel and perpendicular polarization states are assumed for the calculation of the reflectance spectra, which is particularly relevant for off-normal incidence.

2.3 Simulations of the reflected color

Tristimulus values X, Y, and Z are directly inferred from the simulated reflectance spectra, $R(\lambda)$, as:

$$X = \frac{\int I(\lambda)R(\lambda)\bar{x}(\lambda)d\lambda}{\int I(\lambda)\bar{y}(\lambda)d\lambda}, \quad \text{(eq. 1a)}$$

$$Y = \frac{\int I(\lambda)R(\lambda)\bar{y}(\lambda)d\lambda}{\int I(\lambda)\bar{y}(\lambda)d\lambda}, \quad \text{(eq. 1b)}$$

$$Z = \frac{\int I(\lambda)R(\lambda)\bar{z}(\lambda)d\lambda}{\int I(\lambda)\bar{y}(\lambda)d\lambda}, \quad \text{(eq. 1c)}$$

where $\bar{x}(\lambda)$, $\bar{y}(\lambda)$, and $\bar{z}(\lambda)$ are the color matching functions describing the average chromatic response of a 2° observer and $I(\lambda)$ is the spectral irradiance of the light source, in this work considered to be the D65 standard illuminant that describes the spectral power distribution of natural daylight on a clear day around noon.

These tristimulus values can be further converted into coordinates in different color spaces, by applying the corresponding transformation equations [15]. In this work, conversion to both the L*a*b* coordinates defined by the International Commission on Illumination (CIE) using D65 illuminant as white point, and to sRGB coordinates is carried out to calculate color differences and to plot the generated colors, respectively.

3 RESULTS

3.1 Periodic multilayer structures

The most commonly used designs to achieve structural coloration in 1-dimensional multilayer structures explore the conditions for constructive interference of the reflected light happening in a distributed Bragg reflector (DBR). In this case, the thicknesses (d) of the composing materials are set to $d = N\lambda / 4n(\lambda)$, where N is an odd integer number, λ is the design wavelength of the DBR and $n(\lambda)$ is the refractive index of the material at the chosen design wavelength [16,17].

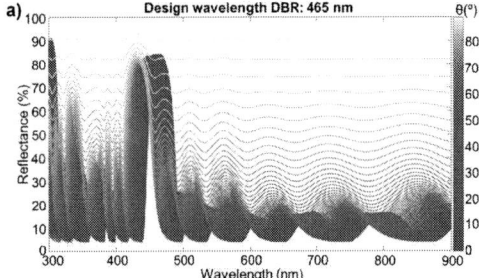

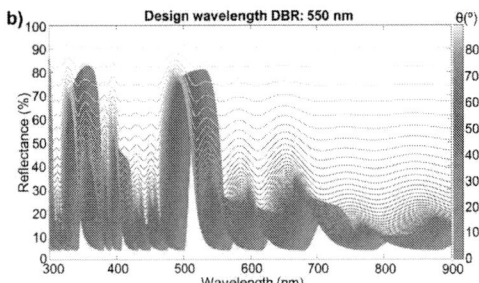

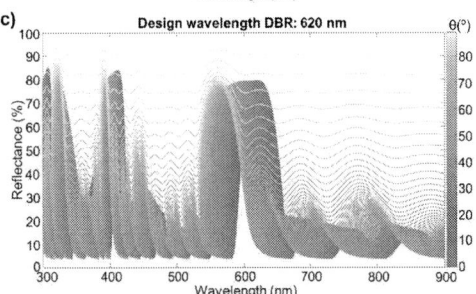

Figure 3: Reflectance spectra and corresponding color in the sRGB color space as a function of the incident angle for 7-layer periodic structures with design wavelengths in the a) blue, b) green, and c) red.

As shown in Figure 3, the resulting periodic structures designed following this approach reflect very efficiently the light in the vicinity of the wavelength of interest, when the total number of thin layers is as low as 7, resulting in quite vivid and saturated reflected colors. For design wavelengths corresponding to blue (465 nm) and green (550 nm) light, the reflected colors are consistent with the desired ones, with some angular dependency (iridescence) originating due to the increase in effective optical path inside the layers when the angle of incidence of the light increases. However, when the design wavelength is further redshifted, this iridescence increases, the main reflectance peak enlarges, and higher-order interference peaks start to appear at wavelengths around 400 nm, as illustrated in Figure 3c for a DBR structure designed to highly reflect the light at 620 nm. As the wavelengths of these higher-order peaks correspond to blue light to which the human eye is already sensitive, the resulting color reflected at normal incidence ends up being pinkish instead of red. Therefore, despite the promising potential of these simpler periodic structures to produce blue and green structural colors, they are unable to realize several red color hues. For this reason, it becomes crucial to develop strategies to break the symmetry of periodic DBRs in a controlled way, to achieve structural coloration with a higher color gamut.

3.2 Beyond periodic structures

Aperiodicity in the thin layer structure appears to be essential to produce a broader range of structural colors. Hence, this section will focus on two distinct approaches where the periodicity of DBRs will be broken to allow for a larger color gamut to be reached.

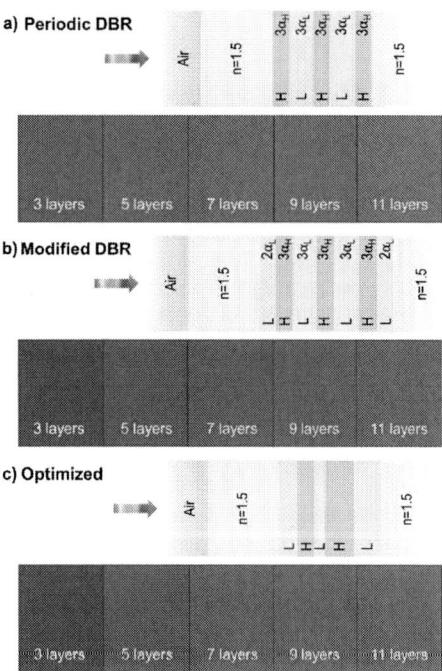

Figure 4: Color reflected at normal incidence, as a function of the total number of dielectric layers, for the three different configurations considered in this work to achieve a structural red color: a) periodic DBR, b) modified DBR, and c) optimized multilayer structure. In the diagrams of the layer structure, H and L refer to the high and low refractive index materials and α is defined as $\alpha = \lambda / 4n(\lambda)$, with $\lambda = 620$ nm.

The first approach consists in modifying the DBR structure by introducing two additional layers of the low refractive index material: one in between the thick glass substrate and the remaining thin layers of the stack and the other one right after the multilayer structure, in contact with the final medium (Figure 4b) [18,19]. By creating a smoother transition in refractive index, unwanted oscillations in the reflectance spectra should be reduced, provided that the thicknesses of these two new layers fulfill the condition that N is now an even integer (as they are placed in between a higher and a lower refractive index material), thus breaking the periodicity of the previous DBR configuration. Indeed, as clearly shown in Figure 5 (blue curve), for a 7-layer structure having this modified DBR configuration, the undesired oscillations in the reflectance spectra are significantly reduced when compared to use of an equivalent periodic structure. The strong secondary peak observed around 400 nm, however, was reduced but not completely removed. For that reason, despite the improvements observed in comparison to the periodic DBR, the structural colors obtained, especially for a higher number of dielectric layers, still present some traces of pink. This is evidenced both in the color maps of Figure 4b and in the a*b* color coordinates represented in Figure 6.

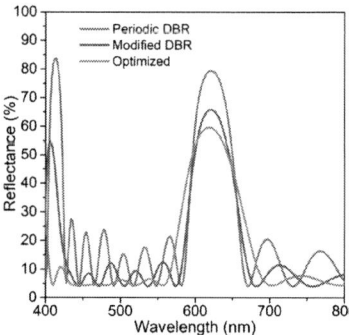

Figure 5: Spectral reflectance at normal incidence for three multilayer structures with a total of 7 layers and with the different layer configurations considered in this work.

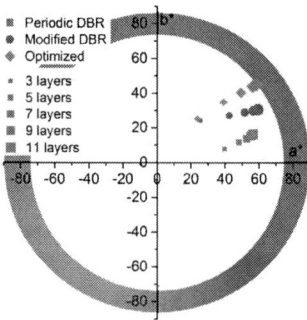

Figure 6: Representation in the a*b* plane of the structural colors reflected by each of the multilayer configurations considered in this work.

To overcome this limitation, a novel approach was explored in this work, in which an optimization algorithm was employed in combination with the transfer matrix method to solve the inverse problem of determining the optimal layer thicknesses that minimize an appropriately defined objective function [20,21], in this case the color difference between a user-defined target red color and the structural color reflected from the multilayer structure. During this procedure, the dielectric materials were fixed to the ones considered for the former two configurations, but the thicknesses of all layers were allowed to freely vary between 0 nm and 350 nm, without any restrictions. For that reason, the resulting configurations, herein designated optimized, consist of nontrivial nonperiodic combinations of layer thicknesses that allow to go beyond the more traditional DBR structures to minimize the unwanted reflectance oscillations and to completely eliminate the higher-order peak at 400 nm, without sacrificing the main reflectance feature in the spectral region corresponding to the red light (Figure 5, green curve). As a result, the structural colors obtained when using this inverse design technique are free from any pink tones, thus approaching the desired red color, as illustrated in Figure 4c. Besides, the representation of the a*b* color coordinates of Figure 6 indicates that an increase in the number of thin layers, allowing for a higher tunability of the layered structure, results in configurations producing more vibrant structural red colors, as opposed to the former two cases where saturation seems to be reached with a low number of dielectrics layers. This allows to anticipate that this methodology will be extremely powerful to design structural colored PV modules with a higher color gamut, therefore extending the application range of BIPV.

4 CONCLUSIONS

The present work explored the use of 1-dimensional thin multilayer structures that take advantage of constructive interference effects to produce different structural colors for application in BIPV. In particular, three different configurations were herein addressed, with varying degrees of periodicity, which included a fully periodic DBR structure, a semi-periodic modified DBR, and a nontrivial optimized configuration with flexible nonperiodic layer thicknesses.

Despite their simplicity and satisfactory performance when designed to reflect blue and green light, periodic DBRs revealed some severe drawbacks that hinder the production of structural red hues. As the periodicity of this configuration started to be broken and layer thicknesses started to vary in a freely way, the unwanted oscillations and secondary peaks in the reflectance spectra were minimized, giving rise to more pure structural red colors. Indeed, it was proven that the use of an optimization algorithm aiming to minimize the color difference between the one reflected by the multilayer structure and a user-specified target color is crucial to achieve structural coloration with broader color gamut.

This proof-of-concept is intended to pave the way for a more extensive use of these optimization algorithms in the design of structural colored PV modules. As a follow-up of this work, future efforts should focus on comparing different material combinations, on reducing iridescence by considering off-normal incidence, and on assessing the effects of including the silicon PV in the simulations (both regarding the influence of the PV cell in terms of the reflectance spectra - and color - of the complete device and how the presence of the multilayer structure affects the light transmitted to and absorbed in the PV).

5 ACKNOWLEDGEMENTS

This work was funded by EUDP as part of the "ColorFoil" project under grant 64022-1027.

6 REFERENCES

[1] SolarPower Europe, "Global Market Outlook for Solar Power 2024-2028", (2024).
[2] H. Lee, H. J. Song, "Current status and perspective of colored photovoltaic modules", Wiley Interdisciplinary Reviews: Energy and Environment 10 (2021) e403.
[3] N. Martín-Chivelet, K. Kapsis, H. R. Wilson, V. Delisle, R. Yang, L. Olivieri, J. Polo, J. Eisenlohr, B. Roy, L. Maturi, G. Otnes, M. Dallapiccola, W. M. P. Upalakshi Wijeratne, "Building-Integrated Photovoltaic (BIPV) products and systems: A review of energy-related behavior", Energy and Buildings 262 (2022) 111998.
[4] C. Ballif, L. E. Perret-Aebi, S. Lufkin, E. Rey, "Integrated thinking for photovoltaics in buildings", Nature Energy 3 (2018) 438.
[5] Q. Bao, T. Honda, S. El Ferik, M. M. Shaukat, M. C. Yang, "Understanding the role of visual appeal in consumer preference for residential solar panels", Renewable Energy 113 (2017) 1569.
[6] A. Røyset, T. Kolås, B. P. Jelle, "Coloured building integrated photovoltaics: Influence on energy efficiency", Energy and Buildings 208 (2020) 109623.
[7] M. Pelle, E. Lucchi, L. Maturi, A. Astigarraga, F. Causone, "Coloured BIPV Technologies: Methodological and Experimental Assessment for Architecturally Sensitive Areas", Energies 13 (2020) 4506.
[8] E. Pascual-San José, A. Sánchez-Díaz, M. Stella, E. Martínez-Ferrero, M. I. Alonso, M. Campoy-Quiles, "Comparing the potential of different strategies for colour tuning in thin film photovoltaic technologies", Science and Technology of Advanced Materials 19 (2018) 823.
[9] T. Jesper Jacobsson, J. P. Correa-Baena, M. Pazoki, M. Saliba, K. Schenk, M. Grätzel, A. Hagfeldt, "Exploration of the compositional space for mixed lead halogen perovskites for high efficiency solar cells", Energy and Environmental Science 9 (2016) 1706.
[10] M. Mittag, C. Kutter, M. Ebert, H. R. Wilson, U. Eitner, "Power loss through decorative elements in the front glazing of BIPV modules", Proceedings 33rd European Photovoltaic Solar Energy Conference and Exhibition (2017).
[11] A. Soman, A. Antony, "Colored solar cells with spectrally selective photonic crystal reflectors for application in building integrated photovoltaics", Solar Energy 181 (2019) 1.
[12] C. O. Ramírez Quiroz, C. Bronnbauer, I. Levchuk, Y. Hou, C. J. Brabec, K. Forberich, "Coloring Semitransparent Perovskite Solar Cells via Dielectric Mirrors", ACS Nano 10 (2016) 5104.
[13] E. Centurioni, "Generalized matrix method for calculation of internal light energy flux in mixed coherent and incoherent multilayers", Applied Optics 44 (**2005**) 7532.
[14] C. L. Mitsas, D. I. Siapkas, "Generalized matrix method for analysis of coherent and incoherent reflectance and transmittance of multilayer structures with rough surfaces, interfaces, and finite substrates", Applied Optics 34 (1995) 1678.

[15] Daniel Malacara, "Color Vision and Colorimetry: Theory and Applications", SPIE (2011).

[16] T. He, T. Ma, B. Bläsi, Z. Li, S. Li, Y. Chen, "Design of periodic dielectric multilayer thin films for colorizing PV panels", Solar Energy 278 (2024) 112655.

[17] A. Røyset, T. Kolås, Ø. Nordseth, C. C. You, "Optical interference coatings for coloured building integrated photovoltaic modules: Predicting and optimising visual properties and efficiency", Energy and Buildings 298 (2023) 113517.

[18] B. Blasi, T. Kroyer, T. Kuhn, O. Hohn, "B. Blasi, T. Kroyer, T. Kuhn, O. Hohn, IEEE J Photovolt 2021, 11, 1305", IEEE Journal of Photovoltaics 11 (2021) 1305.

[19] S. Lee, G. Y. Yoo, B. Kim, M. K. Kim, C. Kim, S. Y. Park, H. C. Yoon, W. Kim, B. K. Min, Y. R. Do, "RGB-Colored $Cu(In,Ga)(S,Se)_2$ Thin-Film Solar Cells with Minimal Efficiency Loss Using Narrow-Bandwidth Stopband Nano-Multilayered Filters", ACS Applied Materials and Interfaces 11 (2019) 9994.

[20] C. G. Ferreira, G. Martínez-Denegri, M. Kramarenko, J. Toudert, J. Martorell, "Light Recycling Using Perovskite Solar Cells in a Half-Cylinder Photonic Plate for an Energy Efficient Broadband Polarized Light Emission", Advanced Photonics Research 2 (2021) 2100077.

[21] C. G. Ferreira, C. Sansierra, F. Bernal-Texca, M. Zhang, C. Ros, J. Martorell, "Bias-Free Solar-to-Hydrogen Conversion in a $BiVO_4$/PM6:Y6 Compact Tandem with Optically Balanced Light Absorption", Energy and Environmental Materials 7 (2023) e12679.

41st European Photovoltaic Solar Energy Conference and Exhibition

Urban Energy Systems Laboratory, Empa

Qiuxian Li*, Natasa Vulic, Hanmin Cai, Philipp Heer

*currently at Building Physics and Sustainable Design, KU Leuven, Ghent, Belgium

UESL

COMPARATIVE ANALYSIS OF INDIVIDUAL AND COLLECTIVE PV INTEGRATION STRATEGIES FOR A RESIDENTIAL NEIGHBORHOOD

Problem statement

Photovoltaics (PV) on buildings are expected to play a significant role in decarbonizing the building sector, which contributes to one third of global carbon emissions. Increased shares of PV in the building sector poses a challenge for the low-voltage grid to absorb surplus generation and may exceed grid hosting capacity. Cooperative structures, such as private self-consumption communities, may offer one way towards a coordinated PV and electrical storage integration strategy. **This work aims to investigate the impacts of the cooperative structure on system design (positioning and capacity of PV systems), operation (grid interaction metrics), and lifecycle costs and emissions.** The approach would help identify potential inefficiencies linked to the lack of coordination between individual prosumers in the design stage, which could support decision-makers in developing suitable incentives for cooperative PV integration strategies aimed at reducing them.

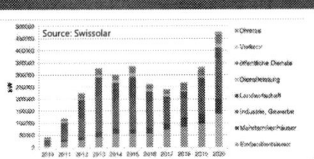

Methodology

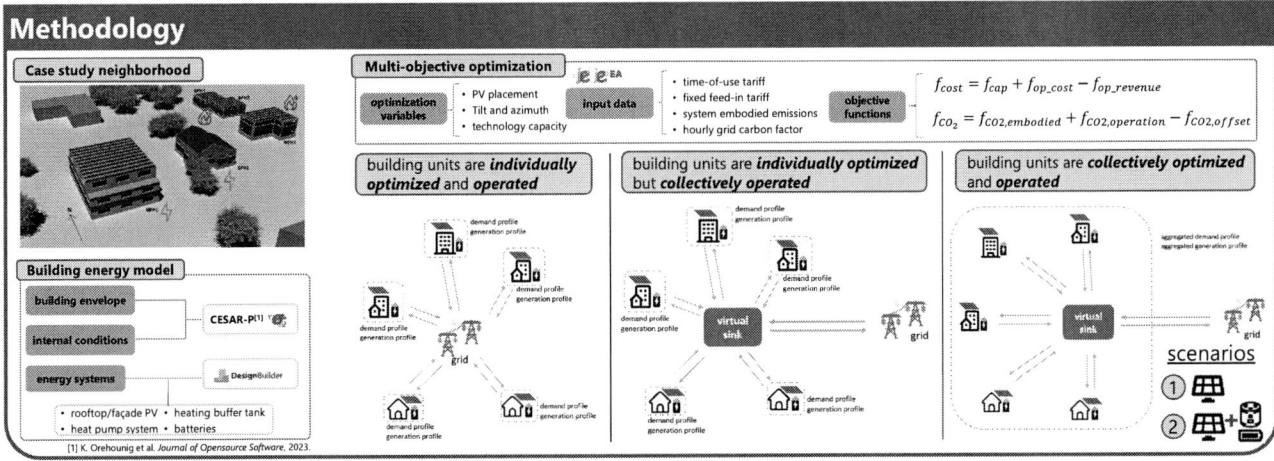

$$f_{cost} = f_{cap} + f_{op\_cost} - f_{op\_revenue}$$

$$f_{CO_2} = f_{CO2,embodied} + f_{CO2,operation} - f_{CO2,offset}$$

[1] K. Orehounig et al. *Journal of Opensource Software*, 2023.

Results

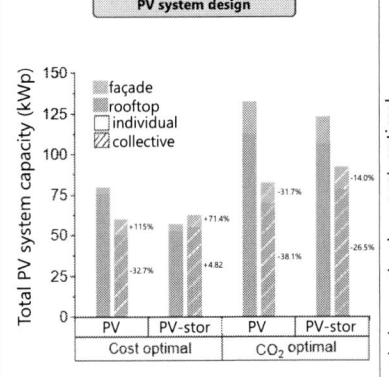

Cost optimal collective:
Smaller rooftop coverage in favor of *larger façade PV*; with therm. storage, higher rooftop coverage.

Emission optimal collective:
Higher rooftop tilt (not shown) in favor of *smaller façade coverage.*

Batteries are *not chosen* in any of the scenarios.

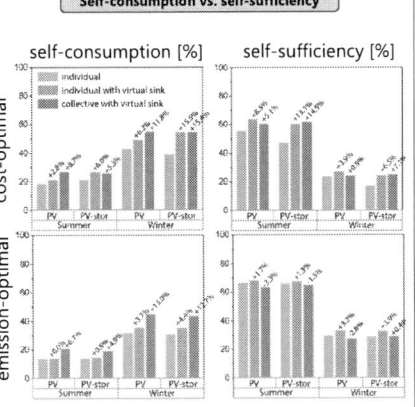

Self-consumption vs. self-sufficiency:
Improved self-consumption and self-sufficiency with the introduction of a virtual sink.

Higher self-consumption in the collectively optimized cases *with limited impact on self-sufficiency.*

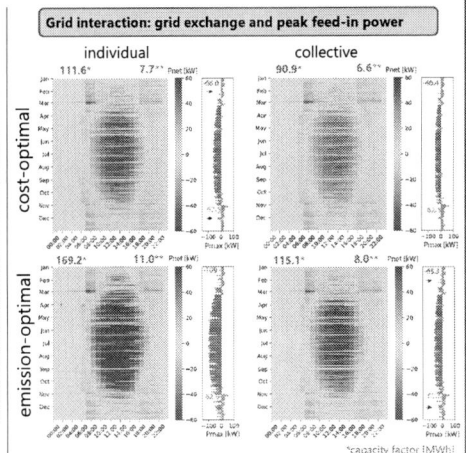

Individually vs. collectively optimized:
Individually optimized scenarios consistently have *higher peak feed-in power.* Grid interaction appears similar for the individually cost-optimal and collectively emission-optimal cases, with *differences in system design* and *winter feed-in.*

Summary

• The presented scenarios highlight the **impact** of **individual** and **collective self-consumption** on **PV system design** and **operation.**

• **Collectively optimized PV systems** tend to be **smaller** than individually optimized ones, with **trade-offs** between **cost** and **emission** objectives influencing design choices, though **batteries** are not selected in any scenario.

• Collectively optimized systems benefit in terms of **improved self-consumption** with a relatively **small impact on self-sufficiency.**

• Individually optimized systems show **higher peak grid feed-in power** than collectively optimized ones, especially for **emission-focused** designs, with policies targeting individual **carbon footprint** potentially causing **oversized systems** and **grid issues.**

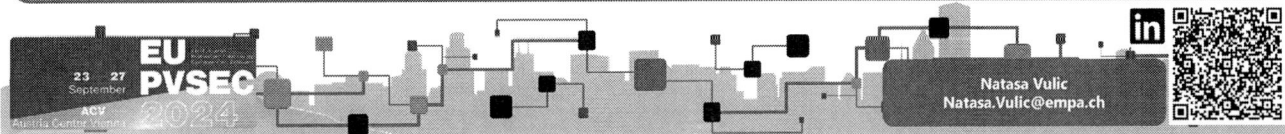

Natasa Vulic
Natasa.Vulic@empa.ch

41st European Photovoltaic Solar Energy Conference and Exhibition

Modelling framework for optimizing hybrid photovoltaic-thermal systems in combination with seasonal heat storage

Z. Ul-Abdin*, A. van Rossum, D. M. Aguilera, D. N. Kanawala, **O. Isabella**, and R. Santbergen

Photovoltaic Materials and Devices Laboratory, Delft University of Technology, The Netherlands (*Contact: Z.ulabdin@.tudelft.nl)

Motivation

- In summer, excess solar supply leads to solar panels being switched off.
- In winter, limited solar energy forces reliance on natural gas for heating.
- A solution is to couple solar collectors [1] with seasonal aquifer thermal energy storage (ATES). Figure adapted from [2].

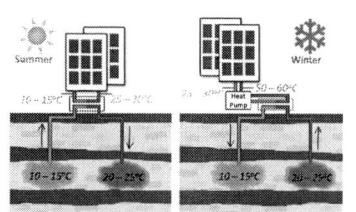

- Extending **lifespan** of PV modules through cooling [3].
- Address literature gap on occupant behavior and building characteristics through the development of a heat demand model.

Heating model

- Developed using **thermal resistance network**, incorporating building geometry, thermal properties and insulation.
- User can adjust thermal properties to simulate different household types.

High-energy vs low-energy consumers
Inside temperature (18 – 24°C)
Heating demand (almost doubled)

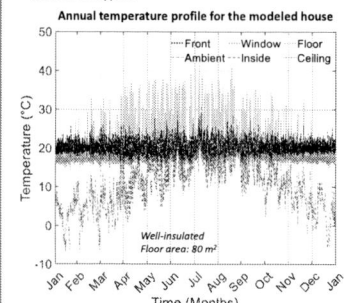

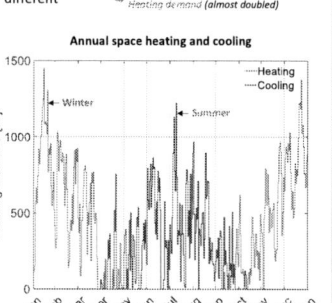

Modelling framework

- Numerical **models** are developed to simulate and evaluate heat storage solutions, considering heat loss mechanisms:
  - Conduction
  - Natural and external convection
  - Radiation
- Modeled **heat storage solutions**, categorizing hours as excess or deficit based on modules output versus demand:
  - Domestic storage tank (**short-term**)
  - ATES-Doublet (**long-term**)
- **Seasonal storage** integrated with PVT and in-built heat exchanger.
  - Aquifers shape assumed as perfectly cylindrical.
  - Heat exchanger effectiveness remained 40% year-round.
  - Homogenous temperature across aquifers depth.
- Models are integrated into the **PVMD Toolbox** [4] which can already predict the energy yield of PV systems.
- Developed a **dynamic model** of a PVT using energy conservation principle.

Water-based PVT collector

Collector area: ≈ 5 m²

Short-term solution: Domestic storage tank

- Single apartment equipped with **two modules**.
- Heat storage tank, radius 0.45 m and height 0.60 m.
- To minimize heat losses, storage insulation measures 0.10 m.
- Conduction loss, significantly impacting overall efficiency, with **combined losses** totaling approximately 0.30 MJ per hour.
- **Small heat storage** tank of 0.38 m³ is **effective for short-term** energy storage.

Domestic storage tank integrated with PVTs

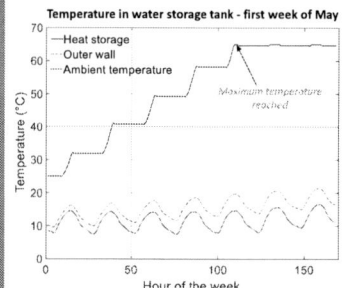

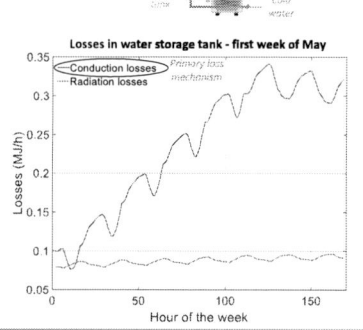

Long-term solution: Aquifer thermal energy storage

- 100 apartments, each equipped with three PVT modules.
  - Similar PVT module is used in both storage cases.
- Climate data affects **module performance**.
- Heat storage system extend the **lifespan** of the module through **effective cooling**.
- Seasonal storage system has **sufficient capacity** for the whole year.
- The **levelized cost** of thermal energy is approximately €0.12/kWh .
  - It is total annual cost divided by thermal demand delivered.

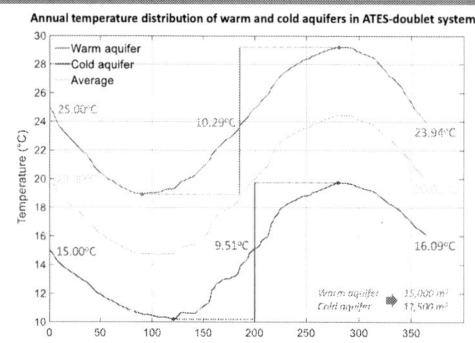

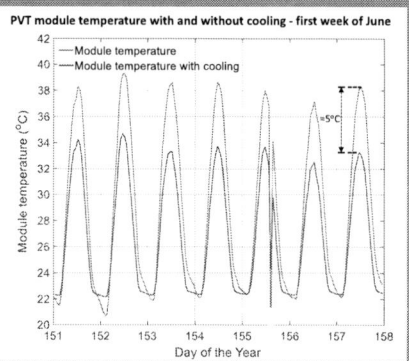

Conclusions

- Dynamic model developed for **PVT collector** generating both solar electricity and heat.
- 0.38 m³ heat storage tank is **not suitable** for seasonal energy storage.
- ATES system for 100 homes is an **effective seasonal heat storage** with limited heat loss.
  - Offer **long-term sustainability** and efficiency for urban energy systems.
- Cooling of PV cells has added advantage of **increased efficiency** and lifetime.

References

[1] S. Kranz, et al., **2015**, World Geothermal Congress, 4.
[2] M. Blomendal and T. N. Olsthoorn, **2018**, Adv. in Geosci., 45, 85-103.
[3] P. Royo, et al., **2016**, Energy, 101, 174-189.
[4] M. Vogt, et al., **2022**, Sol. Energy Mat. Sol. Cells, 247, 111944.

Acknowledgements

This work was carried out as a part of Simply Positive project. The Dutch part is supported by the RVO (the Netherlands Enterprise Agency).

Photovoltaic Materials and Devices Laboratory

www.pvmd.ewi.tudelft.nl

41st European Photovoltaic Solar Energy Conference and Exhibition

PERFORMANCE ASSESSMENT OF NOVEL SOLAR ENERGY SYSTEMS FOR AGED NEIGHBOURHOODS AND BUILDINGS IN DUTCH CITIES

Edward Otoo *[1], Guang Hu[1],
Roel C. G. M. Loonen [2], Angèle H.M. E. Reinders [1]
[1] Energy Technology Group, Eindhoven University of Technology, 5600 MB Eindhoven, the Netherlands
[2] Building Physics and Services, Eindhoven University of Technology, 5600 MB Eindhoven, the Netherlands
*E-mail: e.otoo@tue.nl

ABSTRACT: This paper presents a position paper in the framework of the HERITAGE project of the Dutch 4TU Federation. The HERITAGE, 'Heat Robustness in Relation to Ageing Cities' is an innovative project initiated by the Dutch 4TU Federation that aims to address urban overheating in Dutch cities through innovative strategies, including integration of solar energy systems to enhance the overall quality of urban life. This position paper is related to a PhD project that explores the use of photovoltaic (PV) systems in the built environment as a method for urban heat mitigation. Also, a brief literature review is presented, and suggestions for research to be executed in the coming years. Among the primary areas of focus are the development of integrated designs for solar energy systems in buildings and the evaluation of the operational performance of new solar technologies.

Keywords: Solar energy systems, Photovoltaic systems, Energy performance, Urban sustainability, Indoor thermal comfort

1 INTRODUCTION

In the effort to address the pressing issue of global climate change, which stands as a significant contemporary challenge, the Dutch 4TU Federation has initiated an innovative project known as 'Heat Robustness in Relation to Ageing Cities' (HERITAGE). The general aim of this research programme is to control overheating in Dutch cities, thus fostering a more pleasant living environment. The research programme will develop a high-tech sensing and design system aiming at the detection, reduction and prevention of heat stress occurring in built environments that existed before the eighties in Dutch urban cities. This will be achieved through continuous monitoring and innovative socio-technical solutions. Heat stress at high resolution will be detected and forecasted to provide solutions, through urban design that links energy transition, housing, climate adaptation and digitalisation. According to [1], about 50% of the Dutch urban areas experience seven days of noticeable heat stress annually during the summer period. In Rotterdam, for instance, the heat stress is noticed for 15 days [1,2] (i.e. 4.1% of the days in a year) exceeding the critical threshold of 27.7 °C, and surpassing 32.2 °C for two days (i.e. 0.5% of the days in a year) [2]. Heat stress can lead to health risks, reduced comfort, and higher energy demand. Elderly persons, children and those with pre-existing health conditions are mostly vulnerable to heat stress [1].

Within the "HERITAGE" project, one of the key sub-projects is the implementation of solar photovoltaic (PV) energy in aged urban built settings. The research is focused on developing innovative urban strategies that integrate solar energy technologies such as roof-tops, urban canopies and façade panels, which contribute to climate change effect. The application of PV modules on building surfaces can alter the surroundings' thermal energy balance through several ways: Compared to conventional building envelopes, PV modules have a lower albedo, leading to a higher proportion of absorbed radiation. Part of this absorbed energy is converted into electricity, thereby lowering the temperature of the PV modules and decreasing longwave radiation emissions. Also, due to the lower heat storage capacity of PV modules, they quickly dissipate heat more than traditional materials. The PV arrays add an extra layer to the energy balance, dissipating heat through radiation and convection while providing shading and reducing nighttime cooling by decreasing the sky view factor of the surface beneath them [3]. Over the past years, the Netherlands has made considerable progress in increasing their cumulative PV capacity. At the end of 2023 over 24 GWp PV systems were installed in the Netherlands, with about 43% in residential applications [4]. This significant progress in solar energy deployment in the built environment has made it a crucial area of study, for both energy performance and urban sustainability. **Figure 2** shows some of the typical existing PV technologies in the Netherlands, particularly in the built environment. However, the integration of solar energy systems tailored to improving the comfort of the citizens' lives, particularly in ageing neighbourhoods against urban heating due to climate change presents unique a challenge.

The main focus of this research is to evaluate the applicability, effectiveness and efficiency of novel solar PV energy technologies to further improve their energy performance and contribution to the reduction of the urban heat effect of city areas in the Netherlands that existed before the eighties.

An advanced incorporation of solar energy systems into the historical settlements of Dutch cities can improve the citizens' quality of life in many ways, including lower energy costs [5] and enhanced indoor thermal comfort [3], especially in areas that are highly sensitive to climate effects. It also offers an opportunity for the historical buildings to be equally compared with the modern structures in terms of energy performance and indoor thermal comfort. Furthermore, other cities outside the Dutch settlements facing similar problems could also take precedence from this model and create a more conducive environment for the occupants when it is implemented.

Therefore, this position paper presents suggestions for research to be embarked on in the framework of a PhD project in the coming years (2024 to 2027).

Figure 1: Aerial scenic view of central Amsterdam, the Netherlands. Image courtesy: dreamstime.com

2 LITERATURE REVIEW

2.1 Introduction

Ongoing climate change and urbanization present societies with challenges related to environmental quality, energy management and public health [1]. In this literature review, energy performance of buildings as well as the performance and design optimization of solar energy systems concerning environmental impact, spectral response and energy efficiency, which stand as key indicators that influence the effectiveness of the novel solar energy systems is presented. This review is aimed at gaining knowledge and understanding of the energy performance of buildings and solar PV system performance and its integration into the built environment.

2.2 Related previous studies

Many studies have been conducted on the energy performance of buildings, solar energy integration into buildings and performance optimization of solar PV systems. In this paper, we report eleven publications found by a search in Scopus, ScienceDirect, Web of Science and IEEE explore using the following search terms: Building integrated solar energy system, building energy performance, spectral irradiance and effect of climate change on buildings. Examples of these papers include Krugten et al. [5] who unveiled the role of historical homes and their alterations in improving urban sustainability. The key findings of the paper revealed that historical dwellings predominantly fall under lower energy labels (i.e. F and G), meaning poor energy performance. It was concluded in this study that historical and post-war dwellings have the potential to increase energy performance from a lower energy label (F and G) to a higher energy label (e.g. B, A, etc.) through insulation and renewable energy sources. Manso-Burgos et al. [6] also analysed the energy performance of building stock in València and revealed that the building stock in València, particularly in low-income peripheral regions is dominated by E and G rates. This represents poor energy performance according to their energy labels. It was also mentioned that this poor energy performance was as a result of outdated building standards during the city's expansion in the 1960s and 1970s. Hamdy et al. [7] studied the impact of climate change on the overheating risk in dwellings in the Dutch settlement. According to this study, the findings show that even though most Dutch home types can effectively suppress the effects of global warming, poorly ventilated homes are vulnerable to overheating, particularly if windows lack solar protection. Load matching in residential buildings through the use of thermal energy storage has been presented by Miglioli et al. [8]. According to the study, correctly sized thermal storage with PV and heat pumps can optimize energy usage by up to 40%, reduce grid demand by 30% and CO_2 also by 20-35%. It can also increase economic benefits by 15-25%.

In relation to building applied and building integrated solar energy systems, Khan et al. [9] revealed in their study on the local warming potential of urban rooftop photovoltaic solar panels in cities that solar PV panels can have a considerable contribution related to energy production in an urban setting, but they could also cause an increase in the local temperature. The ambient temperature in summer rises by up to 1.4 °C, whilst surface temperature also increases by up to 2.3 °C. Meng et al. [10] investigate the performance variability between identical rooftop PV systems in residential communities in the Netherlands. Findings show that the PV systems exhibited an annual average performance ratio of 0.71 with a standard deviation of 0.04. According to the research, the Energy Performance Index (EPI) revealed that conventional models overestimate performance by 5.7% annually. It was further established that solar PV system performance inconsistency is significantly greater in real-world conditions due to site-specific factors including geographical location, shading and system age. Perez et al. [11] assessed the lifecycle of façade BIPV. The study revealed that the Energy Payback Time for the As-is is 0.81 years with an Energy Return on Investment (EROI) value of 34.6, while the realistic scenario which includes wafer processing indicates 3.81 years with an EROI of 7.2. The greenhouse gas emissions in the as-is scenario and realistic scenario are -1,578 kg CO_2-eq and 9,329 kg CO_2-eq, respectively. The average solar radiation received by the Solaire BIPV system was 766 kWh/m²/yr with a potential total incident radiation of 822 kWh/m²/yr without shading. The annual average energy production over four years is 5,689 kWh/year, with a Performance Ratio of 64.4%. "Performance analysis of building integrated with photovoltaic thermal system and traditional rooftop photovoltaic system" has been studied by Mishra et al. [12]. According to the findings, the performance of a building-integrated semi-transparent photovoltaic thermal (BiSPVT) system is enhanced by water cooling (10.7 kWh typical day) and thereby outperforms the rooftop semi-transparent photovoltaic (RtSPV) system (8.7 kWh typical day).

One major influencing variable that greatly affects solar PV system performance is the distribution of spectral irradiance [13]. Studies were conducted on real-time spectral variation and its impact on solar PV system performance. For example, Mutiara et al [14] evaluate spectral distribution in the Netherlands concerning various PV technologies' performance. According to the study, the lowest and the highest mean Average Photon Energy (APE) values were 1.80 and 1.93 eV. The organic solar cells yield the highest spectral gain with an SF (Spectral Factor) of 1.01 compared to the other PV technologies, such as c-Si with an SF of 0.966. Spectral irradiance variation in northern Europe has been explored by Paudyal et al [13] using the APE index. The main finding shows that Grimstad and Norway indicate the largest variation in monthly APE value, ranging from 1.82 eV to 1.93 eV between January and July, with an annual average APE value of 1.90 eV. Merklingen and Germany show the smallest APE variation, ranging from 1.86 eV to 1.88 eV, between March and July, with an overall annual average

APE value of 1.86 eV. Witteck et al. [15] studied the effects of spectral on the energy harvesting efficiency of Two-and-Four-Terminal Tandem PV. The research is based on experimental measurements and computational modelling. The study concludes that for a current-matched device, the yearly spectral variation reduces energy harvesting efficiency by 2%rel under STC. After accounting for extra reasonable losses in the 4T arrangement, the energy-harvesting efficiency of the 2T devices is shown to be either equal to or greater. For current matching, deviations in the top cell bandgap are more than 0.1 eV, accounting for a reduced energy harvesting efficiency of more than 5%rel for the 2T tandem device.

These studies have laid the foundation for understanding and identifying the knowledge gap in the performance of solar energy systems in built environments as a whole and are expected to drive the progress of this study.

Figure 2: Typical existing solar PV systems in the built environment in the Netherlands

3 RESEARCH PLAN

3.1 Objective of the study

The main objective of this PhD research study is to assess the performance of the novel solar energy system to further improve the energy performance and indoor thermal comfort of ageing-built urban areas in the Netherlands, considering factors such as geographical locations, weather variables and system configurations.

3.2 Research questions

The following specific research questions are expected to be addressed in this study:

1. What could be proper integrated designs for solar energy systems in buildings given the demand, local (micro-) climate and building geometries?
2. What is the operational performance of the novel solar energy systems, considering system materials, weather conditions, geographical location and user interactions?
3. How do the local environmental parameters interact with the performances of the solar energy system configurations?
4. Is the optimum configuration for current conditions across various climates the same as the future?

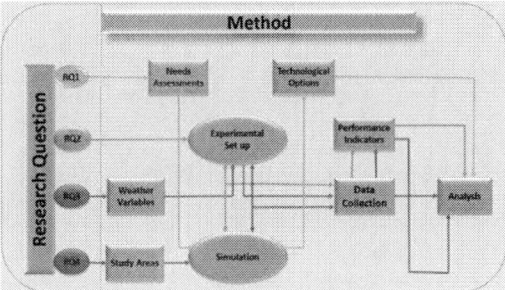

Figure 3: Summary of the proposed method for the study

3.3 Methodology

The general methodology for this study is based on field data collection, simulation and analysis as well as validation in real-life evaluation. This will be done through system monitoring in combination with user interactions. The key procedures involved in addressing the specific research questions are summarized in **Figure 3**, and applied to specific cases in Eindhoven, Delft, Enschede, Wageningen, Rotterdam and Amsterdam. The Aerial scenic view of part of central Amsterdam and its map with buildings' age are shown in **Figure 1** and **4**, respectively.

Figure 4: Map of buildings' age in Amsterdam, with darker orange colour for ageing and blue for new buildings. Image courtesy: reddit.com

ACKNOWLEDGMENT

This work is a PhD project with the Energy Technology Group of the Eindhoven University of Technology (TU/e), funded by the HERITAGE project under the Dutch 4TU Federation.

REFERENCES

[1] R. A. W. Albers *et al.*, "Overview of challenges and achievements in the climate adaptation of cities and in the Climate Proof Cities program," Jan. 01, 2015, *Elsevier Ltd.* doi: 10.1016/j.buildenv.2014.09.006.

[2] G. Steeneveld, S. Koopmans, and L. van Hove, "Quantifying the Urban Heat Island Effect in the Netherlands by Exploring Observations by Hobby Meteorologists," 2011.

[3] E. Fassbender, S. Pytlik, J. Rott, and C. Hemmerle, "Impacts of Rooftop Photovoltaics on the Urban Thermal Microclimate: Metrological

Investigations," *Buildings*, vol. 13, no. 9, Sep. 2023, doi: 10.3390/buildings13092339.

[4] Nationaal Solar Trendrapport 2024 ©Dutch New Energy Research, "Nationaal Solar Trendrapport 2024," https://www.solarsolutions.nl/trendrapport/, 2024.

[5] L. T. F. van Krugten, L. M. C. Hermans, L. C. Havinga, A. R. Pereira Roders, and H. L. Schellen, "Raising the energy performance of historical dwellings," *Management of Environmental Quality: An International Journal*, vol. 27, no. 6, pp. 740–755, 2016, doi: 10.1108/MEQ-09-2015-0180.

[6] Manso-Burgos, D. Ribó-Pérez, J. Van As, C. Montagud-Montalvá, and R. Royo-Pastor, "Diagnosis of the building stock using Energy Performance Certificates for urban energy planning in Mediterranean compact cities. Case of study: The city of València in Spain.," *Energy Conversion and Management: X*, vol. 20, Oct. 2023, doi: 10.1016/j.ecmx.2023.100450.

[7] M. Hamdy, S. Carlucci, P. J. Hoes, and J. L. M. Hensen, "The impact of climate change on the overheating risk in dwellings—A Dutch case study," *Build Environ*, vol. 122, pp. 307–323, Sep. 2017, doi: 10.1016/j.buildenv.2017.06.031.

[8] A. Miglioli, C. D. Pero, F. Leonforte, and N. Aste, "Load matching in residential buildings through the use of thermal energy storage," in *2019 46th IEEE Photovoltaic Specialists Conference*, pp. 272–279.

[9] A. Khan and M. Santamouris, "On the local warming potential of urban rooftop photovoltaic solar panels in cities," *Sci Rep*, vol. 13, no. 1, Dec. 2023, doi: 10.1038/s41598-023-40280-9.

[10] B. Meng, R. C. G. M. Loonen, and J. L. M. Hensen, "Performance variability and implications for yield prediction of rooftop PV systems – Analysis of 246 identical systems," *Appl Energy*, vol. 322, Sep. 2022, doi: 10.1016/j.apenergy.2022.119550.

[11] M. J. R. Perez and V. Fthenakis, "A LIFECYCLE ASSESSMENT OF FAÇADE BIPV IN NEW YORK," in *37rd IEEE Photovoltaic Specialists Conference*, [IEEE], 2011.

[12] G. K. Mishra, T. S. Bhatti, and Tiwari G.N., "Performance analysis of building integrated with photovoltaic thermal system and traditional rooftop photovoltaic system," in *Proceedings of 2017 IEEE International Conference on Technological Advancements in Power and Energy*, IEEE, 2017.

[13] B. R. Paudyal *et al.*, "Analysis of spectral irradiance variation in northern Europe using average photon energy distributions," *Renew Energy*, vol. 224, Apr. 2024, doi: 10.1016/j.renene.2024.120057.

[14] Mutiara l., Pegels K., and Reinders R., "Evaluation of Spectrally Distributed Irradiance in the Netherlands Regarding the Energy Performance of Various PV technologies," *2015 IEEE 42nd Photovoltaic Specialist Conference (PVSC) 14-19 June 2015, New Orleans, LA*, 2015.

[15] R. Witteck, J. F. Geisz, E. L. Warren, and W. E. McMahon, "Spectral Effects on the Energy Harvesting Efficiency of Two- and Four-Terminal Tandem Photovoltaics," *Solar RRL*, vol. 8, no. 3, Feb. 2024, doi: 10.1002/solr.202300782.

41st European Photovoltaic Solar Energy Conference and Exhibition

This presentation was selected by the Sc. Committee of the EU PVSEC 2024 for submission of a full paper to one of the EU PVSEC's collaborating peer-reviewed journals.

STEEL FRAMING/STRUCTURE AS A SOLUTION TO SUPPORT BIPV COMPETITIVENESS

Simon BODDAERT[1,*], Jean-Pierre REYAL[2], Michel DERNIS[3], Philippe ALAMY[4]

[1] CSTB, Energy & Environment Direction, France
[2] SEMPERSTYL Technologies, France
[3] ATRIUM DATA, France
[4] ENERBIM, France

* E-mail to: simon.boddaert@cstb.fr; phone: +33 (0)680 58 1001

ABSTRACT: The development of BIPV solutions makes it possible to respond to the energy and environmental challenges of 2050 (IEA roadmap and EU roadmap), while offering solutions adapted to the requirements of construction codes (CPR n° 305/2011). Current definitions aim to maximize energy production from standard solutions, dictated by the module's dimensions, without integrating, from the design phases, sustainability, safety, and reduction of footprint impacts notions. The SOLARSTYL solution aims to define and validate an innovative BIPV solution that meets these challenges, while offering an advanced aesthetic integration solution, without harming energy production. A global approach combining a specific support solution must and the use of a component widely used in construction, thin rolled steel, may prove to be an approach that makes it possible to respond to these challenges, while aiming for strong competitiveness, based on a controlled carbon footprint and much greater durability to traditional BIPV solutions. This article presents the development methodology of the SOLARSTYL project, which aims to develop a BIPV solution for facade and roof, on a thin steel base with high added value, integrating the manufacturing and experimental validation stages, including the creation a digital tool to optimize the sizing of solutions.
Keywords: BIPV, BIM, Steel, Competitiveness.

1 OBJECTIVES

The use of thin sheets of stainless steel, or more specifically ferritic steel, or a galvanized steel with high corrosion resistance such as Magnélis®, are the materials chosen for the development of an innovative BIPV solution. SOLARSTYL project aims for a high threshold of durability, associated with an innovative implementation solution, allowing high reliability, while limiting labor times. As defined, this integration solution must make it possible to achieve a favorable LCA, while meeting electrical safety requirements (IEC 61215 – 61730 – 63092) and meeting the requirements of the construction of the CPR. The impact of integrating electrical conversion elements directly into the structure of the BIPV solution will also be investigated to guarantee optimal safety and minimize assembly time. All these elements will be studied and validated in the specifically work methodology developed to be compliant with underlined objectives which constitutes the SOLARSTYL project.

In this R&D project we aim to achieve many different targets, by assessing first added values of SOLARSTYL solution, by providing second a digital tool support allowing a widespread dissemination. To correctly address these targets, durability and reliability are pivotal points of this work and expected outcomes are well identified with the support of project partners as:

- Assess STEEL added values (€, CO_2, kWh, Y),
- Fulfill building codes and regulations,
- Compliant with PV/BIPV standards,
- Manufacturing process validation,
- Digital twin tool from early stage to speed up value chain.

SOLARSTYL solution will address PV / BIPV manufacturers as a reliable alternative to standard materials used in PV manufacturing, but also providing a

high-end Add-on in Fab lines. Results harvested during this three years project, experimentation, demo site, monitoring and standard tests, will allow to provide a BIPV relevant alternative for decision markers / planners.

2 AIM and APPROACH

The first investigations made it possible to determine the technical feasibility of a sheet folding solution to replace standard aluminum frames (Figure 1). A proof-of-concept installations are currently in operation, but many aspects have to be investigated to reach an industrial reality. The upscaling to the industrial level must be driven by the achievement of performance results and above all the confirmation of standard compliancy.

Figure 1: Frame drawings and final display

The firsts installations carried out in real test conditions confirmed feasibility and reliability of such approach (Figure 2) and overcoming many technical barriers and reducing significantly manual operations, allowing to reduce work duration on site and possible defaults.

To support from early stages the development of this project and ensure next upscaling steps, the use of a digital tools is validated. Over conception and fine tuning industrial conception, digital design makes it possible to define the technical details necessary for the creation of prototypes and the firsts attempts of mounting systems on demo site and support implementation of KPI allowing to express added values of SOLARSTYL project up to standard compliance and over the time operating conditions.

Figure 2: First SOLARSTYL real demo case.

A work methodology has been implemented in this project to capitalize as best as possible from every learning curve and optimize project achievement. The use of BIMSOLAR toll developed by ENERBIM will support this work.

3 MAIN METHODOLOGY

3.1 Flowchart

To correctly drive this work, the BIM base development is used as a central point allowing to collect data, and to integer additional data providing from different studies (Figure 3). The carbon footprint is a key point development of this project and data collected, material used, and manufacturing details will be integrated, as an example, over the flow to fit with objective to reduce drastically CO_2. Other parameters could be implemented during the work progress is they are relevant for the project.

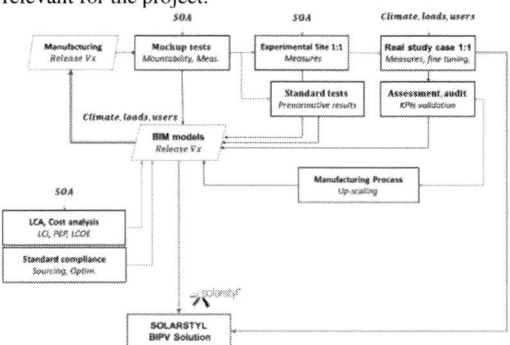

Figure 3: Flowchart of SOLARSTYL project.

The current flowchart version would be implanted or modified according to results and thresholds achieved during the first stages of this project.

3.2 BIM models

To allow a dynamic implementation of work progress and to validate work stages ENERBIM tool will be fine-tuned to fit with project expectation. Over the low carbon footprint material to challenge with current PV manufacturing process, the data collected and validation stages with real study case will be implemented. This

process is highly strategic to demonstrate results, but this crossed validation allows to validate numerical models and calculations before with in operation results before a real case implementation at real scale (demo site)

Feedback loop for solution implementation and dynamic enhancement will be active option allowed by the use of BIMSOLAR. From building description (Figure 4) to detailed results (Figure 5). The digital tool will allow validation of any new development.

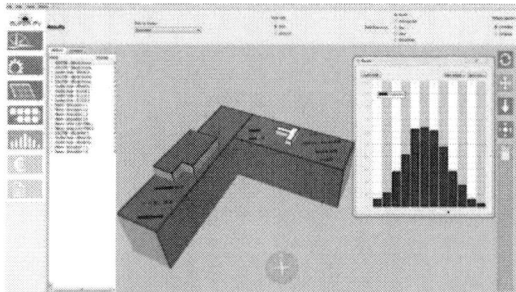

Figure 4: ENERBIM – BIM Display of tool developed to implement SOLARSTYL element in a numerical twin.

Another point addressed with BIM model is the dynamic validation by models/measure validation process used in this project. Indeed, feedback loops will also be used to assess accuracy of PV models and interactions with building needs or objectives in order to validate SOLARSTYL solution as a whole, not only as a PV solution without tied ling with building (as a complete multifunctional BIPV component).

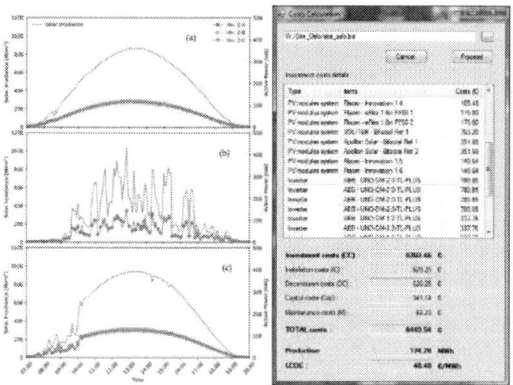

Figure 5: ENERBIM – Results display window and declaration parameters window.

With the large functionalities of BIM base, optimization of every progress stage could be implemented – from bespoke to ready-to-install components. Moreover, the implementation of KPI developed by CSTB, thought different research project as IEA PVPS T15 and INCREASE EU project, straight in the tool will allow to validate technical and industrial choices, and will feed parameters and indicators to support dissemination results.

Another implementation expected with the use of BIM is the over the flow upgrade of the solution filled by monitoring and audits performed during the project. Models/measures permanent comparison will be used to validate project progresses.

3.3 Monitoring and Experimentation

To validate in the field and in operating condition, preliminary experimentation will be carried out on CSTB experimental platforms and then a scale 1:1 mockup will be implemented. A specific monitoring system with dedicated sensors for weather data and energy measurement will by installed. All this process will consider the electrical performance and thermal behavior of SOLARSTYL BIPV solution, by placing the notions of maintenance/replacement and time savings at the center of development concerns.

The data will be implemented by ENERBIM for the creation of an automated sizing and calculation tool for energy and environmental contribution at the building scale.

The pre-validated solution will be then implemented on an ATRIUM DATA demonstration site, which will provide feedback which will make it possible to correct, if necessary, the SOLARSTYL solution to offer a large volume production solution.

4 ONGOING WORKS

Four works are in ongoing stage to address project objectives. As seen on the Figure 3, each work will validate project progresses.

4.1 Industrialization

SOLARSTYL project aims to suggest an innovative framing solution with relevant added values, in terms of cost and time savings. That directly addresses industrialization capabilities and the implementation of this process directly on a factory line. Upscaling from $1m^2$ samples to large modules size to fit with façade expectations is the final goal of the project which is really challenging in the current trend of massive uniformization of the market. To replace metal extrusion process by a steel folding one is to be proved during this project, but based on already achieved results that seems really promising. The Figure 6 present the first-generation automated framing machine allowing to replace aluminum by a folded steel sheet with the same mechanical performances.

Figure 6: Semperstyl Fab Line V1 with hand made loading and visual check – Up to $1m^2$

Complete fully automated line will allow to change patterns and module size as thickness of PV modules allowing to accept double and triple glazing solutions, fitting with BIPV products for façade. To confirm achievements a specific testing plan is developed in order to test bespoke elements according to building and IEC expectations or requirements to validate the range of fab line capabilities. Here is a non-exhaustive list of tests that will be performed during the project, in addition of IEC pre-normative tests carried out by CERTISOLIS.

4.2 Test plan

To reach aims project, a broad test plan has been elaborated to assess as best as possible the SOLARSTYL system to check compliancy with building codes (CPR), IEC standards and thermal regulation to address the French Market. On the other hand, additional investigations will be carried out to assess and manage risks related to PV integration in the built environment, as the multifunctionalities.

To address comfort and thermal regulation the Solar Heat Gain Coefficient (G) (Figure 7) and insulation value (U) will be characterized for each SOLARSTYL framing version/module. Acoustic attenuation could be assessed according to type of façade system implemented.

Figure 7: SHGC setup for G value characterization

For risk management compatibilities, we a PV system is installed in the built environment two main topics have to be addressed and dedicated test performed in accordance. For Fire characterization SBI and ignitability tests will be carried out (Figure 8) thank the support of PV manufacturers allowing to produce bespoke elements in accordance with standard requirements. Accelerating aging tests carried out with WEATHEROMETER will allow to assess impact of aging on ignitability to validate materials selection.

With the same expectations, wind load tests will be carried out at CSTB wind tunnel device with scale 1:1 mockup to assess mechanical resistance and watertightness of SOLARSTYL solution. The Figure 9 show an experimental setup to performed combined stressed Rain/Wind Test. Over initial results, mechanical behavior will be studied to evaluate capabilities of such system to be installed in high wind areas. IEC and electric risk hazards compliancy (cable routing, ground connection, CC mode and integrated inverter in the frame (embedded) will be evaluated by CERTISOLIS.

Figure 8: Example of SBI fire test for material reaction

Auto-plug solution and inverter solution imbedded in the steel frame will have to be tested with specific fatigue protocols and stress tests / accelerated aging to check of the viability of such solution in current standard framework.

Figure 9: Wind tunnel for weather resistance (wind, rain; load)

In addition of all these pre-normative or normative tests, large investigation will be carried out on experimental platforms to assess results in real operating conditions. The figure 10 illustrate one CSTB mockup allowing to test in real test condition BIPV prototypes and apply different stress test scenarios.

4.3 KPIs and Technical assessment
To evaluate and generate explicit outcomes, CSTB has developed specific KPIs allowing to assess combined functionalities of building components.

Figure 10: Outdoor mockups and experimental setup (with shading device)

Based on work carried out in IEA PVPS T15 and T12 and then in BIPVBOOST and INCREASE European projects, time saving / human resources saving will be assessed parameter among Energy, Cost, Aesthetical or Environmental indicators. These elements will support the submission of a French technical assessment by the project partners to fit with French regulation and insurance requirement.

5 CONCLUSION AND PERSPECTIVES

Preliminary studies carried out are very promising and first LCA investigation underline the large CO_2 reduction. Even if the fab Line V1 is ready to be upgraded many challenges are facing the project consortium. CSTB will provide as large as possible support, beside its contribution in the BIM Base compatibility.

One major point is to validate is the scalability up to mega modules of the manufacturing line. This topic will be addressed first to avoid any bottleneck. Then capabilities to manage bespoke ready and offsite optimization will be addressed to assess the range of capabilities allowed by the manufacturing line.

All tests carried out will provide all fail/pass results allowing to bring all security and standard fulfillment to validate industrial needs. Competitiveness of such solution will be validated thank the use KPI and commercial maturity and mass production interests could be estimated. An yearly progress meeting will be carried out to follow work progress.

7 ACKNOWLEDGMENTS

The authors of the article would like to thank the ADEME funding scheme through DEMOTASE – FRANCE 2030 grant as well as support of SEMPERSTYL technologies and ENERBIM as co-founders.

CERTISOLIS is involved in this study on standard compliance and assessment with IEC requirements. CSTB laboratories are involved in compliance with French building requirements, safety, and reliability expectations.

41st European Photovoltaic Solar Energy Conference and Exhibition

ADVANCED PV AND THERMAL MODELING FOR A FEASIBLE AND EFFICIENT BAPV-T SYSTEM DESIGN AND EVALUATION

Iñaki Cornago[a], Mikel Ezquer[a], Patxi Sorbet[a], Alicia Kalms[a], Gonzalo Diarce[b], Olatz Irulegi[c] and Fritz Zaversky[a]

[a]National Renewable Energy Centre (CENER), 31621 Sarriguren, Spain
[b]University of the Basque Country (UPV/EHU), ENEDI Research Group, 48013 Bilbao, Spain
[c]University of the Basque Country (UPV/EHU), CAVIAR Research Group, 20018 San Sebastian, Spain

icornago@cener.com

CENER — NATIONAL RENEWABLE ENERGY CENTRE

4BV.4.13

 EU PVSEC

1. INTRODUCTION

- **PV-T represents a promising technology** in urban environments: limited space + high local thermal and electrical energy demands.

- This study proposes a **straightforward BAPV-T system**, by forming an air ventilation channel, which implies minimal adjustments during PV modules installation and low complexity of the equipment and control set-up.

- A simulation **model** has been developed and validated **to design the system and evaluate its performance.**

2. PV-T MODEL AND EXPERIMENTAL VALIDATION

Comprehensive transient thermal model using the Modelica framework

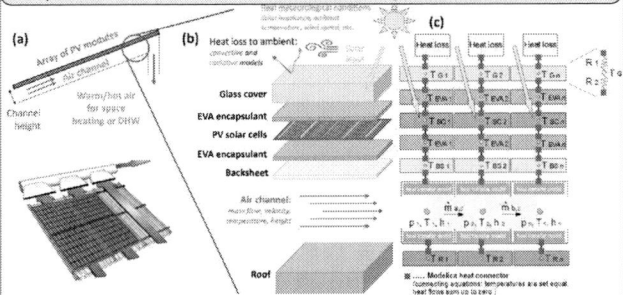

Proposed BAPV-T concept | Considered parameters/layers | Designed 1-D conductive model

Modelling validation though a 3-month experimental campaign

- **3 PV modules** (same model) monitored at **CENER's rooftop:**
  - M1: Air ventilation channel at rear side
  - M2: Standard in-field configuration
  - M3: Thermally insulated rear side

- All 3 modules electrically **biased** at their MPP

- Multiple **variables** continuously monitored: global irradiance, ambient temp., wind speed, module's temp. in different areas, temp. and air velocity in the channel, module's power, etc.

- **Thermography images** periodically captured

  ➡ Deep analysis shows **strong agreement** between simulated results and experimental data, even with highly fluctuating conditions

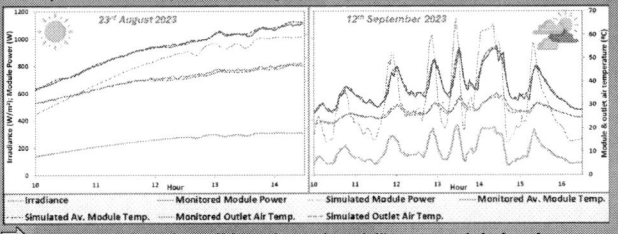

➡ Experimental results **validate** thermal modelling at **module level**

3. REAL CASE STUDY

General simulations conditions

- **IWER building** (Pamplona, Spain), **South-east** oriented roof with 19° tilt.

- 5 in-portrait HJT **PV modules** per channel **(8.9 m x 1.0 m).**

- Typical Meteorological Year **(TMY)** of Pamplona used as meteo-input.

Preliminary simulation results: Influence of channel height & air velocity

Annual generated PV Energy improvement due to channel (%) (*)

Air velocity (m/s)				
5.0	4.3%	4.5%	4.6%	4.9%
2.5	3.2%	3.6%	3.5%	3.6%
1.0	1.9%	2.2%	2.3%	2.3%
0.5	1.1%	1.4%	1.5%	1.5%
	5	10	18.5	30

Channel height (cm)

Average Air temperature increase through channel (°C) (*)

Air velocity (m/s)				
5.0	15.1	7.9	4.3	2.6
2.5	24.1	13.1	7.2	4.4
1.0	24.4	22.0	12.5	7.7
0.5	42.7	28.5	17.0	10.6
	5	10	18.5	30

Channel height (cm)

(*) Relative variation values compared to the case of rear-side insulated modules

⬆ Air Velocity ➡ ⬆ PV production
⬆ Channel height ➡ ⬆ PV production / ≈ PV production

⬆ Air Velocity ➡ ⬇ Air temperature
⬆ Channel height ➡ ⬇ Air temperature

Advanced simulation results: Different fan-control strategies

Channel height constant at 10 cm

Simulated modules temp. for three air flows during a period of high irradiance

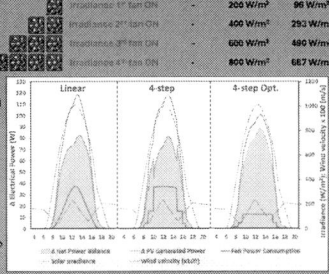

IWER building — 3D building model

Comparison different constant forced air flows

- Increasing forced air velocity significantly reduces module temperature (i.e. down to -19°C with constant 5 m/s).

3 different ventilation fan-control strategies

- **Linear control:** Air flow proportional to irradiance level
  ➡ difficult to implement.

- **4-step control:** 4 low-power fans in **ON/OFF** operation
  ➡ number of working fans depends on equivalent steps of irradiance.

- **Optimized 4-step control:** Control-parameters values are optimized (maximizing **net energy balance**) by applying a differential evolution algorithm.

  Net energy balance = PV generation - fan consumption

Control strategy	Linear	4-step	4-step opt.
Max. Air velocity	3.00 m/s	3.00 m/s	2.11 m/s
Irradiance 1st fan ON		200 W/m²	96 W/m²
Irradiance 2nd fan ON		400 W/m²	293 W/m²
Irradiance 3rd fan ON		600 W/m²	490 W/m²
Irradiance 4th fan ON		800 W/m²	687 W/m²

➡ Settings from iterative optimization process allow **improving the net energy balance.**

Annual simulated results for three fan control strategies proposed (relative to standard rooftop installation)

Temporal evolution of solar irradiance, wind speed, and primary energy-related simulated results, during a clear sunny summer day for the three forced-ventilation control strategies

4. CONCLUSIONS

- **Easily implementable system proposed, which forms a narrow air ventilation channel** between backside of PV modules and top side of roof.

- Model developed and experimentally validated demonstrating **huge potential** for assessing the system feasibility across diverse locations.

- Real case study, three fan control strategies assessed: ⬆ **2.3%** in annual net energy balance; ⬆ **8.1°C** (avg.) in air temperature at the channel exit; ⬇ **31.6°C** (max.) in operating module temperature.

- **Potential extension of PV modules lifespan** due to reduction in operating temperature.

- **Future work:** 1 - further **optimization of control strategy** (wind speed, seasonal dependency); 2 - assessment of **potential uses of warm air** (space heating, DHW); 3 - **economical analysis** of the proposed solutions (cost-effectiveness through the whole lifetime).

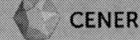 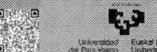 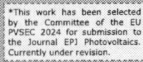 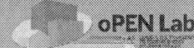

This work has been selected by the Committee of the EU PVSEC 2024 for submission to the Journal EPJ Photovoltaics. Currently under revision.

41st European Photovoltaic Solar Energy Conference and Exhibition

PV ON GREEN ROOFS. TWO YEARS OF COMPARATIVE MEASUREMENT DATA FROM VARIOUS SYSTEM CONCEPTS, SUPPLEMENTED BY SIMULATION RESULTS AND GENERAL CONSIDERATIONS

Markus Klenk[1], Roger Glarner[1], Selina Pfyffer[1], Hartmut Nussbaumer[1], Stephan Brenneisen[2] and Andreas Dreisiebner[3]

[1] Zurich University of Applied Science, School of Engineering, Institute of Energy Systems and Fluid Engineering, Technikumstrasse 9, 8401 Winterthur, Switzerland, E-Mail: markus.klenk@zhaw.ch
[2] ZHAW School of Life Sciences and Facility Management, Institute of Natural Resource Sciences, Switzerland
[3] A777 Gartengestaltung, Switzerland

ABSTRACT: Under Swiss conditions, the optimal use of PV on roof surfaces is just as important as the greening of these surfaces, but the combination of the two involves conflicting objectives. In the project, an energy green roof was realized in which the focus is on the optimal implementation of both approaches. The aim is to show potential customers options for an overall package and to enable a well-founded assessment of the yield. Vertically installed bifacial systems were compared to conventional systems on green roofs. The substructure and substrate structure proved to be easy to install and operate and can be purchased in this form. Substrate thickness and water retention lead to a comparatively strong growth without the need for irrigation, which can nevertheless be easily maintained thanks to the vertical modules. The project can be seen as pioneering with regard to green roofs with high water retention and yet sunny site conditions (vertical PV), which was also demonstrated in the biodiversity studies.
Keywords: flat roof, green roof, bifacial, vertical, bifacial, ground cover ratio

1 INTRODUCTION

Currently, large-scale ground-mounted systems with bifacial modules are being realized worldwide. With regard to land consumption, however, such systems make little sense in Switzerland, apart from a few exceptions. Nevertheless, there are also bifacial applications that are particularly well suited to Swiss conditions. Large flat roofs on shopping centers, industrial or administrative buildings and residential complexes are particularly suitable for PV applications and are being implemented on an increasingly large scale [1] [2].

The installed photovoltaic system supplies energy during the day, exactly when it is needed, at least in commercial buildings. In combination with bifacial modules, there is also a synergy effect when bright, highly reflective surfaces with a high albedo are used, which are favorable for achieving good yields. Even without installed PV systems, the use of white roofs is suggested to reduce the building temperature (and the ambient temperature in the building's surroundings "urban heating") [3]. This can reduce the temperature in the building or the necessary cooling capacity and increase the yield of bifacial systems.

However, also green roofs are suitable with regard to reducing warming, as they also have other positive effects [4] [5]. In addition to providing retreats for plants and insects, they improve air quality and serve to retain water during heavy rainfall [6]. The evaporation of stored rainwater leads to a cooling effect, which radiates into the building (temperature reduction of 2-4 °C) as well as into the surrounding air, thus reducing the effect of the urban heat island. These advantages mean that green roofs are not only realized for idealistic reasons but are increasingly being prescribed or at least promoted by the authorities [7]. In almost all Swiss cities with more than 50,000 inhabitants, green roofs are mandatory for new buildings [5] [8], albeit with varying restrictions. Today, green roofs are central measures in the climate adaptation strategies of cities.

PV systems can be installed on green roofs [8] [9], although the use of the roof as a green space and PV system often results in a conflict of objectives. To optimize the PV-yield, the area is often densely covered with modules, which in turn makes accessibility for maintenance and mowing work more difficult. However, these are particularly important for green roofs to prevent the modules from being shaded by vegetation.

Bifacial modules in particular offer new and additional possibilities for combining green roofs and PV use. Due to their double-sided light sensitivity, they allow a wider range of possible orientations than is possible with monofacial, single-sided light-sensitive modules [10].

Figure 1: The vegetation on green roofs can lead to obviously disadvantageous shading of the PV system. In turn, maintenance work is made more difficult by dense coverage with PV modules.

The ZHAW already had PV/green roof activities and is generally involved with bifacial modules and systems [8] [9] [11] [10] [12]. In Winterthur, a project was realized together with the Solarspar association on the roof of a retirement home in which vertically elevated, 20-cell special modules were installed on a green roof [11] [13] [14]. In this case, the vertical design addressed the conflicting goals of green roofs and photovoltaics. The area covered by modules was significantly reduced and accessibility for maintenance and mowing work was improved. The narrow design also reduces the wind load, which allows for a less massive substructure and reduces visibility. Due to the reduced height, the range of the shading is scaled down. In concrete terms, this means that a system with large, 60-cell modules and three meters spacing between the rows can be replaced by narrow 20-cell laminates with one meter spacing (for the same nominal system output).

In principle, vertically installed bifacial modules have the potential for a high increase in yield compared to monofacial modules with the same nominal output (measured from the front). The conditions must be considered carefully, as mutual shading and the total amount of reflected light from the ground must be taken into account. [15] [16] [17].

2 SYSTEMS UNDER INVESTIGATION AND MEASUREMENT RESULTS

2.1 Motivation for the project "Mattenbach"

The preceding project [11, 13, 14] was met with a very positive response, from regional media and researchers to inquiries from international interested parties for corresponding commercial approaches.

The use of vertical, bifacial systems is widely known in the industry, especially due to the similar application in the field of agro-photovoltaics. However, these differ from the system described here in that they usually have a significantly lower GCR. Potential users are deterred by the general uncertainty of the yield estimation due to the described shading issue and the lower predictive accuracy of yield simulations for vertical bifacial PV systems. A further comparative investigation of the yield, with an improved situation compared to the project described, is therefore obviously of interest. This also applies to the direct comparison of different types of installation, to better assess the cost/benefit ratio.

The overall aim of this follow-up project is therefore to design and test a practical package for combining green roofs and PV. Potential customers should be offered tested components and also receive a well-founded estimate of the expected yield.

2.2 Lay-out of the green roof and the PV system

The PV test system was installed on the roof of the "old print shop" in Mattenbach (Winterthur). The total installed capacity is 93 kWp. Three module types from the manufacturer Hevel were used in different configurations, see diagram in Figure 3.

- 34 pcs. HJT à 330 Wp ; 60 cell monofacial. Portrait ; Butterfly, 15° tilt. 1.67 m x 1 m
- 113 pcs. à 390 Wp bifacial; 72 cell. Vertical and Butterfly (15° tilt). 2 m x 1 m
- 148 pcs. à 255 Wp bifacial; 48 cell. Vertical. 2 m x 0.69 m

The same solar cells are installed in the bifacial modules (48 and 72 cells), they differ only in the number of 24 strings, which are arranged in two rows of 12 cells each. The monofacial modules (60 cells with three 20-cell strings) are equipped with heterojunction cells, which are connected via SmartWire. Two modules at a time are connected via a readable optimizer.

In addition to the different substructures (vertical, portrait and butterfly), half of the roof was covered with two different types of substrates, which also differ in terms of brightness. The northern half of the roof was covered with the darker substrate type. Monocrystalline 60-cells in

portrait and butterfly configuration were installed at the end of the northern half of the roof. At the other end of the roof, 72-cell bifacial modules (as for vertical) were installed on butterfly substructures.

Figure 2: The vertical modules are a good match for the structure of the old commercial building with its vertical struts.

In the previous project individual modules were measured very precisely with great effort. During the course of the year, however, shading by the surroundings could not be avoided anywhere. Accordingly, the yield data had to be elaborately processed to be able to estimate meaningful annual yields, which in turn led to uncertainties. As a result, a great deal of effort was made to obtain data that was nevertheless subject to uncertainties.

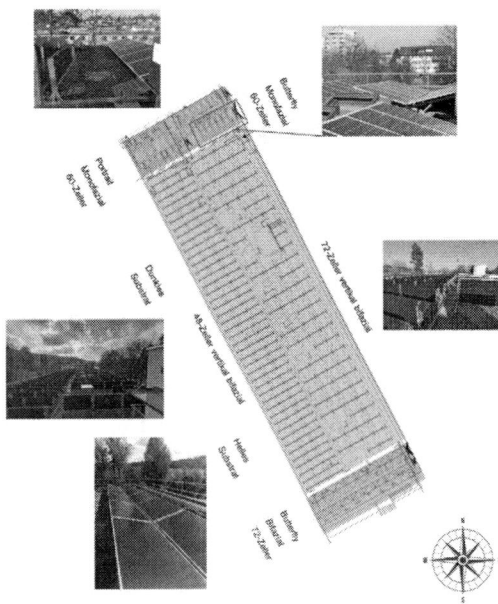

Figure 3: Due to the orientation along the building geometry the test installation is rotated by 28° compared to the south orientation.

In this follow-up project, the yields are now read out using Solaredge elements, which are less accurate and have a lower time resolution. Nevertheless, they enable a locally resolved view of the energy green roof over long periods of time. In addition, the effects of combined solar energy systems and green roofs on biodiversity are to be investigated.

The PV system was to be aligned with the roof geometry, not least for visual reasons. The architect (Zollinger Architekten Winterthur) found the vertical modules, which protrude over the edge of the roof, to be a good match for the structure of the old commercial building with its vertical struts (Fig. 2).

The orientation along the building geometry means that the test installation is rotated by 28° compared to the south orientation, see Figure 3. This means that the monofacial installation is not oriented to the south, and that the vertical segments of the system therefore have rather a south/north- than an east/west-orientation.

In the previous project "Eichgut retirement home" in Winterthur, very narrow solar modules with only two rows of modules were used. On the one hand, this was favorable due to the lower wind load, but it also resulted in lower visibility. In this project, larger modules in the form of bifacial 72-cell and 48-cell modules are now being used for the vertical subsystems to take into account typical industrial module dimensions and to test an intermediate format with the 48-cell modules. Larger formats are more favorable in terms of the system price, as the number of substructures required is reduced. The use of typical industrial formats, such as the 72-cell bifacial modules used here, is obviously preferable for cost reason.

Mutual shading is very pronounced with vertically installed modules, as described above, and is particularly dependent on the spacing of the module rows. Systems consisting of "taller" vertically mounted modules must have larger row spacing in order to achieve the same percentage shading values as narrower designs. In this test system, bifacial modules with 72 (6 x 12 strings) and 48 (4 x 12 strings) cells were installed vertically in a "landscape cape" orientation. The spacing between the large modules was just under 2 m, corresponding to a GCR of 50 %, with a distance to the ground of 40 cm. For the narrower modules, a row spacing of 1.25 m was selected, corresponding to a GCR of 55 %. The area coverage is therefore relatively higher by 10 % than in the case of the large modules. The distance between the lower edges of the modules and the ground is 30 cm.

For the modules at the narrow ends of the roof (monofacial south and butterfly, as well as bifacial butterfly), the determination of a GCR is more difficult due to the conditions (few rows, edge areas, numerous superstructures). However, especially in the monofacial systems with a flat angle of attack of 15°, the GCR does not play a significant role in terms of yield anyway, but it is included in the consideration of the area-related specific yield. The GCR was set at 75 % for these systems, which is a typical value for corresponding PV systems and also roughly corresponds to the conditions in the project. The same GCR of 75 % was used for the bifacial butterfly system, although the slightly different module size results in a slightly different specific line per m².

The modules of the vertical subsystems are installed with substructures from the manufacturer ZinCo AG, (Fig. 4) which are offered now as standard for use on corresponding green energy roofs in the future. Standard modules can be used which leads to lower costs and makes this variant more economically attractive. Other manufacturers now also offer mountings for vertical installation, e.g. SOLYCO with a differing anchoring in the ground.

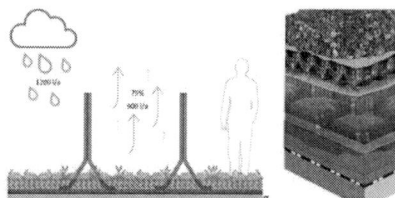

Figure 4: Schematic of the ZinCo substructures, their anchoring and structure of the substrate structure.

With the Zinco system, a good substrate structure is created with 12.5 cm, which has a high storage capacity. This is good for maximizing the retention effect and also stores moisture in liquid reservoirs for dry conditions. Withered planting can thus be avoided, although the costs are higher, and the maximum permissible roof load must also be taken into account. This was not a problem with the roof of the old print shop, the roof load was 160 kg/m², although 500 kg/m² would have been possible. Due to the mass of the substrate structure, absorbing the high wind forces, especially from the vertically mounted PV modules, is not a problem.

A lighter and a darker substrate material (Weiss + Appetito) were deliberately selected for comparison. This was based on experience from the previous project, in which the greening only partially covered the roof surface at times, which means that the substrate's reflexion capacity is of importance. On the other hand, the substrate influences the growth of the greening itself, must therefore also match the seed and represents a cost factor. 190 tons (170 m³) of substrate material and 20 tons (12 m³) of gravel were used.

Figure 5: A lighter and a darker substrate material were deliberately selected for comparison.

Seeds (UFA) that are well suited to the conditions on an energy roof were selected for planting. This means that low-growing, heat- and drought-resistant plants are sown. Two different seed mixtures were sown on the green roof.

One of the mixtures consisted of predominantly hard-leaved plants with a tendency to be "silver-colored". This lighter coloration of the leaves also indicates adaptations in nature that use the albedo effect to reduce the heating of the leaf surfaces and thus also the loss of water through evaporation. However, the reflected spectra of plants fundamentally depends on numerous factors and cannot be derived from the color of the plant. It can be assumed that the vegetation will change considerably over time due to influx, but adapted vegetation will last longer than unsuitable plants. In addition to the actual greening, 15 biodiversity islands were created, particularly in the form of dead wood to attract wild bees. A pond was also created on a side roof without PV.

Figure 6: Deadwood island for the colonization of wild bees and detailed view of the green roof with low-growing plants.

2.3 Yield data

The data presented in this paper focuses on photovoltaics. For data on greening, biodiversity, and water retention, please refer to the project's final report.

A self-developed evaluation tool was used for displaying the results of the PV system with temporal and spatial resolution. To take the different efficiencies of the modules into account, the specific yield (kWh/kWp) on the one hand and the area-related specific yield (kWh/kWp per m²) on the other were examined.

Annual specific yields (Feb. 2022 - Jan. 2023) of the first year for the respective systems (see Fig. 3) are depicted in Figure 7. The numbers over the bars represent the mean values of the respective systems, the error bars indicate the lowest and highest individual value. Fluctuations are differing due to the inhomogeneous shading conditions.

The four bars in the center with lower intensities represent the vertical systems. The systems with the highest specific yield are the conventional bifacial butterfly-, the monofacial "south" – and the monofacial butterfly-system. The differences between the vertical systems are mainly assigned to the frames of the large modules. The smaller modules are frameless and show slightly better performance, even though their ground cover ratio of 0.55 is about 10 % higher than those of the larger ones (0.5). In the first year when the planting was only weakly developed an effect of the substrate color (see Fig. 5) could be observed with lower intensities for the dark substrate. In the later months, the differences were leveled out by the planting and in the second year no differences were observed.

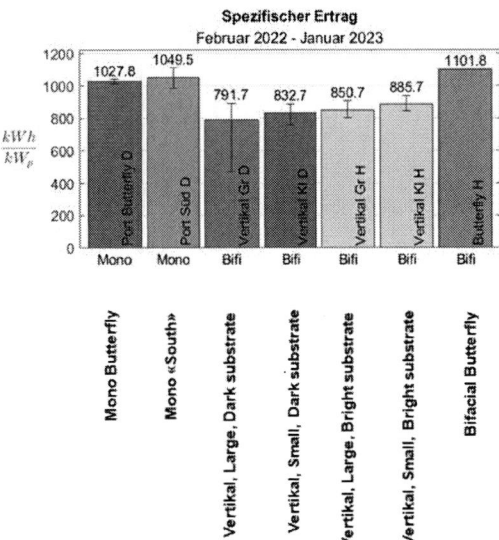

Figure 7 Annual specific yield (kWh/kWp) for the respective PV-systems. The four bars in the center with lower intensities represent the vertical systems.

The plot of the specific yield (kWh/kWp) per month over the course of the year in Figure 8 shows a clear separation between the vertically installed PV systems and those with a flat tilt angle in the summer months, with the details of the system types playing a comparatively subordinate role. In the months from March to August, the specific yield of the vertical installations is significantly lower than that of the more conventional systems, with a maximum difference in midsummer. Furthermore, there is a qualitative division into two segments of very similar duration over the course of the year, with no significant difference between the systems between mid-September and the end of March.

Figure 9 illustrates the seasonal differences and their influence on the total annual yield. Three exemplary chosen months (August, October and December) are shown below in Figure 9 with the same scaling.

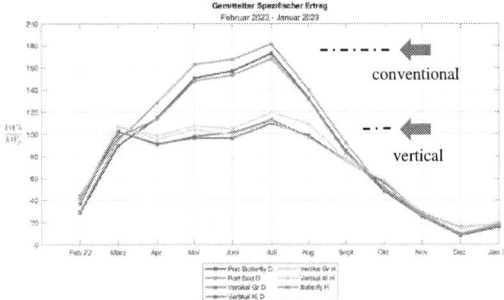

Figure 8: The yield over the course of a year shows a very distinct separation between conventional and vertical systems (see also Fig. 7).

41st European Photovoltaic Solar Energy Conference and Exhibition

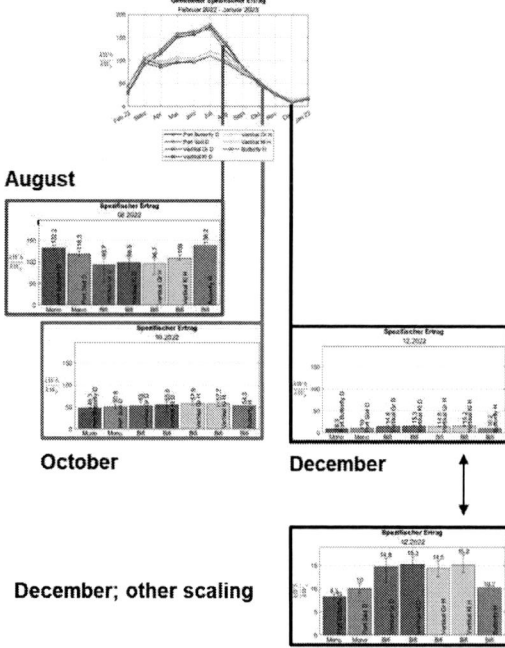

Figure 9: Seasonal specific yield (kWh/kWp) shown for the months August, October, and December.

A representation of the month of December alone in a different scale, as shown in the lower right-hand corner of Figure 9, could give the impression that vertically installed modules are particularly effective in winter, which is true in relative terms, but does not consider the low total annual yield. From an economic point of view, it should be noted that the economic yield of vertical installations is more favorable when electricity prices rise in winter or fall due to oversupply in summer. This is particularly true if dynamic prices and feed-in tariffs are also considered in the daily profile.

The advantage of vertical systems in winter is generally due to the winter conditions of irradiation, but also the lack of snow cover. The advantage of the classical systems in winter was less pronounced in the second year (2023, Fig. 10) in which more snow fell in the Winterthur region and remained longer than in previous years. On the other hand, this advantage, as well as the high albedo effect of the snow, hardly comes into play due to the low radiation intensity on the low-lying Winterthur region in the winter months. It must be emphasized that this is a site-specific assessment that this will not apply to alpine regions with less cover and more snow.

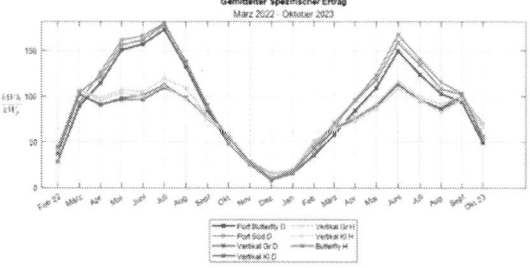

Figure 10: 2022 and 2023 in comparison. The advantage of the classical systems was less pronounced in the second year with more winterly conditions.

Regarding planning, the simulation of projected yields is also important. As part of the project, it was shown that both two simulation tools used, PVSyst and PVcase (ground-mount), are well suited for mapping both conventional and classic systems (Fig. 11).

The weather file used for 2023 contains 80.8 % of measured weather data from Winterthur and 19.2 % of data from PVGis, which originate from the years 2005-2020 and were used to compensate for downtimes in weather data collection. Since mainly the winter months data with lower total contribution are incomplete it can be expected that the irradiation conditions should be largely correctly reflected.

By considering close shading, it was possible to achieve annual yield estimates that correspond well with the measured values, even for vertical systems. The two tools each show characteristic trends, but these have only a minor impact on the annual yield. At monthly resolution (not shown here), the simulations deviate more strongly from the measured values in months with low solar altitudes.

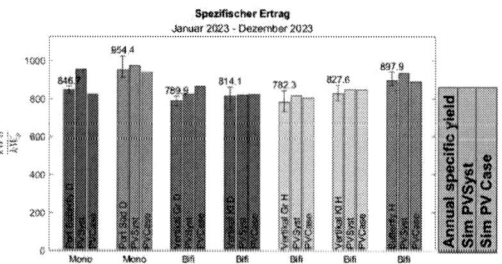

Figure 11: Simulation of the annual specific yield (kWh/kWp) in 2023 (left bar) with PVSyst (center bar) and PV Case (right bar) for the respective systems.

In addition to properties that cannot inherently be captured by the simulation tools (frames, shading details, etc.), there are also other properties of the systems where the simulation tools reach their limits. For example, it was found that the optimizers used in certain shading scenarios were operated at significantly less favorable operating points than those simulated by the software. Particularly in months with heavy shading, it is difficult to estimate the effects that cause deviations in the simulation. It is important to note that the shading environment in this project was adjusted in several simulation runs until it reproduced reality as accurately as possible. This is not possible for an initial yield estimate of a new project, whereby the relevance of the level of detail increases with the complexity of the scenario.

3 DISCUSSION AND SUMMARY

In the preceding project on the retirement home in Winterthur, a low GCR of approx. 37 % was deliberately implemented to minimize losses in the specific yield. In fact, the PV system achieved a specific yield of 942 kWh/kWp over one year, which is close to typical values of approx. 1000 kWh/kWp for east/west-facing modules with a low tilt angle. This means that there are hardly any higher losses in the specific yield, but the total installed output in kWh/m² is limited.

In the Mattenbach project the vertical modules with GCRs of 50 % and 55 % are much more densely arranged.

1405

As expected, the specific yields of the vertical systems described in the project report are also lower than those of the flat-installed conventional systems. Depending on the selection of the individual system types for direct comparison, 83-95 % of the yield of the conventional systems were achieved with the vertical systems in 2023, compared to only 71- 86 % in 2022 see also Fig. 10). On average, a roughly 20 % lower yield can be assumed. The variations are due to the irradiation conditions in the years, which have different effects on the systems.

With tighter staggering, the total installed output and the absolute yield per roof area are increased with a reduced specific yield. With the GCRs realized in Mattenbach, the area-related yields are also on average around 20 % lower than those of a south-facing monofacial system with the same GCR. The difference to flat installed east/west systems is somewhat smaller. However, the GCR only plays a subordinate role for the latter systems with flat tilt angles in terms of yield.

The output per installed Wp, the specific energy yield (kWh/kWp), will always be lower for vertically installed modules in a dense arrangement than for typical systems with a low or medium tilt angle. However, it is obvious that the question of whether it makes sense to install a vertical bifacial system on a green roof cannot be reduced to an optimized specific energy yield. If this is the goal, it would be best to install only a few modules with an optimized southern orientation or almost free-standing bifacial modules in a vertical position. However, the total output and the achievable total energy yield would then be very limited.

The specific energy yield (kWh/kWp) is of particular interest when module prices are a dominant cost factor. However, due to the steady fall in module prices in recent decades, systems with lower specific energy yields have become increasingly interesting. For example, system concepts with almost 100 % GCR with flat tilt angles and east/west orientation are now the most common type of installation on large flat roofs, where the absolute energy yield is maximized, even if the specific yields are lower. Other examples are all systems with orientations that deviate from the previously dominant south-facing orientation or most BIPV systems, especially those with color-faced front sides. In all these examples, the optimally achievable specific energy yield is sacrificed to achieve other goals, in this case optimized total energy yields or façade utilization.

Similar conditions apply to vertical modules on green roofs. It is known that the specific energy yield decreases due to mutual shading and that the albedo on green roofs is comparatively low. On the other hand, this approach enables 100 % green roofs and is advantageous for maintenance. Furthermore, a dense arrangement of vertical modules can be considered as an option to maximize the total energy yield for a given roof area while maintaining the character of a green roof. Both is illustrated in Figure 12 and 13. It is important to emphasize that it is a non-technical consideration to what extent the benefits outweigh the additional costs due to the larger number of modules in this design variant.

If only a limited proportion of the roof area is covered with conventional systems, then one can speak of a combination of PV and green roof. In this regard it should be noted that according to the regulations [18, 19, 20] in some cities it is mandatory to green all flat roof area that is not used as a walkable terrace in an ecologically valuable manner in all zones, even where solar systems are installed. According to this ordinance, solar energy systems and green roofs must be combined since 2015, which means that they are not spatially separated but arranged on top of each other.

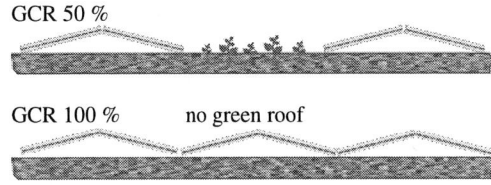

a) Modules at low tilt angles on the ground

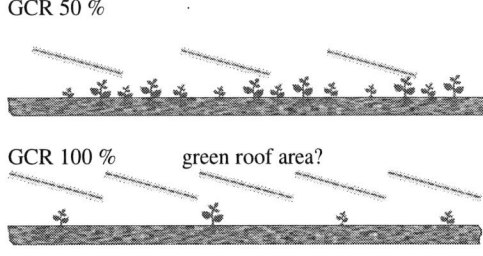

b) Modules at moderate tilt angles and height

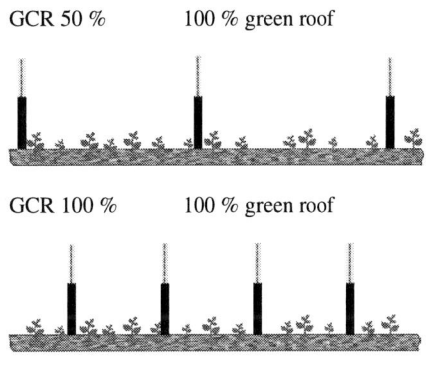

c) Vertical bifacial modules

Figure. 12: PV systems and green roofs with GCRs of 50 % and 100 % respectively.

Referring to the examples in Figure 12, the area under the modules can in principle also be greened and is accessible for maintenance. This is possible for the vertical modules but only to a limited extent for the systems in Figure 12 a) and 12 b). The accessibility of the flat conventional systems can be improved by an increased installation height but will be lowered by increased GCR.

Above a certain GCR, no real greening under the modules and no easy maintenance access will be possible, even with systems especially with heavy vegetation such as the green roof examined here with a high substrate thickness. Such a high substrate thickness is recommended and partly also mandatory now for new buildings to obtain an ecologically valuable greening, due to the better water retention compared to thinner substrates. Heavy vegetation on the other hand leads to the well-known problems regarding shading of the PV modules and maintenance.

Figure. 13: Classic systems with a flat tilt angle are arranged on the right at the edge of the Mattenbach roof, vertical PV systems on the left. The vertical systems maintain the green roof character despite the PV and allow easy access for maintenance.

The City of Zurich's checklist "Green roofs and solar systems" [19, 20], which does not include vertical PV systems but only conventional systems, addresses these conditions. It recommends reducing the substrate thickness in the area in front of the modules to reduce vegetation, while increasing the thickness in other areas to maintain the quality of the green roof and water retention. From a row spacing of 50 cm, it is recommended to use only a continuous thin substrate to avoid high growth, although this document explicitly points out that in this case it is no longer an ecologically valuable greening. Even before this point is reached, one can no longer really speak of a green roof without there being a clear limit. For vertical systems a better ecologically valuable greening is possible, however at increased cost.

In Mattenbach, vertical systems with an area utilization factor of around 50 % were investigated. Compared to a classic, flat PV system with the same area utilization factor on a gravel roof, the costs of the solar green roof as implemented in the project are

+ 20-25 % investment for butterfly on green roof, vs. conventional flat PV on gravel roof
+ 30-35 % investment for vertical on green roof, vs. conventional PV, flat on gravel roof, with 70-95 % of the yields of conventional systems, depending on the year and the comparison system selected.

Combining the costs of PV and green roofs makes sense in this cost-consideration, as the substructures are adapted for use on the green roof.

In addition to the investment costs for the PV components themselves and the yields generated, maintenance plays a key role in the economic viability of the system. The cost of maintaining the vegetation is around CHF 1,000 per year, which is around 20-30 % more than standard maintenance/control for a PV system on a gravel roof.

The project showed that the type of PV installation also determines the vegetation, whereby the vertical modules in combination with the high substrate thickness and water retention lead to increased biomass and a different composition of the greenery.

At the same time, it was shown that the height growth

of the plants in the vertical PV installations can be combined well with efficient maintenance. Only in a few areas was it necessary to cut back the vegetation and, if necessary, simple equipment could be used. A research project is currently underway with the aim of developing a suitable robotic mower to carry out mowing work automatically, which should significantly reduce maintenance costs. Integrated cameras should then also partially eliminate the need for inspections. Also, even more densely arranged vertical PV systems with up to 100% GCR will be investigated and realized in ongoing and future projects.

4 ACKNOWLEDMENT

This research is supported by the Swiss Federal Office of Energy in the following projects:

- Development and comparative test of a complete solution for bifacial PV systems on green roofs, SI/502213-01
- PV systems on flat roofs with highest energy yields per area, SI/502309-01

5 REFERENCES

[1] pv magazine USA, Nov. 11, 2019. https://pv-magazine-usa.com/2019/11/11/big-banker-thinks-worlds-largest-rooftop-bifacial-solar-panel-install-is-worth-it/ (accessed Jan. 13, 2020).

[2] Sunpreme, "Commercial Roof – Sunpreme." http://sunpreme.com/commercial-roof/ (ac-cessed Jan. 13, 2020).

[3] Fred Pearce, "Urban Heat: Can White Roofs Help Cool World's Warming Cities?," Yale Environment360, Mar. 07, 2018. https://e360.yale.edu/features/urban-heat-can-white-roofs-help-cool-the-worlds-warming-cities (accessed Jan. 13, 2020).

[4] Stadt Zürich, "Dachbegrünung - Stadt Zürich." https://www.stadt-zuerich.ch/ted/de/index/gsz/beratung-und-wissen/wohn-und-arbeitsumfeld/dachbegruenungen.html (accessed Jan. 14, 2020).

[5] M. S. und S. Häne, "Mit grünen Dächern gegen die Hitze," Tages-Anzeiger, Jul. 09, 2015.

[6] V. Azeñas et al., "Thermal regulation capacity of a green roof system in the mediterranean region: The effects of vegetation and irrigation level," Energy Build., vol. 164, pp. 226–238, Apr. 2018, doi: 10.1016/j.enbuild.2018.01.010.

[7] Jackie Snow, "Green Roofs Take Root Around the World," National Geographic News, Oct. 27, 2016. https://www.nationalgeographic.com/news/2016/10/san-francisco-green-roof-law/ (ac-cessed Jan. 13, 2020).

[8] Stephan Brenneisen, "Herausforderung Gründach – Chancen und Risiken für den Betrieb der PV-Anlage," presented at the ERFA Photovoltaik Dachdichtigkeit und Gründach, Swisssolar, Uzwil,

Sep. 26, 2018, Accessed: Jan. 13, 2020. [Online]. Available: https://www.swissolar.ch/fileadmin/user_upload/Ta gungen/ERFA_2018/180926_ERFA-Uzwil_3_Brenneisen.pdf.

[9] Stephan Brenneisen, "Naturschutz auf Dachbegrünungen in Verbindung mit Solaranlagen," Sep. 20, 2016.

https://web.archive.org/web/20160920102605/http://ww w.stadtgaertnerei.bs.ch/dms/stadtgaertnerei/downlo ad/der-eigene-garten/dach_solar.pdf (accessed Jan. 13, 2020).

[10] H. Nussbaumer et al., "Accuracy of simulated data for bifacial systems with varying tilt an-gles and share of diffuse radiation," Sol. Energy, vol. 197, pp. 6–21, Feb. 2020, doi: 10.1016/j.solener.2019.12.071.

[11] T. Baumann, H. Nussbaumer, M. Klenk, A. Dreisiebner, F. Carigiet, and F. Baumgartner, "Photovoltaic systems with vertically mounted bifacial PV modules in combination with green roofs," Sol. Energy, vol. 190, pp. 139–146, Sep. 2019, doi: 10.1016/j.solener.2019.08.014.

[12] D. Berrian, J. Libal, M. Klenk, H. Nussbaumer, and R. Kopecek, "Performance of Bifacial PV Arrays With Fixed Tilt and Horizontal Single-Axis Tracking: Comparison of Simulated and Measured Data," IEEE J. Photovolt., vol. 9, no. 6, pp. 1583–1589, Nov. 2019, doi: 10.1109/JPHOTOV.2019.2924394.

[13] T. Baumann, "Vertikale Solarpaneele – Gründach und Solaranlage müssen intelligent kombiniert sein," Emw Energ. Markt Wettbew., no. 6, Dec. 2018, [Online]. Available: https://www.zhaw.ch/storage/engineering/institute-zen-tren/iefe/PDFs/emw_Vertikale_Solarpaneele_Gr% C3%BCndach_und_Solaranlage_m%C3%BCssen_ intelligent_kombiniert_sein.pdf.

[14] T. Baumann, "Senkrechte Solarpanels sollen Leistung glätten," Spektrum Gebäude Technik, pp. 50–51, Feb. 2018.

[15] M. R. Khan, A. Hanna, X. Sun, and M. A. Alam, "Vertical bifacial solar farms: Physics, de-sign, and global optimization," Appl. Energy, vol. 206, pp. 240–248, Nov. 2017, doi: 10.1016/j.apenergy.2017.08.042.

[16] S. Guo, T. M. Walsh, and M. Peters, "Vertically mounted bifacial photovoltaic modules: A global analysis," Energy, vol. 61, pp. 447–454, Nov. 2013, doi: 10.1016/j.energy.2013.08.040.

[17] H. Nussbaumer et al., "PV Installations Based on Vertically Mounted Bifacial Modules Eval-uation of Energy Yield and Shading Effects," 31st Eur. Photovolt. Sol. Energy Conf. Exhib. 2037-2041, 2015, doi: 10.4229/eupvsec20152015-5av.6.34.

[18] I. Sutter, B. Tschander, Anpassung an den Klimawandel / Naturschutz, ZUP Nr. 97, pp. 35-38, Juli 2020

[19] Vorgaben Dachbegrünung (Checkliste), Stadt Zürich, Amt für Hochbauten, Lindenhofstrasse 21, Postfach 8021 Zürich, 2020

[20] Checkliste Dachbegrünungen und Solaranlage, Stadt Zürich, Grün Stadt Zürich, Beatenplatz 2, 8001 Zürich, 2020

41st European Photovoltaic Solar Energy Conference and Exhibition

PV FAÇADES > 30 M - FIRE PREVENTION GUIDELINES ON HIGH-RISE BUILDINGS

Urs Muntwyler, Eva Schüpbach
Dr. Schüpbach & Muntwyler GmbH, Hopfenrain 7, 3007 Bern, Switzerland
Phone : +41 (0)79 864 00 84, E-Mail: urs_muntwyler@gmx.ch

ABSTRACT: PV façade installations at existing or new high-rise buildings are challenging in many ways. One of the challenges is that façade elements of PV modules still do not meet the Swiss fire protection requirements. This constitutes a particular threat in case of fire on high-rise buildings. Hence, implementation projects of tall PV façades at existing or new high-rise buildings are not approved by building insurance companies. To remedy this gap, fire prevention guidelines for PV façades > 30 m were developed. The pioneering project was initiated by the state insurance company of Bern («Gebäudeversicherung Bern», GVB), FAMBAU housing cooperative and PV expert Muntwyler. The work was carried out jointly with an interdisciplinary team of fire protection experts, PV façade and PV system experts, and specialists on electrical engineering in Switzerland. The fire prevention goals in the guidelines, to be met by means of structural and constructive measures, were tested on tall PV façades on a 70 m tower in Bern-West. The PV façades on this high-rise building also served as a demonstrator, and now offer an opportunity for the quality-control of the installed PV façades over their entire lifetime, including regular inspection with IR-drones.
Keywords: PV façades, tall buildings, fire protection guidelines, building insurance, Gebäudeversicherung Bern

1 INTRODUCTION AND MOTIVATION

The Swiss Energy Strategy 2050 [1] aims to phase out the five nuclear power plants and to decarbonize the Swiss energy system by 2050. In this context, roughly 40 TWh electricity (more than 50% of the current electricity needs in Switzerland) must be replaced by renewable energy and by energy efficiency. Hence, additional electricity producers from new renewable energies are required. This will be, for most part, photovoltaics (PV), with current PV systems primarily being installed on roofs and infrastructure [2].

PV façades could achieve 10-40% of the PV electricity production, albeit the current market is far away from this goal. One of the obstacles is that façade elements of PV modules still do not meet the fire protection requirements of category RF 1 (resistance du feu, RF). This is due to the construction of the PV modules, comprising several layers, including plastic films. In addition, there are wires, and possibly optimization electronics, which may both be a source of fire or ignition and could potentially contribute to the fire load. In case of fire on a high-rise building with PV façades, the height is an additional challenge for the fire brigade to reach out to a fire.

To address these obstacles, and to avoid simply prohibiting PV façades > 30m, «Gebäudeversicherung Bern» (GVB), the state insurance company of Berne, Switzerland, has developed guidelines for the fire prevention of tall PV façades. These guidelines were prepared jointly with an interdisciplinary team of fire protection experts, PV façade and PV system experts, and specialists on electrical engineering. The GVB guidelines [3] aim at facilitating the approval process, as the construction of PV façades > 30 m on existing and new high-rise buildings requires approval by a building insurance company prior to implementation.

With the release and publication of the GVB guidelines in 2023 and promoted by simultaneous efforts of other building insurance companies (e.g., Zurich), the PV façade sector in Switzerland gained momentum in 2023, and the market has risen sharply. PV façades realised in Switzerland have increased from 118 PV façades in 2021 (0.6% of the total installed PV capacity), to 156 PV façades with 4 MWp (0.36% of PV capacity) in 2022 and to 661 PV façades with 8 MWp (0,5% of PV

capacity) in 2023. In the first quarter of 2024, there were already 484 PV façades with 6.7 MWp (1,5% of PV capacity). The original aims of the GVP guidelines on fire protection verification of PV façades > 30 m at high-rise buildings have hence essentially already been exceeded.

Another highlight is the involvement of the FAMBAU housing cooperative in the project. FAMBAU currently renovates two high-rise (70 m) buildings in Bern-West, including the installation of tall PV façades. The Holenacker 65 building (Fig. 1) is renovated in 2023 / 2024 and the Holenacker 85 building in 2024 / 2025.

Figure 1: One of the three PV façades on the 70 m high-rise building at Holenacker 65 (Bern-West) by FAMBAU. Source: Muntwyler.

The 70 m tower building at Holenacker 65 served as a real-life example to test and demonstrate all technical recommendations developed in the GVB guidelines. It now also offers an opportunity for the long-term quality-control of the installed tall PV façades over their entire lifetime, which includes regular inspection with IR-drones (Fig. 2).

Figure 2: IR-drone inspection of PV façades on 70 m tower by FAMBAU in Bern-West. Source: Muntwyler.

The new GVB guidelines only address PV façades > 30 m. They describe a verification procedure that is based on fire protection goals and the requirements for the fire protection regulations. Since PV modules do not meet the requirements of the fire behaviour group RF 1 (résistance du feu), measures must be undertaken that meet an equivalent safety level within the VKF-BSN, Art. 12. Due to the verification procedure, PV façades are assigned to quality assurance level 3 at GVB (details see guidelines).

2 THE PROCESS OF GUIDELINE PREPARATION

2.1 Introduction

Until the end of 2022, there was no convergence of fire protection targets for PV façades on high-rise buildings in Switzerland. Therefore, the first step in the development of new guidelines by GVB was the definition of the protection objectives. Based on these objectives, the technical and organizational framework conditions could be determined and described.

Other organisations active in the field of fire protection in Switzerland (e.g., Association of state fire insurance companies, VKF) soon learned and recognized that the GVB guidelines will have a decisive influence on the issue. VKF hence became active and invited all the important players on a round table where it was quickly agreed on that the fire prevention goals for tall PV façades are not only important for the canton of Bern, but for entire Switzerland. Shortly thereafter, VKF formed a working group where it was decided that the fire prevention goals to be developed by GVB need to include all buildings and are not only valid for high-rise buildings.

2.2 Definition of building categories

In the GVB guidelines, buildings hence are divided into three system categories depending on the use of the building, the design, the materialization and the exterior wall construction.

System category 1: Buildings with fire resistance in the façade, extinguishing system concept or window strips of at least 1.3 m height and a variety of technical measures. Allowance due to an argumentative proof is possible.

System category 2: This category includes (i) systems in which the fire protection concept of the high-rise building does not have any relevant deviations from the VKF fire protection regulations and (ii) systems that are not system category 1. Verification only with fire test or mathematical proof (simulation).

System category 3: This system category includes accommodation facilities, outer walls of vertical escape routes with windows without fire resistance. Proof excluded.

Proof of the achievement of the protective measures is recorded in the evidence report. This is required by GVB's fire protection department to assess the equivalence of the achievement of the protection objective. Once the measures have been approved, the assessment of the plausibility of the concepts and the evidence will be reviewed as part of the implementation. These are then accepted by the building insurance company at the trade.

2.3 From participatory consultation to publication

The draft of the GVB guidelines was presented to about 100 experts from the construction and insurance industry, fire prevention representatives, and delegates of associations in two workshops in 2023. Participants were demonstrated all technical details of the guidelines in an excursion to the PV façades on the 70 m high-rise building Holenacker 65 in Bern-West (see Fig. 1 and Fig. 2).

After the exchange and feed-back, refinements were carried out on the draft, and thanks to the additional effort of the GVB fire prevention team, the new guidelines for the planning of high PV façades, including fire prevention specifications, were completed on schedule and are now available in German and French [3].

2.4 Contents of the GVB guidelines

The new GVB fire prevention guidelines (2023) for tall PV façades are a novelty and require that PV façades on high-rise buildings must be made of non-combustible materials. To date, there are no PV modules on the market that are made entirely of non-combustible building materials. This means that PV systems installed on façades at high-rise buildings must provide evidence of an equivalent level of safety based on Art. 12 of the fire protection standard. Equivalence must be proven by means of a verification procedure, either with technical and scientific arguments or with a fire test (see Section 2.2).

To be able to guarantee effective verification, protection goals are necessary. On 27 September 2023, the following protection goals for high-rise buildings were published in the GVB guidelines:

Protective goal «Fire Flashover»: In the event of a fire, the fire must not be transmitted via the outer wall over more than two floors above the fire floor before the fire brigade extinguishes the fire (protected property: personal protection).

Protective goal «Exterior Wall Cladding System»: A fire in the exterior wall cladding system may only spread independently to the next floor level after ignition of the exterior wall cladding system in a vertical direction (protected asset: building protection and personal protection). The function of the vertical escape route must not be impaired (protected asset: personal protection). The exterior wall cladding system must be constructed in such a way that the fire brigade does not have to intervene from the outside (protected asset: building protection).

2.5 Impacts of GVB guidelines publication

The GVB guidelines [3] triggered great public interest in the topic of PV façades and their fire prevention throughout Switzerland and propelled many discussions and meetings. Since the publication of the GVB guidelines, the PV façade market in Switzerland has thus shown pleasing growth from a very low level. However, many efforts are still needed before the volumes targeted within the framework of the Swiss Energy Strategy 2050 [1] are achieved (see Section 1).

The GVB guidelines also enhanced the dynamics surrounding the approval of PV façades by state insurance companies (e.g., in the canton of Zurich). Both, the ensemble of these developments and related public pressure have led to the publication (in 2023) of a so-called «transition paper» by the Swiss association of the solar industry (Swissolar). The Swissolar «transition paper» is based on the GVB guidelines, and the verification / approval procedure for tall PV façades > 30 formulated therein. To ensure a uniform overall framework in Switzerland for the verification procedure for tall PV façades > 30 m, GVB has withdrawn its guidelines and refers all interested parties to Swissolar in the future. Swissolar, in turn, announced a follow-up «state-of-the-art» report on the issue, to be published by the end of 2024.

3 FIRE PREVENTION OF TALL PV FAÇADES

In parallel to the development of the GVB guidelines in 2023, PV façades were mounted on a high-rise building in Bern-West (Fig. 1 and 2). Installation of tall PV façades on two 70 m blocks at Holenacker 65 and 85 (in Bern-West) is a project by FAMBAU (second largest housing company in Switzerland).

FAMBAU was one of the initiators of and served as one of the external experts in the GVB fire prevention guidelines project for PV façades > 30m. The high-rise building at Holenacker 65 in Bern-West was a real-life example to assess technical recommendations developed in the GVP guidelines (including, e.g. fire tests on substructures) and was thus established as a demonstrator.

Today, all PV façades on the Holenacker 65 high-rise building are installed and put into operation electrically. After the construction of the PV façades, the operating phase follows. The focus at Holenacker 65 is now on the concept development for the long-term monitoring of the PV façades over their lifetime. Various monitoring methods are evaluated; on 13 August 2024, monitoring was carried out with an IR-drone (Fig. 2).

The tall PV façades > 30 on the high-rise building at Holenacker 65 in Bern-West fully comply with the required measures for the verification (GVB guidelines on the fire prevention of PV façades > 30 m). Data transfer is operational. Voltage, current, power and energy yield are continuously recorded and documented. Error and fault records are transmitted directly to a specialist for assessment.

In the system documentation, it is demonstrated in a simple and understandable way, which measures need to be initiated in case of disturbances (e.g. discoloration of modules, deviations from target values, or automatic shutdowns). The measures include information of a specialist for further assessment up to the immediate complete shutdown of the entire PV system.

The second high-rise building, Holenacker 85, will be renovated by FAMBAU in 2024-25. As with Holenacker 65, tall PV façades that technically meet the specifications of the GVB guidelines will be installed.

4 SUMMARY AND OUTLOOK

Efforts have been undertaken in Switzerland to develop a regulated procedure (guidelines) for the approval process of PV façades > 30 m, prior to their construction on new and existing high-rise buildings, by state insurance companies. The guidelines, published in 2023, are a novelty, have remedied a gap and successfully moved the issue of PV façades in Switzerland forward.

The development of the new guidelines for the proof of fire protection of high PV façades > 30 m was initiated by GVB, «Gebäudeversicherung Bern», with FAMBAU (Swiss building cooperative) and PV expert Muntwyler, and conducted with an interdisciplinary working group, evaluated with about 100 experts in two workshops and demonstrated on PV façades on a high-rise building (70 m tower) in Bern-West (Switzerland) that served to test all technical recommendations in the guidelines. With the GVB guidelines, the realization of PV façades in Switzerland, and especially of tall PV façades > 30 m, is now simpler, though still very challenging.

The guidelines sparked a great interest among building state insurance companies in Switzerland and the public. The Swiss association of the solar industry (Swissolar) hence now follows-up the pioneering work by GVB and experts and will concentrate on compiling a verification procedure on fire protection goals and the requirements for fire protection regulations for entire Switzerland.

It is believed that these concerted endeavors in Switzerland on the fire prevention of tall PV façades will further foster the market penetration of PV façades and broaden their avenue towards an exploitation of their full potential.

5 REFERENCES

[1] Swiss Energy Strategy 2050
https://www.bfe.admin.ch/bfe/en/home/policy/energy-strategy-2050.html/

[2] R. Rechsteiner, R. Meier, U. Muntwyler, F. Nipkow, Th. Nordmann (2021), Die Energiewende im Warte-saal, Zocher & peter verlag kmg, 266 p.

[3] Gebäudeversicherung Bern, GVB (2023) Photovoltaic systems on high-rise façades - guidelines for fire protection verification, Ittigen, Switzerland.

ACKNOWLEDGEMENTS

We highly appreciate the collaboration with Gebäudeversicherung Bern, GVB and warmly thank FAMBAU Wohnbaugenossenschaft for the interesting cooperation. The technical input and comments provided during two project workshops by organisations such as the cantonal fire insurance association VKF, various building insurance companies and players in the façade construction industry are greatly acknowledged. The GVB project on guidelines for the fire prevention of tall PV façades is financially supported by the Swiss Federal Office of Energy, SFOE (P+D project).

SEMI-TRANSPARENT CIGS THIN-FILM PV MODULES

Peter Borowski[1], Thomas Schutt[2], Julian Röder[1], Maik Schubert[1], Martin Hillmann[2], Kristian Herath[2], Subarna Sapkota[2],
Volker Speer[2], Marko Stölzel[1], Rene Reichel[2], Thomas Dalibor[1]
[1]AVANCIS GmbH, Otto-Hahn-Ring 6, 81739 München, Germany
[2]AVANCIS GmbH, Solarstr. 3, 04860 Torgau, Germany
Phone: +49(0) 89 219620 464, Email: peter.borowski@avancis.de

ABSTRACT: By ablation of tiny spots of photovoltaic CIGS thin films, semi-transparent PV modules are produced with a homogeneous appearance to the human eye. Various degrees of transparency are presented, variations in moduling are being discussed and the resulting modules are characterised in terms of electrical yield and optical properties. Potential applications for these aesthetically appealing semi-transparent PV modules are shown, including the prototype of a photovoltaic window.

Keywords: semitransparency, BIPV, CIGS, thin film, window

1 INTRODUCTION

In contrast to conventional photovoltaics (PV), in which ideally all incoming photons are being transformed into electricity, semi-transparent photovoltaic elements are designed to let a certain percentage of light transmit from the outside to the inside. Table I lists a few application cases in which semi-transparent photovoltaics is desirable, particularly in building-integrated photovoltaics (BIPV).

Application	Purpose of semi-transparency
window, skylight	let in light; prevent excessive heating
canopy, patio roof	let through some light but attenuate full sunshine
car park / traffic way roofing	save on lighting
greenhouse / Agri-PV	attenuate full sunshine

Table I: Potential applications of semi-transparent photovoltaics.

Windows in present-day buildings with extensively glazed facades, e.g., often are semi-transparent by means of a coating ("low-e coating" [1]) that blocks and reflects a certain amount of the spectrum of the sunlight to reduce heat influx into the building and therefore cooling capacity

required. Aesthetically appealing semi-transparent PV can open up areas for electricity generation – particularly in the built environment – that are not available with only non-transparent (opaque) PV.

Different realisations of semi-transparent PV exist, each with their own advantages and disadvantages. Figure 1 depicts three approaches to semi-transparent PV (for further technologies see, e.g., [4] or [5]). In the present work, we focus on type 3 as shown in Figure 1, realised on thin-film PV modules based on the semiconductor CIGS (copper indium gallium (sulphur) selenide). Particularly in the built environment, we assess the presented solution as aesthetically superior compared to type 1 (Figure 1). The strong dark / light patterns generated by type 1 under direct irradiance can be disturbing for an inhabitant or distracting, e.g., for a driver under a roofed motorway. Realisations of type 2 exist, e.g., using organic PV, letting the visible part of the spectrum through and turning (typically) the infrared part of the spectrum into electricity. The underlying photovoltaic material for type 2 realisations, however, is lacking proof of longevity and sufficiently high photon conversion efficiency. The type 3 realisation presented in the current work is based on a trusted and tested commercial PV technology with proven long-term stability, good photon conversion efficiency and superior aesthetics [6][7][8][9].

In the following, we describe the production of the semi-transparent modules, their characterisation in terms

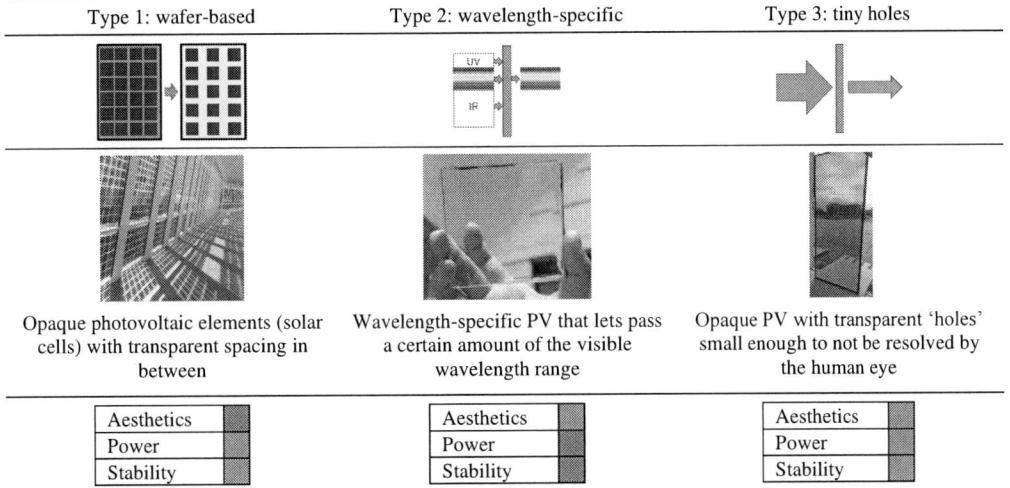

Figure 1: Realisations of semi-transparent photovoltaics. The image on the far right shows a semi-transparent CIGS PV module by manufacturer AVANCIS. For sources of the other images see [2], for an example of type 2 [3]. The bottom row rates the three types according to three criteria aesthetic of appearance, electrical power and stability/longevity.

of optical and electrical properties and finish by suggesting a few potential applications.

2 MAKING THIN-FILM PV MODULES SEMI-TRANSPARENT

2.1 Ablation of thin-film layer stack

Following type 3 as displayed in Figure 1, the AVANCIS CIGS circuits (coated on glass) are made semi-transparent by ablating small areas of the full (opaque) stack of the thin-film layers by means of a laser. With the ablated structures small enough, for the human eye, the PV circuits as well as laminates (modules) then appear as a homogeneous grey filter, since the underlying structure cannot be resolved when viewed from a certain distance. Typical dimensions of the ablated circles or rectangles are in the range of 30µm to 1mm and the degree of transparency can be controlled by varying size and distances of the ablated spots. Figure 2 shows a microscopy image of three CIGS circuits after ablation with a circle-shaped picosecond laser.

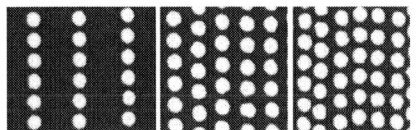

Figure 2: Microscopic image of three CIGS circuits after spot-wise ablation by laser. Optical transparency is increasing from left to right.

Ablation was done with a picosecond laser applied from the glass side of the PV circuit. The very short duration of exposure to the laser of the semi-conductive layers together with the chosen direction of incidence ensures a sudden and clear removal of the full stack of coated layers. Longer laser pulses lead to melting and subsequent condensation of some of the material with a risk of forming electrically conductive shunt paths within the layer stack on the edges of the ablated area.

For the ablated structures, both circles and rectangles were tested. The pattern (orientation) of the ablated structures can be varied, with effect on the optical properties of the resulting semi-transparent PV module. For a monolithically integrated thin-film module, it is, however, important to ensure that the ratio between transparent and absorbing areas is approximately constant perpendicular to the individual cell stripes of the module for all interconnected cells. From the schemes depicted in Figure 3, the theoretical transparency T of the circuit is

$$T_c = \frac{\pi d^2}{4(d + p_1)(d + p_2)} \cdot 100\%$$

for circle-shaped spots with diameter d, and

$$T_s = \frac{d^2}{(d + p_1)(d + p_2)} \cdot 100\%$$

for square-shaped spots with edge length d.

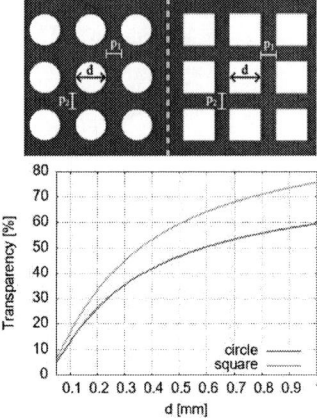

Figure 3: Top: Scheme for the calculation of transparency with ablation spots in the form of circles (diameter d) and squares (edge length d). Bottom: Calculated theoretical transparencies for these patterns with $p_1 = p_2 = 0.1$mm.

To ensure current conduction across the whole module, it is important to avoid uninterrupted ablation within a solar cell (the areas of ablation must not overlap, i.e., a minimum value for p_2 must be ensured). In the direction perpendicular to the cells, a minimum distance p_1 is favourable to avoid a stripe-like appearance of the ablation pattern. As an example, we have chosen $p_1 = p_2 = 0.1$mm for the graph shown in the bottom panel of Figure 3 and plot the resulting theoretical transparency over diameter/edge length of the ablated spots. Transparencies between 10 and 60% (circles) and 75% (squares) can be achieved with this setting, while keeping the dimension of the ablated spots below 1 mm. For the circle pattern, higher transparencies can be reached by arranging the spots in a hexagonal pattern. A hexagonal pattern, however, comes with the danger of overlapping ablation of adjacent rows of spots, which can lead to a fully interrupted current path and therefore non-functioning PV modules, particularly for production-scale laser processes and industrial cycle times of ablation. In a PV module (glass-glass laminate), the actual transmission will be lower as the theoretical values depicted here due to absorption and reflection losses in the glasses and lamination foil as well as non-optimal ablation of the opaque layers.

2.2 Moduling

After ablation of the spots, the semi-transparent circuits are processed to functioning PV modules no different than regular thin-film PV modules, i.e., the circuits are laminated to glass-glass modules and receive electrical contacts and mounting structures with established processes and materials. Figure 4 shows examples of fully functional semi-transparent CIGS thin-film PV modules as produced by AVANCIS in its production site in Torgau, Germany, based on the commercial BIPV module SKALA.

Depending on the type of application of the semi-transparent PV modules, the appearance and features of them can be adjusted, using the same underlying CIGS circuit. For good see-through properties, the junction boxes should not be placed in the semi-transparent area of the module. Junction boxes with a small cross section are available for this purpose that are placed on the glass edge of the glass-glass laminate. For efficient and visually non-disturbing cabling, the junction boxes can be placed on opposing sides on the short edges of the PV module. By varying the type of front glass, the see-through properties of the semi-transparent PV module can be altered from clear to frosted (see Figure 4) or a certain color impression can be achieved by combination with the AVANCIS Colour Technology (ACT) [9]. Choosing the front glass thick enough (starting with 4mm), test requirements of overhead glazing in certain jurisdictions are met and the PV modules become stable to walk on. The mounting of the semi-transparent PV modules should ideally be such that a clear and slim visual impression results, which for the AVANCIS modules with their high aspect ratio can be achieved by a linearly supported mounting along the long edge or by conventional clamping.

Figure 4: Prototype examples of semi-transparent CIGS thin-film PV modules as produced in AVANCIS' production site in Torgau, Germany. While the top and middle image show the front side of the modules, the bottom image looks at the rear (inner) side of the semi-transparent modules. The bottom images show two variations of front glass for clear and frosted see-through properties.

3 CHARACTERIZATION

3.1 Transparency

As discussed in section 2.1, the degree of transparency of the circuit can be varied by adjusting the shape, size and the distance of the ablated patterns. In the final module, further properties like transmission of the front glass, rear glass and lamination foil, optical interfaces between these parts or anti-reflective coating influence the transparency. The optical transmission of the final module / glass-glass laminate was measured with a commercial haze meter as well as by measuring the short circuit current of a small (~1cm^2) solar cell under a sun simulator both with and without the semi-transparent PV module in front of the cell. In Figure 5, these values are plotted on the x-axis. The y-axis shows the short circuit current of the semi-transparent PV modules relative to the short circuit current of the respective opaque PV modules with standard encapsulation. An ideal semi-transparent PV module of type 3 (Figure 1) loses exactly the current that would have

been generated by the ablated photovoltaic material which, in turn, should be a direct measure for the optical transparency (dotted line in Figure 5).

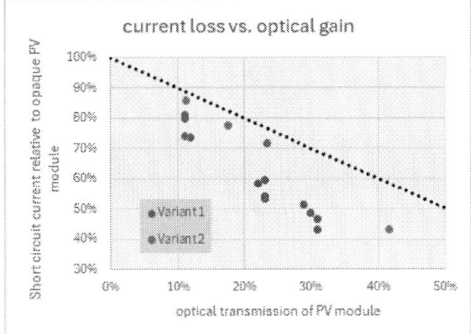

Figure 5: Short circuit current of the semi-transparent PV modules relative to those of an opaque module plotted over their optical transmission. The dotted line indicates an ideal semi-transparent PV module for which the relative loss in current equals the ratio of light that passes through the module.

Shown in Figure 5 are two variants with differing laser processes applied. One reason for the discrepancy of variant 1 to the ideal case is the presence of light-absorbing material that does not contribute to photovoltaic current generation. Microscopy reveals an opaque ring around the fully ablated spot (Figure 6), a region in which the CIGS absorber was ablated by the laser but the metallic back electrode remained on the rear glass. The effect of such a ring increases with increasing spot density (i.e., transmission). The ablation of variant 1 was done on regular production machinery, not optimized for this specific task. For variant 2 as shown in Figure 5, the laser process was changed compared to variant 1, leading to a better current loss to optical gain ratio.

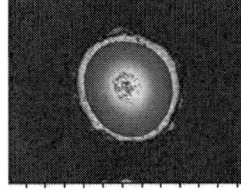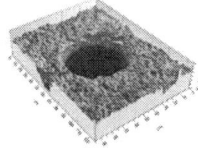

Figure 6: Microscopy image of one spot ablated by a picosecond laser (left panel). The right panel shows an image of a similar spot from confocal microscopy.

3.2 Optical properties

An important characteristic of semi-transparent PV, particularly for clear window applications, is the quality of the optical transmission, i.e., its see-through properties. With the regular patterns of ablated spots, diffraction of the transmitted light can occur when the ablated patterns are small. In conjunction with the optical interfaces present in a glass-foil-glass laminate, double images can result as well as colour effects. Choosing ablation patterns with dimensions large enough, these effects can be diminished. However, ablation spots of size 0.5 to 1mm are easily resolved by the human eye from close distances. The type of application will determine the ideal compromise between good see-through properties, homogeneous look and speed of manufacturing. Further potential for optimization of the visual appearance exists for choosing front and rear glasses as well as lamination foil with optimized optical properties. Also, changing ablation patterns, e.g., to more random structures, can further increase the see-through quality of the semi-transparent PV. Figure 7 shows an image of two lab-scale semi-transparent PV modules as seen from the rear (inner) side.

Figure 7: Two lab-scale (10x10cm²) semi-transparent PV modules with ablated patterns in dimensions above 0.5mm.

3.3 Electrical power

Under full sun irradiance (STC), electrical power generation of up to 100 W/m² was realized, of course depending on the degree of transparency. A few of the semi-transparent modules were installed into a PV system in Torgau, Germany and have been operating since more than a year. One module is installed into an outdoor monitoring setup at the same site with IV and meteorological data been recorded every few seconds. Behind this module, an irradiance sensor is installed and comparing its data to the irradiance into the module's plane, optical transparency during operation is obtained. Figure 8 shows this data for two days, together with the performance ratio of the semi-transparent module. From the data of the irradiance sensors, an optical transparency of about 27% is concluded for a large range of angels of irradiance.

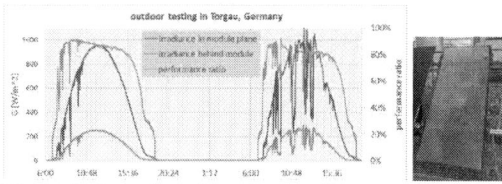

Figure 8: Recordings from two irradiance sensors, one in the plane of a semi-transparent PV module and one behind (see right panel) during two days in September 2023. Also shown is the performance ratio of this module during these two days (right vertical axis).

The left panel of Figure 9 shows the monthly performance ratio of the same module during a duration of 13 months. The values are lower than PR values of opaque PV modules of the same technology but within expectations for this first batch of prototypes and stable during the duration shown (with typical deviations during the winter months). Three other semi-transparent PV modules were measured indoors on a sun simulator at standard test conditions in certain intervals between outdoor operation in a PV system. The data is shown in the right panel of Figure 9 and indicates stable electrical power during the time period of 14 months.

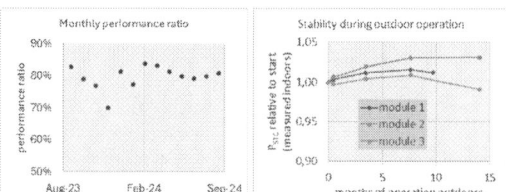

Figure 9: Left: Monthly performance ratio of the same modules as shown in Figure 8. Right: Electrical power at STC (relative to power at start of the deployment) as measured indoor on a pulsed solar simulator of three semi-transparent CIGS thin-film modules that are operating in a PV system in Torgau, Germany.

4 APPLICATIONS

Aesthetically appealing semi-transparent PV modules can cater many potential application cases, particularly in building integration (Figure 10):

- Windows
- Skylights
- Shading elements in facades, curtain walls or protruding roofs
- Greenhouses
- Agri-PV
- Terrace roof / canopy
- Car park / Motorway roofing
- Balcony railings
- Winter gardens
- Fences
- …

Some of these applications (windows, greenhouses) are not meaningful with opaque PV and others (skylights, terrace roof, winter gardens) are more attractive with a homogeneously semi-transparent PV element compared to a wafer-based semi-transparency (cf. Figure 1), by avoiding the strong dark / light patterns under direct irradiance. Such kind of dark / light pattern can be disturbing for an inhabitant or distracting, e.g., for a driver under a roofed motorway thereby endangering the social acceptance of these PV solutions. The – locally varying – requirements for overhead glazing must be met for some of the listed applications.

Figure 10: Renderings of potential applications of homogeneous semi-transparent thin-film PV modules: balcony railing (left) and a canopy (right).

As a showcase and using the CIGS-based technology as described, we manufactured a semi-transparent photovoltaic window as shown in Figure 11. The PV element is integrated as one pane of a standard double glazing thermally insulating window, which in turn is fit into a typical wooden window frame with window sash. For the electrical connection of the PV module, a socket and plug are integrated into the wooden frame.

Figure 11: A double-glazing window including a typical wooden frame incorporating a semi-transparent CIGS thin-film PV module.

5 SUMMARY

CIGS thin-film PV offers a platform for homogenous 'true' semi-transparent PV modules that overcome an important shortcoming of the currently dominating technology for this particular PV application. This, in turn, opens up new areas for PV deployment, particularly in the built environment and BIPV. CIGS thin-film PV can be the basis for homogeneous PV windows, in which the photovoltaic absorption not only generates electrical energy but also is beneficial for the energy demand of the building, contributing in two ways to its sustainability [10]. For a sustainable building product, important parameters like the U- and g-values must be determined, depending on the degree of transparency, the encapsulation method or the incorporation of the photovoltaic element into a more complex product.

6 ACKNOWLEDGEMENTS

The presented work would not have been possible without the members of the structuring and backend teams of AVANCIS in Torgau, Germany. The authors thank Stefan Bergfeld from Bergfeld Lasertech GmbH for processing some samples.

7 REFERENCES

[1] J.Aguilar-Santana et al., "Review on window-glazing technologies and future prospects", Int. J. of Low-Carbon Tech. 15(1) 112-120 (2020).

[2] Images taken from SMA Solar Technology (left) and Michigan State University (middle).

[3] Y.Li et al., "Color-neutral, semitransparent organic photovoltaics for power window applications", PNAS 117 21147 (2020).

[4] K.Chan et al., "Review and Potential Development of Solar Window Technology and Feasibility Study in Urban Area, Hong Kong as Case Study", IOP Conf. Ser.: Earth Environ. Sci. 588 022036 (2020).

[5] J.Sun and J.Jasieniak, "Semi-transparent solar cells", J.Phys.D: Appl.Phys. 50 093001 (2017).

[6] H.Elanzeery et al., "Beyond 20% World Record Efficiency for Thin-Film Solar Modules", IEEE J. of Photovoltaics 14(1) 107-115 (2023).

[7] T.Dalibor et al., "World Record Efficiency Cd-Free CIGS Modules and Their Advances for Large-Scale Production", Proc. of the 2023 MENA-SC (2023).

[8] J.Palm et al., "BIPV Modules: Critical Requirements and Customization in Manufacturing", Proc. of the 7th WCPEC (2018).

[9] T.Dalibor et al., "Cu(In,Ga)(Se,S)$_2$ thin-film technology - aspects of historical development, current status and future prospects", Int. J. of Appl. Glass Science, to be published.

[10] V.Wheeler et al., "Photovoltaic windows cut energy use and CO_2 emissions by 40% in highly glazed buildings", One Earth 5 1271 (2022).

41st European Photovoltaic Solar Energy Conference and Exhibition

Assessing photovoltaic-thermal system performance across diverse climates: An economic and environmental comparative analysis

Z. Ul-Abdin*, **O. Isabella**, and R. Santbergen

Photovoltaic Materials and Devices Laboratory, Delft University of Technology, The Netherlands (*Contact: Z.ulabdin@.tudelft.nl)

Motivation

- Photovoltaic-thermal (PVT) systems generate both **electrical** and **thermal energy simultaneously** [1], enhancing efficiency and longevity of PV cells.
- Introduction of **dynamic models** for air-based and bi-fluid based PVT collectors.
- Highlighting PVT collector adaptability to different **climate zones** [2], ensuring effective solar energy utilization in diverse regions.
- These collectors significantly **reduce CO_2 emissions** compared to traditional PV modules, demonstrating superior **environmental performance**.
- Assessing the **economic feasibility**.

PVT collector basic principle [3]

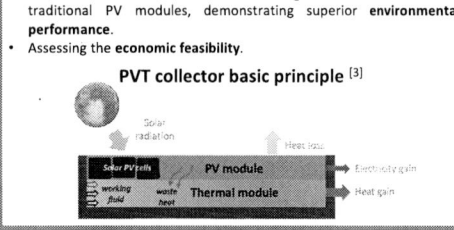

Modelling PVT collectors

- Developed **dynamic models** for different PVT collectors [3].
- Determined governing equations using **energy conservation principle**.
  - Applied at each component of the collector.
  - Considering **heat exchanges** between the layers.
- Several **assumptions** were considered.
  - Homogeneous temperature for each component.
  - Neglecting edge and bottom losses.
  - Overlooking pressure losses and partial shading.

Unglazed air-based PVT

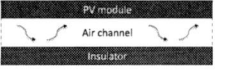

Glazed air-based PVT

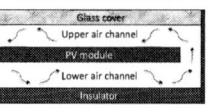

Dual channel air-based PVT

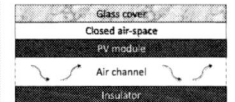

Bi-fluid PVT

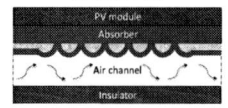

Energetic performance analysis

- Six cities representing **different climate zones** were chosen from northern hemisphere.
- Each corresponds to specific climate zone based on **temperature-precipitation** and **irradiation** [2].
- Weather data from **Amsterdam** as input for monthly comparison.
  - Monthly irradiation varies from 15 kWh/m² (December) to 158 kWh/m² (June), giving an **annual irradiation of 1018 kWh/m²**.
  - Monthly **average temperature** varies between **3.3 °C** (January) and **18.6 °C** (July).

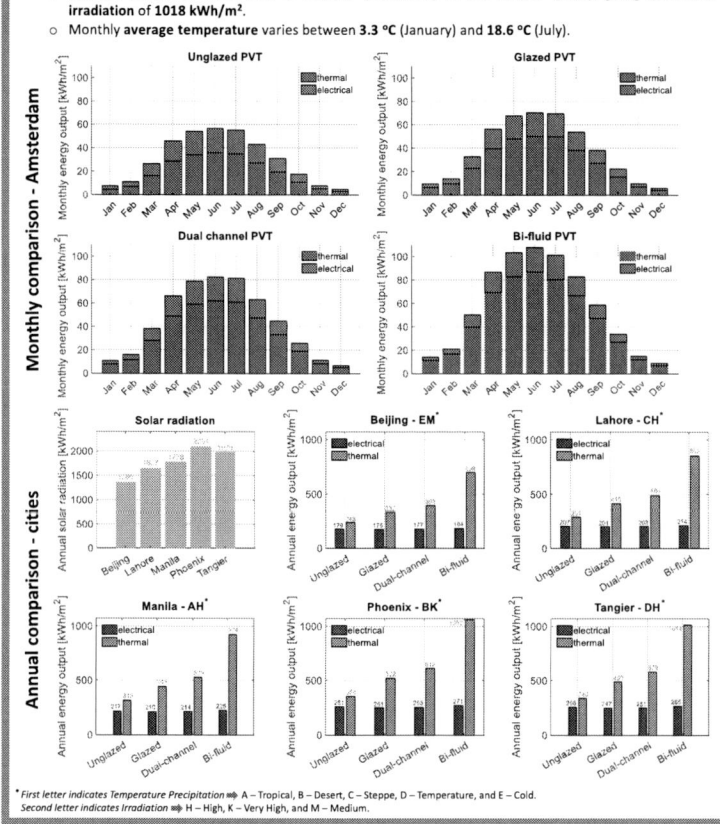

First letter indicates Temperature Precipitation ⇒ A – Tropical, B – Desert, C – Steppe, D – Temperature, and E – Cold.
Second letter indicates Irradiation ⇒ H – High, K – Very High, and M – Medium.

Economic and environmental analysis

- Levelized cost of electricity (LCOE) is calculated by **combining** annual electrical and thermal yield [4].
- Environmental analysis involves modules with a **surface area** of approximately 0.50 m².
- Total electrical equivalent energy production is calculated using **thermal power plant conversion factor** of 0.38 [5].

Comparison of LCOE

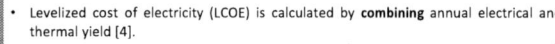

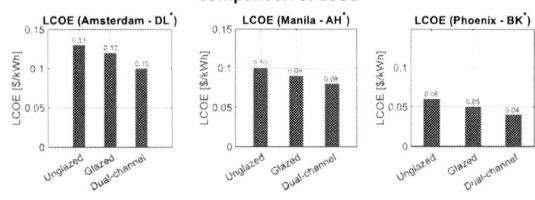

Annual comparison of CO_2 mitigation

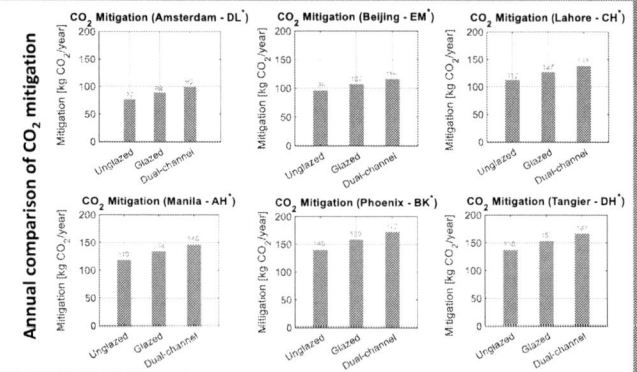

First letter indicates Temperature Precipitation ⇒ A – Tropical, B – Desert, C – Steppe, D – Temperature, and E – Cold.
Second letter indicates Irradiation ⇒ H – High, K – Very High, and M – Medium.

Conclusions

- Unglazed PVTs are **optimal for PV cooling**, while dual-channel PVT **excel in air heating**.
- Using two different fluids simultaneously **boosts thermal output** and effectively cools PV cells.
- Lower LCOE for dual-channel PVT is due to the additional thermal energy.
- PV unit **avoids emissions** at a minimum of 47 kgCO₂/year, while PVTs avoid a maximum of 140, 159 and 172 kgCO₂/year.

References

[1] Z. Ul Abdin, & A. Rachid, **2021**, Energies, 14(4), 1205.
[2] J. Ascencio-Vásquez, et al., **2019**, Sol. Energy, 191, 672-685.
[3] Z. Ul-Abdin, et al., **2024**, Sol. Energy, 276, 112687.
[4] J. Dijkstra, **2024**, MSc thesis, TU Delft.
[5] W. He, et al., **2006**, Appl. Energy, 83(3), 199-210.

Acknowledgements

This work was carried out as a part of Simply Positive project. The Dutch part is supported by the RVO (the Netherlands Enterprise Agency).

Photovoltaic Materials and Devices Laboratory

41st European Photovoltaic Solar Energy Conference and Exhibition

A STRATEGIC APPROACH TO ENABLE LARGE-SCALE PHOTOVOLTAIC ENERGY SYSTEMS DEPLOYMENT IN URBAN AREAS

Joyce Arthllan Oliveira de Sousa *[1, 2, 3], Martin Thebault [1, 2], Lamia Berrah [1, 3]

✉ (joyce-arthllan.oliveira-de-sousa@univ-smb.fr)

[1] Solar Academy Graduate School, Université Savoie Mont Blanc, France
[2] LOCIE, Université Savoie Mont Blanc, CNRS UMR5271, F- 73376 Le Bourget du Lac, France
[3] LISTIC, Université Savoie Mont Blanc, F- 74944 Annecy-le-Vieux, France

ABSTRACT: Deploying large-scale photovoltaic (PV) systems is essential to advancing urban energy transition; however, there are unique challenges that necessitate structured, strategic approaches. This paper introduces a conceptual, actor-oriented framework to support large-scale PV deployment in urban areas. The framework was developed through a Systematic Literature Review (SLR) that aimed to address the research question, "What approach can enable the large-scale deployment of photovoltaic energy systems in urban areas?"— grounded in a recognized research gap. Through this review, four primary actor groups were identified as influential in the deployment process, whether by contributing to, affecting, or being impacted by large-scale PV implementation: (1) Decision-makers, (2) Directly Affected Actors, (3) Indirectly Affected Actors, and (4) Interested Parties. Additionally, six key layers of influence emerged as critical to PV deployment, encompassing (i) External Context, (ii) Internal Context, (iii) Assessment & Planning, (iv) BIPV Technology Advancements, (v) Large-Scale Deployment, and (vi) Urban Energy System Integration. Together, these layers and their 32 influencing factors shape an "urban ecosystem" foundational to effective PV deployment. The findings of the SLR provided the basis for developing a strategic, actor-oriented framework, designed to support territories aiming to undertake extensive PV energy system deployments and contribute to urban sustainability.

Keywords: photovoltaic energy systems, large-scale deployment, energy transition, urban planning, conceptual framework

1 INTRODUCTION

Integrating renewable energy systems, particularly photovoltaic (PV) systems, into existing building infrastructure is instrumental in enhancing energy security and minimizing reliance on distant, disruption-prone energy sources [1]. However, setting achievable and sustainable PV deployment goals across various territorial locations continues to present substantial challenges.

1.1 Research Gap

While the potential of photovoltaic (PV) energy systems in urban areas is well-recognized, their large-scale deployment remains constrained by the absence of strategic approaches that comprehensively address societal, environmental, and techno-economic imperatives [2].

PV systems, due to their scalability and cost-effectiveness, present a viable solution for sustainable urban energy production, particularly when building infrastructure, such as rooftops and façades, is repurposed for energy generation. This transformation could shift urban areas from being major consumers of energy and sources of emissions—currently accounting for 78% of global energy consumption and over 60% of greenhouse gas emissions, despite occupying less than 2% of the Earth's surface—towards becoming key contributors to clean energy generation and emissions reduction [3].

Despite the recognized potential of photovoltaic (PV) systems, the literature lacks tailored, strategic approaches encapsulated within practical frameworks to support their large-scale deployment in urban contexts.

While existing studies broadly explore renewable energy adoption, a clear gap exists in approaches specifically designed to enable the large-scale deployment of PV systems. This study focuses particularly on the promising, yet underutilized potential of solar energy

harnessed through PV systems in urban areas, where PV could play a significant role in advancing the urban energy transition and accelerating urban decarbonization [4].

1.2 Justification

The rapid urbanization projected by the United Nations, with 68% of the global population expected to reside in cities by 2050, underscores the urgent need for sustainable energy solutions to meet rising energy demands and address environmental challenges in urban areas [5].

Photovoltaic (PV) energy systems represent a viable option, given their potential to integrate seamlessly into existing urban infrastructure and provide a cost-effective, scalable means of generating clean energy directly within cities [6].

This study is therefore warranted by the critical need to address these challenges and proposes a strategic approach to enable the large-scale deployment of PV energy systems in urban areas.

1.3 Research Question

In an effort to guide the study's exploration and identify the layers of influence and the influencing factors in enabling the large-scale deployment of photovoltaic (PV) energy systems in urban areas [7], the following research question has been formulated:

- What approach can enable the large-scale deployment of photovoltaic energy systems in urban areas?

1.4 Hypothesis

In light of the identified research gap and the research question concerning the lack of tailored approaches to enable the large-scale deployment of PV energy systems in urban areas, we propose the following hypotheses:

In deductive research, it is hypothesized that a strategic approach specifically designed to facilitate the large-scale deployment of photovoltaic energy systems in urban areas will lead to:

- Increased adoption rates of PV systems within urban areas;
- A consequent reduction in greenhouse gas emissions associated with urban energy consumption.

Conversely, in inductive research, it is hypothesized that an empirical exploration of the proposed approach's usability in enabling large-scale PV system deployment in urban areas will yield insights into:

- The feasibility and effectiveness of the proposed approach in addressing the identified research gap;
- The scalability and adaptability of the approach to various urban contexts.

2 STATE-OF-ART

2.1 PV deployment projections

The Large-scale The role of photovoltaic (PV) systems in the global energy transition is projected to be transformative, particularly under the framework provided by the International Energy Agency's (IEA) Net Zero by 2050 roadmap [8].

In 2023, global cumulative solar PV capacity amounted to 1,624 gigawatts, with roughly 447 gigawatts of new PV capacity installed in that same year [9].

According to this roadmap, solar PV will be essential to decarbonizing the electricity sector, with global solar PV capacity expected to increase more than twentyfold by 2050.

Rooftop PV installations are projected to be a critical component of this growth, providing a distributed and resilient source of renewable electricity that aligns well with urban energy needs and the spatial constraints within densely populated areas.

The International Renewable Energy Agency (IRENA) reinforces this vision in its report on the future of solar PV, emphasizing that rooftop PV systems will play a pivotal role in achieving global renewable energy targets. IRENA projects that by 2050, renewables, including rooftop solar, could supply up to 86% of global power demand [10].

Achieving this goal will necessitate a substantial increase in rooftop PV installations, especially in urban areas where rooftops provide some of the most feasible sites for solar deployment due to limited land availability.

This approach not only optimizes space but also integrates renewable energy production directly into cities, reducing transmission losses and enhancing local energy resilience [11].

However, meeting these ambitious targets requires a multifaceted approach, including robust policy support, favorable regulatory frameworks, and targeted financial incentives to encourage adoption.

IRENA highlights the importance of local decision-making in facilitating PV deployment, as municipalities and city planners must work to harmonize PV integration with existing urban landscapes and infrastructure. Such policies may involve zoning adaptations, incentives for building-integrated PV, and streamlined permitting processes that enable quicker and more cost-effective installation of PV systems on urban rooftops.

Furthermore, as these projections underscore the necessity for widespread PV adoption, particularly at the building level, they also indicate the importance of advancing PV technology to improve efficiency, affordability, and adaptability to various urban conditions.

Efforts to scale up rooftop PV deployments will need to be accompanied by advancements in grid infrastructure to handle the variable nature of solar power and ensure that surplus energy generated by rooftop systems can be efficiently managed, stored, or redistributed.

2.2 Addressing Spatial Challenges

The Large-scale PV deployment in urban areas faces several challenges, including limited space in dense cityscapes and the complexities of setting achievable deployment goals that account for the specific characteristics of each territory.

However, a paradigm shift in perspective could transform our understanding of cities' roles in the renewable energy transition.

By viewing existing building infrastructure as a resource, utilizing rooftops and other surfaces for PV installations, urban areas can support substantial PV capacity without encroaching on land reserved for agriculture or leisure [12].

Strategic identification and utilization of suitable sites for PV installations can minimize conflicts with public legislation, which often supports renewable adoption but must also address urban priorities involving land use, solar irradiance optimization, and environmental impact [13].

In this context, developing strategic approaches that balance urban energy transition with critical considerations for sustainable city planning is essential for urban leaders responsible for energy transition projects and planning [14].

2.3 Data Availability

Effective assessment of PV capacity in urban areas necessitates extensive data on existing building infrastructure and characteristics [15].

Geographic Information Science (GIScience) facilitates this process by supplying critical data for modeling the spatiotemporal distribution of solar potential across 3D urban surfaces within a territory. Such capabilities are instrumental in advancing sustainable cityscapes optimized for PV deployment [16].

Moreover, it is well-established that solar cadastres are essential not only for the large-scale deployment of photovoltaic (PV) energy systems but also for their ongoing management [17].

However, not all territories have access to these data sources, as developing solar cadastres entails significant time and financial investment.

Nonetheless, the value of solar cadastres is undeniable; they provide critical geospatial data that supports strategic analysis of urban areas for PV deployment, facilitating informed decision-making.

In this study, we identify three key dimensions commonly addressed within a solar cadastre, typically grounded in GIS data: (i) rooftop eligibility, which assesses factors such as solar irradiance, heritage protection status, and structural load capacity [18]; (ii) effective surface area available for PV installations, identifying feasible deployment sites [19]; and (iii) grid infrastructure capacity, evaluating the city's ability to accommodate surplus energy generated by rooftop PV systems [20].

Access to clearly documented solar datasets and open-source tools remains essential for the research community to assess PV potential in a transparent and reproducible manner.

Historically, economically favored territories have had the resources to develop and utilize solar cadastres, whereas regions with potentially higher solar potential have been unable to pursue similar efforts due to financial constraints [21].

Consequently, there is an urgent need to disseminate methodologies that support territories both in developing their own solar datasets and in improving existing cadastres [22]. Such advancements provide critical data to inform urban decision-making for optimized PV deployment, ongoing system maintenance, and end-of-life planning.

Large-scale PV deployment implies future requirements for substantial recycling, reuse, and replacement efforts, which demand careful planning today.

Leveraging the key insights from existing solar cadastres can help reduce both the time and financial investments needed for developing new ones, providing a foundational basis for territories seeking to advance their solar data capabilities and thus strengthen their contributions to the urban energy transition.

Large-scale PV deployment implies future requirements for substantial recycling, reuse, and replacement efforts, which demand careful planning by now [23].

Leveraging the key insights from existing solar cadastres can help reduce both the time and financial investments required for developing new ones, providing a foundational basis for territories seeking to enhance their solar data capabilities and strengthen their role in the urban energy transition [24].

In this context, developing strategic frameworks that address key aspects related not only to data assessment but also to scalability across diverse territories is essential [25].

Such frameworks should enable regions committed to energy transition initiatives to systematically select suitable rooftop infrastructure for PV installations, thereby advancing the large-scale deployment of PV systems in urban areas.

3 METHODOLOGY

To address the research gap and respond to the research question posed in this study, a Systematic Literature Review (SLR) was conducted using several scientific databases, including Elsevier/ScienceDirect, Springer Link, IEEE, ACM, and Solar Cadastres Worldwide. A set of 27 keyword clusters were strategically combined using Boolean operators (AND/OR) to refine (AND) or expand (OR) search results. This approach yielded 212 relevant scientific publications, each presenting concepts pertinent to addressing the main concern of this study.

Among the analyzed concepts, three Multi-Scale

Actor Systems were identified as central frameworks: a) Multi-Level Governance (MLG), b) Socio-Technical Systems (STS), and c) Actor-Network Theory (ANT) [26], [27] and [28].

Together, these frameworks may highlight the importance of layered governance structures and the interconnected actors involved in PV deployment across varying scales.

In addition to MLG, STS, and ANT, Urban Metabolism was examined for potential alignment in guiding the study [29].

These frameworks collectively informed the study's approach to understanding the dynamics governing large-scale PV deployment in urban areas and helped shape the review's focus to effectively address the research question.

The principles of Multi-Level Governance (MLG) illustrate how decision-making processes unfold across different governance layers, engaging public, private, and civil society actors. This framework underscores the importance of interactions between external and internal contexts in shaping PV deployment outcomes.

The Socio-Technical Systems (STS) theory, which highlights the co-evolution of technology and society, emphasizes the critical role of Assessment and Planning in PV deployment. STS suggests that integrating large-scale PV systems requires consideration of social, economic, and political shifts accompanying technological advancements.

The Actor-Network Theory (ANT) provides a dynamic view of the network of urban actors—such as decision-makers, directly affected stakeholders, and indirectly affected groups—who interact within a broader network. ANT also revealed how this network, rather than remaining static, adapts as actors may collaborate, influence one another, and respond to challenges encountered in PV system deployment.

The Urban Metabolism (UM) offers a complementary perspective by viewing cities as dynamic systems where energy, resources, and actors interact in a continuous exchange. This approach underscores the importance of analyzing urban areas as integrated environments where material flows—such as energy, water, and waste—circulate within the city. For large-scale PV deployment, UM provides insight into how urban areas can be optimized as self-sustaining energy ecosystems.

This perspective reinforces the importance of adaptive, cross-sectoral strategies in PV deployment, aligning environmental goals with urban planning requirements.

Together with MLG, STS, and ANT, the UM approach forms a multi-faceted conceptual foundation for assessing PV deployment within urban environments, promoting an actor-oriented approach that addresses technical, social, and environmental dimensions critical to urban energy transition.

In Figure 1, the main findings from the SLR are summarized, identifying four primary actor groups as influential in the deployment process, whether by contributing to, affecting, or being impacted by large-scale PV implementation: (1) Decision-makers, (2) Directly Affected Actors, (3) Indirectly Affected Actors, and (4) Interested Parties.

Additionally, six key layers of influence emerged as critical to PV deployment, encompassing (i) External Context, (ii) Internal Context, (iii) Assessment & Planning, (iv) BIPV Technology Advancements, (v) Large-Scale Deployment, and (vi) Urban Energy System

Integration. Together, these layers and their 15 influencing factors shape an "urban ecosystem" foundational to effective PV deployment.

3.1 Urban Actors

This study identifies four primary categories of urban actors whose roles and interests are essential for developing a deployment approach that is inclusive and responsive to societal needs.

Recognizing and understanding these actors' roles is critical, as each group holds distinct interests, responsibilities, and levels of influence, which can either facilitate or impede the successful deployment of photovoltaic (PV) energy systems in urban environments. Based on a systematic literature review (SLR), the four key categories of actors are as follows:

A1. Decision-Makers

This group includes public authorities responsible for the urban energy transition, energy regulators, policymakers, householders, building owners, and tenants. These actors possess direct influence over PV system deployment and play a fundamental role in shaping policies and regulatory frameworks that support sustainable urban energy initiatives [30].

A2. Directly Affected Actors

This category encompasses urban residents, building owners, PV panel and component manufacturers, PV installers, maintenance professionals, utilities, and other energy providers. These stakeholders are directly impacted by PV deployment processes and outcomes, with interests tied to both the practical aspects of system installation and ongoing operational efficiencies [31].

A3. Indirectly Affected Actors

This group includes local urban energy consumers, cross-border citizens, local businesses, industries, and organizations. While these actors may not engage directly in PV deployment, they are affected by the broader urban energy landscape shaped by PV systems and contribute indirectly through energy demand and market support [32].

A4. Interested Parties

This category consists of vocational training institutions, environmental advocacy groups, and investors. These actors play supportive yet influential roles by advocating for sustainability, fostering skilled labor, and providing financial investment to advance PV technology integration within urban energy systems [33].

3.2 Layers of Influence

The analysis of frequency occurrences across six major Layers of Influence reveals insights into the relative importance of various factors within each layer and highlights which dimensions have the highest and lowest impact on overall outcomes.

L.1. External Context (37 occurrences)

External factors significantly shape outcomes, pointing to the importance of aligning local initiatives with global standards and trends. Key factors include:

- International Compliance (12 occurrences) and Global Conjuncture (10 occurrences), highlighting the need for alignment with international regulations and awareness of global economic and environmental trends.
- Environmental Demands (8 occurrences) and Market Dynamics (7 occurrences) emphasize adapting to environmental pressures and market fluctuations as external influences that impact strategic planning.

L.2. Internal Context (19 occurrences)

Internal factors appear less influential overall, suggesting they may play a supporting role rather than being primary drivers in this context. Key factors include:

- Objectives Alignment (8 occurrences) and Readiness to Adapt (6 occurrences), indicating the value of organizational cohesion and adaptability.
- Local Autonomy (5 occurrences) suggests that localized decision-making may be somewhat less critical in the broader scheme.

L.3. Assessment & Planning (31 occurrences)

The assessment and planning also emerge as essential elements, crucial for making informed, data-driven decisions. Key factors within this layer include:

- Territorial Strategy (16 occurrences) reflects the importance of spatial planning to maximize resource efficiency.
- Reliable Support Information (15 occurrences) highlights the need for dependable, high-quality data to guide planning processes.

L.4. BIPV Technology (31 occurrences)

The Building-Integrated Photovoltaics (BIPV) Technology layer is also influential, pointing to the role of innovative technology integration in advancing energy efficiency. Influencing factors include:

- Energy Equity (16 occurrences), underscoring the importance of ensuring fair and widespread access to BIPV solutions.
- Deployment Dynamics (15 occurrences), which highlights the pace and scalability challenges inherent to the deployment of BIPV systems.

L.5. Large-Scale Deployment (51 occurrences)

Large-Scale Deployment stands out with the highest total occurrences, indicating its critical role in influencing outcomes. Within this layer:

- PV Integration (27 occurrences) is the most frequently noted factor, underscoring the essential nature of seamless integration of photovoltaic (PV) systems at scale to enhance energy output and accessibility.

- Spatio-Temporal Initiatives (24 occurrences) closely follow, suggesting that coordinated, time-sensitive deployment strategies are vital for achieving large-scale deployment goals.

L.6. Urban Energy Systems (43 occurrences)

The Urban Energy Systems layer also has a prominent impact, emphasizing the importance of well-structured urban policies and infrastructure to support sustainable energy systems. Within this layer:

- Drivers (23 occurrences) are key factors promoting system development, reflecting the critical role of incentives, supportive policies, and innovation.
- Barriers (20 occurrences), while slightly less frequent, reveal the challenges and resistance that may need to be addressed to ensure successful integration and long-term sustainability.

3.3 Actor-Oriented Framework

The Actor-Oriented Framework presented in Figure 2 has been developed to support the large-scale deployment of PV energy systems in urban areas by integrating insights from the Multi-Scale Actor Systems—MLG, STS, ANT, and UM. This integration enables a comprehensive multi-actor analysis.

Through this framework, the distinct roles of various urban actors within the complex ecosystem of PV deployment are systematically addressed, providing a structured approach to navigating the challenges inherent to large-scale implementation of these energy systems.

This framework organizes the study around both layers of influence, each defined by specific influencing factors, and the unique urban actors who operate within and across these layers.

Each actor group—decision-makers, directly and indirectly affected stakeholders, and interested parties—engages with the layers in distinct ways that reflect their particular levels of influence within the urban energy landscape.

This dual focus on both the structural dimensions (layers) and dynamic interactions (actors) ensures that the framework comprehensively addresses both the systemic and actor-driven aspects critical to effective PV deployment.

The framework is visualized as a metaphorical mountain, with "hiking pathways" symbolizing the adaptive movement—both ascents and descents—required within the Layers of Influence.

Urban actors are positioned within this interconnected "urban ecosystem," fostering collaboration essential for advancing the energy transition.

This framework underscores that effective deployment requires not only cross-layer coordination but also an adaptable, actor-oriented approach that aligns with each actor's specific interactions within a given urban ecosystem.

3.4 Validation and Implications of the Framework

The multi-actor framework developed in this study was validated through an in-depth case study conducted in the Greater Geneva Area (GGA), offering a comprehensive assessment of its applicability in a real-world setting. This case study applied and refined the framework, particularly in addressing the "Assessment and Planning" layer with emphasis on the "Reliable Support Information" influencing factor. To this end, regulatory constraints and urban dynamics were integrated into the analysis, with special attention given to heritage protection levels and building categories within the territory.

The analysis of PV deployment potential across different macro-regions within the GGA, followed by evaluation across building categories, demonstrated the framework's capacity to generate reliable support information in terms of assessment of PV capacity within the territory.

This information enables urban planners to allocate resources effectively and prioritize efforts to support the large-scale deployment of PV energy systems.

The approach aligns with the multi-layered needs of urban territories, thus strengthening the framework's utility in practical planning contexts.

In bridging theoretical planning with actionable implementation, the framework further offers a scalable model suitable for replication across regions with diverse socio-economic, regulatory, and spatial characteristics.

This framework introduces a strategic approach for urban areas aiming to achieve sustainable energy transitions, significantly enhancing decision-making capabilities.

3.5 the FrameworkFuture Directions

Building on the validation through the Greater Geneva Area case study, this framework sets a foundation for future research and practical applications aimed at large-scale PV deployment across diverse urban areas.

To explore its full potential, further studies could examine all the framework's layers of influencing to varied regulatory landscapes, different climatic zones, and distinct socio-economic conditions.

Assessing the framework in contrasting urban contexts will help refine its components and ensure its robustness for wide-scale urban energy transitions.

As scalability remains a critical consideration for the framework's application. While it has been tailored to address the complexities of urban areas, modifications may be required to accommodate suburban or rural settings where building density, infrastructure, and regulatory oversight differ.

Expanding this framework's application across broader geographic regions could yield valuable insights, enhancing its role as a strategic approach, as regions with limited resources or constrained energy policies could highlight potential adaptations to support flexible and cost-effective PV deployment in underserved areas.

Furthermore, integrating high-resolution spatial data, such as GIS data that includes not only roof area availability but also façade surfaces and idle areas capable of hosting PV energy systems without interfering with recreational or agricultural land use, can significantly enhance PV potential assessment and, consequently, the framework's overall utility.

This comprehensive approach ensures that the framework not only addresses current urban planning needs but also adapts to evolving regulatory and technological advancements, positioning it as a long-term asset for sustainable urban energy development.

41st European Photovoltaic Solar Energy Conference and Exhibition

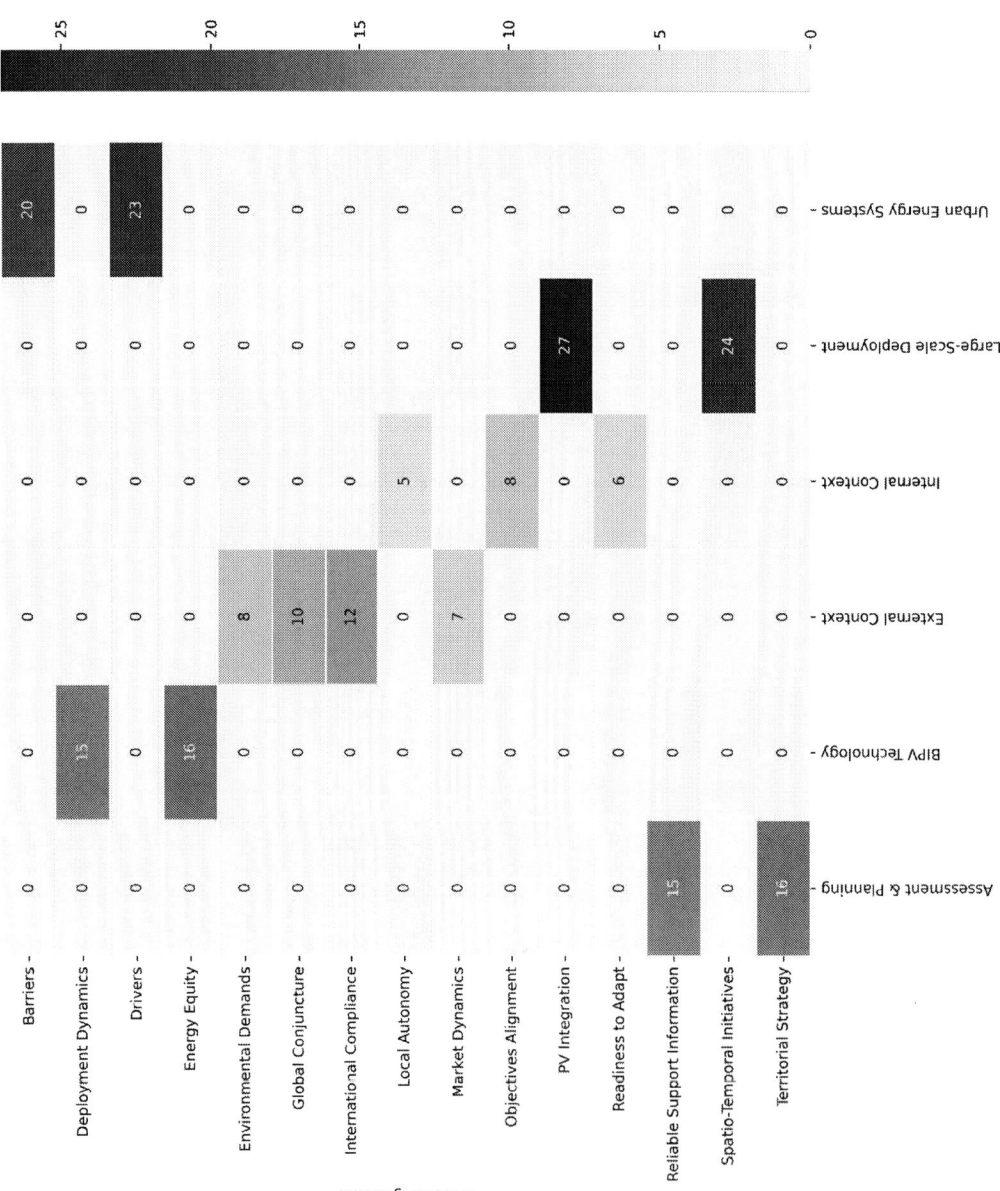

Figure 1. Heatmap of Influencing Factors Across Layers of Influence for Large-Scale PV Deployment in Urban Areas

41st European Photovoltaic Solar Energy Conference and Exhibition

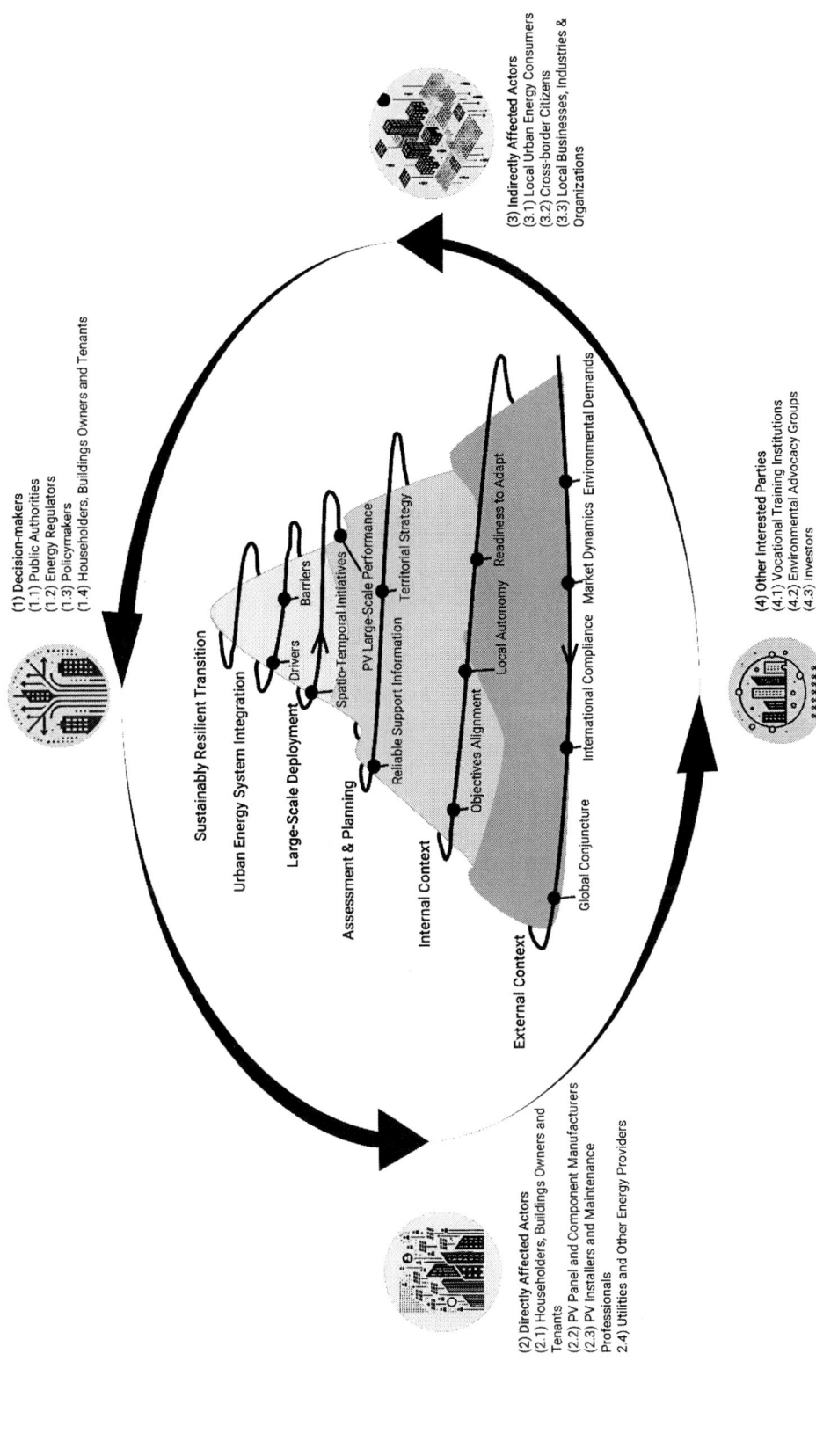

Figure 2. Actor-Oriented Framework for Large-Scale Photovoltaic Deployment

41st European Photovoltaic Solar Energy Conference and Exhibition

4 CASE STUDY

A case study in the Greater Geneva Area has been conducted to demonstrate the application of the conceptual framework proposed in this paper, specifically highlighting its potential to support large-scale PV energy system deployment within a defined territory. This case study emphasizes the "Assessment & Planning Layer," with a particular focus on the "Reliable Support Information" influencing factor.

The study's objective is to quantify the PV capacity across each macro-region and identify the building categories with the highest potential for PV system deployment. By pinpointing priority areas, this approach aims to provide clear guidance for optimizing large-scale PV integration within the region.

4.1 Context

The chose territory is divided into nine territorial planning zones and consists of a total of 209 municipalities. This includes 45 communes in the Canton of Geneva, 47 in the District of Nyon, and 117 communes in the French cross-border region. Of the French communes, 39 are located in the Department of Ain and 78 in the Department of Haute-Savoie.

These zones form an integral part of regional planning in this cross-border region.

Despite differences in political systems and legislation, these territories share a common goal: enabling the large-scale deployment of photovoltaic (PV) energy systems. This shared vision highlights the importance of cross-border cooperation to create a unified strategy for accelerating the adoption of renewable energy systems.

The vision of these territories lies in their planning usage to manage and organize land and resources to achieve sustainable and balanced development across their territory.

It is important to highlight that all of them encourage the expansion of PV energy systems deployment in their territory but there is no consensus in how to enable it among the territories.

The Swiss Canton and the Swiss District have developed Master Plans to guide their sustainable goals by 2050. The Canton of Geneva's Master Plan focuses on three primary areas:

- Enhancing energy efficiency within buildings and promoting renewable energy sources;
- Optimizing mobility in both private and public sectors;
- Refining systems and processes within industrial and service sectors.

The Canton of Geneva's Master Plan includes a significant focus on photovoltaic (PV) energy systems. The canton aims to reach 350 MWp by 2030 and 1000 MWp by 2050. This target not only directs the analysis of necessary MWp installations within the region but also the need to evaluate the feasibility of such goals.

The SCOTs (Le Schéma de Cohérence Territoriale, stands for Territorial Coherence Plan in english) for Genevois, Pays de Gex, Pays Bellegardien, Arve et Salève, Pays Rochois, Faucigny-Glières, and Chablais are responsible for urban planning across the French cross-border region. Although they support PV energy system deployment, highlighting the need to promote the development of photovoltaic solar energy on the rooftops of tertiary buildings by 2030 and continuing its development until 2050, there are no clearly established short-, medium-, and long-term goals for the territory.

In Nyon, solar photovoltaics are identified as the primary source of local renewable electric energy, making its development a significant priority for the city in the context of high electrical consumption pressures. The plan in Nyon is to achieve 18% of the built environment covered by photovoltaics by 2030.

The Greater Geneva Area (GGA) spans three macro regions, each with distinct socio-political boundaries: The Canton of Geneva (GE), the Canton of Vaud (VD), and the French cross-border region (F). A detailed assessment of photovoltaic (PV) capacity potential was performed across these regions by examining all available rooftops surfaces through GIS-data collected in the territorial solar cadaster.

4.2 PV Capacity Scenarios

In the first scenario, depicted in Figure 3a, the cumulative photovoltaic potential is assessed across all rooftops within the Greater Geneva Area (GGA), without restrictions based on suitability class.

This unfiltered analysis provides a baseline for understanding the maximum theoretical capacity potential of rooftop PV installations in the area:

- French cross-border region (F)

The region exhibits the highest total potential capacity, with a projected PV capacity of 2,821 MWp and a total available rooftop area of 12.82 km² distributed across 118,449 buildings.

- Canton of Geneva (GE)

This region closely follows with 2,753 MWp potential, covering 12.51 km² of rooftop space across 108,261 buildings.

- Canton of Vaud (VD)

VD contributes a lesser but significant capacity potential of 597 MWp, covering 2.71 km² of rooftop area from 23,204 buildings.

These results indicate that the GGA as a whole possesses substantial rooftop area, potentially able to host large-scale PV installations. The higher building counts in regions F and GE support greater capacity potentials, illustrating the importance of urban density and building availability in PV capacity projections.

In the second scenario, depicted in Figure 3b, we exclude buildings classified as CLASSE 3, which are deemed strictly protected and are thus unsuitable for rooftop PV installations. This filtered dataset yields a more realistic scenario by aligning with regulatory and architectural preservation constraints.

- French cross-border region (F)

The potential PV capacity decreases to 2,548 MWp over 11.58 km², covering 81,334 buildings. The reduction in building count and area signals the considerable impact of architectural regulations in limiting PV capacity within protected areas.

- Canton of Geneva (GE)

Capacity drops to 2,342 MWp, with available rooftop area reduced to 10.65 km² across 54,738 buildings. This decrease underscores the prevalence of protected buildings in GE, highlighting that regulatory constraints substantially affect the exploitable PV potential.

- Canton of Vaud (VD)

With fewer protected buildings than GE and F, VD's capacity potential drops moderately to 485 MWp over 2.20 km² across 11,338 buildings.

This scenario demonstrates the impact of architectural and historical preservation on PV capacity. Regions with higher counts of strictly protected buildings, such as GE, see notable reductions in PV potential when these buildings are excluded, reflecting the regulatory challenges that limit PV deployment despite theoretical capacity.

The third scenario, depicted in Figure 3c, focuses exclusively on buildings classified as CLASSE 1, which are deemed highly suitable for PV systems. This class provides a focused view of buildings that not only permit PV installation but also offer the most favorable conditions for maximizing PV efficiency and output.

- French cross-border region (F)

PV capacity drops significantly to 1,019 MWp with 4.63 km² of suitable rooftop area across 34,559 buildings. The considerable reduction reflects the stringent filtering, which leaves only buildings that provide the highest suitability and yield for PV installations.

- Canton of Geneva (GE)

Capacity is reduced to 916 MWp with 4.16 km² across 17,173 buildings. This scenario underscores the reduction from initial unfiltered capacity, reflecting the high standard of suitability (CLASSE 1) required for optimal PV installation.

- Canton of Vaud (VD)

VD's capacity falls to 257 MWp over 1.17 km² from 3,904 buildings, underscoring the limited number of highly suitable buildings in this region compared to F and GE.

The final scenario highlights a core group of highly suitable buildings, suggesting a robust but selective pathway for PV deployment in the GGA.

The significant reduction in building count and capacity potential underlines the rigor of suitability standards in optimizing PV output efficiency. This focus on CLASSE 1 buildings provides a realistic view of feasible and high-yield installations that align with current architectural and regulatory standards while maximizing energy yield.

4.3 Building Classification

Table I: Building Type according to SIA 380/1:2009 standard

Building Category	SIA Code
Collective housing	1
Individual housing	2
Administration	3
School	4
Shops	5
Restaurant	6
Meeting places	7
Hospital	8
Industry and crafts	9
Warehouse	10
Sports	11
Indoor swimming pool	12

In Table I, an in-depth analysis of PV capacity within the GGA compares two scenarios: one considering all existing buildings (refer to Figure 3a) and another limited to only suitable and moderately suitable buildings (refer to Figure 3b). This comparative assessment provides more insights into the potential for photovoltaic (PV) deployment across the buildings in this territory.

In both scenarios, collective housing consistently emerges as the largest contributor, representing over 22% of the total capacity in each deployment scenario. This finding underscores the central role of collective housing as a primary target for energy deployment strategies aimed at maximizing PV capacity.

The top three contributors—collective housing, industry and crafts, and individual housing—account for more than 60% of the total capacity across all deployment scenarios, further emphasizing their importance in deployment plans. These categories should be prioritized to achieve the most substantial energy gains in both scenarios.

Warehouse also plays a significant role in both datasets, bringing the cumulative capacity to over 77% once it is included. The strong performance of this category suggests that industrial and storage facilities offer significant potential for PV deployment, reinforcing their relevance in future energy strategies.

Although smaller building categories, such as sports, meeting places, hospital, and administration, contribute less to the overall capacity, their inclusion remains valuable. Collectively, these categories push the cumulative total beyond 90%, demonstrating their importance in contributing to the overall PV capacity. However, the smallest contributors—restaurant and indoor swimming pool—have only a marginal impact, each accounting for less than 1% of the total capacity in both scenarios.

The inclusion of additional heritage-protected buildings (CLASSE 3) in the full dataset results in a slight increase in total capacities but does not significantly alter the overall distribution trends.

The priority building types remain consistent, with collective housing, industry and crafts, and individual housing continuing to be the most impactful categories for PV deployment.

4.3 Further Consideration

Currently, strictly protected buildings are deemed unsuitable for PV installations, largely due to aesthetic constraints associated with heritage preservation.

However, as PV technologies evolve and become more compatible with heritage architecture, the criteria for suitability could shift.

This potential shift underscores the importance of including these buildings in PV capacity assessments, as it anticipates their possible contribution to the region's overall PV capacity in the future.

A parallel challenge involves assessing the structural capacity of rooftops, which traditionally requires individual evaluations.

With the advent of lightweight PV technologies, previously unsuitable building infrastructures could become feasible candidates for PV installations.

Should regulations adapt to accommodate such advancements, and lighter PV solutions become widely available, previously restricted buildings could be targeted for retrofit initiatives, leveraging their PV potential to support local electricity generation.

41st European Photovoltaic Solar Energy Conference and Exhibition

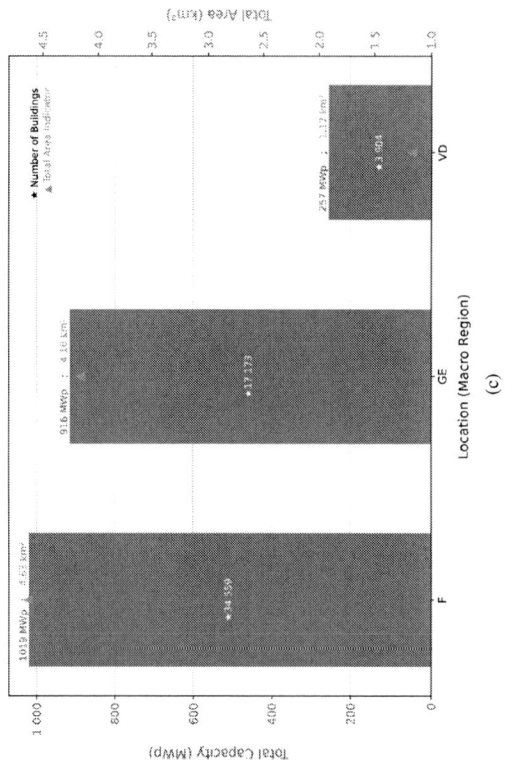

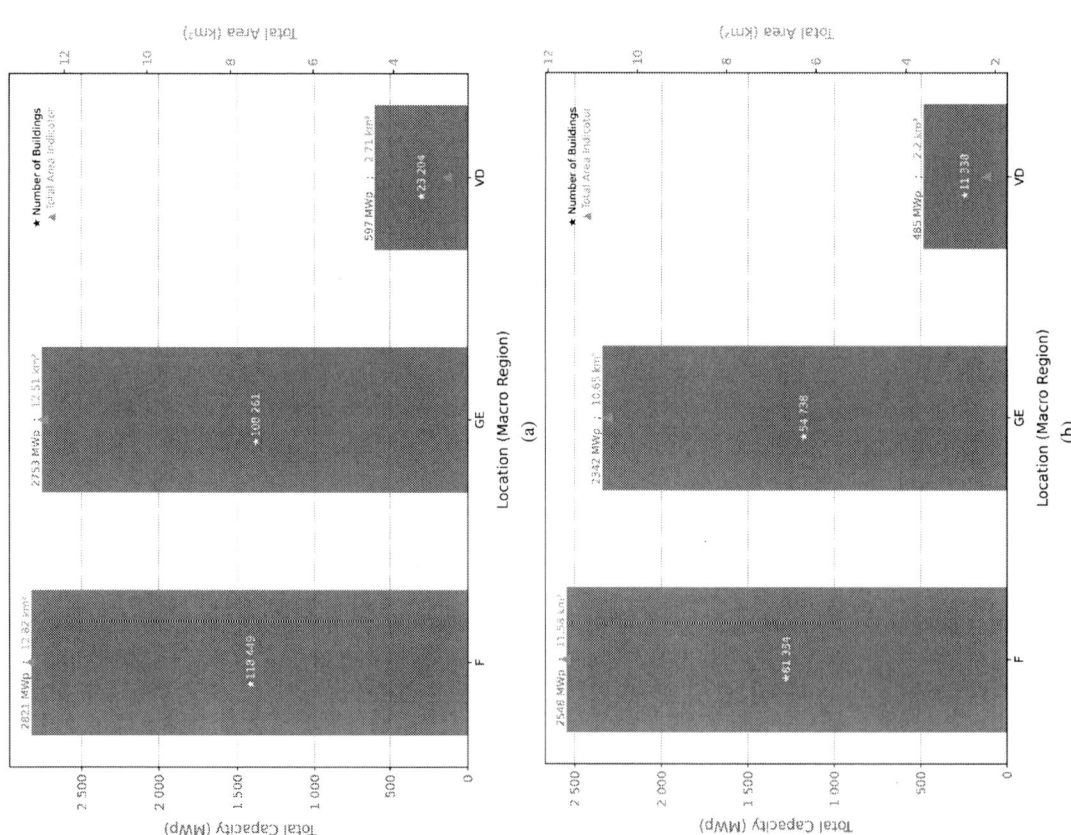

Figure 3. PV Potential in the Greater Geneva Area: (a) Including All Buildings; (b) Excluding Strictly Protected Buildings (Class 3); and (c) Considering Only Buildings Classified as Highly Suitable for PV Systems (Class 1).

5 CONCLUSION

In conclusion, this study presents a strategic, actor-oriented framework to enable the large-scale deployment of photovoltaic (PV) systems within urban areas, with its applicability validated through a comprehensive case study of the Greater Geneva Area (GGA).

The framework, which incorporates insights from key urban actors and multiple layers of influence, provides a structured approach for navigating the complex socio-technical landscape of urban PV deployment.

The GGA case study illustrates the practical impact of this framework by revealing that certain building categories—especially collective housing, industrial facilities, and warehouses—contribute significantly to PV capacity, emphasizing their potential as focal points for strategic deployment.

The case study further highlights the influence of regulatory constraints on deployment potential. For instance, buildings with heritage protections were excluded from certain scenarios, reducing PV capacity within all macro regions due to preservation demands.

However, these findings also suggest potential for future flexibility, as technological advances in PV systems could allow heritage buildings to be retrofitted, thereby unlocking additional capacity without compromising architectural integrity.

This tailored, multi-actor framework, with its defined layers of influence and respective factors, not only addresses the operational complexities of large-scale PV energy system deployment in urban areas but also evaluates the impact of regulatory frameworks on deployment feasibility. Additionally, it serves as a replicable model for application in other territories aiming to prioritize high-potential building types, streamline regulatory alignment, and optimize urban energy transition strategies in support of broader decarbonization goals.

ACKNOWLEDGMENT

This work has been supported by the French National Research Agency, through the Investments for Future Program (ref. ANR-18-EURE-0016 – Solar Academy). This work is also supported by the USMB

REFERENCES

[1] A. Sharifi and Y. Yamagata, "Principles and criteria for assessing urban energy resilience: A literature review," *Renew. Sustain. Energy Rev.*, vol. 60, pp. 1654–1677, Jul. 2016, doi: 10.1016/j.rser.2016.03.028.

[2] S. Freitas, T. Santos, and M. C. Brito, "Impact of large scale PV deployment in the sizing of urban distribution transformers," *Renew. Energy*, vol. 119, pp. 767–776, Apr. 2018, doi: 10.1016/j.renene.2017.10.096.

[3] U. Nations, "Generating power," United Nations. Accessed: Apr. 29, 2024. [Online]. Available: https://www.un.org/en/climatechange/climate-solutions/cities-pollution

[4] O. de Q. F. Araújo, I. B. Boa Morte, C. L. T. Borges, C. R. V. Morgado, and J. L. de Medeiros, "Beyond clean and affordable transition pathways: A review of issues and strategies to sustainable energy supply," *Int. J. Electr.*

Power Energy Syst., vol. 155, p. 109544, Jan. 2024, doi: 10.1016/j.ijepes.2023.109544.

[5] D. M. Kammen and D. A. Sunter, "City-integrated renewable energy for urban sustainability," *Science*, vol. 352, no. 6288, pp. 922–928, May 2016, doi: 10.1126/science.aad9302.

[6] A. A. A. Gassar and S. H. Cha, "Review of geographic information systems-based rooftop solar photovoltaic potential estimation approaches at urban scales," *Appl. Energy*, vol. 291, p. 116817, Jun. 2021, doi: 10.1016/j.apenergy.2021.116817.

[7] S. R. Shakeel, H. Yousaf, M. Irfan, and A. Rajala, "Solar PV adoption at household level: Insights based on a systematic literature review," *Energy Strategy Rev.*, vol. 50, p. 101178, Nov. 2023, doi: 10.1016/j.esr.2023.101178.

[8] IEA, "Net Zero by 2050," IEA, Paris, 2021. [Online]. Available: https://www.iea.org/reports/net-zero-by-2050, Licence: CC BY 4.0

[9] "Global cumulative installed solar PV capacity 2023," Statista. Accessed: Oct. 09, 2024. [Online]. Available: https://www.statista.com/statistics/280220/global-cumulative-installed-solar-pv-capacity/

[10] "Global energy transformation: A roadmap to 2050 (2019 edition)." Accessed: Oct. 09, 2024. [Online]. Available: https://www.irena.org/publications/2019/Apr/Global-energy-transformation-A-roadmap-to-2050-2019Edition

[11] V. Kapsalis *et al.*, "Critical assessment of large-scale rooftop photovoltaics deployment in the global urban environment," *Renew. Sustain. Energy Rev.*, vol. 189, p. 114005, Jan. 2024, doi: 10.1016/j.rser.2023.114005.

[12] O. Bayulgen, "Localizing the energy transition: Town-level political and socio-economic drivers of clean energy in the United States," *Energy Res. Soc. Sci.*, vol. 62, p. 101376, Apr. 2020, doi: 10.1016/j.erss.2019.101376.

[13] M. Mendizabal, O. Heidrich, E. Feliu, G. García-Blanco, and A. Mendizabal, "Stimulating urban transition and transformation to achieve sustainable and resilient cities," *Renew. Sustain. Energy Rev.*, vol. 94, pp. 410–418, Oct. 2018, doi: 10.1016/j.rser.2018.06.003.

[14] M. van Staden, "The Sustainable Energy Transition Cities and Local Governments in Focus," in *Accelerating the Transition to a 100% Renewable Energy Era*, T. S. Uyar, Ed., Cham: Springer International Publishing, 2020, pp. 155–168. doi: 10.1007/978-3-030-40738-4_7.

[15] K. Mainzer, S. Killinger, R. McKenna, and W. Fichtner, "Assessment of rooftop photovoltaic potentials at the urban level using publicly available geodata and image recognition techniques," *Sol. Energy*, vol. 155, pp. 561–573, Oct. 2017, doi: 10.1016/j.solener.2017.06.065.

[16] R. Zhu *et al.*, "GIScience can facilitate the development of solar cities for energy transition," *Adv. Appl. Energy*, vol. 10, p. 100129, Jun. 2023, doi: 10.1016/j.adapen.2023.100129.

[17] G. Desthieux *et al.*, "Solar Cadaster of Geneva: A Decision Support System for Sustainable Energy Management," in *From Science to Society*, B. Otjacques, P. Hitzelberger, S. Naumann, and V. Wohlgemuth, Eds., in Progress in IS. Cham: Springer International Publishing, 2018, pp. 129–137. doi: 10.1007/978-3-319-65687-8_12.

[18] A. C. Lemay, S. Wagner, and B. P. Rand, "Current status and future potential of rooftop solar adoption in the United States," *Energy Policy*, vol. 177, p. 113571, Jun. 2023, doi: 10.1016/j.enpol.2023.113571.

[19] S. Izquierdo, M. Rodrigues, and N. Fueyo, "A method for estimating the geographical distribution of the available roof surface area for large-scale photovoltaic energy-potential evaluations," *Sol. Energy*, vol. 82, no. 10, pp. 929–939, Oct. 2008, doi: 10.1016/j.solener.2008.03.007.

[20] I. Santiago *et al.*, "Assessment of generation capacity and economic viability of photovoltaic systems on urban buildings in southern Spain: A socioeconomic, technological, and regulatory analysis," *Renew. Sustain. Energy Rev.*, vol. 203, p. 114741, Oct. 2024, doi: 10.1016/j.rser.2024.114741.

[21] K. Bódis, I. Kougias, A. Jäger-Waldau, N. Taylor, and S. Szabó, "A high-resolution geospatial assessment of the rooftop solar photovoltaic potential in the European Union," *Renew. Sustain. Energy Rev.*, vol. 114, p. 109309, Oct. 2019, doi: 10.1016/j.rser.2019.109309.

[22] D. Gawley and P. McKenzie, "Investigating the suitability of GIS and remotely-sensed datasets for photovoltaic modelling on building rooftops," *Energy Build.*, vol. 265, p. 112083, Jun. 2022, doi: 10.1016/j.enbuild.2022.112083.

[23] M. K. H. Rabaia, C. Semeraro, and A.-G. Olabi, "Recent progress towards photovoltaics' circular economy," *J. Clean. Prod.*, vol. 373, p. 133864, Nov. 2022, doi: 10.1016/j.jclepro.2022.133864.

[24] L. Bergamasco and P. Asinari, "Scalable methodology for the photovoltaic solar energy potential assessment based on available roof surface area: Application to Piedmont Region (Italy)," *Sol. Energy*, vol. 85, no. 5, pp. 1041–1055, May 2011, doi: 10.1016/j.solener.2011.02.022.

[25] R. Pueblas, P. Kuckertz, J. M. Weinand, L. Kotzur, and D. Stolten, "ETHOS.PASSION: An open-source workflow for rooftop photovoltaic potential assessments from satellite imagery," *Sol. Energy*, vol. 265, p. 112094, Nov. 2023, doi: 10.1016/j.solener.2023.112094.

[26] L. L. B. Lazaro, R. S. Soares, C. Bermann, F. M. A. Collaço, L. L. Giatti, and S. Abram, "Energy transition in Brazil: Is there a role for multilevel governance in a centralized energy regime?," *Energy Res. Soc. Sci.*, vol. 85, p. 102404, Mar. 2022, doi: 10.1016/j.erss.2021.102404.

[27] S. Sareen and H. Haarstad, "Bridging socio-technical and justice aspects of sustainable energy transitions," *Appl. Energy*, vol. 228, pp. 624–632, Oct. 2018, doi: 10.1016/j.apenergy.2018.06.104.

[28] H. Vallecha and P. Bhola, "Sustainability and replicability framework: Actor network theory based critical case analysis of renewable community energy projects in India," *Renew. Sustain. Energy Rev.*, vol. 108, pp. 194–208, Jul. 2019, doi: 10.1016/j.rser.2019.03.053.

[29] D. Rúa, M. Castaneda, S. Zapata, and I. Dyner, "Simulating the efficient diffusion of photovoltaics in Bogotá: An urban metabolism approach," *Energy*, vol. 195, p. 117048, Mar. 2020, doi: 10.1016/j.energy.2020.117048.

[30] C. Nolden, J. Barnes, and J. Nicholls, "Community energy business model evolution: A review of solar photovoltaic developments in England," *Renew. Sustain. Energy Rev.*, vol. 122, p. 109722, Apr. 2020, doi: 10.1016/j.rser.2020.109722.

[31] K. Reindl and J. Palm, "Installing PV: Barriers and enablers experienced by non-residential property owners," *Renew. Sustain. Energy Rev.*, vol. 141, p. 110829, May 2021, doi: 10.1016/j.rser.2021.110829.

[32] M. M. Jackson, J. I. Lewis, and X. Zhang, "A green expansion: China's role in the global deployment and transfer of solar photovoltaic technology," *Energy Sustain. Dev.*, vol. 60, pp. 90–101, Feb. 2021, doi: 10.1016/j.esd.2020.12.006.

[33] D. Noll, C. Dawes, and V. Rai, "Solar Community Organizations and active peer effects in the adoption of residential PV," *Energy Policy*, vol. 67, pp. 330–343, Apr. 2014, doi: 10.1016/j.enpol.2013.12.050.

[34] M. Thebault, G. Desthieux, R. Castello, and L. Berrah, "Large-scale evaluation of the suitability of buildings for photovoltaic integration: Case study in Greater Geneva," *Appl. Energy*, vol. 316, p. 119127, Jun. 2022, doi: 10.1016/j.apenergy.2022.119127.

41st European Photovoltaic Solar Energy Conference and Exhibition

Performance assessment of colorful BIPV façade in Norway

Junjie Zhu[1*] and Jørgen Young [2]

[1]Department of Solar Energy Material and Technology, Institute for Energy Technology, 2007 Kjeller, Norway
[2]Isola Solar AS, Faret 22, 3271 Larvik, Norway

ABSTRACT: A demostration of colorful BIPV facade was installed in Porsgrunn, Norway, featuring an array of colors, including black, red, yellow, white, light gray, and dark gray. In comparison to the black module, the power output of the colorful modules exhibited a range of 15-45% reduction depending on the chosen color. Analyzing the one-year operational data spanning from February 1st, 2023, to January 31th, 2024, it becomes apparent that power production closely correlates with the specified module power, with minor fluctuations that may be attributed to temperature variations throughout the facade. Notably, significant temperature differences were observed from the bottom to the top of the facade, with the primary variations arising from irradiation and/or ventilation rather than the specific color chosen for the modules.

1 INTRODUCTION

Solar energy has been the major renewable energy source, playing a pivotal role in steering societies towards a zero-emission future [1]. Building Integrated Photovoltaics (BIPV) as a building element integrating with photovoltaic to produce the electricity, is always attractive. Tiles, façade are the most BIPV application scenario. In the Nodic country such as Norway, marked by high latitudes and long winter session with heavy snow, BIPV facades offer a significant advantage.

The integration of PV technology with architectural exteriors presents both challenges and opportunities, particularly in addressing environmental demands while enhancing aesthetic appeal within the neighborhood [2]. The color in the PV modules can be achieved by several solution, for example employing colored glass, colored encapsulants, and coatings. It adds a vibrant visual dimension to solar installations. However, this enhancement on aesthetic comes at a cost: not only on the extra material used, but also the power loss due to the optical mismatch to the solar cells [3]. Typically, the power loss can be from few percent to over 60% depending on the color and the color module fabrication process.

Understanding the intricate relationship between color and performance is crucial. Although the color dependence power in the PV module has been measured, the production from a façade on site, especially from the Nordic climate would give the architects and the end users a clear guidance on their color selection and/or decision. In this study, a year-long production of colorful BIPV facades (with different colors) in Porsgrunn, Norway was investigated. By closely monitoring production outputs and temperature fluctuations, this research aims to unravel the nuanced dynamics shaping the efficacy of colorful PV solutions in real-world settings.

2 EXPERIMENTAL DETAILS

High efficiency PERC solar cells in M2 size were used in the BIPV modules. The color of module was obtained by integrating a PQCMTM interlayer between the front encapsulant and the cells. 4mm glass was used on both sides. Figure 1 illustrates the structure of the colorful BIPV module. Seven distinct colors were selected, and three different module sizes were

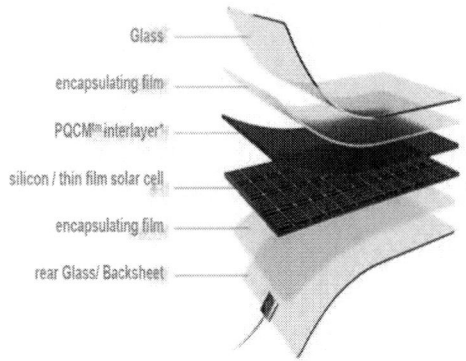

Figure. 1 The schematics of the colorful BIPV

Table 1, the details of the colorful BIPV modules

No	Color		Size (mm*mm)	P_max (Wp)
1	Black		1223*1180	228
2			1180*1002	210
3			1180*580	105
4	White		1180*1002	125
5			1180*580	60
6	Yellow		1180*580	65
7	Red		1180*580	65
8	Brown		1180*580	80
9	Light gray		1180*580	80
10	Dark gray		1180*580	90

manufactured, as detailed in Table 1. Notably, the power output of the modules varied significantly for the same size but different colors. For instance, with the module size as 1180x580mm^2, the power output ranged from

---

[*] Email: Junjie.zhu@ife.no

1430

60Wp to 105Wp, depending on the color. Compared to the black module, which was used as a reference, dark gray module exhibited a slight decrease in power, approximately 15% lower. Light gray and brown modules showed similar power losses, with approximately 25% reduction, while yellow and red modules experienced a larger decrease, approximately 40%. Due to its higher reflectivity, white modules showed the greatest power loss, ranging from 43-45%.

The demonstration façade with those colorful BIPV modules was installed in Porsgrunn, Norway (59.1225, 9.7048), facing southeast, as shown in Figure 2. In this demonstration system, each panel was equipped with an optimizer before being connected to the inverter, allowing for intensive monitoring of the colorful façade module performance. Both the inverter and optimizer were sourced from SolarEdge. Additionally, to gain further insights into performance analysis, seven temperature sensors were mounted under the PV panels, while one reference temperature sensor was directly mounted on the wall, the positions are indicated in Figure 2 as well.

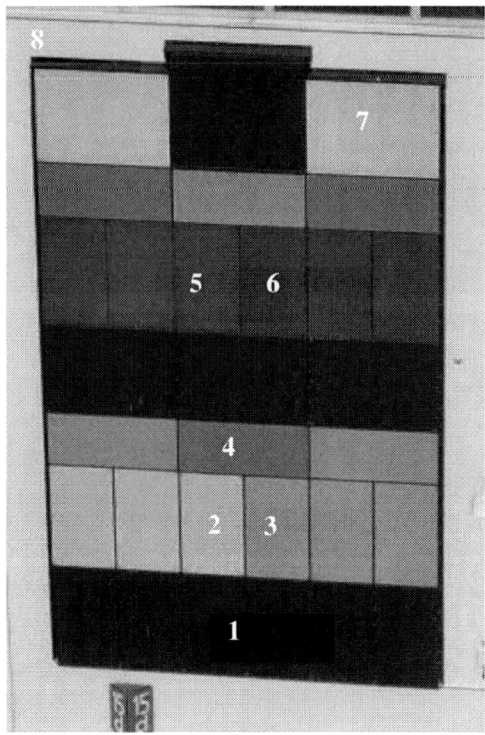

Figure 2, the picture of demostration facade with colorful BIPV modules

3 REASULTS

Given the variation in size among the colorful PV modules, production data was normalized to per square meter (m²), based on the modules with the same size at 1180*580 mm². Figure 3(a) shows the average yearly production per m² for these colorful modules, with different colors, from Feb. 1st, 2023, to Jan. 31st, 2024. The average yearly production ranged from 98.03kwh/m² to 199.47 kwh/m² depending on the color.

The production losses for each color closely mirrored the differences in module power. For example, dark gray modules, with a production of 162.44 kWh/m², experienced a nearly 20% reduction compared to black modules. This decrease is slightly higher than the 15% nominal power difference between dark gray and black modules. Similarly, light gray and brown modules showed approximately 27% loss, while yellow and red modules exhibited around 40% reduction. Notably, the white modules produced nearly half as much energy as the black modules. Figure 3(b) also provides the average monthly production, showing that. Interesting, despite the peak production occurring during summer, the modules continued generating electricity throughout the winter, when roof-mounted systems typically become inactive due to snow coverage [4]. This is the big advantage for BIPV façade in Norway.

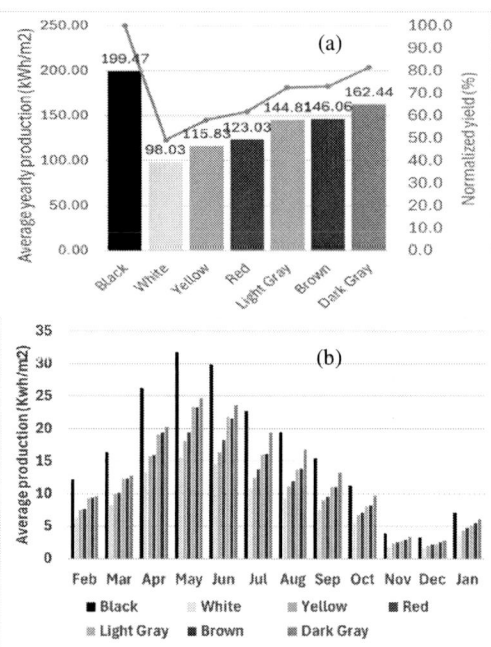

Figure 3, Yearly (a) and monthly (b) average power production per color

The slightly higher production losses, compared to the nominal power differences, could be attributed to increased temperatures in the BIPV modules. Figure 4(a) presents temperature data from different sensors recorded on June 11th, 2023, a typical sunny summer day in Norway. The analysis revealed minimal temperature variation between the different colors at similar heights, but significant differences were observed at varying heights. At midday, when temperatures reached the peak, the sensors from 1 to 7 recorded the following values: 46.5°C, 46.5°C, 49.6°C, 52.6°C, 57°C, 57°C, 57°C, and 35°C (see Table II). Interestingly, the white panels were 3°C cooler than the yellow panels at the same height, while the red and brown panels showed identical temperatures. From the bottom to the top of the BIPV facade, noticeable temperature differences were observed, which likely due to variations in irradiation and/or ventilation, with minimal impact from color differences. Production data from June 11th (Figure 4b) correlated with the temperature logs, showing that the colorful panels

[4] Andenæs E, Jelle B P, Ramlo K, et al. Solar Energy, 159 (2018) 318.

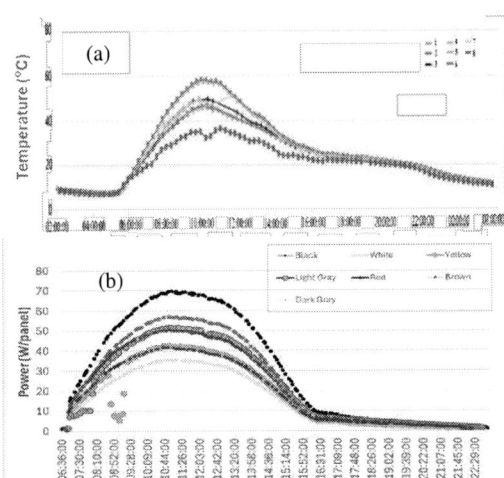

Figure 4, The temperature from different sensors (a) and the power of different colorful BIPV module (b) on June 11th, 2023

experienced slightly lower production due to temperature effects. However, further research is required to fully understand the multiple factors influencing production.

4 CONCLUDITION

A demonstration of a colorful BIPV façade was installed in Porsgrunn, Norway. The one-year operational data highlights the complex relationship between color, temperature, and power output in the colorful BIPV modules. Power production closely followed the nominal module power, with losses ranging from 15% to 45% depending on the color. Darker colors, such as dark gray, exhibited smaller losses, while lighter colors, particularly white, experienced the most significant reductions in power. Temperature variations across the façade, especially from bottom to top, also impacted production, with irradiation and ventilation being the primary factors rather than the color itself. The slightly higher production losses compared to nominal power differences suggest that temperature effects, particularly in certain modules, played a key role.

Overall, while color offers an aesthetic advantage in BIPV systems, it comes with reduced efficiency, highlighting the importance of careful color selection in practical applications. Further research is needed to fully understand the factors affecting performance, especially under different environmental conditions.

Reference

[1] Debbarma, Mary, K. Sudhakar, and Prashant Baredar. *Resource-Efficient Technologies* 3.3 (2017) 263.
[2] Eder G, Peharz G, Trattnig R, et al. Report IEA-PVPS T15-07 (2019).
[3] Saretta E, Bonomo P, Frontini F. 35th European Photovoltaic Solar Energy Conference, (2018), 1472

Implementing strain relief for improved reliability of BIPV modules built on aluminum façade elements

Wiebke Wirtz, Kevin Meyer, Susanne Blankemeyer, Thomas Daschinger, Henning Schulte-Huxel
Institute for Solar Energy Research Hamelin (ISFH), Germany

Motivation

* Building-integrated PV (BIPV) modules combining common materials from building industry and PV

* Increased mismatch of thermal expansion coefficients if using aluminum instead of glass
  → increased thermal stress

* Investigation of strain relief for improvement of reliability

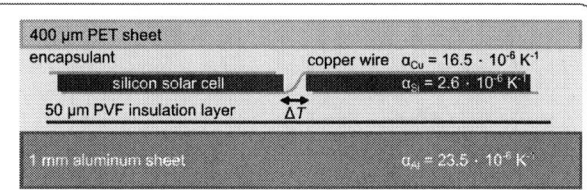

400 μm PET sheet
encapsulant
copper wire $\alpha_{Cu} = 16.5 \cdot 10^{-6}\,K^{-1}$
silicon solar cell $\alpha_{Si} = 2.6 \cdot 10^{-6}\,K^{-1}$
50 μm PVF insulation layer ΔT
1 mm aluminum sheet $\alpha_{Al} = 23.5 \cdot 10^{-6}\,K^{-1}$

Method

* Horizontal crimps (one per cell interconnection) in round interconnection wire as strain relief

* Crimps with smooth curvatures shaped with crimping pliers and tempered for 1 h at 150 °C afterwards

* Crimped part of the wire on rear side or between cells

Experimental

* Test modules laminated on aluminum sheets

* Characterization by current-voltage (*IV*) measurements and electroluminescence (EL)

* Thermal cycling (TC) according to IEC 61215 standard

Results

Strain relief by horizontal crimps

* Horizontal crimps improve module stability in TC

* Less contact degradation visible in EL image of test module with crimps → Fill factor is main reason

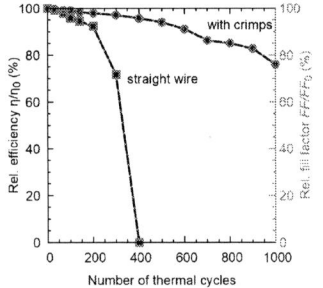

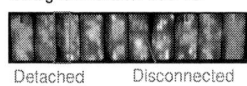
Straight wire after 200 TC
Detached solder pads   Disconnected interconnector

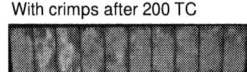
With crimps after 200 TC

Positioning and tempering of crimps

* Negligible influence of crimp position on module reliability

* Small reliability benefit from tempering

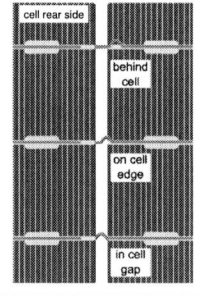
cell rear side
behind cell
on cell edge
in cell gap

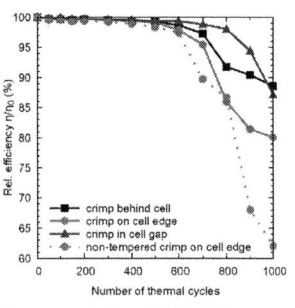

Conclusion

* Strong degradation of BIPV modules built on aluminum façade elements in thermal cycling

* Horizontal crimps in interconnection wires make modules overachieve thermal cycling criterium from IEC 61215

* Small reliability benefit from tempering the crimped wires does not justify the effort of the process

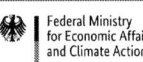
Federal Ministry for Economic Affairs and Climate Action

The authors appreciate the funding by the German State of Lower Saxony and the German Federal Ministry for Economic Affairs and Climate Action within the research project AluPV (Contract no. 03EN1069A).

Industry partner:
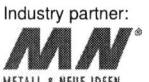
METALL & NEUE IDEEN.

Institute for Solar Energy Research Hamelin (ISFH)
Am Ohrberg 1, 31860 Emmerthal, Phone +49(0)5151 999-303, Email wirtz@isfh.de, Internet www.isfh.de

41st European Photovoltaic Solar Energy Conference and Exhibition

CONIPHER BIPV FACADES: DESIGN AND PERFORMANCE PREDICTION

4BV.4.29

Ya Brigitte Assoa1*, Philippe Thony1, Emmanuel Schmitt2, Olivier Bizzini3, Stephane Gelibert3, Vincent Bressy4, Olivier Wiss1, Alexandre Plissonnier1, Zeina Hamam1

1 CEA, LITEN, Le Bourget du Lac, France;* ya-brigitte.assoa@cea.fr; 2 Vicat, L'Isle-d'Abeau, France; 3 ARaymond, Saint-Egrève, France, 4 Workspaces-architecture, France

CONTEXT AND OBJECTIVES

In the framework of the CONIPHER Life project (**CONcrete Insulation PHotovoltaic Envelope for deep Renovation**), an innovative flexible and efficient photovoltaic cladding panel for facade renovation has been developed. This element is composed of bifacial monocrystalline heterojunction (HET) photovoltaic (PV) modules comprising a thin open air gap and an insulated fiber-concrete reflective frame at the rear side. It is fixed on the facade using optimized fastening nuts from ARaymond. Main results of optimization and demonstration first, during more than a year on the FACT experimental building of CEA INES at Le Bourget du Lac and finally, on a 2-storey building of Vicat at Montalieu in France, are presented.

DESCRIPTION AND MONITORING OF THE CONIPHER BIPV FACADE

- Since April 2017, **Proofs of concept** (Prototypes 1 and 2) on the southern facade of a FACT test room : Integration of 8 PV modules (57 Wp each one; 456 Wp, 6 m²) and a year of weather, thermal, electrical, energy (heat pumps) and air measurements (reference (ref): test room with non-insulated concrete southern facade).

- Since July 2023, **Pilot** on the Northeastern (NE) facade of the building at Montalieu: integration of 24 optimized PV modules (88 Wp each one; 2112 Wp, 15.4 m²).

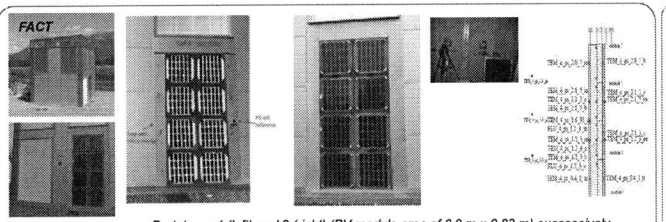

Reference wall and Prototype 2

Prototypes 1 (left) and 2 (right) (PV module area of 0.9 m x 0.83 m) successively integrated on the southern facade of a test room of FACT and their instrumentation

Final design of CONIPHER PV panel (PV module area of 0.595 m x 1.195 m)

CONIPHER Pilot (before and after) at Montalieu and its instrumentation

MAIN PROJECT RESULTS

Impact of CONIPHER solution on the PV field electrical performance, the heat pumps energy consumption for building heating and cooling (Assoa et al., 2021), and on the CO2 emission (life cycle analysis (LCA)) and cost analysis.

➢ Envelope thermal resistance (ref : 5 m².K/W) : (**4.46 ± 0.05) m².K/W (FACT)**.

➢ Additional electrical energy production with bifaciality : **22.5% (FACT)**.

➢ PV Electrical energy production : **322.1 kWh (South - FACT) / 611.03 kWh (NE - Montalieu)**.

➢ Yearly Primary energy savings heating and cooling (ref : 60%) : **68% (FACT)**

➢ Yearly Electrical energy savings for heating and cooling: **509.3 kWh (FACT) / 177585 kWh (Montalieu)**.

➢ Yearly PV self-consumption for heating and cooling (ref : 60%) : **from 22.6% (South-oriented 456 Wp PV field) to 75.7% (14.2 Wp one) (FACT)**.

➢ Reduction of greenhouse gas emission (ref : 75%) : **71% (FACT)**.

➢ Component recyclability in existing industries : **100%**.

➢ Estimated time for return on investment of the 2112 Wp BIPV facade of the pilot (**Montalieu**) (ref : 10 years) : **11.1 years (NE) / 4.8 years (South)**.

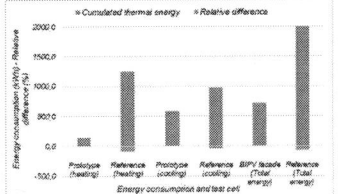

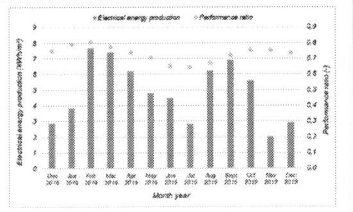

Estimation (with PVGIS and extrapolation from FACT measured data) of the electrical energy production after micro-inverter (AC) of three planned BIPV facades of the pilot at Montalieu (SE: Southeast; SW: Southwest). Differences are mainly due to hypothesis on thermal and orientation losses and weather data.

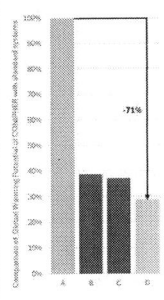

Building at Montalieu	South-West	South-East	North-East (already installed one)	Total BIPV installation
PV field Nominal Power (front face, only) (Wp)	704	2992	2112	5808
Yearly Incident Solar Energy (provided by PVGIS) (kWh/m²)	1037	1014	472	
Estimated Yearly Total Electrical Energy Production (Eac) (kWh) (BIPV in PVGIS)	643	2670	878	4191
Eac extrapolated from the measured yearly performance ratio of the south oriented Prototype 2 and related weather data on FACT, and correction rates for orientation modification (deduced from PVGIS) (kWh)	470	2018	611	3099

CONCLUSION AND FURTHER WORKS

- Main objectives have been reached although some technical issues and challenges.
- Levelized cost of electricity (LCOE) of nearly 0.07874 €/kWh in 2023 is consistent with the average value for a BIPV envelope in Europe of 0.09 €/kWh in 2021.
- As outlooks, all facades of the pilot will be equipped and dissemination activities will be performed over Europe according to industrial partners business plans.

Assoa et al., 2021, Study of a building integrated bifacial photovoltaic façade. Solar Energy, vol. 227, pp. 497-515.

For further information, please contact : emmanuel.schmitt@vicat.fr https://www.life-conipher.eu/

41st European Photovoltaic Solar Energy Conference and Exhibition

THE POTENTIAL OF PLUG&PLAY PV SYSTEMS IN SWITZERLAND

Jan Remund[1], Anne-Kathrin Weber[1], Lukas Meyer[1], David Joss[2], Christof Bucher[2], Theo Zwahlen[2]
[1] Meteotest AG, Fabrikstrasse 14, CH-3012 Bern, Switzerland, jan.remund@meteotest.ch
[2] BFH, Jlcoweg 1, 3400 Burgdorf, Switzerland, david.joss@bfh.ch

ABSTRACT: Plug-in photovoltaic systems for balconies are becoming increasingly popular. This paper analyses the energy potential for such systems in Switzerland. Based on the known potential for facade PV systems, a randomized sample of different building categories is used to investigate whether balconies are present and what their photovoltaic potential is. The potential is then extrapolated to Switzerland's building stock.
In this project, supported by the Swiss Federal Office of Energy, the authors propose to clarify the relevant issues regarding Plug & Play PV systems in Switzerland. Part of this project is the calculation of the energy potential of such systems in Switzerland. The basis of the work is the solar cadastral data (www.sonnenfassade.ch) of Switzerland.
In total 950 facades of the Sonnenfassade.ch dataset were used to assess the suitability for a Plug & Play PV system and determine the potential. This test dataset is representative in terms of house types, sizes and regions. To reach the 950 test examples, 1300 facades were randomly chosen and analyzed by humans regarding the presence and size of balconies with help of publicly available images and maps. 27% of the 1300 facades couldn't be analyzed due to insufficient imagery. The analysis showed, that 60% of the chosen facades don't have any balconies. The example dataset has been extrapolated to all facades of Switzerland.
The areas usable for Plug & Play PV systems on balconies have been defined based on the typical size of such installations (P_{DC_STC}: 400 W on a balcony with approx. 1x2 m and a module efficiency of 20%). Three options were considered: 1. P_{DC_STC}: 800 W with limitation to 600 W AC inverter nominal power; 2. P_{DC_STC}: 1200 W with limitation to 600 W AC inverter nominal power; 3. P_{DC_STC}: 1200 W DC with no AC limitation. The potential is 894 GWh, 1004 GWh and 1040 GWh respectively.

Keywords: Solar Radiation, Satellite Data, PV Potential, Plug & Play PV Systems, Mini PV, Balcony PV

1 INTRODUCTION

Plug-in photovoltaic systems for balconies, also known as Plug & Play PV systems are becoming increasingly popular in Europe. Especially in Germany [1] the growth accelerated lately mainly by lowering bureaucratic thresholds to install such systems. Due to this boom recognizable in Switzerland also, the question arises, how big the potential for such systems might be.

In this project, supported by the Swiss Federal Office of Energy, a group with representatives from the scientific community, the PV and grid industries, and regulatory bodies propose to clarify the relevant issues regarding Plug & Play PV systems for Switzerland.

In a first part of this project the potential of such systems in Switzerland has been calculated.

In Switzerland the solar potentials of all facades and roofs of all existing buildings are well known. They have been calculated and published (www.sonnendach.ch, www.sonnenfassade.ch) by the Swiss Federal Office of Energy. They are updated every 6-7 years.

The total theoretical potential energy production of all facades is 50 TWh (the yearly electricity consumption in Switzerland is about 60 TWh) [2, 3]. But this also includes doors, windows, balconies, and other areas, which are hardly usable for PV. Based on a short analysis the realizable potential was reduced by a factor of 3 to 17 TWh [4].

To our knowledge a detailed analysis of Plug & Play PV systems hasn't yet been done anywhere, although knowing the potential would be crucial regarding future regulations. In 2024 a short analysis of the share of advertised rental apartments in combination with global horizontal irradiance has been made in Germany [5]. However, this does neither show the share of the buildings with balconies nor show the potential irradiance or PV production within the facades.

2 METHOD
2.1 Sample dataset

Based on the known potential for facade PV systems, a randomized sample for different building categories is used to investigate whether balconies are present and how large their photovoltaic potential is. The potential is then extrapolated to Switzerland's building stock.

The existing exceptional database of Sonnenfassade.ch was used as the basis for this evaluation. Initially, a sample dataset of 1000 randomized facades was defined by an automatic algorithm as a starting point. First, a subset of facades was selected according to the following factors: irradiation (above 600 kWh/m²y), usage (residential buildings), area (above 50 m²) and number of floors (more than 1 floor) (Figure 4). Within this selection, the sample is representative of all facades in Switzerland regarding the usage (residential only, mainly residential, and mixed), number of floors, year of construction, facade area, irradiation, region (Canton) and population density.

The facades have been analyzed by humans with the help of Swiss Topographic maps, Google Street View, Google Earth and Apple maps in order to assess their suitability for the potential study.

In the end the sample was enlarged to 1300 facades. The reason to use more than the initial 1000 facades selected in the beginning was that not all of them could be analyzed due to the lack of imagery of the facade available. To determine the parameters of the facades, an image must be available online. Finally, 950 out of 1300 facades (73%) could be examined. It is not known whether the non-inspected buildings differ systematically from the inspected buildings in terms of the characteristics analysed. Approximately 60% of these facades did not have any balconies – a higher percentage than assumed.

The following parameters have been evaluated for each of the 950 sample facades:

Number of balconies, sum of length of balconies (assuming that a balcony is always high enough to place a

PV module (1x2 m area)), number of flats (min and max estimation), potential number of modules on all balconies (min. and max. estimation). The latter two have been evaluated for the three options calculated (see chapter 2.2).

For each facade the production per area [kWh/m² y] (PV$_{rel}$) was available based on Sonnenfassade.ch. This potential included the shading of the surrounding environment (trees, buildings, and topography) as well as the shading of the building itself. Figure 1 shows an example of a facade with many balconies.

Figure 1: Example building facade with 20 balconies, a summed balcony length of 100 m and a maximum of 40 modules (for max. 2 or 3 modules per apartment). Source: Google Maps.

2.2 Calculation of potential production on each sample facade

Based on the aforementioned figures the potential PV production of the modules situated on the balconies have been estimated. In this process, the areas usable for Plug & Play PV systems on balconies have been defined based on the current regulations and typical size of such installations. For the modules a nominal DC power of 400 W at STC was assumed with a module efficiency of 20% and a needed area of 1x2 m. Apart from the limit of 600 W per meter circuit, no other requirements for the electrical installation and spatial planning aspects were taken into account in this analysis.

Only modules in landscape orientation with 90° tilt angle applied to the outer side of balconies have been considered. This was done due to the upscaling method based on the known potential of the underlying facades dataset.

The following **three options have been modelled**:
1. 800 W P$_{DC\_STC}$ system (2 modules) with limitation to 600 W AC inverter nominal power
2. 1200 W P$_{DC\_STC}$ system (3 modules) with limitation to 600 W AC inverter nominal power
3. 1200 W P$_{DC\_STC}$ (3 modules) with no AC limitation

The PV production of the Plug & Play modules (PV$_{pp}$) was calculated as follows:

$$PV_{pp} = PV_{rel} \cdot Area_{pp}$$

Where

$$Area_{pp} = 2 \cdot \frac{Max.nr + Min.nr\ modules}{2}$$

The average number of potentially mountable modules was applied.

Figure 2 shows the facade of a typical small house in Switzerland where the potential for installations of Plug & Play PV systems has been analyzed for.

Figure 2: Example building facade with 1 balcony, a summed balcony length of 7 m and a maximum of 2 resp. 3 modules. Source: Google Maps. The potential production calculated for option 1 was 605 kWh/year, 794 kWh/year for option 2 and 908 kWh/year for option 3. Source: Google Street View.

Additionally, the share of eventually curtailed energy was calculated for each of the three options. This was done based on a typical year for the irradiation in the city of Berne with one minute resolution based on Meteonorm software. Different azimuths of the facades have been applied based on the shares of azimuths found in an analysis over the filtered facade dataset as shown in Figure 3. The average of a south facade and a facade with an azimuth of 120° (60° deviation from the South) was applied as curtailment factor, as this average represents the found distribution well.

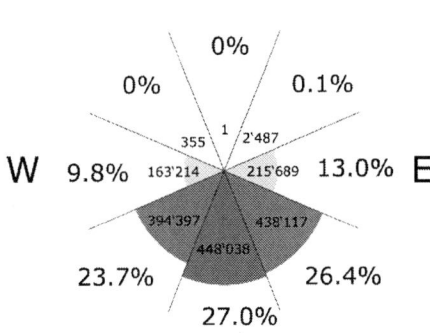

Figure 3: Found distribution of the azimuths of all facades.

This resulted in the following shares for the three options:
1. 800 W P$_{DC\_STC}$ / 600 W AC limit: 1.45% of the annual energy is curtailed.
2. 1200 W P$_{DC\_STC}$ / 600D W AC limit: 12.7% of the annual energy is curtailed.
3. 1200 W P$_{DC\_STC}$ / no AC limit: no curtailment

As the share in option 1 is close to zero and lower than the assumed uncertainties no curtailment was applied for the results of this option. The curtailment for option 2 is lowest in summertime and shows small peaks during spring and autumn. Average monthly curtailment shares have been applied to the original (not curtailed) data to model the final production.

To test the robustness of the approach different weightings to the extrapolation have been applied. A mixture on the buildings based on usage and year of building, number of floors, and usage as well as on usage, number of floors, and year of building and on Cantons alone have been applied. All four weighting options lead to comparable results. The weighting based on usage and year leads to results very close to the results based without weighting.

3 RESULTS

The potential of all the referencing facades in the subset selected initially by the factors irradiation, usage, area, and number of floors sums up on a national scale to 21.611 TWh / year (Figure 4).

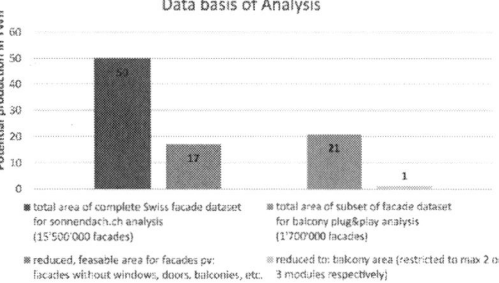

Figure 4: Overview of the data basis and the resulting potentials. For the Plug & Play analysis a subset of facades of the total Swiss facade dataset was used and applied to the percentage of balcony areas taking into account the restrictions of Plug & Play regulations.

The share of the summed production of the Plug & Play systems divided by the summed production of the whole facades subset is 4.1% for option 1. Applying this percentage to the Swiss facade potential, a total of 895 GWh is calculated (Table 1).

Table 1: Results of the estimation of the potential yearly production in Switzerland for the three options chosen.

	Option 1	Option 2	Option 3
Nr. modules	2	3	3
DC	800 W	1200 W	1200 W
Max. AC	600 W	600 W	no limit
Share of curtailed energy	0%	12.7%	0%
Potential production	895 GWh	1002 GWh	1048 GWh
Winter half year share	337 GWh (37.65 %)	379 GWh (37.82 %)	393 GWh (37.5 %)
Summer half year share	558 GWh (62.35 %)	623 GWh (62.18 %)	1009 GWh (62.5 %)

In case of option 2 a total of 1002 GWh was calculated (4.6%). For option 3 a total of 1048 GWh resulted (4.9%).

The production during the winter half-year summed up to 337, 379 and 393 GWh respectively

Option 2 and 3 with 3 modules resulted in higher production – but not as much as assumed based on the number of modules alone. The main reason is that for many balconies only 2 modules can be attached due to limited space. Like this the theoretical curtailment share of 12.7% has a relatively small impact on the potential.

Figure 5 shows the monthly distribution of the PV

production for options 1- 3. In all three options the winter energy share is around 37 %.

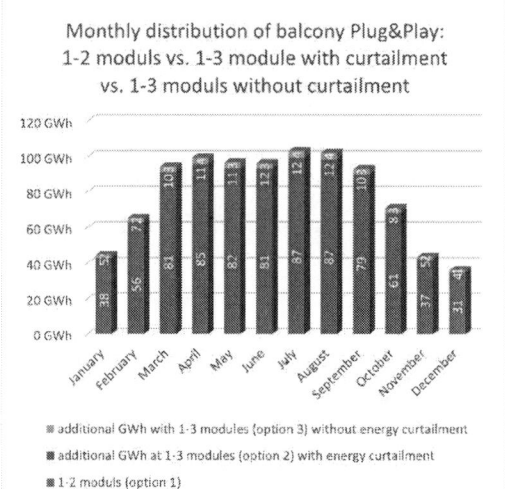

Figure 5: Monthly distribution of balcony PV Plug & Play Potential with all three options

4 CONCLUSIONS

The potential of Plug & Play PV systems in Switzerland was calculated to 900-1000 GWh. In comparison to the total roof potential of 50 TWh [4] on roofs and 17 TWh on facades respectively, this figure is rather small, but still significant. Adding a well distributed 1 TWh would nevertheless be useful, especially producing around 0.37 TWh electricity during winter half year. Additionally, the advantage of this potential is that these systems can be built by anyone. No trained specialists are required. They therefore contribute to a broader and more democratic support of the energy transition. In addition, they are not in direct competition with roof and facade systems, as they do not affect their potential.

The uncertainty of our approach is quite high even considering that the basis – the solar cadaster of Switzerland – is unique regarding accuracy and details. An analysis conducted by humans includes uncertainties as well. The visibility of many facades is limited. The needed balcony dimensions could only be partially measured. The facades are mostly only visible from the side facing the roads. This may induce bias in the extrapolation.

The focus of this paper is on the application of Plug & Play PV systems specifically on balconies. However, the possible use of such systems extends beyond the confines of balcony installations (Figure 6). Within the built environment, numerous surfaces are exposed to sunlight, presenting opportunities for PV system deployment.

Yet, the high costs associated with conventional electrical installations, including safety certifications, often render PV installations on these surfaces economically unviable. Nonetheless, the adoption of Plug & Play PV systems for a low-threshold feed-in, which circumvents the need for extensive electrical work, holds promise for substantial cost reductions at the end customer scale.

Figure 6: Examples of non balcony Plug & Play PV-Systems, whose potential is not included in the present potential study. Source: Solarblitz.ch

Potential additional surfaces for Plug & Play PV installations include bicycle shelters, carports, fences, playground shelters, sunshades, and weather protection. Applying simple scale analysis for such systems with 0.5 – 1 kW scaled up for 2 million buildings a total installed DC power P_{DC_STC} of 1-2 GW could be added. Yet, the energy potential so such non-balcony Plug & Play PV-Systems has not been analyzed in detail but is roughly estimated with additional about 0.7-1.5 TWh a year based on the average specific yield of 725 GWh/MW/a calculated out of the statistics for the year 2023 ($E_{2023,\,PVtot}$: 4624 GWh, $P_{DC\,STC\,tot}$: 6375 MW) [6].

5 REFERENCES

[1] EE-News: 16.05.2024: https://www.ee-news.ch/de/article/53672/boom-der-balkonkraftwerk-geringe-verschattung-und-ein-hoher-eigenverbrauch-sind-die-wichtigsten-voraussetzungen-fur-die-rentabilitat

[2] Remund, J., Albrecht, S., & Stickelberger, D., 2019. Das Schweizer PV-Potenzial basierend auf jedem Gebäude. Photovoltaik Symposium Bad Staffelstein. http://www.sonnendach.ch/

[3] Swiss Federal Office of Energy, Media communication, 2019: https://www.bfe.admin.ch/bfe/de/home/news-und-medien/medienmitteilungen/mm-test.msg-id-74641.html#:~:text=Auf%20Basis%20dieser%20Dat en%20sch%C3%A4tzt,Terawattstunden%20(TWh)% 20pro%20Jahr.

[4] e4plus, 2019: Sonnendach.ch und Sonnenfassade.ch: Berechnung von Potenzialen in Gemeinden Bericht im Auftrag des Bundesamts für Energie; Abrufbar unter: https://www.bfe.admin.ch/bfe/de/home/versorgung/st atistik-und-geodaten/geoinformation/geodaten/solar/solarenergie -eignung-fassade.html

[5] RealEstatePilot, 28.5.2024: https://realestatepilot.com/balkonkraftwerke-in-deutschland/

[6] "Statistik Sonnenenergie 2023", Swissolar, 12.07.2024: https://www.swissolar.ch/de/news/detail/statistik-sonnenenergie-2023-nochmals-ueber-50-prozent-marktwachstum-60956

6 ACKNOWLEDGEMENTS

Our thanks go to Swiss Federal Office of Energy, which funded the work (contract nr. SI/502662-1).

INTEGRATION OF TRANSPARENT PHOTOVOLTAIC PANELS INTO BUILDINGS

Nilşah Özar*[1] and Müjde Altın[2]

[1] Department of Architecture, The Graduate School of Natural and Applies Sciences, Dokuz Eylül University, Türkiye
[2] Department of Architecture, Faculty of Architecture, Dokuz Eylül University, Türkiye
nilsahozar@gmail.com / *mujde.altin@deu.edu.tr

ABSTRACT: The integration of photovoltaic panels, which are used to provide the energy demanded by buildings, is an important field of study today. Today, the use of photovoltaics on roofs is common. However, the facade area is greater than the roof area. Therefore, more energy can be produced when photovoltaic panels are used on facades. However, architects prefer to use aesthetic structural elements on facades. Therefore, transparent photovoltaic panels that provide an aesthetic appearance are needed.In addition to generating energy, transparent photovoltaic panels can be produced transparent or coloured with different transparency values. Thus, they give identity to the building and raise awareness in this way. The aim of this study is to examine the transparent photovoltaic technologies that enable the integration of photovoltaic panels into buildings, beyond its use just on the roof.In addition, different areas of application of transparent photovoltaic panels and the opportunities they offer to architecture were evaluated. The relationship between transparency value, aesthetic concern and efficiency value were analysed through case studies. Comparison and analysis method was used according to the transparency values of the case studies. With this study, the opportunities offered by transparent photovoltaic panels to architecture are explained and it is aimed to be a guide for architects.

Keywords: transparent photovoltaic, transparency, building integrated photovoltaic

1 INTRODUCTION

According to ES-SO (European Solar Shading Organization) data; buildings cause 40% of Europe's energy consumption and 36% of greenhouse gas emissions. Countries implement different policies to ensure that the energy needs of new buildings and the greenhouse gas emissions caused by them are within a certain limit value. However, considering that the current building energy consumption constitutes approximately 30-40% of the total energy consumption of the society, the integration of renewable energy sources into the existing building stock is an important study.

The solar cells that make up photovoltaic panels are elements that are formed by combining two semiconductors and convert sunlight into electricity. When light falls on a surface, some of it is reflected, some of it passes through the material, and another part is absorbed by the atoms of the material. The energy of the absorbed photon passes to the electron. The PV cell directs electrons in one direction, which creates a current. In short, electricity is formed during the passage of electrons from one semiconductor to another. This phenomenon is called the photovoltaic effect.

The aim of this study is to examine the transparent photovoltaic technologies that enable the integration of photovoltaic panels into buildings, beyond its use just on the roof. For this purpose, firstly, the concept of transparency in photovoltaic technologies were analysed. The relationship between transparency value-aesthetic value-efficiency of the building sample in each technology was analyzed through the using the comparison method.

Transparent photovoltaics have low efficiency and their development is still ongoing. However, the reason why they are especially wanted to be used in buildings today is their ability to produce energy while providing an aesthetic appearance.

2 WHAT'S TRANSPARENCY?

Transparency is a physical property that allows light to pass through without interruption. In the study, the issue of transparency in photovoltaic panels was examined in two different ways.

a. In first generation silicon-based technologies, transparent panels are obtained by leaving gaps between solar cells and having different transparency values of the front glass.

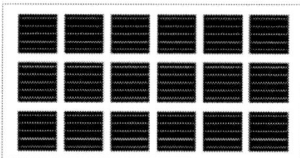

Figure 1: Monocrystalline panel with gaps between cells

b. In other photovoltaic technologies (second and third generation), the solar cells themselves are transparent due to the arrangement of atoms and electrons.

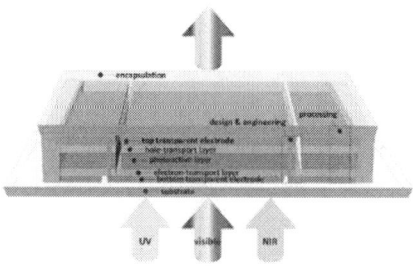

Figure 2: Photons in the non-visible part of the wavelength are absorbed, while visible light is allowed to pass [1]

When an electron has an energy gap equal to a photon energy it absorbs the photon and moves to a higher energy level, in this case, little light passes through that material, which makes that material opaque. In a transparent material the energy gap of the electron is higher than the photons, so the electrons will not be able to use the photons' energy, and light will pass through, which will make the material translucent [2].

The basic principle of the photovoltaic effect is to absorb sunlight. However, in transparent photovoltaics, the purpose is to generate energy by allowing light to pass through. For this reason, in photovoltaics; When allowing the passage of rays with visible wavelengths, materials that provide absorption of photons present in the invisible part of the wavelength should be used.

3 PHOTOVOLTAIC GENERATION

It is possible to classified solar cells into generations with developing technologies. Today, silicon-based technologies belonging to the 1st generation are widespread; thin film, organic, perovskite, dye-sensitized cells are used. Photovoltaic technologies are classified into 3 generations in the study.

3.1 1st Generation Technologies
First-generation photovoltaic cells are usually based on crystalline silicon. Crystalline silicon is the most commonly used material in photovoltaic module manufacturing. It is possible to change the values such as color and opacity of monocrystalline and polycrystalline cell panels in 1st generation technologies. However, if these values are changed, there will be a decrease in cell yields.

The monocrystalline and polycrystalline cells themselves in 1st generation technologies are opaque. However, it is possible to obtain transparent panels of different values by making the windshield in the system component of the panels transparent and leaving gaps between the cells. The windshield can be produced in different transparency values and different colors. In this system module assembly, the colored layer can be used in different positions as follows.

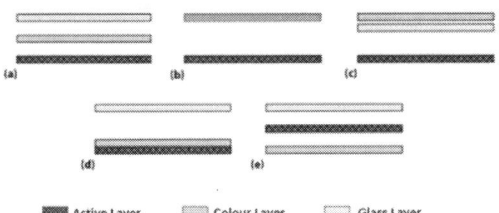

Figure 3: Use of colored layer in different positions

- a- Colored interlayer
- b- Tinted coated windshield
- c- Colored layer on the top of the windshield
- d- Coated active layer
- e- Colored backsheet

Monocrystalline cells are used in the sample office building in Lithuania. Sunlight reaches the interior spaces because there are gaps between the cells. The facade has 90 glass panels with approximately 5000 photovoltaic cells. The estimated energy production is 100 MW per year.

Figure 4: Glassbel office using monocrystalline cell photovoltaic glass panels (https://glassbel.com)

3.2 2nd Generation Technologies
Second generation photovoltaic technologies are thin film cells. Since the active layer of thin films is opaque to light, the only way to achieve transparency can be achieved by reducing the thickness of the active layer to allow light to pass through the glass encapsulating layers [3].

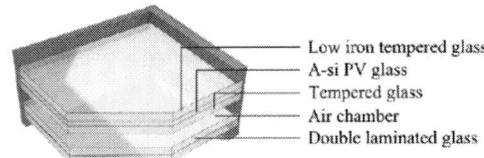

Low iron tempered glass
A-si PV glass
Tempered glass
Air chamber
Double laminated glass

Figure 4: Transparent thin film photovoltaic panel [4]

In order to examine the effect of transparency on the power of thin film photovoltaics, two products with the same length, width, thickness and cell type were examined. When the values are examined, it is observed that the nominal power of the photovoltaic module with 20% transparency and colour usage decreases from 46 Wp to 23 Wp.

Table I: Comparison of two different thin film panels adapted from a manufacturer

	0% Transparency	20% Transparency
Cell Type	a-Si Thin Film	a-Si Thin Film
Front Glass	4mm	4mm(colour)
Nominal Peak Power	46 Wp	23 Wp

Thin film photovoltaic modules with an area of 1000 m2 were applied to the facade of the R&D center in Dubai. Amorphous silicon thin film cells were used in the building and 5 different colors were used: orange, yellow and three shades of green. Semi-transparent photovoltaic panels fixed on the secondary steel construction added to the facade allow 30% sunlight to pass through. A different harmony is created on the facade by the consecutive combination of colors.

1440

Figure 5: Dewa R&D Lab using colored translucent thin film photovoltaic panels (https://www.stantec.com/en)

3.3 3rd Generation Technologies

Organic photovoltaic cells, dye-sensitized solar cells and perovskite cells are third-generation photovoltaic technologies. Lightweight, flexibility, versatility, sustainability of materials, cost-effectiveness, etc. make third-generation photovoltaic technologies the preferred choice for building integration over other photovoltaic technologies.

3.3.1 Organic Photovoltaic Technology (OPV)

The semiconducting materials that compose the light-absorbing layer are the core of an OPV module and directly impact efficiency and transparency [1]. In organic solar cells, ITO (indium tin oxide) or FTO (fluorinated tin oxide) is used as the base material conductive electrode.

OPVs of different colors with the same transparency and physical properties were compared. While blue, green and gray panels have the maximum power at the same value, it is observed that this value is lower in the red panel.

Table 2: Comparison of different color OPVs adapted from a manufacturing company

	Maximum Power
Blue	35-45 W
Green	35-45 W
Red	20-30W
Grey	35-45 W

The media facade consumes as much energy as the photovoltaic modules. Thanks to the energy produced, the light from the LED elements passes through the semi-transparent modules and creates a dynamic facade. The solar-powered media façade consists of 11,608 nodes, combining 10,680 organic photovoltaic modules on the supporting structure and double-sided LEDs embedded within.

Figure 6: Novartis Pavilion using organic photovoltaic modules (https://www.basel.com/en/)

3.3.2 Dye Sensitized Solar Cells (DSSC or DSC)

In dye-sensitive solar cells, the basic principle is to generate electricity by illuminating an organic dye. The transparent surface must have a high rate of sunlight transmission.. When the panel structure is examined, there are active sun strips and transparent gaps between the strips. Energy-generating solar strips can be created in different colors.

Figure 7: Dye-sensitized solar panels produced in different colors [3]

The efficiency of dye-sensitized solar cells depends on the dye of choice. The dye must be well absorbed so that it absorbs sunlight well and increases the efficiency by working with that many electrons.

Figure 8: SwissTech Congress Center using dye-sensitized photovoltaic modules (https://juanbisquert.wordpress.com/)

The Swiss Tech Congress Center in Switzerland is the first large-scale prototype to use dye-sensitized solar cell technology. The semi-transparent colored cells are integrated into a 300 m² area on the building's facade. The building uses 1,500 modules in red, green and orange colors. A dynamic interior space has been created thanks to the semi-transparent panels applied to the facade. Although the panels have high light transmittance, the annual production of the facade is estimated at 2,000 kWh.

3.3.3 Perovskite Technology (PSC)

A semitransparent perovskite panel; It consists of a transparent absorber, a transparent carrier, transparent electrodes and transport layers [5]. As stated by Yang et al. (2016) in their study, the efficiency of perovskite solar cells with four different compositions, which they determined as dark gray, gray, dark red and yellow, was observed as 12.76%, 6.84%, 4.12% and 3.53%, respectively. As a result, the efficiency decreases as the color tone is lightened.

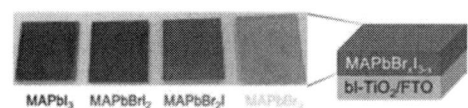

Figure 8: Perovskite cells with different compositions and colors [6]

The office building in Poland is the first commercial scale project to integrate perovskite technology into a facade, with the aim of proving its applicability. 52 colored panels with dimensions of 1.3 x 0.9 m were used on the facade. The photovoltaic modules applied are aimed to meet the energy needs of the offices in the building for 8 hours.

Figure 9: Skanska Office using perovskite photovoltaic modules (https://www.skanska.pl/)

4   CONCLUSION

In the study, photovoltaic cell technologies with transparent panels are analysed through case studies. In this context, the relationship between transparency and efficiency is analysed through examples. However, aesthetic images were obtained by using colours with different transparency values.

Transparency in cell technologies can be achieved in different ways. The degree of transparency in photovoltaic cells, and accordingly whether the color used is in light or dark tones, affects the efficiency. Photovoltaic panels with different transparency can produce clean energy while giving aesthetic value and identity to the building. Transparent panels, whose active function is to generate energy, reduce the cooling need of the building as they reduce the amount of sunlight coming into the interior.

It is seen that panels belonging to the traditional 1st generation cell technology are used more frequently in the sector. The reason for this situation is that they are cheaper and the efficiency of transparent panels is lower. However, with the use of transparent panels, photovoltaics have gone beyond its use on land or on the roof. As a result building facades have become energy generating surfaces.

In the case studies examined, it is seen that colour is also used in transparent panels. The degree of transparency in the use of colour affects the yield. For example, there is a difference in efficiency between a red colour close to opaque and a red panel close to transparent because of the fact that as the panels become opaque, they absorb more sunlight.

Semi-transparent photovoltaic panels are used as glass on the facade or as a facade element placed in a secondary structure, as they allow light to pass through. In addition, these panels create a movable design by coloring the façade, as in the examples. This also adds richness to the interior design by creating light-shadow games. In addition, the integration of transparent panels into the façade is an important development in making existing structures energy-efficient.

Thanks to such applications in public buildings, the awareness of users increases. These integrations reflect and reveal the concept of sustainable environment and energy-efficient building to the society in terms of aesthetics. The designer is free to design the façade with different combinations of transparency and colors. Such innovative product studies strengthen the relationship between energy and architecture.

5   References

[1] Burgues-Ceballos, I., Lucera, L., Tiwana, P., Ocytko, K., Tan, L. W., Kowalski, S., ... & Morse, G. (2021). Transparent organic photovoltaics: A strategic niche to advance commercialization. Joule, 5(9), 2261-2272.

[2] Husain, A. A., Hasan, W. Z. W., Shafie, S., Hamidon, M. N., & Pandey, S. S. (2018). A review of transparent solar photovoltaic technologies. Renewable and sustainable energy reviews, 94, 779-791.

[3] Bunoti, T. (2022). Assessment Of Transparent Luminescent Solar Concentrators For Building Integrated Photovoltaics (Master's thesis, Middle East Technical University).

[4] Li, Q., Li, T., Kutlu, A., & Zanelli, A. (2024). Life cycle cost analysis and life cycle assessment of ETFE cushion integrated transparent organic/perovskite solar cells: Comparison with PV glazing skylight. Journal of Building Engineering, 87, 109140.

[5] Bing, J., Caro, L. G., Talathi, H. P., Chang, N. L., Mckenzie, D. R., & Ho-Baillie, A. W. (2022). Perovskite solar cells for building integrated photovoltaics—Glazing applications. Joule, 6(7), 1446-1474.

[6] Cui, D., Yang, Z., Yang, D., Ren, X., Liu, Y., Wei, Q., ... & Liu, S. (2016). Color-tuned perovskite films prepared for efficient solar cell applications. The Journal of Physical Chemistry C, 120(1), 42-47.

41st European Photovoltaic Solar Energy Conference and Exhibition

Integrating FIDES Reliability Prediction into Building-Integrated Photovoltaic Systems

KU LEUVEN

F. Poormohammadi[1 2*], M. Deckers[1 2], J. Driesen[1 2]

1. KU Leuven, Department of Electrical Engineering (ESAT) - ELECTA, Kasteelpark Arenberg 10, 3001 Leuven, Belgium
2. EnergyVille, Thor Park 8310, 3600 Genk, Belgium

Motivation

The growing adoption of **Building-Integrated Photovoltaic (BIPV)** systems demands robust **reliability assessments** for long-term performance. **Module-Level Power Electronics (MLPE)** are particularly vulnerable to **environmental stresses** such as temperature fluctuations and humidity. This study uses the **FIDES reliability prediction** methodology to evaluate **failure rates** in real-world conditions, focusing on thermal and environmental impacts.

Methodology

The **FIDES** reliability prediction methodology is used to evaluate the reliability of **Module-Level Power Electronics (MLPE)** in BIPV systems. The method breaks down the system's operating life into phases, each contributing to the overall **failure rate (λ)**:

$$\lambda = \prod_{PM} \prod_{Process} \lambda_{Phy}$$

Each phase's **failure rate** is calculated based on environmental stresses, such as temperature and humidity.

$$\lambda_{Phy} = \sum_{i}^{Phases} \left[\frac{t_{annual}}{8760} \right]_i \prod_i \lambda_i$$

Overstress factors ($\prod_i$) include **thermal**, **electrical**, and **mechanical** stresses, allowing for precise failure rate predictions under real-world conditions. By using data from sensors monitoring temperature, humidity, and irradiance, the **FIDES model** predicts how these stresses affect the **MLPE components** over time.

Sensor Data and Mission Profile

Sensors were installed across various layers of the BIPV system to capture **irradiance**, **relative humidity**, and **temperatures** recorded at different locations within the installation (e.g., inner wall of the cavity, cavity, and PV back panel).

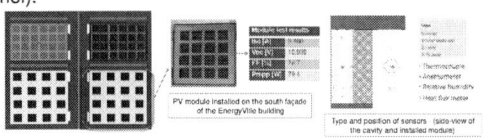

PV module installed on the south façade of the EnergyVille building

Type and position of sensors (side-view of the cavity and installed module)

Data were collected over the course of one year, from **July 1, 2022, to June 30, 2023.**

Mission profiles for three representative days are shown for illustration, highlighting different seasonal conditions and operational stresses.

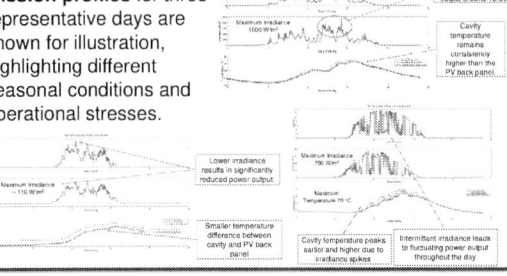

Results and Analysis

To apply the FIDES methodology, the mission profile is divided into **10 power levels**, corresponding to the system's operational phases throughout the day. A representative **normalized P_{mpp} curve** is shown in the figure.

For each power level, the table summarizes key parameters including average temperature, maximum temperature, temperature difference, and humidity.

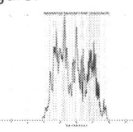

- **Low Power (10%-30%):** Over **2,000 hours** at moderate temperatures (up to **30.67°C**) with **4,886 cycles** at 30%. Thermal stress was minimal with small temperature differences.
- **Mid-High Power (40%-90%):** As power rose, temperatures peaked at **35.76°C** (90% power), with increased thermal stress and cycling (**2,946 cycles** at 90%).
- **Full Power & Dormant:** At **100%** power, temperatures hit **43.34°C** with **1,847 cycles**, while the dormant phase saw **5,032 hours** with significant cooling, but slower thermal impacts.

Conclusion and Outlook

- The **mission profile** of the MLPE boost converter was defined based on real-world environmental data, with ten power levels representing different operational conditions throughout the year.
- The **thermal stress** has been identified as the primary contributor to component failures.

Future work will focus on:

- Detailed **thermal simulations** to model heat dissipation across MLPE components.
- **Electrical simulations** to assess current and voltage stresses during various operational phases.
- Final failure rate and lifetime predictions for MLPE components using the FIDES methodology, once the thermal and electrical simulations are complete. These next steps will provide the foundation for a more comprehensive reliability evaluation of MLPE in BIPV systems, ensuring better system design and longevity.

Learn More About the DAPPER Project:
This research is part of the DAPPER project, aimed at making Building-Integrated Photovoltaics (BIPV) more predictable, reliable, and traceable. The project focuses on enhancing the performance and integration of solar energy systems in buildings. Scan the QR code to explore more about the DAPPER project.

Main References:
[1] De Leon-Aldaco, Susana Estefany, Hugo Calleja, and Jesús Aguayo Alquicira. "Reliability and mission profiles of photovoltaic systems: A FIDES approach." IEEE Transactions on Power Electronics 30.5 (2014): 2578-2586.
[2] Bakeer, Abualkasim, Andrii Chub, and Yanfeng Shen. "Reliability evaluation of isolated buck-boost DC-DC series resonant converter." IEEE Open Journal of Power Electronics 3 (2022): 131-141.

FAST HORIZON ALGORITHM
case of Integrated PV

EUPVSEC, 2024 | Evgenii Sovetkin   Andreas Gerber   Bart E. Pieters |
IMD-3 Photovoltaics Institute, Forschungszentrum Jülich GmbH, Germany

Member of the Helmholtz Association

Outline

Horizon computation problem
    Topography data
    Horizon definition
    Computation time
    Approximate approaches
    Optimising sampling distribution

Algorithm evaluation
    VIPV: impact of orientation
    Maximum sampling distance

High- and low-resolution topography data

- Left: high-resolution DSM (LiDAR based, 0.25 m/pixel).
  Right: low-resolution DEM (satellite based, ASTER, 30 m/pixel).

- Topography plays a primary role in losses for integtated PV applications:
  VIPV: 10–50% losses, depending on road type.

Horizon definition

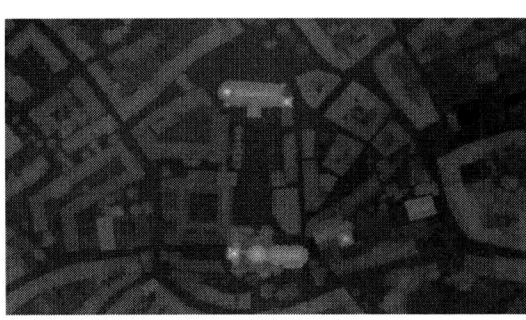

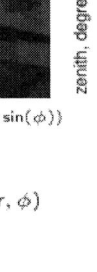

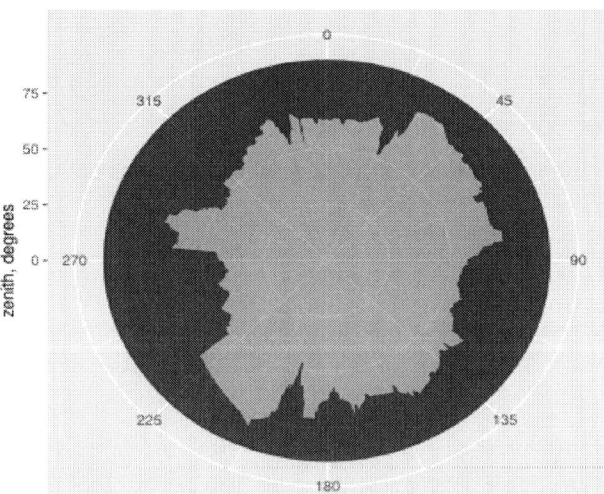

azimuth, degrees

For (x, y) observer location, o observer elevation, T topography raster

$$\xi_o(r, \phi) := T(x + r\cos(\phi), y + r\sin(\phi)) - T(x, y) - o,$$

$$h(\phi) := \max_{r>0} \arctan\left(\frac{\xi_o(r, \phi)}{r}\right), \quad \phi \in [0, 2\pi).$$

- horizon: $h : [0, 2\pi) \to [0, \pi/2]$.
- precise horizon: use all pixels in T.
- approximate horizon: use a subset of pixels in T.

Computation time

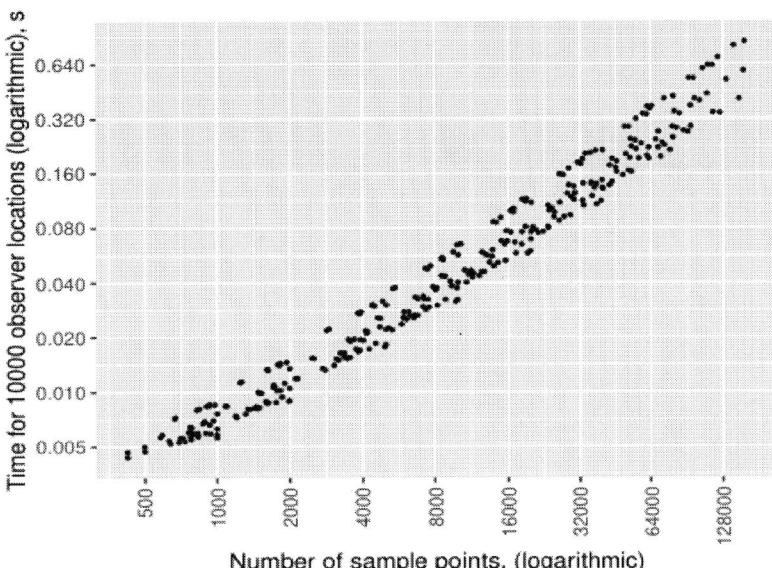

- number of topography sample points correlates with computation time.
- precise algorithm is extremely expensive (e.g. with 40 Mpixel images).

Approximate approaches

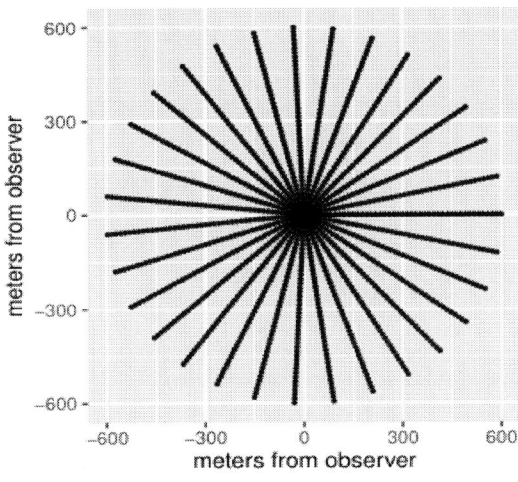

Old method: "rays" sampling

Proposed method: quasirandom sampling

- number of topography sampling points influence time.
- the distribution of points affects the accuracy of the method.
- sample more points near the observer, fewer further away.
- How to optimise the sampling strategy?

Topography data used

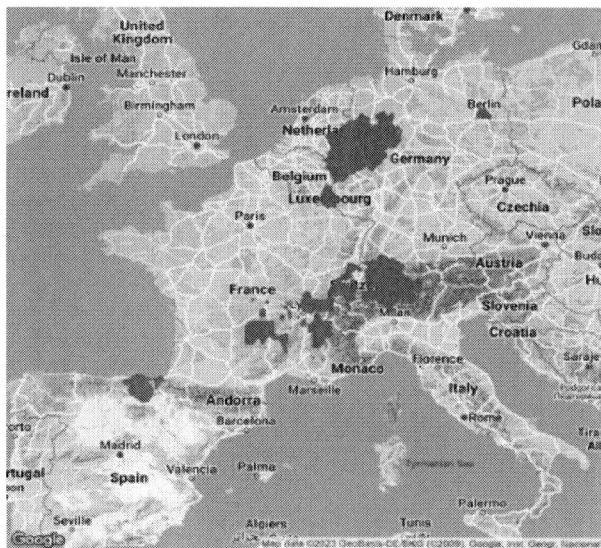 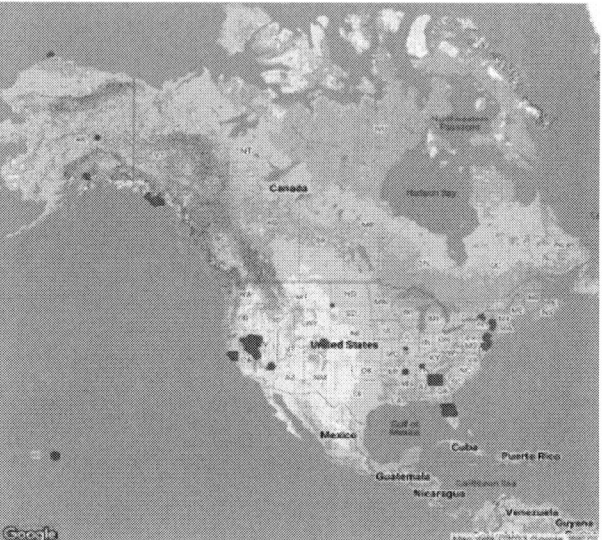

$300\,000\,\mathrm{km}^2$ topography data throughout USA and Europe

- 2 million random locations for topography sampling strategy optimisation.

Data-driven optimisation

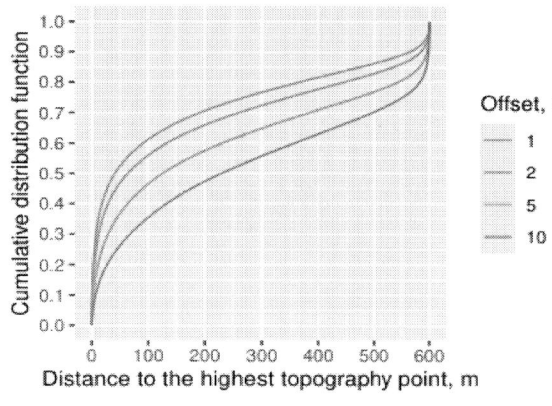

High-resolution topography

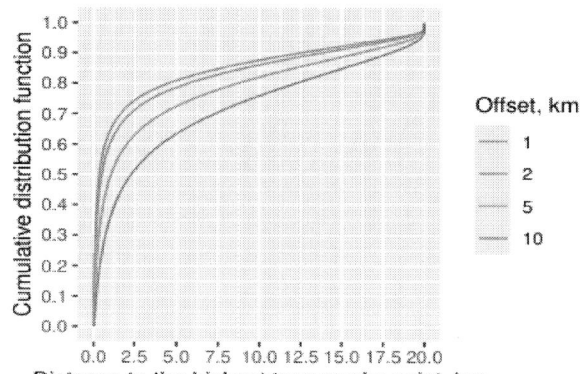

Low-resolution topography

- Estimated cumulative distribution function for ρ_o, o observer height

$$\rho_o(\phi) := \mathrm{argmax}_{r>0} \arctan\left(\frac{\xi_o(r,\phi)}{r}\right), \quad \phi \in [0, 2\pi).$$

- 40% of cases maximum of topography is achieved further away than $100\,\mathrm{m}$ from observer at $1\,\mathrm{m}$ elevation. Do objects further away influence irradiation?

Algorithm evaluation criteria. Relative irradiation. Sky model.

- 11-years annual irradiation
- Perez All-Weather sky
- relative irradiation: irradiation ratio with and without topography
- relative irradiation is dimensionless $\in [0, 1]$
- one minus: losses due to topography

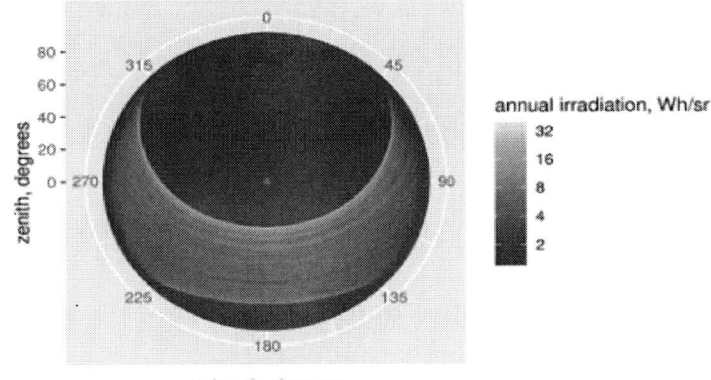

Performance metric:

- RMSE of relative irradiation computed with precise and approximate horizons.

VIPV simulations

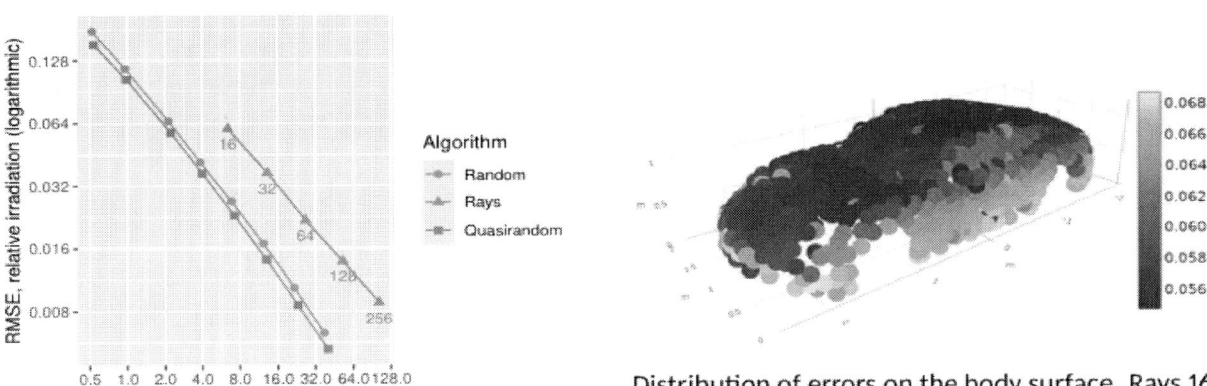

Distribution of errors on the body surface. Rays 16 horizon approximation

Algorithms performance averaged over body surface and locations

- place vehicle on random streets (using OpenStreetMaps).
- quasirandom property is important: random IID sampling reduces performance by 18%.
- RMSE on the vehicle body for "Rays 16": small variation.
- Why VIPV: horizon computation speed matters most here.

How far should one sample topography?

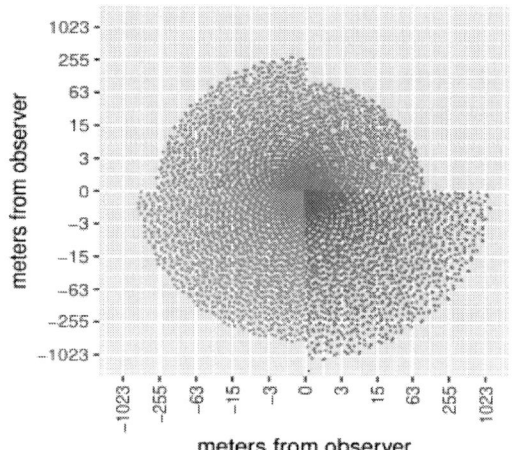

Use truncated Weibull distribution on $[0, M]$, with equal number of points within each disk.

Weibull parameters choice: resembling optimal data-driven distribution.

- $D_{M,1.9\,\text{km}}^{\text{LIDAR}}$ relative irradiation difference computed using approximate horizon computed by sampling at most M and 1.9 km of high-resolution topography.
- $D_{M,120\,\text{km}}^{\text{ASTER}}$ same inference for low-resolution topography.

Right tail of $D_{M,1.9\,\text{km}}^{\text{LIDAR}}$ and $D_{M,120\,\text{km}}^{\text{ASTER}}$

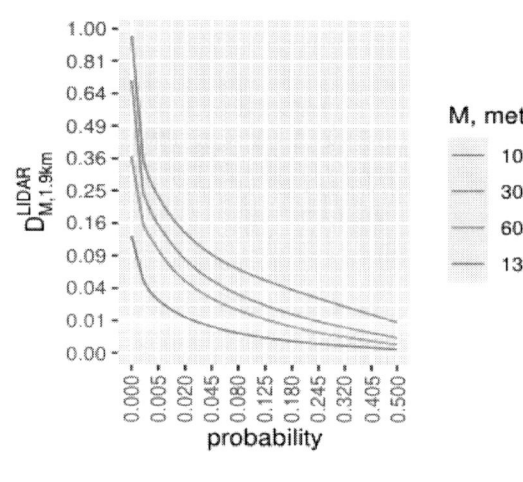

High-resolution topography

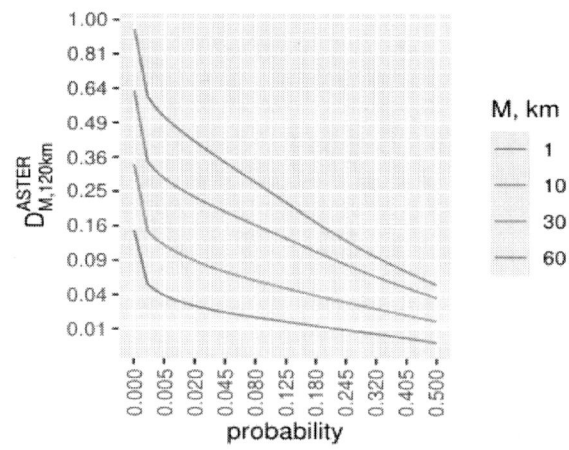

Low-resolution topography

- in 4% of cases the error $D_{100\,\text{m},1.9\,\text{km}}^{\text{LIDAR}} > 0.1$, i.e. 10% overestimation in annual irradiation.
- Which locations have large errors?

Areas with large errors

- Red: simulated locations. Blue: locations $D^{\mathrm{LIDAR}}_{100\,\mathrm{m},1.9\,\mathrm{km}} > 0.1$.
- objects must be sufficiently large enough to have an impact on light and further away than $M = 100$ m to affect the $D^{\mathrm{LIDAR}}_{100\,\mathrm{m},1.9\,\mathrm{km}}$ value.

Take home message

- fast horizon algorithm: quasirandom sampling achieves accurate horizon approximations faster.
- maximum topography sampling distance: matters for some locations. One must use a combination of high- and low-resolution topography in hilly regions.

Papers, software:

- technical, implementation: E. Sovetkin, A. Gerber, and B. E. Pieters. "Improving horizon computation algorithm with quasirandom sequences". In: International Journal of Geographical Information Science (2024). DOI: 10.1080/13658816.2024.2408751
- VIPV, BIPV impact: E. Sovetkin, A. Gerber, and B. E. Pieters. "Fast Horizon Approximation: Impacts on Integrated Photovoltaics Irradiation Simulations". In: Solar RRL (2024). DOI: 10.1002/solr.202400474
- large-scale VIPV: E. Sovetkin, A. Gerber, and B. E. Pieters. "Non-uniformity of irradiation distribution on vehicles' bodies". In: Progress in Photovoltaics: Research and Applications (2024). under review. DOI: 10.22541/au.172114508.82017107/v1
- implementation: B. E. Pieters, E. Sovetkin, and M. Gordon. SSDP: Simple Sky Dome Projector. Version "dev-next". Dec. 2, 2023. URL: github.com/IEK-5/SSDP

poster 4CV.1.21: irradiation modelling with OpenStreetMaps data.

41st European Photovoltaic Solar Energy Conference and Exhibition

This presentation was selected by the Sc. Committee of the EU PVSEC 2024 for submission of a full paper to one of the EU PVSEC's collaborating peer-reviewed journals.

GLOBAL PATTERNS OF SOLAR RESOURCE SHORT-TERM VARIABILITY BASED ON SOLARGIS TIME SERIES DATA

Juraj Betak, Martin Opatovsky, Konstantin Rosina, Marcel Suri
Solargis s.r.o.
Bottova 2A, 811 09 Bratislava, Slovakia

ABSTRACT: The era of modern meteorological satellites provides new opportunities for analysis of variability derived from satellite-based solar resource data. Combining satellite missions together makes it possible to analyse variability on a global scale. This study provides information on the occurrence and magnitude of the short-term solar resource variability, globally. Our approach is based on statistical processing of a 10-minute and 15-minute time series of Global Horizontal Irradiance (GHI), derived from the Solargis satellite-to-irradiance model data from five satellite missions covering the globe. The analysis is performed for a period of the recent 5 years (2019 − 2023, the era of modern meteorological satellites) in a grid with 4-km nominal spatial resolution. The variability statistics are communicated in the form of graphs and maps. Short term solar resource variability (solar ramp rates) shows strong seasonal and geographical patterns. We analyse the typical size of solar ramp rates and patterns in their temporal and geographical spread. The main advantage of the presented approach is understanding of short-term variability and its changing patterns globally, across regions and seasons, especially where availability of solar meteorological measurements is very limited. Information about the variability is important for planning and development of electrical grids and photovoltaic (PV) projects.
Keywords: solar resource ramps, solar resource variability, sub-hourly variability, global map of variability

1 INTRODUCTION

Evaluation of the short-term variability (intermittency) of solar resource has been explored by several authors in the past. Most often, it is presented in statistical terms, optimally derived from (sub)minute high-quality solar resource ground measurements, or carefully validated PV power production data. However, such analysis is typically valid only locally.

Extrapolation of the variability statistics has rarely been attempted by utilisation of satellite-based solar resource data, cloud-specific products, or climate model data. The studies that utilize these data sources are typically focused on one region only [1], [2], [3]. The era of modern meteorological satellites provides opportunities not available years ago: increased temporal and spatial resolution, improved positional geometry, more spectral bands for enhanced identification of cloud characteristics. Combining satellite missions together makes it possible to analyse variability in a global coverage.

In this study we utilize the solar resource data from the Solargis satellite-based model to derive global indicators of solar resource short-term variability. The global indicators allow comparison or correlation of the intermittency among locations and regions globally. The results should be useful to grid utilities, and PV power plant developers and operators can develop approaches to mitigate the impacts of solar resource intermittency.

2 METHODOLOGY

There are two main drivers of the short-term solar resource variability [4]:
a) solar geometry, well documented by formulae,
b) clouds.
Forming or moving clouds cause abrupt changes in solar irradiance within minutes or even seconds, which is also recognized as "solar power ramps" or "ramp rates" [5]. Depending on the type, clouds are capable of consuming a part of the direct irradiance and transforming the rest into diffuse irradiance. Specifically, moving scattered (intermittent) clouds induce frequent up and down solar resource changes. Developing massive storm clouds or moving frontal clouds with a low base often cause abrupt reduction of solar irradiance. For these reasons, photovoltaic (PV) power generation is sometimes referred to as a variable resource, as it is not stable in time and space. Intermittency of the solar resource poses challenges to the grid stability.

In this study, the short-term variability of solar resource data is based on a statistical analysis of time series of Global Horizontal Irradiance (GHI), calculated by the Solargis satellite-based solar model. Data from meteorological satellites, atmospheric models, solar geometry, and terrain are the main inputs into the solar model [6]. One of the model output parameters is a grid of harmonised GHI time series data, available globally.

A subset of the recent 5 years of data (the era of modern meteorological satellites), with nominal spatial resolution of 2 arcmin (nominally 4 km) is considered in this study. The calculation is performed separately for the regions of five satellite missions for the land area between latitudes 60°N and 60°S, with the following calculation step, related to the time step of data acquisition:

- *10-minute* for Americas, Australia, East Asia, and Pacific, covered by the GOES and HIMAWARI satellites, operated by NASA and JMA, respectively
- *15-minute* for Europe, Africa, and West Asia, covered by Meteosat MSG satellites, operated by EUMETSAT.

The source data is then processed in five steps, described below. The processing of the data is performed separately for each satellite region, and the global dataset is merged together in the last step.

Firstly, the solar noon for each day of the year is calculated as a global gridded dataset. An adapted PSA sun position algorithm is used to perform the calculations.

Secondly, Solargis GHI time series data is subset to ±3 hours from the solar noon. This is the time of the day with a potential for high-magnitude solar power ramp occurrences. In the morning or evening hours GHI has lower magnitude due to low sun angles. For this reason, these time slots are filtered out from further processing (Figure 1).

1451

41st European Photovoltaic Solar Energy Conference and Exhibition

Thirdly, the differences between the consecutive time slots are calculated. The calculation is done for each day of the month, season, or year for every grid cell, globally (Figure 2). In this way, a time series of GHI differences is calculated, covering 6 hours every day within the considered time period (2019 – 2023). This global gridded dataset is the basis for the statistical evaluation of the short-term variability.

Fourthly, relevant statistics are calculated to describe different aspects of the short-term variability. The calculated statistics include standard deviation of the GHI difference time series, extreme percentile values (P90, P99), and counts of exceedance occurrences of certain thresholds.

Lastly, the data for the five satellite regions is merged into a single seamless global dataset, and the relevant statistical array data are selected and presented as cartographical outputs. The cartographical outputs can then be investigated and interpreted to understand the regional characteristics of global short-term variability of solar resource data.

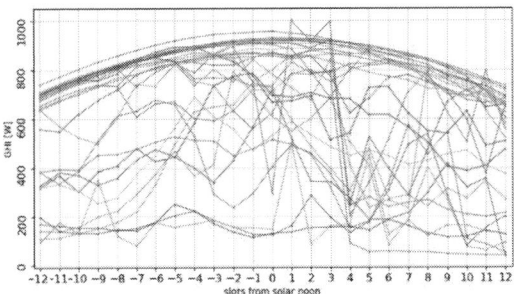

Figure 1: Example of the sub-setting of GHI time series data at a single location: GHI 15-minute time series for ±3 hours from the solar noon for all days in June 2023 in Vienna

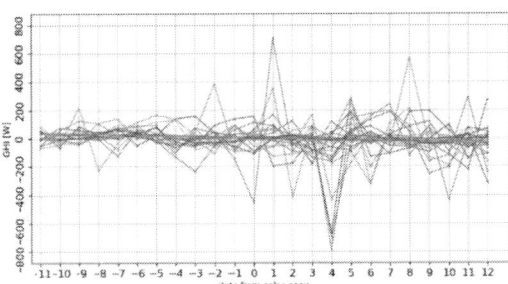

Figure 2: Calculated differences from consecutive 15-minute time slots for all 31 days in June 2023 for the same location as in Figure 1

2.1 Limitations of data from satellite-based solar models compared to ground measurements

It is evident that the variability of solar resource can be large under specific synoptic scenarios, as seen in high-frequency good quality ground measurements. Magnitude and occurrence of ramps in ground-based pyranometer measurements at 1-minute time step can be much higher (even multiple times) compared to the variability observed in time series data with 10- or 15-minute time step (from a pyranometer or satellite-based models). An example of this difference is shown in Figure 3, comparing 1-minute

ground measurements, and 10-minute, 15-minute, and hourly data derived from Solargis time series.

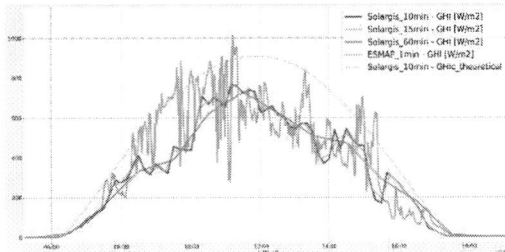

Figure 3: Screenshots from Solargis Analyst software. Comparison of GHI time series of different temporal resolutions for a location in Central Highlands, Vietnam; 1-min derived from ground data measured from ESMAP solar meteorological stations network, station VNCEH [7], [8]. 10-minute, 15-minute and hourly derived from Solargis Time Series

On the other hand, the variability of the power output of a utility-scale PV power plant will be lower than that in pyranometer measurements. This is due to the so-called spatial averaging effect. As the power plant integrates solar resource over its whole area into the power output, the variability of the output will represent the average irradiation over the whole plant. The effect is stronger the larger the powerplant is.

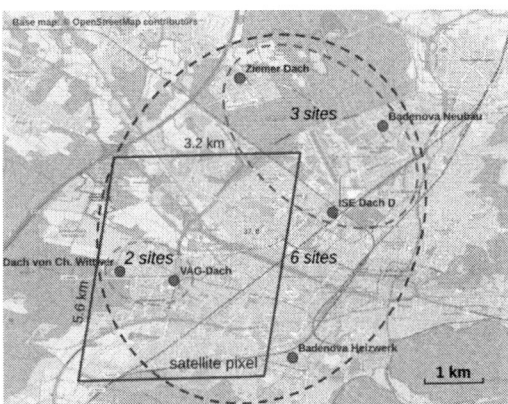

Figure 4: Distribution of the pyranometers in Freiburg (data from [9]), their grouping for the variability analysis, and size of a satellite pixel. (Map background source: © OpenStreetMap contributors)

An analysis of ground-measured data [9] from 6 pyranometers located in Freiburg, Germany demonstrates the spatial averaging effect. The distribution of the sites is shown in Figure 4. As shown in Figure 5, the variability of the individual point measurements with 1-minute time resolution is significantly larger than that of the 15-minute Solargis satellite model time series of GHI. However, when the measured GHI is averaged for 2, 3, and all 6 pyranometers, the resultant variability of the 1-minute time series is comparable to the variability of the 15-minute Solargis data. The variability is measured as the number of ramps over a certain size in one year of data. The distance between the 2 and 3 grouped sites is approximately 1 km and 3 km respectively, which is comparable to the size of a utility-scale PV power plant. Noting the limited geographical context, the analysis indicates that the variability derived from the 15-minute

1452

satellite-based model GHI data is a practical proxy for assessing the variability of the power output of a typical large-scale PV power plant.

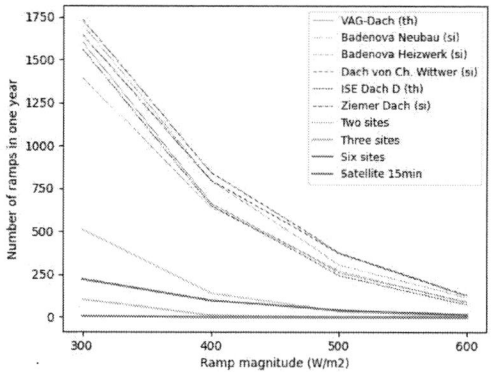

Figure 5: Variability of the measured 1-minute time series data [9], aggregations of the measured data, and Solargis 15-minute satellite model time series

3   RESULTS

In this chapter the resultant maps are analysed and the observed patterns described to understand the characteristics of short-term variability of solar resource in world regions.

3.1   Patterns of solar resource variability

Figure 6 shows the standard deviation of the differences between the consecutive GHI timestamps in the recent five years. The maps are split according to seasons (Dec – Feb, Mar – May, Jun – Aug, Sep – Nov) to identify regional patterns. We provide a short analysis of the likely sources of the observed patterns, however, a more detailed work is needed to provide conclusive explanations.

High variability in all seasons is observed in regions of equatorial Africa, Indonesia, Papua New Guinea, in a belt from Northern Amazonia, through Central American countries, up to the Yucatan peninsula in Mexico. Similar ramp magnitudes are found in the northeast coast of Madagascar. These areas are consistent with occurrence of broken clouds in the tropical regions, which are typically the main cause of solar resource variability.

High variability in two or three seasons, low in one season is observed in regions of Southern Africa, Australia and over large territories of Latin America. These areas experience more stable solar resource during their winter period.

Medium variability with seasonal spikes, is observed e.g., in South India, Southeast Somalia, coastal parts of Kenya, Yemen, Western Europe, or Northwest Iran. Many phenomena with regional impact could be the cause of these patterns, including monsoon and the occurrence of the rainy season in the tropics.

Zonal character of variability, includes interesting patterns in North America or Europe, requiring deeper analysis to offer proper interpretation

Low variability in all seasons, observed in the southeast part of the Sahara Desert, Northeast China (most of Xinjiang province) or in the central part of continental Russia. These areas have either arid or strongly continental climate, which could be expected to lead to low occurrence of weather variations leading to cloud formation.

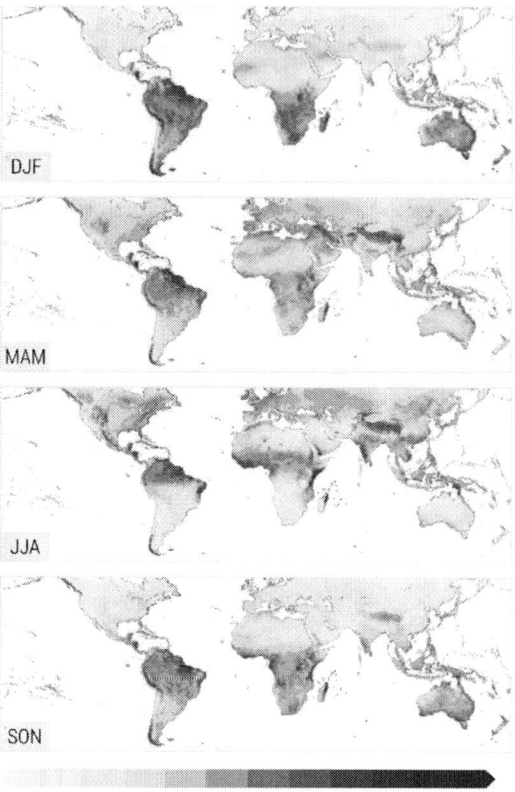

Figure 6: Seasonal indicator of the short-term solar resource variability, calculated as standard deviation from Solargis 10-minute and 15-minute GHI ramps calculated for a subset ±3 hours around the solar noon for a period 2019 to 2023.

3.2   Occurrence of large solar ramps

Figure 7 presents the average annual number of solar ramp rates exceeding defined thresholds, calculated from the Solargis time series. We summarise the number of occurrences in a typical year, when a GHI ramp is larger than 500 W/m², 400 W/m² or 300 W/m², within a 10-minute or 15-minute interval (interval depends on the satellite region).

From this perspective we identify the most apparent solar resource ramps in the northeast coastal regions of Latin America, and some parts of Central America. Here we find areas where the solar resource ramp (rise or drop) of the magnitude 500 W/m² is documented more than 360 times in an average year (or once a day on average), and the 300 W/m² ramp far beyond 1000 times in an typical year (on average 3 to 4 times a day).

The other regions with frequent solar ramps include coastal regions of Australia (with the exception of the West coast), most of the islands in Southeast Asia, high mountains and surrounding plateaus in central Asia, northeast coast of Madagascar, southeast coastal regions of Somalia, and coastal areas in the Gulf of Guinea. Low occurrence of the abrupt solar power ramps is seen in Northeast Africa, central North Asia (Northeast China, Russia) or Northwest Canada.

The patterns correlate with the statistics presented in Figure 6. Some regions deserve further analysis on a regional scale, e.g. we see strong regional variations in China, Australia, USA, but also other regions.

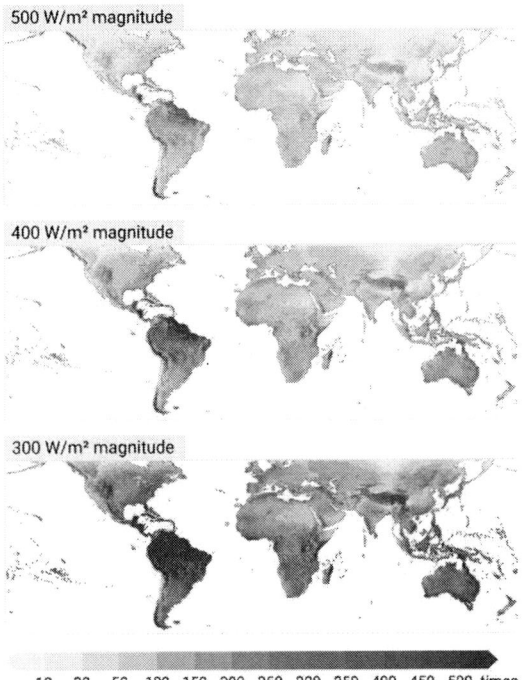

Figure 7: Average annual number of GHI ramps calculated from Solargis 10-minute and 15-minute time series consecutive differences during the day-time ±3 hours around the solar noon in the period 2019 to 2023, exceeding the magnitude 500 W/m^2, 400 W/m^2 and 300 W/m^2

4   DISCUSSION

Solar resource ramps, determined by changing and moving clouds, pose a challenge for management of electric grids and smooth PV electricity supply. In this study we calculated metrics for evaluation of short-term solar resource variability based on Solargis satellite model-based data, and produced world maps of these indicators to study the patterns of the short-term variability. We classify world regions based on the magnitude and occurrence GHI short-term variability, and show the variation of the solar ramps over seasons.

The data and maps can be helpful for strategic planning of electrical grids and PV assets. Specifically, in the high variability areas, the impact of PV power generation on the grid should be carefully analysed, supported by high-quality ground measurements. Adequate dispersion of the PV power plants in a region enables geographical smoothing of the aggregated PV power generation curve. However, to enable this geographical smoothing effect, transmission capacity in the region must be sufficient to support the necessary balancing power flows. Areas with high intermittency are also the best candidates for development of short-term energy storage, capable of providing the balancing power and smoothing the PV power generation curve

Furthermore, the forecast of PV power, but also historical solar models, has a higher uncertainty in the regions during the seasons with high short-term solar

resource variability. Predictability of solar irradiance specifically in regions with frequent intermittent clouds is much more complex than in the regions with prevailing clear-sky or overcast days. This should be reflected in the uncertainty of forecast or historical data, and considered by the users of the data in their further interpretation, including financial models.

Ongoing improvements in satellite-based models and computing infrastructure will make it possible to process data with very high frequency (2.5- and 5-minute frequency of data acquisition) and higher grid resolution (starting at 500 metres). New solar resource models being designed for time series data with high time resolution (1 to 15 minutes) have the potential to support an improved analysis of short-term solar resource variability. In regions with frequent occurrence of large solar ramps, this will make it possible to prepare better technical designs of hybrid PV systems and make the operation of electrical grids more robust.

5   REFERENCES

[1] T. Watanabe, Y. Oishi, and T. Y. Nakajima, "Characterization of surface solar-irradiance variability using cloud properties based on satellite observations," *Sol. Energy*, vol. 140, pp. 83–92, Dec. 2016, doi: 10.1016/j.solener.2016.10.049.

[2] M. Lave, R. J. Broderick, and M. J. Reno, "Solar variability zones: Satellite-derived zones that represent high-frequency ground variability," *Sol. Energy*, vol. 151, pp. 119–128, Jul. 2017, doi: 10.1016/j.solener.2017.05.005.

[3] E. W. Luiz, F. R. Martins, A. R. Gonçalves, and E. B. Pereira, "Analysis of intra-day solar irradiance variability in different Brazilian climate zones," *Sol. Energy*, vol. 167, pp. 210–219, Jun. 2018, doi: 10.1016/j.solener.2018.04.005.

[4] R. Perez *et al.*, "Spatial and Temporal Variability of Solar Energy," *Found. Trends® Renew. Energy*, vol. 1, no. 1, pp. 1–44, Jul. 2016, doi: 10.1561/2700000006.

[5] J. Stein, C. Hansen, and M. Reno, "The Variability Index: A New and Novel Metric for Quantifying Irradiance and PV Output Variability," presented at the World Renewable Energy Forum, May 2012.

[6] R. Perez, T. Cebecauer, and M. Šúri, "Semi-Empirical Satellite Models," in *Solar Energy Forecasting and Resource Assessment*, J. Kleissl, Ed., Boston: Academic Press, 2013, pp. 21–48. doi: 10.1016/B978-0-12-397177-7.00002-4.

[7] World Bank Group, "Vietnam - Solar Radiation Measurement Data - ENERGYDATA.INFO." https://energydata.info/dataset/vietnam-solar-radiation-measurement-data, 2017. Accessed: Sep. 03, 2024. [CSV]. Available: https://energydata.info/dataset/vietnam-solar-radiation-measurement-data

[8] World Bank Group, "Global Solar Atlas." Accessed: Jan. 20, 2024. [Online]. Available: https://globalsolaratlas.info/map?c=12.7535,107.87 61,8&s=12.7535,107.8761&m=solar

[9] N. Straub, W. Herzberg, A. Dittmann, and E. Lorenz, "Blending of a novel all sky imager model with persistence and a satellite based model for high-resolution irradiance nowcasting," *Sol. Energy*, vol. 269, p. 112319, Feb. 2024, doi: 10.1016/j.solener.2024.112319.

6 ACKNOWLEDGEMENTS

The research was partially funded by European Union's Horizon 2020 research and innovation programme under grant agreement No. 953016 SERENDI-PV (https://serendipv.eu/).

41st European Photovoltaic Solar Energy Conference and Exhibition

This presentation was selected by the Sc. Committee of the EU PVSEC 2024 for submission of a full paper to one of the EU PVSEC's collaborating peer-reviewed journals.

CAN DEEP LEARNING REPLACE CLOUD MOTION VECTORS?

[1]Nils Straub, [1]Steffen Karalus, [1]Wiebke Herzberg, [1]Elke Lorenz
[1]Fraunhofer Institute for Solar Energy Systems ISE,
Heidenhofstr. 2, 79110 Freiburg, Germany

ABSTRACT: Satellite-based (SAT) methods are widely used to forecast surface solar irradiance up to several hours ahead. This study applies the established Heliosat method to derive irradiance from Meteosat Second Generation images. As a key parameter the cloud index (CI) is derived from these images, quantifying the impact of clouds on surface solar irradiance. Conventional SAT-methods utilize cloud motion vectors (CMVs) from consecutive CI-images to predict future cloud conditions and subsequently retrieve irradiance. In this study, we introduce HelioNet, a UNet-like neural network designed to predict future CI situations directly from sequences of preceding CI-images. We benchmark forecasts of two HelioNet configurations against CMV and persistence over a full year (2022), with lead times (LT) up to 4 hours. HelioNet$_{15min}$ recursively generates forecasts at 15-minute resolution. HelioNet$_{hybrid}$ begins with forecasts at 15-minute resolution for $LT \leq 45\ min$, then uses a 45 minute-resolved model to forecast all remaining LT-steps. HelioNet$_{15min}$ achieves root mean square error (RMSE) improvements of up to 15% over the CMV model within the first hour on image level. HelioNet$_{hybrid}$ shows superior performance for all LT across all metrics considered, with an average RMSE improvement of >11% on image and >6% on irradiance level.
Keywords: satellite, forecasting, solar irradiance, cloud motion vectors, convolutional neural network

1 INTRODUCTION

Balancing out network load and generation becomes increasingly challenging for electricity grids with a high share of renewable energies. The cause for this is the variable nature of renewable sources, particularly wind and solar energy. Irradiance forecasting makes an important contribution to integrating significant shares of photovoltaics (PV) into electricity grids. Solar geometry introduces diurnal and seasonal variations that can be modelled with relatively little effort and high accuracy. The occurrence of clouds is the main cause for uncertainty in forecasts of surface solar irradiance. Satellite-based (SAT) forecasts are widely used for short-term predictions with inta-hour resolution (up to ~4 hours) [1]. The Heliosat method thereby is a popular approach to retrieve irradiance from satellite images by estimating cloud reflectivity via a cloud index (CI) derived from the broadband visible channel [2].

Conventional SAT methods rely on methods like block-matching or optical flow [3] to compute a field of cloud motion vectors (CMV), describing cloud movement as a two-dimensional vector field [1].

In this work we introduce HelioNet, a convolutional neural network (CNN) with UNet architecture [4], designed to predict future cloud states from sequences of preceding CI-images. This model can be used replace the commonly used CMV-based approach for cloud prediction. HelioNet aims to overcome some of the limitations of CMV-based methods by capturing more temporal context to model more complex non-static cloud movement patterns. Opposed to a CMV-based approach, even forecasting the formation and dissipation of clouds could be possible if loomed in the preceding image sequence. This could significantly reduce forecasting errors, lowering grid integration costs for solar energy, and enhancing grid stability.

The model is trained on sequences of four consecutive CI-images to predict the subsequent one, using a total of 5 years of training data.

A recursive scheme is deployed on the trained model to compute forecasts up to 4 hours ahead. Forecasts from HelioNet are benchmarked against persistence and our current CMV-based method over the year 2022. We directly evaluate forecasted CI-images on image level. Subsequently we validate forecasts of global horizontal irradiance (GHI) derived from these images against measurements from 18 measuring stations across Germany.

In recent years, convolutional neural networks (CNNs) have gained traction in meteorology. In short-term precipitation nowcasting they have outperformed conventional models by using sequences of radar scans to predict future precipitation fields [5–7]. Research has also explored CNN applications for satellite imagery. Works related to our study include Kellerhals et al. [8] and Nielsen et al. [9], who employed a convGRU and convLSTM respectively to forecast surface solar irradiance using patch-based approaches.

In this study, we adopt the UNet architecture, known for its robustness and efficiency with limited training data, to create a single compact model for the entire forecasting domain. This approach reduces computational costs and minimizes border effects, allowing for recursive predictions with a temporal resolution of 15 minutes across arbitrary forecasting horizons.

2 DATASET

In this study we benchmark a UNet-based SAT-forecasting approach against CMV and persistence. For training and validation satellite images and irradiance measurements are deployed.

2.1 Image Dataset

Our satellite-based method utilizes images from the geostationary Meteosat Second Generation (MSG) satellite 10. Equipped with the Spinning Enhanced Visible and InfraRed Imager (SEVIRI), it captures data across 12 spectral channels from 0.3 μm (visible) to 14 μm (mid-infrared). Our method exploits images of the high-resolution broadband visible channel (HRV, 0.6–0.9 μm) measuring reflected irradiance from the Earth's surface. Images are acquired every 15 minutes at a nadir resolution of 1 km x 1 km. Due to the Earth's curvature, resolution decreases with increasing distance from the subsatellite point. From these images we use a crop of 1530x2301 pixels that approximately covers Europe (Europe section

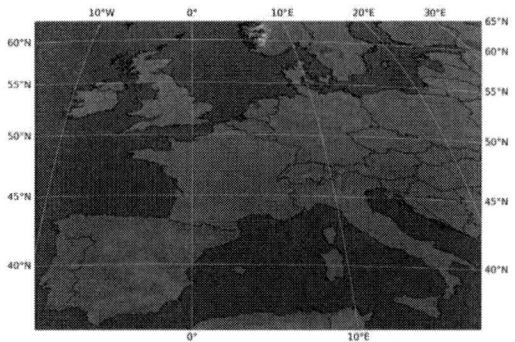

Figure 1: Ground albedo image of the HRV image crop used within this study (referred to as Europe section). Plot was created using Cartopy [10].

in the following, **Figure 1**). We use images from 2017 to 2022, with the first five years for training and 2022 for validation. Since CI is retrieved from the visual spectrum, no valid CI can be computed before sunrise. Dark or only partially illuminated images (night images) with any pixel corresponding to a solar zenith angle of $\Theta_{sun} \geq 85°$ therefore filtered out from the dataset (night filtering).

2.2 Irradiance Measurements

GHI measurements from an irradiance network of 18 stations across Germany, operated by the Deutscher Wetterdienst (DWD), are used to validate our HelioNet GHI forecasts (**Figure 2**). These stations provide GHI data as 10-minute averages. For preprocessing, we first apply quality control flags from the DWD. The GHI measurements are then transformed into 15-minute averages to align with satellite image frequency. This involves converting GHI to a clear-sky index (k*), defined as the ratio of actual GHI to the expected clear-sky GHI [1]:

$$k^* = \frac{GHI}{GHI_{Clear}} \qquad (1)$$

Ten-minute k*-values are upsampled to minute resolution, converted back to GHI (equation (1)) and averaged to 15-minute resolution.

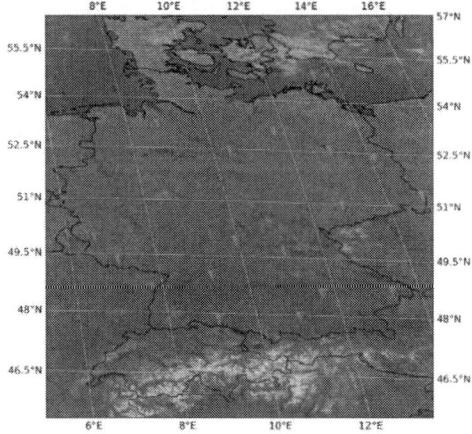

Figure 2: Ground albedo image with DWD irradiance measuring stations used in this study, the red markers denote station positions.

3 METHODOLOGY

Current SAT methods often integrate techniques for inferring clouds and irradiance from satellite data with models that forecast future cloud conditions using cloud motion vectors (CMV). We utilize an enhanced version of the original Heliosat method [11] based on Hammer et al. [12] which first derives cloud index (CI) images from MSG high-resolution visible images and then infers irradiance from forecasted CI-images. In this work we investigate the replacement of CMV by HelioNet to forecast future CI-images (**Figure 3**).

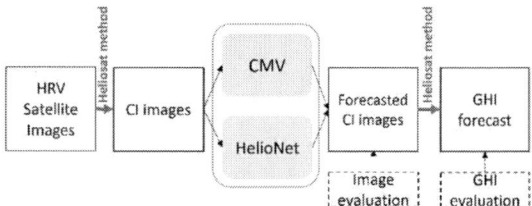

Figure 3: SAT-forecasting scheme, from HRV-images to GHI. HelioNet or CMV can be used to forecast CI-images. The remaining steps are part of the Heliosat method. We evaluate forecasted images and GHI.

3.1 Cloud index and irradiance retrieval

The cloud index (CI) is computed using data from the high-resolution visible (HRV) channel, which measures solar radiation reflected by the Earth (**Figure 4**). The Earth's surface, with its comparatively low albedo, appears dark in satellite images, while clouds, with high albedo, appear bright. CI methods exploit these albedo differences for cloud detection.

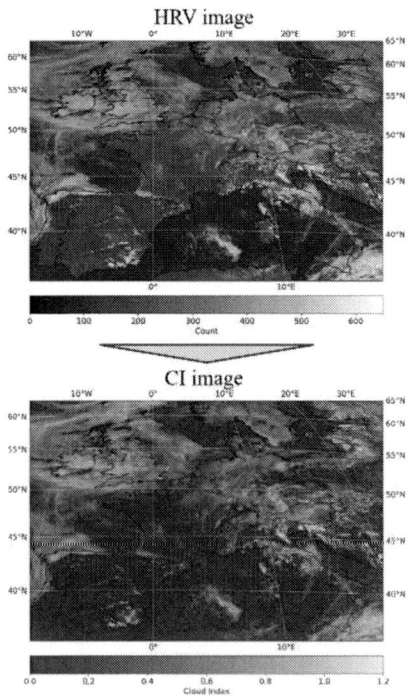

Figure 4: Example CI-image retrieved from raw HRV image from 2022-05-01 12:00h, Europe section.

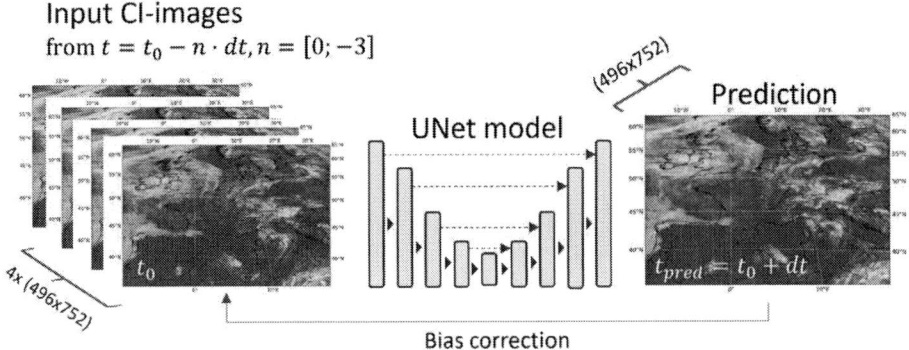

Figure 1: Basic HelioNet application scheme. After each forecasting step the current prediction is bias corrected and recursively used as input for subsequent step.

To derive CI from raw satellite images, pixel radiometer counts are first corrected and normalized by the cosine of the solar zenith angle, yielding reflectance values (ρ) for each pixel. CI is defined as the ratio of the signal from clouds to the mixed signals from clouds (ρ_{cloud}) and the ground (ρ_{ground}), calculated as:

$$CI = \frac{\rho - \rho_{ground}}{\rho_{cloud} - \rho_{ground}} \qquad (2)$$

Ground albedo is determined from sequences of satellite images and updated monthly to account for seasonal variations. Cloud albedo is derived from histograms of reflectance values. More details on deriving CI from HRV images can be found in [12, 13].

To infer global horizontal irradiance (GHI), CI is first linked to the clear sky index (k*) (equation 1). In a first-order approximation, k* is related to CI as (k* = 1 - CI), with adjustments for overcast and variable cloud conditions. To derive 15-minute average GHI values from instantaneous CI pixel values, we integrate CI values from two consecutive images and neighboring pixels. Finally, GHI is computed from k* (equation 1) using the clear-sky model from Dumortier [14] and the turbidity model from Bourges [15].

3.2 Forecasting Models

This section describes HelioNet and our CMV reference forecasting model. Besides CMV we added persistence as naïve baseline model. For CI-image forecasting we define persistence as the most recent CI observation. Regarding GHI we apply persistence of clear-sky index (k*) at the position of the respective measuring station.

3.2.1 CMV

CMV methods analyze consecutive satellite or CI-images to derive a vector field representing cloud motion. This vector field is then applied to the latest CI observation to create a forecast, based on the assumption that cloud development primarily results from horizontal advection over timescales up to a few hours ahead. A major limitation of CMV is its inability to model cloud formation and dissipation [1].

Our CMV model adapts the method from Kühnert et al. [13]. After converting the raw satellite image into a CI-image, we use the DeepFlow algorithm [16] to obtain a CMV field from two consecutive images. This CMV field is, corresponding to the respective LT, repeatedly applied to the latest CI-image. Finally, smoothing filters are applied to the forecasted CI-images to reduce forecasting errors by eliminating randomly varying small-scall

structures [13, 17]. Thereby smoothing increases with proceeding LT.

3.2.2 HelioNet

This study explores the use of UNet for forecasting future CI-images (**Figure 5**). We apply the original architecture from Ronneberger et al. [4] but add padding after each convolutional layer. In reference to the Heliosat method we call our model HelioNet. CI-images are normalized and downscaled to 496x752 pixels before being used within HelioNet. Note, that for the evaluation forecasted CI-images are upscaled again and validated against full resolution observations. The input consists of four consecutive CI-images to forecast the next CI-image. Using the same temporal spacing (dt) between input and forecasted images allows for recursive forecasting. As a slight positive bias was found, that accumulates during recursive application, a dynamic bias correction factor is deployed after each iteration.

We compare two configurations of HelioNet. HelioNet$_{15min}$ recursively forecasts at lead time (LT) 15 min throughout the entire interval. HelioNet$_{hybrid}$ utilizes a 15-minute model to predict the first three steps (up to 45 minutes), after which a 45-minute model is employed to forecast all remaining steps. This configuration allows to forecast with 15 min resolution using a 45 min model. It combines the benefits of a highly resolved model with the longer temporal context of a 45-min model.

3.3 Forecast specifications

For training and evaluation image dataset is night filtered as described in section 2.1. Image data was split into five years for training (2017-2021) and one year for testing (2022). Training utilized all remaining CI-image sequences within the training period with a basetime frequency (starting time of a new forecast run) of 15 Minutes.

For evaluation in 2022, forecast runs where computed between 6 AM and 2 PM, with a basetime frequency of 15 min, temporal resolution of 15 min and a maximum forecasting horizon of 4 hours.

3.4 Error metrics

For the evaluation on image and GHI level standard error metrics [18] are used. The error between an observation x_i and the corresponding prediction y_i is given by:

$$\epsilon_i = y_i - x_i \qquad (3)$$

41st European Photovoltaic Solar Energy Conference and Exhibition

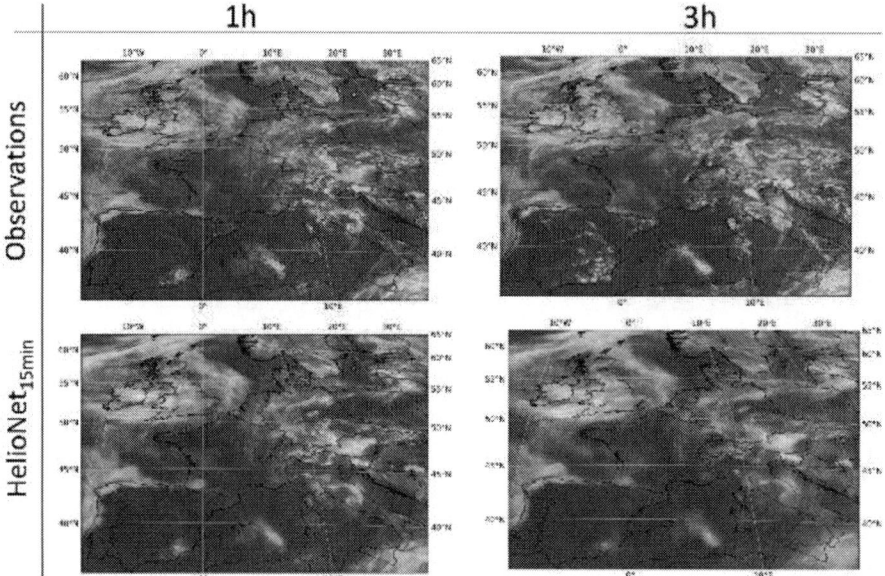

Figure 2. Example forecast run from different models with two selected LT-steps (Basetime=2022-05-01 T10:00)

We evaluate the root mean squared error (RMSE) and mean absolute error (MAE) and their normalized counterparts ($RMSE_{rel}$ and MAE_{rel}):

$$RMSE_{rel} = \frac{100\%}{\bar{x}} \cdot RMSE = \frac{100\%}{\bar{x}} \cdot \sqrt{\frac{1}{n}\sum_{i=0}^{n}\epsilon_i^2} \quad (4)$$

$$MAE_{rel} = \frac{100\%}{\bar{x}} \cdot MAE = \frac{100\%}{\bar{x}} \cdot \frac{1}{n}\sum_{i=0}^{n}|\epsilon_i| \quad (5)$$

For image evaluation $\bar{x}$ is calculated as the mean of the corresponding observed CI-image. To evaluate GHI we compute error metrics across all measuring stations (**Figure 2**) combined. We compute $\bar{x}$ for each LT step individually as the mean GHI of all measuring stations, to account for the diurnal irradiance variation.

Skill score denotes the relative improvement over a trivial reference method with regard to a certain metric (e.g. RMSE). With a non-trivial method as reference this metric is referred to as improvement score [1]. It is calculated from the forecast score S_{fc} in the respective metric, the score of the reference forecast S_{ref} and the score of a prefect forecast S_{perf} (e.g. $S_{perf,RMSE} = 0$):

$$S = \frac{S_{ref} - S_{fc}}{S_{ref} - S_{perf}} \quad (6)$$

A positive score denotes an improvement, negative deterioration compared to the reference method, with a maximum value of 1 implying a perfect forecast ($S_{fc} = S_{perf}$).

4 RESULTS

4.1 CI forecast evaluation

An example forecast of HelioNet15min at two selected LTs is shown in **Figure 6**. The forecasts accurately resemble the CI observations, but show increased blurring for longer LT. During training the model learns blurring as a strategy to minimize its loss. This characteristic of CNNs and specifically UNet-based models is reported in related

work as well [6, 7]. For CMV forecasts blurring is applied as a deliberate postprocessing step to reduce forecasting error as described in section 3.2.1.

Results of the CI-image evaluation are shown in **Figure 7** (a) to (d). Clouds entering from areas outside the image cannot be forecasted by any model. To minimize the impact of such border effects, we limit the evaluation to an inner crop of 850x1621 pixels. HelioNet and CMV significantly outperform persistence (**Figure 7** (a) and (c)). RMSE and MAE increase for all methods with increasing LT as expected.

HelioNet15min shows substantially reduced RMSE and MAE compared to our reference CMV method with $S_{RMSE} > 10\%$ for $LT \leq 90\ min$. Its benefit over CMV drops quickly until for $LT \geq 180\ min$ it is outperformed by CMV in RMSE and MAE. One possible cause for this performance drop is the excessive recursive use of HelioNet15min. Each forecasting step introduces errors and artefacts, which accumulate over each recursive iteration. After only four steps it runs purely on synthetic input.

For $LT \leq 45\ min$ HelioNet_hybrid is identical to HelioNet15min as described in section 3.2.2. To forecast all LT-steps beyond the 45-min model is used. This configuration was found to be superior to any individual model tested (including a 45-min model) in preliminary experiments. HelioNet_hybrid shows superior RMSE and MAE compared to our CMV model for all LT with average improvement scores of $\overline{S_{RMSE}} \approx 11\%$ and $\overline{S_{MAE}} \approx 9\%$ (not shown in plot). The input of HelioNet_hybrid extends further into the past, making model it suitable to forecast cloud development on longer time scales. Additionally, this allows for the inclusion of real CI-images in the input for an extended period. Other than HelioNet15min, HelioNet_hybrid relies solely on synthetic input only for $LT \geq 2:15h$. All models exhibit a slight negative bias $|MBE_{rel,max}| < 5\%$ that increases with LT (**Figure 7** (d)).

1459

41st European Photovoltaic Solar Energy Conference and Exhibition

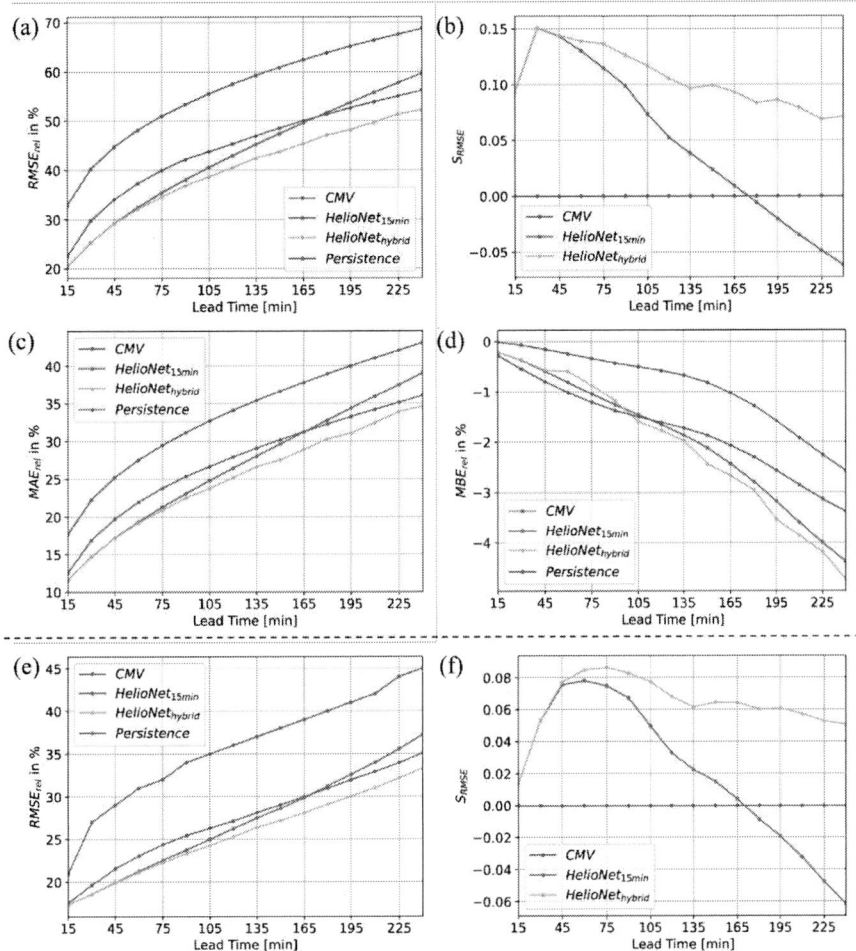

Figure 3. Error metrics over LT of CI-image evaluation (a) relative RMSE, (b) RMSE improvement score over CMV model, (c) relative MAE and (d) relative Bias. GHI evaluation, (e) relative RMSE across all stations and (f) RMSE improvement score.

4.2 GHI evaluation

Relative RMSE and RMSE improvement of our GHI evaluation across all measuring stations (**Figure 2**) are shown in **Figure 7** (e) and (f). As we had similar findings regarding MAE this metric is left out here.

All in all, we find the same trends regarding model performances as in the CI-image evaluation, however with approximately only about half of the improvements seen at image level. HelioNet$_{15min}$ achieves a maximum of $S_{RMSE,max} = 8\%$, while HelioNet$_{hybrid}$ shows an average improvement score of ~6%. This discrepancy may arise from additional averaging between consecutive forecasted CI-images for irradiance retrieval (section 3.1), which diminishes differences between forecast methods. Furthermore, the image-level evaluation covered all of Europe, while the GHI evaluation was limited to stations in Germany, suggesting that improvements in image forecasting may not be uniformly distributed.

5 SUMMARY AND CONCLUSION

Forecasting of future cloud situations is a crucial step in SAT-forecasting of solar irradiance. In this study we used the Heliosat method to derive CI-images from raw satellite images and solar irradiance from forecasted CI-images. To forecast future CI-images HelioNet was introduced, a UNet-based approach that can be used as replacement conventional CMV methods. Two HelioNet configurations were investigated and benchmarked against persistence and our current CMV-based approach with forecast LTs up to 4h ahead. HelioNet uses a sequence of four consecutive CI-images to forecast the subsequent one. Longer LTs are forecasted using a recursive forecasting scheme. HelioNet$_{15min}$ runs on the native temporal resolution of satellite images (i.e. 15 min). HelioNet$_{hybrid}$ combines a 15 min model for $LT \leq 45\ min$ with a 45 min model for all LTs beyond. Both configurations were trained on night filtered sequences of CI-images from 2017 to 2021.

For model validation we computed forecast runs every 15 min between 6h and 14h throughout the entire year 2022 with a temporal resolution of 15 Min up to 4h ahead, likewise excluding night images. On image level we evaluated forecasted CI-images against CI observations on a $850\ x\ 1621$ pixels image section that roughly covers Europe. Based on our CI forecasts we computed full GHI forecasts and validated them against measurements from 18 sites measuring, run and quality controlled by the German Weather service DWD.

HelioNet$_{15min}$ is well-suited for short LT (up to $\approx$ 90 min) with maximum improvement scores of $S_{RMSE,max}(CI) > 15\%$ and $S_{MAE,max}(CI) > 13\%$ on

1460

image level and $S_{RMSE,max}(GHI) > 8\%$ on GHI level. Its performance deterioration on LTs beyond is possibly caused by its excessive recursive use.

HelioNet$_{hybrid}$ shows the best performance for all $LT \geq 1h$ with average improvement scores of $\overline{S_{RMSE}(CI)} \approx 11\%$ and $\overline{S_{MAE}(CI)} \approx 9\%$ on image level and $\overline{S_{RMSE}(GHI)} \approx 6\%$ on irradiance level. Its input extends further back in the past, making it more suitable for forecasting cloud development beyond the immediate short-term.

We consider our current version of HelioNet an important step towards CNN-based SAT-forecasting with further improvement potential. In future work we want to explore more refined ways of using and combining models of different temporal resolutions and further CNN architectures.

6 ACKNOWLEDGEMENTS

The authors express their gratitude towards the German Federal Environmental Foundation (DBU) for supporting this work with a fellowship.

7 REFERENCES

[1] Sengupta M, Habte A, Wilbert S, Gueymard C, Remund J. Best Practices Handbook for the Collection and Use of Solar Resource Data for Solar Energy Applications: Third Edition. Golden, USA: National Renewable Energy Lab. (NREL); 2021 NREL/TP-5D00-77635.

[2] Hammer A, Heinemann D, Lorenz E, Lückehe B. Short-term forecasting of solar radiation: A statistical approach using satellite data. Solar Energy 1999; 67: 139–50 [https://doi.org/10.1016/S0038-092X(00)00038-4.]

[3] Aicardi D, Musé P, Alonso-Suárez R. A comparison of satellite cloud motion vectors techniques to forecast intra-day hourly solar global horizontal irradiation. Solar Energy 2022; 233: 46–60 [https://doi.org/10.1016/j.solener.2021.12.066]

[4] Ronneberger O, Fischer P, Brox T. U-Net: Convolutional Networks for Biomedical Image Segmentation; 2015 May 18.

[5] Xingjian Shi, Zhourong Chen, Hao Wang, Dit-Yan Yeung, Wai-kin Wong, Wang-chun WOO. Convolutional LSTM Network: A Machine Learning Approach for Precipitation Nowcasting 2015.

[6] Ayzel G, Scheffer T, Heistermann M. RainNet v1.0: a convolutional neural network for radar-based precipitation nowcasting. Geosci. Model Dev. 2020; 13(6): 2631–44 [https://doi.org/10.5194/gmd-13-2631-2020]

[7] Ravuri S, Lenc K, Willson M, et al. Skilful precipitation nowcasting using deep generative models of radar. Nature 2021; 597(7878): 672–7 [https://doi.org/10.1038/s41586-021-03854-z][PMID: 34588668]

[8] Kellerhals SA, Leeuw F de, Rodriguez Rivero C. Cloud Nowcasting with Structure-Preserving Convolutional Gated Recurrent Units. Atmosphere 2022; 13(10) [https://doi.org/10.3390/atmos13101632]

[9] Nielsen AH, Iosifidis A, Karstoft H. IrradianceNet: Spatiotemporal deep learning model for satellite-derived solar irradiance short-term forecasting. Solar Energy 2021; 228: 659–69 [https://doi.org/10.1016/j.solener.2021.09.073]

[10] Office M. Cartopy: a cartographic python library with a Matplotlib interface. Exeter, Devon; 2010 - 2015. Available from: URL: https://scitools.org.uk/cartopy.

[11] Cano D, Monget JM, Albuisson M, Guillard H, Regas N, Wald L. A method for the determination of the global solar radiation from meteorological satellite data. Solar Energy 1986; 37(1): 31–9 [https://doi.org/10.1016/0038-092X(86)90104-0]

[12] Hammer A, Heinemann D, Hoyer C, et al. Solar energy assessment using remote sensing technologies. Remote Sensing of Environment 2003; 86(3): 423–32 [https://doi.org/10.1016/S0034-4257(03)00083-X]

[13] Kühnert J, Lorenz E, Heinemann D. Satellite-Based Irradiance and Power Forecasting for the German Energy Market. In: Kleissl J, editor. Solar Energy Forecasting and Resource Assessment. Elsevier 2013; 267–97.

[14] Dumortier D. Modelling global and diffuse horizontal irradiances under cloudless skies with different turbidities: Daylight II, JOU2-CT92-0144,Final Report Vol. 2; 1995.

[15] Bourges B, editor. Climatic data handbook for Europe. Dordrecht: Kluwer Acad. Publ; 1992.

[16] Weinzaepfel P, Revaud J, Harchaoui Z, Schmid C. DeepFlow: Large displacement optical flow with deep matching. ICCV - IEEE International Conference on Computer Vision 2013: 1385-1392.

[17] Lorenz E, Hammer A, Heinemann D. Short term forecasting of solar radiation based on satellite data. In: Short term forecasting of solar radiation based on satellite data; 2004; 841–8.

[18] Yang D, Alessandrini S, Antonanzas J, et al. Verification of deterministic solar forecasts. Solar Energy 2020; 210: 20–37 [https://doi.org/10.1016/j.solener.2020.04.019]

AUTHOR INDEX

Abanda, Amaia 1335
Abdallah, Amir A. 591, 878, 1156
Abdelrahim, Mohamed 1156
Abdin, Zain U. 1391, 1957
Abermann, Stephan 428
Abrahão, Raphael 1528
Acevedo, Maria I. D. 610
Ackermann, Jörg 345
Adam, Zoltan 209
Adier, Marie 25
Adner, David 422
Adothu, Baloji 601, 604, 623, 666, 772, 878, 988, 1168
Adrian, Adrian 466
Aernouts, Tom 323, 494
Aghaei, Mohammadreza 1293, 1297
Aguilera, David M. 1391
Aguirre, Arantxa 315
Aguirre, Aranzazu 323, 494
Aguirre, Miguel 960
Ahnood, Arman 169
Ahuja, Suraj 539
Aiello, Andreas 1782
Aïssa, Brahim 139, 645, 1050, 1156, 1245
Aizpurua, Jon 511
Akiyama, Hidefumi 277
Al-Ahmed, Amir 600
Alam, Muhammad A. 878
Alamy, Philippe 1396
Albadwawi, Omar 330
Al-Bajjali, Saif 2040
Alberts, Vivian 330, 601, 604, 623, 666, 878, 1168
Albinius, Niklas 1348
Albrecht, Steve 399, 466
Albuquerque, Daniel 1304, 1675
Alcocer, Kilian 397
Alet, Pierre-Jean 1548
Algaidy, Sari 163
Algergawy, Alsayed 1085
Alghamdi, Mohammed A. 600
Alheloo, Ahmad 601, 604
Ali, Amjad 600
Alkhatib, Hasan 345
Allagiannis, Christos 651
Allebé, Christophe 182
Allegre, Jules 397
Almeida, José C. 1752, 2021
Almeida, Marcelo P. 1752, 1763, 2021
Almheiri, Ali 601, 604

Almosni, Samy 422
Alonso, Victor 758, 974
Alonso-García, Carmen 1651
Alskaif, Tarek 1354
Altin, Müjde 1439
Alujevic, Neven 1691
Alvarez, Jose M. 1301
Alvarez-Brito, Eduardo 135
Álvarez-Pérez, Guillem 379
Aly, Shahzada P. 623, 988, 1168
Alzahrani, Atif S. 600
Alzate, Juan 2068
Amalu, Emeka H. 695
Anagha, E. R. 619
Anamiati, Gaetana 1124, 1141
Anaya, Julián 974
Anchorena, Oscar 1369
Anderlini, Alessandro 702
Andersen, Nanna L. 632, 1199, 1385
Anderson, Kenrick F. 404, 437
Anderson, Kevin 1089
Andersson, Robin 1799
Andrade, Nathianne M. 1556
Aninat, Remi 943
Antretter, Thomas 627
Antwis, Luke 189
Apostoleris, Harry 772
Appel, Tjade 1476
Arampatzis, Ioannis 651
Ariolli, Daniela 2200
Arslan, Meriç Ç. 225
Arslan, Meric C. 683
Arunagiri, Lingeswaran 446
Asa'A, Shu-Ngwa 1709
Asaa, Shu-Ngwa 1596
Ascencio-Vásquez, Julián 1089, 1885
Asiri, Abdullah M. 800
Asker, Osama 600
Assaid, El M. 776
Aßmann, Nicole 1
Assoa, Ya-Brigitte 1434
Auer, Johann 2033
Avasthi, Sushobhan 414
Axisa, Matthew 1040
Aydemir, Umut 418
Aydogdu, Yildirim 683
Azizi, Ferozan 1828, 2077
Azkona, Nekane 201

Azzopardi, Brian	743, 1305, 1309, 2134, 2145
Azzopardi, Carmel	2134
Babin, Markus	632, 780, 966, 1199, 1385
Baccar, Dorra	125
Bachour, Dunia	1524, 1664
Bagci, Aliihsan	632
Bai, Xueqi	7
Bakovasilis, Apostolos	929, 1519
Balafoutis, Athanasios T.	1304
Balchada, Henrique	1770
Baldacci, Jacopo	1769
Ballif, Christophe	182, 349, 626, 862, 1862
Baloch, Ahmer A. B.	330
Balucani, Marco	411
Bamberger, Evelyn	1234
Bangsund, Audun	1995
Bansal, Nitin K.	429
Baptista, Fátima	1651
Barchi, Grazia	1812, 1921
Bardizza, Giorgio	490, 739, 995, 1046, 1065
Barraza, Rodrigo	562, 584, 618
Barretta, Chiara	684, 809, 892
Barrio, Rocío	163
Barrionuevo, Bruno	1304
Barrou, Alexis	1862
Barth, Vincent	953
Bartholomäus, Martin	1287
Bartolo, Brian	1305, 1309, 2134
Bartsch, Jonas	226
Basler, Felix	1544, 1620
Battisti, Kurt	1782
Baumann, Linus	1079
Baumann, Ulrike	106
Baumgartner, Franz P.	1079, 1501
Bayo, Araceli H.	2145
Bazkir, Özcan	1046
Beaucarne, Guy	539
Behrendt, Julian	63
Beinert, Andreas	527, 665, 1620
Beinert, Angelika	702
Bejat, Timea	940
Belawadi, Aditya G.	812
Belkilani, Kaouther	1776
Bellacicco, Sophie	1496
Bellenda, Giovanni	995
Benatto, Gisele A. D. R.	1241, 1317
Bendfeld, Jörg	563
Bengoechea, Jaione	703, 960, 2052
Benick, Jan	40
Benítez-Fernandez, Rafael	163
Benito, Veronica B.	1156, 1245
Berenguier, Baptiste	397
Beresneviciute, Raminta	295
Bermudez-Benito, Veronica	1664
Berrah, Lamia	1418
Berrian, Djaber	1719
Berson, Solenn	397
Berthet-Rayne, Quentin	1245
Berwind, Matthew	1501
Betak, Juraj	1451
Bhattacharjee, Ankur	2005
Bhoraskar, Akshay	1616
Biezemans, Anne	943
Binani, Ashish	1616, 1729
Bivour, Martin	40, 363
Bizzini, Olivier	1434
Blanc, Philippe	2057
Blankemeyer, Susanne	1433
Bleicher, Friedrich	1704, 1828
Blieske, Ulf	505, 2046, 2139
Blum, Niklas	1466
Boccardi, Roberto	632
Böck, Leonhard	924
Boddaert, Simon	1396
Bodeux, Romain	792
Bogdanov, Dmitrii	1790
Bokalic, Matevž	398, 738, 2138
Bolding, Jons	193
Bolink, Henk J.	50
Bonilla, Ruy S.	110
Bonomo, Pierluigi	1187
Borchert, Juliane	315
Borgers, Tom	101, 929, 1647
Boro, Binita	445
Borowski, Peter	1412, 1534
Borrello, Cosimo	1590, 2027
Borz, Giovanni	1596, 1663
Bosch, Elina	1584, 1682, 1715, 1903
Bosco, Giacomo	1596
Bosman, Johan	943
Bosone, Martina	2129
Bothe, Karsten	35
Boudellioua, Abdelaziz	155
Bouguerra, Sara	854, 1638
Boutov, Dmitri	1705
Bouttemy, Muriel	397
Braga, M.	427, 1029, 1293, 1297, 2009
Brahim, Sarra B.	1085
Braid, Jennifer L.	886
Brand, Andreas	135, 226
Brand, Thorsten	466
Brandstätter, Andreas	684
Brastel, Alexis	940
Braun, Christian	1544

Brecl, Kristijan................................398, 738
Breitenbücher, Marian 101
Brendel, Rolf 13, 35
Breniaux, Edouard 2205
Brenneisen, Stephan 1401
Bressy, Vincent 1434
Breyer, Christian 1790
Bridel-Bertomeu, Agnes 1740
Brivio, Elisabetta 2056
Bruckner, Helmut................................1957
Bruggeman, M. 113
Bruhwyler, Roxane 1606
Bua, Letizia....................................2205
Buceta, Alicia..............................960, 2052
Bucher, Christof..............571, 721, 1435, 1501
Buchholz, Florian 30, 101, 175
Buddgård, Jonas.................................751
Bueno, Bruno 1172
Buffolo, Matteo.................................417
Buijsch, Frans O................................892
Bulkin, Pavel 334
Bunge, Lisa...............................1658, 1675
Bunme, Pawita 1626
Burgers, Antonius R.............................1729
Burkhardt, Daniel 57
Burri, Matthias 721
Busto, Chiara 2205
Busuttil, Daniel................................2145
Byford, Brandon 886
Caballero, David 1663
Cabarrocas, Pere R. I...........................334
Caccivio, Mauro 721, 862, 1009, 1286, 1704
Cai, Hanmin.....................................1390
Cai, Yalun 110
Cai, Yanbo......................................346
Cal, Silvia.....................................960
Caldarelli, Antonio.............................302
Cambarau, Werther 511
Camus, Christian................................466
Candan, Mücahid 1118
Candelise, Chiara...............................2205
Caneva, Silvia..................................2205
Cano, Francisco J...............................511
Cantisano, Jose 1301
Cao, Fangfang...................................800
Cao, Rono.......................................539
Cappelle, Jan...............................709, 1514
Carbone, Rosario 1590, 2027
Cardoso, Andressa D. S..........................1610
Caria, Alessandro...............................417
Carpintero, Luis A. 758
Carr, Anna J....................................1616

Carrillo, Rafael E. 1548
Carroy, Perrine 50
Carvalho, Paulo C. M............................1354
Case, Christopher 447
Castellazzi, Luca...............................1183
Castillo, Juan D. D. 285
Catipovic, Ivan.................................1691
Cattaneo, Gianluca..............................1862
Catthoor, Francky...............................1519
Caudevilla, D. 163
Cavaco, Afonso..................................2138
Cavalcante, Danielle B..........................1556
Cebria, Maria...................................1687
Celi, Edoardo 583
Celik, Duygu....................................2205
Centazzo, Massimo 106
Cereceda, Eneko 201
Cermák, Jan.....................................1476
Cesar, Kay 1729
Cesenia, Eduardo M. 2145
Cester, Andrea 417
Champault, Lisa.................................398
Chan, Catherine 834
Chandra, Amreesh 445
Chang, Han-Chen.............................148, 222
Chapaneri, Kaushal 988, 1055, 1168
Chapon, Julien 1245
Chasparis, Georgios 2015
Chatzipanagi, Anatoli 1894
Chemnitzer, Rene 25
Chen, Angela 404
Chen, Jin-Cheng 222
Chen, Kexun.....................................189
Chen, Ran.......................................834
Chen, Sung-Yu 148
Cheng, Cheng-Liang 148
Chhapia, Gaurang................................1719
Chianese, Domenico..........................1286, 1331
Chiesa, Samuele 1331
Cho, Dae-Hyung 338
Chopard, Jérôme 1496
Choulat, Patrick................................1647
Chowdhury, Gofran...........1516, 1663, 1988
Chozas, Sofia 50
Christiansen, Silke 323
Christöfl, Petra............................570, 892
Chueh, Wei-Lo 144
Chung, Yong-Duck...............................338
Ciesla, Alison 834
Clement, Florian..........89, 155, 164, 226
Clochard, Laurent...............................110
Clyncke, Jan....................................1849

Colberts, Fallon 1638
Coletti, Gianluca 834
Colin, Hervè 1369
Collares-Pereira, Manuel 1687
Collave, Claudia G. 2040
Colvin, Dylan J. 826
Comak, Mertcan 175
Congouleris, Nicolas 1304
Cooper, Emma 886
Çorak, Merve 122, 683
Cordeiro, Diogo 1675
Cornago, Iñaki 1400
Cornaro, Cristina 1020
Cornella, Alessia 2205
Corre, Pierre-Yves 25
Correia, Joana 1696
Corti, Paolo 1187
Coskun, Özlem 35, 101, 213
Cosme, Damien 1245
Costa, Francis 1647
Couderc, Romain 792, 995
Cox, Joel D. 632, 1385
Crawley, Dru B. 2149
Creon, Laura 25
Cristóbal, Ana B. 2138
Cros, Stephane 428
Crozier, Nicole M. 2062
Cuaresma, Jesús 1279
Çubukçu, Mete 1118
Cueli, Ana B. 428, 960
Culot, Dominique 539
Curtis, Taylor L. 1841
Cusenza, Maria A. 2044, 2056
D'Agostino, Delia 1183, 2149
D'Arco, Luigi 702
Daenen, Michaël 854, 1638, 1709
Dahle, Arne 35, 101
Dalibor, Thomas 1412, 1534
Danelli, Andrea 2044, 2056
Danovitch, David 250
Das, Gourab 149
Daschinger, Thomas 1433
Daßler, David 1132
Datas, Alejandro 384
Daume, Darwin 773, 1314
De Biasio, Martin 1825
De Brabandere, Karel 1089
De Castro, C. 758, 974, 983
De Cook, Nicolas 1606
De Jong, Richard 1565, 1576, 1638, 1709
De L'Epine, Mélodie 101, 1305, 1309, 1715, 1906, 2134
De Luca, Daniela 302

De Oliveira, Aline K. V. 1029, 2009
De Rose, Angela 499, 527
De Santi, Carlo 417
De Seoane, Jose M. V. 1682
De Sousa, Joyce A. O. 1418
De Vries, Hindrik 193, 408
Debije, Michael 299
Deckers, Elke 1638
Deckers, Martijn 1443
Deckx, Julien 1089
Defrenne, Nicolas 2068
Del Prado, Alvaro 163
Del Ser, Javier 1335
Delbeke, Oscar 1634
Demant, Matthias 57, 63
Dembélé, Kassiogé 334
Demicoli, Marija 1040
Demir, Melisa 225
Demofonti, Giuseppe 1596
Denafas, Julius 101
Deniz, Esref 1806
Dennenmoser, Martin 1085
Depauw, Valerie 315
Dernis, Michel 1396
Desai, Umang 626
Despeisse, Matthieu 1862
Desrues, Thibaut 50
Devoto, Ignacia 545
Devoto, M. Ignacia 330
Dewallef, Stefan 1647
Di Carlo, Aldo 411, 417, 486
Di Gennaro, Emiliano 302
Di Giusto, Fabio 1638
Di Napoli, Annalisa 2072
Di Sabatino, Marisa 1669
Diarce, Gonzalo 1400
Diaz, Javier 428
Diestel, Christian 57
Dietrich, Andreas 1132
Dijksterhuis, Jakob J. 834
Dilmac, Umran 683
Dippell, Torsten 13
Dirubio, Christopher 826
Dittmann, Sebastian 1606
Dittmar, Hanna 2205
Djeukeu, Ivanol J. 327
Dkhil, Sadok B. 345
Döblinger, Markus 273
Donoso, Jose 1906
Dörn, Markus 1782
Dorn, Silke 35
Dos Santos, Jeremias 1675

Dou, Qizheng 1647
Dozio, Gian C. 1331
Dreisiebner, Andreas 1401, 1736
Driesen, Johan 1443, 1634
Driesse, Anton 1214, 1976
Droudakis, Alexandros I. 2124
Du, Keming 135
Duan, Lian 2099
Duarte, Dorivaldo 1687
Duarte, Sebastian 163
Duck, Benjamin C. 404, 437
Duerinckx, Filip 101, 260, 1647
Duffy, Noel W. 404, 437
Dullweber, Thorsten 35, 101, 106
Duman, Hatice 213, 225
Dunlop, Ewan 999
Dupon, Olivier 1565, 1709
Dupuis, Julien 519, 792
Durand, Salomé 1733
Durusoy, Beyza 422
Dutykh, Denys 1552
Dyson, Paul J. 800
Dzurnák, Branislav 294
Eberlein, Dirk 527
Ebert, Matthias 1132
Ebner, Rita 323, 1305, 1309
Echeverria, Iván G. 618
Eckert, Jonas 155
Eckerter, Sascha 578, 1758
Ecoffey, Serge 250
Eder, Gabriele 694, 966, 1704, 1825, 1828, 2077, 2200
Eggers, Jan-Bleicke 1172
Ehsan, Ali 2145
Eikelboom, Erik 101
Einhaus, Roland 684
Eiternick, Stefan 607
Ekins-Daukes, Nicholas 263
El Ainaoui, Khadija 776
El Mrabet, Yasmine 776
Elamri, Yassin 1496
Element, Adrian 404, 437
Elhamaoui, Said 776
Eliassi, Mojtaba 1988
Ellmann, Martin H. 1178
Emanuel, Gernot 514
Ensslen, Frank 812, 1172
Erber, Alexander 686
Eriksen, Erling W. 1480
Eroglu, Sertaç 122
Eskandari, Aref 1293, 1297
Esmaeilzadeh, Maryam 309
Esmaielpour, Hamidreza 273

Essbai, Soha 743
Estarlich, Pau 285
Esteras, Miguel 1335
Estola, Pirjo 1799
Estrada, Esther L. 384
Ezquer, Mikel 1400
Fabel, Yann 1466
Fabris, Francesca 101
Fabunmi, Oluwagbemiga A. 695
Faes, Antonin 626, 738
Fano, Vanesa 201
Faramarzi, Seyed M. S. 712
Farias-Basulto, Guillermo 393
Farmakis, Filippos V. 2124
Farneda, Rüdiger 330
Farooq, Umar 302
Fath, Moritz 1853
Fath, Peter 76, 1853, 2152
Fava, Luís 1989
Fedrizzi, Maria C. 1763
Feichtner, Markus 1704, 1782
Feldbacher, Sonja 534, 626, 1828, 2077
Feldhof, Anne-Maren 2139
Felipe, Inmaculada C. 898
Fell, Andreas 63
Fernandes, Cláudia 1675
Fernandez, Ana M. 1894
Ferreira, Catarina G. 632, 1385
Fialho, Luís 1651, 1658, 1675, 1687, 1696, 1989, 2138
Fidalgo, Ignacio 898
Figgis, Benjamin W. 1156, 1245
Figueiredo, Gilberto 1752, 2021
Filho, Eduardo S. 1118
Finley, Jonathan 273
Fischer, Marie 1885, 2192
Fki, Rania 2205
Fledderus, Henri 943
Flouchi, Imane 776
Fokuhl, Esther 786
Foles, Ana 1989
Fonseca, Luiz 1528
Fontanot, Thommaso 323
Fooladgar, Ehsan 1799
Formiga, João 1304
Franchi, Norman 1491
Fredj, Donia 345
Frégnaux, Mathieu 397
Friesen, Gabi 721, 862, 995, 1009, 1046, 1286, 1704
Froebel, Jens 552
Frontini, Francesco 1187
Frossard, Pascal 1548
Fuchs, Ida 1273, 1995

Fujii, Masayuki 676
Fumey, Damien 1496
Funahashi, Ryoji 277
Gabbadi, Prashanth 585, 1055
Gagliano, Antonio 1341
Gagnaire, Dimitri 1733
Gaiddon, Bruno 1733
Gaisberger, Lukas 2015
Galán, María I. R. 1610
Galarza, Alejandra 2068, 2099
Gallego-Castillo, Cristobal 2183
Gallmetzer, Sandra 1095, 1225
Gamel, Mansur 289
Gao, Feng 446
Gao, Qi 490, 739
Garabetian, Thomas 2205
García, Fernando 1279
Garcia, Ignacio B. 345
Garcia, José C. 1033
Garcia, Juan L. 878
Garcia, Kévin 1160
García-Hemme, Eric 163
García-Hernansanz, Rodgar 163
Garcin, Jean 1496
Gardeski, Matthew 826
Garín, Moisés 289
Gassner, Anika 694, 1704, 1825, 1828, 2077, 2200
Gatin, Inno 1691
Gaudino, Eliana 302
Gaulding, Ashley 1841
Gayot, Felix 308
Gazbour, Nouha 2205
Gdula, Lukáš 294
Gebhardt, Paul 786, 911
Geerligs, L. J. 113
Geier, Dieter 684
Gelibert, Stephane 1434
Geml, Fabian 1, 7
Genoe, Jan 712
Georghiou, George E. 323, 494, 1209, 1264
Georgilakis, Pavlos 1519
Gerber, Alexander 2138
Gerber, Andreas 1347, 1444, 1543
Gevaerts, Veronique 943
Ghennioui, Abdellatif 776
Ghidesi, Giancarlo 2084
Gioia, Ferdinando 1590, 2027
Girardi, Pierpaolo 2044, 2056
Glarner, Roger 1401
Glatz-Reichenbach, Joachim 610
Glaubitz, Anika 505
Glunz, Stefan W. 327, 363

Glunz, Stefan 40, 176
Göbel, Alexander 158
Godoy-Perez, G. 163
Gohil, Hardik 76, 768
Gok, Abdulkerim 684
Golab, Antonia 2033
Gölboylu, Selin C. 225
Golroodbari, Sara 1501
Golubev, Timofey 1322
Gombás, Z. 209
Gomes, Amanda M. F. 2009
Gomes, João 943
Gonzalez, Alejandra C. 2057
González, Miguel Á. 974, 983
González-Díaz, Benjamín 432
González-Francés, Diego 758, 974, 983
González-Pérez, Sara 432
Goraya, Baljeet S. 2152
Gorchs, Gil 1596
Gordon, Ivan 2205
Gordon, Michael 1543
Gostein, Michael 1245
Gottschalg, Ralph 596, 660, 687, 878, 1005, 1576, 1606
Gou, Yangyang 800
Gouabault, Anaïs 2068
Govaerts, Jonathan 929, 1647
Goverts, Martina 1835
Gracia-Amillo, Ana 2134
Graeber, Dietmar 1776
Graeber, Robin 2046
Grand, Pierre-Philippe 2068
Grasel, Bernhard 686
Gréau, David 1733
Gregory, Geoffrey 106
Greulich, Johannes 63, 164, 176
Grigalevicius, Saulius 295
Groen, Niels 1596, 1663
Grommes, Eva-Maria 2139
Groß, Claudine 466
Großer, Stephan 535, 933
Grosser, Stephan 607
Grübel, Benjamin 499, 527
Grünsteidl, Stefan 1534
Grüttner, Sven 505
Gry, Johannes 40
Guastella, Salvatore 1108
Guerra, Gerardo 1124, 1141
Guerra, Walter 1596
Gueymard, Christian A. 1596
Guidetti, Giulia 2084
Guillemoles, Jean F. 379
Guillevin, Nicolas 101

Guillon, Sebastien 1462
Guštin, Matej 2138
Gutjahr, Astrid 113, 834
Guzman, Francisco 562
Ha, Duy-Long 1220
Haberstroh, René 135, 155, 164
Hacke, Peter 826
Hadadian, Mahboubeh 309
Hadipour, Afshin 323
Hadjipanayi, Maria 323, 494
Hädrich, Ingrid 702, 812
Haedrich, Ingrid 786
Hagendorf, Christian 422
Hahn, Giso 1, 7
Hallensleben, Carina 545
Halm, Andreas 330, 514, 545, 610, 898
Hamada, Toshiyuki 676
Hamam, Zeina 1434
Hameed, Mohammed A. 1576
Hameiri, Ziv 308, 761, 980, 1104
Hamon, Gwenaelle 250
Hanifi, Hamed 552
Hansen, Per-Anders 132, 172
Hanser, Mario 40
Haque, Faiazul 437
Harder, Nils-Peter 1033
Harnisch, Martina 534
Harrison, Samuel 101
Hashem, Ahmad 596, 1005
Hasselblatt, Charlotte 665
Hategan, Sergiu M. 717
Hauer, Martin 1782
Haug, Franz-Josef 182
Haunschild, Jonas 57, 214
Havasi, Gergely 209
Hayez, Valérie 539
Heer, Philipp 1390
Heim, Manuel 1313
Heinonen, Aleksi 1361
Heinrich, Martin 1620
Heinzle, Nino 1264
Heitmann, Johannes 205
Helbig, Matthias 545
Helfer, Eric 570, 627
Hennig, Carsten 1132
Hensel, Andreas 1966
Herath, Kristian 1412
Herceg, Sina 1885, 2192
Herguth, Axel 1, 7
Hermle, Martin 40
Hernandez, Guillermo O. 2200
Hernández, Juan M. 511

Herr, Cornelius 665, 1620
Herrero, Carmen M. R. 345
Herrmann, Werner 490, 739, 995, 1046
Herteleer, Bert 709, 1089, 1514
Herz, Magnus 1065, 1089
Herzberg, Wiebke 1456, 1476
Hess, Donat 571, 721
Hessler-Wyser, Aïcha 182
Heupl, L. 570
Heydarian, Maryamsadat 363
Heydarian, Mina 315
Higueruela, Francisco R. F. 1610
Hillmann, Martin 1412
Hiltebrand, Roger 1736, 1747
Hitchcock, Will 1770
Hitte, Vincent 1496
Hoex, Bram 878
Hoffman, Hannah 2095
Hoffmann, Erik 106
Hofmann, Marc 158
Hofmann, Rene 1961
Holappa, Ville 430
Holder, Emma 404, 437
Holland, Nicolas 1544
Hoppe, Georg 135
Horn, Jonas 327
Horta, Pedro 1658, 1675, 1687, 1696, 1989
Hoß, Jan 30, 175
Hsieh, Hsin-Hsin 328
Hu, Guang 1392
Huang, Chih-Jeng 148
Huang, Meixian 86
Huang, Shujuan 481
Huang, Ying-Yuan 148, 219, 222
Huerta, Hugo 1361, 1969, 2001, 2102
Hughes, David J. 695
Hurni, Julien 182
Hut, Anouk 1988
Hüttl, Bernd 773, 1509
Huyeng, Jonas 89, 158, 164, 226
Hwang, Tae-Ha 338
Iannibelli, Elena 486
Ibrahim, Nabeel 1055
Idlbi, Basem 1369, 1776
Ilse, Klemens 591
Imbuluzqueta, Gorka 511
Infante, Paulo 1658
Irulegi, Olatz 1400
Isabella, Olindo 1391, 1417, 1957
Isaev, Nabi 273
Isasi, Telmo 201
Ishikura, Norio 676

Iwaki, Koshiro 549
Jääskeläinen, Jaakko 2087
Jachmann, Joseph 1509
Jäckel, Bengt...878
Jadaud, Cyril...334
Jadot, Emmanuel...539
Jaeckel, Bengt.............535, 552, 591, 607, 666, 685, 755, 772
Jaeger-Waldau, Arnulf 1894
Jäger, Philip 106
Jager, Wander...2205
Jahani, Babak 1476
Jahn, Mike 19
Jahn, Ulrike..1501, 1909
Jain, Sachin 1664
Jang, Juhee 1269, 1328
Jänkälä, Matti...1375
Jansen, Mark 315
Jasielec, Jerzy J. 1969
Jason, Daniel...1365
Jay, Frédéric 792
Jeangros, Quentin 349, 398, 892
Jeon, Joonyoung 716
Ji, Jingjia..86
Jiang, Hongxu..346
Jiang, Yongjie..800
Jiang, Zonghan..660
Jiménez-Castillo, Gabino...............................1341
Jimeno, Juan C...201
Jo, Sangmin 2160
Job, Enzo..812
John, Jim J. 878, 988, 1168
Johnson, Erik V...334
Jokikyyny, Tommi..530
Jolivet, Raphaël..1859
Jones, Tim W.......................................404, 437
Jooß, Wolfgang 76, 149, 2152
Jooss, Wolfgang 768, 1853
Joseph, Christopher D....................................1620
Joseph, Daniel C. 527
Joss, David 571, 1435
Jošt, Marko 329, 399
Jost, Norman...886
Jouanneau, Corentin 250
Jouttijärvi, Sami.............1361, 1375, 1969, 2102, 2119
Juana, Luis 1654
Juillion, Perrine..1496
Julien, Arthur 379
Jurado, Juan M..1610
Juso, Hiroyuki..254
Kaaya, Ismail.............854, 1565, 1576, 1596, 1638, 1709, 1885
Kadota, Naoki 1194
Kaiser, Martin 1544

Kaizuka, Izumi 1906
Kakoulaki, Georgia 1894, 2108
Kalaghichi, Saman S. 30, 175
Kalliojärvi, Heidi 1258
Kalms, Alicia..1400
Kamide, Kenji 277
Kammerlander, Christoph 342
Kamphues, Joshua 7
Kamppinen, Aleksi..............................303, 530
Kanawala, D. N..1391
Kang, Mangu 338
Kankanamge, Dilshika H. 2087
Kapeller, Rudolf..2205
Karalus, Steffen...............................1456, 1476
Karatepe, Engin...1806
Karhu, Juha A..1361
Kari, Thøger.........................636, 1241, 1317
Karimipour, Massoud 285
Karrenbrock, Anne 2139
Karttunen, Lauri 1969, 2102
Kathan, Johannes 2033
Katsikogiannis, Alexandros...........................1596
Kaufmann, Kai..1132
Kazacos, Duarte..1118
Kazantzidis, Andreas 1466
Kazem, Hussein A...878
Keding, Roman 226
Keiner, Dominik 1790
Kenney, Kayla 539
Kenny, Robert 1894
Kentsch, Ulrich 189
Kerekes, Krisztián 1758
Kern, Jonas...205
Kern, Melanie 2205
Kester, Josco 101
Khan, Firoz 600
Khan, Muhammad..106
Khan, Nabeel..106
Khenkin, Mark 399
Khodr, A. 345
Kikelj, Miha 349
Kim, Changki 2160
Kim, Chongmin................................1269, 1328
Kim, Jinyoung 2160
Kim, Ju-Hee 716
Kim, Kihwan 338
Kim, Moonyong 834
Kim, Rina 338
Kim, Sedong 1775
Kim, Yong H. 716
Kim, Yongil 2160
Kirch, Jochen 755

Kirchhof, Jörg	690
Kirkil, Gökhan	2205
Kishore, Ravi	1647
Kitamura, Ibuki	676
Kizukuri, Rihoko	545
Kladas, Anastasios	709, 1514
Klaus, Daniel	911
Kleinhans, Alexander	1544
Klengel, Robert	1132
Klenk, Markus	1079, 1401, 1736, 1747
Kluska, Sven	19, 135, 155, 164, 226
Klute, Carola	1132
Kobayashi, Nobusato	1194
Koblmüller, Gregor	273
Kobor, Diouma	441
Koduvelikulathu, Lejo	149, 175
Koepge, Ringo	535, 685, 933
Koester, Lukas	1225
Kohn, Norbert	158
Kohno, Tohru	1253, 1953, 1982
Kojima, Nobuaki	254
Kolås, Tore	1381
Könen, Stefanie	2139
Kopp, Nils	545
Korevaar, Marc A. N.	1249
Korkmaz, Güven	213
Korsós, Ferenc	209
Korte, Lars	466
Kossen, Eric J.	113
Kouame, Konan	250
Kousounadis-Knousen, Markos	1519
Kraft, Achim	499
Kraft, Leonard	1132
Kraft, Thomas M.	430
Krähmer, Sabrina	1776
Kräling, Ulli	786
Krammer, Anna	302
Krauter, Stefan	563, 1790, 1932
Krc, Janez	1146
Krieg, Katrin	19, 164
Krishna, Anurag	494
Krisztián, David	209
Kroon, Jan	101
Krucaite, Gintare	295
Kuan, Ta-Ming	144
Kubicek, Bernhard	1305, 1309, 1347, 2134, 2169
Kuhn, Tilmann E.	1172
Kühne, Marcel	892
Kühnert, Jan	1476
Kulhavy, Lukas	125
Kulkarni, Shrikrishna V.	619
Kumano, Kengo	1953, 1982

Kumar, Avinash	76
Kumar, Saravana	57, 214
Kumar, Shubham	1357
Kumar, Sudarshan	1559, 1724
Kumar, Yogesh	585
Kunze, Philipp	63
Kuo, Cheng-Wen	144
Kuraoka, Akihiro	1194
Kurtovic, Enita	545
Kuruganti, Vaibhav	30
Kurumundayil, Leslie L.	63
Kuypers, Ando	943
Kuznetsova, Daria	295
Kyranaki, Nikoleta	854, 1638
Lachowicz, Agata	182
Lacombe, Marie	2068
Lagast, Karel	1514
Laget, Hannes	1647
Lamminaho, Jani	632, 1385
Lampa, Josefin	1799
Landa, Margot	940
Landberg, Lars	1124, 1141
Landes, Dieter	773, 1509
Lang, Margit	570, 627
Lang, Xiting	800
Lange, Gerrit	35
Lanzetta, Ciro	1769
Lappalainen, Kari	1258, 1937, 1977
Larionova, Yevgeniya	35
Lauwaert, Johan	260
Lawrie, Linda K.	2149
Le Rouzo, Judikaël	345
Le, Philip	1638
Lebeau, Frederic	1606
Lee, Chun-Wei	144
Lee, Woo-Jung	338
Lefillastre, Paul	519
Leimgruber, Fabian	2169
Leiva, Amanda M.	50
Leloux, Jonathan	1369, 1596
Lemaitre, Noëlla	397
Leow, Shin W.	995
Leyden, M.	466
Li, Fang	826
Li, Minghui	800
Li, Qiuxian	1390
Li, Yong	404, 437
Li, You-An	219
Li, Yung-Chih	144
Li, Zhuofeng	25
Liang, Tian S.	1187
Lim, Soyoung	338

Lin, Chun-Ping	148, 219, 222
Lin, Shih-Chieh	144
Lin, Yi-Ping	222
Linares, Ana	960
Lindahl, Johan	1903
Linder, Johannes	1719
Lindfors, Anders V.	1361
Lindh, Mattias	1799
Lindig, Sascha	1089
Linke, Jonathan	30, 101, 175
Linsenmeyer, Aswin	755
Lipovšek, Benjamin	349, 399
Lira-Cantu, Mónica	285
Liu, Anyao	25
Liu, Fei	346
Liu, Xirui	800
Liu, Zhipeng	86
Livera, Andreas	1209, 1264
Lizana, Fernando F.	2092
Lizin, Sebastien	2205
Llarena, Elena	432, 960
Lohmüller, Elmar	158, 176
Lohmüller, Sabrina	176
Lomeri, Hamed J.	854, 929
Long, Yean-San	328
Loonen, Roel C. G. M.	1392
López, Gema	289
Lopez-Garcia, Juan	1156, 1245, 1664
Lopez-Velasco, Gerardo	1496
Lorenz, Andreas	226
Lorenz, Elke	1456, 1476, 1544
Lossen, Jan	30, 175
Louwen, Atse	1095, 1225, 1812, 1873, 2205
Lu, Yibo	86
Lübke, Maximilian	1491
Lukinskas, Povilas	101
Luo, Bin	1647
Lyubenova, Teodora S.	999
Maarouf, F.	226
Maaroufi, Hamza	1065
Macarulla, Marcel	1596
Macdonald, Daniel	25
Macdonald, James	1596
Macé, Philippe	101, 1584, 1682, 1715
Mack, Sebastian	164
Mader, Patrick	578, 1758
Madsen, Morten	632, 1385
Maduta, Carmen	1183
Maeda, Kengo	1194
Mahmood, Aysha	636, 1241, 1317
Mahmood, Farrukh I.	826
Maixner, Andreas	552

Makhfudz, Imam	273
Makrides, George	1209, 1264
Malcorps, Philippe	1516
Malguth, Enno	466
Malik, Stephanie	1132
Mamykin, Sergii	295
Mandorlo, Fabien	953
Manito, Alex R. A.	1752, 1763, 2021
Manshanden, Petra	315, 834
Mansour, Djamel E.	911
Mansour, Ridha B.	600
Mansouri, Mathieu	1733
Manzolini, Giampaolo	1921
Marangis, D.	1209, 1264
Marchand, Mathilde	1859
Marcotte, Médérick	250
Marechal, Philippe	1220
Margeat, O.	345
Maria, Enrico D.	1663
Markert, Jochen	786, 812
Markvart, Tom	294
Marrero, Asier M.	428
Marstein, Erik S.	1089
Marteau, Batiste	50
Martín, Isidro	289
Martínez, Mario	746, 1301, 1365
Mártinez, Oscar	758, 974, 983
Martulli, Alessandro	2205
Masmitjà, Gerard	285
Masson, Gaëtan	1584, 1682, 1715, 1903, 1906
Masuda, Atsushi	523, 549
Mateos, Yeray	201
Matheron, Muriel	50
Mathiak, G.	585, 601, 604, 623, 772, 878, 988, 1055, 1168
Matic, Gašper	398
Maticiuc, Natalia	428
Matos, Pedro	1989
Matsumura, Yoko	277
Matteocci, Fabio	417
Maugeri, Giosué	1108
Mazzucchelli, Paola	2205
McCleland, Jacqueline L. C.	2062
Meddahi, Amar	1462
Medina, Eduardo R.	511
Medina, Ismael	999
Medjoubi, Karim	379
Meereboer, Martijn	101
Mehler, Melanie	1
Meinhart, Lisa	684
Meixner, Michael	327
Melgar, David	1118, 1369
Mellone, Celeste	2084

Meneghesso, Gaudenzio 417
Meneghini, Matteo 417
Mercade-Ruiz, Pau 1124
Mercaldo, Lucia V. 428
Mermoud, André 1740
Mertens, Verena 35, 101, 106
Merz, Rainer 578, 1758
Messmer, Christoph 363
Messmer, Marius 164
Messmer, Tobias 101, 514
Meuret, Youri 1647
Meusel, Manuel 53
Meyer, Fabian 135
Meyer, Imke 1770
Meyer, Kevin 1433
Meyer, Lukas 1435
Meza, Carlos 660, 1305, 1309, 1606, 2134
Meza, Cristian V. 363
Miaskiewicz, Aleksandra 466
Midtgård, Ole-Morten 1273
Miech, Juri 7
Mielich, Niko 89
Miettunen, Kati 303, 309, 530, 1361, 1375, 1969, 2102, 2119
Mignonac, Alexandre 1305, 1309
Mihailetchi, Valentin 30, 101
Miklic, Žiga 1146
Mikulic, Antonio 1691
Milimonfared, Jafar 1293, 1297
Min, Byungsul 13
Minuto, Alessandro 583
Mirza, Mark 585
Misfeld, Heidrun 1476
Mittag, Max 920, 2095
Mittal, Ankit 323, 428, 743
Mizuno, Hidenori 1626
Mizushima, Io 342
Mjøs, Øyvind 132
Mo, Alvin 834
Mockeviciute-Azzopardi, Austeja 2134
Mofakhami, Eeva 940
Moine, Gérard 1733
Möller, Marius C. 1932
Mollier, Stéphane 1369
Molto, Cecile 826
Monokroussos, Christos 490, 739, 1005
Montes, Carlos 432
Montes-Romero, Jesus 1264
Moradpoor, Iraj 2109
Mordvinkin, Anton 680, 685, 687, 933
Moreda, Guillermo 1651
Moreda, G.-P. 1654
Moretón, Rodrigo 1160, 1369

Morisset, Audrey 182
Morlier, Arnaud 854, 1565, 1638, 1709
Morozova, Olga 189
Mortazavifar, S. L. 596, 1005
Moschner, Jens 854, 1634, 1647
Mosel, Frank 125
Moser, David 1020, 1095, 1225, 1565, 1596, 1663, 1812, 1873, 1909, 1921, 2200
Motiwala, Saurabh 1559, 1724
Mühlich, Mona 1172
Muka, Eni 197
Mukherjee, Srijani 1552
Müller, Björn 1118
Müller, Matthias 205
Muñoz, Delfina 50, 1909
Muñoz-García, Miguel-Ángel 1651, 1654
Muñoz-Rodriguez, Francisco J. 1341
Muntwyler, Urs 1409, 1679
Murillo, Asier 2052
Musto, Marilena 302, 2072
Muthusamy, Arumugham 1055
Mütter, Gerhard 1966
Myhre, Stine F. 2078
Naas, Tyke 132
Nagel, Henning 40
Nägele, Andreas 514
Nair, Jishnu R. 687
Najafi, Mehrdad 408
Najah, Mohamed 250
Nakajima, Akihiko 1194
Nakamura, Kyotaro 254
Nanno, Ikuo 676
Naspolini, Helena 1029
Nasser, Hisham 19, 197
Nasti, Giuseppe 428
Navarro, Valentina 584
Nazeeruddin, Mohammad K. 800
Naziri, Pouriya 418
Ndioukane, Rémi 441
Nedaei, Amir 1293, 1297
Nekarda, Jan 135, 226
Nemitz, Wolfgang 780
Neuhaus, Dirk H. 920, 2095
Neuhaus, Holger 527
Neumaier, Lukas 1825
Ney, Mylana 250
Nguyen, Hieu T. 25
Nguyen, Nathalie 50
Niederhofer, Stefan 1961
Nieto, María B. 1651
Nigl, Thomas 2077
Nikam, Maitheli 1663

Nikbakht, Hafez 486
Nikitina, Veronika 924
Niskanen, Johannes...1969
Nitzel, Damon..1249
Nizamov, Rustem...309
Noack, Philipp 13
Noels, Serge 1849
Nold, Sebastian..2152
Nordseth, Ørnulf 1381
Norton, Matthew...494
Nour, Christine A. 519
Nouri, Bijan 1466
Ntsala, Palisa G. 2062
Nussbaumer, Hartmut 1401, 1736, 1747
Nyberg, Mikael 309
Nygård, Magnus M. 1480
Oberbeck, Lars.............................. 1859, 2068, 2099
Ocaña, Luis 432
Ochoa, Lluvia 1245
Oh, Jaewon 826
Oh, Sujeong 1269, 1328
Ohdaira, Keisuke 523, 549, 656
Ohshita, Yoshio 254
Oke, Shinichiro 676
Olea, J. 163
Oliosi, Michele 1740
Oliveira, Helena 1658
Olofsson, Arvid 1799
Oozeki, Takashi 846, 1626
Opatovsky, Martin 1451
Oreski, G....534, 570, 626, 627, 684, 809, 892, 966, 2077, 2200
Ortega, Eneko 201
Ortega, Pablo 285
Ortiz, Hugo S. 1606
Osama, Amr 1341
Otaegi, Alona 201
Otoo, Edward 1392
Otto, Nicolas 393
Ouaras, Karim 334
Ourinson, Daniel 164, 226
Oyarzun, Aritz L. 2134
Öz, Aksel Kaan 911
Ozaki, Ryo 254
Özar, Nilsah 1439
Özkalay, Ebrar 862, 1009, 1046, 1704
Paesa, Marta C. 1638
Paez, Pablo S. E. 1095
Palitzsch, Wolfram 101
Palonen, Heikki...530
Pampin, Janire 201
Panchabikesan, Karthik 2149
Pander, Matthias 535, 552, 591, 666, 685, 755, 933

Pandurangan, Karthikeyan 486
Panduri, Fabio 721
Pang, Yongxin 695
Panhuysen, Markus...773, 1314
Paraskeva, Vasiliki...323, 494
Parida, Bhaskar 330
Parion, Jonathan...260, 363
Parisi, Maria L...486
Parker, Danny S. 2149
Parlayan, Onur 527
Parra, Vicente 1279
Parvin, P. 1293, 1297
Passaro, Marcello 2205
Pastor, D. 163
Patel, Dharm...1132
Patha, Andreas...2033
Patton, Daniel J. C. 1279
Paul, Mrittika 445
Paulescu, Marius 717
Pauli, Eva 1476
Paviet-Salomon, Bertrand 182, 349, 1862
Pawar, Vani 414
Payne, David N. R. 481
Pearce, Phoebe 263
Peche, Rene 2045
Pechmann, Sabrina 323
Peibst, Robby 13
Peighambardoust, Naeimeh S... 418
Penas, André 1682, 1715
Peng, Meilin 86
Peratikos, Elias 494
Pereira, Sara 1687
Perelman, Antoine 953
Peres, Paula 25
Perez, Alba 1663
Perez, Inaki 1770
Perez-Astudillo, Daniel.......................................1524, 1664
Perez-Lopez, Paula...1859, 2057
Perrin, Marielle 1733
Persello, Severine 1496
Pervan, Nikolina...534, 626
Peter, Christoph... 30
Petersons, Karlis...632, 1385
Petersson, Anna M. 1799
Petro, Julia...570, 627
Petzschmann, Jonas 1313
Pfau, Charlotte 651
Pfeiffer, Oliver 505, 2046
Pfyffer, Selina.......................................1401, 1736, 1747
Phang, Sieu P... 25
Philipp, Daniel 786, 812
Piazzi, Antonio 1769

Pierce, Benjamin G. 886
Pierro, Marco 1020
Pieters, Bart E. 1347, 1444, 1543
Pignatelli, Angelo 1596
Pilat, Eric 1160, 1369
Pillai, Dhanup 1245, 1664
Piluso, Pierre...940
Pingel, Sebastian 57, 226
Pinto, Cristina 703, 2052
Pirc, Matija 329
Pirelli, Barbara..2129
Pires, Anelise M. 427
Pirot-Berson, Lucie 792
Pittalis, Marco 2108
Plaza, C. 1584, 1682, 1715
Plessing, Lukas 780
Plissonnier, Alexandre 1434
Polacchi, Cristina 1873
Polo, Jesus 1214
Polzin, Jana..19, 40
Ponomarenko, Anna...2040
Poormohammadi, Fereshteh 1443
Poortmans, Jef 260, 712, 929, 1647, 1709
Popplow, Laura...2139
Porter, Jennifer................................... 1663, 1770
Porwal, Shivam...429
Poskela, Aapo 309, 530
Pospischil, Maximilian 101
Poulsen, Peter B............... 632, 636, 1287, 1317, 1385
Prasad, Manjunath 101
Pravettoni, Mauro 995
Preis, Pirmin 149, 175
Preisig, Janis 1747
Preu, Ralf 158, 176, 226, 2152
Protti, Alexander 920
Puel, Jean B. 379
Puglisi, Lisandro 1304
Puigdollers, Joaquim...285
Purohit, Ishan 1559, 1724
Puthiyapurayil, Aafra S. 623
Qin, Yusen..86
Qiu, Zhiheng 800
Queiroz, Isadora M. 1029
Queiroz, Rodrigo S. 1556
Radfar, Behrad 189
Radhakrishnan, Hariharsudan S. 260, 363, 929, 1647, 1709
Radzevicius, Aurimas 101
Rahdan, Parisa 2183
Rajan, S. Prithivi 1596
Ramesh, Santhosh.................................260, 363, 494
Ramos-Fuentes, Isaac A. 1496
Ramspeck, Klaus 327

Randle-Boggis, Richard J.....................................1669
Ransome, Steve..1264
Ranta, Samuli.....................................1361, 1969, 2001, 2102
Ratnasingham, S. R. 408
Rau, Björn 1348
Raval, Mehul.....................................76, 149, 768, 1853
Rebohle, Lars 163
Rebollo, Míguel Á. G. 758
Reekmans, Bart 929, 1647
Rehman, Abdul.. 250
Reichel, Christian................................920, 2095
Reichel, Rene 1412
Reichle, Julian 76, 2152
Reijners, Frits 299
Rein, Stefan.....................................57, 63, 214
Reinders, Angèle H. M. E. 299, 1392
Reise, Christian 1072
Reisecker, Volker................................570, 627
Reiser, Elisabeth 966
Remec, Marko 399
Remund, Jan 1435
Ren, Jinlei 25
Rende, Fedele 1782
Rennhofer, Marcus 743, 1305, 1309, 1961
Rentsch, Jochen 2152
Reyal, Jean-Pierre 1396
Reyes, Valentina A. 610, 618
Rezaei-Hartmann, Nasim 466
Riaño, Sandra 1335
Richter, Armin 40
Riechelmann, Stefan 765, 995, 1046
Riehle, Tim... 924
Rigaud, Eric 2057
Riise, Heine N. 1480, 2078
Rillo, Sergio D. A. 441
Rinio, Markus... 751
Ripke, M. 35
Rist, Tobias 812
Rivas, Jose 746, 1301, 1365
Rivera, Gerard 289
Rivera, Mariella 1072
Rizzi, Stefano 583
Robledo, Jesus... 1596
Röder, Julian 1412
Rodríguez, Osbel A. 2205
Rodríguez-Conde, Sofía.................746, 1279, 1301, 1365
Rodríguez-Lucas, Delia.. 1654
Roessler, Florian.. 135
Roig, Irma 1596
Romer, Pascal.....................................665, 812, 1620
Rosen, Isaac 101
Rosenberg, Eva.. 2078

Roshchina, Nina 295
Rosina, Konstantin 1451
Roß, Marcel 466
Rossetti, Andrea 1108
Rößler, Torsten 924
Rougieux, Fiacre 110
Röver, Ingo 101
Røyset, Arne 1381
Rozanov, Konstantin 2040
Rudolph, Dominik 514
Ruiz, Alfonso L. 1610
Ruiz, Pau M. 1141
Ruiz, Sonia 285
Rummelhoff, Stian 1995
Russo, Roberto 302, 2072
Rüther, Ricardo 427, 1029, 2009
Saegebarth, Kai 1085
Sahli, Florent 349
Saidi-Chalopin, Elika 1733
Saint-Cast, Pierre 176
Sakakibara, Reyu 182
Sakib, Syed N. 481
Sakuma, Jun 277
Salari, Majid 751
Salis, Fabio 2084
Salperwyck, Christophe 1369
Sals, Sem 892
Salvador, Michael 878
Samara, Ayman 645
San Andrés, E. 163
Sanchez, Antonio 562
Sanchez, Hugo 660, 1005
Sánchez-Calvo, Raúl 1654
Sánchez-Friera, Paula 1909
Santamaria, Rodrigo D. P. 636, 1241, 1317
Santbergen, Rudi 1391, 1417, 1957
Santos, J. V. Oliveira 519
Santos, Jose D. 1335
Santos, Leticia D. O. 1354
Sapkota, Subarna 1412
Saucedo, Edgardo 285
Saugues, François 1733
Savin, Hele 189
Savisalo, Tuukka 101
Saviuc, Iolanda 2108
Saw, Min H. 995
Scerri, Kenneth 1305, 1309
Schak, Matthias 607
Schebek, Liselotte 1885, 2192
Scheer, Roland 1576
Scheler, Florian 399
Schenck, Catherina 2062
Schenk, Paul 755
Scherret, Jaqueline 1782
Schifferegger, Raffael 694
Schill, Christian 1118, 1369, 1544
Schimanke, Sabrina 35
Schlatmann, Rutger 393, 399, 1348
Schmid, Alexandra 1046
Schmidt, Jan 35, 193
Schmiga, Christian 155
Schmitt, Emmanuel 1434
Schmitz, Jurriaan 214
Schnaus, Dominik 1466
Schneider, Astrid 1782
Schneider, Jale 135
Schneider, Simon 1957
Schneiders, Thorsten 2139
Schön, Jonas 363
Schönau, Elisabeth 773
Schönau, Maximilian 773, 1314
Schram, Wouter L. 1917
Schranz, Christian 1782
Schube, J. 19, 226
Schubert, Maik 1412
Schubert, Martin C. 363
Schubnel, Baptiste 1548
Schuesler, Daniel 687
Schüler, Andreas 302
Schüler, Marc A. 1620
Schulte-Huxel, Henning 13, 1433
Schultz, Christof 393
Schulz, Philip 397
Schulze, Achim 773, 1314, 1509
Schulz-Ruhtenberg, Malte 89
Schüpbach, Eva 1409, 1679
Schüsler, Daniel 680
Schutt, Thomas 1412
Schwabl, Daniel 2077
Schwarz, Andreas 2046
Schweigstill, Tadeo 89
Sciullo, Alessandro 2205
Scognamiglio, Alessandra 2084
Scroppo, Sofia 2139
Seentakath, Afra 585, 1055
Segura, Oriol 285
Seick, Cinja 1596
Seigneur, Hubert 826
Sen, Nesrin T. 35
Senaud, Laurie-Lou 349
Seppälä, Simeon 2109
Sergio, Lucas A. Z. 427
Serra, Filipe 1675
Serra, João M. 1705

Sgouridis, Sgouris	772
Sha, Nithin	585, 1055
Shakiba, Ali	761, 1104
Sharma, Ashish K.	1559, 1724
Sharma, Bhumika	414
Sharma, Deepak	169
Sharma, Rama	761, 980
Sharma, Ruchi K.	169
Shimokata, Eiko	523
Shimpo, Shuntaro	656
Shin, Donghyeop	338
Shiradkar, Narendra S.	878
Shiradkar, Narendra	619
Shirazi, Elham	1917
Shochet, Ofer	101
Short, Michael	695
Silva, François	334
Silva, José	1658, 1675, 1687, 1696, 1989
Silva, Lucas T.	1556
Simeunovic, Jelena	1548
Singh, Ojas	886
Singh, Trilok	429, 445
Sinicropi, Adalgisa	486
Sinopoli, Alessandro	139
Siquera, Tales	552
Sivaramakrishnan, Hariharsudan	101
Škorjanc, Viktor	466
Slooff-Hoek, Lenneke	1616
Smertenko, Petro	295
Smith, Ligia	1841
Smith, Ryan	826
Sobajima, Yasushi	523, 549
Soeiro, André	1675
Sohani, Ali	1020
Solofra, Nate	1249
Solórzano, Jorge	1160
Søndenå, Rune	132, 172
Sondoqah, Mousa	1095, 1225, 1812
Sorbet, Patxi	1400
Sourd, Francis	1496
Souren, Floor	193, 408
Souza, Francisco A. A.	1354
Sovetkin, Evgenii	1347, 1444, 1543
Spagnolo, Sofia	2056
Spataru, Sergiu V.	636, 1241, 1287, 1317
Späth, Martin	315
Spätlich, Sarah	106
Speer, Volker	1412
Spera, Fabian	2046
Sraisth	768, 911
Srivastava, Sanjay K.	169
Stagno, Luciano M.	1040

Stalmans, Lieven	1647
Stannowski, Bernd	892
Starke, A.	53
Stefanelli, Maurizio	411, 486
Stegemann, Bert	393
Steinlechner, Sebastian	2033
Stellbogen, Dirk	1313
Stensborg, Jan F.	632
Stenzig, Laura	765, 1046
Stepec, Murielle	1220
Stieldorf, Karin	1782
Stierstorfer, J.	101, 2205
Stoicescu, Liviu	1241
Stokkan, Gaute	1669
Stölzel, Marko	1412
Sträter, Hendrik	1046
Straub, Nils	1456
Strazzullo, Paolo	302, 2072
Strömberg, Rich	1849
Stuckelberger, Josua	25
Stueve, William	1245
Suarez, Sergio	746, 1301, 1365
Subbarao, P. M. V.	1357
Subeh, Mosab	1050, 1156
Subramaniam, Sownder	260
Suemitsu, Issei	1953, 1982
Sugimoto, Hiroki	559
Suhonen, Riikka	430
Sulca, Kabir P.	758, 974, 983
Suri, Marcel	1451
Sutariya, Mahesh	1528
Sutkuviene, Simona	295
Sutterlueti, Juergen	1264
Suvarn, Shashank	1055
Svedjeholm, Maria	1799
Syri, Sanna	2087, 2102, 2109, 2119
Szabó, Sandor	1894
Taheri, Nabi	1769
Takamoto, Tatsuya	254
Takashima, Takumi	1626
Talvi, Micke	1977
Tamizhmani, Govindasamy	826
Tanahashi, Katsuto	277
Tanahashi, Tadanori	846
Tang, Peter T.	342
Taylor, Nigel	1894
Taylor, Stephen	257
Ternes, Simon	411
Terrados, Cristian	758, 974, 983
Tervo, Seela	2102, 2119
Tessmann, Christopher	164
Tettenborn, Tuuli	534

Teymouri, Arastoo 834
Thalheimer, Martin 1596
Thawanyavitchajit, Chisanupong 1770
Thebault, Martin 1418
Theelen, Mirjam 943, 1835, 1873
Theristis, Marios 1089
Thomassen, Bent..132
Thony, Philippe 1434
Thorsteinsson, Sune 632, 1199, 1385
Tierney, Paul 110
Tilly, Eric 694
Timò, Gianluca 583
Timofte, Tudor 610
Tina, Giuseppe M. 1341
Tomšic, Špela..329, 399
Topic, Marko.............329, 349, 398, 399, 738, 1146, 2138
Tormena, Noah 417
Torrens, Arnau 285
Torres, Ignacio 163
Torres, Pedro 1752, 2021
Tous, Loic 1647
Trattnig, Roman 780
Traunmüller, Wolfgang 2015
Treberspurg, Christoph 1782
Treberspurg, Martin 1782
Trivellin, Nicola 417
Truong, Thein N..25
Tsai, Min-An 328
Tsanakas, Ioannis 1160, 1552, 2200
Tsanakas, John A. 1220
Tsemekidi-Tzeiranaki, Sofia 1183
Tsoulka, Polyxeni 397
Tsunoda, Jun 1253, 1953, 1982
Tu, Huynh T. C. 656
Tucci, Mauro 1769
Tulinski, Lona 1747
Tune, Daniel 101, 330, 545
Tuomiranta, Arttu 1462
Turala, Artur 250
Turan, Rasit...19, 197
Turek, Marko 53, 422, 607, 651
Turri, Evelyn 1225
Tzinoglou, George I.......................................2124
Übermasser, Stefan 2169
Ugranli, Faruk 1806
Újvári, Gusztáv 323, 743
Ul-Abdin, Zain 1417
Ulbikas, Juras..101
Ulbrich, Carolin 399, 1348
Umeda, Kazuhiko 1194
Unger, Eva 393
Urata, Tomoyuki 277

Urban, Harald..1782
Urbina, Antonio...2052
Uygun, Berkay 19
Vähänissi, Ville 189
Vaidya, Haresh 1528
Valckenborg, Roland 1178
Valdivia, Patricio 618
Valdivia-Lefort, Patricio.................562, 584, 2092
Valencia, Felipe 1501
Valle, Benoît 1496
Valoti, Flavio 995
Van Aken, Bas................................834, 1729
Van De Water, Oscar 1616
Van Den Storme, Guy 929
Van Den Storme, Manuel 929
Van Der Heide, Arvid 712, 854, 995, 1709, 1849, 2200
Van Der Ploeg, Bas 1943
Van Der Vleuten, Maarten 943
Van Dyck, Rik 101, 929
Van Dyk, Ernest E. 2062
Van Gijlswijk, René 1616
Van Overstraeten, Julien 1715
Van Rossum, Aron 1391
Van Sark, Wilfried 1943, 1976
Vandamme, Nicolas..2068
Vanel, Jean-Charles 334
Vanhanen, Tuomas 101
Varjopuro, Julianna 530
Varney, Valérie 2139
Vasquez, Pia 50
Vavilkin, Tatjana 1647
Veettil, Binesh P. 481
Veihelmann, Tobias 1491
Veneri, Paola D. 428
Ventosinos, Federico 50
Verdeil, Olivier 1733
Verkou, Maarten 1957
Vermang, Bart 260
Veronese, Elisa 1921
Vesce, Luigi..411, 486
Vespermann, Merle 1476
Viani, Lucas 1279
Vicente, Diogo 1705
Victoria, Marta 2138, 2183
Vidal, Nerea 1687
Vieira, Bruno J. 1763
Vilela, Jonathan 1301
Villa, Simona.......................................1178, 1835
Villoslada, Daniel..................................746, 1365
Vincent, Robin 1740
Viriyaroj, Bergpob 1375
Virtuani, Alessandro 862, 1862

Vitale, Simone	1369
Vito, Domenico	2129
Vogt, Aaron	89
Voirol, Alexandre	1234
Völler, Steve	1669
Voronko, Yuliya	694, 966
Voroshazi, Eszter	953
Voz, Cristobal	285
Vrielinck, Henk	260
Vuillon, Laurent	1552
Vulic, Natasa	1390
Vyalih, Irina	632, 1385
Wagenmann, Dirk	158
Wagner, Enno	327
Wagner-Mohnsen, Hannes	205
Waldau, Arnulf J.	1906
Wambach, Karsten	2045, 2200
Wang, Deliang	346
Wang, Feng	446
Wang, Guangwei	346
Wang, Jiali	25
Wang, Li	834
Wang, Lu	86, 1859
Wang, Shuo	1249, 1361, 2001
Wang, Xiawa	335
Wang, Yichun	7
Waschl, Alfred	1782
Watrin, Lise	334
Wattenberg, Bianca	13
Weber, Anne-Kathrin	1435
Weber, Juergen W.	308
Weber, Thomas	743
Wehnert, Danny	1132
Weiermair, A.	570
Weinert, Nicolas	1
Weinrich, Frank	765, 995
Weiß, Karl-Anders	809, 1885, 2192
Wellens, Christine	911
Wendt, Michael	680, 687
Wernke, Luka	1348
Wessel, Patrick	680
Westerberg, Amelia O.	1903, 1906
Widler, Adrian	1079
Wienands, Karl	330, 545
Wienberg, Robin	214
Wilbert, Stefan	1466
Willers, Guido	651
Wilson, Gregory J.	404, 437
Winkelmann, Jan	1966
Winter, Michael	35, 193
Winter, Stefan	765, 995, 1046
Wirtz, Wiebke	1433

Wiss, Olivier	1434
Wittmer, Bruno	1740
Woehler, Wilkin	63
Woernhoer, Alexandra	63
Wöhrle, Nico	57
Wojciechowski, Konrad	422
Wolf, Andreas	164
Wong, George	2099
Woodhouse, Michael	2152
Wörnhör, Alexandra	57
Wright, Brendan	761, 980, 1104
Wright, James	110
Wu, Bang-Hao	148
Wu, Haodong	800
Wu, Li-Guo	144
Wu, Yu	113
Xiao, Chuanxiao	800
Xu, Frank	1005
Xu, Wenhao	490, 739
Yagci, Selim	505
Yamaguchi, Akira	1194
Yamaguchi, Masafumi	254
Yde, Leif	632
Ye, Jichun	800
Yeh, Fan-Hsuan	328
Yi, Kai	346
Yildirim, Nurhayat	122
Yordanov, Georgi H.	712, 1647
You, Chang C.	1381
Young, Jørgen	1430
Yu, Cheng-Yeh	144
Yu, Mingzhe	110
Yun, Changyeol	2160
Yurrita, Naiara	511
Zaimi, Mhammed	776
Zamarro, Fernando L.	441
Zanoni, Enrico	417
Zardetto, Valerio	315
Zaror, Yasmin	101
Zarzalejo, Luis F.	1466
Zaversky, Fritz	1400
Zech, Tobias	1476
Zehndorfer, Jakob	780
Zekri, Atef	1050
Zeman, Miro	1957
Zenteno, Franciso J. P.	163
Zerafa, Steve	1305, 1309, 2134
Zeyen, Elisabeth	2183
Zhang, Junchuan	800
Zhang, Shipei	335
Zhang, Yating	490, 739
Zhang, Yi	800

Zheng, Zhiwen 308
Zhou, Guohua 86
Zhu, Junjie 172, 1430
Zhu, Xitong 299
Zhu, Yan.. 308
Zilles, Roberto 1752, 1763, 2021
Zimmermann, Andreas......................... 342, 534
Zubillaga, Oihana 511
Zugasti, Eugenia 428, 1305, 1309, 2052
Züll, Laura 2139
Zult, Michiel 1616
Zwahlen, Theo 1435

WIP – Renewable Energies
Sylvensteinstr. 2
81369 Munchen
Germany

ISBN 979-8-3313-1538-2